Włodzisław Duch Janusz Kacprzyk
Erkki Oja Sławomir Zadrożny (Eds.)

Artificial Neural Networks: Formal Models and Their Applications – ICANN 2005

15th International Conference
Warsaw, Poland, September 11-15, 2005
Proceedings, Part II

Springer

Volume Editors

Włodzisław Duch
Nicolaus Copernicus University
Department of Informatics
ul. Grudziądzka 5, 87-100 Toruń, Poland
and
Nanyang Technological University
School of Computer Engineering
Division of Computer Science
Blk N4-02c-114, Nanyang Avenue, Singapore, 639798
E-mail: wduch@phys.uni.torun.pl

Janusz Kacprzyk
Sławomir Zadrożny
Polish Academy of Sciences
Systems Research Institute
ul. Newelska 6, 01–447 Warsaw, Poland
E-mail: {kacprzyk,zadrozny}@ibspan.waw.pl

Erkki Oja
Helsinki University of Technology
Laboratory of Computer and Information Science
P.O. Box 5400, 02015 Hut, Finland
E-mail: erkki.oja@hut.fi

Library of Congress Control Number: 2005931603

CR Subject Classification (1998): F.1, I.2, I.5, I.4, G.3, J.3, C.2.1, C.1.3

ISSN 0302-9743
ISBN-10 3-540-28755-8 Springer Berlin Heidelberg New York
ISBN-13 978-3-540-28755-1 Springer Berlin Heidelberg New York

Springer is a part of Springer Science+Business Media

springeronline.com

Printed in Germany

Typesetting: Camera-ready by author, data conversion by Scientific Publishing Services, Chennai, India
Printed on acid-free paper SPIN: 11550907 06/3142 5 4 3 2 1 0

Lecture Notes in Computer Science 3697

Preface

This volume is the first part of the two-volume proceedings of the International Conference on Artificial Neural Networks (ICANN 2005), held on September 11–15, 2005 in Warsaw, Poland, with several accompanying workshops held on September 15, 2005 at the Nicolaus Copernicus University, Toruń, Poland.

The ICANN conference is an annual meeting organized by the European Neural Network Society in cooperation with the International Neural Network Society, the Japanese Neural Network Society, and the IEEE Computational Intelligence Society. It is the premier European event covering all topics concerned with neural networks and related areas. The ICANN series of conferences was initiated in 1991 and soon became the major European gathering for experts in those fields.

In 2005 the ICANN conference was organized by the Systems Research Institute, Polish Academy of Sciences, Warsaw, Poland, and the Nicolaus Copernicus University, Toruń, Poland.

From over 600 papers submitted to the regular sessions and some 10 special conference sessions, the International Program Committee selected – after a thorough peer-review process – about 270 papers for publication. The large number of papers accepted is certainly a proof of the vitality and attractiveness of the field of artificial neural networks, but it also shows a strong interest in the ICANN conferences. Because of their reputation as high-level conferences, the ICANN conferences rarely receive papers of a poor quality and thus their rejection rate may be not as high as that of some other conferences. A large number of accepted papers meant that we had to publish the proceedings in two volumes. Papers presented at the post-conference workshops will be published separately.

The first of these volumes, *Artificial Neural Networks: Biological Inspirations*, is primarily concerned with issues related to models of biological functions, spiking neurons, understanding real brain processes, development of cognitive powers, and inspiration from such models for the development and application of artificial neural networks in information technologies, modeling perception and other biological processes. This volume covers dynamical models of single spiking neurons, their assemblies, population coding, models of neocortex, cerebellum and subcortical brain structures, brain–computer interfaces, and also the development of associative memories, natural language processing and other higher cognitive processes in human beings and other living organisms. Papers on self-organizing maps, evolutionary processes, and cooperative biological behavior, with some applications, are also included. Natural perception, computer vision, recognition and detection of faces and other natural patterns, and sound and speech signal analysis are the topics of many contributions in this volume. Some papers on bioinformatics, bioengineering and biomedical applications are also included in this volume.

The second volume, *Artificial Neural Networks: Formal Models and Their Applications*, is mainly concerned with new paradigms, architectures and formal models of artificial neural networks that can provide efficient tools and techniques to model a great array of non-trivial real-world problems. All areas that are of interest to the neural network community are covered, although many computational algorithms discussed in this volume are only remotely inspired by neural networks. A perennial question that the editors and reviewers always face is: how to define the boundary or the limits of a field? What should still be classified as an artificial neural network and what should be left out as a general algorithm that is presented in the network form? There are no clear-cut answers to these questions. Support vector machines and kernel-based methods are well established at neural network conferences although their connections with neural networks are only of a historical interest. Computational learning theory, approximation theory, stochastic optimization and other branches of statistics and mathematics are also of interest to many neural network experts. Thus, instead of asking: Is this still a neural method?, we have rather adopted a policy of accepting all high-quality papers that could be of interest to the neural network community.

A considerable part of the second volume is devoted to learning in its many forms, such as unsupervised and supervised learning, reinforcement learning, Bayesian learning, inductive learning, ensemble learning, and their applications. Many papers are devoted to the important topics in classification and clustering, data fusion from various sources, applications to systems modeling, decision making, optimization, control, prediction and forecasting, speech and text analysis and processing, multimedia systems, applications to various games, and other topics. A section on knowledge extraction from neural networks shows that such models are not always opaque, black boxes. A few papers present also algorithms for fuzzy rule extraction using neural approaches. Descriptions of several hardware implementations of different neural algorithms are also included. Altogether this volume presents a variety of theoretical results and applications covering most areas that the neural network community may be interested in.

We would like to thank, first of all, Ms. Magdalena Gola and Ms. Anna Wilbik for their great contribution in the preparation of the proceedings. Moreover, Ms. Magdalena Gola, Ms. Anna Wilbik, and Ms. Krystyna Warzywoda, with her team, deserve our sincere thanks for their help in the organization of the conference. Finally, we wish to thank Mr. Alfred Hofmann, Ms. Anna Kramer and Ms. Ursula Barth from Springer for their help and collaboration in this demanding publication project.

July 2005 W. Duch, J. Kacprzyk, E. Oja, S. Zadrożny

Table of Contents – Part II

Formal Models and Their Applications

New Neural Network Models

Supervised Learning Algorithms

Ensemble-Based Learning

Unsupervised Learning

Recurrent Neural Networks

Reinforcement Learning

Bayesian Approaches to Learning

Learning Theory

Artificial Neural Networks for System Modeling, Decision Making, Optimalization and Control

Special Session: Knowledge Extraction from Neural Networks
Organizer and Chair: D. A. Elizondo

Temporal Data Analysis, Prediction and Forecasting

Support Vector Machines and Kernel-Based Methods

Soft Computing Methods for Data Representation, Analysis and Processing

Special Session: Data Fusion for Industrial, Medical and Environmental Applications

Organizers and Chairs: D. Mandic, D. Obradovic

Special Session: Non-linear Predictive Models for Speech Processing

Organizers and Chairs: M. Chetouani, M. Faundez-Zanuy, B. Gas, A. Hussain

Special Session: Intelligent Multimedia and Semantics

Organizers and Chairs: Y. Avrithis, S. Kollias

Applications to Natural Language Proceesing

Various Applications

Special Session: Computational Intelligence in Games
Organizer and Chair: J. Mańdziuk

Issues in Hardware Implementation

Table of Contents – Part I

Biological Inspirations

Modeling the Brain and Cognitive Functions

Special Session: The Development of Cognitive Powers in Embodied Systems

Spiking Neural Networks

Models of Biological Functions

Special Session: Projects in the Area of NeuroIT

Evolutionary and Other Biological Inspirations

Self-organizing Maps and Their Applications

Computer Vision

Face Recognition and Detection

Sound and Speech Recognition

Bioinformatics

Biomedical Applications

Special Session: Information-Theoretic Concepts in Biomedical Data Analysis

Neuro-fuzzy Kolmogorov's Network

Yevgeniy Bodyanskiy[1], Yevgen Gorshkov[2],
Vitaliy Kolodyazhniy[3], and Valeriya Poyedyntseva[4]

[1,2,3] Control Systems Research Laboratory,
Kharkiv National University of Radioelectronics,
14, Lenin Av., Kharkiv, 61166, Ukraine
bodya@kture.kharkov.ua, ye.gorshkov@gmail.com
kolodyazhniy@ukr.net
[4] Kharkiv National Automobile and Highway University,
25, Petrovskiy St., Kharkiv, 61002, Ukraine
poyedyntseva@gmx.net

Abstract. A new computationally efficient learning algorithm for a hybrid system called further *Neuro-Fuzzy Kolmogorov's Network* (NFKN) is proposed. The NFKN is based on and is the development of the previously proposed neural and fuzzy systems using the famous superposition theorem by A.N. Kolmogorov (KST). The network consists of two layers of neo-fuzzy neurons (NFNs) and is linear in both the hidden and output layer parameters, so it can be trained with very fast and simple procedures. The validity of theoretical results and the advantages of the NFKN in comparison with other techniques are confirmed by experiments.

1 Introduction

According to the Kolmogorov's superposition theorem (KST) [1], *any* continuous function of d variables can be *exactly* represented by superposition of continuous functions of one variable and addition:

$$f(x_1,\ldots,x_d)=\sum_{l=1}^{2d+1} g_l\left[\sum_{i=1}^{d}\psi_{l,i}(x_i)\right], \tag{1}$$

where $g_l(\bullet)$ and $\psi_{l,i}(\bullet)$ are some continuous univariate functions, and $\psi_{l,i}(\bullet)$ are independent of f. Aside from the exact representation, the KST can be used as the basis for the construction of parsimonious universal approximators, and has thus attracted the attention of many researchers in the field of soft computing.

Hecht-Nielsen was the first to propose a neural network implementation of KST [2], but did not consider how such a network can be constructed. Computational aspects of approximate version of KST were studied by Sprecher [3, 4] and Kůrková [5]. Igelnik and Parikh [6] proposed the use of spline functions for the construction of Kolmogorov's approximation. Yam *et al* [7] proposed the multi-resolution approach to fuzzy control, based on the KST, and proved that the KST representation can be realized by a two-stage rule base, but did not demonstrate how such a rule base could be created from data. Lopez-Gomez and Hirota developed the Fuzzy Functional Link

W. Duch et al. (Eds.): ICANN 2005, LNCS 3697, pp. 1 – 6, 2005.

Network (FFLN) [8] based on the fuzzy extension of the Kolmogorov's theorem. The FFLN is trained via fuzzy delta rule, whose convergence can be quite slow. In [9, 10], a novel KST-based universal approximator called Fuzzy Kolmogorov's Network (FKN) with simple structure and training procedure with high rate of convergence was proposed. However, this training algorithm may require a large number of computations in the problems of high dimensionality. In this paper we propose an efficient and computationally simple learning algorithm, whose complexity depends linearly on the dimensionality of the input space.

2 Network Architecture

The NFKN is comprised of two layers of neo-fuzzy neurons (NFNs) [11] and is described by the following equations:

$$\hat{f}(x_1,\ldots,x_d)=\sum_{l=1}^{n} f_l^{[2]}(o^{[1,l]}),\quad o^{[1,l]}=\sum_{i=1}^{d} f_i^{[1,l]}(x_i),\quad l=1,\ldots,n\,, \tag{2}$$

where n is the number of hidden layer neurons, $f_l^{[2]}(o^{[1,l]})$ is the l-th nonlinear synapse in the output layer, $o^{[1,l]}$ is the output of the l-th NFN in the hidden layer, $f_i^{[1,l]}(x_i)$ is the i-th nonlinear synapse of the l-th NFN in the hidden layer.

The equations for the hidden and output layer synapses are

$$f_i^{[1,l]}(x_i)=\sum_{h=1}^{m_1}\mu_{i,h}^{[1]}(x_i)w_{i,h}^{[1,l]},\quad f_l^{[2]}(o^{[1,l]})=\sum_{j=1}^{m_2}\mu_{l,j}^{[2]}(o^{[1,l]})w_{l,j}^{[2]}\,, \tag{3}$$
$$l=1,\ldots,n\,,\quad i=1,\ldots,d\,,$$

where m_1 and m_2 is the number of membership functions (MFs) per input in the hidden and output layers respectively, $\mu_{i,h}^{[1]}(x_i)$ and $\mu_{l,j}^{[2]}(o^{[1,l]})$ are the MFs, $w_{i,h}^{[1,l]}$ and $w_{l,j}^{[2]}$ are tunable weights. We assume that the MFs are triangular and equidistantly spaced over the range of each NFN input. The parameters of the MFs are not tuned.

Nonlinear synapse is a single input-single output fuzzy inference system with crisp consequents, and is thus a universal approximator [12] of univariate functions. It can provide a piecewise-linear approximation of any functions $g_l(\bullet)$ and $\psi_{l,i}(\bullet)$ in (1). So the NFKN, in turn, can approximate any function $f(x_1,\ldots,x_d)$.

The output of the NFKN is computed as the result of two-stage fuzzy inference:

$$\hat{y}=\sum_{l=1}^{n}\sum_{j=1}^{m_2}\mu_{l,j}^{[2]}\left[\sum_{i=1}^{d}\sum_{h=1}^{m_1}\mu_{i,h}^{[1]}(x_i)w_{i,h}^{[1,l]}\right]w_{l,j}^{[2]}\,. \tag{4}$$

The description (4) corresponds to the following two-level fuzzy rule base:

$$\text{IF } x_i \text{ IS } X_{i,h} \text{ THEN } o^{[1,1]}=w_{i,h}^{[1,1]}d \text{ AND}\ldots\text{AND } o^{[1,n]}=w_{i,h}^{[1,n]}d\,, \tag{5}$$
$$i=1,\ldots,d\,,\quad h=1,\ldots,m_1\,,$$

$$\text{IF } o^{[1,l]} \text{ IS } O_{l,j} \text{ THEN } \hat{y} = w_{l,j}^{[2]} n, \quad l = 1,\ldots,n, \quad j = 1,\ldots,m_2. \tag{6}$$

3 Learning Algorithm

The weights of the NFKN are determined by means of a batch-training algorithm as described below. A training set containing N samples is used. The minimized error function is

$$E(t) = \sum_{k=1}^{N} [y(k) - \hat{y}(t,k)]^2 = \left[Y - \hat{Y}(t)\right]^T \left[Y - \hat{Y}(t)\right], \tag{7}$$

where $Y = \left[y(1),\ldots,y(N)\right]^T$ is the vector of target values, and $\hat{Y}(t) = \left[\hat{y}(t,1),\ldots,\hat{y}(t,N)\right]^T$ is the vector of network outputs at epoch t.

Since the nonlinear synapses (3) are linear in parameters, we can employ direct linear least squares (LS) optimization for the estimation of the output layer weights. To formulate the LS problem for the output layer, re-write (4) as

$$\hat{y} = W^{[2]^T} \varphi^{[2]}(o^{[1]}), \quad W^{[2]} = \left[w_{1,1}^{[2]}, w_{1,2}^{[2]}, \ldots, w_{n,m_2}^{[2]}\right]^T, \tag{8}$$
$$\varphi^{[2]}(o^{[1]}) = \left[\mu_{1,1}^{[2]}(o^{[1,1]}), \mu_{1,2}^{[2]}(o^{[1,1]}), \ldots, \mu_{n,m_2}^{[2]}(o^{[1,n]})\right]^T,$$

and introduce the following regressor matrix of dimensionality $N \times d \cdot m_1 \cdot n$:

$$\Phi^{[2]} = \left[\varphi^{[2]}(o^{[1]}(1)), \ldots, \varphi^{[2]}(o^{[1]}(N))\right]^T. \tag{9}$$

Then the LS solution is [9, 10]:

$$W^{[2]} = \left(\Phi^{[2]^T} \Phi^{[2]}\right)^{-1} \Phi^{[2]^T} Y^{[2]}, \quad Y^{[2]} = Y. \tag{10}$$

The solution (10) is not unique when the matrix $\Phi^{[2]^T}\Phi^{[2]}$ is singular. To avoid this, instead of (10) at every epoch t we will find

$$W^{[2]}(t) = \left(\Phi^{[2]^T}(t)\Phi^{[2]}(t) + \eta I\right)^{-1} \Phi^{[2]^T}(t) Y^{[2]}(t), \tag{11}$$

where η is the regularization parameter with typical value $\eta = 10^{-5}$.

Introducing the regressor matrix of the hidden layer $\Phi^{[1]} = \left[\varphi^{[1]}(x(1)), \ldots, \varphi^{[1]}(x(N))\right]^T$, we can now obtain the expression for the gradient of the error function with respect to the hidden layer weights at the epoch t:

$$\nabla_{W^{[1]}} E(t) = -\Phi^{[1]T} \left[Y - \hat{Y}(t) \right], \tag{12}$$

and then use the well-known gradient-based technique to update these weights:

$$W^{[1]}(t+1) = W^{[1]}(t) - \gamma(t) \frac{\nabla_{W^{[1]}} E(t)}{\left\| \nabla_{W^{[1]}} E(t) \right\|}, \tag{13}$$

where $\gamma(t)$ is the adjustable learning rate, and

$$W^{[1]} = \left[w_{1,1}^{[1,1]}, w_{1,2}^{[1,1]}, \ldots, w_{d,m_1}^{[1,1]}, \ldots, w_{d,m_1}^{[1,n]} \right]^T,$$

$$\varphi^{[1]}(x) = \left[\varphi_{1,1}^{[1,1]}(x_1), \varphi_{1,2}^{[1,1]}(x_1), \ldots, \varphi_{d,m_1}^{[1,1]}(x_d), \ldots, \varphi_{d,m_1}^{[1,n]}(x_d) \right]^T, \tag{14}$$

$$\varphi_{i,h}^{[1,l]}(x_i) = a_l^{[2]}(o^{[1,l]}) \mu_{i,h}^{[1,l]}(x_i),$$

and $a_l^{[2]}(o^{[1,l]})$ is determined as proposed in [9, 10]

$$a_l^{[2]}(o^{[1,l]}) = \frac{w_{l,p+1}^{[2]} - w_{l,p}^{[2]}}{c_{l,p+1}^{[2]} - c_{l,p}^{[2]}}, \tag{15}$$

where $w_{l,p}^{[2]}$ and $c_{l,p}^{[2]}$ are the weight and center of the p-th MF in the l-th synapse of the output layer neuron, respectively. The MFs in an NFN are chosen such that only two adjacent MFs p and p+1 fire at a time [11].

Thus, the NFKN is trained via a two-stage optimization procedure without any nonlinear operations, similar to the ANFIS learning rule for the Sugeno-type fuzzy inference systems [13]. In the forward pass, the output layer weights are calculated. In the backward pass, the hidden layer weights are calculated.

The hidden layer weights are initialized deterministically as proposed in [9, 10]:

$$w_{h,i}^{[1,l]} = \exp\left\{ -\frac{i[m_1(l-1)+h-1]}{d(m_1 n-1)} \right\}, h = 1,\ldots,m_1, i = 1,\ldots,d, l = 1,\ldots,n. \tag{16}$$

4 Simulation Results

To verify the theoretical results and compare the performance of the proposed network to the known approaches, we have carried out two experiments: two spirals classification [14, 13] and Mackey-Glass time series prediction [15].

In the first experiment, the goal was to classify the given input coordinates as belonging to one of the two spirals. The NFKN contained 8 neurons in the hidden layer with 6 MFs per input, 1 neuron in the output layer with 8 MFs per synapse. The results were compared with those obtained with the FKN [9], a two-hidden layer MLP trained with the Levenberg-Marquardt algorithm, the Neuro-Fuzzy Classifier (NFC)

[13], and the cascade correlation (CC) learning architecture [14]. The experiment for the MLP was repeated 10 times (each time with a different random initialization). Because of the deterministic initialization according to (16), the experiments for the FKN and NFKN were not repeated.

As is shown in Table 1, the NFKN reached 0 classification errors after 13 epoch of training, and thus outperformed all the compared approaches, including the FKN.

Table 1. Results of two spirals classification

Network	Parameters	Epochs	Runs	Errors min	Errors max	Errors average
NFKN	160	13	1	0	0	0
FKN	160	17	1	0	0	0
MLP	151	200	10	0	12	4
NFC	446	200	1	0	0	0
CC	N/A (12-19 hidden units)	1700	100	0	0	0

The Mackey-Glass time series was generated by the equation [15]

$$\frac{dy(t)}{dt} = \frac{0.2\, y(t-\tau)}{1+y^{10}(t-\tau)} - 0.1\, y(t) \tag{17}$$

for $t = 0,\ldots,1200$ with the initial condition $y(0) = 1.2$ and delay $\tau = 17$. The values $y(t-18), y(t-12), y(t-6)$, and $y(t)$ were used to predict $y(t+6)$. From the generated data, 500 values for $t = 118,\ldots,617$ were used as the training data set, and the next 500 for $t = 618,\ldots,1117$ as the checking data set.

The NFKN used for prediction had 4 inputs, 9 neurons in the hidden layer with 3 MFs per input, and 1 neuron in the output layer with 5 MFs per synapse (153 adjustable parameters altogether). It demonstrated similar performance as the FKN with the same structure [10]. Both networks were trained for 50 epochs.

Root mean squared error on the training and checking sets (trnRMSE and chkRMSE) was used to estimate the accuracy of predictions (see Table 2).

Table 2. Results of Mackey-Glass time series prediction

Network	Parameters	Epochs	trnRMSE	chkRMSE
FKN	153	50	0.0028291	0.004645
NFKN	153	50	0.0036211	0.0056408

Providing roughly similar performance as the FKN, the NFKN requires much less computations as it does not require matrix inversion for the tuning of the hidden layer.

5 Conclusion

In the paper, a new simple and efficient training algorithm for the NFKN was proposed. The NFKN contains the neo-fuzzy neurons in both the hidden and output layer and is not affected by the curse of dimensionality because of its two-level structure. The use of the neo-fuzzy neurons enabled us to develop fast training procedures for all the parameters in the NFKN.

References

1. Kolmogorov, A.N.: On the representation of continuous functions of many variables by superposition of continuous functions of one variable and addition. Dokl. Akad. Nauk SSSR **114** (1957) 953-956
2. Hecht-Nielsen, R: Kolmogorov's mapping neural network existence theorem. Proc. IEEE Int. Conf. on Neural Networks, San Diego, CA, Vol. 3 (1987) 11-14
3. Sprecher, D.A.: A numerical implementation of Kolmogorov's superpositions. Neural Networks **9** (1996) 765-772
4. Sprecher, D.A.: A numerical implementation of Kolmogorov's superpositions II. Neural Networks **10** (1997) 447–457
5. Kůrková, V.: Kolmogorov's theorem is relevant. Neural Computation 3 (1991) 617-622
6. Igelnik, B., and Parikh, N.: Kolmogorov's spline network. IEEE Transactions on Neural Networks **14** (2003) 725-733
7. Yam, Y., Nguyen, H. T., and Kreinovich, V.: Multi-resolution techniques in the rules-based intelligent control systems: a universal approximation result. Proc. 14th IEEE Int. Symp. on Intelligent Control/Intelligent Systems and Semiotics ISIC/ISAS'99, Cambridge, Massachusetts, September 15-17 (1999) 213-218
8. Lopez-Gomez, A., Yoshida, S., Hirota, K.: Fuzzy functional link network and its application to the representation of the extended Kolmogorov theorem. International Journal of Fuzzy Systems **4** (2002) 690-695
9. Kolodyazhniy, V., and Bodyanskiy, Ye.: Fuzzy Kolmogorov's Network. Proc. 8th Int. Conf. on Knowledge-Based Intelligent Information and Engineering Systems (KES 2004), Wellington, New Zealand, September 20-25, Part II (2004) 764-771
10. Kolodyazhniy, V., Bodyanskiy, Ye, and Otto, P.: Universal Approximator Employing Neo-Fuzzy Neurons. Proc. 8th Fuzzy Days, Dortmund, Germany, Sep. 29 – Oct. 1 (2004) CD-ROM
11. Yamakawa, T., Uchino, E., Miki, T., and Kusanagi, H.: A neo fuzzy neuron and its applications to system identification and prediction of the system behavior. Proc. 2nd Int. Conf. on Fuzzy Logic and Neural Networks "IIZUKA-92", Iizuka, Japan (1992) 477-483
12. Kosko, B.: Fuzzy systems as universal approximators. Proc. 1st IEEE Int. Conf. on Fuzzy Systems, San Diego, CA (1992) 1153-1162
13. Jang, J.-S. R.: Neuro-Fuzzy Modeling: Architectures, Analyses and Applications. PhD Thesis. Department of Electrical Engineering and Computer Science, University of California, Berkeley (1992)
14. Fahlman, S.E., Lebiere, C.: The Cascade-Correlation Learning Architecture. In: Touretzky, D. S. (Ed.): Advances in Neural Information Processing Systems, Morgan Kaufmann, San Mateo, Denver (1990) 524-532
15. Mackey, M. C., and Glass, L.: Oscillation and chaos in physiological control systems. Science **197** (1977) 287-289

A Neural Network Model for Inter-problem Adaptive Online Time Allocation

Matteo Gagliolo[1] and Jürgen Schmidhuber[1,2]

[1] IDSIA, Galleria 2, 6928 Manno-Lugano, Switzerland
[2] TU Munich, Boltzmannstr. 3, 85748 Garching, München, Germany
{matteo, juergen}@idsia.ch

Abstract. One aim of Meta-learning techniques is to minimize the time needed for problem solving, and the effort of parameter hand-tuning, by automating algorithm selection. The predictive model of algorithm performance needed for task often requires long training times. We address the problem in an *online* fashion, running multiple algorithms in parallel on a sequence of tasks, continually updating their relative priorities according to a neural model that maps their current state to the expected time to the solution. The model itself is updated at the end of each task, based on the actual performance of each algorithm. Censored sampling allows us to train the model effectively, without need of additional exploration after each task's solution. We present a preliminary experiment in which this new *inter*-problem technique learns to outperform a previously proposed *intra*-problem heuristic.

1 Problem Statement

A typical machine learning scenario involves a (possibly inexperienced) practitioner trying to cope with a set of problems, that could be solved, in principle, using one element of a set of available algorithms. While most users still solve such dilemmas by trial and error, or by blindly applying some unquestioned rule-of-thumb, the steadily growing area of *Meta-Learning* [1] research is devoted to automating this process. Apart from a few notable exceptions (e.g. [2,3,4,5], see [6], of which we adopt the notation and terminology, for a commented bibliography), most existing techniques amount to the selection of a single candidate solver (e.g. *Algorithm recommendation* [7]), or a small subset of the available algorithms to be run in parallel with the same priority (e.g. *Algorithm portfolio selection* [8]). This approach usually requires a long training phase, which can be prohibitive if the algorithms at hand are computationally expensive; it also assumes that the algorithm runtimes can be predicted *offline*, based on problem features, and do not exhibit large fluctuations. In more complex cases, where the difficulty of the problems cannot be precisely predicted a priori, a more robust approach would be to run the candidate solvers in parallel, adapting their priorities *online* according to their actual performance. We termed this *Adaptive Online Time Allocation* (AOTA) in [6], in which we further distinguish between *intra*-problem AOTA, where the prediction of algorithm performance is made according to some heuristic based on a-priori knowledge about the algorithm's behavior; and *inter*-problem AOTA, in which a time allocation strategy is learned by collecting experience on a sequence of tasks.

W. Duch et al. (Eds.): ICANN 2005, LNCS 3697, pp. 7–12, 2005.

In this work we present an *inter*-problem approach for training a parametric model of algorithm runtimes, and give an example of how this model can be used to allocate time online, comparing its performance with the simple *intra*-problem heuristic from [6].

2 A Parametric Model for Inter-problem AOTA

Consider a finite algorithm set A containing n algorithms a_i, $i \in I = \{1, \ldots, n\}$, applied to the solution of the same problem and running according to some time allocation procedure. Let t_i be the time spent on a_i; $\mathbf{x}_i$ a feature vector, possibly including information about the current problem, the algorithm a_i itself (e.g. its kind, the values of its parameters), and its current state d_i; $H_i = \{(\mathbf{x}_i^{(r)}, t_i^{(r)}), r = 0, \ldots, h_i\}$ a set of collected samples of these pairs; $H = \cup_{i \in I} H_i$ the historic experience set relative to the entire A.

In order to allocate machine time efficiently, we would like to map each pair in each H_i to the time τ_i still left before a_i reaches the solution. If we are allowed to learn such mapping by solving a sequence of related tasks, we can, for a successful algorithm a_i that solved the problem at time $t_i^{(h_i)}$, *a posteriori* evaluate the correct $\tau_i^{(r)} = t_i^{(h_i)} - t_i^{(r)}$ for each pair $(\mathbf{x}_i^{(r)}, t_i^{(r)})$ in H_i. In a first tentative experiment, that led to poor results, these values were used as targets to learn a regression from pairs $(\mathbf{x}, t)$ to residual time values τ. The main problem with this approach is which τ values to choose as targets for the *unsuccessful* algorithms. Assigning them heuristically would penalize with high τ values algorithms that were stopped on the point of solving the task, or give incorrectly low values to algorithms that cannot solve it; obtaining more exact targets τ by running more algorithms until the end would increase the overhead.

The alternative we present here is inspired by *censored sampling* for lifetime distribution estimation [9], and consists in learning a parametric model $g(\tau|\mathbf{x}_i, t_i; \mathbf{w})$ of the conditional probability density function (pdf) of the residual time τ. To see how the model can be trained, imagine we continue the time allocation for a while after the first algorithm solves the current task, such that we end up having one or more successful algorithms a_i, with indices $i \in I_s \subseteq I$, for whose H_i the correct targets $\tau_i^{(r)}$ can be evaluated as above. Assuming each $\tau_i^{(r)}$ to be the outcome of an independent experiment, including t in $\mathbf{x}$ to ease notation, if $p(\mathbf{x})$ is the (unknown) pdf of the $\mathbf{x}_i^{(r)}$ we can write the likelihood of H_i as

$$\mathfrak{L}_{i \in I_s}(H_i) = \prod_{r=0}^{h_i - 1} g(\tau_i^{(r)}|\mathbf{x}_i^{(r)}; \mathbf{w}) p(\mathbf{x}_i^{(r)}) \quad (1)$$

For the unsuccessful algorithms, the final time value $t_i^{(h_i)}$ recorded in H_i is a lower bound on the unknown, and possibly infinite, time to solve the problem, and so are the $\tau_i^{(r)}$, so to obtain the likelihood we have to integrate (1)

$$\mathfrak{L}_{i \notin I_s}(H_i) = \prod_{r=0}^{h_i - 1} [1 - G(\tau_i^{(r)}|\mathbf{x}_i^{(r)}; \mathbf{w})] p(\mathbf{x}_i^{(r)}) \quad (2)$$

where $G(\tau|\mathbf{x};\mathbf{w}) = \int_0^\tau g(\xi|\mathbf{x};\mathbf{w})d\xi$ is the conditional cumulative distribution function (cdf) corresponding to g.

We can then search the value of $\mathbf{w}$ that maximizes $\mathfrak{L}(H) = \prod_{i \in I} \mathfrak{L}(H_i)$, or, in a Bayesian approach, maximize the posterior $p(\mathbf{w}|H) \propto \mathfrak{L}(H|\mathbf{w})p(\mathbf{w})$. Note that in both cases the logarithm of these quantities can be maximized, and terms not in $\mathbf{w}$ can be dropped.

To prevent overfitting, and to force the model to have a realistic shape, we can use some known parametric lifetime model, such as a Weibull distribution [9], with pdf $g(\tau|\mathbf{x},t;\mathbf{w}) = \lambda^\beta \beta \tau^{\beta-1} e^{-(\lambda\tau)^\beta}$ and express the dependency on $\mathbf{x}$ and $\mathbf{w}$ in its two parameters $\lambda = \lambda(\mathbf{x};\mathbf{w}), \beta = \beta(\mathbf{x};\mathbf{w})$. In the example we present here, these will be the two outputs of a feed-forward neural network, which will be trained by back-propagation minimizing the negative logarithm of $\mathfrak{L}(H)$, whose derivatives are easily obtainable, in a fashion that is commonly used for modelling conditional distributions (see e.g. [10], par 6.4).

From the time allocation perspective, one advantage of this approach is that it allows to learn also from the unsuccessful algorithms, suffering less from the trade-off between the accuracy of the learned model, and the time spent on learning it.

3 An Example Application

If the estimated model g was the correct one, the time allocation task would be trivial, as we could allocate all resources to the expected fastest algorithm, i.e., the one with lower expected run time $\int_0^{+\infty} \tau g(\tau|\mathbf{x})d\tau$, periodically re-checking which algorithm is to be selected given the current states $\{\mathbf{x}_i\}$. In practice, however, the predictive power of the model depends on the how the current task compares to the ones solved so far, so trusting it completely would be too risky. In preliminary experiments, we adopted a time allocation technique similar to the one in ([6]), slicing machine time in small intervals ΔT, and sharing each ΔT among elements of A according to a distribution $P_A = \{p_i\}$; the latter is updated at each step based on the current model g, which is re-trained at the end of each task on the whole history H collected so far, as follows:

```
for each problem r
   while (r not solved)
      update {τ_i} based on current g and current {x_i}:
         τ_i = ∫_0^{+∞} τ g(τ|x_i) dτ
      update P_A = {p_i} based on {τ_i}
      for each i = 1..n
         run a_i for a time p_i ΔT
         update x_i
      end
   end
   update H
   update g maximizing 𝔏(H)
end
```

To model g we used an Extreme Value distribution[1] on the logarithms of time values, with parameters $\eta(\mathbf{x};\mathbf{w})$ and $\delta(\mathbf{x};\mathbf{w})$ being the two outputs of a feedforward neural network, with two separate hidden layers of 32 units each, whose weights are obtained by minimizing the negative logarithm of the Bayesian posterior $p(\mathbf{w}|H)$ obtained in Sect. 2, using 20% of the current history H as a validation set, and a Cauchy distribution $p(w) = 1/1 + w^2$ as a prior.

At each cycle of the time allocation, the current expected time τ_i to the solution is evaluated for each a_i from $g(\tau|\mathbf{x}_i;\mathbf{w})$; these values are ranked in ascending order, and the current time slice is allocated proportionally to $(\frac{\log(m+1-j)}{\log(m)})^{-r_i}$, r_i being the current rank of a_i, m the total number of tasks, j the index of current task (from 1 to m). In this way the distribution of time is uniform during the first task (when the model is still untrained), and tends through the task sequence to a sharing pattern in which the expected fastest solver gets half of the current time slice, the second one gets one quarter, and so on. We ran some preliminary tests, using the algorithm set A_3 from [6], a set of 76 simple generational Genetic Algorithms [11], differing in population size ($2^i, i = 1..19$), mutation rate (0 or $0.7/L$, L being the genome length) and crossover operator (uniform or one-point, with rate 0.5 in both cases). We applied these solvers to a sequence of artificial deceptive problems, such as the "trap" described in [3], consisting of n copies of an m-bit trap function: each m-bit block of a bitstring of length nm gives a fitness contribution of m if all its bits are 1, and of $m - q$ if $q < m$ bits are 1. We generated a sequence of 21 different problems, varying the genome length from 30 to 96 and the size of the deceptive block from 2 to 4. The problems were first sorted by genome length, then by block size, such that the resulting sequence is roughly sorted by difficulty (see Table 1). The feature vector $\mathbf{x}$ included two problem features (genome length and block size), the algorithm parameters, the current best and average fitness values, together with their last variation and their current trend, the time spent and its last increment, for a total of 13 inputs.

We compared the presented inter-problem AOTA with the intra-problem AOTA$_{ga}$, the most competitive from [6], in which the $\{\tau_i\}$ were heuristically estimated based on a simple linear extrapolation of the learning curve. In figure 1 we show the significant improvement over AOTA$_{ga}$, which by itself already greatly reduces computation time with respect to a brute-force approach.

4 Conclusions and Future Work

The purpose of this work was to show that a parametric model of algorithm performance can be learned and used to allocate time efficiently, without requiring a long training phase. Thanks to the model, the system was able to learn the bits of a-priori knowledge that we had to pre-wire in the *intra*-problem AOTA$_{ga}$: for example, the fact that increases in the average fitness are an indicator of potentially good performance. Along the sequence of tasks, the model gradually became more reliable, and NN-AOTA was

[1] If τ is Weibull distributed, $l = \log\tau$ has Extreme Value distribution $g(l) = \frac{1}{\delta}e^{\{[(l-\eta)/\delta]-e^{(l-\eta)/\delta}\}}$, with parameters $\delta = 1/\beta$, $\eta = -\log\lambda$. The distribution of the logarithm of residual times was used to learn a common model for a set of tasks whose solution times have different orders of magnitude.

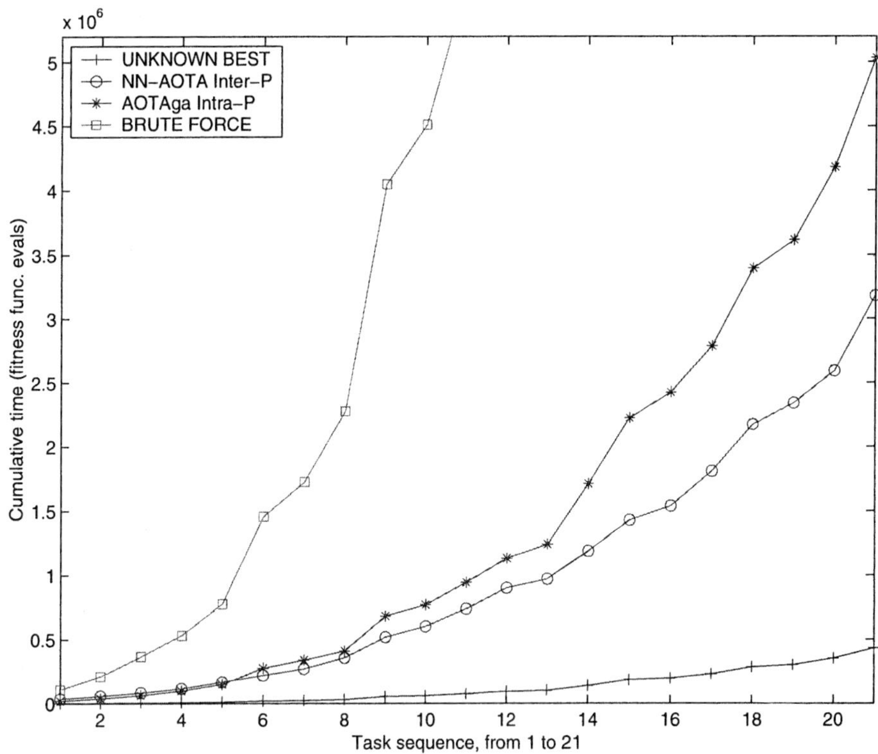

Fig. 1. A comparison between the presented method, labeled NN-AOTA Inter-P, and the intra-problem AOTA$_{ga}$, on a sequence of 21 tasks. Also shown are the the performances of the (a priori unknown, different for each problem and for each random seed) fastest solver of the set (which would be the performance of an ideal AOTA with foresight), labeled UNKNOWN BEST, and the estimated performance of a brute force approach (running all the algorithms in parallel until one solves the problem), labeled BRUTE FORCE, which leaves the figure and completes the task sequence at time 3.3×10^7. The cumulative time spent on the sequence of tasks, i.e. the total time spent in solving the current and all previous tasks, is plotted against current task index. Time is measured in fitness function evaluations; values shown are upper 95% confidence limits calculated on 20 runs.

Table 1. The 21 trap problems used, each listed with its block size m and number of blocks n

	m	n		m	n		m	n
1)	2	15	8)	3	16	15)	4	18
2)	3	8	9)	4	12	16)	2	40
3)	4	6	10)	2	30	17)	3	28
4)	2	20	11)	3	20	18)	4	21
5)	3	12	12)	4	15	19)	2	45
6)	4	9	13)	2	35	20)	3	32
7)	2	25	14)	3	24	21)	4	24

finally able to outperform AOTA$_{ga}$. In spite of the size of the network used, the obtained model is not very accurate, due to the variety of the algorithms behavior on the different tasks; still, it is discriminative enough to be used to rank the algorithms according to their expected runtimes.

The neural network can be replaced by any parametric model whose learning algorithm is based on gradient descent: in future work, we plan to test a more complex mixture model [12], in order to obtain more accurate predictions, and even better performances.

As the obtained model is continuous, and can give predictions also before starting the algorithms (i.e. for $t_i = 0$), it could in principle be used to adapt also the algorithm set A to the current task, guiding the choice of a set of promising points in parameter space.

Acknowledgements. This work was supported by SNF grant 16R1GSMLR1.

References

1. Vilalta, R., Drissi, Y.: A perspective view and survey of meta-learning. Artif. Intell. Rev. **18** (2002) 77–95
2. Schmidhuber, J., Zhao, J., Wiering, M.: Shifting inductive bias with success-story algorithm, adaptive Levin search, and incremental self-improvement. Machine Learning **28** (1997) 105–130 — Based on: Simple principles of metalearning. TR IDSIA-69-96, 1996.
3. Harick, G.R., Lobo, F.G.: A parameter-less genetic algorithm. In Banzhaf, W., Daida, J., Eiben, A.E., Garzon, M.H., Honavar, V., Jakiela, M., Smith, R.E., eds.: Proceedings of the Genetic and Evolutionary Computation Conference. Volume 2., Orlando, Florida, USA, Morgan Kaufmann (1999) 1867
4. Lagoudakis, M.G., Littman, M.L.: Algorithm selection using reinforcement learning. In: Proc. 17th International Conf. on Machine Learning, Morgan Kaufmann, San Francisco, CA (2000) 511–518
5. Horvitz, E., Ruan, Y., Gomes, C.P., Kautz, H.A., Selman, B., Chickering, D.M.: A bayesian approach to tackling hard computational problems. In: UAI '01: Proceedings of the 17th Conference in Uncertainty in Artificial Intelligence, San Francisco, CA, USA, Morgan Kaufmann Publishers Inc. (2001) 235–244
6. Gagliolo, M., Zhumatiy, V., Schmidhuber, J.: Adaptive online time allocation to search algorithms. In Boulicaut, J.F., Esposito, F., Giannotti, F., Pedreschi, D., eds.: Machine Learning: ECML 2004. Proceedings of the 15th European Conference on Machine Learning, Pisa, Italy, September 20-24, 2004, Springer (2004) 134–143 — Extended tech. report available at `http://www.idsia.ch/idsiareport/IDSIA-23-04.ps.gz`.
7. Fürnkranz, J., Petrak, J., Brazdil, P., Soares, C.: On the use of fast subsampling estimates for algorithm recommendation. Technical Report TR-2002-36, Österreichisches Forschungsinstitut für Artificial Intelligence, Wien (2002)
8. Gomes, C.P., Selman, B.: Algorithm portfolios. Artificial Intelligence **126** (2001) 43–62
9. Nelson, W.: Applied Life Data Analysis. John Wiley, New York (1982)
10. Bishop, C.M.: Neural networks for pattern recognition. Oxford University Press (1995)
11. Holland, J.H.: Adaptation in Natural and Artificial Systems. University of Michigan Press, Ann Arbor (1975)
12. Jacobs, R.A., Jordan, M.I., Nowlan, S.J., Hinton, G.E.: Adaptive mixtures of local experts. Neural Computation **3** (1991) 79–87

Discriminant Parallel Perceptrons

Ana González, Iván Cantador, and José R. Dorronsoro*

Depto. de Ingeniería Informática and Instituto de Ingeniería del Conocimiento, Universidad Autónoma de Madrid, 28049 Madrid, Spain

Abstract. Parallel perceptrons (PPs), a novel approach to committee machine training requiring minimal communication between outputs and hidden units, allows the construction of efficient and stable nonlinear classifiers. In this work we shall explore how to improve their performance allowing their output weights to have real values, computed by applying Fisher's linear discriminant analysis to the committee machine's perceptron outputs. We shall see that the final performance of the resulting classifiers is comparable to that of the more complex and costlier to train multilayer perceptrons.

1 Introduction

After their heyday in the early sixties, interest in machines made up of Rosenblat's perceptrons greatly decayed. The main reason for this was the lack of suitable training methods: even if perceptron combinations could provide complex decision boundaries, there were not efficient and robust procedures for constructing them. An example of this are the well known committe machines (CM; [4], chapter 6) for 2–class classification problems. They are made up of an odd number H of standard perceptrons, the output of the i–th perceptron $P_i(X)$ over a D–dimensional input pattern X being given by $P_i(X) = s(act_i(X))$ (we assume $x_D = 1$ for bias purposes). Here $s(\cdot)$ denotes the sign function and $act_i(X) = W_i \cdot X$ is the X activation of P_i. The CM output is then $h(X) = s\left(\sum_{i=1}^{H} P_i(X)\right) = s\left(\mathcal{V}(X)\right)$, i.e., the sign of the overall perceptron vote count $\mathcal{V}(X)$. Assuming that each X has a class label $y_X = \pm 1$, X is correctly classified if $y_X h(X) = 1$. If not, CM training applies Rosenblat's rule

$$W_i := W_i + \eta \ y_X X \tag{1}$$

to the smallest number of incorrect perceptrons (this number is $(1 + |\mathcal{V}(X)|)/2$); moreover, this is done for those incorrect perceptrons for which $|act_i(X)|$ is smallest. Although sensible, this training is somewhat unstable and only able to build not too strong classifiers. A simple but powerful variant of classical CM training, the so–called parallel perceptrons (PPs), recently introduced by Auer et al. in [1], allows a very fast construction of more powerful classifiers, with capabilities close to the more complex (and costlier to train) multilayer

* With partial support of Spain's CICyT, projects TIC 01–572, TIN2004–07676.

W. Duch et al. (Eds.): ICANN 2005, LNCS 3697, pp. 13–18, 2005.

perceptrons (MLPs). In PP training, (1) is applied to all wrong perceptrons but the PP key training ingredient is an output stabilization procedure that tries to keep away from 0 the activation $act_i(X)$ of a correct P_i, so that small random changes on X do not cause its being assigned to another class. More precisely, when X is correctly classified, but for a given margin γ and a perceptron P_i we have $0 < y_X act_i(X) < \gamma$, Rosenblatt's rule is essentially again applied in order to push $y_X act_i(X)$ further away from zero. The value of the margin γ is also adjusted dynamically so that most of the correctly classified patterns have activation margins greater than the final γ^* (see section 2). In spite of their very simple structure, PPs do have a universal approximation property and, as shown in [1], provide results in classification and regression problems quite close to those offered by C4.5 decision trees or MLPs.

There is much work being done in computational learning theory to build efficient classifiers based on low complexity information processing methods. This is particularly important for high dimensionality problems, such as those arising in text mining or bioinformatics. As just mentioned, PPs combine simple processing with good performance. A natural way to try to get a richer behavior is to relax their clamping of output weights to 1, allowing these weights to have real values. In fact, usually PP performance does not depend on the number of perceptrons used, 3 being typically good enough. For classification problems, a natural option, that we shall explore in this work, is to use standar linear discriminant analysis to do so. We shall briefly describe in section 2 the training of these discriminant PPs as well as their handling of margins, while in section 3 we will numerically analize their performance over several classification problems, comparing it to that of standard PPs and MLPs. As we shall see, discriminant PPs will give results somewhat better than those of standard PPs and essentially similar to those of MLPs.

2 Discriminant PPs

We discuss first perceptron weight and margin updates. Assume that a set $\mathcal{W} = (W_1, \ldots, W_H)$ of perceptron weights and of Fisher's weights $A = (a_1, \ldots, a_H)^t$ have been computed. The output hypothesis of the resulting discriminant PP is

$$h(X) = s\left(A \cdot (P(X) - \tilde{P})\right) = s\left(\sum_1^H a_i(P_i(X) - \tilde{P}_i)\right),$$

with $\tilde{P} = (\overline{P}_+ + \overline{P}_-)/2$ and $\overline{P}_\pm$ the averages of the perceptron outputs over the positive and negative classes. We assume that the sign of the A vector has been adjusted so that a pattern X is correctly classified if $y_X h(X) = 1$. Now,

$$|(\overline{P}_\pm)_i| \leq \frac{1}{N_\pm} \sum_{X' \in C_\pm} |P_i(X')| = 1.$$

with $N_\pm$ the sizes of the positive and negative classes $C_\pm$. We can expect in fact that $|(\overline{P}_\pm)_i| < 1$ and hence, $|\tilde{P}_i| < 1$ too. Therefore, $y_X a_i(P(X) - \tilde{P}) > 0$

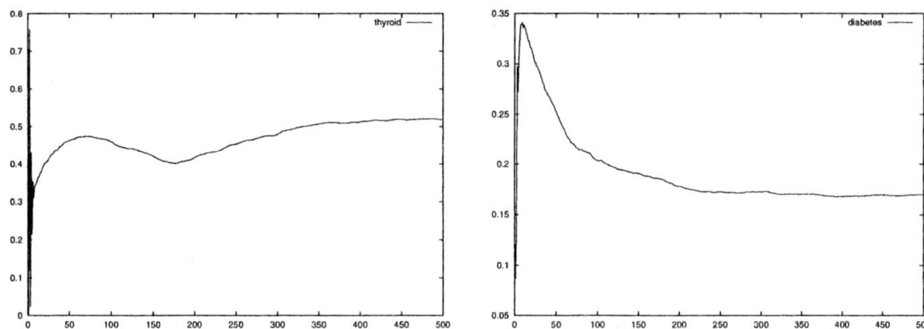

Fig. 1. Margin evolution for the thyroid (left) and diabetes datasets. Values depicted are 10 times 10 fold crossvalidation averages of 500 iteration training runs.

if and only if $y_X a_i P(X) > 0$, and if X is not correctly classified, we should augment $y_X a_i P_i(X)$ over those wrong perceptrons for which $y_X a_i P_i(X) < 0$. This is equivalent to augment $y_X a_i act_i(X) = y_X a_i W_i \cdot X$, which can be simply achieved by using again Rosenblatt's rule (1) adjusted in terms of A:

$$W_i := W_i + \eta s(y_X a_i) X, \tag{2}$$

for then we have

$$y_X a_i (W_i + \eta s(y_X a_i) X) \cdot X = y_X a_i W_i \cdot X + \eta |y_X a_i| |X|^2 > y_X a_i W_i \cdot X.$$

On the other hand, the margin stabilization of discriminant PPs is essentially that of standard PPs. More precisely, if X is correctly classified, $y_X a_i P_i(X) > 0$ and thus $s(y_X a_i) act_i(X) > 0$, which we want to remain > 0 after small X perturbations. For this we may again apply (2) now in the form $W_i := W_i + \lambda\ \eta\ s(y_X a_i) X$ to those correct perceptrons with a too small margin, i.e., those for which $0 < s(y_X a_i) act_i(X) < \gamma$, so that we push $s(y_X a_i) act_i(X)$ further away from zero. The new parameter λ measures the importance given to wide margins. The value of the margin γ is also adjusted dynamically from a starting value γ_0. More precisely, at the beginning of the t–th batch pass, we set $\gamma_t = \gamma_{t-1}$; then, as suggested in [1], if a pattern X is processed correctly, we set $\gamma_t := \gamma_t + 0.25\eta_t$ if all perceptrons P_i that process X correctly also verify $s(y_X a_i) act_i(X) \geq \gamma_{t-1}$, while we set $\gamma_t := \gamma_t - 0.75\eta_t$ if for at least one P_i we have $0 < s(y_X a_i) act_i(X) < \gamma_{t-1}$. In other words, margins are either cautiously increased or substantially decreased. With these (and in fact, other similar) parameter values, γ_t usually has a stable converge to a limit margin γ^* (see figure 1). We normalize the W_i weights after each batch pass so that the margin is meaningful. We also adjust the learning rate as $\eta_t = \eta_0/\sqrt{t}$ after each batch pass, as suggested in [1].

We recall that for 2–class problems, Fisher's discriminants are very simple to construct. In fact, the vector $A = S_T^{-1}(\overline{P}_+ - \overline{P}_-)$ minimizes [2] the ratio $J = s_T/s_B = s_T(A)/s_B(A)$ of the total variance s_T of discriminant PP outputs to their between class variance s_B. However, the total covariance matrix S_T of

Table 1. Input dimensions and training parameters used for the 7 comparison datasets. MLPs were trained by conjugate gradient minimization. The number of hidden units and learning rates were heuristically adjusted to give good training accuracies.

			discr. PPs		PPs		MLPs
Problem set	size pos. %	input dim.	num. hid.	lr. rate	num. hid.	lr. rate	num. hid.
breast cancer	34.5	9	5	0.001	3	0.001	5
diabetes	34.9	7	5	0.01	3	0.001	5
glass	13.6	9	5	0.01	3	0.01	5
heart dis.	46.1	13	5	0.001	5	0.001	5
ionosphere	35.9	33	5	0.001	5	0.0001	7
thyroid	7.4	8	5	0.0005	5	0.001	5
vehicle	25.7	18	10	0.01	5	0.001	5

the perceptrons' outputs is quite likely to be singular (notice that the output space for H perceptrons has just 2^H distinct values). To avoid this, we will take as the output of the perceptron i the value $P_i'(X) = \sigma_\gamma(W_i \cdot X)$, with the ramp function σ_γ taking the values $\sigma_\gamma(t) = s(t)$ if $|t| > \lambda = \min(1, 2\gamma)$ and $\sigma_\gamma(t) = t/\lambda$ when $|t| \leq \lambda$. This makes quite unlikely that S_T will be singular and together with the η and γ updates allows for a fast and quite stable learning convergence. We finally comment on the complexity of this procedure. For D–dimensional inputs and H perceptrons, Rosenblat's rule has an $O(NDH)$ cost. For its part, the S_T covariance matrix computation has an $O(NH^2)$ cost, that dominates the $O(H^3)$ cost of its inversion. While formally similar to the complexity estimates of MLPs, computing times are much smaller for discriminant PPs (and more so for standard PPs), as their weight updates are much simpler. Moreover, as training advances and the number of patterns not classified correctly decreases, so does the number of updates.

3 Numerical Results

We shall compare the performance of discriminant PPs with that of standard PPs and also of multilayer perceptrons (MLPs) over 7 classification problems sets from the well known UCI database; they are listed in table 1, together with the positive class size, their input dimensions and the training parameters used. Some of them (glass, vehicle, thyroid) are multi–class problems; to reduce them to 2–class problems, we are taking as the minority classes the class 1 in the vehicle dataset and the class 7 in the glass problem, and merge in a single class both sick thyroid classes. We refer to the UCI database documentation [3] for more details. In what follows we shall compare the performance of standard and discriminant PPs and also that of standard multilayer perceptrons first in terms of accuracy, that is, the percentage of correctly classified patterns, but also in terms of the value $g = \sqrt{a^+a^-}$, where $a^\pm$ are the accuracies of the positive and negative classes (see [5]). Notice that for sample imbalanced data sets a high

Table 2. Accuracy, g test values and their standard deviations for 7 datasets and different classifier construction procedures. It can be seen that discriminant PPs results are comparable to those of MLPs and both are better than those of standard PPs.

	discr. PPs		PPs		MLPs	
Data set	acc.	g	acc.	g	acc.	g
cancer	96.5 ± 2.2	*96.1 ± 2.2*	**96.6 ± 2.1**	**96.1 ± 2.2**	95.8 ± 1.7	95.5 ± 1.7
diabetes	*75.0 ± 2.4*	**71.9 ± 4.0**	74.2 ± 3.2	68.6 ± 5.3	**76.0 ± 3.1**	*70.4 ± 4.3*
glass	96.9 ± 2.1	**92.1 ± 11.0**	*94.3 ± 2.1*	84.3 ± 11.1	94.0 ± 2.9	*85.3 ± 8.5*
heart d.	**80.0 ± 3.8**	**78.9 ± 3.9**	73.9 ± 3.8	73.8 ± 3.9	*75.2 ± 4.0*	*74.7 ± 4.2*
ionosph.	*84.1 ± 4.6*	**82.2 ± 4.5**	77.0 ± 3.9	74.3 ± 4.1	**84.8 ± 4.2**	*81.4 ± 4.5*
thyroid	**97.9 ± 0.4**	*92.1 ± 1.8*	96.9 ± 0.9	82.5 ± 9.5	*97.6 ± 1.6*	**94.0 ± 4.4**
vehicle	*76.2 ± 0.4*	*70.6 ± 3.5*	74.8 ± 2.5	65.0 ± 5.0	**81.5 ± 4.3**	**74.5 ± 5.8**

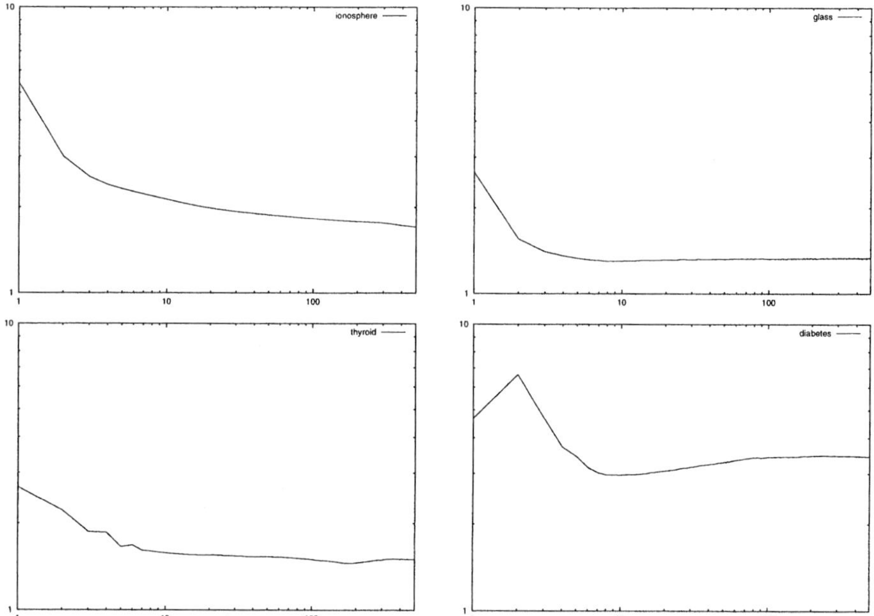

Fig. 2. From top left, clockwise: evolution of Fisher's criterion for the ionosphere, glass, diabetes and thyroid datasets. Values depicted are 10 times 10 fold crossvalidation averages of 500 iteration training runs (all figures in log X and Y scale).

accuracy could be achieved simply by assigning all patterns to the (possibly much larger) negative classes; g gives a more balanced classification performance measure. In all cases, training has been carried out as a batch procedure using 10–times 10–fold cross–validation. Updates (1) and (2) have been applied in standard and discriminant PP training, while conjugate gradient has been used for MLPs. The number of perceptrons in all cases and the initial learning rates for PPs and discriminant PPs for each dataset are described in table 1. Table 2 presents average values of the cross–validation procedure just described (best values in bold face, second best in cursive) for accuracies and g values, together with their standard deviation. As it can be seen, MLPs' accuracy is clearly best in 2 problems, while discriminant PPs give a clear best accuracy in another 2 problems and tie for the best values in the remaining 3. With respecto to g, discriminant PPs' values are clearly best in 4 problems and are tied for the first place in another one; MLPs' g is best in the other 2 problems. The performance of discriminant PPs is thus quite close to that of MLPs and improves on that of standard PPs.

We finish this section by noticing that the weight update (2) only aims to reduce the classification error and to achieve a clear margin, but there is no reason that it should minimize the Fisher criterion J. However, as seen in figure 2, this also happens. The figure depicts in logartihmic X and Y scales the evolution of J for the ionosphere, glass, diabetes and thyroid datasets (clockwise from top left). Values depicted are 10 times 10 fold cross–validation averages of 500 iteration training runs. Although not always monotonic (as in the glass, thyroid and diabetes problems), the overall J behavior is clearly decreasing and it converges.

4 Conclusions

Parallel perceptron training offers a very fast procedure to build good and stable committee machine–like classifiers. In this work we have seen that their classification performance can be improved by allowing their output weights to have real values, obtained by applying Fisher's analysis over the perceptron outputs. The final performance of these discriminant PPs is essentially that of the powerful but costlier to build standard MLPs.

References

1. P. Auer, H. Burgsteiner, W. Maass, *Reducing Communication for Distributed Learning in Neural Networks*, Proceedings of ICANN'2002, Lecture Notes in Computer Science 2415 (2002), 123–128.
2. R. Duda, P. Hart, D. Stork, **Pattern classification** (second edition), Wiley, 2000.
3. P. Murphy, D. Aha, *UCI Repository of Machine Learning Databases*, Tech. Report, University of Califonia, Irvine, 1994.
4. N. Nilsson, **The Mathematical Foundations of Learning Machines**, Morgan Kaufmann, 1990.
5. J.A. Swets, *Measuring the accuracy of diagnostic systems*, Science 240 (1998), 1285–1293.

A Way to Aggregate Multilayer Neural Networks

Maciej Krawczak[1,2]

[1] Systems Research Institute, Polish Academy of Sciences,
Newelska 6, 01-447 Warsaw, Poland
[2] Warsaw School of Information Technology, Newelska 6, 01-447 Warsaw, Poland
krawczak@ibspan.waw.pl

Abstract. In this paper we consider an aggregation way for multilayer neural networks. For this we will use the generalized nets methodology as well as the index matrix operators. The generalized net methodology was developed as a counterpart of Petri nets for modelling discrete event systems. First, a short introduction of these tools is given. Next, three different kinds of neurons aggregation is considered. The application of the index matrix operators allow to developed three different generalized net models. The methodology seems to be a very good tool for knowledge description.

1 Introduction

A multilayer neural network is described by layers and within each layer by a number of neurons. The neurons are connected by weighted links. The network output is strictly related to the presented input, subject to the conditions resulting from the constancy of the structure (the neuron connections), the activation functions as well as the weights.

A neural network can be considered with different degree of aggregation. A case without aggregation means that all layers and all neurons are “visible” for consideration. A second case where neurons within each layer are aggregated and only layers are “visible”. The next case, in which all layers are aggregated, and only inputs and outputs of a neural network are available.

In order to develop these three kinds of aggregation we will apply the methodology of *generalized nets* introduced by K. Atanassov in various works, e.g. [1], [2], [3], [4] and [5].

The generalized nets methodology is defined as an extension of the ordinary Petri nets and their modifications, but in a different way, namely the relation of places, transitions and characteristics of tokens provide for greater modelling possibilities than the individual types of Petri nets.

In the review and bibliography on generalized nets theory and applications of Radeva, Krawczak and Choy in [10] we can find a list 353 scientific works related to the generalized nets.

1.1 Multilayer Neural Networks Structure

The idea of the aggregation will be described in the following way. The network consists of L layers; each layer $l = 0,1,2,...,L$ is composed of $N(l)$ neurons. By $N(0)$

W. Duch et al. (Eds.): ICANN 2005, LNCS 3697, pp. 19 – 24, 2005.

we denote the number of inputs, while by $N(L)$ the number of outputs. The neurons are linked through weighted connections.

The output of the network is strictly related to the presented input, subject to the conditions resulting from the constancy of the structure, the activation functions as well as the weights. The neural network realizes the following mapping:

$$output = NN(input) \tag{1}$$

The neural network consists of neurons described by the activation function as follows

$$x_{pj(l)} = f\left(net_{pj(l)}\right) \tag{2}$$

where

$$net_{pj(l)} = \sum_{i=1}^{N(l-1)} w_{i(l-1)j(l)}\, x_{pi(l-1)} \tag{3}$$

while $x_{pi(l-1)}$ denotes the output of the i-th neuron with respect to the pattern p, $p = 1,2,...,P$, and the weight $w_{i(l-1)j(l)}$ connects the i-th neuron from the $(l-1)$-st layer with the j-th from the l-th layer, $j = 1,2,...,N(l)$, $l = 1,2,...,L$.

The different cases of aggregation determine different streams of information passing through the system.

1.2 Generalized Net Modeling

A generalized net contains *tokens*, which are transferred from place to place. Every token bears some information, which is described by token's *characteristic*, and any token enters the net with an *initial characteristic*. After passing a transition the tokens' characteristics are modified. The places are marked by ○ , and the transitions by | .

The transition has *input* and *output* places, as shown in Fig. 1.

The basic difference between generalized nets and the ordinary Petri nets is the *place – transition relation* (Atanassov, 1991).

Formally, every transition is described by a seven-tuple

$$Z = \langle L', L'', t_1, t_2, r, M, \square \rangle \tag{4}$$

where: $L' = \{l'_1, l'_2, \ldots, l'_m\}$ is a finite, non empty set of the transition's input places, $L'' = \{l''_1, l''_2, \ldots, l''_m\}$ is a finite, non empty set of the transition's output places, t_1 is the current time of the transition's firing, t_2 is the current duration of the transition active state, r is the transition's *condition* determining which tokens will pass the transition, M is an index matrix of the capacities of transition's arcs, $\square$ is an object of a form similar to a Boolean expression, for *true* value the transition becomes active.

The following ordered four-tuple

$$E = \left\langle \left\langle A, \pi_A, \pi_L, c, f, \Theta_1, \Theta_2 \right\rangle, \left\langle K, \pi_k, \Theta_K \right\rangle, \left\langle T, t^0, t^* \right\rangle, \left\langle X, \Phi, b \right\rangle \right\rangle \quad (5)$$

is called *generalized net* if the elements are described as follows: A is a set of transitions, π_A is a function yielding the priorities of the transitions, π_L is a function specifying the priorities of the places, c is a function providing the capacities of the places, f is a function that calculates the truth values of the predicates of the transition's conditions, Θ_1 is a function specifying the next time-moment when a given transition Z can be activated, Θ_2 is a function yielding the duration of the active state of a given transition Z, K is the set of the generalized net's tokens, π_K is a function specifying the priorities of the tokens, Θ_K is a function producing the time-moment when a given token can enter the net, T is the time-moment when the generalized net starts functioning; t^0 is an elementary time-step, related to the fixed (global) time-scale, t^* is the duration of the generalized net functioning, X is the set of all initial characteristics the tokens can receive on entering the net, Φ is a characteristic function that assigns new characteristics to every token when it makes the transfer from an input to an output place of a given transition, b is a function specifying the maximum number of characteristics a given token can receive.

The generalized nets with missing some components are called *reduced generalized nets*.

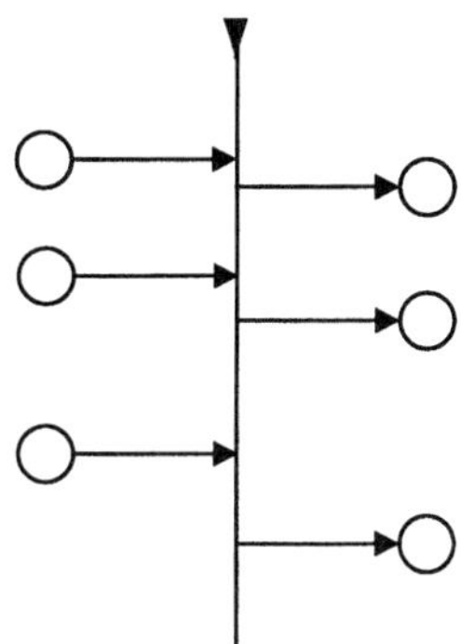

Fig. 1. A generalized net transition

2 Aggregation of MLNN

In this section we will introduce three kinds of MLNN aggregation. Due to the lack of space only the main results will be shown.

2.1 Modeling Without Aggregation

The Generalized net model of this case is shown in Fig. 2.

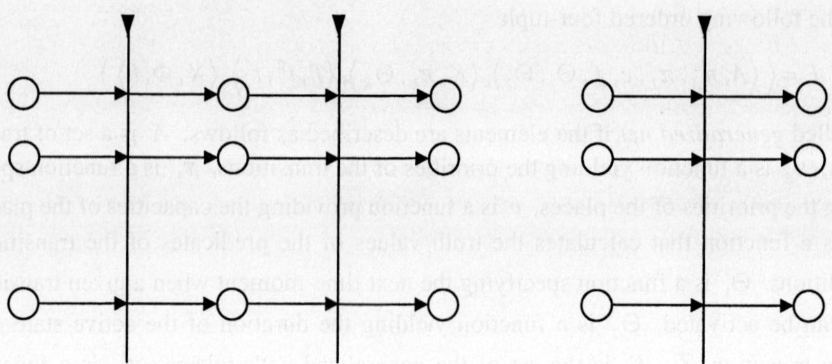

Fig. 2. The Generalized net model of neural network simulation

The model consists of a set of L transitions, each transition has the following form

$$Z_l = \left\langle \left\{ \ddot{x}_{1(l-1)}, \ddot{x}_{2(l-1)}, ..., \ddot{x}_{N(l-1)} \right\}, \left\{ \dddot{x}_{1(l)}, \dddot{x}_{2(l)}, ..., \dddot{x}_{N(l)} \right\}, \tau_l, \tau'_l, r_l, M_l, \Box_l \right\rangle \quad (6)$$

for $l = 1,2,...,L$, where $\left\{ \ddot{x}_{1(l-1)}, \ddot{x}_{2(l-1)}, ..., \ddot{x}_{N(l-1)} \right\}$ - is the set of input places of the l-th transition, $\left\{ \dddot{x}_{1(l)}, \dddot{x}_{2(l)}, ..., \dddot{x}_{N(l)} \right\}$ - is the set of output places of the l-th transition, the rest parameters are described above. The generalized net describing the considered neural network simulation process has the following form:

$$GN = \left\langle \left\langle A, \pi_A, \pi_X, c, g, \Theta_1, \Theta_2 \right\rangle, \left\langle K, \pi_k, \Theta_K \right\rangle, \left\langle T, t^0, t^* \right\rangle, \left\langle Y, \Phi, b \right\rangle \right\rangle \quad (7)$$

where $A = \{Z_1, Z_2, ..., Z_L\}$, and other parameters are similar to the parameters of section 1.2.

2.2 Aggregation of Neurons Within Each Layer

In this case we can obtain two version of aggregation. Here we show only one parallel version in Fig. 3.

Each transition Z_l, $l = 1,2,...,L$, has the following form

$$Z'_l = \left\langle \left\{ \ddot{X}_{(l-1)} \right\}, \left\{ \dddot{X}_{(l)} \right\}, \tau_l, \tau'_l, r_l, M_l, \Box_l \right\rangle \quad (8)$$

The reduced form of the generalized net has the following form:

$$GN = \left\langle \left\langle A, *, \pi_X, c, *, \Theta_1, \Theta_2 \right\rangle, \left\langle K, *, * \right\rangle, \left\langle T, *, t^* \right\rangle, \left\langle Y, \Phi, 1 \right\rangle \right\rangle \quad (9)$$

with the proper components.

In the second version information carrying by tokens is performed sequencially, while in the first one parallel. The number of tokens in the place $\ddot{X}_{(l-1)}$, $l = 1,2,...,L$, corresponds to the number of neurons associated with the l-th layer. In order to introduce a prescribed number of the tokens into each place we need to change the parallelism of signal flows by the sequential flows. In such a case the model of aggregation will become a little bit more complex through introduction of an extra place to each transition.

Version one is characterised by introduction of only one token but with much extended characteristic, due to this fact the information entering each transition is delivered in parallel. For the second version we have introduced the extra places with extra tokens. In this way the information of each token enters each transition in a sequential way.

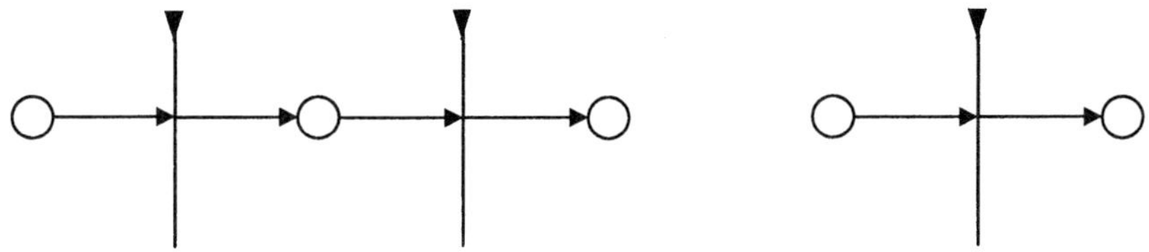

Fig. 3. Neurons aggregation within each layer – parallel processing

2.3 Full Aggregation

In this case we do not distinguish any subsystems (neurons or layers) within the network. We aggregate all transitions Z_l, $l = 1,2,...,L$, in only one transition Z, and all places are aggregated in only three places $\dddot{X}_1, \dddot{X}_2, \dddot{X}_3$ as shown in Fig. 4.

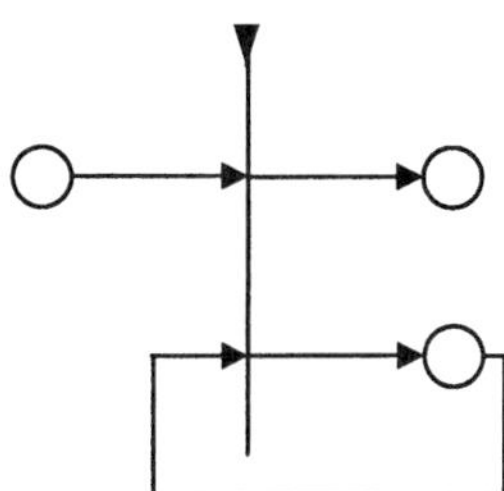

Fig. 4. The Generalized net model of the aggregated neural network

The aggregated model has the following formal description

$$GN = \left\langle \left\langle \{Z\}, *, \pi_X, c, g, \Theta_1, \Theta_2 \right\rangle, \left\langle K, *, \Theta_K \right\rangle, \left\langle T, *, t^* \right\rangle, \left\langle Y, \Phi, b \right\rangle \right\rangle \tag{10}$$

where the transition has the form

$$Z = \left\langle \left\{ \dddot{X}_1, \dddot{X}_3 \right\}, \left\{ \dddot{X}_2, \dddot{X}_3 \right\}, \tau, \tau', r, M, \square \right\rangle \tag{11}$$

3 Conclusions

In this paper we have described the concept of aggregation of multilayer neural networks. We used the generalized nets concept for representing the functioning of the multilayer neural networks.

In somehow informal way we have applied many of the sophisticated tools of the generalized nets theory in order to show different kinds of aggregation.

The Generalized nets theory, as described in [2] by Atanassov, contains many different operations and relations over the transitions, tokens as well as over the characteristics of tokens. The methodology can be used as one possible way to describe knowledge of different kinds of subsystems in order to describe the way of functioning of the whole system.

References

1. Atanassov, K.: On the Concept "Generalized net". AMSE Review. 1, 3 (1984) 39-48
2. Atanassov, K.: Generalized nets. World Scientific, Singapore, New Jersey, London (1991)
3. Atanassov, K. (ed.): Applications of Generalized Net. World Scientific, Singapore, New Jersey, London (1993)
4. Atanassov, K.: Generalized Nets and Systems Theory. „Prof. M. Drinov" Academic Publishing House, Sofia (1997)
5. Atanassov, K.: Generalized Nets in Artificial Intelligence. Vol. 1: Generalized nets and Expert Systems. „Prof. M. Drinov" Academic Publishing House, Sofia (1998)
6. Fine, T. L.: Feedforward Neural Network Methodology. Springer, New York (1999)
7. Hassoun, M. H.: Fundamentals of Artificial Neural Networks. MIT Press (1995)
8. Krawczak, M.: Neural Networks Learning as a Multiobjective Optimal Control Problem. Mathware and Soft Computing. 4, 3 (1997) 195-202.
9. Krawczak, M.: Multilayer Neural Systems and Generalized Net Models. Academic Press EXIT, Warsaw (2003)
10. Radeva, V., Krawczak, M., Choy, E.: Review and Bibliography on Generalized Nets Theory and Applications. Advanced Studies in Contemporary Mathematics. 4, 2 (2002) 173-199.

Generalized Net Models of MLNN Learning Algorithms

Maciej Krawczak[1,2]

[1] Systems Research Institute, Polish Academy of Sciences,
Newelska 6, 01-447 Warsaw, Poland
[2] Warsaw School of Information Technology,
Newelska 6, 01-447 Warsaw, Poland
krawczak@ibspan.waw.pl

Abstract. In this paper we consider generalized net models of learning algorithms for multilayer neural networks. Using the standard backpropagation algorithm we will construct it generalized net model. The methodology seems to be a very good tool for knowledge description of learning algorithms. Next, it will be shown that different learning algorithms have similar knowledge representation – it means very similar generalized net models. The generalized net methodology was developed as a counterpart of Petri nets for modelling discrete event systems. In Appendix, a short introduction is given.

1 Introduction

In the paper we are interested in modelling the learning processes of multilayer neural network using generalized net methodology. Such constructed models of neural networks can describe the changes of the weights between neurons. This model is somehow based on the generalized net model described in: [3], [4], and [5]. In the present case some of absent elements will be involved in the description of the proper characteristic function, which generates the new tokens characteristics. Additionally, the inputs of the neural network are treated also as tokens. Three main parts can be distinguished in the model of the neural network learning. The first part describes the process of simulation or propagation; in the second part the performance index of learning is introduced, while the third part describes the operations that are necessary to change the states of neurons (by changing the connections – i.e. weights).

2 Model of Backpropagation Algorithm

The generalized net model of backpropagation algorithm is shown in Fig. 1. Each neuron or a group of neurons are represented by a token of α-type. These tokens enter the net through the place $\dddot{X}_1$ and have the following initial characteristics $y(\alpha_{i(l)}) = \langle l, i, f_{i(l)} \rangle$ for $i = 1,2,\ldots,N(l)$, $l = 0,1,\ldots,L$, where i - the number of the token (neuron), $f_{1(l)}$ - is an activation function of the i-th neuron associated with the l-th layer of the neural network.

W. Duch et al. (Eds.): ICANN 2005, LNCS 3697, pp. 25 – 30, 2005.

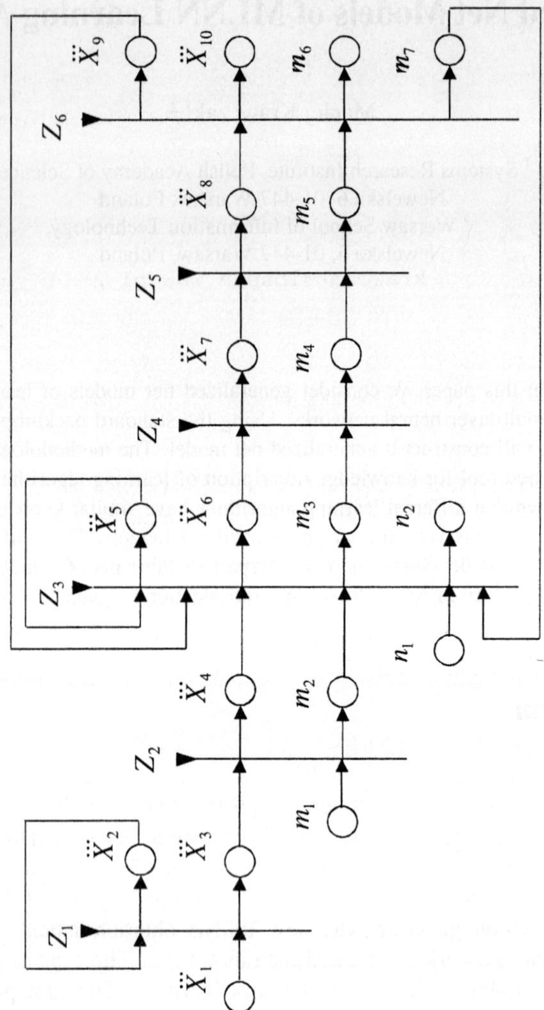

Fig. 1. The generalized net model of the backpropagation algorithm

The basic generalized net model of the backpropagation algorithm contains six transitions.

Every token $\alpha_{i(l)}$, $i = 1,2,...,N(l)$, $l = 0,1,...,L$, is transferred from the place $\dddot{X}_1$ to the place $\dddot{X}_2$ as well as $\dddot{X}_3$ via the transition Z_1. The tokens are transferred sequentially according to increasing indexes $i = 1,2,...,N(l)$ for given $l = 0,1,...,L$, in order to be aggregated with other tokens of the same level l into one new token $\alpha_{(l)}$, representing the whole layer l, according the following conditions of transition Z_1

$$Z_1 = \langle \{\dddot{X}_1, \dddot{X}_2\}, \{\dddot{X}_2, \dddot{X}_3\}, r_1, \vee(\dddot{X}_1, \dddot{X}_2) \rangle \tag{1}$$

where r_1 is the transition's *condition* determining which tokens will pass the transition. Here, we are interested in changing the characteristics of the neurons, therefore the whole neural network is represented by one transition Z_1 with three places $\dddot{X}_1, \dddot{X}_2, \dddot{X}_3$.

The new token $\alpha_{(l)}$ associated with the l-th layer according to the condition (5) is transferred from the place $\dddot{X}_2$ to the place $\dddot{X}_3$, and has the following characteristic for $l = 0,1,2,...,L$, where $[1, N(l)]$ - denotes $N(l)$ tokens (neurons) arranged in a sequence, starting form the first and ending at $N(l)$, associated with the l -th layer, $F_{(l)}$ - is a vector of the activation functions of the neurons associated with the l-th layer of the neural network. In result in the place $\dddot{X}_3$ we obtain L tokens, the representation of the procedure of generating the neural network output.

The second transition Z_2 is devoted to introduction of the performance index of the learning process. This kind of information is associated with the β-type token, which enters the input place m_1 with the following initial characteristic $y(\beta) = \langle E, E_{\max} \rangle$ where E - performance index, $E_{\max}$ - threshold value of the performance index, which must be reached. The transition Z_2 has the following form

$$Z_2 = \langle \{\dddot{X}_3, m_1\}, \{\dddot{X}_4, m_2\}, r_2, \wedge(\dddot{X}_3, m_1) \rangle \tag{2}$$

where r_2 is the transition's *condition* determining which tokens will pass the transition. The token $\alpha_{(l)}$ obtains the following new characteristic in the place $\dddot{X}_4$ $y(\alpha_{(l)}) = \langle l, [1, N(l)], F_{(l)}, \overline{W}_{(l)} \rangle$ for $l = 0,1,2,...,L$, where $[1, N(l)]$ - denotes $N(l)$ tokens (neurons) arranged in a sequence, starting form the first and ending at $N(l)$, associated with the l-th layer, $F_{(l)}$ - is a vector of the activation functions of the neurons associated with the l-th layer of the neural network, $\overline{W}_{(l)}$ - denotes the aggregated initial weights connecting the neurons of the $(l-1)$-st layer with the l-th layer neurons.

The β token obtains the following characteristic in place m_2 $y(\beta) = \langle 0, E_{\max} \rangle$.

The transition Z_3, in which the new tokens of γ-type are introduced. The token γ_p, $p = 1,2...,P$, where p is the number of the training pattern, enters the place n_1 with the initial characteristic $y(\gamma_p) = \langle X_p(0), D_p, p \rangle$, where $X_p(0)$ - the input vector, D_p - the vector of desired network outputs. After the pattern p is applied to the network inputs as $X_p(0)$, the outputs of all layers are calculated sequentially layer by layer. The transition Z_3 describes the process of signal propagation within the neural network

$$Z_3 = \langle \{\dddot{X}_4, \dddot{X}_5, \dddot{X}_9, m_2, m_7, n_1\}, \{\dddot{X}_5, \dddot{X}_6, m_3, n_2\}, r_3, \wedge(\vee(\dddot{X}_4, \dddot{X}_5, \dddot{X}_9), (m_2, m_7), n_1)\rangle \tag{3}$$

where r_2 are the transition's conditions.

In the place $\dddot{X}_5$ the tokens of α-type, $\alpha_{(l)}$, $l = 0,1,2,...,L$, obtain the new characteristics as follows $y(\alpha_{(l)}) = \langle\, l, [1, N(l)], F_{(l)}, W_{(l)}, X_{(l)} \rangle$, where $X_{(l)} = [x_{1(l)}, x_{2(l)}, ..., x_{N(l)}]^T$, $l = 1,2,...,L$, is the vector of outputs of neurons associated with the l-th layer, related to the nominal weights $W_{(l)}$, $l = 1,2,...,L$.

In the place $\dddot{X}_6$ there are tokens with the following characteristics, which contain calculated neuron outputs for the pattern p, $y(\alpha_{(l)}) = \langle l, [1, N(l)], F_{(l)}, \overline{W}_{(l)}, \overline{X}_{p(l)}\rangle$ calculated for the nominal values of the weights $\overline{W}_{(l)}$ and states $\overline{X}_{(l)}$, $l = 1,2,...,L$.

In the place m_3 the token β preserves its characteristic as $y(\beta) = \langle 0, E_{\max}\rangle$, and in the place n_2 the token γ also does not change its characteristic and remains as $y(\gamma_p) = \langle X_p(0), D_p, p\rangle$.

The next transition Z_4 describes the first stage of the estimation and weight adjustment process, which is related to the performance index computation, and has the following form

$$Z_4 = \langle \{\dddot{X}_6, m_3\}, \{\dddot{X}_7, m_4\}, r_4, \wedge(\dddot{X}_6, m_3)\rangle \tag{4}$$

As a result of computations performed within the transition Z_4, the token β obtains the new value of performance index in the place m_4, $y(\beta) = \langle E', E_{\max}\rangle$ where $E' = E + \frac{1}{2}\sum_{j=1}^{N(L)} (d_{p,j} - x_{p,j(L)})^2 = pr_2\langle E, E_{\max}\rangle + \frac{1}{2}\sum_{j=1}^{N(L)} (d_{p,j} - x_{p,j(L)})^2$.

In the place $\dddot{X}_7$ the tokens of α-type do not change their characteristics.

In the next transition

$$Z_5 = \langle \{\dddot{X}_7, m_4\}, \{\dddot{X}_8, m_5\}, r_5, \wedge(\dddot{X}_7, m_4)\rangle \tag{5}$$

the delta factors $\Delta_{p(l)} = [\delta_{p1(l)}, \delta_{p2(l)}, ..., \delta_{pN(l)}]^T$, in backpropagation algorithm are computed for $l = 0,1,2,...,L$.

The tokens of α-type obtain, in the place $\dddot{X}_8$, the following characteristics $y(\alpha_{(l)}) = \langle l, [1, N(l)], F_{(l)}, \overline{W}_{(l)}, \overline{X}_{p(l)}, \Delta_{p(l)}\rangle$. In the place m_5 the token of β-type does not change its characteristic.

The next transition Z_6, describing the process of weight adjustment, has the form

$$Z_6 = \langle \{\dddot{X}_8, m_5\}, \{\dddot{X}_9, \dddot{X}_{10}, m_6, m_7\}, r_6, \wedge(\dddot{X}_8, m_5)\rangle \tag{6}$$

In the place $\dddot{X}_9$ the α-type tokens obtain the new characteristic $y(\alpha_{(l)}) = \langle NN1, l, [1, N(l)], F_{(l)}, W'_{(l)}\rangle$ with updated weight connections

$W'_{(l)} = \left[w'_{1(l)}, w'_{2(l)}, \ldots, w'_{N(l)}\right]^T$, where $w'_{i(l)} = \left[w'_{i(l-1)1(l)}, w'_{i(l-1)2(l)}, \ldots, w'_{i(l-1)N(l)}\right]^T$ are calculated in the following way $\Delta w_{pi(L-1)j(L)} = w'_{pi(L-1)j(L)} - \overline{w}_{pi(L-1)j(L)} = \eta\, \delta_{pj(L)}\, x_{pi(L-1)}$ for $i = 1,2,\ldots,N(L-1)$, $j = 1,2,\ldots,N(L)$,

$$\Delta w_{pi(l-1)j(l)} = w'_{pi(l-1)j(l)} - \overline{w}_{pi(l-1)j(l)} = \eta\, f'_p\left(net_{j(l)}\right) x_{pi(l-1)} \sum_{k(l+1)=1}^{N(l+1)} \delta_{pk(l+1}\, w_{pj(l)k(l+1)}$$

for $i = 1,2,\ldots,N(l-1)$, $j = 1,2,\ldots,N(l)$, and for $l = 1,2,\ldots,L-1$, and replace $\overline{W}[0, L-1] = W'[0, L-1]$, $\overline{X}[1, L] = X'[1, L]$.

In the place m_7 the β token obtains the characteristic $y(\beta) = \langle E, E_{\max} \rangle$, which is not final.

The final values of the weights satisfying the predefined stop condition are denoted by $W^*_{(l)} = pr_5 \langle l, [1, N(l)], F_{(l)}, W'_{(l)} \rangle$, where the characteristics of the α -type tokens in the place $\ddot{X}_{10}$ are described by $y(\alpha_{(l)}) = \langle l, [1, N(l)], F_{(l)}, W'_{(l)} \rangle$ and the β token characteristic in the place m_6 is described by $y(\beta) = \langle NN1, E', E_{\max} \rangle$ while the final value of the performance index is equal $E^* = pr_2 \langle NN1, E', E_{\max} \rangle$.

3 Conclusions

The here developed generalized net model of the backpropagation algorithm describes the main features of the gradient descent based learning algorithms. This model allows for modifying and testing other algorithms by changing a relatively small portion of the generalized net model's formal description.

References

1. Atanassov, K.: Generalized nets. World Scientific, Singapore, New Jersey, London (1991)
2. Atanassov, K. (ed.): Applications of Generalized Net. World Scientific, Singapore, New Jersey, London (1993)
3. Atanassov, K.: Generalized Nets in Artificial Intelligence. Vol. 1: Generalized nets and Expert Systems. „Prof. M. Drinov" Academic Publishing House, Sofia (1998)
4. Krawczak, M.: Multilayer Neural Systems and Generalized Net Models. Academic Press EXIT, Warsaw (2003)
5. Radeva, V., Krawczak, M., Choy, E.: Review and Bibliography on Generalized Nets Theory and Applications. *Advanced Studies in Contemporary Mathematics,* 4, 2 (2002) 173-199.

Appendix

A generalized net contains *tokens*, which are transferred from place to place. Every token bears some information, which is described by token's *characteristic*, and any token enters the net with an *initial characteristic*. After passing a transition the tokens' characteristics are modified. The places are marked by ○, and the transitions by |.

The transition has *input* and *output* places, as shown in Fig. 1. The basic difference between generalized nets and the ordinary Petri nets is the *place – transition relation* (Atanassov, 1991).

Formally, every transition is described by a seven-tuple $Z=\langle L', L'', t_1, t_2, r, M, \square \rangle$ where: $L'=\{l'_1, l'_2, \ldots, l'_m\}$ is a finite, non empty set of the transition's input places, $L''=\{l''_1, l''_2, \ldots, l''_m\}$ is a finite, non empty set of the transition's output places, t_1 is the current time of the transition's firing, t_2 is the current duration of the transition active state, r is the transition's *condition* determining which tokens will pass the transition, M is an index matrix of the capacities of transition's arcs, $\square$ is an object of a form similar to a Boolean expression, for *true* value the transition becomes active.

The following ordered four-tuple:

$E=\langle \langle A, \pi_A, \pi_L, c, f, \Theta_1, \Theta_2 \rangle, \langle K, \pi_k, \Theta_K \rangle, \langle T, t^0, t^* \rangle, \langle X, \Phi, b \rangle \rangle$ is called *generalized net* if the elements are described as follows: A is a set of transitions, π_A is a function yielding the priorities of the transitions, π_L is a function specifying the priorities of the places, c is a function providing the capacities of the places, f is a function that calculates the truth values of the predicates of the transition's conditions, Θ_1 is a function specifying the next time-moment when a given transition Z can be activated, Θ_2 is a function yielding the duration of the active state of a given transition Z, K is the set of the generalized net's tokens, π_K is a function specifying the priorities of the tokens, Θ_K is a function producing the time-moment when a given token can enter the net, T is the time-moment when the generalized net starts functioning; t^0 is an elementary time-step, related to the fixed (global) time-scale, t^* is the duration of the generalized net functioning, X is the set of all initial characteristics the tokens can receive on entering the net, Φ is a characteristic function that assigns new characteristics to every token when it makes the transfer from an input to an output place of a given transition, b is a function specifying the maximum number of characteristics a given token can receive.

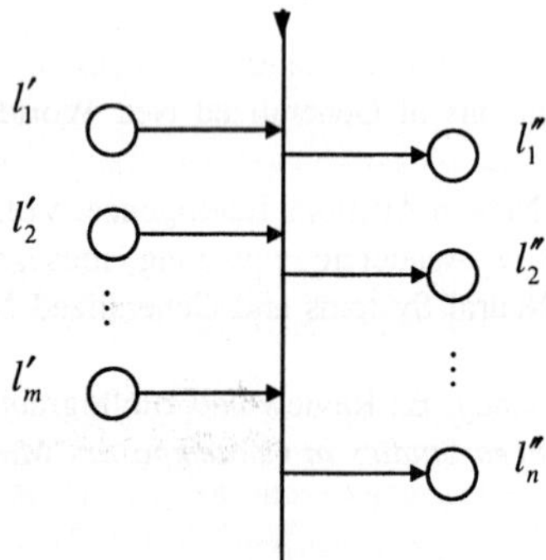

Fig. 2. A generalized net transition

Monotonic Multi-layer Perceptron Networks as Universal Approximators

Bernhard Lang

Siemens AG, Corporate Technology,
Otto-Hahn-Ring 6, D-81730 Munich, Germany
Bernhard.Lang@siemens.com

Abstract. Multi-layer perceptron networks as universal approximators are well-known methods for system identification. For many applications a multi-dimensional mathematical model has to guarantee the monotonicity with respect to one or more inputs. We introduce the MONMLP which fulfils the requirements of monotonicity regarding one or more inputs by constraints in the signs of the weights of the multi-layer perceptron network. The monotonicity of the MONMLP does not depend on the quality of the training because it is guaranteed by its structure. Moreover, it is shown that in spite of its constraints in signs the MONMLP is a universal approximator. As an example for model predictive control we present an application in the steel industry.

1 Introduction

Monotonicity in neural networks is a research topic for several years [1] [2] [3] [4] [5]. Numerous applications in system identification and control require monotonicity for selected input-output relations. The challenge is to fulfil two goals simultaneously: Good approximation properties together with monotonicity for certain inputs. Methods with penalty terms in cost functions are not able to guarantee monotonicity. Piece-wise linear networks like the monotonic network of Sill [4] ensure universal approximation capability and monotonicity for all inputs, but not for some selected ones and their training seems to be demanding. The monotonic multi-layer perceptron network (MONMLP) which is presented here is an approach for multi-dimensional function approximation ensuring monotonicity for selected input-output relations [6] [7][1]. Moreover, we determine the requirements for the network structure regarding universal approximation capabilities.

2 The MONMLP

A fully connected multi-layer perceptron network (MLP) with I inputs, a first hidden layer with H nodes, a second hidden layer with L nodes and a single output is defined by

[1] H. Zhang and Z. Zhang independently introduced the same idea, with a slightly different network structure. We thank the unknown reviewer for his valuable hint.

W. Duch et al. (Eds.): ICANN 2005, LNCS 3697, pp. 31–37, 2005.

$$\hat{y}(\mathbf{x}) = w_b + \underbrace{\sum_{l=1}^{L} w_l \tanh\big(w_{b,l} + \underbrace{\sum_{h=1}^{H} w_{lh} \tanh(\underbrace{w_{b,h} + \sum_{i=1}^{I} w_{hi} x_i}_{\theta_2}))}_{y_{MLP3}}}_{\theta_1} . \tag{1}$$

The MLP ensures a monotonically increasing behavior with respect to the input $x_j \in \mathbf{x}$, if

$$\frac{\partial \hat{y}}{\partial x_j} = \sum_{l=1}^{L} w_l \cdot \underbrace{(1-\theta_1^2)}_{>0} \cdot \sum_{h=1}^{H} w_{lh} \cdot \underbrace{(1-\theta_2^2)}_{>0} \cdot w_{hj} \overset{!}{\geq} 0 \tag{2}$$

The derivative of a hyperbolic tangent is always positive. For this reason, a sufficient condition for a monotonically increasing behavior for the input dimension j (2) is defined as

$$w_l \cdot w_{lh} \cdot w_{hj} \geq 0 \quad \forall \quad l, h \ . \tag{3}$$

To guarantee a monotonically decreasing relation between x_j and $\hat{y}$ we postulate

$$w_l \cdot w_{lh} \cdot w_{hj} \leq 0 \quad \forall \quad l, h \ . \tag{4}$$

Assuming positive signs for the weights between the hidden and output layer and between the hidden layers

$$\boxed{w_l, w_{lh} \overset{!}{>} 0 \quad \forall \quad l, h} \tag{5}$$

simplifies the sufficient conditions for increasing or decreasing monotonicity. Now, constraints on the weights w_{hj} provide sufficient conditions for the monotonicity related to the input dimension j. If all weights related to the input dimension j are greater equal zero then an increasing behavior is guaranteed (6).

$$\boxed{w_{hj} \overset{!}{\geq} 0 \quad \forall \quad h} \tag{6}$$

If all weights w_{hj} are greater than zero then the MLP is strictly increasing with respect to input j. Monotonic decreasing behavior is sufficiently provided by

$$\boxed{w_{hj} \overset{!}{\leq} 0 \quad \forall \quad h} \tag{7}$$

Negative signs for all w_{hj} provide a strictly decreasing behavior. The equations (6) or (7) determine the monotonicity of $\hat{y}$ with respect to one input x_j independently from the other inputs x_i. Applying the constraints for the training of the MONMLP the user is able to define a priori the input dimensions for which he would like to ensure the monotonicity.

3 The MONMLP as Universal Approximator

An MLP is known as a universal approximator [8]. However, the constraints defined above reduce the degrees of freedom for the weights so that the capability to approximate arbitrary, partially monotonic functions has to be shown again.

Firstly, the impact of the conditions defined in (5) is analyzed. The four-layer feed-forward network in (1) is an extension of a three-layer standard MLP y_{MLP3}. Since the three-layer topology is already sufficient for an universal approximator, the extension by a monotonic second hidden layer to a four-layer network respectively additional calculations by hyperbolic tangents and the multiplications with positive weights w_l do not affect this property. Limitations in the sign of the weights w_{lh} can be eliminated by appropriate weights w_{hi} since $\tanh(x) = -\tanh(-x)$. To sum up, a four-layer feed-forward network under the constraints in equation (5) continues to be a universal approximator.

Secondly, the influences of the constraints determined by (6) or (7) are analyzed. The proof of universal approximation is carried out for positive weights w_l, w_{lh}, w_{hi} respectively strictly monotonically increasing networks for all input dimensions i. A monotonic function $m(\mathbf{x})$, $\mathbb{R}^I \rightarrow \mathbb{R}$ [4] can be approximated by a sequence of adjacent, compact areas m_k with $m_k(\mathbf{x}) = const.$, $m_{k-1} < m_k < m_{k+1}$ which are only once cut by hyper-planes $p(\mathbf{x})$ of the form $x_j = const. \forall x_i \neq x_j;\ x_i, x_j \in \mathbb{R}$. In case of steps $m_k - m_{k-1} \rightarrow 0 \,\forall\, k$ an arbitrary approximation accuracy can be achieved. Figure 1 provides an illustration of this description of monotonic functions in case of two inputs x_1 and x_2. Areas m_k, i.e. plateaus and level curves are only once cut by the straight lines $x_1 = const. \forall x_2$ and $x_2 = const. \forall x_1$. The level curves as borders of plateaus consist of piece-wise convex and concave curves.

Figure 2 depicts the case of a single input and a single output. A hyperbolic tangent determines a transition of one plateau to the next one. The maximum gradient is determined by $w_{lh} \cdot w_{hi}$ and the amplitude by $2w_{lh}$. $w_{b,h}/w_{hi}$ and $w_{b,h}$ define the position of a single hyperbolic tangent in the x-y-plane. Adding

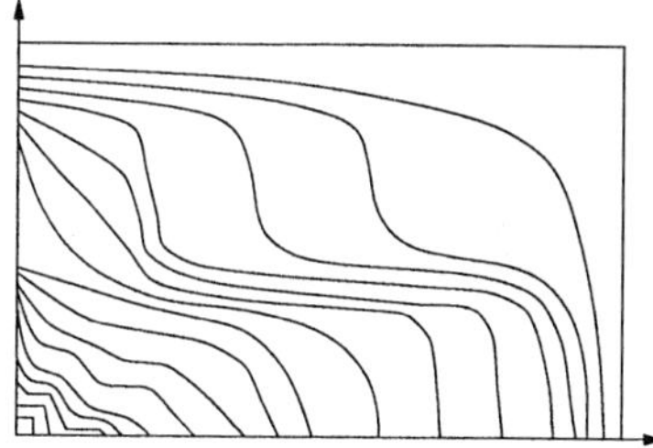

Fig. 1. Level curves for multi-dimensional monotonic functions, here for two inputs

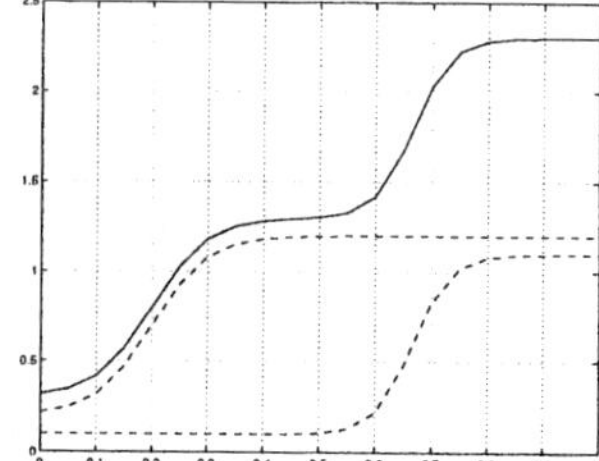

Fig. 2. Approximation of one-dim. monotonic functions by superposition of sigmoids

several hidden nodes corresponds to the superposition of arbitrary increasing transitions among arbitrary plateaus which is another definition of arbitrary increasing monotonic functions.

Considering the multi-dimensional case, a standard three-layer feed-forward network with positive weights (see Fig. 3(a)) provides a convex level curve as border for the lower plateau and a convex level curve as border for the upper plateau. This behavior is due to the addition of two arbitrary hyperbolic tangent functions with positive gradients. Concave level curves, which are also necessary for a universal approximation capability, cannot be created. However, if the three-layer MONMLP is extended by a saturation layer to a four-layer MONMLP then also concave level curves are possible for the lower and upper plateau border (Fig. 3(b) and 3(c)). The output range of this network limited by a single node in the second hidden layer is defined by the maximum amplitude $2w_{lh}$ and the bias w_b. Extending the four-layer MONMLP to more than two hyperbolic tangent in the first hidden layer, arbitrary level curves can be provided (Fig. 4(b)) whereas a three-layer MONMLP still provides convex lower and upper plateaus (Fig. 4(a)). The idea of providing arbitrary level curves by a saturation layer can be applied for more than two inputs: The maximum number of nodes in the first hidden layer

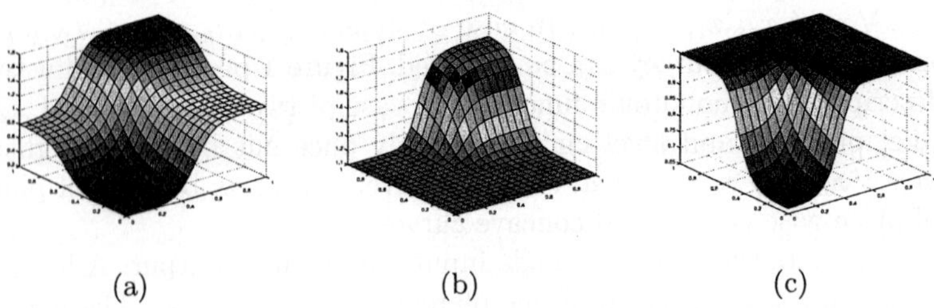

(a) (b) (c)

Fig. 3. Effect of two hidden layers: (a) Addition of two sigmoids for two input dimensions (b) Extension of the network of (a) to a four-layer MONMLP (2 + 1 hidden nodes) (c) The same four-layer MONMLP, but other weights for the saturation node

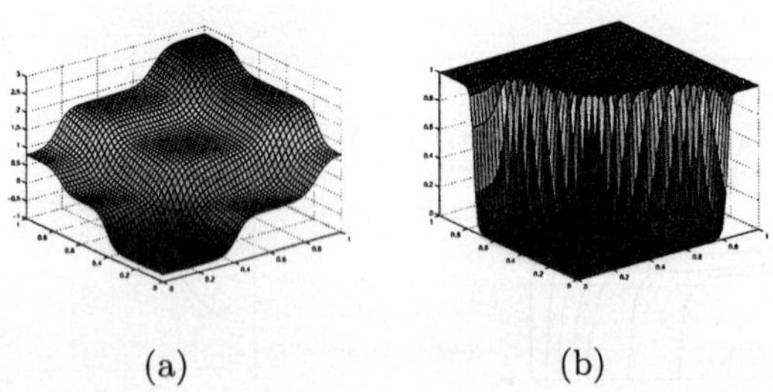

(a) (b)

Fig. 4. Transition between arbitrary plateaus: (a) Three-layer perceptron network with 4 hidden nodes (b) Network of (a) extended to a four-layer MONMLP (4 + 1 hidden nodes)

which are necessary to create a concave or convex line is determined by the input dimension of the network. To sum up, a four-layer MONMLP is able to provide arbitrary monotonic transitions from one plateau to the next one. Extending a four-layer MONMLP by additional hidden nodes enables the concatenation of these arbitrary transitions to an arbitrary monotonic function qed.

Monotonically decreasing behavior for one input x_j can be achieved by multiplying this input by -1 or by applying the constraints of equation (7). If the constraints on the w_{hi} are omitted for some input dimensions $i \neq j$, the relation between these inputs and the output $\hat{y}(\mathbf{x})$ is equivalent to an universal approximator without monotonicity constraints. Any combination with constrained parts of the MONMLP does not affect it, so that the MONMLP is a universal approximator for which a priori monotonic behavior with respect to inputs x_j can be determined.

4 Application in the Steel Industry

Each strip of steel in a hot rolling mill must satisfy customer quality requirements. The analysis of strip properties, however, can only be carried out in the laboratory after the processing. The microstructure monitor of Siemens [9] is able to predict the mechanical properties such as tensile strength or yield point by neural networks. The neural networks predict the expected quality taking into account the current circumstances (alloy components, temperatures, forces, geometry and so on).

In order to not only monitor but also to control such mechanical properties, the model should meet some general rules like 'if the expected quality should be increased then input 1 could be decreased or input 2 could be increased'. We selected a data set of 2429 patterns with 15 inputs and the tensile strength R_m as target for the evaluation of the MONMLP. The following prior knowledge was incorporated into the neural network: An increase in 'carbon content' or 'manganese content' should increase the tensile strength whereas an increase in 'temperature before cooling' should decrease the tensile strength. MONMLPs with and without second hidden layer were trained taking into account the monotonicity constraints. As a benchmark a standard three-layer MLP was selected with the same number of nodes as in the first hidden layer of the MONMLPs. The first 70 percent of the patterns were used for the training and the remaining 30 percent for the validation of the training results.

Table 1. Comparison of MLP and MONMLPs: root-mean-square of approximation errors in normalized units for training and test data

Neural network	MLP	MONMLP		
Nodes in 2nd hidden layer	–	–	1	2
RMSE for training data (normalized units)	84.5	94.3	84.9	84.7
RMSE for test data (normalized units)	100.0	100.4	94.9	93.0

The MONMLP [illegible] the monotonicity [illegible] e.g. [illegible] or noise. Contrary to the MONMLPs, the analysis of the gradients [illegible] for all patterns x did not provide any hint for monotonicity with respect to any input. The question arises if the constraints for the network weights would lead to reduced approximation capabilities. Comparing the approximation performance of the MLP and MONMLPs in Tab. 1 three layers seem to be not sufficient for a MONMLP in case of more than one monotonic input-output relation, but the approximation performance of four-layer MONMLPs is promising. These MONMLPs achieve similar results on the training data and outperform on the test data. The main advantage of the MONMLP, however, is the guaranteed interpolation and extrapolation behavior for regions without or with insufficient training data.

5 Conclusions

The MONMLP is a universal function approximator while ensuring monotonic [illegible] input-output relations [illegible] the MONMLP [illegible] the separation [illegible] not reduce the general approximation capability if a four layer MLP is chosen. Moreover, the reliable incorporation of prior knowledge in MONMLPs can improve the approximation quality in case of insufficient training data which is a common problem of MLP-based approximation tasks. [illegible]

References

1. Abu-Mostafa, Y.S.: Hints. Neural Computation 7 (1995) 639–671
2. Lampinen, J., Selonen, A.: Multilayer perceptron training with inaccurate derivative information. In: Proc. IEEE International Conference on Neural Networks ICNN'95, Volume 5., Perth, West Australia (1995) 2811–2815
3. Sill, J., Abu-Mostafa, Y.S.: Monotonicity hints. In: Advances in Neural Information Processing Systems. Volume 9., Cambridge, MA (1997) 634–640
4. Sill, J.: Monotonic networks. In: Advances in Neural Information Processing Systems. Volume 10., Cambridge, MA (1998) 661–667
5. Rollfinke, R.: Globale Modellierung mit Neuronalen Netzen unter Einhaltung von Nebenbedingungen in den Gradienten. Diploma thesis in cooperation with Siemens AG, Fakultät für Informatik und Automatisierung, TU Ilmenau, Germany (2000)
6. Lang, B.: Können Neuronale Netze monotones Verhalten für bestimmte Dimensionen garantieren? Technical report, Siemens AG, unpublished, Munich, Germany (1999)

Short Term Memories and Forcing the Re-use of Knowledge for Generalization

Laurent Orseau

INSA/IRISA, 35 000 Rennes, France
lorseau@irisa.fr

Abstract. Despite the well-known performances and the theoretical power of neural networks, learning and generalizing are sometimes very difficult. In this article, we investigate how short term memories and forcing the agent to re-use its knowledge on-line can enhance the generalization capabilities. For this purpose, a system is described in a temporal framework, where communication skills are increased, thus enabling the teacher to supervise the way the agent "thinks".

1 Introduction

In a recent work, Marcus et al. [5] tested the learning capabilities of infants when habituated on sequences of events of the type AAB. The purpose was to show that infants are able to extract abstract rules from (positive) examples and could thus generalize to unseen sequences. Abstraction is, here, the fact of knowing that the same letter is repeted. They tested recurrent networks on this same task. Their controversial deduction, e.g. [1], was that neural networks (NNs) are not able to capture the abstraction of the task.

Anyway, it still remains true that NNs are not able to recognize if an input, whichever it is, has been activated more than once. Dominey et al. [2] proposed a Short Term Memory (STM) for improving the abstraction power of NNs and showed that it could then also allow to learn the Marcus et al. task [3]. In this article, their definition of the STM is refined to show that the usefulness of such tools is not limited to "mere" abstraction but also enhances the overall generalization power. A Tapped Delay Neural Network (TDNN) [7,4] is provided with an external loop to allow the agent to "hear" what it produces. Communication skills are thus augmented and the teacher can then force the agent to re-use its knowledge. A classification task is proposed in this temporal framework, not in order to make time series prediction, but rather to use the dynamical properties of the external loop and of the STMs for generalization.

2 Short Term Memories

Dominey et al. [2] introduced their STM in the Temporal Recurrent Network in a psychologic context. The aim was to model the fact that humans could learn surface (the explicit sequence of events) and abstract structures, but could only

W. Duch et al. (Eds.): ICANN 2005, LNCS 3697, pp. 39–44, 2005.

completed by this action. At each time step, the action a predicting the best positive reinforcement for the next time step, or no action if a does not exist, is then selected:

$$a = arg_{i \in I_S}^{\max} Net(i) \text{ if } \max_{i \in I_S} Net(i) > 0, \phi \text{ otherwise} \ , \qquad (2)$$

where I_S is the set of input symbols and $Net(i)$ is the output of the network when only the ith symbol is active. This is the usual action selection method of reinforcement learning system [8]. The STMs are not actions and can be considered as totaly internal tools. When the agent wrongly produces nothing, the teacher tells him the right answer and gives him a reward (positive reinforcement).

For example, suppose the agent generally receives a reward after the sequence ABC. Then, if the sequence AB is presented on its inputs, it searches for an action that completes the sequence and answers C.

Or, for the repetition task, when the C symbol is presented, only the action C activates M_1 (see (1)) and predicts a reward, so C is chosen.

4.2 External Loop and Forcing the Re-use of Knowledge

We now want the agent to be able to "hear" what it produces, so that it can auto-stimulate itself by producing and hearing sequences that could have been produced by the teacher. A loop is thus added around the TDNN and the action produced by the agent at one time step is also provided to its inputs at the next time step. We call this an external loop because it can be thought of as a property of the environment and the learning algorithm does not need to take it into account. Hence, the system is not really a recurrent network (cf. Fig. 1).

Of course, this can generate ambiguities if the teacher's and the agent's symbols are superposed. But, as in any communication protocol, this is not meant to happen and the teacher must be aware of it. In fact, if the agent produces an action when it should not have, the teacher punishes it.

Therefore, for the repetition task, not only will the agent choose C because it would activate M_1, but the inputs are also fed with this action, actually really activating M_1.

The teacher can now explicitely, on-line, make the agent re-use its knowledge. The agent can produce sequences of symbols, which will be reintroduced on its inputs via the loop, enabling it to "think aloud".

Suppose the agent has learned to answer E after the sequence CD, C after AB and D after ABC. If the teacher provides the sequence AB, the agent answers C and thus hears C. ABC being now an input sequence, the agent automatically answers D and hears D. The agent produced the sequence CD, which then auto-stimulates itself to answer E. Yet, an infinite behavior can be produced, for example if the agent learns to produce AB after the sequence AB, continually auto-stimulating itself. In this case, all actions are constants but STMs will emulate variables.

Interpolation Mechanism of Functional Networks*

Yong-Quan Zhou[1,2] and Li-Cheng Jiao[1]

[1] Institute for Intelligence Information Processing, Xidian University, Xi' an, Shanxi 710071, China
zhou_yongquan@163.com
[2] College of Computer and Information Science, Guangxi University for Nationalities, Nanning, Guangxi 530006, China

Abstract. In this paper, the interpolation mechanism of functional networks is discussed. And a kind of three layers Functional networks with single input unit and single output unit and four layers functional networks with double input units and single output unit is designed, a learning algorithm for function approximation is based on minimizing a sum of squares with a unique minimum has been proposed, which can respectively approximate a given one-variable continuous function and a given two-variable continuous function satisfying given precision. Finally, several given examples show that the interpolation method is effective and practical.

1 Introduction

One of the important problems in digital signal processing is a reconstruction of the continuous time signal $F(t)$ from its samples $F(nT)$ [1]. This problem is known as interpolation described by the following equation:

$$F(x) = \sum_{i=1}^{N} w_i \phi_i(x) \tag{1}$$

where w_i are coefficients of interpolation and $\phi_i(x)$ are basis interpolation functions.

If a controlled process is difficult to access, the problem of interpolation becomes complicated. In this case it is difficult to satisfy Nyquist Sampling rate condition and signal is down sampled [1]. In the real time signal process system application of the conventional interpolator based on Lagrange polynomial (parabolic or cubic splines) and orthogonal functions [2] do not satisfy limited time processing requirement related with the generation of functions $\phi_i(x)$, operation of multiplication etc. in accordance to eq. (1).

Interpolation discrete time signal $F(nX)$ by polynomial filtering (Butterworth, Tchebyschev filters) cause a shape distortion. Independent of an original

* This work was supported by NSF of China.

W. Duch et al. (Eds.): ICANN 2005, LNCS 3697, pp. 45–51, 2005.

signal pattern, the result of interpolation is sharply defined at the nodes and it is exponentially varied between the nodes of interpolation [3]. To satisfy time limiting and precision specifications in this paper operation of interpolation is realized using functional network. In this paper, two kinds of Functional networks are designed, which can approximate continuous function satisfying given precision. Finally, several given examples show that the interpolation method is effective and practical.

2 Functional Networks

Castillo et al. [4,5,6] present functional networks as an extension of artificial neural networks. Unlike neural networks, in these networks there are no weights associated with the links connecting neurons, and the internal neuron functions are not fixed but learnable. These functions are not arbitrary, but subject to strong constraints to satisfy the compatibility conditions imposed by the existence of multiple links going from the last input layer to the same output units. In fact, writing the values of the output units in different forms, by considering these different links, a system of functional equation is obtained. When this system is solved, the number of degree of freedom of these initially multidimensional functions is considerably reduced. In learning the resulting functions, a method based on minimizing a least squares error function is used, which, unlike the functions used in neural networks has a single minimum. Our definition is simple but rigorous: a functional network is a network in which the weights of the neurons are substituted by a set of basis functions. For example, in Fig.1 a neural network and its equivalent functional network are shown. Note that weights are subsumed by the neural functions.

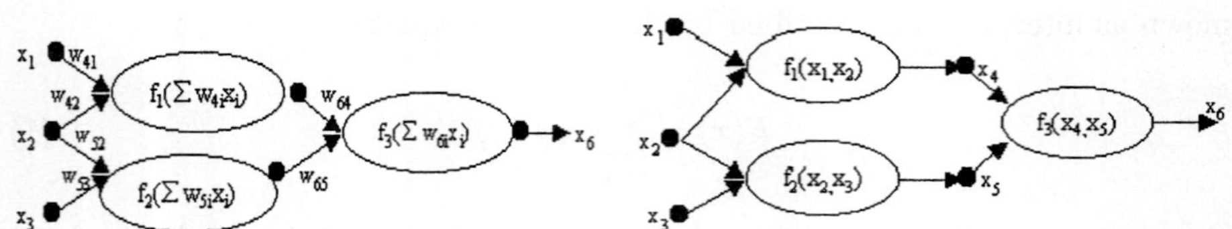

Fig. 1. (a) A neural network (b) A functional network

where f_1, f_2 and f_3 are functional neuron functions in Fig 1 (b).

3 Interpolation Mechanism of Functional Networks

3.1 One-Variable Interpolation Function Model of Functional Network

The one-variable interpolation functional network consists of the following elements (Fig.2): (1) a layer of input units. This first layer accepts input signals.

(2) Two layers of processing units that evaluate a set of input signals and delivers a set of output signals (f_i). (3) A set of directed links. Indicating the signal flow direction. (4) A layers of output units. This is the last layers, and contains the output signals.

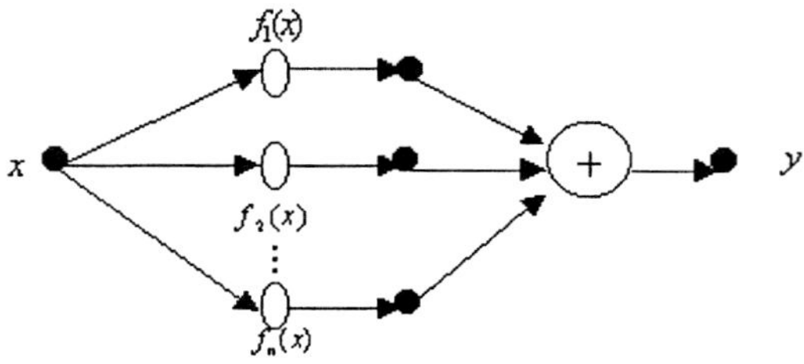

Fig. 2. A functional network for interpolation one-variable function $\hat{F}(x)$

Fig.2 show a functional network corresponding to the functional equation:

$$y = \hat{F}(x) = \sum_{i=1}^{N} f_i(x) \tag{2}$$

Learning interpolation function $F(x)$ is equivalent to learning function $\hat{F}(x)$, and learning function $\hat{F}(x)$ is equivalent to learning the neuron functions $f_i(x)$. To this end, we can approximate $f_i(x): \ i = 1, \ 2, \ldots, \ N$ by

$$\hat{f}_i(x) = \sum_{j=1}^{M} w_{ij}\phi_{ij}(x) \tag{3}$$

Replace this in (2) we get:

$$y = \hat{F}(x) = \sum_{i=1}^{N}\sum_{j=1}^{M} w_{ij}\phi_{ij}(x) \tag{4}$$

where $\{\phi_{ij}(x): \ j = 1, \ 2, \ldots, \ M\}$ is a set of given one-variable interpolation basis of functions capable of approximation interpolation function to the desired accuracy.

3.2 Two-Variable Interpolation Function Model of Functional Network

The two-variable interpolation function model of functional network consists of the following elements (see Fig.3): (1) a layer of input units. This first layer accepts input signals. (2) Three layers of processing units that evaluate a set of input signals and delivers a set of output signals (f_i). (3) A set of directed links

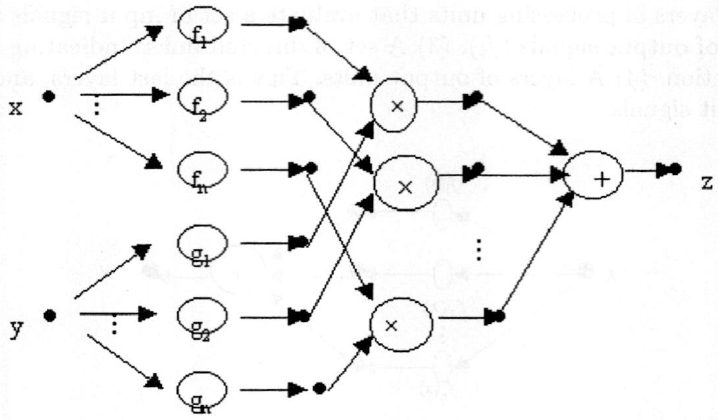

Fig. 3. A functional network for interpolation two-variable function $F(x, y)$

indicating the signal flow direction. (4) A layer of output units. This is the last layer, and contains the output signals.

Consider the following equation:

$$z = F(x, y) = \sum_{i=1}^{n} f_i(x) g_i(y) \tag{5}$$

we now approximate the functions $\{f_i\}$, $\{g_i\}$, $i = 1, 2, \ldots, n$ by using a linear combination of the known functions of a given interpolation basis functions $\Phi_i(x) = \{\phi_{i1}(x), \phi_{i2}(x), \ldots, \phi_{im_{1i}}(x)\}$, $\Psi_i(x) = \{\varphi_{i1}(x), \varphi_{i2}(x), \ldots, \varphi_{im_{2i}}(x)\}$ are two sets of linearly independent functions, then, we get:

$$f_i(x) = \sum_{j=1}^{m_{1i}} a_{ij}\phi_{ij}(x) \qquad g_i(y) = \sum_{k=1}^{m_{2i}} b_{ik}\varphi_{ik}(y) \tag{6}$$

Replace this in (5) we get:

$$z = F(x, y) = \sum_{i=1}^{r} \sum_{j=r+1}^{n} c_{ij}\phi_i(x)\varphi_j(y) \tag{7}$$

where c_{ij} are parameters, and $\phi_i(x)$, $\varphi_i(y)$ are given interpolation basis functions capable of approximation two-variable function $\hat{F}(x, y)$ to the desired accuracy.

4 Interpolation Functional Networks Learning Algorithm

In this section, only we consider the two-variable interpolation functional networks (see Fig.3) learning algorithm. In this case, the problem of learning the polynomial functional network associated with (7) does not involves auxiliary functional conditions and, therefore, a simples least squares method allows us obtained the optimal coefficients c_{ij} from the available data consisting of triplets $\{(x_{0i}, x_{1i}, x_{2i}) \mid i = 1,\ 2, \ldots,\ n\}$, where x_{0i}, x_{1i} and x_{2i} refer to z, x and y, respectively. Then, the error can be measured by:

$$e_i = x_{i0} - \sum_{i=1}^{r} \sum_{j=r+1}^{n} c_{ij} \phi_i(x_{1i}) \varphi_j(x_{2i}) \tag{8}$$

$i = 1, \ldots,\ n$. Training patterns. Thus, to find the optimum coefficients we minimize the sum of square errors:

$$E = \sum_{k=1}^{n} e_k^2 \tag{9}$$

In this case, the parameters are not constrained by extra conditions, so the minimum can be obtained by solving the following system of linear equations, where the unknowns are the coefficients c_{ij}:

$$\frac{\partial E}{\partial c_{pq}} = 2 \sum_{k=1}^{n} e_k \phi_p(x_{1k}) \varphi_q(x_{2k}) = 0; \qquad p = 1, \ldots, r; \qquad q = 1, 2, \ldots, r - s \tag{10}$$

5 Experimental Results

To estimate the performance of approximate, the Root Mean Square Error (RMSE) defined as:

$$RMSE = \sqrt{\frac{1}{r} \sum_{p=1}^{r} \parallel b_p - \hat{b}_p \parallel^2} \tag{11}$$

where $\hat{b}_p$ is the network output, and the norm function $\| \cdot \|$ reduces to the usual absolute value function $| \cdot |$.

Example 1. Consider two-dimensional function approximation problem

$$F : R^2 \to R \qquad z = F(x, y) = 1 + x - y - x^2 y$$

Suppose we are given the data set consisting of the triplets shown in Table 1. where x_0, x_1 and x_2 refer to z, x and y, respectively. Then, by using a structure of the double-input and single output functional network with the learning algorithm, only can we discuss two cases as follows:

1) First, we consider the polynomial interpolation basis functions family $\phi(x_1) = \{1, x_1, x_1^2\}$, $\varphi(x_2) = \{1, x_2, x_2^2\}$ for the neuron function f_1, g_1. We obtain interpolation function $F(x, y)$ of approximation function $\hat{F}(x, y)$.

$$\begin{aligned} z = \hat{F}(x, y) &= 0.997608 + 1.00908x - 0.0105989x^2 - 0.983605y - 0.0686143xy \\ &\quad - 0.928585x^2 y - 0.0166604y^2 + 0.0730798xy^2 - 0.0776385x^2 y^2 \end{aligned}$$

Thus, we get $RMSE = 0$.

Table 1. Triplets $\{x_0, x_1, x_2\}$ for two-variable interpolation function $z = F(x, y)$

x_0	x_1	x_2	x_0	x_1	x_2
0.880	0.384	0.439	1.072	0.136	0.062
1.117	0.976	0.440	1.369	0.801	0.263
0.299	0.309	0.922	1.449	0.457	0.006
0.873	0.725	0.558	1.214	0.377	0.143
0.359	0.449	0.907	0.604	0.818	0.727
0.935	0.058	0.122	0.669	0.799	0.689
0.681	0.673	0.682	0.760	0.663	0.627
1.336	0.697	0.242	1.228	0.862	0.364
1.019	0.387	0.320	0.989	0.405	0.357
0.556	0.662	0.762	0.814	0.028	0.214

2) Second, we consider the polynomial interpolation basis functions family $\phi(x_1) = \{1, x_1, x_1^2\}, \varphi(x_2) = \{1, x_2\}$ for the neuron function f_1, g_1. We obtain the two-variable interpolation function $F(x, y)$ of approximation function $\hat{F}(x, y)$.

$$z = \hat{F}(x, y) = 0.999241 + 1.00004x + 0.000994901x^2 - 0.996202y \\ -0.00108513xy - 0.993437x^2y$$

Thus, we also get $RMSE = 0$.

Example 2. Consider two-dimensional function approximation problem

$$F : I^2 \rightarrow [-1, 1] \qquad F(x, y) = -sin(\pi \cdot x) \cdot cos(\pi \cdot y)$$

Suppose we are given the data set consisting of the triplets shown in Table 2 and we are interested in estimating a representative functional network model of the form (7).

Table 2. Triplets $\{z, x, y\}$ for two-variable interpolation function $z = F(x, y)$

x	y	$F(x,y)$
0.949605	0.564963	0.0319533
0.709702	0.296671	-0.471442
0.636216	0.138527	-0.825018
0.64929	0.260153	-0.610314
0.554334	0.596855	0.29525
0.315312	0.0471869	-0.827172
0.729573	0.879708	0.698005
0.530368	0.360128	-0.423481
0.900224	0.21243	-0.242192
0.743644	0.0738989	-0.701738

Let interpolation basis functions: $\phi(x) = [1, sin(\pi \cdot x)]$, $\varphi = [1, cos(\pi \cdot y)]$, we can obtain the two-variable approximation interpolation function $\hat{F}(x, y)$.

$$\hat{F}(x, y) = -1 sin(\pi \cdot x) \cdot cos(\pi \cdot y)$$

Thus, we get $RMSE = 0$.

6 Conclusions

Two kinds of interpolation functional networks are designed, A learning algorithm for function approximation is based on minimizing a sum of squares with a unique minimum is proposed, and the learning of parameters of the functional networks is carried out by the solving linear equations, which can respectively approximate continuous function satisfying given precision. Finally, several given examples show that the interpolation method is effective and practical.

References

1. Fred J Taylor. Principles of Signals and Systems. New York:McGraw-Hill,1994.
2. John H. Mathews. Numerical Methods.NJ:Prentice-Hall,1992.
3. Fakhreddin M. Communication System for The Plant with Difficult Of Access. Intentional Conference on "Components and Electronics systems", Algeria,1991.
4. Enrique Castillo. Functional Networks. Neural Processing Letters, 1998,7:151 159
5. Enrique Castillo, Jose Manuel Gutierrez. A Comparison Between Functional Networks and Artificial Neural Networks. In: Proceedings of the IASTED International Conference on Artificial Intelligence and Soft Computing.1998:439-442.
6. Enrique Castillo, Angel Cobo, Jose Manuel Gutierrez and Rose Eva Pruneda, Functional Networks with Applications. Kluwer Academic Publishers,1999.

Neural Network Topology Optimization

Mohammed Attik[1,2], Laurent Bougrain[1], and Frédéric Alexandre[1]

[1] LORIA/INRIA-Lorraine, Campus Scientifique - BP 239 - 54506, Vandœuvre-lès-Nancy Cedex - France
attik@loria.fr
http://www.loria.fr/~attik
[2] BRGM, 3 av Claude Guillemin - BP 6009 - 45060 Orléans Cedex 2 - France

Abstract. The determination of the optimal architecture of a supervised neural network is an important and a difficult task. The classical neural network topology optimization methods select weight(s) or unit(s) from the architecture in order to give a high performance of a learning algorithm. However, all existing topology optimization methods do not guarantee to obtain the optimal solution. In this work, we propose a hybrid approach which combines variable selection method and classical optimization method in order to improve optimization topology solution. The proposed approach suggests to identify the relevant subset of variables which gives a good classification performance in the first step and then to apply a classical topology optimization method to eliminate unnecessary hidden units or weights. A comparison of our approach to classical techniques for architecture optimization is given.

1 Introduction

Supervised neural networks are widely used in various areas like pattern recognition, marketing, geology or telecommunications. They are well adapted to prediction using information of databases. Generally, we optimize the architecture of these techniques before any use. The objectives of neural network topology optimization can be numerous: improving the prediction performance, resolving the over-training problem, providing a faster predictor, providing a better understanding of the underlying process that generates data (facilitating extraction of rules) and reducing the time and the cost of collecting and transforming data.

In the last years, several neural network topology optimization methods have been proposed [1,2]. Generally, these methods are grouped in three classes: (i) *Network Growing Techniques:* the algorithms of this class start with a small network and add units or connections until an adequate performance level is obtained; (ii) *Network Pruning Techniques:* the algorithms of this class start with a fully trained large network and then attempt to remove some of redundant weights and/or units; (iii) *Regularization Techniques:* the algorithms of this class add a penalty term to the network error function when training the network which allows to reduce the number of units and weights in a network.

W. Duch et al. (Eds.): ICANN 2005, LNCS 3697, pp. 53–58, 2005.

These topology optimization methods give a sub-optimal architecture of a supervised neural network. The goal of this work is to maximize reduction of network complexity, i.e. removal of all the unnecessary units and weights. It is known that the neural network architecture size strongly depends on the number of variables (input units) and some of these variables might contain highly correlated information or even irrelevant information. *It is important to begin with the selection of variables before finding irrelevant weights or hidden units in the network structure.* The goal of variable selection methods is to find the subset of variables which gives a high classification performance by elimination the unuseless variables or variables which present redundancy. Irrelevant variables are defined as variables not having any influence on the classification performance. Contrary to relevant variables which are variables that have an influence on the classification performance. For a discussion of relevance *vs.* usefulness and definitions of the various notions of relevance, see the review articles of [3,4,5]. The variable selection methods can be subdivided into filter methods and wrapper methods. The main difference is that the wrapper method makes use of the classifier, while the filter method does not. The wrapper approach is clearly more accurate.

With a small subset size of variables resulting from variable selection task, compared to original number of variables, we have inevitably reduced the network complexity. The next section describe in detail our approach which recommend to apply variable selection before applying optimization topology technique. In section 3 an illustration of this approach using Optimal Brain Surgeon (OBS) variants for variable selection and architecture optimization is presented. Finally, some conclusions are drawn in section 4.

2 Variables-Weights Neural Network Topology Optimization

Our hybrid approach is based on the divide-and-conquer principle. It combines variable selection method and classical topology optimization method i.e. it suggests to identify the relevant subset of variables in the first step and then to apply a classical topology optimization method to eliminate unnecessary hidden units or weights. We called this approach *Variables-Weights Neural Network Topology Optimization (VW-NNTO).*

When applying the variable selection method to select a relevant subset of variables, we have a panoply of methods which can be used for selection task. According to used method, the solution presents: only the relevant subset of variables or the subset with the topology architecture. When applying a classical topology optimization method, we have also a panoply of methods which can be used for optimization task by selecting the weights and/or hidden units. The choice of the used method can takes in consideration the solution type obtained by variable selection method. In the case of the solution with the subset of variables and topology architecture, several questions can be raised, it is good to keep: (1) the same architecture? (2) the same training parameters? (3) the values of the weights?

Practically, irrelevance variable does not imply that it should not be in the relevant variable subset [3,4]. This step will allows also refine the results obtained by the selection methods.

3 Illustration: OBS Techniques

3.1 Techniques Describtion

Several heuristic methods based on computing the saliency (also termed sensitivity) have been proposed: Optimal Brain Damage (OBD) [6] and Optimal Brain Surgeon (OBS) [7]. These methods are known as pruning methods. The principle of these techniques is: the weight with the smallest saliency will generate the smallest error variation if it is removed. These techniques considered a network trained to a local minimum in error by using the Taylor function. The functional Taylor series of the error with respect to weights is:

$$\delta E = \sum_i \frac{\partial E}{\partial w_i}\delta w_i + \frac{1}{2}\sum_i \frac{\partial^2 E}{\partial w_i^2}(\delta w_i)^2 + \frac{1}{2}\sum_{i,j\neq i} \frac{\partial^2 E}{\partial w_i \partial w_j}\delta w_i \delta w_j + O(\|\delta W\|^3) \quad (1)$$

A well trained network implies that the first term in (Eq. 1) will be zero because E is at a minimum. When the perturbations are small, the last term will be negligible.

Optimal Brain Surgeon (OBS) was introduced by Hassibi et al. [7,8]. In these works the saliency of a weight is measured by approximating (Eq. 1):

$$\delta E = \frac{1}{2}\sum_i \sum_j h_{ij}\delta w_i \delta w_j = \frac{1}{2}\delta \mathbf{w}^T.\mathbf{H}.\delta \mathbf{w} \quad (2)$$

it is possible to update the magnitude of all weights in the network by:

$$\delta \mathbf{w} = -\frac{w_q}{[\mathbf{H}^{-1}]_{qq}}\mathbf{H}^{-1}.\mathbf{e}_q \quad (3)$$

and the saliency of the weight q is given by:

$$L_q = \frac{1}{2}\frac{w_q^2}{[\mathbf{H}^{-1}]_{qq}} \quad (4)$$

where $\mathbf{e}_q$ is the unit vector in weight space corresponding to (scalar) weight w_q.

Optimal Brain Surgeon has inspired some methods for feature selection such as Generalized Optimal Brain Surgeon (G-OBS), Unit-Optimal Brain Surgeon (Unit-OBS) and Flexible-Optimal Brain Surgeon (F-OBS).

Stahlberger and Riedmiller [9] proposed to the OBS's users, a calculation, called Generalized Optimal Brain Surgeon (G-OBS) which can update every weights when deleting a subset of m weights in a single step. The solution is given by:

$$\Delta E = \frac{1}{2}\mathbf{w}^T M(M^T H^{-1} M)^{-1} M^T \mathbf{w} \tag{5}$$

$$\Delta \mathbf{w} = -H^{-1} M(M^T H^{-1} M)^{-1} M^T \mathbf{w} \tag{6}$$

where M is the selection matrix and $q_1, q_2, \cdots, q_m$ are the indices of the weights that will be removed.

This calculation presents a combinatory calculation in order to know which weights should be deleted. An implementation for this method is proposed in [10] which defines the subset of connections by the smallest saliencies.

Unit-OBS was proposed by Stahlberger and Riedmiller [9]; it is a variable selection algorithm, which computes, using the calculation G-OBS, which input unit will generate the smallest increase of error if it is removed (Eq. 5).

The F-OBS has been proposed in [10], where its particularity is to remove connections only between the input layer and the hidden layer.

OBS variants are less efficiency on large networks [10,11] because are severely influenced by the cumulated noise generated by Hessian matrix calculation on the weights. They require only one training for all pruning. An alternative is proposed to re-applied these techniques in several times to resolve this problem.

3.2 Experiments

We use the first Monk's problem of the proposed approach to optimize the topology of Multilayer perceptron (MLP). This well-known problem (See [12]) requires the learning agent to identify (true or false) friendly robots based on six nominal attributes. The attributes are head_shape (round, square, octagon), body_shape (round, square, octagon), is_smiling (yes, no), holding (sword, balloon, flag), jacket_color (red, yellow, green, blue) and has_tie (yes, no). The "true" concept for this problem is (head_shape = body_shape) or (jacket_color = red). The training dataset contains 124 examples and the validation and test datasets contains 432 examples.

To forecast the class according to the 17 input values (one per nominal value coded as 1 or -1 if the characteristic is true or false), the multilayer perceptron starts with 3 hidden neurons containing a hyperbolic tangent activation function. This number of hidden neurons allows a satisfactory representation able to solve this discrimination problem. The total number of weights for this fully connected network (including a bias) is 58. This value will be compared to the remaining weights after pruning. The performance in classification is equal to 100% according to the confusion matrix. For each method, we build 500 MLPs by varying: the training/validating subset decomposition, the initialization of the weights and the order of the pattern presentation.

We select two values as measures of the performance for optimization: the number of preserved weights (see Figures 2,4,6) and the number of preserved variables (see Figures 1,3,5). According to all obtained histograms in the figures, we notice that: (i) There is a certain compromise between the number of variables and the number of weights for all methods; (ii) The hybrid techniques give good results compared to the simple optimization methods OBS.

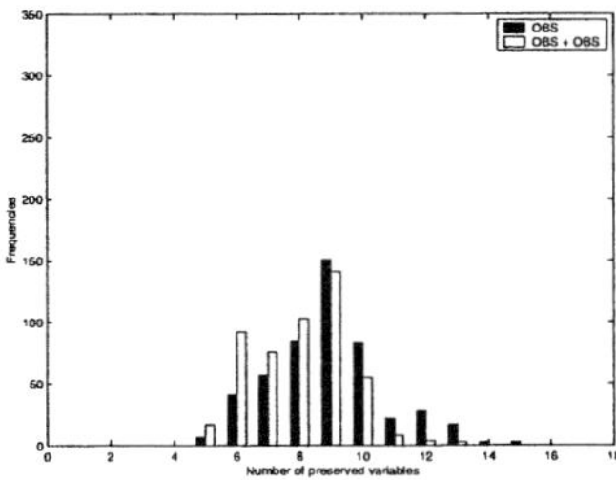

Fig. 1. OBS : variables distribution

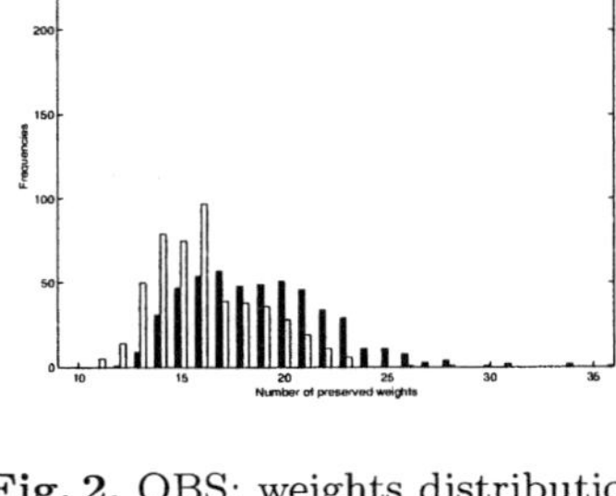

Fig. 2. OBS: weights distribution

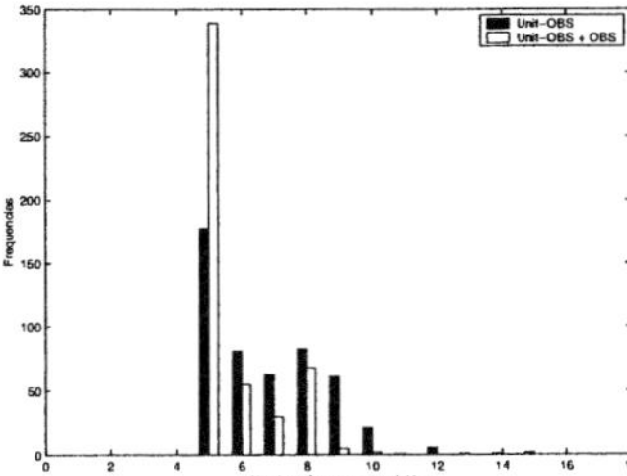

Fig. 3. Unit-OBS + OBS: variables distribution

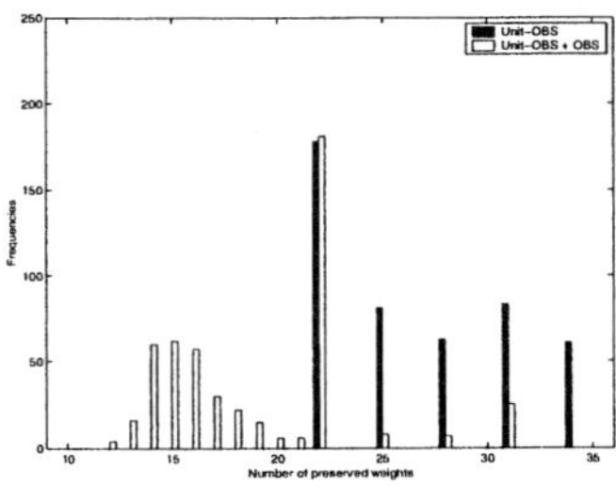

Fig. 4. Unit-OBS + OBS: weights distribution

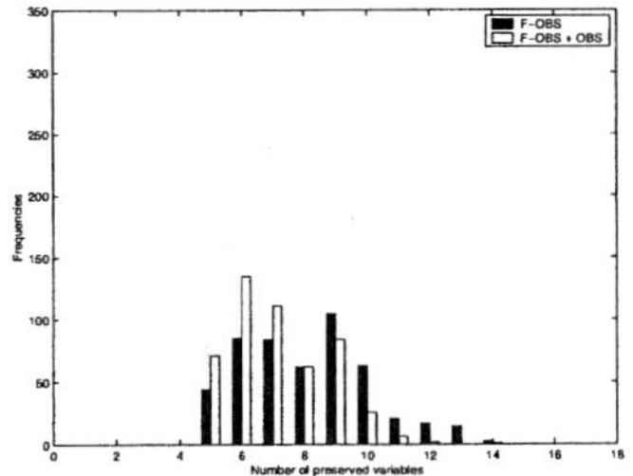

Fig. 5. F-OBS + OBS: variables distribution

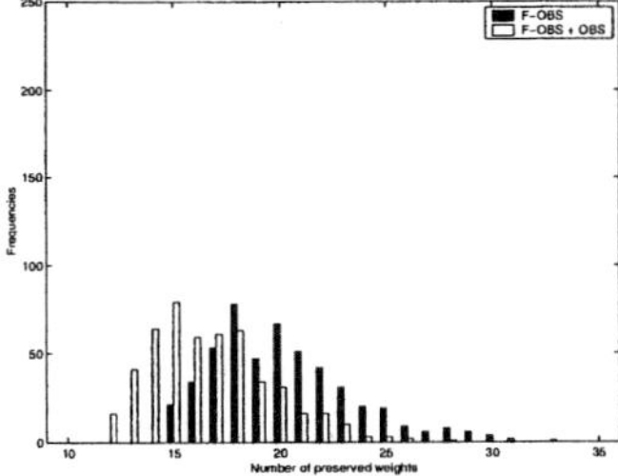

Fig. 6. F-OBS + OBS: weights distribution

4 Conclusion

We have proposed a hybrid approach for architecture optimization combining variable selection methods and classical topology optimization technique. In the first step we have proposed to select all relevant features or a subset of relevant features and then to apply classical optimization architecture until there is no more weight or hidden unit to eliminate. We have presented an illustration of this approach by using the Optimal Brain Surgeon variants specialized in variable selection and architecture optimization.

Acknowledgment

We would like to thank Abdessamad Imine for his helpful remarks.

References

1. Bishop, C.M.: Neural Networks for Pattern Recognition. Clarendon Press, Oxford (1995)
2. Reed, R.: Pruning algorithms — A survey. IEEE Transactions on Neural Networks **4** (1993) 740–746
3. John, G.H., Kohavi, R., Pfleger, K.: Irrelevant features and the subset selection problem. In: International Conference on Machine Learning. (1994) 121–129 Journal version in AIJ, available at http://citeseer.nj.nec.com/13663.html.
4. Kohavi, R., John, G.H.: Wrappers for feature subset selection. Artificial Intelligence **97** (1997) 273–324
5. Blum, A., Langley, P.: Selection of relevant features and examples in machine learning. Artificial Intelligence **97** (1997) 245–271
6. Le Cun, Y., Denker, J.S., Solla, S.A.: Optimal brain damage. In Touretzky, D.S., ed.: Advances in Neural Information Processing Systems: Proceedings of the 1989 Conference, San Mateo, CA, Morgan-Kaufmann (1990) 598–605
7. Hassibi, B., Stork, D.G.: Second order derivatives for network pruning: Optimal brain surgeon. In Hanson, S.J., Cowan, J.D., Giles, C.L., eds.: Advances in Neural Information Processing Systems. Volume 5., Morgan Kaufmann, San Mateo, CA (1993) 164–171
8. Hassibi, B., Stork, D.G., Wolff, G.: Optimal brain surgeon: Extensions and performance comparison. In Cowan, J.D., Tesauro, G., Alspector, J., eds.: Advances in Neural Information Processing Systems. Volume 6., Morgan Kaufmann Publishers, Inc. (1994) 263–270
9. Stahlberger, A., Riedmiller, M.: Fast network pruning and feature extraction by using the unit-OBS algorithm. In Mozer, M.C., Jordan, M.I., Petsche, T., eds.: Advances in Neural Information Processing Systems. Volume 9., The MIT Press (1997) 655
10. Attik, M., Bougrain, L., Alexandre, F.: Optimal brain surgeon variants for feature selection. In: International Joint Conference on Neural Networks - IJCNN'04, Budapest, Hungary. (2004)
11. Attik, M., Bougrain, L., Alexandre, F.: Optimal brain surgeon variants for optimization. In: 16h European Conference on Artificial Intelligence - ECAI'04, Valencia, Spain. (2004)
12. Thrun, S.B., Bala, J., Bloedorn, E., Bratko, I., Cestnik, B., Cheng, J., Jong, K.D., Džeroski, S., Fahlman, S.E., Fisher, D., Hamann, R., Kaufman, K., Keller, S., Kononenko, I., Kreuziger, J., Michalski, R.S., Mitchell, T., Pachowicz, P., Reich, Y., Vafaie, H., de Welde, W.V., Wenzel, W., Wnek, J., Zhang, J.: The MONK's problems: A performance comparison of different learning algorithms. Technical Report CS-91-197, Carnegie Mellon University, Pittsburgh, PA (1991)

Rough Sets-Based Recursive Learning Algorithm for Radial Basis Function Networks

Yevgeniy Bodyanskiy[1], Yevgen Gorshkov[1], Vitaliy Kolodyazhniy[1], and Irina Pliss[1]

Control Systems Research Laboratory,
Kharkiv National University of Radioelectronics,
14, Lenin Av., Kharkiv, 61166, Ukraine
{bodya, pliss}@kture.kharkov.ua, ye.gorshkov@gmail.com,
kolodyazhniy@ukr.net

Abstract. A recursive learning algorithm based on the rough sets approach to parameter estimation for radial basis function neural networks is proposed. The algorithm is intended for the pattern recognition and classification problems. It can also be applied to neuro control, identification, and emulation.

1 Introduction

Neural networks techniques play an essential role in data mining and intelligent data processing tasks. They can be applied a variety of problems like nonlinear systems identification, time series forecasting, filtration, adaptive control, pattern recognition, technical diagnostics, etc.

Radial basis function networks (RBFNs) are known as networks with locally-tuned processing units and possess universal approximation capabilities. The RBFN consist of two information processing layers, in contrast to the multilayer perceptrons, include only linear synaptic weights of the output layer providing desired performance to the nonlinear input-output mapping.

In the general case, the RBFN contains n inputs and m outputs and performs the following nonlinear mapping: $y_j = F_j(x) = w_{j0} + \sum_{i=1}^{h} w_{ji}\phi_i(x) = w_j^T \phi(x)$, where y_j is the j-th network output signal ($j = 1, 2, \ldots, m$), $F_j(x)$ is a nonlinear mapping of the input vector $x = (x_1, x_2, \ldots, x_n)^T$ into the j-th output, w_{ji} represent the adjustable synaptic weights, and $\phi_i(x)$ denote a radial basis or kernel functions, $w_j = (w_{j0}, w_{j1}, \ldots, w_{jh})^T$, $\phi(x) = (1, \phi_1(x), \phi_2(x), \ldots, \phi_h(x))^T$.

The only hidden layer of the networks of this type performs a nonlinear mapping of the input vector space $\mathbb{R}^n$ into the space $\mathbb{R}^h$ of a larger dimension ($h \gg n$). The output layer consists of a set of adaptive linear associators, and calculates the network response $y = (y_1, y_2 \ldots, y_m)^T$ to the input signal x.

The most widely used learning algorithms for the RBFNs are based on the quadratic criteria of the error function. These methods range from the simple one-step Widrow-Hoff learning algorithm to the exponentially weighted least-squares method:

W. Duch et al. (Eds.): ICANN 2005, LNCS 3697, pp. 59–65, 2005.

$$\begin{cases} w_j(k) = w_j(k-1) + \dfrac{P_\phi(k-1)(d_j(k) - w_j^T(k-1)\phi(x(k)))}{\alpha + \phi^T(x(k))P_\phi(k-1)\phi(x(k))}\phi(x(k)), \\ P_\phi(k) = \dfrac{1}{\alpha}\left(P_\phi(k-1) - \dfrac{P_\phi(k-1)\phi(x(k))\phi^T(x(k))P_\phi(k-1)}{\alpha + \phi^T(x(k))P_\phi(k-1)\phi(x(k))}\right), \end{cases}$$

where $k = 1, 2, \ldots$ denotes discrete time, $0 < \eta < 2$ is a scalar parameter, which determines the convergence of the learning process, $d_j(k)$ is the target signal, and $0 < \alpha < 1$ is a forgetting factor.

Estimates of the synaptic weights obtained with these algorithms have a clear statistical meaning. The weights also reach their optimal values if both the useful information and disturbances are stochastic signals generated by normal distributions. The learning algorithms based on non-quadratic criteria are still connected with a certain distribution law, and have a spatial estimate represented by the mean value of the distribution.

It is natural that the application of the statistical criteria to the non-stochastic information processing tasks (e.g. dynamic reconstruction of chaotic signals [1,2,3]) will not provide reasonable accuracy. In this case we could assume that both the useful signals $x(k)$, $d_j(k)$ (here $d_j(k)$ is the training signal) and unobserved disturbances $\zeta(k)$ lie within a bounded range. Moreover, these signals might have a regular, chaotic or synthetic nature (e.g. noise). It is also obvious that even for the optimal values of the synaptic weights w_j^* an exact equality $y_j(k) = d_j(k)$ at the network output cannot be achieved during the learning process. The optimal values w_j^* can only define a range of compatible values [4,5]

$$d_j(k) - r(k) \le w_j^{*T}\phi(x(k)) \le d_j(k) + r(k), \tag{1}$$

with $r(k)$ denoting the bounds of the disturbances: $|\zeta(k)| \le r(k)$.

It could be readily seen that the inequality (1) defines two hyperplanes in the synaptic space, which are the bounds for $w_j(k)$ values. The sequence $d_j(1), d_j(2), \ldots, d_j(N)$ of the training signals generates N pairs of such hyperplanes. Denote by $D_j(N)$ the intersection of the sets of synaptic weights compatible with all observations (i.e. for all k). $D_j(N)$ is usually called a polytope. All points of $D_j(N)$ are equivalent, viz. we cannot distinguish the best synaptic weight vector value. In this case the result of the learning procedure will not be a spatial estimate, but an interval one. However, for convenience the center of the polytope $D_j(N)$ could be chosen as an estimate, in a manner, similar to the defuzzification procedures used in the fuzzy inference systems.

The approach discussed above is called a set-membership approach to parameter estimation and is quite popular in the tasks of identification of the control objects and systems.

One of the possible approaches to find the set of synaptic weights lies in solving the linear system of N inequalities (1). However, as the number of vertices of the polytope $D_j(N)$ increases much faster than $k = 1, 2, \ldots, N, \ldots$, this approach seems to be quite ineffective in a numerical implementation.

An alternative approach consists in the approximation of the polytope $D_j(k)$ by ellipsoids

$$L_j(k): \ (w_j^* - w_j(k))^T P_j^{-1}(k)(w_j^* - w_j(k)) \leq 1, \tag{2}$$

with the centers at $w_j(k)$ and the positive definite covariance matrices $P_j(k)$, which are tuned during the learning process to reach the most accurate approximation of $D_j(k)$ provided by $L_j(k)$. Since $w_j(k)$ and $P_j(k)$ contain $(h+1) + (h+2)(h+1)/2$ adjustable parameters, the use of the ellipsoidal approximation is more effective than solving the linear system defined by a polytope.

One can see that the polytope $D_j(k)$ is actually a rough-set [6], and the ellipsoid $L_j(k)$ containing all the elements of $D_j(k)$ is its upper approximation. Notice also that the rough-sets approach could be applied to the learning of the multilayer neural networks (see [7]).

The approach developed in this paper is based on the ideas of F. Schweppe [4] and consists in the concept that the ellipsoid $L_j(k)$ must contain all the possible values of the synaptic weights, belonging to the intersection of $L_j(k-1)$ (an ellipsoid, built at the time $k-1$) with the area $G_j(k)$, lying between two hyperplanes defined by the current k-th observation (1).

Since the intersection of $L_j(k-1)$ and $G_j(k)$ generally is not an ellipsoid, we need to determine the values of the parameters $w_j(k)$ and $P_j(k)$ giving the best approximation of the intersection which is provided by $L_j(k)$. From the equations (1) and (2) it can be readily seen that the desired parameters values are determined by the set of inequalities:

$$\begin{cases} (w_j^* - w_j(k-1))^T P_j^{-1}(k-1)(w_j^* - w_j(k-1)) \leq 1, \\ r^{-2}(k)(d_j(k) - w_j^{*T}\phi(x(k)))^2 \leq 1, \end{cases} \tag{3}$$

or for a nonnegative $\rho_j(k)$:

$$\begin{aligned} &(w_j^* - w_j(k-1))^T P_j^{-1}(k-1)(w_j^* - w_j(k-1)) + \\ &\qquad + \rho_j(k) r^{-2}(d_j(k) - w_j^{*T}\phi(x(k)))^2 \leq 1 + \rho_j(k). \end{aligned} \tag{4}$$

Denoting the error vector of determining the synaptic weight by $\tilde{w}_j(k) = w_j^* - w_j(k)$, after simple but tedious transformations of the quadratic form in the left side of (4), we could obtain the Fogel-Huang's algorithm [8] most popular in ellipsoidal estimation:

$$\begin{cases} w_j(k) = w_j(k-1) + \rho_j(k) r^{-2}(k) \tilde{P}_j(k)(d_j(k) - w_j^T(k-1)\phi(x(k)))\phi(x(k)), \\ \tilde{P}_j(k-1) = P_j(k-1) - \dfrac{\rho_j(k) r^{-2}(k) P_j(k-1)\phi(x(k))\phi^T(x(k))P_j(k-1)}{1 + \rho_j(k) r^{-2}(k)\phi^T(x(k))P_j(k-1)\phi(x(k))}, \\ P_j(k) = \tilde{P}_j(k-1)\left(1 + \rho_j(k) - \dfrac{\rho_j(k)(d_j(k) - w_j^T(k-1)\phi(x(k)))^2}{r^2(k) + \rho_j(k)\phi^T(x(k))P_j(k-1)\phi(x(k))}\right). \end{cases} \tag{5}$$

The procedure (5) contains undefined parameter $\rho_j(k)$, which should be chosen to minimize the volume of the ellipsoid $L_j(k)$ approximating an intersection of $L_j(k-1)$ and $G_j(k)$, that is in fact similar to the D-criterion of optimality used in experimental design theory. This can be formulated as the minimization of the function:

$$\det P_j(k) = \left(1 + \rho_j(k) - \frac{\rho_j(k)e_j^2(k)}{r^2(k) + \rho_j(k)\phi^T(x(k))P_j(k-1)\phi(x(k))}\right)^{h+1} \cdot$$
$$\cdot \left(1 - \frac{\rho_j(k)\phi^T(x(k))P_j(k-1)\phi(x(k))}{r^2(k) + \rho_j(k)\phi^T(x(k))P_j(k-1)\phi(x(k))}\right) \det P_j(k-1) \quad (6)$$

(where $e_j(k) = d_j(k) - w_j^T(k-1)\phi(x(k))$ is a learning error of the neural network on the j-th output) or, equivalently, as solving the differential equation:

$$\partial \det P_j(k)/\partial \rho_j = 0. \quad (7)$$

It is quite obvious that (7) does not have an explicit analytic solution. To determine the value of the parameter $\rho_j(k)$, one-dimensional global minimization procedures or other numerical algorithms of finding real nonnegative roots can be applied to the equations (6) or (7) respectively. However, these methods are not applicable in real-time learning because of their low computational performance.

2 The Learning Algorithm

In order to overcome the difficulties noted above let us introduce a scalar $\gamma_j(k)$ satisfying the following conditions [9]: $D_j^{-1}(k) = \gamma_j(k)P_j^{-1}(k)$, $D_j(k) = \gamma_j^{-1}(k)$ $P_j(k)$, $\gamma_j(k) > 0$, and then rewrite equations (3), (4) in the form

$$\begin{cases} (w_j^* - w_j(k-1))^T D_j^{-1}(k-1)(w_j^* - w_j(k-1)) \le \gamma_j(k-1), \\ r^{-2}(k)(d_j(k) - w_j^{*T}\phi(x(k)))^2 \le 1, \end{cases}$$

and $(w_j^* - w_j(k-1))^T D_j^{-1}(k-1)(w_j^* - w_j(k-1)) + \rho_j(k)r^{-2}(k)(d_j(k) - w_j^{*T}\phi(x(k)))^2 \le \gamma_j(k-1) + \rho_j(k)$, respectively.

Performing further a sequence of transformations, we obtain an algorithm:

$$\begin{cases} w_j(k) = w_j(k-1) + \dfrac{\delta_j(k)e_j(k)D_j(k-1)\phi(x(k))}{1 + \delta(k)\phi^T(x(k))D_j(k-1)\phi(x(k))}, \\ D_j(k) = D_j(k-1) - \delta_j(k)\dfrac{D_j(k-1)\phi(x(k))\phi^T(x(k))D_j(k-1)}{1 + \delta(k)\phi^T(x(k))D_j(k-1)\phi(x(k))}, \end{cases} \quad (8)$$

(where $\delta_j(k) = \rho_j(k)r^{-2}(k)$) which structurally coincides with the Hägglund's algorithm [10] minimizing the criterion: $E_j(k) = \sum_{p=0}^{k} \delta_j(p)e_j^2(p)$. It should be noted that unlike the Hägglund's algorithm, the procedure (8) contains two free parameters $\gamma_j(k)$ and $\delta_j(k)$, which fully determine the character of the learning process.

To find $\gamma_j(k)$, let us write an inequality: $\tilde{w}_j^T(k)D_j^{-1}(k)\tilde{w}_j(k) + \delta_j(k)e_j^2(k) - \delta_j^2(k)e_j^2(k)\phi^T(x(k))D_j(k)\phi(x(k)) \le \gamma_j(k-1) + \delta_j(k)r^2(k)$, whence it is easy to obtain: $\tilde{w}_j^T(k)D_j^{-1}(k)\tilde{w}_j(k) \leq \gamma_j(k-1) + \delta_j(k)r^2(k) - (\delta_j(k)e_j^2(k))/(1 + \delta_j(k)\phi^T(x(k))D_j(k-1) \cdot \phi(x(k))) = \gamma(k)$. Now the learning algorithm can be written by a set of recursive equations:

$$\begin{cases} w_j(k) = w_j(k-1) + \dfrac{\delta_j(k)e_j(k)D_j(k-1)\phi(x(k))}{1+\delta(k)\phi^T(x(k))D_j(k-1)\phi(x(k))}, \\ D_j(k) = D_j(k-1) - \delta_j(k)\dfrac{D(k-1)\phi(x(k))\phi^T(x(k))D_j(k-1)}{1+\delta_j(k)\phi^T(x(k))D_j(k-1)\phi(x(k))}, \\ \gamma_j(k) = \gamma_j(k-1) + \delta_j(k)r^2(k) - \dfrac{\delta_j(k)e_j^2(k)}{1+\delta_j(k)\phi^T(x(k))D_j(k-1)\phi(x(k))}. \end{cases} \tag{9}$$

In order to find the value of the parameter $\delta(k)$, providing the convergence to the algorithm (9), consider a process of error decreasing as $\|\tilde{w}_j(k)\|^2_{D_j^{-1}(k)}$. The following estimation for the $\delta_j(k)$ can be obtained:

$$0 < \delta_j(k) \leq (e_j(k)r^{-2}(k) - 1)/(\phi^T(x(k))D_j(k-1)\phi(x(k))). \tag{10}$$

Hence the algorithm (9) adjusts the weights $w_j(k)$ till the following inequality holds $e_j^2(k) \geq r^2(k)$, i.e. it converges to the domain bounded by $r(k)$. If the latter condition is not satisfied, the current observation is ignored as it falls into the dead-zone of the algorithm.

3 Experiments

The goal of our experiments is to compare the performance of the proposed learning algorithm with the standard recursive least-squares (RLSE) method on a set of well-known benchmarks from the UCI and PROBEN1 [11] repositories.

All tests were performed 100 times each and averaged results for the standard RBFN with RLSE learning and the proposed algorithm are presented in Table 1. Average mean class error (MCE) values are given in percents of the training and checking data sets respectively. Each training pass was performed on the same data and clustering results for both RLSE and the proposed algorithm. In order to achieve the convergence of the proposed algorithm, it was executed in a batch mode for several epochs (5 in our case). All input data sets were preliminarily normalized on a unit hypercube. The subtractive clustering procedure was used to find the centers and radii of the neurons in the hidden layer. The "iris" and "wine" data sets from the UCI repository were randomly divided into the training and checking sets with 70% to 30% ratio respectively.

To determine the values of the $\delta_j(k)$ parameters we used the standard exponential-decay procedure: $\tilde{\delta}_j(k) = (1-1/\tau)(\tilde{\delta}_j(k-1) - \tilde{\delta}_\infty) + \tilde{\delta}_\infty$ (here $\tilde{\delta}_\infty$ denotes the lower bound of the $\tilde{\delta}_j(k)$ values), subject to constraints (10) at every learning step.

The following parameter values were used for all the data sets: $r(k) = 0.25$ for all k; $w_{ji}(0) = 0$, $\delta_j(0) = 1$, $\gamma_j(0) = 1$, $P_j(0) = \theta I_{h+1}$ (here I_{h+1} is the

Table 1. Experimental results for the standard RLSE and the proposed learning algorithms

Data set	RLSE		**The proposed algorithm**		$mean\{h\}$
	$mean\{MCE_{tr}\}$	$mean\{MCE_{ts}\}$	$mean\{MCE_{tr}\}$	$mean\{MCE_{ts}\}$	
iris	3.54	4.09	3.49	4.06	7.29
wine	2.77	4.21	2.10	3.19	8.89
cancer1	4.57	2.30	4.19	0.98	16
cancer2	4.19	5.17	3.49	4.60	14
cancer3	4.00	4.60	3.83	3.22	15
horse1	30.77	29.67	29.66	27.29	26
horse2	21.61	34.07	20.42	33.16	42
horse3	31.50	35.16	30.53	33.67	28

identity matrix of the dimension $(h+1)$, and $\theta = 10000$) for all $j = 1, \ldots, m$, $i = 1, \ldots, h$; $\tilde{\delta}_\infty = 0.1$, $\tau = 0.25$.

It should be noted that introducing a learning procedure for the radii of the neurons in the hidden layer, better classification results may be achieved. However, this should not influence significantly the advantage of one of the learning algorithms over another.

4 Conclusion

The proposed algorithm is quite simple in its computational realization, involving only simple matrix operations like summation, multiplication, and division by a scalar. It provides convergence of the adjustable weights to the ellipsoids of the minimal volume, which contain the optimal parameters. This algorithm does not require solving of the additional resource-consuming optimization or root finding problems. During the learning process and accumulation of information the algorithm gradually takes the form of the weighted recursive least squares method, which is quite popular in the neural networks learning tasks.

References

1. S. Abe. *Neural Networks and Fuzzy Systems.* Kluwer Academic Publishers, Boston, 1997.
2. S. Haykin. *Neural Networks. A Comprehensive Foundation.* Prentice Hall, Inc., Upper Saddle River, N.Y., 1999.
3. Da Ruan, editor. *Intelligent Hybrid Systems: Fuzzy Logic, Neural Networks, and Genetic Algorithms*, Boston, 1997. Kluwer Academic Publishers.
4. F. C. Schweppe. *Uncertain Dynamic Systems.* Prentice Hall, Englewood Cliffs, N.Y., 1973.
5. J. P. Norton. *An Introduction to Identification.* Academic Press Inc., London, 1986.
6. Z. Pawlak. Rough sets present state and further prospects. In *Proc. Int. Workshop on Rough Sets and Soft Computing*, pages 72–76, San Jose, California, 1994.

7. R. Yasdi. Combining rough sets learning and neural learning method to deal with uncertain and imprecise information. *Neurocomputing*, 7:61–84, 1995.
8. E. Fogel and Y. F. Huang. On the value of information in system identification — bounded noise case. *Automatics*, 18(2):229–238, 1982.
9. M. Halwass. "Least-Squares"-Modificationen in Gegenwart begrenzter Stoerungen. *MSR*, 33(8):351–355, 1990.
10. J. Hägglund. Recursive identification of slowly time-varying parameters. In *Proc. IFAC/IFORS Symp. on Identification and System Parameters Estimation*, pages 1137–1142, York, UK, 1985.
11. L. Prechelt. Proben1 a set of neural network problems and benchmarking rules. Technical Report 21/94, Fakultät für Informatik, Universität Karlsruhe, Sep. 1994.

Support Vector Neural Training

Włodzisław Duch

Department of Informatics, Nicolaus Copernicus University, Grudziądzka 5, Toruń, Poland, and School of Computer Engineering, Nanyang Technological University, Singapore

Abstract. SVM learning strategy based on progressive reduction of the number of training vectors is used for MLP training. Threshold for acceptance of useful vectors for training is dynamically adjusted during learning, leading to a small number of support vectors near decision borders and higher accuracy of the final solutions. Two problems for which neural networks have previously failed to provide good results are presented to illustrate the usefulness of this approach.

1 Introduction

In the final part of the multi-layer perceptron (MLP) training presentation of most vectors has no influence on the network parameters. Support Vector Machines (SVMs, [1]) progressively remove such vectors from the training procedure. This approach contributes not only to the increased speed near the end of training, but also to the higher accuracy that is finally achieved. If the margin between the hyperplane and the vectors from two classes is small the long tails of sigmoidal output function contributing to the error function may shift the position of the MLP hyperplane, and although the mean square error will decrease the number of classification errors may increase. Such behavior has been observed [2] after initialization of the MLP network with parameters obtained from the linear discrimination analysis (LDA).

Selection of training vectors near the border is a form of active learning [3] in which the training algorithm has an influence on what part of the inputs space the information comes from. Sensitivity analysis in respect to input perturbations has been used [4] to visualize and analyze decision borders. Another way to select border vectors is to use distances between vectors of different classes [5] but for large databases this is costly, scaling with a square of the number of vectors, and it does not include the information about possible influence of a given vector on the network training.

In this paper perhaps the simplest, but it seems that so far little explored, approach to active learning is investigated. Following SVM approach all data is initially used and progressively vectors that do not contribute much to the training process are removed. This number of support vectors used in the final stages of training is small, contributing to the precision and speed of learning. The algorithm and properties of this approach are described in the next section. In the third section an illustration of the selection process is shown on a version of XOR data and in section four two applications where MLPs have previously failed to provide good answers are presented. The paper ends with a few conclusions.

2 Active Learning by Dynamic Selection of Training Vectors

The Support Vector Neural Training (SVNT) algorithm selects those training vectors in each iteration that may support the training process. The neural network $F_k =$

W. Duch et al. (Eds.): ICANN 2005, LNCS 3697, pp. 67–72, 2005.

$M_k(\mathbf{X};\mathbf{W})$, for $k = 1 \ldots K$ outputs, is a vector mapping $F_k \in [0,1]$ that depends on parameters $\mathbf{W}$. These parameters are updated after presentation of the training data $\mathcal{T}$, depending on the difference between the target output values and the achieved network output $Y_k - M_k(\mathbf{X};\mathbf{W})$:

$$\Delta W_{ij} = -\eta \frac{\partial E(\mathbf{W})}{\partial W_{ij}} = \eta \sum_{k=1}^{K} (Y_k - M_k(\mathbf{X};\mathbf{W})) \frac{\partial M_k(\mathbf{X};\mathbf{W})}{\partial W_{ij}} \tag{1}$$

If the difference $\varepsilon(\mathbf{X}) = \max_k |Y_k - M_k(\mathbf{X};\mathbf{W})|$ is suffciently small $\mathbf{X}$ vector will have negligible influence on the training process (a more costly alternative is to check the gradient). A forward propagation step before each training epoch selects support vectors close to the decision borders for which $\varepsilon(\mathbf{X}) > \varepsilon_{\min}$ that should be included in the current training set. The number of SVs decreases because the backpropagation training leads to growing weights, and therefore slopes of sigmoidal functions become effectively steeper. This comes from the fact that $\sigma(\mathbf{W} \cdot \mathbf{X} - \theta) = \sigma(\beta(\mathbf{W}' \cdot \mathbf{X} - \theta'))$, where $\beta = ||\mathbf{W}||$, $\mathbf{W}' = \mathbf{W}/||\mathbf{W}||$ and $\theta' = \theta/||\mathbf{W}||$. The network provides sharper decision borders, leaving fewer support vectors that have $\varepsilon(\mathbf{X}) > \varepsilon_{\min}$.

For noisy data or for strongly overlapping clusters the number of vectors selected in each training epoch may change rapidly. To avoid such oscillations in the first training epoch $\varepsilon_{\min} = 0$ and all training vectors are used. Then $\varepsilon_{\min}$ is increased by $\Delta\varepsilon$ (0.01 is usually a good choice) after every epoch in which the accuracy has been increased. If the number of currently selected SVs increases after the next selection $\varepsilon_{\min}$ should be set slightly lower to stabilize the iterative process, therefore it is decreased by $\Delta\varepsilon$. To reduce oscillations $\Delta\varepsilon$ is then decreased (dividing it by 1.2 was a good empirical choice).

This works well if the number of mislabeled patterns, outliers, or noisy patterns giving large errors is not too large, but it is clear that such patterns may end up as support vectors even though their contribution to gradients is close to zero. This problem is solved by excluding from training also the patterns that give very large errors, keeping only those for which $\min_k |Y_k - M_k(\mathbf{X};\mathbf{W})| > 1 - \varepsilon_{\min}$ has been used. The $\varepsilon_{\min}$ threshold determines the margin of the size $1 - 2\varepsilon_{\min}$, centered at 1/2, for all output values. This may be translated into a margin around the decision surface near selected SVs $\mathbf{X}$, using the curvature of the sigmoidal functions in this area.

The SVNT algorithm selects best support vectors for training, cleaning the data at the same time. The algorithm proceeds as follows:

1. Initialize the network parameters $\mathbf{W}$, set $\Delta\varepsilon = 0.01$, $\varepsilon_{\min} = 0$, set $SV = \mathcal{T}$.
   Until no improvement is found in the last N_{last} iterations do
2. Optimize network parameters for N_{opt} steps on SV data.
3. Run feedforward step on $\mathcal{T}$ to determine overall accuracy and errors,
   take $SV = \{\mathbf{X}|\varepsilon(\mathbf{X}) \in [\varepsilon_{\min}, 1 - \varepsilon_{\min}]\}$.
4. If the accuracy increases:
   compare current network with the previous best one, choose the better one as the current best (take lower MSE if the number of errors is identical);
   increase $\varepsilon_{\min} = \varepsilon_{\min} + \Delta\varepsilon$.

5. If the number of support vectors $|SV|$ increases
   decrease $\varepsilon_{\min} = \varepsilon_{\min} - \Delta\varepsilon$;
   decrease $\Delta\varepsilon = \Delta\varepsilon/1.2$.

MSE error should always be determined on the whole data set. Variants of this algorithm may include changes of basic parameters, more sophisticated schedules of parameter changes, for example increase of the number of iterations N_{opt} between support vector selections that may speed up the final convergence, tolerate small increase of the number of SVs, and many others. Of course various thresholds may also be built in the backpropagation procedure itself, but in this paper only the simplest solution that does not require changes to the network optimization procedure is investigated.

In the implementation used in our experiments the optimization procedure runs for a small number of iterations (typically $N_{\mathrm{opt}} = 2 - 50$), and then selection of support vectors is performed. After the number of selected vectors stabilizes the number of iterations performed between reductions may be increased, or a stopping criterion for optimization may be used to converge to the particular solution for reduced data. After subsequent reduction of the training set another solution may be found. Thus the algorithm explores various solutions in the parameter space, keeping the core support vectors and changing small percentage of these vectors after every restart. Convergence at the later stages may therefore be far from monotonic, and it is worthwhile to wait for some number of iterations before stopping. Network with the lowest number of the training errors, and among those with the lowest MSE error, is returned at the end and used on the test data.

3 Numerical Experiments

Implementation of the SVNT algorithm has been made using Scaled Conjugated Gradient (SCG) optimization procedure from the Netlab package [6]. In order to see that SVNT algorithm is indeed capable of selecting correct support vectors noisy version of XOR data has been created, with Gaussian clouds around the corners and two additional vectors per cluster defining decision borders. Correct support vectors have been selected by the SVM algorithm implemented in the GhostMiner data mining package [7] using Gaussian and quadratic kernels. Parameter tuning (determined automatically using crossvalidation) was necessary for Gaussian kernel, but some kernels (linear, cubical) could not find the optimal solution within the range of parameters investigated. SVNT algorithm with batch selection of support vectors after every 5 iterations and $\Delta\varepsilon = 0.01$ converges to optimal solution in almost all runs. The number of support vectors is rapidly reduced to 8 (Fig.1).

Satellite Image data [8] consists of the multi-spectral values of pixels in the 3x3 neighborhoods in a small section (82x100) of an image taken by the Landsat Multi-Spectral Scanner. The intensities are one-byte numbers (0-255), the training set includes 4435 samples and the test set 2000 samples. Central pixel in each neighborhood is assigned to one of the six classes: red soil (1072), cotton crop (479), grey soil (961), damp grey soil (415), soil with vegetation stubble (470), and very damp grey soil (1038 training samples). This dataset is rather difficult to classify, with strong overlaps between some classes. Test set accuracies reported in the Statlog project [9] ranged between

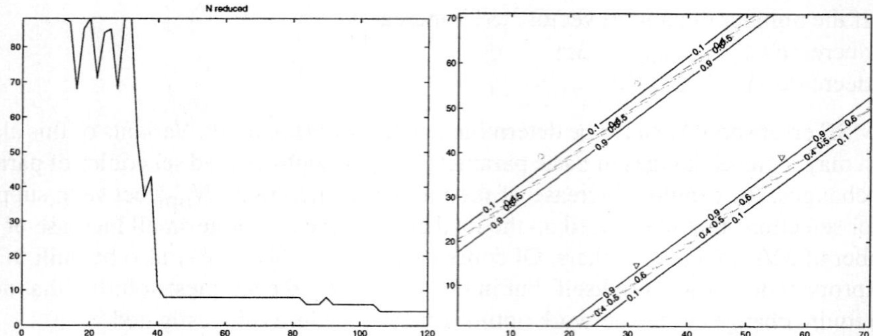

Fig. 1. Typical run on the noisy XOR data; left, the number of support vectors decreasing during iterations; right, the final solution with only support vectors shown

71% (Naive Bayes method) to 91% for the k-Nearest Neighbor classifier. Accuracy of MLPs was at the level of 86%, and RBFs at 88% on the test set.

This dataset has been re-analyzed with a number of methods available in the Ghostminer package [7]. Best results (Table 1) were achieved with the k-Nearest Neighbors classifier with small k (optimal k=3), suggesting that rather complex decision borders are needed. Creating a network with 36 hidden nodes, one per each output, and adding a regularization term with $\alpha = 0.5$ to avoid overfitting led to a very good results. The maximum variance of the hidden node responses reached 0.85, but for 5 nodes it was of the order of 0.001, and for another 4 nodes it was below 0.1, showing that such network is slightly too large, but most neurons are fully utilized. SVNT algorithm has achieved best results using a fraction of all training vectors. Large classes 1 and 5 may be reduced to 10% without any degradation of accuracy, while some classes retain almost all their vectors. With regularization at $\alpha = 0.5$ level no overfitting is observed, the training results are well correlated with the test results – the number of training errors goes below the number of test errors without any increase in the latter. Strong overlap of classes for this dataset requires relatively large number of support vectors – the final number of support vectors was 2075, still rather large.

Hypothyroid dataset [8] has been for a long time a challenge to neural networks. This data has been collected during two years of real medical screening tests for thyroid

Table 1. Results for the Satellite Image data

System and parameters	Train accuracy	Test accuracy
SVNT MLP, 36 nodes, $\alpha = 0.5$	96.5	91.3
kNN, k=3, Manhattan	–	90.9
SVM Gaussian kernel (optimized)	91.6	88.4
RBF, Statlog result	88.9	87.9
MLP, Statlog result	88.8	86.1
C4.5 tree	96.0	85.0

Table 2. Results for the hypothyroid dataset

Method	% train	% test	Ref.
C-MLP2LN rules	99.89	99.36	[10]
MLP+SCG, 4 neurons	99.81	99.24	this work
MLP+SCG, 4 neurons, 67 SV	99.95	99.01	this work
MLP+SCG, 4 neurons, 45 SV	100.0	98.92	this work
Cascade correlation	100.0	98.5	[11]
MLP+backprop	99.60	98.5	[11]
SVM Gaussian kernel	99.76	98.4	[10]

diseases and therefore contains mostly healthy cases. Among 3772 cases there are only 93 cases of primary hypothyroid, and 191 of compensated hypothyroid, the remaining 3488 cases are healthy cases. 3428 cases are provided for testing, with similar class distribution. 21 attributes (15 binary, 6 continuous) are given, but only two of the binary attributes (on thyroxine, and thyroid surgery) contain useful information, therefore the number of attributes has been reduced to eight. MLPs trained using all tricks of the trade (local learning rates, genetic optimization), and the cascade correlation algorithm reach only 98.5% percent accuracy on the test set [11]. Best results (Table 2) are obtained with 4 optimized logical rules, or very simple decision trees [10]. This shows the need of logical, sharp decision borders that may only be provided with a small network with large weights (at least 4 neurons are needed).

SVM algorithm with 8 features (results with all 21 features were significantly worse) and Gaussian kernels after optimization achieved best results for $C = 1000$ and bias$= 0.05$, found by crossvalidation on the training set. The total number of support vectors was 391 and the test result was 98.4%. Experiments with SVNT algorithm showed that with automatic determination of ε zero errors may be reached on the training set repeating SCG optimization for 100 iterations, followed by training set reduction. A rather flat error curve has been reached after 27400 iterations, with only two errors on the training, 34 errors on the test set. The ε has stabilized at 0.048 and only 67 support vectors were selected. If the training is continued further MSE drops very slowly, and after 43500 iterations zero training errors is reached, with 37 test errors and 45 support vectors (7, 21, and 17 respectively, from the three classes). The final weights grow to quite high values, of the order of 100, showing that this problem has an inherent logical structure [10]. Convergence is quite slow because little is gained by a very large increase of network weights. The number of iterations may be large, but with less than 100 vectors for training and a very small network time to do such calculations is quite short. Significant improvement over previously published neural network results has been achieved. These results represent probably the limit of accuracy that may be achieved for this data set.

4 Discussion and Conclusions

The SVNT algorithm gave significant improvements in classification of unbalanced data. In case of the hypothyroid problem the total number of SVs was amazingly small (Table 2) and their class distribution was well balanced. SVs are useful in defining class boundaries in one-class learning problems [12] and SVNT may easily compete with SVM in generating them. Depending on the rejection threshold, MSE, the number of errors and the number of selected vectors may oscillate. Large oscillations are damped by the dynamic rejection threshold in the SVNT algorithm, but small oscillations may actually be useful. Reduction of the traing set introduces a stochastic element to the training, pushing the system out of the local minima. Comparing this approach to the algorithms based on evolutionary or on other global optimization algorithms it is clear that at the initial stages convergence is fast and that all good models need to share roughly the same characteristics. Near the end of the training gradients are small and wider exploration of the parameter space is worthwhile to fine tune the decision borders. This is exactly what SVNT does.

As all gradient-based training algorithms SVNT may sometimes fail. More empirical tests are needed to evaluate it, but the ability to find minimal number of support vectors, handle multiclass problems with strong class overlaps, and excellent results on the unbalanced data demonstrated in this paper, combined with simple modifications to the standard backpropagation algorithms in batch and on-line learning needed to achieve this, encourage further experimentation.

References

1. B. Schölkopf, and A.J. Smola, *Learning with Kernels. Support Vector Machines, Regularization, Optimization, and Beyond.* Cambridge, MA: MIT Press, 2001.
2. W. Duch, R. Adamczak, and N. Jankowski, "Initialization and optimization of multilayered perceptrons". Proc. 3rd Conf. on Neural Networks and Their Applications, Kule, Poland, pp. 105-110, 1997.
3. D. Cohn, L. Atlas, and R. Ladner, "Improving Generalization with Active Learning," Machine Learning, vol. 15, pp. 201-221, 1994.
4. A. P. Engelbrecht, "Sensitivity Analysis for Selective Learning by Feedforward Neural Networks," Fundamenta Informaticae, vol. 45(1), pp. 295-328, 2001.
5. W. Duch, "Similarity based methods: a general framework for classification, approximation and association," Control and Cybernetics, vol. 29 (4), pp. 937-968, 2000.
6. I. Nabnay, and C. Bishop, NETLAB software, Aston University, Birmingham, UK, 1997. http://www.ncrg.aston.ac.uk/netlab/
7. W. Duch, N. Jankowski, K. Grąbczewski, A. Naud, and R. Adamczak, Ghostminer software, http://www.fqspl.com.pl/ghostminer/
8. C.L, Blake, and C.J. Merz, UCI Repository of machine learning databases, University of California, Irvine, http://www.ics.uci.edu/ ~mlearn/MLRepository.html
9. D. Michie, D.J. Spiegelhalter, and C.C. Taylor, *Machine Learning, Neural and Statistical Classification.* London, UK: Ellis Horwood, 1994.
10. W. Duch, R. Adamczak, and K. Grąbczewski, "A New Methodology of Extraction, Optimization and Application of Crisp and Fuzzy Logical Rules," *IEEE Transactions on Neural Networks*, vol. 12, pp. 277–306, 2001.
11. W. Schiffman, M. Joost, and R. Werner, "Comparison of optimized backpropagation algorithms". Proc. of European Symp. on Artificial Neural Networks, Brussels 1993, pp. 97-104.
12. D.M.J. Tax and R.P.W. Duin, "Uniform Object Generation for Optimizing One-class Classifiers", Journal of Machine Learning Research Vol. 2, pp. 155-173, 2001.

Evolutionary Algorithms for Real-Time Artificial Neural Network Training

Ananda Jagadeesan, Grant Maxwell, and Christopher MacLeod

School of Engineering, The Robert Gordon University,
Schoolhill, Aberdeen, UK

Abstract. This paper reports on experiments investigating the use of Evolutionary Algorithms to train Artificial Neural Networks in real time. A simulated legged mobile robot was used as a test bed in the experiments. Since the algorithm is designed to be used with a physical robot, the population size was one and the recombination operator was not used. The algorithm is therefore rather similar to the original Evolutionary Strategies concept. The idea is that such an algorithm could eventually be used to alter the locomotive performance of the robot on different terrain types. Results are presented showing the effect of various algorithm parameters on system performance.

1 Introduction

Artificial Neural Networks (ANNs) are normally trained before use, off-line. Online learning, in complex Neural Networks - for example, those designed for pattern classification - has proved difficult and complex solutions like Adaptive Resonance Theory [1] have had to be applied. This is because learning new patterns effectively means altering the State Landscape of the network.

However, in some cases - for example, in many Control Systems - On-Line Learning is possible because, in contrast with Pattern Recognition, the network parameters may only have to change gradually as the controlled system changes. As an example, consider a legged mobile robot walking across a series of different ground types (for example, sandy, rocky or boggy). Obviously, in this case, for optimum locomotive efficiency, the robot will have to alter its gait pattern in response to the conditions underfoot. It may have to shorten its stride, for example, when moving from a hard to a sandy surface. One possible way to achieve this is for the robot's leg parameters (such as stride length, etc) to be under the control of a series of learned or pre-programmed gait patterns. However, this has the disadvantage of complexity and inflexibility. This latter point is illustrated when we consider the situation which might occur if the robot meets a surface for which it has not been prepared. Since it has no way of finding a suitable gait pattern, it has to lumber on with the "best guess".

The idea behind the Real-Time Evolutionary Algorithm (RTEA) is to constantly alter the robot's locomotive algorithms by a small random amount, evaluating corresponding changes to the fitness function (in the first approximation,

W. Duch et al. (Eds.): ICANN 2005, LNCS 3697, pp. 73–78, 2005.

the efficiency of walking). Mutations which cause beneficial changes in fitness are kept; those which make the situation worse are discarded. In this way, the robot's control system is constantly seeking a better solution to the walking problem and will move towards such a solution, even when the robot moves onto a different surface.

Artificial Neural Networks were used to control the robot's legs, rather than direct control (for example, an algorithm generating a rhythmical step pattern), so that lessons learned from the experience could be applied more generally to other neural network controlled systems. Such networks are commonly used to control a variety of different mechanical and mechatronic systems.

2 Robotic Test System

It was decided to use a quadruped robotic system as a test bed for the neural network since it represented a generalised control system and also it has been used successfully several times in the past. The system is similar to that used by Muthuraman [2] and McMinn [3] for work on Evolutionary ANNs. Due to space restrictions here, only a brief overview of the system is given. The leg model used is a simulated linear servo mechanism with two active degrees of freedom, as shown in Fig. 1. The leg advances forward by one unit on receiving a positive pulse for one clock cycle and backward on receiving a similar negative going pulse. A similar arrangement is used to lift and lower the leg. The robot is driven forward when the legs are in contact with the ground and moving in the correct direction, the distance moved being equal to leg units moved under these conditions.

The neurons used in the network produce a positive followed by a negative pulse to drive the leg as shown in Fig. 2. Four evolvable parameters (D_1, D_2, T_{on1} and T_{on2}) are associated with each neuron; these are trained by the real-time EA.

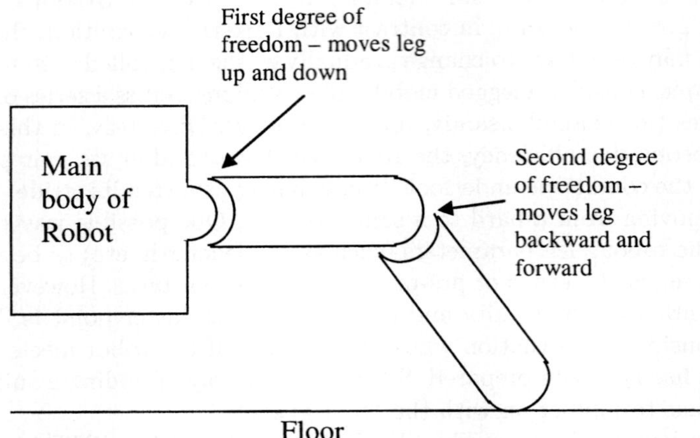

Fig. 1. Model of Robot Leg (Front View)

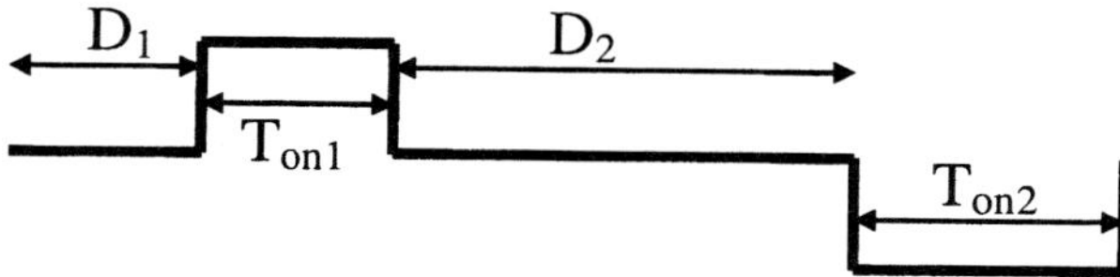

Fig. 2. Neuron Output

The fitness function of the robot is a programmable combination of three factors - Stability, Fuel Consumption and Distance Travelled. The weighting between these can be changed in order that the system can develop different gaits. Stability takes into account the inclination of the robot and how many feet it has on the ground over its walking cycle. Fuel Consumption is related to the powered phase of leg use. Finally, Distance Travelled is measured, as indicated above, using the leg movements.

3 Neural Network Design

Several common network topologies and neuron types were tried at the start of the project. From these experiments, two important principles were discovered.

Firstly, neuron types with built in threshold functions (for example, Threshold Perceptron types) performed poorly. This was because either many small or a single large mutation is required to overcome the threshold and this causes large jumps in the network's fitness function, rather then small adjustments, making the global minima difficult to find. Therefore, for success, it is important to choose a unit which alters its behaviour gradually in order to avoid this problem.

Secondly, the network topology is equally important. Fully connected networks perform badly because mutations in one side of the network affect the other side. In the case of the legged robot used in the experiments, this meant that changes in one system of legs caused changes in an unrelated system and so the network fitness tended not to increase. Of course, it is often useful to provide some connection between different synchronised networks within a system like this and an effective network was one which took its overall timing from a master clock. Networks with recurrent connections performed poorly for similar reasons. The network topology is shown in Fig. 3. The outputs are taken from the child neurons (marked C); two are required for each leg (one for each degree of freedom).

4 Real-Time Evolutionary Algorithm Parameters

The parameters of the RTEA were tested to establish their effect on system performance. The first to be tested was the effect of mutating all the parameters in the system compared with mutating only one at a time. Figure 4 shows the

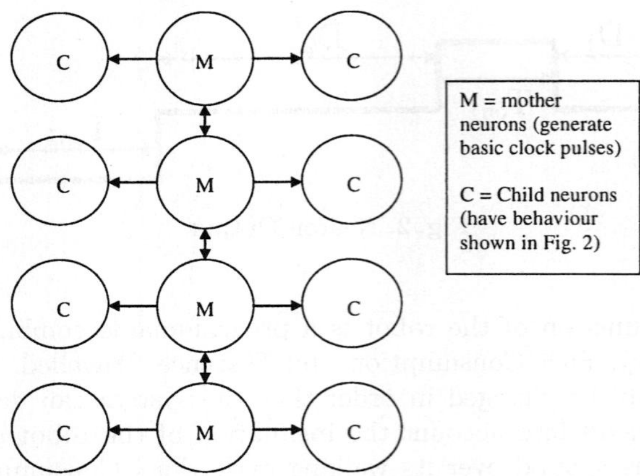

Fig. 3. Network Topology

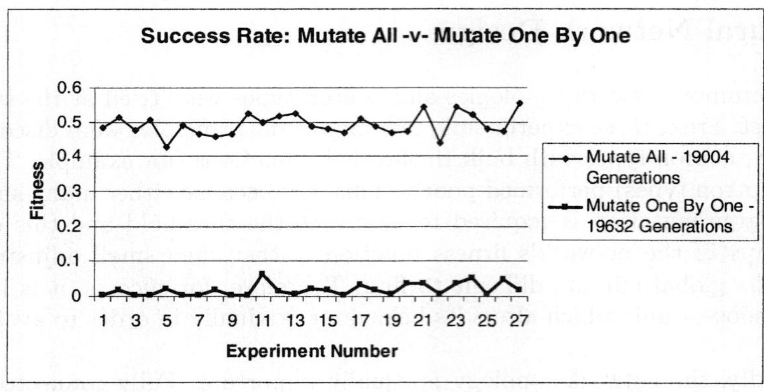

Fig. 4. The effect of mutating each parameter singly or in groups

effect of mutating all the parameters compared with the fitness profile of only mutating one at a time. This experiment was done over a wide variety of system setups and parameter variation (over 2000 experiments representing all possible variations of system setup). Each point on the graph is the average of 81 such experiments. We might expect this effect to be even greater if we used mutation with an expectation value of zero [4]; however, this was not tested. It can be seen from the graph that mutating all of the parameters is more successful.

The effect of choosing a uniform versus a normal (Gaussian) distribution [4] is shown in Fig. 5. In this case the normally distributed numbers proved more effective in most cases. As before, the graph shows the effect over the whole range of system setup.

When performing these experiments, we have not allowed error increases to escape from local minima (which some algorithms such as simulated annealing

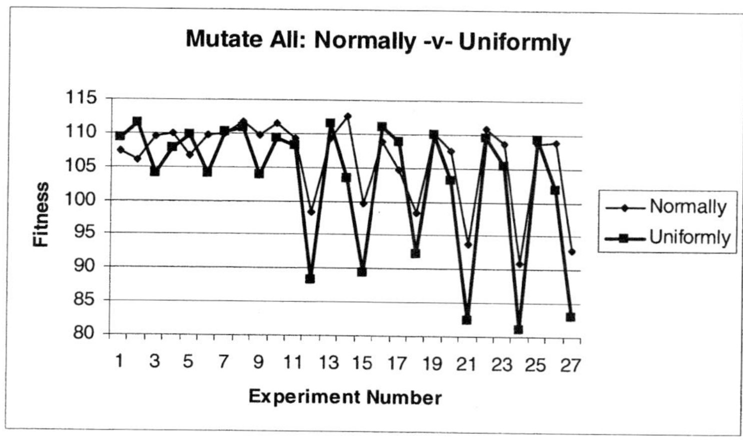

Fig. 5. The effect of Gaussian and Uniform mutation

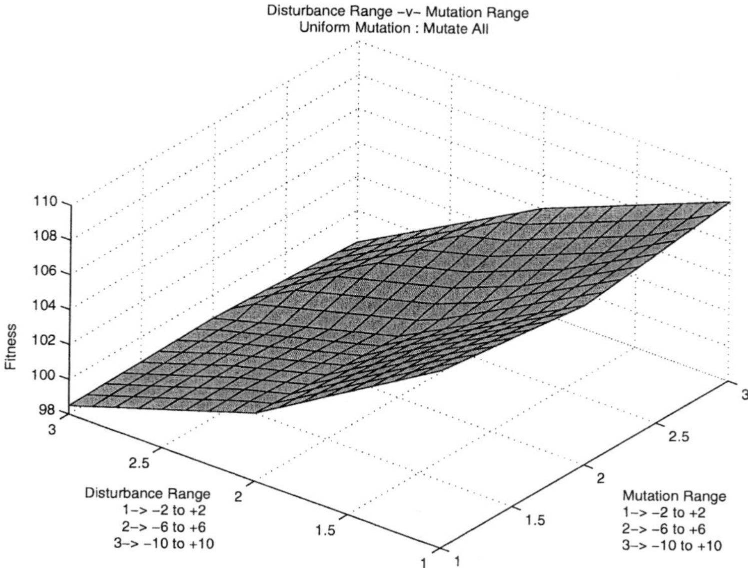

Fig. 6. Relationship between average mutation size and distance from equilibrium

do allow). The reason for this is that, in the case of this system, we are operating quite close to equilibrium. Should the system have to change gait (for example, from a walk to a trot), then this facility would have to be utilised.

In many such systems, mutation size is controlled by distance from equilibrium; examples include simulated annealing, where an artificial temperature controls mutation size [5], and Evolutionary Strategies with its one in five rule [4].

Tests were therefore conducted to establish how distance from equilibrium interacted with mutation size. It was found that these generalisations do not necessarily apply; Fig. 6 demonstrates this. In this graph disturbance range means distance from equilibrium (higher is further) and mutation rate is the average size of the mutation. Again, this is averaged over all system parameters.

5 Conclusion

The conclusions which were drawn from these experiments were that Real-Time Evolutionary Algorithms can optimise Neural Networks on-line in certain applications. Such applications are principally those in which small changes are needed to "tune" the performance of the system and include many types of control system.

In such systems the type of neuron unit and network topology types are critical. The neuron unit should be one where output changes gradually with input and the network topology should be a feedforward type where unnecessary lateral connections are suppressed.

The experiments confirm that many of the recommendations for off-line optimization using EAs also apply on-line. These include that normal distributions and variable mutation rates are effective.

References

1. G. A. Carpenter and S. Grossberg, ART - 2: self organisation of stable category recognition codes for analog input patterns, *Applied Optics*, 26:4919-4930, 1987.
2. S. Muthuraman, G. M. Maxwell and C. MacLeod, The Evolution of Modular Artificial Neural Networks for Legged Robot Control. In *proceedings of the International Conference on Neural Networks and Neural Information Processing* (ICANN/ICONIP 2003), 488-495, Istanbul (Turkey), June 2003.
3. D. McMinn, C. MacLeod and G. M. Maxwell, Evolutionary Artificial Neural Networks for Quadruped Locomotion. In *proceedings of the International Conference on Neural Networks* (ICANN 2002), 789-794, Madrid (Spain), August 2002.
4. H.-P. Schwefel. *Evolution and Optimum Seeking*, Wiley, New York, 1995.
5. P. P. C. Yip and Y. Pao, Growing Neural Networks using Guided Evolutionary Simulated Annealing. In *proceedings of the International Conference on Evolutionary Programming*, 17-20, Istanbul (Turkey), June 1994.

Developing Measurement Selection Strategy for Neural Network Models

Przemysław Prętki and Marcin Witczak

Institute of Control and Computation Engineering,
University of Zielona Góra,
ul. Podgórna 50, 65–246 Zielona Góra, Poland
{P.Pretki, M.Witczak}@issi.uz.zgora.pl

Abstract. The paper deals with an application of the theory of optimum experimental design to the problem of selecting the data set for developing neural models. Another objective is to show that neural network trained with the samples obtained according to D-optimum design is endowed with less parameters uncertainty what allows to obtain more reliable tool for modelling purposes.

1 Introduction

In recent years neural networks have been willingly used in many fields of researcher activities.A still increasing popularity of neural networks results from their capability of modelling a large class of natural and artificial phenomena which cannot be successfully dealt with classical parametric models. In the literature devoted to neural networks it is very often assumed that it is possible to obtain a neural network with an arbitrary small uncertainty. In this paper, authors assume that the confidence measure is associated with a neural network - variance of the predicted output. Thereby the problem of minimization neural network uncertainty can be solved by utilizing the well-known concepts of experimental design theory [1]. The main purpose of this paper is to develop an effective approach that makes it possible to design an artificial neural network with possibly small parameter uncertainty which, undoubtedly increases its reliability in many applications.

The paper is organized as follows. In Section 2 basic definitions and terminology is given. Section 3 presents important properties regarding Optimum Experimental Design (OED) applied for neural networks. These properties are very useful while applying OED in practice. Finally, the last part is devoted to an illustrative example regarding the DAMADICS benchmark [2].

2 Preliminaries

2.1 Structure of a Neural Network

Let us consider the standard single-hidden layer feed-forward network (SLFN) with n_h non-linear neurons and a single linear output neuron. The choice of such

W. Duch et al. (Eds.): ICANN 2005, LNCS 3697, pp. 79–84, 2005.

structure is not an accidental one since it has been proved that this kind of neural networks can learn any finite set $\{(\boldsymbol{u}_k, y_k) : \boldsymbol{u}_k \in \mathbb{R}^{n_u}, y_k \in \mathbb{R}, k = 1, 2, ...N\}$ with an arbitrarily small error. The input-output mapping of the network is defined by:

$$y_{m,k} = \boldsymbol{g}\left(\boldsymbol{P}^{(n)}\overline{\boldsymbol{u}}_k\right)^T \boldsymbol{p}^{(l)} = f(\overline{\boldsymbol{u}}_k, \boldsymbol{p}), \tag{1}$$

where $y_{m,k} \in \mathbb{R}$ stands for the model output, $\boldsymbol{g}(\cdot) = [g_1(\cdot), \ldots, g_{n_h}(\cdot), 1]^T$, where $g_i(\cdot) = g(\cdot)$ is a non-linear differentiable activation function,

$$\boldsymbol{p} = [(\boldsymbol{p}^{(l)})^T, \boldsymbol{p}^{(n)}(1)^T, \ldots, \boldsymbol{p}^{(n)}(n_h)^T]^T \in \mathbb{R}^{n_p} \tag{2}$$

represents the parameters (weights) of the model which enter linearly and non-linearly to the $f(\cdot)$ respectively. Moreover, $\overline{\boldsymbol{u}}_k \in \mathbb{R}^{n_r+1}$, $\overline{\boldsymbol{u}}_k = [u_{1,k}, \ldots, u_{n_r,k}, 1]^T$ where $u_{i,k}$, $i = 1, \ldots, n_r$ are the system inputs.

2.2 Parameter Estimation

Ones the neural network structure has been selected, the next step to build a model of a real system is to obtain the best value $\hat{\boldsymbol{p}}$ of its the parameters. Let us assume that the system output satisfies the following equality:

$$y_k^s = y_{m,k} + \varepsilon_k = f(\overline{\boldsymbol{u}}_k, \boldsymbol{p}) + \varepsilon_k, \tag{3}$$

where ε is i.i.d and $\mathcal{N}(0, \sigma^2)$. Among many popular estimation techniques one of them, seems to be of great importance - namely, well known maximum likelihood estimation [6]. The most important properties of maximum likelihood estimators is that they are asymptotically Gaussian and unbiased i.e. the distribution of $\hat{\boldsymbol{p}}$ tends to $\mathcal{N}(\boldsymbol{p}^*, F^{-1}(\boldsymbol{p}^*))$, where $\boldsymbol{p}^*$ stands for the value of true parameters. In the presence of additive normal noise (see (3)) the parameters of neural network can be estimated with the equality criterion:

$$\hat{\boldsymbol{p}} = \arg\min_{\boldsymbol{p} \in \mathbb{R}^{n_p}} ||y^s(\Xi) - f(\Xi, \boldsymbol{p})||_2^2 \tag{4}$$

where $||\cdot||_2$ is the Euclidean norm and Ξ stands for experimental conditions. To solve the optimization problem (4) many algorithms can be successful used, where the Levenberg-Marquardt method is the most popular one. Moreover, Walter and Pronzato [5] shown that appropriate experimental conditions Ξ may remove all suboptimal local minima, so that local optimization algorithms can be used to find the best neural network parameters.

2.3 Parameter Uncertainty

The main problem in estimating parameters of a neural network lies in the fact that the training data are typically affected by noise and disturbances. It causes the presence of the so-called parameter uncertainty, which means that the neural network output should be viewed as a probabilistic one. In the case of the regression problem defined in (3), the feasible parameter set can be defined according to the following formulae [6]:

$$\mathbb{P} = \left\{ \boldsymbol{p} \in \mathbb{R}^{n_p} \Big| \sum_{i=1}^{n_t} (y_i^s - f(\overline{\boldsymbol{u}}_k, \boldsymbol{p}))^2 \leq \sigma^2 \chi^2_{\alpha,n_t} \right\} \tag{5}$$

where χ^2_{α,n_t} is the Chi-square distribution quantile. It can be shown [1,6] that the difference between neural network and the modelled system $z_k = y_k^s - y_{m,k}$ is given by:

$$|z_k| \leq t^{\alpha/2}_{n_t - n_p} \hat{\sigma} \left(1 + \boldsymbol{r}_k^T \boldsymbol{F}^{-1} \boldsymbol{r}_k\right)^{1/2}. \tag{6}$$

n_t denotes the number of input-output measurements used for parameter estimation, $t^{\alpha/2}_{n_t-n_p}$ is the t-Student distribution quantile, $\hat{\sigma}$ is the standard deviation estimate, $\boldsymbol{r}_k = \left.\frac{\partial f(\overline{\boldsymbol{u}}_k, \boldsymbol{p})}{\partial \boldsymbol{p}}\right|_{\boldsymbol{p}=\boldsymbol{p}^*}$, and $\boldsymbol{F}^{-1}$ is the inverse of Fisher information matrix defined as :

$$\boldsymbol{F} = \sum_{i=1}^{n_t} \boldsymbol{r}_i \boldsymbol{r}_i^T. \tag{7}$$

It is easy to see that the length of the confidence interval (6) is strongly related with the Fisher information matrix (7) that depends on the experimental conditions $\Xi = [\boldsymbol{u}_1, \ldots, \boldsymbol{u}_{n_t}]$. Thus, optimal experimental conditions can be found by choosing $\boldsymbol{u}_i$, $i = 1, \ldots, n_t$, so as to minimize some scalar function $\phi(\cdot)$ of (7). Such a function can be defined in several different ways [1], while probably the most popular is the D-optimality criterion:

$$\phi(\boldsymbol{F}) = \det \boldsymbol{F}, \tag{8}$$

A D-optimum design minimizes the volume of the confidence ellipsoid approximating the feasible parameter set 5. Moreover, from the Kiefer-Wolfowitz equivalence theorem [1], it follows that the variance of the predicted models output can be minimized by determining of experimental conditions according to D-optimum criterion, assuming that the continuous experimental design [1] is used. Another reason which speaks in favor of using this criterium is its popularity, what makes that in the literature one can find several, very effective algorithms developed for the D-optimum design e.g. DETMAX or Wynn-Fedorov [1],[6].

3 D-Optimum Experimental Design for Neural Networks

3.1 Partially Nonlinear Model

One of the main difficulties associated with the application of an experimental design theory to neural networks is the dependency of the optimal design on the model parameters. It can be proved [7] that the experimental design for a general structure (1) is independent of the parameters that enter linearly into (1). In consequence the process of minimizing the determinant of $\boldsymbol{F}^{-1}$ with respect to $\boldsymbol{u}$ is independent of the linear parameters $\boldsymbol{p}^l$. This means that at least a rough estimate of $\boldsymbol{P}^{(n)}$ is required to solve the experimental design problem. Of course, it can be obtained with any training method for feed-forward neural networks, and then the specialized algorithms for D-optimum experimental design can be applied [1,6].

3.2 Fisher Information Matrix in a Neural Network Regression

The Fisher information matrix $\boldsymbol{F}$ of (1) may be singular for some parameter configurations, and in such cases it is impossible to obtain its inverse $\boldsymbol{F}^{-1}$ that is necessary to calculate (6) as well as to utilize the specialized algorithms for obtaining the D-optimum experimental design [1,6]. Fukumizu [3] established the conditions under which $\boldsymbol{F}$ is singular and developed the procedure that can be used for removing the redundant neurons what guarantees that information matrix is positive defined. Unfortunately, Fukumizu's theorem have strictly theoretical meaning as in most practical situations FIM would be close to singular but not singular in an exact sense. In such a case FIM should be regularized with the help of the methods proposed in [1]. In this paper we propose to use a completely different approach that is more appropriate for the models (1). As has already been mentioned, process of minimizing the determinant of $\boldsymbol{F}$ with respect to $\boldsymbol{u}$ is independent of the linear parameters $\boldsymbol{p}^l$. This means that $\boldsymbol{p}^l$ can be set arbitrarily. This implies that it can be employed as an elegant tool for controlling the value of the determinant (7). Thus, before the iterative algorithm for finding the D-optimum design is started (e.g. the Wynn-Fedorov algorithm [1,6] that is used in this work), it is necessary to select $\boldsymbol{p}^l$ so as to ensure that the matrix $\boldsymbol{F}$ is far from a singular one.

4 Experimental Simulations

In order to show that the proposed approach effectively leads to the decrease of the neural network parameter uncertainty, let us consider the following example. The problem is to develop a neural network that can be used as a model of the valve plant actuator in the sugar factory in Lublin, Poland. The data for the normal operating mode have been collected by actuator simulator developed with MATLAB Simulink (worked out in DAMADICS [1] project). The actuator itself can be considered as a two-input and one-output system:

$$\boldsymbol{u} = (U_1, U_2) \qquad y = F \tag{9}$$

where U_1 is the control signal, U_2 is the pressure at the inlet of the valve and the F is the juice flow at the outlet of the valve (for more details see [2]). On the basis

[1] Research Training Network funded by the European Commission under the 5th Framework Programme.

of a number of trail-and-error experiments a suitable number of hidden neurons has been established $n_h = 12$. In the preliminary experiment $\boldsymbol{u}_k, k = 1, \ldots, n_t = 68$ were obtained in such a way so as to equally divide the input space. Then the Levenberg-Marquardt algorithm [6] was employed for parameter estimation. Next, in order to prevent the fact that FIM is almost a singular matrix it was appropriately transformed with the use of linear parameters $\boldsymbol{p}_l$. Based on the obtained estimates the Wynn-Fedorov algorithm [1] was employed to obtain the D-optimum experimental design and then the parameter estimation process was repeated once again. Figure 1 presents the variance function $\boldsymbol{r}_k^T F^{-1} \boldsymbol{r}_k$ for the network obtained with the application of optimal experimental design OED (a) and the network obtained without it (b). Moreover in Fig. 2 residuals and its bounds (determined with $1 - \alpha = 0.95$) can be seen for both networks. It can be observed that the use of OED results in a network with smaller uncertainty.

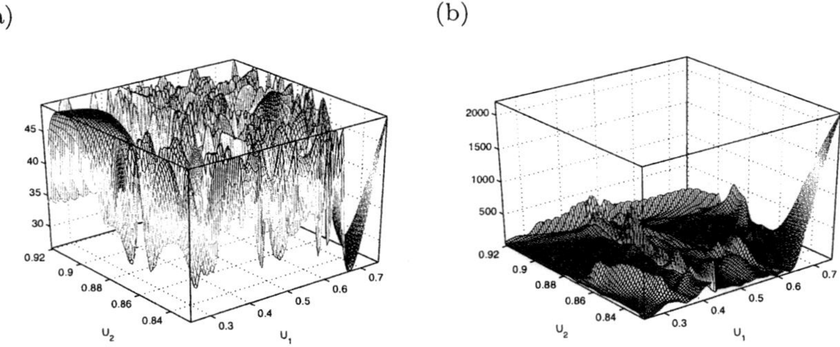

Fig. 1. Variance function of the neural network trained with samples chosen according to (a) D-optimal design and (b) in the way to uniformly divide input space

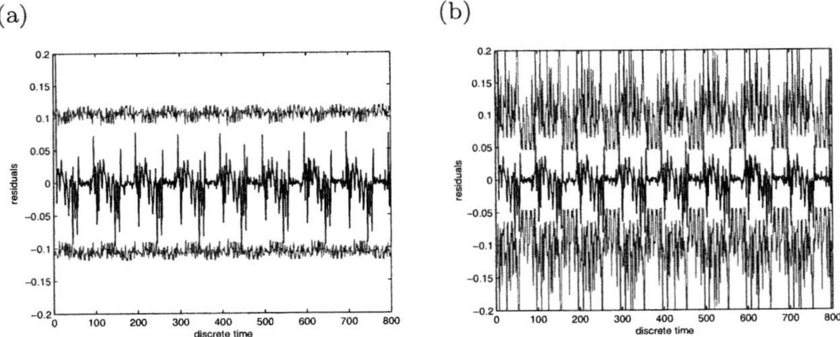

Fig. 2. Difference between output of the real system and its neural model along with bounds for a network trained with samples chosen according to (a) D-optimal design and (b) in the way to uniformly divide input space

5 Concluding Remarks

The paper deals with the problem of minimization parameters uncertainty of neural networks by means of utilizing optimum experimental design theory. One of the main advantages of the proposed approach is that it is possible to obtain a neural model which more accurately describes a real system. This is a crucial subject in many industrial applications where such systems are used for fault detection purposes. Moreover, with the use of the proposed approach it is possible to minimize the size of a training set, what is invaluable in applications for which process of collecting training samples is very expensive or time consuming. Besides, an important property of independence of the design of the parameters that enter linearly into the neural network was revealed. This property was also employed to prevent the singularity of FIM. This is especially important from the point of view of the Wynn-Fedorov algorithm because it evaluates FIM in order to obtain the D-optimum design. The proposed approach was tested on the DAMADICS benchmark problem. The experimental results show that the approach makes it possible to obtain a suitably accurate mathematical description of the system with small model uncertainty.

Acknowledgment

The work was supported by State Committee for Scientific Research in Poland (KBN) under the grant 4T11A01425 *Modelling and identification of non-linear dynamic systems in robust diagnostics.*

References

1. Atkinson A.C. and Donev A.N. (1992): *Optimum Experimental Designs.* – New York: Oxford University Press.
2. DAMADICS (2004): Website of the Research Training Network *DAMADICS: Development and Application of Methods for Actuator Diagnosis in Industrial Control Systems.* `http://diag.mchtr.pw.edu.pl/damadics/`.
3. Fukumizu K. (1996): *A regularity condition of the information matrix of a multilayer perceptron network.* – Neural Networks, Vol.9, No.5, pp.871–879.
4. Huang G.B. and Haroon A.B. (1998): *Upper Bounds on the Number of Hidden Neurons in Feedfoward Networks with Aribitrary Bounded Nonlinear Activation Functions.* – IEEE Trans. Neural Networks, Vol.9, No.1, pp.224–228.
5. Pronzato L. and Walter E. (2001): *Eliminating Suboptimal Local Minimizers in Nonlinear Parametr Estimation.* – Technometrics, Novenberg 2001, vol. 43, no. 4.
6. Walter E. and Pronzato L. (1997): *Identification of Parametric Models from Experimental Data.* – London: Springer.
7. Witczak M. and Prętki P. (2004): *Proc. Methods of Artificial Intelligence: AI–METH Series*, November 17-19, 2004, Gliwice, Poland, pp. 159-161.

Nonlinear Regression with Piecewise Affine Models Based on RBFN

Masaru Sakamoto, Dong Duo, Yoshihiro Hashimoto, and Toshiaki Itoh

Department of Systems Engineering, Nagoya Institute of Technology,
Gokiso-cho, Showa-ku, Nagoya, 466-8555, Japan

Abstract. In this paper, a modeling method of high dimensional piecewise affine models is proposed. Because the model interpolates the outputs at the orthogonal grid points in the input space, the shape of the piecewise affine model is easily understood. The interpolation is realized by a RBFN, whose function is defined with max-min functions. By increasing the number of RBFs, the capability to express nonlinearity can be improved. In this paper, an algorithm to determine the number and locations of RBFs is proposed.

1 Introduction

In moving towards the production of high added value chemicals, changes in operating conditions become larger and more frequent and increases in the strength of nonlinear characteristics of the dynamics of the plant occur. In some processes, switching of dynamics such as follows occurs. If temperature increases to a certain value, some reaction occurs. Manipulation has saturation.

Mixed logical dynamical system [BM1] can deal with switching of dynamics. The system equations including switching can be described with mixed (1,0) integer inequalities. If the nonlinearity of the switched dynamics was approximated by a piecewise affine model, model predictive control problem of every nonlinear system could be described with mixed (1,0) integer linear programming.

For approximation of nonlinear systems with piecewise linear models, several methods have been proposed such as hinging sigmoids [HH1], clustering and local affine models [CC1][CC2][FT1][KT2] and Max-Min propagation neural networks [EP1]. In the models, an affine function is defined for each partition and the partition is defined by the combination of linear functions. If the dimension of the input space is high, it is very difficult to understand the boundaries.

Hashimoto et. al. [Ha1] proposed a piecewise affine modeling using a radial basis function network (RBFN). Each RBF is located on a grid point of orthogonal coordinate of the input space. The output is generated by linear combination of heights of RBFs. It is equal to the result from interpolating of tops of RBFs. Therefore, the space of the piecewise affine model is easily understood. The reliability of a local model can be estimated based on the number of data and variance of estimation errors in its local area. Because the model was proposed for on-line modeling and control, recursive least square method was adopted to

W. Duch et al. (Eds.): ICANN 2005, LNCS 3697, pp. 85–90, 2005.

calculate weights of RBFNs. The results of RLS deteriorated when the number of RBFs were increased to improve the expression capability.

In this paper, an off-line algorithm to approximate nonlinear system equation is discussed. In the next section, the RBFN proposed in [Ha1] is illustrated. In the third section, the modeling procedure of the RBFN is proposed. By using a numerical example, the effectiveness of proposed method is illustrated in comparison with the result of the previous work.

2 RBFN for Piecewise Affine Models

The RBFN for piecewise affine model has a popular structure shown in Fig.1.

$$\hat{y}(x) = \sum_{n_1=1}^{N_1} \cdots \sum_{n_L=1}^{N_L} \omega_{[n1,\cdots,n_L]} \cdot a_{[n_1,\ldots,n_L]}(\mathbf{x}) \tag{1}$$

Its characteristics are RBFs' shape and locations. Each RBF is located on a grid point of orthogonal coordinate in the input space. The output of each RBF $a_{[n_1,\ldots,n_L]}(\mathbf{x})$ is defined by the metric from the grid point $D(\mathbf{X}, [n_1, \ldots, n_L]^T)$, where L is the dimension of the input space, n_l indicates the grid number of the lth axis of the input space and $\mathbf{X}$ is a scaled vector of the input vector $\mathbf{x}$.

$$a_{[n_1,\ldots,n_L]}(\mathbf{x}) = 1 - D(\mathbf{X}, [n_1, \ldots, n_L]^T) \tag{2}$$

The scaling defined in Eq. (3) is executed to make the intervals of grids unity, where $gr_{[n_1,\ldots,n_l,\ldots,n_L]}$ indicate the coordinate of the grid in the input space.

$$\begin{array}{ll} \textit{if} & gr_{[n_1,\ldots,n_l,\ldots,n_L]} \le x_l < gr_{[n_1,\ldots,n_l+1,\ldots,n_L]} \\ \textit{then} & X_l = n_l + \dfrac{x_l - gr_{[n_1,\ldots,n_l,\ldots,n_L]}}{gr_{[n_1,\ldots,n_l+1,\ldots,n_L]} - gr_{[n_1,\ldots,n_l,\ldots,n_L]}} \end{array} \tag{3}$$

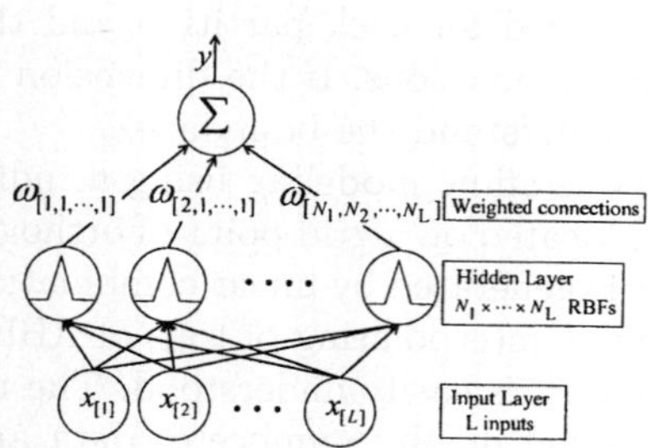

Fig. 1. RBFN structure

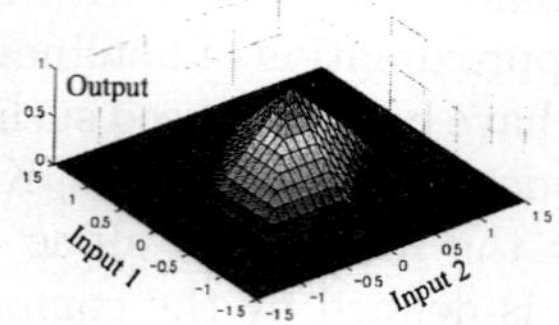

Fig. 2. RBF for 2-inputs

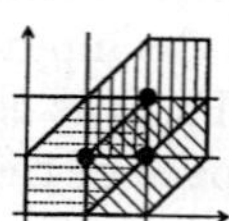

Fig. 3. Intersection of RBFs

The metric in the scaled space is defined in Eq. (4) by using max-min functions.

$$D(\mathbf{X},\mathbf{Xo}) = \min\left(\max_{l\in\{1,\dots,L\}}(X_l - Xo_l, 0) - \min_{l\in\{1,\dots,L\}}(X_l - Xo_l, 0), 1\right) \tag{4}$$

This equation can be applied to any dimensional systems. The shape of a RBF in 2-input systems is a six-sided pyramid as shown in Fig. 2. An Intersection with neighboring three RBFs in 2-input system is shown in Fig. 3. By using the metric defined in Eq. (4) and locating RBFs on orthogonal grids, the linear combinations of RBFs generate affine planes which interpolating the tops of neighboring RBFs. In 1-input systems, the affine planes are line segments between two tops. They are triangles in 2-input systems and are trigonal pyramids in 3-input systems. The RBFN's weight parameter $\omega_{[n_1,\dots,n_L]}$ is equal to the height of the RBF on the $[n_1,\dots,n_L]$-th grid point. The shape of the piecewise affine model is easily understood.

The weights ω can be determined by the least-square method. A new performance index to determine the weights is proposed as shown in Eq.(5), where K is the number of data.

$$\begin{aligned} P.I. &= \sum_{k=1}^{K}(\hat{y}(x[k]) - y[k])^2 \\ &+\rho\cdot\sum_{n_1=1}^{N_1}\cdots\sum_{n_L=1}^{N_L}\sum_{l=1}^{L}\Big(\frac{\omega_{[n1,\cdots,n_L]} - \omega_{[n1-\delta 1l,\cdots,n_L-\delta Ll]}}{gr_{[n1,\cdots,n_L]} - gr_{[n1-\delta 1l,\cdots,n_L-\delta Ll]}} \\ &-\frac{\omega_{[n1+\delta 1l,\cdots,n_L+\delta Ll]} - \omega_{[n1,\cdots,n_L]}}{gr_{[n1+\delta 1l,\cdots,n_L+\delta Ll]} - gr_{[n1,\cdots,n_L]}}\Big)^2 \\ &\rightarrow \min_{\omega_{[n_1,\cdots,n_L]}} \end{aligned}$$

$$\textit{where} \quad \delta ij = \begin{cases} 1\cdots(i=j) \\ 0\cdots(i\neq j) \end{cases} \tag{5}$$

The first term indicates the magnitude of the estimation errors. The second term's aim is smoothing.

3 Incrementation of RBFs to Improve Approximation

Approximation capability can be improved by increasing the number of RBFs. Because the proposed RBFN utilizes the orthogonal coordinate, the number of RBFs are increased by dividing axes of the input space. In this section, it is discussed to divide the axes effectively.

1. **(Initial model)**
   A RBF is put on the each end point of the input space. Dots on the corners in Fig.4 show the center points of RBFs in the initial model. Weighting parameters are calculated by Eq.(5).

2. **(Assess of the deviations from the model)**
For every input datum $\mathbf{x}[k]$, the estimated out $\hat{y}(\mathbf{x}[k])$ is calculated and memorized. The variance of approximation errors around each grid point is estimated by calculating $\sigma_{[n_1,\ldots,n_L]}$ defined in Eq. (6). Select the grid point, whose σ^2 is maximum.

$$\sigma^2_{[n_1,\ldots,n_L]} = \frac{1}{n_{[n_1,\ldots,n_L]}} \sum_{k=1}^{K} \{(\hat{y}(x_k - y_k)^2 \cdot a_{[n_1,\ldots,n_L]}(x_k)\} \tag{6}$$

$$\text{where} \quad n_{[n_1,\ldots,n_L]} = \sum_{k=1}^{K} a_{[n_1,\ldots,n_L]} \tag{7}$$

If the number of data around the grid n is small, the grid was omitted from the estimation of σ.
3. **(Temporary divide of neighborhood of the max point)**
Every grid line around the selected point is divided into 5 sections as shown in Fig.5. RBFs are added and rearranged. Because the calculation load of least square method is heavy, the model parameters are not calculated. By using memorized data $\hat{y}(\mathbf{x}[k]) - y(\mathbf{x}[k])$ and new RBFs, Bias of approximation errors $\epsilon_{[n_1,\ldots,n_L]}$ defined in Eq. (8) is calculated at each grid point.

$$\epsilon_{[n_1,\ldots,n_L]} = \sum_{k=1}^{K} \{(\hat{y}(\mathbf{x}[k]) - y_k) \cdot a_{[n_1,\ldots,n_L]}(\mathbf{x}[k])\} \tag{8}$$

4. **(Select the dividing line)**
For each dividing line (hyper plane), $\epsilon_{[n_1,\ldots,n_L]}$s are summed up. Choose two lines(hyper planes), whose absolute values of the sum of $\epsilon_{[n_1,\ldots,n_L]}$s are top two, as shown in Fig. 6.
In our previous work [Ha1], only variance σ was utilized to estimate the approximation errors. The grid line whose end-point has the largest σ was selected for divide. The advantage of ϵ is ability to check biases of errors. Another disadvantage of the previous method is that it didn't the distribution of errors in intermediate regions.
5. **(Update RBFN and estimation errors)**
For the new grids on the two added lines, RBFs are reassigned. Weighting parameters are calculated by Eq.(5) and estimation error for every datum is memorized.

The procedure from 2 to 5 is iterated. The maximum value of $\sigma_{[n_1,\cdots,n_L]}$ in Eq.(6) can be utilized as an index to terminate the iteration.

4 Numerical Example

In this section, the performance of this modeling is illustrated in comparison with our previous work [Ha1]. As an example nonlinear function, the function defined in Eq. (9) is utilized.

$$y = \sin(1.5 * \pi * x_2) - 0.1 * x_1 \tag{9}$$

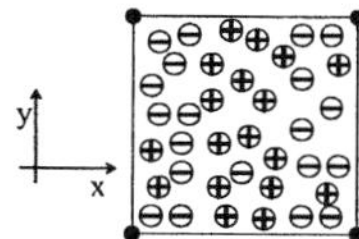

Fig. 4. Errors of Initial model

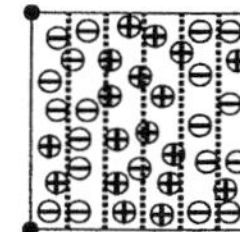
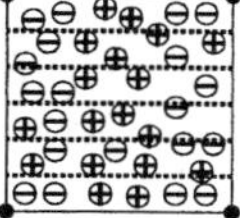

Fig. 5. Temporary divide of axes

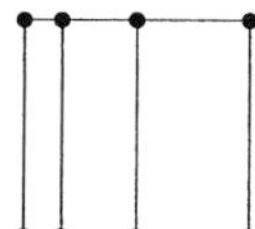

Fig. 6. Select the dividing line

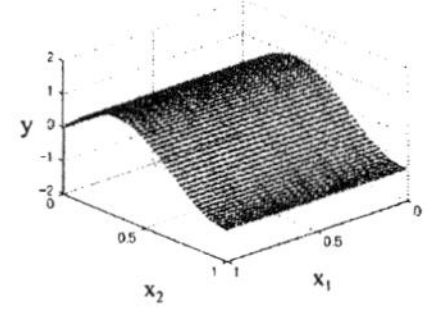

Fig. 7. Original nonlinear function

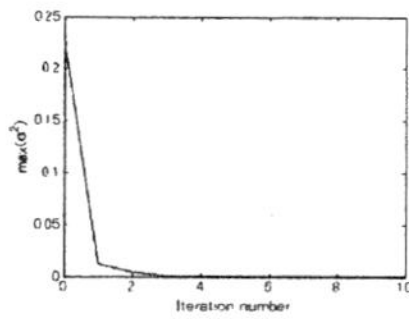

Fig. 8. Maximum variances in iteration

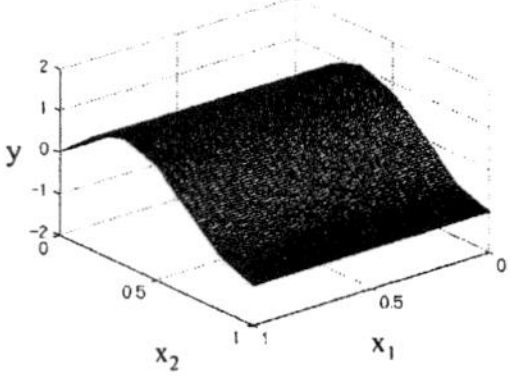

Fig. 9. Result after 3rd division iteration

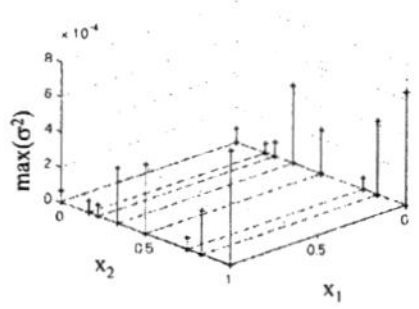

Fig. 10. Variances of RBFs

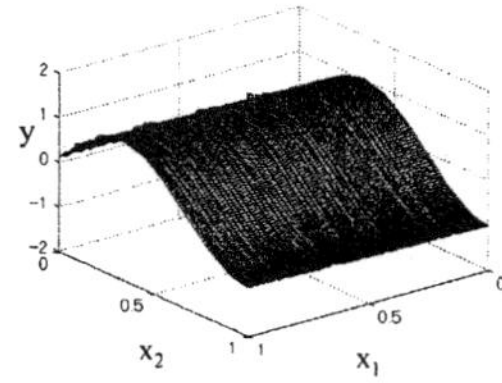

Fig. 11. Result of previous method

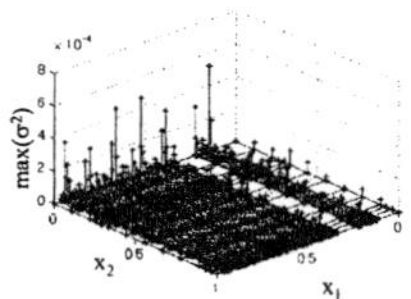

Fig. 12. σ^2 by previous method

Fig. 7 shows the original function. 2500 data, which are equally-spaced discrete points in the input space, are utilized for approximation. For the value of smoothing factor ρ in Eq. (5), 1.5 was selected in this example. It is shown in Fig. 8 that the approximation errors become almost zeros after the division iteration is executed three times. Fig. 9 shows the approximation result after third division iteration. It is similar to the original function in Fig. 7. Fig. 10 shows the grid points of its RBFs and heights of bars on them show the values σ^2. The heights are almost zeros. Fig. 11 shows the approximation result of our previous method. Fig. 12 shows the grid points of its RBFs. Because the nonlinear function in Eq. (9) has linear dependence with x_1, x_1 axis is not necessary to be dived. While the proposed method divided only x_2 axis and only 16 RBFs make approximate almost perfect, the previous one divided x_1 and the approximation result using much more RBFs is inferior. The advantage of the proposed method to estimate the biases of errors in intermediate regions is illustrated.

5 Conclusion

In this paper, a multidimensional piecewise affine modeling method based on a Radial Basis Function Network was proposed. Because this model interpolates the values at orthogonal grid points, the shape of the piecewise affine model is easily understood. The algorithm to improve approximation capability with small number of RBFs was introduced. This modeling technique can be applied to higher dimension systems. The reliability of the model outputs can be estimated. This piecewise affine model is hoped to be applied to many nonlinear problems.

References

[HH1] Hush, R. D., B. Horne: Efficient Algorithms for Function Approximation with Piecewise Linear Sigmoidal Networks. IEEE Trans. Neural Networks. **9(6)** (1998) 1129–1141

[CC1] Kahlert, C., C. O. Chua: A generalized canonical piecewise linear representation. IEEE Trans. Cirsuit Syst. **37** (1990) 373–382

[CC2] Kahlert, C., C. O. Chua: The complete canonical piecewise linear representation-Part 1 The geometry of the domain space. IEEE Trans. Cirsuit Syst. **39** (1992) 222–236

[FT1] Ferrari-Trecate, G., M. Muselli, D.Liberati, M.Morari: A Comparison of Piecewise-Linear Model Descriptions. ACC Arlilgton **5** (2001) 3521-3526

[EP1] Estevez, P. A.: Max-Min Propagation Neural Networks: Representation Capabilities, Learning Algorithms and Evolutionary Structuring. Dr thesis of Tokyo University.

[KT2] Kevenaar, T. A. M., D. M. W. Leenaerts: A Comparison of Piecewise-Linear Model Descriptions. IEEE Trans. Cirsuit Syst. **39(12)** (1992) 222–236

[Ha1] Hashimoto, Y. et al.: Piecewise linear controller improving its own reliability. Journal of process control. **6** (1996) 129–136

[BM1] Bemporad,A., M.Morari: Control of systems integrating logic, dynamics, and constraints. Automatica. **35** (1999) 407–427

Batch-Sequential Algorithm for Neural Networks Trained with Entropic Criteria*

Jorge M. Santos[1,3], Joaquim Marques de Sá[1], and Luís A. Alexandre[2]

[1] INEB - Instituto de Engenharia Biomédica
[2] IT - Networks and Multimedia Group, Covilhã
[3] Instituto Superior de Engenharia do Porto, Portugal
jms@isep.ipp.pt, jmsa@fe.up.pt, lfbaa@di.ubi.pt

Abstract. The use of entropy as a cost function in the neural network learning phase usually implies that, in the back-propagation algorithm, the training is done in batch mode. Apart from the higher complexity of the algorithm in batch mode, we know that this approach has some limitations over the sequential mode. In this paper we present a way of combining both modes when using entropic criteria. We present some experiments that validates the proposed method and we also show some comparisons of this proposed method with the single batch mode algorithm.

1 Introduction

In our previous work we introduced the use of Entropy as cost function in the learning process of Multi Layer Perceptrons (MLP) for classification [1]. This method computes the entropy of the error between the output of the neural network and the desired targets as the function to be minimized. The entropy is obtained using probability density estimation with the Parzen window method which implies the use of all available samples to estimate its value. This fact forces the use of the batch mode in the Back-propagation algorithm limiting the use of, in some cases most appropriate, sequential mode. To overcome this limitation we propose a new approach that combines these to modes (the batch and the sequential) to try to use their mutual advantages. What we call the batch-sequential mode divides, in each epoch, the training set in several groups and sequentially presents each one to the learning algorithm to perform the appropriate weight updating.

The next section of this work introduces the Error Entropy Minimization Algorithm and several optimizations to achieve a faster convergence by manipulating the smoothing parameter and the learning rates. Section 3 presents the new batch-sequential algorithm and section 4 several experiments that show the applicability of the proposed method. In the final section we conclude with some discussion of the paper.

* This work was supported by the Portuguese Fundação para a Ciência e Tecnologia(project POSI/EIA/56918/2004).

W. Duch et al. (Eds.): ICANN 2005, LNCS 3697, pp. 91–96, 2005.

2 The EEM Algorithm

The use of entropy and related concepts in learning systems is well known. The Error Entropy Minimization concept was introduced in [2] and, using the same approach, we introduced in [1] the Error Entropy Minimization Algorithm for Neural Network Classification. This algorithm uses the entropy of the error between the output of the neural network and the desired targets as the function to be minimized in the training phase of the neural network. Despite the fact that we use Renyi's Quadratic Entropy we may as well use other kinds of entropy measures, like Shannon's entropy, as in [4].

Let $y = \{y_i\} \in \mathbb{R}^m$, $i = 1, ..., N$, be a set of samples from the output vector $Y \in \mathbb{R}^m$ of a mapping $\mathbb{R}^n \mapsto \mathbb{R}^m : Y = g(w, X)$, where w is a set of neural network weights, X is the input vector and m is the dimensionality of the output vector. Let $d = d_i \in \{-1, 1\}^m$ be the desired targets and $e_i = d_i - y_i$ the error for each data sample. In order to compute the Renyi's Quadratic Entropy of e we use the Parzen window probability density function (pdf) estimation using Gaussian kernel with zero mean and unitary covariance matrix, $G(e, I) = \frac{1}{(2\pi)^{\frac{m}{2}}} exp(-\frac{1}{2}e^T e)$. This method estimates the pdf as

$$f(e) = \frac{1}{Nh^m} \sum_{i=1}^{N} G(\frac{e - e_i}{h}) \tag{1}$$

where h is the bandwidth or smoothing parameter.

Renyi's Quadratic Entropy of the error can be estimated, applying the integration of gaussian kernels [5], by

$$\begin{aligned} \hat{H}_{R2}(e) &= -\log \left[\frac{1}{N^2 h^{2m-1}} \sum_{i=1}^{N} \sum_{j=1}^{N} G(\frac{e_i - e_j}{h}, 2I) \right] \\ &= -\log V(e) \end{aligned} \tag{2}$$

The gradient of $V(e)$ for each sample i is:

$$F_i = -\frac{1}{2Nh^{2m+1}} \sum_{j=1}^{N} G(\frac{e_i - e_j}{h}, 2I)(e_i - e_j) \tag{3}$$

The update of the neural network weights is performed using the backpropagation algorithm with $\Delta w = -\eta \frac{\partial V}{\partial w}$.

One of the first difficulties in estimating the entropy is to find the best value for h in pdf estimation. In our first experiments with this entropic approach the value of the smoothing parameter was experimentally selected. Latter, we have developed a formula to obtain an appropriate value for h, as a function of the number of samples and the dimensionality of the neural network output (related with the number of classes in the classification problem). This formula, proposed in [6]

$$h_{op} = 25\sqrt{\frac{m}{N}} \tag{4}$$

gives much higher values than those formulas usually proposed for probability density function estimation and gives very good results for the EEM algorithm.

3 Batch-Sequential Algorithm

The Batch-Sequential algorithm tries to combine the two methods applied in the back propagation learning algorithm: the *sequential* mode, also referenced as on-line or stochastic, where the update is made for each sample of the training set, and the *batch* mode, where the update is performed after the presentation of all samples of the training set.

We know that, the estimated pdf approximates the true pdf as $N \to \infty$ but, in the EEM algorithm, we only need to compute the entropy and its gradient; we do not need to estimate the probability density function of e. This is a relevant fact because, in the gradient descent method, more important than computing with extreme precision the gradient is to get accurately its direction. Also, the computation with extreme accuracy of the probability density function causes the entropy to have high variability. This fact could lead to the occurrence of local minima. The sequential mode updating of the weights leads to a sample by sample *stochastic* search in the weight space implying that becomes less likely for the back-propagation algorithm to be trapped in local minima [7]. However, we still need some samples to estimate the entropy what limits the use of the sequential mode. Other advantage of the sequential mode occurs when there are some redundancy in the training set. The batch mode also presents some advantages: the gradient vector is estimated with more accuracy guarantying the convergence to, at least, a local minima and the algorithm is more easily parallelized than using sequential mode.

In order to make use of the advantages of both modes and also to speedup the algorithm, we developed a batch-sequential algorithm consisting of the splitting of the training set in several groups that are presented to the algorithm in a sequential way. In each group we apply the batch mode.

Let $\{Ts\}$ be the training set of a given data set and $\{Ts_j\}$ the subsets obtained by randomly dividing Ts in several groups with an equal number of samples, such as

$$\#Ts = n + \sum_{j=1}^{L} \#Ts_j \tag{5}$$

where L is the number of subsets and n the remainder. Leaving, in each epoch, some samples out of the learning process (when $n \neq 0$) is not significant because those samples will most likely be included in the next epoch. The partition of the training set in subsets being performed in a random way reduces the probability of the algorithm getting trapped in local minima. The subsets are sequentially presented to the learning algorithm, that applies to each one, in batch mode, the respective computation and subsequent weight update. The pseudo code for the Error Entropy Minimization Batch-Sequential algorithm (EEM-BS) is presented in Table 1.

Table 1. Pseudo-code for the EEM-BS Algorithm

```
For k:=1 to number of epochs
    Create L subsets of Ts
    For j:=1 to L
        - Compute the error entropy gradient of Ts_j applying formula 3
        - Perform weight update
    End For
End For
```

One of the advantages in using the batch-sequential algorithm is the decrease of the algorithm complexity. The complexity of the original EEM algorithm, due to formulas 2 and 3, is $O(Ts^2)$. We clearly see that, for large training sets, the algorithm is highly time consuming. With the EEM-BS algorithm the complexity is proportional to:

$$L\left(\frac{Ts}{L}\right)^2 \tag{6}$$

Therefore, the complexity ratio of both algorithms is:

$$\frac{Ts^2}{L(\frac{Ts}{L})^2} = L \tag{7}$$

which means that, in terms of computational processing time, we achieve a reduction proportional to L. For a complete experiment, similar to the one presented in the next section with the data set "Olive", we reduce the processing time from about 30 to 6 minutes in our machine.

The number of subsets, L, is determined by the size of the data set. If, in a given problem, the training set has a large number of data samples, we can use a higher number of subsets than if we have a small training set. We recommend the division of the training set in a number of subsets with a number of samples not less than 40, even though we had some good results with less elements.

In order to perform the experiments with the batch-sequential algorithm, we tried to use the optimization proposed in [3], the EEM-VLR. This optimization is based on the use of a global variable learning rate (VLR) during the training phase, as a function of the entropy value in consecutive iterations. Since this optimization compares H_{R2} of a certain epoch with the same value in the previous one, we could not use it because, in each epoch, we use different sets of samples and, by this simple fact, we would have different values of H_{R2}. To overcome this limitation we implemented a similar process, also using variable learning

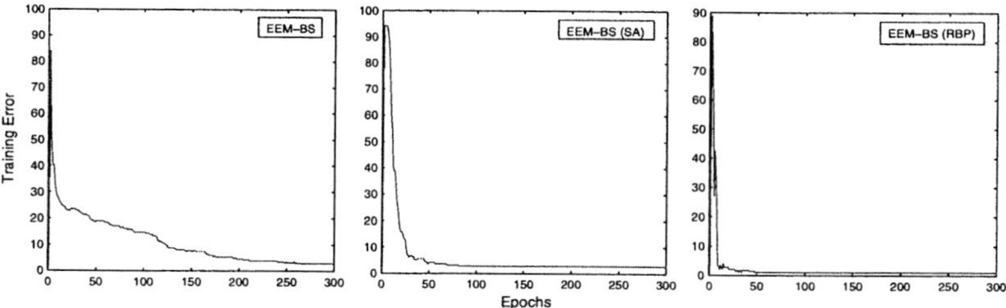

Fig. 1. Training Error for the EEM-BS and the two optimizations

rate, but this time, the variation of the learning rate is done for each neural network weight by comparing the respective gradient in consecutive iterations (EEM-BS(SA)). This approach was already used in back-propagation with MSE [8]. We also used, for the same purpose of speeding up the convergence, the combination of this implementation with the resilient back-propagation, achieving very good results (EEM-BS(RBP)). Examples of the training phase for the three different methods, with data set "Olive", are depicted in Fig.1.

4 Experiments

In order to establish the validity of the proposed algorithm we performed several classification experiments, comparing the results obtained with the EEM-BS algorithm and with the simple EEM-VLR algorithm. The characteristics of the data sets used in the experiments are summarized in Table 2.

In all experiments we used (I, n_h, m) MLP's, where I is the number of input neurons, n_h is the number of neurons in the hidden layer and m is the number of output neurons. We applied the cross-validation method using half of the data for training and half for testing. The experiments for each data set were performed varying the number of neurons in the hidden layer, the number of subsets used and the number of epochs. Each result is the mean error of 20 repetitions. In Table 3 we only present the best results for each experiment with 4 and 8 subsets for the EEM-VLR and the EEM-BS algorithms.

The results of EEM-BS algorithm are, in some cases, even better than those of EEM-VLR. The complexity of the neural networks for each experiment is very

Table 2. Data sets used for the experiments.

Data set	*# Samples*	*# Features*	*# Classes*
Ionosphere	351	33	2
Olive	572	8	9
Wdbc	569	30	2
Wine	178	13	3

Table 3. Results for EEM-VLR and EEM-BS (Tpe: Time per epoch $\times 10^{-3}$ sec.)

Ionosphere	*Error (Std)*	*L*	*n_h*	*Epochs*	*Tpe*	**Olive**	*Error (Std)*	*L*	*n_h*	*Epochs*	*Tpe*
EEM-VLR	12.06 (1.11)	-	12	40	16.7	*EEM-VLR*	5.04 (0.53)	-	25	200	77.7
EEM-BS	12.00 (1.22)	4	16	80	6.4	*EEM-BS*	5.17 (0.51)	4	30	140	17.6
EEM-BS	12.22 (1.14)	8	16	60	4.8	*EEM-BS*	5.24 (0.70)	8	20	180	12.8
Wdbc	*Error (Std)*	*L*	*n_h*	*Epochs*	*Tpe*	**Wine**	*Error (Std)*	*L*	*n_h*	*Epochs*	*Tpe*
EEM-VLR	2.33 (0.37)	-	4	40	38.7	*EEM-VLR*	1.83 (0.83)	-	14	40	5.8
EEM-BS	2.31 (0.35)	5	10	60	13.6	*EEM-BS*	1.88 (0.80)	4	16	60	3.2
EEM-BS	2.35 (0.48)	8	10	40	9.6	*EEM-BS*	1.88 (0.86)	8	16	60	2.5

similar for both algorithms what reenforces the validity of the proposed method. Since the best results were obtained with different neural network complexity we present in column *Tpe* the processing time per epoch for each algorithm.

5 Conclusions

We presented, in this paper, a way of combining the sequential and batch modes when using entropic criteria in the learning phase, taking profit of the advantages of both methods. We show, using experiments, that this is a valid approach that can be used to speed-up the training phase, maintaining a good performance.

References

1. Jorge M. Santos, Luis A. Alexandre, and Joaquim Marques de Sá. The Error Entropy Minimization Algorithm for Neural Network Classification. *Int. Conf. on Recent Advances in Soft Computing*, pages 92–97, 2004.
2. D. Erdogmus and J. Prncipe. An error-entropy minimization algorithm for supervised training of nonlinear adaptive systems. *Trans. On Signal Processing*, 50(7):1780–1786, 2002.
3. Jorge M. Santos, Joaquim Marques de Sá, Luis A. Alexandre, and Fernando Sereno. Optimization of the Error Entropy Minimization Algorithm for Neural Network Classification. In C.H.Dagli, A. L. Buczak, D. L. Enke, M. J. Embrechts, and O. Ersoy, editors, *Intelligent Engineering Systems Through Artificial Neural Networks*, volume 14, pages 81–86. ASME Press, 2004.
4. Luis M. Silva, Joaquim Marques de Sá, and Luis A. Alexandre. Neural Network Classification using Shannon's Entropy. In *European Symposium on Artificial Neural Networks (Accepted for publication)*, 2005.
5. D. Xu and J. Princpe. Training mlps layer-by-layer with the information potential. In *Intl. Joint Conf. on Neural Networks*, pages 1716–1720, 1999.
6. Jorge M. Santos, Joaquim Marques de Sá, and Luis A. Alexandre. Neural Networks Trained with the EEM Algorithm: Tuning the Smoothing Parameter. In *6th WSEAS Int. Conf. on Neural Networks, (accepted)*, 2005.
7. Simon Haykin. *Neural Networks: A Comprehensive Foundation.* Prentice Hall, New Jersey, 2 edition, 1999.
8. F. Silva and L. Almeida. Speeding up backpropagation. In Eckmiller R., editor, *Advanced Neural Computers*, pages 151–158, 1990.

Multiresponse Sparse Regression with Application to Multidimensional Scaling

Timo Similä and Jarkko Tikka

Helsinki University of Technology, Laboratory of Computer and Information Science, P.O. Box 5400, FI-02015 HUT, Finland
timo.simila@hut.fi
tikka@mail.cis.hut.fi

Abstract. Sparse regression is the problem of selecting a parsimonious subset of all available regressors for an efficient prediction of a target variable. We consider a general setting in which both the target and regressors may be multivariate. The regressors are selected by a forward selection procedure that extends the Least Angle Regression algorithm. Instead of the common practice of estimating each target variable individually, our proposed method chooses sequentially those regressors that allow, on average, the best predictions of all the target variables. We illustrate the procedure by an experiment with artificial data. The method is also applied to the task of selecting relevant pixels from images in multidimensional scaling of handwritten digits.

1 Introduction

Many practical data analysis tasks, for instance in chemistry [1], involve a need to predict several target variables using a set of regressors. Various approaches have been proposed to regression with a multivariate target. The target variables are often predicted separately using techniques like Ordinary Least Squares (OLS) or Ridge Regression [2]. An extension to multivariate prediction is the Curds and Whey procedure [3], which aims to take advantage of the correlational structure among the target variables. Latent variable models form another class with the same goal including methods like Reduced Rank, Canonical Correlation, Principal Components and Partial Least Squares Regression [4].

Prediction accuracy for novel observations depends on the complexity of the model. We consider only linear models, where the prediction accuracy is traditionally controlled by shrinking the regression coefficients toward zero [5,6]. In the latent variable approach the data are projected onto a smaller subspace in which the model is fitted. This helps with the curse of dimensionality but the prediction still depends on all of the regressors. On the contrary, sparse regression aims to select a relevant subset of all available regressors. Many automatic methods exist for the subset search including forward selection, backward elimination and various combinations of them. Least Angle Regression (LARS) [7] is a recently introduced algorithm that combines forward selection and shrinkage.

The current research in sparse regression is mainly focused on estimating a univariate target. We propose a Multiresponse Sparse Regression (MRSR) algo-

W. Duch et al. (Eds.): ICANN 2005, LNCS 3697, pp. 97–102, 2005.

rithm, which is an extension of the LARS algorithm. Our method adds those regressors sequentially to the model, which allow the most accurate predictions averaged over all the target variables. This allows to assess the average importance of the regressors in the multitarget setting. We illustrate the MRSR algorithm by artificially generated data and also apply it in a discriminative projection of images representing handwritten digits.

2 Multiresponse Sparse Regression

Suppose that the targets are denoted by an $n \times p$ matrix $\boldsymbol{T} = [\boldsymbol{t}_1 \cdots \boldsymbol{t}_p]$ and the regressors are denoted by an $n \times m$ matrix $\boldsymbol{X} = [\boldsymbol{x}_1 \cdots \boldsymbol{x}_m]$. The MRSR algorithm adds sequentially active regressors to the model

$$\boldsymbol{Y}^k = \boldsymbol{X}\boldsymbol{W}^k \tag{1}$$

such that the $n \times p$ matrix $\boldsymbol{Y}^k = [\boldsymbol{y}_1^k \cdots \boldsymbol{y}_p^k]$ models the targets $\boldsymbol{T}$ appropriately. The $m \times p$ weight matrix $\boldsymbol{W}^k$ includes k nonzero rows in the beginning of the kth step. Each step introduces a new nonzero row and, thus, a new regressor to the model. In the case $p = 1$ MRSR coincides with the LARS algorithm. This makes MRSR rather an extension than an improvement of LARS.

Set $k = 0$, initialize all elements of $\boldsymbol{Y}^0$ and $\boldsymbol{W}^0$ to zero, and normalize both $\boldsymbol{T}$ and $\boldsymbol{X}$ to zero mean. The columns of $\boldsymbol{T}$ and the columns of $\boldsymbol{X}$ should also have the same scales, which may differ between the matrices. Define a cumulative correlation between the jth regressor $\boldsymbol{x}_j$ and the current residuals

$$c_j^k = ||(\boldsymbol{T} - \boldsymbol{Y}^k)^T \boldsymbol{x}_j||_1 = \sum_{i=1}^{p} |(\boldsymbol{t}_i - \boldsymbol{y}_i^k)^T \boldsymbol{x}_j|. \tag{2}$$

The criterion measures the sum of absolute correlations between the residuals and the regressor over all p target variables in the beginning of the kth step. Let the maximum cumulative correlation be denoted by $c_{\max}^k$ and the group of regressors that satisfy the maximum by $\mathcal{A}$, or formally

$$c_{\max}^k = \max_j \{c_j^k\}, \quad \mathcal{A} = \{j \mid c_j^k = c_{\max}^k\}. \tag{3}$$

Collect the regressors that belong to $\mathcal{A}$ as an $n \times |\mathcal{A}|$ matrix $\boldsymbol{X}_{\mathcal{A}} = [\cdots \boldsymbol{x}_j \cdots]_{j \in \mathcal{A}}$ and calculate an OLS estimate

$$\bar{\boldsymbol{Y}}^{k+1} = \boldsymbol{X}_{\mathcal{A}} (\boldsymbol{X}_{\mathcal{A}}^T \boldsymbol{X}_{\mathcal{A}})^{-1} \boldsymbol{X}_{\mathcal{A}}^T \boldsymbol{T}. \tag{4}$$

Note that the OLS estimate involves $k + 1$ regressors at the kth step.

Greedy forward selection adds regressors based on (2) and the OLS estimate (4) is used. However, we define a less greedy algorithm by moving from the MRSR estimate $\boldsymbol{Y}^k$ toward the OLS estimate $\bar{\boldsymbol{Y}}^{k+1}$, i.e. in the direction $\boldsymbol{U}^k = \bar{\boldsymbol{Y}}^{k+1} - \boldsymbol{Y}^k$, but we will not reach it. The largest step possible is taken in the direction of $\boldsymbol{U}^k$ until some $\boldsymbol{x}_j$, where $j \notin \mathcal{A}$, has as large cumulative correlation with the current residuals as the already added regressors [7]. The MRSR estimate $\boldsymbol{Y}^k$ is updated

$$\boldsymbol{Y}^{k+1} = \boldsymbol{Y}^k + \gamma^k(\bar{\boldsymbol{Y}}^{k+1} - \boldsymbol{Y}^k). \quad (5)$$

In order to make the update, we need to calculate the correct step size γ^k. The cumulative correlations c_j^{k+1} may be obtained by substituting (5) into (2). According to (4), we may write $\boldsymbol{X}_{\mathcal{A}}^T(\bar{\boldsymbol{Y}}^{k+1} - \boldsymbol{Y}^k) = \boldsymbol{X}_{\mathcal{A}}^T(\boldsymbol{T} - \boldsymbol{Y}^k)$. This gives the cumulative correlations in the next step as a function of γ

$$c_j^{k+1}(\gamma) = |1-\gamma| c_{\max}^k \text{ for all } j \in \mathcal{A} \quad (6)$$

$$c_j^{k+1}(\gamma) = ||\boldsymbol{a}_j^k - \gamma \boldsymbol{b}_j^k||_1 \text{ for all } j \notin \mathcal{A}, \quad (7)$$

where $\boldsymbol{a}_j^k = (\boldsymbol{T} - \boldsymbol{Y}^k)^T \boldsymbol{x}_j$ and $\boldsymbol{b}_j^k = (\bar{\boldsymbol{Y}}^{k+1} - \boldsymbol{Y}^k)^T \boldsymbol{x}_j$. A new regressor with index $j \notin \mathcal{A}$ will enter the model when (6) and (7) are equal. This happens if the step size is taken from the set

$$\Gamma_j = \left\{ \frac{c_{\max}^k + \boldsymbol{s}^T \boldsymbol{a}_j^k}{c_{\max}^k + \boldsymbol{s}^T \boldsymbol{b}_j^k} \right\}_{\boldsymbol{s} \in \mathcal{S}}, \quad (8)$$

where $\mathcal{S}$ is the set of all 2^p sign vectors of size $p \times 1$, i.e. the elements of $\boldsymbol{s}$ may be either 1 or -1. The correct choice is the smallest of such positive step sizes that introduces a new regressor

$$\gamma^k = \min\{\gamma \mid \gamma \geq 0 \text{ and } \gamma \in \Gamma_j \text{ for some } j \notin \mathcal{A}\}, \quad (9)$$

which completes the update rule (5).

The weight matrix, which satisfies (5) and (1) may be updated

$$\boldsymbol{W}^{k+1} = (1-\gamma^k)\, \boldsymbol{W}^k + \gamma^k\, \bar{\boldsymbol{W}}^{k+1}, \quad (10)$$

where $\bar{\boldsymbol{W}}^{k+1}$ is an $m \times p$ sparse matrix. Its nonzero rows, which are indexed by $j \in \mathcal{A}$, contain the corresponding rows of the OLS parameters $(\boldsymbol{X}_{\mathcal{A}}^T \boldsymbol{X}_{\mathcal{A}})^{-1} \boldsymbol{X}_{\mathcal{A}}^T \boldsymbol{T}$. The parameters of the selected regressors are shrunk according to (10) and the rest are kept at zero during the steps $k = 0, \ldots, m-2$. The last step coincides with the OLS parameters. The selection of the final model from m possibilities is based on prediction accuracy for novel data.

3 Multidimensional Scaling

Multidimensional scaling (MDS) [8] is a collection of techniques for exploratory data analysis that visualize proximity relations of objects as points in a low-dimensional Euclidean feature space. Proximities are represented as pairwise dissimilarity values δ_{ij}. We concentrate on the Sammon criterion [9]

$$E(\boldsymbol{Y}) = \sum_{i=1}^{n} \sum_{j>i} \alpha_{ij} (||\hat{\boldsymbol{y}}_i - \hat{\boldsymbol{y}}_j||_2 - \delta_{ij})^2. \quad (11)$$

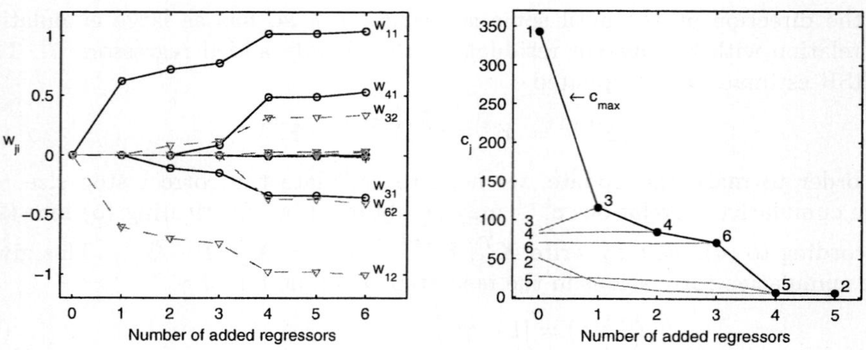

Fig. 1. Results for the artificial data: (*left*) Estimates of the weights w_{ji} and (*right*) cumulative correlations c_j as a function of the number of regressors in the model

Normalization coefficients $\alpha_{ij} = 2/(n(n-1)\delta_{ij})$ put focus on similar objects. The vector $\hat{\boldsymbol{y}}_i$ is the ith row of an $n \times p$ feature configuration matrix $\boldsymbol{Y}$.

Differing from the ordinary Sammon mapping, we are not only seeking for $\boldsymbol{Y}$ that minimizes the error (11), but also the parameterized transformation from the data space to the feature space that generates $\boldsymbol{Y}$ as a function of an $n \times m$ matrix $\boldsymbol{X}$. More specifically, we define $\boldsymbol{Y}$ as a linear combination of some relevant columns of $\boldsymbol{X}$. Next, a gradient descent procedure for such a minimization of (11) is outlined by modifying the Shadow Targets algorithm [10].

Make an initial guess $\boldsymbol{Y}^0$ and set the learning rate parameter η^0 to a small positive value. The estimated targets at each of the following iterations are

$$\boldsymbol{T}^{\ell+1} = \boldsymbol{Y}^{\ell} - \eta^{\ell} \frac{\partial E(\boldsymbol{Y}^{\ell})}{\partial \boldsymbol{Y}}. \tag{12}$$

Calculate $\boldsymbol{W}^{\ell+1}$ by feeding $\boldsymbol{T}^{\ell+1}$ and $\boldsymbol{X}$ to the MRSR algorithm and update $\boldsymbol{Y}^{\ell+1} = \boldsymbol{X}\boldsymbol{W}^{\ell+1}$. As suggested in [10], set $\eta^{\ell+1} = 0.1\eta^{\ell}$ if error (11) has increased from the previous iteration, and otherwise set $\eta^{\ell+1} = 1.2\eta^{\ell}$.

The only difference between the original Shadow Targets algorithm is the way in which the weights $\boldsymbol{W}^{\ell+1}$ are calculated. MRSR replaces the calculation of OLS parameters $(\boldsymbol{X}^T\boldsymbol{X})^{-1}\boldsymbol{X}^T\boldsymbol{T}^{\ell+1}$. This allows us to control the sparsity of the solution. The number of nonzero rows in $\boldsymbol{W}^{\ell+1}$ depends on the number of steps we perform in the MRSR algorithm.

4 Experiments

To illustrate the MRSR algorithm, we generated artificial data from the setting $\boldsymbol{T} = \boldsymbol{X}\boldsymbol{W} + \boldsymbol{E}$, where the elements of a 200×6 matrix $\boldsymbol{X}$ are independently distributed according to the Gaussian distribution $N(0,1)$, the elements of a 200×2 matrix $\boldsymbol{E}$ according to $N(0, 0.35^2)$, and the weights are set to

$$\boldsymbol{W}^T = \begin{bmatrix} 1 & 0 & -1/3 & 1/2 & 0 & 0 \\ -1 & 0 & 1/3 & 0 & 0 & -2/5 \end{bmatrix}.$$

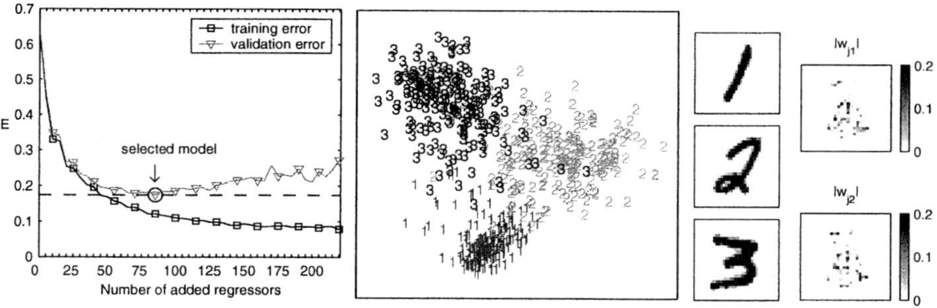

Fig. 2. Results for the digits data: (*left*) Training and validation errors as a function of the number of regressors in the model. (*middle*) Projection of the test set. (*right*) Example images from the test set and images illustrating the weights w_{ji}.

Fig. 1 shows results of MRSR analysis of the artificial data. The regressors are added to the model in the order 1, 3, 4, 6, 5, 2 and each addition decreases the maximum cumulative correlation between the regressors and residuals. The apparently most important regressor $\boldsymbol{x}_1$ is added first and the two useless regressors $\boldsymbol{x}_2$ and $\boldsymbol{x}_5$ last. Importantly, $\boldsymbol{x}_3$ enters the model before $\boldsymbol{x}_4$ and $\boldsymbol{x}_6$, because it is overall more relevant. However, $\boldsymbol{x}_4$ ($\boldsymbol{x}_6$) would enter before $\boldsymbol{x}_3$ if the first (second) target was estimated individually using the LARS algorithm.

The second experiment studies images of handwritten digits 1, 2, 3 with 28×28 resolution from the MNIST database. An image is represented as a row of $\boldsymbol{X}$, which consists of grayscale values of 784 pixels between zero and one. We constructed randomly three distinct data sets: a training set with 100, validation set with 200, and test set with 200 samples per digit. The aim is to form a model that produces a discriminative projection of the images onto two dimensions by a linear combination of relevant pixels. Pairwise dissimilarities are calculated using a discriminative kernel [11]. A within digit dissimilarity is $\delta_{ij} = 1 - \exp(-||\hat{\boldsymbol{x}}_i - \hat{\boldsymbol{x}}_j||^2/\beta)$ and a between digit dissimilarity is $\delta_{ij} = 2 - \exp(-||\hat{\boldsymbol{x}}_i - \hat{\boldsymbol{x}}_j||^2/\beta)$, where $\hat{\boldsymbol{x}}_i$ denotes the ith image. The parameter β controls discrimination and we found a visually suitable value $\beta = 150$.

Fig. 2 displays results for the digits data. The left panel shows the best training set error of MDS starting from 100 random initializations $\boldsymbol{Y}^0$ and the corresponding validation error as a function of the number of effective pixels in the model. The validation error is at the minimum when the model uses 85 pixels. The middle panel shows a projection of test images obtained by this model and the right panel illustrates sparsity of the estimated weights w_{ji}. The selected group of about 11% of the pixels is apparently enough to form a successful linear projection of novel images.

5 Conclusions

We have presented the MRSR algorithm for forward selection of regressors in the estimation of a multivariate target using a linearly parameterized model. The algorithm is based on the LARS algorithm, which is designed for a univariate target. MRSR adds regressors one by one to the model such that the added regressor always correlates most of all with the current residuals. The order in which the regressors enter the model reflects their importance. Sparsity places focus on relevant regressors and makes the results more interpretable. Moreover, sparsity coupled with shrinkage helps to avoid overfitting.

We used the proposed algorithm in an illustrative experiment with artificially generated data. In another experiment we studied images of handwritten digits. The algorithm fitted a MDS model that allows a discriminative projection of the images onto two dimensions. The experiment combines the two major categories of dimensionality reduction methods: input selection and input transformation.

LARS is closely connected to the Lasso estimator [6,7]. As such, MRSR does not implement a multiresponse Lasso, which constraints the ℓ_1-norm of the weight matrix. MRSR updates whole rows of the matrix instead of its individual elements. However, the connection might emerge by modifying the constraint structure of Lasso. Another subject of future research could be basis function selection for linear neural networks.

References

1. Burnham, A.J., MacGregor, J.F., Viveros, R.: Latent Variable Multivariate Regression Modeling. Chemometrics and Intelligent Laboratory Systems **48** (1999) 167-180
2. Hoerl, A.E., Kennard, R.W.: Ridge Regression: Biased Estimation for Nonorthogonal Problems. Technometrics **12** (1970) 55-67
3. Breiman, L., Friedman, J.H.: Predicting Multivariate Responses in Multivariate Regression. Journal of the Royal Statistical Society. Series B **59** (1997) 3-54
4. Abraham, B., Merola, G.: Dimensionality Reduction Approach to Multivariate Prediction. Computational Statistics & Data Analysis **48** (2005) 5-16
5. Copas, J.: Regression, Prediction and Shrinkage. Journal of the Royal Statistical Society. Series B **45** (1983) 311-354
6. Tibshirani, R.: Regression Shrinkage and Selection via the Lasso. Journal of the Royal Statistical Society. Series B **58** (1996) 267-288
7. Efron, B., Hastie, T., Johnstone, I., Tibshirani, R.: Least Angle Regression. Annals of Statistics **32** (2004) 407-499
8. Cox, T.F., Cox, M.A.A.: Multidimensional Scaling. Monographs on Statistics and Applied Probability 88. Chapman & Hall (2001)
9. Sammon, J.W.: A Nonlinear Mapping for Data Structure Analysis. IEEE Transactions on Computers **C-18** (1969) 401-409
10. Tipping, M.E., Lowe, D.: Shadow Targets: A Novel Algorithm for Topographic Projections by Radial Basis Functions. Neurocomputing **19** (1998) 211-222
11. Zhang, Z.: Learning Metrics via Discriminant Kernels and Multidimensional Scaling: Toward Expected Euclidean Representation. In: International Conference on Machine Learning. (2003) 872-879

Training Neural Networks Using Taguchi Methods: Overcoming Interaction Problems

Alagappan Viswanathan, Christopher MacLeod, Grant Maxwell, and Sashank Kalidindi

School of Engineering, The Robert Gordon University, Schoolhill, Aberdeen, UK

Abstract. Taguchi Methods (and other orthogonal arrays) may be used to train small Artificial Neural Networks very quickly in a variety of tasks. These include, importantly, Control Systems. Previous experimental work has shown that they could be successfully used to train single layer networks with no difficulty. However, interaction between layers precluded the successful reliable training of multi-layered networks. This paper describes a number of successful strategies which may be used to overcome this problem and demonstrates the ability of such networks to learn non-linear mappings.

1 Introduction

A number of papers have outlined the use of Taguchi and other Orthogonal arrays to train Artificial Neural Networks (ANNs). The idea was originated by C. MacLeod in 1994 [1] at the Robert Gordon University and implemented in practice by his student Geva Dror in an MSc project [2]. Another group at the JPL research centre in NASA also developed the same idea independently at around the same time [3].

The technique works by trying a series of different weight combinations on the network and then using Taguchi's techniques [4] to interpolate the best combination of these. A detailed description is not presented here due to space restrictions and the fact that the method is explained fully in several of the references [1,3,5]. The American team added to the basic technique by proposing an iterative approach [3].

The technique can operate very quickly to set network weights and it has been suggested that this could be used in "disaster control" networks [5] where the ANN takes over control from a damaged system (for example, an aircraft with compromised control surfaces).

The problem with the technique is that Taguchi Methods only operate well on variables which do not interact. In the case of a two layer ANN, the weights in the initial layer interact strongly with those of the second layer (that is, if you change the first layer weights, those in the second layer must also change if the error is to stay the same). This meant that, although it was occasionally possible to get a two layer network to train, more often than not, it did not.

W. Duch et al. (Eds.): ICANN 2005, LNCS 3697, pp. 103–108, 2005.

Maxwell suggested some possible ways around this problem [5] but these did not work reliably in all cases. However, the methods outlined below do show good results when used with two layer networks. They do this by treating the neurons, rather then the individual weights, as the basic units of the network.

2 Coding the State of Each Neuron

Since interaction between weights in different layers is the cause of the problem described above, one possible way around this is to have each variable used in the Orthogonal Array (OA) correspond to the state of a particular neuron [6]. For example, an array with eight levels could be used to code a two input neuron as shown in Fig. 1. The possible combinations of weights are shown in table 1. These are then used in the OA which represents the overall network.

As with the original method [1], the weights are quantised. As noted in the references [5], although standard texts on the Taguchi method [4] give only simple arrays, it is possible to generate others using standard formulae [7,8].

When this method is used, it does give a better error reduction than applying the standard method to a two-layer network. However, when compared against the full-factorial results for the same problem, although the error reduction is generally good, it is not as high as theoretically possible. The reason for this is neuron to neuron interaction replacing layer by layer interaction as a problem.

One can see this if one considers the structure of the experimental arrays. Consider a very simple example of the middle two experiments in a L4 array: 1 2 2 and 2 1 2. We can see that both these experiments correspond to the same network (although the neurons are in a different order) as shown in Figs. 2(a) and 2(b). This means that, when we calculate the best state for a particular

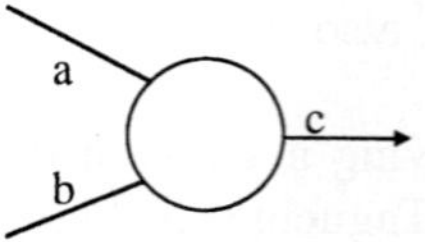

Fig. 1. Using levels to code neuron states

Table 1. List of all weight combinations

Level	Weight a	Weight b	Weight c
1	-1	-1	-1
2	-1	-1	+1
3	-1	+1	-1
4	-1	+1	+1
5	+1	-1	-1
6	+1	-1	+1
7	+1	+1	-1
8	+1	+1	+1

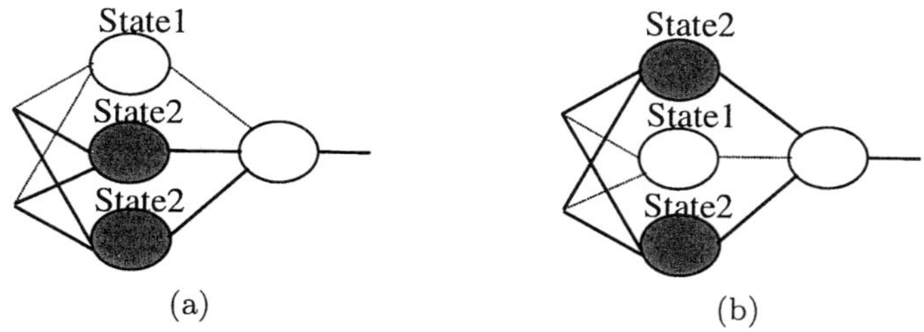

Fig. 2. Two different combinations of neuron weights which give the same result

neuron, the system cannot differentiate between states 1 and 2 for a particular neuron as the table is symmetric.

It is possible that this problem might be overcome by allocating different states to the same level in different neurons or by using interaction columns in the tables; however, this has not been tested.

The advantage of this approach is that reasonably good (although not optimum) results can be achieved. Its disadvantage is that large tables are required (as the size of the network increases) because the size of the table is proportional to the number of neurons.

3 Neuron by Neuron Training

A more successful technique is to train each neuron one by one. This does allow the error to fall to its lowest theoretical limit. It works as shown in Fig. 3. In network (a), the weights of the neuron are set using an orthogonal array in the usual way. These weights are then fixed and not altered during the rest of the training. Next, as in (b), a new neuron is added and its weights trained; the first neuron's output is used in the calculation of the error. Finally, the third neuron is added and the process is repeated, again using the first two neurons' outputs in the error calculation.

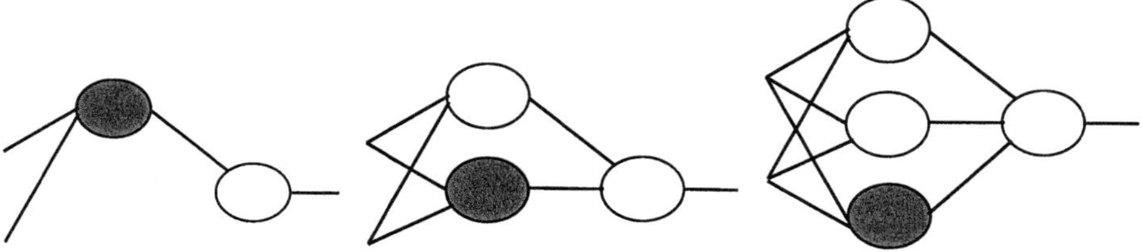

(a) First neuron's weights are trained

(b) Second neuron added and trained. The first neuron's output is also used to calculate the error.

(c) Third neuron added and trained. Previous neurons used in calculation of error.

Fig. 3. Neuron by Neuron training

Apart from the guarantee of reaching a low error (within the limits that the quantised weights apply), this method also has the advantage that the orthogonal arrays used are small (because their size is proportional to the number of weights associated with an individual neuron, not the whole network). The disadvantage is that each succeeding neuron is refining the output and so the initial neuron has the greatest affect and each successive addition has less. In effect, the method acts rather like the iterative method discussed earlier [3]. Indeed, one can tailor the succeeding weights to refine the output in a similar way. When the network reaches the correct number of neurons required to solve the problem, the overall error will not continue to fall.

4 Power Series Neurons

Power series neurons are a refinement of perceptron types and are discussed in previous papers [9]. They allow a single neuron unit to fulfil any differential function as shown in Fig. 4. It was shown in earlier work [5,9] that power series neurons can be trained using these methods.

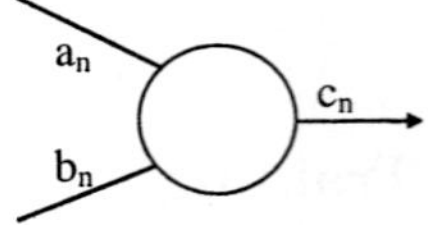

Fig. 4. A power series neuron

$$c = f((a_0.i_1 + b_0.i_2) + (a_1.i_1^2 + b_1.i_2^2) + (a_2.i_1^3 + b_2.i_2^3) + \ldots) \quad (1)$$

Where $f(x)$ is the activation function of x (typically a sigmoid), i_n^x is the xth power of nth input and a_y, b_y are the weights a and b associated with the $(y+1)^{th}$ power of the inputs.

Such a neuron can fulfil complex mathematical functions without having to resort to many layers. Although it is not capable in this form of separating discrete areas of input space, this is not necessary for fulfilling many functions required of control systems. It is possible to combine the power series and neuron-by-neuron approaches. Each additional power series neuron provides a new classifier.

5 Results

The training methods were shown to work by testing them on some non-linear mappings. These were a sigmoid, reverse sigmoid and gaussian. In this case the first method (coding the state of the neuron as a level) was used. The network consists of two inputs, one of which is held constant (at one unit);

the other is varied as shown on the x axis. There are nine hidden layer and one output neuron. The array used is 64 rows of 8 levels coded as shown in Fig. 1 and generated using Owen's data [8]. The results are shown in Figs. 5, 6 and 7.

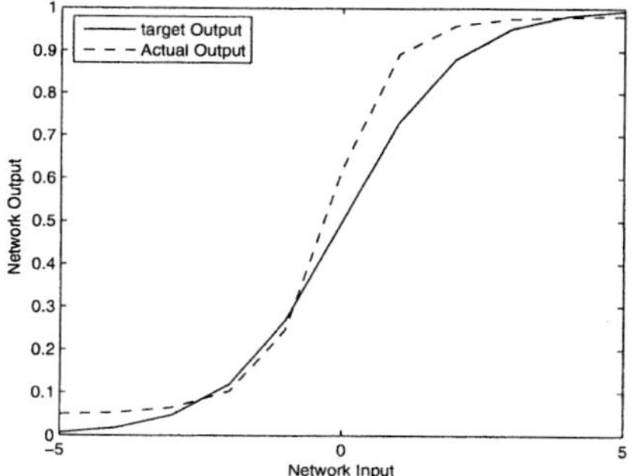

Fig. 5. Sigmoid Test

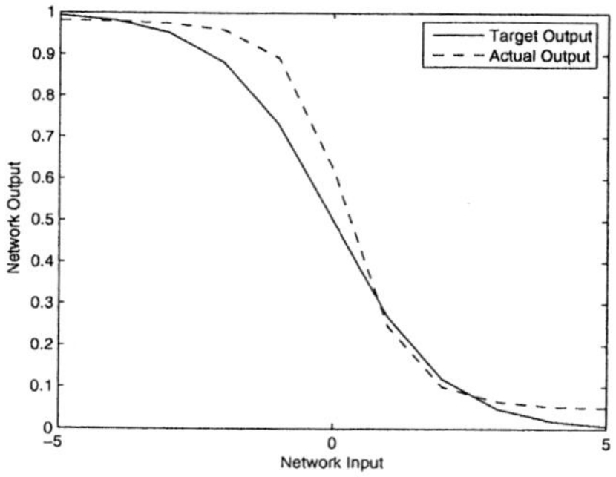

Fig. 6. Reverse Sigmoid Test

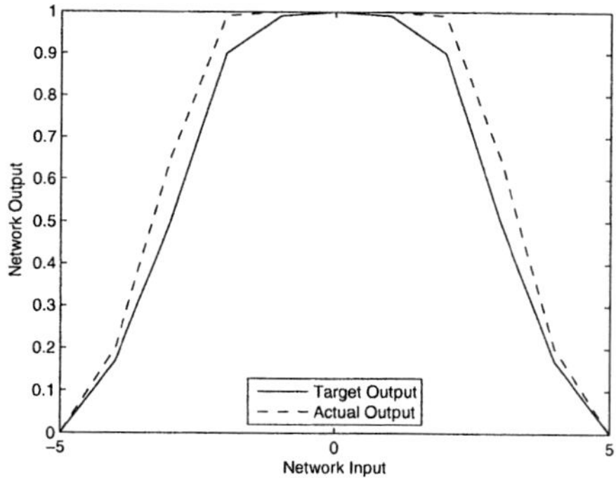

Fig. 7. Gaussian Test

6 Conclusion

Using Orthogonal or Taguchi arrays to train neural networks is a promising technique. It allows relatively small networks to be trained very quickly and simply. This makes it an ideal technique for some specialised applications in control, particularly in the stabilisation of systems which may have undergone some change which makes their system model difficult or impossible to determine.

The method, however, suffers from the problem of interaction which means that it is difficult to apply to multi-layer networks. However, it is possible to overcome this problem. The three techniques discussed above are examples of how this may be done, the last two being particularly effective.

References

1. C. MacLeod, G. Dror and G. M. Maxwell, Training Artificial Neural Networks Using Taguchi Methods, *AI Review*, (13) 3, 177-184, Kluwer, 1999.
2. G. Dror, Training Neural Networks Using Taguchi Methods, MSc thesis, 1995, The Robert Gordon University.
3. A. Stoica, J. Blosiu, Neural Learning Using Orthogonal Arrays. In proceedings of the *International Symposium on Intelligent Systems*, Reggio (Italy), 418-423, IOS Press, Amsterdam, 1997.
4. R. K. Roy, *A Primer on the Taguchi Method*, Wiley, New York, 1990.
5. G. Maxwell and C. MacLeod, Using Taguchi Methods To Train Artificial Neural Networks In Pattern Recognition, Control And Evolutionary Applications. In proceedings of the *International Conference on Neural Information Processing*, (ICONIP 2002), vol 1, 301-305, Singapore, 2002.
6. A. Viswanathan, Using Orthogonal Arrays to Train Artificial Neural Networks, MPhil Thesis, forthcoming.
7. A. Dey, *Orthogonal Fractional Factorial Designs*, Wiley, New York, 1985.
8. A. Owen, Orthogonal Arrays for Computer Experiments, Integration and Visualization, Statistica Sinica 2, 439-452, 1992.
9. N. Capanni, C. MacLeod, G. Maxwell, An Approach to Evolvable Neural Functionality. In proceedings of the *International Conference on Artificial Neural Networks and Neural Information Processing*, (ICANN/ICONIP 2003), Istanbul (Turkey), 220-223.

A Global-Local Artificial Neural Network with Application to Wave Overtopping Prediction

David Wedge, David Ingram, David McLean, Clive Mingham, and Zuhair Bandar

Department of Computing and Mathematics, Manchester Metropolitan University, John Dalton building, Chester Street, Manchester, M1 5GD, United Kingdom

Abstract. We present a hybrid Radial Basis Function (RBF) - sigmoid neural network with a three-step training algorithm that utilises both global search and gradient descent training. We test the effectiveness of our method using four synthetic datasets and demonstrate its use in wave overtopping prediction. It is shown that the hybrid architecture is often superior to architectures containing neurons of a single type in several ways: lower errors are often achievable using fewer hidden neurons and with less need for regularisation. Our Global-Local Artificial Neural Network (GL-ANN) is also seen to compare favourably with both Perceptron Radial Basis Net (PRBFN) and Regression Tree RBFs.

1 Introduction

Multi-Layer Perceptron (MLP) and RBF networks have complementary properties. While both are theoretically capable of approximating a function to arbitrary accuracy using a single hidden layer, their operation is quite different [1]. MLP networks have a fixed architecture and are usually trained using a variant of gradient descent. MLP networks invariably incorporate neurons with sigmoid activation functions. Their response therefore varies across the whole input space and weight training is affected by all training points. RBF networks, on the other hand, are most commonly created using a constructive algorithm. Gradient descent training is usually replaced by deterministic, global methods such as Forward Selection of Centres with Orthogonal Least Squares (FS-OLS). Unlike sigmoid neurons, RBF neurons generally respond strongly only to inputs within a local region [2]. RBF training is usually quicker, since training methods often involve the solving of linear equations. However RBF networks are usually larger, partly offsetting the improvement in computational efficiency.

We present a hybrid network that combines the global approximation capabilities of MLP networks with the local approximation capabilities of RBF networks. The hybrid structure of our network is reflected in a hybrid training algorithm that combines gradient descent with forward selection. It is tested using 4 synthetic datasets and comparisons are made with alternative architectures and training methods, including PRBFN and RT-RBF [3,4]. Our network is then applied to the real-world problem of wave overtopping prediction.

2 Global-Local Artificial Neural Networks

A hybrid ANN containing both sigmoidal and radial neurons may have the advantages of both RBF and MLP ANNs, i.e. fast training and compact networks with

W. Duch et al. (Eds.): ICANN 2005, LNCS 3697, pp. 109 – 114, 2005.

good generalisation. Our approach may be compared with those of PRBFN and RT-RBF. They both cluster the data and choose a neuron that approximates the local function within each cluster [3,4]. In the case of PRBFN each neuron may be either sigmoidal or RBF, leading to a hybrid network. We approximate on a global level first using a MLP and then add RBF neurons using FS-OLS, in order to add local detail to the approximating function. For this reason we call our network a Global-Local Artificial Neural Network (GL-ANN). The advantages of our approach are:

- There is no need to cluster the data prior to training. This gives more flexibility to the FS-OLS process and avoids possible problems when clustering reflects the distribution of the available data rather than the underlying functionality.
- All phases of training take into account all of the training data.
- Unlike pure RBF networks and PRBFNs, our networks do not require regularisation. The MLP created in the first phase of training has a moderating effect on the selection and training of RBF neurons added subsequently.

3 Training Method

In order to achieve rapid training, the Levenberg-Marquardt (L-M) method is used to train the MLP networks. Fixed numbers of sigmoidal neurons are used in a single hidden layer and the output neuron has a linear activation function. Initial weights are set to small random values at the start of training and inputs and outputs undergo a linear transformation to give them a range of [-0.8, 0.8].

RBF neurons are then added using FS to choose RBF centres from the training data. Symmetrical radial functions with fixed widths are employed. After each addition the output weights from both sigmoidal and RBF neurons are adjusted using OLS minimisation. We have introduced a modification to the traditional OLS method to allow for the presence of non-radial neurons while maintaining the computational efficiency of the method [2].

Finally all weights, including hidden layer weights and RBF steepnesses, are optimised using L-M training. We have found that this step is valuable in creating networks that are compact and have strong generalisation capability.

Using this algorithm a series of networks with different architectures may be created. Their performance is then assessed using unseen test data. In each case the data is sampled several times to determine the training - test split and averages are taken over all runs. 10 runs are used for the benchmark tests and 30 for the wave overtopping data.

4 Benchmark Tests

4.1 Method and Datasets

Four benchmark tests are employed. They are all function approximation tasks using synthetic data. The first function is

$$f(x)=\sin(12x) . \quad (1)$$

with x randomly selected from [0,1]. The second function is the 2D sine wave

$$f(x)=0.8\sin(x_1/4)\sin(x_2/2) . \quad (2)$$

with x_1= [0,10] and x_2=[-5,5]. The third function is a simulated alternating current used by Friedman in the evaluation of multivariate adaptive regression splines (MARS) [5]. It is given by

$$Z(R,\omega,L,C)=\sqrt{(R^2+(\omega L-1/\omega C)^2)} . \quad (3)$$

where Z is the impedance, R the resistance, ω the angular frequency, L the inductance and C the capacitance of the circuit. The input ranges are R=[0,100], ω=[40π,560π], L=[0,1] and C=[1×10^{-6},11×10^{-6}]. The fourth function is the Hermite polynomial,

$$f(x)=1+(1-x+2x^2)\exp(-x^2) . \quad (4)$$

with x randomly selected from [-4,4]. This function was first used by Mackay [6]. Gaussian noise is added to all training outputs, with a standard deviation (s.d.) of 0.1 except for Friedman's dataset which has noise with s.d. of 175. Clean data is used for test purposes, except for the 1D sine data which uses test data containing added noise with s.d. of 0.1. The latter dataset therefore has a minimum test MSE of 0.01.

4.2 Results and Discussion

Mean MSE results averaged over 10 runs for MLP, FS-OLS and GL-ANN networks are given in Table 1. In the case of the third dataset we follow Friedman [5] in dividing the MSE by the variance of the test data.

Table 1. Mean square errors for 4 benchmark datasets

	1D sine	2D sine	Friedman	MacKay
MLP (L-M)	1.75e-2	1.28e-3	1.02e-1	2.14e-3
FS-OLS	1.19e-2	0.95e-3	1.55e-1	1.41e-3
GL-ANN	1.20e-2	1.11e-3	0.98e-1	1.30e-3

GL-ANN does better than the pure networks with the complex datasets (Friedman and Mackay). With the 1D sine data GL-ANN and FS-OLS achieve comparable results. The sine 2D dataset gives best results using a pure RBF network (FS-OLS).

Table 2 gives the number of neurons used in the most successful networks. In each case S, R and T refer to the number of sigmoid, RBF and total neurons in the hidden layer, respectively. For the first two datasets, the GL-ANN imitates the RBF networks, using just 1 sigmoid neuron. The GL-ANN uses just 3 hidden neurons to reproduce Mackay's function and 6 for Friedman's. These results show that the GL-ANN is parsimonious in its use of hidden neurons.

Table 2. Number of hidden layer neurons for 4 benchmark datasets

	1D sine			2D sine			Friedman			MacKay		
	S	R	T	S	R	T	S	R	T	S	R	T
MLP(L-M)	12	0	12	5	0	5	5	0	5	14	0	14
FS-OLS	0	6	6	0	13	13	0	43	43	0	7	7
GL-ANN	1	6	7	1	16	17	3	3	6	1	2	3

Finally, the introduction of regularisation into OLS training of GL-ANNs does not improve the errors of the most effective networks. The initial MLP appears to have a moderating effect on the weights of RBF neurons added subsequently.

4.3 Comparison with PRBFN and RT-RBF

Table 3 shows the test MSEs for PRBFN and RT-RBF. For all but the 1D sine data these errors are higher than those produced by GL-ANN. Further, the results quoted for the first dataset are below the minimum error achievable, given the noisy test data. We believe that these results cannot therefore be taken at face value.

Table 3. MSEs for PRBFN and Regression Tree RBF on 4 benchmark tests

	1D sine	2D sine	Friedman	MacKay
PRBFN	0.66e-2	1.28e-3	1.50e-1	1.50e-3
RT-RBF	0.88e-2	2.28e-3	1.12e-1	1.52e-3
Minimum error	1.00e-2	-	-	-

5 Wave Overtopping

5.1 Introduction

Much research has been conducted into predicting overtopping at sea-walls during storm events. One approach is to use scale models in laboratories, [7], but this is expensive and time-consuming. An alternative is to numerically model a particular seawall configuration and sea-state, e.g. [8]. However, accurate simulation requires a detailed knowledge of both the geometry of the seawall and sea conditions. Results may therefore be applied only to individual scenarios. As part of the European CLASH project [9] a large overtopping database has been compiled, of sufficient size to allow ANN training. We have used this database to test our GL-ANNs.

5.2 Data Characteristics and Pre-processing

10 input parameters are selected for training primarily on the basis of information content [10]. 7 of these describe the structure of the seawall in question, including the height (crest freeboard) of the wall, R_0. The remaining 3 are the wave period T_0, water depth and angle of wave attack. The single output is the logarithm of the wave overtopping rate, $\ln(q_0)$. It is known that $\ln(q_0)$ is related to R_0 and T_0 by the

approximate relationship of equation (5), where A and B are determined by the remaining parameters but usually vary slowly across the input space [7]. However, there are regions in which there are larger variations in q_0, due to phenomena such as impacting waves.

$$\ln(q_0) \cong A - BR_0/T_0 \,. \tag{5}$$

A hybrid network could be well-suited to this data, since the MLP network may represent the global relationship in equation (5) well, leaving the RBF neurons to identify local variations in the function.

6 Wave Overtopping Results and Discussion

The GL-ANN and RBF networks give lower MSEs than the MLP network, but require more hidden neurons (Table 4). The GL-ANNs are superior to the pure RBF

Table 3. Mean Square Errors and Hidden Layer Sizes using overtopping data

	Average test MSE	Hidden layer size
MLP (L-M)	1.16e-2	14
FS-OLS	1.04e-2	195
GL-ANN	0.94e-2	80

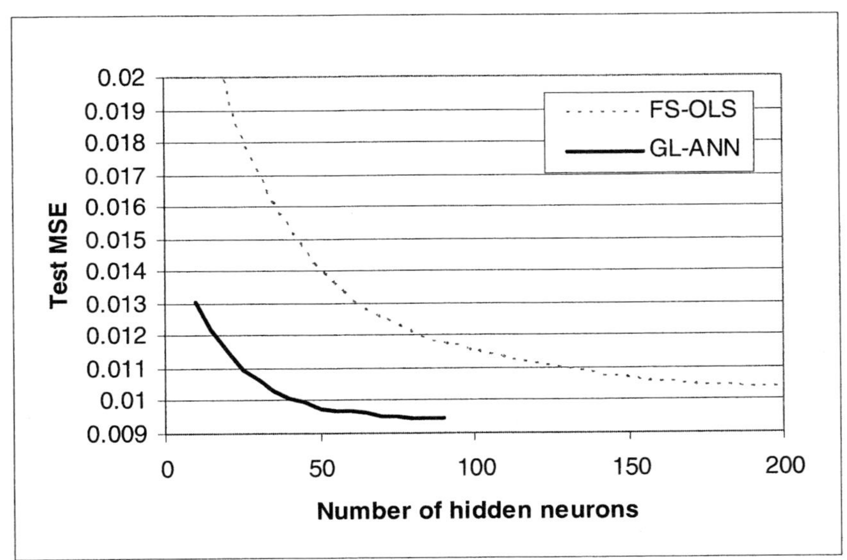

Fig. 1. Test errors for given hidden layer sizes with FS-OLS and GL-ANN

networks in terms of both MSE and hidden layer size. As neurons are added, the errors of both networks decrease before reaching a minimum (Fig 1). However, the GL-ANN networks require considerably fewer neurons to achieve a given test error. They also give errors comparable to those from numerical simulation, even though the latter is specific to a particular structure and sea-state.

These results suggest that GL-ANNs may be particularly suited to high-dimensional, noisy data. Further investigation of the strengths and weaknesses of GL-ANNs with different types of data is required.

7 Conclusions

It has been shown that GL-ANNs have strong generalisation capabilities, particularly with noisy, high-dimensional data. They compare favourably with traditional MLP and RBF networks, as well as with variable width RBF networks, exemplified here by RT-RBF. The training algorithm used by GL-ANN also seems more effective than the early clustering method used by PRBFNs. When compared to pure RBFs, GL-ANNs are more compact and have less need of regularisation.

References

1. Haykin S. Neural Networks: A Comprehensive Foundation. 2nd edn. Prentice Hall (1999)
2. Moody J. Darken C. Fast learning in networks of locally-tuned processing units. *Neural Computation*, 1: 281-294 (1989)
3. Cohen S., Intrator N. A Hybrid Projection-based and Radial Basis Function Architecture. *Pattern Analysis and Applications*, 5: 113-120 (2002)
4. Orr M. Introduction to Radial Basis Function Networks. Available from http://www.anc.ed.ac.uk/~mjo/rbf.html (1996)
5. Friedman J., Multivariate adaptive regression splines. *Annals of Statistics,* 19:1-141, 1991
6. Mackay D., Bayesian Interpolation. *Neural Computation*, 4(3): 415-447 (1992)
7. Besley P. Overtopping of Seawalls: Design and Assessment Manual, Environment Agency R&D Technical Report W178 (1999).
8. Shiach J., Mingham C., Ingram D., Bruce T. The applicability of the shallow water equations for modelling violent wave overtopping. *Coastal Engineering*, 51:1-15 (2004)
9. http://www.clash-eu.org
10. D. Wedge, D. Ingram, C. Mingham, D. McLean, Z. Bandar. Neural Network Architectures and Overtopping Predictions. Submitted to *Maritime Engineering* (2004)

Learning with Ensemble of Linear Perceptrons

Pitoyo Hartono[1] and Shuji Hashimoto[2]

[1] Future University-Hakodate, Kamedanakanocho 116-2 Hakodate City, Japan
hartono@fun.ac.jp
[2] Waseda University, Ohkubo3-4-1 Shinjuku-ku, Tokyo 169-8555, Japan
shuji@waseda.jp

Abstract. In this paper we introduce a model of ensemble of linear perceptrons. The objective of the ensemble is to automatically divide the feature space into several regions and assign one ensemble member into each region and training the member to develop an expertise within the region. Utilizing the proposed ensemble model, the learning difficulty of each member can be reduced, thus achieving faster learning while guaranteeing the overall performance.

1 Introduction

Recently, several models of neural networks ensemble have been proposed [1,2,3], with the objective of achieving a higher generalization performance compared to the singular neural network. Some of the ensembles, represented by Boosting [4] and Mixture of Experts [5], proposed mechanisms to divide the learning space into a number of sub-spaces and assign each sub-space into one of the ensemble's member, hence the learning burden of each member is significantly reduced, leading to a better overall performance. In this paper, we proposed an algorithm that effectively divides the learning space in a linear manner, and assign the classification task of each sub-space to a linear perceptron [6] that can be rapidly trained. The objective of this algorithm is to achieve linear decomposition of nonlinear problems through an automatic divide and conquer approach utilizing ensemble of linear perceptrons. In addition to the ordinary output neurons, each linear perceptron in the proposed model also has an additional neuron in its output layer. The additional neuron is called "confident neuron", and produced an output that indicates the "confidence level" of the perceptron with regard to its ordinary output. An output of the perceptron which has a high confidence level can be considered as a reliable output, while an output with low confidence level is unreliable one. The proposed ensemble is equipped with a competitive mechanism for learning space division based on the confidence levels and at the same time to train each member to perform in the given sub learning space.

The linearity of each member also enables us to analyze the division of the problem space that can be useful in understanding the structure of the problems and also to analyze the overall performance of the ensemble.

2 Ensemble of Linear Perceptron

The Ensemble of Linear Perceptron (ELP) consists of several linear perceptrons (called members), each with an additional output neuron that indicates the confidence level as shown in Fig. 1.

W. Duch et al. (Eds.): ICANN 2005, LNCS 3697, pp. 115–120, 2005.

The ordinary output of each member is shown as follows.

$$O_j^i = \sum_{k=1}^{N_{in}} w_{kj}^i I_k + \theta_j^i \tag{1}$$

Where O_j^i is the j-th output neuron of the i-th member, while w_{kj}^i is the connection weight between the k-th input neuron and the j-th output neuron in the i-th member and θ_j^i is the threshold for the j-th output neuron in the i-th member.

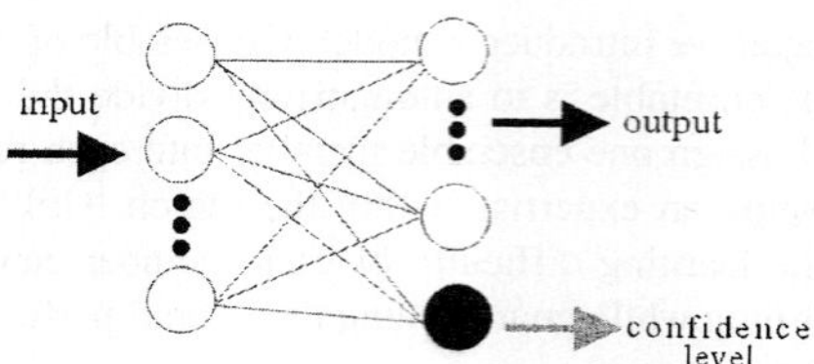

Fig. 1. Ensemble's Member

The output of the confidence neuron can be written as follows.

$$O_c^i = \sum_{k=1}^{N_{in}} v_k^i I_k + \theta_c^i \tag{2}$$

Where O_c^i is the output of the confidence neuron in the i-th member, v_k^i is the connection weight between the k-th input neuron to the confidence neuron in the i-th member, and θ_c^i is the threshold of the confidence neuron in the i-th member.

As illustrated in Fig. 2, in the running phase, the output of the member with the highest confidence value is taken is as the ensemble's output, O^{ens} as follows.

$$\begin{aligned} O^{ens} &= O^w \\ w &= \arg\max_i \{O_c^i\} \end{aligned} \tag{3}$$

In the training phase only the winning member (a member with the highest confidence level) is allowed to modify its connections weights between the ordinary output neurons and the input neurons, while the connection weights between the input neuron and ordinary output neurons for the rest of members remain unchanged. The weights corrections are executed as follows.

$$\begin{aligned} W^w(t+1) &= W^w(t) - \eta \frac{\partial E(t)}{\partial W^w(t)} \\ W^i(t+1) &= W^i(t) \ \ for\ i \neq w \\ E(t) &= \left\| O^w(t) - D(t) \right\|^2 \end{aligned} \tag{4}$$

Where D(t) indicates the teacher signal.

However all members are required to modify the connection weights between their input neurons and confidence neurons as follows.

$$
\begin{aligned}
V_c^i(t+1) &= V_c^i(t) - \eta \frac{\partial E_c^i(t)}{\partial V_c^i(t)} \\
E_c^i(t) &= \left\| O_c^i(t) - C^i(t) \right\|^2 \\
C^i(t) &= 1 - \frac{\exp(\alpha \left\| O_c^i(t) - O_c^w(t) \right\|^2)}{\sum_{j=1}^{N} \exp(\alpha \left\| O_c^j(t) - O_c^w(t) \right\|^2)}
\end{aligned} \tag{5}
$$

Where V_c^i is the connection weight vector between the input neurons and the confidence neuron in the *i*-th member. The correction of the weights leading to the confidence neuron is illustrated in Fig. 3.

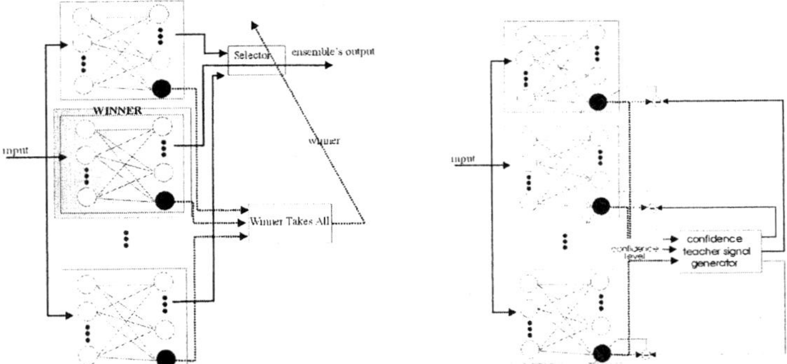

Fig. 2. Running Phase **Fig. 3.** Confidence Level Training

The learning mechanism triggers competition among the members to take charge of a certain region in the learning space. Because initially the connection weights of the members were randomly set, we can expect some bias in the performance of the members regarding different regions in the learning space. The training mechanism assures that a member with higher confidence level with regard to a region will perform better than other members regarding that region and at the same time produce a better confidence level, while other members' confidence will be reduced. Thus, the learning space is divided into subspaces where a member will develop an expertise upon a particular subspace.

3 Experiments

In the first experiment, we trained the ELP with XOR problem, which cannot be solved with a single linear perceptron [6]. In this experiment we use ELP with two members, whose training rates were uniformly set to 0.1 and α in Equation 5 is set to

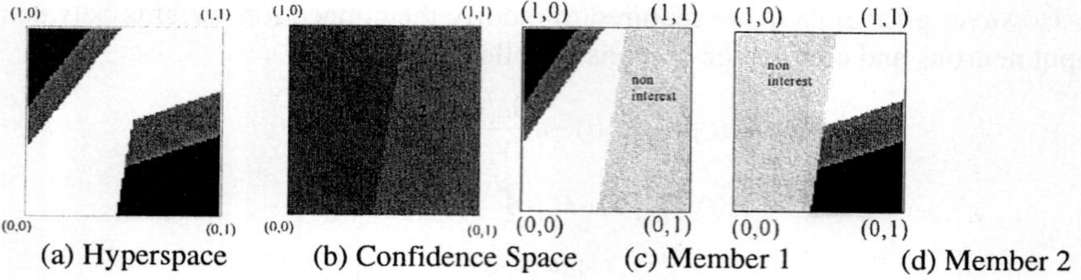

(a) Hyperspace (b) Confidence Space (c) Member 1 (d) Member 2

Fig. 4. Hyperspace of ELP

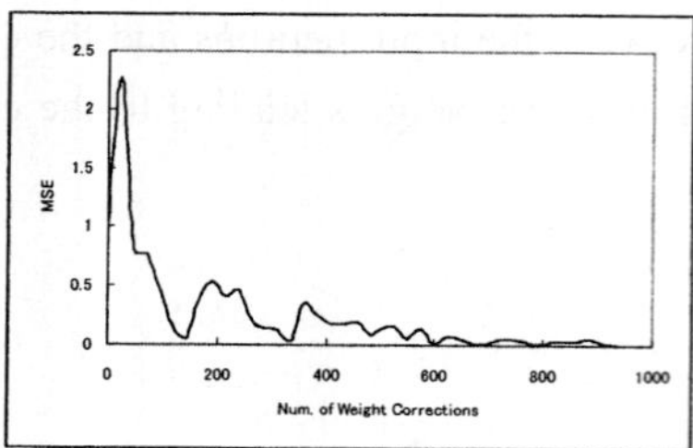

Fig. 5. Learning Curve of ELP

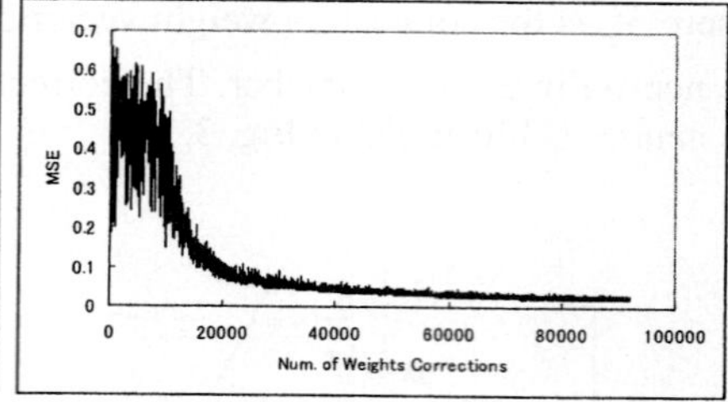

Fig. 6. Learning Curve of ELP

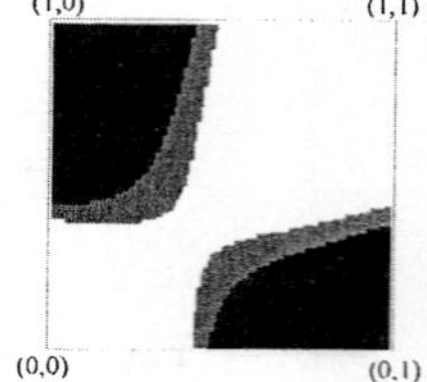

Fig. 7. Hyperspace (MLP)

Figure 4(a) shows the hyperspace of ELP, where the nonlinear problem is divided into two linearly separable problems. The black shows the regions that are classified as 1, and white shows the region of 0, while the gray regions are ambiguous where the outputs are in the range of 0.4 and 0.6. Figure 4(b) indicates the confidence space of ELP, where the region marked "1" is a region where the confidence level of Member 1 exceeds that of Member 2, while the region marked "2" indicates where the confidence level of Member 2 exceeds that of Member 1. Figs. 4(c) and 4(d) show the hyperspace of Member 1 and Member 2, respectively. From Fig. 4, it is obvious that ELP have the ability solve nonlinear problem through linear decomposition.

We compared the performance of the ELP to that of MLP [7] with 3 hidden neurons. The learning rate for MLP is set to 0.3. Fig.6 (a) shows the hyperspace formed by the MLP, where the learning space is nonlinearly classified. We also compared the learning performances between ELP and MLP, where we calculate the number of weight corrections. For ELP, for each training iteration, the connection weights of between the input neurons and output neurons of the winning member and the connection weights between confidence neuron and the input neurons of all the members are corrected, hence the number of weight corrections, C_{ELP} is as follows.

$$C_{ELP} = N_{in} \cdot (N_{out} + M) \tag{6}$$

where N_{in}, N_{out} and M are the number of input neurons, the number of output neurons and the number of members in the ensemble, respectively.

The number of the weight corrections in MLP, C_{MLP} is calculated as follows.

$$C_{MLP} = N_{hid} \cdot (N_{in} + N_{out}) \tag{7}$$

N_{hid} shows the number of hidden neurons in the MLP.

From Fig. 5 and Fig. 6 (b), we can see that the ELP can achieve the same classification performance with significantly lower number of weight corrections.

In the second experiment we trained ELP, with two members with Breast Cancer Classification Problem [8][9]. In this problem the task of neural network is to classify, a 9 dimensional input into two classes. In this experiment, the parameter settings for the ELP are the same as the previous experiment, while for MLP we set 5 hidden neurons. Comparing Fig. 8 and Fig. 9, we can see that ELP achieves similar performance to MLP with significantly less number of weights corrections.

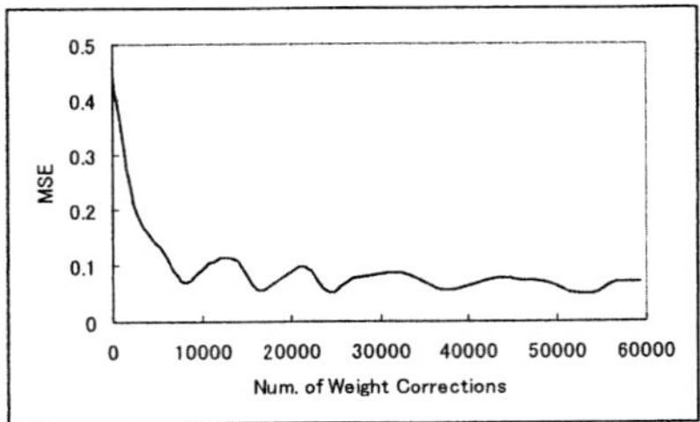

Fig. 8. Learning Curve of ELP

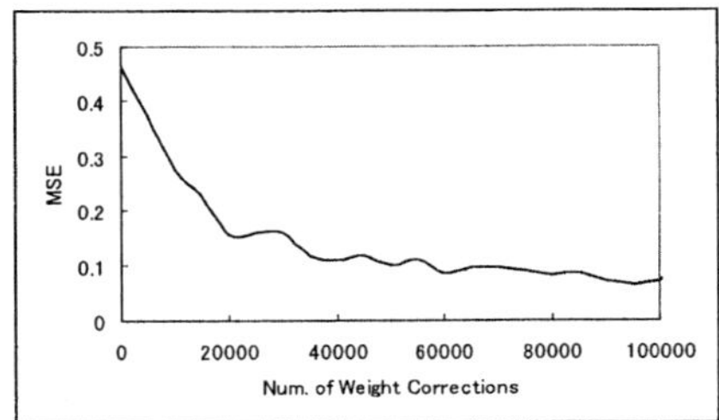

Fig. 9. Learning Curve of MLP

In the third experiment we trained the ELP with the 3-classed Iris Problem, in which it is known one classis linearly inseparable from the rest of the classes. The performance of 3-membered ELP is shown in Fig. 10, while the performance of MLP with 5 hidden neurons is shown in Fig. 11.

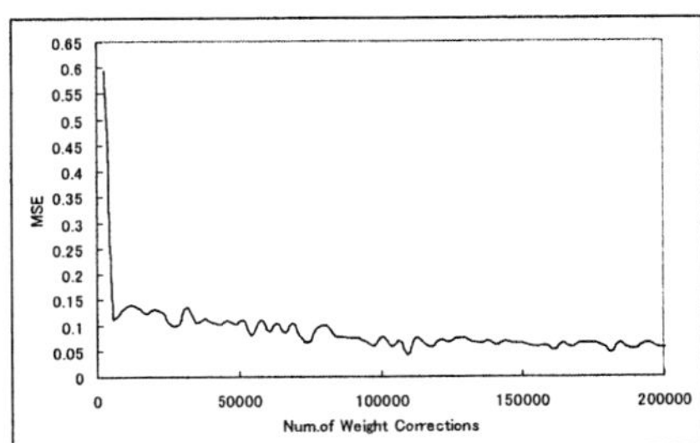

Fig. 10. Learning Curve of ELP

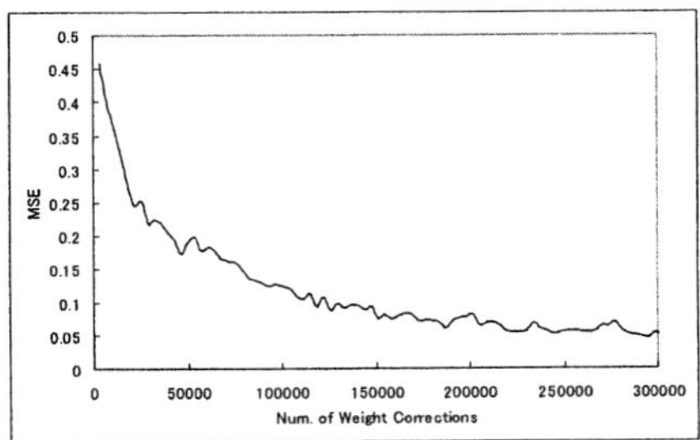

Fig. 11. Learning Curve of MLP

4 Conclusion and Future Works

In this study, we propose an ensemble of linear perceptron that can automatically divide the problem space in linear manner and assign one of the ensemble members to the sub-problem space. This division of problem space is achieved based on the confidence level of each member, in which each member is only responsible to perform in the region where its confidence is the highest, hence the learning burden of each member can be significantly lessen. Because the members are linear perceptrons, overall ELP learns significantly faster than MLP, because the number of connection weights to be corrected is significantly less than MLP, while the performances are similar. In the future, we consider to develop a more efficient competitive mechanism so that the uniqueness of each member expertise is enforced. We also consider to develop the proposed ELP for efficient Boosting mechanism.

References

1. Baxt, W.: Improving Accuracy of Artificial Neural Network using Multiple Differently Trained Networks. Neural Computation Vol. 4 (1992) 108–121.
2. Sharkey, A.: On Combining Artificial Neural Nets. Connection Science, Vol. 9, Nos. 3 and 4 (1996) 299-313.
3. Hashem, S.: Optimal Linear Combination of Neural Networks. Neural Networks, Vol. 10, No.4 (1996) 559-614.
4. Freund, Y.: Boosting a weak learning algorithm by Majority. Information and Computation, Vol. 121, No.2, (1995), 256-285.
5. Jacobs, R., Jordan, M., Nowlan, S., and Hinton, G: Adaptive Mixture of Local Experts. Neural Computation, Vol. 3 (1991) 79-87.
6. Minsky, M., and Papert, S.: Perceptron, The MIT Press (1969).
7. Rumelhart, D.E., and McClelland, J.: Learning Internal Representation by Error Propagation. Parallel Distributed Processing, Vol.1 MIT Press (1984) 318-362.
8. Mangasarian, O.L., Wolberg, W.H., Cancer Diagnosis via linear programming, *SIAM News*, Vol. 23, No. 5, (1990) 1-18.
9. UCI Machine Learning Repository, http://www.ics.uci.edu/~mlearn/MLRepository.html

Combination Methods for Ensembles of RBFs[1]

Carlos Hernández-Espinosa, Joaquín Torres-Sospedra,
and Mercedes Fernández-Redondo

Universidad Jaume I. Dept. de Ingeniería y Ciencia de los Computadores,
Avda Vicente Sos Baynat s/n. 12071 Castellon, Spain
{redondo, espinosa}@icc.uji.es

Abstract. Building an ensemble of classifiers is a useful way to improve the performance. In the case of neural networks the bibliography has centered on the use of Multilayer Feedforward (MF). However, there are other interesting networks like Radial Basis Functions (RBF) that can be used as elements of the ensemble. In a previous paper we presented results of different methods to build the ensemble of RBF. The results showed that the best method is in general the *Simple Ensemble*. The combination method used in that research was averaging. In this paper we present results of fourteen different combination methods for a simple ensemble of RBF. The best methods are Borda Count, Weighted Average and Majority Voting.

1 Introduction

The most important property of a neural network (NN) is the generalization capability. One method to increase this capability with respect to a single NN consist on training an ensemble of NNs, i.e., to train a set of NNs with different weight initialization or properties and combine the outputs in a suitable manner.

In the field of ensemble design, the two key factors to design an ensemble are how to train the individual networks and how to combine the different outputs.

It seems clear from the bibliography that this procedure generally increases the generalization capability in the case of the NN Multilayer Feedforward (MF) [1,2].

However, in the field of NNs there are other networks besides MF, and traditionally the use of ensembles of NNs has restricted to the use of MF.

Another useful network which is quite used in applications is Radial Basis Functions (RBF). This network can also be trained by gradient descent [3] and it can be also an element of an ensemble.

In [4], we obtain the first results on ensembles of RBF, we presented a comparison of different methods to build the ensemble and we concluded that the "Simple Ensemble" was the most appropriate. The combination method was averaging.

In this paper we present results of different combination methods for the case of a "simple ensemble" of RBFs. The number of combination methods analyzed is fourteen. With these results we can have a hint to select the appropriate combination method and improve the performance of RBFs ensembles.

---

[1] This research was supported by the project MAPACI TIC2002-02273 of CICYT in Spain.

W. Duch et al. (Eds.): ICANN 2005, LNCS 3697, pp. 121 – 126, 2005.

2 Theory

In this section, first we briefly review the basic concepts of RBFs networks and after that we review the different methods of combining the outputs of the ensemble.

2.1 RBF Networks with Gradient Descent Training

A RBF has two layers of networks. The first layer is composed of neurons with a Gaussian transfer function and the second layer has neurons with a linear transfer function. The output of a RBF network can be calculated with equation 1.

$$F_k(x) = \sum_{q=1}^{Q} w_q^k \cdot \exp\left(-\sum_{n=1}^{N} \left(C_{q,n}^k - X_n\right)^2 \Big/ \left(\sigma_q^k\right)^2 \right) \tag{1}$$

Where $C_{q,n}{}^k$ are the centers of the Gaussian units, $\sigma_q{}^k$ control the width of the Gaussian functions and $w_q{}^k$ are the weights among the Gaussian units and the output units.

The equations for adaptation of centers and weights are the following [3].

$$\Delta w_q^k = \eta \cdot \varepsilon_k \cdot \exp\left(-\sum_{n=1}^{N} \left(C_{q,n}^k - X_n\right)^2 \Big/ (\sigma)^2 \right) \tag{2}$$

Where η is the step size and ε_k is the difference between the target and the output and the equation for the adaptation of the centers is number 3.

$$\Delta C_q = \eta \cdot (X_k - C_q) \cdot \frac{2}{\sigma} \cdot \exp\left(-\sum_{n=1}^{N} \left(C_{q,n}^k - X_n\right)^2 \Big/ (\sigma)^2 \right) \cdot \sum_{k=1}^{n_o} \varepsilon_k \cdot w_q^k \tag{3}$$

2.2 Ensemble Combination Methods

Average: This approach simply averages the individual classifier outputs.

Majority Vote: The correct class is the one most often chosen by the classifiers.

Winner Takes All (WTA): In this method, the class with overall maximum value in all the classifiers is selected.

Borda Count: For any class q, the Borda count is the sum of the number of classes ranked below q by each classifier. If $B_j(q)$ is the number of classes ranked below the class q by the jth classifier, then the Borda count for class q is in the equation 4.

$$B(q) = \sum_{j=1}^{L} B_j(q) \tag{4}$$

Bayesian Combination: This combination method was proposed in reference [5]. According to this reference a belief value that the pattern x belongs to class i can be approximated by the following equation.

$$Bel(i) = \frac{\prod_{k=1}^{L} P(x \in q_i \mid \lambda_k(x) = j_k)}{\sum_{i=1}^{Q} \prod_{k=1}^{L} P(x \in q_i \mid \lambda_k(x) = j_k)}, \quad 1 \le i \le Q \tag{5}$$

Where the conditional probability that sample x actually belongs to class i, given the classifier k assign it to class j ($\lambda_k(x)=j_k$) is estimated from the confusion matrix [6].

Weighted Average: This method introduces weights to the outputs of the different networks prior to averaging [7]. The weights try to minimize the difference between the output of the ensemble and the "desired or true" output.

Choquet Integral: This method is based in the fuzzy integral and the Choquet integral. The method is complex and a full description can be found in reference [6].

Combination by Fuzzy Integral with Data Dependent Densities (Int. DD): It is another method based on the fuzzy integral and the Choquet integral. But in this case [6], prior to the application of the method it is performed a partition of the input space in regions by k-means clustering or frequency sensitive learning.

Weighted Average with Data Dependent weights (W.Ave DD): This method is the weighted average described above. But in this case, a partition of the space is performed by using k-means clustering and the weights are calculated for each partition.

BADD Defuzzification Strategy: It is another combination method based on fuzzy logic concepts. The method is complex and the description can be found in [6].

Zimmermann's Compensatory Operator: This combination method is based in the Zimmermann's compensatory operator [8]. The method is described in [6].

Dynamically Averaged Networks (DAN), version 1 and 2: It is proposed in reference [9]. In this method instead of choosing static weights derived from the network performance on a sample of the input space, we allow the weights to adjust to be proportional to the certainties of the respective network output.

Nash Vote: In this method [10] each net assigns a number between zero and one for each output. The maximum of the product of the values of the nets is the winner.

3 Experimental Results

We have applied the 20 ensemble methods to 9 different classification problems. They are from the UCI repository and their names are Balance Scale (Bala), Cylinders Bands (Band), Liver Disorders (Bupa), Credit Approval (Credit), Glass Identification (Glass), Heart Disease (Heart), the Monk's Prob. (Monk 1, Monk 2) and Voting Records (Vote).

We have constructed ensembles of 3 and 9 networks. We repeated the process of training ten times for different partitions of data in training, cross-validation and test. With this procedure we can obtain a mean performance of the ensemble for each database (the mean of the ten ensembles) and an error in the performance calculated by standard error theory. We have used the error, but this measure is related to the standard deviation as equation 6 and to the confidence interval with equation 7.

$$\sigma = \sqrt{n} \cdot Error \tag{6}$$

Where n is the number of experiments performed to obtain the mean, which is 10.

$$Confidence\ Interval = \left(\overline{X} - z_{\alpha/2} \cdot Error,\ \overline{X} + z_{\alpha/2} \cdot Error \right) \tag{7}$$

Where $z_{\alpha/2}$ is obtained from the following probability of the normal distribution.

$$P\left(\overline{X} - z_{\alpha/2} \frac{\sigma}{\sqrt{n}} \le \mu \le \overline{X} + z_{\alpha/2} \frac{\sigma}{\sqrt{n}} \right) = 1 - \alpha \tag{8}$$

The results of the performance are in table 1 and 2.

As table 1, 2 show the improvement by the use of an ensemble depends of the problem. We get an improvement in Bupa, Credit, Glass, Heart, Monk1 and Vote.

The results of tables 1 and 2 show that the improvement of training nine networks (instead of three) is low. The best alternative might be an ensemble of three networks.

Comparing the results of the different combination methods of tables 1 and 2, we can see that the differences are low. The largest difference between simple average and other method is around 0.7% in the problem Heart.

However, we should point out that the computational cost of any combination method is very low. So the selection of an appropriate combination method allows an improvement without extra cost.

Table 1. Results for the ensemble of three networks

	Bala	Band	Bupa	Credit	Glass
Single Network	90.2 ± 0.5	74.0 ± 1.1	70.1 ± 1.1	86.0 ± 0.8	93.0 ± 0.6
Average	89.7 ± 0.7	73.8 ± 1.2	71.9 ± 1.1	87.2 ± 0.5	93.2 ± 1.0
Majority V.	89.9 ± 0.7	74.4 ± 1.2	72.0 ± 1.0	87.1 ± 0.6	93.2 ± 1.0
WTA	90.1 ± 0.8	72.9 ± 1.1	71.1 ± 1.2	87.2 ± 0.6	93.2 ± 1.0
Borda	89.8 ± 0.7	74.4 ± 1.2	72.0 ± 1.0	87.1 ± 0.6	93.2 ± 1.0
Bayesian	89.9 ± 0.7	74.4 ± 1.2	72.0 ± 1.0	87.1 ± 0.6	93.2 ± 1.0
W. Average	89.9 ± 0.7	72.9 ± 1.5	72.4 ± 1.2	87.2 ± 0.5	93.0 ± 1.2
Choquet	89.9 ± 0.7	73.1 ± 1.1	71.4 ± 1.0	86.9 ± 0.6	93.2 ± 1.0
Int. DD	89.9 ± 0.7	72.9 ± 1.2	71.9 ± 0.9	86.9 ± 0.6	93.0 ± 0.9
W. Ave DD	89.7 ± 0.7	74.2 ± 1.0	71.9 ± 1.1	87.2 ± 0.5	93.2 ± 1.0
BADD	89.7 ± 0.7	73.8 ± 1.2	71.9 ± 1.1	87.2 ± 0.5	93.2 ± 1.0
Zimmermann	65 ± 5	63 ± 5	62 ± 3	70 ± 5	87.2 ± 1.5
DAN	89.8 ± 0.8	73.5 ± 1.3	71.9 ± 1.0	87.3 ± 0.5	93.6 ± 1.1
DAN version 2	89.9 ± 0.8	73.1 ± 1.3	71.7 ± 1.0	87.2 ± 0.5	93.8 ± 1.1
Nash Vote	89.7 ± 0.7	74.0 ± 1.2	72.3 ± 1.1	87.2 ± 0.5	93.2 ± 1.0

Table 1. *(Continued).*

	Heart	Monk 1	Monk 2	Vote
Single Network	82.0 ± 1.0	98.5 ± 0.5	91.3 ± 0.7	95.4 ± 0.5
Average	83.9 ± 1.6	99.6 ± 0.4	91.5 ± 1.2	96.3 ± 0.7
Majority V.	84.6 ± 1.5	99.6 ± 0.4	90.9 ± 1.1	96.4 ± 0.6
WTA	83.9 ± 1.6	99.6 ± 0.4	91.4 ± 1.3	96.3 ± 0.7
Borda	84.6 ± 1.5	99.6 ± 0.4	90.9 ± 1.1	96.4 ± 0.6
Bayesian	84.6 ± 1.5	99.4 ± 0.4	90.1 ± 1.1	96.4 ± 0.6
W. Average	83.6 ± 1.6	99.8 ± 0.3	92.0 ± 1.2	96.3 ± 0.6
Choquet	83.6 ± 1.6	99.6 ± 0.4	91.5 ± 1.2	96.3 ± 0.7
Int. DD	83.6 ± 1.4	99.6 ± 0.4	91.1 ± 1.3	96.3 ± 0.7
W. Ave DD	83.9 ± 1.6	99.6 ± 0.4	92.0 ± 1.2	96.3 ± 0.7
BADD	83.9 ± 1.6	99.6 ± 0.4	91.5 ± 1.2	96.3 ± 0.7
Zimmermann	75 ± 4	90 ± 5	82 ± 3	92 ± 3
DAN	83.6 ± 1.3	99.5 ± 0.4	90.8 ± 1.3	96.0 ± 0.6
DAN version 2	83.4 ± 1.5	99.6 ± 0.4	90.6 ± 1.4	96.1 ± 0.6
Nash Vote	84.1 ± 1.7	99.6 ± 0.4	91.6 ± 1.1	96.3 ± 0.7

Table 2. Results for the ensemble of nine networks

	Bala	Band	Bupa	Credit	Glass
Single Network	90.2 ± 0.5	74.0 ± 1.1	70.1 ± 1.1	86.0 ± 0.8	93.0 ± 0.6
Average	89.7 ± 0.7	73.3 ± 1.4	72.4 ± 1.2	87.2 ± 0.5	93.0 ± 1.0
Majority V.	89.7 ± 0.7	74.0 ± 1.5	72.1 ± 1.1	87.1 ± 0.5	93.2 ± 1.0
WTA	89.8 ± 0.8	73.6 ± 1.7	72.0 ± 1.3	87.2 ± 0.5	93.4 ± 1.0
Borda	89.6 ± 0.7	74.0 ± 1.5	72.1 ± 1.1	87.1 ± 0.5	93.2 ± 1.0
Bayesian	90.2 ± 0.7	74.2 ± 1.5	72.3 ± 1.1	87.2 ± 0.5	92.6 ± 1.0
W. Average	89.5 ± 0.7	73.1 ± 1.6	71.6 ± 1.3	86.9 ± 0.5	92.8 ± 1.2
Choquet	89.8 ± 0.8	74.0 ± 1.5	72.0 ± 1.4	87.3 ± 0.5	93.4 ± 1.0
Int. DD	89.8 ± 0.8	74.0 ± 1.5	72.1 ± 1.5	87.2 ± 0.5	93.4 ± 1.0
W. Ave DD	89.7 ± 0.7	73.3 ± 1.4	72.3 ± 1.2	87.2 ± 0.5	93.0 ± 1.0
BADD	89.7 ± 0.7	73.3 ± 1.4	72.4 ± 1.2	87.2 ± 0.5	93.0 ± 1.0
Zimmermann	69 ± 5	66 ± 3	62 ± 4	75 ± 5	80 ± 3
DAN	89.8 ± 0.8	72.7 ± 1.7	71.6 ± 1.3	87.2 ± 0.5	92.8 ± 1.0
DAN version 2	89.8 ± 0.8	73.3 ± 1.7	71.4 ± 1.3	87.3 ± 0.5	92.8 ± 1.0
Nash Vote	89.6 ± 0.7	73.1 ± 1.4	72.6 ± 1.2	87.2 ± 0.5	93.0 ± 1.0

Table 2. *(Continued).*

	Heart	Monk 1	Monk 2	Vote
Single Network	82.0 ± 1.0	98.5 ± 0.5	91.3 ± 0.7	95.4 ± 0.5
Average	83.9 ± 1.5	99.6 ± 0.4	91.4 ± 1.2	96.3 ± 0.7
Majority V.	84.6 ± 1.6	99.6 ± 0.4	91.5 ± 1.2	96.4 ± 0.6
WTA	83.6 ± 1.7	99.8 ± 0.3	90.8 ± 1.2	96.0 ± 0.6
Borda	84.6 ± 1.6	99.6 ± 0.4	91.5 ± 1.2	96.4 ± 0.6
Bayesian	84.6 ± 1.6	99.5 ± 0.3	90.9 ± 1.1	96.4 ± 0.6
W. Average	83.7 ± 1.4	99.8 ± 0.3	91.8 ± 1.4	96.6 ± 0.7
Choquet	83.4 ± 1.6	99.6 ± 0.4	90.6 ± 1.2	96.0 ± 0.6
Int. DD	83.4 ± 1.6	99.6 ± 0.4	90.8 ± 1.2	96.1 ± 0.6
W. Ave DD	83.9 ± 1.5	99.6 ± 0.4	91.3 ± 1.2	96.3 ± 0.7
BADD	83.9 ± 1.5	99.6 ± 0.4	91.4 ± 1.2	96.3 ± 0.7
Zimmermann	73 ± 4	92 ± 2	76 ± 5	81 ± 6
DAN	84.4 ± 1.7	99.4 ± 0.4	88.9 ± 1.6	96.0 ± 0.6
DAN version 2	84.1 ± 1.8	99.5 ± 0.4	88.9 ± 1.5	96.0 ± 0.6
Nash Vote	83.9 ± 1.5	99.6 ± 0.4	91.5 ± 1.1	96.3 ± 0.7

Table 3. Relative Performance with respect to a Single Network

	Three Networks	Nine Networks
Average	13,07	12,64
Majority V.	13,24	13,93
WTA	12,26	12,63
Borda	13,24	13,84
Bayesian	11,39	12,91
W. Average	13,89	13,62
Choquet	12,21	11,70
Int. DD	11,50	12,10
W. Ave DD	13,86	12,42
BADD	13,07	12,64
Zimmermann	-118,01	-168,30
DAN	10,94	6,46
DAN version 2	12,00	7,45
Nash Vote	13,58	12,68

To appreciate the results more clearly, we have also calculated the percentage of error reduction of the ensemble with respect to a single network. We have used equation 9 for this calculation.

$$PorError_{reduction} = 100 \cdot \frac{PorError_{single\ network} - PorError_{ensemble}}{PorError_{single\ network}} \tag{9}$$

This new measurement is just a simple way to perform a ranking of the performance, but the results are not statistical significant. A correct way to obtain statistical significant results might be to use the normal percentage and a McNemar's test with Bonferroni correction, this will be addressed in future researches.

Also, this new measurement is relative and we can calculate a mean value across all databases. The result is in table 3 for the case of three and nine networks.

According to the values of the mean performance of error reduction, the best performing methods are Majority Vote, Borda Count and Weighted Average.

4 Conclusions

In this paper we have presented experimental results of 14 different methods to combine the outputs of an ensemble of RBF nets, using 10 databases. We trained ensembles of 3 and 9 nets. The results showed that the improvement in performance from three to nine networks in the ensemble is usually low. Taking into account the computational cost, an ensemble of three networks might be the best alternative. The differences among the different combination methods are low, but we can obtain a extra performance. Finally, we have obtained the mean percentage of error reduction over all databases. According to their values the best performing methods are Majority Vote, Borda Count and Weighted Average.

References

1. Tumer, K., Ghosh, J., "Error correlation and error reduction in ensemble classifiers", Connection Science, vol. 8, nos. 3 & 4, pp. 385-404, 1996.
2. Raviv, Y., Intrator, N., "Bootstrapping with Noise: An Effective Regularization Technique", Connection Science, vol. 8, no. 3 & 4, pp. 355-372, 1996.
3. Karayiannis, N.B., Randolph-Gips, M.M., "On the Construction and Training of Reformulated Radial Basis Function Neural Networks", IEEE Trans. On Neural Networks, Vol.14, no. 4, pp. 835-846, 2003.
4. Hernandez-Espinosa, Carlos, Fernandez-Redondo, Mercedes, Torres-Sospedra, Joaquin, "Ensembles of RBFs Trained by Gradient Descent", Internacional Symposium on Neural Networks, Lecture Notes in Computer Science, vol. 3174, pp. 223-228, 2004.
5. Xu, L., Krzyzak, A., SUen, C.Y., "Methods of combining multiple classifiers and their applications to handwriting recognition", IEEE Trans. On Systems, Man and Cybernetics, vol. 22, no. 3, pp. 418-435, 1992.
6. Verikas, A., Lipnickas, A., Malmqvist, K., Bacauskiene, M., Gelzinis, A., "Soft combination of neural classifiers: A comparative study", Pattern Recognition Letters, vol. 20, pp. 429-444, 1999.
7. Krogh, A., Vedelsby, J., "Neural network ensembles, cross validation, and active learning", Advances in Neural Information Processing Systems 7, pp. 231-238, 1995.
8. Zymmermann, H.J., Zysno, P., "Decision and evaluations by hierarchical aggregation of information", Fuzzy Sets and Systems, vol. 10, no. 3, pp. 243-260, 1984.
9. Jimenez, D., "Dynamically Weighted Ensemble Neural Network for Classification", IEEE World Congress on Computational Intelligence, vol. 1, pp. 753-756, 1998.
10. Wanas, N.M., Kamel, M.S., "Decision Fusion in Neural Network Ensembles", International Joint Conference on Neural Networks, vol. 4, pp. 2952-2957, 2001.

Ensemble Techniques for Credibility Estimation of GAME Models

Pavel Kordík and Miroslav Šnorek

Department of Computer Science and Engineering, Czech Technical University, Prague, Czech Republic
{kordikp, snorek}@fel.cvut.cz

Abstract. When a real world system is described either by means of mathematical model or by any soft computing method the most important is to find out whether the model is of good quality, and for which configuration of input features the model is credible. Traditional methods restrict the credibility of model to areas of training data presence. These approaches are ineffective when non-relevant or redundant input features are present in the modeled system and for non-uniformly distributed data. Even for simple models, it is often hard to find out how credible the output is for any input vector. We propose a novel approach based on ensemble techniques that allows to estimate credibility of models. We experimentally derived an equation to estimate the credibility of models generated by Group of Adaptive Models Evolution (GAME) method for any configuration of input features.

1 Introduction

If we build a black-box model of a real world system, we would like to know when we can trust the model. The very first information about the quality of our model we can get from its error on the training set[1]. The quality of models is usually evaluated on a testing data set. The disadvantage of this approach is that the quality of the model is tested just in few testing points (testing data). We cannot get any information about the model's quality for input configurations that are out of testing data. This is a problem particularly for multivariate real world systems (data are sparse or non-uniformly distributed) and for systems where some inputs are of low relevance. For such systems we can evolve inductive models using the GAME [2] method. These models are valid even for configurations of inputs that are very far from training and testing data (eg. values of irrelevant inputs can be out of areas of data presence). For these models we need more sophisticated technique to evaluate their quality and so estimate their credibility even for areas, where data about the system behavior are absent. Such technique is introduced in this paper.

[1] The data set describing a real system is usually split into two subsets: training set and the testing set. The model is built on the training data set and its performance is evaluated on the testing data set.

W. Duch et al. (Eds.): ICANN 2005, LNCS 3697, pp. 127–132, 2005.

At first, we shortly describe the process of GAME models evolution. In the next section, we mention ensemble techniques our approach is based on. Then we experimentally derive the equation that estimates the credibility of GAME models for artificial and real world data. At last, we apply the derived theory to the uncertainty signaling for visual knowledge mining.

2 Group of Adaptive Models Evolution (GAME)

Our work proceeds from the theory of inductive models construction commonly known as Group Method of Data Handling (GMDH) that was originally introduced by A.G. Ivachknenko in 1966 [1]. Where the traditional modeling methods

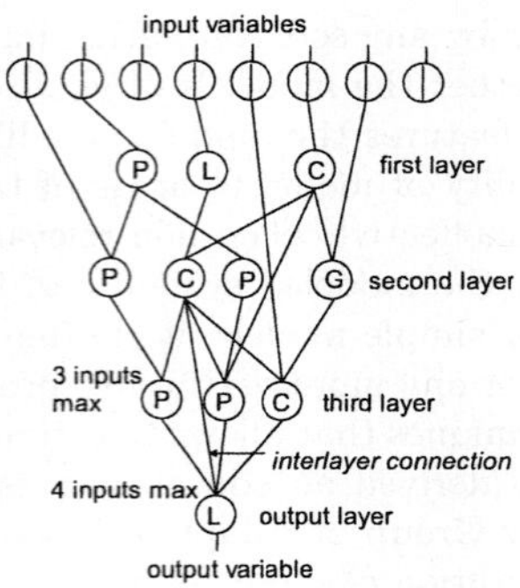

Fig. 1. The example of the GAME network. Network evolved on the training data consisting of units with suitable transfer function (P-percepton unit optimized by backpropagation algorithm, L-linear transfer unit and C-polynomial transfer unit, both optimized by Quasi Newton method).

(eg. MLP neural network) fail due to the "curse of dimensionality" phenomenon, the inductive methods are capable to build reliable models. The problem is decomposed into small subtasks. At first, the information from most important inputs is analyzed in the subspace of low dimensionality, later the abstracted information is combined to get a global knowledge of the system variables relationship. Figure 1 shows the example of inductive model (GAME network). It is constructed layer by layer during the learning stage from units that transfer information feedforwardly form inputs to the output. The coefficients of units' transfer functions are estimated using the training data set describing the modeled system. Units within single model can be of several types (hybrid model) - their transfer function can be linear(L), polynomial(C), logistic(S), exponential(E), small multilayer perceptron network(P), etc. Each type of unit has its own learning algorithm for coefficients estimation. The niching genetic algorithm is employed in each layer to choose suitable units. Which types of units are selected during the evolution to make up the model depends on the nature of modeled data. More information about inductive modeling can be found in [2].

3 Ensemble Techniques

Ensemble techniques [6] are based on the idea that a collection of a finite number of models (eg. neural networks) is trained for the same task. Neural network ensemble [5] is a learning paradigm where a group of neural networks is trained for the same task. It originates from Hansen and Salamons work [3], which shows that ensembling a number of neural networks can significantly improve the generalization ability. To create the ensemble of neural networks, the most prevailing approaches are Bagging and Boosting. Bagging is based on bootstrap sampling [5]. It generates several training sets from the original training set and then trains a component neural network from each of those training sets. Our Group of Adaptive Models Evolution method generates the ensemble of models (GAME networks) by the Bagging approach. By using the ensemble, instead of single GAME model, we can improve the accuracy of modelling. But this is not only advantage of using ensembles. There is a highly interesting information encoded in the ensemble behaviour. It is the information about credibility of member models.

4 GAME Models' Credibility Estimation

Following experiments explore the relationship between the dispersion of models' responses in the GAME ensemble and the credibility of models. Given a training data set L and testing data set T, suppose that $(x_1, x_2, ..., x_m, y)$ is a single testing vector from T, where $x_1...x_m$ are input values and y is the corresponding output value. Let G is an ensemble of n GAME models evolved on L using the Bagging technique [5]. When we apply values $x_1, x_2, ..., x_m$ to the input of each model, we receive models' outputs $y'_1, y'_2, ...y'_n$. Ideally, all responses would match the required output ($y'_1 = y'_2 = ... = y'_n = y$). This can be valid just for certain areas of the input space that are well described by L and just for data without noise. In most cases models' outputs will differ from the ideal value y. Figure 2

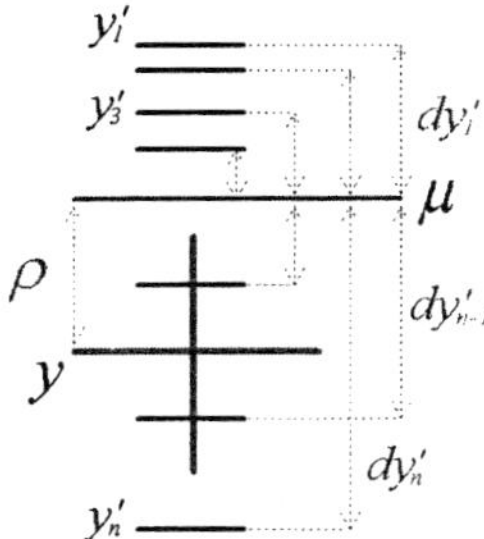

Fig. 2. Responses of GAME models for a testing vector lying in the area insufficiently described by the training data set

illustrates a case when a testing vector is lying in the area insufficiently defined by L. Responses of GAME models $y'_1, y'_2, ..., y'_n$ significantly differ. The mean of responses is defined as $\mu = \frac{1}{n}\sum_{i=1}^{n} y'_i$, the distance of ith model from the mean is $dy'_i = \mu - y'_i$ and the distance of the mean response from required output $\rho = \mu - y$ (see Figure 2). We observed that there may be a relationship between $dy'_1, dy'_2, ..., dy'_n$ and ρ. If we could express this relationship, we would be able to compute for any input vector not just the estimate of the output value (μ), but also the interval $\langle \mu + \rho', \mu - \rho' \rangle$ where the real output should lie with a certain significant probability.

4.1 Credibility Estimation - Artificial Data

We designed an artificial data set to explore this relationship by means of inductive modeling. We generated 14 random training vectors (x_1, x_2, y) in the range $x_1, x_2 \in \langle 0.1, 0.9 \rangle$ and 200 testing vectors in the range $x_1, x_2 \in \langle 0, 1 \rangle$, $y = \frac{1}{2} sinh(x_1 - x_2) + x_1^2(x_2 - 0.5)$. Then using the training data and the Bagging scheme we evolved n inductive models G by the GAME method. The

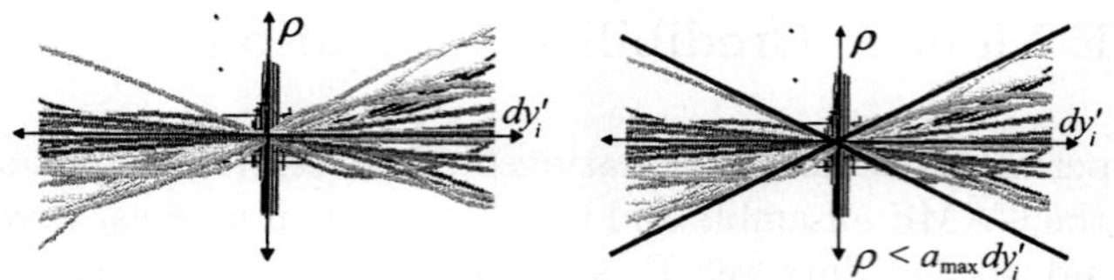

Fig. 3. The dependence of ρ on dy'_i is linear for artificial data without noise

density of the training data in the input space is low, therefore there are several testing vectors far from training vectors. For these vectors responses of GAME models considerably differ (similarly to the situation depicted on the Figure 2). For each testing vector, we computed $dy'_1, dy'_2, ..., dy'_n$ and ρ. This data $(x_1 = dy'_1, ..., x_n = dy'_n, y = \rho)$ we used to train a GAME model D to explore the relationship between dy'_i and ρ. In the Figure 3 there are responses of the model D for input vectors lying on dimension axes of the input space. Each curve express the sensitivity of ρ to the change of one particular input whereas other inputs are zero. We can see that with growing deviation of model G_i from the required value y, the estimate of ρ given by the model D increases with a linear trend. There exist a coefficient a_{max} that limits the maximal slope of linear dependence of ρ on dy'_i. If we present an input vector[2] to models from G, we can approximately limit (upper bound) the maximal deviation of their mean response from the real output as $\rho \leq \frac{a_{max}}{n}\sum_{i=1}^{n} |dy'_i|$.

[2] We show relationship between dy'_i and ρ just on dimension axes of the input space(Fig.3), but the linear relationship was observed for whole input space (inputs are independent), therefore the derived equation can be considered valid for any input vector.

4.2 Credibility Estimation - Real World Data

We repeated the same experiment with a real world data set (mandarin tree water consumption data - see [2] for description). Again, GAME model D was trained on distances of GAME models from μ on the testing data (2500 vectors). Figure 4 shows the sensitivity of D to the deviation of each GAME model from

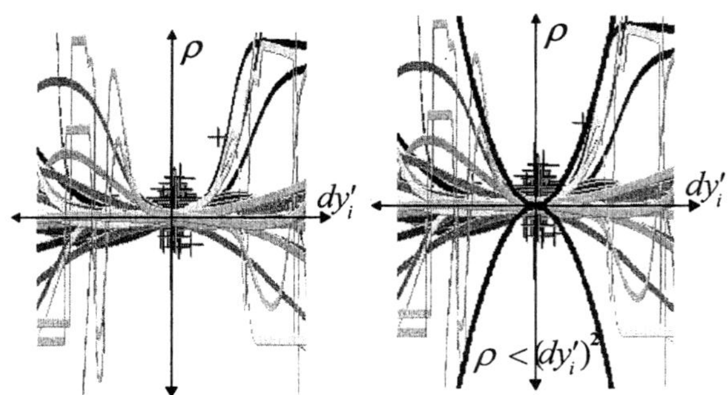

Fig. 4. The dependence of ρ on dy_i' is quadratic for real world data

the mean response. Contrary to the artificial data set, the dependence of ρ on dy_i' is quadratic. We can approximately limit the maximal error of models' mean response for this real word data set as $\rho \leq \frac{a_{max}}{n} \sum_{i=1}^{n} (dy_i')^2$, so the credibility models is inversely proportional to the size of this interval.

5 An Application: Uncertainty Signaling for Visual Knowledge Mining

The "Building data set" is frequently used for benchmarking modeling methods [4]. It consists of more than three thousand measurements inside the building (hot water ($wbhw$), cold water (wbc), energy consumption (wbe)) for the specific weather conditions outside (temperature (temp), humidity of the air (humid), solar radiation (solar), wind strength (wind)). We excluded the information about time of measurement. On this data set we evolved nine GAME models for each output (wbc, $wbhw$, wbe). Figure 5 *left* shows the relationship of the cold water consumption variable on the temperature outside the building under conditions given by the other weather variables. The relationship of the variables can be clearly seen within the area of models' compromise response. We have insufficient information for modeling in the areas where models' responses differ. By the y thickness of dark background we signal the uncertainty of the models' response for particular values of inputs. It is computed according to the equation derived above for real world data $\langle y_{wbc} - \frac{1}{n} \sum_{i=1}^{n} (dy_i')^2, y_{wbc} + \frac{1}{n} \sum_{i=1}^{n} (dy_i')^2 \rangle$. The real value of the output variable should be in this interval with a significant degree of probability. In Figure 5 *right* we show that the level of uncertainty for

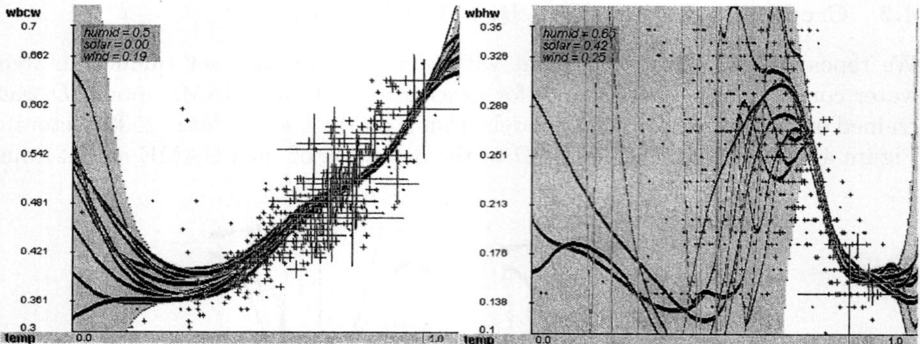

Fig. 5. GAME models of the cold water (left) and hot water (right) consumption variable. *Left*: Models are not credible for low temperature of the air. With increasing temperature, the consumption of cold water grows. *Right*: When its too hot outside, the consumption of hot water drops down. Nothing else is clear - we need more data or include more relevant inputs.

wbhw is significantly higher than for *wbc*. For this specific conditions (values of humid, solar, wind) our GAME models are credible just in thin area where the consumption of hot water drops down.

6 Conclusions

We presented the method for credibility estimation of GAME models. Proposed approach can be also used for many similar algorithms generating black-box models. The main problem of black-box models is that user does not know when one can trust the model. Proposed techniques allow estimating the credibility of the model's response, reducing the risk of misinterpretation.

References

1. Madala, Ivakhnenko,*Inductive Learning Algorithm for Complex System Modelling*, CRC Press, Boca Raton, 1994.
2. Kordík,*Group of Adaptive Models Evolution*, Technical Report DCSE-DTP-2005-07, CTU Prague 2005.
3. L.K. Hansen, P. Salamon, *Neural network ensembles*, IEEE Trans. Pattern Anal. Machine Intelligence 12 (10) (1990) 9931001.
4. Prechelt, L.,*A Set of Neural Network Benchmark Problems and Rules*, Technical Report 21/94, Karlsruhe, Germany 1994.
5. Zhi-Hua Zhou , Jianxin Wu, Wei Tang: *Ensembling neural networks: Many could be better than all*, Artificial Intelligence 137 (2002) p. 239263
6. Gavin Brown: *Diversity in Neural Network Ensembles*, Ph.D. thesis, The University of Birmingham, January 2004

Combination Methods for Ensembles of MF[1]

Joaquín Torres-Sospedra, Mercedes Fernández-Redondo,
and Carlos Hernández-Espinosa[1]

Universidad Jaume I. Dept. de Ingeniería y Ciencia de los Computadores,
Avda Vicente Sos Baynat s/n. 12071 Castellon, Spain
{espinosa, redondo}@icc.uji.es

Abstract. As shown in the bibliography, training an ensemble of networks is an interesting way to improve the performance. The two key factors to design an ensemble are how to train the individual networks and how to combine the different outputs of the nets. In this paper, we focus on the combination methods. We study the performance of fourteen different combination methods for ensembles of the type "simple ensemble" (SE) and "decorrelated" (DECO). In the case of the "SE" and low number of networks in the ensemble, the method Zimmermann gets the best performance. When the number of networks is in the range of 9 and 20 the weighted average is the best alternative. Finally, in the case of the ensemble "DECO" the best performing method is averaging.

1 Introduction

The most important property of a neural network (NN) is the generalization. One technique to increase the generalization capability with respect to a single NN consist on training an ensemble of NNs, i.e., to train a set of neural network with different weight initialization or properties and combine the outputs in a suitable manner.

It is clear from the bibliography that this procedure in general increases the generalization capability [1,2] for the case of Multilayer Feedforward and other classifiers.

The two key factors to design an ensemble are how to train the individual networks and how to combine the different outputs to give a single output.

Among the methods of training the individual networks there are several alternatives. Our research group has performed a comparison [3], which shows that the best performing method is called "Decorrelated" (DECO). It is also shown that the "simple ensemble" (SE) provides a reasonable performance with a lower computing cost.

In the other aspect, (the combination methods) there are also several different methods in the bibliography, and we can also find a comparison in paper [4]. In [4], they conclude that the combination by the weighted average with data dependent weights is the best method. However, the comparison lacks of two problems. The method CVC was used and the results were obtained in only four databases.

In this paper, we present a comparison of 14 different methods of combining the outputs for a total of ten databases. So these results are more complete. Furthermore, we present results for two methods of building the ensemble: "SE" and "DECO".

---

[1] This research was supported by the project MAPACI TIC2002-02273 of CICYT in Spain.

W. Duch et al. (Eds.): ICANN 2005, LNCS 3697, pp. 133 – 138, 2005.

2 Theory

In this section, we briefly review the methods of combination, "DECO" and "SE".

2.1 Simple Ensemble and Decorrelated

Simple Ensemble: A simple ensemble can be constructed by training different networks with the same training set, but different random weight initialization.

Decorrelated (Deco): This ensemble method was proposed in [5]. It consists on introducing a penalty added to the usual error function of Backpropagation. The penalty term for network *j* is:

$$Penalty = \lambda \cdot d(i,j)(y - f_i) \cdot (y - f_j) \tag{1}$$

Where λ determines the strength of the penalty term and should be found by trial and error, *y* is the target of the training pattern and f_i and f_j are the outputs of networks number *i* and *j* in the ensemble. The term *d(i,j)* is in equation 2.

$$d(i,j) = \begin{cases} 1, & if\ i = j-1 \\ 0, & otherwise \end{cases} \tag{2}$$

2.2 Combination Methods

Average: This approach simply averages the individual classifier outputs. The output yielding the maximum values is chosen as the correct class.

Majority Vote: Each classifier provides a vote to a class, given by its highest output. The correct class is the one most often voted by the classifiers.

Winner Takes All (WTA): In this method, the class with overall maximum output across all classifier and outputs is selected as the correct class.

Borda Count:. For any class *q*, the Borda count is the sum of the number of classes ranked below *q* by each classifier. If $B_j(q)$ is the number of classes ranked below the class *q* by the *j*th classifier, then the Borda count for class *q* is in the equation 3.

$$B(q) = \sum_{j=1}^{L} B_j(q) \tag{3}$$

Bayesian Combination: This combination method was proposed in reference [6]. According to this reference a belief value that the pattern *x* belongs to class *i* can be approximated by the following equation.

$$Bel(i) = \frac{\prod_{k=1}^{L} P(x \in q_i \mid \lambda_k(x) = j_k)}{\sum_{i=1}^{Q} \prod_{k=1}^{L} P(x \in q_i \mid \lambda_k(x) = j_k)}, \quad 1 \le i \le Q \tag{4}$$

Where the probability that sample *x* actually belongs to class *i*, given that classifier *k* assign it to class *j* ($\lambda_k(x)=j_k$) can be estimated from the confusion matrix [4].

Weighted Average: This method introduces weights to the outputs of the different networks prior to averaging. The weights try to minimize the difference between the output of the ensemble and the "desired or true" output. The weights can be estimated from the error correlation matrix. A description of the method can be found in [7].

Choquet Integral: This method is based in the fuzzy integral and the Choquet integral. The method is complex and a full description can be found in reference [4].

Combination by Fuzzy Integral with Data Dependent Densities (Int. DD): It is another method based on the fuzzy integral and the Choquet integral [4]. But in this case, prior to the application of the method it is performed a partition of the input space in regions by k-means clustering or frequency sensitive learning.

Weighted Average with Data Dependent Weights (W.Ave DD): This method is the weighted average described above. But in this case, a partition of the space is performed by using k-means clustering and the weights are calculated for each partition. We have a different combination scheme for the different partitions of the space.

BADD Defuzzification Strategy: It is another combination method based on fuzzy logic concepts. The method is complex and the description can be found in [4].

Zimmermann's Compensatory Operator: This method is based in the Zimmermann's compensatory operator described in [8]. The description can be found in [4].

Dynamically Averaged Networks (DAN), Version 1 and 2: It is proposed in [9]. In this method instead of choosing static weights derived from the NN performance on a sample of the input space, we allow the weights to adjust to be proportional to the certainties of the respective network output.

Nash Vote: In this method each voter assigns a number between zero and one for each candidate output. The product of the voter's values is compared for all candidates. The higher is the winner. The method is reviewed in reference [10].

3 Experimental Results

We have applied the fourteen combination methods in ten different problems. They are from the UCI repository and their names are Cardiac Arrhythmia (Aritm), Dermatology (Derma), Protein Location Sites (Ecoli), Solar Flares (Flare), Image Segmentation (Image), Johns Hopkins University Ionosphere (Ionos), Pima Indians Diabetes (Pima), Haberman's survival (Survi), Vowel Recognition (Vowel) and Wisconsin Breast Cancer (Wdbc).

We have constructed ensembles of 3, 9, 20 and 40 networks. We repeated the process of training an ensemble ten times for different partitions of data in training, cross-validation and test. With this procedure we can obtain a mean performance of the ensemble for each database (the mean of the ten ensembles) and an error in the performance. We have used the error, but this measure is related to the standard deviation as equation 5 and to the confidence interval with equation 6.

$$\sigma = \sqrt{n} \cdot Error \tag{5}$$

Where n is the number of experiments performed to obtain the mean, which is 10.

$$Confidence\ Interval = \left(\overline{X} - z_{\alpha/2} \cdot Error,\ \overline{X} + z_{\alpha/2} \cdot Error \right) \tag{6}$$

Where $z_{\alpha/2}$ is obtained from the following probability of the normal distribution.

$$P\left(\overline{X} - z_{\alpha/2} \frac{\sigma}{\sqrt{n}} \leq \mu \leq \overline{X} + z_{\alpha/2} \frac{\sigma}{\sqrt{n}} \right) = 1 - \alpha \tag{7}$$

The results of the performance for an ensemble of the type "SE" and 3 networks are in table 1. For the ensemble of type "DECO" the results are in table 3 also for the case of three NNs. The rest of results are omitted by the lack of space.

As the results show, the difference in performance among the different combination methods is low. For example, for the case of table 1 ("SE", three networks) the largest difference between two combination methods is only 1.9% in the case of Ionos.

However, the computational cost of the different combination methods is very low and it can be worthy to select an appropriate combination method.

Table 1. Results of combination methods for the ensemble of three nets, "Simple Ensemble"

	ARITM	DERMA	ECOLI	FLARE	IMAGEN
Single Network	75.6 ± 0.7	96.7 ± 0.4	84.4 ± 0.7	82.1 ± 0.3	96.3 ± 0.2
Average	73.5 ± 1.1	97.2 ± 0.7	86.6 ± 0.8	81.8 ± 0.5	96.5 ± 0.2
Majority V.	73.1 ± 1.0	96.9 ± 0.8	86.0 ± 0.9	81.5 ± 0.5	96.2 ± 0.3
WTA	73.6 ± 1.0	97.2 ± 0.7	86.3 ± 0.9	81.7 ± 0.5	96.4 ± 0.2
Borda	73.1 ± 1.0	97.0 ± 0.7	86.5 ± 0.8	81.5 ± 0.5	95.70 ± 0.2
Bayesian	73.6 ± 0.9	96.9 ± 0.8	86.3 ± 0.9	81.5 ± 0.5	96.3 ± 0.3
W. Average	73.0 ± 0.9	96.3 ± 0.7	85.9 ± 0.9	81.3 ± 0.6	96.7 ± 0.3
Choquet	74.1 ± 1.1	97.2 ± 0.7	86.3 ± 0.9	81.7 ± 0.5	96.3 ± 0.2
Int. DD	74.1 ± 1.1	97.2 ± 0.7	85.9 ± 0.7	81.8 ± 0.5	96.3 ± 0.2
W. Ave DD	73.5 ± 1.1	97.2 ± 0.7	86.6 ± 0.8	81.8 ± 0.5	96.5 ± 0.2
BADD	73.5 ± 1.1	97.2 ± 0.7	86.6 ± 0.8	81.8 ± 0.5	96.5 ± 0.2
Zimmermann	74.7 ± 1.4	97.3 ± 0.7	86.0 ± 1.2	81.5 ± 0.6	96.6 ± 0.3
DAN	73.2 ± 1.1	96.9 ± 0.6	85.7 ± 1.0	81.4 ± 0.6	95.7 ± 0.2
DAN version 2	73.2 ± 1.1	96.9 ± 0.6	85.4 ± 0.9	81.4 ± 0.6	95.7 ± 0.2
Nash Vote	73.5 ± 1.1	97.3 ± 0.7	86.6 ± 0.8	81.8 ± 0.5	95.8 ± 0.2

	IONOS	PIMA	SURVI	VOWEL	WDBC
Single Network	87.9 ± 0.7	76.7 ± 0.6	74.2 ± 0.8	83.4 ± 0.6	97.4 ± 0.3
Average	91.1 ± 1.1	75.9 ± 1.2	74.3 ± 1.3	88.0 ± 1.0	96.9 ± 0.5
Majority V.	91.3± 1.0	75.9 ± 1.3	74.4 ± 1.4	86.9 ± 0.9	96.9 ± 0.5
WTA	91.1 ± 1.1	75.9 ± 1.2	73.9 ± 1.4	86.7 ± 0.8	96.9 ± 0.5
Borda	91.3± 1.0	75.9 ± 1.3	74.4 ± 1.4	85.9 ± 1.0	96.9 ± 0.5
Bayesian	91.4 ± 1.1	75.8 ± 1.3	74.3 ± 1.4	86.4 ± 1.0	96.9 ± 0.5
W. Average	91.3± 0.9	75.3± 1.3	74.1 ± 1.3	87.7 ± 1.0	96.9 ± 0.5
Choquet	91.3± 1.1	76.1 ± 1.2	74.1 ± 1.4	86.4 ± 0.7	96.9 ± 0.5
Int. DD	91.1 ± 1.1	75.6 ± 1.3	74.1 ± 1.3	86.3 ± 0.7	96.9 ± 0.5
W. Ave DD	91.1 ± 1.1	75.9 ± 1.2	74.3 ± 1.3	88.0 ± 0.9	96.9 ± 0.5
BADD	91.1 ± 1.1	75.9 ± 1.2	74.3 ± 1.3	88.0 ± 0.9	96.9 ± 0.5
Zimmermann	91.9± 1.1	76.0 ± 1.0	74.3 ± 1.3	87.8 ± 1.0	96.9 ± 0.5
DAN	90.0 ± 1.2	75.9± 1.2	74.4 ± 1.4	84.6 ± 1.2	96.9 ± 0.5
DAN version 2	90.0 ± 1.2	75.9± 1.2	74.4 ± 1.4	84.5 ± 1.2	96.9 ± 0.5
Nash Vote	91.3± 1.2	75.9 ± 1.2	74.3± 1.3	86.2 ± 1.0	96.9 ± 0.5

Table 2. Results of the combination methods for the ensemble of three nets, "Decorrelated"

	ARITM	DERMA	ECOLI	FLARE	IMAGEN
Single Network	75.6 ± 0.7	96.7 ± 0.4	84.4 ± 0.7	82.1 ± 0.3	96.3 ± 0.2
Average	74.9 ± 1.3	97.2 ± 0.7	86.6 ± 0.6	81.7 ± 0.4	96.7 ± 0.3
Majority V.	74.9 ± 1.1	97.5 ± 0.6	86.3 ± 0.7	81.4 ± 0.5	96.5 ± 0.3
WTA	74.9 ± 1.1	97.0 ± 0.7	86.0 ± 0.6	81.7 ± 0.5	96.7 ± 0.2
Borda	74.9 ± 1.1	97.0 ± 0.9	86.6 ± 0.8	81.4 ± 0.5	96.0 ± 0.4
Bayesian	74.8 ± 1.2	96.9 ± 0.8	86.8 ± 0.5	81.3 ± 0.5	96.7 ± 0.3
W. Average	74.4 ± 1.2	97.3 ± 0.5	86.6 ± 0.7	81.6 ± 0.5	96.7 ± 0.3
Choquet	74.9 ± 1.2	97.0 ± 0.7	86.0 ± 0.6	81.7 ± 0.5	96.58 ± 0.19
Int. DD	74.9 ± 1.2	97.0 ± 0.7	86.2 ± 0.8	81.5 ± 0.4	96.56 ± 0.19
W. Ave DD	75.1 ± 1.3	97.2 ± 0.7	86.6 ± 0.7	81.5 ± 0.4	96.7 ± 0.3
BADD	74.9 ± 1.3	97.2 ± 0.7	86.6 ± 0.6	81.7 ± 0.4	96.7 ± 0.3
Zimmermann	74.6 ± 1.2	97.3 ± 0.5	86.0 ± 0.9	81.4 ± 0.6	96.5 ± 0.3
DAN	72.9 ± 1.1	96.8 ± 1.1	85.2 ± 0.7	81.3 ± 0.5	95.9 ± 0.3
DAN version 2	72.9 ± 1.1	96.6 ± 1.2	85.0 ± 0.8	81.3 ± 0.5	96.0 ± 0.3
Nash Vote	75.3 ± 1.3	97.2 ± 0.7	86.3 ± 0.5	81.6 ± 0.5	96.2 ± 0.3

	IONOS	PIMA	SURVI	VOWEL	WDBC
Single Network	87.9 ± 0.7	76.7 ± 0.6	74.2 ± 0.8	83.4 ± 0.6	97.4 ± 0.3
Average	90.9 ± 0.9	76.4 ± 1.2	74.6 ± 1.5	91.5 ± 0.6	97.0 ± 0.5
Majority V.	90.7 ± 1.2	75.8 ± 1.1	74.1 ± 1.5	89.4 ± 0.5	97.0 ± 0.5
WTA	91.4 ± 0.9	76.1 ± 1.1	74.6 ± 1.5	91.2 ± 0.7	96.8 ± 0.4
Borda	90.7 ± 1.2	75.8 ± 1.1	74.1 ± 1.5	87.8 ± 0.8	97.0 ± 0.5
Bayesian	92.3 ± 1.0	75.7 ± 1.1	73.8 ± 1.3	88.0 ± 0.4	97.0 ± 0.5
W. Average	91.6 ± 0.8	76.1 ± 0.9	73.4 ± 1.2	91.1 ± 0.4	96.9 ± 0.4
Choquet	91.1 ± 1.0	75.9 ± 1.1	74.8 ± 1.3	90.1 ± 0.5	96.7 ± 0.4
Int. DD	91.0 ± 0.9	75.5 ± 1.1	74.6 ± 1.3	90.0 ± 0.5	96.7 ± 0.4
W. Ave DD	91.1 ± 1.0	76.4 ± 1.1	74.6 ± 1.5	91.7 ± 0.6	97.0 ± 0.5
BADD	90.9 ± 0.9	76.4 ± 1.2	74.6 ± 1.5	91.5 ± 0.6	97.0 ± 0.5
Zimmermann	91.4 ± 1.0	76.6 ± 1.0	74.1 ± 1.3	90.6 ± 0.6	96.7 ± 0.4
DAN	89.9 ± 1.2	75.2 ± 1.0	74.4 ± 1.4	85.8 ± 0.7	97.1 ± 0.4
DAN version 2	89.9 ± 1.2	75.2 ± 1.0	74.4 ± 1.4	86.3 ± 0.8	97.1 ± 0.4
Nash Vote	91.0 ± 0.9	76.4 ± 1.2	74.6 ± 1.5	88.1 ± 0.8	96.9 ± 0.4

We can obtain further conclusions and insights in the performance of the different methods by calculating the percentage of error reduction of the ensemble with respect to a single network. We have used equation 5 for this calculation.

$$PorError_{reduction} = 100 \cdot \frac{PorError_{single\ network} - PorError_{ensemble}}{PorError_{single\ network}} \quad (5)$$

The value of the percentage of error reduction ranges from 0%, where there is no improvement by the use of a particular ensemble method to 100%. There can also be negative values when the performance of the ensemble is worse than the single net.

This new measurement is just a simple way to perform a ranking of the performance, but the results are not statistical significant. A correct way to obtain statistical significant results might be to use the normal percentage and a McNemar's test with Bonferroni correction, this will be addressed in future researches.

Furthermore, we can calculate the mean performance of error reduction across all databases. This value is in table 3 for the case "SE".

As the results of table 3 show the average is quite appropriate for SE. It provides the best performance for the case of 40 nets. However, the method Zimmermann gets the best results and should be used for ensembles of low number of networks (3 nets). There is another method that should be taken into account. Weighted average gets the best performance when an intermediate number of networks is used (9 and 20).

In the case of "DECO" the best performing methods are clearly the simple average and BADD over a wide spectrum in the number of networks in the ensemble.

Table 3. Mean percentage of error reduction for the different ensembles, "Simple ensemble"

	Average	Majority V.	WTA	Borda	Bayesian	W. Average	Choquet
Ensem. 3 Nets	5,49	2,73	4,16	1,55	3,20	2,22	4,35
Ensem. 9 Nets	8,25	6,61	6,01	4,63	-0,52	9,77	4,87
Ensem. 20 Nets	8,13	7,52	6,65	5,35	-9,05	10,82	
Ensem. 40 Nets	9,73	8,11	6,14	6,39	-16,58	6,38	

	Int. DD	W. Ave DD	BADD	Zimmermann	DAN	DAN 2	Nash
Ensem. 3 Nets	3,74	5,54	5,49	6,80	-1,27	-1,50	3,30
Ensem. 9 Nets	3,75	8,52	8,25	9,18	1,72	1,15	5,13
Ensem. 20 Nets			8,13	4,98	-1,38	-1,46	6,34
Ensem. 40 Nets			9,73	-16,36	-0,71	-0,83	7,08

4 Conclusions

In this paper, we have focused in the different alternatives of ensemble combination methods. We have performed experiment with a total of fourteen different combination methods for ensembles of the type "simple ensemble" and "decorrelated". The experiments are performed with ten different databases. The ensembles are trained with 3, 9, 20 and 40 networks. The results show that in the case of the "simple ensemble" and low number of networks, the method Zimmermann gets the best performance. When the number of networks is in greater the weighted average is the best alternative. Finally, in the case of the ensemble "decorrelated" the best performing method is averaging over a wide spectrum of networks in the ensemble.

References

1. Tumer, K., Ghosh, J., "Error correlation and error reduction in ensemble classifiers", Connection Science, vol. 8, nos. 3 & 4, pp. 385-404, 1996.
2. Raviv, Y., Intrator, N., "Bootstrapping with Noise: An Effective Regularization Technique", Connection Science, vol. 8, no. 3 & 4, pp. 355-372, 1996.
3. Fernandez-Redondo, Mercedes, Hernández-Espinosa, Carlos, Torres-Sospedra, Joaquín, "Classification by Multilayer Feedforward ensembles", Internacional Symposium on Neural Networks, Lecture Notes in Computer Science, Vol. 3173, pp. 852-857, 2004.
4. Verikas, A., Lipnickas, A., Malmqvist, K., Bacauskiene, M., Gelzinis, A., "Soft Combination of neural classifiers: A comparative study", Pattern Recognition Letters, Vol. 20, pp 429-444, 1999.
5. Rosen, B., "Ensemble Learning Using Decorrelated Neural Networks", Connection Science, vol. 8, no. 3 & 4, pp. 373-383, 1996.
6. Xu, L., Krzyzak, A., SUen, C.Y., "Methods of combining multiple classifiers and their applications to handwriting recognition", IEEE Trans. On Systems, Man and Cybernetics, vol. 22, no. 3, pp. 418-435, 1992.
7. Krogh, A., Vedelsby, J., "Neural network ensembles, cross validation, and active learning", Advances in Neural Information Processing Systems 7, pp. 231-238, 1995.
8. Zymmermann, H.J., Zysno, P., "Decision and evaluations by hierarchical aggregation of information", Fuzzy Sets and Systems, vol. 10, no. 3, pp. 243-260, 1984.
9. Jimenez, D., "Dynamically Weighted Ensemble Neural Network for Classification", IEEE World Congress on Computational Intelligence, vol. 1, pp. 753-756, 1998.
10. Wanas, N.M., Kamel, M.S., "Decision Fusion in Neural Network Ensembles", International Joint Conference on Neural Networks, vol. 4, pp. 2952-2957, 2001.

New Results on Ensembles of Multilayer Feedforward[1]

Joaquín Torres-Sospedra, Carlos Hernández-Espinosa, and Mercedes Fernández-Redondo

Universidad Jaume I. Dept. de Ingeniería y Ciencia de los Computadores, Avda Vicente Sos Baynat s/n. 12071 Castellon, Spain
{espinosa, redondo}@icc.uji.es

Abstract. As shown in the bibliography, training an ensemble of networks is an interesting way to improve the performance. However there are several methods to construct the ensemble. In this paper we present some new results in a comparison of twenty different methods. We have trained ensembles of 3, 9, 20 and 40 networks to show results in a wide spectrum of values. The results show that the improvement in performance above 9 networks in the ensemble depends on the method but it is usually low. Also, the best method for a ensemble of 3 networks is called "Decorrelated" and uses a penalty term in the usual Backpropagation function to decorrelate the networks outputs in the ensemble. For the case of 9 and 20 networks the best method is conservative boosting. And finally for 40 networks the best method is Cels.

1 Introduction

The most important property of a neural network (NN) is the generalization.

One technique to increase the generalization capability with respect to a single NN consist on training an ensemble of NN, i.e., to train a set of NNs with different weight initialization or properties and combine the outputs in a suitable manner.

It is clear from the bibliography that this procedure in general increases the generalization capability [1,2].

The two key factors to design an ensemble are how to train the individual networks and how to combine the different outputs to give a single output.

Among the methods of combining the outputs, the two most popular are *voting* and *output averaging* [3]. In this paper we will normally use *output averaging*.

In the other aspect, nowadays, there are several different methods in the bibliography to train the individual networks and construct the ensemble [1-3].

However, there is a lack of comparison among the different methods.

One comparison can be found in [4], it is a previous work developed by our research group. In paper [4], eleven different methods are compared.

Now, we present more complete results by including nine new methods, so we increase the number of methods in the comparison to a total of twenty. The empirical results are quite interesting, one of the new methods analyzed in this paper seems to have the best performance in several situations.

[1] This research was supported by the project MAPACI TIC2002-02273 of CICYT in Spain.

W. Duch et al. (Eds.): ICANN 2005, LNCS 3697, pp. 139 – 144, 2005.

2 Theory

In this section we briefly review the new nine ensemble methods introduced in this paper for comparison. The description of the rest of methods can be found in [4].

CVC version 2: The version 2 of CVC included in this paper is used in reference [5]. The data for training and cross-validation is jointed in one set and with this jointed set the usual division of CVC is performed. In this case, one subset is omitted for each network and the omitted subset is used for cross-validation.

Aveboost: Aveboost is the abbreviation of Average Boosting. This method was propossed in reference [6] as a variation of Adaboost. In Adaboost, it is calculated a probability for each pattern of being included in the training set for the following network. In this case a weighted adaptation of the probabilities is performed.

TCA, Total Corrective Adaboost: It was also proposed in [6] and it is another variation of Adaboost. In this case the calculation of the probability distribution for each network is treated as an optimization problem and an iterative process is performed.

Aggressive Boosting: Aggressive Boosting is a variation of Adaboost. It is reviewed in [7]. In this case it is used a common step to modify the probabilities of a pattern for being included in the next training set.

Conservative Boosting: It is another variation of Adaboost reviewed in [7]. In this case the probability of the well classified patterns is decreased and the probability of wrong classified patterns is kept unchanged.

ArcX4: It is another variation of Boosting, it was proposed and studied in reference [8]. The method selects training patterns according to a distribution, and the probability of the pattern depend on the number of times the pattern was not correctly classified by the previous networks. The combination procedure proposed in the reference is the mean average. In our experiments we have used this procedure and also voting.

EENCL Evolutionary Ensemble with Negative Correlation: This method is proposed in reference [9]. The ensemble is build as a population of a genetic algorithm. The fitness function is selected to consider the precision in the classification of the individual networks and also to penalize the correlation among the different networks in the ensemble. Two variations of the method are used, EENCL UG and MG.

3 Experimental Results

We have applied the twenty ensemble methods to ten different classification problems. They are from the UCI repository of machine learning databases. Their names are Cardiac Arrhythmia Database (Aritm), Dermatology Database (Derma), Protein Location Sites (Ecoli), Solar Flares Database (Flare), Image Segmentation Database (Image), Johns Hopkins University Ionosphere Database (Ionos), Pima Indians Diabetes (Pima), Haberman's survival data (Survi), Vowel Recognition (Vowel) and Wisconsin Breast Cancer Database (Wdbc).

We have constructed ensembles of a wide number of networks in the ensemble, 3, 9, 20 and 40. We repeated the process of training each ensemble ten times for different partitions of data in training, cross-validation and test. With this procedure we can obtain a mean performance of the ensemble for each database (the mean of the ten ensembles) and an error in the performance calculated by standard error theory. We have used the error, but this measure is related to the standard deviation as equation 1 and to the confidence interval with equation 2.

$$\sigma = \sqrt{n} \cdot Error \tag{1}$$

Where n is the number of experiments performed to obtain the mean, which is 10.

$$Confidence\ Interval = \left(\overline{X} - z_{\alpha/2} \cdot Error,\ \overline{X} + z_{\alpha/2} \cdot Error \right) \tag{2}$$

Where $z_{\alpha/2}$ is obtained from the following probability of the normal distribution.

$$P\left(\overline{X} - z_{\alpha/2} \cdot \frac{\sigma}{\sqrt{n}} \le \mu \le \overline{X} + z_{\alpha/2} \cdot \frac{\sigma}{\sqrt{n}} \right) = 1 - \alpha \tag{3}$$

The results are in table 1 for the case of ensembles of 3 networks and in table 2 for 9. We omit the results of 20 and 40 networks by the lack of space.

By comparing the results of table 1, and 2 with the results of a single network we can see that the improvement by the use of the ensemble methods depends clearly on the problem. For example in databases Aritm, Flare, Pima and Wdbc there is not a clear improvement. In the rest of databases there is an improvement; perhaps the most important one is in database Vowel.

There is, however, one exception in the performance of the method Evol. This method did not work well in our experiments. In the original reference the method was tested only in database Heart.

Now, we can compare the results of tables 1 and 2 for ensembles of different number of networks. We can see that the results are in general similar and the improvement of training an increasing number of networks, for example 20 and 40, is in general low. Taking into account the computational cost, we can say that the best alternative for an application is an ensemble of three or nine networks.

We have also calculated the percentage of error reduction of the ensemble with respect to a single network. We have used equation 4 for this calculation.

$$PorError_{reduction} = 100 \cdot \frac{PorError_{single\ network} - PorError_{ensemble}}{PorError_{single\ network}} \tag{4}$$

The value of the percentage of error reduction ranges from 0%, where there is no improvement by the use of a particular ensemble method to 100%. There can also be negative values when the performance of the ensemble is worse than the single net.

This new measurement is just a simple way to perform a ranking of the performance, but the results are not statistical significant. A correct way to obtain statistical significant results might be to use the normal percentage and a McNemar's test with Bonferroni correction, this will be addressed in future researches.

Table 1. Results for the ensemble of three networks

	ARITM	DERMA	ECOLI	FLARE	IMAGEN
Single Net.	75.6 ± 0.7	96.7 ± 0.4	84.4 ± 0.7	82.1 ± 0.3	96.3 ± 0.2
Adaboost	71.8 ± 1.8	98.0 ± 0.5	85.9 ± 1.2	81.7 ± 0.6	96.8 ± 0.2
Bagging	74.7 ± 1.6	97.5 ± 0.6	86.3 ± 1.1	81.9 ± 0.6	96.6 ± 0.3
Bag_Noise	75.5 ± 1.1	97.6 ± 0.7	87.5 ± 1.0	82.2 ± 0.4	93.4 ± 0.4
Boosting	74.4 ± 1.2	97.3 ± 0.6	86.8 ± 0.6	81.7 ± 0.4	95.0 ± 0.4
Cels_m	73.4 ± 1.3	97.7 ± 0.6	86.2 ± 0.8	81.2 ± 0.5	96.82 ± 0.15
CVC	74.0 ± 1.0	97.3 ± 0.7	86.8 ± 0.8	82.7 ± 0.5	96.4 ± 0.2
Decorrelated	74.9 ± 1.3	97.2 ± 0.7	86.6 ± 0.6	81.7 ± 0.4	96.7 ± 0.3
Decorrelated2	73.9 ± 1.0	97.6 ± 0.7	87.2 ± 0.9	81.6 ± 0.4	96.7 ± 0.3
Evol	65.4 ± 1.4	57 ± 5	57 ± 5	80.7 ± 0.7	77 ± 5
Ola	74.7 ± 1.4	91.4 ± 1.5	82.4 ± 1.4	81.1 ± 0.4	95.6 ± 0.3
CVC version 2	76.1 ± 1.6	98.0 ± 0.3	86.8 ± 0.9	82.5 ± 0.6	96.9 ± 0.3
AveBoost	73.4 ± 1.3	97.6 ± 0.7	85.3 ± 1.0	81.8 ± 0.8	96.8± 0.2
TCA	70.7 ± 1.9	96.1 ± 0.6	85.4 ± 1.3	81.9 ± 0.7	94.8 ± 0.5
ArcX4	75.4 ± 0.8	97.8 ± 0.5	85.3 ± 1.1	78.3± 0.9	96.6 ± 0.2
ArcX4 Voting	73.0 ± 0.8	97.0 ± 0.5	85.7 ± 1.1	80.6 ± 0.9	96.5 ± 0.2
Aggressive B	72.3 ± 1.9	97.0 ± 0.5	85.7 ± 1.4	81.9 ± 0.9	96.6 ± 0.3
Conservative B	74.8 ± 1.3	96.9 ± 0.8	85.4 ± 1.3	82.1 ± 1.0	96.5 ± 0.3
EENCL UG	71 ± 2	96.8 ± 0.9	86.6 ± 1.2	81.4 ± 0.8	96.3 ± 0.2
EENCL MG	74.5 ± 1.3	97.2 ± 0.8	86.6 ± 1.2	81.9 ± 0.5	96.0 ± 0.2
Simple Ens.	73.4 ± 1.0	97.2 ± 0.7	86.6 ± 0.8	81.8 ± 0.5	96.5 ± 0.2

Table 1. *(Continued).*

	IONOS	PIMA	SURVI	VOWEL	WDBC
Single Net.	87.9 ± 0.7	76.7 ± 0.6	74.2 ± 0.8	83.4 ± 0.6	97.4 ± 0.3
Adaboost	88.3 ± 1.3	75.7 ± 1.0	75.4 ± 1.6	88.43 ± 0.9	95.7 ± 0.6
Bagging	90.7 ± 0.9	76.9 ± 0.8	74.2 ± 1.1	87.4 ± 0.7	96.9 ± 0.4
Bag_Noise	92.4 ± 0.9	76.2 ± 1.0	74.6 ± 0.7	84.4 ± 1.0	96.3 ± 0.6
Boosting	88.9 ± 1.4	75.7 ± 0.7	74.1 ± 1.0	85.7 ± 0.7	97.0 ± 0.4
Cels_m	91.9 ± 1.0	76.0 ± 1.4	73.4 ± 1.3	91.1 ± 0.7	97.0 ±0.4
CVC	87.7 ± 1.3	76.0 ± 1.1	74.1 ± 1.4	89.0 ± 1.0	97.4 ± 0.3
Decorrelated	90.9 ± 0.9	76.4 ± 1.2	74.6 ± 1.5	91.5 ± 0.6	97.0 ± 0.5
Decorrelated2	90.6 ± 1.0	75.7 ± 1.1	74.3 ± 1.4	90.3 ± 0.4	97.0 ± 0.5
Evol	83.4 ± 1.9	66.3 ± 1.2	74.3 ± 0.6	77.5 ± 1.7	94.4 ± 0.9
Ola	90.7 ± 1.4	69.2 ± 1.6	75.2 ± 0.9	83.2 ± 1.1	94.2 ± 0.7
CVC version 2	89.7 ± 1.4	76.8 ± 1.0	74.1 ± 1.2	89.8 ± 0.9	96.7 ± 0.3
AveBoost	89.4 ± 1.3	76.5 ± 1.1	75.1 ± 1.2	88.1 ± 1.0	95.6 ± 0.5
TCA	87.9 ± 1.2	75.4 ± 0.8	73.0 ± 1.5	87.5 ± 1.1	91 ± 4
ArcX4	89.4 ± 1.0	76.0 ± 0.8	68 ± 2	90.8 ± 0.9	96.3 ± 0.6
ArcX4 Voting	89.0 ± 1.0	76.3 ± 0.8	74 ± 2	86.2 ± 0.9	96.1 ± 0.6
Aggressive B	90.3 ± 0.9	74.3 ± 1.5	73.8 ± 1.5	86.9 ± 1.2	96.6 ± 0.6
Conservative B	89.4 ± 1.0	75.6 ± 1.2	75.6 ± 1.1	88.8 ± 1.1	97.0 ± 0.6
EENCL UG	93.0 ± 1.0	74.7 ± 1.0	73.9 ± 1.2	87.2 ± 0.8	96.2 ± 0.4
EENCL MG	93.7 ± 0.9	75.3 ± 1.0	73.9 ± 0.8	87.4 ± 0.7	96.4 ± 0.5
Simple Ens.	91.1 ± 1.1	75.9 ± 1.2	74.3 ± 1.3	88.0 ± 0.9	96.9 ± 0.5

Furthermore we can calculate the mean performance of error reduction across all databases this value is in table 4 for ensembles of 3, 9, 20 and 40 nets. According to this global measurement *Ola*, *Evol* and *BagNoise* performs worse than the *Simple Ensemble*. The best methods are *Bagging*, *Cels*, *Decorrelated, Decorrelated2* and *Conservative Boosting*.

The best methods for 3 nets in the ensemble are *Cels* and *Decorrelated*, the best method for the case of 9 and 20 nets is *Conservative Boosting* and the best method for the case of 40 networks is *Cels* but *Conservative Boosting* is also good.

Table 2. Results for the Ensemble of nine networks

	ARITM	DERMA	ECOLI	FLARE	IMAGEN
Adaboost	73.2 ± 1.6	97.3 ± 0.5	84.7 ± 1.4	81.1 ± 0.7	97.3 ± 0.3
Bagging	75.9 ± 1.7	97.7 ± 0.6	87.2 ± 1.0	82.4 ± 0.6	96.7 ± 0.3
Bag_Noise	75.4 ± 1.2	97.0 ± 0.7	87.2 ± 0.8	82.4 ± 0.5	93.4 ± 0.3
Cels_m	74.8 ± 1.3	97.3 ± 0.6	86.2 ± 0.8	81.7 ± 0.4	96.6 ± 0.2
CVC	74.8 ± 1.3	97.6 ± 0.6	87.1 ± 1.0	81.9 ± 0.6	96.6 ± 0.2
Decorrelated	76.1 ± 1.0	97.6 ± 0.7	87.2 ± 0.7	81.6 ± 0.6	96.9 ± 0.2
Decorrelated2	73.9 ± 1.1	97.6 ± 0.7	87.8 ± 0.7	81.7 ± 0.4	96.84 ± 0.18
Evol	65.9 ± 1.9	54 ± 6	57 ± 5	80.6 ± 0.8	67 ± 4
Ola	72.5 ± 1.0	86.7 ±1.7	83.5 ± 1.3	80.8 ± 0.4	96.1 ± 0.2
CVC version 2	76.1 ± 1.6	98.0 ± 0.3	86.8 ± 0.9	82.5 ± 0.6	96.9 ± 0.3
AveBoost	73.4 ± 1.3	97.6 ± 0.7	85.3 ± 1.0	81.8 ± 0.8	96.8 ± 0.2
TCA	70.7 ± 1.9	96.1 ± 0.5	85.4 ± 1.3	81.9 ± 0.7	94.8 ± 0.5
ArcX4	75.4 ± 0.8	97.8 ± 0.5	85.3 ± 1.1	78.3 ± 0.9	96.6 ± 0.2
ArcX4 Voting	73.3 ± 0.8	97.6 ± 0.5	84.9 ± 1.1	80.1 ± 0.9	97.2 ± 0.2
Aggressive B	72.3 ± 1.9	97.0 ± 0.5	85.7 ± 1.4	81.9 ± 0.9	96.6 ± 0.3
Conservative B	74.8 ± 1.3	96.9 ± 0.8	85.4 ± 1.3	82.1 ± 1.0	96.5 ± 0.3
EENCL UG	71 ± 2	96.8 ± 0.9	86.6 ± 1.2	81.4 ± 0.8	96.3 ± 0.2
EENCL MG	74.5 ± 1.3	97.2 ± 0.8	86.6 ± 1.2	81.9 ± 0.5	96.0 ± 0.2
Simple Ens	73.8 ± 1.1	97.5 ± 0.7	86.9 ± 0.8	81.6 ± 0.4	96.7 ± 0.3

Table 2. *(Continued).*

	IONOS	PIMA	SURVI	VOWEL	WDBC
Adaboost	89.4 ± 0.8	75.5 ± 0.9	74.3 ± 1.4	94.8 ± 0.7	95.7 ± 0.7
Bagging	90.1 ± 1.1	76.6 ± 0.9	74.4 ± 1.5	90.8 ± 0.7	97.3 ± 0.4
Bag_Noise	93.3 ± 0.6	75.9 ± 0.9	74.8 ± 0.7	85.7 ± 0.9	95.9 ± 0.5
Cels_m	91.9 ± 1.0	75.9 ± 1.4	73.4 ± 1.2	92.7 ± 0.7	96.8 ± 0.5
CVC	89.6 ± 1.2	76.9 ± 1.1	75.2 ± 1.5	90.9 ± 0.7	96.5 ± 0.5
Decorrelated	90.7 ± 1.0	76.0 ± 1.1	73.9 ± 1.3	92.8 ± 0.7	97.0 ± 0.5
Decorrelated2	90.4 ± 1.0	76.0 ± 1.0	73.8 ± 1.3	92.6 ± 0.5	97.0 ± 0.5
Evol	77 ± 3	66.1 ± 0.7	74.8 ± 0.7	61 ± 4	87.2 ± 1.6
Ola	90.9 ± 1.7	73.8 ± 0.8	74.8 ± 0.8	88.1 ± 0.8	95.5 ± 0.6
CVC version 2	89.7 ± 1.4	76.8 ± 1.0	74.1 ± 1.2	89.8 ± 0.9	96.7 ± 0.3
AveBoost	89.4 ± 1.3	76.5 ± 1.1	75.1 ± 1.2	88.1 ± 1.0	95.6 ± 0.5
TCA	87.9 ± 1.2	75.4 ± 0.8	73.0 ± 1.5	87.5 ± 1.1	91 ± 4
ArcX4	89.4 ± 1.0	76.0 ± 0.8	68 ± 2	90.8 ± 0.9	96.3 ± 0.6
ArcX4 Voting	91.3 ± 1.0	76.3 ± 0.8	73.9 ± 1.0	94.6 ± 0.9	96.6 ± 0.6
Aggressive B	90.3 ± 0.9	74.3 ± 1.5	73.8 ± 1.5	86.9 ± 1.2	96.6 ± 0.6
Conservative B	89.4 ± 1.0	75.6 1.2	75.6 ± 1.1	88.8 ± 1.1	97.0 ± 0.6
EENCL UG	93.0 ± 1.0	74.7 ± 1.0	73.9 ± 1.2	87.2 ± 0.8	96.2 ± 0.4
EENCL MG	93.7 ± 0.9	75.3 ± 1.0	73.9 ± 0.8	87.4 ± 0.7	96.4 ± 0.5
Simple Ens	90.3 ± 1.1	75.9 ± 1.2	74.2 ± 1.3	91.0 ± 0.5	96.9 ± 0.5

Table 3. Mean percentage of error reduction for the different ensembles

	Ensemble 3 Nets	Ensemble 9 Nets	Ensemble 20 Nets	Ensemble 40 Nets
Adaboost	1.33	4.26	9.38	12.21
Bagging	6.86	12.12	13.36	12.63
Bag_Noise	-3.08	-5.08	-3.26	-3.05
Boosting	-0.67	---	---	---
Cels_m	9.98	9.18	10.86	14.43
CVC	6.18	7.76	10.12	6.48
Decorrelated	9.34	12.09	12.61	12.35
Decorrelated2	9.09	11.06	12.16	12.10
Evol	-218.23	-297.01	-375.36	-404.81
Ola	-33.11	-36.43	-52.53	-47.39
CVC version 2	10.25	10.02	7.57	7.49
AveBoost	1.13	10.46	9.38	10.79
TCA	-9.71	-25.22	-43.98	-53.65
ArcX4	1.21	2.85	7.85	10.05
ArcX4 Voting	-2.08	9.73	10.76	11.14
Aggressive B	1.22	7.34	13.03	13.54
Conservative B	4.45	13.07	14.8	14.11
EENCL UG	0.21	-3.23	-3.59	1.10
EENCL MG	3.96	1.52	2.84	7.89
Simple Ens	5.89	8.39	8.09	9.72

So, we can conclude that if the number of networks is low it seems that the best methods are *Cels* and *Decorrelated* and if the number of network is high the best method is in general *Conservative Boosting*.

Also in table 3, we can see the effect of increasing the number of networks in the ensemble. There are several methods (*Adaboost, Cels, ArcX4, ArcX4 Voting, Aggressive Boosting* and *Conservative Boosting*) where the performance seems to increase slightly with the number of networks in the ensemble. But other methods does not increase the performance beyond 9 or 20 networks in the ensemble.

4 Conclusions

In this paper we have presented experimental results of twenty different methods to construct an ensemble of networks, using ten different databases. We trained ensembles of 3, 9, 20 and 40 networks. The results showed that in general the improvement by the use of the ensemble methods depends on the database. Also the improvement in performance from three or nine networks in the ensemble to a higher number of networks it is usually low. Taking into account the computational cost, an ensemble of nine networks may be the best alternative. Finally, we have obtained the mean percentage of error reduction over all databases. According to this measurement the best methods are *Bagging*, *Cels*, *Decorrelated, Decorrelated2* and *Conservative Boosting*. The best method for 3 networks in the ensemble is *Cels*, the best method for the case of 9 and 20 nets is *Conservative Boosting* and the best method for 40 is *Cels* but the performance of *Conservative Boosting* is also good.

References

1. Tumer, K., Ghosh, J., "Error correlation and error reduction in ensemble classifiers", Connection Science, vol. 8, nos. 3 & 4, pp. 385-404, 1996.
2. Raviv, Y., Intrator, N., "Bootstrapping with Noise: An Effective Regularization Technique", Connection Science, vol. 8, no. 3 & 4, pp. 355-372, 1996.
3. Drucker, H., Cortes, C., Jackel, D., et alt., "Boosting and Other Ensemble Methods", Neural Computation, vol. 6, pp. 1289-1301, 1994.
4. Fernandez-Redondo, Mercedes, Hernández-Espinosa, Carlos, Torres-Sospedra, Joaquín, "Classification by Multilayer Feedforward ensembles", Internacional Symposium on Neural Networks, Lecture Notes in Computer Science, Vol. 3173, pp. 852-857, 2004.
5. Verikas, A., Lipnickas, A., Malmqvist, K., Bacauskiene, M., Gelzinis, A., "Soft Combination of neural classifiers: A comparative study", Pattern Recognition Letters, Vol. 20, pp 429-444, 1999.
6. Oza, N.C., "Boosting with Averaged Weight Vectors", Multiple Classifier Systems, Lecture Notes in Computer Science, vol. 2709, pp. 15-24, 2003.
7. Kuncheva, L.I., "Error Bounds for Aggressive and Conservative Adaboost", Multiple Classifier Systems, Lecture Notes in Computer Science, vol. 2709, pp. 25-34, 2003.
8. Breiman, L., "Arcing Classifiers", Annals of Statistic, vol. 26, no. 3, pp. 801-849, 1998.
9. Liu, Y., Yao, X., Higuchi, T., "Evolutionary Ensembles with Negative Correlation Learning", IEEE Trans. On Evolutionary Computation, vol. 4, no. 4, pp. 380-387, 2000.

On Variations of Power Iteration

Seungjin Choi

Department of Computer Science,
Pohang University of Science and Technology,
San 31 Hyoja-dong, Nam-gu, Pohang 790-784, Korea
seungjin@postech.ac.kr

Abstract. The power iteration is a classical method for computing the eigenvector associated with the largest eigenvalue of a matrix. The subspace iteration is an extension of the power iteration where the subspace spanned by n largest eigenvectors of a matrix, is determined. The natural power iteration is an exemplary instance of the subspace iteration, providing a general framework for many principal subspace algorithms. In this paper we present variations of the natural power iteration, where n largest eigenvectors of a symmetric matrix without rotation ambiguity are determined, whereas the subspace iteration or the natural power iteration finds an invariant subspace (consisting of rotated eigenvectors). The resulting method is referred to as *constrained natural power iteration* and its fixed point analysis is given. Numerical experiments confirm the validity of our algorithm.

1 Introduction

A symmetric eigenvalue problem where the eigenvectors of a symmetric matrix are required to be computed, is a fundamental problem encountered in a variety of applications involving the spectral decomposition. The power iteration is a classical and the simplest method for computing the eigenvector with the largest modulus. The subspace iteration is a natural generalization of the power iteration, where the subspace spanned by n largest eigenvectors of a matrix, is determined.

The natural power iteration [1] is an exemplary instance of the subspace iteration, that was investigated mainly for principal subspace analysis. In this paper we present variations of the natural power iteration and show that its fixed point is the n largest eigenvectors of a symmetric matrix up to a sign ambiguity, whereas the natural power iteration just finds a principal subspace (i.e., arbitrarily rotated eigenvectors). The resulting algorithm is referred to as *constrained natural power iteration*. Numerical experiments confirm the validity of our algorithm.

2 Natural Power Iteration

The power iteration is a classical method which finds the largest eigenvector (associated with the largest eigenvalue) of a matrix $\boldsymbol{C} \in \mathbb{R}^{m \times m}$ [2]. Given a

W. Duch et al. (Eds.): ICANN 2005, LNCS 3697, pp. 145–150, 2005.

symmetric matrix $\boldsymbol{C} \in \mathbb{R}^{m \times m}$ (hence its eigenvalues are real), the power iteration starts from a nonzero vector $\boldsymbol{w}(0)$ and iteratively updates $\boldsymbol{w}(t)$ by

$$\widetilde{\boldsymbol{w}}(t+1) = \boldsymbol{C}\boldsymbol{w}(t), \tag{1}$$

$$\boldsymbol{w}(t+1) = \frac{\widetilde{\boldsymbol{w}}(t+1)}{\|\widetilde{\boldsymbol{w}}(t+1)\|_2}, \tag{2}$$

where $\|\cdot\|_2$ represents Euclidean norm. Combining (1) and (2) leads to the updating rule which has the form

$$\boldsymbol{w}(t+1) = \boldsymbol{C}\boldsymbol{w}(t)\left[\boldsymbol{w}^T(t)\boldsymbol{C}^2\boldsymbol{w}(t)\right]^{-\frac{1}{2}}. \tag{3}$$

Assume that $\boldsymbol{C}$ has an unique eigenvalue of maximum modulus λ_1 associated with the leading eigenvector $\boldsymbol{u}_1$. Then the power iteration (3) leads $\boldsymbol{w}(t)$ to converge to $\boldsymbol{u}_1$.

The subspace iteration [3] is a direct generalization of the power iteration, for computing several eigenvectors of $\boldsymbol{C}$. Starting from $\boldsymbol{W}(0) \in \mathbb{R}^{m \times n}$, the subspace iteration updates $\boldsymbol{W}(t)$ by

$$\boldsymbol{W}(t+1) = \boldsymbol{C}\boldsymbol{W}(t). \tag{4}$$

The space spanned by $\boldsymbol{W}(t)$ converges to invariant subspace determined by n largest eigenvectors of $\boldsymbol{C}$, provided that $|\lambda_n| > |\lambda_{n+1}|$ [3]. As in the power iteration, the subspace iteration requires the normalization or orthogonalization.

The subspace iteration

$$\widetilde{\boldsymbol{W}}(t+1) = \boldsymbol{C}\boldsymbol{W}(t), \tag{5}$$

followed by an orthogonalization

$$\boldsymbol{W}(t+1) = \widetilde{\boldsymbol{W}}(t+1)\left[\widetilde{\boldsymbol{W}}^T(t+1)\widetilde{\boldsymbol{W}}(t+1)\right]^{-\frac{1}{2}}, \tag{6}$$

leads to

$$\boldsymbol{W}(t+1) = \underbrace{\boldsymbol{C}\boldsymbol{W}(t)}_{\text{power term}} \underbrace{\left[\boldsymbol{W}^T(t)\boldsymbol{C}^2\boldsymbol{W}(t)\right]^{-\frac{1}{2}}}_{\text{normalizer}}, \tag{7}$$

which is known as the *natural power iteration* proposed in [1].

Denote the eigendecomposition of the symmetric matrix $\boldsymbol{C} \in \mathbb{R}^{m \times m}$ of rank $r(> n)$ as

$$\boldsymbol{C} = \left[\boldsymbol{U}_1\, \boldsymbol{U}_2\right] \begin{bmatrix} \boldsymbol{\Lambda}_1 & 0 \\ 0 & \boldsymbol{\Lambda}_2 \end{bmatrix} \left[\boldsymbol{U}_1\, \boldsymbol{U}_2\right]^T, \tag{8}$$

where $\boldsymbol{U}_1 \in \mathbb{R}^{m \times n}$ contains n largest eigenvectors, $\boldsymbol{U}_2 \in \mathbb{R}^{m \times (m-n)}$ consists of the rest of eigenvectors, and associated eigenvalues are in $\boldsymbol{\Lambda}_1$, $\boldsymbol{\Lambda}_2$ with $|\lambda_1| > |\lambda_2| > \cdots > |\lambda_m|$. The key result in regards to the natural power iteration is summarized in the following theorem

Theorem 1 (Y. Hua *et al.* [1]). *The weight matrix $\boldsymbol{W}(t) \in \mathbb{R}^{m \times n}$ in the natural power iteration (10) globally and exponentially converges to $\boldsymbol{W} = \boldsymbol{U}_1 \boldsymbol{Q}$ where $\boldsymbol{Q} \in \mathbb{R}^{n \times n}$ is an arbitrary orthogonal matrix, provided that the nth and $(n+1)$th eigenvalues of $\boldsymbol{C}$ are distinct and the initial weight matrix $\boldsymbol{W}(0)$ meets a mild condition, saying that there exists a nonsingular matrix $\boldsymbol{L} \in \mathbb{R}^{(m-n) \times n}$ such that $\boldsymbol{U}_2^T \boldsymbol{W}(0) = \boldsymbol{L}\boldsymbol{U}_1^T \boldsymbol{W}(0)$ for a randomly chosen $\boldsymbol{W}(0)$.*

The natural power iteration was mainly studied for principal subspace analysis where $\boldsymbol{C} = E\{\boldsymbol{x}(t)\boldsymbol{x}(t)\}$ is the covariance matrix of m-dimensional stationary vector sequences, $\boldsymbol{x}(t)$, with zero mean. In such a case, the matrix $\boldsymbol{C}$ is symmetric as well as positive semidefinite. For the case of principal subspace analysis, the weight vector $\boldsymbol{W}(t)$ of the natural power iteration (7) converges to n principal arbitrary rotated eigenvectors of $\boldsymbol{C}$. A variety of algorithms, including Oja's subspace rule [4], PAST [5], OPAST [6], can be viewed as the implementations of the natural power iteration [1]. However, all these algorithms belong to the principal subspace method where arbitrarily rotated eigenvectors are determined, unless the deflation method was used to extract principal components one by one. Next section describes a simple variation of the natural power iteration, incorporating the upper-triangularization operator into the normalizer in (7). This variation is referred to as a *constrained natural power iteration.* It is shown here that a fixed point of the constrained natural power iteration is $\boldsymbol{W} = \boldsymbol{U}_1$ (up to a sign ambiguity). Thus, the constrained natural power iteration computes the exact eigenvectors of a given symmetric matrix, whereas the natural power method finds a principal subspace.

3 Constrained Natural Power Iteration

We impose a constraint in the normalization term in the natural power method (7), through an upper-triangularization operator $\mathcal{U}_T[\cdot]$ which sets all elements of its matrix argument that are below the diagonal to zero, i.e., $\mathcal{U}_T[\boldsymbol{Y}]$ for an arbitrary matrix $\boldsymbol{Y} \in \mathbb{R}^{n \times n}$ gives

$$\mathcal{U}_T[y_{ij}] = \begin{cases} 0 & \text{if } i > j \\ y_{ij} & \text{if } i \leq j \end{cases}, \tag{9}$$

where y_{ij} is the (i,j)-element of $\boldsymbol{Y}$. The constrained natural power iteration updates the weight matrix by

$$\boldsymbol{W}(t+1) = \boldsymbol{C}\boldsymbol{W}(t) \left\{ \mathcal{U}_T \left[\boldsymbol{W}^T(t) \boldsymbol{C}^2 \boldsymbol{W}(t) \right] \right\}^{-\frac{1}{2}}. \tag{10}$$

Only difference between the constrained natural power iteration (10) and the natural power method (7) lies in the presence of $\mathcal{U}_T$ in the normalization term. As will be shown below, the operator $\mathcal{U}_T$ leads the algorithm (10) to find exact principal eigenvectors of $\boldsymbol{C}$ up to a sign ambiguity under mild conditions that are generally required for power iteration. That is, the fixed point of (10) satisfies $\boldsymbol{W} = \boldsymbol{U}_1 \mathring{\boldsymbol{I}}$ where $\mathring{\boldsymbol{I}}$ is a diagonal matrix with its diagonal entries being 1 or -1, whereas the fixed point of (7) is $\boldsymbol{U}_1 \boldsymbol{Q}$ for an arbitrary orthogonal matrix $\boldsymbol{Q}$.

Theorem 2. *The fixed point* $\boldsymbol{W}$ *of the constrained natural power iteration (10) satisfies* $\boldsymbol{W} = \boldsymbol{U}_1 \overset{\circ}{\boldsymbol{I}}$*, under the same conditions as Theorem 1.*

Proof. We define $\boldsymbol{\Phi}(t) = \boldsymbol{U}_1^T \boldsymbol{W}(t)$ and $\boldsymbol{\Omega}(t) = \boldsymbol{U}_2^T \boldsymbol{W}(t)$. With this definition, pre-multiplying both sides of (10) by $[\boldsymbol{U}_1 \, \boldsymbol{U}_2]^T$ leads to

$$\begin{bmatrix} \boldsymbol{\Phi}(t+1) \\ \boldsymbol{\Omega}(t+1) \end{bmatrix} = \begin{bmatrix} \boldsymbol{\Lambda}_1 & 0 \\ 0 & \boldsymbol{\Lambda}_2 \end{bmatrix} \begin{bmatrix} \boldsymbol{\Phi}(t) \\ \boldsymbol{\Omega}(t) \end{bmatrix} \boldsymbol{Z}(t), \tag{11}$$

where

$$\boldsymbol{Z}(t) = \left\{ \mathcal{U}_T \left[\boldsymbol{\Phi}^T(t) \boldsymbol{\Lambda}_1^2 \boldsymbol{\Phi}(t) + \boldsymbol{\Omega}^T(t) \boldsymbol{\Lambda}_2^2 \boldsymbol{\Omega}(t) \right] \right\}^{-\frac{1}{2}}. \tag{12}$$

As in the convergence proof of the natural power iteration in [1], one can show that $\boldsymbol{\Omega}(t)$ goes to zero. Assume that $\boldsymbol{\Phi}(0) \in \mathbb{R}^{n \times n}$ is a nonsingular matrix, then it implies that $\boldsymbol{\Omega}(0) = \boldsymbol{L}\boldsymbol{\Phi}(0)$ for some matrix $\boldsymbol{L}$. Then it follows from (11) that we can write

$$\boldsymbol{\Omega}(t) = \boldsymbol{\Lambda}_2^t \boldsymbol{L} \boldsymbol{\Lambda}_1^{-t} \boldsymbol{\Phi}(t). \tag{13}$$

The assumption that first n eigenvalues of $\boldsymbol{C}$ are strictly larger than the others, together with (13), implies that $\boldsymbol{\Omega}(t)$ converges to zero and is asymptotically in the order of $|\lambda_{n+1}/\lambda_n|^t$ where $|\lambda_n|$ and $|\lambda_{n+1}|$ ($< |\lambda_n|$) are nth and $(n+1)$th largest eigenvalues of $\boldsymbol{C}$.

Taking into account that $\boldsymbol{\Omega}(t)$ goes to zero, the fixed point $\boldsymbol{\Phi}$ of (11) satisfies

$$\boldsymbol{\Phi} \left\{ \mathcal{U}_T \left[\boldsymbol{\Phi}^T \boldsymbol{\Lambda}_1^2 \boldsymbol{\Phi} \right] \right\}^{\frac{1}{2}} = \boldsymbol{\Lambda}_1 \boldsymbol{\Phi}. \tag{14}$$

Note that $\boldsymbol{\Lambda}_1$ is a diagonal matrix with diagonal entries λ_i for $i = 1, \ldots, n$. Thus, one can easily see that $\boldsymbol{\Phi}$ is the eigenvector matrix of $\left\{ \mathcal{U}_T \left[\boldsymbol{\Phi}^T \boldsymbol{\Lambda}_1^2 \boldsymbol{\Phi} \right] \right\}^{\frac{1}{2}}$ with associated eigenvalues in $\boldsymbol{\Lambda}_1$. Note that the eigenvalues of an upper-triangular matrix are the diagonal elements. Then it follows from (14) that we have a set of equations

$$\left(\varphi_i^T \boldsymbol{\Lambda}_1^2 \varphi_i \right)^{\frac{1}{2}} = \lambda_i, \quad i = 1, \ldots, n. \tag{15}$$

where φ_i is the ith column vector of $\boldsymbol{\Phi}$, i.e., $\boldsymbol{\Phi} = [\varphi_1 \, \varphi_2 \, \cdots \, \varphi_n]$. We can re-write (15) as

$$\sum_{i=1}^{n} \lambda_i^2 \varphi_{ij}^2 = \lambda_j^2, \quad j = 1, \ldots, n, \tag{16}$$

where φ_{ij} is the (i,j)-element of $\boldsymbol{\Phi}$. Assume $n \leq \operatorname{rank}(\boldsymbol{C})$, then $\lambda_i \neq 0$, $i = 1, \ldots, n$. For non-zero λ_i, the only $\boldsymbol{\Phi}$ satisfying (16) is $\boldsymbol{\Phi} = \overset{\circ}{\boldsymbol{I}}$. Therefore, $\boldsymbol{W} =$

$\boldsymbol{U}_1 \mathring{\boldsymbol{I}}$, implying that the fixed point of (10) is the true eigenvector matrix $\boldsymbol{U}_1$ up to a sign ambiguity. ■

Based on the result in Theorem 2, we can consider a variation of the constrained natural power iteration (10), described by

$$\boldsymbol{W}(t+1) = \boldsymbol{C}\boldsymbol{W}(t)\left\{\mathcal{U}_T\left[\boldsymbol{W}^T(t)\boldsymbol{C}^2\boldsymbol{W}(t)\right]\right\}^{-1}. \tag{17}$$

Following Theorem 2, one can easily see that the weight matrix $\boldsymbol{W}(t)$ in (17) also converges to the scaled eigenvector matrix of $\boldsymbol{C}$. Algorithms (10) and (17) have a difference in their normalizers. The matrix inverse requires less complexity, compared to the square-root-inverse of a matrix, although (17) finds scaled eigenvectors.

4 Numerical Experiments

Two simple numerical examples are shown in order to verify that the weight matrix $\boldsymbol{W}(t)$ converges to true eigenvectors of a given symmetric matrix $\boldsymbol{C}$. The first experiment was carried out with a symmetric matrix $\boldsymbol{C} \in \mathbb{R}^{5\times 5}$ whose eigenvalues are $2.48, -2.18, 1.20, -0.50, 0.34$. Fig. 1 (a) shows the the evolution of $|\boldsymbol{w}_i^T \boldsymbol{u}_i|$ for $i = 1, 2, 3$, where $\boldsymbol{w}_i$ is the ith column vector of $\boldsymbol{W}$ and $\boldsymbol{u}_i$ are true eigenvectors computed by SVD in Matlab.

The second experiment is related to principal component analysis. We generated 100-dimensional data vectors of length 1000, $\boldsymbol{x}(t) \in \mathbb{R}^{100}$, $t = 1, \ldots, 1000$, through linearly transforming 5-dimensional Gaussian vectors, $\boldsymbol{s}(t) \in \mathbb{R}^5$, with zero mean and unit variance, i.e., $\boldsymbol{x}(t) = \boldsymbol{A}\boldsymbol{s}(t)$ where $\boldsymbol{A} \in \mathbb{R}^{100\times 5}$ and its elements were randomly drawn from Gaussian distribution. We applied the constrained natural power iteration (10) with a weight matrix $\boldsymbol{W}(t) \in \mathbb{R}^{100\times 3}$ to estimate first 3 eigenvectors of $\boldsymbol{C} = \frac{1}{1000}\sum_{t=1}^{1000} \boldsymbol{x}(t)\boldsymbol{x}^T(t)$. Fig. 1 (b) shows the evolution of $|\boldsymbol{w}_i^T \boldsymbol{u}_i|$ for $i = 1, 2, 3$, where $\boldsymbol{u}_i$ are true eigenvectors computed by SVD in Matlab.

5 Discussions

We have presented the constrained natural power iteration and have shown that its fixed point corresponded to the exact eigenvectors of a given symmetric matrix, up to sign ambiguity. Its slight variation was also discussed. Numerical experiments confirmed that the constrained natural power iteration successfully first n eigenvectors of $\boldsymbol{C}$. The constrained natural power iteration will be useful, especially for the case where a few eigenvectors are required to be determined from very high-dimensional data. Constrained natural power iteration could be viewed as a recognition model counterpart of the generative model-based methods in [7,8] where EM optimization were used. The constrained natural power iteration has an advantage over EM algorithms in [7,8], in the sense that the former involves a single-step updating whereas the latter needs two-step updating (E and M steps), although both share a similar spirit.

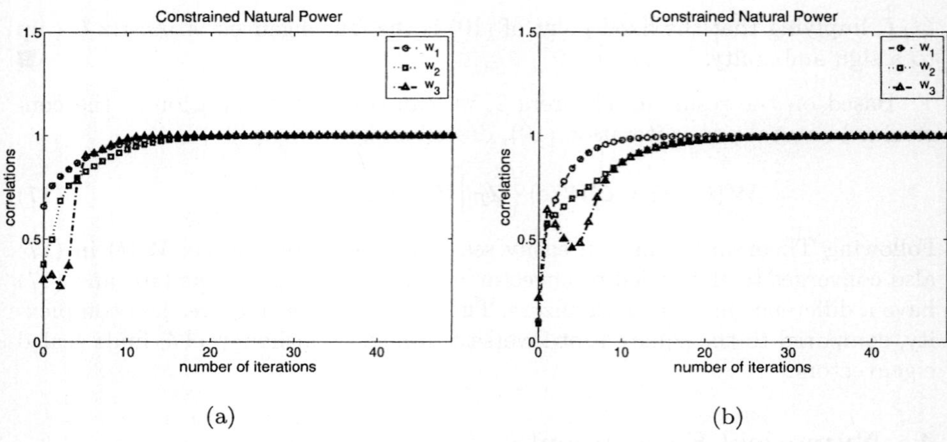

Fig. 1. Convergence of $\boldsymbol{W} = [\boldsymbol{w}_1\, \boldsymbol{w}_2\, \boldsymbol{w}_3]$ in the constrained natural power iteration, is shown in terms of the absolute values of the inner product between these weight vectors and first three true eigenvectors of $\boldsymbol{C}$: (a) experiment 1; (b) experiment 2

Acknowledgments

This work was supported by ITEP Brain Neuroinformatics program, KISTEP International Cooperative Research Program, Systems Bio-Dynamics Research Center under NCRC progrm, KOSEF 2000-2-20500-009-5, and Basic Research Fund in POSTECH.

References

1. Hua, Y., Xiang, Y., Chen, T., Abed-Meraim, K., Miao, Y.: A new look at the power method for fast subspace tracking. Digital Signal Processing **9** (1999) 297–314
2. Golub, G.H., Loan, C.F.V.: Matrix Computations. 2 edn. Johns Hopkins (1993)
3. Heath, M.T.: Scientific Computing: An Introductory Survey. 2 edn. McGraw-Hill, New York (2002)
4. Oja, E.: Neural networks, principal component analysis, and subspaces. International Journal of Neural Systems **1** (1989) 61–68
5. Yang, B.: Projection approximation subsapce tracking. IEEE Trans. Signal Processing **43** (1995) 95–107
6. Abed-Meraim, K., Chkeif, A., Hua, Y.: Fast orthonormal PAST algorithm. IEEE Signal Processing Letters **7** (2000) 60–62
7. Ahn, J.H., Oh, J.H.: A constrained EM algorithm for principal component analysis. Neural Computation **15** (2003) 57–65
8. Ahn, J.H., Choi, S., Oh, J.H.: A new way of PCA: Integrated-squared-error and EM algorithms. In: Proc. IEEE Int'l Conf. Acoustics, Speech, and Signal Processing, Montreal, Canada (2004)

Linear Dimension Reduction Based on the Fourth-Order Cumulant Tensor

M. Kawanabe

Fraunhofer FIRST.IDA, Kekuléstr. 7, 12439 Berlin, Germany

Abstract. In high dimensional data analysis, finding non-Gaussian components is an important preprocessing step for efficient information processing. By modifying the contrast function of JADE algorithm for *Independent Component Analysis*, we propose a new *linear dimension reduction* method to identify the non-Gaussian subspace based on the fourth-order cumulant tensor. A numerical study demonstrates the validity of our method and its usefulness for extracting sub-Gaussian structures.

1 Introduction

Recently enormous amount of data with a huge number of features have been stored and are necessary to be analyzed. In most real-world applications, the 'signal' or 'information' is typically contained only in a low-dimensional subspace of the high-dimensional data, thus dimensionality reduction is a useful preprocessing for further data analysis. Here we make an assumption on the data: the high-dimensional data includes low-dimensional non-Gaussian components ('signal') and the other components are Gaussian noise. Under this modeling assumption, therefore, the task is to recover the relevant non-Gaussian components. Once such components are identified and extracted, it can be applied for various tasks in the data analysis process, say, data visualization, clustering, denoising or classification.

If the number of Gaussian components is *at most one* and all the non-Gaussian components are mutually independent, *Independent Component Analysis (ICA)* techniques [3,6] can be applied to identify the non-Gaussian subspace. Unfortunately, however, this is often a too strict assumption on the data. On the other hand, we treat here more complicated non-Gaussian structures as *Projection Pursuit (PP)* algorithms [4,5,6]. In fact, PP methods can extract non-Gaussian components in a general setting, i.e., the number of Gaussian components can be more than one and the non-Gaussian components can be dependent.

In this paper, we will propose a new approach to identifying the non-Gaussian subspace based on the fourth-order cumulant tensor. JADE algorithm [1] minimizes the sum of squared fourth-order *cross* cumulants for extracting *independent* components. It is important to remark that as far as we know, JADE has no theoretical guarantee for non-Gaussian structures except the independent model. Therefore, we will introduce a different contrast function which measures, roughly speaking, the total squared cumulants in the subspaces. The optimization of this contrast function leads a simple iterative

W. Duch et al. (Eds.): ICANN 2005, LNCS 3697, pp. 151–156, 2005.

scheme of the eigenvalue decomposition. We will also present results of numerical experiments with synthetic data, in order to show validity of our method. In particular, it can extract sub-Gaussian structures quite well.

2 The Model and Cumulant Tensor

Suppose $\{\boldsymbol{x}_i\}_{i=1}^n$ are i.i.d. samples in $\mathbb{R}^d$ drawn from an unknown distribution with density $f(\boldsymbol{x})$, which is expressed as

$$f(\boldsymbol{x}) = g(B_N \boldsymbol{x}) \phi_{\boldsymbol{\theta},\Gamma}(\boldsymbol{x}), \tag{1}$$

where B_N is an unknown linear mapping from $\mathbb{R}^d$ to another space $\mathbb{R}^m$ with $m \leq d$, g is an unknown function on $\mathbb{R}^m$, and $\phi_{\boldsymbol{\theta},\Gamma}$ is a Gaussian density with unknown mean $\boldsymbol{\theta}$ and unknown covariance matrix Γ. Note that the general semiparametric model (1) includes as particular cases both the pure parametric ($m = 0$) and pure non-parametric ($m = d$) models. We effectively consider an intermediate case where d is large and m is rather small. In this paper, we assume the effective dimension m to be known. In order to simplify the coming analysis, we will actually impose the following two assumptions: **(A1)** $\mathrm{E}_f[\boldsymbol{x}] = \int \boldsymbol{x} f(\boldsymbol{x}) d\boldsymbol{x} = \boldsymbol{0}$; **(A2)** $\boldsymbol{\theta} = 0$.

Our goal is to estimate the m-dimensional *non-Gaussian subspace*

$$\mathcal{I} = \mathrm{Ker}(B_N)^{\perp} = \mathrm{Range}(B_N^{\top})$$

from the samples $\{\boldsymbol{x}_i\}_{i=1}^n$ and to project out the "Gaussian part" of the data. Note that we do *not* estimate the nuisance parameters Γ and g at all.

At first, we remark that the semiparametric model (2) can be translated into a linear mixing model (3) which is more familiar in signal processing.

Lemma 1. *Suppose that data $\boldsymbol{x} \in \mathbb{R}^d$ have a density of the form*

$$f(\boldsymbol{x}) = g(B_N \boldsymbol{x}) \phi_{\boldsymbol{0},\Gamma}(\boldsymbol{x}), \tag{2}$$

where B_N is an $m \times d$ matrix, g is a function on $\mathbb{R}^m$ and $\phi_{\boldsymbol{0},\Gamma}$ is the density of $N(\boldsymbol{0}, \Gamma)$. Then, it can be expressed as a linear mixing model,

$$\boldsymbol{x} = A_N \boldsymbol{s}_N + A_G \boldsymbol{s}_G, \tag{3}$$

where $(B_N^{\top}, B_G^{\top})^{\top} = (A_N, A_G)^{-1}$ and $A_G^{\top} \Gamma^{-1} A_N = 0$. Furthermore $\boldsymbol{s}_N$ and $\boldsymbol{s}_G$ are independent and $\boldsymbol{s}_G$ is Gaussian distributed.

The fourth-order cumulant tensor

$$\begin{aligned} &\mathrm{cum}(x_i, x_j, x_k, x_l) \\ &:= \mathrm{E}[x_i x_j x_k x_l] - \mathrm{E}[x_i x_j]\mathrm{E}[x_k x_l] - \mathrm{E}[x_i x_k]\mathrm{E}[x_j x_l] - \mathrm{E}[x_i x_l]\mathrm{E}[x_j x_k] \end{aligned}$$

is used in JADE algorithm for extracting independent components. From the linear model representation (3), we can show that many components of the cumulant tensor of the factors $(\boldsymbol{s}_N^{\top}, \boldsymbol{s}_G^{\top})$ take 0.

Lemma 2. *Suppose the linear model (3) holds. If* $\boldsymbol{s}_N = (s_1, \ldots, s_m)^\top$ *and* $\boldsymbol{s}_G = (s_{m+1}, \ldots, s_d)^\top$ *are independent and* $\boldsymbol{s}_G$ *is Gaussian distributed,*

$$\operatorname{cum}(s_i, s_j, s_k, s_l) = 0, \tag{4}$$

unless $1 \leq i, j, k, l \leq m$.

The whitening transformation $\boldsymbol{z} = V^{-1/2}\boldsymbol{x}$ is often used as a preprocessing in ICA, where $V = \operatorname{Cov}[\boldsymbol{x}]$. Let us define the matrices

$$W_N := B_N V^{1/2}, \qquad W_G := B_G V^{1/2},$$

which are the linear transformations from the sphered data to the factors $\boldsymbol{s} = (\boldsymbol{s}_N^\top, \boldsymbol{s}_G^\top)^\top$. We remark that the non-Gaussian index space can be expressed as

$$\mathcal{I} = \operatorname{Range}(B_N^\top) = V^{-1/2}\operatorname{Range}(W_N^\top).$$

and therefore, it is enough to estimate the matrix W_N. Without loss of generality, we can assume that $\operatorname{Cov}[\boldsymbol{s}] = I$. Then, $(W_N^\top, W_G^\top)$ becomes an orthogonal matrix.

In JADE algorithm for ICA, the linear operator

$$\{Q(M)\}_{ij} := \sum_{k,l=1}^{d} q_{ijkl} M_{kl}$$

from a $d \times d$ matrix M to a $d \times d$ matrix is considered, where $q_{ijkl} = \operatorname{cum}(z_i, z_j, z_k, z_l)$ is the fourth-order cumulant tensor of the sphered data $\boldsymbol{z}$. In contrast to ICA, we can prove that the linear operator $Q(M)$ has at most $\frac{m(m+1)}{2}$ non-zero eigenvalues under our model assumption. Furthermore, the corresponding eigen matrices take the form as $W_N^\top \tilde{M} W_N$, where $\tilde{M}$ is an $m \times m$ symmetric matrix. Therefore, in principle, the non-Gaussian subspace can be estimated by solving the eigenvalue problem of the linear operator $Q(M)$. However, calculating eigenvalues of a $d^2 \times d^2$ matrix is computationally heavy, if d is large.

3 Contrast Function and Algorithm

Instead of solving the $d^2 \times d^2$ eigenvalue problem, we will introduce a contrast function which is inspired by that of JADE algorithm. JADE extracts the independent components by maximizing the contrast function

$$\mathcal{L}_{\mathrm{JADE}}(W) = \sum_{i=1}^{d} \sum_{k,l=1}^{d} |\operatorname{cum}(y_i, y_i, y_k, y_l)|^2 \tag{5}$$

where $\boldsymbol{y} = W\boldsymbol{z}$ is a linear transformation of the sphered data by an orthogonal matrix W. In contrast to ICA, we want to get an m-dimensional vector $\boldsymbol{y}_N = (y_1, \ldots, y_m)^\top =$

$W_N z$ by an $m \times d$ matrix W_N for dimension reduction with the property $W_N W_N^\top = I$. Here we propose the following contrast function

$$\mathcal{L}(W_N) = \sum_{k,l=1}^{d} \sum_{i',j'=1}^{m} |\mathrm{cum}(y_{i'}, y_{j'}, z_k, z_l)|^2$$

$$= \sum_{k,l=1}^{d} \sum_{i',j'=1}^{m} \left\{ \sum_{i,j=1}^{d} w_{i'i} w_{j'j} q_{ijkl} \right\}^2 = \sum_{k,l=1}^{d} \|W_N Q^{(kl)} W_N^\top\|_{\mathrm{Fro}}^2, \quad (6)$$

where $Q^{(kl)} = (Q_{ij}^{(kl)}) = (q_{ijkl})$ for each $1 \le k, l \le d$ and $\| \cdot \|_{\mathrm{Fro}}^2$ is Frobenius norm of matrices.

Theorem 1. *The objective function $\mathcal{L}(W_N)$ takes maximum, when W_N is equal to the true matrix W_N^*.*

Now, let us take the derivative of the criterion $\mathcal{L}$ with the orthonormal constraints

$$\sum_{k,l=1}^{d} \|W_N Q^{(kl)} W_N^\top\|_{\mathrm{Fro}}^2 - 2\mathrm{tr}(W_N W_N^\top - I_m)\Lambda$$

with respect to W_N, where Λ is an $m \times m$ symmetric matrix (Lagrange multipliers). Then, we get

$$W_N \sum_{k,l=1}^{d} Q^{(kl)} W_N^\top W_N Q^{(kl)} = \Lambda W_N. \quad (7)$$

We can assume Λ is diagonal without loss of generality, because UW_N is also a solution of the optimization problem for any orthogonal matrix $U \in O(m)$, and so we can fix U arbitrary. The equation (7) reminds us the eigenvalue of $\sum_{k,l=1}^{d} Q^{(kl)} W_N^\top W_N Q^{(kl)}$, once W_N in this matrix is fixed. Therefore, we propose the following iterative scheme.

Algorithm

1. Sphere the data $\{\boldsymbol{x}_i\}_{i=1}^n$. Let $\widehat{\boldsymbol{z}}_i = \widehat{V}^{-1/2}\boldsymbol{x}_i$, where $\widehat{V} = \widehat{\mathrm{Cov}}[\boldsymbol{x}]$
2. Calculate the fourth-order cumulant tensor from the sphered data $\{\widehat{\boldsymbol{z}}_i\}_{i=1}^n$.
$$\widehat{q}_{ijkl} = \widehat{\mathrm{cum}}(\widehat{z}_i, \widehat{z}_j, \widehat{z}_k, \widehat{z}_l)$$
3. Compute m eigen vectors with largest absolute eigenvalues.
$$W_N^{(0)} \sum_{k,l=1}^{d} \widehat{Q}^{(kl)} = \Lambda W_N^{(0)}$$
4. Solve the following eigenvalue problem until the matrix $W_N^{(t)}$ converges.
$$W_N^{(t+1)} \sum_{k,l=1}^{d} \widehat{Q}^{(kl)} \{W_N^{(t)}\}^\top W_N^{(t)} \widehat{Q}^{(kl)} = \Lambda W_N^{(t+1)}$$

The symbols $\widehat{\mathrm{Cov}}$ and $\widehat{\mathrm{cum}}$ denote the sample covariance and the sample cumulant, respectively.

4 Numerical Experiments

For testing our algorithm, we performed numerical experiments using various synthetic data used in Blanchard et al.[7]. Each data set includes 1000 samples in 10 dimension. Each sample consists of 8-dimensional independent standard Gaussian and 2 non-Gaussian components as follows.

(A) Simple: 2-dimensional independent Gaussian mixtures with density of each component given by $\frac{1}{2}\phi_{-3,1}(x) + \frac{1}{2}\phi_{3,1}(x)$.
(B) Dependent super-Gaussian: 2-dimensional isotropic distribution with density proportional to $\exp(-\|x\|)$.
(C) Dependent sub-Gaussian: 2-dimensional isotropic uniform with constant positive density for $\|x\| \leq 1$ and 0 otherwise.
(D) Dependent super- and sub-Gaussian: 1-dimensional Laplacian with density proportional to $\exp(-|x_{Lap}|)$ and 1-dimensional dependent uniform $U(c, c+1)$, where $c = 0$ for $|x_{Lap}| \leq \log 2$ and $c = -1$ otherwise.

The profiles of the density functions of the non-Gaussian components in the above data sets are described in Figure 1. The mean and standard deviation of samples are normalized to zero and one in a component-wise manner.

Besides the proposed algorithm, we applied for reference the following three methods in the experiments: PPs with 'pow3' or 'tanh' index[1] (denoted by PP3 and PPt, respectively) and JADE. We remark that the purpose of these experiments is not comparing our methods to the others, but checking its validity. To avoid local optima, additionally 9 runs from random initial matrices were also carried out and the optimum among these 10 solutions were chosen. Figure 2 shows boxplots of the error criterion

$$\mathcal{E}(\widehat{\mathcal{I}}, \mathcal{I}) = \frac{1}{m}\|(I_d - P_{\mathcal{I}})P_{\widehat{\mathcal{I}}}\|^2_{\mathrm{Fro}}, \tag{8}$$

obtained from 100 runs, where $P_{\mathcal{I}}$ (resp. $P_{\widehat{\mathcal{I}}}$) is the projection matrix onto the true non-Gaussian subspace $\mathcal{I}$ (resp. the estimated one $\widehat{\mathcal{I}}$).

Although we did not prove theoretically, JADE could find the non-Gaussian subspace $\mathcal{I}$ in all these examples. The performance of the proposed algorithm was essentially same as JADE for data (B) and (D) which contain super-Gaussian structures. On the other hand, it outperformed JADE for data (A) and (C) only with sub-Gaussian structures.

5 Conclusions

In this paper, we proposes a new *linear* method to identify the non-Gaussian subspace based on the fourth-order cumulant tensor. We also checked the validity of our method by numerical experiments. In particular, the proposed method works well in extracting sub-Gaussian structures. Although JADE is designed to extract independent components, in our examples JADE could estimate the non-Gaussian subspace $\mathcal{I}$. However,

[1] We used the deflation mode of the FastICA code [6] as an implementation of PP. 'pow3' means with the kurtosis based index while 'tanh' means with a robust index for heavy-tailed data.

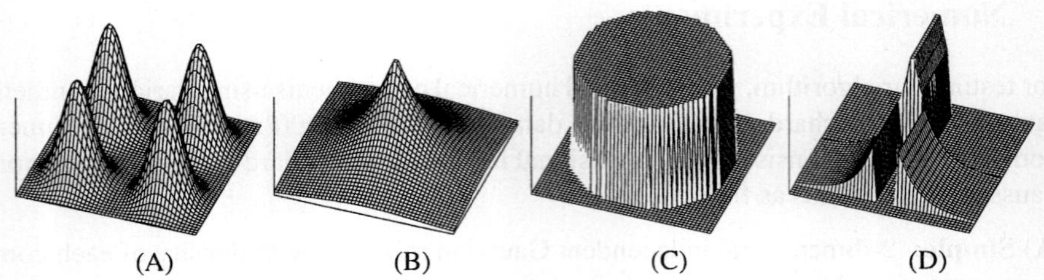

Fig. 1. Densities of non-Gaussian components. The datasets are: (a) 2D independent Gaussian mixtures, (b) 2D isotropic super-Gaussian, (c) 2D isotropic uniform and (d) dependent 1D Laplacian + 1D uniform

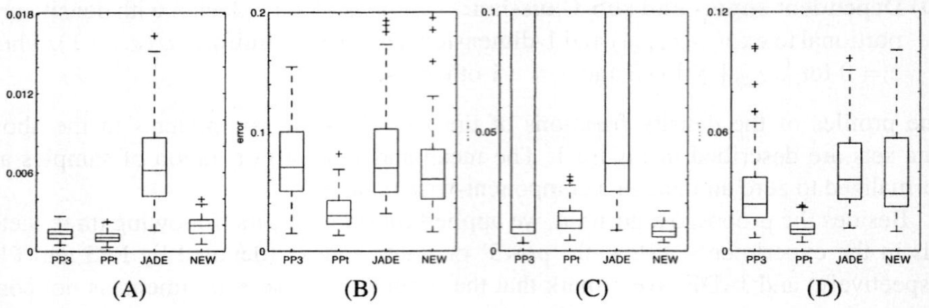

Fig. 2. Boxplots of the error criterion $\mathcal{E}(\widehat{\mathcal{I}}, \mathcal{I})$. Algorithms are PP3, PPt, JADE and the NEW method (from left to right)

our method has at least two advantages over JADE: (i) better performance for sub-Gaussian data sets, and (ii) a theoretical guarantee in the general setting. Further research should be done to prove global consistency of JADE in our model assumption.

Acknowledgement. We acknowledge Prof. J.-F. Cardoso for important suggestions, Prof. K.-R. Müller, Dr. M. Sugiyama and Dr. G. Blanchard for valuable discussions.

References

1. J.-F. Cardoso and A. Souloumiac. Blind beamforming for non Gaussian signals. *IEE Proceedings-F*, 140(6):362–370, 1993.
2. J.-F. Cardoso. Multidimensional independent component analysis In *Proceedings of ICASSP '98*, Seattle, WA, 1998.
3. P. Comon. Independent component analysis—a new concept? *Signal Processing*, 36:287–314, 1994.
4. J.H. Friedman and J.W. Tukey. A projection pursuit algorithm for exploratory data analysis. *IEEE Transactions on Computers*, 23(9):881–890, 1975.
5. P.J. Huber. Projection pursuit. *The Annals of Statistics*, 13:435–475, 1985.
6. A. Hyvarinen, J. Karhunen and E. Oja. *Independent Component Analysis*. Wiley, 2001.
7. G. Blanchard, M. Sugiyama, M. Kawanabe, V. Spokoiny and K.-R. Müller In search of non-Gaussian components of a high-dimensional distribution. submitted to *JMLR*.

On Spectral Basis Selection for Single Channel Polyphonic Music Separation

Minje Kim and Seungjin Choi

Department of Computer Science,
Pohang University of Science and Technology,
San 31 Hyoja-dong, Nam-gu, Pohang 790-784, Korea
{minjekim, seungjin}@postech.ac.kr

Abstract. In this paper we present a method of separating musical instrument sound sources from their monaural mixture, where we take the harmonic structure of music into account and use the sparseness and the overlapping NMF to select representative spectral basis vectors which are used to reconstruct unmixed sound. A method of spectral basis selection is illustrated and experimental results with monaural instantaneous mixtures of voice/cello and saxophone/viola, are shown to confirm the validity of our proposed method.

1 Introduction

The nonnegative matrix factorization (NMF) [1] or its extended version, nonnegative matrix deconvolution (NMD) [2], was shown to be useful in polyphonic music description [3], in the extraction of multiple music sound sources [2], and in general sound classification [4]. On the other hand, a method based on multiple cause models and sparse coding was successfully applied to automatic music transcription [5]. Some of these methods regard each note as a source, which might be appropriate for music transcription and work for source separation in a very limited case.

In this paper we present a method for single channel polyphonic music separation, the main idea of which is to select a few representative spectral basis vectors using the sparseness and the overlapping NMF [6], which are used to reconstruct unmixed sound signals. We assume that the structure of harmonics of a musical instrument approximately remains the same, even if it is played at different pitches. This view allows us to reconstruct original sound using only a few representative spectral basis, through the overlapping NMF. We illustrate a method of spectral basis selection from the spectrogram of mixed sound and show how these basis vectors are used to restore unmixed sound. Promising results with monaural instantaneous mixtures of voice/cello and saxophone/viola, are shown to confirm the validity of our proposed method.

2 Overlapping NMF

Nonnegative matrix factorization (NMF) is a simple but efficient factorization method for decomposing multivariate data into a linear combination of basis

W. Duch et al. (Eds.): ICANN 2005, LNCS 3697, pp. 157–162, 2005.

vectors with nonnegativity constraints for both basis and encoding matrix [1]. Given a nonnegative data matrix $\boldsymbol{V} \in \mathbb{R}^{m \times N}$ (where $V_{ij} \geq 0$), NMF seeks a factorization

$$\boldsymbol{V} \approx \boldsymbol{W}\boldsymbol{H}, \tag{1}$$

where $\boldsymbol{W} \in \mathbb{R}^{m \times n}$ $(n \leq m)$ contains nonnegative basis vectors in its columns and $\boldsymbol{H} \in \mathbb{R}^{n \times N}$ represents the nonnegative encoding variable matrix. Appropriate objective functions and associated multiplicative updating algorithms for NMF can be found in [7].

The overlapping NMF is an interesting extension of the original NMF, where transform-invariant representation and a sparseness constraint are incorporated with NMF [6]. Some of basis vectors computed by NMF could correspond to the transformed versions of a single representative basis vector. The basic idea of the overlapping NMF is to find transformation-invariant basis vectors such that fewer number of basis vectors could reconstruct observed data. Given a set of transformation matrices, $\mathcal{T} = \left\{\boldsymbol{T}^{(1)}, \boldsymbol{T}^{(2)}, \ldots, \boldsymbol{T}^{(K)}\right\}$, the overlapping NMF finds a nonnegative basis matrix $\boldsymbol{W}$ and a set of nonnegative encoding matrix $\left\{\boldsymbol{H}^{(k)}\right\}$ (for $k = 1, \ldots, K$) which minimizes

$$\mathcal{J}(\boldsymbol{W}, \boldsymbol{H}) = \frac{1}{2}\left\|\boldsymbol{V} - \sum_{k=1}^{K} \boldsymbol{T}^{(k)}\boldsymbol{W}\boldsymbol{H}^{(k)}\right\|_F^2, \tag{2}$$

where $\|\cdot\|_F$ represents Frobenious norm. As in [7], the multiplicative updating rules for the overlapping NMF were derived in [6], which are summarized below.

Algorithm Outline: Overlapping NMF [6].

Step 1 Calculate the reconstruction: $\boldsymbol{R} = \sum_{k=1}^{K} \boldsymbol{T}^{(k)}\boldsymbol{W}\boldsymbol{H}^{(k)}$.

Step 2 Update the encoding matrix by

$$\boldsymbol{H}^{(k)} \leftarrow \boldsymbol{H}^{(k)} \odot \frac{\boldsymbol{W}^T \left[\boldsymbol{T}^{(k)}\right]^T \boldsymbol{V}}{\boldsymbol{W}^T \left[\boldsymbol{T}^{(k)}\right]^T \boldsymbol{R}}, \quad k = 1, \ldots, K, \tag{3}$$

where $\odot$ denotes the Hadamard product and the division is carried out in an element-wise fashion.

Step 3 Calculate the reconstruction $\boldsymbol{R}$ again using the encoding matrix $\boldsymbol{H}^{(k)}$ updated in Step 2, as in Step 1.

Step 4 Update the basis matrix by

$$\boldsymbol{W} \leftarrow \boldsymbol{W} \odot \frac{\sum_{k=1}^{K} \left[\boldsymbol{T}^{(k)}\right]^T \boldsymbol{V} \left[\boldsymbol{H}^{(k)}\right]^T}{\sum_{k=1}^{K} \left[\boldsymbol{T}^{(k)}\right]^T \boldsymbol{R} \left[\boldsymbol{H}^{(k)}\right]^T}. \tag{4}$$

---

3 Spectral Basis Selection

The goal of spectral basis selection is to choose a few representative vectors from $\boldsymbol{V} = [\boldsymbol{v}_1 \cdots \boldsymbol{v}_N]$ where $\boldsymbol{V}$ is the data matrix associated with the spectrogram of mixed sound. In other words, each column vector of $\boldsymbol{V}$ corresponds to the power spectrum of the mixed sound at time $t = 1, \ldots, N$. Selected representative vectors are fixed as basis vectors that are used to learn an associated encoding matrix through the overlapping NMF with the sparseness constraint, in order to reconstruct unmixed sound.

Our spectral basis selection method consists of two parts, which is summarized in Table 1. The first part is to select several candidate vectors from $\boldsymbol{V}$ using a sparseness measure and a clustering technique. We use the sparseness measure proposed by Hoyer [8], described by

$$\text{sparseness}(\boldsymbol{v}) = \frac{\sqrt{m} - (\sum |v_i|)/\sqrt{\sum v_i^2}}{\sqrt{m} - 1}, \tag{5}$$

where v_i is the ith element of the m-dimensional vector $\boldsymbol{v}$.

Table 1. Spectral basis selection procedure

```
Calculate the sparseness value of every input vector, v_t, using (5);
Normalize every input vector;
repeat until the number candidates < threshold or all input vectors are eliminated
  Select a candidate with the highest sparseness value;
  Estimate the fundamental frequency bin for each input vector;
  Align each input vector such that its frequency bin location is the same as the candidate;
  Calculate Euclidean distances between the candidate and every input vector;
  Cluster input vectors using Euclidean distances;
  Eliminate input vectors in the cluster which the candidate belongs to;
end (repeat)
repeat for every possible combination of candidates
  Set all candidate vectors as input vectors;
  Select a combination of candidates;
  Learn a encoding matrix, through the overlapping NMF,
    with fixing these selected candidates as basis vectors;
  Compute the reconstruction error of the overlapping NMF;
end (repeat)
Select the combination of candidates with the lowest reconstruction error;
```

The first part of our spectral basis selection method starts with choosing a candidate vector that has the largest sparseness values (see Fig. 1 (d)). Then we estimate the location of fundamental frequency bin for each input vector, which corresponds to the lowest frequency bin above the mean value. Each input vector is aligned to the candidate vector such that the fundamental frequency bin appears at the same location as the candidate vector. Euclidean distances between these aligned input vectors and the candidate vectors are calculated and a hierarchical clustering method (or any other clustering methods) is applied to eliminate whatever vectors belong to a group which the candidate vector belongs to. This procedure is repeated until we choose a pre-specified number of candidate vectors. Increasing this pre-specified number provides more feasible candidate vectors, however, the computational complexity in the second part increases. The repetition procedure produces several candidates, some of which are expected to represent a original musical instrument sound in such a way that a set of vertically-shifted basis restores the original sound.

The second part of our method is devoted for the final selection of representative spectral basis vectors from candidates obtained in the first part. Candidate vectors are regarded as input vectors for the overlapping NMF. For every possible combination of candidates (for the case of 2 sources, 2 out of the number of candidates), we learn an associated encoding matrix with selected candidates fixed as basis vectors, and calculate the reconstruction error. Final representative spectral basis vectors are the ones which give the lowest reconstruction error.

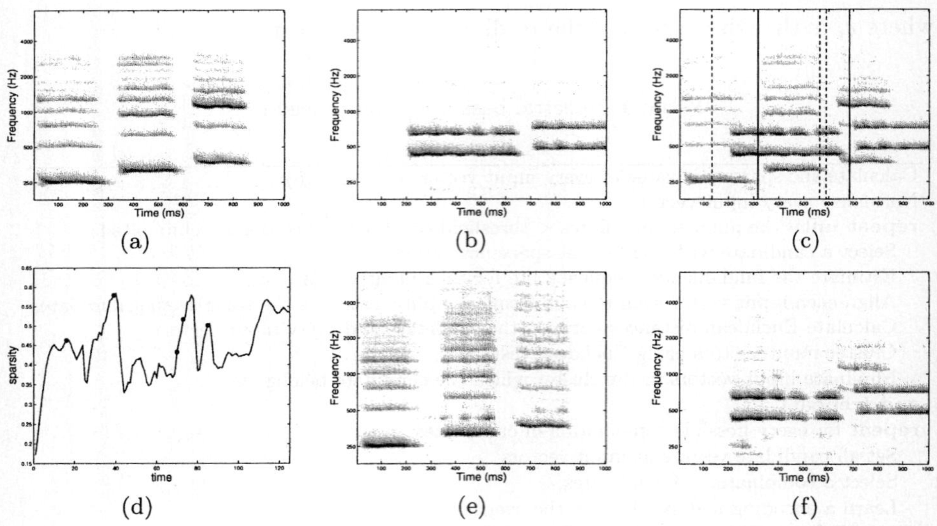

Fig. 1. Spectrograms of original sound of voice and a single string of a cello are shown in (a) and (b), respectively. Horizontal bars reflect the structure of harmonics. One can see that every note is the vertically-shifted version of each other if their musical instrument sources are the same. Monaural mixture of voice and cello is shown in (c) where 5 candidate vectors selected by our algorithm in Table 1 are denoted by dotted or solid vertical lines. Two solid lines represent final representative spectral basis vectors which give the smallest reconstruction error in the overlapping NMF. Each of these two basis vectors is a representative one for voice and a string of cello. Associated sparseness values are shown in (d) where black dots on a graph are associated with the candidate vectors. Unmixed sound is shown in (e) and (f) for voice and cello, respectively.

4 Numerical Experiments

We present two simulation results for monaural instantaneous mixtures of: (1) voice and cello; (2) saxophone and viola. We apply our spectral basis selection method with the overlapping NMF to these two data sets. Experimental results are shown in Fig. 1 and 2 where figure captions describe detailed results.

The set of transformation matrices, $\mathcal{T}$, that we used, is

$$\mathcal{T} = \left\{ \boldsymbol{T}^{(k)} \mid \boldsymbol{T}^{(k)} = \overset{\overset{k-m}{\longmapsto}}{\boldsymbol{I}}, \quad 1 \leq k \leq 2m-1 \right\}, \tag{6}$$

where $\boldsymbol{I} \in \mathbb{R}^{m \times m}$ is the identity matrix and $\overset{\overset{j}{\longmapsto}}{\boldsymbol{I}}$ leads to the shift-up or shift-down of row vectors of $\boldsymbol{I}$ by j, if j is positive or negative, respectively. After shift-up or -down, empty elements are zero-padded.

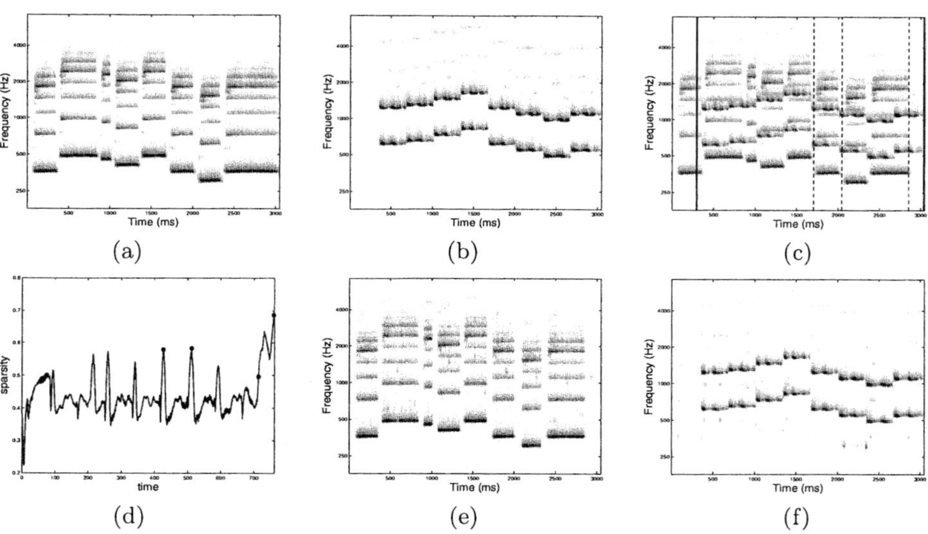

Fig. 2. Spectrograms of original sound of saxophone and viola are shown in (a) and (b), respectively. Every note is artificially generated by changing the frequency of a real sample sound, so that the spectral character of each instrument is constant in all the variations of notes. Monaural mixture is shown in (c) where 5 selected candidate vectors are denoted by vertical lines. Each of two solid lines among them represents final representative spectral basis vector of each instrument. Associated sparseness values are shown in (d) where black dots associated with the candidate vectors are marked. Unmixed sound is shown in (e) and (f) for saxophone and viola, respectively.

For the case where $m = 3$ and $k = 2$, $\boldsymbol{T}^{(2)}$ and $\boldsymbol{T}^{(5)}$ are defined as

$$\boldsymbol{T}^{(2)} = \overset{2-3}{\overleftrightarrow{\boldsymbol{I}}} = \begin{bmatrix} 0 & 0 & 0 \\ 1 & 0 & 0 \\ 0 & 1 & 0 \end{bmatrix}, \quad \boldsymbol{T}^{(5)} = \overset{5-3}{\overleftrightarrow{\boldsymbol{I}}} = \begin{bmatrix} 0 & 0 & 1 \\ 0 & 0 & 0 \\ 0 & 0 & 0 \end{bmatrix}. \tag{7}$$

Multiplying a vector by these transformation matrices, leads to a set of vertically-shifted vectors.

5 Discussions

We have presented a method of spectral basis selection for single channel polyphonic music separation, where the harmonics, sparseness, clustering, and the overlapping NMF were used. Rather than learning spectral basis vectors from the data, our approach is to select a few representative spectral vectors among given data and fix them as basis vectors to learn associated encoding variables through the overlapping NMF, in order to restore unmixed sound. The success of our approach lies in the assumption that the distinguished timbre of a given musical instrument can be expressed by a transform-invariant time-frequency representation, even though their pitches are varying. A string instrument has multiple distinguished harmonic structures. In such a case, it is reasonable to assign a basis vector for each string.

Acknowledgments. This work was supported by ITEP Brain Neuroinformatics Program and ETRI.

References

1. Lee, D.D., Seung, H.S.: Learning the parts of objects by non-negative matrix factorization. Nature **401** (1999) 788–791
2. Smaragdis, P.: Non-negative matrix factor deconvolution: Extraction of multiple sound sources from monophonic inputs. In: Proc. Int'l Conf. Independent Component Analysis and Blind Signal Separation, Granada, Spain (2004) 494–499
3. Smaragdis, P., Brown, J.C.: Non-negative matrix factorization for polyphonic music transcription. In: Proc. IEEE Workshop on Applications of Signal Processing to Audio and Acoustics, New Paltz, NY (2003) 177–180
4. Cho, Y.C., Choi, S.: Nonnegative features of spectro-temporal sounds for classfication. Pattern Recognition Letters **26** (2005) 1327–1336
5. Plumbley, M.D., Abdallah, S.A., Bello, J.P., Davies, M.E., Monti, G., Sandler, M.B.: Automatic transcription and audio source separation. Cybernetics and Systems (2002) 603–627
6. Eggert, J., Wersing, H., Körner, E.: Transformation-invariant representation and NMF. In: Proc. Int'l Joint Conf. Neural Networks. (2004)
7. Lee, D.D., Seung, H.S.: Algorithms for non-negative matrix factorization. In: Advances in Neural Information Processing Systems. Volume 13., MIT Press (2001)
8. Hoyer, P.O.: Non-negative matrix factorization with sparseness constraints. Journal of Machine Learning Research **5** (2004) 1457–1469

Independent Subspace Analysis Using k-Nearest Neighborhood Distances

Barnabás Póczos and András Lőrincz

Department of Information Systems, Eötvös Loránd University,
Research Group on Intelligent Information Systems,
Hungarian Academy of Sciences, Budapest, Hungary
pbarn@cs.elte.hu, andras.lorincz@elte.hu
http://nipg.inf.elte.hu

Abstract. A novel algorithm called independent subspace analysis (ISA) is introduced to estimate independent subspaces. The algorithm solves the ISA problem by estimating multi-dimensional differential entropies. Two variants are examined, both of them utilize distances between the k-nearest neighbors of the sample points. Numerical simulations demonstrate the usefulness of the algorithms.

1 Introduction

Independent component analysis (ICA) [1,2] aims at recovering linearly mixed and unknown sources. It is generally assumed by ICA algorithms that all sources are one-dimensional and independent. This assumption, however, may be too restrictive in practice. In more realistic situations not all, but only some groups of sources may be independent with some dependence within the subspaces. Traditional ICA algorithms form a subset of Independent Subspace Analysis (ISA) problems. These problems were first treated in [3]. Application areas include data analysis of EEG-fMRI measurements [4], the modelling of cell properties in the visual stream [5,6], and so on. Despite the efforts that have been made to develop ISA algorithms [3,7,4,8], there are still several open theoretical problems. Certain approaches use the 2-dimensional Edgeworth expansion [4], which leads to sophisticated equations. Furthermore, they have not been extended to 3 or more dimensions. Another suggestion on solving the ISA problems is to start with ICA and then to permute the columns of the mixing matrix to find the best ISA estimation [3]. This case has not been worked out and permutations may not be general enough.

Here, a particular objective function is proposed for solving ISA problems. Multi-dimensional differential entropies will be estimated. The estimation is based on the distances of the k-nearest neighbor sample points.

The paper is built as follows: Section 2 gives an overview of the ISA problem. The objective function will be derived in Section 3. Section 4 deals with the estimations of the entropy. Section 5 will introduce some efficient procedures to minimize the objective function by using Jacobi-rotations. Numerical simulations are presented in Section 6. Conclusions are drawn in the last section.

W. Duch et al. (Eds.): ICANN 2005, LNCS 3697, pp. 163–168, 2005.

2 The ISA Model

Assume that we have d of m-dimensional independent sources. Let these sources be denoted by $\boldsymbol{y}^1, \ldots, \boldsymbol{y}^d$, where $\boldsymbol{y}^i \in \mathbb{R}^m$. Let us introduce $\boldsymbol{y} = [(\boldsymbol{y}^1)^T, \ldots, (\boldsymbol{y}^d)^T]^T \in \mathbb{R}^{dm}$, where superscript T denotes transposition. Assume that these sources are hidden and the observed quantity is $\boldsymbol{x} = \boldsymbol{A}\boldsymbol{y}$, where $\boldsymbol{A} \in \mathbb{R}^{dm \times dm}$. The task is to recover the hidden source $\boldsymbol{y}$ and mixing matrix $\boldsymbol{A}$ from the observed data $\boldsymbol{x} \in \mathbb{R}^{dm}$. In the ISA model we assume that $\boldsymbol{y}^i \in \mathbb{R}^m$ is independent of $\boldsymbol{y}^j \in \mathbb{R}^m$ $(i \neq j)$. The special case of $m = 1$ corresponds to the ICA model. Assume further that both the sources and the observations are whitened, i.e., $E\boldsymbol{y} = E\boldsymbol{x} = \boldsymbol{0}$, and $E\{\boldsymbol{y}\boldsymbol{y}^T\} = E\{\boldsymbol{x}\boldsymbol{x}^T\} = \boldsymbol{I}_{md}$, where $\boldsymbol{I}_n$ is the n dimensional identity matrix an E denotes the expectation operation.

By construction, for ICA sources y^i can be recovered only up to signs and can be ordered arbitrarily. The ISA model has similar, but more general symmetry properties: sources $\boldsymbol{y}^i$ can be permuted and are undetermined up to m dimensional orthogonal transformations. In other words, if $\boldsymbol{C} \in \mathbb{R}^{m \times m}$ is orthogonal then one can not determine from the measured data $\boldsymbol{x}$ whether the original signal was $\boldsymbol{y}^i$, or if it was $\boldsymbol{C}\boldsymbol{y}^i$. Because, for whitened signals, one may assume without the loss of generality that $\boldsymbol{I}_{md} = \boldsymbol{A}\boldsymbol{A}^T$, thus it is satisfactory to restrict the search for the mixing matrix $\boldsymbol{A}$ and for its inverse separation matrix $\boldsymbol{W}$ to orthogonal matrices.

3 The ISA Objective

We introduce the ISA objective subject to constraints $\boldsymbol{W}^T\boldsymbol{W} = \boldsymbol{I}_{md}$. The separation matrix $\boldsymbol{W}$ is one of the global minima of this ISA objective function. Let $I(\boldsymbol{y}^1, \ldots, \boldsymbol{y}^d)$ denote the mutual information between vectors $\boldsymbol{y}^1, \ldots, \boldsymbol{y}^d \in \mathbb{R}^m$. Further, let $H(\boldsymbol{y})$ denote the joint entropy of vector-valued stochastic variable $\boldsymbol{y}$. Let $\boldsymbol{y} = \boldsymbol{W}\boldsymbol{x}$. Then

$$I(\boldsymbol{y}^1, \boldsymbol{y}^2, \ldots, \boldsymbol{y}^d) = -H(\boldsymbol{x}) + \log|\boldsymbol{W}| + H(\boldsymbol{y}^1) + \ldots + H(\boldsymbol{y}^d) \tag{1}$$

Our task is to minimize (1). $H(\boldsymbol{x})$ is constant and $\boldsymbol{W}^T\boldsymbol{W} = \boldsymbol{I}$. Consequently, $\log|\boldsymbol{W}| = 0$ and the minimization of (1) is equivalent to the minimization of

$$J(\boldsymbol{W}) \doteq H(\boldsymbol{y}^1) + \ldots + H(\boldsymbol{y}^d). \tag{2}$$

4 Multi-dimensional Entropy Estimation

Two estimations of entropy $H(\boldsymbol{y}^i)$ will be described below. For stochastic variable y of distribution f, Rényi's α-entropy is defined as

$$H_\alpha \doteq \frac{1}{1-\alpha} \log \int f^\alpha(x) \mathrm{d}x \tag{3}$$

It is known that $\lim\limits_{\alpha \to 1} H_\alpha = -\int f(x) \log f(x) \mathrm{d}x$, that is, the limit $\alpha \to 1$ corresponds to the Shannon-entropy.

Let $\{\boldsymbol{y}^i(1),\ldots,\boldsymbol{y}^i(n)\}$ be an i.i.d. sample set from distribution $\boldsymbol{y}^i$. Let $\mathcal{N}^i_{k,j}$ be the k nearest neighbors of $\boldsymbol{y}^i(j)$ in this sample set. Let us set α equal to $(m-\gamma)/m$. Then under mild assumptions, the Beadword-Halton-Hammersley theorem holds [9,10]:

$$\frac{1}{1-\alpha}\log\left(\frac{1}{n^\alpha}\sum_{j=1}^{n}\sum_{\boldsymbol{z}\in\mathcal{N}^i_{k,j}}\|\boldsymbol{z}-\boldsymbol{y}^i(j)\|^\gamma\right)\to H_\alpha(\boldsymbol{y}^i)+c,\ \text{as } n\to\infty, \tag{4}$$

where $\|\cdot\|$ denotes the Euclidean-norm and c is a constant. This estimation is asymptotically unbiased, strongly consistent [9], and in the limit $\alpha\to 1$ it is an estimation of the Shannon-entropy. Trivially, $\alpha\to 1$ if $\gamma\to 0$. Computation of the limit $\gamma\to 0$ is, however, troublesome and, instead, a small γ value can be used in practice. The limit $n\to\infty$ also remains an approximation.

Observe that in the ISA problem the task is not to compute the *value* of the entropy for J, but only to compute the *argument*, where the objective function takes its minimum value. Thus (4) can be modified freely by any arbitrary monotone increasing transformation. This transformation will not modify the ISA solution. Therefore the following estimation can be applied:

$$\hat{H}^1(\boldsymbol{y}^i)\doteq\sum_{j=1}^{n}\sum_{\boldsymbol{z}\in\mathcal{N}^i_{k,j}}\|\boldsymbol{z}-\boldsymbol{y}^i(j)\|^\gamma \tag{5}$$

Another estimation can be derived by means of the theorems of Kozahenko and Leonenko [11], which suggest to use 1-nearest neighbor entropy estimation [12]:

$$\hat{H}(\boldsymbol{y}^i)=\frac{1}{n}\sum_{j=1}^{n}\sum_{\boldsymbol{z}\in\mathcal{N}^i_{1,j}}\log(n\|\boldsymbol{z}-\boldsymbol{y}^i(j)\|)+\ln(2)+C_E \tag{6}$$

where C_E is the Euler constant $(=-\int_0^\infty e^{-t}\ln(t)dt)$. Under certain assumptions, the mean-square consistency of this estimation holds for any dimension m [11].

Because (i) arbitrary monotone increasing transformation of the objective is possible, and (ii) it might be useful to take into account more than one neighbors in practice, we propose the following k-nearest neighbor estimation:

$$\hat{H}^2(\boldsymbol{y}^i)\doteq\sum_{j=1}^{n}\sum_{\boldsymbol{z}\in\mathcal{N}^i_{k,j}}\log(\|\boldsymbol{z}-\boldsymbol{y}^i(j)\|) \tag{7}$$

5 Optimization and Error Measurement

It is useful to execute an ICA preprocessing step for ISA and to minimize $\sum_{j=1}^{d}\sum_{i=1}^{m}H(y_i^j)$, because (2) can be written as

$$J(\boldsymbol{W})=\sum_{j=1}^{d}\sum_{i=1}^{m}H(y_i^j)-\sum_{j=1}^{d}I(y_1^j,\ldots,y_m^j), \tag{8}$$

In this case the minimization of (2) is equivalent to the maximization of the mutual entropies ($I(y_1^j, \ldots, y_m^j)$ for all j) *within* the subspaces, which can be performed by means of Jacobi-rotations. Let $1 \le p < q \le md$, let the Jacobi-rotation matrix belonging to angle θ be denoted by $\boldsymbol{G}(p, q, \theta) \in \mathbb{R}^{md \times md}$. This is an identity matrix $\boldsymbol{I}_{md}$ with 4 modified elements, which are as follows: $\boldsymbol{G}(p, q, \theta)_{pp} = \boldsymbol{G}(p, q, \theta)_{qq} = \cos(\theta)$, $\boldsymbol{G}(p, q, \theta)_{qp} = -\boldsymbol{G}(p, q, \theta)_{pq} = \sin(\theta)$. Then, multiplying vector $\boldsymbol{y} \in \mathbb{R}^{md}$ by matrix $\boldsymbol{G}(p, q, \theta)$ from the left, will rotate the pth and the qth elements of vector $\boldsymbol{y}$. We will perform the optimization for each single $\theta \in [-\pi/2, \pi/2]$ ($\forall\ 1 \le p < q \le md$) with p and q belonging to different $m \times m$ blocks. 1D global search will be executed in each 1D optimization. An iteration cycle consists of $m^2 d(d-1)/2$ steps of 1-dimensional optimization tasks, which is much less demanding than the exhaustive search for the optimal rotation in $\mathbb{R}^{md \times md}$. Cycles will be repeated until convergence is achieved. The algorithm does not warrant to reach a global minimum, unless random rotations of the $m \times m$ blocks are carefully applied between the cycles. Such rotations were not used in the experiments.

Note also that if the ISA algorithm works properly, then the product of the estimated separation matrix $\boldsymbol{W}$ and the original mixing matrix $\boldsymbol{A}$ produces a permutation matrix made of $m \times m$ blocks. The distance of $\boldsymbol{WA}$ and the block-permutation matrix is measured by using a generalization of the Amari-distance [13]. Let b_{ij} denote the sum of the absolute values of elements at the intersection of the $i(m-1)+1, \ldots, im$ rows and the $j(m-1)+1, \ldots, jm$ columns of matrix $\boldsymbol{WA}$. Then the generalized Amari-distance $\rho(\boldsymbol{A}, \boldsymbol{W})$ is defined as follows:

$$\rho(\boldsymbol{A}, \boldsymbol{W}) \doteq \frac{1}{2d} \sum_{i=1}^{d} \left(\frac{\sum_{j=1}^{d} |b_{ij}|}{\max_j |b_{ij}|} - 1 \right) + \frac{1}{2d} \sum_{j=1}^{d} \left(\frac{\sum_{i=1}^{d} |b_{ij}|}{\max_i |b_{ij}|} - 1 \right) \ge 0 \qquad (9)$$

Clearly, $\rho(\boldsymbol{A}, \boldsymbol{W}) = 0$ iff $\boldsymbol{WA}$ is a permutation matrix of $m \times m$ blocks.

6 Numerical Simulations

In this section, we demonstrate that the estimations $\hat{H}^1$ and $\hat{H}^2$ can solve ISA problems. In two different cases, 10 (6) of 2 (3) dimensional independent sources were used. Neither of these sources could be separated linearly in the 2 (3) dimensional subspaces. For the sake of visualization, the sources formed simple 2 (3) dimensional curves and 2 (3) dimensional points of these curves were sampled independently with different distributions. The generated inputs (5000 in number) were whitened. These 2 (3) dimensional sources are depicted in Fig. 1(a)[Fig. 1(d)]. The sources were then mixed by 20×20 (18×18) real random matrix and a 20 (18) dimensional mixture was received. This high dimensional space is hard to illustrate and we show the appropriately mixed 10 (6) pieces of 2 (3) dimensional projections. See Fig. 1(b)[Fig. 1(e)]. Optimizations using the estimations $\hat{H}^1$ and $\hat{H}^2$ with $k = 20$ neighbors and $\gamma = 0.01$ values were executed. Separated sources are depicted in Fig. 1(c)[Fig. 1(f)]. Both algorithm could recover the subspaces encapsulating the original sources, apart from the order of the subspaces and

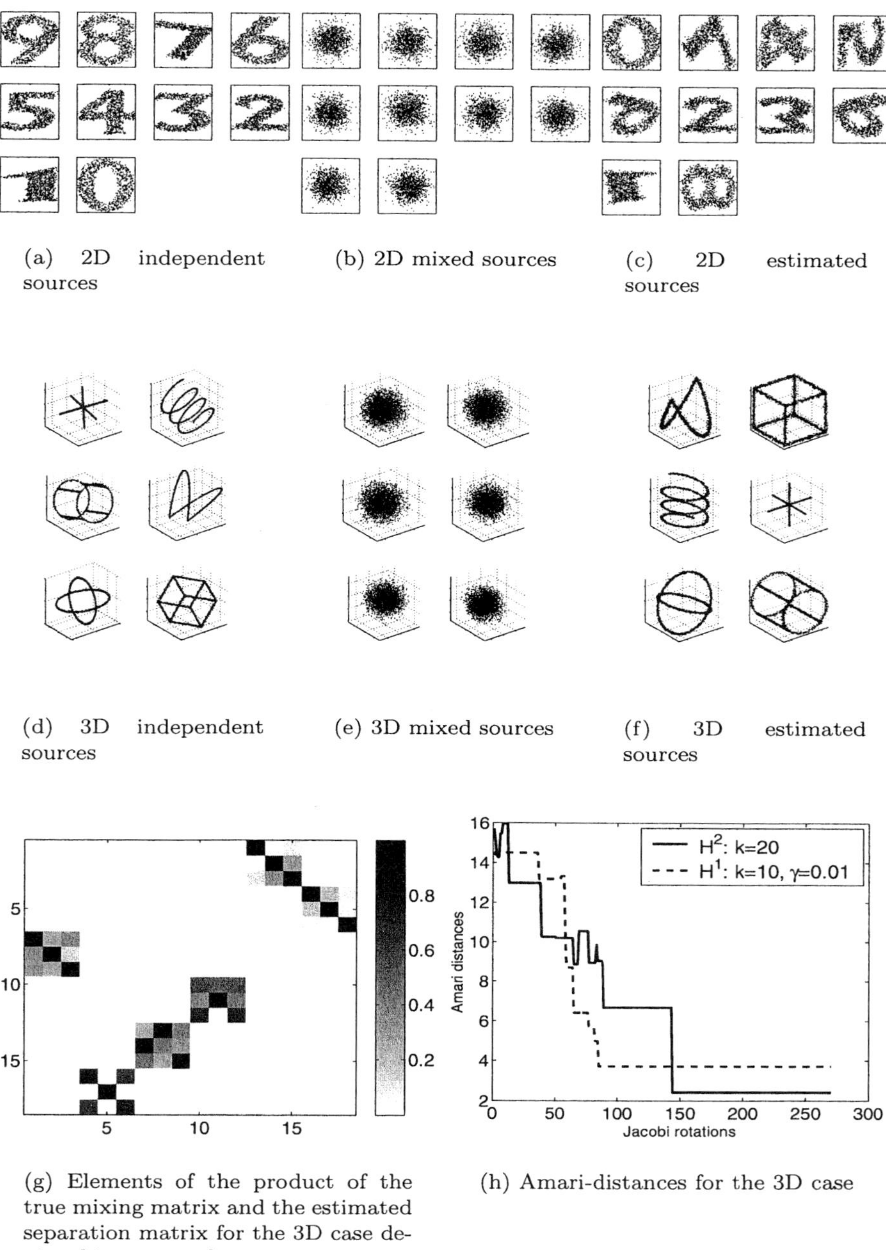

(a) 2D independent sources

(b) 2D mixed sources

(c) 2D estimated sources

(d) 3D independent sources

(e) 3D mixed sources

(f) 3D estimated sources

(g) Elements of the product of the true mixing matrix and the estimated separation matrix for the 3D case depicted in gray scale

(h) Amari-distances for the 3D case

Fig. 1. ISA results for simple curves. The algorithm properly estimates the subspaces of the independent sources (a-f). The product of the true mixing matrix and the estimated separation matrix forms a permutation matrix made of $m \times m$ blocks (g). (h): Amari distances for $\hat{H}^1$, and $\hat{H}^2$.

the directions within the subspaces, as it was expected. This is demonstrated in Fig. 1(g), which depicts the absolute values of the product of the true mixing matrix and the estimated separation matrix in gray scale for the 3D case. The result is a permutation matrix of $m \times m$ sized blocks. Amari-errors for estimations $\hat{H}^1$ and $\hat{H}^2$ are shown in Fig. 1(h). Both approximations have proved to be appropriate, the Amari-distance decreased for both cases.

7 Conclusions

We have introduced two objectives to solve the ISA problem. ICA algorithm has been applied as a preprocessing step to speed up convergence. The algorithms have used series of one-dimensional Jacobi rotations during the course of the optimization. ICA preprocessing and 1D optimization have resulted in fast and efficient minimization of the joint dependencies. Illustrative examples have been used to demonstrate the efficiency of our approach.

References

1. Jutten, C., Herault, J.: Blind separation of sources, part 1: an adaptive algorithm based on neuromimetic architecture. Signal Processing **24** (1991) 1–10
2. Comon, P.: Independent Component Analysis, a new concept? Signal Processing, Elsevier **36** (1994) 287–314 Special issue on Higher-Order Statistics.
3. Cardoso, J.: Multidimensional independent component analysis. In: Proc. of Int. Conf. on Acoust. Speech and Signal Processing, Seattle, WA. (1998)
4. Akaho, S., Kiuchi, Y., Umeyama, S.: Mica: Multimodal independent component analysis. in Proc. of IJCNN'99 (1999)
5. Hyvärinen, A., Hoyer., P.O.: Emergence of topography and complex cell properties from natural images using extensions of ica. In: Proc. of NIPS'99. (2000) 827–833
6. Hyvärinen, A., Hoyer, P.: Emergence of phase and shift invariant features by decomposition of natural images into independent feature subspaces. Neural Computation **7** (2000) 1705–1720
7. Vollgraf, R., Obermayer, K.: Multi dimensional ica to separate correlated sources. In: Proc. of NIPS'00. (2001) 993–1000
8. Bach, F.R., Jordan, M.I.: Finding clusters in independent component analysis. In: Proc. of Fourth International Symposium on Independent Component Analysis and Blind Signal Separation. (2003)
9. Yukich, J.E.: Probability Theory of Classical Euclidean Optimization Problems. Volume 1675 of Lecture Notes in Mathematics. Springer-Verlag, Berlin (1998)
10. Costa, J.A., Hero, A.O.: Manifold learning using k-nearest neighbor graphs. In: Proc. of Int. Conf. on Acoust. Speech and Signal Processing, Montreal, Canada. (2004)
11. Kozachenko, L.F., Leonenko, N.N.: Sample estimate of entropy of a random vector. Problems of Information Transmission **23** (1987) 95–101
12. Beirlant, J., Dudewicz, E.J., Györfi, L., van der Meulen, E.C.: Nonparametric entropy estimation: An overview. International Journal of Mathematical and Statistical Sciences **6** (1997) 17–39
13. Amari, S., Cichocki, A., Yang, H.: A new learning algorithm for blind source separation. In: Proc of. NIPS'95. (1996) 757–763

Study of the Behavior of a New Boosting Algorithm for Recurrent Neural Networks

Mohammad Assaad, Romuald Boné, and Hubert Cardot

Université François-Rabelais de Tours,
Laboratoire d'Informatique (EA 2101),
64, avenue Jean Portalis, 37200 Tours, France
mohammad.assaad@etu.univ-tours.fr
{romuald.bone, hubert.cardot}@univ-tours.fr

Abstract. We present an algorithm for improving the accuracy of recurrent neural networks (RNNs) for time series forecasting. The improvement is achieved by combining a large number of RNNs, each of them is generated by training on a different set of examples. This algorithm is based on the boosting algorithm and allows concentrating the training on difficult examples but, unlike the original algorithm, by taking into account all the available examples. We study the behavior of our method applied on three time series of reference with three loss functions and with different values of a parameter. We compare the performances obtained with other regression methods.

1 Introduction

The reliable prediction of future values of real-valued time series has many important applications ranging from ecological modeling to dynamic system control passing by finance and marketing. Generally the characteristics of the phenomenon which generate the series are unknown. The most usually adopted approach to consider the future values $\hat{x}(t+1)$ consists in using a function which takes as input the recent history of the time series. Using a time window of fixed size proves however to be limiting in many applications.

Ideally, for a given problem, the size of the time window should adapt to the context. This can be accomplished by employing recurrent neural networks (RNNs) learned by a gradient-based algorithm [1]. To improve the obtained results, we may use a combination of models to obtain a more precise estimate than the one obtained by a single model. In the boosting algorithm, the possible small gain a "weak" model can bring compared to random estimate is boosted by the sequential construction of several such models, which concentrate progressively on the difficult examples of the original training set. The boosting [2] [3] [4] works by sequentially applying a classification algorithm to re-weighted versions of the training data, and then taking a weighted majority vote of the sequence of classifiers thus produced. Freund and Schapire in [3] outline their ideas for applying the Adaboost algorithm to regression problems; they presented the Adaboost.R algorithm that attacks the regression problem by reducing it to a classification problem.

W. Duch et al. (Eds.): ICANN 2005, LNCS 3697, pp. 169 – 174, 2005.

Recently, a new approach to regressor boosting as residual-fitting was developed [7] [8]. Instead of being trained on a different sample of the same training set, as in previous boosting algorithms, a regressor is trained on a new training set having different target values (e.g. the residual error of the sum of the previous regressors). Before presenting our algorithm, let us mention the few existing applications of boosting to time series modelling. In [9] a boosting method belonging to the family of boosting algorithms presented in [2] is applied to the classification of phonemes. The learners employed are RNNs, and the authors are the first to notice the implications the internal memory of the RNNs has on the boosting algorithm. A similar type of boosting algorithm is used in [10] for the prediction of a benchmark time series, but with MLPs as regressors. It constructs triplets of learners and shows that the median of the three regressors has a smaller error than the individual regressors.

In this paper we focus on the definition of a boosting algorithm for improving the prediction performance of RNNs. Our algorithm is defined in section 2. Section 3 is devoted to the study of the behavior of this algorithm applied on three different benchmarks.

2 Recurrent Neural Networks with Boosting Algorithm

The boosting algorithm employed should comply with the restrictions imposed by the general context of application. In our case, it must be able to work well when a

Table 1. The boosting algorithm proposed for regression with recurrent neural networks

1. Initialize the weights for the examples: $D_1(q)=1/Q$, and Q, the number of training examples. Put the iteration counter at 0: $n=0$
2. Iterate
   (a) increment n. Learn with BPTT [1] a RNN h_n by using the entire training set and by weighting the squared error computed for example q with $D_n(q)$, the weight of example q for the iteration n;
   (b) update the weights of the examples:
      (i) compute $L_n(q)$ for every $q=1,\cdots,Q$ according to the loss function :
      $L_n^{linear}(q)=\left|y_q^{(n)}(x_q)-y_q\right|/S_n$, $L_n^{squared}(q)=\left|y_q^{(n)}(x_q)-y_q\right|^2/S_n^2$
      $L_n^{exponential}(q)=1-\exp\left(-\left|y_q^{(n)}(x_q)-y_q\right|/S_n\right)$, with
      $S_n=\sup_q\left|y_q^{(n)}(x_q)-y_q\right|$;
      (ii) compute $\varepsilon_n=\sum_{q=1}^{Q}D_n(q)L_n(q)$ and $\alpha_n=(1-\varepsilon_n)/\varepsilon_n$;
      (iii) the weights of the examples become (Z_n is a normalizing constant)
      $D_{n+1}(q)=\dfrac{1+k\cdot p_{n+1}(q)}{Q+k}$ with $p_{n+1}(q)=\dfrac{D_n(q)\alpha_n^{(L_n(q)-1)}}{Z_n}$ until $\varepsilon_n<0.5$.
3. Combine RNNs by using the weighted median.

limited amount of data is available and accept RNNs as regressors. We followed the generic algorithm of [5]. We had to decide which loss function to use for the regressors, how to update the distribution on the training set and how to combine the resulting regressors. Our updates are based on the suggestion in [6], but we apply a linear transformation to the weights before employing them (see the definition of $D_{n+1}(q)$ in the Table 1) in order to prevent the RNNs from simply ignoring the easier examples for problems similar to the sunspots dataset. Then, instead of sampling with replacement according to the updated distribution, we prefer to weight the error computed for each example (thus using all the data points) at the output of the RNN with the distribution value corresponding to the example.

3 Experimental Results

This set of experiments was carried out in order to explore the performance of the constructive algorithm and to study the influence of the parameter k on its behavior. The boosting algorithm described was used with the learning algorithm Back-Propagation Through Time (BPTT [1]) and evaluated on the sunspots time series and two Mackey-Glass time series (MG17 and MG30). In a previous paper [11], we gave some basic results on the first two series but with only one simulation for each test.

We will now come back to a more detailed study of the behavior of the algorithm, providing average results and standard deviation which have been determined after 5 trial runs for each configuration: (linear, squared or exponential loss functions; value of the parameter k). We'll also evoke extension of the range of values of k , which were due to the noticeably different results found on average. The error criterion used was the normalized mean squared error (NMSE).

The employed architectures had a single input neuron, a single linear output neuron, a bias unit and a fully recurrent hidden layer composed of neurons with tanh activation functions. For the sunspots dataset, we tested RNNs having 12 neurons in the hidden layer, and for the Mackey-Glass dataset 7 neurons. We compared the results given by our algorithm to other results in the literature (see [12] for more details).

The sunspots dataset contains the yearly number of dark spots on the sun from 1700 to 1979. The time series has a pseudo-period of 10 to 11 years. It is common practice to use as the training set the data from 1700 to 1920 and to evaluate the performance of the model on two sets, 1921-1955 (test1) and 1956-1979 (test2). Test2 is considered to be more difficult because it has a larger variance. The Mackey-Glass time-series are generated by the following nonlinear differential equation:

$$\frac{dx(t)}{dt} = -0.1 \cdot x(t) + \frac{0.2 \cdot x \cdot (t-\tau)}{1 + x^{10} \cdot (t-\tau)} \tag{1}$$

We consider here $\tau = 17$ (MG17) and $\tau = 30$ (MG30), the values which are usually retained. The data generated with $x(t) = 0.9$ for $0 \le t \le \tau$, is then sampled with a period of 6. We use the first 500 values of this series for the learning set and the next 100 values for the test set.

Table 2. Best results (NMSE*10^3)

Model	Sunspots1	Sunspots2
TAR	97	280
MLP	86	350
IIR MLP	97	436
RNN/BPTT	84	300
DRNN1	91	273
DRNN2	93	246
EBPTT	78	227
CBPTT	92	251
Boost.(sq., 5)	78	250
Boost.(lin., 10)	80	270

Model	MG17	MG30
Linear	269	324
Polynomial	11.2	39.8
RBF	10.7	25.1
MLP	10	31.6
FIR MLP	4.9	16.2
TDFFN	0.8	_
DRNN	4.7	7.6
RNN/BPTT	0.23	0.89
EBPTT	0.13	0.05
CBPTT	0.14	0.73
Boosting (lin.,150)	0.13	0.45
Boosting (sq., 100)	0.15	0.41

Table 3. Best mean results (NMSE*10^3)

Model	Sunspots1	Sunspots2
RNN/BPTT	102	371
EBPTT	92	308
CBPTT	94	281
Boost.(sq., 20)	90	296
Boost.(lin., 10)	80	314

Model	MG17	MG30
RNN/BPTT	4.4	13
EBPTT	0.62	1.84
CBPTT	1.6	2.5
Boosting (sq., 100)	0.16	0.45
Boosting (sq., 200)	0.18	0.45

Table 4. Standard deviations*10^3

Model	Sunspots1	Sunspots2
RNN/BPTT	2.3	10
EBPTT	0.9	4.6
CBPTT	6.4	33.8
Boost.(sq., 5)	9.4	34.1
Boost.(lin., 10)	1.7	25.5

Model	MG17	MG30
RNN/BPTT	3.8	41
EBPTT	0,074	0.84
CBPTT	2.0	1.6
Boosting (sq., 100)	0.016	0.028
Boosting (sq., 200)	0.025	0.014

The number of weak learners in our experiments depends on the series and loss function used. For the sunspots series our algorithm develop around 9 weak learners with the linear and quadratic loss functions, and 36 weak learners with the exponential function. For the two series MG, we obtain around 26 networks with the linear function, 37 with the quadratic function and between 46 and 50 for the exponential function. 50 is the maximal number of networks we allow.

The table 2 shows the best NMSE obtained by various models on the two test sets of sunspots and on the test sets of MG 17 and MG 30. To obtain the best results in this table, we choose the normalised best results from the 5 trials for each set of parameters (k , loss function). The best results reported in table 2 were obtained with ($k=5$, square loss function) for the sunspots, with ($k=150$, linear loss function) for MG17 and with ($k=100$, square loss function) for MG30. These tables also show that the best results obtained with our boosting algorithm are close to the best results reported in the literature (a detailed description of the mentioned models in the tables can be found in [12]).

To obtain the best mean results in the table 3, we take the normalised mean results of the 5 trials of each set of parameters, and then we choose the best results.

Table 4 gives the standard deviations of the NMSE obtained in our experiments; this information was not available for all the models in the literature. We generally find that the standard deviation is about 10% of the average. For the sunspots time series (figure 1), the average error decreases from $k=5$ to $k=20$ and increases until $k=150$. The average error for MG17 on figure 2 decreases to $k=150$ (the result is similar for MG30, the optimal value being $k=200$).

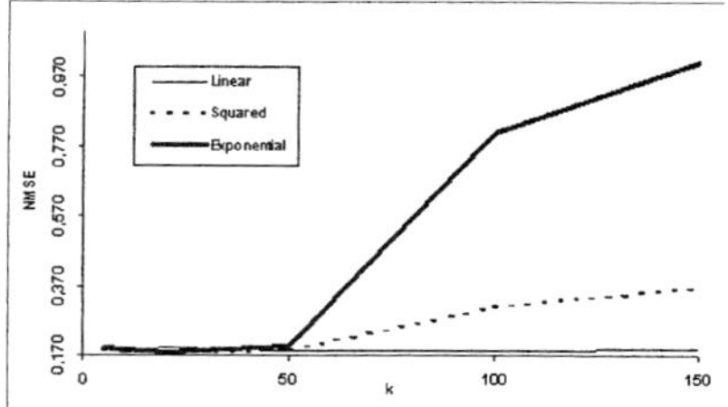

Fig. 1. NMSE of the weighted median for different values of k for sunspots

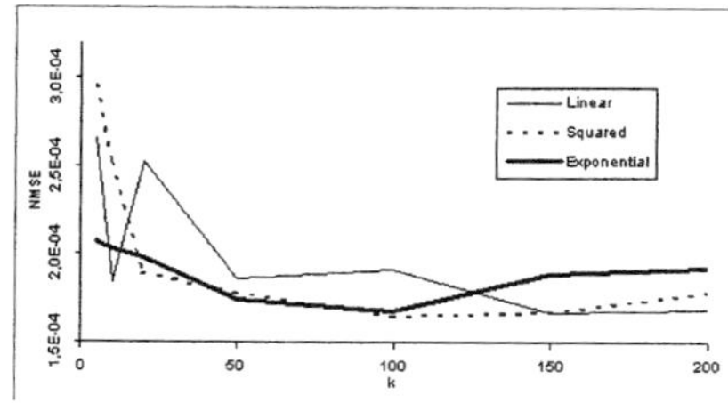

Fig. 2. NMSE of the weighted median for different values of k for MG17

To better understand why performance depends on k and why the behavior on the two datasets is different, we can notice that when $k=0$, $D_{n+1}(q)=1/Q$, and when $k>>0$, $D_{n+1}(q)\approx p_{n+1}(q)$. The relations used to calculate the values of $p_{n+1}(q)$ are those of [6] for updating the distribution and can be very close to 0 for the easier examples. If $k>>0$ the RNNs are in reality trained on the examples having big errors. These examples are very few and rather isolated for the sunspots dataset. The RNNs learn then poorly on this dataset. For low k , the examples have almost equal weights and boosting brings little improvement. We are currently trying to identify a simple method for adjusting k according to the distribution of the errors of the first learner. The experiments we performed up to now do not allow us to distinguish between the linear and the squared loss functions for the updating of the example weights, but show that the exponential loss function is not suited to our algorithm.

4 Conclusion and Future Work

We adapted boosting to the problem of learning time-dependencies in sequential data for predicting future values, using recurrent neural networks as "weak" regressors. The experimental results we obtained show that the boosting algorithm actually improves upon the performance comparatively with the use of only one RNN. We compared our results with the results obtained from other combination methods. This results obtained with our boosting algorithm are close to the best results reported in the literature.

The evaluation on multi-step-ahead prediction is one of our further works on this algorithm. We are also searching for more rigorous accounts for the various choices incorporated in this boosting algorithm.

References

1. Rumelhart, D.E., Hinton, G.E., Williams, R.J.: Learning Internal Representations by Error Propagation. In: Parallel Distributed Processing: Explorations in the Microstructure of Cognition. MIT Press (1986) 318-362
2. Schapire, R.E.: The Strength of Weak Learnability. Machine Learning 5 (1990) 197-227
3. Freund, Y., Schapire, R.E.: A Decision-Theoretic Generalization of On-Line Learning and an Application to Boosting. Journal of Computer and System Sciences 55 (1997) 119-139
4. Ridgeway, G., Madigan, D., Richardson, T.: Boosting Methodology for Regression Problems. Artificial Intelligence and Statistics (1999) 152-161
5. Freund, Y.: Boosting a Weak Learning Algorithm by Majority. Workshop on Computational Learning Theory (1990) 202-216
6. Drucker, H.: Boosting Using Neural Nets. In: Sharkey, A. (ed.): Combining Artificial Neural Nets: Ensemble and Modular Learning. Springer (1999) 51-77
7. Mason, L., Baxter, J., Bartlett, P.L., Frean, M.: Functional gradient techniques for combining hypotheses. In: Smola, A.J., *et al.* (eds.): Advances in Large Margin Classifiers. MIT Press (1999) 221-247
8. Duffy, N., Helmbold, D.: Boosting Methods for Regression. Machine Learning 47 (2002) 153-200
9. Cook, G.D., Robinson, A.J.: Boosting the Performance of Connectionist Large Vocabulary Speech Recognition. International Conference in Spoken Language Processing (1996) 1305-1308
10. Avnimelech, R., Intrator, N.: Boosting Regression Estimators. Neural Computation 11 (1999) 491-513
11. Boné, R., Assaad, M., Crucianu, M.: Boosting Recurrent Neural Networks for Time Series Prediction. In: Pearson, D.W., Steele, N.C., Albrecht, R.F. (eds.): Artificial Neural Networks and Genetic Algorithms. Springer (2003) 18-22
12. Boné, R., Crucianu, M., Asselin de Beauville, J.-P.: Learning Long-Term Dependencies by the Selective Addition of Time-Delayed Connections to Recurrent Neural Networks. NeuroComputing 48 (2002) 251-266

Time Delay Learning by Gradient Descent in Recurrent Neural Networks

Romuald Boné and Hubert Cardot

Université François-Rabelais de Tours,
Laboratoire d'Informatique (EA 2101),
64, avenue Jean Portalis, 37200 Tours, France
{romuald.bone, hubert.cardot}@univ-tours.fr

Abstract. Recurrent Neural Networks (RNNs) possess an implicit internal memory and are well adapted for time series forecasting. Unfortunately, the gradient descent algorithms which are commonly used for their training have two main weaknesses: the slowness and the difficulty of dealing with long-term dependencies in time series. Adding well chosen connections with time delays to the RNNs often reduces learning times and allows gradient descent algorithms to find better solutions. In this article, we demonstrate that the principle of time delay learning by gradient descent, although efficient for feed-forward neural networks and theoretically adaptable to RNNs, shown itself to be difficult to use in this latter case.

1 Introduction

Recurrent Neural Networks (RNNs), which possess an internal memory owing to cycles in their connection graph, have interesting universal approximation capabilities for sequential problems (see [1] for example), comparable to the Feed-forward Neural Network (FNNs) ones for static problems. Most of the time they are associated to gradient based learning algorithms, requiring more computing time than the gradient based algorithms for FNNs with the same number of parameters. Moreover, they have difficulties in dealing with long-term dependencies in the data [2] [3].

An alternative is to use globally feed-forward architectures. They share the characteristic of having been initially elaborated for using the error gradient back-propagation of FNNs (some of which have an adapted version today [4]). Hence the Locally Recurrent Globally Feed-forward Networks [5] introduce particular neurons, with local feedback loops. In the most general form, these neurons feature delays in inputs as well as in their loops. All these architectures remain limited: hidden neurons are mutually independent and therefore, cannot pick up some complex behaviors which require the collaboration of several neurons of the hidden layer. In order to overcome this problem, a certain number of recurrent architectures have been suggested (see [3] for a presentation). It has been shown that in practice the use of delay connections in these networks gives rise to a reduction in learning time [6] as well as improving the taking into account of long term dependencies [3] [7]. The resulting network is named Time Delay Recurrent Neural Networks (TDRNN). In this case,

W. Duch et al. (Eds.): ICANN 2005, LNCS 3697, pp. 175 – 180, 2005.

unless to apply an algorithm for selective addition of connections with time delays [7], which improve forecasting performance capacity but at the cost of increasing computations, the networks finally retained are often oversized and use meta-connections with consecutive delay connections, also named Finite Impulse Response (FIR) connections or, if they contain loops, Infinite Impulse Response (IIR) connections [5].

Thus, the solution could be found in the learning of the connection delays themselves. [8] (see also [9]) have suggested, for a FNN that associate a delay to each connection, an algorithm based on the gradient which simultaneously adjusts weights and delays. We propose to adapt this technique to recurrent architecture.

2 Learning of the Delays

Considering a RNN in which two values are associated to each connection from a neuron *j* to a neuron *i*. These two values are an usual weight w_{ij} of the signal and a delay τ_{ij} which is a real value indicating the needed time for the signal to propagate through the connection. Note that this parameter is not the same as the maximal order of a FIR connection: indeed, when we consider a connection of delay τ_{ij}, we do not have simultaneously $\tau_{ij}-1$ connections with integer delays between 1 and τ_{ij}. The neuron output $s_i(t)$ is given by:

$$s_i(t)=f_i(net_i(t-1)) \text{ with } net_i(t-1)=\sum_{j\in Pred(i)} w_{ij}s_j(t-\tau_{ij}-1) \tag{1}$$

The values $s_j(t-\tau_{ij}-1)$ are obtained by applying a linear interpolation between the two nearest whole numbers of the delay τ_{ij} [8]. The set $Pred(i)$ contains, for each neuron *i*, the index of the incoming neurons $Pred(i)=\{j\in N \mid \exists(w_{ij},\tau_{ij})\}$. Likewise, we have defined the successors of a neuron *i*: $Succ(i)=\{j\in N \mid \exists(w_{ji},\tau_{ji})\}$.

We have adapted the BPTT algorithm [10] to this architecture with a simultaneous learning of weights and delays of the connections, inspired from [8]. Central idea of BPTT algorithm is to unfold in time the original RNN to obtain a l-layer feed-forward neural network. This allows applying the well-known back-propagation learning algorithm.

The variation of a delay τ_{ij} can be compute as the sum of the variations of this parameter copies corresponding to the times from t_1 to t_l. Then we add this variation to all copies of τ_{ij}. We will only give here the demonstration of the learning of the delays as the weight one can easily be deducted from it.

We note $T(t)$ the set of neuron indices which have a desired output at time *t*, $d_p(t)$ the desired output of neuron *p* at this time and $\Delta\tau_{ij}(\tau)$ the copy of τ_{ij} for $t=\tau$ in the unfold in time neural net which is virtually constructed with BPTT [10]. $\lceil . \rceil$ is the operator of upward roundness.

We apply a back-propagation of the gradient of the mean quadratic error $E(t_1,t_l)$ which is defined as the sum of the instantaneous errors $e(t)$ from t_1 to t_l:

$$E(t_1,t_l)=\sum_{t=t_1}^{t_l} e(t)=\sum_{t=t_1}^{t_l}\frac{1}{2}\sum_{p\in T(t)}\left(d_p(t)-s_p(t)\right)^2 \tag{2}$$

$$\Delta\tau_{ij}(t_1,t_l-1)=-\lambda\frac{\partial E(t_1,t_l)}{\partial\tau_{ij}}=\sum_{\tau=t_1+\lceil\tau_{ij}\rceil}^{t_l-1}\Delta\tau_{ij}(\tau)=-\lambda\sum_{\tau=t_1+\lceil\tau_{ij}\rceil}^{t_l-1}\frac{\partial E(t_1,t_l)}{\partial\tau_{ij}(\tau)} \tag{3}$$

We can write $\partial E(t_1,t_l)/\partial\tau_{ij}(\tau)=\partial E(t_1,t_l)/\partial net_i(\tau)\bullet\partial net_i(\tau)/\partial\tau_{ij}(\tau)$. With a first order approximation, $\partial net_i(\tau)/\partial\tau_{ij}(\tau)\approx w_{ij}\left(s_j(\tau-\tau_{ij}-1)-s_j(\tau-\tau_{ij})\right)$. We expand $\partial E(t_1,t_l)/\partial net_i(\tau)$:

$$\frac{\partial E(t_1,t_l)}{\partial net_i(\tau)}=\frac{\partial E(t_1,t_l)}{\partial s_i(\tau+1)}\frac{\partial s_i(\tau+1)}{\partial net_i(\tau)}\text{ with }\frac{\partial s_i(\tau+1)}{\partial net_i(\tau)}=f'(net_i(\tau)) \tag{4}$$

If neuron i belongs to the last layer ($\tau=t_l-1$):

$$\frac{\partial E(t_1,t_l)}{\partial s_i(\tau+1)}=\frac{\partial e(t_l)}{\partial s_i(t_l)}=\underset{i\in T(\tau+1)}{\delta}(s_i(t_l)-d_i(t_l)) \tag{5}$$

where $\underset{i\in T(\tau+1)}{\delta}=1$ if $i\in T(\tau+1)$ and 0 otherwise.

If neuron i belongs to the preceding layers:

$$\frac{\partial E(t_1,t_l)}{\partial s_i(\tau+1)}=\frac{\partial e(\tau+1)}{\partial s_i(\tau+1)}+\sum_{j\in Succ(i)}\left(\frac{\partial E(t_1,t_l)}{\partial net_j(\tau+\tau_{ji}+1)}\frac{\partial net_j(\tau+\tau_{ji}+1)}{\partial s_i(\tau+1)}\right) \tag{6}$$

As $\partial net_j(\tau+\tau_{ji}+1)/\partial s_i(\tau+1)=w_{ji}(\tau+1)$, we obtain the final relations to learn the delay associated to each connection:

$$\Delta\tau_{ij}(t_1,t_l-1)=-\lambda\sum_{\tau=t_1+k}^{t_l-1}\frac{\partial E(t_1,t_l)}{\partial net_i(\tau)}w_{ij}\left(s_j(\tau-\tau_{ij}-1)-s_j(\tau-\tau_{ij})\right) \tag{7}$$

with for $\tau=t_l-1$

$$\frac{\partial E(t_1,t_l)}{\partial net_i(\tau)}=\underset{i\in T(\tau+1)}{\delta}(s_i(t_l)-d_i(t_l))f_i'(net_i(\tau)) \tag{8}$$

and for $t_1\le\tau<t_l-1$

$$\frac{\partial E(t_1,t_l)}{\partial net_i(\tau)}=\left[\begin{array}{l}\underset{i\in T(\tau+1)}{\delta}(s_i(\tau+1)-d_i(\tau+1))\\+\displaystyle\sum_{j\in Succ(i)}\frac{\partial E(t_1,t_l)}{\partial net_j(\tau+\tau_{ji}+1)}w_{ji}(\tau+1)\end{array}\right]f_i'(net_i(\tau)) \tag{9}$$

3 Experiments

The experimental results we present here concern univariate regression, but our algorithm is not limited to such problems. We applied our algorithm to RNNs having an input neuron, a linear output neuron, a bias unit and a fully recurrent hidden layer composed of neurons with a symmetric sigmoid as activation function. 10 experiments were performed for every architecture, by randomly initializing the weights in $[-0.3, 0.3]$ and the delays in $[0, 10]$. We employed two kinds of datasets: a natural one (sunspots [11]) and a synthetic one (Mackey-Glass [12]). The error criterion used was the normalized mean squared error (NMSE).

Table 1. Results (NMSE) obtained by various models on the sunspots time series

Model	Test1	Test2
Carbon Copy	0.427	0.966
TAR	0.097	0.280
MLP	0.086	0.350
IIR MLP	0.097	0.436
RNN / BPTT	0.084	0.300
DRNN1	0.091	0.273
DRNN2	0.093	0.246
RNN / CBPTT	0.092	0.251
RNN / EBPTT	0.078	0.227
Our algorithm	0.081	0.261

Table 2. Results (NMSE) obtained by various models on the MG17 time series

Model	NMSE
Polynomial	$11.2 \cdot 10^{-3}$
RBF	$10.7 \cdot 10^{-3}$
MLP	$10 \cdot 10^{-3}$
FIR MLP	$4.9 \cdot 10^{-3}$
TDFNN [8]	$0.8 \cdot 10^{-3}$
DRNN	$4.7 \cdot 10^{-3}$
RNN/BPTT	$0.23 \cdot 10^{-3}$
RNN/CBPTT	$0.14 \cdot 10^{-3}$
RNN/EBPTT	$0.13 \cdot 10^{-3}$
Our algorithm	$0.15 \cdot 10^{-3}$

The Sunspots dataset contains the yearly number of dark spots on the sun from 1700 to 1979. It is common practice to use as the training set the data from 1700 to 1920 and to evaluate the performance of the model on two sets, 1921-1955 (test1) and 1956-1979 (test2). Test2 is considered to be more difficult because it has a larger variance and is non-stationary. Table 1 compares the results obtained by various models applied to the two test sets of this benchmark (see [7] [13] for more details). Table 1 shows that the result obtained by our algorithm on Test1 is close to the best results reported in the literature. The result for Test2 is close to the results obtained by networks with fixed delays (DRNN). Our results correspond to a three hidden neurons network.

The Mackey-Glass time series [12] are generated by the following non-linear differential equation:

$$\frac{dx(t)}{dt} = -0.1x(t) + \frac{0.2x(t-\tau)}{1 + x^{10}(t-\tau)} \tag{10}$$

We consider here $\tau = 17$ (MG17), the value which is usually retained. The resulting time series exhibits then a chaotic behavior. The data is generated and then sampled with a period of 6, according to the common practice (see e.g. [14]). We use the first 500 values for the learning set and the next 100 values for the test set. Table 2 gives the results obtained with various models on the MG17 benchmark. The Time Delay FNN in [8] has a single input, 20 neurons in the hidden layer and one output neuron. The DRNN [15] has FIR connections of order 4 between the input and the hidden layer, FIR connections of order 2 between the 4 to 7 hidden neurons, and simple connections to the output neuron. The results reported for CBPTT were obtained with RNNs having 6 hidden neurons, with 10 time-delayed connections. EBPTT gives the best results for the MG17 dataset with the same network. Several additional models applied to MG17 can be found in [16], some of them with better results than the mentioned models but obtained from a different dataset (number of data, sampling, ...)

The experiments show an occasionally unstable behavior of our algorithm: some learning attempts being soon blocked with high values of error. The internal state of the network (the set of neuron outputs belonging to the hidden layer) happens to be very sensitive to delay variation. The choice of the two learning steps, either for the weights or for connection delays, require a very precise tuning. We can mention that the value ranges of the two parameters are very different, typical values for the delays being above one. It is worth noticing that our results are not as good as those of the EBPTT algorithm which adds connections to a RNN without using a method directly based on an error gradient calculation. Despite those remarks, our algorithm has already performances near the best of other architectures and it seems to be promising. We are still working on his amelioration.

4 Conclusion

A recent type of neuron whose connections have a real value was adapted to recurrent networks. A new dedicated learning algorithm has been presented. The best results obtained from two forecast problems are encouraging but demonstrate a tricky configuration and an unstable behavior of the algorithm.

The architecture seems to bring additional power but to overcome the mentioned limitations, it remains to study an alternative to the use of the gradient calculation for the learning of the delays. A stochastic version is presently under consideration

Moreover, good results were obtained with the two constructive algorithms CBPTT and EBPTT thanks to the addition of time-delayed connections to RNNs. Those RNNs can simultaneously contain, between neurons, an usual connection and a delayed one. This allows us to consider an improvement of performances if we apply our algorithm on such architectures.

Acknowledgements

The authors would like to thank Richard Duro from the University of Coruña (Spain) for the interesting discussion related to the use of connections with real time delays in FNNs. They also thank Alexandre Chriqui, who spent four months in their lab, for his help in the programming task.

References

1. Jin, L., Nikiforuk, N., Gupta, M.M.: Uniform Approximation of Nonlinear Dynamic Systems Using Dynamic Neural Networks. International Conference on Artificial Neural Networks (1995) 191-196
2. Bengio, Y., Simard, P., Frasconi, P.: Learning Long-Term Dependencies with Gradient Descent is Difficult. IEEE Transactions on Neural Networks 5 (1994) 157-166
3. Lin, T., Horne, B.G., Tino, P., Giles, C.L.: Learning Long-Term Dependencies in NARX Recurrent Neural Networks. IEEE Transactions on Neural Networks 7 (1996) 13-29
4. Campolucci, P., Uncini, A., Piazza, F., Rao, B.D.: On-Line Learning Algorithms for Locally Recurrent Neural Networks. IEEE Transactions on Neural Networks 10 (1999) 253-271
5. Tsoi, A.C., Back, A.D.: Locally Recurrent Globally Feedforward Networks: A Critical Review of Architectures. IEEE Transactions on Neural Networks 5 (1994) 229-239
6. Guignot, J., Gallinari, P.: Recurrent Neural Networks with Delays. International Conference on Artificial Neural Networks (1994) 389-392
7. Boné, R., Crucianu, M., Asselin de Beauville, J.-P.: Learning Long-Term Dependencies by the Selective Addition of Time-Delayed Connections to Recurrent Neural Networks. NeuroComputing 48 (2002) 251-266
8. Duro, R.J., Santos Reyes, J.: Discrete-Time Backpropagation for Training Synaptic Delay-Based Artificial Neural Networks. IEEE Transactions on Neural Networks 10 (1999) 779-789
9. Pearlmutter, B.A.: Dynamic Recurrent Neural Networks. Research Report CMU-CS-90-196, Carnegie Mellon University School of Computer Science (1990)
10. Rumelhart, D.E., Hinton, G.E., Williams, R.J.: Learning Internal Representations by Error Propagation. In: Rumelhart, D.E., McClelland, J. (eds.): Parallel Distributed Processing: Explorations in the Microstructure of Cognition. MIT Press (1986) 318-362
11. Weigend, A.S., Huberman, B.A., Rumelhart, D.E.: Predicting the Future: A Connectionist Approach. International Journal of Neural Systems 1 (1990) 193-209
12. Mackey, M., Glass, L.: Oscillations and chaos in physiological control systems. Science (1977) 197-287
13. Boné, R., Crucianu, M., Verley, G., Asselin de Beauville, J.-P.: A Bounded Exploration Approach to Constructive Algorithms for Recurrent Neural Networks. International Joint Conference on Neural Networks (2000) 6 p.
14. Back, A., Wan, E.A., Lawrence, S., Tsoi, A.C.: A Unifying View of some Training Algorithms for Multilayer Perceptrons with FIR Filter Synapses. Neural Networks for Signal Processing IV (1994) 146-154
15. Aussem, A.: Nonlinear Modeling of Chaotic Processes with Dynamical Recurrent Neural Networks. Neural Networks and Their Applications (1998) 425-433
16. Gers, F., Eck, D., Schmidhuber, J.: Applying LSTM to Time Series Predictable Through Time-Window Approaches. Int. Conference on Artificial Neural Networks (2001) 669-675

Representation and Identification Method of Finite State Automata by Recurrent High-Order Neural Networks

Yasuaki Kuroe

Center for Information Science, Kyoto Institute of Technology,
Matsugasaki, Sakyo-ku, Kyoto 606-8585, Japan
kuroe@dj.kit.ac.jp

Abstract. This paper presents a new architecture of neural networks for representing deterministic finite state automata. The proposed model is a class of high-order recurrent neural networks. It is capable of representing FSA with the network size being smaller than the existing models proposed so far. We also propose an identification method of FSA from a given set of input and output data by training the proposed model of neural networks.

1 Introduction

The problem of representing and learning finite state automata (FSA) with artificial neural networks has attracted a great deal of interest recently. Several models of artificial neural networks for representing and learning FSA have been proposed and their computational capabilities have been investigated [1,2,3,4,5].

In recent years, there have been increasing research interests of hybrid control systems, in which controlled objects are continuous dynamical systems and controllers are implemented as discrete event systems. One of the most familiar model representations of discrete event systems is a model representation by using FSA. It has been strongly desired to develop an identification method of unknown FSA from given input and output data. One of the promising approaches to the problem is to develop a method by using neural networks. The problem that comes out first in the approach is to investigate what architectures of neural networks are suitable for identification of FSA.

High-order neural networks (HONNs) which allow high-order nonlinear connections among neurons have been recognized to possess higher capability of nonlinear functions representations than the conventional sigmoidal neural networks. It is expected, therefore, that high-order neural networks having recurrent connections, recurrent high-order neural networks (RHONNs), possess excellent approximations capability of nonlinear dynamical systems [7,8].

We have already proposed an architecture of neural networks, which is a class of recurrent neural networks (RNNs), for representing deterministic FSA [6]. The proposed model is capable of representing FSA with the network size being smaller than the existing models proposed so far. In this paper, we propose

W. Duch et al. (Eds.): ICANN 2005, LNCS 3697, pp. 181–189, 2005.

another architecture of neural networks, which is a class of RHONNs. We also propose an identification method of FSA from a given set of input and output data by training the proposed neural networks. It is shown that the proposed neural networks are easier to train because of less number of learning parameters.

2 Finite State Automata

In this paper we consider finite state automata (FSA) M defined by

$$M = (Q, q_0, \Sigma, \Delta, \delta, \varphi) \tag{1}$$

where Q is the set of state symbols: $Q = \{q_1, q_2, \cdots, q_r\}$, r is the number of state symbols, $q_0 \in Q$ is the initial state, Σ is the set of input symbols: $\Sigma = \{i_1, i_2, \cdots, i_m\}$, m is the number of input symbols, Δ is the set of output symbols: $\Delta = \{o_1, o_2, \cdots, o_l\}$, l is the number of output symbols, δ: $Q \times \Sigma \to Q$ is the sate transition function and φ: $Q \times \Sigma \to \Delta$ is the output function.

We suppose that the FSA M operates at unit time intervals. Letting $i(t) \in \Sigma$, $o(t) \in \Delta$ and $q(t) \in Q$ be the input symbol, output symbol and state symbol at time t, respectively, then the FSA M is described by the discrete dynamical system of the form:

$$M: \; q(t+1) = \delta(q(t), i(t)), \; q(0) = q_0, \quad o(t) = \varphi(q(t), i(t)) \tag{2}$$

The objective of this paper is to discuss the problem of neural-network representation and identification of the FSA described by (1) or (2). We first discuss architecture of neural networks for representing the FSA.

3 Recurrent High-Order Neural Networks

We consider a general class of RHONNs in order to represent FSA. Figure 1 shows the schematic diagram of RHONN models. The network consists of neuron units, connection units, external inputs and external outputs. Two types of neurons are considered, dynamic neurons and static neurons. All the neurons and the external inputs are connected through the connection units which allow high-order nonlinear interactions.

Let N_d, N_s, M, L and K be the numbers of the dynamic neurons, the static neurons, the high-order units, the external inputs and the external outputs existing in the network, respectively. The mathematical model of the dynamic neurons is given by

$$v_i^d(t+1) = \sum_{m=1}^{M} w_{im}^d z_m^d(t) + \theta_i^d, \quad v_i^d(0) = v_{i0}^d, \qquad h_i^d(t) = S(v_i^d(t)) \tag{3}$$

$i = 1, 2, \cdots, N_d$, where $z_m^d(t)$, $v_i^d(t)$, $h_i^d(t)$ and θ_i^d are the input, the state, the output and threshold value of the i-th dynamic neuron, respectively, and w_{im}^d

is the connection weight from the m-th input to the i-th dynamic neuron, and $S(\cdot)$ is a nonlinear output function such as threshold or sigmoidal function. The mathematical model of the static neurons is given by the following equations.

$$u_i^s(t) = \sum_{m=1}^{M} w_{im}^s z_m(t) + \theta_i^s, \quad h_i^s(t) = S(u_i^s(t)), \quad i = 1, 2, \cdots, N_s \tag{4}$$

where z_m^s, u_i^s, h_i^s and θ_i^s are the input, the state, the output and the threshold value of the i-th static neuron, respectively, and w_{im}^s is the weight from the m-th input to the i-th static neuron. The way of connections among the neurons and the external inputs is determined by the connection units (high-order units). Let $\boldsymbol{s}$ be a $N_d + N_s + L$ dimensional vector defined by $\boldsymbol{s} = [h_1^d, h_2^d, \cdots, h_{N_d}^d, h_1^s, h_2^s, \cdots, h_{N_s}^s, I_1, I_2, \cdots, I_L]^T$ where $I_1, I_2, \cdots, I_L$ are the external inputs. The model of the connection units is given by

$$z_m(t) = G_m(\boldsymbol{s}(t)), \qquad (m = 1, 2, \cdots, M) \tag{5}$$

where $G_m(\boldsymbol{s}(t))$ is defined by $G_m := \prod_{j \in J_m} s_j^{d_j(m)}$. In the above equation z_m is the output of the m-th high-order unit, J_m is a subset of the index set $\{1, 2, \cdots, N_d + N_s + L\}$ and $d_j(m)$ are nonnegative integers. Note that z_m is given to the static neurons and dynamic neurons as their inputs as shown in (3) and (4). It can be seen from (3),(4) and (5) that the connections among the neurons and the external inputs allow not only linear combinations but also high-order product combinations. The external outputs O_k are expressed by

$$O_k(t) = \sum_{i=1}^{N_d} \delta_{ki}^d h_i^d(t) + \sum_{i=1}^{N_s} \delta_{ki}^s h_i^s(t) \qquad (k = 1, 2, \cdots, K) \tag{6}$$

where δ_{ki}^d and δ_{ki}^s take values 1 or 0. If the output of the i-th dynamic (static) neuron is connected to the k-th external output, $\delta_{ki}^d = 1$ ($\delta_{ki}^s = 1$), otherwise $\delta_{ki}^d = 0$ ($\delta_{ki}^s = 0$).

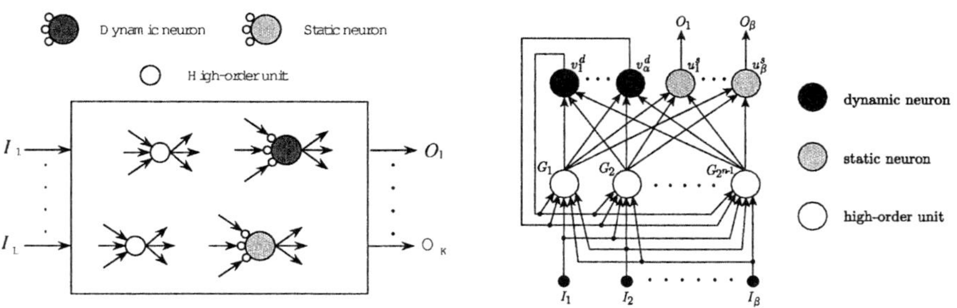

Fig. 1. Recurrent high-order neural networks

Fig. 2. Recurrent high-order neural networks representing FSA

4 Neural Network Architectures for Representing FSA

4.1 Recurrent Second-Order Neural Networks for Representing FSA

There have been done several works on the representation of FSA with neural networks or on investigation of relationship between neural network architectures and FSA. A typical representative of neural network architectures for representing FSA is a class of second-order neural networks.

Let each state symbol q_i be expressed by r dimensional unit basis vector, that is $q_1 = (1, 0, \cdots, 0)$, $q_2 = (0, 1, \cdots, 0)$, $\cdots$, $q_r = (0, 0, \cdots, 1)$. Similarly let each input symbol i_i and each output symbol o_i be expressed by m and l dimensional unit basis vectors, respectively. With the use of these expression the functions δ and φ in (2) can be represented by the product of the state $q(t)$ and input $i(t)$. Then (2) can be rewritten as follows.

$$q_i(t+1) = \sum_{l=1}^{m} i_l(t) \sum_{j=1}^{r} a_{ij}^l q_j(t), \quad o_k(t) = \sum_{l=1}^{L} i_l(t) \sum_{j=1}^{N} c_{kj}^l q_j(t) \tag{7}$$

where a_{ij} and c_{kj} are parameters. From this equation, we can construct a neural network which can represent FSA as follows. Consider the RHONNs described by (3), (4), (5) and (6) where $N_d = r, N_s = l, M = r \times l$, $L = \beta$, $K = m$ and s is defined by $s = [h_1^d, h_2^d, \cdots, h_{N_d}^d, I_1, I_2, \cdots, I_L]^T$. Furthermore, in the high order units (5), $J_m, m = 1, 2, \cdots M$ are defined by $J_1 = \{1, r+1\}, J_2 = \{2, r+1\}, \cdots J_r = \{r, r+1\}, J_{r+1} = \{1, r+2\}, J_{r+2} = \{2, r+2\}, \cdots, J_{2r} = \{r, r+2\}, \cdots, J_{rl-1} = \{r-1, r+l\}, J_{rl} = \{r, r+l\}$ and $d_j(m)$ are chosen as $d_j(m) = 1, m = 1, 2, \cdots M$. In the external outputs (6) δ_{ki}^d and δ_{ki}^s are chosen as $\delta_{ki}^d = 0, k = 1, 2, \cdots, K, i = 1, 2, \cdots, N_d$ and $\delta_{ki}^s = 1$, for $i = k$, $\delta_{ki}^s = 0$ for $i \neq k, k = 1, 2, \cdots, K$. The networks thus constructed are called recurrent second-order neural networks (R2ONNs), because they only contain the second-order (product) connections in their high-order units. It can be shown that the R2ONNs can represent any FSA exactly by letting the states $v_i^d(t)$ of the dynamic neurons, the external inputs I_i and the external outputs O_k be corresponding to the states $q_j(t)$, the inputs $i_l(t)$ and the outputs $o_k(t)$ of FSA (7). [3] and [5] proposed similar architectures of neural networks, recurrent second-order neural networks and show their capability of representing FSA.

4.2 Recurrent High-Order Neural Networks for Representing FSA

In the R2ONNs obtained in §4.1 or in [3] and [5], each state of FSA is represented by assigning one neuron individually. Then, as the number of states of a target FSA increases, the number of neurons required for representing the FSA increases, which makes it difficult to identify the FSA. We propose a new architecture of neural networks for representing FSA.

We encode all the state symbols q_i, input symbols i_i and output symbols o_i of FSA as binary variables. Then $q(t)$, $i(t)$ and $o(t)$ in (2) can be expressed as [6]:

$q(t) = (s_1(t), s_2(t), \cdots, s_\alpha(t))$ where $s_i(t) \in \{0,1\}$, $i(t) = (x_1(t), x_2(t), \cdots, x_\beta(t))$ where $x_i(t) \in \{0,1\}$ and $o(t) = (y_1(t), y_2(t), \cdots, y_\gamma(t))$ where $y_i(t) \in \{0,1\}$. α, β and γ are natural numbers, which are determined depending on r, m and l, respectively, that is, α is the minimum natural number satisfying $r \le 2^\alpha$, β is the minimum natural number satisfying $m \le 2^\beta$ and γ is the minimum natural number satisfying $l \le 2^\gamma$. By using these representations, we can transform (2) into

$$M : \begin{cases} s_i(t+1) = \delta_i(s_1(t), \cdots, s_\alpha(t), x_1(t), \cdots, x_\beta(t)) & (i = 1, 2, \cdots, \alpha) \\ y_i(t) = \varphi_i(s_1(t), \cdots, s_\alpha(t), x_1(t), \cdots, x_\beta(t)) & (i = 1, 2, \cdots, \gamma) \end{cases} \tag{8}$$

where δ_i and φ_i are Boolean functions: $\delta_i : \{0,1\}^{\alpha+\beta} \to \{0,1\}$ and $\varphi_i : \{0,1\}^{\alpha+\beta} \to \{0,1\}$. It is well known that any Boolean function can be expanded into one of the canonical forms. We represent the Boolean functions δ_i and φ_i in the principal disjunctive canonical form. For simplicity, we introduce new variables z_i, $i = 1, 2, \cdots, n$ $(n = \alpha + \beta)$ defined by $z_1 = s_1, z_2 = s_2, \cdots, z_\alpha = s_\alpha, z_{\alpha+1} = x_1, z_{\alpha+2} = x_2, \cdots, z_n = x_\beta$. Let $Z_1(t), Z_2(t), \cdots, Z_{2^n}(t)$ be defined by $Z_1(t) = z_1(t) \wedge z_2(t) \wedge \cdots \wedge z_{n-1}(t) \wedge z_n(t)$, $Z_2(t) = \bar{z}_1(t) \wedge z_2(t) \wedge \cdots \wedge z_{n-1}(t) \wedge z_n(t)$, $\cdots$, $Z_{2^n}(t) = \bar{z}_1(t) \wedge \bar{z}_2(t) \wedge \cdots \wedge \bar{z}_{n-1}(t) \wedge \bar{z}_n(t)$ which are called the fundamental products of z_1,z_2, $\cdots$, and z_n. We can rewrite (8) as follows.

$$M : \ s_i(t+1) = \bigvee_{j=1}^{2^n} a_{ij} Z_j(t), \qquad y_i(t) = \bigvee_{j=1}^{2^n} b_{ij} Z_j(t) \tag{9}$$

where a_{ij} and b_{ij} are the coefficients of the Boolean functions δ_i and φ_i represented in the principal disjunctive canonical form and they take the values '1' or '0'.

We now discuss the expression of logical operations in (9). Let 'true = 1' and 'false = 0' and define the function $H(\cdot)$ by $H(x) = 1$ for $x \ge 0$ and $H(x) = 0$ for $x < 0$. Then the logical product: $y = x_1 \wedge x_2 \wedge \cdots \wedge x_k$ is given by $y = x_1 x_2 \cdots x_k$, the logical sum: $y = x_1 \vee x_2 \vee \cdots \vee x_k$ is given by $y = H(x_1 + x_2 + \cdots + x_k - \gamma)$ $0 < \gamma \le 1$ and the not: $y = \bar{x}$ is given by $y = 1 - x$.

By using these expressions of the logical operations, (9) can be transformed into the following equations without logical operations.

$$M : \begin{cases} s_i(t+1) = H(\sum_{j=1}^{2^n-1} a^*_{ij} z^*_j(t) + a_{i2^n} - \gamma^q_i), s_i(0) = s_{i0}, \quad i = 1, 2, \cdots, \alpha \\ y_i(t) = H(\sum_{j=1}^{2^n-1} b^*_{ij} z^*_j(t) + b_{i2^n} - \gamma^y_i), \quad i = 1, 2, \cdots, \gamma \end{cases} \tag{10}$$

where $z^*_1(t) = z_1(t) z_2(t) z_3(t) \cdots z_n(t)$, $z^*_2(t) = z_2(t) z_3(t) \cdots z_n(t)$, $z^*_3(t) = z_1(t)$ $z_3(t) \cdots z_n(t)$, $\cdots$, $z^*_{2^n-1}(t) = z_n$. a^*_{ij} and b^*_{ij} are integers, and γ^q_i and γ^y_i are real numbers satisfying $0 < \gamma^q_i \le 1$ and $0 < \gamma^q_i \le 1$.

We now propose a new architecture of neural networks for representing FSA M. The neural network is constructed based on the expression (10) as shown in Fig. 2. Consider the RHONNs described by (3), (4), (5), and (6) where we let

$N_d = \alpha$, $N_s = \gamma$, $M = 2^n - 1$ $(n = \alpha + \beta)$, $L = \beta$, $K = \gamma$ and $S(x) = H(x)$. In the high order units (5), $\boldsymbol{s}$ is defined by $\boldsymbol{s} = [h_1^d, h_2^d, \cdots, h_{N_d}^d, I_1, I_2, \cdots, I_L]^T$. For each z_m^* in (5), we define an index set J_m^* consisting of indexes of the elements of the monomial, that is, $J_1^* = \{1, 2, 3, \cdots, n\}$, $J_2^* = \{2, 3, \cdots, n\}$, $\cdots$, $J_M^* = \{n\}$ and we let $J_m = J_m^*, 1, 2, \cdots, M$ and $d_j(m) = 1, m = 1, 2, \cdots M$. In the external outputs (6), δ_{ki}^d and δ_{ki}^s are chosen as $\delta_{ki}^d = 0, k = 1, 2, \cdots, K, i = 1, 2, \cdots, N_d$ and $\delta_{ki}^s = 1$, for $i = k$, $\delta_{ki}^s = 0$ for $i \neq k, k = 1, 2, \cdots, K$. Note that the RHONNs thus constructed contain the at most n-th order product connections in their high order units. It can be shown that the RHONNs can represent any FSA, by using the expression (10) and letting the states $v_i^d(t)$ of the dynamic neurons, the external inputs I_i and the external outputs O_k be corresponding to the states $q_j(t)$, the inputs $i_l(t)$ and the outputs $o_k(t)$ of FSA.

5 Identification of FSA

5.1 Identification Method

In this section we discuss the identification problem of unknown FSA by using the proposed architecture of neural networks, that is, the RHONNs. We formulate the identification problem of FSA as follows. Given a set of data of input and output sequences of a target FSA, determine values of the parameters of the neural network such that its input and output relation becomes equal to that of the FSA. In the identification problem of FSA it is natural to make the following assumptions.

A1. The set of state symbols and the initial states of FSA are unknown.
A2. The state transition function δ and output function φ of FSA are unknown.
A3. The sets of input symbols Σ and output symbols Δ of FSA are known.
A4. A set of data of input sequences $\{i(t)\}$ and the corresponding output sequences $\{o(t)\}$ are available.

For the identification we can construct the RHONN in the manner discussed in the previous section. Note that, β and γ can be uniquely determined since the number of the input states and the output states of a target FSA are known, on the other hand, α cannot be determined since the number of the states of the FSA is not known. That is to say, the number of the static neurons $N_s = \gamma$, the number of external inputs $L = \beta$ and the number of the external outputs $K = \gamma$ in the RHONN can be determined uniquely, but the number dynamic neurons $N_d = \alpha$ and the number of the high-order units $M = 2^n - 1, (n = \alpha + \beta)$ cannot be uniquely determined. We provide the number of dynamic neurons (so the number of the high order units) large enough so that $N_d \geq \alpha$ of the target FSA.

Define the performance index by $J = \frac{1}{2} \sum_{t=0}^{t_f} \sum_{k=1}^{K} |o_k(t) - O_k(t)|^2$ where $O_k(t)$, $t = 0, 1, 2, \cdots, t_f$ are the output sequences of the RHONN when input sequences $\{i(t), t = 0, 1, 2, \cdots, t_f\}$ are given to its external inputs: $I_l(t) = i_l(t)$. The problem now is reduced to a learning problem of neural networks, that is to finding the value the parameters of the RHONN which minimize the performance index J. Note that the nonlinear function $S(x)(= H(x))$ in the RHNN is not differentiable, which implies that the gradient-based algorithms such as the steepest descent method cannot be applied to the optimization problem. We replace the nonlinear function of each neuron by a smooth sigmoidal function which can approximate it with reasonable accuracy and utilize the gradient-based algorithms. The main problem associated with these algorithms is the computation of the gradients $\partial J/\partial \boldsymbol{\omega}$. The gradients can be calculated by deriving adjoint neural networks of the RHONNs [8].

Note also that the initial values of the dynamic neurons $v_i^d(0)$ can not be given a priori because of the assumption that the initial states of FSA are unknown. Therefore we choose as the learning parameters not only the connection weights and the threshold values but also the initial states of the dynamic neurons. In the RHONN, the learning parameters are w_{im}^d, θ_i^d, w_{im}^s, θ_i^s and $v_i^d(0)$, the total number of which is $2^{(\alpha+\beta)} \times (\alpha+\beta)+\alpha$. In the R2ONN in §4.1, the learning parameters are w_{im}^d, w_{im}^s and $v_i^d(0)$, the total number of which is $rm(r+l)+r$. For example, consider FSA with $|\Sigma| = 2$ and $|\Delta| = 2$. The number of parameters in the RHONN is 9 for $|Q| = 2$, 26 for $|Q| = 3$ and 26 for $|Q| = 4$, on the other hand, that in the R2ONN is 24 for $|Q| = 2$, 33 for $|Q| = 3$ and 60 for $|Q| = 4$. It can be seen that the identifications with the RHONN require less number of parameters, which becomes more remarkable as the number of the states of FSA increases.

5.2 Numerical Experiment of Identification

Here we present experimental results of identification of FSA by the proposed RHONNs. We have performed identification for three examples of FSA. The first one is a simple FSA whose state transition diagram is shown in Fig. 3. This FSA accepts the sequences consisting of only '1'. The number of state symbols of the FSA is one and $\Sigma = \Delta = \{0,1\}$. The second example is a FSA whose state transition diagram is shown in Fig. 4, which accepts the sequence $(10)^*$. The number of the state symbols of the FSA is three and $\Sigma = \Delta = \{0,1\}$. The third one is a FSA whose state transition diagram is shown in Fig. 5. This FSA accepts the sequences which do not include '000'. The number of state symbols of the FSA is four and $\Sigma = \Delta = \{0,1\}$.

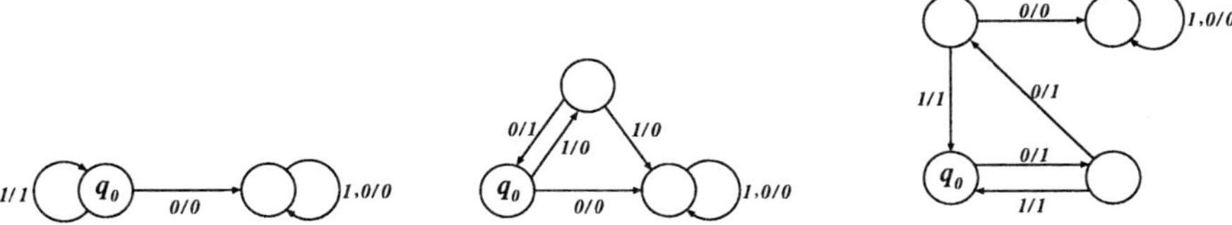

Fig. 3. Example 1: FSA accepting the sequences consisting of only '1'

Fig. 4. Example 2: FSA accepting the sequence $(10)^*$

Fig. 5. Example 3: FSA accepting the sequences that do not include '000'

For identification of these FSAs, we construct the RHONNs in the manner discussed in §4. In all these examples we can choose $L = \beta = 1$ and $K = \gamma = 1$. We have performed the identification experiments several times by changing the number of dynamic neurons $N_d = \alpha$ (so the number of the high order units $M = 2^n - 1, (n = \alpha + \beta)$). In the experiments we use the nonlinear function $S(x) = \frac{1}{1+\exp(-50x)}$ and utilize the quasi-Newton algorithm to search values of the network parameters which minimize the performance index. It is known that FSA with r state symbols is uniquely identified by using all input sequences of length $2r - 1$. We train the RNONN's by using all sequences of length $2r - 1$ as learning data. For each experiment by the RHONN with given α, several learning (optimization) trials have been done by changing the initial values of the learning parameters. Table 1 shows the minimum values of the performance index J among these trials. For each experiment it has been checked that the target FSA is successfully identified with the RHONNs. For comparison, similar identification experiments have been done by using the R2ONNs. The results are shown in Table 2. For each experiment it has been checked that the target FSA is successfully identified by the neural networks. Note that the RHONN with $N_d = 3$ can represent FSA with $r \leq 8$, on the other hand, the R2ONN with $N_d = 5$ can represent FSA with $r \leq 5$. It can be seen from Tables 1 and 2 that the performance of identification with the RHONNs is superior or equivalent to that with the R2ONNs, with the RHONNs being smaller network size.

Table 1. Results of Identification of FSA by RHONN

the number of dynamic neurons in RHONN	the minimum value of J		
	Example 1	Example 2	Example 3
$N_d = 1$	2.7734×10^{-20}		
$N_d = 2$	3.7408×10^{-25}	40294×10^{-17}	9.9975×10^{-17}
$N_d = 3$	1.0122×10^{-15}	1.4543×10^{-9}	3.2567×10^{-10}

Table 2. Results of Identification of FSA by R2ONN

the number of dynamic neurons in R2ONN	the minimum value of J		
	Example 1	Example 2	Example 3
$N_d = 2$	4.3528×10^{-22}		
$N_d = 3$	5.4690×10^{-18}	2.0818×10^{-15}	
$N_d = 4$	3.6260×10^{-17}	1.3158×10^{-14}	1.9092×10^{-14}
$N_d = 5$	1.7123×10^{-10}	2.1379×10^{-15}	1.5884×10^{-15}

6 Conclusions

This paper presented a new architecture of neural networks, RHONNs, for representing deterministic FSA. They consist of two types of neurons, static neurons and dynamic neurons, and high order connection units. The proposed RHONN models are capable of learning and identifying FSA with the network size being smaller than the existing models. It should be noted that the proposed identification method of FSA is efficient and easy to implement because of less number of learning parameters.

References

1. M. L. Minsky:"Computation:Finite and Infinite Machines", Prentice-Hall, New York, 1967.
2. N. Alon, A. K. Dewdney and T. J. Ott: "Efficient Simulation of Automata by Neural Nets", Journal of Association for Computing Machinery, vol.38, No.2, April, pp.495-514, 1991.
3. C. L. Giles, C. B. Miller, D. Chen, H. H. Chen, G. Z. Sun and Y. C. Lee: "Learning and Extracting Finite State Automata with Second-Order Recurrent Neural Networks", Neural Computation, 4, pp.393-405, 1992.
4. Z. Zegn, R. M. Goodman and P. Smyth: "Learning Finite State Machines with Self-Clustering Recurrent Networks", Neural Computation, 5, pp.976-990, 1993.
5. C. L. Giles, D. Chen, G. Sun, H. Chen, Y. Lee and W. Goudreau: "Constructive Learning of Recurrent Neural Networks: Limitations of Recurrent Casade Correlation and a Simple Solution", IEEE Trans. on Neural Networks, Vol.6, No.4, pp.829-836, July, 1995.
6. Y. Kuroe: Representation and Identification of Finite State Automata by Recurrent Neural Networks; Neural Information Processing - ICONIP 2004 Proc., N. R. Pal et. al.(Eds.), Lecture Notes in Computer Science, 3316, pp.261-268, Springer, 2004
7. E. B. Kosmatopoulos, M. M. Polycarpou, M. A. Christodoulou, and P. A. Ioannos, "High-Order Neural Network Structures for Identification of Dynamical Systems" *IEEE Trans. Neural Networks*, vol.6, no.2, pp.422-431, Mar. 1995.
8. Y. Kuroe, H. Ikeda and T. Mori: Identification of Nonlinear Dynamical Systems by Recurrent High-Order Neural Networks; Proc. of IEEE International Conference on Systems, Man, and Cybernetics, Vol.1, pp.70-75, 1997.

Global Stability Conditions of Locally Recurrent Neural Networks*

Krzysztof Patan, Józef Korbicz, and Przemysław Prętki

Institute of Control and Computation Engineering,
University of Zielona Góra
{k.patan, j.korbicz, p.pretki}@issi.uz.zgora.pl

Abstract. The paper deals with a discrete-time recurrent neural network designed with dynamic neural models. Dynamics is reproduced within each single neuron, hence the considered network is a locally recurrent globally feed-forward. In the paper, conditions for global stability of the considered neural network are derived using the pole placement and Lyapunov second method.

1 Introduction

Structures of discrete-time recurrent neural networks have been proved useful in modelling and identification of dynamic processes. However, simulation results report that networks with locally recurrent architectures perform better and converges faster that fully recurrent ones [1,2]. The paper is devoted to the so called Locally Recurrent Globally Feed-forward (LRGF) networks [1,3]. Such networks have a feed-forward multi-layer architecture and their dynamic properties are obtained using a specific kind of neuron models. The neural networks considered are composed of neuron models with an Infinite Impulse Response (IIR) filter [3,4]. Such networks can also be advantageous in terms of stability of learning. In the paper, the global stability of the LRGF networks are derived by using the pole placement for a network with a single hidden layer and by using the global stability theorem of Lyapunov for a network with two hidden layers.

2 Dynamic Neural Networks

The topology of the neural network considered is analogous to the multi-layered feed-forward one and dynamics is reproduced by so called dynamic neuron models. Such neural networks are called locally recurrent globally feed-forward. Dynamic properties of the model are achieved by introducing an Infinite Impulse Response (IIR) filter into a neuron structure. Thus, the states of the i-th neuron in the network can be described by the following state equation:

$$\boldsymbol{x}_i(k+1) = \boldsymbol{A}_i \boldsymbol{x}_i(k) + \boldsymbol{W}_i \boldsymbol{u}(k) \quad (1)$$

* This work was supported by the State Committee for Scientific Research in Poland (KBN) under the grant No. 4T11A01425

W. Duch et al. (Eds.): ICANN 2005, LNCS 3697, pp. 191–196, 2005.

where $\boldsymbol{x}_i(k) = \in \mathbb{R}^r$ is the state vector, $\boldsymbol{W}_i = \mathbf{1}\boldsymbol{w}_i$ is the weight matrix ($\boldsymbol{w}_i \in \mathbb{R}^n$, $\mathbf{1} \in \mathbb{R}^r$ is the vector of ones), $\boldsymbol{u}(k) \in \mathbb{R}^n$ is the input vector, n is the number of inputs, and the state matrix $\boldsymbol{A}_i$ has the form:

$$\boldsymbol{A}_i = \begin{bmatrix} -a_{1_i} & -a_{2_i} & \dots & -a_{r-1_i} & -a_{r_i} \\ 1 & 0 & \dots & 0 & 0 \\ 0 & 1 & \dots & 0 & 0 \\ \vdots & \vdots & \ddots & \vdots & \vdots \\ 0 & 0 & \dots & 1 & 0 \end{bmatrix} \tag{2}$$

Finally, the neuron output is described by:

$$y_i(k) = f\big(g_{2_i}(\boldsymbol{b}_i\boldsymbol{x}_i(k) + \boldsymbol{d}_i\boldsymbol{u}(k)) - g_{1_i}g_{2_i}\big) \tag{3}$$

where $f(\cdot)$ is a non-linear activation function, $\boldsymbol{b}_i = [b_{1_i} \dots b_{r_i}]$ is the vector of feed-forward filter parameters, $\boldsymbol{d}_i = [b_{0_i}w_1 \dots b_{0_i}w_n]$, g_{1_i} and g_{2_i} are the bias and slope of the activation function, respectively.

2.1 State-Space Representation of the Network

One hidden layer network. Let us consider a discrete-time neural network with n inputs and m outputs. A neural model with one hidden layer consisting of v neurons with IIR filters of the order r is described by the following formulae:

$$\begin{cases} \boldsymbol{x}(k+1) = \boldsymbol{A}\boldsymbol{x}(k) + \boldsymbol{W}\boldsymbol{u}(k) \\ \boldsymbol{y}(k) = \boldsymbol{C}\boldsymbol{f}(\boldsymbol{G}_2(\boldsymbol{B}\boldsymbol{x}(k) + \boldsymbol{D}\boldsymbol{u}(k)) - \boldsymbol{G}_2\boldsymbol{g}_1)^T \end{cases} \tag{4}$$

where $N = v \times r$ represents a number of model states, $\boldsymbol{x} \in \mathbb{R}^N$ is the state vector, $\boldsymbol{u} \in \mathbb{R}^n$, $\boldsymbol{y} \in \mathbb{R}^m$ – input and output vectors, respectively, $\boldsymbol{A} \in \mathbb{R}^{N\times N}$ is the block diagonal state matrix ($\mathrm{diag}(\boldsymbol{A}) = [\boldsymbol{A}_1, \dots, \boldsymbol{A}_v]$), $\boldsymbol{W} \in \mathbb{R}^{N\times n}$ and $\boldsymbol{C} \in \mathbb{R}^{m\times v}$ are input and output matrices, respectively, $\boldsymbol{B} \in \mathbb{R}^{v\times N}$ is block diagonal matrix of feed-forward filter parameters ($\mathrm{diag}(\boldsymbol{B}) = [\boldsymbol{b}_1, \dots, \boldsymbol{b}_v]$), $\boldsymbol{D} \in \mathbb{R}^{v\times n}$ is the transfer matrix ($\boldsymbol{D} = [b_{0_1}\boldsymbol{w}_1^T, \dots b_{0_v}\boldsymbol{w}_v^T]^T$), $\boldsymbol{g}_1 = [g_{1_1} \dots g_{1_v}]^T$ denotes the vector of biases, $\boldsymbol{G}_2 \in \mathbb{R}^{v\times v}$ is the diagonal matrix of slope parameters ($\mathrm{diag}(\boldsymbol{G}_2) = [g_{2_1} \dots g_{2_v}]$) and $\boldsymbol{f} : \mathbb{R}^v \to \mathbb{R}^v$ is nonlinear vector valued function.

Two hidden layers network. Let us consider a discrete-time dynamic neural network with n inputs and m outputs. A neural model with two hidden layers with v_1 neurons in the first layer and v_2 neurons in the second layer, each neuron consists of r-th order IIR filter, is described as follows:

$$\begin{cases} \boldsymbol{x}(k+1) = \boldsymbol{g}\left(\boldsymbol{x}(k), \boldsymbol{u}(k)\right) \\ \boldsymbol{y}(k) = \boldsymbol{h}\left(\boldsymbol{x}(k), \boldsymbol{u}(k)\right) \end{cases} \tag{5}$$

where $\boldsymbol{g}$, $\boldsymbol{h}$ are nonlinear functions. Taking into account layered topology of the network, one can decompose the state vector as follows $\boldsymbol{x}(k) = [\boldsymbol{x}^1(k) \quad \boldsymbol{x}^2(k)]^T$, where $\boldsymbol{x}^1(k) \in \mathbb{R}^{N_1}$ ($N_1 = v_1 \times r$) represents states of the first layer, and $\boldsymbol{x}^2(k) \in$

$\mathbb{R}^{N_2}$ $(N_2 = v_2 \times r)$ represents states of the second layer, the state equation can be rewritten in the form:

$$\boldsymbol{x}^1(k+1) = \boldsymbol{A}^1\boldsymbol{x}^1(k) + \boldsymbol{W}^1\boldsymbol{u}(k) \tag{6a}$$

$$\boldsymbol{x}^2(k+1) = \boldsymbol{A}^2\boldsymbol{x}^2(k) + \boldsymbol{W}^2\boldsymbol{f}\big(\boldsymbol{G}_2^1(\boldsymbol{B}^1\boldsymbol{x}^1(k) + \boldsymbol{D}^1\boldsymbol{u}(k)) - \boldsymbol{G}_2^1\boldsymbol{g}_1^1)\big) \tag{6b}$$

where $\boldsymbol{u} \in \mathbb{R}^n$, $\boldsymbol{y} \in \mathbb{R}^m$ are inputs and outputs, respectively, matrices $\boldsymbol{A}^1 \in \mathbb{R}^{N_1 \times N_1}$, $\boldsymbol{B}^1 \in \mathbb{R}^{v_1 \times N_1}$, $\boldsymbol{W}^1 \in \mathbb{R}^{N_1 \times n}$, $\boldsymbol{D}^1 \in \mathbb{R}^{v_2 \times n}$, $\boldsymbol{g}_1^1 \in \mathbb{R}^{v_1}$, $\boldsymbol{G}_2^1 \in \mathbb{R}^{v_1 \times v_1}$ have the form analogous to matrices describing network with one hidden layer, $\boldsymbol{A}^2 \in \mathbb{R}^{N_2 \times N_2}$ is the block diagonal state martix of the second layer $(\mathrm{diag}(\boldsymbol{A}^2) = [\boldsymbol{A}_1^2, \ldots, \boldsymbol{A}_{v_2}^2])$, $\boldsymbol{W}^2 \in \mathbb{R}^{N_2 \times v_1}$ is the weight matrix between the first and second hidden layers. Finally, output of the model is represented by the equation:

$$\boldsymbol{y}(k) = \boldsymbol{C}^2\boldsymbol{f}\big(\boldsymbol{G}_2^2(\boldsymbol{B}^2\boldsymbol{x}^2(k) + \boldsymbol{D}^2\boldsymbol{f}\big(\boldsymbol{G}_2^1(\boldsymbol{B}^1\boldsymbol{x}^1(k) + \boldsymbol{D}^1\boldsymbol{u}(k)) - \boldsymbol{G}_2^1\boldsymbol{g}_1^1) - \boldsymbol{G}_2^2\boldsymbol{g}_1^2\big) \tag{7}$$

where $\boldsymbol{C}^2 \in \mathbb{R}^{m \times v_2}$ is the output matrix, $\boldsymbol{B}^2 \in \mathbb{R}^{v_2 \times N_2}$ is the block diagonal matrix of the second layer feed-forward filter parameters, $\boldsymbol{D}^2 \in \mathbb{R}^{v_2 \times v_1}$ is the transfer matrix of the second layer, $\boldsymbol{g}_1^2 \in \mathbb{R}^{v_2}$ is the vector of the second layer biases, $\boldsymbol{G}_2^2 \in \mathbb{R}^{v_2 \times v_2}$ represents the diagonal matrix of the second layer activation function slope parameters. Matrices $\boldsymbol{B}^2$, $\boldsymbol{D}^2$, $\boldsymbol{g}_1^2$ and $\boldsymbol{G}_2^2$ have the form analogous to matrices of the first hidden layer.

3 Stability Analysis

One hidden layer network. It is well known that linear discrete time system is stable iff all its poles z_i are inside unit circle:

$$\forall i \quad |z_i| < 1 \tag{8}$$

As one can see in (4), the state equation in linear. Thus, the system (4) is stable iff roots of the characteristic equation satisfy (8). In a general case, a characteristic equation can has quite complex form and analytical computation of roots can be extremely difficult. In a considered case hovewer, the state equation is linear with the block-diagonal matrix $\boldsymbol{A}$, what makes the consideration about stability relatively easier. The determinant of a block-diagonal matrix can be represented as a product of determinants of matrices placed on the main diagonal. Using this property, the characteristic equation of (4) can be rewritten in the following way:

$$\prod_{i=1}^{v} \det(\boldsymbol{A}_i - \boldsymbol{I}z_i) = 0 \tag{9}$$

where $\boldsymbol{I} \in \mathbb{R}^{r \times r}$ is the identity matrix, z_i represents the poles of the i-th neuron. Thus, from (9) one can determine poles of (4) solving a set of equations:

$$\forall i \quad \det(\boldsymbol{A}_i - \boldsymbol{I}z_i) = 0 \tag{10}$$

From above analysis one can conclude that poles of the i-th subsystem (i-th dynamic neuron) can be calculated separately. Finally, it can be stated that *If all neurons in the networks are stable, the whole neural network model is stable.* If during training the poles will be kept inside unit circle, the stability of the neural model will be guaranteed. The main problem now is how one can elaborate a method of keeping poles inside unit circle during neural network training. This problem can be solved by deriving a region of admissible values of filter parameters [5].

Two hidden layers network. In this case the state equation has a nonlinear form. From decomposed states equation (6) it is clearly seen, that states of the first layer of the network are independent on the states of the second layer, and have a linear form (6a). States of the second layer are described by nonlinearity (6b). Let us observe once again that states of a neuron of the second layer are independent on the states of other neurons the same layer. Follows this fact, (6b) can be rewritten as follows:

$$\boldsymbol{x}_i^2(k+1)=\boldsymbol{A}_i^2\boldsymbol{x}_i^2(k)+\boldsymbol{W}_i^2\boldsymbol{f}\big(\boldsymbol{G}_2^1(\boldsymbol{B}^1\boldsymbol{x}^1(k)+\boldsymbol{D}^1\boldsymbol{u}(k)-\boldsymbol{g}_1^1)\big) \text{ for } i=1,\ldots,v_2 \quad (11)$$

Let $\boldsymbol{\Psi} = \boldsymbol{G}_2^1\boldsymbol{B}^1$ and $\boldsymbol{s}_1 = \boldsymbol{G}_2^1\boldsymbol{D}^1\boldsymbol{u}(k) - \boldsymbol{G}_2^1\boldsymbol{g}_2^1$ where $\boldsymbol{s}_1$ can be regarded as threshold or fixed input, then (11) takes the form

$$\boldsymbol{x}_i^2(k+1) = \boldsymbol{A}_i^2\boldsymbol{x}_i^2(k) + \boldsymbol{W}_i^2\boldsymbol{f}\big(\boldsymbol{\Psi}\boldsymbol{x}^1(k) + \boldsymbol{s}_1\big) \quad (12)$$

using a linear transformation $\boldsymbol{y}^1(k) = \boldsymbol{\Psi}\boldsymbol{x}^1(k) + \boldsymbol{s}_1$, and $\boldsymbol{y}^2(k) = \boldsymbol{x}_i^2(k)$ one obtain an equivalent system:

$$\begin{cases} \boldsymbol{y}^1(k+1) = & \boldsymbol{\Psi}\boldsymbol{A}^1\boldsymbol{\Psi}^-\boldsymbol{y}^1 - \boldsymbol{\Psi}\boldsymbol{A}^1\boldsymbol{\Psi}^-\boldsymbol{s}_1 + \boldsymbol{s}_2 \\ \boldsymbol{y}^2(k+1) = & \boldsymbol{A}_i^2\boldsymbol{y}^2(k) + \boldsymbol{W}_i^2\boldsymbol{f}(\boldsymbol{y}^1(k)) \end{cases} \quad (13)$$

where $\boldsymbol{\Psi}^-$ is a pseudoinverse of the matrix $\boldsymbol{\Psi}$ (e.g. in a Moore-Penrose sense) and $\boldsymbol{s}_2 = \boldsymbol{W}^1\boldsymbol{u}(k)$ is the threshold or fixed input. Let $\boldsymbol{y}^* = [\boldsymbol{y}^{1*}\ \boldsymbol{y}^{2*}]^T$ be an equilibrium point of the (13) ($\boldsymbol{y}^* = \boldsymbol{f}(\boldsymbol{y}^*)$). Introducing an equivalent coordinate transformation $\boldsymbol{z}(k) = \boldsymbol{y}(k) - \boldsymbol{y}^*(k)$, the system (13) can be transformed to the form:

$$\begin{cases} \boldsymbol{z}^1(k+1) = & \boldsymbol{\Psi}\boldsymbol{A}^1\boldsymbol{\Psi}^-\boldsymbol{z}^1(k) \\ \boldsymbol{z}^2(k+1) = & \boldsymbol{A}_i^2\boldsymbol{z}^2(k) + \boldsymbol{W}_i^2\boldsymbol{\sigma}(\boldsymbol{z}^1(k)) \end{cases} \quad (14)$$

where $\boldsymbol{\sigma}(\boldsymbol{z}(k)) = \boldsymbol{f}(\boldsymbol{z}(k)+\boldsymbol{y}^*(k)) - \boldsymbol{f}(\boldsymbol{y}^*(k))$. Substituting $\boldsymbol{z}(k) = [\boldsymbol{z}^1(k)\,\boldsymbol{z}^2(k)]^T$ finally one obtain

$$\boldsymbol{z}(k+1) = \mathcal{A}_i\boldsymbol{z}(k) + \mathcal{W}_i\boldsymbol{\sigma}(\boldsymbol{z}(k)) \quad (15)$$

where

$$\mathcal{A}_i = \begin{bmatrix} \boldsymbol{\Psi}\boldsymbol{A}^1\boldsymbol{\Psi}^- & 0 \\ 0 & \boldsymbol{A}_i^2 \end{bmatrix} \quad \mathcal{W}_i = \begin{bmatrix} 0 & 0 \\ \boldsymbol{W}_i^2 & 0 \end{bmatrix} \quad (16)$$

In order to obtain stability conditions for the system (15), one use the second Lyapunov method [6].

Let $\boldsymbol{V}(\boldsymbol{z}) = \|\boldsymbol{z}\|$ be a Lyapunov function for the system (15). This function is positive definite with minimum at $\boldsymbol{x}(k) = 0$. The difference along the trajectory of the system is given as follows:

$$\begin{aligned} \Delta \boldsymbol{V}(\boldsymbol{z}(k)) &= \|\boldsymbol{z}(k+1)\| - \|\boldsymbol{z}(k)\| = \|\mathcal{A}_i \boldsymbol{z}(k) + \mathcal{W}_i \boldsymbol{\sigma}(\boldsymbol{z}(k))\| - \|\boldsymbol{z}(k)\| \\ &\leqslant \|\mathcal{A}_i \boldsymbol{z}(k)\| + \|\mathcal{W}_i \boldsymbol{\sigma}(\boldsymbol{z}(k))\| - \|\boldsymbol{z}(k)\| \end{aligned} \tag{17}$$

Using a property of an activation function $\|\boldsymbol{\sigma}(\boldsymbol{z}(k))\| \leqslant \|\boldsymbol{z}(k)\|$ [6], (17) can be expressed as follows:

$$\Delta \boldsymbol{V}(\boldsymbol{z}(k)) \leqslant \|\mathcal{A}_i\| \|\boldsymbol{z}(k)\| + \|\mathcal{W}_i\| \|\boldsymbol{z}(k)\| - \|\boldsymbol{z}(k)\| \leqslant (\|\mathcal{A}_i\| + \|\mathcal{W}_i\| - 1) \|\boldsymbol{z}(k)\| \tag{18}$$

From (18) one can see that if

$$\|\mathcal{A}_i\| + \|\mathcal{W}_i\| < 1 \tag{19}$$

then $\Delta \boldsymbol{V}(\boldsymbol{z}(k))$ is negative definite and system (15) is globally asymptotically stable. It is easy to see that this result can be expanded to the entire network. Thus, the system (6) is globally asymptotically stable if conditions

$$\|\mathcal{A}_i\| + \|\mathcal{W}_i\| < 1 \quad \text{for } i = 1, \ldots, v_2 \tag{20}$$

are satisfied.

4 Experiment

The main objective of the experiment is to show that stability of the network may have a crucial influence on training quality. The network with a single hidden layer consisting of 3 dynamic neurons with second order IIR filters and hyperbolic tangent activation functions has been trained off-line by using the SPSA method [3] for 500 iterations using 100 learning patterns. The process to be identified is a proces described in Experiment 4 in the outstanding paper of Narendra and Parthasarathy [7]. The results of training are presented in Fig. 1. Results for the unstable model are depicted in Fig.1 a), c) and e) whilst results for the stable one in Fig. 1 b), d) and f), respectively. As one can observe, training in both cases is convergent Fig.1 e) and f). However, the first model is unstable (states are divergent - Fig. 1 a)) and generalization properties are very poor (Fig. 1 c)). In turn, states of the stable neural model are depicted in Fig. 1 b) and testing of this network in Fig. 1 d). For this network modelling results are much better. This simple experiment shows, that the stability problem is of a crucial importance and should be taken into account during training, otherwise the obtained model may be improper.

5 Concluding Remarks

The paper presents the stability conditions for locally recurrent globally feed-forward neural networks composed of dynamic neuron models with IIR filters.

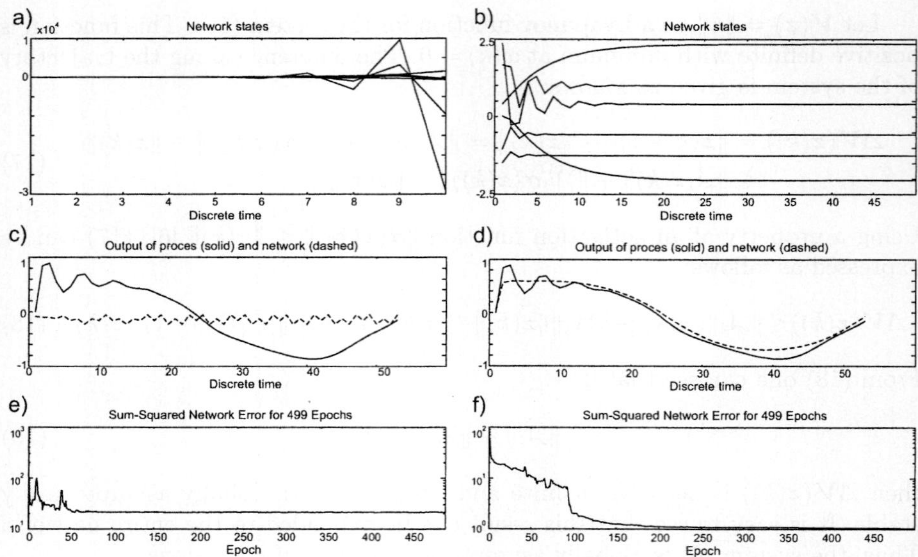

Fig. 1. Result of the experiment

For networks with single hidden layer the stability considerations rely on the classical stability theory for linear systems. In the case of networks with two hidden layers the stability conditions were derived using the second Lyapunov method. The further work will focus on extension of achieved results in order to obtain feasible regions of the network parameters and on this basis elaborating a training procedure which guarantee the stability of the network during training.

References

1. Tsoi, A.C., Back, A.D.: Locally recurrent globally feedforward networks: A critical review of architectures. IEEE Transactions on Neural Networks **5** (1994) 229–239
2. Gori, M., Bengio, Y., Mori, R.D.: BPS: A learning algorithm for capturing the dynamic nature of spech. In: International Joint Conference on Neural Networks. Volume II. (1989) 417–423
3. Patan, K., Parisini, T.: Identification of neural dynamic models for fault detection and isolation: the case of a real sugar evaporation process. Journal of Process Control **15** (2005) 67–79
4. Korbicz, J., Patan, K., Obuchowicz, A.: Dynamic neural networks for process modelling in fault detection and isolation systems. International Journal of Applied Mathematics and Computer Science **9** (1999) 519–546
5. Patan, K.: Training of the dynamic neural networks via constrained optimization. In: Proc. IEEE Int. Joint Conference on Neural Networks, IJCNN 2004, Budapest, Hungary. (2004) published on CD-ROM.
6. Gupta, M.M., Jin, L., Homma, N.: Static and Dynamic Neural Neetworks. From Fundamentals to Advanced Theory. John Wiley & Sons, New Jersey (2003)
7. Narendra, K.S., Parthasarathy, K.: Identification and control of dynamical systems using neural networks. IEEE Transactions on Neural Networks **1** (1990) 12–18

An Agent-Based PLA for the Cascade Correlation Learning Architecture

Ireneusz Czarnowski and Piotr Jędrzejowicz

Department of Information Systems, Gdynia Maritime University,
Morska 83, 81-225 Gdynia, Poland

Abstract. The paper proposes an implementation of the agent-based population learning algorithm (PLA) within the cascade correlation (CC) learning architecture. The first step of the CC procedure uses a standard learning algorithm. It is suggested that using the agent-based PLA as such an algorithm could improve efficiency of the approach. The paper gives a short overview of both - the CC algorithm and PLA, and then explains main features of the proposed agent-based PLA implementation. The approach is evaluated experimentally.

1 Introduction

In order to obtain good generalization performance when training a neural network, it must have the right size. Networks that are too small cannot represent the required function, while networks that are too large are prone to overfitting. Apart from choosing the network architecture one has also to decide on learning algorithm to be used for the network training. Commonly used gradient-based algorithms face a danger of being caught in a local optimum. Besides, larger networks require a lot of computational time. Since choosing an architecture and a learning algorithm are interrelated, it seems only logical to consider them simultaneously. The first who came up with such an idea were Fahlman and Lebiere proposing the cascade correlation learning architecture [3], which in fact allows for both i.e. building a network and training its neurons.

In this paper we propose developing a neural network architecture using the cascade correlation (CC) procedure and applying the agent-based population learning algorithm (PLA) as an embedded learning algorithm.

The paper is organized as follows. Sections 2 and 3 contain a short review of the CC procedure and the PLA, respectively. Section 4 presents, in a more detailed manner, the suggested agent-based PLA implementation. Section 5 gives an overview of the computational experiment results. The experiment has been carried with a view to validate the approach. Finally, in Section 5 some conclusions are drawn and directions of further research are suggested.

2 The Cascade Correlation Algorithm

The Cascade Correlation approach is a simple and powerful method for training a neural network [3]. It belongs to a class of the supervised learning algo-

W. Duch et al. (Eds.): ICANN 2005, LNCS 3697, pp. 197–202, 2005.

rithms. The CC network determines its own size and topology starting with an input/output layer and building a minimal multi-layer network by creating its own hidden layer. The procedure is iterative and based on recruiting new units according to the residual approximation error. Iterations involve:

- Training and adding hidden units one by one to tackle new tasks, hence *"Cascade "*,
- Maximizing the residual error. *"Correlation"* between the new units and its output is maximized,
- Input weights going into the new hidden unit become frozen (fixed).

CC combines two ideas - cascade architecture where hidden units are added one at a time and frozen, and learning algorithm which trains and installs new hidden units. The CC algorithm starts with a minimal network consisting of an input and output layer. The network is trained with a learning algorithm (i.e. gradient descent or some metaheuristic-based algorithm like simulating annealing) until no significant error reduction can be observed. Hidden units are added one by one to the network which is connected by all input units and to every pre-existing hidden unit with all input weights of the hidden unit becoming frozen. The cycle is repeated until desired performance is reached or no further improvement is possible.

The CC has been used successfully in many real-life applications although its potential problem is convergence to suboptimal local optimum solutions in the correlation phase of network training [5].

3 Population Learning Algorithm

Population Learning Algorithm, introduced in [2], in contrast to many other population-based methods, has been inspired by analogy to a social phenomenon rather than to evolutionary processes.

In the PLA an individual represents a coded solution of the considered problem. Initially, a number of individuals, known as the initial population, is generated. Once the initial population has been generated, individuals enter the first learning stage. It involves applying some, possibly basic and elementary, improvement schemes. The improved individuals are then evaluated and better ones pass to a subsequent stage. A strategy of selecting better or more promising individuals must be defined and applied. In the following stages the whole cycle is repeated. Individuals are subject to improvement and learning, either individually or through information exchange, and the selected ones are again promoted to a higher stage with the remaining ones dropped-out from the process. At the final stage the remaining individuals are reviewed and the best represents a solution to the problem at hand.

The PLA can be also run as a parallel algorithm. The parallel PLA requires a scheme for grouping individuals at various stages, and the respective rules for carrying parallel computations. The parallel implementation can be based on information exchange between the individuals or groups of individuals during the learning process. Potential benefits of the parallel PLA include a reduction of the computation time and improvement of efficiency through better exploration of the solution space and more effective avoidance of the local optima traps. Even more flexibility offers the agent-based PLA, where multiple agents execute population learning schemes using identical or different learning and improvement procedures and each taking care of a population of individuals. Agents act in parallel and independently. A supervisory agent can be added to supervise and coordinate information exchange between learning agents.

4 Agent-Based PLA Implementation

The paper investigates using an implementation of the agent-based population-learning algorithm to train artificial neural networks with a CC architecture. The approach has been motivated by promising results obtained when applying the PLA to training MLP networks [1]. It was observed that using the PLA as a training algorithm has some important advantages over standard gradient-based methods. First of all, the PLA is less likely to get trapped in a local optimum since it is operating on a set of solutions and not on a single solution. Second, using the PLA does not place any constraints on the neuron transfer function, whereas for standard back propagation methods this function must be differentiable. Third, using the PLA does not require backward propagation of an error signal. Hence, network topology does not impose any restrictions on the PLA implementation. Advantages of the PLA are counter measured by one significant drawback. Obtaining high quality results requires an initial population of a substantial size, which, in turn, results in an excessive computation time. To overcome the problem an agent-based approach with a set of training agents managing a set of relatively small populations of solutions is suggested. Each training agent is using an identical PLA scheme which has been implemented basing on the following assumptions:

- An individual is a vector of real numbers from the predefined interval, each representing a value of weight of the respective link between neurons in the considered neural network,
- The initial population of individuals is generated randomly,
- There are five learning/improvement procedures - standard mutation, local search, non-uniform mutation, gradient mutation and application of the gradient adjustment operator. Number of iterations for each procedure is set by the respective agent,
- There is a common selection criterion for all learning stages. At each stage, individuals with fitness below the current average are rejected.

In addition to training agents there is also a supervisory agent performing two main tasks. The first is the construction of the neural network architecture and the second - is the coordination of information exchange between training agents.

Supervisor role is to initiate a neural network consisting of input and output layers. The number of neurons in each layer is equal to the number of attributes characterizing data elements from the training set. Within the first learning cycle, such a network is trained to minimize the classification error. Training is based on the earlier described PLA scheme. Next, candidate neurons are independently trained by learning agents and the respective values of their input weights are established using again the same PLA structure. The number of candidate neurons determines the number of learning agents used. This time, however, optimization criterion is the value of the covariance calculated for each candidate neuron. Candidate with the highest covariance value is selected and added to the network. Following it, values of all inputs leading to neurons in the output layer are established. To do it agents use only the gradient adjustment procedure, which is the last one from the earlier proposed PLA. The initial population of individuals is then generated by producing multiple copies from a template which in this case is a vector of network weights produced at the previous stage. Population members are then randomly diversified. Weights of inputs to the output layer neurons are generated randomly according to the uniform distribution from the interval $(0, -K)$, where K is the maximum covariance obtained at the previous stage. The method of initializing values of the hidden unit input connections is adopted from [5].

Training neural network of the CC architecture is performed by the Cascade-PLA which is an agent based software environment managing and performing the above described training cycles consisting of candidate neurons training and determining values of input weights for the output layer neurons in accordance with the above described procedure, until some stopping criterion is met. General idea of the Cascade-PLA is shown in Fig. 1.

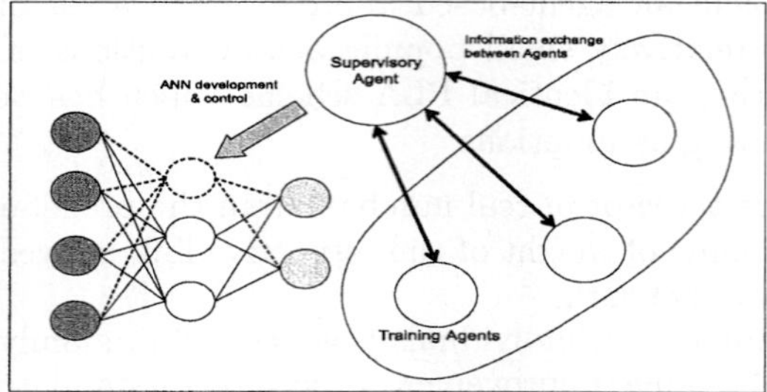

Fig. 1. General idea of the Cascade-PLA multi-agent environment

As it can be seen the proposed solution belongs to the class where each agent manages a group of individuals. An agent is represented by a code, which is executed in the parallel environment and the multi-agent system is implemented as a set of interacting and co-operating training agents. Interaction and co-operation is controlled by the supervisory agent. Supervisory agent manages communication flow, sets work parameters of training agents, initiates execution of tasks by training agents, receives solution values at each computation stage from each agent, compares solutions and chooses the best one, sends best solutions to agents, stops computations and decides on adding a hidden layer neuron to the network under development. A training agent executes training procedures set by the supervisory agent, sends solution values to the supervisory agent and modifies the population of solutions by replacing the worst individual by currently the best one.

5 Computational Experiment Results

The proposed environment has been used to train the CC architecture neural networks applied to solving two popular benchmarking classification problems - Cleveland heart disease (303 instances, 13 attributes, 2 classes) and Credit approval (690, 15, 2). Both datasets were obtained from UCI Machine Learning Repository [4]. Experiments have been preceded by data cleaning and data normalisation procedures.

Computational experiment carried to evaluate the proposed implementation has been based on the correct classification ratio for the 10-cross-validation approach. Each benchmarking problem has been solved 30 times and the reported values of the quality measures have been averaged over all runs. The resulting accuracy of classification for the Credit problem was 85.1% with 18 ± 11 hidden neurons and 87.6% with 10 ± 2 hidden units for the Heart problem. In Table 1 these results are compared with other reported in the literature.

Table 1. Cascade-PLA versus the reported accuracy of training algorithms (source for the best reported: http://www.phys.uni.torun.pl/kmk/projects/datasets.html and http://www.phys.uni.torun.pl/kmk/projects/datasets-stat.html)

Approach	Credit		Heart	
	Literature reported accuracy (%)	Accuracy of cascade-PLA (%)	Literature reported accuracy (%)	Accuracy of cascade-PLA (%)
MLP+PLA	86.6 [1]	85.1	86.5 [1]	87.6
MLP+BP	84.6		81.3	
C 4.5	85.5		77.8	
CART	85.5		80.8	
kNN	86.4		85.1	

To better evaluate the agent-based Cascade-PLA it has been decided to develop the Cascade Correlation network architecture trained by a standard back propagation algorithm (CC). Additionally, a non-cascade MLP network trained by a standard back propagation algorithm (BackPro) has been also developed. Its architecture has been determined by a trial and error procedure. The results obtained by applying these algorithms using identical hardware platform are shown in Table 2.

All compared algorithms assume a sigmoid activation function with $\beta = 1$. The Cascade-PLA employed 5 learning agent each managing 10 individuals and

Table 2. Cascade-PLA versus the computational experiments results of CC and Back-Pro

Problem	Accuracy (%)			Training time (sec.)		
	cascade-PLA	CC	BackPro	cascade-PLA	CC	BackPro
Credit	85.1	82.6	81.2	246	377	580
Heart	87.6	81.8	76.4	108	87	332

executing 60 training procedure iterations. The CC algorithm generated results with 7 ± 2 hidden neurons for Heart problem and 12 ± 4 hidden neuron units for Credit problem. In case of BackPro the network structure was (15, 15, 1) and (13, 13, 1) for Credit and Heart problems, respectively. BackPro stopped after 5000 epochs.

6 Conclusion

It can be concluded that the proposed agent-based population learning algorithm is a successful and, possibly, a competitive tool as applied to training the cascade correlation architecture neural networks. Future research should concentrate on finding critical features of the proposed agent-based environment and on developing and implementing within such an environment even more effective training procedures.

References

1. Czarnowski, I., P. Jędrzejowicz: An Approach to Artificial Neural Network Training. In: Bramer, M., Preece, A., Coenen, F., ed., "Research and Development in Intelligent Systems XIX", Springer, London (2003) 149–162
2. Jędrzejowicz, P.: Social Learning Algorithm as a Tool for Solving Some Difficult Scheduling Problems. Foundation of Computing and Decision Sciences, **24** (1999) 51–66
3. Fahlman, S.E., Lebiere, C.: The Cascade-Correlation Learning Architecture. Technical Report CMU-CS-90-100, School of Computer Science, Carnegie Mellon University, Pittsburgh (1990)
4. Merz, C.J., Murphy, P.M.: UCI Repository of Machine Learning Databases [http://www.ics.uci.edu/∼ mlearn/MLRepository.html.] Irvine, CA: University of California, Department of Information and Computer Science (1998)
5. Potter, M.A.: A Genetic Cascade-Correlation Learning Algorithm. In: Proceedings of COGANN 92, International Workshop on Combinations of Genetic Algorithms and Neural networks, IEEE Computer Society Press (1992) 123–133

Dual Memory Model for Using Pre-existing Knowledge in Reinforcement Learning Tasks

Kary Främling

Helsinki University of Technology, PL 5500, FI-02015 TKK, Finland
Kary.Framling@hut.fi

Abstract. Reinforcement learning agents explore their environment in order to collect reward that allows them to learn what actions are good or bad in what situations. The exploration is performed using a policy that has to keep a balance between getting more information about the environment and exploiting what is already known about it. This paper presents a method for guiding exploration by pre-existing knowledge expressed as heuristic rules. A dual memory model is used where the value function is stored in long-term memory while the heuristic rules for guiding exploration act on the weights in a short-term memory. Experimental results from a grid task illustrate that exploration is significantly improved when appropriate heuristic rules are available.

1 Introduction

Supervised and unsupervised learning construct models that represent training samples given by a "teacher" as well as possible. Reinforcement learning (RL) methods differ from these learning methods because the RL agent has to explore its environment by itself and collect training samples. Initially the agent has to take actions without knowing how good or bad they are, which may be known only much later when a reward signal is provided. The agent takes actions following a policy, usually referred to by the symbol π that should make it explore the environment efficiently enough to learn at least a near-optimal policy.

This paper presents a new method for using pre-existing knowledge for state-space exploration. It uses a dual memory model where a so-called long-term memory is used for action-value learning and a short-term memory is used for modifying action selection by heuristic rules. Experimental results with simple heuristic rules show that it is possible to converge to a good policy with less exploration than with some well-known methods.

After this introduction, the most relevant RL methods to the scope of this paper are described in Section 2. Section 3 presents methods to guide exploration, while Section 4 shows comparative test results for grid world tasks, followed by conclusions.

2 Reinforcement Learning Principles

One of the main domains treated by RL is Markov Decision Processes (MDPs). A (finite) MDP is a tuple $M=(S,A,T,R)$, where: S is a finite set of states; $A = \{a_1, \ldots, a_k\}$

W. Duch et al. (Eds.): ICANN 2005, LNCS 3697, pp. 203 – 208, 2005.

is a set of $k \geq 2$ actions; $T = [P_{sa}(\cdot) \mid s \in S, a \in A\}$ are the next-state transition probabilities, with $P_{sa}(s')$ giving the probability of transitioning to state s' upon taking action a in state s; and R specifies the reward values given in different states $s \in S$. RL methods try to learn a value function that allows them to predict future reward from any state when following a given action selection policy π. Value functions are either *state-values* (i.e. value of a state) or *action-values* (i.e. value of taking an action in a given state). Action-values are denoted $Q(s,a)$, where $a \in A$.

The currently most popular RL methods are so-called *temporal difference* (TD) methods. *Q-learning* is a TD control algorithm that updates action values according to

$$\Delta Q(s_t, a_t) = \beta \left[r_{t+1} + \gamma \max_a Q(s_{t+1}, a) - Q(s_t, a_t) \right] e_{t+1}(s) \tag{1}$$

where $Q(s_t, a_t)$ is the value of action a in state s at time t, β is a learning rate, r_{t+1} is the immediate reward and γ is a discount factor that determines to what degree future rewards affect the value. The max-operator signifies the greatest action value in state s_{t+1}. $e_{t+1}(s)$ is an eligibility trace that allows rewards to be propagated back to preceding states.

Exploration of the state space is performed by *undirected* or *directed exploration* methods [3]. Undirected exploration methods do not use any task-specific information, e.g. *ε-greedy exploration* (take greedy action with probability (1-ε) and an arbitrary action with probability ε) and *Boltzmann* action selection (selection probabilities proportional to action values). Directed exploration methods use task-specific knowledge for guiding exploration in such a way that the state space would be explored more efficiently. *Counter-based* methods direct exploration towards states that were visited least frequently in the past, *recency-based exploration* prefers states that were visited least recently while other directed exploration methods use confidence estimates for the current value function [2]. The well-known technique *optimistic initial values* uses initial value function estimates that are bigger than the expected ones, therefore privileging unexplored actions (and states). Initializing action values to zero and giving negative reward for every action are often used to implement this technique [1].

3 SLAP Reinforcement and the BIMM Network

This section describes the *bi-memory model* (BIMM) that uses a *short-term memory* (STM) for controlling exploration and a *long-term memory* (LTM) for action-value learning. Both memories use linear function approximation by the one-layer Adaline artificial neural net (ANN) [4] but any function approximator could be used for both STM and LTM. Weights are stored in a two-dimensional matrix of size $P \times N$, where P is the number of actions and N is the number of state variables. If only one state variable is allowed to be one while all others are zero this representation is identical to the lookup-table representation usually used in discrete RL tasks. A major advantage of using a function approximator approach over lookup-tables is that they can handle continuous-valued state variables. BIMM outputs are calculated according to

$$a_j(s) = K_0 \sum_{i=1}^{N} ltw_{i,j} s_i + K_1 \sum_{i=1}^{N} stw_{i,j} s_i \tag{2}$$

where $a_j(s)$ is the estimated action-value and K_0 and K_I are positive constants that control the balance between exploration and exploitation as in [2]. $ltw_{i,j}$ is the LTM weight and $stw_{i,j}$ is the STM weight for action neuron j and input i. Q-learning or some other action-value learning method can be used to update LTM weights by replacing Q with ltw in equation (1). STM is an exploration bonus that decreases or increases the probability of an action being selected in a given state and whose influence on action selection is determined by the value of K_I. In the tests performed in this paper, action probabilities are only decreased by the *SLAP* (Set Lower Action Priority) method according to heuristic rules (see test section). SLAP updates STM weights using the Widrow-Hoff rule with the target value

$$a_j'(s) = a_{min}(s) - margin \tag{3}$$

where $a_{min}(s)$ is the smallest $a_j(s)$ value in state s[1]. Only STM weights are modified by the Widrow-Hoff rule, which becomes

$$stw_{i,j}^{new} = stw_{i,j} + \alpha(a_j' - a_j)s_i \ . \tag{4}$$

where α is the learning rate The new activation value is

$$\begin{aligned} a_j^{new}(s) &= K_0\sum_{i=1}^{N} ltw_{i,j}s_i + K_1\sum_{i=1}^{N} stw_{i,j}^{new}s_i = \\ K_0\sum_{i=1}^{N} ltw_{i,j}s_i &+ K_1\sum_{i=1}^{N} s_i\left(stw_{i,j} + \alpha(a_j' - a_j)s_i\right) = a_j(s) + \alpha\sum_{i=1}^{N} s_i^2 K_1(a_j' - a_j) \end{aligned} \tag{5}$$

where we can see that setting α to $1/(\Sigma s_i^2 K_I)$ guarantees that a_j^{new} will become a_j' in state s after SLAPing action j. This is a generalization of the well-known Normalized Least Mean Squares (NLMS) method, so from now on α will systematically refer to α_{norm} in BIMM. In stochastic tasks, α should be inferior to $1/(\Sigma s_i^2 K_I)$ because even the optimal action may not always be successful, so immediately making it the least attractive is not a good idea. A general algorithm for using SLAP in a learning task is given in Fig. 1.

```
Initialize parameters
REPEAT (for each episode)
  s ← initial state of episode
  REPEAT (for each step in episode)
    a  ←  action given by π for s
    Take action a, observe next state s'
    SLAP "undesired" actions
    Update action-value function in LTM
    s ← s'
```

Fig. 1. General algorithm for using SLAP in typical RL task

[1] The *margin* should have a "small" value that ensures that an action that is repeatedly SLAPed will eventually have the lowest action value. 0.1 has been used in all tests.

3.3 Increasing Exploration

As shown for the experimental tasks in section 4, heuristic rules can make exploration faster but they may not guarantee a sufficient exploration of the state space. Using an undirected exploration method in addition to the heuristic rules can compensate for this. In the tests reported in this paper, STM weights have been initialized to random values in the interval [0,1) while LTM weights are initialized to zero. Therefore actions will be selected in a random order as long as LTM weights remain zero and no action has been SLAPed. When LTM weights become none-zero, the degree of randomness depends on the value of K_l. Setting K_l to a small value will give no or little randomness (10^{-6} has been used in the tests of section 4) while a greater value will give more randomness. In deterministic tasks it is easy to show that setting $K_l = 1.0$ and always SLAPing the greedy action from equation (2) will perform a depth-first search of the state space.

4 Experimental Results

This section compares different methods on a typical maze task (Fig. 2) with four possible actions. Both deterministic and stochastic state transition rates 0.2 and 0.5 (the probability of another direction being taken than the intended one) are tested.

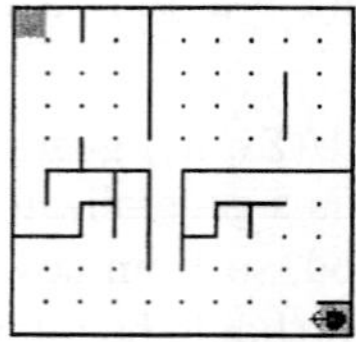

Fig. 2. Maze. Agent is in start state (lower right corner), terminal state in upper left corner

The compared methods are: **1)** ***Q***: ε-greedy exploration, zero initial Q-values, $r = 1.0$ at terminal state and zero for all other state transitions; **2)** ***OIV***: optimistic initial values, zero initial Q-values, $r = 0.0$ at terminal state and $r = -1.0$ for all other state transitions; **3)** ***BIMM***: zero initial Q-values (LTM weights), $r = 1.0$ at goal and zero for all other state transitions and **4)** ***CTRB***: counter-based exploration, zero initial Q-values, $r = 1.0$ at goal and zero for all other state transitions. Q-learning without eligibility traces is used by all methods for action-value learning. Learning parameters are indicated in Table 1. In deterministic tasks $\beta = 1$ is always the best learning rate for Q-learning [1].

The counter-based exploration bonus is implemented as

$$\delta_1(s,a) = \begin{cases} 1.0 - \dfrac{cnt(s,a) - cnt(s)_{\min}}{cnt(s)_{\max} - cnt(s)_{\min}} & \text{if } cnt(s)_{\max} - cnt(s)_{\min} > 0 \\ 0.0 & \text{if } cnt(s)_{\max} - cnt(s)_{\min} = 0 \end{cases} \tag{6}$$

where $cnt(s,a)$ is the counter for action a in state s and $cnt(s)_{min}$ and $cnt(s)_{max}$ are the smallest and greatest counter values for all actions in state s. Counter values and

BIMM STM weights are reset before beginning a new episode. SLAP was used according to the following rules when entering a new state and before taking the next action: 1) SLAP the "inverse" action and 2) if the new state is already visited during the episode, SLAP action with the biggest value $a_j(s)$ in equation (2) for the new state. For BIMM, using ε-greedy exploration in the deterministic task converged to a better policy than when using a "high" K_I-value (e.g. 0.1). In the stochastic tasks such supplementary exploration was not needed. Parameter values are indicated in Table 1.

Table 1. Parameter values used in grid world tests. $K_I = 10^{-6}$ for BIMM in all tasks. All parameters not indicated here were set to zero.

Agent	***Q***, ***γ=0.95***		***OIV***		***BIMM, γ=0.95***			***CTRB, γ=0.95***	
Task	β	ε	β	γ	α	β	ε	β	K_I
Deterministic	1	0.1	1	1	0.1	1	0.1	1	0.1
Stochastic, 0.2	0.1	0.1	0.5	0.95	0.2	0.5	0.0	0.5	0.01
Stochastic, 0.5	0.1	0.1	0.5	0.95	0.1	0.5	0.0	0.5	0.01

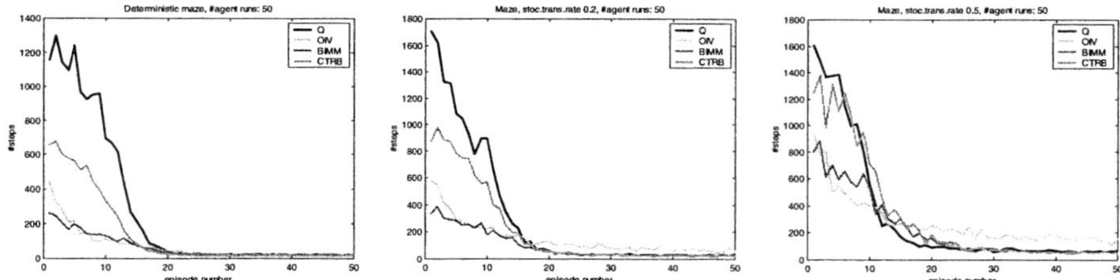

Fig. 3. Maze results as average number of steps per episode. Stochastic transition rate and the number of agent runs used for calculating the average number of steps are indicated at the top of each graph.

All agents performed 250 episodes. After 200 episodes actions were selected greedily using the learned action-values by setting ε and K_I to zero for all agents. From Fig. 3 it is clear that BIMM converges towards a "good" policy after less exploration than Q- and CTRB-agents. Only the OIV agent can compete in the beginning of exploration but it converges very slowly. With stochastic transitions OIV fails to converge due to the increased probability of cycles during exploration, which causes the action-values to be continually modified due to the negative step reward.

As indicated by the first column in Table 2, on the first episode BIMM agents reach the terminal state faster than all other agents. BIMM also represents the best compromise between how good the "converged" policy is and how much exploration is needed. BIMM has the best converged policy or is very close to it in all tasks while using the smallest total number of steps in both stochastic tasks.

Table 2. Maze results. Numbers are averages from 50 agents runs, indicated as Q/OIV/BIMM/CTRB. Steps with "converged" policy are averages of episodes 241-250. STR: stochastic transition rate.

Task	Steps first episode	Steps conv. policy	Total nbr. of steps
Deterministic	1160/438/**263**/654	18.0/18.0/18.0/18.2	17700/**7070**/7380/10600
STR 0.2	1700/571/**335**/871	**24.0**/26.7/24.4/25.1	20300/13600/**9730**/15600
STR 0.5	1610/937/**798**/1250	45.2/50.7/**44.3**/46.1	26100/26400/**21300**/25800

5 Conclusions

The results show that with appropriate heuristic rules for guiding exploration, BIMM and SLAP can improve exploration both in deterministic and stochastic environments. BIMM agents only use their own general domain knowledge, which makes them interesting compared with methods like reward shaping that usually require a priori knowledge about the task itself and easily lead to sub-optimal solutions.
Even though only a grid world task is used in this paper, BIMM and SLAP are also applicable to tasks involving continuous-valued state variables. Such tasks are a subject of current and future research.

References

1. Kaelbling, L.P., Littman, M.L., Moore, A.W.: Reinforcement Learning: A Survey. Journal of Artificial Intelligence Research, Vol. 4 (1996) 237-285
2. Ratitch, B., Precup, D.: Using MDP Characteristics to Guide Exploration in Reinforcement Learning. In: Lecture Notes in Computer Science, Vol. 2837, Springer-Verlag, Heidelberg (2003) 313-324
3. Thrun, S.B.: The role of exploration in learning control. In: DA White & DA Sofge (eds.): Handbook of Intelligent Control: Neural, Fuzzy and Adaptive Approaches. New York, Van Nostrand Reinhold (1992)
4. Widrow, B., Hoff, M.E.: Adaptive switching circuits. In: 1960 WESCON Convention record Part IV, Institute of Radio Engineers, New York (1960) 96-104

Stochastic Processes for Return Maximization in Reinforcement Learning

Kazunori Iwata[1], Hideaki Sakai[2], and Kazushi Ikeda[2]

[1] Faculty of Information Sciences, Hiroshima City University, Hiroshima, 731-3194, Japan
kiwata@im.hiroshima-cu.ac.jp
[2] Graduate School of Informatics, Kyoto University, Kyoto, 606-8501, Japan

Abstract. In the framework of reinforcement learning, an agent learns an optimal policy via return maximization, not via the instructed choices by a supervisor. The framework is in general formulated as an ergodic Markov decision process and is designed by tuning some parameters of the action-selection strategy so that the learning process eventually becomes almost stationary. In this paper, we examine a theoretical class of more general processes such that the agent can achieve return maximization by considering the asymptotic equipartition property of such processes. As a result, we show several necessary conditions that the agent and the environment have to satisfy for possible return maximization.

1 Introduction

Reinforcement learning (RL) [1, 2] is an effective framework to comprehensively describe a decision-making process that consists of interactions between an agent and an environment. One of the outstanding features of the framework is that the agent learns an optimal policy via return maximization (RM), not via the right choices indicated by a supervisor. The RL process is usually formulated as an ergodic Markov decision process (MDP) [1, Section 3] [2, Section 3.6], and is designed by tuning some parameters of the action-selection (AS) strategy [1, Section 2.2] [2, Sections 2.2 and 2.3] so that the learning process eventually becomes almost stationary. This leads us to the question of whether we indeed need the Markov property, stationary property, and ergodicity for RM, and further, to the question of what conditions are the least needed for RM in arbitrary general processes. The aim of this paper is to shed further light onto a theoretical class of processes so that RM can be achieved by showing several conditions for the agent and the environment. Considering such a class is meaningful, in particular, when we apply the framework to practical applications such as mechanical robot-learning in the real world, because some of the processes of the applications are not strictly Markovian, stationary, and/or ergodic. The results derived later enable us to judge whether the RL framework is suitable for the environment of a treating application. They also provide the necessary conditions that the policy of the agent has to satisfy for RM when it is suitable.

The organization of this paper is as follows. In Section 2 we extend the framework of RL to treat an arbitrary general process and introduce key properties to explore the processes. Section 3 describes the main results of this paper. Finally, we give conclusions in Section 4.

W. Duch et al. (Eds.): ICANN 2005, LNCS 3697, pp. 209–214, 2005.

2 Extension of Reinforcement Learning Framework

We concentrate on the discrete-time single-agent decision process with discrete states, actions, and rewards in this paper. We denote the finite set of states, actions, and rewards by $\mathcal{S} \triangleq \{s_1, s_2, \ldots, s_I\}$, $\mathcal{A} \triangleq \{a_1, a_2, \ldots, a_J\}$, and $\Re_0 \triangleq \{r_1, r_2, \ldots, r_K\} \subset \Re$. For notational simplicity, let $\mathcal{Z} \triangleq \mathcal{S} \times \mathcal{A} \times \Re_0$. Let $s(t)$, $a(t)$, and $r(t)$ be the stochastic variables of state, action, and reward at time step t, respectively. The MDP is the most popular process in the field of RL to describe the framework of interactions between an agent and the environment. The processes are determined by the two probability distributions; the agent's policy $\{P(a(t) = a_j | s(t) = s_i)\}$ and the state-transition probabilities $\{P(s(t+1) = s_{i'}, r(t+1) = r_k | s(t) = s_i, a(t) = a_j)\}$ of the environment, and are actually time-varying because the agent usually improves the policy by updating both the estimates of the value-function [2, Section 3.7] and the parameter of the AS strategy after every state-transition. However, to guarantee the convergence of the estimates to the expected ones and to simplify the analysis of the process, we often assume that the MDP satisfies ergodicity and that the time-evolutions of the estimates and the parameter are sufficiently slow so that we can deal with the MDP as a stationary one [3, Chapter 2] [4,5]. In this section, we extend the framework based on the MDP to treat an arbitrary general process.

The agent improves the policy via RM by observing each element of the empirical sequence, $s(1), a(1), s(2), r(2), a(2), \ldots$, which is generated one-by-one from an arbitrary general process. For example, consider non-Markovian, non-stationary, and/or non-ergodic processes. Now we consider the empirical sequence of n time steps,

$$s(1), a(1), s(2), r(2), a(2), \ldots, s(n), r(n), a(n), r(n+1).$$

For notational convenience, let $r(n+1) = r(1)$, and let $\boldsymbol{x} = \{s(t), a(t), r(t)\}_{t=1}^{n}$ denote the empirical sequence of n time steps, hereafter. Let us describe a probability measure under which an empirical sequence occurs. The probability measure does not needed to satisfy any constraints. Since the policy of the agent is determined by the estimates of the value-function and the parameters of the action-selection strategy, the probability measure depends on just the time-evolution of the two factors. We use $Q_{ij}^{(t)}$ to express the estimate of the value-function with respect to $(s_i, a_j) \in \mathcal{S} \times \mathcal{A}$ at time step t. We define an event $\omega \triangleq \{\theta_t | t = 1, 2, \ldots, n\} \in \Omega_n$, where $\theta_t \triangleq (\boldsymbol{Q}_t, \beta_t)$ is a pair of an $I \times J$ matrix $\boldsymbol{Q}_t \triangleq \left\{Q_{ij}^{(t)}\right\}$ and the parameter β_t of the softmax method [2, Section 2.3] at time step t. The definition of event for other AS strategies is virtually the same. Let Ω_n denote the entire set of possible events, in other words, the sample space of events. Let y be an arbitrary probability measure of the measurable space of Ω_n such that $\int_{\Omega_n} dy(\omega) = 1$ holds (alternatively, written as $\int_{\Omega_n} y(d\omega) = 1$). We assume that for all n the probability $P_{\mathcal{Z}_\omega^n}(\boldsymbol{x})$ for $\boldsymbol{x} \in \mathcal{Z}^n$ is a measurable function of ω. To put it simply, $P_{\mathcal{Z}_\omega^n}(\boldsymbol{x})$ implies the probability that an empirical sequence $\boldsymbol{x} \in \mathcal{Z}^n$ occurs under a history ω up to the time step n. Then, the probability $P_{\mathcal{Z}^n}(\boldsymbol{x})$ for $\boldsymbol{x} \in \mathcal{Z}^n$ is given by a mixed source, defined as

$$P_{\mathcal{Z}^n}(\boldsymbol{x}) = \int_{\Omega_n} P_{\mathcal{Z}_\omega^n}(\boldsymbol{x}) dy(\omega). \tag{1}$$

This is the probability that should be considered in order to analyze general processes of RL. Henceforth, we call a process given by (1) a general process. The entropy of $P_{\mathcal{Z}^n}(\boldsymbol{x})$ is written as

$$\mathrm{H}(P_{\mathcal{Z}^n}) = \sum_{\boldsymbol{x} \in \mathcal{Z}^n} P_{\mathcal{Z}^n}(\boldsymbol{x}) \log \frac{1}{P_{\mathcal{Z}^n}(\boldsymbol{x})}. \tag{2}$$

Also, the entropy $\mathrm{H}(P_{\mathcal{Z}^n_\omega})$ with respect to $P_{\mathcal{Z}^n_\omega}(\boldsymbol{x})$ for any ω is described in the same way.

Remark 1 (Stationary and ergodic process). If the process is stationary and ergodic, then the entropy rate $\lim_{n\to\infty} \mathrm{H}(P_{\mathcal{Z}^n})/n$ has a fixed point and it satisfies

$$\lim_{n\to\infty} \frac{1}{n} \log \frac{1}{P_{\mathcal{Z}^n}(\boldsymbol{x})} = \lim_{n\to\infty} \frac{1}{n} \mathrm{H}(P_{\mathcal{Z}^n}), \tag{3}$$

with probability one. Obviously, this equation similarly holds when the process is asymptotically mean stationary, defined in [6]. For proofs, see [7].

Let us introduce the following notions that play a fundamental role in later discussions.

Definition 1 (Limit superior and inferior in probability [8]). *For an arbitrary sequence of real-valued random variables* $\{Z_n\}_{n=1}^{\infty}$,

$$\operatorname*{p\text{-}limsup}_{n\to\infty} Z_n \triangleq \inf\left\{ b \,\middle|\, \lim_{n\to\infty} \Pr(Z_n > b) = 0 \right\}, \tag{4}$$

$$\operatorname*{p\text{-}liminf}_{n\to\infty} Z_n \triangleq \sup\left\{ b \,\middle|\, \lim_{n\to\infty} \Pr(Z_n < b) = 0 \right\}. \tag{5}$$

Using these notions, we define

$$\overline{\mathrm{H}}(P_{\mathcal{Z}^\infty}) \triangleq \operatorname*{p\text{-}limsup}_{n\to\infty} \frac{1}{n} \log \frac{1}{P_{\mathcal{Z}^n}(\boldsymbol{x})}, \tag{6}$$

and

$$\underline{\mathrm{H}}(P_{\mathcal{Z}^\infty}) \triangleq \operatorname*{p\text{-}liminf}_{n\to\infty} \frac{1}{n} \log \frac{1}{P_{\mathcal{Z}^n}(\boldsymbol{x})}. \tag{7}$$

In the same manner, we define $\overline{\mathrm{H}}(P_{\mathcal{Z}^\infty_\omega})$ and $\underline{\mathrm{H}}(P_{\mathcal{Z}^\infty_\omega})$ for any ω.

Example 1 (General process). In a general process, the quantity

$$\lim_{n\to\infty} \frac{1}{n} \log \frac{1}{P_{\mathcal{Z}^n}(\boldsymbol{x})}, \tag{8}$$

has a spectrum that ranges between $\overline{\mathrm{H}}(P_{\mathcal{Z}^\infty})$ and $\underline{\mathrm{H}}(P_{\mathcal{Z}^\infty})$ (but in general does not have a fixed point). For example, consider

$$P_{\mathcal{Z}^n}(\boldsymbol{x}) = P_{\mathcal{Z}^n_{\omega_1}}(\boldsymbol{x}) y(\omega_1) + P_{\mathcal{Z}^n_{\omega_2}}(\boldsymbol{x}) y(\omega_2), \quad \text{where} \quad y(\omega_1) + y(\omega_2) = 1, \tag{9}$$

where $P_{\mathcal{Z}^n_{\omega_1}}$ and $P_{\mathcal{Z}^n_{\omega_2}}$ are also the probability measures of a general process. Then, the probability density of (8) given by (9) is illustrated in Figure 1. The probability density of $\frac{1}{n} \log \frac{1}{P_{\mathcal{Z}^n}(\boldsymbol{x})}$ is called the entropy spectrum [8, Section 1.3].

3 Main Results

First, we introduce the following key properties to explore learning mechanisms in general processes.

Definition 2 (Strong converse property [8]). *If the equation*

$$\underline{\mathrm{H}}(P_{\mathcal{Z}^\infty}) = \overline{\mathrm{H}}(P_{\mathcal{Z}^\infty}), \tag{10}$$

holds, this is called the strong converse property.

Definition 3 (Asymptotic equipartition property [9]). *If for all $\delta > 0$, as $n \to \infty$*

$$P(B_n(\delta)) \to 0, \qquad P(S_n(\delta)) \to 0, \tag{11}$$

where the subsets of atypically big and atypically small probability masses are denoted by

$$B_n(\delta) = \{\boldsymbol{x} \in \mathcal{Z}^n | P_{\mathcal{Z}^n}(\boldsymbol{x}) \geq \exp(-(1-\delta)\mathrm{H}(P_{\mathcal{Z}^n}))\}, \tag{12}$$
$$S_n(\delta) = \{\boldsymbol{x} \in \mathcal{Z}^n | P_{\mathcal{Z}^n}(\boldsymbol{x}) \leq \exp(-(1+\delta)\mathrm{H}(P_{\mathcal{Z}^n}))\}, \tag{13}$$

respectively, then this is called the asymptotic equipartition property (AEP).

The strong converse property and the AEP are equivalent under a certain assumption. In fact, the following theorem holds.

Theorem 1 ([9]). *If $\mathcal{Z}$ is finite and $\lim_{n\to\infty} \mathrm{H}(P_{\mathcal{Z}^n})/n$ exists and is positive, Definition 2 is equivalent to Definition 3.*

Now we are in a position to clearly describe the situation of RM. We give the definition of RM in the same way as [5].

Definition 4 (Return Maximization). *We denote a proper subset of best sequences by*

$$\mathcal{X}_n^\dagger \subseteq \{\boldsymbol{x} \in \mathcal{Z}^n |\ \textit{the empirical distribution of } \boldsymbol{x} \textit{ yields a maximal expected return.}\}. \tag{14}$$

Also, we define the typical set as

$$C_n(\delta) \triangleq \{\boldsymbol{x} \in \mathcal{Z}^n | \boldsymbol{x} \notin B_n(\delta), \boldsymbol{x} \notin S_n(\delta)\}, \tag{15}$$

for all δ. Then, RM means that $\mathcal{X}_n^\dagger \subset C_n(\delta)$ and then $\mathcal{X}_n^\dagger = C_n(\delta)$ as $n \to \infty$.

Note that $P(C_n(\delta)) \to 1$ in probability as $n \to \infty$. In short, RM corresponds to the situation where the typical set $C_n(\delta)$ includes the subset $\mathcal{X}_n^\dagger$ by a learning algorithm as $n \to \infty$ and the AEP holds for the empirical sequences; otherwise the subset cannot have an arbitrarily high probability. Note that the learning algorithm consists of methods to update the estimates of the value-function such as temporal difference method and to tune the parameter of AS strategies. Figure 2 shows the relationship between the AEP and RM. Incidentally, if the definition of RM does not require giving probability one for the subset, it allows a pointless learning algorithm that sometimes obtains the

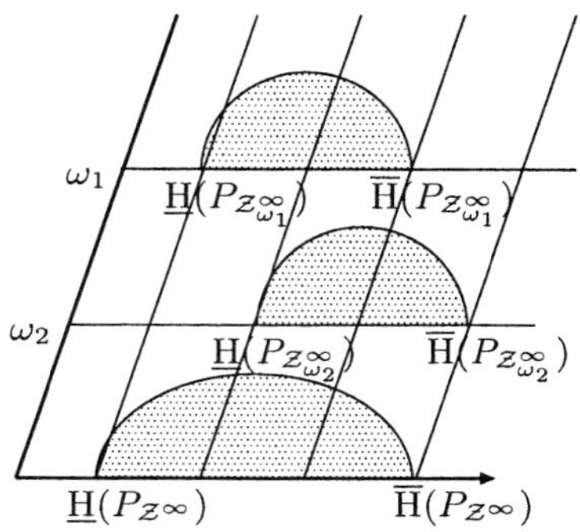

Fig. 1. Entropy spectrum when $n \to \infty$. The horizontal axis expresses the value of $\lim_{n\to\infty} \frac{1}{n} \log \frac{1}{P_{\mathcal{Z}^n}(\boldsymbol{x})}$.

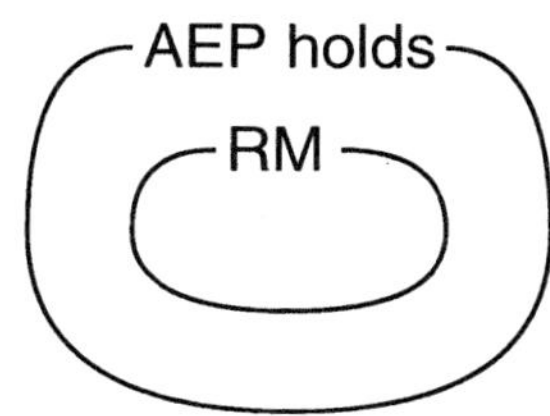

Fig. 2. Relationship between the AEP and RM in a general process. The AEP is a necessary condition for RM.

optimal policy unexpectedly and by luck, but does not guarantee the achievement of the optimal policy mathematically even under proper conditions. Hence, RM should be defined such that the probability of the subset goes to one in probability.

We show that RM can be performed with a positive probability in general processes by examining what conditions are required to establish the AEP in general processes. The following definitions are necessary condition of the learning process of the agent.

Definition 5 (Coincidence in inf- and sup-entropy spectrum). *If for all $\omega,\ \omega' \in \Omega_\infty$ such that $y(\omega) > 0$ and $y(\omega') > 0$, the following equations*

$$\overline{\mathrm{H}}(P_{\mathcal{Z}^\infty_\omega}) = \overline{\mathrm{H}}(P_{\mathcal{Z}^\infty_{\omega'}}), \qquad \underline{\mathrm{H}}(P_{\mathcal{Z}^\infty_\omega}) = \underline{\mathrm{H}}(P_{\mathcal{Z}^\infty_{\omega'}}), \tag{16}$$

are satisfied, then this is called the coincidence in inf- and sup-entropy spectrum.

The processes that satisfy Definition 5 are in a much wider class of processes than stationary and ergodic ones. Then, the following definition is the condition for the behavior of the environment.

Definition 6 (Strong converse property for ω). *If for all $\omega \in \Omega_\infty$ such that $y(\omega) > 0$, the equation*

$$\underline{\mathrm{H}}(P_{\mathcal{Z}^\infty_\omega}) = \overline{\mathrm{H}}(P_{\mathcal{Z}^\infty_\omega}), \tag{17}$$

holds, then this is called the strong converse property for ω.

Definition 6 allows a wider class of environments than stationary environments; for example, asymptotically mean stationary environments. Intuitively speaking, Definition 6 holds in such environments that the probability density of $\frac{1}{n} \log \frac{1}{P_{\mathcal{Z}^n_\omega}(\boldsymbol{x})}$ oscillates with respect to n but converges to a fixed point as $n \to \infty$.

Theorem 2 (Direct and converse theorems). *Assume that $\mathcal{Z}$ is finite and $\lim_{n\to\infty} \mathrm{H}(P_{\mathcal{Z}^n})/n$ exists and is positive. Then, if Definitions 5 and 6 are satisfied, the AEP in Definition 3 holds. Conversely, if the AEP in Definition 3 holds, then Definitions 5 and 6 are satisfied. Therefore, when $\mathcal{Z}$ is finite and $\lim_{n\to\infty} \mathrm{H}(P_{\mathcal{Z}^n})/n$ exists and is positive, the AEP in Definition 3 is equivalent to Definitions 5 and 6.*

Proof. First, we prove the direct part of Theorem 2. From [8, Chapter 1],

$$\overline{\mathrm{H}}(P_{Z^\infty}) = \sup_{\omega \in \Omega_\infty : y(\omega) > 0} \overline{\mathrm{H}}(P_{Z^\infty_\omega}), \tag{18}$$

$$\underline{\mathrm{H}}(P_{Z^\infty}) = \inf_{\omega \in \Omega_\infty : y(\omega) > 0} \underline{\mathrm{H}}(P_{Z^\infty_\omega}). \tag{19}$$

By Definition 5, for every $\omega \in \Omega_\infty$ such that $y(\omega) > 0$ these equations become

$$\overline{\mathrm{H}}(P_{Z^\infty}) = \overline{\mathrm{H}}(P_{Z^\infty_\omega}), \qquad \underline{\mathrm{H}}(P_{Z^\infty}) = \underline{\mathrm{H}}(P_{Z^\infty_\omega}), \tag{20}$$

respectively. Then, from Definition 6, Definition 2 holds. Therefore, if Definitions 5 and 6 are satisfied, Definition 3 holds because of Theorem 1.

Next, we prove the converse part of Theorem 2. Recall that (19) and (18) hold. Since $\underline{\mathrm{H}}(P_{Z^\infty_\omega})$ and $\overline{\mathrm{H}}(P_{Z^\infty_\omega})$ exist between $\underline{\mathrm{H}}(P_{Z^\infty})$ and $\overline{\mathrm{H}}(P_{Z^\infty})$ for any $\omega \in \Omega_\infty$ such that $y(\omega) > 0$, Definition 2 yields

$$\overline{\mathrm{H}}(P_{Z^\infty}) = \overline{\mathrm{H}}(P_{Z^\infty_\omega}) = \overline{\mathrm{H}}(P_{Z^\infty_{\omega'}}) = \underline{\mathrm{H}}(P_{Z^\infty_{\omega'}}) = \underline{\mathrm{H}}(P_{Z^\infty_\omega}) = \underline{\mathrm{H}}(P_{Z^\infty}). \tag{21}$$

Hence, if Definition 3 holds, Definitions 5 and 6 are satisfied. ∎

4 Conclusions

The framework of RL has been extended to treat an arbitrary general process. We discussed that RM can be performed with a positive probability in such processes that Definitions 5 and 6 hold. Such processes are in general non-Markovian, non-stationary, and/or non-ergodic. Also, these definitions show the conditions necessary for the policy, and for the state-transition probabilities of the environment.

References

1. Kaelbling, L.P., Littman, M.L., Moore, A.W.: Reinforcement learning: A survey. Journal of Artificial Intelligence Research **4** (1996) 237–285
2. Sutton, R.S., Barto, A.G.: Reinforcement Learning: An Introduction. Adaptive Computation and Machine Learning. MIT Press, Cambridge, MA (1998)
3. Kushner, H.J., Yin, G.G.: Stochastic Approximation Algorithms and Applications. Volume 35 of Applications of Mathematics. Springer-Verlag, New York (1997)
4. Iwata, K., Ikeda, K., Sakai, H.: A new criterion using information gain for action selection strategy in reinforcement learning. IEEE Transactions on Neural Networks **15** (2004) 792–799
5. Iwata, K., Ikeda, K., Sakai, H.: The asymptotic equipartition property in reinforcement learning and its relation to return maximization. Neural Networks (in press)
6. Gray, R.M., Kieffer, J.C.: Asymptotically mean stationary measures. The Annals of Probability **8** (1980) 962–973
7. Barron, A.R.: The strong ergodic theorem for densities: Generalized Shannon-Mcmillan-Breiman theorem. The Annals of Probability **13** (1985) 1292–1303
8. Han, T.S.: Information-Spectrum Methods in Information Theory. Volume 50 of Applications of mathematics. Springer (2003)
9. Verdú, S., Han, T.S.: The role of the asymptotic equipartition property in noiseless source coding. IEEE Transactions on Information Theory **43** (1997) 847–857

Maximizing the Ratio of Information to Its Cost in Information Theoretic Competitive Learning

Ryotaro Kamimura and Sachiko Aida-Hyugaji

Information Science Laboratory, Tokai University,
1117 Kitakaname Hiratsuka Kanagawa 259-1292, Japan
ryo@cc.u-tokai.ac.jp

Abstract. In this paper, we introduce costs in the framework of information maximization and try to maximize the ratio of information to its associated cost. We have shown that competitive learning is realized by maximizing mutual information between input patterns and competitive units. One shortcoming of the method is that maximizing information does not necessarily produce representations faithful to input patterns. Information maximizing primarily focuses on some parts of input patterns used to distinguish between patterns. Thus, we introduce the ratio of information to its cost that represents distance between input patterns and connection weights. By minimizing the ratio, final connection weights reflect well input patterns. We applied unsupervised information maximization to a voting attitude problem and supervised learning to a chemical data analysis. Experimental results confirmed that by minimizing the ratio, the cost is decreased with better generalization performance.

Keywords: mutual information maximization, competitive learning, winner-take-all, Gaussian, generalization, ratio, cost, supervised learning.

1 Introduction

In this paper, we introduce costs in the framework of information maximization. The new method can increase information content while minimizing the corresponding cost. The new method can contribute to neural computing from two perspectives: (1) this is a new type of information-theoretic competitive learning in which competition is realized by maximizing mutual information between input patterns and competitive units and (2) the ratio of information to the cost is maximized to produce faithful representations.

First, this is a new type of information-theoretic method to realize competition. Information-theoretic methods have been applied to competitive learning and self – organizing maps. For example, van Hulle [1] attempted to use entropy maximization for realizing equiprobabilistic outputs and to solve the fundamental problems of competitive learning such as dead neurons and dependency on initial conditions [2], [3], [4], [5], [6], [7], [8], [9], [10]. On the other hand, Linsker tried to use a more direct method to maximize mutual information. He assumed that living systems should preserve as much information as possible in every stage of processing. However, it seems to us that his method could not give clear rules to maximize mutual information, and their validity may be confined to simple artificial data [11], [12], [13]. Our method here

W. Duch et al. (Eds.): ICANN 2005, LNCS 3697, pp. 215–222, 2005.

proposed is one that maximizes mutual information directly [14], [15], [16], [17]. Contray to Linsker's formulation, our method is simple and powerful enough to be applied to practical problems. In our method, competitive unit outputs are computed directly by using a Gaussian function of distance between input patterns and connection weights. Then, information is directly maximized by using the ordinary update rules.

Second, we introduce the ratio of information to the cost. A cost function is introduced in our framework of information maximization to produce representations more faithful to input patterns. We have observed that connection weights by conventional competitive learning and our information-theoretic method are sometimes different from each other. By examining carefully a mechanism of information maximization, we find that information maximization focuses upon some parts of input patterns that are necessary to distinguish between patterns. On the other hand, conventional competitive learning imitates input patterns as much as possible. Information maximization focuses on some parts of input patterns at the expense of imitating input patterns. At this stage, to incorporate the property of conventional competitive learning in the framework of information maximization, we introduce a cost function that measures how much connection weights are similar to input patterns. The cost is actually average distance between input patterns and connection weights. By minimizing the cost, connection weights become similar to input patterns. For minimizing the cost and maximizing information, we introduce the ratio of information to the cost, and we try to maximize this ratio. Thus, it is possible to maximize information, while keeping representations faithful to input patterns.

2 Cost Minimization and Information Maximization

2.1 Unsupervised Information Maximization

We consider information content stored in competitive unit activation patterns. For this purpose, let us define information to be stored in a neural system. Information stored in a system is represented by decrease in uncertainty [18]. Uncertainty decrease, that is, information I, is defined by

$$I = -\sum_{\forall j} p(j) \log p(j) + \sum_{\forall s} \sum_{\forall j} p(s) p(j \mid s) \log p(j \mid s), \qquad (1)$$

where $p(j)$, $p(s)$ and $p(j|s)$ denote the probability of firing of the jth unit, the probability of the sth input pattern and the conditional probability of firing of the jth unit, given the sth input pattern, respectively. Let us define a cost function

$$C = \sum_{\forall s} p(s) \sum_{\forall j} p(j \mid s) C_j^s, \qquad (2)$$

where C_j^s is a cost of the jth unit for the sth input pattern. Thus, we must maximize the ratio R of information to the cost

$$R = \frac{-\sum_{\forall j} p(j) \log p(j) + \sum_{\forall s} \sum_{\forall j} p(s) p(j \mid s) \log p(j \mid s)}{\sum_{\forall s} p(s) \sum_{\forall j} p(j \mid s) C_j^s}. \qquad (3)$$

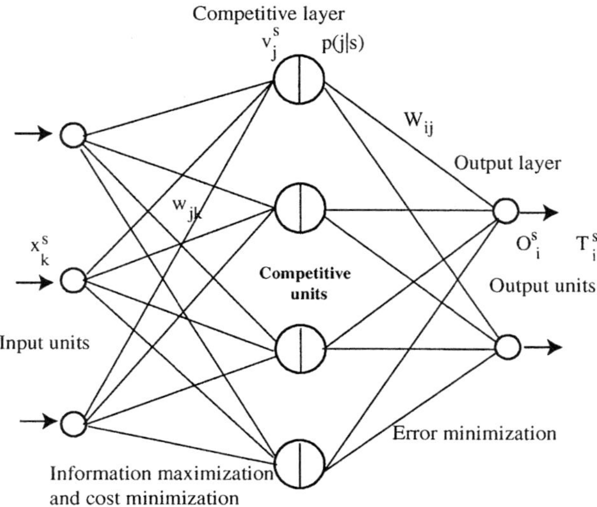

Fig. 1. A network architecture for information maximization

Let us present update rules to maximize the ratio in neural networks. As shown in Figure 1, a network is composed of input units x_k^s and competitive units v_j^s. The jth competitive unit receives a net input from input units, and an output from the jth competitive unit can be computed by

$$v_j^s = \exp\left(-\frac{\sum_{k=1}^{L}(x_k^s - w_{jk})^2}{2\sigma^2} \right), \tag{4}$$

where L is the number of input units, w_{jk} denote connections from the kth input unit to the jth competitive unit and σ controls the width of the Gaussian function. The output is increased as connection weights are closer to input patterns. The conditional probability $p(j \mid s)$ is computed by

$$p(j \mid s) = \frac{v_j^s}{\sum_{m=1}^{M} v_m^s}, \tag{5}$$

where M denotes the number of competitive units. Since input patterns are supposed to be given uniformly to networks, the probability of firing of the jth competitive unit is computed by

$$p(j) = \frac{1}{S}\sum_{s=1}^{S} p(j \mid s). \tag{6}$$

Information I is computed by

$$I = -\sum_{j=1}^{M} p(j)\log p(j) + \frac{1}{S}\sum_{s=1}^{S}\sum_{j=1}^{M} p(j \mid s)\log p(j \mid s). \tag{7}$$

A cost function is computed by

$$C = \frac{1}{S} \sum_{s=1}^{S} \sum_{j=1}^{M} p(j \mid s) \sum_{k=1}^{L} (x_k^s - w_{jk})^2. \tag{8}$$

Thus, we must maximize the following function:

$$R = \frac{-\sum_{j=1}^{M} p(j) \log p(j) + S^{-1} \sum_{s=1}^{S} \sum_{j=1}^{M} p(j \mid s) \log p(j \mid s)}{S^{-1} \sum_{s=1}^{S} \sum_{j=1}^{M} p(j \mid s) \sum_{k=1}^{L} (x_k^s - w_{jk})^2}. \tag{9}$$

Differentiating information with respect to input-competitive connections w_{jk}, we have

$$\begin{aligned}
\Delta w_{jk} &= -\frac{\beta}{SC\sigma^2} \sum_{s=1}^{S} \left(\log p(j) - \sum_{m=1}^{M} p(m \mid s) \log p(m) \right) p(j \mid s)(x_k^s - w_{jk}) \\
&+ \frac{\beta}{SC\sigma^2} \sum_{s=1}^{S} \left(\log p(j \mid s) - \sum_{m=1}^{M} p(m \mid s) \log p(m \mid s) \right) p(j \mid s)(x_k^s - w_{jk}) \\
&- \frac{\beta I}{S\sigma^2 C^2} \left(\sum_{k=1}^{L} (x_k^s - w_{jk})^2 - \sum_{m=1}^{M} p(m \mid s) \sum_{k=1}^{L} (x_k^s - w_{jk})^2 \right) \cdot \\
& p(j \mid s)(x_k^s - w_{jk}) + \frac{2\beta I}{SC^2} \sum_{s=1}^{S} p(j \mid s)(x_k^s - w_{jk}),
\end{aligned} \tag{10}$$

where β is the learning parameter.

2.2 Extension to Supervised Learning

Unsupervised information maximization can easily be transformed into supervised learning by adding an output layer. In the output layer, errors between targets and outputs are minimized. The outputs from the output layer are computed by

$$O_i^s = \sum_{j=1}^{M} W_{ij} p(j|s), \tag{11}$$

where W_{ij} denote connection weights from the jth competitive unit to the ith output unit. Errors between targets and outputs can be computed by

$$E = \frac{1}{2} \sum_{s=1}^{S} \sum_{i=1}^{N} (T_i^s - O_i^s)^2, \tag{12}$$

where T_i^s denote targets for output units O_i^s and N is the number of output units. This linear equation is directly solved by using the pseudo-inverse of the matrices of competitive unit outputs. Following the standard matrix notation, we suppose that $\mathbf{W}$ and $\mathbf{T}$ denote the matrices of connection weights and targets and $\mathbf{P}^\dagger$ shows the pseudo-inverse of the matrix of competitive unit activations. Then, we can obtain final connection weights by $\mathbf{W} = \mathbf{P}^\dagger \mathbf{T}$.

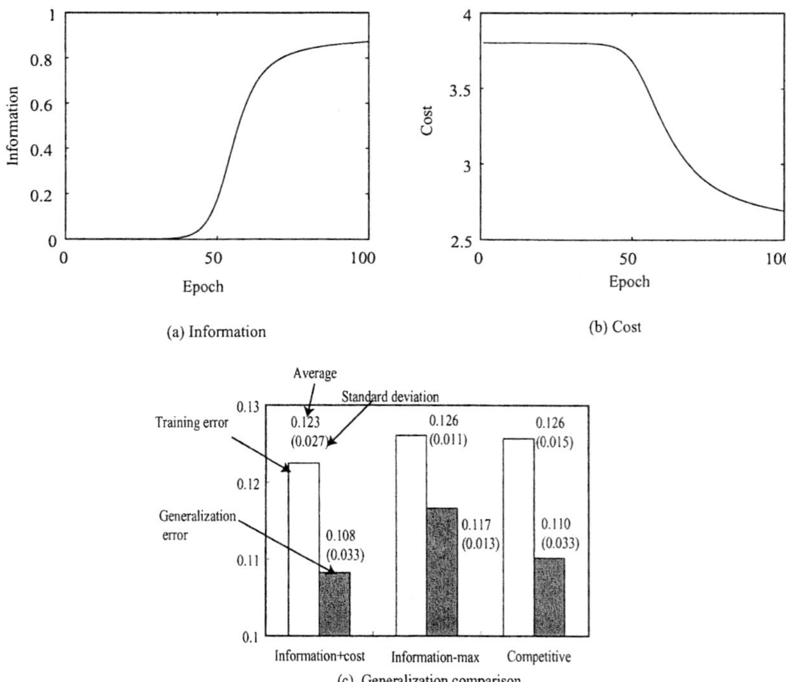

Fig. 2. Experimental results for the voting attitude problem: Figure (a), (b) and (c) show information as a function of the number of epochs, the cost as a function of the number of epochs and error comparison by three methods, respectively

3 Voting Attitude

In this experiment, we used the voting attitude data from the machine learning data base[1]. The data set includes votes for each of the U.S. House of Representatives Congressmen on the 16 key votes. The number of input patterns was 435. The half of this data was training patterns, and the remaining one was used as a test data set. The number of input units is 16, and the number of competitive unit is two.

Figure 2(a) shows information as a function of the number of epochs by information maximization with costs. Information is rapidly increased from around the 50th epoch and reaches the ninety percent of a maximum point. Figure 2(b) shows the cost as a function of the number of epochs. The cost is also decreased rapidly from around the 50th epoch. Figure 2(c) shows training and generalization errors by three methods[2]. By

[1] http://www1.ics.uci.edu/ mlearn/MLRepository.html

[2] We did experiments ten times with ten different initial conditions and ten different sets of training and test data, and averaged the results. The results were obtained when the generalization error was the lowest. Competitive learning was performed by using Matlab package with default parameter values except the number of epochs.

using standard competitive learning, the training error is 0.126, and the generalization error is 0.110. By maximizing information, we have the same training errors, but the generalization error is slightly increased to 0.117. By information maximization with costs, the training error is 0.123, but the generalization error is the lowest, that is, 0.108. These results show that information can be maximized and simultaneously the cost can be decreased significantly and that generalization performance could be improved by introducing the cost.

4 Chemical Data Analysis

In this experiment, we extend our method to supervised learning by adding an output layer as shown in Figure 1. We employed electron affinity of each atoms, calculated solvation free energy (dG), and estimated hydrophobic property (LogP) as an input data set of each unit to predict cancer drug resistance ratio [19] that was measured by biochemical experiments. The numbers of input and output unit are nine and one, respectively. Because the number of input patterns is just ten, we used the one-leave-out cross validation. As already mentioned in the previous section, we directly solved linear equations for the output layer to obtain connection weights to output units.

Figure 3 shows generalization errors by information maximization, information maximization with costs and by standard BP. As can be seen in the figure, generalization errors are overwhelmingly large by standard BP for any number of competitive units. The best generalization performance can be obtained by information maximization with costs when the number of competitive units is 34.

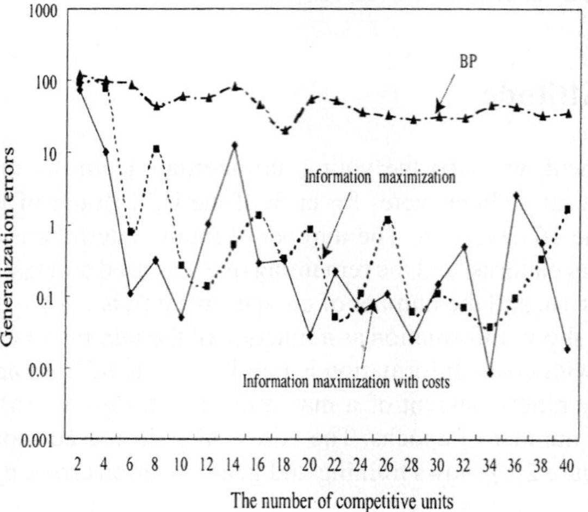

Fig. 3. Generalization as a function of the number of epochs by three methods: information maximization, information maximization with costs and BP

5 Conclusion

In this paper, we have introduced the ratio of information to its cost in the framework of information maximization. By minimizing the ratio, information can produce more faithful representations with better generalization performance. Though our method has shown fairly good experimental results, we can point out several problems for the present method to be more practically applicable. First, we maximize the ratio of information to its cost. However, it is better to evaluate more exactly relations between information and its cost. Second, we have shown that generalization errors are significantly improved by our method. However, relations between information and generalization remain obscure at the present stage of research. We need to examine the relations more explicitly. Third, as mentioned in the main text, the parameter δ controls the Gaussian width, and is directly related to a process of information maximization. Thus, more subtle analysis of the relations should be needed. Though some improvement must be needed for practical applications, it is certain that our new method can open up a new perspective in neural networks.

References

1. M. Marc and M. V. Hulle, *Faithful representations and topographic maps.* New York: John Wiley and Sons, Inc, 2000.
2. S. Grossberg, "Competitive learning: From interactive activation to adaptive resonance," *Cognitive Science*, vol. 11, pp. 23–63, 1987.
3. D. E. Rumelhart and D. Zipser, "Feature discovery by competitive learning," in *Parallel Distributed Processing* (D. E. Rumelhart and G. E. H. et al., eds.), vol. 1, pp. 151–193, Cambridge: MIT Press, 1986.
4. D. E. Rumelhart and J. L. McClelland, "On learning the past tenses of English verbs," in *Parallel Distributed Processing* (D. E. Rumelhart, G. E. Hinton, and R. J. Williams, eds.), vol. 2, pp. 216–271, Cambrige: MIT Press, 1986.
5. S. Grossberg, "Competitive learning: from interactive activation to adaptive resonance," *Cognitive Science*, vol. 11, pp. 23–63, 1987.
6. D. DeSieno, "Adding a conscience to competitive learning," in *Proceedings of IEEE International Conference on Neural Networks*, (San Diego), pp. 117–124, IEEE, 1988.
7. S. C. Ahalt, A. K. Krishnamurthy, P. Chen, and D. E. Melton, "Competitive learning algorithms for vector quantization," *Neural Networks*, vol. 3, pp. 277–290, 1990.
8. L. Xu, "Rival penalized competitive learning for clustering analysis, RBF net, and curve detection," *IEEE Transaction on Neural Networks*, vol. 4, no. 4, pp. 636–649, 1993.
9. A. Luk and S. Lien, "Properties of the generalized lotto-type competitive learning," in *Proceedings of International conference on neural information processing*, (San Mateo: CA), pp. 1180–1185, Morgan Kaufmann Publishers, 2000.
10. M. M. V. Hulle, "The formation of topographic maps that maximize the average mutual information of the output responses to noiseless input signals," *Neural Computation*, vol. 9, no. 3, pp. 595–606, 1997.
11. R. Linsker, "Self-organization in a perceptual network," *Computer*, vol. 21, pp. 105–117, 1988.
12. R. Linsker, "How to generate ordered maps by maximizing the mutual information between input and output," *Neural Computation*, vol. 1, pp. 402–411, 1989.

13. R. Linsker, "Local synaptic rules suffice to maximize mutual information in a linear network," *Neural Computation*, vol. 4, pp. 691–702, 1992.
14. R. Kamimura, T. Kamimura, and T. R. Shultz, "Information theoretic competitive learning and linguistic rule acquistion," *Transactions of the Japanese Society for Artificial Intelligence*, vol. 16, no. 2, pp. 287–298, 2001.
15. R. Kamimura, T. Kamimura, and O. Uchida, "Flexible feature discovery and structural information," *Connection Science*, vol. 13, no. 4, pp. 323–347, 2001.
16. R. Kamimura, T. Kamimura, and H. Takeuchi, "Greedy information acquisition algorithm: A new information theoretic approach to dynamic information acquisition in neural networks," *Connection Science*, vol. 14, no. 2, pp. 137–162, 2002.
17. R. Kamimura, "Progressive feature extraction by greedy network-growing algorithm," *Complex Systems*, vol. 14, no. 2, pp. 127–153, 2003.
18. L. L. Gatlin, *Information Theory and Living Systems*. Columbia University Press, 1972.
19. M. Yoshikawa, Y. Ikegami, S. Hayasaka, K. Ishii, A. Ito, K. Sano, T. Suzuki, T. Togawa, H. Yoshida, H. Soda, M. Oka, S. Kohno, S. Sawada, T. Ishikawa, and S. Tanabe, "Novel camtothecin analogues that circumvent abcg2-associated drug resistance in human tumor," *Int. J. Cancer*, vol. 110, pp. 921–927, 2004.

Completely Self-referential Optimal Reinforcement Learners

Jürgen Schmidhuber

IDSIA, Galleria 2, 6928 Manno (Lugano), Switzerland
TU Munich, Boltzmannstr. 3, 85748 Garching, München, Germany
juergen@idsia.ch
http://www.idsia.ch/~juergen

Abstract. We present the first class of mathematically rigorous, general, fully self-referential, self-improving, optimal reinforcement learning systems. Such a system rewrites any part of its own code as soon as it has found a proof that the rewrite is *useful,* where the problem-dependent *utility function* and the hardware and the entire initial code are described by axioms encoded in an initial proof searcher which is also part of the initial code. The searcher systematically and efficiently tests computable *proof techniques* (programs whose outputs are proofs) until it finds a provably useful, computable self-rewrite. We show that such a self-rewrite is globally optimal—no local maxima!—since the code first had to prove that it is not useful to continue the proof search for alternative self-rewrites. Unlike previous *non*-self-referential methods based on hardwired proof searchers, ours not only boasts an optimal *order* of complexity but can optimally reduce any slowdowns hidden by the $O()$-notation, provided the utility of such speed-ups is provable at all.

1 Introduction and Outline

Traditional reinforcement learning (RL) algorithms [6] are hardwired. They are designed to improve some limited type of policy through experience, but are not part of the modifiable policy, and cannot improve themselves. Humans are needed to create new / better RL algorithms and to prove their usefulness under appropriate assumptions.

Let us eliminate the restrictive need for human effort in the most general way possible, leaving all the work including the proof search to a system that can rewrite and improve itself in arbitrary computable ways and in a most efficient fashion. To attack this "Grand Problem of Artificial Intelligence," we introduce a novel class of optimal, fully self-referential [3] general problem solvers called *Gödel machines* [11,10]. They are universal RL systems that interact with some (partially observable) environment and can in principle modify themselves without essential limits besides the limits of computability. Their initial RL algorithm is not hardwired; it can completely rewrite itself, but only if a proof searcher embedded within the initial algorithm can first prove that the rewrite is useful, given a formalized utility function reflecting expected rewards and computation

W. Duch et al. (Eds.): ICANN 2005, LNCS 3697, pp. 223–233, 2005.

time. We will see that self-rewrites due to this approach are actually *globally optimal* (Theorem 1, Section 4), relative to Gödel's well-known fundamental restrictions of provability [3]. These restrictions should not worry us; if there is no proof of some self-rewrite's utility, then humans cannot do much either.

The initial proof searcher is $O()$-optimal (has an optimal order of complexity) in the sense of Theorem 2, Section 5. Unlike Hutter's hardwired systems [5] (Section 2), however, a Gödel machine can further speed up its proof searcher to meet *arbitrary* formalizable notions of optimality beyond those expressible in the $O()$-notation. Our approach yields the first theoretically sound, fully self-referential, optimal, general reinforcement learners.

Outline. Section 2 presents basic concepts, relation to previous work, and limitations, Section 3 the essential details of a self-referential axiomatic system, Section 4 the Global Optimality Theorem 1, and Section 5 the $O()$-optimal (Theorem 2) initial proof searcher.

2 Basic Overview and Relation to Previous Work and Limitations

Notation and Set-Up. Unless stated otherwise or obvious, throughout the paper newly introduced variables and functions are assumed to cover the range implicit in the context. B denotes the binary alphabet $\{0, 1\}$, B^* the set of possible bitstrings over B, $l(q)$ denotes the number of bits in a bitstring q; q_n the n-th bit of q; λ the empty string (where $l(\lambda) = 0$); $q_{m:n} = \lambda$ if $m > n$ and $q_m q_{m+1} \ldots q_n$ otherwise (where $q_0 := q_{0:0} := \lambda$).

Our hardware (e.g., a universal or space-bounded Turing machine or the abstract model of a personal computer) has a single life which consists of discrete cycles or time steps $t = 1, 2, \ldots$. Its total lifetime T may or may not be known in advance. In what follows, the value of any time-varying variable Q at time t will be denoted by $Q(t)$.

During each cycle our hardware executes an elementary operation which affects its variable state $s \in \mathcal{S} \subset \mathcal{B}^*$ and possibly also the variable environmental state $Env \in \mathcal{E}$. (Here we need not yet specify the problem-dependent set $\mathcal{E}$). There is a hardwired state transition function $F : \mathcal{S} \times \mathcal{E} \rightarrow \mathcal{S}$. For $t > 1$, $s(t) = F(s(t-1), Env(t-1))$ is the state at a point where the hardware operation of cycle $t-1$ is finished, but the one of t has not started yet. $Env(t)$ may depend on past output actions encoded in $s(t-1)$ and is simultaneously updated or (probabilistically) computed by the possibly reactive environment.

At any given time t ($1 \leq t \leq T$) the goal is to maximize future success or *utility*. A typical *"value to go"* utility function (to be maximized) is of the form $u(s, Env) : \mathcal{S} \times \mathcal{E} \rightarrow \mathcal{R}$, where $\mathcal{R}$ is the set of real numbers:

$$u(s, Env) = E_\mu \left[\sum_{\tau=time}^{T} r(\tau) \;\middle|\; s, Env \right], \tag{1}$$

where $r(t)$ is a real-valued reward input (encoded within $s(t)$) at time t, $E_\mu(\cdot \mid \cdot)$ denotes the conditional expectation operator with respect to some possibly unknown distribution μ from a set M of possible distributions (M reflects whatever is known about the possibly probabilistic reactions of the environment), and the above-mentioned $time = time(s)$ is a function of state s which uniquely identifies the current cycle. Note that we take into account the possibility of extending the expected lifespan through appropriate actions.

Basic Idea. Our machine becomes a self-referential [3] *Gödel machine* by loading it with a machine-dependent, particular form of self-modifying code p. The initial code $p(1)$ at time step 1 includes a (typically sub-optimal) problem solving subroutine for interacting with the environment, such as Q-learning [6], and a general proof searcher subroutine (Section 5) that systematically makes pairs *(switchprog, proof)* until it finds a *proof* of a target theorem which essentially states: *'the immediate rewrite of* p *through current program* switchprog *on the given machine implies higher utility than leaving* p *as is'*. Then it executes *switchprog*, which may completely rewrite p, including the proof searcher. Section 3 will explain details of the necessary initial axiomatic system $\mathcal{A}$ encoded in $p(1)$.

The **Global Optimality Theorem** (Theorem 1, Section 4) shows this self-improvement strategy is not greedy: since the utility of *'leaving p as is'* implicitly evaluates all possible alternative *switchprog*s which an unmodified p might find later, we obtain a globally optimal self-change—the *current switchprog* represents the best of all possible relevant self-changes, relative to the given resource limitations and initial proof search strategy.

Proof Techniques and an $O()$-optimal Initial Proof Searcher. Section 5 will present an $O()$-optimal initialization of the proof searcher, that is, one with an optimal *order* of complexity (Theorem 2). Still, there will remain a lot of room for self-improvement hidden by the $O()$-notation. The searcher uses an online extension of *Universal Search* [7] to systematically test *online proof techniques*, which are proof-generating programs that may read parts of state s (similarly, mathematicians are often more interested in proof techniques than in theorems). To prove target theorems as above, proof techniques may invoke special instructions for generating axioms and applying inference rules to prolong the current *proof* by theorems. Here an axiomatic system $\mathcal{A}$ encoded in $p(1)$ includes axioms describing **(a)** how any instruction invoked by a program running on the given hardware will change the machine's state s (including instruction pointers etc.) from one step to the next (such that proof techniques can reason about the effects of any program including the proof searcher), **(b)** the initial program $p(1)$ itself (Section 3 will show that this is possible without introducing circularity), **(c)** stochastic environmental properties, **(d)** the formal utility function u, e.g., equation (1). The evaluation of utility automatically takes into account computational costs of all actions including proof search.

Hutter's Previous Work. Hutter's non-self-referential but still $O()$-optimal *'fastest' algorithm for all well-defined problems* HSEARCH [4] uses a *hardwired* brute force proof searcher. Assume discrete input/output domains X/Y, a formal problem specification $f : X \to Y$ (say, a functional description of how integers are decomposed into their prime factors), and a particular $x \in X$ (say, an integer to be factorized). HSEARCH orders all proofs of an appropriate axiomatic system by size to find programs q that for all $z \in X$ provably compute $f(z)$ within time bound $t_q(z)$. Simultaneously it spends most of its time on executing the q with the best currently proven time bound $t_q(x)$. It turns out that HSEARCH is as fast as the *fastest* algorithm that provably computes $f(z)$ for all $z \in X$, save for a constant factor smaller than $1 + \epsilon$ (arbitrary $\epsilon > 0$) and an f-specific but x-independent additive constant [4]. This constant may be enormous though.

Hutter's AIXI*(t,l)* [5] is related. In discrete cycle $k = 1, 2, 3, \ldots$ of AIXI*(t,l)*'s lifetime, action $y(k)$ results in perception $x(k)$ and reward $r(k)$, where all quantities may depend on the complete history. Using a universal computer such as a Turing machine, AIXI*(t,l)* needs an initial offline setup phase (prior to interaction with the environment) to examine all proofs of length at most L, filtering out those that identify programs (of maximal size l and maximal runtime t per cycle) which not only could interact with the environment but which for all possible interaction histories also correctly predict a lower bound of their own expected future reward. In cycle k, AIXI*(t,l)* then runs all programs identified in the setup phase (at most 2^l), finds the one with highest self-rating, and executes its corresponding action. The problem-independent setup time (where almost all of the work is done) is $O(L \cdot 2^L)$. The online time per cycle is $O(t \cdot 2^l)$. Both are constant but typically huge.

Advantages and Novelty of the Gödel Machine. There are major differences between the Gödel machine and Hutter's HSEARCH [4] and AIXI*(t,l)* [5], including:

1. The theorem provers of HSEARCH and AIXI*(t,l)* are hardwired, non-self-referential, unmodifiable meta-algorithms that cannot improve themselves. That is, they will always suffer from the same huge constant slowdowns (typically $\gg 10^{1000}$) buried in the $O()$-notation. But there is nothing in principle that prevents our truly self-referential code from proving and exploiting drastic reductions of such constants, in the best possible way that provably constitutes an improvement, if there is any.
2. The demonstration of the $O()$-optimality of HSEARCH and AIXI*(t,l)* depends on a clever allocation of computation time to some of their unmodifiable meta- algorithms. Our Global Optimality Theorem (Theorem 1, Section 4), however, is justified through a quite different type of reasoning which indeed exploits and crucially depends on the fact that there is no unmodifiable software at all, and that the proof searcher itself is readable and modifiable and can be improved. This is also the reason why its self-improvements can be more than merely $O()$-optimal.
3. HSEARCH uses a "trick" of proving more than is necessary which also disappears in the sometimes quite misleading $O()$-notation: it wastes time on finding programs that provably compute $f(z)$ for all $z \in X$ even when the current $f(x) (x \in X)$ is the only object of interest. A Gödel machine, however, needs to prove only what is relevant to its goal formalized by u. For example, the general u of eq. (1) completely ignores the limited concept

of $O()$-optimality, but instead formalizes a stronger type of optimality that does not ignore huge constants just because they are constant.

4. Both the Gödel machine and AIXI*(t,l)* can maximize expected reward (HSEARCH cannot). But the Gödel machine is more flexible as we may plug in *any* type of formalizable utility function (e.g., *worst case* reward), and unlike AIXI*(t,l)* it does not require an enumerable environmental distribution.

Limitations. The fundamental limitations are closely related to those first identified by Gödel's celebrated paper on self-referential formulae [3]. Any formal system that encompasses arithmetics (or ZFC etc) is either flawed or allows for unprovable but true statements. Hence even a Gödel machine with unlimited computational resources must ignore those self-improvements whose effectiveness it cannot prove, e.g., for lack of sufficiently powerful axioms in $\mathcal{A}$. In particular, one can construct pathological examples of environments and utility functions that make it impossible for the machine to ever prove a target theorem. Compare Blum's speed-up theorem [1] based on certain incomputable predicates. Similarly, a realistic Gödel machine with limited resources cannot profit from self-improvements whose usefulness it cannot prove within its time and space constraints. Nevertheless, unlike previous methods, it can in principle exploit at least the *provably* good speed-ups of *any* part of its initial software, including those parts responsible for huge (but problem class-independent) slowdowns ignored by the earlier approaches [5].

3 Essential Details of One Representative Gödel Machine

Theorem proving requires an axiom scheme yielding an enumerable set of axioms of a formal logic system $\mathcal{A}$ whose formulas and theorems are symbol strings over some finite alphabet that may include traditional symbols of logic (such as $\rightarrow, \wedge, =, (,), \forall, \exists, \ldots, c_1, c_2, \ldots, f_1, f_2, \ldots$), probability theory (such as $E(\cdot)$, the expectation operator), arithmetics ($+, -, /, =, \sum, <, \ldots$), string manipulation (in particular, symbols for representing any part of state s at any time, such as $s_{7:88}(5555)$). A proof is a sequence of theorems, each either an axiom or inferred from previous theorems by applying one of the inference rules such as *modus ponens* combined with *unification*, e.g., [2].

The remainder of this paper will omit standard knowledge to be found in any proof theory textbook. Instead of listing *all* axioms of a particular $\mathcal{A}$, we will focus on the novel and critical details: how to overcome problems with self-reference and how to deal with the potentially delicate online generation of proofs that talk about and affect the currently running proof generator itself.

Proof Techniques. Brute force proof searchers (used in Hutter's AIXI*(t,l)* and HSEARCH) systematically generate all proofs in order of their sizes. To produce a certain proof, this takes time exponential in proof size. Instead our $O()$-optimal $p(1)$ will produce many proofs with low algorithmic complexity [7] much more quickly. It systematically tests (see Section 5) *proof techniques* written in universal language $\mathcal{L}$ implemented within $p(1)$. A proof technique is composed of instructions that allow any part of s to be read, such as inputs x, or the code of $p(1)$. It may write on s^p, a part of s reserved for temporary results. It also may rewrite *switchprog*, and produce an incrementally growing proof placed in the string variable *proof* stored somewhere in s. *proof* and s^p are reset to the empty string at the beginning of each new proof technique test. Apart from standard arithmetic and function-defining instructions [9] that modify s^p, the programming language $\mathcal{L}$ includes special instructions for prolonging the current *proof* by correct theorems, for setting *switchprog*, and for checking whether a provably optimal p-modifying program was found and should be executed now. Certain long proofs can be produced by short proof techniques.

The nature of the five *proof*-modifying instructions below (there are no others) makes it impossible to insert an incorrect theorem into *proof*, thus trivializing proof verification:

1. **get-axiom(n)** takes as argument an integer n computed by a prefix of the currently tested proof technique with the help of arithmetic instructions such as those used in previous work [9]. Then it appends the n-th axiom (if it exists, according to the axiom scheme below) as a theorem to the current theorem sequence in *proof*. The initial axiom scheme encodes:
   (a) **Hardware axioms** describing the hardware, formally specifying how certain components of s (other than environmental inputs) may change from one cycle to the next. For example, the following axiom could describe how some 64-bit hardware's instruction pointer stored in $s_{1:64}$ is continually increased by 64 as long as there is no overflow and the value of s_{65} does not indicate that a jump to some other address should take place:

$$(\forall t \forall n : [(n < 2^{64} - 1) \wedge (n > 0) \wedge (t > 1) \wedge (t < T) \wedge (string2num(s_{1:64}(t)) = n)$$

$$\wedge (s_{65}(t) = \text{`0'})] \rightarrow (string2num(s_{1:64}(t+1)) = n + 1))$$

   Here the semantics of used symbols such as '(' and '>' and '→' (implies) are the traditional ones, while '*string2num*' symbolizes a function translating bitstrings into numbers. It is clear that any abstract hardware model can be fully axiomatized in a similar way.
   (b) **Reward axioms** defining the computational costs of any hardware instruction, and physical costs of output actions (e.g., control signals encoded in $s(t)$). Related axioms assign values to certain input events (encoded in s) representing reward or punishment (e.g., when a Gödel machine-controlled robot bumps into an obstacle). Additional axioms define the total value of the Gödel machine's life as a scalar-valued function of all rewards (e.g., their sum) and costs experienced between cycles 1 and T, etc.
   (c) **Environment axioms** restricting the way the environment will produce new inputs (encoded within certain substrings of s) in reaction to sequences of outputs encoded in s. For example, it may be known

in advance that the environment is sampled from an unknown probability distribution that is *computable,* given the previous history [12,5]. Or, more restrictively, the environment may be some unknown but deterministic computer program sampled from the Speed Prior [8] which assigns low probability to environments that are hard to compute by any method. Or the interface to the environment is Markovian, that is, the current input always uniquely identifies the environmental state [6]. Even more restrictively, the environment may evolve in completely predictable fashion known in advance. All such prior assumptions are perfectly formalizable in an appropriate $\mathcal{A}$ (otherwise we could not write scientific papers about them).

(d) **Uncertainty axioms; string manipulation axioms:** Standard axioms for arithmetics and calculus and probability theory and statistics and string manipulation that (in conjunction with the environment axioms) allow for constructing proofs concerning (possibly uncertain) properties of future values of $s(t)$ as well as bounds on expected remaining lifetime / costs / rewards, given some time τ and certain hypothetical values for components of $s(\tau)$ etc.

(e) **Initial state axioms:** Information about how to reconstruct the initial state $s(1)$ or parts thereof, such that the proof searcher can build proofs including axioms of the type

$$(s_{\mathbf{m:n}}(1) = \mathbf{z}), \ e.g.: \ (s_{7:9}(1) = \text{`010'}).$$

Here and in the remainder of the paper we use bold font in formulas to indicate syntactic place holders (such as **m,n,z**) for symbol strings representing variables (such as m,n,z) whose semantics are explained in the text (in the present context z is the bitstring $s_{m:n}(1)$).

Note that it is **no fundamental problem** to fully encode both the hardware description *and* the initial hardware-describing p within p itself. To see this, observe that some software may include a program that can print the software.

(f) **Utility axioms** describing the overall goal in the form of utility function u; e.g., equation (1).

2. **apply-rule(k, m, n)** takes as arguments the index k (if it exists) of an inference rule such as *modus ponens* (stored in a list of possible inference rules encoded within $p(1)$) and the indices m, n of two previously proven theorems (numbered in order of their creation) in the current *proof.* If applicable, the corresponding inference rule is applied to the addressed theorems and the resulting theorem appended to *proof.* Otherwise the currently tested proof technique is interrupted. This ensures that *proof* is never fed with invalid proofs.
3. **set-switchprog(m,n)** replaces *switchprog* by $s^p_{m:n}$, provided that $s^p_{m:n}$ is indeed a non-empty substring of s^p, the storage writable by proof techniques.
4. ♠ **state2theorem(m, n)** takes two integer arguments m, n and tries to transform the current contents of $s_{m:n}$ into a theorem of the form

$$(s_{\mathbf{m:n}}(\mathbf{t_1}) = \mathbf{z}), \ e.g.: \ (s_{6:9}(7775555) = \text{`1001'}),$$

where t_1 represents a time measured (by checking *time*) shortly after *state2theorem* was invoked, and z the bistring $s_{m:n}(t_1)$ (recall the special case $t_1 = 1$ of Item 1e). So we accept the time-labeled current observable contents of any part of s as a theorem that does not have to be proven in an alternative way from, say, the initial state $s(1)$, because the computation so far has already demonstrated that the theorem is true.

We must avoid inconsistent results through parts of s that change while being read. For example, the present value of a quickly changing instruction pointer IP *(continually updated by the hardware) may not be directly observable without changing the observable itself: the execution of the reading subroutine itself will already modify* IP *many times. So* state2theorem *first uses some fixed protocol to check whether the current* $s_{m:n}$ *is readable at all or whether it might change if it were read by the remaining code of* state2theorem. *If so, or if* m, n, *are not in the proper range, then the instruction has no further effect.*

5. ♠ **check()** verifies whether the goal of the proof search has been reached. First it tests whether the last theorem (if any) in *proof* has the form of a **target theorem**. A target theorem states that given the *current* axiomatized utility function u (Item 1f), the utility of a switch from p to the current *switchprog* would be higher than the utility of continuing the execution of p (which would keep searching for alternative *switchprogs*). Target theorems are symbol strings (encoded in s) of the (decoded) form

$$(u[s(\mathbf{t_1}) \oplus (\mathit{switchbit}(\mathbf{t_1}) = \text{`1'}), \mathit{Env}(\mathbf{t_1})] > u[s(\mathbf{t_1}) \oplus (\mathit{switchbit}(\mathbf{t_1}) = \text{`0'}), \mathit{Env}(\mathbf{t_1})]) \quad (2)$$

where the variable t_1 (represented by syntactic place holder $\mathbf{t_1}$) stands for a time step, while all other symbols belong to the alphabet of the theorem-proving calculus, that is, the set of possible target theorems is parameterized only by t_1. Here the calculus should permit the notation $s(\mathbf{t_1}) \oplus (\mathit{switchbit}(\mathbf{t_1}) = \text{`}b\text{'})$ as a shortcut for the state obtained when we replace $\mathit{switchbit}(t_1)$, the true value of the variable bit *switchbit* (encoded in s) at time t_1, by $b \in \{0, 1\}$. This will facilitate the formulation of theorems that compare values conditioned on various alternative hypothetical properties of $s(t_1)$. (Note that $s(t_1)$ may be only partially known by the current proof technique even in environments where $s(t_1)$ and $\mathit{switchbit}(t_1)$ are fully predetermined for all valid t_1.)

The purpose of introducing t_1 is to deal with hardware-specific temporal delays that may be involved in checking and switching—it may take a significant amount of time to match abstract symbol strings found during proof search to the Gödel machine's real current state. If a target theorem has been found, *check()* uses a simple prewired subroutine to check whether there is enough time left to set variable *switchbit* (originally 0) to 1 before the continually increasing *time* will equal t_1. If this subroutine returns a negative result, *check()* exits. Otherwise it sets $\mathit{switchbit} := 1$ (there is no other way of changing *switchbit*). Then it repeatedly tests *time* until $\mathit{time} > t_1$, to make sure the condition of formula (2) was fulfilled at t_1. Then it transfers control to *switchprog* (there is no other way of calling *switchprog*). The *switchprog* may subsequently rewrite all parts of s, excluding hardware-reserved parts such as *time* and x, but including p.

Note that a proof technique does not necessarily have to compute the true expected utilities of switching and not switching—it just needs to determine which is higher. For example, it may be easy to prove that speeding up a subroutine of the proof searcher by a factor of 2 will certainly be worth the negligible (compared to lifetime T) time needed to execute the subroutine-changing algorithm, no matter what is the precise utility of the switch.

The axiomatic system $\mathcal{A}$ is a defining parameter of a given Gödel machine. Clearly, $\mathcal{A}$ must be strong enough to permit proofs of target theorems. In particular, the theory of uncertainty axioms (Item 1d) must be sufficiently rich. This is no fundamental problem: we simply insert all traditional axioms of probability theory.

4 Global Optimality Theorem

Intuitively, at any given time p should execute some self-modification algorithm only if it is the 'best' of all possible self-modifications, given the utility function, which typically depends on available resources, such as storage size and remaining lifetime. At first glance, however, target theorem (2) seems to implicitly talk about just one single modification algorithm, namely, *switchprog*(t_1) as set by the systematic proof searcher at time t_1. Isn't this type of local search greedy? Couldn't it lead to a local optimum instead of a global one? No, it cannot, according to the global optimality theorem:

Theorem 1 (Globally Optimal Self-Changes, given u and $\mathcal{A}$ encoded in p). *Given any formalizable utility function u (Item 1f), and assuming consistency of the underlying formal system $\mathcal{A}$, any self-change of p obtained through execution of some program* switchprog *identified through the proof of a target theorem (2) is globally optimal in the following sense: the utility of starting the execution of the present* switchprog *is higher than the utility of waiting for the proof searcher to produce an alternative* switchprog *later.*

Proof. Target theorem (2) implicitly talks about all the other *switchprogs* that the proof searcher could produce in the future. To see this, consider the two alternatives of the binary decision: (1) either execute the current *switchprog* (set *switchbit* $= 1$), or (2) keep searching for *proofs* and *switchprogs* (set *switchbit* $= 0$) until the systematic searcher comes up with an even better *switchprog*. Obviously the second alternative concerns all (possibly infinitely many) potential *switchprogs* to be considered later. That is, if the current *switchprog* were not the 'best', then the proof searcher would not be able to prove that setting *switchbit* and executing *switchprog* will cause higher expected reward than discarding *switchprog*, assuming consistency of $\mathcal{A}$. *Q.E.D.*

5 Bias-Optimal Proof Search (BIOPS)

Here we construct a $p(1)$ that is $O()$-optimal in a certain limited sense to be described below, but still might be improved as it is not necessarily optimal in the sense of the given u (for example, the u of equation (1) neither mentions nor cares for $O()$-optimality). Our Bias-Optimal Proof Search (BIOPS) is essentially an application of Universal Search [7] to proof search. Practical extensions [9] of Universal Search have been applied to *offline* program search tasks where the program inputs are fixed such that the same program always produces the same results. In our *online* setting, however, BIOPS has to take into account that the same proof technique started at different times may yield different proofs, as it may read parts of s (e.g., inputs) that change as the machine's life proceeds.

BIOPS starts with a probability distribution P (the initial bias) on the proof techniques w that one can write in $\mathcal{L}$, e.g., $P(w) = K^{-l(w)}$ for programs composed from K possible instructions [7]. BIOPS is *near-bias-optimal* [9] in the sense that it will not spend much more time on any proof technique than it deserves, according to its probabilistic bias, namely, not much more than its probability times the total search time:

Method 51 (BIOPS). *In phase* $(i = 1, 2, 3, \ldots)$ DO*:*

FOR *all self-delimiting [7] proof techniques* $w \in \mathcal{L}$ *satisfying* $P(w) \geq 2^{-i}$ DO*:*

1. *Run* w *until halt or error (such as division by zero) or* $2^i P(w)$ *steps consumed.*
2. *Undo the effects of* w *on* s^p *(does not cost significantly more time than executing* w*).*

A proof technique w can interrupt Method 51 only by invoking instruction *check()* (Item 5), which may transfer control to *switchprog* (which possibly even will delete or rewrite Method 51). Since the initial p runs on the formalized hardware, and since proof techniques tested by p can read p and other parts of s, they can produce proofs concerning the (expected) performance of p and BIOPS itself. Method 51 at least has the optimal *order* of computational complexity in the following sense.

Theorem 2. *If independently of variable* time(s) *some unknown fast proof technique* w *would require at most* $f(k)$ *steps to produce a proof of difficulty measure* k *(an integer depending on the nature of the task to be solved), then Method 51 will need at most* $O(f(k))$ *steps.*

Proof. It is easy to see that Method 51 will need at most $O(f(k)/P(w)) = O(f(k))$ steps—the constant factor $1/P(w)$ does not depend on k. *Q.E.D.*

Note again, however, that the proofs themselves may concern quite different, arbitrary formalizable notions of optimality (stronger than those expressible in the $O()$-notation) embodied by the given, problem-specific, formalized utility function u. This may provoke useful, constant-affecting rewrites of the initial proof searcher despite its limited (yet popular and widely used) notion of $O()$-optimality.

6 Conclusion

The initial software $p(1)$ of our machine runs an initial problem solver (e.g., one of Hutter's approaches [5] which have at least an optimal *order* of complexity). Simultaneously, it runs an $O()$-optimal initial proof searcher using an online variant of Universal Search to test *proof techniques*, which are programs able to compute proofs concerning the system's own future performance, based on an axiomatic system $\mathcal{A}$ encoded in $p(1)$, describing a formal *utility* function u, the hardware and $p(1)$ itself. If there is no provably good, globally optimal way of rewriting $p(1)$ at all, then humans will not find one either. But if there is one, then $p(1)$ itself can find and exploit it. This approach yields the first class of theoretically sound, fully self-referential, optimal, general RL machines.

After the theoretical analysis above, one practical question remains: to build a particular, especially practical Gödel machine with small initial constant overhead, which generally useful theorems should one add as axioms to $\mathcal{A}$ (as initial bias) such that the initial searcher does not have to prove them from scratch?

References

1. M. Blum. On effective procedures for speeding up algorithms. *Journal of the ACM*, 18(2):290–305, 1971.
2. M. C. Fitting. *First-Order Logic and Automated Theorem Proving*. Graduate Texts in Computer Science. Springer-Verlag, Berlin, 2nd edition, 1996.
3. K. Gödel. Über formal unentscheidbare Sätze der Principia Mathematica und verwandter Systeme I. *Monatshefte für Mathematik und Physik*, 38:173–198, 1931.
4. M. Hutter. The fastest and shortest algorithm for all well-defined problems. *International Journal of Foundations of Computer Science*, 13(3):431–443, 2002. (On J. Schmidhuber's SNF grant 20-61847).
5. M. Hutter. *Universal Artificial Intelligence: Sequential Decisions based on Algorithmic Probability*. Springer, Berlin, 2004. (On J. Schmidhuber's SNF grant 20-61847).
6. L. P. Kaelbling, M. L. Littman, and A. W. Moore. Reinforcement learning: a survey. *Journal of AI research*, 4:237–285, 1996.
7. L. A. Levin. Randomness conservation inequalities: Information and independence in mathematical theories. *Information and Control*, 61:15–37, 1984.
8. J. Schmidhuber. The Speed Prior: a new simplicity measure yielding near-optimal computable predictions. In J. Kivinen and R. H. Sloan, editors, *Proceedings of the 15th Annual Conference on Computational Learning Theory (COLT 2002)*, Lecture Notes in Artificial Intelligence, pages 216–228. Springer, Sydney, Australia, 2002.
9. J. Schmidhuber. Optimal ordered problem solver. *Machine Learning*, 54:211–254, 2004.
10. J. Schmidhuber. Gödel machines: fully self-referential optimal universal problem solvers. In B. Goertzel and C. Pennachin, editors, *Artificial General Intelligence*. Springer Verlag, in press, 2005.
11. J. Schmidhuber. Gödel machines: Towards a technical justification of consciousness. In D. Kudenko, D. Kazakov, and E. Alonso, editors, *Adaptive Agents and Multi-Agent Systems III (LNCS 3394)*, pages 1–23. Springer Verlag, 2005.
12. R. J. Solomonoff. Complexity-based induction systems. *IEEE Transactions on Information Theory*, IT-24(5):422–432, 1978.

Model Selection Under Covariate Shift*

Masashi Sugiyama[1] and Klaus-Robert Müller[2]

[1] Tokyo Institute of Technology, Tokyo, Japan
sugi@cs.titech.ac.jp
http://sugiyama-www.cs.titech.ac.jp/~sugi/

[2] Fraunhofer FIRST.IDA, Berlin, and University of Potsdam, Potsdam, Germany
klaus@first.fhg.de
http://ida.first.fraunhofer.de/~klaus/

Abstract. A common assumption in supervised learning is that the training and test input points follow the same probability distribution. However, this assumption is not fulfilled, e.g., in interpolation, extrapolation, or active learning scenarios. The violation of this assumption—known as the covariate shift—causes a heavy bias in standard generalization error estimation schemes such as cross-validation and thus they result in poor model selection. In this paper, we therefore propose an alternative estimator of the generalization error. Under covariate shift, the proposed generalization error estimator is unbiased if the learning target function is included in the model at hand and it is asymptotically unbiased in general. Experimental results show that model selection with the proposed generalization error estimator is compared favorably to cross-validation in extrapolation.

1 Introduction

Let us consider a regression problem of estimating an unknown function $f(\boldsymbol{x})$ from training examples $\{(\boldsymbol{x}_i, y_i) \mid y_i = f(\boldsymbol{x}_i) + \epsilon_i\}_{i=1}^{n}$, where $\{\epsilon_i\}_{i=1}^{n}$ are i.i.d. random noise with mean zero and unknown variance σ^2. Using a linear regression model

$$\hat{f}(\boldsymbol{x}) = \sum_{i=1}^{p} \alpha_i \varphi_i(\boldsymbol{x}), \tag{1}$$

where $\{\varphi_i(\boldsymbol{x})\}_{i=1}^{p}$ are fixed linearly independent functions and $\boldsymbol{\alpha} = (\alpha_1, \alpha_2, \ldots, \alpha_p)^\top$ are parameters, we would like to learn the parameter $\boldsymbol{\alpha}$ such that the squared test error expected over all test input points (or the *generalization error*) is minimized. Suppose the test input points independently follow a probability distribution with density $p_t(\boldsymbol{x})$ (> 0). Then the generalization error is expressed as

$$J = \int \left(\hat{f}(\boldsymbol{x}) - f(\boldsymbol{x})\right)^2 p_t(\boldsymbol{x}) d\boldsymbol{x}. \tag{2}$$

* The authors would like to thank Dr. Motoaki Kawanabe and Dr. Gilles Blanchard for their valuable comments. We acknowledge the Alexander von Humboldt Foundation and from the PASCAL Network of Excellence (EU #506778) for financial support.

W. Duch et al. (Eds.): ICANN 2005, LNCS 3697, pp. 235–240, 2005.

A common assumption in this supervised learning is that the training *input* points $\{\boldsymbol{x}_i\}_{i=1}^n$ independently follow the *same* probability distribution as the test input points [4]. However, this assumption is not fulfilled, for example, in *interpolation* or *extrapolation* scenarios: only few (or no) training input points exist in the regions of interest, implying that the test distribution is significantly different from the training distribution. *Active learning* also corresponds to such cases because the locations of training input points are designed by users while test input points are provided from the environment [1]. The situation where the training and test distributions are different is referred to as the situation under the *covariate shift* [3] or the *sample selection bias* [2]. Let $p_x(\boldsymbol{x})$ (> 0) be the probability density function of training input points $\{\boldsymbol{x}_i\}_{i=1}^n$. An example of an extrapolation problem where $p_x(\boldsymbol{x}) \neq p_t(\boldsymbol{x})$ is illustrated in Figure 1.

When $p_x(\boldsymbol{x}) \neq p_t(\boldsymbol{x})$, two difficulties arise in a learning process. The first difficulty is parameter learning. The ordinary least-squares learning, given by

$$\min_{\boldsymbol{\alpha}} \left[\sum_{i=1}^{n} \left(\hat{f}(\boldsymbol{x}_i) - y_i \right)^2 \right], \tag{3}$$

tries to fit the data well in the region with high training data density. This implies that the prediction can be inaccurate if the region with high test data density has low training data density. Theoretically, it is known that when the training and test distributions are different and the true function is *unrealizable* (i.e., the learning target function is not included in the model at hand), least-squares learning is no longer *consistent* (i.e., the learned parameter does not converge to the optimal one even when the number of training examples goes to infinity). This problem can be overcome by using a least-squares learning weighted by the *ratio* of test and training data densities[1] [3].

$$\min_{\boldsymbol{\alpha}} \left[\sum_{i=1}^{n} \frac{p_t(\boldsymbol{x}_i)}{p_x(\boldsymbol{x}_i)} \left(\hat{f}(\boldsymbol{x}_i) - y_i \right)^2 \right]. \tag{4}$$

A key idea of this weighted version is that the training data density is adjusted to the test data density by the density ratio, which is similar in spirit to *importance sampling.* Although the consistency becomes guaranteed by this modification, the weighted least-squares learning tends to have large variance. Indeed, it is no longer *asymptotically efficient* even when the noise is Gaussian. Therefore, in practical situations with finite samples, a stabilized estimator, e.g.,

$$\min_{\boldsymbol{\alpha}} \left[\sum_{i=1}^{n} \left(\frac{p_t(\boldsymbol{x}_i)}{p_x(\boldsymbol{x}_i)} \right)^{\lambda} \left(\hat{f}(\boldsymbol{x}_i) - y_i \right)^2 \right] \quad \text{for } 0 \leq \lambda \leq 1 \tag{5}$$

would give more accurate estimates. The learned parameter $\hat{\boldsymbol{\alpha}}_\lambda$ obtained by the weighted least-squares learning (5) is given by $\hat{\boldsymbol{\alpha}}_\lambda = \boldsymbol{L}_\lambda \boldsymbol{y}$, where $\boldsymbol{L}_\lambda = (\boldsymbol{X}^\top \boldsymbol{D}^\lambda \boldsymbol{X})^{-1} \boldsymbol{X}^\top \boldsymbol{D}^\lambda$, $\boldsymbol{X}_{i,j} = \varphi_j(\boldsymbol{x}_i)$, $\boldsymbol{D}$ is the diagonal matrix with the i-th diagonal element $p_t(\boldsymbol{x}_i)/p_x(\boldsymbol{x}_i)$, and $\boldsymbol{y} = (y_1, y_2, \ldots, y_n)^\top$. Note that $\lambda = 0$

[1] In theory, we assume that $p_x(\boldsymbol{x})$ and $p_t(\boldsymbol{x})$ are known. Later in experiments, they are estimated from the data and we evaluate the practical usefulness of the theory.

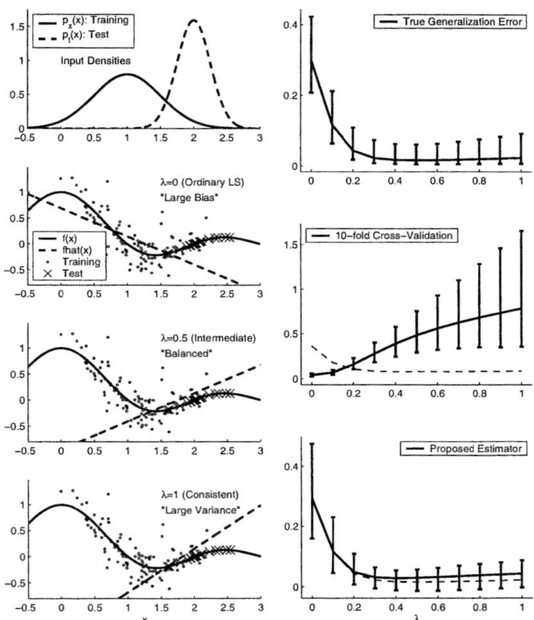

Fig. 1. An illustrative example of extrapolation by fitting a linear function $\hat{f}(\boldsymbol{x}) = \alpha_1 + \alpha_2 x$. [Left column]: The top graph depicts the probability density functions of the training and test input points, $p_x(x)$ and $p_t(x)$. In the bottom three graphs, the learning target function $f(x)$ is drawn by the solid line, the noisy training examples are plotted with o's, a learned function $\hat{f}(x)$ is drawn by the dashed line, and the (noiseless) test examples are plotted with $\times$'s. Three different learned functions are obtained by weighted least-squares learning with different tuning parameter λ. $\lambda = 0$ corresponds to the ordinary least-squares learning (small variance but large bias), while $\lambda = 1$ gives a consistent estimate (small bias but large variance). With finite samples, an intermediate λ, say $\lambda = 0.5$, often provides better results. [Right column]: The top graph depicts the mean and standard deviation of the generalization error over 300 independent trials, as a function of λ. The middle and bottom graphs depict the means and standard deviations of the estimated generalization error obtained by the standard 10-fold cross-validation (10CV) and the proposed method. The dotted lines are the mean of the true generalization error. 10CV is heavily biased because of $p_x(x) \neq p_t(x)$, while the proposed estimator is almost unbiased with reasonably small variance.

corresponds to the ordinary least-squares learning (3), while $\lambda = 1$ corresponds to consistent weighted least-squares learning (4). Thus, the parameter learning problem is now relocated to the model selection problem of choosing λ.

However, the second difficulty when $p_x(\boldsymbol{x}) \neq p_t(\boldsymbol{x})$ is model selection itself. Standard unbiased generalization error estimation schemes such as cross-validation are heavily biased, because the generalization error is over-estimated in the high training data density region and it is under-estimated in the high test data density region.

In this paper, we therefore propose a *new* generalization error estimator. Under covariate shift, the proposed estimator is proved to be exactly unbiased with finite samples in realizable cases and asymptotically unbiased in general. Furthermore, the proposed generalization error estimator is shown to be able to accurately estimate the *difference* of the generalization error, which is a useful property in model selection.

For simplicity, we focus on the problem of choosing the tuning parameter λ in the following. Note, however, that the proposed theory can be easily extended to general model selection of choosing basis functions or regularization constant.

2 A New Generalization Error Estimator

Let us decompose the learning target function $f(\boldsymbol{x})$ into $f(\boldsymbol{x}) = g(\boldsymbol{x}) + r(\boldsymbol{x})$, where $g(\boldsymbol{x})$ is the orthogonal projection of $f(\boldsymbol{x})$ onto the span of $\{\varphi_i(\boldsymbol{x})\}_{i=1}^p$ and the residual $r(\boldsymbol{x})$ is orthogonal to $\{\varphi_i(\boldsymbol{x})\}_{i=1}^p$, i.e., $\int r(\boldsymbol{x})\varphi_i(\boldsymbol{x})p_t(\boldsymbol{x})d\boldsymbol{x} = 0$. Since $g(\boldsymbol{x})$ is included in the span of $\{\varphi_i(\boldsymbol{x})\}_{i=1}^p$, it is expressed by $g(\boldsymbol{x}) = \sum_{i=1}^p \alpha_i^* \varphi_i(\boldsymbol{x})$, where $\boldsymbol{\alpha}^* = (\alpha_1^*, \alpha_2^*, \ldots, \alpha_p^*)^\top$ are unknown optimal parameters.

Let $\boldsymbol{U}$ be a p-dimensional matrix with the (i,j)-th element $\boldsymbol{U}_{i,j} = \int \varphi_i(\boldsymbol{x})\varphi_j(\boldsymbol{x})p_t(\boldsymbol{x})d\boldsymbol{x}$, which is assumed to be accessible in the current setting. Then the generalization error J is expressed as

$$J(\lambda) = \int \hat{f}_\lambda(\boldsymbol{x})^2 p_t(\boldsymbol{x})d\boldsymbol{x} - 2\int \hat{f}_\lambda(\boldsymbol{x})f(\boldsymbol{x})p_t(\boldsymbol{x})d\boldsymbol{x} + \int f(\boldsymbol{x})^2 p_t(\boldsymbol{x})d\boldsymbol{x}$$
$$= \langle \boldsymbol{U}\hat{\boldsymbol{\alpha}}_\lambda, \hat{\boldsymbol{\alpha}}_\lambda\rangle - 2\langle \boldsymbol{U}\hat{\boldsymbol{\alpha}}_\lambda, \boldsymbol{\alpha}^*\rangle + C, \tag{6}$$

where $C = \int f(\boldsymbol{x})^2 p_t(\boldsymbol{x})d\boldsymbol{x}$. In Eq.(6), the first term $\langle \boldsymbol{U}\hat{\boldsymbol{\alpha}}_\lambda, \hat{\boldsymbol{\alpha}}_\lambda\rangle$ is accessible and the third term C does not depend on λ. Therefore, we focus on estimating the second term "$-2\langle \boldsymbol{U}\hat{\boldsymbol{\alpha}}_\lambda, \boldsymbol{\alpha}^*\rangle$".

Hypothetically, let us suppose that the following two quantities are available.

(i) A matrix $\boldsymbol{L}_u$ which gives a linear unbiased estimator of the unknown true parameter $\boldsymbol{\alpha}^*$: $\mathbb{E}_\epsilon \boldsymbol{L}_u \boldsymbol{y} = \boldsymbol{\alpha}^*$, where $\mathbb{E}_\epsilon$ denotes the expectation over the noise $\{\epsilon_i\}_{i=1}^n$.

(ii) An unbiased estimator σ_u^2 of the noise variance σ^2: $\mathbb{E}_\epsilon \sigma_u^2 = \sigma^2$.

Note that $\boldsymbol{L}_u$ does not depend on $\boldsymbol{L}_\lambda$. Then it holds that

$$\mathbb{E}_\epsilon\langle \boldsymbol{U}\hat{\boldsymbol{\alpha}}_\lambda, \boldsymbol{\alpha}^*\rangle = \langle \mathbb{E}_\epsilon \boldsymbol{U}\boldsymbol{L}_\lambda\boldsymbol{y}, \mathbb{E}_\epsilon \boldsymbol{L}_u\boldsymbol{y}\rangle = \mathbb{E}_\epsilon[\langle \boldsymbol{U}\boldsymbol{L}_\lambda\boldsymbol{y}, \boldsymbol{L}_u\boldsymbol{y}\rangle - \sigma_u^2 \mathrm{tr}(\boldsymbol{U}\boldsymbol{L}_\lambda\boldsymbol{L}_u^\top)], \tag{7}$$

which implies that we can construct an unbiased estimator of $\mathbb{E}_\epsilon\langle \boldsymbol{U}\hat{\boldsymbol{\alpha}}_\lambda, \boldsymbol{\alpha}^*\rangle$ if $\boldsymbol{L}_u$ and σ_u^2 are available. However, in general, neither $\boldsymbol{L}_u$ nor σ_u^2 may be available. So we use the following approximations instead:

$$\widehat{\boldsymbol{L}}_u = (\boldsymbol{X}^\top\boldsymbol{D}\boldsymbol{X})^{-1}\boldsymbol{X}^\top\boldsymbol{D} \quad \text{and} \quad \widehat{\sigma_u^2} = \|\boldsymbol{G}\boldsymbol{y}\|^2/\mathrm{tr}(\boldsymbol{G}), \tag{8}$$

where $\boldsymbol{G} = \boldsymbol{I} - \boldsymbol{X}(\boldsymbol{X}^\top\boldsymbol{X})^{-1}\boldsymbol{X}^\top$. Actually, $\widehat{\boldsymbol{L}}_u$ corresponds to Eq.(4), which implies that $\widehat{\boldsymbol{L}}_u$ exactly fulfills the requirement (i) in realizable cases and asymptotically satisfies it in general [3]. On the other hand, it is known that the above $\widehat{\sigma_u^2}$ exactly fulfills the requirement (ii) in realizable cases [1]. Although, in unrealizable cases, $\widehat{\sigma_u^2}$ does not satisfy the requirement (ii) even asymptotically, it turns out that the asymptotic unbiasedness of $\widehat{\sigma_u^2}$ is not needed in the following.

Based on the above discussion, we define the following estimator $\hat{J}$ of the generalization error J.

$$\hat{J}(\lambda) = \langle \boldsymbol{U}\boldsymbol{L}_\lambda\boldsymbol{y}, \boldsymbol{L}_\lambda\boldsymbol{y}\rangle - 2\langle \boldsymbol{U}\boldsymbol{L}_\lambda\boldsymbol{y}, \widehat{\boldsymbol{L}}_u\boldsymbol{y}\rangle + 2\widehat{\sigma_u^2}\mathrm{tr}(\boldsymbol{U}\boldsymbol{L}_\lambda\widehat{\boldsymbol{L}}_u^\top). \tag{9}$$

Let B_ϵ be the bias of $\hat{J}$: $B_\epsilon = \mathbb{E}_\epsilon[\hat{J} - J] + C$. Then we have the following theorem (proof is omitted because of lack of space).

Theorem 1 *If $r(\boldsymbol{x}_i) = 0$ for $i = 1, 2, \ldots, n$, $B_\epsilon = 0$. If $\delta = \max\{|r(\boldsymbol{x}_i)|\}_{i=1}^n$ is sufficiently small, $B_\epsilon = \mathcal{O}(\delta)$. If n is sufficiently large, $B_\epsilon = \mathcal{O}_p(n^{-\frac{1}{2}})$.*

This theorem implies that, except for the constant C, $\hat{J}$ is exactly unbiased if $f(\boldsymbol{x})$ is strictly realizable, it is almost unbiased if $f(\boldsymbol{x})$ is almost realizable, and it is asymptotically unbiased in general. We can also prove that the above $\hat{J}$ can estimate the *difference* of the generalization error among different models. However, because of lack of space, we omit the detail.

3 Numerical Examples

Figure 1 shows the numerical results of an illustrative extrapolation problem. The curves in the right column show that the proposed estimator gives almost unbiased estimates of the generalization error with reasonably small variance (note that the target function is not realizable in this case).

We also applied the proposed method to *Abalone* data set available from the UCI repository. It is a collection of 4177 samples, each of which consists of 8 input variables (physical measurements of abalones) and 1 output variable (the age of abalones). The first input variable is qualitative (male/female/infant) so it was ignored, and the other input variables were normalized to $[0, 1]$ for convenience. From the population, we randomly sampled n abalones for training and 100 abalones for testing. Here, we considered a biased sampling: the sampling of the 4-th input variable (weight of abalones) has negative bias for training and positive bias for testing. That is, the weight of training abalones tends to be small while that for the test abalones tends to be large. We used multi-dimensional linear basis functions for learning. Here we suppose that the test input points are known (i.e., the setting corresponds to transductive inference [4]) and the density functions $p_x(\boldsymbol{x})$ and $p_t(\boldsymbol{x})$ were estimated from the training input points and test input points, respectively, using a kernel density estimation method.

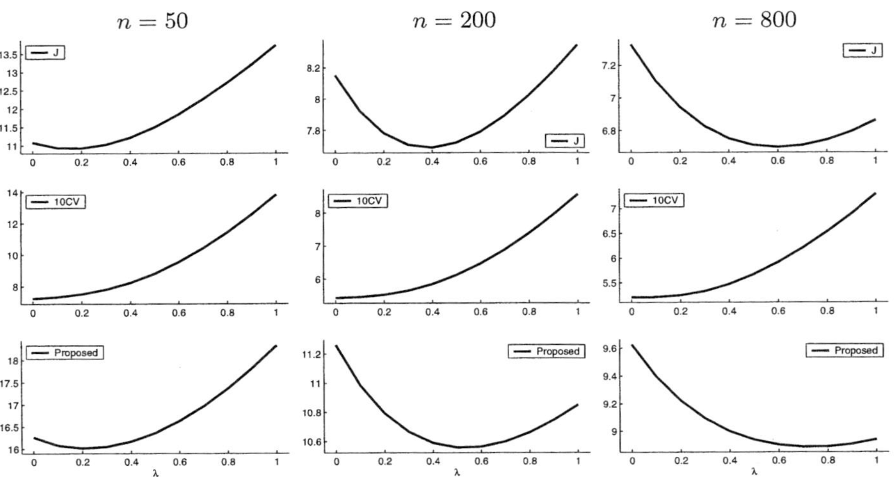

Fig. 2. Extrapolation of the 4-th variable in the Abalone dataset. The mean of each method is described. Each column corresponds to each n.

Table 1. Extrapolation of the 4-th variable (left) or the 6-th variable (right) in the Abalone dataset. The mean and standard deviation of the test error obtained with each method are described. The better method and comparable one by the t-test at the significance level 5% are described with boldface.

n	$\hat{J}$	10CV
50	$\mathbf{11.67 \pm 5.74}$	$\mathbf{10.88 \pm 5.05}$
200	$\mathbf{7.95 \pm 2.15}$	$\mathbf{8.06 \pm 1.91}$
800	$\mathbf{6.77 \pm 1.40}$	7.23 ± 1.37

n	$\hat{J}$	10CV
50	$\mathbf{10.67 \pm 6.19}$	$\mathbf{10.15 \pm 4.95}$
200	$\mathbf{7.31 \pm 2.24}$	$\mathbf{7.42 \pm 1.81}$
800	$\mathbf{6.20 \pm 1.33}$	6.68 ± 1.25

Figure 2 depicts the mean values of each method over 300 trials for $n = 50, 200$, and 800. The error bars are omitted because they were excessive and deteriorated the graphs. Note that the true generalization error J is calculated using the test examples. The proposed $\hat{J}$ seems to give reasonably good curves and its minimum roughly agrees with the minimum of the true test error. On the other hand, irrespective of n, the minimizer of 10CV tend to be small.

We chose the tuning parameter λ by each method and estimated the age of the test abalones by using the chosen λ. The mean squared test error for all test abalones were calculated, and this procedure was repeated 300 times. The mean and standard deviation of the test error of each method are described in the left half of Table 1. It shows that $\hat{J}$ and 10CV work comparably for $n = 50, 200$, while $\hat{J}$ outperforms 10CV for $n = 800$. Hence, the proposed method overall compares favorably to 10CV.

We also carried out similar simulations when the sampling of the 6-th input variable (weight of gut after bleeding) is biased. The results described in the right half of Table 1 showed similar trends to the previous ones.

4 Conclusions

In this paper, we proposed a new generalization error estimator under covariate shift. The proposed estimator is shown to be unbiased with finite samples in realizable cases and asymptotically unbiased in general. Experimental results showed that model selection with the proposed generalization error estimator is compared favorably to the standard cross-validation in extrapolation scenarios.

References

1. V. V. Fedorov. *Theory of Optimal Experiments.* Academic Press, New York, 1972.
2. J. J. Heckman. Sample selection bias as a specification error. *Econometrica*, 47(1):153–162, 1979.
3. H. Shimodaira. Improving predictive inference under covariate shift by weighting the log-likelihood function. *Journal of Statistical Planning and Inference*, 90(2):227–244, 2000.
4. V. N. Vapnik. *Statistical Learning Theory.* John Wiley & Sons, Inc., New York, 1998.

Smooth Performance Landscapes of the Variational Bayesian Approach

Zhuo Gao* and K.Y. Michael Wong

Hong Kong University of Science and Technology, Hong Kong, China
zhuogao@bnu.edu.cn
phkywong@ust.hk

Abstract. We consider the practical advantage of the Bayesian approach over maximum *a posteriori* methods in its ability to smoothen the landscape of generalization performance measures in the space of hyperparameters, which is vitally important for determining the optimal hyperparameters. The variational method is used to approximate the intractable distribution. Using the leave-one-out error of support vector regression as an example, we demonstrate a further advantage of this method in the analytical estimation of the leave-one-out error, without doing the cross-validation. Comparing our theory with the simulations on both artificial (the "sinc" function) and benchmark (the Boston Housing) data sets, we get a good agreement.

1 Introduction

Bayesian approaches provide a unified framework for probabilistic inference [1]. Compared with other heuristic or maximum *a posteriori* (MAP) approaches, they make explicit the models of priors and noises underlying the data, thus avoiding the overfitting of data and facilitating the natural development in analyses and model selection [2]. However, since Bayesian approaches analyse a *distribution* of models explaining the data, in contrast to MAP approaches which only considers the most likely one, the computational intractability increases. To deal with this problem, variational methods have become popular recently [3].

In this paper, we consider a practical advantage of the Bayesian approach over MAP methods, namely, its ability to smoothen the landscape of generalization performance measures in the space of hyperparameters. This smoothness is vitally important when the hyperparameters are tuned to search for their optimal choices. The generalization performance measures may refer to the marginal likelihood, cross-validation error, bootstrap error, or leave-one-out (LOO) error. Since these measures are estimated empirically from the example sets of finite sizes, changes in the relative weights of individual examples often roughen the landscape of the hyperparameters. This problem is particularly relevant to the recently popular Support Vector Machines [4].

* Permanent address: Physics, Beijing Normal University, Beijing 100875, China.

W. Duch et al. (Eds.): ICANN 2005, LNCS 3697, pp. 241–246, 2005.

On the other hand, the Bayesian formulation considers a *distribution* of models, whose relative weights can be controlled by a *temperature* hyperparameter. At a "low temperature", one recovers the MAP formulation, while "high temperature" corresponds to uniform distributions of the models. Hence, when the temperature increases, the landscape of the performance measures smoothens. The variational approach is used to overcome the increased intractability.

We will mention an additional advantage of the variational Bayesian approach, namely, the reduced computational complexity through analytical estimation of the generalization performance measures [5,6]. Consider the example of Support Vector Machines. Some empirical suggestions for model selection have been given [7,8]. Theoretical bounds on the risk were also used [9,10,11]. However, the general applicability of these methods remain to be studied. In practice, one of the commonly used methods is resampling, such as cross-validation [12,13], but it is usually computationally expensive. Using the variational Bayesian approach described in this paper, the LOO error can be estimated analytically, without having to go through the cumbersome process of actual LOO validation. Furthermore, the formulation of the LOO error is simpler than the variational replica approach of the bootstrap error [14]. We compare our analytical approximations with computer simulations for both artificial and benchmark datasets. The results show good agreement.

2 The Bayesian Model and the Variational Method

In the Bayesian approach [1], we consider a probabilistic model explaining a dataset D. The model is described by a set of parameters θ, whose prior distribution is $P(\theta)$. The process of generating the dataset from the parameters is described by the likelihood $P(D|\theta)$. For a dataset of N independent training examples D_i $(i = 1, \cdots, N)$, the likelihood can be factorized as $P(D|\theta) = \prod_{i=1}^{N} P(D_i|\theta)$. According to Bayes' theory, the posterior probability of the model θ is then $P(\theta|D) = P(D|\theta)P(\theta)/P(D)$, where $P(D) = \int d\theta\, P(D|\theta)P(\theta)$.

We consider prior and the likelihood distributions expressed in the exponential form, $P(\theta) \propto \exp[-\beta V(\theta)]$ and $P(D_i|\theta) \propto \exp[-\beta U(D_i, \theta)]$, where β is an inverse temperature parameter. Then, the posterior probability of the model is $P(\theta|D) = \exp(-\beta H)/Z$, where $H = V(\theta) + \sum_i U(D_i, \theta)$ is the regularized risk function, and $Z \equiv P(D) = \int d\theta \exp(-\beta H)$ is the marginal likelihood. So, maximizing the posterior with respect to θ is equivalent to minimizing H. The Bayesian average of an arbitrary quantity A is then $\langle A\rangle = \int d\theta\, A \exp(-\beta H)/Z$. In statistical mechanics, the marginal likelihood is called the partition function, and its logarithm is related to the free energy by $F = -\ln Z/\beta$.

In general, the computation of the marginal likelihood is intractable. So, the variational approach is considered. We approximate the posterior probability by a tractable distribution, $P_0(\theta) = \exp(-\beta H_0)/Z_0$. Then, the free energy can be written as $F = -\ln Z_0/\beta - \ln\langle\exp[-\beta(H - H_0)]\rangle_0/\beta$, where $\langle\cdots\rangle_0$ denotes average over the distribution P_0. Since $\langle\exp x\rangle \geq \exp\langle x\rangle$, we have $F \leq -\ln Z_0/\beta + \langle H - H_0\rangle_0 \equiv F_{\rm var}$, Our objective is then to find an approximated H_0 such that the variational free energy F_{var} is minimized in a family of tractable distributions, hoping that the optimized model can capture the essence of the true model.

In the following, we will use the support vector regression (SVR) as an example, and introduce a mean-field approximation as the candidate distribution. The probabilistic model of SVR deals with a fixed set of data $D = \{\mathbf{x}_i, y_i\}_{i=1}^N$, where $\mathbf{x}_i$ is the input vector of the ith example, and y_i is the output. The input $\mathbf{x}$ is first mapped into a feature space, which usually has a higher dimension than the input space. We denote the map as $\mathbf{g}(\mathbf{x})$. Then the linear model in the feature space is given by $f(\mathbf{x}) = \mathbf{w} \cdot \mathbf{g}(\mathbf{x}) + b$, where $\mathbf{w}$ is the *weight* vector and b is the *bias*. Thus, the set of parameters $\{\theta\}$ of the model refers to $\{\mathbf{w}, b\}$.

In SVR, V and U in the exponential arguments of the prior $P(\mathbf{w}, b)$ and the likelihood $P(y_i|\mathbf{x}_i; \mathbf{w}, b)$ take the forms $V(\mathbf{w}, b) = w^2/2$ and $U(\mathbf{x}, y, f) = C \max(0, |f - y| - \epsilon)$ respectively, where $U(\mathbf{x}, y, f)$ is the linear ϵ-insensitive loss function. The regularized risk function is then $H = w^2/2 + \sum_{i=1}^N U(\mathbf{x}_i, y_i, f(\mathbf{x}_i))$. Adopting the variational approach, we propose a mean-field variational distribution decribed by $H_0 = w^2/2 + \sum_{i=1}^N U_0(\mathbf{x}_i, y_i, f(\mathbf{x}_i))$, where $U_0(\mathbf{x}_i, y_i, f(\mathbf{x}_i)) = \hat{q}_i[\mathbf{w} \cdot \mathbf{g}(\mathbf{x}_i)]^2/2 - \hat{r}_i[\mathbf{w} \cdot \mathbf{g}(\mathbf{x}_j)] + \hat{s}b^2/2 - \hat{t}b$. In contrast with the usual quadratic expression of H_0 in terms of the biased fields $f(\mathbf{x}_i)$ [14], we have omitted the cross-terms between $\mathbf{w} \cdot \mathbf{g}(\mathbf{x}_i)$ and b. This simplifies the analysis considerably, and the results are invariant in the limit when the bias is self-averaging. Minimizing F_{var} leads us to a set of variational equations involving the variational parameters $\hat{q}_i$, $\hat{r}_i$, $\hat{s}$ and $\hat{t}$, and the distribution $P_0(\mathbf{w}, b)$ is completely specified. The estimates of the data, described by the moments $R_i \equiv \langle \mathbf{w} \cdot \mathbf{g}_i \rangle_0$, $Q_{ij} \equiv \langle [\mathbf{w} \cdot \mathbf{g}(\mathbf{x}_i)][\mathbf{w} \cdot \mathbf{g}(\mathbf{x}_j)] \rangle_0 - R_i R_j$, $B \equiv \langle b \rangle_0$, and $S \equiv \langle b^2 \rangle_0 - B^2$ can be expressed in terms of the variational parameters [15]. In particular, we find

$$\langle f(\mathbf{x}_i) \rangle_0 = \sum_j K_{ij} \alpha_j + B, \quad \alpha_i \equiv - \int DX \, U'(\mathbf{x}_i, y_i, R_i + B + X\sqrt{Q_{ii} + S}), \quad (1)$$

where $DX \equiv dX \exp(-X^2/2)/\sqrt{2\pi}$ is the Gaussian measure, and the prime in U' represents the derivative of U with respect to X. $K_{ij} \equiv K(\mathbf{x}_i, \mathbf{x}_j) \equiv \mathbf{g}(\mathbf{x}_i) \cdot \mathbf{g}(\mathbf{x}_j)$ is the kernel. For an arbitrary input $\mathbf{x}$, the regression estimate has the form $\langle f(\mathbf{x}) \rangle_0 = \langle \mathbf{w} \cdot \mathbf{g}(\mathbf{x}) + b \rangle_0 = \sum_{i=1}^N \alpha_i K(\mathbf{x}_i, \mathbf{x}) + B$.

It is instructive to compare our approach with the standard SVR, which can be obtained in a MAP approach [4]. There, the regression problem is to find a function $\bar{f}(\mathbf{x})$ that maximizes the posterior. After dual transformation, the solution reads $\bar{f}(\mathbf{x}) = \sum_{i=1}^N \bar{\alpha}_i K(\mathbf{x}_i, \mathbf{x}) + \bar{B}$, where the dual variables $\bar{\alpha}_i$ satisfy $-C \leq \bar{\alpha}_i \leq C$ and $\sum_i \bar{\alpha}_i = 0$. The example i is called a *support vector* (*non-support vector*) if $\bar{\alpha}_i \neq 0$ ($\bar{\alpha}_i = 0$). For support vectors, we have $|\bar{f}(\mathbf{x}_i) - y_i| \geq \epsilon$, while for non-support vectors $|\bar{f}(\mathbf{x}_i) - y_i| < \epsilon$. This result is identical to that of the Bayesian approach, when the temperature β^{-1} approaches zero, causing the variance $Q_{ii} + S$ of the Gaussian noise in α_i to vanish.

Figure 1 shows the regression results of both the standard maximum-likelihood SVR and our variational Bayesian method. Here, we use an artificial data set, in which 500 examples are generated from $\text{sinc}(x) = (\sin \pi x)/(\pi x)$, and a

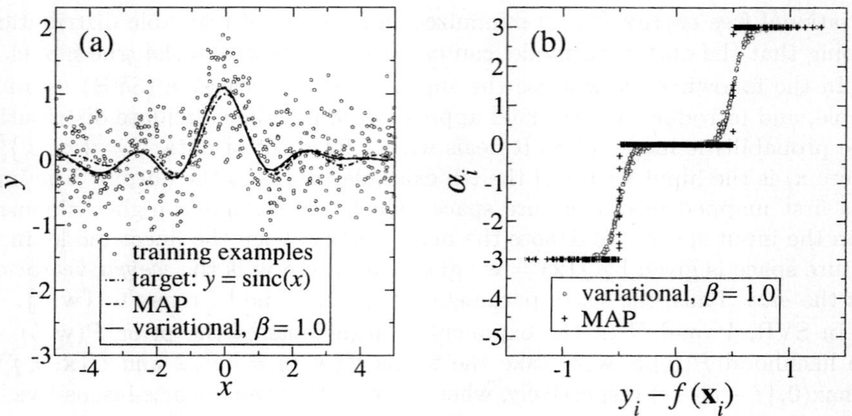

Fig. 1. Comparison of variational approach with the MAP approach. (a) The regression results. (b) The relation between the parameters α_i and the estimated function $f(\mathbf{x}_i)$. The parameters are $C = 3.0$, $\epsilon = 0.5$, and $l = 1.0$.

Gaussian noise with mean 0 and standard deviation 0.5 is added to the output. The radial-basis-function (RBF) kernel, $K(\mathbf{x}, \mathbf{x}') = \exp(-||\mathbf{x}-\mathbf{x}'||^2/2l^2)$, is used for the regression. The two regression curves in Fig. 1(a) are very close to each other. So, the variational Bayesian method is a very good approximation of the standard SVR. Figure 1(b) illustrates the relation between α_i and the estimated function $f(\mathbf{x}_i)$. The smoothened corners of the variational Bayesian result indicate the effects of Gaussian noise inherent in the Bayesian distribution.

LOO error ε_{LOO} is an unbiased estimate of the generalization error. It is the average error of examples, obtained by removing an example from the training data one at a time, and measuring the error of that example. Hence we have $\varepsilon_{LOO} \equiv \sum_{i=1}^{N}[\langle f(\mathbf{x}_i)\rangle^{\backslash i} - y_i]^2/N$, where the superscript $\backslash i$ represents expressions evaluated when example i is removed from the training data. In the variational Bayesian approach, we have $\langle f(\mathbf{x}_i)\rangle_0^{\backslash i} = \langle[\mathbf{w}\cdot\mathbf{g}(\mathbf{x}_i)+b]\mathrm{e}^{\beta U_0(f(\mathbf{x}_i))}\rangle_0/\langle\mathrm{e}^{\beta U_0(f(\mathbf{x}_i))}\rangle_0$ $= \langle f(\mathbf{x}_i)\rangle_0 - \alpha_i/[(\beta Q_{ii})^{-1} - u_{ii}]$, where $u_{ii} = \int DXU''(\mathbf{x}_i, y_i, R_i + B + X\sqrt{Q_{ii}+S})$. We note that this result is similar to that in [5], which applies to the MAP solution, whereas the present result extends to the Bayesian approach. So, the estimated LOO error becomes

$$\varepsilon_{LOO}^{\mathrm{est}} = \frac{1}{N}\sum_{i=1}^{N}[\langle f(\mathbf{x}_i)\rangle_0^{\backslash i} - y_i]^2 = \frac{1}{N}\sum_{i=1}^{N}\left[\langle f(\mathbf{x}_i)\rangle_0 - \frac{\alpha_i}{(\beta Q_{ii})^{-1} - u_{ii}} - y_i\right]^2 . \quad (2)$$

This shows that the LOO error can be estimated without having to go through the Bayesian learning process for leaving out each example.

In order to test our theory, both artificial ("sinc" function) and benchmark (Boston Housing) data sets are used. The Boston Housing data set has 506 examples and the input dimension is 13. Before we do the test, each component of the input vectors is normalized to zero mean and unit variance. The choice

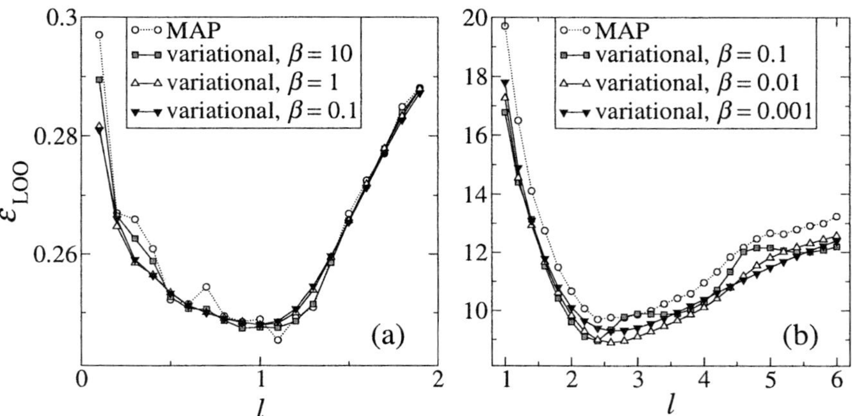

Fig. 2. The dependences of the leave-one-out error on the kernel width l for (a) the "sinc" data set, and (b) the Boston Housing data set. The other hyperparameters are $C = 3.0$ and $\epsilon = 0.5$ in (a), and $C = 100.0$ and $\epsilon = 2.0$ in (b).

of the parameter C follows those of [7,8]. While the results are sensitive to the choice of the parameter ϵ, we have fixed its value to make the demonstration of the principles more transparent. For both data sets, we use the RBF kernel and calculate the LOO error for different kernel widths l.

In Fig. 2, we present the estimated LOO error calculated by Eq. (2) as a function of the kernel width l, and the simulation curve of the standard SVR by cross-validation. For the latter, the curve is rough due to the leaping of the number of support vectors when the hyperparameter is changed. The roughness of the curve shows that it is easy to get trapped at local minima in the search for the optimal value of l. In contrast, the theoretical curves are much smoother for sufficiently high temperature. Furthermore, we find that the theoretical LOO estimate can be varied by tuning β. When β decreases, the rough behavior of the estimate smooths out, and the profile is rather insensitive to the choice of β.

3 Conclusions

We have presented a method to estimate the LOO averages of SVRs using a variational Bayesian approach. We take advantage of the Bayesian framework to get the smoothness of the error curves, and use the variational approach to estimate the LOO averages analytically. The comparison between our theory and the numerical experiments shows that our method works very well and fast for both yielding a regression estimate close to the standard SVR and finding the optimal hyperparameters with minimum generalization error.

There are other techniques for searching the optimal generalization performance, such as maximizing the marginal likelihood. As long as these techniques are based on data sets of finite size, they will be prone to the roughness problem as described here. Nevertheless, we find that the variation approach continue to produce good results [15].

The remaining issue is the determination of the maximum value of β above which the roughness characteristic of the standard SVR reappears. This is a topic for further study. Nevertheless, we have shown that as long as β is sufficiently small, the LOO error remains smooth.

Acknowledgements. We thank M. Opper and J. Kwok for meaningful discussions and encouragement. This work is supported by the Research Grant Council of Hong Kong (HKUST6062/02P and DAG04/05.SC25).

References

1. MacKay, D.J.C. (2003). *Information Theory, Inference, and Learning Algorithms.* Cambridge: Cambridge University Press.
2. Seeger, M. (2000). Bayesian model selection for support vector machines, Gaussian processes and other kernel classifiers. In S. A. Solla, T. K. Leen, and K.-R. Müller (eds.), *Advances in Neural Information Processing Systems 12*, pp. 603–609. Cambridge, MA: MIT Press.
3. Opper, M. & Saad, D. eds. (2001). *Advanced Mean Field Methods: Theory and Practice.* Cambridge, MA: MIT Press. For reviews on the variational method, see the chapters by T. Jaakola and Z. Ghahramani & M. J. Beal. For reviews on the mean-field approximation, see the chapters by M. Opper and O. Winther.
4. Vapnik, V.N. (1995). *The Nature of Statistical Learning Theory.* New York: Springer-Verlag.
5. Opper, M. & Winther, O. (1999). Gaussian Processes and SVM: Mean Field Results and Leave-One-Out Estimator. *Advances in Large Margin Classifiers*, Smola, Schölkopf and Schuurmans, eds. Cambridge, MA: MIT Press.
6. Wong, K.Y.M. & Li, F. (2000). Fast Parameter Estimation Using Green's Functions. In Z. Ghahramani (ed.), *Advances in Neural Information Processing Systems 14*, pp. 535–542. Cambridge, MA: MIT Press.
7. Cherkassky, V. & Ma, Y (2004). Practical Selection of SVM Parameters and Noise Estimation for SVM Regression. *Neural Networks* **17**:113–126.
8. Chalimourda, A., Schölkopf, B. & Smola, A.J. (2004). Experimentally Optimal ν in Support Vector Regression for Different Noise Models and Parameter Settings. *Neural Networks* **17**:127–141.
9. Vapnik, V. (1998). *Statistical Learning Theory.* New York: Wiley.
10. Cristianini, N. & Shawe-Taylor, J. (2000). *An Introduction to Support Vector Machines.* Cambridge: Cambridge University Press.
11. Chapelle, A. & Vapnik, V. (2000). Model Selection for Support Vector Machines. In S. A. Solla, T. K. Leen, and K.-R. Müller (eds.), *Advances in Neural Information Processing Systems 12*, pp. 230–236. Cambridge, MA: MIT Press.
12. Cherkassky, V. & Mülier, F. (1998). *Learning From Data: Concepts, Theory, and Methods.* New York: Wiley.
13. Schölkopf, B., Burges, J. & Smola, A. (1999). *Advances in Kernel Methods: Support Vector Machines, Regularization, and Beyond.* Cambridge, MA: MIT Press.
14. Malzahn, D. & Opper, M. (2003). A Statistical Mechanics Approach to Approximate Analytical Bootstrap Averages. In S. Becker, S. Thrun and K. Obermayer (eds.), *Advances in Neural Information Processing Systems 15*, pp. 327–334. Cambridge, MA: MIT Press.
15. Gao, Z. and Wong, K. Y. M., in preparation (2005).

Jacobi Alternative to Bayesian Evidence Maximization in Diffusion Filtering

Ramūnas Girdziušas and Jorma Laaksonen

Helsinki University of Technology, Laboratory of Computer and Information Science, P.O. Box 5400, FI-02015 HUT, Espoo, Finland
{Ramunas.Girdziusas, Jorma.Laaksonen}@hut.fi

Abstract. Nonlinear diffusion filtering presents a way to define and iterate Gaussian process regression so that large variance noise can be efficiently filtered from observations of size n in m iterations by performing approximately $O(mn)$ number of multiplications, while at the same time preserving the edges of the signal. Experimental evidence indicates that the optimal stopping time exist and the steady state solutions obtained by setting m to an arbitrarily large number are suboptimal. This work discusses the Bayesian evidence criterion, gives an interpretation to its basic components and proposes an alternative, simple optimal stopping method. A synthetic large-scale example indicates the usefulness of the proposed stopping criterion.

1 Introduction

Regression by means of nonlinear diffusion filtering [10] is fast and scales well with increasing input dimensionality. It becomes an especially useful alternative to neural networks and kernel machines when filtering long discontinuous oscillatory signals hidden in noise. The key feature of the diffusion filtering is that the optimal solution emerges long before the steady state. However, sensible probabilistic criteria for choosing the optimal stopping time are lacking.

This problem can be partially circumvented by putting nonlinear diffusion filtering into Bayesian Gaussian process (GP) regression framework [5]. This work further develops a kernel-based method for edge-preserving filtering when the number of discontinuities and the approximate intervals of their locations are not available. This comes in contrast to various extensions of Bayesian GP regression whose kernels have explicit parametric forms and where the locations of discontinuities are considered as hyperparameters [4]. We show how the nonlinear diffusion filtering defines the Gaussian process kernel in a data-dependent way, state several interpretations of the Bayesian evidence in the case of dynamic GP regression model, and examine a particular stopping criterion suitable for large-scale GP regression.

For this purpose, we first give a brief summary of the GP approach to diffusion filtering in Section 2. Several interpretations of the Bayesian evidence and the connection to Jacobi's conjugate point theory are briefly stated in Section 3. A large-scale numerical example with a discussion concludes our study in Section 4.

W. Duch et al. (Eds.): ICANN 2005, LNCS 3697, pp. 247–252, 2005.

2 Nonlinear Diffusion as Gaussian Process Regression

Consider that we are given n noisy scalar observations $y(\mathbf{x}_1), \ldots, y(\mathbf{x}_n)$ which are specified at locations $\mathbf{x}_i \in \mathbb{R}^d$. Let the observations represent a true signal which has Heaviside-type discontinuities and is corrupted by an additive Gaussian noise of variance comparable to the range of the signal.

A variety of practical ways to solve the problem of recovering the signal from its noisy observations can be seen in GP regression models equipped with the Bayesian evidence maximization [3]. If we gather noisy observations into a single vector $\mathbf{y} \in \mathbb{R}^n$, then the corresponding solution vector $\mathbf{u} \in \mathbb{R}^n$ at the same spatial locations can be found according to

$$\mathbf{u}^* = \arg\min_{\mathbf{u}} \big(\underbrace{\frac{1}{\theta_0^*}||\mathbf{y}-\mathbf{u}||^2}_{\text{'loglikelihood'}} + \underbrace{\mathbf{u}^T\mathbf{K}_{\theta^*}^{-1}\mathbf{u}}_{\text{'logprior'}} \big) = \underbrace{\mathbf{K}_{\theta^*}(\mathbf{K}_{\theta^*} + \theta_0^*\mathbf{I})^{-1}}_{\text{'posterior filter'}}\mathbf{y}, \tag{1}$$

$$\boldsymbol{\theta}^* = \arg\min_{\boldsymbol{\theta}} \big(\underbrace{\frac{1}{\theta_0}[||\mathbf{y}-\mathbf{u}^*||^2 + \mathbf{u}^T(\mathbf{y}-\mathbf{u}^*)]}_{\text{'log of best fit likelihood'}} + \underbrace{\ln[(2\pi\theta_0)^n \det(\mathbf{K}_\theta + \theta_0\mathbf{I})]}_{\text{'log of Occam factor'}} \big). \tag{2}$$

The symmetric positive-definite matrix $\mathbf{K}_\theta$ is usually postulated via a kernel (covariance) function $[\mathbf{K}_\theta]_{ij} = k_\theta(\mathbf{x}_i, \mathbf{x}_j)$ which amplifies slowly varying components of the signal. The kernel depends on a few hyperparameters $\boldsymbol{\theta}$. In the case of the regression with a single optimal hyperparameter vector it can be chosen to maximize the Bayesian evidence criterion whose doubled negative logarithm is given by Eq. (2). We note that in the GP regression literature the logarithm of the best fit likelihood is usually written in an equivalent form $\mathbf{y}^T(\mathbf{K}_\theta + \theta_0)^{-1}\mathbf{y}$.

The problem is that Gaussian, Brownian motion and alike kernels blur the edges of a signal. In a one-dimensional case, the two-sided exponential kernel would be a much more efficient candidate as it would result in a tridiagonal inverse $\mathbf{K}_\theta^{-1}$ [8], but the problem of the edge-preserving filtering demands kernels with discontinuities [4]. In general, it is hard to choose a parametric form of a kernel so that the location of a discontinuity, viewed as a hyperparameter, would have a unimodal posterior density. In other words, Eq. (2) is plagued to possess multiple minima. However, we believe the problem of the edge-preserving filtering can be solved with a single optimal hyperparameter regression within Bayesian evidence maximization.

A general principle of the edge-preserving filtering can be informally stated as 'iteratively average where a current estimate of a signal's gradient is small'. In the case of a one-dimensional domain $x \in \Omega \equiv [0, 1]$ this idea can be implemented by solving the regularization problem

$$u_k^* = \arg\inf_u \big(\int_\Omega [2\tau g(u_{k-1})(\partial_x u)^2 + u^2 + (u - u_{k-1})^2] dx \big) \ \text{s.t. bound. cond.} \tag{3}$$

for u_k, $k = 0, 1, \ldots, m$, where $u_k \equiv u(x, k\tau)$ and $u(x, 0) \equiv y$. Eq. (3) represents an implicit Euler time stepping of the nonlinear diffusion filtering with time increment τ [10,1,5]. Aside from its obvious classical meaning, Eq. (3) has yet another, rather difficult to see, interpretation. When defined on an unbounded domain $x \in \mathbb{R}^1$ with $g(u_{k-1}) = \text{const}$ and the constraint $u > 0$, it minimally transports u_k to u_{k-1} while maximizing the differential entropy of u viewed as a density function [7].

The weighting function should give higher penalties to larger absolute values of derivative estimates [10]:

$$g(u) \equiv 1 - e^{-c(\frac{\partial_x u_\sigma}{\lambda})^{-s}} . \tag{4}$$

Here the time-dependent signal u is passed through a Gaussian filter resulting in the signal u_σ, whose spatial derivative's value is denoted by $\partial_x u_\sigma$. The constant c can always be chosen beforehand so that the diffusion of the original noisy signal y takes place only in low derivative regions where $|\partial_x u_\sigma| < \lambda$ [1]. The even number $s \geq 2$ denotes the sharpness of the nonlinearity.

An implementation example of Eq. (3) can be written for the discrete domain $x = 0, h, \ldots, 1$ by approximating the derivatives with their first order finite differences. If we assume reflecting boundary conditions $\partial_n u|_\Omega = 0$ and define the tridiagonal matrix

$$[\mathbf{B}_k]_{ij} = \begin{cases} 1 + \frac{\tau}{h}(g_{i-1} + g_{i+1}) & \text{if } |i-j| = 0, \\ -\frac{\tau}{h}(g_i) & \text{if } |i-j| = 1, \end{cases} \tag{5}$$

then Eq. (3) in the discrete space reduces to the matrix-vector product $\mathbf{u}_k = \mathbf{B}_k^{-1}\mathbf{u}_{k-1}$ with $\mathbf{u}_0 = \mathbf{y}$. Tridiagonal matrices can be inverted by applying Thomas algorithm [10]. Such a process is very efficient as it requires only $O(n)$ number of multiplications to perform a single inversion. This comes in contrast to the inversion in Eq. (1), which in the case of a GP model with Gaussian kernel requires $O(n^2) \ldots O(n^3)$ multiplications. The structure of the matrix $\mathbf{B}_k$ depends on the boundary conditions, while the values of its elements depend on the discrete space and time step sizes h and τ. Eq. (5) introduces $o(h)$ approximation whereas the stability of the time stepping according to Eq. (3) holds for all step sizes $\tau > 0$.

In spite of efficiency, nonlinear diffusion lacks systematic criteria in choosing the nonlinear diffusivity parameters λ and σ and, most importantly, the termination step number m. This problem can be partially solved if we consider diffusion filtering as a GP regression with the covariance matrix [5]

$$\mathbf{K}_\theta \equiv \theta_0^2(\mathbf{B}_1 \cdot \mathbf{B}_2 \cdots \mathbf{B}_m - \mathbf{I})^{-1}. \tag{6}$$

The matrix $\mathbf{K}_\theta$ determines spatial covariance between any two points of the nonlinear diffusion outcome and in the case of nonlinear diffusivity, presented in Eq. (4), attains the form of step-like functions [5]. This GP model is suitable to large-scale applications. Given that hyperparameters are known, only storage and inversion of tridiagonal matrices are required to perform the regression. At the same time, the covariance matrix of the final diffusion result is not restricted to lie in the subspace of the tridiagonal inverses.

3 Optimal Stopping with Jacobi Criterion

Given the hyperparameters $\boldsymbol{\theta}$, solving Eq. (1) with the covariance matrix defined by Eq. (6) is a very efficient process. However, this does not hold for the logevidence criterion stated in Eq. (2). In what follows, we attempt to alleviate this difficulty when determining the optimal diffusion stopping iteration m. Let us denote the shortcut $\mathbf{B}_{1:k} = \mathbf{B}_1 \cdot \mathbf{B}_2 \cdots \mathbf{B}_k$ and rewrite Eq. (2) for the stopping iteration m:

$$m = \arg\min_k \Big(\underbrace{\ln \Big((2\pi\theta_0)^n \sqrt{\frac{|\mathbf{B}_{1:k}|}{|\mathbf{B}_{1:k} - \mathbf{I}|}} \Big)}_{\text{`compressibility'}} + \underbrace{\frac{1}{\theta_0} \mathbf{u}_k^T \mathbf{B}_{1:k} \overbrace{(\mathbf{y} - \mathbf{u}_k)}^{\text{'noise'}}}_{\text{`temporal decorrelation'}} \Big). \tag{7}$$

As can be seen, logevidence of the 'dynamic GP' model comprises two terms. The determinants in the first term can be viewed as functions whose argument is the observation vector $\mathbf{y}$. When integrated over any initial region of uncertainty in the space $\mathbf{y} \in \mathbb{R}^n$, they represent uncertainty volumes at time k, driven by diffusion Eq. (3) with the norm term u^2 and without it. A starting point for a more detailed analysis could be the application of the Abel-Liouville-Jacobi-Ostrogradskii idenity [6], but these determinants are hard to estimate when $n > 10^3$ because the sparsity of the matrix $\mathbf{B}_{1:k}$ decreases with each iteration: $\mathbf{B}_1$ is tridiagonal, $\mathbf{B}_2$ will be pentadiagonal, etc. If the matrix $\mathbf{B}$ were diagonal, the second term, namely the best-fit likelihood, would equal to zero whenever the optimal model output is orthogonal to the 'noise term' $\mathbf{y} - \mathbf{u}_m$. In the case of a small variance θ_0, this criterion can indeed be close to the heuristic stopping based on decorrelation of the model output and noise, clf. [5].

As it is difficult to implement the criterion in Eq. (7) in the large-scale problem, it becomes rather handy to consider that $\mathbf{u}_{k+1}$ is always 'less noisy' than $\mathbf{u}_k$, and view a single iteration of Eq. (3) as a temporal GP regression with $\mathbf{K}_\theta$ defined as in Eq. (6) with $\mathbf{B} \equiv \mathbf{B}_k$. In this case, both terms are easier computable. The Jacobi theory on conjugate points in the calculus of variations yields the following, rather surprising, result [2]:

$$\ln\det(\mathbf{B}_k) - \ln\det(\mathbf{B}_k - \mathbf{I}) = \ln z(1) - \ln \tilde{z}(1), \tag{8}$$

where $z(1)$ solves a complete Jacobi equation of the functional in Eq. (3):

$$\partial_x[\tau g \partial_x z] = z, \;\; z(0) = 0, \;\; \partial_x z(0) = 1, \tag{9}$$

and $\tilde{z}(1)$ is a solution which excludes the quadratic term u^2 in Eq. (3):

$$\partial_x[\tau g \partial_x \tilde{z}] = 0, \;\; \tilde{z}(0) = 0, \;\; \partial_x \tilde{z}(0) = 1. \tag{10}$$

Due to the space limits we exclude derivation. Essential steps in obtaining this result can be found in [2,9]. Intuitively, the functions $z(x)$ and $\tilde{z}(x)$ measure the ellipticity of the functional in Eq. (3) with and without the term u^2. Eqs. (9) and (10) then indicate that when performing evidence maximizing iteration of the nonlinear diffusion filtering, 'temporal model complexity' is the difference between the ellipticity of a complete functional in Eq. (3) and its degenerate counterpart which does not constrain the Euclidean norm of the filtered signal. Eqs. (9) and (10) reduce to non-integrable in quadratures Riccati equations, but solving a one-dimensional ordinary differential equation numerically does not present difficulties in the case of large n [6].

The importance of this result to practice lies in the following observation: the 'compressibility' term in Eq. (7), when evaluated by using $\mathbf{B}_k$ rather than $\mathbf{B}_{1:k}$, always decreases, and the time it reaches the steady state correlates with the optimal stopping time. We call the expression in Eq. (8) the Jacobi criterion and will test it in the next section.

Notice that the maximal value of the best-fit likelihood in the case of a single iteration with $p=$ const is attained when the frequency spectrum of the model output is orthogonal to the noise spectrum. In other words, the tridiagonal matrix $\mathbf{B} = \mathbf{B}_k$ is diagonal in the discrete cosine transform basis [8].

4 Numerical Example and Discussion

We create a numerical example from a telecommunication setting, where a discontinuous signal is distorted with additive Gaussian noise of very large variance $\theta_0^2 = 1$. Due its discontinuous nature, the signal can still be reliably recovered. Fig. 1 shows the noisy observations and the filtered signal, which closely matches its true counterpart. This result was obtained by employing the Jacobi criterion to optimally stop the diffusion filtering, which happened after four iterations. The result also supports that the implicit time stepping with very large step sizes can be used to filter signals without distorting the diffusion outcome. Fig. 2 indicates that the time when the criterion $\ln z(1) - \ln \tilde{z}(1)$ reaches its steady state corre-

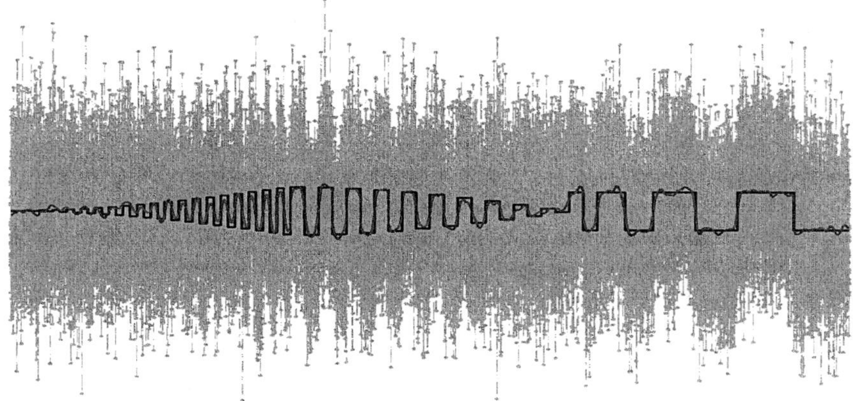

Fig. 1. Restoring a discontinuous signal of 30000 samples distorted by very large values of additive Gaussian noise. The result of the optimally stopped nonlinear diffusion filtering closely matches the true signal.

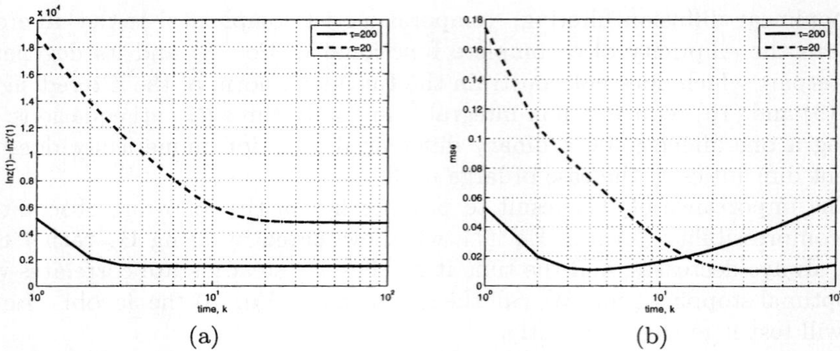

(a) (b)

Fig. 2. Time evolution of (a) the Jacobi criterion $\ln z(1) - \ln \tilde{z}(1)$, and (b) the mean of squared errors between the true signal and the diffusion outcome. Notice that the Jacobi criterion optimally stops the diffusion independently of the time step size τ.

lates with the time when the mean squared error between the true and filtered signals reaches its minimum.

In conclusion, we would like to note that in case when the number of discontinuities and their approximate location intervals are known, various kernel functions can be explicitly constructed and the discontinuous regression is likely to outperform the nonlinear diffusion, especially at larger noise variances. However, multiple maxima in the Bayesian evidence criterion would be unavoidable even if such explicit knowledge were available.

References

1. Frederico D'Almeida. Nonlinear diffusion toolbox. MATLAB Central.
2. I. M. Gelfand and S. V. Fomin. *Calculus of Variations.* Prentice-Hall, 1963.
3. M.N. Gibbs. *Bayesian Gaussian Processes for Regression and Classification.* Ph.d. thesis, Cambridge University, 1997.
4. I. Gijbels. *Recent Advances and Trends in Nonparametric Statistics*, chapter Inference for nonsmooth regression curves and surfaces using kernel-based methods, pages 183–202. Elsevier Science Ltd, 2003.
5. R. Girdziušas and J. Laaksonen. Gaussian process of nonlinear diffusion filtering. IJCNN, 2005.
6. E. Hairer, S.P. Nørsett, and G. Wanner. *Solving Ordinary Differential Equations I. Nonstiff Problems.* Springer, 1993.
7. R. Jordan, D. Kinderlehrer, and F. Otto. The variational formulation of the Focker–Planck equation. *SIAM J. Math. Anal.*, 29(1):1–17, 1998.
8. A. Kavčić and J.M.F. Moura. Matrices with banded inverses: inversion algorithms and factorization of gauss–markov processes. *IEEE Trans. on Information Theory*, 46(4):1495–1509, July 2000.
9. H. Kawasaki. A conjugate points theory for a nonlinear programming problem. *SIAM J. Control Optim.*, 40(1):54–63, 2001.
10. J. Weickert, B. M. ter Haar Romeny, and M. A.Viergever. Efficient and reliable schemes for nonlinear diffusion filtering. *IEEE Trans. on Image Processing*, 7(3):398–410, March 1998.

Bayesian Learning of Neural Networks Adapted to Changes of Prior Probabilities

Yoshifusa Ito[1], Cidambi Srinivasan[2], and Hiroyuki Izumi[3]

[1,3] Department of Information and Policy Studies, Aichi-Gakuin University, Nisshin, Aichi-ken, 470-0195 Japan
ito@psis.aichi-gakuin.ac.jp
[2] Department of Statistics, University of Kentucky, Lexington, Kentucky 40506, USA
srini@ms.uky.edu

Abstract. We treat Bayesian neural networks adapted to changes in the ratio of prior probabilities of the categries. If an ordinary Bayesian neural network is equipped with $m-1$ additional input units, it can learn simultaneously m distinct discriminant functions which correspond to the m different ratios of the prior probabilities.

1 Introduction

We propose an algorithm for Bayesian learning of neural networks which can learn several discriminant functions simultaneously. It is useful when the ratio of prior probabilities changes depending on the situation but the state-conditioned probability distributions are not changed. For simplicity, we present the details in the case of two categories and m situations in this paper. The algorithm, however, can be extended to multiple categories. The point of this paper is to remark that the discriminant functions are differ only by constants in the distinct situations. If an ordinary Bayesian neural network is equipped with $m-1$ additional linear input units, it can learn all the m distinct discriminant functions. The main part of the network learns the discriminant function at one of the situations, and the additional part modifies it for other $m-1$ situations.

The proposed neural network has a relatively minimal number of units. Though Funahashi's network for the case of normal state-conditioned distributions is considered as a Bayesian neural network having rather a small number of units [2], our network has half the number of hidden layer units in comparison to his. Our network has direct connections between the input units and the output unit.

The phrase "Bayesian learning" may be somewhat confusing. Unlike [7], our theory is based on the traditional theory on Bayesian neural networks [2,4,5,6,8]. The simple case of the approximation theory in [3] is used. Our terminology is mainly drawn from [1]. Two simple simulations are presented in Section 5 to show how the algorithm works.

W. Duch et al. (Eds.): ICANN 2005, LNCS 3697, pp. 253–259, 2005.

2 Discriminant Functions

Suppose that there are two categories $\Theta = \{\theta_1, \theta_2\}$ and several situations $\Gamma = \{\gamma_1, \cdots, \gamma_m\}$ of the source of observables $x \in \mathbf{R}^d$. Further suppose that the ratio of the prior probabilities $P_k(\theta_1)/P_k(\theta_2)$, $1 \leq k \leq m$, depends on the situation but the state-conditioned probability distributions $P(x|\theta_i)$ are common. We treat the cases where $P(x|\theta_i)$, $i = 1, 2$, are one of familiar probability distributions of the exponential family such as the normal, binomial, multinomial, Poisson, exponential and other distributions. Then, the log ratio $\log \frac{p(x|\theta_1)}{p(x|\theta_2)}$ is a polynomial of low degree. Hence, the log ratios of the posterior probabilities

$$g_k(x) = \log \frac{P_k(\theta_1|x)}{P_k(\theta_2|x)} = \log \frac{P_k(\theta_1)}{P_k(\theta_2)} + \log \frac{p(x|\theta_1)}{p(x|\theta_2)}, \qquad k = 1, \cdots, m, \tag{1}$$

are also polynomials of low degree. It is well know that the log ratio can be a dicriminant function because $g_k(x) > 0$ implies $P_k(\theta_1|x) > P_k(\theta_2|x)$. If $P(x|\theta_i)$ are normal then the log ratio is a quadratic function, and if it is some of other distributions mentioned above the log ratio is a linear function.

In the case of the two categories, the posterior probability at the state γ_k is

$$P_k(\theta_1|x) = \frac{P_k(\theta_1|x)}{P_k(\theta_1|x) + P_k(\theta_2|x)} = \sigma\left(\log \frac{P_k(\theta_1|x)}{P_k(\theta_2|x)}\right) = \sigma(g_k(x)), \tag{2}$$

where $\sigma(t) = (1 + e^{-t})^{-1}$. Hence, if the inner potential of the output unit can approximate polynomials of low degree and its activation function is the logistic function σ, its outputs can approximate the posterior probability [2]. The posterior probability $P_k(\theta_1|x) = \sigma(g_k(x))$ can also be a discriminant function as the logistic function is strictly monotone increasing.

If the $p(x|\theta_i)$ are the multinomial distribution $\frac{n!}{x_1! \cdots x_d!} p_{i1}^{x_1} \cdots p_{id}^{x_d}$, the log ratio (2) is a linear function

$$\log \frac{P_k(\theta_1)}{P_k(\theta_2)} + \log \frac{p_{11}}{p_{21}} \cdot x_1 + \cdots + \log \frac{p_{d1}}{p_{d2}} \cdot x_d. \tag{3}$$

If the $p(x|\theta_i)$ are normal $N(\mu_i, \Sigma_i)$ with distinct covariance matrices then the log ratio (2) is a quadratic function

$$\log \frac{P_k(\theta_1)}{P_k(\theta_2)} - \frac{1}{2} \log \frac{|\Sigma_1|}{|\Sigma_2|} - \frac{1}{2}\{(x-\mu_1)^t \Sigma_1^{-1}(x-\mu_1) - (x-\mu_2)^t \Sigma_2^{-1}(x-\mu_2)\}. \tag{4}$$

3 Bayesian Neural Network

The training set is a sequence of triplets $(x, \gamma, \theta) \in \mathbf{R}^d \times \Gamma \times \Theta$. The trained neural network receives pairs (x, γ). Let $F(x, \gamma, w)$ denote the output of the neural network, where w is the weight vector, and let $\xi(x, \gamma, \theta)$ be a function on $\mathbf{R}^d \times \Gamma \times \Theta$. Let $E[\xi(x, \gamma, \cdot)|x]$ and $V[\xi(x, \gamma, \cdot)|x]$ be the conditional expectation and variance of $\xi(x, \gamma, \theta)$. Set $p_k(x) = \Sigma_i P_k(\theta_i) p(x|\theta_i)$. Let q_k be the probability of the situation γ_k: $\Sigma_k q_k = 1$. The proof of the proposition below is a modification of that for the simpler case in [8].

Proposition 1. Set

$$\mathcal{E}(w) = \sum_{k=1}^{m} q_k \int_{\mathbf{R}^d} \sum_{j=1}^{2} (F(x, \gamma_k, w) - \xi(x, \gamma_k, \theta_j))^2 P_k(\theta_j) p(x|\theta_j) dx. \tag{5}$$

Then,

$$\mathcal{E}(w) = \sum_{k=1}^{m} q_k \int_{\mathbf{R}^d} (F(x, \gamma_k, w) - E[\xi(x, \gamma_k, \cdot)|x])^2 p_k(x) dx$$

$$+ \sum_{k=1}^{m} q_k \int_{\mathbf{R}^d} V[\xi(x, \gamma_k, \cdot)|x] p_k(x) dx. \tag{6}$$

If $\xi(x, \gamma_k, \theta_1) = 1$ and $\xi(x, \gamma_k, \theta_2) = 0$, then $E[\xi(x, \gamma_k, \cdot)|x] = P_k(\theta_1|x)$. Hence, when $\mathcal{E}(w)$ is minimized, the output $F(x, \gamma_k, w)$ is expected to approximate $P_k(\theta_1|x)$. Accordingly, learning of the network is carried out by minimizing

$$\mathcal{E}_n(w) = \frac{1}{n} \sum_{t=1}^{n} (F(x^{(t)}, \gamma^{(t)}, w) - \xi(x^{(t)}, \gamma^{(t)}, \theta^{(t)}))^2. \tag{7}$$

This method of training has actually been stated by several authors [2,4,5,6,8,9].

4 Construction of Neural Network

We show below that a neural network having a small number of units, when ideally trained, can approximate the discriminant function $\sigma(g_k(x))$ receiving a pair (x, γ). For the proofs of the lemma and corollary below, the inclusions $t^n \in L^p(\mathbf{R}, \nu)$ and $|x|^2 \in L^p(\mathbf{R}^d, \nu)$ are essential respectively [3,5].

Lemma 2. Let ν be a probability measure on $\mathbf{R}$. If $t^n \in L^p(\mathbf{R}, \nu)$, $1 \leq p < \infty$, $\phi \in C^n(\mathbf{R})$ and $\phi^{(i)}, 0 \leq i \leq n$, are bounded, then, for any $\varepsilon > 0$, there exists $\delta > 0$ for which

$$\| \frac{1}{n!} \phi^{(n)}(0) t^n - \frac{1}{\delta^n} \phi(\delta t) - \sum_{i=0}^{n-1} \phi^{(i)}(0)(\delta t)^i \|_{L^p(\mathbf{R},\nu)} < \varepsilon. \tag{8}$$

Corollary 3. Let ν and Q be a probability measure and a quadratic form on $\mathbf{R}^d$. If $|x|^2 \in L^p(\mathbf{R}^d, \nu)$, $1 \leq p < \infty$, $\phi \in C^2(\mathbf{R})$ and $\phi^{(j)}$, $j = 0, 1, 2$, are bounded, then, for any $\varepsilon > 0$, there exist vectors $w_i \in \mathbf{R}^d$ and constants a_i, $0 \leq i \leq d$, for which

$$\|Q(x) - \bar{Q}(x)\|_{L^p(\mathbf{R}^d,\nu)} < \varepsilon, \tag{9}$$

where

$$\bar{Q}(x) = \sum_{i=1}^{d} a_i \phi(w_i \cdot x) + w_0 \cdot x + a_0. \tag{10}$$

If $g_m(x)$ defined by (1) is linear as in the case of (3), it is equale to

$$w \cdot x + a_0, \qquad w \in \mathbf{R}^d, \quad a_0 \in \mathbf{R}. \tag{11}$$

with constants optimally adjusted. This can be exactly realized by the sum of the outputs of the main part units (large circles) of the network in Fig.1a. If $g_m(x)$ is quadratic as in the case of (4), then it can be approximated by (10) in the sense of $L^p(\mathbf{R}^d, \nu)$ which can be realized by the sum of the outputs of the main part units (large circles) of the network in Fig.1b.

The neural networks have additional parts (small circles) respectively. Its role is to approximate the differences $u(k) = g_k(x) - g_m(x)$, $k = 1, \cdots, m$. The information on the situation is fed into the additional part as a vector $(s_1, \cdots, s_{m-1})$. In the k-th situation, $k < m$, $s_k = 1$ and other elements are zero, and in the m-th situation the entire vector is zero. If the connection weight between the k-th additional input unit and the output unit is equal to the difference $g_k(x) - g_m(x)$, then the sum (the inner potential of the output unit) of the inputs to the output unit from the two parts approximates the discriminant functions:

$$g_k(x) = g_m(x) + u(k) = \log \frac{P_k(\theta_1|x)}{P_k(\theta_2|x)}, \quad k = 1, \cdots, m-1. \tag{12}$$

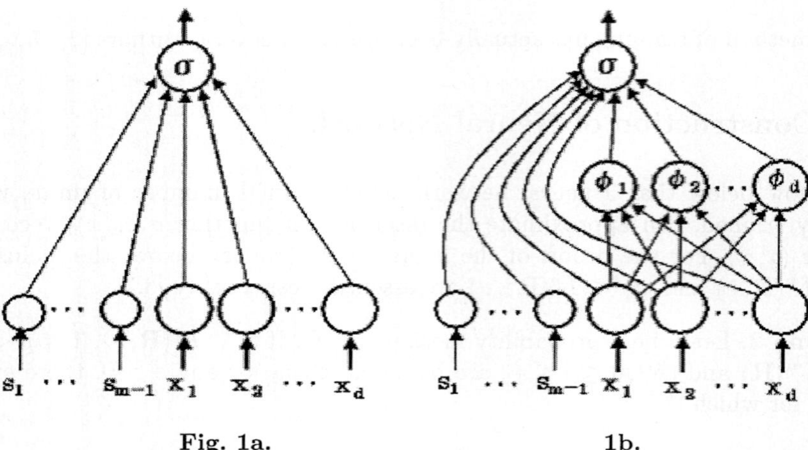

Fig. 1a. **1b.**

5 Training and Simulations

Since the training is carried out by (7), the activation function of the output unit must be the logistic function. In accordance with (12), we divide the inner potential of the output unit into two components $G(x, w)$ and $U(k, w)$. They are the inputs from the main and additional parts of the network respectively. Then,

$$\mathcal{E}_n(w) = \frac{1}{n}\sum_{t=1}^{n}\{\sigma(G(x^{(t)}, w) + U(k^{(t)}, w)) - \xi(x^{(t)}, k^{(t)}, \theta_j^{(t)})\}^2. \tag{13}$$

It can be seen from this equation that the main part of the network, (responsible for $G(x^{(t)}, w)$, is trained on each arrival of a training triplet $(x^{(t)}, k^{(t)}, \theta_j^{(t)})$, but the connection weights of the additional part (responsible for $U(x^{(t)}, w)$ are trained separately.

We present two simple examples of simulations to show that the theoretically obtained algorithm works well. To this end, simple examples may be enough. The examples used for illustrations involve only two categories with two situations. The activation functions of the output unit and the hidden layer units are the logistic function and the input units are linear. The network was trained with a sequence of 1000 triplets (x, γ, θ). It is a mixture of triplets generated in the two situations having even probabilistic chances of generating a signal. Fig. 2a, 3a are the state-conditioned probabilities, Fig. 2b, 3b are the exact posterior probabilities $P_k(\theta_1|x) = \sigma(g_k(x))$, $k = 1, 2$, of the category 1 theoretically obtained and Fig. 2c,3c are their approximations obtained by simulations. If we artificially adjust the connection weights the discriminant functions can be approximated

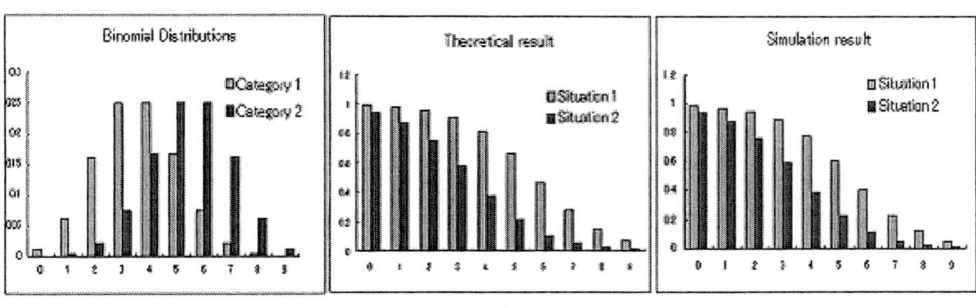

Fig. 2a. **2b.** **2c.**

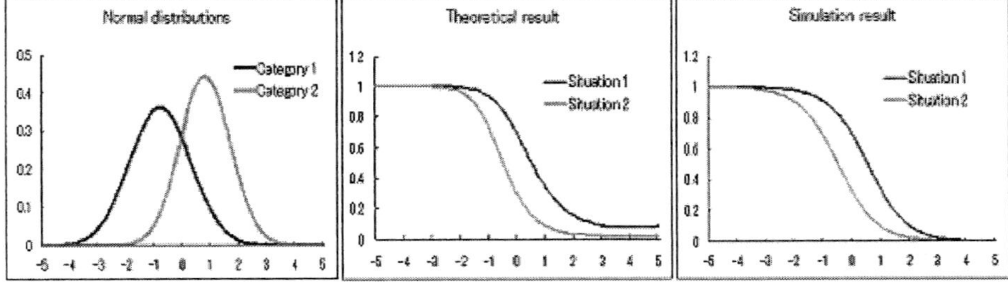

Fig. 3a. **3b.** **3c.**

with any level of accuracy. The prior probabilities of the first and second categories are 0.7 and 0.3 in the situation 1, and 0.3 and 0.7 in the situation 2. The capabilities of the networks were tested by distinct 1000 signals (x, γ).

In the first example, the state-conditioned probability distributions are binomial, $B(9, 0.4)$ and $B(9, 0.6)$, as illustrated in Fig. 2a. This is the case where the log ratio of the posterior probabilities is linear. The discriminant functions obtained by simulations are slightly distinct from those theoretically obtained. But the classification capability of the trained network was exactly the same as the theoretical discriminant functions $g_k(x)$. Since the observable are integer-valued, a small deviation of the approximated discriminant function does not affect the decision.

In the second example, the state-conditioned probability distributions are normal, $N(-0.8, 1.2)$ and $N(0.8, 0.8)$, as illustrated in Fig. 3a. In this case the log ratio is a quadratic function. The training sequence was constructed randomly but it was a little biased. Among 1000 observables, there was only one x greater than 3, though theoretically 4 observables are expected. The approximations of the discriminant functions obtained by simulations are also slightly deviated from theoretical ones for x, $|x| > 2$. This is because the density of training signals is low in that domain. In spite of this deviation, the decisions of categories for 970 observables coincide with the theoretical decision. The deviation of the curve in the area $|x| > 2$ does not necessarily imply a great statistical L^2 error, because the probability density in the domain is low.

6 Discussions

Thus a single Bayesian neural network, with a simple additional structure, can be used in several situations. The discriminant function is divided into two parts, $g_k(x) = g_m(x) + u(k)$. The main part of the network is trained to approximate $g_m(x)$ with the whole sequence of teaching signals, and the respective units of the additional part to approximate the constants $u(k)$, $k = 1, \cdots, m-1$, respectively with the small numbers of signals generated at the respective situations. Since training of the main part is harder, this method of learning is reasonable. We can replace the additional input $(s_1, \cdots, s_{m-1})$ by a scalar input, say $k \in \{1, \cdots, m\}$, using an additional structure which can be trained to realize the function $u(k)$ with $u(m) = 0$. The idea of this paper may be extended to the case where the conditional distributions depend on the situation, if they can be separated into a common main part and the individual simple parts respectively.

This paper is written assuming that the number of the categories is two. However, using the method stated in [5], the number of categories can be increased with any restriction. The number of the units of the neural network is the minimum or close to the minimum. In the case of statistical learning, the generalization capability of the network is particularly important. Too many units adversely affect the generalization capability. Our network with minimal number of units may be more suitable for statistical learning.

References

[1] R.O. Duda and P.E. Hart, Pattern classification and scene analysis, Joh Wiley & Sons, New York, 1973.

[2] Funahashi, K., "Multilayer neural networks and Bayes decision theory", Neural Networks, 11, 209-213, 1998.

[3] Y. Ito, "Simultaneous L^p -approximations of polynomials and derivatives on $\mathbf{R}^d$ and their applications to neural networks", (in preparation)

[4] Y. Ito and C. Srinivasan. "Multicategory Bayesian decision using a three-layer neural network", in Proceedings of ICANN/ICONIP 2003, 253-261, 2003.

[5] Y. Ito and C. Srinivasan. "Bayesian decision theory on three-layer neural networks", Neurocomputing, vol. 63, 209-228, 2005.

[6] R.M. Neal, "Bayesian learning for neural networks", Springer, 1996.

[7] M.D.Richard and R.P. Lipmann,"Neural network classifiers estimate Bayesian a posteriori probabilities", Neural Computation, vol. 3, pp461-483, 1991.

[8] M.D.Ruck, S. Rogers, M. Kabrisky, H. Oxley, B. Sutter, "The multilayer perceptron as approximator to a Bayes optimal discriminant function", IEEE Transactions on Neural Networks, vol. 1, pp296-298, 1990.

[9] H. White "Learning in artificial neural networks: A statistical perseptive". Neural Computation, vol. 1, pp425-464, 1989.

A New Method of Learning Bayesian Networks Structures from Incomplete Data

Xiaolin Li[1], Xiangdong He[2], and Senmiao Yuan[1]

[1] College of Computer Science and Technology, Jilin University,
Changchun 130012, China
lixl@email.jlu.edu.cn
yuansenmiao@hotmail.com
[2] VAS of China Operations, Vanda Group,
Changchun 130012, China
hexd_163@163.com

Abstract. This paper describes a new data mining algorithm to learn Bayesian networks structures from incomplete data based on extended Evolutionary programming (EP) method and the Minimum Description Length (MDL) principle. This problem is characterized by a huge solution space with a highly multimodal landscape. The algorithm presents fitness function based on expectation, which converts incomplete data to complete data utilizing current best structure of evolutionary process. The algorithm adopts a strategy to alleviate the undulate phenomenon. Aiming at preventing and overcoming premature convergence, the algorithm combines the niche technology into the selection mechanism of EP. In addition, our algorithm, like some previous work, does not need to have a complete variable ordering as input. The experimental results illustrate that our algorithm can learn a good structure from incomplete data.

1 Introduction

The Bayesian belief network is a powerful knowledge representation and reasoning tool under conditions of uncertainty. Recently, learning the Bayesian network from a database has drawn noticeable attention of researchers in the field of artificial intelligence. To this end, researchers developed many algorithms to induct a Bayesian network from a given database [1], [2], [3], [4], [5], [6].

Very recently, researchers have begun to tackle the problem of learning the network from incomplete data. A major stumbling block in this research is that when in closed form expressions do not exist for the scoring metric used to evaluate network structures. This has led many researchers down the path of estimating the score using parametric approaches such as the expectation-maximization (EM) algorithm [7]. However, it has been noted [7] that the search landscape is large and multimodal, and deterministic search algorithms find local optima. An obvious choice to combat the problem is to use a stochastic search method.

This paper developed a new data mining algorithm to learn Bayesian networks structures from incomplete data based on extended Evolutionary Programming (EP) method and the Minimum Description Length (MDL) principle. The algorithm presents fitness function by using expectation, which converts incomplete data to com-

W. Duch et al. (Eds.): ICANN 2005, LNCS 3697, pp. 261 – 266, 2005.

plete data utilizing current best structure of evolutionary process. The algorithm adopts a strategy to alleviate the undulate phenomenon. Another important characteristic of our algorithm is that, in order to preventing and overcoming premature convergence, we combine the niche technology [8] into the selection mechanism of EP. Furthermore, our algorithm, like some previous work, does not need to impose the restriction of having a complete variable ordering as input.

We'll begin by briefly introducing Bayesian network and MDL principle. Next we will introduce the extended EP method. In section 4, we will describe the algorithm based on the extended EP method and the MDL metric. In the end, we will conduct a series of experiments to demonstrate the performance of our algorithm and sum up the whole paper in section 5 and 6, respectively.

2 Bayesian Networks and MDL Metric

2.1 Bayesian Networks

A Bayesian network is a directed acyclic graph (DAG), nodes of which are labeled with variables and conditional probability tables of the node variable given its parents in the graph. The joint probability distribution (JPD) is then expressed by the formula:

$$P(x_1,\cdots\cdots,x_n) = \prod_{i=1\cdots n} P(x_i \mid \pi(x_i)) \tag{1}$$

where $\pi(x_i)$ is the configuration of X_i's parent node set $\Pi(X_i)$.

2.2 The MDL Metric

The MDL metric [9] is derived from information theory and incorporates the MDL principle. With the composition of the description length for network structure and the description length for data, the MDL metric tries to balance between model accuracy and complexity. Using the metric, a better network would have a smaller score. Similar to other metrics, the MDL score for a Bayesian network, S, is decomposable and could be written as in equation 2. The MDL score of the network is simply the summation of the MDL score of $\Pi(X_i)$ of every node X_i in the network.

$$MDL(S) = \sum_i MDL(X_i, \Pi(X_i)) \tag{2}$$

According to the resolvability of the MDL metric, equation 2 can be written when we learn Bayesian networks form complete data as follow:

$$MDL(S) = N\sum_{i=1}^{N} \sum_{X_i,\Pi(X_i)} P(X_i,\Pi(X_i)) \log P(X_i,\Pi(X_i)) \tag{3}$$

$$-\sum_{i=1}^{N} \frac{\log N}{2} \| \Pi(X_i) \| (\| X_i \| -1)$$

Where N is population size.

The problem of learning Bayesian networks from incomplete data is much more difficult than for learning from complete data because the MDL metric no longer resolves into formula 3.

3 The Extended EP Method

Although EP was first proposed as an evolutionary algorithm to artificial intelligence, it has been recently applied to many numerical and combinatorial optimization problems successfully.

EP applies mutation operators. We can know by father analyzing EP that the mutation operation modifies aspects of the parent randomly according to a statistical distribution. So the mutation operation can not only provide evolutionary chance for the individuals, but also produce the undulate phenomenon. The algorithm adopts a strategy to alleviate the undulate phenomenon. We examine the offspring after finishing mutation operation. If the fitness of the offspring individual is not as good as the parent, we can consider that the evolutionary process produce a serious undulate phenomenon. Then, the parent will instead of the offspring to compete.

A niche is a stable sub-population of competing prototypes sharing the same environmental resources. The niche technology is a mature、effective and important method of keeping population variety. So the niche technology can prevent and overcome premature convergence effectively. The niche technology has three selection methods, which are preselection, crowding and sharing.

We combine the niche technology (crowding) into the selection mechanism of EP and put the new selection process in force to keep population variety and prevent premature convergence after learning the characteristics of the niche technology.

Define the standard Euclidean distance for the two random individuals $(x_1, \eta_1), (x_2, \eta_2)$,

$$S(x_1, x_2) = \sqrt{\frac{1}{n}\sum_{k=1}^{n}\left|\frac{x_1(k) - x_2(k)}{b_k - a_k}\right|^2} \tag{4}$$

The selection method of niche-EP is as follows. For each selected individual, q individuals compare with it. If the standard Euclidean distance S for the selected individual and the certain individual of the q individuals is less than the constant $dist$, the two individuals do not compare with each other. Then for the certain individual, it loses a "win". Those individuals which are correspondingly dense are reduced the times of "win" and the livability in the offspring by using niche-EP. This method has no influence on other individuals. It can keep multi-niche and population variety. So it can also prevent premature convergence.

4 Learning Bayesian Network from Incomplete Data

The algorithm we propose is shown below.

1. Set to 0.
2. Create an initial population, Pop(t), of PS random DAGs. The initial population size is PS.
3. Convert incomplete data to complete data utilizing a DAG of the initial population randomly
4. Each DAG in the population Pop(t) is evaluated using the MDL metric.
5. While t is smaller than the maximum number of generations G
   a) Each DAG in Pop(t) produces one offspring by performing mutation operations. If the offspring has cycles, assign a poor fitness. If choices of set of edges exist, we randomly pick one choice.
   b) The DAG in Pop(t) and all new offspring are stored in the intermediate population Pop'(t).The size of Pop'(t) is 2*PS. If the fitness of the offspring individual is not as good as the parent, the parent will instead of the offspring to compete.
   c) Conduct a number of pair-wise competitions over all DAGs in Pop'(t). Then perform selection operations according to Section 3.
   d) Select PS DAGs with the highest scores from Pop'(t) and store them in the new population Pop(t+1).
   e) Remain the best individual and convert incomplete data to complete data.
   f) Increase t by 1
6. Return the DAG with lowest MDL metric found in any generation of a run as the result of the algorithm.

5 Experimental Results and Analyses

We have conducted a number of experiments to evaluate the performance of our algorithm. The learning algorithms take the data set only as input. The data set is derived from ALARM network (http://www.norsys.com/netlib/alarm.htm).

Firstly, we generate 5,000 cases from this structure and learn a Bayesian network from the data set ten times. Then we select the most perfect network structure as the final structure. We also compare our algorithm with a classical GA algorithm. The algorithms run without missing data. The MDL metric of the original network structures for the ALARM data sets of 5,000 cases is 81,219.74.

The population size PS is 30 and the maximum number of generations is 5,000. We employ our learning algorithm to solve the ALARM problem. The value of q is set to be 5. We also implemented a classical GA to learning the ALARM network. The one-point crossover and mutation operations of classical GA are used. The crossover probability p_c is 0.9 and the mutation probability p_m is 0.01. The MDL metric for our learning algorithm and the classical GA are delineated in Figure 1.

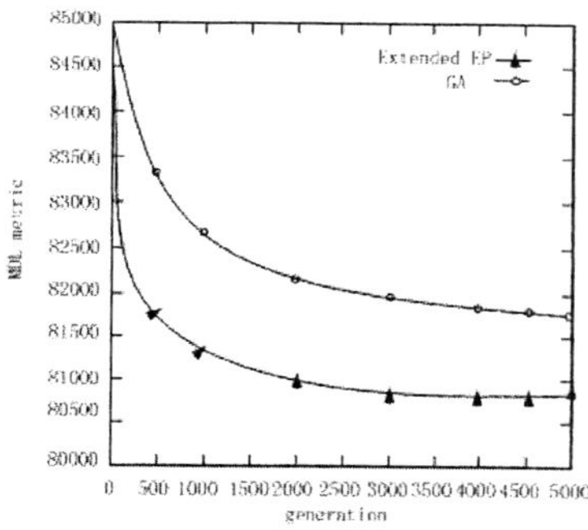

Fig. 1. The MDL metric for the ALARM network

From Figure 1, we see that the value of the average of the MDL metric for extended EP is 81258.9 and the value of the average of the MDL metric for the GA is 8,1789.4. We find our learning algorithm evolves good Bayesian network structures at an average generation of 4187.5. The GA obtains the solutions at an average generation of 4495.4. Thus, we can conclude that our learning algorithm finds better network structures at earlier generations than the GA does. Our algorithm can also prevent and overcome the premature convergence.

Our algorithm generates 1000, 10000 cases from the original network for training and testing. The algorithm runs with 10%, 20%, 30%, and 40% missing data. The experiment runs ten times for each level of missing data. Using the best network from each run we calculate the log loss. The log loss is a commonly used metric appropriate for probabilistic learning algorithms. Figure 2 shows the comparison of log loss between our algorithm and Friedman (1998b) [10].

As can be seen from figure 2, the algorithm finds better predictive networks at 10%, 20%, 30%, and 40% missing data than Friedman (1998b) does.

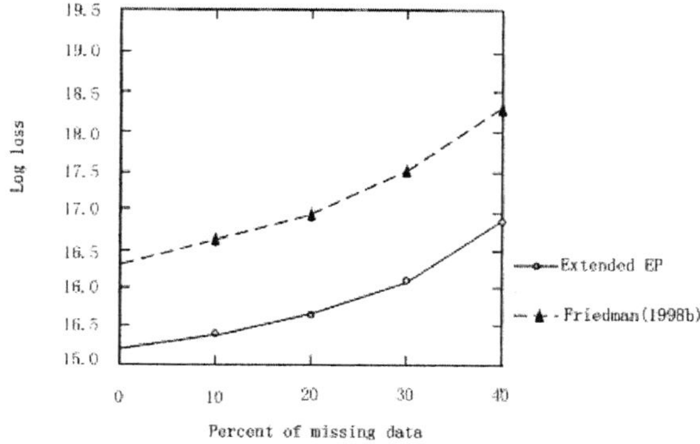

Fig. 2. The Comparison of log loss

6 Conclusions

In this paper we describe a novel evolutionary algorithm for learning Bayesian networks from incomplete data. This problem is extremely difficult for deterministic algorithms and is characterized by a large, multi-dimensional, multi-modal search space. The experimental results show that our learning algorithm can learn a good structure from incomplete data.

References

1. J Suzuki. A construction of Bayesian networks from databases based on a MDL scheme, Proc of the 9th Confon Uncertainty in Artificial Intelligence. San Mateo, CA: Morgan Kaufmann, (1993) 266~273,
2. Y Xiang, S K M Wong. Learning conditional independence relations from a probabilistic model, Department of Computer Science, University of Regina, CA, (1994) Tech Rep: CS-94-03,.
3. D Heckerman. Learning Bayesian network: The combination of knowledge and statistic data, Machine Learning, Vol. 20, No. 2, (1995), 197～243,.
4. Cheng J, Greiner R, Kelly J. Learning Bayesian networks from data: An efficient algorithm based on information theory, Artificial Intelligence. Vol.137, No.1-2, (2002) 43-90.
5. Lam, W. and Bacchus, F., Learning Bayesian belief networks: An algorithm based on the MDL principle, Computational Intelligence, Vol 10, No.4, 1994.
6. P. Larranaga, M. Poza, Y. Yurramendi, R. Murga, and C. Kuijpers, Structure Learning of Bayesian Network by Genetic Algorithms: A Performance Analysis of Control Parameters, IEEE Trans. Pattern Analysis and Machine Intelligence, vol.
7. Friedman, N. (1998a). The Bayesian Structural EM Algorithm. Proceedings of the Fourteenth Conference on Uncertainty in Artificial Intelligence, Madison, WI, Morgan Kaufmann Publishers.
8. Mahfound S.W., Crowding and Preselection Revisited, Parallel Problem Solving from Nature – II, pp. 27-36, 1992.
9. W. Lam and F. Bacchus. Learning Bayesian belief networks: an algorithm based on the MDL principle, Computational Intelligence, Vol.10, No.4, pp: 269–293, 1994.
10. Friedman, N. (1998b). Learning Belief Networks in the Presence of Missing Values and Hidden Variables. Fourteenth International Conference on Machine Learning (ICML-97), Vanderbilt University, Morgan Kaufmann Publishers.

Bayesian Hierarchical Ordinal Regression

Ulrich Paquet, Sean Holden, and Andrew Naish-Guzman

Computer Laboratory, University of Cambridge, Cambridge CB3 0FD, UK
{ulrich.paquet, sean.holden, andrew.naish-guzman}@cl.cam.ac.uk

Abstract. We present a Bayesian approach to ordinal regression. Our model is based on a hierarchical mixture of experts model and performs a soft partitioning of the input space into different ranks, such that the order of the ranks is preserved. Experimental results on benchmark data sets show a comparable performance to support vector machine and Gaussian process methods.

1 Introduction

Many applications in Machine Learning require the prediction of ordered categories, and thereby ask of us to bridge the gap between regression and classification problems. *Ordinal regression*, or ranking, often arise when a judgment of preference is made. In collaborative filtering, for example, we seek to predict a consumer's rating of a novel item on an ordinal scale such as *good* > *average* > *bad*, using past ratings of similar items. The problem shares properties with classification since the targets are discrete and finite, but also with regression estimation by the existence of an ordering in the target space.

In this paper we adopt a Bayesian approach to the ordinal regression problem, based on the *hierarchical mixture of experts* (HME) model (Jordan & Jacobs, 1994; Waterhouse et al. 1996). The HME model consists of a hierarchy of 'experts', where each expert models some data-generating process on a subset of the data. We simplify each expert to an indicator function, such that an expert is responsible for labeling a pattern with a certain rank on a subset of the input space. The ordering of the targets is imposed by a left-to-right assignment of ranks to experts in a binary HME tree.

2 Learning From Examples

We are given a data set $\mathcal{D}$ of independent and identically distributed examples of real-valued input vectors $\mathbf{X} = \{\mathbf{x}_n\}_{n=1}^N$ and corresponding targets $\mathbf{y} = \{y_n\}_{n=1}^N$. The targets come from a space $\mathcal{Y}$ consisting of a finite number of ranks, $\mathcal{Y} = \{1, \dots, R\}_>$. The subscript $>$ denotes that there is an ordering between the ranks, and can be interpreted as 'preferred to'. For simplicity we use integers to indicate the ordered set of ranks, but any labels will do. Given a new example $\mathbf{x}_*$ and the observed data, we wish to determine the probability distribution of its rank, $P(y_* = r|\mathbf{x}_*, \mathcal{D})$.

W. Duch et al. (Eds.): ICANN 2005, LNCS 3697, pp. 267–272, 2005.

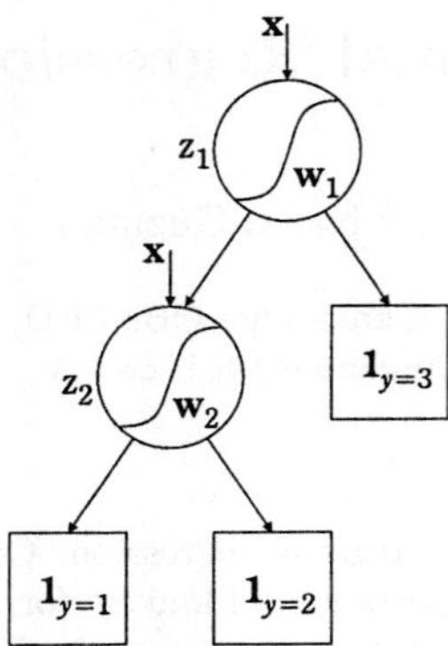

Fig. 1. A binary mixture of experts tree for ordinal regression. The expert (leaf) nodes are indicator functions, each responsible for labeling one possible rank. Here $\mathbf{1}_A$ is one if A is true, and zero otherwise. The gating nodes indicate the probability of following the left—or conversely right—branch down the tree to a rank. The structure of the HME tree, with a left-to-right assignment of ranks to the 'experts', encapsulates the ordinal regression problem.

3 Hierarchical Mixture of Experts for Ordinal Regression

We formulate the distribution of the ordinal target variables with a binary mixture of experts tree. Figure 1 illustrates such a tree, where the leaves, called 'experts', are component distributions of the targets. The non-leaf nodes, called 'gates', form coefficients that mix the experts. Each gate is conditioned on an input variable and indicates the probability of following its left—or conversely right—branch down the tree; consequently the gates perform a soft partitioning of the input space. This soft partitioning is used as our ordinal regression model.

We associate a binary variable z_i with each gate, and set it to one if the left branch is followed from the ith gate. The parameters of the model are the real-valued weight vectors of the gates, which we indicate with $\mathbf{W} = \{\mathbf{w}_i\}_{i=1}^{I}$. The experts are labeled with discrete labels $1, \ldots, R$, and we require the experts to be indicator functions. Hence, given expert r, the probability that it labeled $(\mathbf{x}, y)$ is one if $y = r$, and zero otherwise. With a left-to-right assignment of ranks to the experts, the structure of the HME tree and the resulting partitioning of the input space impose a natural ordering on the targets. In this paper we restrict ourselves to complete binary trees, although a more judicious choice of tree structure, based on evidence maximization, can be made.

The probability of y having rank r, given $\mathbf{x}$, is equal to the probability that expert r was responsible for generating the target. Equivalently it is equal to the probability of correctly setting the binary indicator variables z_i to form a path from the root to the rth 'expert',

$$P(y = r|\mathbf{x}, \mathbf{W}) = \prod_{i:\text{root}\to r} P(z_i|\mathbf{x}, \mathbf{w}_i). \quad (1)$$

We use notation $i : \text{root} \to r$ to indicate that the product is taken over the gates on the unique path from the root to the rth expert, and note that summing (1) over all ranks give unity. By defining $\sigma(a) = 1/(1 + e^{-a})$, the probability of following the left branch from the ith gate is

$$P(z_i = 1|\mathbf{x}, \mathbf{w}_i) = \sigma(\mathbf{w}_i^{\top}\mathbf{x}).$$

Throughout this paper, we implicitly augment input vectors with a bias clamped at 1. From (1), the likelihood of observing the entire data set is

$$P(\mathcal{D}|\mathbf{W}) \equiv P(\mathbf{y}|\mathbf{X},\mathbf{W}) = \prod_{n=1}^{N} \prod_{i:\text{root}\to y_n} P(z_{in}|\mathbf{x}_n,\mathbf{w}_i). \tag{2}$$

3.1 The Posterior

A probabilistic formulation—often prone to overfitting, as in the familiar case of supervised learning—can be found by maximizing the likelihood (2) with respect to the model parameters $\mathbf{W}$. We rather use the usual Bayesian approach of making predictions by computing the expected value of $P(y_* = r|\mathbf{x}_*,\mathbf{W})$ for a new example $\mathbf{x}_*$ with respect to the posterior distribution of $\mathbf{W}$. For the purpose of obtaining this posterior distribution from Bayes' theorem, we place a Gaussian prior on each gate's parameter vector,

$$p(\mathbf{w}_i|\alpha_i) = \left(\frac{\alpha_i}{2\pi}\right)^{d/2} \exp\left\{-\frac{\alpha_i}{2}\mathbf{w}_i^\top\mathbf{w}_i\right\},$$

and combine it with the likelihood (2), normalized by the evidence. The hyperparameter α_i controls the width of the prior.

The weight vector of gate i, conditioned on the observed data, is independent of the parameters of the other gates, and only dependent on the examples that were labeled by its left and right subtrees. As a notational convenience, let $\mathcal{T}_i$ indicate the set of experts that are leaves in the subtree with gate i as root. Define $\mathcal{D}_i$ to be the subset of examples associated with $\mathcal{T}_i$. From Bayes' theorem, the posterior distribution of each gate's parameters is

$$p(\mathbf{w}_i|\mathcal{D}_i,\alpha_i) = \frac{P(\mathcal{D}_i|\mathbf{w}_i)p(\mathbf{w}_i|\alpha_i)}{p(\mathcal{D}_i|\alpha_i)} \tag{3}$$

$$\propto \prod_{n:y_n\in\mathcal{T}_i} \sigma(\mathbf{w}_i^\top\mathbf{x}_n)^{z_{in}}(1-\sigma(\mathbf{w}_i^\top\mathbf{x}_n))^{1-z_{in}} \exp\left\{-\frac{\alpha_i}{2}\mathbf{w}_i^\top\mathbf{w}_i\right\}. \tag{4}$$

The full posterior is simply the product over all individual gate posterior distributions, $p(\mathbf{W}|\mathcal{D},\boldsymbol{\alpha}) = \prod_{i=1}^{I} p(\mathbf{w}_i|\mathcal{D}_i,\alpha_i)$.[1]

3.2 Inference

To determine the rank of a new example $\mathbf{x}_*$, we marginalize over the posterior distribution of the weights, given the observed data:

$$P(y_* = r|\mathbf{x}_*,\mathcal{D},\boldsymbol{\alpha}) = \int P(y_* = r|\mathbf{x}_*,\mathbf{W})\, p(\mathbf{W}|\mathcal{D},\boldsymbol{\alpha})\, d\mathbf{W}$$

$$= \prod_{i:\text{root}\to r} \int P(z_i|\mathbf{x}_*,\mathbf{w}_i)\, p(\mathbf{w}_i|\mathcal{D}_i,\alpha_i)\, d\mathbf{w}_i. \tag{5}$$

[1] Ideally we want $p(\mathbf{W}|\mathcal{D}) = \int p(\mathbf{W}|\mathcal{D},\boldsymbol{\alpha})p(\boldsymbol{\alpha}|\mathcal{D})\, d\boldsymbol{\alpha}$, a matter that we shall touch on in Section 3.3.

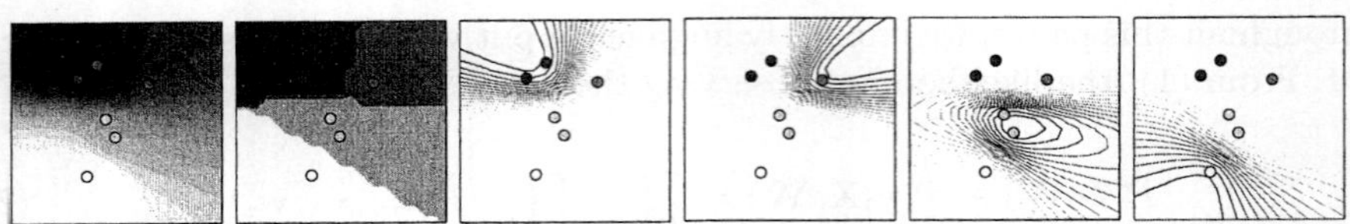

Fig. 2. An example showing four ranks. Shown from left to right is the expected rank; most probable rank; posterior probabilities of ranks 1 to 4.

Figure 2 illustrates a toy problem with four ranks, and the respective posterior probabilities of each rank.

It is not possible to perform the integration in (5) analytically, so we make a Laplace approximation (MacKay, 1992) to each $p(\mathbf{w}_i|\mathcal{D}_i,\alpha_i)$. Laplace's method involves a quadratic approximation of the log-posterior around its mode: the negative logarithm of the posterior (3) is maximized over $\mathbf{w}_i$ to give the most probable weight vector $\mathbf{w}_{\mathrm{MP}_i}$. We find $\mathbf{w}_{\mathrm{MP}_i}$ by setting the first derivative of $-\ln p(\mathbf{w}_i|\mathcal{D}_i,\alpha_i)$ to zero and solving with a standard Newton-Raphson method. The second-order Taylor expansion of $-\ln p(\mathbf{w}_i|\mathcal{D}_i,\alpha_i)$ around its maximum $\mathbf{w}_{\mathrm{MP}_i}$ allows us to approximate the posterior with a Gaussian distribution with mean $\mathbf{w}_{\mathrm{MP}_i}$ and variance-covariance matrix $\mathbf{A}_i^{-1}$. Here $\mathbf{A}_i$ is the Hessian, the matrix of second derivatives $-\nabla^2 \ln p(\mathbf{w}_i|\mathcal{D}_i,\alpha_i)$ evaluated at the most probable parameter values $\mathbf{w}_{\mathrm{MP}_i}$. This leads to an approximation of (5) with

$$P(y_* = r|\mathbf{x}_*,\mathcal{D},\boldsymbol{\alpha}) \simeq \prod_{i:\mathrm{root}\to r} \int P(z_i|\mathbf{x}_*,\mathbf{w}_i)\,\mathsf{Normal}(\mathbf{w}_i;\mathbf{w}_{\mathrm{MP}_i},\mathbf{A}_i^{-1})\,d\mathbf{w}_i. \quad (6)$$

The probability $P(z_i = 1|\mathbf{x}_*,\mathbf{w}_i) = \sigma(\mathbf{w}_i^\top\mathbf{x}_*)$ has a linear dependence on the weight parameter through the scalar $a_i = \mathbf{w}_i^\top\mathbf{x}_*$, and hence the dimensionality of the integral can be reduced by finding the probability density $p(a_i|\mathbf{x}_*,\mathcal{D}) = 1/\sqrt{2\pi s_i^2}\cdot\exp\{-(a_i - a_{\mathrm{MP}_i})^2/2s_i^2\}$ with the mean and variance given by $a_{\mathrm{MP}_i} = \mathbf{w}_{\mathrm{MP}_i}^\top\mathbf{x}_*$ and $s_i^2 = \mathbf{x}_*^\top\mathbf{A}_i^{-1}\mathbf{x}_*$ respectively. The marginalized output, where each of the integrals in the product (6) is effectively $P(z_i|\mathbf{x}_*,\mathcal{D}_i,\alpha_i)$, is therefore

$$P(z_i = 1|\mathbf{x}_*,\mathcal{D}_i,\alpha_i) = \psi(a_{\mathrm{MP}_i}, s_i^2) \equiv \int \sigma(a_i)\,\mathsf{Normal}(a_i; a_{\mathrm{MP}_i}, s_i^2)\,da_i.$$

The integral of a sigmoid times a Gaussian is approximated by $\psi(a_{\mathrm{MP}_i}, s_i^2) \simeq \sigma(\kappa(s_i^2)\cdot a_{\mathrm{MP}_i})$, with $\kappa(s_i^2) = 1/\sqrt{1+\pi s_i^2/8}$ (MacKay, 1992), so that we make a final prediction with

$$P(y_* = r|\mathbf{x}_*,\mathcal{D},\boldsymbol{\alpha}) \simeq \prod_{i:\mathrm{root}\to r} \sigma(\kappa(s_i^2)\cdot a_{\mathrm{MP}_i}).$$

3.3 Finding Values for Hyperparameters $\boldsymbol{\alpha}$

The preferred Bayesian treatment for hyperparameters such as $\boldsymbol{\alpha}$ is to integrate them out of any predictions with $p(\mathbf{w}_i|\mathcal{D}_i) = \int p(\mathbf{w}_i|\mathcal{D}_i,\alpha_i)p(\alpha_i|\mathcal{D}_i)\,d\alpha_i$. We will assume rather that the hyperparameter posterior $p(\alpha_i|\mathcal{D}_i)$ is sharply peaked around its most probable value α_{MP_i}, so that $p(\mathbf{w}_i|\mathcal{D}_i) \simeq p(\mathbf{w}_i|\mathcal{D}_i, \alpha_{\mathrm{MP}_i})$. The hyperparameters which maximize the posterior $p(\alpha_i|\mathcal{D}_i)$ need to be found; by assuming a non-informative hyperprior over α_i, this task amounts to maximizing the likelihood term (or *evidence*: the denominator in (3)). The log of the evidence as a function of α_i is $\ln p(\mathcal{D}_i|\alpha_i) = \frac{d}{2}\ln\alpha_i - \frac{\alpha_i}{2}\mathbf{w}_{\mathrm{MP}_i}^\top \mathbf{w}_{\mathrm{MP}_i} - \frac{1}{2}\ln|\mathbf{A}_i| + c$. Following MacKay (1992), maximizing the log-evidence with respect to α_i leads to $\alpha_{\mathrm{MP}_i} = (d - \alpha_i \mathsf{Trace}(\mathbf{A}_i^{-1}))/\mathbf{w}_{\mathrm{MP}_i}^\top \mathbf{w}_{\mathrm{MP}_i}$, which we use as a re-estimation formula for α_i. The Hessian and most probable weights are recomputed, and the process repeated until convergence of α_i.

3.4 Nonlinear Decision Boundaries

Nonlinearity is introduced to the model with a fixed set of basis functions, and we replace $\mathbf{w}^\top\mathbf{x}$ by $\sum_{m=1}^{M} w_m \phi_m(\mathbf{x}) = \mathbf{w}^\top \boldsymbol{\phi}(\mathbf{x})$. For simplicity, we let the basis functions be shared over all the gates. For practical results, we use radial basis functions, $\phi_m(\mathbf{x}) = \exp\{-\frac{1}{2h^2}\|\mathbf{x} - \boldsymbol{\mu}_m\|^2\}$, and keep one basis function fixed at unity (the bias). The basis function centres are set by a k-means clustering on each rank. The M basis functions used in each gate are the collection of all basis functions over the ranks. The width h of the basis functions is set to twice the average spacing between the cluster centres. We defer other methods of implementing the gates to Sec. 5.

4 Experimental Results

The proposed HME approach to ordinal regression was evaluated on benchmark data sets from Chu & Ghahramani (2004), who have discretized the targets from the data sets, normally used for metric regression, into 5 and 10 ordinal ranks using equal-length binning. The data were partitioned into training and test sets, with a repartitioning performed 20 times on each data set.[2]

We evaluate the accuracy by taking the most likely rank as the predicted rank $\hat{y}_n$, and comparing it to the true rank y_n. If there are N' elements in the test set, the *mean zero-one error* averages the number of incorrect predictions with $\frac{1}{N'}\sum_{n=1}^{N'} \mathbf{1}_{\hat{y}_n \neq y_n}$. For the nonlinear case we added 10 basis functions per rank to the set of basis functions used. Table 1 shows the averages over 20 trials, along with the standard deviation. The first three columns are taken from Chu & Ghahramani (2004), who have compared Gaussian processes with Gaussian basis functions to the support vector machine (SVM) approach of Shashua & Levin (2003). Both a MAP estimation with Laplace approximation (MAP) and Expectation Propagation algorithm with variational bound (EP) was used as inference techniques to implement the Gaussian process. The HME model with both linear and nonlinear gates gives comparable performance.

[2] The datasets and partitions are downloadable from www.gatsby.ucl.ac.uk/~chuwei/ordinalregression.html.

Table 1. The test results of five algorithms. The data sets used, with (attributes, training instances, test instances), are **Di.** Diabetes (2, 30, 13); **Py.** Pyrimidines (27, 50, 24); **Tr.** Triazines (60, 100, 86); **Wi.** Wisconsin Breast Cancer (32, 130, 64); **St.** Stocks Domain (9, 600, 350); **Ab.** Abalone (8, 1000, 3177).

	Mean zero-one error (5 equal-length bins)				
Data	SVM	GP (MAP)	GP (EP)	HME (linear)	HME (nonlinear)
Di.	57.31±12.09%	54.23±13.78%	54.23±13.78%	51.54±6.16%	57.69±15.28%
Py.	41.46±8.49%	39.79±7.21%	36.46±6.47%	46.25±8.32%	47.71±8.16%
Tr.	54.19±1.48%	52.91±2.15%	52.62±2.66%	56.80±8.50%	55.12±4.55%
Wi.	70.78±3.73%	65.00±4.71%	65.16±4.65%	74.61±4.83%	68.36±2.91%
St.	10.81±1.70%	11.99±2.34%	12.00±2.06%	19.26±1.80%	14.43±2.16%
Ab.	21.58±0.32%	21.50±0.22%	21.56±0.36%	21.91±0.30%	21.91±0.30%
	Mean zero-one error (10 equal-length bins)				
Di.	90.38±7.00%	83.46±5.73%	83.08±5.91%	76.54±7.27%	80.77±9.50%
Py.	59.37±7.63%	55.42±8.01%	54.38±7.70%	64.79±8.60%	60.83±9.21%
Tr.	67.91±3.63%	63.72±4.34%	64.01±3.78%	68.37±5.65%	69.30±4.37%
Wi.	85.86±3.78%	78.52±3.58%	78.52±3.51%	88.75±4.11%	79.53±4.53%
St.	17.79±2.23%	19.90±1.72%	19.44±1.91%	32.00±3.82%	23.87±2.24%
Ab.	44.32±1.46%	42.60±0.91%	42.27±0.46%	43.14±0.52%	42.56±1.27%

5 Conclusion and Future Work

We have described a novel Bayesian approach to ordinal regression, based on a hierarchical mixture of experts tree. The model was made analytically tractable with a Laplace approximation to the parameter posterior: future work will involve using Markov-chain Monte Carlo methods to average (integrate) predictions over the posterior distribution. The gates can equally well be impemented with Gaussian processes, a matter worthy of investigation.

Ulrich Paquet is supported by a Commonwealth Scholarship, and expresses his thanks to the Commonwealth Scholarship Commission. Andrew Naish-Guzman holds a Millennium Scholarship.

References

Chu, W. & Ghahramani, Z. (2004) Gaussian processes for ordinal regression. Technical report, Gatsby Computational Neuroscience Unit, University College London.

Jordan, M. I. & Jacobs, R. A. (1994). Hierarchical mixtures of experts and the EM algorithm. *Neural Computation, 6*, 181–214.

MacKay, D. J. C. (1992). The evidence framework applied to classification networks. *Neural Computation, 4*(5), 698–714.

Shashua, A. & Levin, A. (2003). Ranking with large margin principle: two approaches. In *Advances in Neural Information Processing Systems 15*, (pp. 937–944). MIT Press

Waterhouse, S. R., MacKay, D. J. C., & Robinson, A. J. (1996). Bayesian methods for mixtures of experts. In *Advances in Neural Information Processing Systems 8*, (pp. 351–357). MIT Press.

Traffic Flow Forecasting Using a Spatio-temporal Bayesian Network Predictor

Shiliang Sun, Changshui Zhang, and Yi Zhang

State Key Laboratory of Intelligent Technology and Systems,
Department of Automation, Tsinghua University, Beijing, China 100084
sunsl02@mails.tsinghua.edu.cn
{zcs, zhyi}@mail.tsinghua.edu.cn

Abstract. A novel predictor for traffic flow forecasting, namely spatio-temporal Bayesian network predictor, is proposed. Unlike existing methods, our approach incorporates all the spatial and temporal information available in a transportation network to carry our traffic flow forecasting of the current site. The Pearson correlation coefficient is adopted to rank the input variables (traffic flows) for prediction, and the best-first strategy is employed to select a subset as the cause nodes of a Bayesian network. Given the derived cause nodes and the corresponding effect node in the spatio-temporal Bayesian network, a Gaussian Mixture Model is applied to describe the statistical relationship between the input and output. Finally, traffic flow forecasting is performed under the criterion of Minimum Mean Square Error (M.M.S.E.). Experimental results with the urban vehicular flow data of Beijing demonstrate the effectiveness of our presented spatio-temporal Bayesian network predictor.

1 Introduction

Short-term traffic flow forecasting, which is to determine the traffic volume in the next time interval usually in the range of five minutes to half an hour, is an important issue for the application of Intelligent Transportation Systems (ITS) [1]. Up to the present, some approaches ranging from simple to elaborate on this theme were proposed including those based on neural network approaches, time series models, Kalman filter theory, simulation models, non-parametric regression, fuzzy-neural approach, layered models, and Markov Chain models [1]~[8]. Although these methods have alleviated difficulties in traffic flow modelling and forecasting to some extent, from a careful review we can still find a problem, that is, most of them have not made good use of spatial information from the viewpoint of networks to analyze the trends of the object site. Though Chang et al utilized the data from other roadways to make judgmental adjustments, the information was still not used to its full potential [9]. Yin et al developed a fuzzy-neural model to predict the traffic flows in an urban street network whereas it only utilized the upstream flows in the current time interval to forecast the selected downstream flow in the next interval [7].

W. Duch et al. (Eds.): ICANN 2005, LNCS 3697, pp. 273–278, 2005.

The main contribution of this paper is that we proposed an original spatio-temporal Bayesian network predictor, which combines the available spatial information with temporal information in a transportation network to implement traffic flow modelling and forecasting. The motivation of our approach is very intuitive. Although many sites may be located at different even distant parts of a transportation network, there exist some common sources influencing their own traffic flows. Some of the distributed sources include shopping centers, home communities, car parks, etc. People's activities around these sources usually obey some consistent laws in a long time period, such as the usually common working hours. To our opinion, these hidden sources imply some information useful for traffic flow forecasting in different sites. Therefore, construct a causal model (Bayesian network) among different sites for traffic flow forecasting is reasonable. This paper covers how to use the information from a whole transportation network to design feasible spatio-temporal Bayesian networks and carry our traffic flow forecasting of the object sites. Encouraging experimental results with real-world data show that our approach is rather effective for traffic flow forecasting.

2 Methodology

In a transportation network, there are usually a lot of sites (road links) related or informative to the traffic flow of the current site from the standpoint of causal Bayesian networks. However, using all the related links as input variables (cause nodes) would involve much irrelevance, redundancy and would be prohibitive for computation. Consequently, a variable selection procedure is of great demand. Up to date many variable selection algorithms include variable ranking as a principal or auxiliary selection mechanism because of its simplicity, scalability, and good empirical success [10]. In this article, we also adopt the variable ranking mechanism, and the Pearson correlation coefficient is used as the specific ranking criterion defined for individual variables.

2.1 Variable Ranking and Cause Node Selection

Variable ranking can be regarded as a filter method: it is a preprocessing step, independent of the choice of the predictor [11]. Still, under certain independence or orthogonality assumptions, it may be optimal with respect to a given predictor [10]. Even when variable ranking is not optimal, it may be preferable to other variable subset selection methods because of its computational and statistical scalability [12]. This is also the motivation of our using the best-first search strategy to select the most relevant traffic flows from the ranking result as final cause nodes of a Bayesian network.

Consider a set of m samples $\{x_k, y_k\}(k = 1, ..., m)$ consisting of n input variable $x_{k,i}(i = 1, ..., n)$ and one output variable y_k. Variable ranking makes use of a scoring function $S(i)$ computed from the value $x_{k,i}$ and $y_k(k = 1, ..., m)$. By convention, we assume that a high score is indicative of a valuable variable and that we sort variables in decreasing order of $S(i)$. Furthermore, let X_i denote the random variable corresponding to the ith component of input vector x, and Y

denote the random variable of which the outcome y is a realization. The Pearson correlation coefficient between X_i and Y can be estimated by:

$$R(i) = \frac{\sum_{k=1}^{m}(x_{k,i} - \overline{x}_i)(y_k - \overline{y})}{\sqrt{\sum_{k=1}^{m}(x_{k,i} - \overline{x}_i)^2 \sum_{k=1}^{m}(y_k - \overline{y})^2}} \quad (1)$$

where the bar notation stands for an average over the index k [10].

In this article, we use the norm $|R(i)|$ as a variable ranking criterion. After the variable ranking stage, a variable selection (cause node selection) process is adopted to determine the final cause nodes (input variables) for predict the effect node (output). Here we use the best-first search strategy to find the cause nodes as the first several variables in the ranking list because of its fastness, simplicity and empirical effectiveness.

2.2 Flow Chart for Traffic Flow Forecasting

Given the derived cause nodes and the effect node in a Bayesian network, we utilize the Gaussian Mixture Model (GMM) and the Competitive EM (CEM) algorithm to approximate their joint probability distribution. Then we can obtain the optimum prediction formulation as an analytic solution under the M.M.S.E. criterion. For details about the GMM, CEM algorithm and the prediction formulation, please refer to articles [8][13][14].

Now we describe the flow chart of our approach for traffic flow forecasting. First the data set is divided into two parts, one serving as training set for input variable (cause node) selection and parameter learning, and the other test set. The flow chart can be given as follows: 1) Choose an object road site whose traffic flow should be forecasted (effect node) and collect all the available traffic flows in a traffic network as the original input variables; 2) Compute the Pearson correlation coefficients between the object traffic flow (effect node) and the input variables on the training set with different time lags respectively, and then select several most related variables in the ranking list as the final cause nodes of the spatio-temporal Bayesian network; 3) Derive the optimum prediction formulation using GMM and CEM algorithm detailed in articles [8][14]; 4) Implement forecasting on the test set using the derived formulation.

Conveniently, the flow chart can be largely reduced and for real-time utility when forecasting a new traffic flow, because the cause node selection and the prediction formulation need only be computed one time based on the historical traffic flows (learning stage), and thus can be derived in advance.

3 Experiments

The field data analyzed in this paper is the vehicle flow rates of discrete time series recorded every 15 minutes along many road links by the UTC/SCOOT system in Traffic Management Bureau of Beijing, whose unit is vehicles per hour (veh/hr). Fig. 1 depicts a real patch used to verify our proposed predictor. The

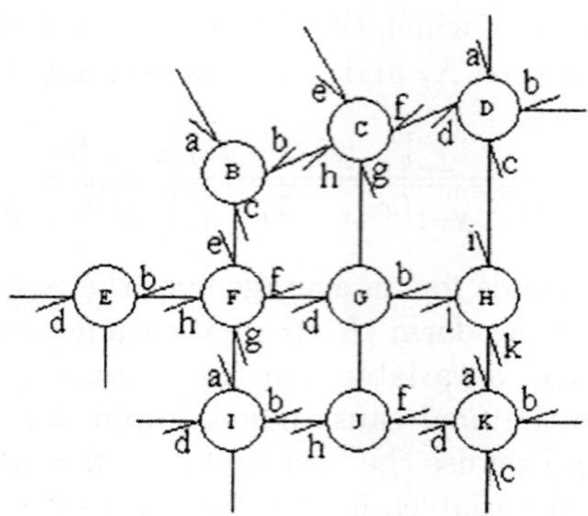

Fig. 1. The analyzed transportation network

raw data for utility are of 25 days and totally 2400 sample points taken from March, 2002. To validate our approach, the frist 2112 points (training set) are employed to carry out input cause node selection and to learn parameters of the spatio-temporal Bayesian network, and the rest (test set) are employed to test the forecasting performance.

In addition, we utilize the the Random Walk method and Markov Chain method as base lines to evaluate our presented approach [8]. Random Walk is a classical method for traffic flow forecasting. Its core idea is to forecast the current value using its last value, and can be formulated as:

$$\hat{x}(t+1) = x(t) . \tag{2}$$

Markov Chain method models traffic flow as a high order Markov chain. It has shown great merits over several other approaches for traffic flow forecasting [8]. In this paper the joint probability distribution for the Markov Chain method is also approximated by the GMM whose parameters are estimated through CEM algorithm. The number of input variables is also taken as 4 (same as in [8]) for each object site in our approach. This entire configuration is to make an equitable comparison as much as possible. Now the only difference between our Bayesian network predictor and the Markov Chain method is that we utilize the whole spatial and temporal information in a transportation network to forecast while the latter only uses the temporal information of the object site.

We take road link Gd as an example to show our approach. Gd represents the vehicle flow from upstream link F to downstream link G. All the available traffic flows which may be informative to forecast Gd in the transportation network includes $\{Ba, Bb, Bc, Ce, Cf, ..., Ka, Kb, Kc, Kd\}$. Considering the time factor, to forecasting the traffic flow $Gd(t)$ (effect node), we need judge the above sites with different time indices, such as $\{Ba(t-1), Ba(t-2), ..., Ba(t-d)\}, \{Bb(t-1), Bb(t-2), ..., Bb(t-d)\}$, etc. In this paper, d is taken as 100 empirically. Four most correlated traffic flows which are selected with the correlation variable ranking criterion and the best-first strategy for five different object traffic flows and the corresponding correlation coefficient values are listed in Table 1.

Table 1. Four most correlated traffic flows for five object traffic flows

Object traffic flows	Strongly correlated traffic flows (cause nodes)			
$Ch(t)$	$Bc(t-1)$	$Hl(t-1)$	$Ch(t-1)$	$Fe(t-3)$
	0.971	0.968	0.967	0.966
$Dd(t)$	$Dd(t-1)$	$Ch(t-1)$	$Bc(t-1)$	$Hl(t-2)$
	0.963	0.961	0.959	0.959
$Fe(t)$	$Fe(t-1)$	$Ba(t-1)$	$Fe(t-2)$	$Fe(t-96)$
	0.983	0.978	0.964	0.961
$Gd(t)$	$Gd(t-1)$	$Fh(t-1)$	$Hl(t-1)$	$Fe(t-1)$
	0.967	0.962	0.962	0.957
$Ka(t)$	$Hi(t-1)$	$Cf(t-1)$	$Ka(t-1)$	$Bb(t-2)$
	0.967	0.967	0.967	0.966

Table 2. A performance comparison of three methods for short-term traffic flow forecasting of five different road links

Methods	Ch	Dd	Fe	Gd	Ka
Random Walk	79.85	70.99	157.60	177.57	99.20
Markov Chain	68.51	66.15	122.65	151.31	80.46
Spatio-Temporal Bayesian Network	*65.95*	*57.46*	*115.07*	*141.37*	*73.02*

With the selected cause nodes (input traffic flows), we can approximate the joint probability distribution between the input and output with GMM, then derive the optimum prediction formulation for road link *Gd*. In addition, we also conducted experiments on four other traffic flows. Table 2 gives the forecasting errors denoted by Root Mean Square Error (RMSE) of all the five road links through Random Walk method, Markov Chain method and our predictor. In the same column of Table 2, the smaller RMSE corresponds to the better forecasting accuracy. From the experimental results, we can find the significant improvements of forecasting capability brought by the spatio-temporal Bayesian network predictor which integrates both spatial and temporal information for forecasting.

4 Conclusions and Future Work

In this paper, we successfully combine the whole spatial with temporal information available in a transportation network to carry out short-term traffic flow forecasting. Experiments show that distant road links in a transportation network can have high correlation coefficients, and this relevance can be employed for traffic flow forecasting. This knowledge would greatly broaden people's traditional knowledge about transportation networks and the transportation forecasting research. Many existing methods can be illuminated and further developed on the scale of a transportation network. In the future, how to extend the presented spatio-temporal Bayesian network predictor to forecast traffic flows in case of incomplete data would be a valuable direction.

Acknowledgements

The authors are grateful to the anonymous reviewers for giving valuable remarks. This work was supported by Project 60475001 of the National Natural Science Foundation of China.

References

1. William, B.M.: Modeling and Forecasting Vehicular Traffic Flow as a Seasonal Stochastic Time Series Process. Doctoral Dissertation. University of Virginia, Charlottesville (1999)
2. Yu, E.S., Chen, C.Y.R.: Traffic Prediction Using Neural Networks. Proceedings of IEEE Global Telecommunications Conference, Vol. 2 (1993) 991-995
3. Moorthy, C.K., Ratcliffe, B.G.: Short Term Traffic Forecasting Using Time Series Methods. Transportation Planning and Technology, Vol. 12 (1988) 45-56
4. Okutani, I., Stephanedes, Y.J.: Dynamic Prediction of Traffic Volume through Kalman Filter Theory. Transportation Research, Part B, Vol. 18B (1984) 1-11
5. Chrobok, R., Wahle, J., Schreckenberg, M.: Traffic Forecast Using Simulations of Large Scale Networks. Proceedings of IEEE Intelligent Transportation Systems Conference (2001) 434-439
6. Davis, G.A., Nihan, N.L.: Non-Parametric Regression and Short-Term Freeway Traffic Forecasting. Journal of Transportation Engineering (1991) 178-188
7. Yin, H.B., Wong, S.C., Xu, J.M., Wong, C.K.: Urban Traffic Flow Prediction Using a Fuzzy-Neural Approach. Transportation Research, Part C, Vol. 10 (2002), 85-98
8. Yu, G.Q., Hu, J.M., Zhang, C.S., Zhuang, L.K., Song J.Y.: Short-Term Traffic Flow Forecasting Based on Markov Chain Model. Proceedings of IEEE Intelligent Vehicles Symposium (2003) 208 - 212
9. Chang, S.C., Kim, R.S., Kim, S.J., Ahn, B.H.: Traffic-Flow Forecasting Using a 3-Stage Model. Proceedings of IEEE Intelligent Vehicle Symposium (2000) 451-456
10. Guyon, I., Elisseeff, A.: An Introduction to Variable and Feature Selection. Journal of Machine Learning Research, Vol. 3 (2003) 1157-1182
11. Kohavi, R., John, M.: Wrappers for Feature Selection. Artificial Intelligence (1997) 273-324
12. Hastie, T., Tibshirani, R., Friedman, J.: The Elements of Statistical Learning. Springer Series in Statistics. Springer, New York (2001)
13. Zhang, B.B., Zhang, C.S., Yi, X.: Competitive EM Algorithm for Finite Mixture Models. Pattern Recognition, Vol. 37 (2004) 131-144
14. Sun, S.L., Zhang, C.S., Yu, G.Q., Lu, N.J., Xiao, F.: Bayesian Network Methods for Traffic Flow Forecasting with Incomplete Data. ECML 2004, Lecture Notes In Artificial Intelligence, Vol. 3201. Springer-Verlag, Berlin Heidelberg (2004) 419-428

Manifold Constrained Variational Mixtures

Cédric Archambeau* and Michel Verleysen**

Machine Learning Group - Université catholique de Louvain,
Place du Levant 3, B-1348 Louvain-la-Neuve, Belgium
{archambeau, verleysen}@dice.ucl.ac.be

Abstract. In many data mining applications, the data manifold is of lower dimension than the dimension of the input space. In this paper, it is proposed to take advantage of this additional information in the frame of variational mixtures. The responsibilities computed in the VBE step are constrained according to a discrepancy measure between the Euclidean and the geodesic distance. The methodology is applied to variational Gaussian mixtures as a particular case and outperforms the standard approach, as well as Parzen windows, on both artificial and real data.

1 Introduction

Finite mixture models [1] are commonly used for clustering purposes and modeling unknown densities. Part of their success is due to the fact that their parameters can be computed in an elegant way by the expectation-maximization algorithm (EM) [2]. Unfortunately, it is well known that mixture models suffer from an inherent drawback. EM maximizes iteratively the data log-likelihood, which is an ill-posed problem that can lead to severe overfitting; maximizing the likelihood may result in setting infinite probability mass on a single data point.

Among others, the variational Bayesian framework was introduced in order to avoid this problem [3]. In variational Bayes (VB) a factorized approximation of the joint posterior of the latent variables and the model parameters is used in order to compute a variational lower bound on the marginal data likelihood. In addition, VB allows determining the optimal number of components in the mixture by comparing the value of this variational lower bound. In [4] a variant was proposed to perform automatic model selection.

Recently, manifold Parzen [5] was introduced in order to improve nonparametric density estimation when the data is lying on a manifold of lower dimensionality than the one of the input space. In this paper, a related technique for variational mixtures is proposed by constraining the responsibilities according to the mismatch between the Euclidean and the geodesic distance. The key idea is to favor the directions along the manifold when estimating the unknown density, rather than wasting valuable density mass in directions perpendicular to the manifold orientation. The approach is applied to VB Gaussian mixtures as a particular case. Manifold constrained variational Gaussian mixtures (VB-MFGM) are compared experimentally to standard VB-FGM and standard Parzen.

* C.A. is supported by the European Commission project IST-2000-25145.

** M.V. is a Senior Research Associate of Belgian National Fund for Scientific Research.

W. Duch et al. (Eds.): ICANN 2005, LNCS 3697, pp. 279–284, 2005.

2 Variational Bayes for Mixtures Models

Let $X = \{\mathbf{x}_n\}_{n=1}^N$ be an i.i.d. sample, $Z = \{\mathbf{z}_n\}_{n=1}^N$ the latent variables associated to X and $\Theta = \{\boldsymbol{\theta}_m\}_{m=1}^M$ the model parameters, M being the number of mixture components. Finite mixture models are latent variable models in the sense that we do not know by which component a data sample was generated. We may thus associate to each data sample $\mathbf{x}_n$ a binary latent vector $\mathbf{z}_n$, with latent variables $z_{nm} \in \{0,1\}$ that indicate which component has generated $\mathbf{x}_n$ ($z_{nm} = 1$ if $\mathbf{x}_n$ was generated by component m and 0 otherwise).

In Bayesian learning, both the latent variables Z and the model parameters Θ are treated as random variables. The quantity of interest is the marginal data likelihood, also called incomplete likelihood (i.e. of the observed variables only). For a fixed model structure $\mathcal{H}_M$, it is obtained by integrating out the nuisance parameters Z and Θ:

$$p(X|\mathcal{H}_M) = \int_\Theta \int_Z p(X, Z, \Theta|\mathcal{H}_M) dZ d\Theta \ . \tag{1}$$

This quantity is usually untractable. However, for any arbitrary density $q(Z,\Theta)$ a lower bound on $p(X|\mathcal{H}_M)$ can be found using Jensen's inequality:

$$\log p(X|\mathcal{H}_M) \geq \int_\Theta \int_Z q(Z,\Theta) \log \frac{p(X,Z,\Theta|\mathcal{H}_M)}{q(Z,\Theta)} dZ d\Theta \tag{2}$$

$$= \log p\left(X|\mathcal{H}_M\right) - \mathrm{KL}\left[q(Z,\Theta)||p\left(Z,\Theta|X,\mathcal{H}_M\right)\right] \ , \tag{3}$$

where KL[·] is the Kullback-Leibler (KL) divergence. It is easily seen from (3) that the equality holds when $q(Z,\Theta)$ is equal to the joint posterior $p(Z,\Theta|X,\mathcal{H}_M)$.

In VB, the variational posterior approximates the joint posterior by assuming the latent variables and the parameters are independent:

$$p(Z,\Theta|X,\mathcal{H}_M) \approx q(Z,\Theta) = q(Z)q(\Theta) \ . \tag{4}$$

By assuming this factorization, the lower bound (2) on the marginal likelihood is tractable and the gap between both can be minimized by minimizing the KL divergence between the true and the variational posterior. Setting the derivatives of KL with respect to $q(Z)$ and $q(\Theta)$ to zero results in an EM-like scheme [6]:

$$\textbf{VBE step}: \quad q(Z) \propto \exp\left(\mathrm{E}_\Theta\{\log p(X,Z|\Theta,\mathcal{H}_M)\}\right) \ . \tag{5}$$

$$\textbf{VBM step}: \quad q(\Theta) \propto p(\Theta|\mathcal{H}_M) \exp\left(\mathrm{E}_Z\{\log p(X,Z|\Theta,\mathcal{H}_M)\}\right) \ . \tag{6}$$

In these equations $\mathrm{E}_\Theta\{\cdot\}$ and $\mathrm{E}_Z\{\cdot\}$ denote respectively the expectation with respect to Θ and Z, and $p(X,Z|\Theta,\mathcal{H}_M)$ is the complete likelihood (i.e. of the observed and unobserved variables). Note also that the prior $p(\Theta|\mathcal{H}_m)$ on the parameters appears in (6). If we choose $p(\Theta|\mathcal{H}_m)$ conjugate[1] to the exponential family, learning in the VB framework consists then simply in updating the parameters of the prior to the parameters of the posterior.

---

[1] The prior $p(\Theta)$ is said to be conjugate to $r(\Theta)$ if the posterior $q(\Theta)$ is of the same form as $p(\Theta)$, that is $q(\Theta) \propto p(\Theta)r(\Theta)$. In (6), $r(\Theta)$ is of the exponential family.

Since sample X is i.i.d., the posterior $q(Z)$ factorizes to $\prod_n q(\mathbf{z}_n)$. Furthermore, in the case of mixture models $q(\mathbf{z}_n)$ factorizes as well, such that $q(Z) = \prod_n \prod_m q(z_{nm})$. The resulting VBE step for mixture modes is:

$$q(z_{nm}) \propto \exp\left(\mathrm{E}_{\boldsymbol{\theta}_m}\{\log p(\mathbf{x}_n, \mathbf{z}_n|\boldsymbol{\theta}_m, \mathcal{H}_M)\}\right) \ , \tag{7}$$

where $\mathrm{E}_{\boldsymbol{\theta}_m}\{\cdot\}$ is the expectation with respect to $\boldsymbol{\theta}_m$. As in EM, the quantities computed in the VBE step are the responsibilities, each of them being proportional to the posterior probability of having a component m when $\mathbf{x}_n$ is observed.

3 Manifold Constrained Mixtures Models

Nonlinear data projection techniques [7,8] aim at finding the lower dimensional data manifold (if any) embedded in the input space and at unfolding it. A central concept is the geodesic distance, which is measured along the manifold and not through the embedding space, akin the Euclidean distance. The geodesic distance thus takes the intrinsic geometrical structure of the data into account.

Data Manifold. Consider two data points $\mathbf{x}_i$ and $\mathbf{x}_j$ on the multidimensional manifold $\mathcal{M}$ of lower dimensionality than the one embedding space. The geodesic distance between $\mathbf{x}_i$ and $\mathbf{x}_j$ is defined as the minimal arc length in $\mathcal{M}$ connecting both data samples. In practice, such a minimization is untractable. However, geodesic distances can easily be approximated by graph distances [9]. The problem of minimizing the arc length between two data samples lying on $\mathcal{M}$ reduces to the problem of minimizing the length of path (i.e. broken line) between these samples, while passing through a number of other data points of $\mathcal{M}$. In order to follow the manifold, only the smallest jumps between successive samples are permitted. This can be achieved by using either the K-rule, or the ϵ-rule. The former allows jumping to the K nearest neighbors. The latter allows jumping to samples lying inside a ball of radius ϵ centered on them. Below, we only consider the K-rule as the choice for ϵ is more difficult in practice.

The data and the set of allowed jumps constitute a weighted graph, the vertices being the data, the edges the allowed jumps and the edge labels the Euclidean distance between the corresponding vertices. In order to be a distance, the path length (i.e. the sum of successive jumps) must satisfy the properties of non-negativity, symmetry and triangular inequality. The first and the third are satisfied by construction. Symmetry is ensured when the graph is undirected. For the K-rule, this is gained by adding edges as follows: if $\mathbf{x}_j$ belongs to the K nearest neighbors of $\mathbf{x}_i$, but $\mathbf{x}_i$ is not a neighbor of $\mathbf{x}_j$ then the corresponding edge is added. Besides, extra edges are added to the graph in order to avoid disconnected parts. For this purpose a minimum spanning tree [10] is used.

The only remaining problem for constructing the distance matrix of the weighted undirected graph is to compute the shortest path between all data samples. This is done by repeatedly applying Dijkstra's algorithm [11], which computes the shortest path between a source vertex and all other vertices in a weighted graph provided the labels are non-negative (which is here the case).

Manifold Constrained VBE step. Let us denote the Euclidean and graph distances (i.e. approximate geodesic distances) between sample $\mathbf{x}_n$ and the component center $\boldsymbol{\mu}_m$ by $\delta^e(\mathbf{x}_n, \boldsymbol{\mu}_m)$ and $\delta^g(\mathbf{x}_n, \boldsymbol{\mu}_m)$ respectively. The exponential distribution $\mathcal{E}(y|\eta, \zeta) = \zeta^{-1} \exp\{-(y-\eta)/\zeta\}$ is suitable to measure the discrepancy between both distances by setting η to $\delta^e(\mathbf{x}_n, \boldsymbol{\mu}_m)^2$ and y to $\delta^g(\mathbf{x}_n, \boldsymbol{\mu}_m)^2$, since $\delta^e(\mathbf{x}_n, \boldsymbol{\mu}_m) \leq \delta^g(\mathbf{x}_n, \boldsymbol{\mu}_m)$. The manifold constrained responsibilities are obtained by penalizing the complete likelihood by the resulting discrepancy:

$$\begin{aligned} q'(z_{nm}) &\propto \exp\left(\mathrm{E}_{\boldsymbol{\theta}_m}\{\log p(\mathbf{x}_n, \mathbf{z}_n|\boldsymbol{\theta}_m, \mathcal{H}_M)\mathcal{E}(\delta^g(\mathbf{x}_n, \boldsymbol{\mu}_m)^2|\delta^e(\mathbf{x}_n, \boldsymbol{\mu}_m)^2, \zeta = 1)\}\right) \\ &\approx q(z_{nm}) \exp\left(\delta^e(\mathbf{x}_n, \boldsymbol{\alpha}_m)^2 - \delta^g(\mathbf{x}_n, \boldsymbol{\alpha}_m)^2\right) , \end{aligned} \tag{8}$$

where it is assumed that the variance of $\boldsymbol{\mu}_m$ is small and $\boldsymbol{\alpha}_m = \mathrm{E}_{\boldsymbol{\theta}_m}\{\boldsymbol{\mu}_m\}$. Choosing ζ equal to 1 leaves the responsibility unchanged if both distances are identical. However, when the mismatch increases, $q'(z_{nm})$ decreases, which means that it is less likely that $\mathbf{x}_n$ was generated by m because the corresponding geodesic distance is large compared to the Euclidean distance. This results in a weaker responsibility, reducing the influence of $\mathbf{x}_n$ when updating the variational posterior of the parameters of m in the VBM step.

4 Manifold Constrained Variational Gaussian Mixtures

In this section, the manifold constrained variational Bayes machinery is applied to the Gaussian mixture case. A finite Gaussian mixture (FGM) [1] is a linear combination of M multivariate Gaussian distributions with means $\{\boldsymbol{\mu}_m\}_{m=1}^M$ and covariance matrices $\{\boldsymbol{\Sigma}_m\}_{m=1}^M$: $\hat{p}(\mathbf{x}) = \sum_{m=1}^M \pi_m \mathcal{N}(\mathbf{x}|\boldsymbol{\mu}_m, \boldsymbol{\Sigma}_m)$, with $\mathbf{x} \in \mathbb{R}^d$. The mixing proportions $\{\pi_m\}_{m=1}^M$ are non-negative and must sum to one. Their conjugate prior is a Dirichlet $p(\pi_1, ..., \pi_M) = \mathcal{D}(\pi_1, ..., \pi_M|\kappa_0)$ and the conjugate prior on the means and the covariance matrices is a product of Normal-Wisharts $p(\boldsymbol{\mu}_m, \boldsymbol{\Sigma}_m) = \mathcal{N}(\boldsymbol{\mu}_m|\boldsymbol{\alpha}_0, \boldsymbol{\Sigma}_m/\beta_0)\mathcal{W}\left(\boldsymbol{\Sigma}_m^{-1}|\gamma_0, \boldsymbol{\Lambda}_0\right)$. The variational posterior factorizes similarly as the prior and is of the same functional form. The posterior on the mixture proportions $q(\pi_1, ..., \pi_M)$ are jointly Dirichlet $\mathcal{D}(\pi_1, ..., \pi_M|\kappa_1, ..., \kappa_m)$ and the posterior on the means and the covariance matrices $q(\boldsymbol{\mu}_m, \boldsymbol{\Sigma}_m)$ are Normal-Wishart $\mathcal{N}(\boldsymbol{\mu}_m|\boldsymbol{\alpha}_m, \boldsymbol{\Sigma}_m/\beta_m)\mathcal{W}\left(\boldsymbol{\Sigma}_m^{-1}|\gamma_m, \boldsymbol{\Lambda}_m\right)$.

Training Procedure. The parameters of manifold constrained variational Gaussian mixtures (VB-MFGM) can be learnt as follows:

1. Construct the training manifold by the K-rule and compute the associated distance matrix $\delta^g(\mathbf{x}_i, \mathbf{x}_j)$ by Dijkstra's shortest path algorithm.
2. Repeat until convergence:
   Update the distance matrix of the expected component means. Find for each $\boldsymbol{\alpha}_m$ the K nearest training samples $\{\mathbf{x}_k\}_{k=1}^K$ and compute its graph distances to all training data: $\delta^g(\mathbf{x}_n, \boldsymbol{\alpha}_m) = \min_k\{\delta^g(\mathbf{x}_n, \mathbf{x}_k) + \delta^e(\mathbf{x}_k, \boldsymbol{\alpha}_m)\}$.
   VBE step. Compute the manifold constrained responsibilities using (8):

$$\begin{aligned} q'(z_{nm}) \propto \tilde{\pi}_m \tilde{\Lambda}_m^{1/2} \exp\left\{-\frac{\gamma_m}{2}(\mathbf{x}_n - \boldsymbol{\alpha}_m)^{\mathrm{T}} \boldsymbol{\Lambda}_m (\mathbf{x}_n - \boldsymbol{\alpha}_m) - \frac{d}{2\beta_m}\right\} \\ \times \exp\left\{\delta^e(\mathbf{x}_n, \boldsymbol{\alpha}_m)^2 - \delta^g(\mathbf{x}_n, \boldsymbol{\alpha}_m)^2\right\} , \end{aligned} \tag{9}$$

where $\log \tilde{\pi}_m \equiv \psi(\kappa_m) - \psi\left(\sum_m \kappa_m\right)$ and $\log \tilde{\Lambda}_m \equiv \sum_{i=1}^{d} \psi\left(\frac{\gamma_m+1-i}{2}\right) + d \log 2 - \log|\boldsymbol{\Lambda}_m|$, with $\psi(\cdot)$ the digamma function.

VBM step. Update the variational posteriors by first computing the following quantities:

$$\bar{\boldsymbol{\mu}}_m = \frac{\sum_n q'(z_{nm})\mathbf{x}_n}{\sum_n q'(z_{nm})}, \quad \bar{\boldsymbol{\Sigma}}_m = \frac{\sum_n q'(z_{nm}) C(\mathbf{x}_n, \bar{\boldsymbol{\mu}}_m)}{\sum_n q'(z_{nm})}, \quad \bar{\pi}_m = \frac{\sum_n q'(z_{nm})}{N},$$

where $C(\mathbf{x}_n, \bar{\boldsymbol{\mu}}_m) = (\mathbf{x}_n - \bar{\boldsymbol{\mu}}_m)(\mathbf{x}_n - \bar{\boldsymbol{\mu}}_m)^{\mathrm{T}}$. Next, update the parameters of the posteriors:

$$\boldsymbol{\alpha}_m = \frac{N\bar{\pi}_m \bar{\boldsymbol{\mu}}_m + \beta_0 \boldsymbol{\alpha}_0}{\beta_m}, \qquad \beta_m = N\bar{\pi}_m + \beta_0, \qquad \gamma_m = N\bar{\pi}_m + \gamma_0, \quad (10)$$

$$\boldsymbol{\Lambda}_m^{-1} = N\bar{\pi}_m \bar{\boldsymbol{\Sigma}}_m + \frac{N\bar{\pi}_m \beta_0 C(\bar{\boldsymbol{\mu}}_m, \boldsymbol{\alpha}_0)}{\beta_m} + \boldsymbol{\Lambda}_0^{-1}, \qquad \kappa_m = N\bar{\pi}_m + \kappa_0. \quad (11)$$

The computational overhead at each iteration step is limited with respect to standard VB-FGM, as the number of components in the mixture is usually small and updating $\delta^g(\mathbf{x}_n, \boldsymbol{\alpha}_m)$ does not require to recompute $\delta^e(\mathbf{x}_i, \mathbf{x}_j)$.

5 Experimental Results and Conclusion

In order to asses the quality of the density estimators the average negative log-likelihood of the test set $\{\mathbf{x}_q\}_{q=1}^{N_t}$ is used: ANLL $= -\frac{1}{N_t}\sum_{q=1}^{N_t} \log \hat{p}(\mathbf{x}_q)$. In the following, VB-MFGM is compared to standard VB-FGM and standard nonparametric density estimation (Parzen) [12] on artificial and real data.

The first example is presented for illustrative purposes. The data samples are generated from a two dimensional noisy spiral: $\mathbf{x}_1 = 0.04t \sin t + \epsilon_1$ and $\mathbf{x}_2 = 0.04t \cos t + \epsilon_2$, where t follows a uniform $\mathcal{U}(3, 15)$ and $\epsilon_1, \epsilon_2 \sim \mathcal{N}(0, .03)$ is zero-mean Gaussian noise in each direction. The training, validation and test sets have respectively 300, 300 and 10000 samples. The optimal parameters are $M = 15$ and $K = 5$. The estimators are shown in Figure 1. On the one hand, VB-MFGM avoids manifold related local minima in which standard VB-FGM may get trapped into by forcing the expected component centers to move through the training manifold and the covariance matrices to be oriented along it. On the other hand, VB-MFGM clearly produces smoother estimators than Parzen.

In order to asses the performance of VB-MFGM on a real data set, the density of the Abalone[2] data is estimated after normalization. Note that the information regarding the sex is not used. The available data is divided in 2500 training, 500 validation, and 1177 test points. The optimal parameters are $M = 7$ and $K = 20$. The optimal width of the Gaussian kernel in Parzen is 0.17. The ANLL of test set for Parzen windows, VB-FGM and VB-MFGM are respectively 2.49, 0.84 and 0.37. Remark that the improvement of VB-MFGM compared to VB-FGM is statistically significant (the standard error of the ANLL is 0.025).

[2] The Abalone data is available from the UCI Machine Learning repository: htttp://www.ics.uci.edu/ mlearn/MLRepository.html.

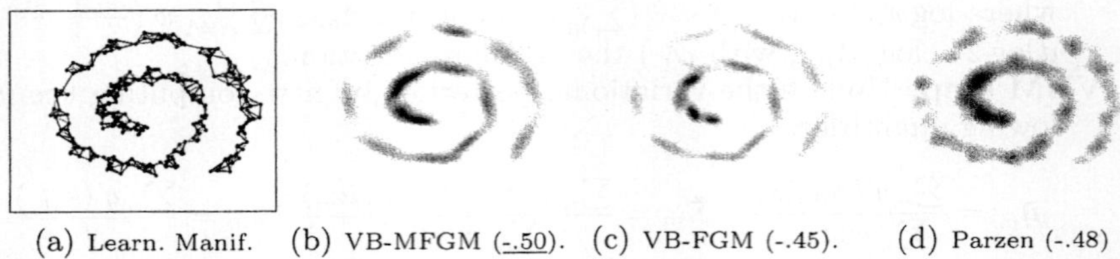

(a) Learn. Manif. (b) VB-MFGM (-.50). (c) VB-FGM (-.45). (d) Parzen (-.48)

Fig. 1. Training manifold of a noisy spiral, as well as the VB-MFGM, the standard VB-FGM and the Parzen window estimator. For each one, the ANLL of the test set is between parentheses (and the best is underlined).

Conclusion. The knowledge that the data is lying on a lower dimensional manifold than the dimension of the embedding space can be exploited in the frame of variational mixtures. By penalizing the complete data likelihood, the responsibilities (VBE step) are biased according to a discrepancy between the Euclidean and the geodesic distance. Following this methodology, manifold constrained variational Gaussian mixtures (VB-MFGM) were constructed. It was demonstrated experimentally that the resulting estimators are superior to standard variational approaches and nonparametric density estimation. In the future, we plan to investigate alternative mismatch measures.

References

1. McLachlan, G.J., Peel, D.: Finite Mixture Models. Wiley, New York, NY. (2000)
2. Dempster, A.P., Laird, N.M., Rubin, D.B.: Maximum likelihood from incomplete data via the EM algorithm (with discussion). J. Roy. Stat. Soc., B **39** (1977) 1–38.
3. Attias, H.: A variational bayesian framework for graphical models. In Solla, S., Leen, T., Mller, K.R., eds.: NIPS 12. MIT Press. (1999)
4. Corduneanu, A., Bishop, C.M.: Variational bayesian model selection for mixture distributions. In Jaakkola, T., Richardson, T., eds.: AISTATS 8, Morgan Kaufmann (2001) 27–34.
5. Vincent, P., Bengio, Y.: Manifold Parzen windows. In S. Becker, S.T., Obermayer, K., eds.: NIPS 15. MIT Press (2003) 825–832.
6. Beal, M.J.: Variational Algorithms for Approximate Bayesian Inference. PhD thesis, University College London (UK). (2003)
7. Lee, J.A., Lendasse, A., Verleysen, M.: Nonlinear projection with curvilinear distances: Isomap versus Curvilinear Distance Analysis. Neucom **57** (2003) 49–76.
8. Tenenbaum, J.B., de Silva, V., Langford, J.C.: A global geometric framework for nonlinear dimensionality reduction. Science **290** (2000) 2319–2323.
9. Bernstein, M., de Silva, V., Langford, J., Tenenbaum, J.: Graph approximations to geodesics on embedded manifolds. Techn. report Stanford University, CA. (2000)
10. West, D.B.: Introduction to Graph Theory. Prentice Hall, Upper Saddle River, NJ. (1996)
11. Dijkstra, E.W.: A note on two problems in connection with graphs. Num. Math. **1** (1959) 269–271.
12. Parzen, E.: On estimation of a probability density function and mode. Ann. Math. Stat. **33** (1962) 1065–1076.

Handwritten Digit Recognition with Nonlinear Fisher Discriminant Analysis*

Pietro Berkes

Institute for Theoretical Biology, Humboldt University Berlin,
Invalidenstraße 43, D - 10115 Berlin, Germany
p.berkes@biologie.hu-berlin.de
http://itb.biologie.hu-berlin.de/~berkes

Abstract. To generalize the Fisher Discriminant Analysis (FDA) algorithm to the case of discriminant functions belonging to a nonlinear, finite dimensional function space $\mathcal{F}$ (Nonlinear FDA or NFDA), it is sufficient to expand the input data by computing the output of a basis of $\mathcal{F}$ when applied to it [1,2,3,4]. The solution to NFDA can then be found like in the linear case by solving a generalized eigenvalue problem on the between- and within-classes covariance matrices (see e.g. [5]). The goal of NFDA is to find linear projections of the expanded data (i.e., nonlinear transformations of the original data) that minimize the variance within a class and maximize the variance between different classes. Such a representation is of course ideal to perform classification. The application of NFDA to pattern recognition is particularly appealing, because for a given input signal and a fixed function space it has no parameters and it is easy to implement and apply. Moreover, given C classes only $C - 1$ projections are relevant [5]. As a consequence, the feature space is very small and the algorithm has low memory requirements and high speed during recognition.

Here we apply NFDA to a handwritten digit recognition problem using the MNIST database, a standard and freely available set of 70,000 handwritten digits (28×28 pixels large), divided into a training set (60,000 digits) and a test set (10,000 digits). Several established pattern recognition methods have been applied to this database by Le Cun et al. [6]. Their paper provides a standard reference work to benchmark new algorithms.

We perform NFDA on spaces of polynomials of a given degree d, whose corresponding basis functions include all monomials up to order d in all input variables. It is clear that the problem quickly becomes intractable because of the high memory requirements. For this reason, the input dimensionality is first reduced by principal component analysis. On the preprocessed data we then apply NFDA by expanding the training patterns in the polynomial space and solving the linear FDA eigenvalue problem. As mentioned above, since we have 10 classes we only need to compute the first 9 eigenvectors. Since the within-class variance is minimized, the patterns belonging to different classes tend to cluster in

* This work has been supported by a grant from the Volkswagen Foundation.

W. Duch et al. (Eds.): ICANN 2005, LNCS 3697, pp. 285–287, 2005.

Table 1. Performance comparison Error rates on test data of various algorithms. All error rates are taken from [6].

METHOD	% ERRORS
Linear classifier	12.0
K-Nearest-Neighbors	5.0
1000 Radial Basis Functions, linear classifier	3.6
Best Back-Propagation NN (3 layers with 500 and 150 hidden units)	2.95
Reduced Set SVM (5 deg. polynomials)	1.0
LeNet-1 (16×16 input)	1.7
LeNet-5	0.95
Tangent Distance (16×16 input)	1.1
Nonlinear Fisher Discriminant Analysis (3 deg. polynomials, 35 input dim)	**1.5**

the feature space when projected on the eigenvectors. For this reason we classify the digits with a simple method such as Gaussian classifiers.

We perform simulations with polynomials of degree 2 to 5. With polynomials of degree 2 the explosion in the dimensionality of the expanded space with increasing number of input dimensions is relatively restricted, so that it is possible to use up to 140 dimensions. With higher order polynomials one has to rely on a smaller number of input dimensions, but since the function space gets larger and includes new nonlinearities, one obtains a remarkable improvement in performance. The best performance is achieved with polynomials of degree 3 and 35 input dimensions, with an error rate of 1.5% on test data. This error rate is comparable to but does not outperform that of the most elaborate algorithms (Table 1). The performance of NFDA is however remarkable considering the simplicity of the method and the fact that it has no a priori knowledge on the problem, in contrast for example to the LeNet-5 algorithm [6] which has been designed specifically for handwritten character recognition. In addition, for recognition, NFDA has to store and compute only 9 functions and has thus small memory requirements and a high recognition speed.

It is also possible to formulate NFDA using the *kernel trick*, in which case one can in principle use function spaces of infinite dimensionality [1,2,3,4]. However, the limiting factor in that formulation is the number of training patterns, which makes it not realistic for this application. The performance of NFDA could be further improved using for example a more problem-specific preprocessing of the patterns (e.g., by increasing the size of the training set with new patterns generated by artificial distortion of the original one), boosting techniques, or mixture of experts with other algorithms [5,6].

Keywords: nonlinear fisher discriminant analysis, pattern recognition, digit recognition, feature extraction.

References

1. Mika, S., Rätsch, G., Weston, J., Schölkopf, B., Müller, K.R.: Fisher discriminant analysis with kernels. In Hu, Y.H., Larsen, J., Wilson, E., Douglas, S., eds.: Neural Networks for Signal Processing. Volume IX. (1999) 41–48 Proceedings of the IEEE Signal Processing Society Workshop.
2. Baudat, G., Anouar, F.: Generalized discriminant analysis using a kernel approach. Neural Computation **12** (2000) 2385–2404
3. Mika, S., Rätsch, G., Weston, J., Schölkopf, B., Smola, A., Müller, K.R.: Invariant feature extraction and classification in kernel spaces. In Solla, S., Leen, T., Müller, K.R., eds.: Advances in Neural Information Processing Systems. Volume 12. (2000) 526–532
4. Mika, S., Smola, A., Schölkopf, B.: An improved training algorithm for kernel Fisher discriminants. In Jaakkola, T., Richardson, T., eds.: Proceedings AISTATS 2001. (2001) 98–104
5. Bishop, C.M.: Neural Networks for Pattern Recognition. Oxford University Press (1995)
6. LeCun, Y., Bottou, L., Bengio, Y., Haffner, P.: Gradient-based learning applied to document recognition. Proceedings of the IEEE **86** (1998) 2278–2324

Separable Data Aggregation in Hierarchical Networks of Formal Neurons[1]

Leon Bobrowski

Faculty of Computer Science, Bialystok Technical University,
Institute of Biocybernetics and Biomedical Engineering, PAS, Warsaw, Poland

Abstract. In this paper we consider principles of such data aggregation in hierarchical networks of formal neurons which allows one to preserve the separability of the categories. The postulate of the categories separation in the layers of formal neurons is examined by means of the concept of clear and mixed dipoles. Dependence of separation of the categories on the feature selection is analysed.

Keywords: Multilayer perceptrons, dipolar neural models, linear separability, feature selection.

1 Introduction

Hierarchical networks of formal neurons are often called multilayer perceptrons [1]. There are still unresolved fundamental questions related to hierarchical networks of formal neurons [2]. For example, the problem of determining the structure (architecture) of such neural networks is still open.

A basic principle of optimising the neural structure of networks can be based on the postulate of separable data aggregation [3]. The separable layer of formal neurons preserves the separation of feature vectors belonging to different learning sets (categories). The ranked and the dipolar strategies of the designing of separable layers of formal neurons have been proposed ([3], [4]). These strategies have been implemented through the minimisation of the convex and piecewise linear (*CPL*) criterion functions.

The problem of aggregation of different streams of data by separated sublayers of formal neurons is considered in this paper. Such data streams could be viewed as generated from different sources and represented as data submatrices in different feature subspaces. Multisource data aggregation could be carried out by applying a dipolar approach in particular feature subspaces.

2 Separability of Learning Sets

Let us assume that the objects' O_j ($j = 1,\ldots,m$) descriptions stored in a given database are represented as the so called feature vectors $\mathbf{x}_j = [x_{j1},\ldots,x_{jn}]^T$, or as points in the *n*-

[1] This work was partially supported by the W/II/1/2005 and SPB-M (COST 282) grants from the Białystok University of Technology and by the 16/St/2005 grant from the Institute of Biocybernetics and Biomedical Engineering PAS.

W. Duch et al. (Eds.): ICANN 2005, LNCS 3697, pp. 289 – 294, 2005.

dimensional feature space $F[n]$. The components (*features*) x_i ($x_i \in F[n]$) of the vector $\mathbf{x}$ are numerical results of a variety of examinations of a given object O. The feature vectors $\mathbf{x}$ can be of mixed, qualitative-quantitative type ($x_i \in \{0,1\}$ or $x_i \in R$).

We assume that the database contains the descriptions $\mathbf{x}_j(k)$ of m objects $O_j(k)$ ($j = 1,......,m$) labelled in accordance with their *category* (*class*) ω_k ($k = 1,....,K$). The learning set C_k contains m_k feature vectors $\mathbf{x}_j(k)$ assigned to the k-th category ω_k

$$C_k = \{\mathbf{x}_j(k)\} \quad (j \in I_k) \tag{1}$$

where I_k is the set of indices j of the feature vectors $\mathbf{x}_j(k)$ belonging to the class ω_k.

Definition 1: The learning sets C_k (1) are *separable* in the feature space $F[n]$, if they are disjoined in this space ($C_k \cap C_{k'} = \varnothing$, if $k \neq k'$). It means that the feature vectors $\mathbf{x}_j(k)$ and $\mathbf{x}_{j'}(k')$ belonging to different learning sets C_k and $C_{k'}$ cannot be equal:

$$(k \neq k') \Rightarrow (\forall j \in I_k) \; and \; (\forall j' \in I_{k'}) \quad \mathbf{x}_j(k) \neq \mathbf{x}_{j'}(k') \tag{2}$$

We are also considering the separation of the sets C_k (1) by the hyperplanes $H(\mathbf{w}_k,\theta_k)$ in the feature space $F[n]$

$$H(\mathbf{w}_k,\theta_k) = \{\mathbf{x}: (\mathbf{w}_k)^T\mathbf{x} = \theta_k\}. \tag{3}$$

where $\mathbf{w}_k = [w_{k1},....,w_{kn}]^T \in R^n$ is the weight vector, $\theta_k \in R^1$ is the threshold, and $(\mathbf{w}_k)^T\mathbf{x}$ is the inner product.

Definition 2: The learning sets (1) are *linearly separable* in the n-dimensional feature space $F[n]$ if each of the sets C_k can be fully separated from the sum of the remaining sets C_i by some hyperplane $H(\mathbf{w}_k,\theta_k)$ (3):

$$(\forall k \in \{1,...,K\}) \; (\exists \, \mathbf{w}_k,\theta_k) \; (\forall \mathbf{x}_j(k) \in C_k) \; (\mathbf{w}_k)^T\mathbf{x}_j(k) > \theta_k. \tag{4}$$
$$\textbf{and} \; (\forall \mathbf{x}_j(k) \in C_i, i \neq k) \; (\mathbf{w}_k)^T\mathbf{x}_j(k) < \theta_k$$

In accordance with the relation (4), all the vectors $\mathbf{x}_j(k)$ belonging to the learning set C_k are situated on the positive side ($(\mathbf{w}_k)^T\mathbf{x}_j(k) > \theta_k$) of the hyperplane $H(\mathbf{w}_k,\theta_k)$ (3) and all the feature vectors $\mathbf{x}_j(i)$ from the remaining sets C_i are situated on the negative side ($(\mathbf{w}_k)^T\mathbf{x}_j(k) < \theta_k$) of this hyperplane.

The linear separability of the learning sets C_k (1) may result from the linear independence of the feature vectors $\mathbf{x}_j(k)$ [5]:

Remark 1: If m feature vectors $\mathbf{x}_j(k)$ constituting the learning sets C_k (1) are linearly independent in a given feature space $F[n]$, then these sets are linearly separable (4).

Remark 2: If m feature vectors $\mathbf{x}_j(k)$ ($j \in I_{ind}$) are linearly independent in the n-dimensional feature space $F[n]$, and $m \leq n$, then there exists at least one such feature subspace $F_k[m\text{-}1]$ of dimension $m - 1$, that the *augmented* feature vectors $\mathbf{y}_j[m] = [\mathbf{x}_j[m\text{-}1]^T,1]^T$ ($\mathbf{x}_j[m\text{-}1] \in F_k[m\text{-}1]$) are linearly independent.

The reduced feature vectors $\mathbf{x}_j[m\text{-}1]$ belonging to the feature subspace $F_k[m\text{-}1]$ ($\mathbf{x}_j[m\text{-}1] \in F_k[m\text{-}1]$) are obtained from the vectors $\mathbf{x}_j(k)$ by neglecting identical ($n - m$ $+1$) features x_i The learning sets $C_k[m\text{-}1]$ (1) composed of the vectors $\mathbf{x}_j[m\text{-}1]$ are linearly separable (4) in the feature subspace $F_k[m\text{-}1]$. A quality of the feature subspaces $F_k[m\text{-}1]$ assuring the linear separability can be determined by taking into account additional properties of the separating hyperplanes $H(\mathbf{w}_k,\theta_k)$ (3). Such an additional property can be directed at separating hyperplanes $H(\mathbf{w}_k,\theta_k)$ (3) with the maximal margin, similarly as it is done in the *Support Vector Machines* approach [6].

The feature selection problem aimed at finding optimal feature subspaces $F_k[m\text{-}1]$ and separating hyperplanes $H(\mathbf{w}_k^*[m\text{-}1],\theta_k^*)$ (3) can be formulated and solved without exhaustive search among feature subsets on the base of minimisation of the convex and piecewise linear (*CPL*) criterion functions [5].

3 Layers of Formal Neurons

The formal neuron $NF(\mathbf{w},\theta)$ can be defined by *the activation function* $r_t(\mathbf{w},\theta;\,\mathbf{x})$

$$r = r_t(\mathbf{w},\theta;\,\mathbf{x}) = \begin{cases} 1 & \textit{if} \quad \mathbf{w}^T\mathbf{x} \geq \theta \\ 0 & \textit{if} \quad \mathbf{w}^T\mathbf{x} < \theta \end{cases} \tag{5}$$

where $\mathbf{w} = [w_1,....,w_n]^T \in R^n$ is the weight vector, $\theta \in R^1$ is the threshold, and r is the output signal.

The layer of L formal neurons $NF(\mathbf{w}_k,\theta_k)$ transforms feature vectors $\mathbf{x}$ into the output vectors $\mathbf{r} = [r_1,.....,r_L]^T$ with L binary components $r_i \in \{0,1\}$ which are determined by the equation $r_i = r_t(\mathbf{w}_i,\theta_i;\mathbf{x})$ (8):

$$\mathbf{r} = \mathbf{r}(\mathbf{W};\mathbf{x}) = [r_t(\mathbf{w}_1,\theta_1;\mathbf{x}),.......,\, r_t(\mathbf{w}_L,\theta_L;\mathbf{x})]^T \tag{6}$$

where $\mathbf{W} = [\mathbf{w}_1^T,\, \theta_1,.......,\, \mathbf{w}_L^T,\, \theta_L]^T$ is the vector of the layer parameters.

The transformed learning sets C'_k (1) are constituted by the vectors $\mathbf{r}_j(k)$

$$C'_k = \{\mathbf{r}_j(k)\} \quad (j \in I_k) \tag{7}$$

where $\mathbf{r}_j(k) = \mathbf{r}(\mathbf{W};\mathbf{x}_j(k))$ (6).

We are examining such properties of the transformation (6) which assure the separability (2) or the linear separability (4) of the transformed sets C'_k (7). Such a property can be achieved through the separation of dipoles [3].

Definition 3: A pair of different feature vectors $(\mathbf{x}_j(k),\mathbf{x}_{j'}(k'))$ $(\mathbf{x}_j(k) \neq \mathbf{x}_{j'}(k'))$ constitutes a *mixed dipole* if and only if these vectors belong to different classes ω_k $(k \neq k')$. Similarly, a pair of different feature vectors from the same class ω_k constitutes the *clear dipole* $(\mathbf{x}_j(k),\mathbf{x}_{j'}(k))$.

Definition 4: The formal neuron $NF(\mathbf{w}_k,\theta_k)$ (5) separates the dipole $(\mathbf{x}_j(k),\mathbf{x}_{j'}(k'))$ if *only one* feature vector $(\mathbf{x}_j(k),\mathbf{x}_{j'}(k'))$ from this pair activates $(r_k = r_t(\mathbf{w}_k,\theta_k;\mathbf{x}) = 1)$ this neuron.

Lemma 1: The necessary and sufficient conditions for separability (*Def.* 1) of the transformed sets C'_1 (7) by the layer (7) is the separation of each mixed dipole $(\mathbf{x}_j(k),\mathbf{x}_{j'}(k'))$ by at least one neuron $NF(\mathbf{w}_k,\theta_k)$ of this layer. [3].

In order to increase the generality of the designed neural layers, the following postulates have been proposed and implemented as a multistage procedure [3].

Dipolar Designing Postulate: The hyperplane $H(\mathbf{w}_l,\theta_l)$ (3) designed during the l-th stage should separate the highest possible number of mixed and still undivided dipoles $(\mathbf{x}_j(k),\mathbf{x}_{j'}(k'))$ and, at the same time, the lowest possible number of the clear dipoles $(\mathbf{x}_j(k),\mathbf{x}_{j'}(k))$ should be divided.

Ranked Designing Postulate: The hyperplane $H(\mathbf{w}_l,\theta_l)$ (3) designed during the l-th stage should separate $((\mathbf{w}_l)^T\mathbf{x}_j(k) > \theta_l)$ as many feature vectors $\mathbf{x}_j(k)$ as possible from one set $C_k[l]$ under the condition that no vector $\mathbf{x}_{j'}(k')$ from the remaining sets $C_{k'}[l]$ is separated.

The symbol $C_k[l]$ used in the above postulate denotes the learning set C_k (1) after neglecting such feature vectors $\mathbf{x}_j(k)$ which have been separated by the hyperplanes $H(\mathbf{w}_l,\theta_l)$ (3) during the previous l - 1 steps. The designing procedure is stopped during the L-th step when all the sets $C_k[L]$ become empty.

Lemma 2: The layer of L formal neurons $NF(\mathbf{w}_l,\theta_l)$ (5) designed in accordance with the ranked postulate produces such sets C'_k (7) which are linearly separable (4). [4]

The procedure of the separable layer designing can be based on a sequence of minimisation of the dipolar or the ranked criterion functions $\Psi_k(\mathbf{w},\theta)$, which belong to the family of the convex and piecewise linear (*CPL*) criterion function [3], [4]. The parameters $(\mathbf{w}_k^*,\theta_k^*)$ constituting the minimum of the function $\Psi_k(\mathbf{w},\theta)$ define the k-th neuron $NF(\mathbf{w}_k^*,\theta_k^*)$ (9) of the layer and the separating hyperplane $H(\mathbf{w}_k^*,\theta_k^*)$ (3). The criterion functions $\Psi_k(\mathbf{w},\theta)$ can be specified in accordance with one of the above postulates.

4 Decomposition of the Feature Space

Let us consider a decomposition of the feature space $F[n]$ into a number L of disjoined feature subspaces $F_i[n_i]$ of dimensionality n_i.

$$F[n] = F_1[n_1] \cup F_2[n_2] \cup ......\cup F_L[n_L] \tag{8}$$

where $n = n_1 + n_2 + ...+ n_l$, and $F_i[n_i] \cap F_{i'}[n_{i'}] = \varnothing$ if $i \neq i'$.

The learning sets C_k (1) can be decomposed in accordance with the above principle.

$$C_k[n_i] = \{\mathbf{x}_j(k[n_i])\} \quad (j \in I_k) \tag{9}$$

where $C_k[n_i]$ is obtained from the learning sets C_k (1) by neglecting such features x_i which do not belong to the subspace $F_i[n_i]$ $(x_i \notin F_i[n_i])$.

The data sets $C_k[n_i]$ can be seen as different representations of the objects $O_j(k)$ (1) in particular feature subspaces $F_i[n_i]$. The reduction of the feature space $F[n]$ into the subspace $F_i[n_i]$ can result in the deprivation of the learning sets C_k (1) or the dipoles separability (2). The separability of the mixed dipole $(\mathbf{x}_j(k[n_i]),\mathbf{x}_{j'}(k'[n_i]))$ in the subspace $F_i[n_i]$ results in the separabilty of the dipole $(\mathbf{x}_j(k),\mathbf{x}_{j'}(k'))$ in the feature space $F[n]$, but there is no reverse implication:

$$(\mathbf{x}_j(k[n_i]) \neq \mathbf{x}_{j'}(k'[n_i])) \Rightarrow (\mathbf{x}_j(k) \neq \mathbf{x}_{j'}(k')) \tag{10}$$

Definition 5: The i-th sublayer of formal neurons $NF_i(\mathbf{w}_k[n_i],\theta_k)$ is separable in the feature subspace $F_i[n_i]$, if and only if each dipole $(\mathbf{x}_j(k[n_i]),\mathbf{x}_{j'}(k'[n_i]))$ $(\mathbf{x}_j(k[n_i]) \neq \mathbf{x}_{j'}(k'[n_i]))$ is separated (*Def.* 4) by at least one neuron $NF_i(\mathbf{w}_k[n_i],\theta_i)$.

Let us consider for a moment the transformation $\mathbf{r}_j(k[n_i]) = \mathbf{r}_i(\mathbf{W}_i;\mathbf{x}_j(k))$ (7) of the vectors $\mathbf{x}_j(k[n_i])$ in the feature subspace $F_i[n_i]$ (9) by a separable sublayer of formal neurons $NF_i(\mathbf{w}_k[n_i],\theta_k)$. As a result, the feature vectors $\mathbf{x}_j(k)$ have the below forms.

$$\mathbf{x}'_j(k) = [\mathbf{x}_j(k[n_1])^T,\ldots\ldots, \mathbf{r}_i(\mathbf{W}_i;\mathbf{x}_j(k))^T, \ldots\ldots\ldots, \mathbf{x}_j(k[n_L])^T]^T \tag{11}$$

The sufficient conditions for separability (2) of the sets $C'_k = \{\mathbf{x}'_j(k)\}$ of the vectors $\mathbf{x}'_j(k)$ (11) can be based on the separability of the neural layer in the subspace $F_i[n_i]$.

5 Separable Data Aggregation

Separable data aggregation could be a primary goal in designing successive layers of a hierarchical network. Data aggregation means here decreasing the number of different feature vectors $\mathbf{r}_j(k)$ (6) resulting from the transformation by the neuronal layer. The transformation (6) aggregates the number of the feature vectors $\mathbf{x}_j(k)$ (1) into one output vector $\mathbf{r}_j$. We assume here that the different indices n and n' $(n \neq n')$ are related to different output vectors $\mathbf{r}_n$ and $\mathbf{r}_{n'}$.

$$(n \neq n') \Rightarrow \mathbf{r}_n \neq \mathbf{r}_{n'} \tag{12}$$

Let us introduce the concept of the *active fields* S_n in order to characterise the aggregation properties [4]. Each active field S_n is a convex polyhedron with the walls defined by the hyperplanes $H(\mathbf{w}_k,\theta_k)$ (3).

Definition 6: The n-th active field S_n of the neural layer (6) is constituted by such feature vectors $\mathbf{x}$ which are transformed (integrated) in one output vector $\mathbf{r}_n$ $(\mathbf{r}_n \neq \mathbf{0})$.

$$S_n = S(\mathbf{W}, \mathbf{r}_n) = \{\mathbf{x}: \mathbf{r}(\mathbf{W}, \mathbf{x}) = \mathbf{r}_n\} \tag{13}$$

Definition 7: The *clear active field* S_n contains the labelled feature vectors $\mathbf{x}_j(k)$ belonging to only one learning set C_k (1). The *mixed active field* S_n contains the labelled feature vectors $\mathbf{x}_j(k)$ belonging to more than one learning set C_k (1).

Remark 3: The layer of L formal neurons $NF(\mathbf{w}_k,\theta_k)$ (9) is *separable* if each active field S_n of this layer is clear and each labelled vector $\mathbf{x}_j(k)$ belongs to some set (13).

The following relation is fulfilled between the active fields $S_n[l]$ (13) in the hierarchical network of two successive layers of formal neurons $NF(\mathbf{w}_k,\theta_k)$ (9) [4]

$$S_n[l] = \bigcup_{n' \in I(l,n)} S_{n'}[l\text{-}1] \qquad (14)$$

where $I(l,n)$ is the set of indices n' of such active fields $S_{n'}[l'\text{-}1]$ of the previous, $(l-1)$-th layer, which are aggregated in one field $S_n[l]$ of the l – th layer.

6 Concluding Remarks

As it follows from the relation (14), an increase in the number of layers in a hierarchical network should result in a summation (aggregation) of some active fields S_n (13) and in a decrease in the number of different output vectors $\mathbf{r}_n$ (12). The principles sketched out in this paper are aimed at designing separable layers of formal neurons with clear active fields $S_n[l]$ (13). Relatively large clear fields $S_n[l]$ give a chance for obtaining a separable layer with large generalisation power. In order to increase generalisation power, a given neural layer should aggregate the possibly greatest number of clear dipoles $(\mathbf{x}_j(k),\mathbf{x}_{j'}(k))$ under the condition of separating a sufficiently high fraction of the mixed dipoles $(\mathbf{x}_j(k),\mathbf{x}_{j'}(k'))$ [3].

Designing neural layers should be related to the problems of feature space decomposition and reduction of dimensionality. If the number of the independent feature vectors $\mathbf{x}_j(k)$ used by the neural sublayer (6) is equal to m, then there is no need for a dimensionality grater than m–1 in the feature space of this layer [5].

Bibliography

1. Rosenblatt F.: *Principles of Neurodynamics*, Spartan Books, Washington 1962
2. Duda O. R. and Hart P. E., Stork D. G.: *Pattern Classification,* J. Wiley, New York 2001
3. Bobrowski L.: "*Piecewise-Linear Classifiers, Formal Neurons and Separability of the Learning Sets*", Proceedings of ICPR'96, pp. 224-228, Wienna, Austria, 1996
4. Bobrowski L.: *Eksploracja danych oparta na wypukłych i odcinkowo-liniowych funkcjach kryterialnych* (*Data mining based on convex and piecewise linear (CPL) criterion functions*) (in Polish), Technical University Białystok, 2005
5. Bobrowski L., Łukaszuk T.: "Selection of the linearly separable feature subsets", pp. 544-549 in: *Artificial Intelligence and Soft Computing* - ICAISC 2004, Eds. Rutkowski L. et al, Springer Lecture Notes in Artificial Intelligence 3070, 2004
6. Vapnik V. N.: *Statistical Learning Theory*, J. Wiley, New York 1998

Induced Weights Artificial Neural Network

Slawomir Golak

Sielsian University of Technology, Electrotechnology Department,
Division of Informatics and Modeling of Technological Processes
sgolak@polsl.pl

Abstract. It is widely believed in the pattern recognition field that the number of examples needed to achieve an acceptable level of generalization ability depends on the number of independent parameters needed to specify the network configuration. The paper presents a neural network for classification of high-dimensional patterns. The network architecture proposed here uses a layer which extracts the global features of patterns. The layer contains neurons whose weights are induced by a neural subnetwork. The method reduces the number of independent parameters describing the layer to the parameters describing the inducing subnetwork.

1 Introduction

The great potential of the neural networks is most frequently used in pattern recognition. The most challenging problem here is achieving the proper generalization. Typical images and time-series are usually large, often with several hundred variables. Fully connected, unrestricted networks do not work well as far as recognizing such large patterns is concerned.

The number of examples needed to achieve an acceptable level of generalization ability is dependent on the intrinsic entropy of the chosen architecture, and can be decreased by reducing the number of independent parameters needed to specify the network configuration. One of the ways to improve generalization is a reduction of the network structure on the base of a pruning algorithm (e.g. the Optimal Brain Damage [7]).

Another deficiency of the fully-connected architectures is that the topology of the inputs is entirely ignored. In fact, images have a strong 2D structure, while time-series have a strong 1D structure. Pixels, or variables, spatially or temporally adjacent are correlated.

The application of a specialized network architecture, instead of a fully-connected net, can reduce the number of free parameters.

There are many papers that propose specialized network architectures for the recognition of large patterns. Convolutional networks, for instance, use the techniques of local receptive fields and shared weights. These networks extract and combine local features of pattern [4,5]. Principal component analysis transforms a number of correlated variables into a smaller number of uncorrelated variables called principal components and is frequently adopted for dimensionality reduction [1].

W. Duch et al. (Eds.): ICANN 2005, LNCS 3697, pp. 295 – 300, 2005.

The idea that has been followed in the IWANN is based on the invention of dynamically-calculated weights, which results in giving the individual neurons the ability to transform large patterns, using only a limited number of parameters describing their connection weights. The proposed network architecture extracts and transforms the global features of the patterns.

2 Network Architecture

The IWANN network makes use of dynamically-calculated connection weights. As a result, the number of parameters describing neural network connections is reduced.

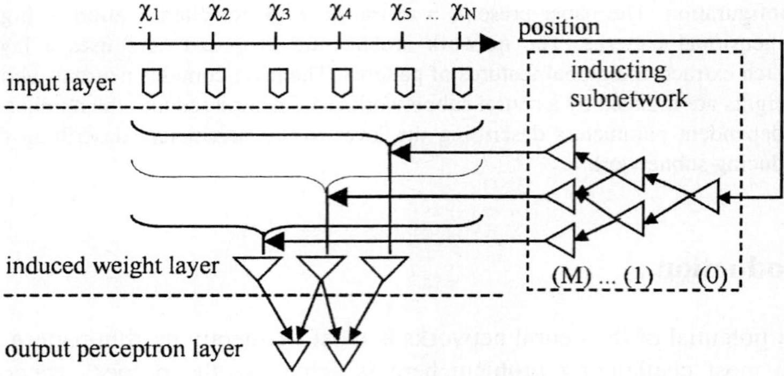

Fig. 1. Induced weights network scheme

The input layer of the network proposed can be one- or multidimensional, and every neuron in this layer is described by its geometric position.

The layer is a data source for the induced weights layer. It contains radial basis neurons which apply the Gaussian transformation function. The input of this radial basis transformation function is the Euclidean distance between the input vector and the vector of weights calculated by the inducing subnetwork, multiplied by the bias.

$$y_i^{(1)} = \varphi\left(b_i^{(1)} \sqrt{\frac{1}{2} \sum_{j=1}^{N^{(0)}} \left(\dot{y}_{ji}^{(M)} - y_j^{(0)} \right)^2} \right) \tag{1}$$

where, $y_i^{(1)}$ output of the i-th neuron in induced weights layer, $\varphi()$ Gaussian transformation function, $b_i^{(1)}$ bias of neuron, $\dot{y}_{ji}^{(M)}$ output of neuron in output (M-th) layer of inducing network equal to the weight of the j-th input of the i-th neuron of the induced layer, and $y_j^{(o)}$ output of j-th input neuron – network input.

The task of the inducing neural subnetwork consists in positioning of high-dimensional centers. The inducing network is a multilayer perceptron:

$$\dot{y}_{ji}^{(\dot{m})} = f\left(\sum_{k=1}^{\dot{N}^{(\dot{m}-1)}} \dot{y}_{jk}^{(\dot{m}-1)} \dot{w}_{ik}^{(\dot{m})}\right) \tag{2}$$

where, $\dot{y}_{ji}^{(\dot{m})}$ output of i-th neuron in the m-th layer for j-th network input, $\dot{w}_{ik}^{(\dot{m})}$ weight of the k-th input of the i-th neuron in m-th layer.

The number of neurons in the inducing network output layer is equal to the number of neurons in the induced weights layer. Geometrical positions of neurons in the input layer are introduced to the input layer of the inducing neural network:

$$\dot{y}_{jk}^{(0)} = \chi_{jk} \tag{3}$$

where, $\dot{y}_{jk}^{(0)}$ value of inducing network k-th input, χ_{jk} - k-th coordinate of j-th input neuron

The inducing network calculates connection weights between every neuron in the induced and input layer by using coordinates' values of the input neurons.

Remaining output layers of the IWANN network are perceptron layers:

$$y_i^{(m)} = f\left(\sum_{k=1}^{N^{(m-1)}} y_k^{(m-1)} w_{ik}^{(m)}\right) \tag{4}$$

where, $y_i^{(m)}$ output i-th neuron from m-th layer (m>1), $\dot{w}_{ik}^{(\dot{m})}$ weight of the k-th input of the i-th neuron from m-th layer.

The similar idea is used in the mixture of experts model, were weights of the gating function depend on the output of the gating network [2,3]. However, in the ME algorithm, the input of gating network are the global network inputs' values and not geometrical coordinates of these inputs. Furthermore, the gating function is a linear or non-linear weight function, while the induced layer uses a distance function.

3 Learning Algorithm

We can express the training error E as a function (5). If the transformation functions used in the network are continuous and differentiable, it is possible to calculate derivatives of this error function. Therefore, the proposed network can trained with the use of gradient learning methods.

$$E = \frac{1}{2}\sum_{i=1}^{N^{(M)}} \left(y_i^{(M)} - d_i\right)^2 \tag{5}$$

where, d_i and $y_i^{(M)}$ are the target and output values for training example.

The gradient of the error function specifies the vector in whose direction the greatest increase in E can be obtained. Our aim is to calculate the partial derivatives of error function for each weight of the network. The algorithm to calculate the error derivatives for perceptron output layers is well-known:

$$\delta_i^{(m)} = \sum_{k=0}^{N^{(m+1)}} \left(\delta_k^{(m+1)} w_{ki}^{(m+1)}\right) f'\left(\sum_{k=1}^{N^{(m-1)}} y_k^{(m-1)} w_{ik}^{(m)}\right) \tag{6}$$

$$\frac{\partial E}{\partial w_{ik}^{(m)}} = \delta_i^{(m)} y_k^{(m-1)} \tag{7}$$

where, $\delta_i^{(m)}$ signal error of i-th neuron in m-th perceptron layer, $\delta_k^{(m+1)}$ signal error of k-th neuron in next layer, $w_{ki}^{(m+1)}$ connection weight between i-th neuron in current layer and k-th neuron in the next layer, $w_{ik}^{(m)}$ weight of the i-th input of the k-th neuron in the current layer, $y_k^{(m-1)}$ output of neuron in the previous layer.

Applying the algorithm, we can obtain signal errors for the neurons of the induced layer, and the derivatives for biases of these neurons:

$$\delta_i^{(1)} = \sum_{k=0}^{N^{(2)}} \left(\delta_k^{(2)} w_{ki}^{(2)}\right) \varphi'\left(b_i^{(1)} \sqrt{\frac{1}{2}\sum_{j=1}^{N^{(0)}} \left(\dot{y}_{ji}^{(\dot{M})} - y_j^{(0)}\right)^2}\right) \tag{8}$$

$$\frac{\partial E}{\partial b_i^{(1)}} = \delta_i^{(1)} \sqrt{\frac{1}{2}\sum_{j=1}^{N^{(1)}} \left(\dot{y}_{ji}^{(\dot{M})} - y_j^{(0)}\right)^2} \tag{9}$$

The values of signal error may be used to calculate errors of neurons in the output layer of the inducing network. They have to be calculated individually for every j-th input of the induced layer:

$$\dot{\delta}_{ji}^{(\dot{M})} = \delta_i^{(1)} b_i^{(1)} \frac{1}{2\sqrt{\frac{1}{2}\sum_{j=1}^{N^{(0)}} \left(\dot{y}_{ji}^{(\dot{M})} - y_j^{(0)}\right)^2}} \left(\dot{y}_{ji}^{(\dot{M})} - y_j^{(0)}\right) f'\left(\sum_{k=1}^{\dot{N}^{(\dot{M}-1)}} \dot{y}_{jk}^{(\dot{M}-1)} \dot{w}_{ik}^{(\dot{M})}\right) \tag{10}$$

Signal errors of neurons in the remaining layers are calculated by using the same backpropagation method as in (6).

The derivatives for the weights of a neuron of the inducing network are the sums of the derivatives calculated for every j-th input of the network:

$$\frac{\partial E}{\partial \dot{w}_{ik}^{(\dot{m})}} = \sum_{j=1}^{N^{(0)}} \dot{\delta}_{ji}^{(\dot{m})} \dot{y}_{jk}^{(\dot{m}-1)} \tag{11}$$

Owing to the evaluation of these derivatives, various gradient-based optimization methods can be utilized (e.g. QuickProp, RPROP).

4 Experiments

The proposed neural network was applied to classification of one- and two-dimensional patterns. The first dataset consisted of 200 artificially generated one-dimensional patterns of size 100. These data were evenly divided into the training and test sets. The dataset contained four classes of patterns. Figure 2a shows examples of patterns in the dataset.

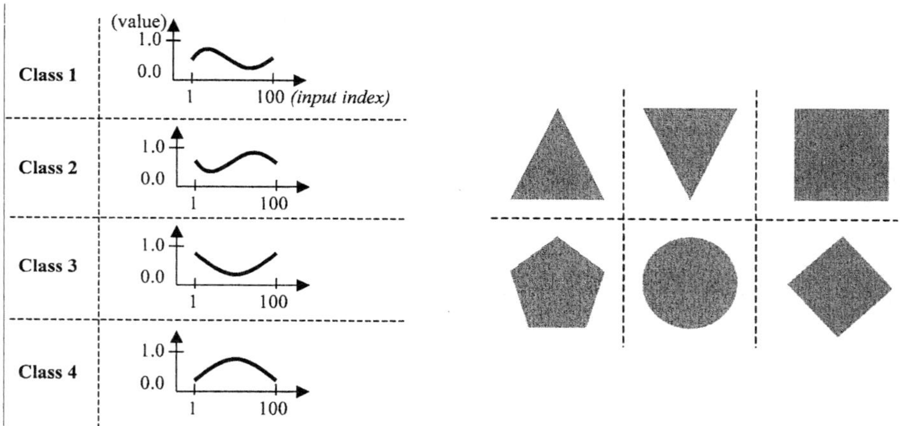

Fig. 2. Classes of patterns in dataset 1 (a) and dataset 2 (b)

The base patterns were randomly scaled (60÷140%), translated horizontally (±25) and vertically (±0.25).

The architecture of the neural network consisted of an induced layer with 4 neurons and an output perceptron layer with 4 neurons. The inducing subnetwork was a multilayer perceptron. This neural network consisted of three layers and the number of neurons in the hidden layer was 10. The error function was minimized by the RPROP method [6]. The training was stopped after 340 epochs. At the end of the training, the mean square error was 0.0022 and the percentage of wrong classifications was below 1%. Results of a classic multilayer perceptron with the optimized structure (100x20x4) were 0.0352 (MSE) and 16% (wrong classifications).

The second dataset consisted of 200 two-dimensional patterns of size 50x50. The patterns were evenly divided into the six classes and into the training and test datasets.

Figure 4b contains examples of patterns in the six classes. The base patterns were randomly transformed by rotation (±5°) and translation of polygons vertexes (±5 pixels). Uniformly distributed random noise was introduced to patterns at the level of 20%.

The network consisted of the induced layer with 10 neurons, the output layer with 6 neurons and the inducing subnetwork with 25 neurons in the hidden layer.

The network was trained by the RPROP method [6]. After 1230 epochs the mean square error was 0.0034 and the percentage of wrong classifications was equal to 3%. Results of the multilayer perceptron (2500x40x6) were 0.1134 and 41%.

5 Conclusions

The experiments described in this paper show that the IWANN network is a suitable model for the classification of large patterns. The presented algorithm may be of great help to the network designers in their time-consuming task of preprocessing the patterns. The method reduces the number of independent parameters of the induced layer to the parameters describing the biases of neurons in this layer and parameters describing the inducing subnetwork. Owing to this, it is possible to obtain satisfactory results of classification for high-dimensional patterns with the help of a limited number of learning examples.

References

1. Diamantaras K. I., Kung S. Y.: Principal Component Neural Networks: Theory and Applications, Wiley, 1996
2. Jacobs R. A., Jordan M. I.: Adaptative Mixture of Local Expert, Neural Computation, v. 3, pp. 79-87, 1991
3. Jordan M. I., Jacobs R.A.: Herarchical mixtures of experts and the EM algorithm, Neural Computation, v. 6, pp. 181-214, 1994
4. LeCun Y., Bengio Y. Convolutional networks for images, speech, and time series, The Handbook of Brain Theory and Neural Networks, pp. 255-258. MIT Press, Cambridge, Massachusetts, 1995
5. LeCun Y., Bottou L., Benigo Y., Haffner P.: Gradient-Based Learning Applied to Document Recognition, Proceedings of the IEEE, v. 86, pp. 2278-2324, 1998
6. Riedmiller M. and Braun H. A direct adaptive method for faster backpropagation learning: The RPROP algorithm. In Proceedings of the IEEE International Conference on Neural Networks 1993 (ICNN 93), 1993
7. Solla S., LeCun Y., Denker J.: Optimal Brain Damage, Advances in Neural Information Processings Systems 2, pp. 598-605, San Mateo, Morgan Kaufmann Publishers Inc., 1990

SoftDoubleMinOver: A Simple Procedure for Maximum Margin Classification

Thomas Martinetz, Kai Labusch, and Daniel Schneegaß

Institute for Neuro- and Bioinformatics,
University of Lübeck,
D-23538 Lübeck, Germany
martinetz@informatik.uni-luebeck.de
http://www.inb.uni-luebeck.de

Abstract. The well-known MinOver algorithm is a simple modification of the perceptron algorithm and provides the maximum margin classifier without a bias in linearly separable two class classification problems. DoubleMinOver as a slight modification of MinOver is introduced, which now includes a bias. It is shown how this simple and iterative procedure can be extended to SoftDoubleMinOver for classification with soft margins and with kernels. On benchmarks the extremely simple SoftDoubleMinOver algorithm achieves the same classification performance with the same computational effort as sophisticated Support-Vector-Machine software.

1 Introduction

The Support-Vector-Machine (SVM) [1], [2] has been applied very successfully and become a standard tool in classification and regression tasks (e.g. [3], [4], [5]). A major drawback, particularly for industrial applications where easy and robust implementation is an issue, is the large Quadratic-Optimization problem which has to be solved. The users have to rely on existing software packages, which are hardly comprehensive and, in some cases at least, error-free. This is in contrast to most Neural Network approaches, where learning has to be simple and incremental almost by definition. The pattern-by-pattern nature of learning in Neural Networks usually leads to simple training procedures which can easily be implemented. It is desirable to have similar training procedures also for the SVM.

Several approaches for obtaining more or less simple incremental training procedures for the SVM have been introduced so far [6], [7], [8], [9]. We want to mention in particular the Kernel-Adatron by Friess, Cristianini, and Campbell [6] and the Sequential-Minimal-Optimization algorithm (SMO) by Platt [7]. The SMO algorithm by Platt is the most widespread iterative training procedure for SVMs. It is fast and robust, but it is still not yet of a pattern-by-pattern nature and not yet as easy to implement as one is used from Neural Network approaches.

Therefore, in this paper we will revisit and extend the MinOver algorithm which was introduced by Krauth and Mézard [10] for constructing synaptic weight matrices of optimal stability in spin-glass models of Neural Networks. It is well-known that MinOver yields the maximum margin hyperplane without a bias in linearly separable classification problems. We will reformulate MinOver. In this reformulation the MinOver

W. Duch et al. (Eds.): ICANN 2005, LNCS 3697, pp. 301–306, 2005.

algorithm is a slight modification of the perceptron learning rule and could hardly be simpler. Then we extend MinOver to DoubleMinOver which remains as simple as MinOver but now provides the maximum margin solution also for linear classifiers with a bias. Then we will show how DoubleMinOver can be extended to SoftDoubleMinOver for classification with soft margins and with kernels. SoftDoubleMinOver yields solutions also in case the classification problem is not linearly separable. Finally, we will present results and comparisons on standard benchmark problems.

2 The DoubleMinOver Algorithm

Given a linearly separable set of patterns $\mathbf{x}_i \in \mathbb{R}^D$, $i = 1, \ldots, N$ with corresponding class labels $y_i \in \{-1, 1\}$. We want to find the hyperplane which separates the patterns of these two classes with maximum margin. The hyperplane for classification is determined by its normal vector $\mathbf{w} \in \mathbb{R}^D$ and its bias $b \in \mathbb{R}$. It achieves a separation of the two classes, if

$$y_i(\mathbf{w}^T \mathbf{x}_i - b) > 0 \qquad \text{for all} \qquad i = 1, \ldots, N$$

is valid. The margin Δ of this separation is given by

$$\Delta(\mathbf{w}, b) = \min_i [y_i(\mathbf{w}^T \mathbf{x}_i - b)/||\mathbf{w}||].$$

Maximum margin classification is given by the $\mathbf{w}_*$, $||\mathbf{w}_*|| = 1$ and b_* for which $\Delta(\mathbf{w}_*, b_*) = \Delta_*$ becomes maximal.

A simple and iterative algorithm which provides the maximum margin classification in linearly separable cases is the well-known MinOver algorithm introduced by [10] in the context of constructing synaptic weight matrices of optimal stability in spin-glass models of Neural Networks. However, it only provides the maximum margin solution if no bias b is included. The MinOver algorithm yields a vector $\mathbf{w}_t$ which converges against the maximum margin solution with increasing number of iterations t. This is valid as long as a full separation, i.e. a $\Delta_* > 0$, exists. The MinOver algorithm works like the perceptron algorithm, with the slight modification that with each training step t the pattern $\mathbf{x}_{\min}(t)$ out of the training set $\mathcal{T} = \{\mathbf{x}_i | i = 1, \ldots, N\}$ with the worst, i.e. the minimum distance (overlap) $y_i \mathbf{w}^T \mathbf{x}_i$ is chosen ($b = 0$). Hence, the name MinOver.

We now modify MinOver such that a bias b can be included. For this purpose we divide $\mathcal{T}$ into the set $\mathcal{T}^+$ of patterns with class label $y_i = +1$ and the set $\mathcal{T}^-$ of patterns with class label $y_i = -1$. Then, instead of looking for the pattern with minimum distance on $\mathcal{T}$, we look for the pattern $\mathbf{x}_{\min^+}(t)$ with minimum distance $y_i(\mathbf{w}^T \mathbf{x}_i - b)$ on $\mathcal{T}^+$ and for the pattern $\mathbf{x}_{\min^-}(t)$ with minimum distance $y_i(\mathbf{w}^T \mathbf{x}_i - b)$ on $\mathcal{T}^-$. Hence, the name DoubleMinOver.

With t_{max} as the number of desired iterations, DoubleMinOver works like follows:

0. Set $t = 0$, choose a t_{max}, and set $\mathbf{w}_{t=0} = 0$.
1. Determine the $\mathbf{x}_{\min^+}(t) \in \mathcal{T}^+$ and the $\mathbf{x}_{\min^-}(t) \in \mathcal{T}^-$ which minimize $y_i \mathbf{w}_t^T \mathbf{x}_i$ on $\mathcal{T}^+$ and $\mathcal{T}^-$, respectively.
2. Set $\mathbf{w}_{t+1} = \mathbf{w}_t + \mathbf{x}_{\min^+}(t) - \mathbf{x}_{\min^-}(t)$.
3. Set $t = t + 1$ and go to 1.) if $t < t_{max}$.
4. Determine $\mathbf{x}_{\min^+}$ and $\mathbf{x}_{\min^-}$ according to 1. and set $b = \frac{1}{2}\, \mathbf{w}_{t_{\max}}^T (\mathbf{x}_{\min^+} + \mathbf{x}_{\min^-})$.

2.1 On the Convergence of DoubleMinOver

For a given $\mathbf{w}$, the margin $\Delta(\mathbf{w}, b)$ is maximized with the $b(\mathbf{w})$ for which the margin to both classes is equal, i.e., for which

$$\min_{\mathbf{x}_i \in \mathcal{T}+} [y_i(\mathbf{w}^T \mathbf{x}_i - b(\mathbf{w}))] = \min_{\mathbf{x}_i \in \mathcal{T}-} [y_i(\mathbf{w}^T \mathbf{x}_i - b(\mathbf{w}))]$$

is valid. This leads to the expression of step 4. for the bias

$$b(\mathbf{w}) = \frac{1}{2}\left(\min_{\mathbf{x}_i \in \mathcal{T}+} y_i \mathbf{w}^T \mathbf{x}_i - \min_{\mathbf{x}_i \in \mathcal{T}-} y_i \mathbf{w}^T \mathbf{x}_i\right) = \frac{1}{2}\left(\mathbf{w}^T \mathbf{x}_{\min^+} + \mathbf{w}^T \mathbf{x}_{\min^-}\right);.$$

We now have to look for the $\mathbf{w}$ which maximizes $\Delta(\mathbf{w}) = \Delta(\mathbf{w}, b(\mathbf{w}))$. We obtain

$$\begin{aligned}\Delta(\mathbf{w}) &= \min_{\mathbf{x}_i \in \mathcal{T}} \frac{y_i(\mathbf{w}^T \mathbf{x}_i - b(\mathbf{w}))}{||\mathbf{w}||} = \min_{\mathbf{x}_i \in \mathcal{T}+} \frac{\mathbf{w}^T \mathbf{x}_i - b(\mathbf{w})}{||\mathbf{w}||} = \min_{\mathbf{x}_i \in \mathcal{T}-} \frac{-\mathbf{w}^T \mathbf{x}_i + b(\mathbf{w})}{||\mathbf{w}||} \\ &= \frac{1}{2}\left(\min_{\mathbf{x}_i \in \mathcal{T}+} \frac{\mathbf{w}^T \mathbf{x}_i}{||\mathbf{w}||} - \max_{\mathbf{x}_i \in \mathcal{T}-} \frac{\mathbf{w}^T \mathbf{x}_i}{||\mathbf{w}||}\right).\end{aligned}$$

With

$$\mathcal{Z} = \left\{\mathbf{z}_{ij} = \mathbf{x}_i - \mathbf{x}_j; |; \forall\,(i,j) : x_i \in \mathcal{T}^+, x_j \in \mathcal{T}^-\right\}$$

we obtain

$$\Delta(\mathbf{w}) = \frac{1}{2} \min_{\mathbf{z}_{ij}} \frac{\mathbf{w}^T \mathbf{z}_{ij}}{||\mathbf{w}||};.$$

In this formulation we can directly apply the $\mathcal{O}(t^{-1/2})$ convergence proofs for MinOver in [10] or [11]. For both methods in fact even a $\mathcal{O}(t^{-1})$ convergence is valid [11].

2.2 DoubleMinOver in Its Dual Formulation and with Kernels

The vector $\mathbf{w}_t$ which determines the dividing hyperplane is given by

$$\mathbf{w}_t = \sum_{\tau=0}^{t-1} (\mathbf{x}_{\min^+}(\tau) - \mathbf{x}_{\min^-}(\tau)) = \sum_{\mathbf{x}_i \in \mathcal{T}} y_i n_i(t) \mathbf{x}_i$$

with $n_i(t) \in \mathbb{N}_0$ as the number of times each $\mathbf{x}_i$ has been used for training up to step t. $\sum_{\mathbf{x}_i \in \mathcal{T}} n_i(t) = 2t$ is valid. In this dual formulation the training step of the DoubleMinOver algorithm simply consists of searching for $\mathbf{x}_{\min^+}(t)$ and $\mathbf{x}_{\min^-}(t)$ and increasing their corresponding n_i by one.

In the dual representation the inner product $\mathbf{w}^T \mathbf{x}$ can be written as

$$\mathbf{w}^T \mathbf{x} = \sum_{\mathbf{x}_i \in \mathcal{T}} y_i n_i \mathbf{x}_i^T \mathbf{x};. \quad (1)$$

If the input patterns $\mathbf{x} \in \mathbb{R}^D$ are transformed into another (usually higher dimensional) feature space by a transformation $\mathbf{\Phi}(\mathbf{x})$, DoubleMinOver can work with the Kernel $K(\mathbf{x}, \mathbf{x}') = \mathbf{\Phi}(\mathbf{x})^T \mathbf{\Phi}(\mathbf{x}')$ instead of the usual inner product. At each step of the algorithm where $\mathbf{w}^T \mathbf{x}_i$ occures one then uses

$$\mathbf{w}^T \mathbf{\Phi}(\mathbf{x}_i) = \sum_{\mathbf{x}_j \in \mathcal{T}} y_j n_j K(\mathbf{x}_j, \mathbf{x}_i) = y_i h(\mathbf{x}_i);. \quad (2)$$

3 SoftDoubleMinOver

So far linear separability of the patterns was required. Since this is not always the case, the concept of a "soft margin" was introduced in [1], [2]. With a soft margin training patterns are allowed to be misclassified for a certain cost. With DoubleMinOver we can easily realize a 2-norm soft margin.

In Cristianini and Shawe-Taylor [12] it is shown that solving the 2-norm soft margin classification problem within a feature space implicitly defined by a kernel $K(\mathbf{x}, \mathbf{x}')$ is equivalent to solving the hard margin problem within a feature space defined by a kernel $\hat{K}(\mathbf{x}, \mathbf{x}')$ for which $\hat{K}(\mathbf{x}_i, \mathbf{x}_j) = K(\mathbf{x}_i, \mathbf{x}_j) + C^{-1}\delta_{ij}$ is valid for each $\mathbf{x}_i, \mathbf{x}_j \in \mathcal{T}$, with δ_{ij} as the Kronecker δ which is 1 for $i = j$ and 0 otherwise. Within the feature space defined by $\hat{K}(\mathbf{x}, \mathbf{x}')$ the training data are linearly separable by construction. The scalar parameter C determines the "hardness" of the margin. The smaller C, the softer the margin. For $C \to \infty$ we obtain the dual formulation of DoubleMinOver (hard margin).

The SoftDoubleMinOver algorithm in its dual formulation and with kernels then works like follows:

0. Set $t = 0$, choose a t_{max}, and set $n_i = 0$ for $i = 1, \ldots, N$.
1. Determine $\mathbf{x}_{\min^+}(t) \in \mathcal{T}^+$ and $\mathbf{x}_{\min^-}(t) \in \mathcal{T}^-$ which minimize

$$\hat{h}(\mathbf{x}_i) = y_i \sum_{\mathbf{x}_j \in \mathcal{T}} y_j n_j \left(K(\mathbf{x}_j, \mathbf{x}_i) + \frac{\delta_{ij}}{C} \right) = \frac{n_i}{C} + h(\mathbf{x}_i)$$

   on $\mathcal{T}^+$ and $\mathcal{T}^-$, respectively.
2. Increase the $n_{\min^+}$ and $n_{\min^-}$ of $x_{\min^+}$ and $x_{\min^-}$ by one, respectively.
3. Set $t = t + 1$ and go to 1.) if $t < t_{max}$.
4. Determine $\mathbf{x}_{\min^+}$ and $\mathbf{x}_{\min^-}$ according to 1. and set $b = \frac{1}{2}\left(\hat{h}(\mathbf{x}_{\min^+}) - \hat{h}(\mathbf{x}_{\min^-})\right)$.

Having determined the n_i and b via SoftDoubleMinOver, the class assignment of a new pattern $\mathbf{x}$ takes place, of course, based on the original kernel. The decision depends on whether

$$\sum_{\mathbf{x}_i \in \mathcal{T}} y_i n_i K(\mathbf{x}_i, \mathbf{x}) - b$$

is larger or smaller than zero.

4 Experimental Results on Benchmark Problems

To validate and compare the performance of SoftDoubleMinOver[1] we tested it on a number of common classification benchmark problems. The classification benchmarks stem from the UCI[2], DELVE[3] and STATLOG[4] [13] collection. We compare our results

[1] A SoftDoubleMinOver package is available at http://www.inb.uni-luebeck.de/maxminover
[2] UCI Repository: `http://www.ics.uci.edu/~mlearn/MLRepository.html`
[3] DELVE Datasets: `http://www.cs.utoronto.ca/~delve/index.html`
[4] STATLOG Datasets: `http://www.niaad.liacc.up.pt/statlog/index.html`

Table 1. Classification results obtained with SoftDoubleMinOver on standard benchmarks. For comparison the results obtained with the C-SVM Implementation of the OSU-SVM Toolbox and those reported in the Fraunhofer benchmark repository are listed.

			C2-SDMO		OSU-SVM		Reference
Benchmark	#TR	#TE	Seconds/Iter.	ERR	Seconds	ERR	ERR_{REF}
banana	400	4900	;0.030/200	11.6 ± 0.83	0.031	10.4 ± 0.46	12.0 ± 0.66
br-cancer	200	77	;0.019/100	27.1 ± 4.96	0.012	28.2 ± 4.62	26.0 ± 4.74
diabetis	468	300	;0.060/300	23.3 ± 1.78	0.065	23.1 ± 1.82	24.0 ± 1.73
fl-solar	666	400	;0.148/300	32.4 ± 1.80	0.229	32.3 ± 1.82	32.0 ± 1.82
german	700	300	;0.142/200	24.1 ± 2.67	0.177	24.0 ± 2.17	24.0 ± 2.07
heart	170	100	;0.010/100	15.5 ± 3.22	0.006	15.2 ± 3.21	16.0 ± 3.26
image	1300	1010	;0.811/2000	13.1 ± 4.33	0.812	9.8 ± 0.62	3.0 ± 0.60
ringnorm	400	7000	;0.030/300	2.6 ± 0.41	0.021	2.5 ± 0.38	1.7 ± 0.12
splice	1000	2175	;0.615/500	16.1 ± 0.65	0.654	14.9 ± 0.78	11.0 ± 0.66
titanic	150	2051	;0.034/1500	22.4 ± 0.96	0.013	22.3 ± 1.04	22.0 ± 1.02
waveform	400	4600	;0.047/300	11.4 ± 0.59	0.045	10.7 ± 0.53	10.0 ± 0.43
thyroid	140	75	;0.004/200	4.2 ± 2.40	0.003	4.1 ± 2.42	4.8 ± 2.19
twonorm	400	7000	;0.057/200	2.4 ± 0.13	0.033	2.4 ± 0.14	3.0 ± 0.23

#Tr : number of training data, #Te : number of test data

with those reported in the SVM-benchmark repository of the Fraunhofer Institute[5] and results we obtained with the C-SVM of the OSU-SVM Matlab Toolbox[6] that is based on SMO [7].

Each result reported in the benchmark repository of the Fraunhofer Institute is based on 100 different partitionings of the respective benchmark problem data into training and test sets (Except for the splice and image benchmark which consist of 20 partitionings). For classification they used the standard C-SVM with RBF-kernels. The reported classification result is the average and standard deviation over all 100 realizations. Each partitioning is available from this repository.

Table 1 lists the average classification errors we obtained with SoftDoubleMinOver and the OSU-SVM on the different benchmark problems. We used the default parameter settings of the OSU-SVM Toolbox. As the Fraunhofer Institute we used RBF-kernels, and we took their kernel widths γ. The C values in SoftDoubleMinOver and the OSU-SVM where chosen such that the minimum error is obtained. On all benchmarks the simple SoftDoubleMinOver is as fast as and achieves results comparable to those of the OSU-SVM and those reported in the Fraunhofer benchmark repository. Only a few training steps are necessary. On the "ringnorm", the "image" and the "splice" benchmark both the OSU-SVM as well as SoftDoubleMinOver are worse than the Fraunhofer reference. By either performing more iterations for SoftDoubleMinOver or tweeking the parameters of the OSU-SVM one can, of course, obtain comparable results for these benchmarks, too.

---

[5] Benchmark Repository: http://ida.first.fraunhofer.de/projects/bench/benchmarks.htm

[6] OSU SVM Classifier Toolbox: `http://www.ece.osu.edu/~maj/osu_svm/`

5 Conclusions

The main purpose of this paper is to present a very simple, incremental algorithm which solves the maximum margin classification problem with or without kernels and with or without a soft margin. SoftDoubleMinOver as an extension of MinOver learns by simply iteratively selecting patterns from the training set. Based on previous work it can be shown that SoftDoubleMinOver converges like $\mathcal{O}(t^{-1})$ to the exact solution, with t as the number of iteration steps. The incremental nature of the algorithm allows one to trade-off the computational time and the precision of the obtained hyperplane. The computational effort increases only linearly with the number of training patterns N. For the similar Kernel-Adatron the computational effort increases with N^2, however, the Kernel-Adatron also converges exponentially with t. SoftDoubleMinOver is advantegoeus particularly for large training sets. In experiments on standard benchmark problems SoftDoubleMinOver achieves a performance comparable to the widespread OSU-SVM which is based on the SMO-algorithm. However, SoftDoubleMinOver as a "three-liner" is much easier to implement, and with its pattern-by-pattern nature it might be a good starting point for a real on-line learning procedure for maximum margin classification.

References

1. Cortes, C., Vapnik, V.: Support-vector networks. Machine Learning **20** (1995) 273–297
2. Vapnik, V.: The Nature of Statistical Learning Theory. Springer-Verlag, New York (1995)
3. LeCun, Y., Jackel, L., Bottou, L., Brunot, A., Cortes, C., Denker, J., Drucker, H., Guyon, I., Muller, U., Sackinger, E., Simard, P., Vapnik, V.: Comparison of learning algorithms for handwritten digit recognition. Int.Conf.on Artificial Neural Networks (1995) 53–60
4. Osuna, E., Freund, R., Girosi, F.: Training support vector machines:an application to face detection. CVPR'97 (1997) 130–136
5. Schölkopf, B.: Support vector learning (1997)
6. Friess, T., Cristianini, N., Campbell, C.: The kernel adatron algorithm: a fast and simple learning procedure for support vector machine. Proc. 15th International Conference on Machine Learning (1998)
7. Platt, J.: Fast Training of Support Vector Machines using Sequential Minimal Optimization. In: Advances in Kernel Methods - Support Vector Learning. MIT Press (1999) 185–208
8. Keerthi, S.S., Shevade, S.K., Bhattacharyya, C., Murthy, K.R.K.: A fast iterative nearest point algorithm for support vector machine classifier design. IEEE-NN **11** (2000) 124–136
9. Li, Y., Long, P.: The relaxed online maximum margin algorithm. Machine Learning **46(1-3)** (2002) 361–387
10. Krauth, W., Mezard, M.: Learning algorithms with optimal stability in neural networks. J.Phys.A **20** (1987) 745–752
11. Martinetz, T.: Minover revisited for incremental support-vector-classification. Lecture Notes in Computer Science **3175** (2004) 187–194
12. Cristianini, N., Shawe-Taylor, J.: Support Vector Machines (and other kernel-based learning methods). Cambridge University Press, Cambridge (2000)
13. King, R., Feng, C., Shutherland, A.: Statlog: comparison of classification algorithms on large real-world problems. Applied Artificial Intelligence **9** (1995) 259–287

On the Explicit Use of Example Weights in the Construction of Classifiers

Andrew Naish-Guzman, Sean Holden, and Ulrich Paquet

Computer Laboratory, University of Cambridge, Cambridge CB3 0FD, UK
{andrew.naish-guzman, sean.holden, ulrich.paquet}@cl.cam.ac.uk

Abstract. We present a novel approach to two-class classification, in which a classifier is parameterised in terms of a distribution over examples. The optimal distribution is determined by the solution of a linear program; it is found experimentally to be highly sparse, and to yield a classifier resistant to noise whose error rates are competitive with the best existing methods.

1 Introduction

Many classification algorithms associate a weight with each element of the training set. In support vector machines, these weights are Lagrange multipliers in a quadratic optimisation problem; when set correctly, they define a separating hyperplane in the kernel-induced feature space (Schölkopf et al. (1999)). The relevance vector machine (Tipping (2001)) places a Gaussian of constant width over every data point and, in a Bayesian setting, assigns a weight to each such basis function. By an explicit assumption on the form of the solution, the distribution of weights is encouraged to be sparse. Boosting methods, in contrast, work iteratively and update the weights in response to each hypothesis chosen by a so-called *weak learner* (Freund and Schapire (1995)). An example's weight is related to the frequency with which it has been misclassified; by appropriate reweighting of the data, boosting algorithms encourage the weak learner to explore advantageous regions of hypothesis space. While studying the behaviour of boosting when applied to a simple weak learner, we observed the approximate convergence of the example weights, and the correlated convergence of the decision boundary. This observation motivated the idea that a *fixed* distribution over examples may be capable of inducing a useful distribution over the class of functions available to the weak learner. In this work, we show how a novel interpretation of example weights may indeed yield a sensible distribution over hypotheses. The optimal weight assignment is given by the solution of a linear program, and we show that the predictions of this distributed classifier may then be evaluated efficiently. Preliminary results indicate our algorithm is stable in noisy conditions, and performs competitively with the best existing methods. It also yields sparse solutions, in that many examples are given weights equal to or very close to zero. The relevance vector machine also exhibits this property, but is computationally more involved than our approach, and shows greater sensitivity to its parameters' settings (in particular, to the width of the Gaussians).

W. Duch et al. (Eds.): ICANN 2005, LNCS 3697, pp. 307–312, 2005.

2 Interpretation of Weights

Let us formalise the problem. We have a data set $D = \{(x_i, y_i)\}_{i=1}^m$, where $x_i \in X$ and $y_i \in Y = \{\pm 1\}$, a distribution over examples d_i, such that $\|\mathbf{d}\|_1 = 1$ and $d_i \geq 0$, and also a basis class of hypotheses $\mathcal{H}$ with an associated measure, allowing us to place over it a distribution $p(h)$. We will find that even for uniform $p(h)$, a novel interpretation of $\mathbf{d}$ has the potential to yield a distribution over $\mathcal{H}$ that corresponds to a complex classifier.

Consider the following scheme for classifying a new point $x \in X$. We draw examples from D according to the probability vector $\mathbf{d}$; for each labelled example (x_i, y_i) selected, we sample a hypothesis from $\mathcal{H}$ according to $p(h : h(x_i) = y_i)$, and evaluate $h(x)$. If we sum many such classifications and normalise the result, the final output will tend to

$$\begin{aligned} F_{\mathbf{d}}(x) &= \sum_{i=1}^m d_i \int \mathbb{I}[h(x_i) = y_i] h(x) dp(h) \\ &= \mathbb{E}_{(x_i,y_i)\sim\mathbf{d}} \left[\mathbb{E}_{h:h(x_i)=y_i}[h(x)]\right] . \end{aligned} \tag{1}$$

Assume that $p(h)$ is symmetric with respect to the two classes; that is, we have $p(h(x) = 1) = p(h(x) = -1)$ for all x. We may now classify x:

$$\begin{aligned} F_{\mathbf{d}}(x) &= \sum_{i=1}^m d_i \int \mathbb{I}[h(x_i) = y_i] h(x) dp(h) \\ &= \sum_{i=1}^m d_i \left(\int \mathbb{I}[h(x) = h(x_i) = y_i] y_i dp(h) - \int \mathbb{I}[h(x) \neq h(x_i) = y_i] y_i dp(h) \right) \\ &= \sum_{i=1}^m d_i y_i \left(\int \mathbb{I}[h(x_i) = y_i] dp(h) - 2 \int \mathbb{I}[h(x) \neq h(x_i) = y_i] dp(h) \right) \\ &= \sum_{i=1}^m d_i y_i \left(\frac{1}{2} - \int \mathbb{I}[h(x) \neq h(x_i)] dp(h) \right) , \end{aligned} \tag{2}$$

where in the last line we have used the symmetry of $p(h)$. We note that in (2), the final bracketed expression is a kernel function; its value is related to the probability that an arbitrary hypothesis drawn from $p(h)$ will have equal sign at x and x_i.

2.1 Assignment of Weights

Write the *margin* of the classifier on each element of the training set as a vector:

$$\begin{aligned} [y_j F_{\mathbf{d}}(x_j)]_{j=1}^m &= \left[y_j \sum_{i=1}^m d_i y_i \left(\frac{1}{2} - \int \mathbb{I}[h(x_j) \neq h(x_i)] p(h) dh \right) \right]_{j=1}^m \\ &= \mathbf{d}^\top \mathsf{Q}, \end{aligned} \tag{3}$$

where

$$\mathsf{Q}_{ij} = y_i y_j \left(\frac{1}{2} - \int \mathbb{I}[h(x_j) \neq h(x_i)] p(h) dh \right).$$

Q is symmetric; we have also $\mathsf{Q}_{ii} = \frac{1}{2}$ for all i. The linear formulation (3) allows us to find suitable weights by solving a linear program. For example, we can choose weights that maximise the minimum margin over the training set:

$$\begin{aligned} &\max_{\mathbf{d},\gamma} \gamma \\ &\text{subject to } y_i F_{\mathbf{d}}(x_i) \geq \gamma \text{ for } i = 1 \ldots m \\ &\qquad d_i \geq 0 \text{ and } \|\mathbf{d}\|_1 = 1. \end{aligned}$$

Alternatively, we can introduce a parameter $C > 0$ and slack variables $\boldsymbol{\xi}$ to allow a small number of misclassifications:

$$\begin{aligned} &\max_{\mathbf{d},\boldsymbol{\xi},\gamma} \gamma - C \sum_{i=1}^{m} \xi_i \\ &\text{subject to } y_i F_{\mathbf{d}}(x_i) \geq \gamma - \xi_i \text{ for } i = 1 \ldots m \\ &\qquad d_i \geq 0 \text{ and } \|\mathbf{d}\|_1 = 1 \\ &\qquad \xi_i \geq 0. \end{aligned} \tag{4}$$

In either case, the final classifier is given by $\text{sgn}(F_{\mathbf{d}}(x))$.

Optimisation with respect to the l_1 norm has been investigated in the context of support vector machines by Bradley and Mangasarian (1998) and Zhu et al. (2003). They both observe that when used as a regulariser, it generally favours sparser solutions than the l_2 norm. However, the latter is prevalent since it is more amenable when applied to infinite feature spaces. In some sense our method borrows from both schemes, by fitting a model which corresponds implicitly to an infinite set of basis functions, but which also allows the sparsity-inducing l_1 constraint on the model coefficients. The nature of this model is such that we cannot form an *arbitrary* finite or infinite linear combination of basis functions; however, we will see that even a coarse approximation is capable of excellent performance.

3 Implementation

To construct the matrix Q, we need to choose $\mathcal{H}$ and $p(h)$, and thus determine

$$\int \mathbb{I}[h(x_i) \neq h(x_j)] dp(h) \qquad \text{for } i, j = 1, 2, \ldots, m. \tag{5}$$

In the following analysis, we restrict our attention to the two-dimensional case, and fix $\mathcal{H}$ to be linear halfspaces in $\mathbb{R}^2$. Extending these concepts to higher dimensions and further classes is deferred to future work.

Without loss of generality, let the mean of the data be at the origin $O = (0, 0)$, and let all training coordinates lie in the region $[-R, R]^2$. All hypotheses $h \in \mathcal{H}$, with the exception of those that pass through the origin, may be paramaterised by a pair $(\mathbf{r}, s) \in (\mathbb{R}^2, \{\pm 1\})$. The coordinate $\mathbf{r}$ indicates the closest point on the line to O, while the sign term s defines the classification of the origin. Let us now define the measure $p(h)$ on $\mathcal{H}$ by placing a uniform distribution over $\mathbf{r}$ in the range $[-R, R]^2$, and assigning equiprobably $s = 1$ or $s = -1$.

In order to calculate (5), we must find the expected proportion of hypotheses discriminating between x_i and x_j. With the preceding assumptions, we may now consider (5) as the volume of parameter space $\mathcal{H}' \subseteq \mathcal{H}$, in which $h(x_i) \neq h(x_j) \Leftrightarrow h \in \mathcal{H}'$. The situation is illustrated in Figure 1. We note that, for a given pair $(\mathbf{r}, s)$, if the hypothesis parameterised by $(\mathbf{r}, s)$ satisfies this property, so also will that parameterised by $(\mathbf{r}, -s)$.

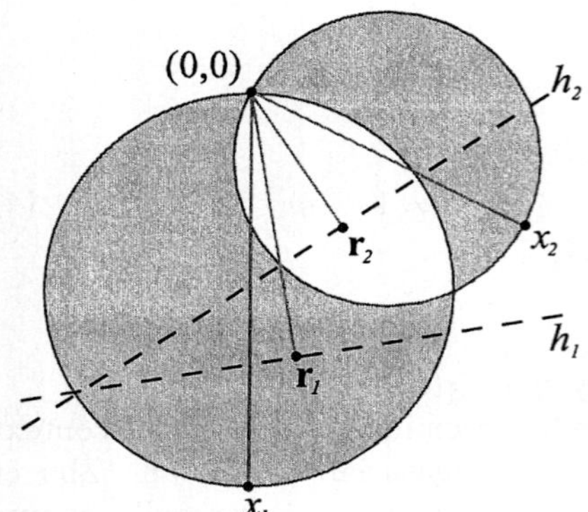

Fig. 1. Visualisation of (5). The shaded region parameterises hypotheses $h \in \mathcal{H}' \subseteq \mathcal{H}$ for which $h(x_1) \neq h(x_2) \Leftrightarrow h \in \mathcal{H}'$. Two hypotheses are shown, h_1 and h_2, parameterised by $\mathbf{r}_1$ and $\mathbf{r}_2$ respectively. Independently of s, $h_1 \in \mathcal{H}'$ discriminates between x_1 and x_2, while $h_2 \notin \mathcal{H}'$ classifies the two examples identically.

For a point $x \in X$, write the circular region parameterising hypotheses that discriminate between x and O as $\bigcirc_x$. Now,

$$\begin{aligned}\int \mathbb{I}[h(x_i) \neq h(x_j)]dp(h) &\propto \left|\bigcirc_{x_i} \setminus \bigcirc_{x_j}\right| + \left|\bigcirc_{x_j} \setminus \bigcirc_{x_i}\right| \\ &= \left|\bigcirc_{x_i}\right| + \left|\bigcirc_{x_j}\right| - 2\left|\bigcirc_{x_i} \cap \bigcirc_{x_j}\right| . \qquad (6)\end{aligned}$$

It can be shown that the area of intersection $\left|\bigcirc_{x_i} \cap \bigcirc_{x_j}\right|$ is given by

$$\frac{1}{2}\left(\|x_i\|^2(\theta_i - \sin\theta_i) + \|x_j\|^2(\theta_j - \sin\theta_j)\right),$$

where θ_i (θ_j) is the angle subtended at the centre of $\bigcirc_{x_i}$ ($\bigcirc_{x_j}$) by radii extending to the two points of intersection. Using (6), define

$$A(x_i, x_j) = \|x_i\|^2 (\pi - \theta_i + \sin\theta_i) + \|x_j\|^2 (\pi - \theta_j + \sin\theta_j),$$

so that

$$\mathsf{Q}_{ij} = y_i y_j \left(\frac{1}{2} - \frac{1}{4R^2} A(x_i, x_j)\right). \qquad (7)$$

4 Results

Results were obtained for two benchmark data sets: Ripley's mixture of Gaussians (Ripley (1996)) consists of 250 training examples and 1000 test examples; the Banana set[1] consists of 100 realisations of 400 training examples and 4900 test examples. Both are two-dimensional. We chose the parameter C on the Ripley set by examining the decision boundary for a variety of choices, and selecting the one with qualitatively best fit; this was found to be $C = 0.009$. For the Banana benchmark, we split the training set into equal subsets for training and validation, to find the optimal $C \in \{0.01, 0.012, \ldots, 0.02\}$. In each case, we used the formulation (7), and set $R = 5$.

On Ripley's set, the test error was 8.6%. This compares favourably with existing methods: using an SVM, Ripley achieved 10.6%, while Tipping's RVM achieved 9.3%.[2] The Bayes rate is around 8%. Over the first ten realisations of the Banana set, our method achieved a mean test error of 10.9%; the support and relevance vector machines obtained error rates of 10.9% and 10.8% respectively.[3] The decision boundary we obtained on the Ripley set is illustrated in Figure 2. The training data are shown, together with surrounding circles, each of whose radii is proportional to the weight of the associated data point. It is interesting to observe that many components of this distribution are equal to or close to zero, and that the heavily weighted examples tend to be some distance from the decision boundary. The SVM solution to this problem used 38 support vectors, while the RVM solution used 4 relevance vectors; our solution places non-zero weight on 8 examples.

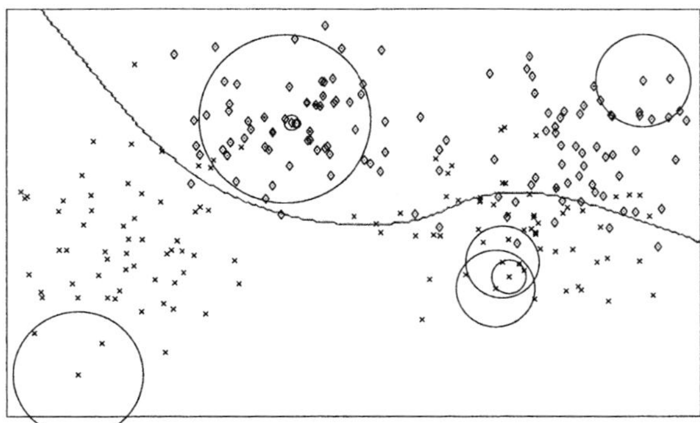

Fig. 2. The decision boundary obtained on Ripley's training set by solving (4) with $C = 0.009$. Data points with non-zero weight assignment have been circled; the radius of the circle is proportional to the example's weight.

[1] Available from `http://ida.first.fhg.de/projects/bench/benchmarks.htm`.
[2] Results from Bishop and Tipping (2003).
[3] Results from Tipping (2001).

5 Conclusions

We have shown how a simple sampling scheme for classification and a novel interpretation of weighted examples induces a distribution over a basis class of hypotheses. We have evaluated the predictions of this distributed classifier for an optimal weighting of the training set, and found these predictions to be resistant to overfitting. Our method has certain advantages: the weight assignment can be determined easily by solving a linear program, with a single parameter defining the degree to which misclassifications are tolerated; the weight vector is experimentally found to be sparse when the solution has not overfit; new classifications are then possible in time $\mathcal{O}(m')$, where $m' \leq m$ is the number of examples in the training set with non-zero weight. We have observed also that our "support vectors" lie away from the decision boundary and tend to be fewer in number than for the standard SVM solution.

The development of a rigorous explanation for our algorithm's strong performance is an area of active research. In particular, we have recently extended the concepts of Section 3 to allow data of arbitrary dimensionality. We hope to present the results of this work in a future publication.

Acknowledgments

Andrew Naish-Guzman is supported by a Millennium Scholarship, and a grant from Gonville and Caius College, Cambridge. Ulrich Paquet is supported by a Commonwealth Scholarship, and expresses his thanks to the Commonwealth Scholarship Commission.

References

Bishop, C. M and Tipping, M. E. (2003) Bayesian regression and classification. In J. Suykens, G. Horvath, S. Basu, C. Micchelli, and J. Vandewalle, *Advances in Learning Theory: Methods, Models and Applications 190.* IOS Press, Amsterdam.

Bradley, P. and Mangasarian, O. (1998). Feature selection via concave minimization and support vector machines. In J. Schavlik, *International Conference on Machine Learning '98.* Morgan Kaufmann.

Freund, Y. and Schapire, R. E. (1995). A decision-theoretic generalization of on-line learning and an application to boosting. *European Conference on Computational Learning Theory*, 23–37.

Ripley, B. D. (1996). *Pattern Recognition and Neural Networks.* Cambridge University Press.

Schölkopf, B., C. J. C. Burges, and A. J. Smola (1999). *Advances in Kernel Methods: Support Vector Learning.* MIT Press.

Tipping, M. E. (2001). Sparse Bayesian learning and the relevance vector machine. *Journal of Machine Learning Research 1*, 211–244.

Zhu, J., Rosset, S., Hastie, T. and Tibshirani, R. (2003) 1-norm support vector machines. *Neural Information Processing Systems 16.*

A First Approach to Solve Classification Problems Based on Functional Networks

Rosa Eva Pruneda[1], Beatriz Lacruz[2], and Cristina Solares[1]

[1] University of Castilla-La Mancha Spain
rosa.pruneda@uclm.es,cristina.solares@uclm.es
[2] University of Zaragoza Spain
lacruz@unizar.es

Abstract. In this paper the ability of the functional networks approach to solve classification problems is explored. Functional networks were introduced by Castillo et al. [1] as an alternative to neural networks. They have the same purpose, but unlike neural networks, neural functions are learned instead of weights, using families of linear independent functions. This is illustrated by applying several models of functional networks to a set of simulated data and to the well-known Iris data and Pima Indian data sets.

1 Introduction

Classification involving two classes often appear in real-life. The problem consists of discovering a function f of a vector of predictors $\mathbf{X} = (X_1, X_2, \ldots, X_k)$ which allows us to classify an individual, such that $X_1 = x_1, X_2 = x_2, \ldots, X_k = x_k$, in one of the two classes, represented by a binary response variable Y.

Classical statistical techniques propose f to be a linear classification function obtained via *logistic regression* or *discriminant analysis*. More general functions are considered when using *generalized additive models*, *neural networks* or other nonparametric methods (see, for example, [3] and the references therein).

In this paper, we propose some nonlinear functions based on functional networks models to determine f. It can be estimated using the appropriate learning procedures usually used in this setting. This approach is introduced in Section 2. In Section 3 we investigate the performance of the proposed method using a simulated and two real-life data sets. Finally, a summary and some concluding remarks are given in Section 4.

2 Functional Networks

In this paper we propose two functional network models to approximate $f(\mathbf{X})$:

1. The Generalized Associativity model which leads to the additive model

$$f(X_1, \ldots, X_k) = h_1(X_1) + h_2(X_2) + \ldots + h_k(X_k), \tag{1}$$

(see Castillo et al. [1] for further details). The corresponding functional network is shown in Figure 1.

W. Duch et al. (Eds.): ICANN 2005, LNCS 3697, pp. 313–318, 2005.

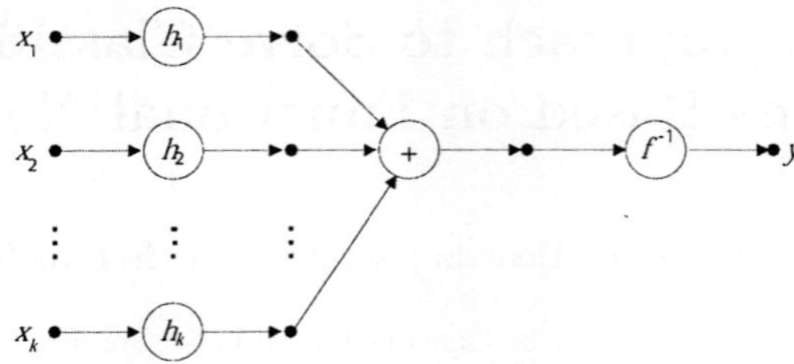

Fig. 1. Functional network representing the additive model

2. The Separable model which considers a more general form for f

$$f(X_1, \ldots, X_k) = \sum_{r_1=1}^{q_1} \ldots \sum_{r_k=1}^{q_k} c_{r_1 \ldots r_k} \phi_{r_1}(X_1) \ldots \phi_{r_k}(X_k), \tag{2}$$

where $c_{r_1 \ldots r_k}$ are unknown parameters and the sets of functions $\Phi_s = \{\phi_{r_s}(X_s),\ r_s = 1, 2, \ldots, q_s\}$, $s = 1, 2, \ldots, k$, are linearly independent. An example of this functional network for $k = 2$ and $q_1 = q_2 = q$ is shown in Figure 2.

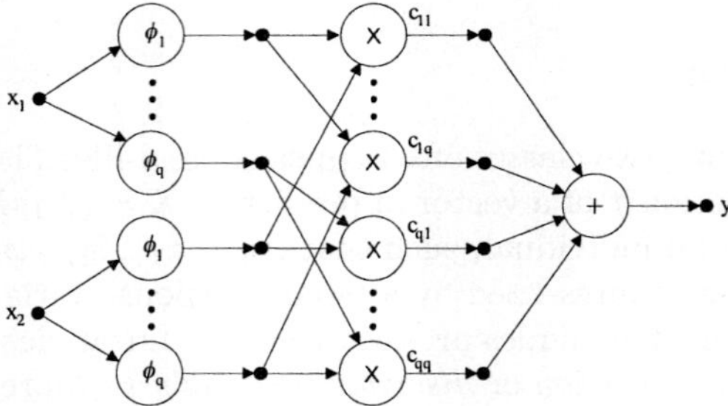

Fig. 2. The functional network for the separable model with $k = 2$ and $q_1 = q_2 = q$

Equations (1) and (2) are functional equations since their unknowns are functions. Their corresponding functional networks are the graphical representations of these functional equations. Note that the graphical structure is very similar to a neural network.

Our problem consists of learning from data

1. $h_1, h_2, \ldots, h_k$ in (1) and
2. $c_{r_1 \ldots r_k}$ in (2).

In order to obtain $h_1, h_2, \ldots, h_k$ in (1), we approximate each $h_i(x_i)$ by a linear combination of linearly independent functions ϕ_{ij}, that is,

$$h_i(x_i) \simeq \sum_{j=1}^{q_i} a_{ij} \phi_{ij}(x_i) \tag{3}$$

and the problem is reduced to estimate the parameters $a_{ij}, \forall i, j$.

The sets of linearly independent functions used in equations (2) and (3) can be some of the following:

1. Polynomial family:
$$\Phi = \{1, x, x^2, \ldots, x^q\} \tag{4}$$
2. Exponential family:
$$\Phi = \{1, e^x, e^{-x}, e^{2x}, e^{-2x}, \ldots, e^{qx}, e^{-qx}\} \tag{5}$$
3. Fourier series family:
$$\Phi = \{1, \sin x, \cos x, \sin(2x), \cos(2x), \ldots, \sin(qx), \cos(qx)\}. \tag{6}$$

Finally, we choose the least squares criterion to estimate the parameters, but the additive model requires to add some constrains to guarantee uniqueness (see Castillo et al. [1] for further details). The main advantage is that least squares criterion leads to solve a linear system of equations in both cases.

3 Examples

In this section we apply the proposed functional networks approach to a set of simulated data and to the well-known Iris data and Diabetes in Pima Indian data sets. In order to analyze the performance of the technique we divide randomly both sets in a training data set and a test data set. The training set is used to learn the classification function and the test set is used to compute test error.

A model selection method is also needed. In both examples we use the exhaustive method, i.e., all possible models are investigated in order to gain some understanding of the performance of our proposed technique.

3.1 Simulated Data

This data set consists of a random sample of size 100 generated using two explanatory variables X_1 and X_2, independent and uniformly distributed, with a classification function

$$f(X_1, X_2) = 0.1 * \exp(X_1) + X_2.$$

Figure 3 shows the scatter plot of X_2 versus X_1 and the class each pair belongs to. The obtained model using the Additive functional network model with the set of linearly independent functions $\Phi = \{1, x, \exp(x), \exp(2x), \exp(3x)\}$ is

$$y = 0.0389 + 0.0893 * \exp(x_1) + 0.9924 * x_2, \tag{7}$$

with training error equal to 0.0360 and test error equal to 0.0166. The well classified observation rates for training and test data sets are 85% and 80%, respectively. Figure 4 shows the training and test data sets against the predicted probabilities.

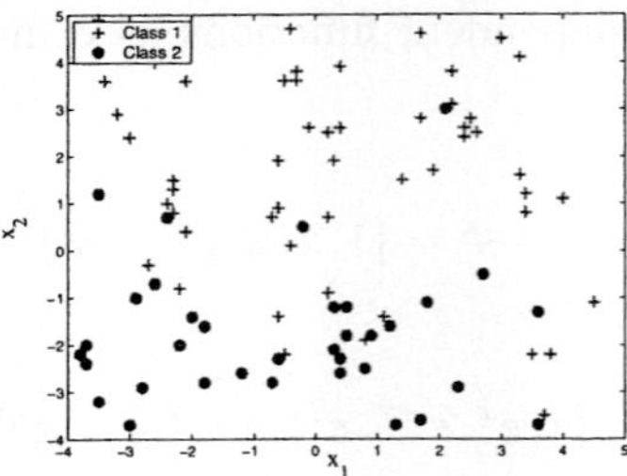

Fig. 3. Simulated data

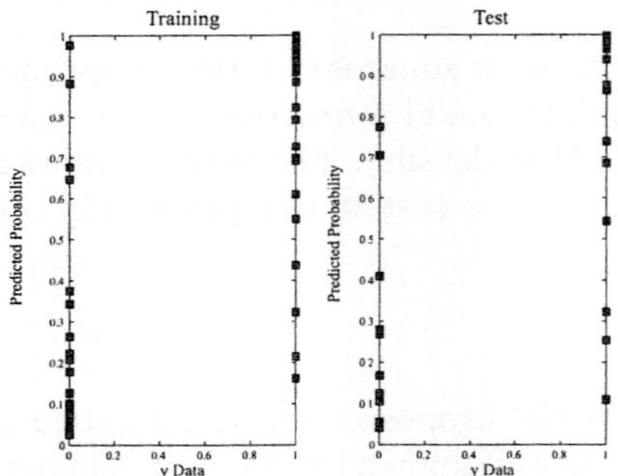

Fig. 4. Training and test data sets against the predicted probabilities

3.2 The Iris Data Set

The well-known Iris data set consists of 150 random samples of flowers from the iris species setosa, versicolor, and virginica, (which is available at the UCI ML Repository database [5]). From each species there are 50 observations for sepal length, sepal width, petal length, and petal width, measured in centimeters.

Even though flowers are classified in three classes in this data set, as a first approximation, we consider a bivariate response variable taking into account if flowers belong to the iris species setosa or not. We know that this class is linearly separable.

In order to approximate the classification function we will use:

1. The Additive functional network model using the polynomial, exponential and Fourier families of linearly independent functions, and
2. The Separable functional network model using the polynomial family.

The training and test errors are showed in Table 1. Since setosa class is separable, the accuracy for both training and test data sets is 100%.

Now, we repeat the same experiment taking into account if flowers belong to the iris species versicolor or not. It is a non separable class, however all the functional network models provide good levels of accuracy for the training and test data sets as is shown in Table 2.

Table 1. Training and test errors for the classification of Iris data set considering the two classes setosa-non setosa

	Training Error	Test Error
Additive Polynomial	0.0058	0.0233
Additive Exponential	0.0182	0.0276
Additive Fourier	0.0092	0.0236
Separable Polynomial	0.0126	0.0227

Table 2. Training and test errors for the classification of Iris data set considering the two classes versicolor-non versicolor

	Tra. Error	Tra. Accuracy	Test Error	Test Accuracy
Additive Polynomial	0.0354	95%	0.1053	94%
Additive Exponential	0.0280	94%	0.1122	92%
Additive Fourier	0.3350	93%	0.1088	94%
Separable Polynomial	0.2552	93%	0.0969	98%

3.3 Diabetes in Pima Indians

A population of women who were at least 21 years old, of Pima Indian heritage and living near Phoenix, Arizona, was tested for diabetes according to World Health Organization criteria. The data were collected by the US National Institute of Diabetes and Digestive and Kidney Diseases. This data set, which is available in the UCI machine-learning database collection [5], is profusely analyzed in Ripley [4] by means of different classical classification techniques. To carry out a comparison between functional networks and the results provide by Ripley [4], we have selected the same training and test sets: a random training set of size 200 and a test set of size 332. Serum insulin variable and incomplete records have been omitted.

The results reported by Ripley [4] show a best rate error around 20% which is attained using standard linear discrimination and simple logistic regression models. A neural network with one hidden gives a error rate of 21%. Other classification tools as multivariate adaptative Regression Splines (MARS) or Projection Pursuit Regression (PPR) give a 23.4% error rate on the test set.

We have applied several functional networks models obtaining similar error rates similar than those provided in [4]. For example, the additive functional network model combining the polynomial and Fourier families of linearly independent functions: $\{1, x, \sin(x)\}$ gives a training error rate of 20.5% and a test error rate of 22.3%. The obtained model is:

$$y = 2.1123 - 0.0231x_1 - 0.0071x_2 - 0.0145x_5 + 0.0574\sin(x_5) - 0.3510x_6 - 0.0096x_7. \quad (8)$$

4 Conclusions

In this paper we propose a new method based on functional networks to solve two-class classification problems. Two nonlinear functional network models are proposed to obtain the classification function. The main advantage of working with this kind of models is that the learning procedure is based on solving a linear system of equations.

On the one hand, the performance of the proposed method is investigated using a simulated example in which the classification function is nonlinear in the variables. The method provides the true classification function and high rates of well classified observations for both training and test data sets.

On the other hand, the well-known Iris data set is used to illustrate the performance of the proposed method when, (1) a separable class (setosa-non setosa), and (2) a non separable class (versicolor-non versicolor) are considered. For the first case, 100% of accuracy is obtained with all of the proposed models. For the second case, good levels of accuracy for both training and test data sets are also provided by all of the proposed models.

Furthermore, we have applied our methods to the Diabetes Indian Pima obtaining similar results that those provided by Ripley when using several well-known classification techniques.

These are some promising preliminar experiences in working with functional networks for classification problems. Much work must be made in order to asses the performance of our method. In the future, we will explore the ability of the models in discovering nonlinear classification functions and a deeper comparison study with more examples will be addressed.

References

1. Castillo, E., Cobo, A., Gutiérrez, J. M. and Pruneda, E. *An Introduction to Functional Networks with Applications.* Kluwer Academic Publishers (1998) New York.
2. Castillo, E. and Ruiz-Cobo, R. *Functional Equations in Science and Engineering.* Marcel Dekker (1992) New York.
3. Hastie, T., Tibshirani, R. and Friedman, J. *The Element of Statistical Learning. Data Mining, Inference and Prediction.* Springer (2001).
4. Ripley, B.D. *Pattern Recognition and Neuronal Networks.* Cambridge University Press (2002).
5. UCI ML Repository Database. http://www.ics.uci.edu/ mlearn/MLSummary.html.

A Description of a Simulation Environment and Neural Architecture for A-Life

Leszek Rybicki

Faculty of Mathematics and Computer Science,
Nicolaus Copernicus University, Torun, Poland

Abstract. This paper describes a project in progress, a modular environment for a-life experiments. The main purpose of the project is to design a neural architecture that would allow artificial creatures (biots) to learn to perform certain simple tasks within the environment, having to deal with only the information they can gather during exploration and semi-random trials. That means that the biots are given no explicit information about their position, distance from surrounding objects or even any measure of progress in a task. Information that a task has been started and either accomplished or failed is to be the only reinforcement passed to the learning process.

1 Introduction

The model we use in the project involves an agent (agents) exploring an unknown environment. When the agent notices an object of interest, a "game" starts and all actions taken by the agent are regarded as moves within the game. If the game is won, e.g. the object of interest happens to be food and the agent succeeds in eating it, the agent remembers the steps that led to success.

This bears certain similarity to how simple organisms explore their surrounding. When exposed to unknown stimuli, they probe the environment and remember the effects of their actions, making their movements more defined and precise, as they depend more on previous experience and less blind trials.

2 The Underlying Idea

As a biot explores its world, input from the sensors comes as a vector of activation values. The biot makes a decision, which is represented as a vector of motor (effector) activation. The two vectors create a state-action pair that is used in the learning process as a description of a single step.

The exploration process needs to be a mixture of random movement (i.e. when no object of interest is within sensor range or the current situation is new to the biot) and repetition of previously remembered movements with some degree of innovation and generalization when sensor input is similar to something the biot had encountered earlier.

Although Boltzmann machines are often criticized for low effectivity, they have certain features that seem to be perfect for this project. An untrained Boltzmann machine, with randomized weights, generates state-action pairs that result

W. Duch et al. (Eds.): ICANN 2005, LNCS 3697, pp. 319–324, 2005.

in a random walk. The weight relaxation procedure can be longer or shorter and depends on a modifiable temperature parameter. This allows for controlling the biots' "temper", making them either slow and thorough or rapid and innovative (or simply dumb). On the other hand, the temperature and relaxation length are extra parameters that have to be balanced.

Boltzmann machines are good for implementing content-addressed memory [3] and their limitation to second-order statistics can be an issue when it comes to more complex tasks, but allows for simplifying (generalizing) data given as a set of state-action vectors into a set of rules, associations between sensors and motors.

3 The Simulated World

The simulated environment the biots inhabit is a two-dimensional area filled with different active (or inactive) objects such as walls or food and the biots themselves, represented by insect-like symbols.

We designed the physics of the simulation ground as a simplified implementation of newtonian physics with collision detection dependent on the type of colliding objects and momentum for movement and rotation of biots.

4 The Biots

The purpose of the biots is to undergo individual development by changing their preferred actions accordingly to information gathered during interaction with the environment and other biots. The simulation allows for differentiating the biots, giving them different anatomies - different sets of sensors, motors and visual appearance, but also setting them to perform different tasks.

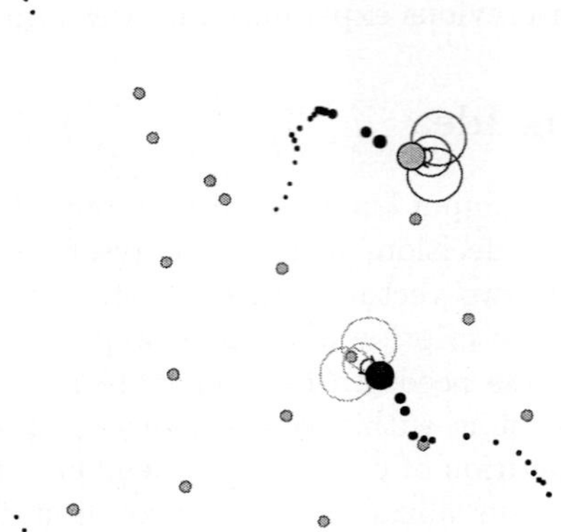

Fig. 1. Screenshot from the project. Two biots competing in a "finding food" task. Sensor ranges are marked with circles. Biots leave trail marks to show their position in time.

4.1 The Sensors

A sensor is an abstract mechanism that can be probed by the biot for its activation. This way, biots can have sensors as simple as constant sensors, having a fixed activation value, or as complex as environment sensors.

Environment sensors are designed to activate when an object (or another biot) is close to their owner. An environment sensor has a finite range (r) and a center point, represented with radial coordinates centered in the center of the biot's body (α, d). The value of the activation depends on the distance of the object from the sensor center point. An environment sensor can be specialized in detecting only one type of objects, e.g. food.

4.2 The Motors

Motors (effectors) can either move the biot (forward or backward) or make it turn. Each motor has a set force (γ) it applies to the biot when fully activated and a minimum neural signal needed to activate it. The overall activation of the motor is a product of the maximum force of the motor (γ) and activation of its assigned neuron (act), provided that $act > act_{min}$, where act_{min} is the minimum activation value needed, zero otherwise. Although the network in the brain is bipolar, it is assumed throughout the implementation that $act_{min} > 0$.

4.3 The Instinct

The role of the instinct is to inform the brain that a game has started or finished and if the game was won or not. It can also dynamically modify certain parameters of the brain, such as relaxation time, temperature or learning rate. We made sure the instinct mechanism had no direct impact on motors and was deprived of any complex calculation, as it would undermine the premise of the project. For example - the instinct mechanism is informed that a food piece is in sight, but not which sensor sensed the piece or even what was the value of activation of that sensor.

4.4 The Brain

The biot brain, as mentioned earlier, is a Boltzmann machine. Its visible layer (or visible subset of neurons) reflects the state-action vector while the hidden layer is used to boost the calculating ability of the network beyond the second-order statistics within the visible layer.

We have dropped traditional Boltzmann Machines in favor of Continuous Boltzmann Machines in early stages of the project. Precise restoration of remembered patterns is less important than the ability for the biot to vary the strength of impulses sent to motors, making the biot's motion more precise and fluent. A continuous decision making unit can also use less sensors, as they too can provide a spectrum of activation values relating to distance, size or number of sensed objects.

Neuron activations change asynchronously, with gaussian noise added to each neuron:

$$s_j = \sigma((\sum_i w_{ij} s_i + N(0,1)\beta) \cdot a_j)$$

where σ is the *tanh* sigmoid function, w_{ij} is the weight between ith and jth neuron, β is a noise ratio factor (relating to temperature in non-continuous models) and a_j is the jth neuron's own transfer function slope.

5 The Learning Process

The learning process takes place only during "game" mode. From the start to the end of a game, vectors composed of sensory input and effector activations (state-action pairs) are being stored in a static array that serves a purpose similar to short-term memory. This way, the array contains a verbatim record of each step taken by the biot between key events - one that had started the game and another one that meant the biot's success or failure.

After a biot wins a game, the contents of the short-term memory are treated as learning data for the Boltzmann Machine. The vectors are not evaluated for relevance to the overall strategy with the exception that the common weight of all vectors within a game linearly depends on the length of the game, favoring shorter games and strategies giving results more quickly.

5.1 The Experiment

The biots were equipped with two long range environmental sensors side by side, dubbed *eyes*, one environment sensor in front, corresponding to the sense of *smell* and a short range sensor directly in front of the biot, which we called *mouth* for more than one reason. All sensors were set to detect food and only food.

The biots entered in-game mode when any of their sensors sensed food and exited the mode when the food was either eaten or lost from within the sensor range. The biots had to learn to "aim" at food, whether approaching it from left or right. Food grew back at the same rate as it was eaten and the pieces formed a line, so that a well trained biot could follow it indefinately.

5.2 Two Learning Algorithms

Two algorithms were tried in different stages of development of the project.

The first one was a mixture of Hebbian learning within the visible layer and a simplified MCD learning[5] between the visible layer and the hidden layer. Gibbs probing was short and slope factors (a_j) were set to be constant during training.

All sensor input was duplicated with a one-step time shift - the input from the previous move was given together with the current input. This method gave visual effects relatively quickly and accurately as biots devised a tendency to aim, stop and jump forward at food. This was most probably due to the Hebbian learning of the visible layer, which allowed the network to 'discover' basic dependencies between sensors and motors: if you see food on the left, turn left. Unfortunately, after some time, biots started to move in circles. This is only understandable. As each food piece that was eaten was immediately replaced by another one somewhere in the ground, it made sense to move in circles, and make jumps every now and then. This strategy is often utilized by mushroom pickers.

At the same time, the biots were losing their ability to aim, missing food many times before finally catching a piece. Whether we can count this as a success of the network in finding a strategy that satisfied the premise of the experiment or an unwanted effect caused by an anomaly is a matter of discussion. Simply speaking, the biots had managed to find a 'cheat mode'.

An interesting phenomenon that accompanied this experiment was that when a biot had found food when moving backwards (which they did quite often at first, before the first learning), it started moving backwards as it was the only way of finding food it 'knew'.

The second algorithm we tried was a full implementation [4] of the MCD rule [2] with no connections within the visible layer, no duplication of sensory input and an extra modification in order to prevent the fore-mentioned phenomena. In order for a biot to eat a food piece, it had to be placed within the range of the mouth sensor. This made the task a little harder, preventing the 'mushroom picker' cheat (any such strategy would be much less efficient than aiming here) and making it less probable for the biot to find food while moving backwards, as backward jumps were much shorter than the size of the biot's body. Also, to provide a comparison, a wandering biot was introduced - a biot with its brain replaced by a generator of random neural activations.

We observed the visual effects of stopping, aiming and eating, but they were no longer as apparent as previously (especially stopping) nor did they appear as quickly. This architecture seemed to be more disturbed by noise as biots could switch back and forth between aiming accurately and completely ignoring food.

When we introduced a quantative measure of effectivity ratio, more phenomena emerged. It soon turned out that the randomly moving biot was quite capable of finding food by pure chance, while the biot depending on the output of its neural network made many wrong moves. The *wanderer* biot would just attack a cluster of food pieces, missing most, but eating a fair amount, while the *eater* biot kept losing points by botching attempts, making sudden turns in the worst possible moments. Although the effectivity of the *eater* tended to be bigger throughout the simulation, the *wanderer* had more successful 'meals' in a unit of time.

Also, the measure, described by the following formula:

$$effectivity = \frac{wins}{wins + loses}$$

turned out to be greatly dependent on seemingly irrelevant factors such as distance of food pieces from each other, precise placement and range of sensors and motor parameters.

A trial-and-error modification of motor parameters allowed increasing the eater biot's effectivity. Most importantly, both biots were given a forward effector that required a strong neural activation, but caused long jumps when activated. The normal forward motor was weakened, threshold was slightly increased for rotation motors, while their strength was decreased.

The new effector configuration slowed down the random biot and increased the smart biot's precision, as it managed to control the jumping effector. In simulations involving three biots capable of learning and one random biot, the smart biots 'win' one in four games (as opposed to around one in six for the random biot), but their speed (as in number of pieces eaten per unit of time) is at least twice the speed of the random biot.

6 Possible Future Development and Conclusion

The project is in progress and subsequent experiments suggest various directions of future development. In earlier stages, a selection scheme was used to determine an optimal placement of sensors for the eating task. We plan to explore the area further - having the biots evolve optimal anatomies and learning parameters, but disallowing genetic programming to interfere with the learning process itself - i.e. each new biot is to learn from its own experience.

Other learning algorithms[6], are being taken into consideration. The problem stated in the project is specific and it is apparent that methods commonly used in training Boltzmann machines as associative memory might not apply here.

Finally, it is a priority to establish a method for training biots to not only pursue targets, but also actively avoid obstacles and danger, expanding the number of possible tasks biots might perform.

Acknowledgments

The development of the project continues thanks to the support of dr Tomasz Schreiber. Some past and future modifications were inspired by discussions with dr Norbert Jankowski, dr Krzysztof Grabczewski and Filip Piekniewski.

References

1. Zhou, Z.-H.; Shen, X.-H.: "Virtual Creatures Controlled by Developmental and Evolutionary CPM Neural Networks". Intelligent Automation and Soft Computing, 2003, 9(1): 23-30, 2003
2. Ackley, D. H.; Hinton, G. E. and Sejnowski, T. J.: "A learning algorithm for Boltzmann machines". Cognitive Science, 9, 147–169., 1985
3. Aarts, E; Korst, J.: "Simulated annealing and Boltzmann machines. A stochastic approach to combinatorial optimization and neural computing" Wiley-Interscience Series in Discrete Mathematics and Optimization. John Wiley & Sons, Ltd., Chichester, 1989
4. Chen, H; Murray, A.: "A continuous restricted boltzmann machine with an implementable training algorithm" IEE Proc. of Vision, Image and Signal Processing, vol. 150, no. 3, pp. 153–158, 2003.
5. Hinton, G.E.: "Training Products of Experts by Minimizing Contrastive Divergence" Neural Computation 14(8) 1771-1800, 2002
6. Sallans, B.; Hinton, G.E.: "Using Free Energies to Represent Q-Values in a Multiagent Reinforcement Learning Task" in T. K. Leen, T. Dietterich and V. Tresp (eds.) Advances in Neural Information Processing Systems 13, The MIT Press, Cambridge, MA., 2001

Neural Network Classifers in Arrears Management

Esther Scheurmann and Chris Matthews

Faculty of Science, Technology & Engineering,
La Trobe University, P.O. Box 199 Bendigo 3552, Victoria, Australia
Phone: +61 3 54447998, Fax: +61 3 54447557
c.matthews@latrobe.edu.au
EssScheurmann@gmail.com

Abstract. The literature suggests that an ensemble of classifiers outperforms a single classifier across a range of classification problems. This paper investigates the application of an ensemble of neural network classifiers to the prediction of potential defaults for a set of personal loan accounts drawn from a medium sized Australian financial institution. The imbalanced nature of the data sets necessitates the implementation of strategies to avoid under learning of the minority class and two such approaches (minority over-sampling and majority under-sampling) were adopted here. The ensemble out performed the single networks irrespective of which strategy was used. The results also compared more than favourably with those reported in the literature for a similar application area.

Keywords: neural network ensembles, minority over-sampling, majority under-sampling, loan default, arrears management.

1 Introduction

Authorised Deposit-Taking Institutions (ADIs) are corporations that are authorised under the Australian Banking Act (1959) to invest and lend money. ADIs include banks, building societies and credit unions. ADIs generate a large part of their revenue through new lending or extension of existing credit facilities as well as investment activities. The work described here focuses on lending, in particular the creation and management of customer personal loan accounts. The development of credit scoring models to aid in loan approval is well established. Traditionally these have been statistically based[9,11] although more recently artificial neural network approaches have attracted some research interest[4,13,15]. However there has been less work in the management of existing accounts. Substantial amounts of money are spent on recovery of defaulted loans, which could be significantly decreased by having the option of tracking a high default risk borrowers' repayment performance. This is sometimes referred to as *arrears* or *collections* management.

This is essentially a classification problem. Loan accounts could be classified as high or low risk depending on the risk of the customer not meeting their

W. Duch et al. (Eds.): ICANN 2005, LNCS 3697, pp. 325–330, 2005.

repayment committments. Multi-layer artificial neural networks can be considered as non-linear classifiers and, given their success in credit scoring, may be of use in identifying high risk accounts. A recent study[2] compared a neural network approach to the prediction of early repayment and loan default with more traditional approaches. The results were promising and suggested that a neural network approach outperformed the traditional approaches, particular for the prediction of early repayment.

The research reported here focuses only loan default and applies an ensemble as well as a single classifier approach. The data used is real life data sourced from a medium sized Australian bank and includes a low proportion of bad accounts. The paper is organised as follows: Section 2 provides a brief overview of ensembles and classifiers, section 3 discusses the data used in more detail and the experiments conducted, and section 4 discusses the experimental results. The paper concludes with a discussion of possible areas for future work that arise from the results presented here.

2 Classifiers and Ensembles

In simplest terms, a classifier divides examples into a number of categories. Classifiers may be trained on a data set and then tested on unseen data to determine their generalisation capabilities. Typically training uses a supervised learning approach i.e the target class is known for both the training and testing data. It has been shown that the use of an ensemble, rather than a single classifier, significantly improves classification performance [5,8,16]. Ensembles are particularly useful for classification problems involving large data sets[3] and can be constructed and combined in various ways[5,14].

Each member of the ensemble could be trained and tested on a subset of the total data set. This approach works well for unstable learning algorithms such as those used by artificial neural networks[5]. Several methods are available for the selection of these subsets. They can simply be selected at random (with or without replacement). The data set could be divided into a series of disjoint subsets and the training sets could be formed by leaving out one or more of the subsets, which might be reserved for testing. In these situations the ensemble members are trained independently of each other[10]. Another approach is to use a boosting algorithm such as the ADABOOST algorithm[6] which builds the ensemble by using datasets formed by focusing on misclassified examples. Ensembles can also be constructed using subsets of the input attributes. This approach is particularly useful when there is some redundancy amongst the inputs. In situations where there are many target classes ensemble members can be constructed using a reduced set. The number of target classes can be reduced combining several together. Whatever methods are choosen for ensemble construction the designer should ensure that there is diversity amongst individual ensemble members.

There are several ways of combining or fusing the decision of each individual classifier into one final ensemble decision. The simplest is to use an unweighted voting system where it is assumed that the relative importance of each individual decision is the same. If this is not the case then appropriate weightings could be introduced. A discussion of the possibilities can be found in [5,14] and examples of ensemble application areas in [1,12,17].

3 Experimental Work

The networks were developed using personal loan accounts created in May 2003. The observation point was 12 months later i.e May 2004. This was considered sufficient time before a realistic assessment of their performance could be made (Fig. 1). The networks were trained to classify whether an account was likely to

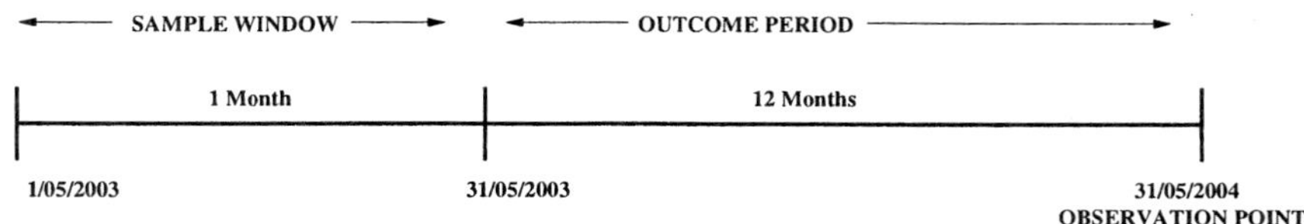

Fig. 1. Data Selection

lapse into arrears or remain healthy. An account was considered *in arrears* (i.e. 'bad') if, at the observation point, the contractual repayment obligations had not been met. Otherwise it was considered *not in arrears*, or 'good'. The data set totalled 1534 accounts consisting of 1471 'good' examples and 63 'bad' examples. The imbalanced nature of the data set was typical across the unsecured loan accounts of the financial institution involved.

23 input attributes were used of which 22 were collected at the time of loan approval and one during the outcome period, reflecting the loan performance. Of these 17 were continuous and 6 discrete. There was no significant correlation between any of the input attributes and the target class except for that collected during the outcome period and even in this case it was weak. The continuous attributes were linearly scaled from 0 to 1 and the discrete attributes were widened and represented as a suitable vector. There was little missing data. There were two target classes. All networks used 46 input neurons and one output neuron. The number of hidden layers and hidden layer neurons varied, depending on the experimental results. The networks were developed using the publically available neural network software *NevProp* and trained using the *quickprop* learning algorithm.

The literature suggests that networks trained on imbalanced data sets of the type used here tend to learn the majority class at the expense of the minority one[7]. A series of preliminary experiments using single networks trained, tested and validated on sets containing a ratio of good to bad examples equal to that in the original data set confirmed this. In arrears management it is important that the classifiers predict well the minority class (i.e. the 'bad' accounts). Several strategies have been suggested to overcome the data imbalance[7] and two (a single *minority over-sampled* network and an ensemble of *majority under-sampled networks*) were used here.

For the minority over-sampled network all majority class examples were retained and the data set was enlarged by sampling each minority class example five times. For each ensemble member all minority class examples were retained and a subset of the majority class, drawn at random, was added. Seven such data sets were created.

In all cases the data sets were subdivided into a training, a testing and a validation set. The proportion of 'good' to 'bad' accounts was 2:1 in each set. Multiple experiments were run to determine the best performing network based on testing set performance, particularly on the classification of 'bad' examples. A validation set was used to provide an estimation of performance on unseen data in the development data set.

4 Experimental Results and Discussion

The training, testing and validation performance of each individual network on the May 2003 – 2004 data is shown in table 1. The minority-oversampled network out performed all individual ensemble members, particularly in the classification of the 'bad' accounts. This is not surprising as the proportion of training and testing examples to the total available examples used during the development of this network was greater than that for the development of each ensemble member.

Table 1. Individual network performance on development (May 2004) data

Ensemble member	Testing % good	Testing % bad	Validation % good	Validation % bad
#1	95	85	88.5	84.6
#2	92.5	80	96	61.5
#3	85	85	57.7	84.6
#4	60	85	88.5	69
#5	85	80	80.8	80
#6	90	80	88.5	80
#7	80	90	65.4	61.5
minority-oversampled network	95	100	93	100

The trained ensemble and minority-oversampled networks were then applied to unseen data viz: personal accounts from June, Nov and Dec 2003 – 2004 (table 2). The proportion of 'good' to 'bad' accounts in these sets was similar to that in the development data set. A simple non-weighted majority voting system was used to determine ensemble performance. The ensemble clearly outperformed the minority-oversampled network in the classification of both 'good' and 'bad' accounts across the three data sets. It also outperformed the average performance of each individual ensemble member. These averages are also shown in the table. These results support the literature observation that the classification performance of an ensemble is superior to that of a single network (in this case that of both a minority-oversampled and a majority-undersampled network)[5,8,16].

The ensemble results also compare more than favourably with those in the analagous part of the study reported in [2]. In this case single networks were used to predict personal loan default after 12 months for a set of accounts from a U.K. financial institution. The minority class (loan default) was over-sampled and the input attributes, although less numerous, were similar to ones used here. The trained network yielded a classification accuracy of 78.8% overall (87.4 % on the good accounts, but only 33 % on the default accounts).

Table 2. Performance of the *ensemble* and the *minority-oversampled* network on unseen data

Observation point	June 2004		Nov 2004		Dec 2004	
	good	bad	good	bad	good	bad
ensemble	97.6	100	89	85	94.3	91.3
minority-oversampled network	83.7	91.7	72.5	63.8	75.7	78.8
ensemble member (average)	(84.8)	(89.9)	(77.7)	(71.8)	(80.7)	(78.8)

5 Conclusion and Future Work

Arrears management involves identifying and tracking high risk customer loan accounts. An ensemble of neural network classifiers shows promise as an accurate classifier for predicting potential personal loan defaults. The results reported here illustrate that ensembles outperform single networks, even when the data set is under or over-sampled. Future work includes the application of these approaches to the construction of systems that investigate the effectiveness of the loan approval process. This may include the identification of rejected loan applications that would possibly not default. Finally the development of single and ensembles of rule based classifiers, in an effort to supply a classification explanation for unsecured lending such as personal loan and credit card accounts, is another possible area for future research.

References

1. M.H.L.B. Abdullah and V. Ganapathy. Neural Network Ensemble for Financial Trend Prediction. In *Proceedings TENCON 2000*, volume 3, pages 157—161, Kuala Lumpar, Malaysia, 2000.
2. Bart Baesens, Tony Van Gestel, Maria Stepanova, and Jan Vanthienen. Neural Network Survival Analysis for Personal Loan Data. In *Proceedings of the Eighth Conference on Credit Scoring and Credit Control (CSCC VIII'2003)*, Edinburgh, Scotland, 2003.

3. Nitesh V. Chawla, Lawrence O. Hall, Kevin Bowyer, and W. Philip Kegelmeyer. Learning Ensembles from Bites: A Scalable and Accurate Approach. *Journal of Machine Learning*, 5:421—451, 2004.
4. V.S. Desai, J.N. Crook, and G.A. Overstreet Jr. A Comparison of Neural Networks and Linear Scoring Models in the credit union environment. *European Journal of Operational Research*, 95:24—37, 1995.
5. Thomas G. Dietterich. Machine-Learning Research: Four Current Directions. *AI Magazine*, 18(4):97—136, 1997.
6. Y. Freund and R. Schapire. Experiments with a new boosting algorithm. In *Proceedings Thirteenth International Conference on Machine Learning*, Bari, Italy, 1996.
7. Hongyu Guo and Herna L. Viktor. Learning from Imbalanced Data Sets with Boosting and Data Generation: The Databoost-IM Approach. *ACM SIGKDD Explorations Newsletter: Special Issue on Learning from Imbalanced Datasets*, 6(1):30—39, June 2004.
8. L.K. Hansen and P. Salamon. Neural Network Ensembles. *IEEE Transactions on Pattern Analysis and Machine Intelligence*, 12(10):993—1001, 1990.
9. E.M Lewis. *An Introduction to Credit Scoring.* The Athena Press, San Rafael, California, 1992.
10. Y. Liu, X. Yao, Q. Zhao, and T. Higuchi. An Experimental Comparison of Ensemble Learning Methods on Decision Boundaries. In *Proceedings 2002 International Joint Conference on Neural Networks, IJCNN '02*, volume 1, pages 221—226, Honolulu, HI USA, 2002.
11. E Mays. *Handbook of Credit Scoring.* Glenlake Publishing, Chicago, 2001.
12. Yair Shimshoni and Nathan Intrator. Classification of Seismic Signals by Integrating Ensembles of Neural Networks. *IEEE Transactions on Signal Processing*, 46(5):1194—1201, 1998.
13. R.P. Srivastva. Automating judgemental decisions using neural network: a model for processing business loan applications. In *Proceedings of the 1992 ACM Conference on Communications*, Kansas City, Missouri, 1992.
14. Nayer M. Wanas and Mohamed S. Kamel. Decision Fusion in Neural Network Ensembles. In *Proceedings 2001 International Joint Conference on Neural Networks, IJCNN '01*, pages 2952—2957, Washington, DC USA, 2001.
15. D. West. Neural Network Credit Scoring Models. *Computer & Operations Research*, 27:1131—1152, 2000.
16. Yunfeng Wu and Juan I. Arribas. Fusing Output Information in Neural Networks: Ensemble Performs Better. In *Proceedings of the 25th Annual Conference of the IEEE EMBS*, pages 2265—2268, Cancum, Mexico, 2003.
17. Xin Yao, Manfred Fischer, and Gavin Brown. Neural Network Ensembles and Their Application to Traffic Flow Prediction in Telecommunications Networks. In *Proceedings 2001 International Joint Conference on Neural Networks, IJCNN '01*, pages 693—698, Washington, DC USA, 2001.

Sequential Classification of Probabilistic Independent Feature Vectors Based on Multilayer Perceptron

Tomasz Walkowiak

Institute of Engineering Cybernetics, Wroclaw University of Technology,
ul. Janiszewskiego 11/17, 50-372 Wroclaw, Poland
Tomasz.Walkowiak@pwr.wroc.pl

Abstract. The paper presents methods of classification based on a sequence of feature vectors extracted from signal generated by the object. The feature vectors are assumed to be probabilistic independent. Each feature vector is separately classified by a multilayer perceptron giving a set of local classification decisions. This set of statistical independent decisions is a base for a global classification rule. The rule is derived from statistical decision theory. According to it, an object belongs to a class for which product of corresponding neural network outputs is the largest. The neural outputs are modified in a way to prevent them vanishing to zero. The performance of the proposed rule was tested in an automatic, text independent, speaker identification task. Achieved results are presented.

1 Introduction

The subject of the paper is an object classification based on signals incoming in time. The real examples of such problem are speaker recognition, recognition of vehicles based on acoustic signals or recognition of vehicles based on seismic signals. The framework of the classification problem with predefined classes is as follows. Certain objects are to be classified as coming from one of a fixed number of classes. Each object gives rise to certain measurements (signal), which could be interpreted as a change of some physical parameter in time. The aim is to classify the object based on measured signal to a given group. One solution of such problem could be a so called long period recognition, which is based on estimating some parameters for long periods of signal. However, we want to focus on the other approach. Like in the classical object recognition we have a pre-processing method which extracts feature vectors from short parts of signal. For each feature vector, representing a short part of signal, we have a classifier - multilayer perceptron [4,11], which generates classification answers. The goal of the paper is to develop the rule which will integrate answers from neural network classifier for n consecutive feature vectors and make a global sequential classification. Moreover, we assume that feature vectors could be treated as realizations of probabilistic independent random variable.

The problem of making a global decision base on a set of local and statistical independent classifications was studied for the long time. In case of pattern recognition it is known as classifier combination (a good review could be found in [10]). The most

W. Duch et al. (Eds.): ICANN 2005, LNCS 3697, pp. 331 – 336, 2005.

common methods includes voting classification [2], the mean rule [7,8], product rule [3], max, min rule [9] or Dempster-Shafer theory [1].

In this paper we will use very similar techniques, but we will apply it to a slightly different task. In most of the applications of classifier combination methods, the classification is based on one feature vector with different kind of classifiers or on different feature representations of the same input signal. However, in the analyzed in this paper independent, speaker recognition task [14] we will have one classifier with consequent inputs for different parts of speech signal.

2 Sequential Decision Schema

Assume that we have some objects from one of 1,...,K classes described by a feature vector $\boldsymbol{x}$ (belonging to the $\Re^d$ space). The process of classification could be understood as taking one of K possible decisions, i.e. the classificatory is a function $c(\boldsymbol{x})$, which has values in set 1...,K. Assuming that 'a posteriori' probabilities of the classes, i.e. conditional probabilities that a given object belongs to a given class under observation $\boldsymbol{x}$, are known (denote it as $\Pr(k|\boldsymbol{x})$), optimal (minimum-error-rate) Bayes theory classifier[4] is based on selection of a class for which the probability $\Pr(k|\boldsymbol{x})$ is the largest. If we assume that the prior probabilities (probabilities that an example is drawn from a given class) are equal, the 'a posteriori' probabilities of the classes could be simply calculated based on probability density function for each class (denote it by $f_k(\boldsymbol{x})$):

$$\Pr(k \mid \boldsymbol{x}) = \frac{f_k(\boldsymbol{x})}{\sum_{i=1}^{K} f_i(\boldsymbol{x})} . \tag{1}$$

Therefore, the Bayes classification rule based on probability density functions is:

$$c(\boldsymbol{x}) = \underset{k \in (1,\ldots,K)}{\arg\max}\, f_k(\boldsymbol{x}) . \tag{2}$$

2.1 Sequential Classification for Probabilistic Independent Feature Vectors

As it was stated in the introduction, we focused in this paper on a case when a sequence of feature vector is probabilistic independent. Therefore, the overall probability density for a given class under a sequence of observations $\boldsymbol{x}_1$,$\boldsymbol{x}_2$,..., $\boldsymbol{x}_n$ could be calculated by multiplying probability densities for each feature vector $\boldsymbol{x}_i$. Therefore, based on (2), the optimal Bayes classifier for the sequence of probabilistic independent feature vector is equal to:

$$c(\boldsymbol{x}_1, \boldsymbol{x}_2 \ldots \boldsymbol{x}_n) = \underset{k \in (1,\ldots,K)}{\arg\max} \prod_{i=1}^{n} f_k(\boldsymbol{x}_i) . \tag{3}$$

2.2 Multilayer Perceptron Multiplication Rule

It is a well known fact, for example in [12], that when a multilayer perceptron is trained in order to minimize the mean square error between the target and the network outputs, it provides after learning estimates of the 'a posteriori' probabilities of the classes. Therefore, the outputs of multilayer perceptron (denote is as $F_k(\boldsymbol{x})$) could substitute the 'a posteriori' probabilities in equation (1) giving a relation between probability density functions for each class and multilayer perceptron outputs:

$$f_k(\boldsymbol{x}) = F_k(\boldsymbol{x}) \cdot \sum_{i=1}^{K} f_i(\boldsymbol{x}) . \tag{4}$$

Substituting it into equation (3) one could achieve:

$$c(\boldsymbol{x}_1, \boldsymbol{x}_2 \ldots \boldsymbol{x}_n) = \underset{k \in (1,\ldots,K)}{\arg\max} \prod_{i=1}^{n} \left(F_k(\boldsymbol{x}_i) \cdot \sum_{l=1}^{K} f_l(\boldsymbol{x}_i) \right) \tag{5}$$

$$= \underset{k \in (1,\ldots,K)}{\arg\max} \left(\prod_{i=1}^{n} F_k(\boldsymbol{x}_i) \cdot \prod_{i=1}^{n} \sum_{l=1}^{K} f_l(\boldsymbol{x}_i) \right)$$

Since the second factor in multiplication inside argmax operator is equal for all k classes, resulting multilayer perceptron sequential classification rule is:

$$c(\boldsymbol{x}_1, \boldsymbol{x}_2 \ldots \boldsymbol{x}_n) = \underset{k \in (1,\ldots,K)}{\arg\max} \prod_{i=1}^{n} F_k(\boldsymbol{x}_i) . \tag{6}$$

3 Speaker Identification Test Bed

Presented in the previous chapter sequential classification rule (6) and its modification (described in next chapters) have been tested on a closed set, text independent, speaker identification task. The population of speakers consists of 15 persons. Each person produced a set of three 10s utterance of text. Text consisted of freely spoken sentences in Polish language [13].

The main assumption of this paper is that feature vectors are probabilistic independent. Therefore, we have used special way of pre-selecting input data [13]. The method is based on selecting only some parts of speech signal (we called it speech events) which are in most cases vowels. Around 95% of speech events lay in the middle of vowels [14]. The sequence of vowels in a freely spoken speech could be treated as a statistical independent, so the feature vectors calculated based on speech signal around these vowels too.

Next, each of selected speech event was preprocessed to achieve feature vectors which are inputs to the multilayer perceptron. The pre-processing was developed on the basis of empirical studies of the human ear. The short-time spectra from a conventional form was converted to perceptual domain. Finally, the logarithm perceptually spectrum was cosine transformed to produce the cepstral coefficients [13]. In performed experiments 14 cepstral coefficients were used to form the feature vector.

4 Classification Results

The multilayer perceptron with 14 input neurons which correspond to cepstral coefficients, 17 hidden layer neurons and 15 output neurons (corresponding to 15 speakers-classes) was used. The number of neurons in the hidden layer was chosen experimentally. Such network produces the best results and enlarging number of neurons in the hidden layer did not give any improvement. The tan-sigmoid was used as a transfer function in the hidden layer and log-sigmoid in the output layer. The network was trained using the Levenberg-Marquardt algorithm [4].

The network gives 60.1% of correct classification for one feature vector for the test set. Next, we used the algorithm (6). The achieved results, i.e. percent of correct recognition, for consecutive features vectors are presented in Table 1, row i. We have found the results not satisfying. As expected, taking into consideration larger number of analyzed feature vectors (i.e. longer text for classification) gives better recognition results, but only for the number of feature vectors not larger then 5.

5 Multilayer Perceptron Summation Rule

The failure of the theoretical multiplication rule (6) in a real sequential recognition task resulted in a search for the other method. In many papers (for example [15]), the sequential recognition using multilayer perception is based on a simple summation rule. According to it, the analyzed sequence of feature vectors belongs to a class for which the sum of corresponding neural outputs is the largest, i.e.:

$$c(\boldsymbol{x}_1, \boldsymbol{x}_2 ... \boldsymbol{x}_n) = \mathop{\arg\max}_{l \in (1,..,K)} \sum_{i=1}^{K} F_l(\boldsymbol{x}_i) \,. \tag{7}$$

The rule is equivalent to a mean rule [8], i.e. a weighted average of the outputs of the individual classifiers or an averaged Bayes classifier [16]. It could be also derived from the assumption that the 'a posteriori' probability for the sequence of feature vector is a mixture of 'a posteriori' probabilities of each feature vector. Therefore, it is similar to mixture model approach of probability density estimation (i.e. [4]). The achieved results are presented in Table 1, row ii.

Table 1. Percent of correct recognition in a function of number of feature vectors for different sequential recognition algorithms: i – multiplication rule (6), ii – summation rule (7), iii - multiplication rule with added threshold (8), iv - multiplication rule with max operator (9)

Rule	Number of consecutive feature vectors									
	2	3	4	5	6	7	8	9	10	11
i	70.2	75.4	76.5	76.6	76.4	75.0	74.3	73.7	71.7	70.1
ii	74.1	82.4	87.1	90.8	92.3	94.6	95.7	96.7	97.7	98.1
iii	75.2	83.3	88.0	91.5	94.0	95.6	96.8	97.8	98.3	98.8
iv	74.4	83.2	88.3	91.6	94.1	95.6	96.7	97.7	98.4	98.8

6 Multiplication Rule with Modified Neuron Outputs

The success of the summing rule, and the failure of the multiplication rule in the reference speaker identification task raises the question why multiplication rule with better theoretical background is not working. We have analyzed the neural network outputs and found that these values sometimes are equal to zero even for neurons corresponding to the correct class. Multiplication of the neuron outputs by zero value in (6) makes the correct recognition impossible. A class for which only one neuron output value for any analyzed feature vector is equal to zero has no chance to be selected as a correct class. Therefore, we propose two modifications of multiplication rule which prevent the neuron outputs vanishing to zero. It could be done simply by adding to all neuron outputs a small constant value (mark it by θ):

$$c(\boldsymbol{x}_1,\boldsymbol{x}_2...\boldsymbol{x}_n) = \underset{l\in(1,..,K)}{\arg\max} \prod_{i=1}^{K} \left(F_l(\boldsymbol{x}_i)+\theta\right). \tag{8}$$

The results of multiplication rule with added threshold in the reference speaker recognition task are presented in Table 1, row iii. The value of threshold was selected experimentally. The achieved results are better then in case of summing rule.

The other way of preventing neuron outputs from vanishing to zero is to use max operator which selects the larger value from the neuron output or some small threshold:

$$c(\boldsymbol{x}_1,\boldsymbol{x}_2...\boldsymbol{x}_n) = \underset{l\in(1,..,K)}{\arg\max} \prod_{i=1}^{K} \max\left(F_l(\boldsymbol{x}_i),\theta\right). \tag{9}$$

Resulting multiplication rule with max operator gives very similar results to the above rule (8). These results could be found in Table 1, row iv.

7 Conclusion and Further Work

We have presented methods of integration of multilayer perceptron outputs to allow the classification of an object based on a sequence of feature vectors. The sequential recognition rule was derived using statistical decision theory (chapter 2) and resulting method, after some heuristic modification (chapter 6), was applied to the speaker identification task. The achieved results are quite good. As expected, taking into consideration larger number of feature vectors (longer text) gives better classification results. The result for 11 consecutive feature vectors, approximately 1.4s of text, is 98.8% of correct classifications, when the result for one feature vector was 60.1%.

The presented sequential rule deduction and achieved experimental results rise several questions. Firstly, there is a need to verify the assumption that the feature vectors are probabilistic independent. We compared the classification results achieved for original sequence of feature vectors with results for the shuffled sequence. They differ in average 1.4%. Therefore, the assumption seems to be justified. However, more formal way, i.e. applying a statistical test for verification independence in time series, is required. We plan to use independence tests based on BDS statistics [5,6]. However, these methods were developed for univariate data and needs some modification to be used for the sequence of multivariate feature vectors.

Secondly, it would very useful to have some methods for setting the optimal values of thresholds in rules described by equations (8) and (9). Not just a simple empirical method - look which values give the best results.

References

1. Al-Ani, A., Deriche, M.: A new technique for combining multiple classifiers using the Dempster-Shafer theory of evidence. Journal of Artificial Intelligence Research **17** (2002) 333-361
2. Bauer, R., Kohavi, R.: An Empirical Comparison of Voting Classification Algorithms: Bagging, Boosting, and Variants, Machine Learning **36** (1999) 105-139
3. Bilmes, J., Kirchhoff, K.: Generalized rules for combination and joint training of classifiers, Pattern Analysis and Applications **6** (2003) 201-211
4. Bishop, Ch. M.: Neural Networks for Pattern Recognition, Clarendon Press, Oxford (1995)
5. Brock, W. A., Dechert, W. D., Scheinkman, J. A. and LeBaron, B.: A test for independence based on the correlation dimension. Econometric Reviews **15** (1996) 197-235
6. Diks, C., Manzan, S.: Tests for Serial Independence and Linearity Based on Correlation Integrals, Studies in Nonlinear Dynamics & Econometrics **6** (2002) no. 2, article 2
7. Hashem, S.: Optimal linear combinations of neural networks. Neural Networks **10** (1997) 599–614
8. Kittler, J., Hataf, M., Duin, R.P.W. and Matas, J.: On combining classifiers. IEEE Transactions on Pattern Analysis and Machine Intelligence **20** (1998) 226–239
9. Kirchhoff, K., Bilmes, J.: Dynamic classifier combination in hybrid speech recognition systems using utterance-level confidence values. Proceedings international conference on acoustic, speech and signal processing (1999) 693–696
10. Kuncheva, L.I.: Combining Pattern Classifiers: Methods and Algorithms. John Wiley, New York (2004)
11. Ripley, B. D.: Pattern Recognition and Neural Networks. University Press, Cambridge (1996)
12. Ruck, D.W., Rogers, S.K., Kabrisky, M., Oxley, M.E. and Suter B.W.: The multilayer perceptron as an approximation to a Bayes optimal discriminant function. IEEE Transactions on Neural Networks **1** (1990) 296-298
13. Walkowiak, T.: Probabilistic Neural Network for Open Set Classification. IV National Conference Neural Networks and their Application, Poland, Zakopane (1999) 232-237
14. Walkowiak, T.: A t-Student based Neural Network for Speaker Recognition. Proceedings of the Fourth International Workshop Neural Networks in Applications, Germany, Magdeburg (1999) 133-139
15. Walkowiak, T., Zamojski, W.: Heuristic sequential classifiers based on multilayer perceptron and probabilistic neural network. 2nd International Conference on Information Technology, Jordan, Amman (2005) 202-206
16. Xu, L., Krzyzak, A., and Suen, C.: Methods of combining multiple classifiers and their applications to handwriting recognition. IEEE Transactions on Systems, Man and Cybernetics **22** (1992) 418–435

Multi-class Pattern Classification Based on a Probabilistic Model of Combining Binary Classifiers

Naoto Yukinawa[1], Shigeyuki Oba[1], Kikuya Kato[2], and Shin Ishii[1]

[1] Nara Institute of Science and Technology,
8916-5 Takayama-cho, Ikoma, Nara 630-0192, Japan
{naoto-yu, shige-o, ishii}@is.naist.jp

[2] Research Institute, Osaka Medical Center for Cancer and Cardiovascular Diseases,
1-3-2 Nakamichi, Higashinari-ku, Osaka 537-8511, Japan
katou-ki@mc.pref.osaka.jp

Abstract. We propose a novel probabilistic model for constructing a multi-class pattern classifier by weighted aggregation of general binary classifiers including one-versus-the-rest, one-versus-one, and others. Our model has a latent variable that represents class membership probabilities, and it is estimated by fitting it to probability estimate outputs of binary classfiers. We apply our method to classification problems of synthetic datasets and a real world dataset of gene expression profiles. We show that our method achieves comparable performance to conventional voting heuristics.

1 Introduction

Pattern classification methods that have been studied in the field of machine learning can be categorized into two. Those of one type have applicability to multi-class classification problems as well as binary classification problems – such as K nearest neighbours method [1], parametric mixture models [2], and Naive Bayes method. On the other hand, those of the other type have been developed in particular for binary classification problems. The most popular one is Support Vector Machine (SVM) [3] which tries to find the optimal hyperplane that separates samples of two classes with a maximum margin.

When applying a method belonging to the latter type to multi-class (M classes) classification problems, we need some devices; the following voting heuristics are frequently used: 1) prepare a set of M binary classifiers, each of which separates one class from the other classes (1–R), then a class is decided by voting the outputs of probability estimates derived by M binary classifiers [4]; and 2) prepare a set of $M(M-1)/2$ binary classifiers, each of which separates one from another class (1–1), then a class is decided by a vote done by them [5][6].

Although the voting methods have weak theoretical background, they demonstrate fairly good performance for problems in which a binary classification subproblem is well performed by a binary classifier like SVM. However, which is

W. Duch et al. (Eds.): ICANN 2005, LNCS 3697, pp. 337–342, 2005.

better to use 1–1 or 1–R is still an unknown problem. A previous study [7] evaluated various methods for multi-class classification problems by using several published datasets of gene expression pattern vectors. They found that SVM-based methods showed overwhelming performance in most cases, but also how to choose a set of binary classifiers was problem-specific.

In this study, we propose a statistical framework for obtaining optimal decision with aggregation of binary classifiers including not only 1–R and 1–1 but also others such as 1–2 and 2–2, in classification problems for more than two classes. Especially when the number of classes is not large, a simple voting procedure like 1–R or 1–1 has no plausibility, and any combination can be considered. To deal with this problem, we propose a probabilistic model for aggregating binary classifiers, in which we assume class membership probabilities of each data point are consistent with set of class membership probabilities of arbitrary binary classifications. This model exhibits a natural voting mechanism by the classifiers.

2 Probabilistic Model of Combining Binary Classifiers

There are N observations $\boldsymbol{L} = \{\boldsymbol{x}^{(n)}, t^{(n)}\}_{1:N}, \boldsymbol{x}^{(n)} \in \mathcal{R}^D, t^{(n)} \in C$, where $\boldsymbol{x}^{(n)}$ and $t^{(n)}$ are the pattern vector and the true class label, respectively, of the n-th sample. $C \equiv \{1, 2, \ldots, M\}$ is a set of M class labels. The objective of a multi-class pattern classification is to predict the class label of an unknown test pattern vector based on the training dataset $\boldsymbol{L}$.

2.1 Unit Binary Classifiers and Class Probability Estimates

We decompose an M-class classification problem into all possible binary classification problems by drawing two subsets of class labels, l and m ($l, m \in \tilde{2}^C, l \cap m = \emptyset$), without overlapping, from the label's power set: $\tilde{2}^C \equiv 2^C - \{\emptyset, C\} = \{\{1\}, \{2\}, \ldots, \{1,2\}, \ldots, \{1,2,3\}, \ldots, \}$. We call a pair of label subsets a "target", represented by $[l|m] \in B$, where B is a set of targets. We also consider some types for the set of targets B. Those are:

Type 1–1 One to one, i.e., $B^{11} = B_{1,1}$
Type 1–R One to the rest, i.e., $B^{1\mathrm{R}} = B_{1,M-1}$
Type 1–A One to a subset in the rest, i.e., $B^{1\mathrm{A}} = \bigcup_{i=1}^{M-1} B_{1,i}$
Type A–A All possible pairs of subsets without overlapping, i.e.,
$$B^{\mathrm{AA}} = \bigcup_{j=1}^{\lfloor M/2 \rfloor} \left(\bigcup_{i=j}^{M-1} B_{j,i} \right)$$

where $B_{j,i} \equiv \{[l|m] \,|\, l, m \in \tilde{2}^C, l \cap m = \emptyset, \#l = j, \#m = i\}$, and $\#l$ and $\#m$ are the numbers of class labels in the subsets l and m, respectively, and $\lfloor \cdot \rfloor$ denotes the floor integer.

Provided that we have a discriminant function $f^{\boldsymbol{L}}_{[l|m]}(\boldsymbol{x}) \in \mathcal{R}$ on a target $[l|m]$, of a binary classification algorithm trained with learning dataset $\boldsymbol{L}$. Let $q_{[l|m]}(\boldsymbol{x}) = \Pr\left(\boldsymbol{t} \in l | f^{\boldsymbol{L}}_{[l|m]}(\boldsymbol{x}), \boldsymbol{t} \in l \cup m\right)$ be the class membership probability after applying a specific method to the discriminant function value; in this study, we use 1-D logistic regression [8] for this conversion.

2.2 Class Probabilities

Let $p_i(\boldsymbol{x}) = \Pr(\boldsymbol{t} = i|\boldsymbol{x}) \in [0,1], \sum_{i \in C} p_i(\boldsymbol{x}) = 1$ describe the membership probability of a pattern vector $\boldsymbol{x}$ to a class label i. We also define the membership probability vector of whole class labels as $\boldsymbol{p}(\boldsymbol{x}) = \{p_i(\boldsymbol{x})\}_{i \in C}$. The membership probability of $\boldsymbol{x}$ to an arbitrary set of class labels $l \in \tilde{2}^C$, $p_l(\boldsymbol{x})$, is given by $p_l(\boldsymbol{x}) = \sum_{i \in l} p_i(\boldsymbol{x})$. If we know $\boldsymbol{p}(\boldsymbol{x})$, the $\boldsymbol{x}$'s class label is decided as $\hat{t} = \text{argmax}_{i \in C}\, p_i(\boldsymbol{x})$; this decision is Bayes optimal if $p_i(\boldsymbol{x})$ gives the posterior probability of the class label i.

2.3 Probabilistic Model of Binary Classifications

In reality, we do not know the true posterior probability $\boldsymbol{p}(\boldsymbol{x})$, but have a set of class membership probabilities $\boldsymbol{q}(\boldsymbol{x}) = \{q_{[l|m]}(\boldsymbol{x})\}_{[l|m] \in B}$, corresponding to the set of binary classifiers, B. The problem here is to set $\boldsymbol{p}(\boldsymbol{x})$ so as to show the best fit to $\boldsymbol{q}(\boldsymbol{x})$. Given a true $\boldsymbol{p}(\boldsymbol{x})$, the true class probability of binary classification on target $[l|m]$, $\pi_{[l|m]}(\boldsymbol{x}) = \Pr(\boldsymbol{t} \in l|\boldsymbol{x}, \boldsymbol{t} \in l \cup m)$ is given by

$$\pi_{[l|m]} = \frac{p_l}{p_l + p_m}. \tag{1}$$

For simplicity, we omit the argument $(\boldsymbol{x})$ in the followings. Then, our objective is to obtain a $\boldsymbol{p}$ so that $\hat{\boldsymbol{\pi}} \equiv \{\hat{\pi}_{[l|m]}\}_{[l|m] \in B}$ shows the best correspondence with $\boldsymbol{q}$. To do this, we use the Kullback-Leibler divergence as similarity measure between $\boldsymbol{q}$ and $\boldsymbol{\pi}(\boldsymbol{p})$, and maximize

$$\begin{aligned} L_0(\boldsymbol{p}) &\equiv -KL(\boldsymbol{q}; \boldsymbol{\pi}(\boldsymbol{p})) \\ &= -\sum_{[l|m] \in B} \left\{ q_{[l|m]} \ln \frac{q_{[l|m]}}{\pi_{[l|m]}} + (1 - q_{[l|m]}) \ln \frac{1 - q_{[l|m]}}{1 - \pi_{[l|m]}} \right\}, \quad \text{w.r.t. } \boldsymbol{p}. \end{aligned} \tag{2}$$

In addition, we introduce a Dirichlet prior $\boldsymbol{p} \sim \text{Dir}(\gamma_0)$ to (2) for regularization where γ_0 is the hyper parameter, then we have a modified objective function:

$$L_1(\boldsymbol{p}) = \sum_{[l|m] \in B} \{q_{[l|m]} \ln p_l + q_{[m|l]} \ln p_m - \ln(p_l + p_m)\} + \sum_{i \in C} \gamma_0 \ln p_i + R, \tag{3}$$

where R is a constant depending on $\boldsymbol{q}$. In the experiments in the later section, we set $\gamma_0 = 1$. The objective function (3) is maximized with respect to $\boldsymbol{p}$ under the condition $\sum_{i \in C} p_i = 1$; this optimization can be performed by the Lagrange method. This model and method are an extension of the Bradley-Terry model for paired (one to one) comparisons [9] and the multi-class classification method by pairwised coupling of probability estimates [5] to for any possible pairs. We call this new probabilistic approach to optimize the membership probability vector $\boldsymbol{p}$ the Optimization of class probabilities (OPT) method.

3 Estimate Class Probabilities Based on Heuristics

To evaluate our proposed method, we use two ways to calculate class probabilities based on existing simple voting heuristics and a modification of it.

Single Summation Method. In the simplest heuristics called Single Summation (SIS) method [4][10], the class probability is given as $p_i = \left\{ \sum_{[l|m] \in B \text{ s.t. } l=i} q_{[l|m]} + \sum_{[l|m] \in B \text{ s.t. } m=i} q_{[m|l]} \right\} / Z$, where Z is a normalization term: $Z = \sum_{i \in C} p_i$. This heuristics sums up probability estimate of binary classification on each target whose subclass has a single class label: $q_{[l|m]}$, where $\#l = 1$ and $\#m = 1$.

Shared Summation Method. We here propose another heuristics called Shared Summation (SHS). In SHS, if a target subclass consists of multiple class labels, the probability estimate output is distributed equally to every class label in the subclass. This allows sets of targets such as B_{22} to join in voting. In contrast, such a target cannot be used in SIS. The class membership probability by SHS is given by $p_i = \left\{ \sum_{[l|m] \in B \text{ s.t. } i \in l} q_{[l|m]} / \#l + \sum_{[l|m] \in B \text{ s.t. } i \in m} q_{[m|l]} / \#m \right\} / Z$, where Z is a normalization term similar to the above.

4 Experiments

4.1 Application to Synthesized Dataset

In order to examine the performance of our OPT method, we first prepared a synthesized dataset; the true distribution was a mixture of four 2-D Gaussians (Fig. 1), consisting of four classes, from which $20 \times 4 = 80$ training data and $40 \times 4 = 160$ test data were generated. We compared the classification performances of each combination of the three voting procedures (OPT, SHS, and SIS) and four types of targets (1–1, 1–R, 1–A and A–A). In this experiment, we used an SVM with a linear kernel $K(\boldsymbol{x}, \boldsymbol{x}') = \boldsymbol{x}^T \boldsymbol{x}'$ as a binary classifier. Table 1 shows

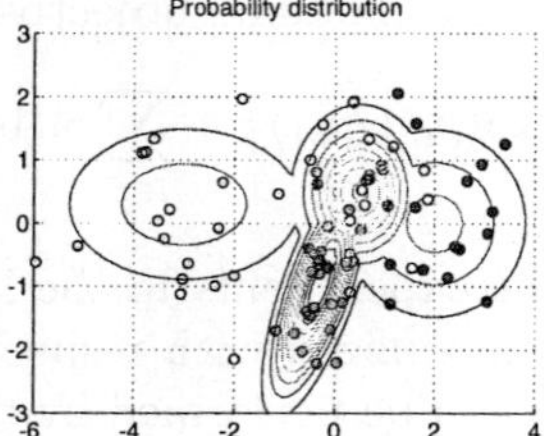

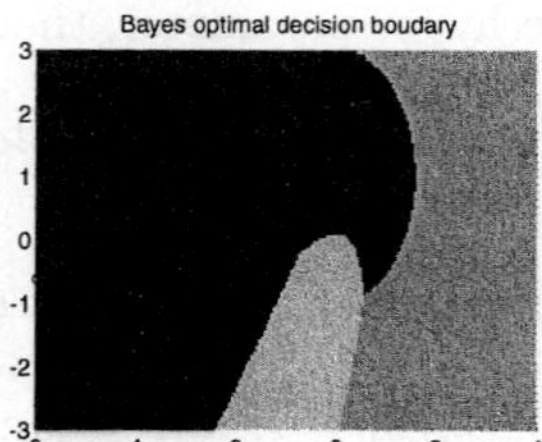

Fig. 1. Synthesized dataset. The left panel represents training data and the true probability distribution. The right panel represents the Bayes optimal decision boundaries of the four classes based on the true distribution.

Table 1. Classification accuracy of the synthesized data

	OPT	SHS	SIS
1–R	0.794	0.787	0.806
1–1	0.819	0.800	0.825
1–A	0.812	0.812	0.812
A–A	0.812	0.812	N/A

the classification result for the test dataset. In each entry, the value represents the classification accuracy.

The binary results show that our OPT method has comparable or even better performance than SHS and SIS, with an exceptional case for 1–R; in 1–R, the high accuracies by SHS and SIS are due to the large number of indeterminate samples. In addition, we can see that the performance of the combination of either of OPT and SHS, and A–A is fairly good; although the A–A combines a lot of classifiers, its performance is better than the conventionally-used 1–R and 1–1, and such a higher-order combination is naturally done by our OPT method.

4.2 Application to a Gene Expression Dataset

As an application to a realistic problem, our method was applied to a tumor classification problem using gene expression profiling data. We used a dataset of gene expressions from four classes (FA, FC, N and PC) of thyroid cancer: 119 training samples consisted of 41 FA, 20 FC, 28 N, and 30 PC, and 49 test samples consisted of 17 FA, 8 FC, 12 N, and 12 PC. Each gene expression was a vector of log-expression ratios of 2,000 genes. We used weighted voting [11][12], a kind of linear discriminator after the gene selection based on the statistical test using signal-to-noise ratio, as a binary classifier, because it has often been used in the field of gene expression analyses. We set the significant level p in the gene selection to 0.001.

Table 2 shows the accuracy by each combination (left part: leave-one-out (LOO) accuracy, right part: test dataset accuracy). From these results, we can see that the accuracies are higher by using a higher-order combination (1–A and A–A) than those by the conventionally-used combination (1–R and 1–1). If we can use classifiers of higher-order combination, the performance does not depend on the combination way, i.e., OPT and SHS show comparable results. SIS may

Table 2. Classification accuracy of the gene expression dataset

	LOO accuracy			test dataset accuracy		
	OPT	SHS	SIS	OPT	SHS	SIS
1–R	0.765	0.765	0.765	0.816	0.816	0.816
1–1	0.782	0.782	0.782	0.816	0.816	0.816
1–A	0.798	0.798	0.798	0.857	0.857	0.857
A–A	0.807	0.807	N/A	0.857	0.857	N/A

be inferior to OPT or SHS when the A–A combination shows the best performance. As for the reason why the results by 1–A and A–A are the same in the independent test case, we consider the classification accuracy has been saturated with 1–A, and additional classifiers in A–A are no more necessary. As a consequence, we have found that there are cases in which a higher-order combination like A–A shows the best performance, and our OPT method shows comparable classification accuracies with the heuristic voting methods, SHS and SIS. SIS is not good, because it cannot deal with higher-order combination (A–A).

5 Conclusion

In this article, we proposed a probabilistic model of binary classifiers for constructing a multi-class classifier, and its estimation method. We showed that our method achieves comparable performance to the heuristics voting methods. We found that there are cases in which higher-order combination of binary classifiers shows the best accuracy, and in such cases, our probabilistic approach to constructing a multi-class classifier exhibits a natural model for the vote by the classifiers. Although the eligibility of each classifier is fixed in this study, its tuning can be done in the framework of probabilistic inference, which is our near future work.

References

1. Duda, R.O., Hart, P.E., Stork, D.G.: Pattern Classification. Wiley Interscience (2000)
2. Oba, S., Sato, M., Ishii, S.: Variational Bayes method for Mixture of Principal Component Analyzers. In: Proceeding for 7th International Conference on Neural Information Processing (ICONIP2000). Volume 2. (2000) 1416–1421
3. V. Vapnik: Statistical Learning Theory. Wiley, NY (1998)
4. B. Schölkopf and C. Burges and V. Vapnik: Extracting support data for a given task. In: Proceedings of the First International Conference on Knowledge Discovery and Data Mining. (1995) 252–257
5. Hastie, T., Tibshirani, R.: Classification by pairwise coupling. In Jordan, M.I., Kearns, M.J., Solla, S.A., eds.: Advances in Neural Information Processing Systems. Volume 10., The MIT Press (1998)
6. B. Schölkopf and C. Burges and A. Smola: Advances in Kernel Methods *Support Vector Learning*. The MIT Press (1999)
7. Li, T., Zhang, C., Ogihara, M.: A comparative study of feature selection and multiclass classification methods for tissue classification based on gene expression. Bioinformatics **20** (2004) 2429–2437
8. Anderson, J.: Logistic discrimination. Biometrika **59** (1972) 19–35
9. Bradley, R.A., Terry, M.E.: Rank Analysis of incomplete block designs, I. The method of paired comparisons. Biometrika **41** (1952) 324–345
10. Tax, D., Duin, R.P.W.: Using two-class classifiers for multi-class classification. In: Proceedings 16th International Conference on Pattern Recognition (ICPR). Volume 2. (2002) 124–127
11. S. Ramaswamy et al:: Multiclass cancer diagnosis using tumor gene expression signatures. Proc Natl Acad Sci U S A **98** (2001) 15149–15154
12. T. R. Golub et al.: Molecular classification of cancer: class discovery and class prediction by gene expression monitoring. Science **286** (1999) 531–537

Evaluating Performance of Random Subspace Classifier on ELENA Classification Database

Dmitry Zhora

Institute of Software Systems, 03187, 40 Glushkov Pr., Kiev, Ukraine
dvz73@bigfoot.com

Abstract. This work describes the model of random subspace classifier and provides benchmarking results on the ELENA database. The classifier uses a coarse coding technique to transform the input real vector into the binary vector of high dimensionality. Thus, class representatives are likely to become linearly separable. Taking into account the training time, recognition time and error rate the RSC network in many cases surpasses well known classification algorithms.

1 Introduction

The random subspace classifier is a generalized version of random threshold classifier, originally suggested in [1]. The model change makes the classifier more competitive when the number of input parameters and training set size increase. An interesting comparative analysis of the coarse coding schemes is provided in [2]. The main advantage of RSC scheme, especially in comparison with one-threshold schemes, is the ability to control the density of binary representation, which allows to regulate informational properties of binary images like correlation, Hamming distance etc. Besides, this scheme allows an effective transformation of multidimensional real data into the binary representation for use e.g. in associative-projective neural networks [3]. The classifier can be considered as a discrete counterpart of the RBF network. The RSC description presented below is used for probabilistic analysis provided in [4].

2 Classifier Architecture

The classifier is designed for the classification of points located within n-dimensional unit cube. Any classification task can be transformed to the classification within a unit cube using the linear transformation. The random subspace classifier, which scheme is shown in the Fig. 1 below, has four neuron layers. The first three layers, i.e. neurons $\mathbf{l}$ and $\mathbf{h}$, layer $\mathbf{A}$ and layer $\mathbf{B}$ make a non-linear transformation of input real vector into the high-dimensional binary vector. The last two neuron layers $\mathbf{B}$ and $\mathbf{C}$ represent usual perceptron with synapse matrix $\mathbf{W}$. Sometimes, the binary layer $\mathbf{B}$ is referred as hidden layer.

Let's designate the number of neurons in the layer $\mathbf{B}$ as N. Each j-th neuron in this layer represents a group of neurons $\mathbf{G}_j$. This group also contains corresponding

W. Duch et al. (Eds.): ICANN 2005, LNCS 3697, pp. 343 – 349, 2005.

afferent neurons in layers $\mathbf{l}$, $\mathbf{h}$ and $\mathbf{A}$, see Fig. 1. Each group of neurons use some η different components of the input real vector, where $1 \le \eta \le n$. The components to be selected are determined by the index $\varphi(i, j)$, where the function φ has values in the range from 1 to n. Here i is an index of a component within the group, it changes from 1 to η, j is a group index and changes from 1 to N.

Let $\mathbf{x} = (x_1, \ldots, x_n)$ is the network input vector. Each component of the input vector with index $\varphi(i, j)$ is submitted to threshold neurons $\mathbf{l}_{ij}$ and $\mathbf{h}_{ij}$. The output signal of neuron $\mathbf{l}_{ij}$ has the value 1 in the case the input signal is greater than the threshold value l_{ij}. In other cases the output of neuron $\mathbf{l}_{ij}$ is equal to 0. Similarly, the output signal of neuron $\mathbf{h}_{ij}$ has the value 1 only in the case it's input signal is not less than h_{ij}. Further, output signal of the neuron $\mathbf{A}_{ij}$ has the value 1 when the neuron $\mathbf{l}_{ij}$ fires and the neuron $\mathbf{h}_{ij}$ has zero output. In all other cases the output of neuron $\mathbf{A}_{ij}$ is 0. In other words, the neuron $\mathbf{A}_{ij}$ fires only in the case the following condition holds $l_{ij} < x_{\varphi(i,j)} < h_{ij}$. The layer $\mathbf{B}$ neurons also may have only 1 or 0 as the output value. The neuron $\mathbf{B}_j$ fires only in the case all afferent neurons $\mathbf{A}_{ij}$ fire, i.e. under condition

$$\forall i \in 1, \ldots, \eta: \; l_{ij} < x_{\varphi(i,j)} < h_{ij}. \tag{1}$$

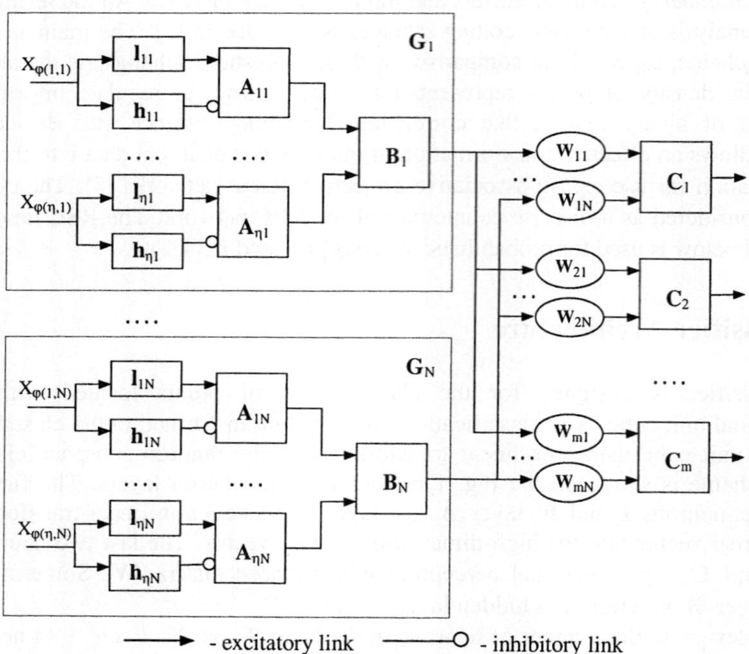

Fig. 1. Random subspace classifier scheme

In fact, the last condition means that the neuron $\mathbf{B}_j$ fires when the input point is located within the space area, limited by hyperplanes $x_{\varphi(i,j)} = l_{ij}$ and $x_{\varphi(i,j)} = h_{ij}$. These hyperplanes are defined for all values of index $i = 1,\ldots,\eta$. The postsynaptic potentials of layer $\mathbf{C}$ neurons are determined according to the following formula:

$$u_k = \sum_{j=1}^{N} W_{kj} B_j . \qquad (2)$$

As soon as the output values of layer $\mathbf{B}$ neurons are calculated, it's possible to obtain the required potentials. The synapse weights W_{kj}, and thereby potentials u_k, are represented by integer values due to the training rules provided below. Finally, the layer $\mathbf{C}$ neuron with maximum postsynaptic potential defines the class, which the classifier refers the input vector to.

3 Generating Classifier Structure

Before training the network, it's needed to generate the classifier structure, which includes subspace index $\varphi(i, j)$ and threshold values l_{ij} and h_{ij} for neurons of the first layer. Originally, the random threshold classifier was suggested as a special case of this model when $\eta = n$ and $\varphi(i, j) = i$. The generalization was done in order to obtain better performance in the case of high-dimensional data. Particularly, when the space dimension is increasing and $\eta = n$, the probability that the layer $\mathbf{B}$ neuron fires is approaching to zero.

Each value $\varphi(i, j)$ is generated as an independent random variable, which can have integer values in the range $1,\ldots,n$ with equal probability. There is a restriction, that the values $\varphi(i, j)$ should be different within the same group, i.e. for the particular value of j. This selection is useful for the applications and is easy to program. Usually, the subspace index values can be considered as independent variables.

The classifier structure also depends on two parameters δ_1 and δ_2, which should be defined in advance, and should satisfy the following conditions: $0 < \delta_2$, $0 \le \delta_1 \le \delta_2$. The threshold values are calculated as

$$l_{ij} = \xi_{ij} - \zeta_{ij}, \; h_{ij} = \xi_{ij} + \zeta_{ij}, \qquad (3)$$

where the center ξ_{ij} is random variable with uniform distribution in the range $[-\delta_2, 1+\delta_2]$, and half-width ζ_{ij} is the random variable with uniform distribution in $[\delta_1, \delta_2]$. It's interesting to compare different approaches for the calculation of the threshold values. All variables ξ_{ij} and ζ_{ij} are independent, hence random variables l_{ij} and h_{ij}, which have different index pairs, are also independent.

The algorithm suggested for the calculation of threshold values in [1] is somewhat different from the described above. Particularly, the following formulas are proposed:

$$l_{ij} = \max(0, \upsilon_{ij} - \omega_{ij}), \; h_{ij} = \min(\upsilon_{ij} + \omega_{ij}, 1). \qquad (4)$$

Here υ_{ij} are independent and uniformly distributed in [0,1], ω_{ij} are independent and uniformly distributed in $[0,\delta_2]$. The approach, suggested in this article, insures that all points in the range [0,1] have equal probability to be located within the interval (l_{ij}, h_{ij}). Thus, the classifier sensitivity is equal for all points of the unit hypercube. At the same time, the presence of threshold values outside the range [0,1] gives the opportunity to classify points outside the hypercube. This feature should be useful for classifier applications. However, in comparison with the algorithm suggested in [1], this approach requires a little more memory for subspace index values $\varphi(i, j)$, assuming the average threshold density is equal in [0,1].

Thus, the classifier architecture depends on four parameters N, δ_1, δ_2 and η. The input and output dimensionalities n and m are defined by the task to be solved. The greater the hidden layer size N the better fitting abilities will be provided by the neural network. Usually this parameter should be selected according to the resource limitations of the computing system. The typical RSC application has tens of thousands neurons in the hidden layer. As shown in [4], for most classifier applications the selection of $\delta_1 = \delta_2 = 0.5$ should be close to optimal. The subspace parameter η should be determined empirically for the particular input data distribution. However, in many cases the selection of η=3,4,5 should be good enough.

4 Network Training

Initially, all synapse weights W_{kj} have zero values. For each vector from the training set, it's needed to find the class, which the classifier refers the input vector to. Reinforcement of connections W_{kj} is done in the case of misclassifications. Let t is an actual class number, and f is a class number, determined by the classifier. In this case, for all values of $j = 1,\ldots,N$, weights should be recalculated according to formulas

$$W'_{tj} = W_{tj} + B_j,\ W'_{fj} = \max(W_{fj} - B_j, 0)\,. \tag{5}$$

Another option is to use the rule "with penalty", where negative weights are allowed and the calculations are as follows:

$$W'_{tj} = W_{tj} + B_j,\ W'_{fj} = W_{fj} - B_j\,. \tag{6}$$

Usually, the first rule provides better success rate on the test set, while the second assures faster convergence. The training procedure is usually repeated until the convergence process is complete, i.e. misclassifications are absent for the whole training set. Sometimes, the training is performed some particular number of epochs. In any case the final synapse matrix $\mathbf{W}$ depends on the order of vectors in the training set.

5 Generating Sensitive Structure

In the basic classifier model the thresholds have a uniform distribution, while the actual input data usually do not. In this case the classifier provides relatively fine presentation for regions with low probability density and relatively poor presentation for regions with a high density. Let's consider one particular component of the input real vector. It is possible to improve RSC performance by generating threshold values with the density proportional to the vector component probability density. Usually the only information about the input set distribution is the input set itself. Let's denote the i-th component of the j-th vector as x_{ij}, and L – the size of the training set. In this case the probability density of the i-th component could be well approximated e.g. by the following function:

$$\psi_i(x) = \sum_{j=1}^{L} \frac{e^{L(x-x_{ij})}}{(1+e^{L(x-x_{ij})})^2}. \tag{7}$$

Such approximation of the probability distribution is similar to the method employed for probabilistic neural networks [5]. The probability function of this distribution is

$$F_i(x) = \int_{-\infty}^{x} \psi_i(\tau)d\tau = \frac{1}{L}\sum_{j=1}^{L} \frac{e^{L(x-x_{ij})}}{1+e^{L(x-x_{ij})}}. \tag{8}$$

As shown in [4], in order to obtain thresholds satisfying the required distribution, it's needed to generate them using the usual procedure in one space X' and then transform values into required space X using the inverse transform function $F^{-1}(x')$. The classifier structure obtained using this transformation will be referred as sensitive.

6 Testing on ELENA Classification Database

Enhanced Learning for Evolutive Neural Architecture project is presented in [6,7]. Each dataset provided with detailed description of properties like number of samples, fractal dimensionality, inertia of principal components, confusion matrix etc.

The RSC network had the following configuration: number of groups $N = 32768$, subspace parameter $\eta = \min(3,n)$, half-width parameters $\delta_1 = \delta_2 = 0.5$. The error rate for each dataset was estimated using the results of five independent experiments. In order to reduce the bias provided by particular sample distribution, each time the dataset was rotated by moving the first 20% of samples to the end of the dataset. The first half of the resulting set was used as the training set, while the second half – as the test set. The classifier had the sensitive structure generated using the training set only.

The datasets *Clouds*, *Concentric* and *Gaussian* were artificially generated, so the probability density distributions and Bayesian decision are available. The real datasets *Iris*, *Phoneme*, *Satimage* and *Texture* were obtained from different application areas and the underlying distributions are not known. The real datasets were preprocessed, in particular, *CR* designates the centering to zero mean and reducing to unit variance, *PCA* – principal component analysis applied to the *CR* dataset, *DFA* – discriminant factorial analysis applied to the *PCA* dataset. The experimental results are presented in the Table 1. The approximate error rates for classifiers KNN, MLP and IRVQ in their best configurations [6,7] are presented for comparison purposes.

Table 1. Error rates and average number of epochs for RSC on ELENA datasets

Dataset	n	m	RSC				KNN	MLP	IRVQ
			Min.	Average	Max.	Epochs	Average		
Clouds	2	2	13.04	14.25	15.44	1696	11.8	12.2	11.7
Concentric	2	2	0.64	1.17	1.60	75.8	1.7	2.8	1.5
Gaussian 2D	2	2	33.56	35.67	37.76	2993	27.4	26.8	27.2
Gaussian 3D	3	2	28.60	29.42	30.60	468.4	22.2	22.4	22.6
Gaussian 4D	4	2	23.44	24.34	24.64	248.6	19.4	19.5	18.5
Gaussian 5D	5	2	19.00	19.79	21.00	171.6	17.7	17.4	15.3
Gaussian 6D	6	2	17.36	17.74	18.36	116.8	16.8	16.4	13.5
Gaussian 7D	7	2	14.36	15.30	16.68	102.0	15.9	15.3	11.5
Gaussian 8D	8	2	12.96	13.60	14.40	80.4	16.1	14.5	9.9
Iris	4	3	1.33	4.53	8.00	5.4	n/a	n/a	n/a
Iris CR	4	3	2.67	4.53	6.67	6.6	4.0	4.3	6.7
Iris PCA	4	3	4.00	8.27	12.00	4.2	n/a	n/a	n/a
Phoneme	5	2	12.47	12.89	13.58	207.2	12.3	n/a	n/a
Phoneme CR	5	2	11.88	12.50	13.40	184.4	12.3	16.3	16.4
Phoneme PCA	5	2	12.66	13.18	14.43	221.8	n/a	n/a	n/a
Satimage	36	6	9.73	10.15	10.60	82.4	n/a	n/a	n/a
Satimage CR	36	6	9.42	9.95	11.04	86.0	9.9	12.3	11.4
Satimage PCA	18	6	11.04	11.77	13.15	42.2	9.6	11.6	11.4
Satimage DFA	5	6	12.15	13.31	13.86	150.0	n/a	11.8	11.4
Texture	40	11	1.60	1.82	2.04	19.0	n/a	n/a	n/a
Texture CR	40	11	1.24	1.84	2.44	20.4	1.9	2.0	3.1
Texture PCA	18	11	0.87	1.29	1.64	11.2	2.0	1.2	3.0
Texture DFA	10	11	0.55	0.88	1.20	10.0	0.4	1.4	0.4

7 Discussion

The RSC network provides good or sometimes the best results on real databases, comparing to the seven classifier models considered in [6,7]. Also RSC gives the best error rate on the *Concentric* dataset. The results for *Clouds* and *Gaussian* datasets, which have highly overlapping probability densities are poor. Despite the Bayesian decision surface is simple, the early stopping on average does not provide better results. However, the relative error rates for *Gaussian* datasets become better with increasing input space dimensionality. Only two classifiers exceed RSC for *Gaussian 8D*. The classifier performance may be improved by adjusting it's configuration parameters for the particular classification task.

References

1. Kussul, E.M., Baidyk, T.N., Lukovich, V.V., Rachkovskij, D.A.: Adaptive high performance classifier based on random threshold neurons. Cybernetics and Systems'94, Ed. R. Trappl. Singapore: World Scientific Publishing Co. Pte. Ltd., (1994) 1687-1695
2. Kussul, E.M., Rachkovskij, D.A., Wunsch, D.C.: The random subspace coarse coding scheme for real-valued vectors. Proc. Int. Joint Conf. Neural Networks 1999, Vol. 1, 450-455

3. Rachkovskij, D.A., Kussul, E.M.: Binding and normalization of binary sparse distributed representations by context-dependent thinning. Neural Computation, Vol. 13, n. 2, (2001) 411-452
4. Zhora, D.V.: Random threshold classifier functioning analysis. Cybernetics and System Analysis, Vol. 3, Kiev (2003) 72-91, in Russian. Available: http://rsc.netfirms.com/
5. Duda, R.O., Hart, P.E., Stork, D.G.: Pattern Classification. 2-nd ed. Wiley Interscience (2000) 654
6. Aviles-Cruz, C., Guérin-Dugué, A., Voz, J.L., Van Cappel, D.: Databases, Enhanced Learning for Evolutive Neural Architecture. Tech. Rep. R3-B1-P, INPG, UCL, TSA (1995) 47. Available: http://www.dice.ucl.ac.be/neural-nets/Research/Projects/ELENA/elena.htm
7. Blayo, F., Cheneval, Y., Guérin-Dugué, A., Chentouf, R., Aviles-Cruz, C., Madrenas, J., Moreno, M., Voz, J.L.: Benchmarks, Enhanced Learning for Evolutive Neural Architecture. Tech. Rep. R3-B4-P, INPG, EERIE, EPFL, UPC, UCL (1995) 114

A New RBF Neural Network Based Non-linear Self-tuning Pole-Zero Placement Controller

Rudwan Abdullah, Amir Hussain, and Ali Zayed

Department of Computing Science and Mathematics,
University of Stirling, FK9 4LA, Scotland
{raa, ahu, asz}@cs.stir.ac.uk

Abstract. In this paper a new self-tuning controller algorithm for non-linear dynamical systems has been derived using the Radial Basis Function Neural Network (RBF). In the proposed controller, the unknown non-linear plant is represented by an equivalent model consisting of a linear time-varying sub-model plus a non-linear sub-model. The parameters of the linear sub-model are identified by a recursive least squares algorithm with a directional forgetting factor, whereas the unknown non-linear sub-model is modelled using the (RBF) network resulting in a new non-linear controller with a generalised minimum variance performance index. In addition, the proposed controller overcomes the shortcomings of other linear designs and provides an adaptive mechanism which ensures that both the closed-loop poles and zeros are placed at their pre-specified positions. Example simulation results using a non-linear plant model demonstrate the effectiveness of the proposed controller.

1 Introduction

During the past three decades, a great deal of attention has been paid to the problem of designing pole-placement controllers and minimum variance controllers[1, 2, 3]. Various self-tuning controllers based on classical pole-placement and minimum variance ideas have been developed and employed in real applications, e.g. [1, 2, 3, 4].

Comparatively, only little attention has been given to zeros since they are considered to be less crucial than poles. Most of the previous discussions on zeros are centred around the choice of the sampling time so that the resulting system is invertible. However, it is important to note that zeros may be used to achieve better set point tracking [5], and they may also help reduce the magnitude of the control action [6].

In this paper a control algorithm is proposed which builds on the works of Zayed *et al.* [7, 8] and Zhu and Warwick [4]. In the proposed controller an unknown non-linear plant is represented by an equivalent model consisting of a linear time-varying sub-model plus a non-linear sub-model. Models of this type have previously been shown to be particularly useful in an adaptive pole-placement based control framework by Zhu and Warwick [4]. In this work, following [7, 8] the non-linear controller is designed to incorporate a zero-pole placement structure, in which the parameters of the linear sub-model are identified by a standard recursive algorithm. The non-linear

W. Duch et al. (Eds.): ICANN 2005, LNCS 3697, pp. 351 – 357, 2005.

sub-model is now detected using a Radial-Basis Function neural network since the RBF network is more capable of implementing arbitrary nonlinear transformations of the input space [9].

The paper is organised as follows: the derivation of the control law is discussed in section 2. In section 3, a simulation case study is carried out in order to demonstrate the effectiveness of the proposed controller. Finally, some concluding remarks are presented in section 4.

2 Derivation of the New Control Law

In deriving the control law it is considered that the plant being investigated can be described by: [4]

$$A(z^{-1})y(t) = z^{-k} B(z^{-1})u(t) + f_{0,t}(Y,U) . \tag{1}$$

Where $y(t)$ and $u(t)$ are respectively the measured output and the control input at the sampling instant t, $k=1$ is the integer-sample dead time of the process, and $f_{0,t}(Y,U)$ is potentially a nonlinear function, where the length of its inputs Y and U depend on the order of the plant model. Therefore the equivalent model is a combination of a linear time varying sub-model plus a non-linear sub-model as shown in figure(1). The polynomials $A(z^{-1})$ and $B(z^{-1})$ are respectively of orders n_a and n_b. The polynomial $A(z^{-1})$ is assumed to be monic. In what follows, the z^{-1} notation will be omitted from the various polynomials to simplify the presentation.

The generalised minimum variance controller minimises the following cost function:

$$J = E\{(Py(t+k) + Qu(t) - Rw(t) - Hf_{0,t}(.,.))^2\} . \tag{2}$$

Where $w(t)$ is the set point and $P(z^{-1})$, $Q(z^{-1})$, $R(z^{-1})$ and $H(z^{-1})$ are user-defined transfer functions in the backward shift operator z^{-1}. $E\{.\}$ is the expectation operator.

The control law which minimises J is [8]:

$$u(t) = \frac{[Rw(t) - (\frac{F'}{P_d})y(t) + (H - E')f_{0,t}(.,.)]}{(E'B + Q)} . \tag{3}$$

Where E' is of order $(k-1)$, and the order of F' is $(n_a + n_{P_d} - 1)$ where $(P = \frac{P_n}{P_d})$ and n_{P_d} is the order of P_d.

The polynomials E' and F' are obtained by solving the least square identity

$$CP_n = AP_d E' + z^{-k} F' . \tag{4}$$

Equation (3) can also be expressed as:

$$u(t)=\frac{[R^{'}w(t)-F^{'}y(t)+H^{'}f_{0,t}(.,.)]}{q}. \tag{5}$$

Where

$$\left.\begin{aligned} R^{'} &= P_d R \\ q &= P_d E^{'} B + Q P_d \\ H^{'} &= P_d (H - E^{'}) \end{aligned}\right\}. \tag{6}$$

$R^{'}$, q and $H^{'}$ are still user defined transfer functions since they depend on the transfer functions R, Q and H, respectively.

We further assume that q can also be expressed as [7, 8]:

$$q=\Delta[\frac{q_n}{q_d}]. \tag{7}$$

2.1 Pole Zero-Placement Design

If we substitute for $u(t)$ given by equation (5) into the process model given by equation (1) and make use of equations (6) and (7), we obtain [7, 8]

$$(q_n\Delta A+z^{-k}q_d BF^{'})y(t)=z^{-k}q_d BR^{'}w(t)+(z^{-k}H^{'}Bq_d+\Delta q_n)f_{0,t}(.,.). \tag{8}$$

It is obvious from equation (8) that at steady state the output $y(t)$ equals the constant reference command $w(t)$ when:

$$R^{'}=F^{'}\Big|_{z=1} \text{ and } H^{'}=\Delta=(1-z^{-1}). \tag{9}$$

If we assume $\overline{B}=q_d B$ and $\overline{A}=\Delta A$, then equation (8) becomes:

$$(q_n\overline{A}+z^{-k}F^{'}\overline{B})y(t)=z^{-k}q_d BR^{'}w(t)+\Delta(z^{-k}Bq_d+q_n)f_{0,t}(.,.). \tag{10}$$

We can now introduce the identity:

$$(q_n\overline{A}+z^{-k}F^{'}\overline{B})=T. \tag{11}$$

Where T and q_d are respectively the desired closed loop poles and zeros, and q_n is the controller polynomial. For equation (11) to have a unique solution, the order of the regulator polynomials and the number of the desired closed loop poles have to be [7, 8]:

$$\left.\begin{aligned} n_f &= n_{\overline{a}}-1 \\ n_{q_n} &= n_{\overline{b}}+k-1 \\ n_T &\le n_{\overline{a}}+n_{\overline{b}}+k-1 \end{aligned}\right\}. \tag{12}$$

Where $n_{\overline{a}}$, $n_{\overline{b}}$, n_{q_d} and n_{q_n} are the orders of the polynomials $\overline{A}$, $\overline{B}$, q_d and q_n, respectively, n_T and k denote the number of desired closed loop poles and the time delay. Also, $n_{\overline{b}} = n_{q_d} + n_b$ and $n_{\overline{a}} = n_a + 1$.

The pole-zero placement can be achieved by assuming that:

$$q = \Delta q_n \text{ and } R' = q_d R_0' = q_d F'(1)[q_d(1)]^{-1} . \tag{13}$$

Where q_d is the desired closed loop zeros.

By using (5), (6) and (13) the closed loop system becomes:

$$(q_n \overline{A} + z^{-k} F' B)y(t) = z^{-k} B q_d R_0' w(t) + \Delta(z^{-k} B + q_n) f_{0,t}(.,.) . \tag{14}$$

Now we can introduce the identity:

$$(q_n \overline{A} + z^{-k} F' B) = T . \tag{15}$$

It is obvious that the order of the polynomial q_n becomes:

$$n_{q_n} = n_b + k - 1 . \tag{16}$$

It is clear from equation (14) and (15) that both closed loop poles and zeros are now at their desired positions.

As can be seen in figure (1), a recursive least squares algorithm is initially used to estimate the parameters A and B (equation (1)) of the linear sub-model. Then an Radial-Basis Function (RBF) neural network is used to approximate the non-linear sub-model $f_{0,t}$.

The desired non-linear output function $\tilde{x}(t)$ is detected as follows:

$$\tilde{x}(t) = y(t) - \tilde{y}(t) = y(t) - \varphi^T(t)\hat{\theta}(t) . \tag{17}$$

where $\hat{\theta}$ is the parameter vector, $\varphi \in \Re^m$ is the data factor and $\tilde{y}(t)$ is the output of the RLS estimator. In equation (17), the output $\tilde{x}(t)$ is considered as the target output for the RBF network, as shown in figure(1) and equation (19). The actual output of the nonlinear RBF-based sub-model $f_{0,t}$ is achieved using a multivariate Gaussian function incorporating Delta rule to update the hidden layer weights. Equations (18-21) illustrate the online learning procedure of the neural network

$$f_{0,t}(.,.) = \frac{1}{1 + \exp[-\beta(\sum_{j=1}^{n} w_j g_j + b)]} . \tag{18}$$

$$\delta_w = \eta g_{j,t} f_{0,t} [1 - f_{0,t}][\tilde{x}(t) - f_{0,t}] . \tag{19}$$

$$w_j(t) = w_j(t-1) + \delta_w . \tag{20}$$

$$g_{0,t}(.,.) = \exp\left(-\sum_{i=1}^{l} \frac{\left\| x_i - c_j^i \right\|^2}{\left(2\sigma_j^i\right)^2} \right). \tag{21}$$

Where w_j is the hidden layer weights, β is the output layer threshold, δ_w is the change in weights, η is the learning rate, x_i is the inputs, c_j^i is the mean, $l = n_a + n_b$, and g_j is the output of the hidden layer. The variance of the Gaussian units σ_j^i is dependent on the input dimension because the RBF inputs are scaled differently [10].

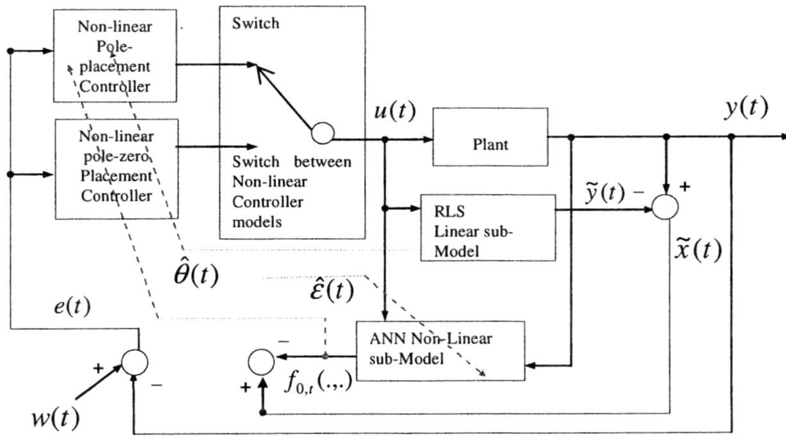

Fig. 1. Control system structure

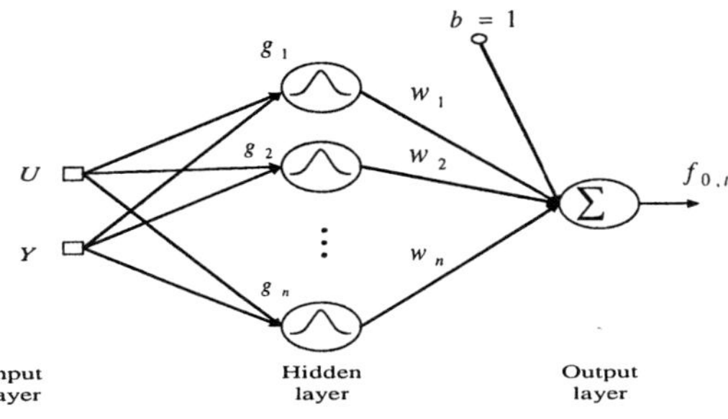

Fig. 2. ANN model to approximate NL function

3 Simulation Results

The objective of this section is to study the performance and the robustness of the closed loop system using the technique proposed in section 2.1. A simulation case study will be carried out in order to demonstrate the ability of the proposed algorithm to locate the closed-loop poles and zeros at their desired locations under set point changes. The simulation example was performed over 600 samples with the set point changing every 100 sampling instants.

Consider the following second-order non-minimum phase and unstable process treated previously by Zhu and Warwick [4]. The set point changes from 10 to 20.

$$y(t) = \frac{1.5\sin[y(t-1)]y(t-1)}{1+y(t-1)^2+y(t-2)^2} + 1.1y(t-1) - 1.2u(t-1) + 2u(t-2)\,. \tag{22}$$

The desired closed loop poles polynomial and zeros were selected as follows: $T = 1-0.65z^{-1}$ and $q_d = 1+0.7z^{-1}$. In order to see clearly the effect of the zeros in the performance of the closed loop system, the controller is arranged to work as either a pole-placement or pole-zero placement controller as follows:

a) From (0th up to 250th sampling time) only the pole placement controller is on-line.
b) The pole-zero placement controller is switched on from (251st to 600th sampling time). The output and the control input are shown in the figures (3a) and (3b).

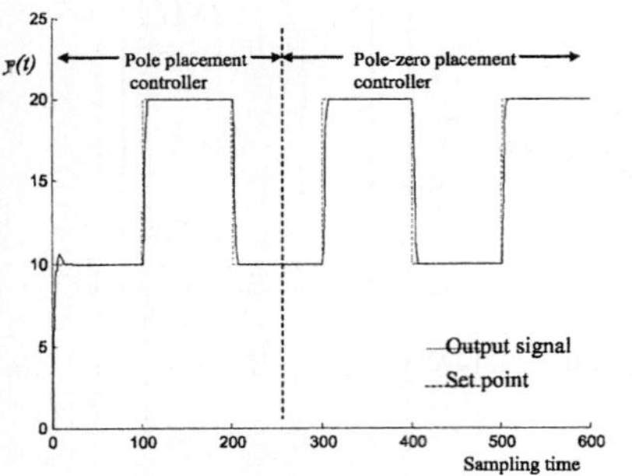

Fig. 3a. The output signal

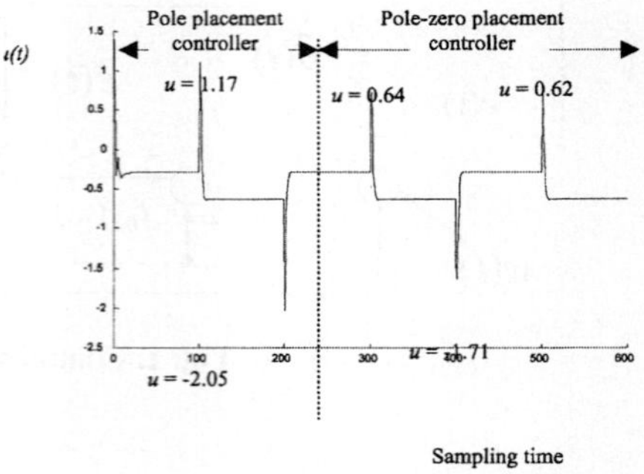

Fig. 3b. The control input signal

It is clear from the figures (3a) and (3b) that, the transient response is shaped by the choice of the polynomial T and ensures steady state error to zero. It is also obvious that excessive control action, which resulted from set-point changes, is tuned after using zero-pole placement from the sampling interval 251, onwards.

4 Conclusions

In this paper, an algorithm to extend the generalised minimum variance stochastic self-tuning controller for non-linear plants has been proposed incorporating the (RBF) network. The unknown non-linear plant is represented by an equivalent model composed of a simple linear sub-model plus a non-linear sub-model. The parameters of the linear sub-model are identified by a standard recursive least squares algorithm, whereas the non-linear sub-model is approximated using RBF neural network. The resulting self-tuning controller provides an adaptive mechanism, which ensures that *both* the closed loop poles and zeros are located at their pre-specified positions. The design was successfully tested on simulated non-linear model. The results presented here indicate that the controller tracks set point changes with the desired speed of response, penalises the excessive control action, and can deal with non-minimum phase systems. The transient response is shaped by the choice of the pole polynomial $T(z^{-1})$, while the zero polynomial $q_d(z^{-1})$ can be used to reduce the magnitude of control action or to achieve better set point tracking [7, 8]. In addition, the controller has the ability to ensure zero steady state error.

In future work, closed loop stability analysis will be carried out and the performance of the RBF based pole-zero placement controller need to be compared with other linear and non-linear neural network based controllers.

References

1. D. W. Clarke and P. J. Gawthrop. "Self-tuning control," *Proc. Inst. Electr. Engineering,* Part D, vol. 126, pp. 633-640, 1979
2. K. J. Astrom and B. Wittenmark. "On self-tuning regulators," *Automatica* , 9, pp. 185-199,1973.
3. A. Y. Allidina and F. M. Hughes, "Generalised self-tuning controller with pole assignment," Proc. Inst. Electr. Engineering, Part D, vol. 127, pp. 13-18, 1980.
4. Q. Zhu, Z. Ma and K. Warwick, "Neural network enhanced generalised minimum variance self-tuning controller for nonlinear discrete-time systems," IEE Proc. Contol theory Appl., vol. 146, pp. 319-326, 1999.
5. A. Zayed, A. Hussain and L. Smith, A New Multivariable Generalised Minimum-variance Stochastic Self-tuning with Pole-zero Placement, International Journal of Control and Intelligent Systems, 32 (1) (2004) 35-44.
6. H. R. Sirisena and F. C. Teng, "Multivariable pole-zero placement self-tuning controller," Int. J. Systems Sci., vol. 17, No.2 pp. 345-352, 1986.
7. A. Zayed, Hussain A and M. Grimble, Novel non-linear Multiple-Controller framework for Complex Systems incorporating a Learning sub-model, invited paper, to appear, Int. J. of Control & Intelligent Systems, Special Issue, 33, (2005).
8. A. Zayed, A. Hussain and L. Smith, A New Non-linear Self-tuning pole-zero placement incorporating Neural Networks, Proceedings of the third International NASO Symposium on Engineering of Intelligent Systems Conference, Malaga Spain, 24-27 Dec 2002, pp 107.paper no 130 in CD.
9. S. Haykin, Neural Networks A Comprehensive Foundation, (Macmillan 1994).
10. K. Passino, Biomimcry for Optimization, Control, and Automation, (Springer 2005).

Using the Levenberg-Marquardt for On-line Training of a Variant System

Fernando Morgado Dias[1], Ana Antunes[1], José Vieira[2], and Alexandre Manuel Mota[3]

[1] Escola Superior de Tecnologia de Setúbal do Instituto Politécnico de Setúbal, Departamento de Engenharia Electrotécnica, Campus do IPS, Estefanilha, 2914-508 Setúbal, Portugal
Tel.: +351 265 790000
{aantunes, fmdias}@est.ips.pt

[2] Escola Superior de Tecnologia de Castelo Branco, Departamento de Engenharia Electrotécnica, Av. Empresário, 6000 Castelo Branco, Portugal
Tel.: +351 272 330300
zevieira@est.ipcb.pt

[3] Departamento de Electrónica e Telecomunicações, Universidade de Aveiro, 3810 - 193 Aveiro Portugal
Tel.: +351 234 370383
alex@det.ua.pt

Abstract. This paper presents an application of the Levenberg-Marquardt algorithm to on-line modelling of a variant system. Because there is no iterative version of the Levenberg-Marquardt algorithm, a batch version is used with a double sliding window and Early Stopping to produce models of a system whose poles change during operation. The models are used in a Internal Model Controller to control the system which is held functioning in the initial phase by a PI controller.

1 Introduction

On-line learning is usually required for time-variant systems. Most real systems vary slowly because of the degradation of their components. Nevertheless, in the present work it has been decided to use an artificial system that has an abrupt change in its characteristics, namely the position of its poles.

On-line learning usually requires an iterative algorithm. For the Levenberg-Marquardt algorithm a true iterative version hasn't been proposed yet, so the present work uses this algorithm in a batch version through the use of a double sliding window with Early Stopping.

The difficulties of implementing the algorithm in an iterative version come from computing the derivatives for the Hessian matrix, inverting it and computing the region for which the approximation contained in the matrix is valid (the trust region).

The solution proposed is tested with a cruise control system with variant characteristics. The location of the pole changed in the middle of the experiment to test the capabilities of adaptation of the solution proposed.

W. Duch et al. (Eds.): ICANN 2005, LNCS 3697, pp. 359–364, 2005.

2 Review of the Algorithm

In this section, a short review of the Levenberg-Marquardt algorithm is done to enable easier perception of the problems found in the on-line implementation. Equation 1 shows the updating rule for the algorithm where x_k is the current iteration, $v(x)$ is the error between the output obtained and the pretended output, $J(x_k)$ is the Jacobian of the system at iteration k and $2.J^T(x_k).J(x_k) + \mu_k I$ is the Hessian matrix approximation used, where I is the identity matrix and μ_k is a value (that can be changed in each iteration) that makes the approximation positive definite and therefore allowing its inversion.

$$\Delta x_k = -\left[2.J^T(x_k).J(x_k) + \mu_k I\right]^{-1} .2.J^T(x_k).v(x_k) \tag{1}$$

The Levenberg-Marquardt algorithm is due to the independent work of both authors in [1] and [2].

The parameter μ_k is the key of the algorithm since it is responsible for stability (when assuring that the Hessian can be inverted) and speed of convergence. It is therefore worth to take a closer look on how to calculate this value.

The modification of the Hessian matrix will only be valid in a neighbourhood of the current iteration. This corresponds to search for the correct update of the next iteration x_{k+1} but restricting this search to $|x - x_k| \leqslant \delta_k$.

There is a relationship between δ_k and μ_k since raising μ_k makes the neighbourhood δ_k diminish [3]. As an exact expression to relate these two parameters is not available, many solutions have been developed.The one used in the present work was proposed by Fletcher [3] and uses the following expression:

$$r_k = \frac{V_N(x_k) - V_N(x_k + p_k)}{V_N(x_k) - L_k(x_k + p_k)} \tag{2}$$

to obtain a measure of the quality of the approximation. Here V_N is the function to be minimized, L_k is the estimate of that value calculated from the Taylor series of second order and p_k is the search direction, in the present situation, the search direction given by the Levenberg-Marquardt algorithm.

The value of r_k is used in the determination of μ_k in an iterative way[3]

3 On-line Version

As pointed out before, the difficulties come from computing the derivatives for the Hessian matrix, inverting this matrix and computing the trust region, the region for which the approximation contained in the calculation of the Hessian matrix is valid.

In the literature, some attempts to build on-line versions can be found, namely the work done by Ngia [4] developing a modified iterative Levenberg-Marquardt algorithm which includes the calculation of the trust region and the work in [5] which implements a Levenberg-Marquardt algorithm in sliding window mode for Radial Basis Functions.

3.1 A Double Sliding Window Approach with Early Stopping

In the present work two sliding windows are used, one for the training set and another for the evaluation set with all the data being collected on-line. The Early Stopping technique [6], [7] is used for avoiding the overfitting problem because it is almost mandatory to employ a technique to avoid overtraining when dealing with systems that are subject to noise. The Early Stopping technique consists of stopping training when the test error starts to increase with the training iterations and was chosen because it has less computational burden then other solutions.

The use of two sliding windows will introduce some difficulties since both data sets will be changing during training and evaluation phases. For these two windows it is necessary to decide their relative position. In order to be able to perform Early Stopping in a valid way, it was decided to place the windows in a way that the new samples will go into the test window and the samples that are removed from the test set will go in to the training set.

The procedure used for the identification of the direct model on-line is represented in figure 1.

Training stars before collecting all the samples for both windows to save some time. For stability the test window has always the same size while the training window is growing in the initial phase.

The windows may not change in each training iteration since all the time between sampling is used for training which may permit several training epochs before a new sample is collected. But each time the composition of the windows is changed the test and training errors will probably be subjected to an immediate change that might be interpreted as an overtraining situation. The Early Stopping technique is here used in conjunction with a measure of the best model that is retained for control. Each time there is a change in the windows the values of the best models (direct and inverse) must be re-evaluated because the previous ones, obtained over a different test set, are no longer valid for a direct comparison.

After each epoch the ANN is evaluated with a test set. The value of the Mean Square Error (MSE) obtained is used to perform Early Stopping and to retain the best models.

The conditions for overtraining and the maximum number of epochs are then verified. If they are true, the Flag, which indicates that the threshold of quality has been reached, will also be verified and if it is on, the training of the inverse model starts, otherwise the models will be reset since new models need to be prepared. Resetting here means that the model's weights are replaced by random values between -1 and 1 as in the initial models. After testing these conditions, if they are both false, the predefined threshold of quality will also be tested and if it has been reached the variable Flag will be set to on. In either case the remaining time of the sampling period is tested to decide if a new epoch is to be performed or if a new sample is to be collected and training is to be performed with this new sample included in the sliding window.

The procedure for the inverse model is very similar and almost the same block diagram could be used to represent it. The on-line training goes on switching from direct to inverse model each time a new model is produced. The main difference between the procedure for direct and inverse model lies in the evaluation step. While the direct model is evaluated with a simple test set, the inverse model is evaluated with a control simulation corresponding to the hybrid Direct/Specialized approach for generating inverse models [8]. During the on-line training the NNSYSID [9] and NNCTRL [10] toolboxes for MATLAB were used.

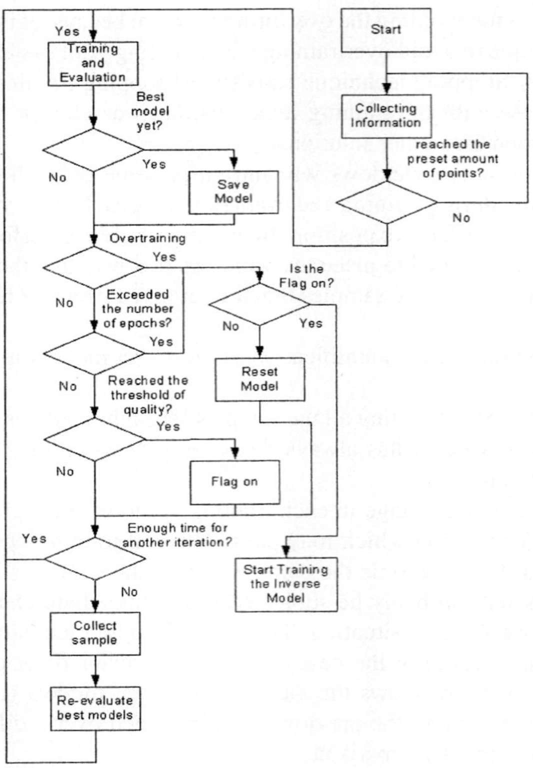

Fig. 1. Block diagram for the identification of a direct model on-line

4 Time Variant System

The time variant system used for this test is a cruise control with a variable pole position according to the following equation:

$$\begin{cases} \frac{0.05}{s+0.05} \; if \; sample \leqslant 500 \\ \frac{0.15}{s+0.15} \; if \; sample > 500 \end{cases} \tag{3}$$

with this change in the pole position a new system is obtained and a new set of models must be prepared. The system used is rather simple but the variance introduced allows testing the functionality of the algorithm proposed.

5 Results

The test sequence is composed of 100 points, the sliding window used for training has a maximum of 200 samples and training starts after 240 samples have been collected.

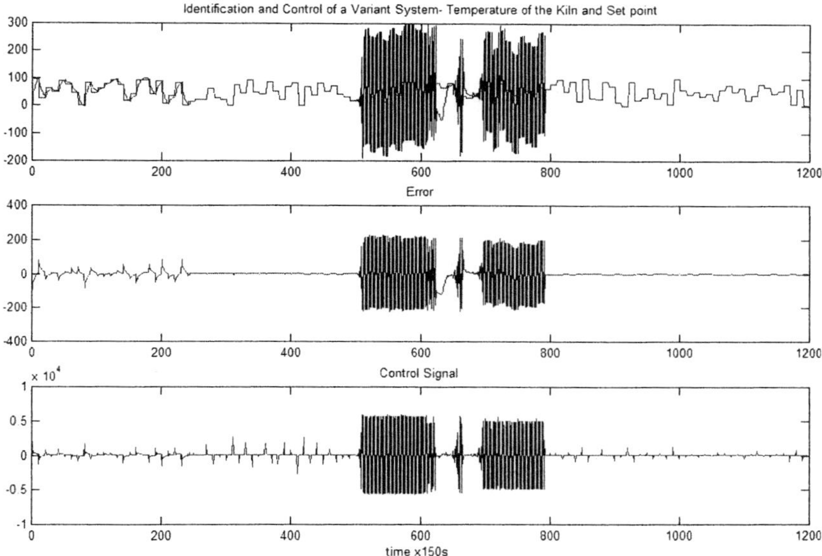

Fig. 2. Result obtained using the IMC control strategy and the proposed on-line learning solution with a random reference

Both direct and inverse models were one hidden layer models with 6 neurons on the hidden layer and one linear output neuron. The direct model has as inputs the past two samples of both the output of the system and the control signal.

The sampling period used was 150 seconds, which allowed performing several epochs of training between each control iteration. During the initial phase of collecting data a Proportional Integral controller (PI) was used in order to keep the system operating within the range of interest. The PI parameters are Kp=0.01 and Ki=0.01. After this initial phase the PI is replaced by an Internal Model Controller (IMC) controller, using the direct and inverse models trained on-line.

The first inverse model is ready at sample 243, that is only 2 samples after the training has started. After the 240 samples have been collected it only took one sampling period to complete the training of the direct model and another sampling period to complete the inverse model. As can be seen from figure 2 after the pole position is changed the closed loop enters a stage of heavy oscillation but as soon as the information about the new characteristics of the system is dominant in the training window(that is at sample 795) the quality of control is re-established.

6 Conclusion

This paper presents on-line identification and control of a pole position varying system using the Levenberg-Marquardt algorithm in a batch version with two sliding windows and Early Stopping.

As shown here, even for a noisy system, for which overtraining is a real problem it is possible to create models on-line of acceptable quality and recover from changes in the system.

The artificial pole position variant system used in this experiment is an extreme situation compared with most real time variant systems, which vary slowly. Nevertheless the successful application of the Levenberg-Marquardt sliding window solution to this situation shows that it will also work for slowly variant systems.

With this artificial time variant system it can be seen that learning is very difficult when the sliding windows contains data from the system previously and after the change. This corresponds to training an ANN with mixed data from two different systems. Once this situation is overcome the models of the new system are rapidly obtained. This problem would not happen for a slow changing system.

The sliding window solution with Early Stopping for the Levenberg-Marquardt algorithm is very interesting since it does not limit the capabilities of the algorithm and overcomes the limitations of application of the traditional solution.

References

1. K. Levenberg, "A method for the solution of certain problems in least squares," *Quart. Appl. Math.*, vol. 2, pp. 164–168, 1944.
2. D. Marquardt, "An algorithm for least -squares estimation of nonlinear parameters," *SIAM J. Appl. Math.*, vol. 11, pp. 431–441, 1963.
3. M. Nørgaard, O. Ravn, N. K. Poulsen, and L. K. Hansen, *Neural Networks for Modelling and Control of Dynamic Systems*, Springer, 2000.
4. Lester S. H. Ngia, *System Modeling Using Basis Functions and Application to Echo Cancelation*, Ph.D. thesis, Department of Signals and Systems School of Electrical and Computer Engineering, Chalmers University of Technology, 2000.
5. P. Ferreira, E. Faria, and A. Ruano, "Neural network models in greenhouse air temperature prediction," *Neurocomputing*, vol. 43, no. 1-4, pp. 51–75, 2002.
6. N. Morgan and H. Bourlard, "Generalization and parameter estimation in feedforward nets: Some experiments.," *Advances in Neural Information Processing Systems, Ed. D.Touretzsky, Morgan Kaufmann*, pp. 630–637, 1990.
7. Jonas Sjöberg, *Non-Linear System Identification with Neural Networks*, Ph.D. thesis, Dept. of Electrical Engineering, Linköping University, Suécia, 1995.
8. Fernando Morgado Dias, Ana Antunes, and Alexandre Mota, "A new hybrid direct/specialized approach for generating inverse neural models," *WSEAS Transactions on Systems*, vol. 3, Issue 4, pp. 1521–1529, 2004.
9. M. Nørgaard, "Neural network based system identification toolbox for use with matlab, version 1.1, technical report," Tech. Rep., Technical University of Denmark, 1996.
10. M. Nørgaard, "Neural network based control system design toolkit for use with matlab, version 1.1, technical report," Tech. Rep., Technical University of Denmark, 1996.

Optimal Control Yields Power Law Behavior

Christian W. Eurich and Klaus Pawelzik

Universität Bremen, Institut für Theoretische Physik,
Otto-Hahn-Allee 1, D-28359 Bremen, Germany
{eurich, pawelzik}@neuro.uni-bremen.de
http://www.neuro.uni-bremen.de/~web/index.php

Abstract. Power law tails can be observed in the statistics of human motor control such as the balancing of a stick at the fingertip. We derive a simple control algorithm that employs optimal parameter estimation based on past observations. The resulting control system self-organizes into a critical regime, whereby the exponents of power law tails do not depend on system parameters. The occurrence of power laws is robust with respect to the introduction of delays and a variation in the length of the memory trace. Our results suggest that multiplicative noise causing scaling behavior may result from optimal control.

1 Introduction

Power law tails, i. e., distributions $p(y) \sim y^{-\delta}$ for large values of y, occur ubiquitously in the statistics of natural and man-made systems. Examples include avalanche sizes and durations of granular matter, the distribution of the magnitudes of earthquakes (Gutenberg-Richter law), firing behavior of neural populations in cortical tissue, and stock-market fluctuations. Scaling behavior has also been identified in human sensorimotor control systems such as the balancing of a stick at the fingertip [2,3] and the visuomotor control of a virtual target on a computer screen [1]. In the case of stick balancing, the occurrence of power law tails has been attributed to on-off intermittency [5] resulting from the existence of multiplicative noise and a fine-tuning of system parameters to a stability boundary [3]. Here, we suggest a simple control mechanism that yields self-organized critical behavior without the need of parameter tuning. Moreover, multiplicative noise turns out to be the result of optimal parameter estimation.

2 The Basic Model

We define our control problem as a discrete random map and employ a maximum likelihood approach to optimize the prediction of the control parameter.

2.1 Derivation of the Control Equations

The uncontrolled system is defined by a one-dimensional linear random map

$$y_{t+1} = \alpha^0 y_t + \beta_t \, , \tag{1}$$

W. Duch et al. (Eds.): ICANN 2005, LNCS 3697, pp. 365–370, 2005.

where the dynamical variable y_t denotes the deviation from some target value at time t ($t = 0, 1, 2, \ldots$). α^0 is a system parameter unknown to the controller; it is assumed to be constant in the following. For $\alpha^0 > 1$, the fixed point at the origin is unstable. $\beta_t \sim \mathcal{N}(0, \sigma^2)$ is a Gaussian random variable describing nonpredictable fluctuations. Its variance $\sigma^2 \equiv$ const. is a second hidden system parameter. Its estimation turned out to be irrelevant for the system under study. In the following, we shall therefore focus on the estimation of α^0.

In unconstrained control, the controller simply minimizes deviations from the target value, i. e., $\langle y^2 \rangle \stackrel{!}{=} \min$. The controller is assumed to know the form of the dynamical equation (1). A control strategy consists in computing an estimate α_t of the parameter α^0 from observations $y_{t+1}, y_t, y_{t-1}, \ldots$ of the system and subtracting the term $\alpha_t y_t$. If control is switched on, (1) has therefore to be replaced by

$$y_{t+1} = (\alpha^0 - \alpha_t) y_t + \beta_t \,. \tag{2}$$

The estimate of the system parameter α_0 is obtained from an optimality principle as follows. In the basic form of the control algorithm, we consider the observation of only two subsequent values y_t and y_{t+1}. Since α^0 is unknown to the controller, it is regarded a variable, and α_{t+1} is taken to be the value α^0 that maximizes the compound density $p(y_{t+1}, y_t | \alpha^0, \alpha_t, \alpha_{t-1}, \ldots)$:

$$\alpha_{t+1} = \underset{\alpha^0}{\operatorname{argmax}}\; p(y_{t+1}, y_t | \alpha^0, \alpha_t, \alpha_{t-1}, \ldots) \,. \tag{3}$$

In other words, the new estimate α_{t+1} is defined to be the control parameter α^0 that is most probable to give rise to the two observations y_t and y_{t+1}, given the past estimates $\alpha_t, \alpha_{t-1}, \ldots$. The compound density can be written as

$$\begin{aligned} p(y_{t+1}, y_t | \alpha^0, \alpha_t, \alpha_{t-1}, \ldots) &= p(y_{t+1} | y_t, \alpha^0, \alpha_t, \alpha_{t-1}, \ldots) p(y_t | \alpha^0, \alpha_t, \alpha_{t-1}, \ldots) \\ &= p(y_{t+1} | y_t, \alpha^0, \alpha_t) p(y_t | \alpha^0, \alpha_t, \alpha_{t-1}, \ldots) \,. \end{aligned} \tag{4}$$

The second equality is due to the Markov property with respect to the control parameter α_t which becomes clear from (2).

Now consider $p(y_{t+1} | y_t, \alpha^0, \alpha_t)$. It is computed with the help of (2) which can be written as

$$y_{t+1} - (\alpha^0 - \alpha_t) y_t = \beta_t \sim \mathcal{N}(0, \sigma^2) \,. \tag{5}$$

If the left hand side is regarded as a function of y_{t+1}, a transformation of variables yields the result $p(y_{t+1} | y_t, \alpha^0, \alpha_t) \sim \mathcal{N}(0, \sigma_t^2)$:

$$p(y_{t+1} | y_t, \alpha^0, \alpha_t) = \frac{1}{\sqrt{2\pi\sigma_t^2}} e^{-\frac{\beta_t^2}{2\sigma_t^2}} = \frac{1}{\sqrt{2\pi\sigma_t^2}} e^{-\frac{(y_{t+1} - (\alpha^0 - \alpha_t) y_t)^2}{2\sigma_t^2}} \,. \tag{6}$$

We now assume that the second term on the right hand side of (4) is independent of α^0: $\partial p(y_t | \alpha^0, \alpha_t, \alpha_{t-1}, \ldots) / \partial \alpha^0 = 0$. This assumption is by no means trivial and will have to be checked later in the analysis of the control system. If it holds, (3) can be replaced by

$$\alpha_{t+1} = \underset{\alpha^0}{\operatorname{argmax}}\; p(y_{t+1} | y_t, \alpha^0, \alpha_t) \,.$$

Inserting (6) yields the result

$$\alpha_{t+1} = \alpha_t + \frac{y_{t+1}}{y_t} . \tag{7}$$

Employing (2), it can be written in the more common form $\alpha_{t+1} = \alpha^0 + \beta_t / y_t$.

2.2 Properties of the Control System

The control system (2,7) is a two-dimensional random map giving rise to a compound density $p(y_t, \alpha_t | \alpha^0)$. Here we shall focus on the behavior of y_t under the iteration of both equations. Figure 1 shows an example of the time series y_t after removal of the transient. The trajectory is irregular. In most cases, deviations of y_t from zero are rather small, with large deviations interspersed. The time series seems to have a self-similar structure: Magnifications of the trajectory yield trajectories of similar appearance (middle and bottom). In particular, fluctuations

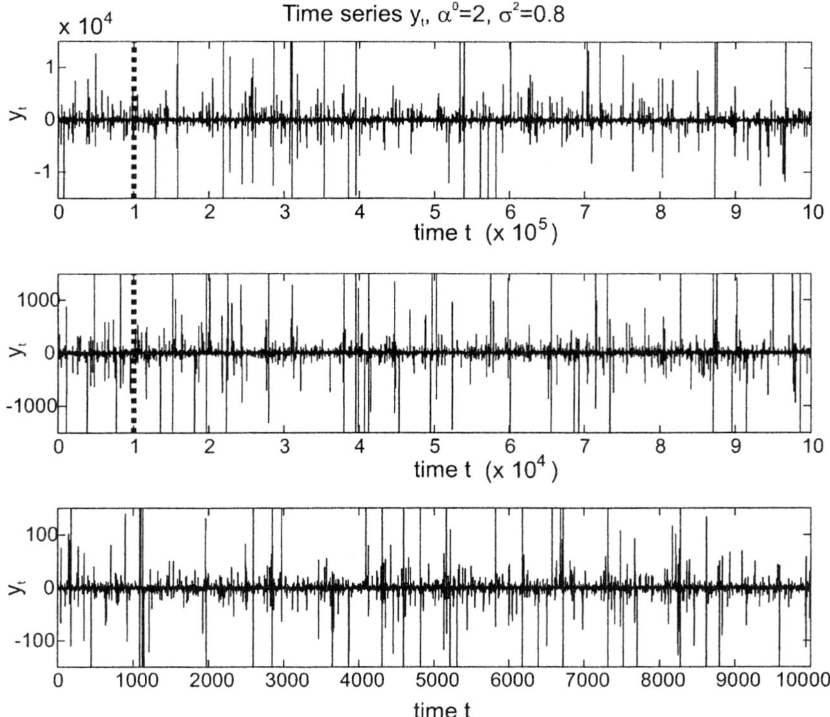

Fig. 1. Solutions y_t of the two-dimensional random map (2, 7) in three different resolutions. Top: 10^6 iterations, middle: 10^5 iterations, bottom: 10^4 iterations. The interval on the left of the vertical dotted lines in the top (middle) figure is expanded in the middle (bottom) figure. In all cases, $\alpha^0 = 2$, $\sigma^2 = 0.8$.

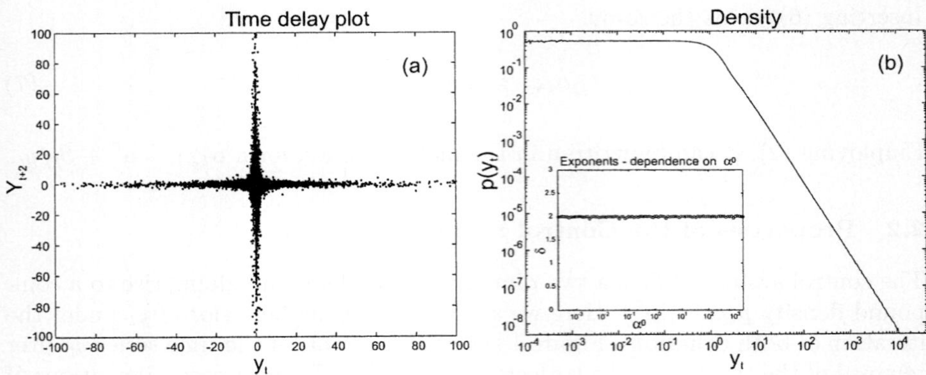

Fig. 2. (a) Time delay plot y_{t+2} vs. y_t. 25000 iterations after the transient. $\alpha^0 = 2$, $\sigma^2 = 0.8$. (b) Marginal density $p(y_t|\alpha^0)$ for $\alpha^0 = 2$. A clear power tail is visible, the exponent of which was estimated to be $\delta \approx -2.0$. $\sigma^2 = 0.8$, 10^8 iterations. The inset shows exponents δ for different values of the true control parameter α^0 as fitted from the marginal densites $p(y_t)$. Note the logarithmic scaling of the abscissa. In all cases, $\sigma^2 = 0.8$, 10^8 iterations for each value of α^0.

with large amplitudes in y_t (and also in α_t) occur because there are no further constraints on the estimated parameter. The existence of large fluctuations becomes clear from the structure of the system equations: a good estimate at time t, i. e., a small value y_t yields a large value α_{t+1} because y_t appears in the denominator on the right hand side of (7). This in turn yields a bad estimate (i. e., a large value) of y_{t+2} via (2). This mechanism can also be seen by eliminating α_t from the system (2,7), resulting in a random map with delay 2,

$$y_{t+2} = \beta_{t+1} - \beta_t \frac{y_{t+1}}{y_t} . \tag{8}$$

The time delay plot y_{t+2} vs. y_t in Fig. 2a corroborates the above explanation: large values of y_t yield small values of y_{t+2}, and only small values of y_t can result in large values of y_{t+2}.

An example for the marginal density $p(y_t|\alpha^0)$ is shown in Fig. 2b. For small values of y_t, the density is constant. For large values, it shows clear power law behavior with an exponent close to 2. Extensive numerical simulations show that the exponent is independent of the system parameter α^0 (see inset of Fig. 2). In fact, the complete density $p(y_t|\alpha^0)$ is invariant with respect to α^0: $\partial p(y_t|\alpha^0)/\partial\alpha^0 \equiv 0$ (data not shown). This can also be seen from (8) where α^0 has dropped out.

The exponents of the power law tails of $p(y_t|\alpha^0)$ are also independent of the variance σ^2 of the distribution of the noise variable β_t. This was tested numerically over 7 orders of magnitude of σ.

The control system and the resulting distributions can be studied analytically by means of the Frobenius-Perron equation which reveals an asymptotic power law exponent $\delta = -2$ of the marginal density $p(y_t)$ [6].

3 Extensions of the Basic Model

The basic model introduced in the previous section exhibits a fundamental property of biological motor control systems, the occurrence of power law tails. An application to realistic systems (such as the balancing of a stick or human postural sway) will only be possible if power law behavior in the model is robust with respect to the introduction of delays, changes in the system dynamics, etc. Here we briefly consider two extensions: delays and memory.

Interaction delays are ubiquitous in motor control (e. g., [7,4,2,3]). They can be introduced in the model simply by replacing (7) by

$$\alpha_{t+n} = \alpha_t + \frac{y_{t+1}}{y_t} \tag{9}$$

which will be referred to a system with delay n. Figure 3a shows the marginal density $p_n(y_t|\alpha^0)$ for the system with delay $n = 10$. Again, a clear power law tail can be identified. More systematic numerical investigations show that the exponents δ_n decrease with n.

A second extension of the basic model considers a longer history of past observations than only y_t and y_{t+1} as employed in (3):

$$\alpha_{t+1} = \underset{\alpha^0}{\operatorname{argmax}}\, p(y_{t+1}, y_t, y_{t-1}, \ldots, y_{t-n}|\alpha^0, \alpha_t, \alpha_{t-1}, \ldots)\,.$$

which will be referred to a system with with memory n. In this case, the control equation reads

$$\alpha_{t+1} = \frac{\sum_{i=0}^{n} \left(y_{t-i+1}y_{t-i} + \alpha_{t-i}y_{t-i}^2\right)}{\sum_{i=0}^{n} y_{t-i}^2}$$

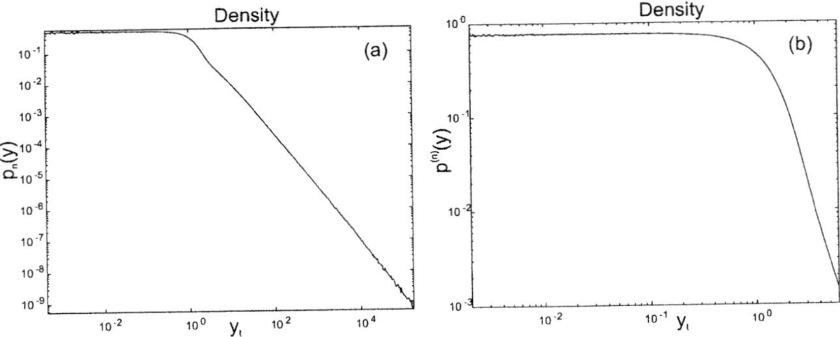

Fig. 3. (a) The marginal density $p_n(y_t)$ for the delayed system with $n = 10$. The densities show clear power law tails. The exponent was estimated to be $\delta_n \approx -1.7$. (b) The marginal density $p^{(n)}(y_t|\alpha^0)$ of a system with memory $n = 2$. The exponents was estimated to be $\delta^{(n)} \approx -3.9$. In both cases, $\alpha^0 = 2$, $\sigma^2 = 0.8$. 10^7 (a) and 10^8 (b) iterations.

instead of (7). Figure 3b gives an example of a marginal density $p^{(n)}(y_t|\alpha^0)$ for $n = 2$. Numerical investigations suggest that for a system with memory n, $p^{(n)}(y_t|\alpha^0)$ has an exponent $\delta^{(n)} = -(n+2)$, independent of the value of α^0.

Finally, the control equation (7) can be equipped with constraints on α_t. For example, a clipping at some value $\alpha^{\max}$ again yields power law tails of $p(y_t)$ that have an exponential cutoff (data not shown).

4 Discussion

We developed a simple control algorithm to explain the occurrence of power law tails in human movement data. As can be seen from (5), two mechanisms may contribute to this behavior: first, the dynamical variable y_t appears in the numerator, resulting in large amplitudes if the controller has previously been successful. Second, the noise term is multiplicative. Multiplicative noise is well known to produce on-off intermittency and power law scaling in random maps [5]. Whereas previous models for motor control explicitly incorporated multiplicative noise [2], in our model it results from a maximum-likelihood parameter estimation of a standard unstable fixed-point situation with additive noise.

So far, the occurrence of power law behavior has proven to be robust with respect to delays and extended memory. The fact that the consideration of a longer history yields faster decaying densities suggests that power laws observed in sensorimotor control result from a compromise between stability and fast adaptation with respect to changes in system parameters. Future work will focus on mechanisms that produce the power spectra and exponents found in human motor control.

References

1. Bormann, R., Cabrera, J. L., Milton, J. G., Eurich, C. W.: Visuomotor tracking on a computer screen - an experimental paradigm to study the dynamics of motor control. Neurocomp. **58–60** (2004) 517–523
2. Cabrera, J. L., Milton, J. G.: On-off intermittency in a human balancing task. Phys. Rev. Lett. **89** (2002) 158702
3. Cabrera, J. L., Bormann, R., Eurich, C. W., Ohira, T., Milton, J. G.: State-dependent noise and human balance control. Fluct. Noise Lett. **4** (2004) L107–L117
4. Eurich, C. W., Milton, J. G.: Noise-induced transitions in human postural sway. Phys. Rev. E **54** (1996) 6681–6684
5. Heagy, J. F., Platt, N., Hammel, S. M.: Characterization of on-off intermittency. Phys. Rev. E **49** (1994) 1140–1150
6. Pawelzik, K., Eurich, C. W.: Self-organized critical control. Submitted
7. Woollacott, M. H., von Hosten, C., Rösblad, B.: Relation between muscle response onset and body segmental movements during postural perturbations in humans. Exp. Brain Res. **72** (1988) 593–604

A NeuroFuzzy Controller for 3D Virtual Centered Navigation in Medical Images of Tubular Structures

Luca Ferrarini, Hans Olofsen, Johan H.C. Reiber, and Faiza Admiraal-Behloul

LKEB - Leiden University Medical Center, The Netherlands
L.Ferrarini@lumc.nl
http://www.lkeb.nl

Abstract. In this paper we address the problem of virtual central navigation in 3D tubular structures. A virtual mobile robot, equipped with a neuro-fuzzy controller, is trained to navigate inside image datasets of tubular structures, keeping a central position; virtual range sensors are used to sense the surrounding walls and to provide input to the controller. Aim of this research is the identification of smooth and continuous central paths which are useful in several medical applications: virtual endoscopy, virtual colonoscopy, virtual angioscopy, virtual bronchoscopy, etc. We fully validated the algorithm on synthetic datasets, and performed successful experiments on a colon dataset.

1 Introduction

Mobile robots often need to keep a central position while moving in corridors: several approaches have been proposed in literature both for holonomic and non-holonomic robots ([1], [2]), but most of them deal with real robots moving on the ground: thus, with a 2D navigation.

Identifying central lines in 3D tubular structures is an important task in several fields; magnetic resonance imaging (MRI) techniques allow the exploration of the inner human body through high resolution multi-slice images; 3D visualization of reconstructed volumes and virtual tours inside organs provide better analyses of the organ's morphology, function and pathology. However, virtual tours (like virtual angioscopy, virtual colonoscopy, virtual bronchoscopy, etc.) need a smooth and unique central line through the organ in order to provide internal views.

Different methods for central path extraction have been proposed in literature: skeleton based approaches (Yeorong et al. [3], Kiraly et al. [4]) usually provide non-smooth and non-unique paths which need post-processing steps (see Deschamps [5]). Paik et al. [6] use thinning techniques to project a surface minimum path in the middle of the structure: it results in smooth paths, but the algorithm is computationally heavy. Haigron et al. [7] proposed a fly-through approach based on active vision for virtual angioscopy: the system is highly automatic but time consuming.

We suggest a new approach to the problem of central path extraction for 3D tubular objects. By using a virtual mobile robot, the smoothness and uniqueness of the final path is granted. While moving through a virtual structure, the robot uses virtual range sensors to analyze its surrounding: results of the analysis are given to a 3D neuro-fuzzy

W. Duch et al. (Eds.): ICANN 2005, LNCS 3697, pp. 371–376, 2005.

controller which continuously adjusts the robot position and orientation, in order to generate a smooth central path through the organ. The 3D controller has a key role in the central navigation, and its design is based on a 2D controller introduced by Ng et al. [1] for car-like vehicles. Real-time interaction with the system is allowed: the user can suspend the exploration and inspect the surrounding in detail. Due to its generality, the method can be applied in different areas where central line extraction is needed: we present an application for virtual colonoscopy.

2 Method

The virtual mobile robot is represented as a flying vehicle with a direction system located on the front: the desired direction is defined in the local system by the two angles ϕ and ψ (figure 1.a). The mobile robot is equipped with range sensors on all its sides: they are simulated as lines which propagate in the environment and return information on detected obstacles or walls. In order to guarantee smooth trajectories, the direction system is limited ([-ϕ_{min},ϕ_{max}], [-ψ_{min},ψ_{max}]), and kinematic constraints for a 3D nonholonomic three-cycle vehicle are applied: we had already used nonholonomic constraints in [8], and we have extended them in this work by considering a 3D movement as a sequence of 2D steps on different planes.

The 3D central navigation problem is solved by separately considering the local xy and xz planes; for each plane we apply a procedure similar to the one proposed in [1], and by merging the results we obtain the final desired direction in 3D. We report the procedure for the xy local plane.

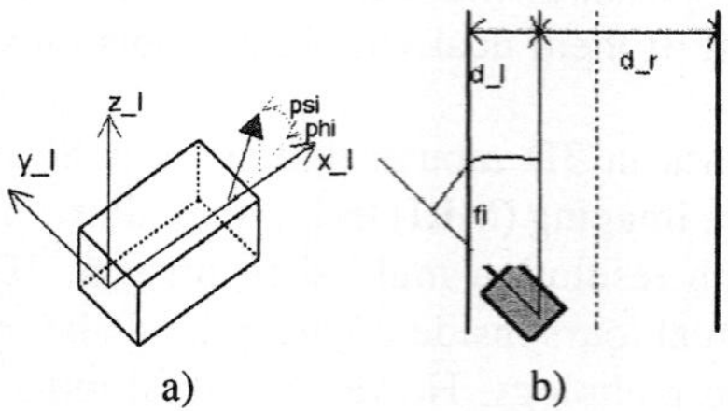

Fig. 1. (a) the robot in 3D and its local coordinate system. The desired direction (red arrow) is identified by two angles; (b) the robot evaluates its position and orientation on its local xy plane.

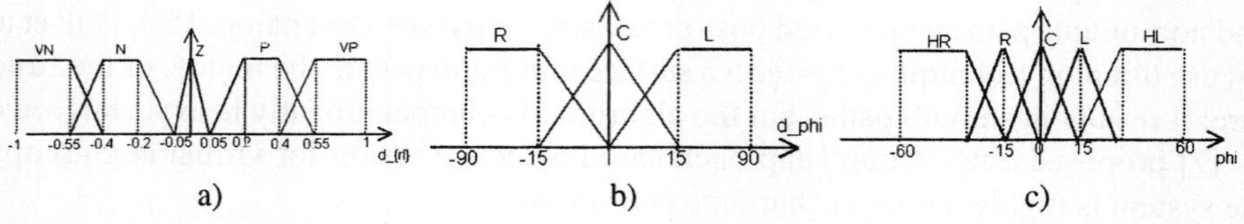

Fig. 2. Membership functions: (a) 5 for d_{rl} (*very negative, negative, central, positive, very positive*); (b) 3 for d_{ϕ} (*right, center, left*); (c) 5 for the output (*strongly right, right, center, left, strongly left*)

The mobile robot, using its range sensors, evaluates its position and orientation (see figure 1.b). The position is evaluated at each step as:

$$d_{rl} = \frac{d_r - d_l}{d_r + d_l}, \quad (1)$$

where d_r and d_l are the distances from the right and left walls; the orientation is defined as d_ϕ and is positive for the left orientation, and negative for the right one.

The two variables d_{rl} and d_ϕ are first fuzzyfied through the membership functions shown in figure 2.a and 2.b.

The output of the fuzzyfication is given to a feed-forward neural network (FFNN) called RNN[1]: the RNN maps the 8 input values onto 5 levels, corresponding to the output membership functions (see figure 2.c) associated with the output variable ϕ (steering angle on xy local plane).

The output of RNN is given to a second FFNN, ORNN[2], whose task is to defuzzyfy the 5 input levels into one crisp output: the desired steering angle. The RNN is trained on a set of basic rules (see table 1), while the ORNN is trained to reproduce the function:

$$O = \frac{\sum_{i=1..5} \omega_i * V_i}{\sum_{i=1..5} V_i} \quad (2)$$

where ω_i is the fuzzy value for the ith output membership function and V_i is the function's center. The ORNN learns a mathematical formula, and it would be right to argue that a neural network is not necessary for this aim; nevertheless, by building the system with neural networks, we end up with a more flexible tool which can be further improved by interaction with users. Particularly, an expert could provide the right central path, and the ORNN would then adapt to the user capabilities rather than to a mathematical formulation of the problem.

By applying the same procedure to the xz local plane, the desired ψ angle is evaluated, and the combination of ϕ and ψ angles gives the desired direction in 3D.

Table 1. Example of a fuzzy rule: the first 8 columns are the input for training RNN, and the last 5 columns are the desired output; if the robot is close to the right wall (d_{rl} is *very negative*) but is already pointing towards the left direction (d_ϕ is *left*), it should keep going toward the left wall (ϕ is *left*)

d_{rl}					d_ϕ			Output ϕ				
vn	n	z	p	vp	l	c	r	hl	l	c	r	hr
1	0	0	0	0	1	0	0	0	1	0	0	0

3 Experiments and Results

The validation for the neuro-fuzzy controller in 3D has been made in synthetic environments; the algorithm has then been tested on a colon dataset.

[1] RNN, Rule Neural Network: 8 input nodes, 20 hidden nodes, 5 output nodes.

[2] ORNN, Output Refinement Neural Network: 5 input nodes, 10 hidden nodes, 1 output nodes.

3.1 Synthetic Environments

The first tests[3] were performed on the tubes shown in figure 3.a-d; for each case, we let the application run 50 times, and for each step of the robot we evaluated the *error variable* as the distance between the robot position and the closest position on the ideal central line. In average, the error amounts to 6% of the diameter; table 2 reports the statistical analysis.

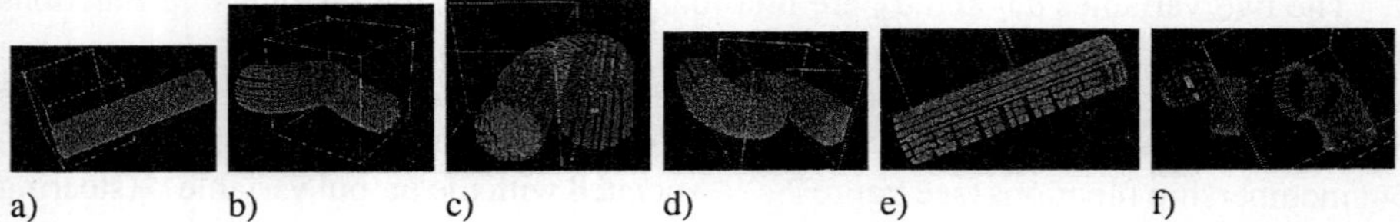

a) b) c) d) e) f)

Fig. 3. (a)-(d) test environments: *straight*, *tube*, *U-shaped*, and *S-shaped*; (e)-(f) *single side slice, straight corridor, 80% of the length removed*, and *entire slice, tube corridor, 80% of the length removed*

Table 2. Statistical analysis for the *error variable* on different shapes; results are in voxels; **N** is the total number of robot's step being considered

	N	μ	σ	95% Conf. Int.	
				Lower	Upper
Straight	1755	.06	.21	.05	.07
Tube	1469	1.59	1.06	1.53	1.64
U-shape	1895	1.42	1.16	1.37	1.47
S-shape	2670	1.61	1.13	1.57	1.65
All	7789	1.21	1.17	1.18	1.24

Table 3. Conf. Int. (95%, Lower, Upper) for missing information (results are in voxels); case I: *Straight, single side*, case II: *Straight, entire*, case III: *Tube, single side*, case IV: *Tube, entire*

	case I		case II		case III		case IV	
	L	U	L	U	L	U	L	U
20%	.12	.16	.13	.17	1.31	1.41	1.46	1.57
40%	.10	.13	.07	.10	1.30	1.40	1.38	1.48
60%	.04	.07	.14	.19	1.33	1.43	1.54	1.64
80%	.05	.07	.14	.19	1.57	1.68	1.82	1.97

In order to test the robustness of our solution, we performed tests on the *straight* and *tube* corridors with missing information; holes were created at regular distance, and the amount of missing data varied from 20% up to 80%; two frameworks were used: *single side slice* (only half side of a tube slice removed) and *entire slice* (see figure 3.e-f). The average error amounts to 9% of the tube diameter: table 3 reports the confidential intervals for the *error variable* in the four scenarios.

[3] Parameters: safety distance=5 units, maximum steering angle=$\pm$30 degrees, tube radius=10 units.

Finally, we tested the neuro-fuzzy controller on tubes of different sizes: we considered *straight tubes* with different constant radius (4, 5, 20, 25 voxels), two *straight tubes* with changing radius, and a *tube* with a radius of 6 units: we obtained an average error of 0.6 voxels over 50 runs per case.

3.2 Medical Dataset

We tested the algorithm on a colon dataset (CT scan, 128 x 128 x 488 voxels of 2.88 x 2.88 x 1.00 mm). The colon is a challenging environment presenting curves with high curvature and changes in diameter: the robot went successfully through it for several runs (different initial positions and orientations). Visual results are shown in figure 4: thanks to the use of a virtual mobile robot and kinematic constraints, the extracted path is always inside the organ (the robot can not pass through the walls), and is always unique, continuous, and smooth. The exploration with visualization is in real time: the robot goes through the colon, approximately 1.3 meters long, in 2 minutes.

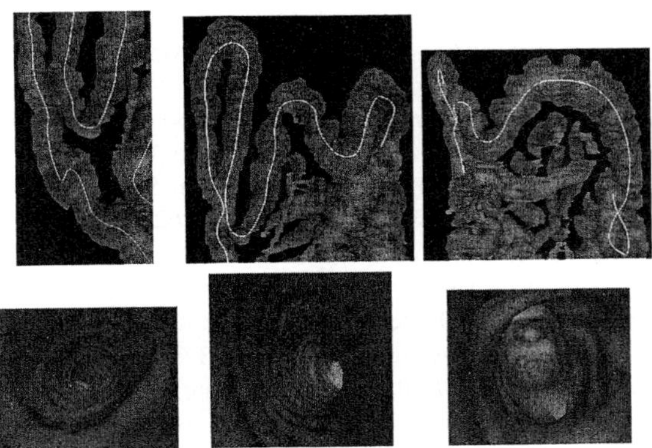

Fig. 4. Three outside views (top row) with central line, and three internal views (bottom row)

4 Conclusion

The main contribution of this paper is the development of a neuro-fuzzy controller for central navigation in 3D tubular structures (central path extraction); by using a non-holonomic mobile robot we guarantee smooth and unique paths, which are useful in several medical applications. We have thoroughly validated the algorithm on synthetic datasets. The experiment performed on the medical dataset resulted in a successful exploration (virtual colonoscopy), based on the identified central path. The navigation is in real time, allowing the user to interact with the environment. Finally, due to its generality, the method is suitable for different applications like virtual angioscopy and virtual bronchoscopy. We are currently investigating the use of extra modules for the robot (like novelty filters) which could highlight abnormal situations (branches, polyps, aneurysms, etc.), helping users in the diagnosis of diseases.

Acknowledgments

This work was supported by the Technology Foundation STW, applied science division of NWO and the technology programme of the Ministry of Economic Affairs, and by Medis medical imaging systems, Leiden, The Netherlands (www.medis.nl).

References

1. K. C. Ng, M. M. Trivedi, *A Neuro-Fuzzy Controller for Mobile Robot Navigation and Multirobot Convoying*, IEEE Transaction on Systems, Man, and Cybernetics-part B:Cybertetics, Vol. 28, NO. 6, Dec. 1998, pp. 829-840
2. R. Braunstingl, P. Sanza, J.M. Ezkerra, *Fuzzy Logic Wall Following of a Mobile Robot Based on the Concept of General Perception*, ICAR 1995, 7th Int. Conf. on Advanced Robotics, Sant Feliu de Guixols, Spain, pp. 367-376
3. Yerong G., Stelts D.R., Jie W., Vining D.J., *Computing the centrerline of a colon: robust and efficient method based on 3D skeletons*, Journal of Computer-Assisted Tomography 23 (5), 1999, pp. 786-794
4. A.P. Kiraly, J. P. Helferty, E.A. Hoffman, G. McLennan, W.E. Higgins, *Three-Dimensional Path Planning for Virtual Bronchoscopy*, IEEE Transaction on Medical Imaging, Nov. 2004, Vol. 23, Issue 11, pp. 1365-1379
5. T. Deschamps, L.D. Cohen, *Fast extraction of minimal paths in 3D images and applications to virtual endoscpy*, Medical Image Analysis, Vol 5, Issue 4, December 2001, pp. 281-299
6. Paik D.S., Beaulieu C.F., Jeffrey R.B., Rubin G.D., Naper S., *Automated flight path planning for virtual endoscopy*, Medical Physics 25 (5), 1998, pp. 629-637
7. P. Haigron, M. E. Bellemare, P. Acorsa, C. Gksu, C. Kulik, K. Rioul, A. Lucas, *Depth-Map-Based Scene Analysis for Active Navigation in Virtual Angioscopy*, IEEE Transaction on Medical Imaging, Nov. 2004, Vol. 23, Issue 11, pp. 1380-1390
8. F. Admiraal-Behloul, B.P.F. Lelieveldt, L. Ferrarini, H. Olofsen, R.J. van der Geest, J.H.C. Reiber, *A Virtual Exploring Mobile Robot for Left Ventricle Contour Tracking*, Proceedings IJCNN 2004, Vol. I, pp. 333-338

Emulating Process Simulators with Learning Systems

Daniel Gillblad, Anders Holst, and Björn Levin

Swedish Institute of Computer Science, Box 1263, SE-164 29 Kista, Sweden
{dgi, aho, blevin}@sics.se

Abstract. We explore the possibility of replacing a process simulator with a learning system. This is motivated in the presented test case setting by a need to speed up a simulator that is to be used in conjunction with an optimisation algorithm to find near optimal process parameters. Here we will discuss the potential problems and difficulties in this application, how to solve them and present the results from a paper mill test case.

1 Introduction

In the process industries there is often a need to find optimal production parameters, for example to reduce energy costs or to improve quality or production speed. Many of the parameter settings are scheduled some time in advance, e.g. to produce necessary amounts of different product qualities. A parameter schedule that is sufficiently near optimal as evaluated by a cost function can possibly be found using an optimiser that iteratively tests different scenarios in e.g. a first principles simulator, i.e. a simulator that tries to mimic the physical properties of the process, gradually converging to an optimal solution. An initial state for the simulator must be retrieved from the actual process, and the final scheme is suggested to the operators as an effective way to control the real process.

Unfortunately, although the simulator in question is faster than real time, it might still not be fast enough. The number of iterations that is required for the optimisation might easily stretch into the thousands, which means that even a relatively fast simulator cannot be used to reach a conclusion before the optimisation horizon is well over. If this is the case, some way to speed up the simulations is critical.

1.1 Learning Systems and Simulators

Let us consider two fundamentally different ways to model process behaviour. One is to build a first principles simulator of some sophistication. It has to consider how e.g. flows, temperatures, pressures, concentrations, etc. varies as material flows through components. The other approach is to approximate the input-output mapping in the process with some mathematical function, *without* considering the actual physical path. This is essentially what is done by a learning system, in which at least a subset of the parameters are estimated from examples.

W. Duch et al. (Eds.): ICANN 2005, LNCS 3697, pp. 377–382, 2005.

If we want to replace the simulator with a learning system we have a choice of either modelling the actual outputs of the simulator, i.e. training the system to map simulator inputs to corresponding outputs, or to associate simulator states to corresponding objective function values. The latter approach is very elegant and could probably yield very good results, but it is highly dependant on the specific outline of the objective function. In our test case it was not possible to try this direct modelling of the objective function, since all data necessary was not available. Instead, the first approach of mapping simulator inputs, consisting of a description of the initial state and a proposed schedule, to corresponding outputs was used.

We also have to make a choice of either using real process data, or to generate data with the simulator. There are benefits and drawbacks with both approaches, but using a simulator is actually very attractive mainly because of two reasons. First, most simulators are not only free of random measuring noise and drifting sensors, they also lack the stochastic nature of real data in the sense that we do not need several samples from identical input states to reliably estimate the mapping. Second, and perhaps more importantly, is the fact that real processes are kept within only a fraction of the total state space by the operators, following best practises known to produce good results. Most states are simply worthless from the process point of view. Nevertheless, the learning system does not know that these states are worthless unless the training data contains examples showing this, and will probably not produce very realistic results when the optimising algorithm moves out of the region covered by training data.

With a simulator we can cover a larger part of the state space in a controlled manner, but the actual generation of this training data now becomes somewhat of a problem. At first, this might seem like an ideal case: We should be able to generate arbitrary amounts of relatively noise free data. Unfortunately, disregarding the effects of finite precision calculations and time resolution in the simulator, there is still one problem remaining: Having a large amount of training data is still effectively useless if it does not reflect the data the learning system will encounter in use. When generating data, we would like to cover as much as possible of all possibly relevant states of the process. Fortunately, this can be rather straightforward using e.g. a statistical modelling approach, as we will describe later for the test case.

2 The Paper Mill Test Case

A simulator of a part of the system at the Jämsänkoski paper mill in Finland was used to test the approach. In the Jämsänkoski mill, thermo mechanical pulp refiners are used to grind wood chips, resulting in pulp that is fed to a number of production lines through a complex system of tanks and filters. There are two separate refiner lines, each consisting of five refiners, and three paper machines, PM4–6. Refiner line one is connected through two production lines to PM4 and PM5, while the second refiner line is only connected to PM6. The second refiner line and production line is largely representative for the whole system and was chosen for the test case. The state of the system is mainly represented by a number of tank levels, and the external control parameters by production rates, refiner schedules and production quality schedules.

The cost function for the optimization problem is constructed so that electricity costs for running the refiners are minimized while maintaining consistency of schedules and tank levels. It can be expressed as $C_{tot} = \frac{C_E}{10^4} + C_C + \frac{C_{\Delta_{tot}}}{4} + \frac{|V_\Delta|}{4}$, where C_{tot} is the total cost, C_E the cost of electricity, C_C and $C_{\Delta_{tot}}$ consistency terms related to the refiner and set points schedules, and V_Δ the difference between desired and actual tank levels at the end of a schedule. For a further explanation of these terms, see [1].

To generate training and validation data from the simulator, we modelled the joint distribution of the "normal" conditions over all inputs to the simulator, and then sampled random input values from this distribution. By "normal" states, we refer not only to the states and conditions the process normally encounters in daily use, but rather conditions that do not produce obviously faulty or unacceptable behaviour. This might seem complicated at first, but there are reasonable assumptions that simplify the procedure considerably. As a first approach, we can consider all inputs to be independent. This means that the joint multivariate distribution is simply the product of all marginal distributions. We only have to describe these marginal distributions over each variable, which simplifies the task significantly. Common parameterisations such as uniform distributions, normal distributions etc. were used, preferably making as few assumptions about the variable and its range as possible. When we e.g. from explicit constraints know that two or more variables are dependant, we model these using a joint distribution over these variables.

For the external control variables the situation is a little bit more difficult, since we need to generate a time series that to at least some degree reflect a real control sequence in the plant. We model these control sequences using Markov processes, each with transfer probabilities that will generate a series with about the same rate of change and mean value as the real control sequences. Great care was taken to assure that the statistical model and its parameters reflected actual plant operations. In total, around ten million samples were generated, representing about six years of operations.

2.1 Test Results

A natural approach to emulate the simulator with a learning system is by step-by-step recursive prediction. However, initial test results using this approach were not encouraging. The state space is large and complex, making the error in the prediction add up quickly and diverging from the true trajectory.

Fortunately, we can actually re-write the data into a form that does not involve time directly. From the cost function we know that we are not interested in intermediate tank levels, but only the final levels at the end of the optimisation horizon. However, we do not want to overflow or underflow any tanks during the time interval, as this disturbs the process and does not produce optimal quality. We also know that apart from the internal control loops, all parameter changes in the process are scheduled. This allows us to re-code the data as *events*, where an event occurs at every change of one of the scheduled external control parameters, i.e. one data point describing the momentary state is generated for each external change to the control parameters. If we assume that the process stays in one state, i.e. that flows or levels are stable or at least monotonous during the whole event or after a shorter stabilisation period, it should be possible to predict the difference in tank levels at the end of an event from initial levels, flows, quality and perhaps event length.

The previously generated raw data was transformed to this event form and three types of learning systems were tested: A Multi-Layer Perceptron trained with backward error propagation in batch mode using sigmoid activation functions [2], a k-Nearest Neighbour model using an euclidean distance measure on normalised inputs [3], and a mixture [4] of Naïve Bayes models [5], one for each production quality and all using normal distributions for all attributes. One separate model was constructed for each tank. The parameters of each model were chosen experimentally to values producing good results on a separate validation data set. This data was also used to perform backwards selection of the input attributes for each model, resulting in 4 to 10 used inputs out of 10 available depending on model and tank. Testing was performed by predicting the difference in a tanks level at the end of an event from the initial level in the beginning of the event on a test data set separate from the training and validation set. The results can be found in table 1. For a detailed description of the models used and the results, see [1].

Table 1. Results from predicting the difference in tank levels. σ denotes the standard deviation of the data, and results are displayed using both the correlation coefficient (ρ) and the root mean square error (RMS) for the multi-layer perceptron (MLP), k-Nearest Neighbour (k-NN), and Naïve Bayes (NB) models.

	Events							Monotonous Events						
		MLP		k-NN		NB			MLP		k-NN		NB	
Tank	σ	ρ	RMS	ρ	RMS	ρ	RMS	σ	ρ	RMS	ρ	RMS	ρ	RMS
2	22.5	0.50	20.3	0.59	18.2	0.52	19.3	37.8	0.89	33.3	0.89	17.4	0.71	26.6
3	16.0	0.70	11.7	0.63	12.4	0.71	11.2	7.5	0.82	5.18	0.81	4.85	0.84	4.29
4	15.9	0.48	16.0	0.64	12.1	0.57	13.2	16.4	0.77	12.5	0.72	12.4	0.53	15.5
5	14.9	0.34	14.0	0.35	14.0	0.35	13.9	15.6	0.44	6.66	0.44	15.1	0.33	15.9
6	15.8	0.61	12.8	0.55	13.2	0.55	13.2	14.2	0.59	8.13	0.57	11.9	0.51	12.8
7	22.0	0.54	18.7	0.57	18.0	0.47	19.3	21.2	0.63	12.8	0.58	16.5	0.44	18.3
8	19.4	0.69	14.1	0.75	12.7	0.54	16.4	20.1	0.76	14.9	0.72	14.5	0.51	17.9

The different models' performances are reasonable similar, but not particularly good. Although there might be a reasonable correlation coefficient in some cases, the square errors are still much too high for the predictions to be seen as very useful. Also note that for optimisation horizons consisting of more than one event, which usually would be the case, these errors will accumulate and make

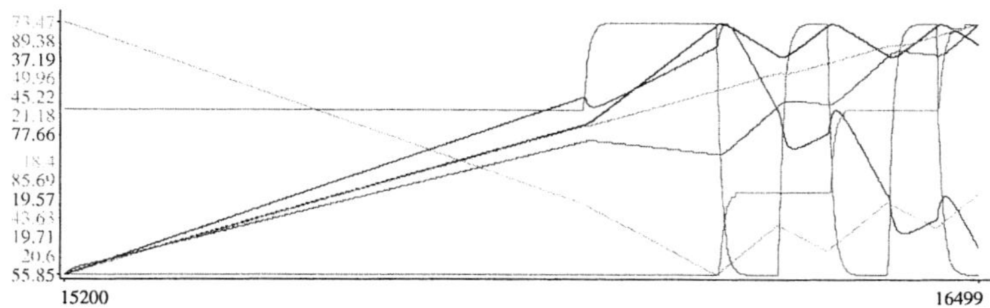

Fig. 1. An example of the characteristic oscillations of tank levels

the predictions unusable. Since all three types of models perform similarly, we can conclude that the problem probably is ill posed.

So why does it actually go wrong? A closer look at data reveals a likely reason for the poor performance, so let us study an example. Figure 1 shows that the tank levels behave very nicely during the first part of the event. Then, one tank overflows and the regulatory systems in the simulator change behaviour. The tank volumes start to oscillate and behave in an unstable manner, the overflow affecting almost the entire tank system. These oscillations in the tank levels are very difficult for a learning system to predict, since it in essence has to learn how to predict the phase, frequency and shape of these oscillations. However, we can now actually try to use this observation to change the representation of the problem in a manner that would make it easier for a learning system to solve.

Overflows or underflows are not desired from the cost functions perspective, which means that we have an opportunity to restrict the state space we train the learning system on to data when this does not occur. The model would of course only be valid when there are no overflows or underflows, but since we in practise predict a derivative for the tank levels we can easily estimate whether we would have an overflow or not in any of the tanks during the event. We transformed the data as earlier, but with the exception that an overflow or an underflow of any tank also constitutes the end of an event, although not the start of a new since the process then is in an unstable state. The models used in testing is as before, and the results can be seen in table 1. An improvement of the results compared to earlier tests was observed, but some tank levels are still very difficult to predict. The main storage line of tanks 2, 3 and 4 show decent results on the test data set, but the tanks with a direct connection to the paper mill (5 and 6) are very difficult to predict accurately. The differences in these tanks are actually usually zero, only occasionally changing drastically when a control loop need to use these tanks for temporary storage. Predicting when and to what degree this happens is very difficult.

The differences between the different models performance is again not that high. Usable predictions could possibly be made for the tanks in the main line and some tanks used for temporary storage. However, if the exact levels of the tanks connected more directly to the process itself are necessary, then there is a question of whether the predictions produced by the models are good enough. The correlation coefficient is definitely very low, but the tanks do not usually fluctuate much, which means that the mean absolute error of the predictions still could be kept rather low.

3 Discussion

The idea of replacing a slow simulator with a faster learning system is certainly attractive. The neural network and Naïve Bayes models are at least 100 to 1000 times faster than the simulator in the test case. However, as the results showed, it is by no means an easy process and not necessarily an effective solution. The generation and representation of data require quite a lot of work, which might easily make it more effective to develop a simpler, faster simulator instead.

It can also be argued that learning systems often are not a suitable solution for approximating a process simulator. The reason is that most "real" processes are described by a system of non-linear differential equations. Such systems will display chaotic behaviour, i.e. small changes in input data are quickly amplified, and lose correlation with the input. The time horizon for accurately predicting the output from input data is likely about as short as the time span within which the non-linear differential equations can be approximated by linear differential equations. However, this might not be a problem if we are not interested in the actual output values after a longer time period, but rather a mean value over a certain time or similar.

Even if we need these actual output values, it might still be possible to reformulate the problem so that it is solvable. It might potentially also be possible to divide the simulator into smaller parts and replacing some or all of these parts with fast learning systems, overcoming the problem of non-linearity for these systems. This division will, unfortunately, require a substantial amount of domain knowledge, as there is no reliable way to perform the division automatically today. Method development might need to get further before we can expect the learning system simulator replacement to become directly viable as an alternative in the general case.

References

1. D. Gillblad et al. Approximating process simulators with learning systems. SICS Technical Report T:2005:03, Swedish Institute of Computer Science, 2004.
2. D.E. Rumelhart and J.L. McClelland. Learning internal representations by error propagation. In *Parallel Distributed Processing,* volume I. M.I.T. Press, Cambridge, MA (1986).
3. T. Cover. Estimation by The Nearest Neighbour rule. In *IEEE Transactions on Information Theory* 14(1) (1968) 50–55.
4. G. McLachlan and D. Peel. *Finite Mixture Models.* Wiley & Sons (2000).
5. I.J. Good. *Probability and the Weighting of Evidence.* Charles Griffin (1950).

Evolving Modular Fast-Weight Networks for Control

Faustino Gomez[1] and Jürgen Schmidhuber[1,2]

[1] IDSIA, Galleria 2, 6928 Manno (Lugano), Switzerland
[2] TU Munich, Boltzmannstr. 3, 85748 Garching, München, Germany
{tino, juergen}@idsia.ch

Abstract. In practice, almost all control systems in use today implement some form of linear control. However, there are many tasks for which conventional control engineering methods are not directly applicable because there is not enough information about how the system should be controlled (i.e. reinforcement learning problems). In this paper, we explore an approach to such problems that evolves *fast-weight* neural networks. These networks, although capable of implementing arbitrary non-linear mappings, can more easily exploit the piecewise linearity inherent in most systems, in order to produce simpler and more comprehensible controllers. The method is tested on 2D mobile robot version of the pole balancing task where the controller must learn to switch between two operating modes, one using a single pole and the other using a jointed pole version that has not before been solved.

1 Introduction

All real-world systems are non-linear to some degree, yet almost all control systems in operation today employ some variant of linear feedback control. The wide applicability of linear methods relies on the fact that most non-linear systems of interest are either nearly linear around some useful operating point or can be decomposed into multiple linear operating regions. Methods such as *gain-scheduling* provide powerful tools to control such systems [1]: first a linear model is built for each operating mode, and then a linear controller (e.g. PID) is designed with parameters (i.e. gains) that are switched by a *scheduler* when the system transitions from one mode to another.

Gain scheduling works well when the mode of the system is observable, and, like all classical approaches, when the appropriate type of strategy is known *a priori*. For very complex tasks, such as those encountered in robotics, the designer often does not know what action should be taken in each system state. One method for solving control tasks under these more general conditions is neuroevolution [2] where a genetic algorithm is used to search the space of neural network controllers by repeatedly recombining the best performing candidates according to the principle of natural selection.

Artificial neural networks can potentially implement *global* non-linear controllers, but ensuring their stability and analyzing their behavior is difficult [3].

W. Duch et al. (Eds.): ICANN 2005, LNCS 3697, pp. 383–389, 2005.

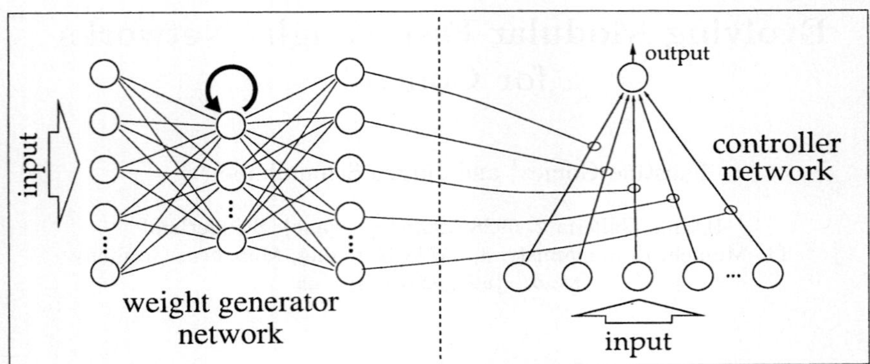

Fig. 1. Fast-Weight Network Module. The figure shows the two components of a Fast-Weight Module. On the left is the *generator* network that is evolved by the GA. This recurrent network receives input from the environment and outputs a set of weight values for the single-layer *controller* network at right. The controller network also receives input from the environment and outputs the control action.

This paper explores a method for evolving a special kind of *fast-weight* neural network that can potentially provide simpler automatically designed controllers. The networks consist of separate modules containing a recurrent neural network that generates weights for a linear controller.

The next section describes the fast-weight network architecture. Section 3 describes the neuroevolution method, Hierarchical Enforced SubPopulation (H-ESP) that is used to evolve the fast-weight networks. Section 4 presents our experimental results in applying the method to a two-mode robot pole balancing task, and section 5 concludes with a brief discussion of the results.

2 Modular Fast-Weight Networks

When neural networks are used to solve tasks in which the output depends on a history of inputs, they usually contain recurrent connections that feed back previous activations. Temporal information is encoded in the form of internal activation patterns (i.e. state) generated by propagating external inputs and previous activations through a fixed set of weights. Another possibility is to have dynamic weights or *fast weights* that can change in value over time. The little work that has been done using this concept has either used fast-weights as a mechanism to provide more robust associative memories [4], or to reduce network learning complexity [5]. Here we use the idea of fast-weights to generate controllers capable of switching easily between linear functions.

Networks are composed of a separate fast-weight module for each output unit. Each module consists of a recurrent *generating network* and a single-layer, feedforward *controller network* (figure 1). The output o_m of module m is:

$$o_m = \delta\left(\sum_{k=1}^{I} x_k \widehat{w}_{km}\right) \tag{1}$$

$$\widehat{w}_{km} = \sum_{j=1}^{H}\left(w_{jk}\ \delta\left(\sum_{i=1}^{I} x_i w_{ij} + \sum_{h=1}^{H} a_h w_{hj}\right)\right) \tag{2}$$

where $x \in \Re^I$ is the external input, $a \in \Re^H$ is the hidden layer activation from the previous time step, w_{ij} is the weight from unit i to unit j in the generating network, $\widehat{w}_{ij}$ is the weight from i to j in the controller network, and δ is the sigmoid function. Equation 1 computes the output of the controller network *after* the generating network has produced the I weights according to equation 2.

This network architecture is theoretically no more powerful than a standard fully recurrent network. The underlying intuition behind its design is that such an architecture will bias the search toward controllers that are potentially simpler and better suited to non-stationary environments characterized by transitions between operating modes. Although a module can generate weights that are a non-linear function of the entire history of inputs, it can easily implement a linear controller, if it is all that is required, by having the generating network output constant values.

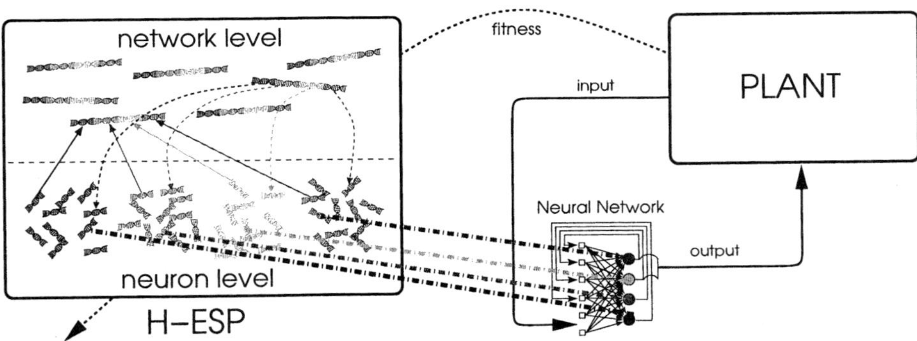

Fig. 2. H-ESP. Evolution occurs at both the level of neurons and of networks. The neuron level (L1) consists of multiple subpopulations of neurons, shown here in different colors. The network level (L2) consists of complete network representations that have either migrated up from below or have been created by recombining networks. During evolution, networks are evaluated in two possible ways: from L2 directly, and from L1 by randomly selecting a neuron from each subpopulation and combining them into a complete network. The dashed lines from the neuron level to the network being evaluated indicate a network formed in this manner. A network from L2 that has higher fitness than any network formed so far in L1, has its neurons copied into their corresponding subpopulations in L1 (shown with the dashed arrows from L2 to L1). A network form in L1 that has higher fitness than the worst L2 network is copied into L2 (the solid arrows from L1 to L2). In this way, the two levels supply each other with new genetic material with which to search in their respective weight spaces.

3 Hierarchical Enforced Subpopulations

Fast-weight networks are evolved using a method introduced in [6] called Hierarchical Enforced SubPopulations (H-ESP). H-ESP searches the space of recurrent networks by evolving at two levels in tandem: the level of network components or *neurons*, and the level of full networks. The neuron level (i.e. plain ESP) searches the space of networks indirectly by sampling the possible networks that can be constructed from the subpopulations of neurons. Network evaluations provide a fitness statistic that is used to produce better neurons that can eventually be combined to form a successful network. Figure 2 shows the basic operation of the algorithm (see [6] for further details).

The network level provides a repository or "hall of fame" of the best networks found so far by the neuron level, and allows H-ESP to search within the space of highly fit neuron combinations in a way that is not possible at the neuron level because it constructs networks at random.

To evolve fast-weight networks each neuron encodes the input, recurrent, and output weights of one of the units of a generating network.

4 Experiments

To evaluate approach, we evolved controllers for a simulated version of the three-wheeled Robertino mobile robot (figure 3a). Each wheel can slide along its rotational axis using six small sub-wheels (figure 3b). This *holonomic drive* enables the robot to change direction without having to rotate. On top of the robot is a vertical pole that is attached to the chassis with a ball joint. The pole can be either a single rigid rod or two rods, one on top of the other, with a ball joint

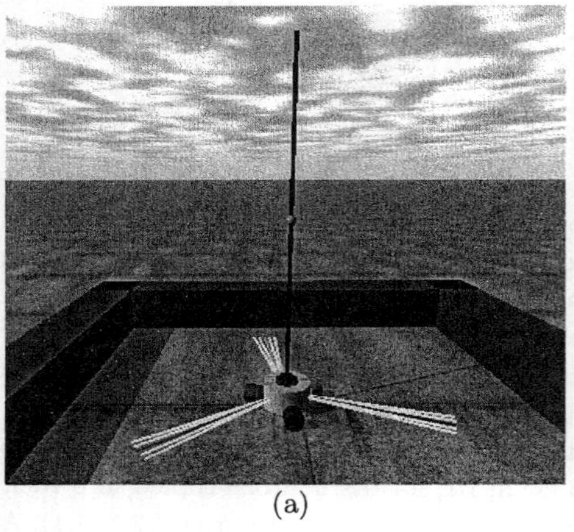
(a)

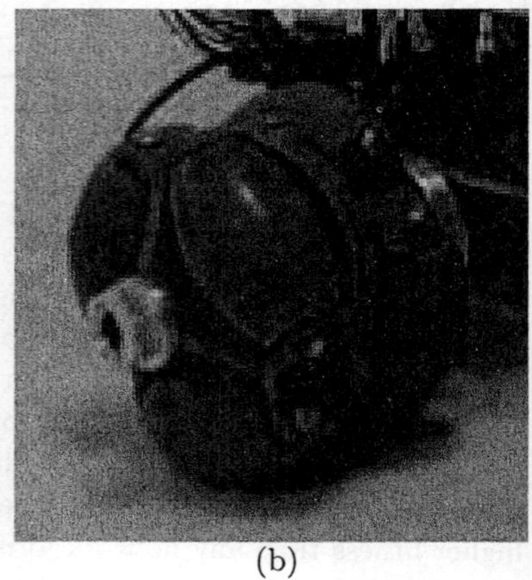
(b)

Fig. 3. Robertino with pole. (a) A snapshot from the Robertino ODE simulation showing the robot in the center of a walled arena. (b) Robertino wheel. The small sub-wheels allow the Robertino to slide as well as roll.

connecting them. The objective is to balance the both types of poles by applying a torque to each of the three wheels. Balancing each type of pole requires a different strategy. Both systems are nearly linear around their unstable equilibrium points (i.e. poles in vertical position), but when the angle of the pole(s) increases they become non-linear, more so in the case of the jointed pole.

Note that unlike the 2-dimensional version of the classic pole balancer [7], the system cannot be controlled by solving the 1-dimensional case and then using two copies of this controller, one for each principle axis. Because the robot can rotate around its vertical axis using 3 wheels spaced 120° from each other, this simple symmetry cannot be exploited. To move in a given direction, the velocity of all 3 wheels must be correctly modulated.

H-ESP was used to evolve networks consisting of three fast-weight modules, one for each wheel as shown in figure 4. The weight generating network of each module had 5 hidden units and 8 inputs scaled to the range [-1.0,1.0]: 3 proximity sensors, the angle of the lower pole in the x-axis θ_x^l and y-axis θ_y^l, the angles for the upper pole θ_x^u, θ_y^u, and the rotation of the chassis; all angles were measured in absolute (global) coordinates. For the single pole mode $\theta_{x,y}^u$ were set to zero. The neuron level subpopulations consisted of 200 neurons, and the network level population of 100 networks. The robot was simulated using the Open Dynamics Engine (www.ode.org) with a 0.01 second integration time.

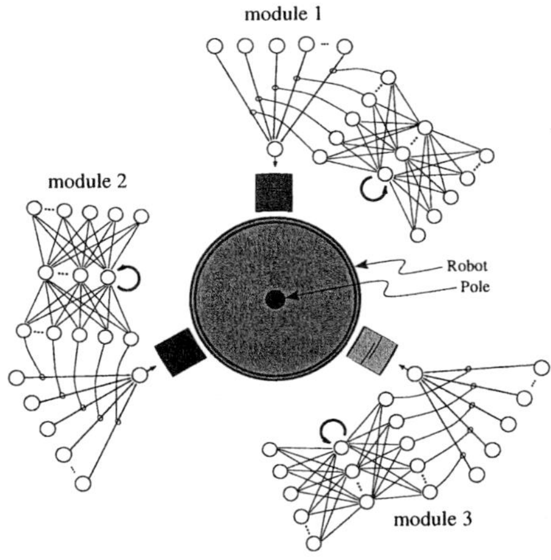

Fig. 4. Control Architecture for the Robertino robot. The Robertino (overhead view) is controlled by of three fast-weight modules, one for each wheel. At each time step, the generator networks produce weights for and activate their respective controller network.

During evaluation the controllers were tested in two trials: one with a single pole of 1.0 meter in length, and one with a jointed pole with two 0.5 meter segments. Each trial starts with the robot sitting in the center of a 1.5×1.5 meter walled arena (see figure 3a) with the pole(s) leaning 6° (0.5°) from vertical in the x direction. Every 0.04 seconds (i.e. 25Hz control frequency) each module outputs the desired angular velocity for its corresponding wheel $[-3.6\pi, 3.6\pi]$, until the pole angle(s) exceed 36°. The fitness of the controller was the number of time steps the pole(s) could be balanced in the shorter of the two trials. The task was considered solved with a fitness of 10 thousand. In order to solve the task, the controller must determine which mode it is in and apply the appropriate strategy for that mode.

4.1 Results

Figure 5 shows the controller network weight values produced during the successful operation of a typical controller for the first 100 time steps of operation, in each mode. The solid curve is for the single pole and the dotted curve is for the jointed pole. For most weights, the difference between the modes occurs at the beginning of the trial when the pole angles are relatively large, and the controller must employ a different strategy to bring the pole(s) into the linear region. Other weights, specifically those in module 3, quickly reach a constant value for the jointed pole mode, and then transition to the same value used for the single pole after about 80 time steps, by which time the jointed pole has been stabilized.

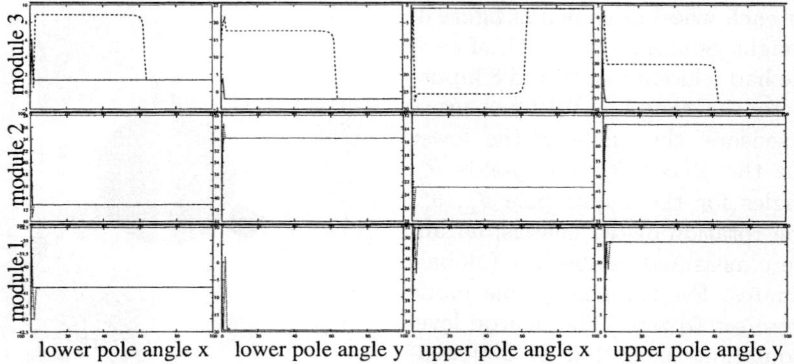

Fig. 5. Fast-weight values during control. Each plot shows the weight value for one of the inputs for the two modes. Each row corresponds to one of the 3 modules. The solid curve is for the single pole, the dotted curve is for the jointed pole.

5 Discussion and Conclusion

The experiments show that fast-weight networks can be evolved to produce relatively simple controllers. The weights produced by the generating networks implement almost piecewise linear controllers. While simpler architectures such as fully recurrent networks can solve each case, we were unable to do so for the two mode problem, and even the jointed pole version by itself could not be solved reliably. Furthermore, with such networks it is often difficult to understand the strategy being implemented. Using fast-weight networks, each period of constant weight values is a linear controller that can be "cut away" from its generating network during testing, leaving a set of simple linear filters that are more amenable to formal control theory analysis.

Future work will apply this approach to bipedal robot walking where it might be possible to implemented controllers for different, potentially non-linear, gait modes by using fast-weight networks.

Acknowledgments

This research was partially funded by CSEM Alpnach and the EU MindRaces project: FP6 511931. We would like to thank Frank Pasemann and Keyan Mahmoud Ghazi-Zahedi of Fraunhofer AIS for the Robertino ODE code.

References

1. Rugh, W., Shamma, J.: A survey of research on gain-scheduling. Automatica (2000) 1401–1425
2. Yao, X.: Evolving artificial neural networks. Proceedings of the IEEE **87** (1999) 1423–1447
3. Kretchmar, R.M.: A Synthesis of Reinforcement Learning and Robust Control Theory. PhD thesis, Department of Computer Science, Colorado State University, Fort Collins, Colorado (2000)
4. Hinton, G.E., Plaut, D.C.: Using fast weights to deblur old memories. In: Proceedings of the Ninth Annual Cognitive Science Society Conference, Hillsdale, NJ, Lawrence Erlbaum Associates (1987) 177–186
5. Schmidhuber, J.: Learning to control fast-weight memories: An alternative to dynamic recurrent networks. Neural Computation **4** (1992) 131–139
6. Gomez, F.J., Schmidhuber, J.: Co-evolving recurrent neurons learn deep memory pomdps. Technical Report 17–04, IDSIA, Lugano, Switzerland (2004)
7. Gomez, F., Miikkulainen, R.: 2-D pole-balancing with recurrent evolutionary networks. In: Proceedings of the International Conference on Artificial Neural Networks, Berlin; New York, Springer-Verlag (1998) 425–430

Topological Derivative and Training Neural Networks for Inverse Problems*

Lidia Jackowska-Strumiłło[1], Jan Sokołowski[2], and Antoni Żochowski[3]

[1] Computer Engineering Department, Technical University of Łódz,
Al. Politechniki 11, 90-924 Łódz, Poland
lidia_js@kis.p.lodz.pl

[2] Institut Elie Cartan, Université Henri Poincaré,
Nancy I, B.P. 239, 54506 Vandoeuvre lès Nancy Cedex, France
sokolows@iecn.u-nancy.fr

[3] Systems Research Institute of the Polish Academy of Sciences,
ul. Newelska 6, 01-447 Warszawa, Poland
zochowsk@ibspan.waw.pl

Abstract. We consider the problem of locating small openings inside the domain of definition of elliptic equation using as the observation data the values of finite number of integral functionals. Application of neural networks requires a great number of training sets. The approximation of these functionals by means of topological derivative allows to generate training data very quickly. The results of computations for $2D$ examples show, that the method allows to determine an approximation of the global solution to the inverse problem, sufficiently closed to the exact solution.

1 Introduction

We consider a given geometrical domain $\Omega \subset \mathbb{R}^2$, e.g. $\Omega = (0,1) \times (0,1)$, which contains a small opening $B_\rho(y) = \{ \, x \mid |x-y| < \rho \, \}$. In such a domain $\Omega_\rho = \Omega \setminus \overline{B_\rho(y)}$ the following model boundary value problems are defined:

$$\begin{gathered} \Delta u^i_\rho = f^i \;\text{ in }\; \Omega_\rho, \\ u^i_\rho = g^i \;\text{ on }\; \Gamma^i_1, \quad \frac{\partial u^i_\rho}{\partial n} = h^i \;\text{ on }\; \Gamma^i_2, \quad \frac{\partial u^i_\rho}{\partial n} = 0 \;\text{ on }\; \Gamma_\rho = \partial B_\rho(y), \end{gathered} \tag{1}$$

with solutions $u^i_\rho \in H^1(\Omega_\rho)$. The superscript "$i$", $i = 1, \ldots, K$, denotes here different sets of of data for which (1) is solved in the same domain Ω_ρ. By u^i we mean the solutions for the same data, but in the domain without opening. We want to locate the position y and radius ρ of the opening. To this end we utilize the observed values of shape functionals, assuming $K > 3$:

* Supported by the grant 4-T11A-015-24 of the State Committee for the Scientific Research of the Republic of Poland.

W. Duch et al. (Eds.): ICANN 2005, LNCS 3697, pp. 391–396, 2005.

$$I_i(\Omega_\rho) = I_i(y,\rho) = \int_{\Omega_\rho} [F(u^i_\rho) + G(\nabla u^i_\rho)]\, d\Omega, \tag{2}$$

with smooth $F(\cdot), G(\cdot,\cdot)$.

As a result the problem boils down to inverting the mapping $\mathcal{G} : R^3 \mapsto R^K$

$$\mathcal{G}(y,\rho) = [I_1(y,\rho), \ldots, I_K(y,\rho)]\,. \tag{3}$$

For the approximation of $\mathcal{G}^{-1}$ we may use artificial neural networks (ANN). However, generation of training data, i.e. the values of $\mathcal{G}(y,\rho)$ for a big number of arguments (y,ρ), requires numerous solutions of the problem (1) in domains of varying geometries, complicated by changing openings. This is very time consuming. The main idea consists in approximating the components of the mapping $\mathcal{G}$ by

$$\hat{I}_i(y,\rho) = I_i(\Omega) + \frac{1}{2}\rho^2 \mathcal{T}_\Omega I_i(y) \tag{4}$$

and inverting the resulting $\hat{\mathcal{G}}$ instead of $\mathcal{G}$. Here $I_i(\Omega)$ denotes, according to (2), the value of the functional I_i for the solution u^i, and $\mathcal{T}_\Omega I_i(y)$, $y \in \Omega$, is a function defined in Ω and called **topological derivative** of the functional I_i.

The notion of the topological derivative has been proposed by the authors and developed in several papers [10,11,12,13,4,14]. It is based on results concerning asymptotic behaviour of solutions to PDE's, see [3,6,8]. Partial inspiration for this work came from [9]. The gain in using (4) follows from the observation that $I_i(\Omega)$ and $\mathcal{T}_\Omega I_i(y)$ must be computed only once for every set of data, i.e. only K – times. Since this requires, as we shall see, only solving (1) in Ω $2K$ times, generating very big training sets for $\hat{\mathcal{G}}$ becomes practicable.

The topological derivative may be applied in a great many other cases, in particular in optimal shape design. We refer the reader to our papers listed in the bibliography for a detailed discussion and more exhaustive references.

2 Topological Derivatives of Shape Functionals

In the present section we recall only the most relevant results concerning topological derivative. Let us fix data and consider the functional

$$I(\Omega_\rho) = I(y,\rho) = \int_{\Omega_\rho} [F(u_\rho) + G(\nabla u_\rho)]\, d\Omega. \tag{5}$$

Using standard notation for Sobolev spaces $H^1_g(\Omega_\rho) = \{\psi \in H^1(\Omega_\rho) | \psi = g \text{ on } \Gamma_1\}$, $H^1_{\Gamma_1}(\Omega_\rho) = \{\psi \in H^1(\Omega_\rho) | \psi = 0 \text{ on } \Gamma_1\}$, the weak solution $u_\rho \in H^1_g(\Omega_\rho)$ satisfies the following integral identity

$$\int_{\Omega_\rho} \nabla u_\rho \cdot \nabla \phi \, d\Omega = \int_{\Gamma_2} h\phi \, dS - \int_{\Omega_\rho} f\phi \, d\Omega, \quad \forall \phi \in H^1_{\Gamma_1}(\Omega_\rho)\,. \tag{6}$$

We define **topological derivative** $\mathcal{T}_\Omega I(y)$ as

$$\mathcal{T}_\Omega I(y) = \lim_{\rho \downarrow 0} \frac{dI(y,\rho)}{d(|B_\rho(y)|)}\,,$$

Then the following theorem holds, [10].

Theorem 1. *The topological derivative of the functional*

$$I(\Omega) = \int_\Omega [F(u) + G(\nabla u)]\, d\Omega$$

is given by the following formula

$$\mathcal{T}I(y) = -\frac{1}{2\pi}\left[2\pi F(u(y)) + g(\nabla u(y)) + 2\pi f(y)v(y) + 4\pi \nabla u(y)\cdot\nabla v(y)\right] ,$$

where $\nabla u(y) = (a,b)^T$,

$$g(\nabla u(y)) = \frac{1}{2\pi}\int_0^{2\pi} G\left(a\sin^2\theta - b\sin\theta\cos\theta, -a\sin\theta\cos\theta + b\cos^2\theta\right)\, d\theta$$

and the adjoint state $v \in H^1_{\Gamma_1}(\Omega)$ *solves the boundary value problem*

$$-\int_\Omega \nabla v\cdot\nabla\phi d\Omega = -\int_\Omega [F'(u)\phi + \nabla G(\nabla u)\cdot\nabla\phi] d\Omega \ , \ \forall\phi\in H^1_{\Gamma_1}(\Omega) \ .$$

This result implies immediately the following asymptotic expansion

$$I(y,\rho) = I(\Omega) + \frac{1}{2}\rho^2 \mathcal{T}_\Omega I_i(y) + o(\rho^2) \tag{7}$$

uniformly for any bounded set of data (in relevant space).

3 Analysis of Inverse Problem

In the present section we discuss the uniqueness of solutions to inverse problems under consideration. Let Ω be the unit square in $\mathbb{R}^2$, $\Omega = (0,1)\times(0,1)$. Its edges will be denoted respectively:

$$\Gamma_1 = [0,1]\times\{1\} \quad \Gamma_2 = \{1\}\times[0,1] \quad \Gamma_3 = [0,1]\times\{0\} \quad \Gamma_4 = \{0\}\times[0,1].$$

Let us assume that Ω contains a small circular hole $B_\rho(y)$. The the question is:

Can we identify the hole $B_\rho(y)$ by means of measurements performed inside Ω as well as on the exterior boundary of Ω?

More precisely, consider the following boundary value problem defined in Ω_ρ:

$$\begin{gathered} \Delta u = 0 \ \text{ in } \ \Omega_\rho \\ u = 1 \ \text{ on } \ \Gamma_1, \quad \frac{\partial u}{\partial n} = 0 \ \text{ on } \ \Gamma_2\cup\Gamma_4\cup\Gamma_\rho. \quad u = 0 \ \text{ on } \ \Gamma_3. \end{gathered} \tag{8}$$

Let us also assume that we solve the same boundary value problem after the rotation of Ω_ρ by an angle of $\frac{\pi}{2}$, that is we may use the notations: $\hat{\Gamma}_1 = \Gamma_4, \hat{\Gamma}_2 = \Gamma_1, \hat{\Gamma}_3 = \Gamma_2, \hat{\Gamma}_4 = \Gamma_1$. Then following [4] we have the result.

Theorem 2. *Let u and $\hat{u}$ be the solutions of these two systems and assume that we are able to compute the following domain and boundary integrals*

$$\int_{\Omega_\rho} x_1^2 u_{/11}\, d\Omega, \quad \int_{\Omega_\rho} x_2^2 u_{/22}\, d\Omega, \quad \int_{\Omega_\rho} x_1 x_2 u_{/12}\, d\Omega, \quad \int_{\Omega_\rho} x_1 u_{/1}\, d\Omega,$$

$$\int_{\Omega_\rho} x_2 u_{/2}\, d\Omega, \quad \int_{\Omega_\rho} |\nabla u|^2 d\Omega, \quad \int_{\Gamma_2} u\, dS, \quad \int_{\Gamma_4} u\, dS,$$

and the same integrals with u replaced by $\hat{u}$. Then, we can determine, in an unique way, the center $y = (y_1, y_2)$ and the radius ρ of the ball $B_\rho(y)$.

4 Numerical Example of Shape Functionals

We consider four boundary value problems defined in the same domain $\Omega = (0,1) \times (0,1)$. It means, that for $i = 1,2,3,4$ we have $\Delta u_i = 0$ in Ω. These problems differ with respect to the boundary conditions. For $i = 1$ they have the form

$$u_1 = 1 \quad \text{on } \{0\} \times \left(\frac{1}{3}, \frac{2}{3}\right); \quad u_1 = 0 \quad \text{on } \{1\} \times (0,1); \quad \frac{\partial u_1}{\partial n} = 0 \quad \text{otherwise.}$$

For $i = 2,3,4$ they are obtained from the above conditions applying the successive rotation by the angle $\pi/2$. The shape functionals $I_j = I_j(\Omega)$ are defined as follows: for $j = 1, \ldots, 12$, $i = 1, \ldots, 4$.

$$I_{\{1+3(i-1)\}} = \int_\Omega u_i^2 d\Omega, \; I_{\{2+3(i-1)\}} = \int_\Omega (u_{i/1})^2 d\Omega, \; I_{\{3+3(i-1)\}} = \int_\Omega (u_{i/2})^2 d\Omega$$

The topological derivatives of shape functionals are obtained from Theorem 1.

The inverse mapping G^{-1}, which allows for identification of inclusion, is difficult to calculate from the mathematical relations and therefore was modeled using ANN's. Similarly as in the classical approach, the inverse mapping G^{-1}, shown in Fig. 1, may be determined unambiguously only when the transformation G from (y_1, y_2, ρ) into $I_1, \ldots, I_{12}$ is one to one. The knowledge about the inverse mapping is stored within the network structure and network connection weights. Values of functionals $I_1, \ldots, I_{12}$ are calculated by the use of topological derivative method for the square with the inclusion are the network input vector. The approximated values of the corresponding inclusion's parameters,

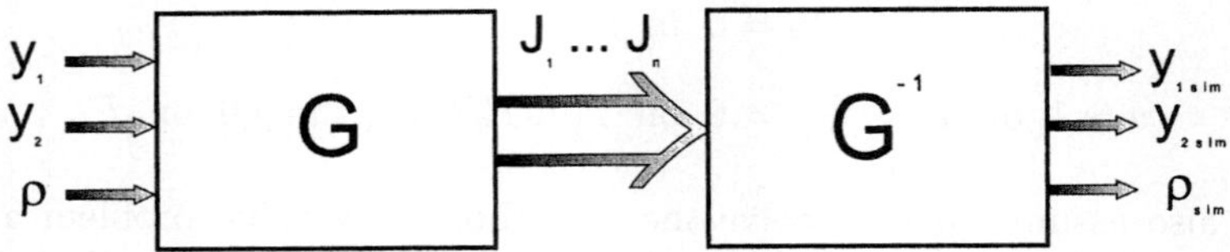

Fig. 1. An inverse mapping problem

such as radius ρ_{sim} and position (y_{1sim}, y_{2sim}), are calculated at the network output. An unknown mapping of the input vector to the output vector is approximated in an iterative neural network training [2]. In our particular problem feed forward MLP network, sum square error cost function and back propagation learning algorithm with Levenberg–Marquardt [2] optimisation method were applied. This algorithm was implemented in MATLAB. Different MLP networks with a single hidden layer were considered and two of them were tested. The network structure (12-18-3) i.e.: twelve inputs, eighteen processing units with a sigmoidal transfer function in the network hidden layer and three linear units in the output layer, comprising 291 weights; and network (12-24-3) with 387 weights. Numerical computations that were based on the topological derivative have provided data both for network training and testing procedures. The training and testing data were computed for different values of inclusion radius, which were changed from 0,05 to 0,2 and for the corresponding values of the inclusion position. Position coordinates were changed in the range $2\rho_i < y_{1i} < 1 - 2\rho_i$, $2\rho_i < y_{2i} < 1 - 2\rho_i$.

From available data sets, 1285 that correspond to the radii [0.05, 0.088, 0.125, 0.16, 0.2] were selected for network training and 205 for radii [0.075, 0.1, 0.18] were selected for network testing. The latter one is required for validation of the network true generalization capabilities. The stopping condition for the learning procedure was the value of sum square error SSE less then 0.02. The network (12-18-3) was trained by the use of Levenberg-Marquardt algorithm in 69 epochs and the network (12-24-3) in 44 epochs. The maximum values of relative errors for both of the tested networks and their training times are given in a Table 1. Increasing the number of neurons in the hidden layer has improved the accuracy by a factor of two, but the learning time has also increased ten times. The maximum relative errors for the network (12-24-3) for three values of ρ are given in Table 2. The largest values of the errors in position identification are observed at the corners of the square.

Conclusions. An example of numerical solution of $2D$ shape inverse problem was presented in the paper. Identification of the position and radius of the small

Table 1. Maximum values of relative errors for $\rho = 0.075$

network structure	δy_1 [%]	δy_2 [%]	$\delta\rho$ [%]	learning time [min]
(12-18-3)	12	12	5	24
(12-24-3)	6	6	3	240

Table 2. Maximum relative errors for the network (12-24-3)

ρ	δy_1 [%]	δy_2 [%]	$\delta\rho$ [%]
0.075	6	6	3
0.1	5	5	3
0.18	2	2	3

inclusion in a square, which is difficult to calculate from the mathematical relations, was computed using ANN's. The presented experiments indicate, that the approach based on using topological derivative for producing training data for neural networks, gives promising results.

References

1. Barron, A.R.: Universal approximation bounds for superpositions of a sigmoidal function, IEEE Transactions on Information Theory, **39** (1993) 930-945.
2. Hagan, M., Menhaj, M.: Training feedforward networks with the Marquardt algorithm, IEEE Trans. on Neural Networks, **5** (1994) 989-993.
3. Il'in, A. M.: Matching of Asymptotic Expansions of Solutions of Boundary Value Problems, Translations of Mathematical Monographs, **102** AMS 1992.
4. Jackowska, L., Sokołowski, J., Żochowski, A., Henrot, A.: On numerical solutions of shape inverse problems Computational Optimization and Applications, —23 (2002) 231-255.
5. Lewiński, T., Sokołowski, J., Żochowski, A.: Justification of the bubble method for the compliance minimization problems of plates and spherical shells CD-Rom, 3rd World Congress of Structural and Multidisciplinary Optimization (WCSMO-3) Buffalo/Niagara Falls, New York, May 17-21, (1999).
6. Nazarov, S. A., Sokołowski, J.: Asymptotic analysis of shape functionals, Les prépublications de l'Institut Élie Cartan 51/2001.
7. Roche, J.R., Sokołowski, J.: Numerical methods for shape identification problems Special issue of Control and Cybernetics: Shape Optimization and Scientific Computations, **5** (1996) 867-894.
8. Schiffer, M., Szegö, G.: Virtual mass and polarization, Transactions of the American Mathematical Society, **67** (1949) 130-205.
9. Shumacher, A.: Topologieoptimierung von Bauteilstrukturen unter Verwendung von Lochpositionierungkriterien, Ph.D. Thesis, Universität–Gesamthochschule–Siegen, Siegen, 1995.
10. Sokołowski, J., Żochowski, A.: On topological derivative in shape optimization, SIAM Journal on Control and Optimization. **37**, No. 4 (1999) 1251-1272.
11. Sokołowski, J., Żochowski, A.: Topological derivative for optimal control problems, Control and Cybernetics, **28**, No. 3 (1999) 611–626.
12. Sokołowski, J., Żochowski, A.: Topological derivatives for elliptic problems, Inverse Problems, **15**, No. 1 (1999) 123–134.
13. Sokołowski, J., Żochowski, A.: Topological derivatives of shape functionals for elasticity systems Mechanics of Structures and Machines **29** (2001) 333-351.
14. Sokołowski, J., Żochowski, A.: Optimality conditions for simultaneous topology and shape optimization, SIAM Journal on Control and Optimization. **42**, No. 4 (2003) 1198-1221.

Application of Domain Neural Network to Optimization Tasks

Boris Kryzhanovsky and Bashir Magomedov

Institute for optical-neural technologies RAS,
Vavilova st. 44/2, 199333, Moscow, Russia
kryzhanovsky@mail.ru, bashir.magomedov@gmail.com
http://www.iont.ru

Abstract. A new model of neural network (the domain model) is proposed. In this model the neurons are joined together into more large groups (domains), and accordingly the updating rule is modified. It is shown that memory capacity grows linearly as function of the domain size. In optimization tasks, this kind of neural network allows one to find more deep local minima of the energy than the standard asynchronous dynamics.

1 Introduction

The dynamics of well-known spin models of neural networks [1,2] consists in aligning of each spin along the direction of the local field. The storage capacity M of such a network is comparatively low: $M \sim N/2\ln N$, where N is the number of neurons. The network storage capacity depends both on the way of organization of interaction between neurons (the architecture of the network) and the way of relaxation into a stable state (the dynamics of the network). However, usually only the possibility to increase the storage capacity by means of changing of the architecture is discussed, and at the same time the standard spin dynamics related to the Hopfield model is used [3-9]. In what follows we would like to show that the storage capacity can be increased noticeably by means of changing of the neural network dynamics.

In cite [10] we proposed a new type of neural network, which was called *the domain neural network*. Its dynamics is defined by overturns of domains. Each domain is a group of strongly coupled spins. Overturn of a domain means the simultaneous changing of orientations of all the k spins constituting the domain. We will show that replacing of the spin dynamics by the domain one leads to k times increase of the storage capacity. Moreover, it will be shown that the domain neural network can be efficiently used in optimization problems. The point is that this model allows us to find minima on the energy surface that are deeper than the ones obtained with the aid of the Hopfield model.

2 Description of the Domain Model

Let us examine a system of N spins, which take the values $s_i = \pm 1$, where $i = 1,2,...,N$. The behavior of the system is described by the Hamiltonian

W. Duch et al. (Eds.): ICANN 2005, LNCS 3697, pp. 397 – 403, 2005.

$$E = -\tfrac{1}{2}\sum_{i=1}^{N} J_{ij} s_i s_j \tag{1}$$

where J_{ij} are matrix elements of the Hebb connection matrix [2],

$$J_{ij} = (1-\delta_{ij})\sum_{m=1}^{M} s_i^{(m)} s_j^{(m)} \tag{2}$$

In Eq.(2) $S_m = (s_1^{(m)},\ s_2^{(m)}\ ,\ ...,\ s_N^{(m)})$, $m = 1,...,M$, are randomized binary patterns. The local field acting on the ith spin is calculated according to the usual rule: $h_i = -\partial H / \partial s_i$.

Let us define the domain neural network. We suppose that the system of N spins is divided into groups each of which contains k spins. Each group is a domain. In the domain the spins are strongly coupled, and when the domain turns over, all the spins in the domain change their signs simultaneously(*). Thus, our system consists of N/k domains. When the state of the system is changing due to overturns of domains only, its dynamics is called *the domain dynamics*. From physical point of view the behavior of the domain network is determined by stability of domains in the local field. The given domain turns over, if as a result the energy of the system decreases. For example, let us examine the first domain, i.e. the group of coupled spins whose numbers are $1 \le r \le k$. To define the domain stability, let us write down its energy (the sum of the energies of all k spins constituting the domain) in the form of two terms. The first term is the intrinsic energy of the domain that is the energy of interaction of the spins of the domain. The second term is the energy of interaction of the given domain with other domains of the system (E_{int}) , i.e. the energy of interaction of the spins belonging to the given domain with spins of all other domains:

$$E_{\text{int}} = -\sum_{r=1}^{k}\sum_{j=k+1}^{N} J_{rj} s_r s_j = -\sum_{\mu=1}^{M}\sum_{r=1}^{k}\sum_{j=k+1}^{N} s_r s_r^{(\mu)} s_j^{(\mu)} s_j \tag{3}$$

Evidently, the domain stability is defined completely by the sign of the interaction energy E_{int} . The value and the sign of the intrinsic energy of the domain are of no importance, since they do not change when the domain turns over. Consequently, the domain dynamics of the network is defined as follows. If at the time t inequality $E_{\text{int}}(t) > 0$ is fulfilled, then the domain turns over at the next step, i.e. it transforms to the state $s_r(t+1) = -s_r(t),\ \forall r = 1,...,k$, with the negative interaction energy $E_{\text{int}}(t+1) < 0$. If $E_{\text{int}}(t) < 0$, then the domain is stable and at the next step its state is the same: $s_r(t+1) = s_r(t),\ \ \forall r = 1,...,k$. Under the described dynamics the energy of the system as a whole decreases, and, consequently, the algorithm converges after finite number of steps. It should be stressed that a domain can overturn even if each of its spins is in the stable state, i.e. each spin is directed along the its own local field.

(*) In contrast to a real domain where all the spins are aligned, the spins of our formal "domain" can have different directions.

3 Recognizing Ability of Domain Model

Let us examine the recognizing ability of the domain neural network. Let $S=(s_1,s_2,...,s_N)$ be the input vector. It is a distorted copy of the pattern S_m. To be concrete, we suppose that not individual spins are distorted, but the domains as a whole: p is the probability that a domain is distorted (all the spins of the domain change their signs); $1-p$ is the probability that the domain of the pattern is not distorted. When the number of domains is sufficiently large, $N>>k$, the probability of correct recognition of the pattern can be obtained using the central limit theorem. For this purpose let us represent $E_{\mathbf{int}}$ in the form $E_{\mathbf{int}}=S+R$, where S is the useful signal (the part of the sum (3) related to the mth pattern) and R is the noise (the contribution to (3) from all other patterns). The analysis of statistical properties of S and R shows [8] that the signal mean value is $S=k(N-k)(1-2p)$. The noise is zero mean and its dispersion is $\sigma^2=k(N-k)M$. With regard to this relations it is easy to obtain the expression for the probability of the error of recognition. In the most interesting limit $\gamma=S/\sigma>>1$ this probability has the form:

$$P=\frac{N}{\sqrt{2\pi}\ \ k\gamma}\exp\left(-\frac{kN}{2M}(1-2p)^2\right) \tag{4}$$

When $k=1$, Eq.(4) represents the known result for the Hopfield model. However, when the size k of the domains increases, the probability of error decreases exponentially. This statement is justified by the results of computer simulations for N=600, M=1200, p=0. In Fig.1 we present the dependence of the value of P on the size of the domain k: experimental data correspond well with the curve plotted according with the formula (4). We see that if the size of domains is small (k<10), the network does not recognize the input patterns. When $k>15\div20$ the domain network recognizes the patterns with confidence, though the value of the loading parameter is sufficiently large ($M/N=2$). For comparison , the Hopfield network ($k=1$) can functionate as an associative memory only if the loading parameter is small $M/N\le 0.14$.

From (4) it follows that for the domain model the estimation of the number of fixed points in the asymptotic limit $N\to\infty$ is

$$\overline{M}=kN/2\ln N \tag{5}$$

We see that the storage capacity $\overline{M}$ of the domain neural network is k times greater than for the Hopfield network. We accentuate that the expression (5) defines the number of fixed points only. The number of distorted patterns, which can be recognized, is less: $M=\overline{M}\ (1-2p)^2$. The results of computer simulations for N=1000 are in good agreement with the plot made with the aid of the formula (5), which is shown in Fig.2. These results confirm the possibility to obtain the loading parameter (M/N) by an order of magnitude greater than for the Hopfield model.

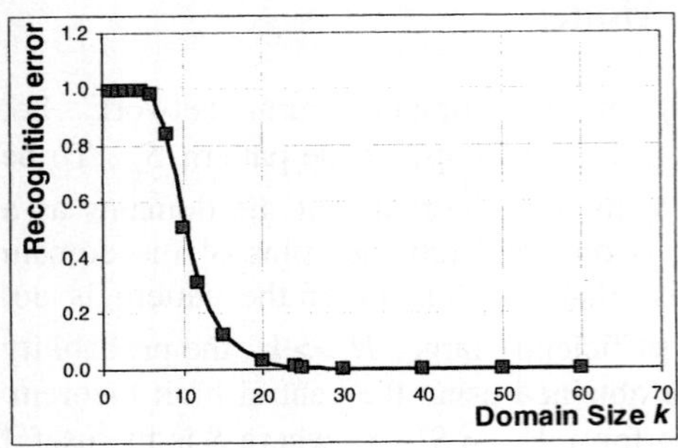

Fig. 1. Dependence of recognition error on domain size

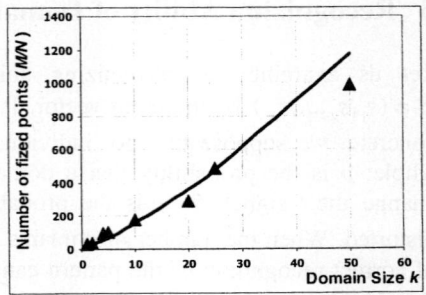

Fig. 2. Dependence of number of fixed points on domain size

We see that the storage capacity of the network can be increased noticeably by means of changing of the network dynamics only. Imperfection of the model lies in its inability to recognize multiplicatively distorted patterns, since no overturn of a domain can improve a distortion of a single bit (a single spin). On the other hand, this model can be used under the most difficult conditions of recognition, when the bits of transmitting information have correlated distortions.

4 Application of Domain Dynamics for Optimization

Let us examine the possibility to apply the domain model for minimization of the energy functional (1) for an arbitrary connection matrix J_{ij}. The functioning of the network can be treated as a motion of a particle along the energy surface (1). In Fig.3 the dotted line is the typical form of the descent along the energy surface under the standard spin dynamics, the solid line corresponds to the domain dynamics. We see that when starting from the point A the domain network gets through shallow minima (they are traps for the Hopfield network) and has the possibility to move towards the

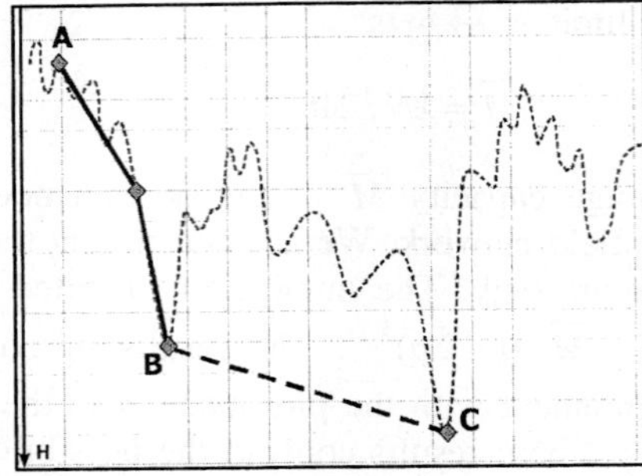

Fig. 3. The motion along the energy surface (1). The dotted line corresponds to the spin dynamics, the solid line corresponds to the domain dynamics.

more deep minimum $\boldsymbol{B}$. Moreover, due to an overturn of a single domain only, the domain network can jump from the point $\boldsymbol{B}$ into the point $\boldsymbol{C}$. The last is impossible in the case of spin dynamics, where such a transition can be done due to tunneling trough the barrier only.

The domain dynamics can be compared with movement of a big rock, which rolls down a mountainside and does not notice shallow defects of the surface. Drawing this analogy, one might expect that moving towards the bottom the rock feels the general slope of the surface. This assumption was verified with the aid of experiments, where for random Hebbian connection matrices with different loading parameters the abovementioned two dynamics were compared.

The description of the experiments. The connection matrices were defined according the generalized Hebb rule. All the patterns S_m with random statistical weights $w_m \in (0,1)$ were used to construct the matrix elements (2). Introducing of statistical weights allowed us to make an intricate energy landscape: local minima of different depth corresponded to different patterns.

We used random binary vectors as starting points. The coordinates of these vectors were combined into domains of a size $k > 1$. Then the domain dynamics was used. When a local minimum was achieved, we set k=1 and, if it was possible, the descending along the landscape was continued according to the spin dynamics. Beginning from the same starting point we used the domain dynamics with different values of the parameter k, including $k = 1$ (the spin dynamics). For different k we fixed the values of energy minima. We found out how frequently the domain network reached more deep or more high minima (E_d) comparing with the Hopfield network (E_h). Furthermore, we estimated the benefit or the loss resulting from using of the domain network. Note, sometimes when using the spin dynamics and the domain dynamics, the values of the reached local minima were equal ($E_d = E_h$). Note, that domains could be chosen arbitrary. But in current experiments, domains were chosen as follows: the first domain – consists of spins s_i, $i = 1,2,..,k$; second domain - s_i, $i = k+1,..,2k$, and so on.

Five different loading parameters were examined: M/N = 0.05, 0.1, 0.2, 0.5, 1. For each value of M/N we used $5\text{x}10^4$ random starting points. The results of one of the experiments (N=100, k=2) averaged over all the starts are given in Table 1. In the first column we write down the probability to find the domain network in the deeper minimum than the Hopfield network. In the second column the probability of relaxation of the both networks in the same minima is presented. In the third column we show the probability of relaxation of the Hopfield network in the deeper minimum.

Thus, comparing with the Hopfield network, we find the domain network in a deeper minimum more frequently. Moreover, when the reached minimum is deeper, for the domain dynamics we get rather large energy benefit $\delta E = 100\% \times (E_d - E_h)/E_h$. In the described experiment for a small loading parameter (M/N=0.05) we had $\delta E \sim 14\%$. The loss of the domain network was an order of magnitude less: $\delta E \sim -2\%$. In different experiments the benefits and the

Table 1.

N=100, *k*=2			
	$E_d < E_h$	$E_d = E_h$	$E_d > E_h$
$M = 5$	0.31	0.47	0.22
$M = 10$	0.42	0.26	0.32
$M = 20$	0.40	0.36	0.24
$M = 50$	0.47	0.07	0.46
$M = 100$	0.54	0.04	0.42

losses of the domain network vary. Sometimes the obtained benefit was even $\delta E = 48\%$. After averaging over *N* experiments, the mean values of the benefit and the loss were $\delta E = 11.3\%$ and $\delta E = -3.4\%$, respectively.

For larger values of the loading parameter, all minima were practically the same. In this case, as it is seen from Table 1, only very rarely we have $E_d = E_h$. However, even in this case, when the energy landscape has no pronounced local minima, the domain network reaches deeper minima more frequently, and the averaged benefit is $\delta E = 3.6\%$.

5 Conclusions

Summing the abovementioned results we state that comparing with the Hopfield network:

- As a rule, the domain network reaches deeper minima more frequently. The frequency depends on the loading parameter and the size of domains.
- For the domain network the benefit always exceeds the loss.
- The domain networks wins not only if there are pronounced local minima, but also when the local minima differ slightly (for example, when $M / N \geq 1$). Our experiments showed that the domain network could fix very small energy differences.

Concluding this article we would like to emphasize that all the advantages of the domain network are obtained by changing the dynamics of the network only, without any modification of its architecture.

Acknowledgment

Authors are grateful to Dr. L.B.Litinskii and to academician A.L.Mikaelyan for helpful discussions during the course of this work. The work was supported by Russian Basic Research Foundation (the project 04-07-90038) and the program "Intellectual computer systems" (the project 2.45).

References

1. Hopfield, J.: Neural Networks and physical systems with emergent collective computational abilities. Proc.Nat.Acad.Sci.USA. v.79 (1982) 2554-2558.
2. Hebb, D.: The Organization of Behavior. New York: Wiley, 1949.
3. Palm, G., Sommer, F.: Information capacity in recurrent McCulloch-Pitts networks with sparsely coded memory states. Network v.3 (1992) 177-186
4. Abu-Mostafa Y. S., St. Jacques J. Information capacity of the Hopfield model. IEEE Transactions on Information Theory vol.31 (4), (1985) 461-64.
5. Cohen M. A., Grossberg S. G. Absolute stability of global pattern formation and parallel memory storage by compatitive neural networks. IEEE Transactions on Systems, Man and Cybernetics v. 13 (1983) 815-26.
6. Grossberg S. The adapptive brain, vol. 1 and 2. Amsterdam: North-Holland. 1987.
7. Amit D.J., Gutfreund H. & Sompolinsky H. Statistical mechanics of neural networks near saturation. Annal of Physics, v.173, (1987) 30-67.
8. Kinzel W. Learning and pattern recognition in spin glass models. 2 Physik B, v. 60 (1985) 205-213.
9. Kryzhanovsky B.V., Mikaelian A.L. An associative memory capable of recognizing strongly correlated patterns. Doklady Mathematics, v.67, No.3 (2003) 455-459.
10. Kryzhanovsky B., Magomedov B., Mikaelian A.: Domain Model of a Neural Network. Doklady Mathematics, v. 71, No. 2 (2005) 310-314.

Eigenvalue Problem Approach to Discrete Minimization

Leonid B. Litinskii

Institute of Optical Neural Technologies Russian Academy of Sciences,
Moscow, Russia
litin@iont.ru

Abstract. The problem of finding of the deepest local minimum of a quadratic functional of binary variables is discussed. Our approach is based on the asynchronous neural dynamics and utilizes the eigenvalues and eigenvectors of the connection matrix. We discuss the role of the largest eigenvalues. We report the results of intensive computer experiments with random matrices of large dimensions $N \sim 10^2 - 10^3$.

1 Introduction

In various applications ([1], [2], [3]) it is necessary to minimize a quadratic functional depending on N binary variables $s_i = \{\pm 1\}$:

$$\min_{\mathbf{s}} \left\{ E(\mathbf{s}) = -\sum_{\mathrm{i,j=1}}^{N} J_{ij} s_i s_j = -(\mathbf{Js}, \mathbf{s}) \right\}, \tag{1}$$

where $(\mathbf{Js}, \mathbf{s})$ is the scalar product of N-dimensional vectors. Vectors $\mathbf{s} = (s_1, s_2, ..., s_N)$ with binary coordinates will be called *configuration vectors*. They define 2^N binary configurations, among which the optimal configuration with regard to the objective function $E(\mathbf{s})$ has to be found. When N increases, the number of different states increases exponentially. Already if $N > 50$, it is almost impossible to solve the problem by means of exhaustive search.

Without loss of generality we consider a symmetric connection matrix $\mathbf{J} = (J_{ij})_{i,j=1}^{N}$ with zero diagonal elements, $J_{ij} = J_{ji},\ J_{ii} = 0,\ \forall i, j$. The objective function $E(\mathbf{s})$ will be called *the energy* of the state.

Usually, one uses the asynchronous dynamic procedure to minimize $E(\mathbf{s})$. According this procedure, if $\mathbf{s}(t) = (s_1(t), s_2(t), ..., s_N(t))$ is the state of the system at the moment t, in the next moment, $t+1$, coordinates of the state change according the rule

$$s_i(t+1) = \mathrm{sign}\left(\sum_{j=1}^{N} J_{ij} s_j(t) \right). \tag{2}$$

It is known that starting from an arbitrary initial state $\mathbf{s}(0)$, after several steps the system (2) gets into the nearest local energy minimum. As a rule, the number of local minima is very large [4]. To solve the problem, it is important to start the system from "a good" initial state, which is located in the basin of attraction of the global minimum.

W. Duch et al. (Eds.): ICANN 2005, LNCS 3697, pp. 405–410, 2005.

We suggest to start the dynamic system (2) from the configuration vectors, which are the nearest to the eigenvectors of the matrix $\mathbf{J}$ corresponding to the largest eigenvalues. The idea is based on evident geometrical argumentation (see below). This approach proved itself for matrices of small dimensions $N \sim 20$, allowing us to find the global energy minimum for statistical assembly of 1500 random matrices with probability 0.97 [5], [6]. The approach has the computational complexity of $O(N^3)$. It is of interest due to its universality.

In this publication we present the computer simulation results for random matrices of large dimensions $N \sim 10^2 - 10^3$. The organization of the paper is as follows. In Sect. 2 we describe the algorithm allowing us to find the deepest local minimum. In Sect. 3 we estimate the computational complexity of the algorithm and present the numerical results for random Gaussian matrices.

2 The Algorithm

The symmetric matrix $\mathbf{J}$ possesses the full set of the eigenvectors $\mathbf{f}^{(i)}$. Let us sort them in descending order with regard to eigenvalues λ_i:

$$\mathbf{J} \cdot \mathbf{f}^{(i)} = \lambda_i \mathbf{f}^{(i)}, \ \lambda_1 \geq \lambda_2 \geq \ldots \geq \lambda_k > 0 > \ldots \lambda_N, (\mathbf{f}^{(i)}, \mathbf{f}^{(j)}) = \delta_{ij}.$$

Since the diagonal elements of the matrix are equal to zero, a part of the eigenvalues is positive, and another part is negative.

The functional (1) can be rewritten as

$$E(\mathbf{s}) = -\left(\lambda_1(\mathbf{s}, \mathbf{f}^{(1)})^2 + \ldots + \lambda_k(\mathbf{s}, \mathbf{f}^{(k)})^2 + \ldots + \lambda_N(\mathbf{s}, \mathbf{f}^{(N)})^2\right). \tag{3}$$

The expression (3), first, allows one to find the lower estimate for the global minimum of the functional (1):

$$E(\mathbf{s}) \geq -\lambda_1 N. \tag{4}$$

Second, from Eq.(3) it is clear that one has to look for the solution of the problem among those configuration vectors $\mathbf{s}$, which have the smallest possible projection onto the eigenvectors corresponding to negative eigenvalues. Indeed, $E(\mathbf{s})$ is proportionate to a weighted sum of squared projections of a configuration vector $\mathbf{s}$ onto eigenvectors $\mathbf{f}^{(i)}$. The weights entering the sum (3) are λ_i. Then, the larger the projection of a configuration vector onto the subspace spanned over *the largest* eigenvectors, the smaller is the value of the functional $E(\mathbf{s})$. Eigenvectors corresponding to some first maximal eigenvalues will be called *the largest* eigenvectors.

To clarify the aforesaid, we analyze the situation when the eigenvalue λ_1 exceeds essentially all the other eigenvalues in modulus (one faces such a situation not so rarely):

$$\lambda_1 >> \mid \lambda_i \mid, \ i = 2, \ldots N.$$

Then in the expression (3) we can restrict ourselves with the first term only:

$$E(\mathbf{s}) \approx -\lambda_1(\mathbf{s}, \mathbf{f}^{(1)})^2.$$

Evidently, in this case the solution $\mathbf{s}^*$ is the configuration vector the nearest to $\mathbf{f}^{(1)}$.

It is very simple to find out this configuration vector: its coordinates are defined by the signs of the coordinates of $\mathbf{f}^{(1)}$:

$$s_i^* = \mathrm{sign}(f_i^{(1)}), i = 1, ..., N \Rightarrow (\mathbf{s}^*, \mathbf{f}^{(1)}) = \sum_{i=1}^{N} |f_i^{(1)}| \geq |(\mathbf{s}, \mathbf{f}^{(1)})| \; \forall \, \mathbf{s}. \quad (5)$$

Thus, if the maximal eigenvalue λ_1 of the matrix $\mathbf{J}$ is very large, the solution of the problem (1) is very simple. Difficulties emerge when there are some eigenvalues comparable with the maximal one. In this situation the contributions of these eigenvalues to the functional (3) compete with each other. It can occur that the global minimum is achieved on the configuration vector, which is the nearest not to the first eigenvector, but to some other large eigenvector. However, from the expression (3) it is evident that the global minimum cannot be achieved on the configuration vector orthogonal to the largest eigenvectors.

This argument defines the main idea of our approach: to find out the deepest local minimum of the functional (1), the dynamic system (2) has to be started from configuration vectors the nearest to the largest eigenvectors. It is very easy to determine these nearest configuration vectors (see Eq.(5)).

3 Experiments with Gaussian Matrices

In this Section we present the results of computer simulations for random symmetric Gaussian matrices of dimensionalities $N = 60$, 100, 200, 300, 400, 500, 750, 1000, 1500, and 2000. Their matrix elements where chosen randomly and independently from the standard Gaussian distribution. In physics matrices of such a kind are used when describing the so called *spin glass*, which is characterized by multiply degenerated global minimum of the energy [2]. In this case the number of local minima is exponentially large. All these minima have approximately the same depth, and distributed with a small dispersion around a mean value. When N increases, the dispersion tends to zero. In other words, there is no a pronounced global minimum, and, consequently, the search of the deepest minimum is especially difficult.

3.1 Computer Simulation Parameters

Because the exact value of the global minimum is unknown, a reasonable way to verify the suggested approach is to compare our deepest local minimum with the deepest local minimum determined with the aid of the random search. For each N we generated 1000 random matrices and using this statistical assembly compared the two aforementioned minimization methods. In addition we took care of both methods being of nearly the same computational complexity.

The computational complexity of our approach is defined by the time required for calculation of eigenvalues and eigenvectors of the connection matrix. In order of magnitude this time is equal to $O(N^3)$. When the system (2) starts from an arbitrary configuration $\mathbf{s}(0)$, about N^2 operations are needed to get into the nearest local minimum. Consequently, we can start the system (2) from N configuration vectors that are the nearest to all the eigenvectors of the matrix $\mathbf{J}$. Then we have to start the dynamic system from the same number N of random configurations. Thus, the computational complexity of the both minimization methods in order of magnitude is $O(N^3)$.

Starting configurations $\mathbf{s}(0)$ that are the nearest to the eigenvectors of the matrix $\mathbf{J}$ will be called *the EIGEN-configurations*; the obtained local minima will be called *the EIGEN-local minima*. Similarly, local minima obtained when starting from RANDOM-configurations will be called *the RANDOM-local minima*.

3.2 Main Results

1. Our experiments showed that the deepest EIGEN-local minima could be obtained when starting the system from the vicinity of the largest eigenvectors.

This is clearly seen from Fig.1, where for N= 100, 500 and 1000 we present histograms of distribution of the numbers of those eigenvectors, starting from whose vicinity leads the system (2) to the deepest EIGEN-local minimum. We would like to remind that numbers of eigenvectors corresponding to the largest eigenvalues are small: 1, 2, 3 and so on. To get the histograms, 1000 random tests were done.

For every dimensionality N we see the leading role of the largest eigenvalues. When N increases, this tendency becomes only stronger. Thus, for $N = 1000$ in the overwhelming majority of the cases the deepest EIGEN-local minimum was achieved when starting from configuration vectors the nearest to 10 largest eigenvectors only.

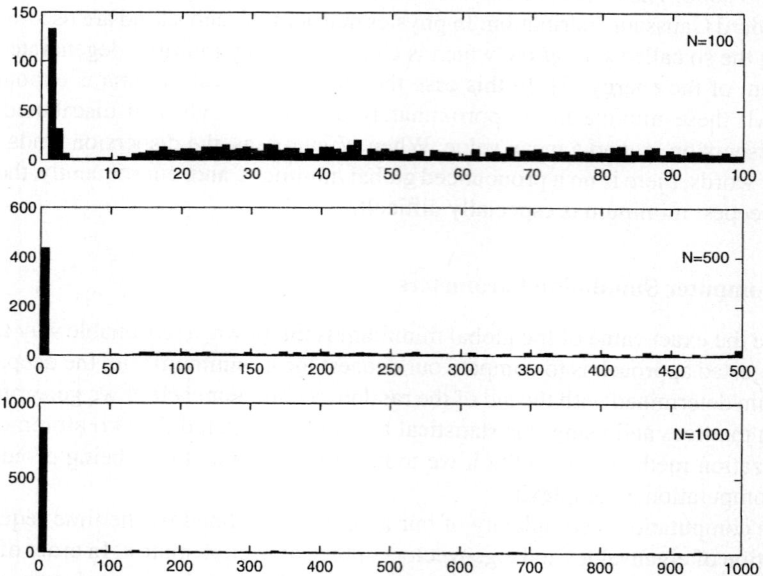

Fig. 1. The numbers of eigenvectors starting from whose vicinity provided the deepest EIGEN-local minimum. The dimensionality of matrices N=100, 500 and 1000.

2. For each matrix the deepest EIGEN-local minimum was compared with the deepest RANDOM-local minimum to define whichever was deeper. Then using 1000 random tests we calculated the probabilities of two events:

i) the probability $p_e = \text{Prob}(\text{EIGEN} < \text{RANDOM})$ to get the deeper local minimum with the aid of our EIGEN-approach;
ii) the probability $p_r = \text{Prob}(\text{RANDOM} < \text{EIGEN})$ to get the inverse result.

Our results are given in the second and third columns of Table 1, respectively. We see that for small values $N \sim 60 - 100$ both minimization approaches are practically identical: the probabilities p_e and p_r are approximately equal. In addition, there is a large probability that both deepest local minima are the same. For $N = 60$ we have $\text{Prob}(\text{EIGEN} = \text{RANDOM}) = 0.63$, and $\text{Prob}(\text{EIGEN} = \text{RANDOM}) = 0.30$ for $N = 100$.

Table 1. Results for two methods of minimization

N	p_e	p_r	Δ_e	Δ_r
60	0.18	0.19	0.14	0.16
100	0.37	0.33	0.27	0.23
200	0.58	0.41	0.47	0.33
300	0.56	0.44	0.52	0.34
400	0.61	0.39	0.57	0.26
500	0.62	0.38	0.59	0.26
750	0.76	0.24	0.74	0.12
1000	0.87	0.13	0.98	0.06
1500	0.97	0.03	1.33	0.01
2000	0.996	0.004	1.64	0.001

However, beginning from $N = 200$ the probability p_e exceeds the probability p_r. When N increases, this superiority increases too. When $N = 2000$ the probability p_e is close to one. Note, as it was mentioned above, for large $N \sim 1000$ the deepest EIGEN-local minimum is reached when starting the system (2) from several largest eigenvectors only.

3. It is important to estimate the worth of the improvement obtained with the aid of the EIGEN-approach comparing with the results of the RANDOM-minimization. To make this estimate we used the following method. Suppose, for the ith matrix, E_e is the deepest EIGEN-local minimum and E_r is the deepest RANDOM-local minimum. We also suppose that $E_e < E_r$. We remind that always $E_e, E_r < 0$. Then for the ith matrix the value

$$\Delta_i = \frac{E_e - E_r}{E_r} \cdot 100 > 0$$

shows (in percentage terms) how low the RANDOM-local minimum can be deepen when using the EIGEN-approach. Averaging the positive Δ_i over 1000 random tests, we obtain the value of Δ_e that characterizes in average the efficiency of the EIGEN-approach,

$$\Delta_e = \frac{1}{1000} \sum_{\Delta_i > 0} \Delta_i.$$

For those matrices, where the inverse inequality $E_r < E_e$ was true, we calculated

$$\delta_i = \frac{E_r - E_e}{E_e} \cdot 100 > 0, \ \Delta_r = \frac{1}{1000} \sum_{\delta_i > 0} \delta_i.$$

The value of Δ_r characterizes in average the efficiency of the RANDOM- minimization compared with the EIGEN-approach (in percentage terms).

The values of Δ_e and Δ_r are given in the fourth and fifth columns of Table 1, respectively. As above, for small $N \sim 60 - 100$ both minimization methods are practically the same. When N increases, the value of Δ_e increases monotonically, and the value of Δ_r decreases.

In other words, the larger N, the more is the possibility to make gains (to reach a deeper local minimum) when using the EIGEN-approach. The fact that the benefit is relatively small, $\sim 1.5\%$ only, is understood easily. The spin glass has no a pronounced global minimum. On the contrary, when $N \to \infty$ there are a lot of local minima of approximately the same depth. In other words, starting from any initial configuration, the system reaches a local minimum whose depth is practically the same as depths of other local minima. The fact that for large N the results of the EIGEN-approach are better is the evidence of its efficiency.

Acknowledgment

It would not be possible to complete this work without stimulative support and help of B.V.Kryzhanovsky and D.V.Vylegzhanin.

The work was supported by the grant of President of Russian Federation (SS-1152) and in part by Russian Basic Research Foundation (the grant 05-07-90049).

References

1. Hertz, J., Krogh, A., Palmer, R.: Introduction to the Theory of Neural Computation. Addison-Wesley (1991)
2. Hartmann, A.K., Rieger, H.: Optimization Algorithms in Physics. Wiley-VCH, Berlin (2001)
3. Smith, Kate A. Neural networks for combinatorial optimization: A review of more than a decade of research. INFORMS Journal on Computing, **11**(1) (1999) 15-34.
4. Joya, G., Atencia, M., Sandoval, F. Hopfield Neural Networks for Optimization: Study of the Different Dynamics. Neurocomputing, **43**(1-4) (2002) 219-237.
5. Kryzhanovsky, B.V., Litinskii, L.B.: Finding a global minimum of one multiextremal functional. Artificial Intelligence #3 (2003) 116-120 (in Russian).
6. Litinskii, L.B., Magomedov, B.M.: Global Minimization of a Qudratic Functional: Neural Networks Approach. Pattern Recognition and Image Analysis **15**(1) (2005) 80-82.

A Neurocomputational Approach to Decision Making and Aging

Rui Mata

Max Planck Institute for Human Development, Lentzeallee 94,
14195 Berlin, Germany
mata@mpib-berlin.mpg.de

Abstract. The adaptive toolbox approach to human rationality analyzes environments and proposes detailed cognitive mechanisms that exploit the structures identified. This paper argues that the posited mechanisms are suitable for implementation as connectionist networks and that this allows (1) integrating behavioral, biological, and information processing levels, an attractive feature of any approach in cognitive science; and (2) addressing developmental issues. These claims are supported by reporting implementations of decision strategies using simple recurrent networks and showing how age differences related to attenuation in cholaminergic modulation can be modeled by lowering the *G* parameter in these networks. This approach is shown to be productive by deriving empirically testable predictions of age differences in decision making tasks.

1 The Adaptive Toolbox Approach

The adaptive toolbox approach is concerned with how simple mechanisms can exploit the information structure of an environment to make effective decisions. Accordingly, it has been shown that, in some environments, simple algorithms perform very well compared to more complex ones [1]. One example of such an algorithm is Take-the-Best (TTB), a simple lexicographic heuristic for multiattribute pair-comparison tasks (e.g., inferring which of two diamonds is more expensive based on a set of cues, such as carat, clarity, and cut). A lexicographic strategy is one that looks up cues in a fixed order, much like we use the alphabetic order arrangement of words when consulting a dictionary. TTB makes decisions by selecting the alternative with the highest value on the cue with highest validity (i.e., the proportion of correct inferences a cue supplies). If the two alternatives have the same value, the cue with the second highest validity is considered, and so forth. It has been shown that TTB cannot be outperformed by a multiple-reason decision strategy like a weighted additive rule (WADD) in a noncompensatory environment, that is, an environment in which one cue cannot be outweighed by any combination of less important cues. WADD is an information-intensive strategy; it creates a weighted value for each cue by multiplying the cue value by its weight and summing over all weighted cue values to arrive at an overall evaluation of an alternative. In comparison, TTB's performance is striking given it has no cue integration process – this success is due to the fit of TTB to the structure of noncompensatory environments.

W. Duch et al. (Eds.): ICANN 2005, LNCS 3697, pp. 411 – 416, 2005.

Several connectionist models of decision making have been proposed [2], [3], [4] and shown to handle complex processes of information integration. This contrasts with the adaptive toolbox approach's emphasis on how the fit between environment and mechanisms allows decision strategies to processs *little* information. Hence, it is understandable that attempts have been made to pit connectionist models against heuristics [2]. Nevertheless, this opposition is not warranted. According to an implementationist perspective to connectionism [5], it is possible to think of networks as implementing aspects of symbol-manipulation. Thus, in principle, one can model decision strategies, such as TTB, as neural networks. Moreover, combining the adaptive toolbox and the connectionism frameworks provides clear benefits. First, the models proposed by the adaptive toolbox approach can inform a connectionst agenda in the domain of inference by supplying plausible algorithms to be modeled. Second, the synergy can help connect different levels of explanation, from the behavioral to the neurological levels. Finally, as proposed here, it can help tackle issues such as ontogenetic change in efficiency of strategy use due to age-related cognitive decline.

1.1 Age-Related Cognitive Decline

Age-related cognitive decline has been studied at different levels. For example, at the behavioral level, researchers have identified age differences in asymptotic performance [6], and complexity cost [7]. At the neurological level, neuroanatomical as well as neurochemical changes due to aging have been reported [8]. Finally, at the information processing level, it has been proposed that there are age related reductions in general processing resources [9].

Although there is a considerable body of work on age-related cognitive decline, little is known about how strategy use in decision making changes with age. A survey of aging and decision making [10] suggests that older adults look up less information and take longer to make a choice than younger adults. Hence, some preliminary evidence exists supporting the idea that age-related cognitive decline impacts strategy use. However, this work did not identify the strategies used or the occurrence of application errors. Thus, it is not known how strategy efficiency changes as a function of age. A principled way of investigating this issue is to consider already detailed models of decision making, associate them with a theory of aging and, subsequently, design empirical studies to test the models' predictions against data. That is the approach taken here.

2 A Connectionist Approach to Decision Making and Aging

Simple recurrent networks (SRN; [11]) were used to implement two decision strategies: TTB, and an evidence accumulation strategy (EAS), which can be thought of as a version of WADD that implies sequential information search. This sequentially feature is common in decision making tasks like pair-comparison ones which are the focus of the reported modeling efforts. SRN are well suited to deal with these tasks because internal states are fed back at every step, which supplies such networks with a memory, and allows them to process information sequentially over time.

The SRN used had 2 input, 20 hidden, 20 context, and 2 output units. Fifty networks with different initial random weights were trained using both TTB-congruent and EAS-congruent target activations. Thus, each network was trained once to implement TTB and once to implement EAS. The number of possible input patterns for training, given all combinations of 5 binary cues for two alternatives, is 1024. However, to insure the networks implemented TTB and EAS, 1/3 of these trials were withheld (341) from the training set to be used after training in a generalization test. The total number of input patterns used to train the networks was, therefore, 683. Each training epoch consisted of supplying as input the same sequence of 683 sets of 5 vectors with 2 cue values each, corresponding to 683 decisions between 2 different objects based on 5 cues.

At the end of each epoch the network weights were updated using a gradient descent backpropagation algorithm [12] with adaptive learning rate and momentum. The networks were trained for 500 epochs. Target activations were constrained to range between 0 and 1. The two versions of each of the 50 networks differed only in terms of the specific targets provided. The crucial difference between the targets were that for the TTB implementations the first object having a positive value on a discriminating cue had an activation of 1, while for the EAS each positive cue value led to a proportion of the maximum possible activation[1]. For example, given 5 cues with equal validities, the contribution of each cue to an object's activation is .2. Thus, EAS targets reflected the principle of information accumulation (see Table 1).

Table 1. Example of mapping between input and activation of alternatives for TTB and EAS

	Input		Targets TTB		Targets EAS	
Cue 1	1	1	0	0	.20	.20
Cue 2	1	0	1	0	.40	.20
Cue 3	0	1	1	0	.40	.40
Cue 4	1	1	1	0	.60	.60
Cue 5	0	1	1	0	.60	.80

After 500 training epochs the mean MSE was considerably small (MSE_{TTB} = .0007; MSE_{EAS} = .0051). However, performance was not perfect: only 13 networks showed perfect performance in the sense that the difference between objects' activations for each decision was in the prescribed direction both when trained with TTB and EAS targets. This subset of networks was further analyzed and it was found that only 2 generalized perfectly to the new 341 trials (MSE < .003). These networks were used to obtain the quantitative predictions presented below regarding strategy efficiency. To obtain a probability that a network would choose a particular option out of the two alternatives a Luce choice rule was employed ($P(A,B) = Activation_A / (Activation_A + Activation_B)$). This rule specifies that the probability of choosing the object

[1] For EAS, the object target activation corresponding to each input cue value was defined by: $value_i \times v_i/\Sigma v$, where $value_i$ represents the object's value on cue i, v_i the validity of that cue, and Σv the sum of all cue validities.

with highest activation tends to .5 as the magnitude of the difference between A and B decreases.

This framework makes the prediction that someone using an evidence-accumulation strategy, will more often make a choice not recommended by that strategy than someone using TTB, which relies on a single piece of information. This is because TTB ensures that the difference between objects with different profiles will be of magnitude 1, while EAS allows differences between 0 and 1, depending on the number of positive cue values. The claim that information intensive strategies are particularly error prone is supported by research showing that people trained to use, for instance, WADD, choose alternatives not prescribed by it more often than when using other, more frugal strategies [13].

2.1 Aging Decision Strategies

Li and colleagues [14] have proposed that deficits in neuromodulatory efficiency due to aging can be conceptualized as noisy information processing and the existence of less distinct neural representations. Specifically, they proposed that deficits in catecholaminergic activation in the prefrontal cortex [8] can be modeled by adjusting the gain (*G*) parameter of the sigmoidal activation function of neural networks (see [15] for the general approach). Li et al. showed that manipulation of the *G* parameter allowed simulating various behavioral findings, for instance, the fact that older adults require more trials to learn paired-associates (i.e., arbitrary word pairs, such as violin-computer; see [14] for more examples). According to this approach, the activation function of neural units takes the form:

$$Activation_t = \frac{1}{1+e^{-(G_t x Input_t + bias)}} \tag{1}$$

where *t* indicates the processing step, *Input* refers to the input supplied to the unit ($Input = \Sigma\, W_{ij} I_j$), and *bias* refers to a negative bias unrelated to aging (usually, $bias = -1$; [15]). Finally, *G* is a random number sampled from a uniform distribution ($G_i \in [G_{min}, G_{max}]$, $G_{min} > 0$; see Fig. 1 for the values of *G* used).

Age-related decline in neuromodulation and its effect on strategy efficiency was modeled by using Equation (1) as the activation function of the output layer of the two successfully trained SRN. Reducing *G* flattens the sigmoidal activation function, lowering units' responsivity (see Figure 1). Quantitative results were obtained by applying the Luce choice rule after a network was presented with the 5 cues concerning each decision for which both TTB and EAS did not have to guess. The outcomes for the two networks were averaged to produce the final results.

In general, sampling *G* from a distribution with a lower mean should produce less pronounced and more variable activations [14], thus making differences between object's activations harder to distinguish by a choice rule. As a consequence, the probability of someone choosing an alternative not recommended by a particular strategy should increase with age. In sum, the relation between aging cognition and decision making was modeled by manipulating the signal to noise ratio of information processing in strategies implemented as neural networks. The following sections present illustrative predictions of this approach concerning strategy efficiency of human decision makers which mirror effects found in other domains, such as memory [14].

Mean Performance. Young and older adults usually differ in terms of mean performance even after considerable periods of training [6]. Figure 1 shows average efficiency of the young and old TTB and EAS networks. As expected, young TTB and EAS perform better than the equivalent old networks.

Complexity Cost. Another robust empirical finding in the aging literature is an age by complexity effect, that is, an increase in the difference between young and older adults with increasing processing demands or task difficulty [7]. Researchers have claimed that lexicographic rules are less computationally demanding than other more information-intensive strategies [1], [13]. Accordingly, an age by complexity effect can be observed in Figure 1: The difference between young and old TTB networks is smaller than the difference between young and old EAS networks (2% vs. 9%). The reason for this is that differences between activations for TTB are on average larger than for EAS due to the pattern of target activations and that older networks suffer considerably with less sparser codes than that of TTB.

The same logic predicts additional complexity cost effects regarding efficiency of EAS: (1) older adults should have more difficulties with increased number of cues than young adults; (2) options with similar profiles should be harder to distinguish for older compared to young decision makers. In both cases, this arises from the fact that the differences between alternative's activations become increasingly small with increased number of cues or similarity which, due to the choice rule, leads to an increase in the probability of choosing the wrong alternative.

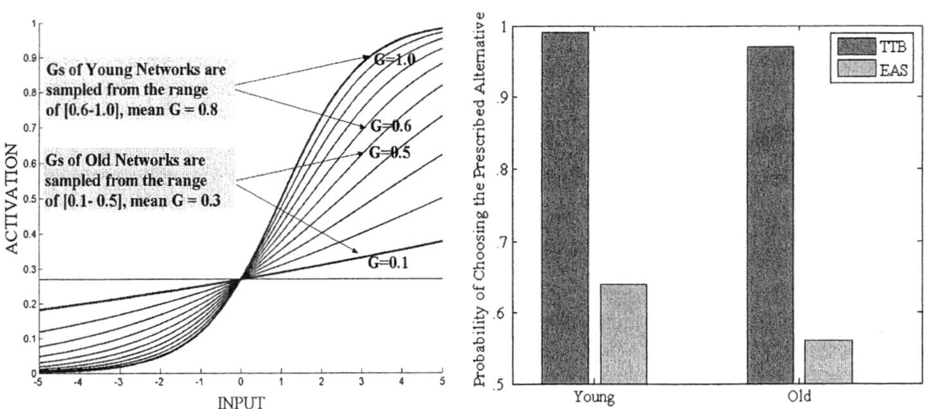

Fig. 1. The S-shaped logistic activation function at different values of G and the probability of choosing the option prescribed by TTB and EAS given different age (G) levels

3 Conclusion

A neurocomputational approach was presented which builds on previous accounts of decision making [1] and formal modeling of aging [14] to predict effects of age-related cognitive decline on the efficiency of decision strategies. However, the results obtained provide only the groundwork for understanding the relation between cognitive aging and decision making; these preliminary modeling efforts will have to be further developed, for example, to include predictions about variability in performance [14]. Nevertheless, these basic results remain empirically testable and will be examined by training young and older adults to use TTB and EAS (see [13] for such a procedure) and analyzing their error patterns as a function of trial difficulty. One additional step that could provide support for the approach would be to estimate G for each participant and relate this person-specific parameter to individual measures of cognitive capacity, such as speed [9].

References

1. Gigerenzer, G., Todd, P. M., & the ABC Research Group (1999). *Simple heuristics that make us smart*. New York: Oxford University Press.
2. Chater, N., Oaksford, M., Nakisa, R. & Redington, M. (2003). Fast, frugal and rational: How rational norms explain behavior. *Organizational Behavior and Human Decision Processes, 90*, 63-86.
3. Roe, R. M., Busemeyer, J. R., & Townsend, J. T. (2001). Multi-alternative decision field theory: A dynamic artificial neural network model of decision making. *Psychological Review, 108*, 370-392.
4. Usher, M., & McClelland, J. L. (2004). Loss aversion and inhibition in dynamical models of multialternative choice. *Psychological Review, 111*, 757-769.
5. Marcus, G. F. (2001). *The algebraic mind: Integrating connectionism and cognitive science*. Cambridge, MA, USA: MIT Press.
6. Baltes, P. B., & Kliegl, R. (1992). Further testing the limits of cognitive plasticity in old age: Negative age differences in a mnemonic skill are robust. *Developmental Psychology, 28*, 121-125.
7. Lair, C. J., Moon, W. H., & Kausler, D. H. (1969). Associative interference in the paired-associate learning of middle-aged and old subjects. *Developmental Psychology, 9*, 548-552.
8. Arnsten, A. F. T. (1998). Catecholamine modulation of prefrontal cortical cognitive function. *Trends in Cognitive Sciences, 2*, 436-447.
9. Salthouse, T. A. (1996). The processing-speed theory of adult age differences in cognition. *Psychological Review, 103*, 403-428.
10. Sanfey, A. G., & Hastie, R. (1999). Judgment and decision making across the adult life span. In D. Park and N. Schwarz (Eds.), *Aging and cognition: A primer*. Philadelphia: Psychology Press.
11. Elman, J. F. (1990). Finding structure in time. *Cognitive Science, 14*, 179-211.
12. Rumelhart, D. E., Hinton, G. E., & Williams, R. J. (1986). Learning representations by back-propagating errors. *Nature, 323*, 533-536.
13. Payne, J. W., Bettman, J. R., & Johnson, E. J. (1993). *The Adaptive Decision Maker*. New York, NY: Cambridge University Press.
14. Li, S.-C., Lindenberger, U., & Sikström, S. (2001). Aging cognition: From neuromodulation to representation. *Trends in Cognitive Sciences, 5*, 479-486.
15. Servan-Schreiber, D., Printz, H., & Cohen, J. D. (1990). A network model of chatecolamine effects: Gain, signal-to-noise ratio, and behavior. *Science, 249*, 892-895.

Comparison of Neural Network Robot Models with Not Inverted and Inverted Inertia Matrix

Jakub Możaryn and Jerzy E. Kurek

Warsaw University of Technology,
Institute of Automatic Control and Robotics,
02-525 Warszawa, ul. św.Andrzeja Boboli 8, Poland
J.Mozaryn@mchtr.pw.edu.pl

Abstract. The mathematical model of an industrial robot is usually described in the form of Lagrange-Euler equations, Newton-Euler equations or generalized d'Alambert equations. However, these equations require the physical parameters of a robot that are difficult to obtain. In this paper, two methods for calculation of a Lagrange-Euler model of robot using neural networks are presented and compared. The proposed network structure is based on an approach where either a not inverted or inverted inertia matrix is calculated. The presented models show good performance for different sets of data.

1 Introduction

An accurate robot mathematical model is useful for the design of advanced robot control systems. Its calculation requires knowledge of exact values of a robot's physical parameters. It is rather difficult to obtain the data without disassembling a robot. One of the methods which enables the identification of the robot mathematical model without a'priori knowledge of these parameters is using the neural networks [3,4,5].

This paper presents identification method of the robot mathematical model using neural networks. Firstly, the robot mathematical model is presented. Then, two neural network models of a robot are described, with not inverted inertia matrix (NIIM), and with inverted inertia matrix (IIM). Next, computer simulations and comparison of the robot neural network models are presented. Finally, concluding remarks are given.

2 Discrete Time Robot Model

The discrete time model of a robot with n degrees of freedom, in the form of the Lagrange-Euler equation [2] can be presented as follows

$$\tau(k) = T_p^{-2} M(k)[q(k+1) - 2q(k) + q(k-1)] + V(k) + G(k) \ , \qquad (1)$$

where $\tau(k) \in R^n$ is a vector of control signals, $q \in R^n$ is a vector of generalized joint coordinates, $M(k) = M(q(k)) = [m_{ij}(k)] \in R^{n \times n}$ is a robot inertia matrix, $G(k) = G(q(k)) = [g_i(k)] \in R^n$ is a vector of gravity loading,

W. Duch et al. (Eds.): ICANN 2005, LNCS 3697, pp. 417–422, 2005.

$V(k) = V(q(k), q(k-1)) = [v_i(k)] \in R^n$ is a vector of Coriolis and centrifugal effects, k is discrete time, $t = kT_p$, T_p is sampling time.

3 Design of the Neural Network Robot Models

In the robot model (1) the unknown nonlinear elements of $M(k), V(k), G(k)$ have to be identified. For their identification neural networks are proposed, because of their ability to approximate the nonlinear multidimensional functions [6]. Two different approaches to design robot neural models can be developed.

In the first approach, with a not inverted inertia matrix, equation (1) can be rewritten as

$$M(k)\gamma(k) + H(k) = \tau(k) \ , \tag{2}$$

where

$$H(k) = [h_i(k)] = V(k) + G(k) \ , \tag{3}$$

$$\gamma(k) = [\gamma_i(k)] = T_p^2[q(k+1) - 2q(k) + q(k-1)] \ . \tag{4}$$

Matrix equation (2) can be presented in the form of n independent equations

$$\sum_{j=1}^{n} m_{ij}(k)\gamma_j(k) + h_i(k) = \tau_i(k), i = 1 \ldots n \ . \tag{5}$$

The structure of neural network should be designed to identify the elements $h_i(k)$ and $m_{ij}(k)$. The inputs of the neural network are $q(k-1), q(k)$, and $\gamma(k)$. The output of the neural network is $\tau_i(k)$. The performance function of the neural network is in the form

$$J_{\mathrm{NIIM}i} = \frac{1}{N}\sum_{k=1}^{N}(\tau_i(k) - \tau_{\mathrm{NN}i}(k))^2 \ , \tag{6}$$

where $\tau_{\mathrm{NN}i}(k)$ is the neural network output and N is the number of training samples.

The neural network model of a robot with a not inverted inertia matrix can be used to calculate the values of $M_{\mathrm{NN}}(k) = [m_{\mathrm{NN}ij}(k)]$ and $H_{\mathrm{NN}}(k) = [h_{\mathrm{NN}i}(k)]$ in (2). Thus, using neural networks generalized coordinates $q_{\mathrm{NN}}(k+1)$ can be calculated based on (2) and (4) as follows

$$q_{\mathrm{NN}}(k+1) = M_{\mathrm{NN}}^{-1}(k)T_p^2[\tau(k) - H_{\mathrm{NN}}(k)] + 2q(k) - q(k-1) \ . \tag{7}$$

In the second approach, with inverted inertia matrix, equation (1) can be presented as follows

$$\overline{M}(k)\tau(k) + P(k) = \gamma(k) \ , \tag{8}$$

where

$$\overline{M}(k) = [\overline{m}_{ij}] = M^{-1}(k) \ , \tag{9}$$

$$P(k) = [p_i(k)] = -M^{-1}(k)[V(k) + G(k)] \ . \tag{10}$$

Each of n independent equations of matrix equation (8) can be written as follows

$$\sum_{j=1}^{n} \overline{m}_{ij}(k)\tau_i(k) + p_i(k) = \gamma_i(k), i = 1 \ldots n \ . \tag{11}$$

In this case, the structure of neural network should be designed to identify the elements $p_i(k)$ and $\overline{m}_{ij}(k)$. The inputs of the neural network are $q(k-1), q(k)$, and $\tau_i(k), i = 1 \ldots n$. The output of the neural network is $\gamma_i(k)$. The performance function of the neural network is in the form

$$J_{\mathrm{IIM}i} = \frac{1}{N} \sum_{k=1}^{N} (\gamma_i(k) - \gamma_{\mathrm{NN}i}(k))^2 \ , \tag{12}$$

The neural network robot model with inverted inertia matrix can be used to calculate the values of $\overline{M}_{\mathrm{NN}}(k) = [\overline{m}_{\mathrm{NN}ij}(k)]$, $P_{\mathrm{NN}}(k) = [p_{\mathrm{NN}i}(k)]$, and $\gamma_{\mathrm{NN}}(k) = [\gamma_{\mathrm{NN}i}(k)]$ in (8). Thus, using neural networks generalized coordinates $q_{\mathrm{NN}}(k+1)$ can be calculated based on (4) as follows

$$q_{\mathrm{NN}}(k+1) = T_p^{-2}\gamma_{\mathrm{NN}}(k) - 2q(k) + q(k-1) \ . \tag{13}$$

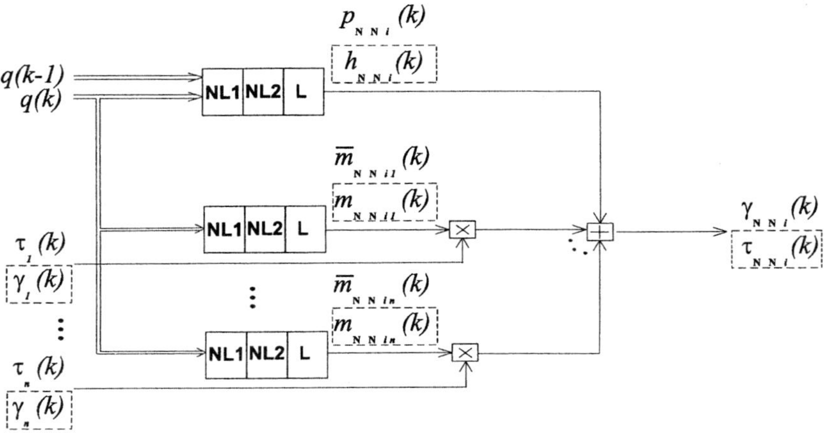

Fig. 1. Structure of the neural network for the robot model identification

For both approaches, the structure of neural network is similar, with different input and output signals, which is presented in Fig.1.

4 Computer Simulations

To compare the presented methods sets of data were generated from the simulation of the robot PUMA 560 [1] with 6 degrees of freedom, revolute joints and

Table 1. Parameters of training ($\varpi_i = 18[\frac{\circ}{s}]$, $\varphi_i = 0[^\circ]$, $i = 1\ldots6$) and testing ($\varpi_i = 18[\frac{\circ}{s}]$, $\varphi_i = 180[^\circ], i = 2,3,5,6$, $\varphi_i = 0[^\circ]$, $i = 1,4$) trajectories

	$q_1[^\circ]$	$q_2[^\circ]$	$q_3[^\circ]$	$q_4[^\circ]$	$q_5[^\circ]$	$q_6[^\circ]$
q_{mini}, *training trajectories*	-20	10	-180	-20	20	-100
q_{maxi}, *training trajectories*	160	45	225	170	100	266
q_{mini}, *testing trajectories*	-20	10	-180	-20	20	-100
q_{maxi}, *testing trajectories*	86	140	30	40	90	120

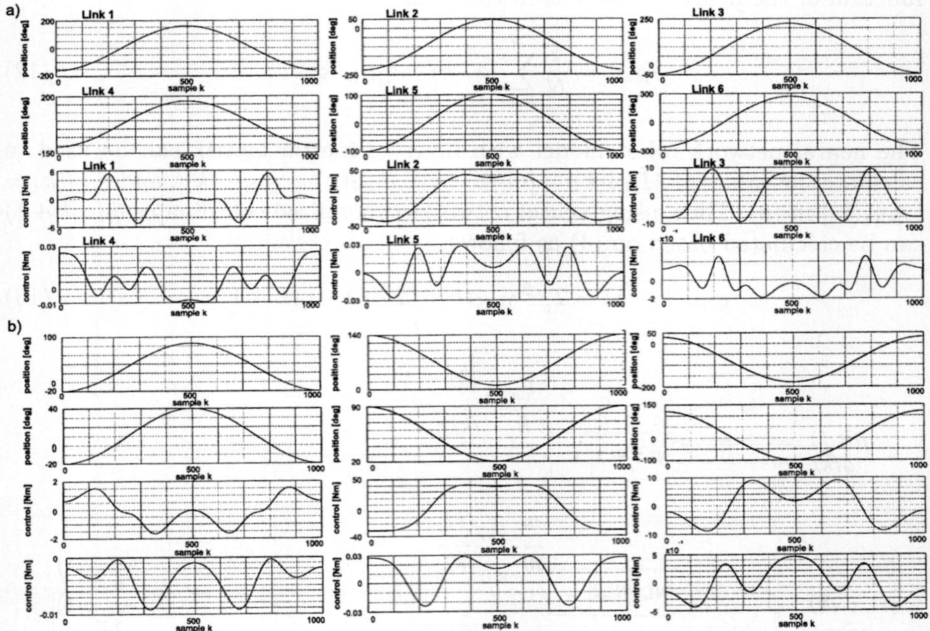

Fig. 2. Trajectories and control signals: a. training data, b. testing data

sliding mode control algorithm. They were used to train and test the proposed neural network models.

The trajectory for i-th joint was set according to the following formula

$$q_i(k) = a_i cos(\varpi_i k T_p + \varphi_i) + \frac{a_i}{2} + q_{\mathrm{mini}}, i = 1\ldots n \ , \tag{14}$$

where $a_i = |q_{\mathrm{maxi}} - q_{\mathrm{mini}}|$ is the amplitude, q_{maxi} is the maximal value of angular position, q_{mini} is the minimal value of angular position, ϖ_i is the angular velocity and φ_i is phase.

The values of q_{maxi}, q_{mini}, ϖ_i, φ_i for the training and testing data are given in Table 1. The trajectories and control signals are presented in Fig.2.

The robot was simulated with given trajectory in time interval $T = 10$[sec], with sampling time $T_p = 0.01$[sec]. Thus, there were 1000 data samples for training and 1000 data samples for testing of the neural models. In all nonlinear layers (NL) the neurons are described by the sigmoidal activation function

$$y = f_{\mathrm{nl}}(x) = \frac{1}{1+e^{-x}} - 1 \ . \tag{15}$$

In linear layers (L) the neurons are described by the linear activation function

$$y = f_{\mathrm{l}}(x) = x \ , \tag{16}$$

where $x = \sum_{i=1}^{L} w_i x_i + b$, L is the number of neuron inputs, w_i is the weight of the i-th input to neuron, x_i is the i-th input to neuron, b is bias.
In both robot neural network models, (IIM) and (NIIM), there were two neurons in each nonlinear layer and one neuron in each linear layer. Models were trained using the the Levenberg - Marquardt method to update weights in all layers [6]. There were 1000 training iterations.

To compare both neural network models, (IIM) and (NIIM), quality indexes were chosen as average absolute error $Q_{\mathrm{av}i}$ and maximum absolute error $Q_{\mathrm{max}i}$ in i-th joint

$$Q_{\mathrm{av}i} = \frac{1}{N} \sum_{k=1}^{N} |e_i(k)|, i = 1 \ldots n \ , \tag{17}$$

$$Q_{\mathrm{max}i} = \max_{k} |e_i(k)|, k = 1 \ldots N, i = 1 \ldots n \ , \tag{18}$$

where errors were calculated as

$$e_i(k) = q_{\mathrm{NN}i}(k) - q_i(k), i = 1 \ldots n \ , \tag{19}$$

k is the number of the data sample, N is the number of all data samples.
The values of quality indexes (17) and (18) for (IIM) and (NIIM) robot neural network models for training and testing data are given in Table 2.
The obtained results show that for the robot neural network model with a not inverted inertia matrix the average errors are generally smaller than for the robot neural network model with an inverted inertia matrix. The main drawbacks

Table 2. Values of the average errors $Q_{\mathrm{av}i}$ and maximum errors $Q_{\mathrm{max}i}$

$Q_{\mathrm{av}i}$	$q_1[°]$	$q_2[°]$	$q_3[°]$	$q_4[°]$	$q_5[°]$	$q_6[°]$
NIIM Model, training	0.080	0.040	0.017	0.115	0.030	0.035
NIIM Model, testing	0.032	0.018	0.090	0.034	0.130	0.049
IIM Model, training	0.360	0.070	0.820	0.380	0.740	0.740
IIM Model, testing	0.210	0.260	0.420	0.120	0.140	0.440
$Q_{\mathrm{max}i}$	$q_1[°]$	$q_2[°]$	$q_3[°]$	$q_4[°]$	$q_5[°]$	$q_6[°]$
NIIM Model, training	0.480	0.210	0.530	1.510	0.130	1.100
NIIM Model, testing	2.800	3.660	15.420	2.890	12.460	8.090
IIM Model, training	0.560	0.110	1.270	0.600	0.250	1.200
IIM Model, testing	0.330	0.410	0.660	0.200	0.220	0.700

of the robot neural network model with a not inverted inertia matrix are the significantly bigger maximal errors and the requirement of inversion of the matrix $M_{\mathrm{NN}}(k)$ to calculate the generalized coordinates. The neural network model with the inverted inertia matrix has better overall performance. The main problem is the difficulty to recognize exactly which equation from the matrix equation (8) is calculated. It is due to the fact that as the output is γ_i, it can be the same for different joints during use of the robot neural network model.

5 Concluding Remarks

In this work the two approaches to design neural networks that can be used for a robot mathematical model identification, have been described. The results obtained during the simulations have shown, that presented robot neural network models have good approximation properties for training and testing data. However, there are a few drawbacks of both methods that should be considered during the design and use of these models. The model with a not inverted inertia matrix can be used for the the identification of the Lagrange-Euler equation coefficients. The model with an inverted inertia matrix has better overall performance. It is also useful in practice, especially for the design of robot control algorithms, where the inverted inertia matrix is needed.

We plan further research to improve the robot neural network models, and to use them in the synthesis of model-based robot control algorithms.

References

1. Corke, P.I., Armstrong-Helouvry, B.: A Meta Study of Puma 560 Dynamics: A Critical Appraisal of Literature Data. Robotica **13** (1995) 253-258
2. Fu, K. S., Gonzalez, R. C., Lee, C. S. G.: Robotics: control, sensing, vision, and inteligence. McGraw-Hill Book Company (1987)
3. Lewis, F. L., Liu, K., Yesildirek, A.: Neural Net Robot Controller with Guaranteed Tracking Performance. IEEE Transactions on Neural Networks **6** (1995) 703-715
4. Możaryn, J., Wildner, C., Kurek, J. E.: Wyznaczanie Parametrów Modelu Robota Przemysłowego Przy Pomocy Sieci Neuronowych (Calculation of Industrial Robot Model Coefficients Using Neural Networks). Proc. XIV Krajowa Konferencja Automatyki KKA 2002 **2** (2002) 675-678 (in polish)
5. Możaryn, J., Kurek, J. E.: Neural Network Robot Model with Not Inverted Inertia Matrix. Proc. 10th IEEE Int. Conf. on Methods and Models in Automation and Robotics MMAR 2004 **2** (2004) 1021-1026
6. Osowski, S.: Sieci Neuronowe do Przetwarzania Informacji (Neural Networks for Information Processing). OWPW Warszawa, Poland (1994) (in polish)

Causal Neural Control of a Latching Ocean Wave Point Absorber

T.R. Mundon[1], A.F. Murray[1], J. Hallam[2], and L.N. Patel[1]

[1] Edinburgh University, The Kings Buildings, Edinburgh, UK
tim.mundon@ed.ac.uk
[2] University of Southern Denmark, Campusvej 55,
DK-5230 Odense M, Denmark

Abstract. A causal neural control strategy is described for a simple "heaving" wave energy converter. It is shown that effective control can be produced over a range of off-resonant frequencies. A latching strategy is investigated, utilising a biologically inspired neural oscillator as the basis for the control.

1 Introduction

It is vital that the energy retrieved from the ocean by a *Wave Energy Converter (WEC)* be maximised and much work has been done in this area [4]. However, due to the inherent unpredictability of the future wave, many advanced techniques that can implement near-optimum transfer require knowledge of the sea state immediately prior to reaching the WEC device. Our system uses a phase locked neural oscillator that tracks and optimises the motions of a simple point absorber WEC. A system with only one degree of freedom, with motion constrained to the vertical direction, is studied as a simplified exemplar of a wider class of WEC's. A biological system, the lamprey [2], provided the inspiration for the artificially evolved neural controller presented here. This system is implemented using a neural network in order to optimise the power generated over a range of frequencies and for a variety of input waveforms. A time domain based system was developed that does not require explicit prior knowledge of the input sea state. This approach effectively solves the equations of motion and the neural equations in parallel as separate, mutually dependent systems.

One of the fundamental requirements for efficiency in WEC's is that the correct phase must be maintained at frequencies away from resonance [1]. The primary method of phase control studied here is "latching" control [5,6,8], which has been shown to yield significant increases in power. Latching control provides a pseudo-resonant system by clamping the device rigid at the extremities of its motion until the wave force has increased to an optimum. The device is then released and thus generates power until reaching the next extremity of excursion. Although latching has so far been difficult to implement in irregular waves we outline a system that points towards a practical latching strategy for irregular waves that does not require explicit knowledge of the future state of the ocean.

W. Duch et al. (Eds.): ICANN 2005, LNCS 3697, pp. 423–429, 2005.

2 The Wave Energy Converter (WEC) Model

Advanced models of ocean wave energy converters can take account of complex wave-body interactions, including non-linear effects in order to model exactly how to absorb maximum energy from their environment. Whilst these models are essential to develop increased overall power capture, their inherent complexity makes them unsuitable for concept-proving experiments such as this. In order to better illustrate the method of control proposed, a point absorber, restricted to a single degree of movement in the vertical plane is considered. Experience with full WEC models suggests that this simplification captures most of the important characteristics of a more complex WEC model.

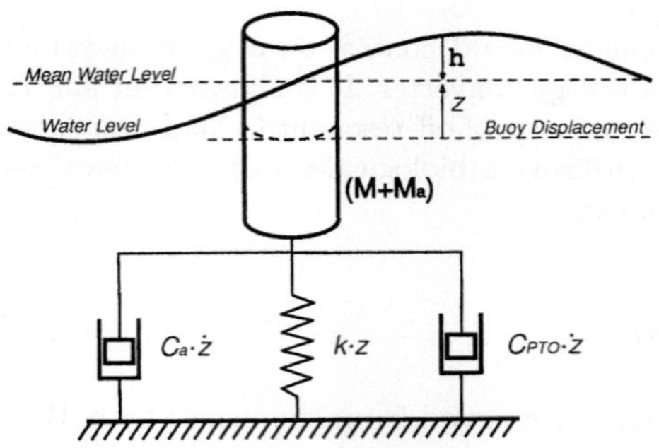

Fig. 1. Mechanical model of heaving buoy

Consider the simple mechanical system described in fig. 1, consisting of a single buoy constrained to oscillate in heave mode (i.e. vertically) only. The float is a cylinder of radius 1.65m and length 5m that is 50% submerged at equilibrium. The power take-off is represented as a damper, C_{PTO}. This simple harmonic oscillator system is described by equation 1.

$$F_{e(t)} = (M + M_a)\ddot{z} + C_a\dot{z} + C_{PTO}\dot{z} + kz \quad (1)$$

$$displacement = \eta - z \quad (2)$$

$$F_{e(t)} = \eta(\pi r^2 \rho_w g) \quad \& \quad kz = z(\pi r^2 \rho_w g) \quad (3)$$

The hydrodynamic co-efficients of added mass (M_a) and added damping (C_a) are defined by the buoy's geometry and are frequency dependent. As we aim to construct a control strategy requiring no foreknowledge of future waves, these have simply been approximated to static values. Although we appreciate that this is not strictly accurate it was found that such an approximation was adequate for this model.

In order to maintain a linear approximation, the force on the buoy is decomposed into the wave force F_e and the spring force kz which are proportional to water level(η) and buoy displacement(z) respectively. This is shown in eqn. 3, where $(\pi r^2 \rho_w g)$ is the restoring force on the cylinder.[3]

As power is the rate of energy absorbed in damper C_{PTO}, and given that from eqn. 1, $F_{(t)} = C_{PTO} \cdot \dot{z}$, it follows that: $P_{(t)} = F_{(t)} \cdot \dot{z} = C_{PTO} \cdot \dot{z}^2$

2.1 Existing Control Methods

Maximum power transfer between the wave and the device will occur when the natural period of these coincides, i.e resonance. Budal [1] found that it is the phase difference of $\frac{\pi}{2}$ between the forcing component and the device displacement that characterises this point of maximum power transfer. Away from resonance this phase difference reduces or increases. However, by varying the value of C_{PTO} in eqn. 1, it is possible to adjust the WEC's response at off-resonant periods, increasing the phase difference so as to optimise the power captured. However this is still an uncontrolled system as C_{PTO} remains constant over many periods, hence is is classed as "optimal real" damping in this paper (see fig. 3). Due to the unpredictable nature of sea waves, increases in off-resonance performance are desirable so that the effective bandwidth can be as wide as possible.

Many approaches to optimising the power developed with WEC's have been proposed and are reviewed in [4]. Latching, the simplest and hence perhaps the most significant of these methods, was initially proposed independently in [6] and [5] and then later in [8]. Latching enforces the correct phase shift between the water level and device displacement by locking the device at the extremities of its oscillatory cycle. This is implemented by locking the device displacement when velocity, $\dot{z} = 0$ and releasing it a certain time T_L later. The optimum power achievable from this method is intrinsically non-continuous, although it can be calculated iteratively [8]. Alternatively, reactive control [4] involves application of forces to the device that are in phase with both displacement and acceleration. Both latching and reactive control can develop significantly improved off-resonance power by using knowledge of the incoming wave.

3 A Neurally-Inspired Solution

In order to appreciate the inspiration for the control method proposed here, we first look at the articulated (snake-like) configuration of some real WEC devices. Generally consisting of two or more floats, it can be seen that these devices generate power through the relative motions of their individual sections.[10]

It can be further seen that this motion is similar to the propulsive anguiliform movement shown in many aquatic organisms. In general, it can be seen that this motion is controlled through a local neural system known as a "Central Pattern Generator" *(CPG)*. One particular vertebrate to demonstrate this locomotion, the lamprey, has a very well documented CPG structure and has been shown [7] to have a surprisingly simple and elegant neural architecture.

The structure of the lamprey CPG consists of a series of neural oscillators *(segments)*, each responsible for controlling a single section of the body. Interconnections between adjacent segments produce a small phase difference such that a travelling wave propagates along the length of the body. This overall motion is subject to modulation via sensory interaction [9].

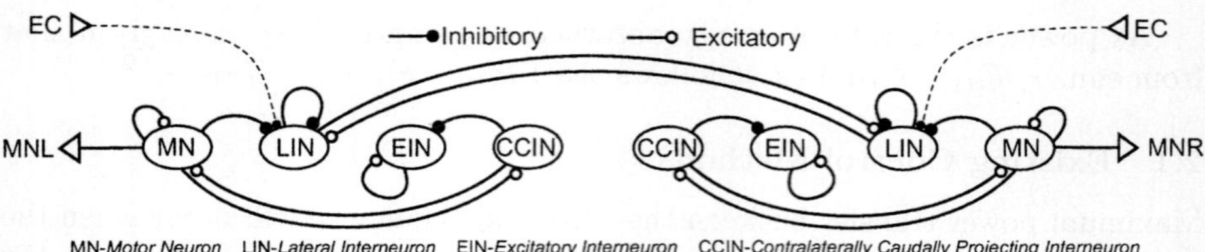

Fig. 2. Evolved controller for the system in fig 1. A fully connected topology was developed. However, for clarity, only weights> 1.7× average synaptic weight are illustrated.

It is proposed that we may be able to apply the lamprey CPG to such an articulated WEC. However in order to test this hypothesis, we must start with a simple system. If we consider that we can approximately model an articulated WEC as a series of point absorbers limited to heave only, whereby the power out-take (C_{PTO}) is between adjacent floats, we can further simplify this and consider the control of a single float as in fig. 1.

This is convenient as it has been shown [2] that a single neural CPG segment, see fig. 2, will oscillate at a natural frequency defined by the weights of the network, modulated by the sensory input. In future work we will show the extension of this to include interconnected buoys as described above and how we can apply a series of interconnected CPG segments as the control.

3.1 Developing a Neural Controller

The mechanical simulation described in section 2 was coupled to the neural network using the buoy velocity ($\dot{z}$) as a sensory feedback input. This allowed the period of oscillation of the CPG network to adapt to that of the WEC device. Furthermore, optimisation of the network weights allows the adjustment of the phase difference range and the bandwidth over which the CPG output *(the neural network output)* and the device displacement could become matched. It was found that an effective latching strategy could be implemented by fixing the buoy displacement at extremity by using $C_{PTO} \ggg optimum$ when $\dot{z} = 0$ [6] and using the CPG output to trigger the release, where $C_{PTO} = optimum$.

The synaptic weights of the network were optimised by a genetic search process. The fitness of each individual was rated using the average power developed over four separate 60 second simulations, each at a different, constant wave period. The Genetic Algorithm (GA) was also given the freedom to evolve the value of $C_{PTO\ (optimum)}$ as it was not clear exactly what value would be appropriate to the evolved solution.

Using a real valued population and taking advantage of the network symmetry, individual chromosomes consisted of 73 variables each of 6dp precision, giving a search space of $10^{73\times6}$. For the results here, three separate evolutions were invoked, each of which converged within 500 generations. The average variation in individual weights between these solutions was less than 2% and the "fitnesses" of the final solutions matched to within 0.2%. The evolved result is shown in fig. 2.

4 Results

The results in fig. 3 show that the evolved strategy was successful over a useful range of frequencies, but it is interesting to note that it was not possible to produce a network that would provide latching control close to the resonant period of 2.9s. This can be explained relatively straightforwardly. If T_L defines the duration of latching, then the phase lag introduced by latching can be described as: $\phi_i = \frac{T_L}{Period}\pi$. For maximum power transfer, the optimal phase difference between the wave and buoy displacement is $\phi = \phi_n + \phi_i = \frac{\pi}{2}$. At resonance however, the natural phase shift of the buoy is $\phi_n = \frac{\pi}{2}$, so the induced additional phase ϕ_i (though latching) should be 0 *at resonance*. Close to resonance however, a lower limit is imposed upon T_L by the neuron model, forcing $\phi > \frac{\pi}{2}$, resulting in sub-optimal power transfer. Under these conditions, since the network phase locks to the buoy displacement rather than to the water level, a feed-forward effect reinforces this error and neural control fails. In theory it would be possible to adjust the neuron coefficients to reduce the minimum value of T_L, ultimately side-stepping the control failure at resonance. However, as the release point must occur *after* the fix point, then due to the discrete nature of the simulation T_L can never equal zero as it is limited by the minimum time-step of the simulation.

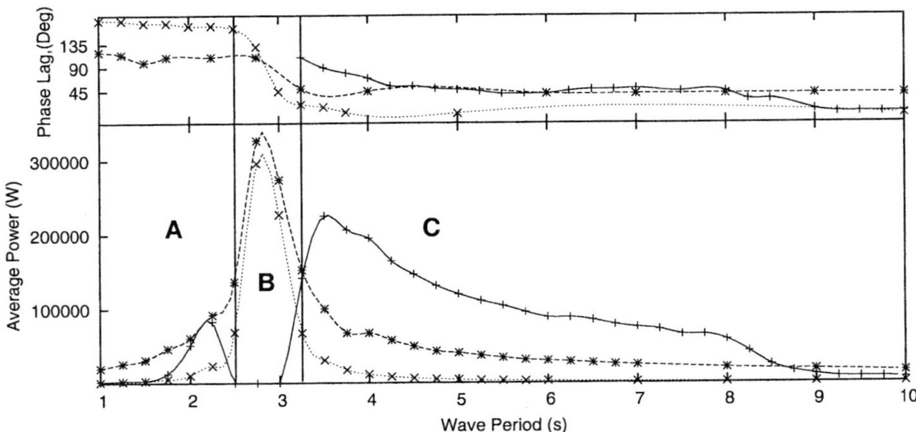

Fig. 3. Phase and power response for regular waves with periods between 1 and 10 seconds. Within region "C" the neural controlled latching strategy (+), can be seen to outperform optimal real damping (∗) and the uncontrolled system (×) over much of this frequency range. Region "B" illustrates areas where a solution could not be evaluated. Results in region "A" are not of interest in this paper.

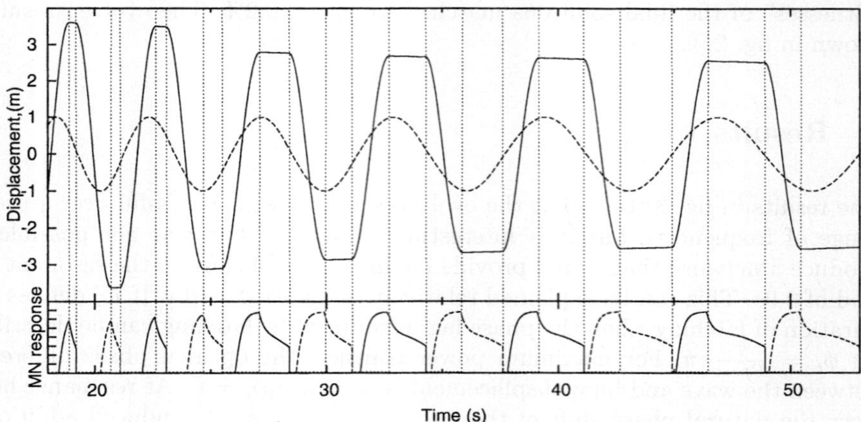

Fig. 4. Response of the system to a changing period from 3 to 10s over a duration of 35s. The top pane shows the wave displacement (dashed) relative to the device displacement (solid), while the lower pane shows the Motor Neuron (MN) output from the network. The vertical lines indicate the latch points for the system. Although these are not conditions experienced in reality, temporal scales for areas of interest are correct.

In order to concentrate on producing a general solution for periods above 3.5s, the network was evolved using four regular wave periods of 3.5, 4, 5.5 & 7 seconds with overall developed power used as the fitness. Over this bandwidth it can be seen to show a significant improvement over optimal real damping.

The GA evolved a single value of C_{PTO} for all periods. However in optimal real damping an ideal value was computed for C_{PTO} at each period and this value increased significantly with the period. From fig. 3 it can be seen that there is a distinct reduction in power from the neural control above its evolved bandwidth (>8s). It was seen that this reduction was due to the value of C_{PTO} being too low at these longer periods. By simply increasing this value for periods 8s< *period* <10s, increases of up to 246% were observed.

The ability of the neural controller to adjust to changing input conditions is clearly illustrated in fig. 4. The sensory input to the network allows the oscillatory period of the network to adapt to the displacement of the buoy and hence to define the latching release points accurately. Even though a practical strategy is implemented, it is possible to see that the ideal $\frac{\pi}{2}$ phase shift is not obtained. This is because future knowledge of the wave is *required* for this optimum condition. However, by using the neural system as we have, we can effectively extrapolate the correct release point from the past knowledge contained within the network. This can be seen in fig. 4 whereby the release point is somewhat too early. Nevertheless, good performance is maintained over an input wave sweeping from 3.5 to 10s.

5 Concluding Remarks

We have shown that "causal" latching control for wave energy converters (WEC), based on a model of a biological (CPG) controller, can be developed using genetic search techniques. This method produces controllers that can outperform optimised real damping over a wide range of sea conditions, with an idiosyncrasy at resonance that can be avoided straightforwardly. Although the wave conditions demonstrated here are far from real conditions, the controller demonstrates a basic ability to adapt accurately to changing input conditions. Extension to testing in realistic-wave conditions will be covered in a future paper.

This work is based upon a simple abstraction of a real WEC device that captures much of the richness of a more complex model and we are optimistic that the promising results reported here will translate into a simple, implementable strategy for control of power out-take in real WEC devices. Furthermore, we speculate that, using naturally-evolved examples of "neural" computing and control structures as a starting point, it may be possible to evolve novel solutions to further hard problems by altering the constraints under which the neural solution is optimised.

References

1. K. Budal and J. Falnes. Interacting point absorbers with controlled motion. In *Power from Sea Waves*, pages 129–142, Edinburgh, UK, 1979.
2. O. Ekeberg. A combined neuronal and mechanical model of fish swimming. *Biological Cybernetics*, pages 363–374, 1993.
3. J. Falnes. *Ocean Waves and Oscillating Systems: linear interactions including wave-energy extraction*. Cambridge University Press, 2002.
4. J. Falnes. Optimum control of oscillation of wave-energy converters. *International Journal of Offshore and Polar Engineering*, 12(2), June 2002.
5. J. Falnes and K. Budal. Wave power conversion by point absorbers. *Norwegian Maritime Research*, (4), 1978.
6. M. J. French. A generalized view of resonant energy transfer. *Journal Mechanical Engineering Science*, 21(4):299–300, 1979.
7. S. Grillner, P. Wallén, and L. Brodin. Neuronal network generating locomotor behaviour in lamprey. *Annu. Rev. Neurosci*, pages 169–199, 1991.
8. R. E. Hoskin and N. K. Nichols. Latching control of a point absorber. In *3rd International Symposium of Wave, Tidal, OTEC and small scale Hydro Energy*, pages 317–329, Brighton, UK, May 1986.
9. P. Wallén and S. Grillner. Central patern generators and their interaction with sensory feedback. In *Proceedings of the American Control Conference*, volume 5, pages 2851–2855, 1997.
10. www.oceanpd.com. Ocean power delivery. Internet Site.

An Off-Policy Natural Policy Gradient Method for a Partial Observable Markov Decision Process

Yutaka Nakamura, Takeshi Mori, and Shin Ishii

Nara Institute of Science and Technology
{yutak-na, tak-mori, ishii}@is.naist.jp

Abstract. There has been a problem called "exploration-exploitation problem" in the field of reinforcement learning. An agent must decide whether to explore a better action which may not necessarily exist, or to exploit many rewards by taking the current best action. In this article, we propose an off-policy reinforcement learning method based on a natural policy gradient learning, as a solution of the exploration-exploitation problem. In our method, the policy gradient is estimated based on a sequence of state-action pairs sampled by performing an arbitrary "behavior policy"; this allows us to deal with the exploration-exploitation problem by handling the generation process of behavior policies. By applying to an autonomous control problem of a three-dimensional cart-pole, we show that our method can realize an optimal control efficiently in a partially observable domain.

1 Introduction

Reinforcement learning (RL) is a machine learning scheme for achieving optimal controls based on trial-and-error, and has been mainly applied to various Markov decision processes (MDPs), in which all state variables of the environment are assumed to be observable [7]. In many real problems, however, there are unobservable factors which affect the dynamics of the environment, and it is then reasonable to model the environment as a partially observable MDP (POMDP) [2]. The optimal policy in a POMDP should be a mapping from the history of observations to an action. The belief state MDP is one possible solution for a POMDP, by means of the belief state, which typically represents a posterior distribution of the current state estimated from the history. It is difficult to solve a belief state MDP, however, because the space of belief states is very large. Furthermore, the model of the environment is necessary to estimate a belief state, while various system identification techniques have been used for modeling the environment in fully observable domains [9].

On the other hand, policy gradient methods, a type of RL, can be applied to both MDPs and POMDPs. In these methods, a policy is defined as a parametric probability distribution (stochastic policy), and the parameter is updated according to the policy gradient; the partial derivative of a performance indicator

W. Duch et al. (Eds.): ICANN 2005, LNCS 3697, pp. 431–436, 2005.

(e.g., the expected reward accumulation) with respect to the policy parameter. Recently, policy gradient methods combined with value learning have been developed, and they attract attention because of their efficiency and stability [8][4].

Because the current observation does not have sufficient information of the state, the performance of a stochastic policy which depends only on the current observation is low. In order to increase the efficiency, a policy gradient method for tuning a policy that possesses an internal state was formerly proposed [1]. The internal state has a dynamics based on a state-space model, and the dynamics is changed by applying an external input such to maximize the accumulated reward. Then, it is expected that the internal state extracts the information essential to the reward maximization from the history of observation-action pairs. However, it was suggested that a policy tends to converge to a local optimum, where the internal state was not used effectively [1].

In order to overcome this problem, it is useful to focus on the "exploration-exploitation problem" which has been studied in the field of RL [3]. Although the aim of RL is to obtain an optimal controller, i.e., to exploit the optimal policy to get rewards as much as possible, such an optimal controller should be explored in the space of possible controllers, because RL is based on trial-and-error in its concept. If a parameter which determines the randomness of the policy is adjusted by the maximization criterion of the performance indicator, it becomes an automatic meta-control of exploration and exploitation. However, the learning of the randomness parameter is not easy; as the learning proceeds, the randomness parameter becomes small to exploit the rewards, but in that case the policy becomes non-ergodic and the estimate of the policy gradient is likely to diverge.

In this article, we propose an off-policy learning method based on the natural policy gradient learning, as a solution of the exploration-exploitation problem. In an off-policy learning method, actions are not generated by the current policy, but by an arbitrary policy (such a policy is called a behavior policy)[6]. Because large variety of behavior policies leads to exploration, and employing a behavior policy similar to the current policy corresponds to exploitation, we can realize an appropriate control of exploration and exploitation by the meta-control of the behavior policy. We apply our proposed method to the acquisition of an automatic control of a three-dimensional cart-pole, and computer simulation shows a good controller possessing an internal state can be obtained even in a partial observation situation.

2 Policy Involving Internal State

The motion of a physical system is expressed as $\dot{\mathbf{x}} = F(\mathbf{x}, \mathbf{u}_x)$, where $\mathbf{x}$ and $\dot{\mathbf{x}}$ denote the state and time derivative, respectively, of the physical system, and $\mathbf{u}_x$ denotes the control signal applied to the physical system. We assume the observation is given by a function of the state $\mathbf{x}$: $\mathbf{X} = G(\mathbf{x})$, where G is an observation function, and the dimensionality of $\mathbf{X}$ is smaller than that of $\mathbf{x}$, i.e., the environment is partially observable.

Controller Possessing An Internal State. In this study, we assume that the output of the controller depends on an internal state $\mathbf{y}$, and its state transition is given by a linear dynamics system (LDS): $\dot{\mathbf{y}} = \mathbf{A}\mathbf{y} + \mathbf{u}_y$, where $\mathbf{u}_y$ is an input to the LDS, and $\mathbf{A}$ denotes the parameter that represents the linear dynamics. The control signal $\mathbf{u}_x$ to the physical system depends on $\mathbf{X}$ and $\mathbf{y}$, and the input to the LDS $\mathbf{u}_y$ is also determined from $\mathbf{X}$ and $\mathbf{y}$. Note that the dynamics of the LDS can be controlled by appropriately controlling its input $\mathbf{u}_y$.

Learning Framework. The dependence of $\mathbf{u}_x$ on $\mathbf{y}$ makes the controller to possess an internal state $\mathbf{y}$. Then, the action selection probability cannot be specified from the state of the physical system $\mathbf{x}$, which leads to a non-stationarity. Although policy gradient methods can be applied to POMDPs, they cannot be applied to training of a non-stationary policy by itself.

In this study, the controller is trained by the same framework as in [1]. The physical system and the internal state are treated as a single dynamical system, which we call LDS-embedded system, and the controller for the LDS-embedded system is trained by a policy-gradient-based RL method. Then, the stochastic policy is conceptually represented as $\pi(\mathbf{s}, \mathbf{u}) \equiv p(\mathbf{u}|\mathbf{s})$, where $\mathbf{s} \equiv \{\mathbf{x}, \mathbf{y}\}$ and $\mathbf{u} \equiv \{\mathbf{u}_x, \mathbf{u}_y\}$ denote the state and the input, respectively, of the LDS-embedded system. In this case, the policy depends only on the state of the controlled object, the LDS-embedded system, and hence becomes stationary. In partially observable domains, a policy depends only on the observation $\mathbf{o} \equiv \{\mathbf{X}, \mathbf{y}\}$.

3 Learning Algorithm

At a discrete time step t, the controller receives the observation $\mathbf{o}(t)$, and outputs a control signal $\mathbf{u}(t)$ according to the stochastic policy π. The LDS-embedded system receives $\mathbf{u}(t)$, and changes its state $\mathbf{s}(t)$ to $\mathbf{s}(t+1)$. Simultaneously, the controller receives an immediate reward $r(\mathbf{s}(t), \mathbf{u}(t))$.

The stochastic policy $\pi_{\boldsymbol{\theta}}$ is assumed to depend only on the observation $\mathbf{o}$, and is defined by $p(\mathbf{u}|\mathbf{o};\boldsymbol{\theta})$, where $\boldsymbol{\theta}$ is the policy parameter which is an n-dimensional vector. We assume that $\pi_{\boldsymbol{\theta}}$ is differentiable with respect to each parameter component θ_i, i.e., $\frac{\partial}{\partial \theta_i}\pi_{\boldsymbol{\theta}}$ exists, and that under any stochastic policy $\pi_{\boldsymbol{\theta}}$, there exists a stationary invariant distribution of states, $D_{\boldsymbol{\theta}}(\mathbf{s})$, which is independent of initial states of the LDS-embedded system.

3.1 Natural Policy Gradient Method

The objective of RL here is to obtain the policy parameter that maximizes the expected reward accumulation defined by $\rho(\boldsymbol{\theta}) \equiv \mathrm{E}_{\boldsymbol{\theta}}\left[\sum_t \gamma^{t-1} r(\mathbf{s}(t), \mathbf{u}(t))\right]$, where $\gamma \in (0, 1]$ is a discount factor. The partial derivative of $\rho(\boldsymbol{\theta})$ with respect to the policy parameter θ_i is calculated by $\frac{\partial \rho(\boldsymbol{\theta})}{\partial \theta_i} = \langle \psi_i(\mathbf{s}, \mathbf{u}) Q_{\boldsymbol{\theta}}(\mathbf{s}, \mathbf{u}) \rangle$, where $\psi_i(\mathbf{s}, \mathbf{u}) \equiv \frac{\partial}{\partial \theta_i} \ln \pi_{\boldsymbol{\theta}}(\mathbf{u}|\mathbf{s})$ and $Q_{\boldsymbol{\theta}}(\mathbf{s}, \mathbf{u})$ denotes the action-value function (Q-function) [8]. $\langle \cdot \rangle$ stands for the expectation with respect to the stationary distribution of state-action pair $(\mathbf{s}, \mathbf{u})$. When the Q-function is approximated by a weighted sum of bases $\boldsymbol{\psi}$: $Q^w_{\boldsymbol{\theta}}(\mathbf{s}, \mathbf{u}) \equiv \sum_i w_i \psi_i(\mathbf{s}, \mathbf{u})$, where $\mathbf{w}$ is the weight vector of the approximate Q-function, the optimal weight in the least square sense: $\tilde{\mathbf{w}} = \arg\min_{\mathbf{w}} \left\langle \left(Q_{\boldsymbol{\theta}}(\mathbf{s},\mathbf{u}) - Q^w_{\boldsymbol{\theta}}(\mathbf{s},\mathbf{u})\right)^2 \right\rangle$, provides the natural policy gradient without introducing any bias, so that the policy parameter can be updated as $\theta_i := \theta_i + \eta_a \tilde{w}_i$ [4].

The Q-function is required to satisfy probabilistically the self-consistency equation for a fixed policy: $Q_{\boldsymbol{\theta}}(\mathbf{s},\mathbf{u}) = r(\mathbf{s},\mathbf{u}) + \gamma V_{\boldsymbol{\theta}}(\mathbf{s}')$, where $\mathbf{s}'$ denotes the state at the next discrete time step. Then, the Q-function is approximated as $r(\mathbf{s},\mathbf{u}) + \gamma \hat{V}_{\boldsymbol{\theta}}(\mathbf{s}')$, where $\hat{V}_{\boldsymbol{\theta}}(\cdot) \equiv \sum_j v_j \phi_j(\cdot)$ is an approximate state-value function, ϕ_i for $i = 1, \ldots, M$ are arbitrary basis functions of state $\mathbf{s}$, and $\mathbf{v}$ is the weight vector. Hence, $\tilde{\mathbf{w}}$ is calculated by using one-step-ahead prediction: $\tilde{\mathbf{w}} = \arg\min_{\mathbf{w}} \left\langle \left(r(\mathbf{s},\mathbf{u}) + \gamma \mathbf{v}\boldsymbol{\phi}(\mathbf{s}')^T - \mathbf{v}\boldsymbol{\phi}(\mathbf{s})^T - \mathbf{w}\boldsymbol{\psi}(\mathbf{s},\mathbf{u})^T\right)^2 \right\rangle$. In this case, the weight vectors, $\mathbf{w}$ and $\mathbf{v}$, are estimated together by the least square method:

$$\mathbf{W} = \langle r\hat{\boldsymbol{\varphi}}^T \rangle \langle \boldsymbol{\varphi}\hat{\boldsymbol{\varphi}}^T \rangle^{-1}, \tag{1}$$

where $\boldsymbol{\varphi}(t) = \begin{pmatrix} \boldsymbol{\psi}(\mathbf{s}(t),\mathbf{u}(t)) \\ \boldsymbol{\phi}(\mathbf{s}(t)) - \gamma\boldsymbol{\phi}(\mathbf{s}(t+1)) \end{pmatrix}$, $\hat{\boldsymbol{\varphi}}(t) = \begin{pmatrix} \boldsymbol{\psi}(\mathbf{s}(t),\mathbf{u}(t)) \\ \boldsymbol{\phi}(\mathbf{s}(t)) \end{pmatrix}$ and $\mathbf{W} \equiv \begin{pmatrix} \mathbf{w} \\ \mathbf{v} \end{pmatrix}$ [5]. Note that $\mathbf{W}$ is the parameter of the value function (critic).

3.2 Importance Sampling Method

In ordinary policy gradient methods, sufficient statistics, $\langle r\hat{\boldsymbol{\varphi}}^T \rangle$ and $\langle \boldsymbol{\varphi}\hat{\boldsymbol{\varphi}}^T \rangle$ in Eq. (1), are calculated using state-action pairs which are sampled by the current policy. In this study, these statistics are instead calculated using state-action pairs which are generated by some behavior policies that differ from the current policy. In this case, unbiased estimators for the expectations can be obtained by an importance sampling method [6].

According to the importance sampling, an expectation $\langle f \rangle$ of a function $f(\mathbf{s},\mathbf{u})$ with respect to the current policy π is calculated by using a sequence of state-action pairs, $h_{\mathrm{B}} = \{(\mathbf{s}(t),\mathbf{u}(t))\}$, which is generated by a behavior policy π_{B}: $\langle f \rangle \approx E_{h_{\mathrm{B}}}\left[\eta \frac{1}{t_e} \sum_{t=1}^{t_e} f(\mathbf{s}(t),\mathbf{u}(t))\right]$, where η is the importance weight for the sequence h_{B}: $\eta = \prod_t \frac{\pi_{\boldsymbol{\theta}}(\mathbf{s}(t),\mathbf{u}(t))}{\pi_{\mathrm{B}}(\mathbf{s}(t),\mathbf{u}(t))}$, and $E_{h_{\mathrm{B}}}[\cdot]$ denotes the expectation with respect to the sample sequence h_{B} [6].

In our experimental setting, after generating a behavior policy π_{B}, the controller interacts with the environment for a certain number of steps in a single 'episode', then the expectation $\langle f \rangle$ is approximated as

$$\langle f \rangle \approx \frac{1}{t_B} \sum_{b=1}^{B} \eta_b \sum_{t=t_{b-1}+1}^{t_b} f(\mathbf{s}(t),\mathbf{u}(t)) = \frac{1}{t_E} \sum_{t=0}^{t_E} \eta(t) f(\mathbf{s}(t),\mathbf{u}(t)), \tag{2}$$

where B denotes the number of episodes, t_b and η_b represent the time at the end of the b-th episode and its importance weight, respectively. In Eq. (2), $\eta(t)$ is fixed at η_b in the b-th episode ($t \in (t_{b-1}, t_b]$).

In order to make the effects of old policies decay, instead of using the simple mean (2), we introduce the weighted mean: $\langle\langle f \rangle\rangle(t_E) = \alpha(t_E) \sum_{t=0}^{t_E}$

$(\prod_{\tau=t+1}^{t_E} \beta(\tau))\eta(t)f(\mathbf{s}(t), \mathbf{u}(t))$, where $\alpha(t_E)$ and $\beta(t) \in (0, 1]$ are the normalization term and the discount factor, respectively. The normalization term $\alpha(t_E)$ is calculated iteratively: $\alpha(t_E) = \alpha(t_E - 1)/(\beta(t_E) + \alpha(t_E - 1))$, thus the weighted mean $\langle\langle f \rangle\rangle(t_E)$ is iteratively calculated by $\langle\langle f \rangle\rangle(t_E) = (1 - \alpha(t_E))\langle\langle f \rangle\rangle(t_E - 1) + \alpha(t_E)\eta(t)f(\mathbf{s}(t_E), \mathbf{u}(t_E))$. These weighted means are updated after a single episode, and $\beta(t)$ is fixed at 1 during the episode.

In this method, behavior policies can be chosen arbitrarily, but appropriate choice will accelerate the learning. In this study, a behavior policy is produced by adding a small randomness $\Delta\boldsymbol{\theta}_{\mathrm{B}}$ to the current policy parameter $\boldsymbol{\theta}$. When the variance of $\Delta\boldsymbol{\theta}_{\mathrm{B}}$ is large, the behavior policy may be different from the current policy so that the exploration happens. When $\Delta\boldsymbol{\theta}_{\mathrm{B}}$ is small, in contrast, a behavior policy becomes close to the current policy and the policy parameter is updated to obtain a larger reward accumulation, namely, the exploitation occurs. This variance is decreased as the learning proceeds, which reminds us of the annealing procedure.

4 Experiment

We applied our off-policy RL scheme based on the natural policy gradient learning to an automatic control problem of a three-dimensional cart-pole depicted in Fig 1(a). The cart and the pole were connected by a ball joint. The state of the cart-pole is represented by coordinates of d_1 and d_2, angles ζ_1 and ζ_2, and their time derivatives $\mathbf{x} = (d_1, \dot{d}_1, \zeta_1, \dot{\zeta}_1, d_1, \dot{d}_2, \zeta_2, \dot{\zeta}_2)$. A control signal to the cart-pole consists of forces applied to the cart in the direction of d_1 and d_2. The controller observes the state of the cart-pole each 0.01 sec. and outputs a control signal.

The aim of RL is to stabilize the cart-pole at the upright position, and for that purpose we defined an immediate reward as $r(\mathbf{s}, \mathbf{u}) = -\sqrt{\zeta_1^2 + \zeta_2^2} - 0.02(\dot{d}_1^2 + \dot{d}_2^2)$. The maximum time period of each RL episode was 20 sec. and if the pole fell down or the cart moved too fast before 20 sec elapsed, the episode was terminated at that time and a large penalty was incurred. The RL proceeded by repeating such learning episodes.

In this experiment, we assumed the angular velocity $\dot{\zeta}_1$ is not observable, and trained a controller which possesses a one-dimensional internal state. We performed ten learning tasks by our proposed (off-policy) method and by the on-policy method ($\Delta\boldsymbol{\theta}_{\mathrm{B}} = 0$). Fig. 1(b) shows a learning curve, the horizontal axis denotes the number of learning episodes, and the vertical axis denotes the

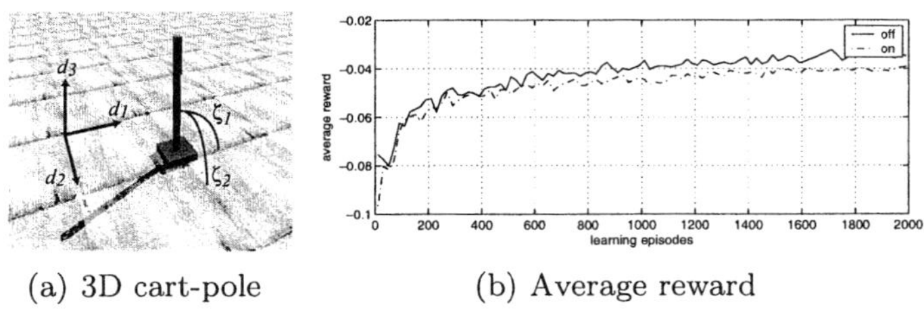

(a) 3D cart-pole (b) Average reward

Fig. 1. Learning of a controller for a 3D cart-pole

average reward. Eight controllers that allow the pole to keep upright position for 20 seconds were obtained by either of the methods. The solid (dashed) line denotes the average reward when the cart-pole was controlled by a controller which was trained by off(on)-policy learning method. This figure shows a good controller can be obtained by our method even in a partially observable situation, and obtained controllers by the off-policy method was better than those by the on-policy method.

5 Conclusion

In this article, we proposed an off-policy learning method based on the natural policy gradient learning. In our method, sufficient statistics in the natural policy gradient were estimated based on state-action pairs sampled by a behavior policy that was different from the current policy. We applied our method to the autonomous control problem of a 3D cart-pole, and computer simulation showed a good controller can be obtained by our method, even in a partially observable domain.

In experiments, behavior policies were produced by adding a noise to the current policy parameter. It is important to develop a more efficient way to appropriately control the exploration-exploitation balance by considering the uncertainly and non-stationarity of the environment. Furthermore, to develop a framework to determine the dimensionality of the internal state automatically is our future work.

References

1. D. Aberdeen and J. Baxter. Scaling internal-state policy-gradient methods for pomdps. In *Proceedings of the International Conference on Machine Learning*, 2002.
2. Douglas Aberdeen. A survey of approximate methods for solving partially observable markov decision processe. Technical report, National ICT Australia, Canberra, Austalia, 2003.
3. S. Ishii, W. Yoshida, and J. Yoshimoto. Control of exploitation-exploration metaparameter in reinforcement learning. *Neural Networks*, 15(4):665–687, 2002.
4. S. Kakade. A natural policy gradient. *In Advances in Neural Information Processing Systems*, 14:1531–1538, 2001.
5. Y. Nakamura, T. Mori, and S. Ishii. In *International conference on parallel problem solving from nature (PPSN VIII)*, pages 972–981, 2004.
6. D. Precup, R. S. Sutton, and S. Dasgupta. Off-policy temporal-difference learning with function approximation. In *Proceedings of the 18th international conference on machine learning.*, pages 417–424, 2001.
7. R. S. Sutton and A. G. Barto. *Reinforcement Learning: An Introduction.* MIT Press, 1998.
8. R. S. Sutton, D. McAllester, S. Singh, and Y. Manour. Policy gradient method for reinforcement learning with function approximation. In *Advances in Neural Information Processing Systems*, volume 12, pages 1057–1063, 2000.
9. Junichiro Yoshimoto, Shin Ishii, and M. Sato. System identification based on online variational bayes method and its application to reinforcement learning. In *in Artificial Neural Networks and Neural Information Processing, LNCS 2714*, pages 123–131, 2003.

A Simplified Forward-Propagation Learning Rule Applied to Adaptive Closed-Loop Control

Yoshihiro Ohama, Naohiro Fukumura, and Yoji Uno

Department of Information and Computer Sciences,
Toyohashi University of Technology,
1-1 Hibarigaoka, Tempaku-cho, Toyohashi, Aichi 441-8580, Japan
{ohama, fkm, uno}@system.tutics.tut.ac.jp

Abstract. In terms of computational neuroscience, several theoretical learning schemes have been proposed to acquire suitable motor controllers in the human brain. The controllers have been classified into a feedforward manner and a feedback manner as inverse models of controlled objects. For learning a feedforward controller, we have proposed a forward-propagation learning (FPL) rule which propagates error "forward" in a multi-layered neural network to solve a credit assignment problem. In the current work, FPL is simplified to realize accurate learning, and to be extended to adaptive feedback control. The suitability of a proposed scheme is confirmed by computer simulation.

1 Introduction

Computational neural researches have suggested the existence of inverse models in the human brain. Investigating how to acquire such inverse models in the motor control, several learning schemes have been proposed to implement the inverse model in artificial multi-layered neural networks. Preparing forward models of controlled objects in advance, supervied learning can be realized[1]. The learning method has been applied to engineering. On the other hand, it was suggested in consideration of physiological structure that feedback-error signal can be used to learn the neural networks without the forward models[2]. Back-propagation (BP) rule[3] is applied to both schemes for learning in the neural networks. However, the back-propagation channels in biogenic neural networks have not been found yet. We have proposed a forward-propagation learning (FPL) rule in which back-propagation errors are not used[4,5]. Approximating the gradient of the learning model to the inverse gradient of the controlled object, FPL performs the inverse model learning based on a Newton-like method. However, FPL might not be able to learn the inverse model in the neural network if a gradient of the inverse model is not bound. Moreover, both feedforward and feedback controllers are required in biological motion tasks, such as optokinetic eye movement response, smooth pursuit, posture control, and so on.

In the current work, FPL is modified for accurate learning in solving a credit assignment problem which inquires how to assign appropriately the observed error to each layer with neural network. The modification is achieved

W. Duch et al. (Eds.): ICANN 2005, LNCS 3697, pp. 437–443, 2005.

by simplifying the estimation of a desired signal in each layer of the network. Furthermore, applying Taylor expansion, we extend FPL to an adaptive closed-loop control system.

2 Forward-Propagation Learning Scheme

FPL for an open-loop control system has been proposed as shown in Fig.1[4]. Given a desired trajectory $\boldsymbol{\theta_d}$, a neural network $\boldsymbol{g}(\boldsymbol{\theta_d};\boldsymbol{w})$ produces motor command $\boldsymbol{\tau}$. Driving a controlled object $\boldsymbol{f}$ by $\boldsymbol{\tau}$, an achieved trajectory $\boldsymbol{\theta} = \boldsymbol{f}(\boldsymbol{\tau})$ is observed. The purpose of the learning is that the trajectory error $\boldsymbol{\Delta\theta} = \boldsymbol{\theta_d} - \boldsymbol{\theta}$ comes to zero by acquiring $\boldsymbol{f}^{-1}$ in $\boldsymbol{g}(\boldsymbol{\theta_d};\boldsymbol{w})$.

In this learning, the desired motor command $\hat{\boldsymbol{\tau}}$ is unknown. Thus, inverse model learning schemes face a problem how $\hat{\boldsymbol{\tau}}$ is determined. FPL can provide a solution of this problem based on a Newton-like method assuming that an open-loop system can be regarded as an approximated identity mapping.

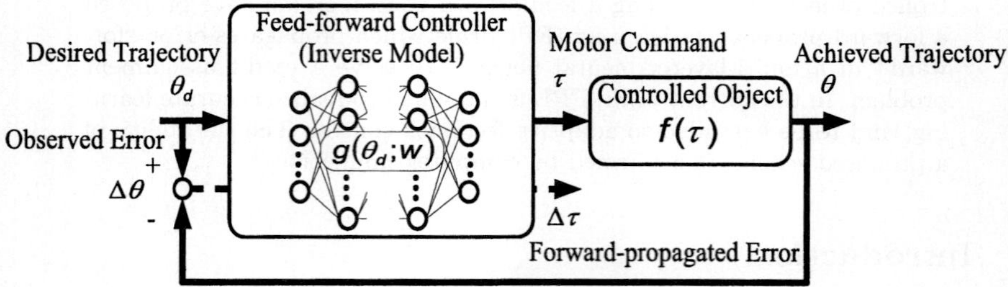

Fig. 1. A forward-propagation learning scheme for a neural inverse model

$$\boldsymbol{f} \circ \boldsymbol{g} \simeq \boldsymbol{I}. \tag{1}$$

Applying a Newton method to find a zero point of $\boldsymbol{\Delta\theta}$ according to $\boldsymbol{\tau}$ on the assumption, $\hat{\boldsymbol{\tau}}$ can be approximated as follows:

$$\begin{aligned} \hat{\boldsymbol{\tau}} &\simeq \boldsymbol{\tau} - \left[\nabla_{\tau}\boldsymbol{\Delta\theta}(\boldsymbol{\tau})\right]^{-1}\boldsymbol{\Delta\theta} \\ &= \boldsymbol{\tau} + \left[\nabla_{\tau}\boldsymbol{f}(\boldsymbol{\tau})\right]^{-1}\boldsymbol{\Delta\theta} \\ &= \boldsymbol{\tau} + \left[\nabla_{\theta}\boldsymbol{f}^{-1}(\boldsymbol{\theta})\right]\boldsymbol{\Delta\theta} \\ &\simeq \boldsymbol{\tau} + \varepsilon\left[\nabla_{\theta_d}^{T}\boldsymbol{g}(\boldsymbol{\theta_d};\boldsymbol{w})\right]\boldsymbol{\Delta\theta}, \end{aligned} \tag{2}$$

where ε is a positive small number in cosideration of nonlinearity. Since $\nabla_{\theta}\boldsymbol{f}^{-1}(\boldsymbol{\theta})$ is approximated by $\varepsilon\nabla_{\theta_d}^{T}\boldsymbol{g}(\boldsymbol{\theta_d};\boldsymbol{w})$, this is one of Newton-like methods. If $\boldsymbol{g}$ is the multi-layered neural network, the error $\boldsymbol{\Delta\tau} = \hat{\boldsymbol{\tau}} - \boldsymbol{\tau}$ must be propagated backward in the network to learn a mapping of $\boldsymbol{f}^{-1}$ by BP rule.

Let us consider a $(L+1)$-layered neural network described by

$$\begin{aligned} \boldsymbol{g}(\boldsymbol{\theta_d};\boldsymbol{w}) &= \boldsymbol{g}^{L}(\boldsymbol{\theta_d};\boldsymbol{w}_L,\cdots,\boldsymbol{w}_1), \\ \boldsymbol{g}^{l}(\boldsymbol{\theta_d};\boldsymbol{w}_l,\cdots,\boldsymbol{w}_1) &= \boldsymbol{g}_l(\boldsymbol{w}_l)\circ\cdots\boldsymbol{g}_1(\boldsymbol{\theta_d};\boldsymbol{w}_1),\ (l=1,2,\cdots,L). \end{aligned}$$

If Eq.(1) is applicable, the Newton-like method cannot be applied to find a zero point of $\boldsymbol{\Delta\theta}$ but be applied to find a zero point of $\boldsymbol{\Delta\tau}_l = \left[\nabla_{\theta_d}^T \boldsymbol{g}^l\right] \boldsymbol{\Delta\theta}$ according to $\boldsymbol{\tau}_l = \boldsymbol{g}^l(\boldsymbol{\theta_d})$ in each layer. Thus, the desired output $\hat{\boldsymbol{\tau}}_l$ which should be calculated by $\boldsymbol{g}_l$ is described as follows:

$$\begin{aligned}\hat{\boldsymbol{\tau}}_l &\simeq \boldsymbol{\tau}_l - \left[\nabla_{\tau_l} \boldsymbol{\Delta\tau}_l\right]^{-1} \boldsymbol{\Delta\tau}_l \\ &\simeq \boldsymbol{\tau}_l + \varepsilon \left[\nabla_{\theta_d}^T \boldsymbol{g}^l\right] \boldsymbol{\Delta\theta}.\end{aligned} \tag{3}$$

Then, parameters $\boldsymbol{w}_l$ can be adjusted by various methods which minimize not only the norm of $\boldsymbol{\Delta\tau}_l$ but also the norm of $\boldsymbol{\Delta\theta}$[6].

$$\hat{\boldsymbol{w}}_l = \arg\min_{w_l} \|\boldsymbol{\Delta\tau}_l\|^2. \tag{4}$$

Since the error signals are only propagated forward to derive $\hat{\boldsymbol{\tau}}_l$ in each layer from $\boldsymbol{\Delta\theta}$, this scheme has been called "forward-propagation learning rule."

Although feedback-error learning (FEL)[2] has been also based on the Newton-like method, $\nabla_\theta \boldsymbol{f}^{-1}(\boldsymbol{\theta})$ has been approximately prepared by the feedback controller. Moreover, FEL uses BP rule to learn the multi-layered neural network according to $\boldsymbol{\Delta\tau} = \hat{\boldsymbol{\tau}} - \boldsymbol{\tau}$ while FPL does not use BP rule.

If a differentiable forward model can be prepared in advance, BP rule can be used to solve the credit assignment problem instead of Newton-like method. Forward and inverse modeling[1] is typical one of such schemes. But the learning schemes based on BP rule require huge iterations, and seem to be improper in terms of neurophysiology.

3 Simplified Forward-Propagation Learning Rule

Since FPL requires the forward-propagated error $\boldsymbol{\Delta\tau}_l$, the neural network must not only acquire the inverse model but also estimate its gradient for the learning. If the function of a true inverse model is discontinuous or its gradient is complicated, FPL might not acquire the inverse model. In order to overcome this serious problem, we revise FPL successfully to estimate the inverse function and its "approximated" gradient.

Applying gradient descent methods to the minimization in Eq.(4), the sign of gradient is mainly used to find a solution. Furthermore, since the activation function of neurons is usually monotonous increased function, the sign of its gradient is always plus. Thus, $\boldsymbol{\Delta\tau}_l$ can be approximated as follows:

$$\begin{aligned}\boldsymbol{\Delta\tau}_l &= \left[\nabla_{\theta_d}^T \boldsymbol{g}^l\right] \boldsymbol{\Delta\theta} \\ &= \nabla \left[\boldsymbol{\sigma} \circ \boldsymbol{w}_l \circ \boldsymbol{\sigma} \circ \cdots \circ \boldsymbol{w}_1(\boldsymbol{\Delta\theta})\right] \\ &\simeq \boldsymbol{W}_L \boldsymbol{W}_{L-1} \cdots \boldsymbol{W}_1 \boldsymbol{\Delta\theta} = \boldsymbol{\Delta}\tilde{\boldsymbol{\tau}}_l,\end{aligned} \tag{5}$$

where $\boldsymbol{\sigma}$ is an activation function of a neuron and $\boldsymbol{W}_l$ is the l-th translation matrix in each linear mapping $\boldsymbol{w}_l$. $\boldsymbol{\Delta}\tilde{\boldsymbol{\tau}}_l$ can be derived from linear translation without calculation of the gradient of non-linear activation.

4 Adaptive Closed-Loop Control by Forward-Propagation Learning Rule

Although FPL has been proposed as a learning scheme that acquires the inverse model serving as a feedforward controller, we extend FPL to adaptive closed-loop control. Let us consider applying Taylor expansion to the neural feedforward controller $\boldsymbol{g}(\boldsymbol{\theta_d}; \boldsymbol{w})$ by perturbation $\boldsymbol{\Delta\theta} = \boldsymbol{\theta_d} - \boldsymbol{\theta}$ in which $\boldsymbol{\theta} = \boldsymbol{f} \circ \boldsymbol{g}(\boldsymbol{\theta_d}; \boldsymbol{w})$. If Eq.(1) is applicable, $\boldsymbol{\Delta\theta}$ is enough small. Then, the higher order terms of Taylor expansion can be neglected.

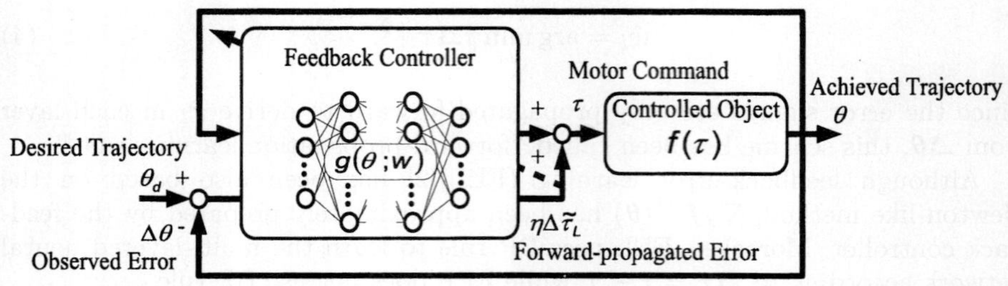

Fig. 2. Adaptive closed-loop control by FPL

$$\boldsymbol{g}(\boldsymbol{\theta_d}; \boldsymbol{w}) \simeq \boldsymbol{g}(\boldsymbol{\theta}; \boldsymbol{w}) + \nabla_{\theta}^{T} \boldsymbol{g}(\boldsymbol{\theta}; \boldsymbol{w}) \boldsymbol{\Delta\theta}. \tag{6}$$

Since the second term in the right hand of Eq.(6) is the same form as $\boldsymbol{\Delta\tau}$, the term can be approximated by simplified forward-propagated error $\boldsymbol{\Delta\tilde{\tau}}_L$ as follows:

$$\boldsymbol{g}(\boldsymbol{\theta_d}; \boldsymbol{w}) \simeq \boldsymbol{g}(\boldsymbol{\theta}; \boldsymbol{w}) + \eta \boldsymbol{\Delta\tilde{\tau}}_L, \tag{7}$$

where η is small positive number in consideration of the approximation. According to Eq.(7), the forward-propagated error $\eta\boldsymbol{\Delta\tilde{\tau}}_L$ should add to the motor command in adaptive closed-loop control by FPL as shown in Fig.2.

Adaptive closed-loop control of FPL is similar to that of FEL[7] since the error signal $\eta\boldsymbol{\Delta\tilde{\tau}}_L$ can be replaced with the signal from feedback controller. However, they differ in the adjustment of parameter $\boldsymbol{w}$; FEL must use BP rule while FPL can use the simplified forward-propagated error $\boldsymbol{\Delta\tilde{\tau}}_l$ in each layer. Naturally, FPL can use a steepest descent method to adjust the parameters without back-propagation signals, since the error signals in each layer are obviously estimated as $\boldsymbol{\Delta\tilde{\tau}}_l$.

5 Simulation of Motor Learning

Here the suitability of the proposed scheme is shown by computer simulation. We examined a learning problem for a 3-layered feedforward neural network in a closed-loop controller of a 2-link arm in x-y plane as shown in Fig.3(a). The dynamics of the 2-link arm can be described by

$$\boldsymbol{\tau} = \boldsymbol{M}(\boldsymbol{q})\ddot{\boldsymbol{q}} + \boldsymbol{h}(\boldsymbol{q}, \dot{\boldsymbol{q}}),$$

where $\boldsymbol{\tau}$ is the drive torque, $\boldsymbol{q}$ is the joint angle, $\boldsymbol{M}$ is the inertia matrix of the arm, and $\boldsymbol{h}$ represents the Coriolis and centrifugal force. $\boldsymbol{M}$ and $\boldsymbol{h}$ were calculated according to the values: $m_1 = 1.59\,\mathrm{kg}, m_2 = 1.44\,\mathrm{kg}, l_1 = 0.35\,\mathrm{m}, l_2 = 0.35\,\mathrm{m}, l_{g1} = 0.18\,\mathrm{m}, l_{g2} = 0.21\,\mathrm{m}, I_1 = 0.0163\,\mathrm{kgm}^2, I_2 = 0.0164\,\mathrm{kgm}^2$. Here, m is the mass of the arm, l is the length of the arm, l_g is the center of gravity and I is the inertia of the arm.

The 3-layered neural network with 6 hidden units $\boldsymbol{g}(\boldsymbol{\theta}; \boldsymbol{w})$ was used for learning the closed-loop controller. The input signal $\boldsymbol{\theta}$ was defined as $\boldsymbol{\theta} = \left[\boldsymbol{q}^T, \dot{\boldsymbol{q}}^T, \ddot{\boldsymbol{q}}^T\right]^T$. The activation functions of input and output layer units were linear but those of hidden units were nonlinear as $\sigma(x) = [1 + \exp(-x)]^{-1}$.

The initial values of $\boldsymbol{W}_1$ were randomly selected in the uniform distribution on (-0.1,0.1). To satisfy Eq.(1) before learning, we prepared a sequence of the torque patterns $\boldsymbol{T} = [\boldsymbol{\tau}(1), \boldsymbol{\tau}(2), \cdots, \boldsymbol{\tau}(500)]^T \in R^{500\times 2}$ and a sequence of the corresponding angular trajectories $\boldsymbol{\Theta} = [\boldsymbol{\theta}(1), \boldsymbol{\theta}(2), \cdots, \boldsymbol{\theta}(500)]^T \in R^{500\times 6}$. Here the torque patterns are selected to be various sine waves. Inputting $\boldsymbol{\Theta}$ to the neural network, the output signal in the hidden layer $\boldsymbol{\Gamma}_1 \in R^{500\times 6}$ was derived. An approximated inverse model was obtained by substituting the solution of the following equation for $\boldsymbol{W}_2$ which was obtained by multiple linear regression.

$$\boldsymbol{T} = \boldsymbol{\Gamma}_1 \boldsymbol{W}_2.$$

The desired trajectory of an end effector was defined as a minimum jerk trajectory from $(0.45, 0.0)$m to $(0.0, 0.4)$m in x-y plane during 1.0 s. This trajectory was sampled at 100 Hz in the angular space to obtain a training sequence $\boldsymbol{\Theta}_d \in R^{100\times 6}$. For updating the connection weights to achieve Eq.(4), we applied a RLS algorithm in bach-mode learning[5]. The learning results of FPL were simulated with the learning parameters $\eta = 0.1$, $\varepsilon = 0.0001$ and the regularization parameter in the RLS algorithm $\delta = 1.0 \times 10^{-6}$ without the forgetting factor[5].

Fig.3(b) shows that the mean square errors of the observed error $\boldsymbol{\Delta\theta}$ and the forward-propagated error $\eta\boldsymbol{\Delta}\tilde{\boldsymbol{\tau}}_L$ by FPL would be decreased as iterations. The desired trajectories are indicated by solid lines, and the achieved trajectories before and after learning are indicated by dash-dotted and dashed lines in Fig.4. Although the achieved trajectories before learning (dash-dotted lines) could not follow the desired trajectories (solid lines), the achieved trajectories after 100 iterations roughly agreed with the desired trajectories. Moreover, the achieved trajectories after 1000 iterations agreed with the desired trajectories. Fig.3(b) and Fig.4 suggest that FPL can acquire a suitable closed-loop controller in the neural network whose ability is good enough to follow the desired trajectory after 1000 iterations.

To compare with other scheme, FEL applied to closed-loop control[7] is also implemented in computer simulation using the same initial neural network in

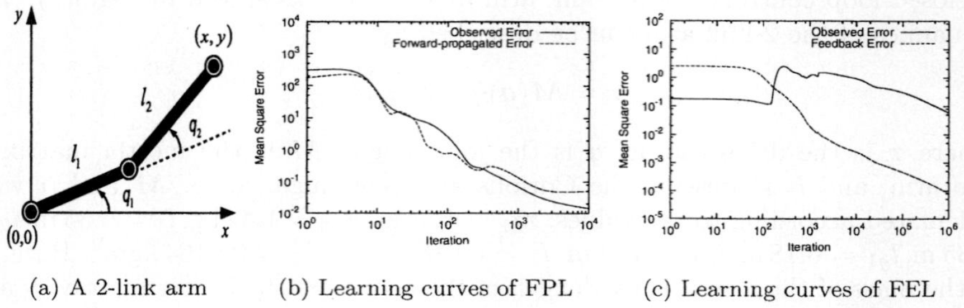

(a) A 2-link arm (b) Learning curves of FPL (c) Learning curves of FEL

Fig. 3. A controlled object and the results of learning control

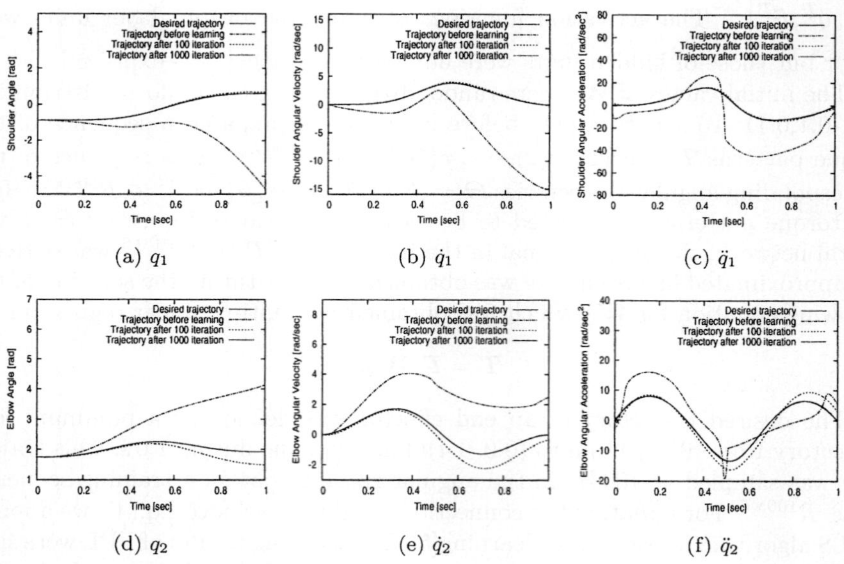

(a) q_1 (b) $\dot{q}_1$ (c) $\ddot{q}_1$

(d) q_2 (e) $\dot{q}_2$ (f) $\ddot{q}_2$

Fig. 4. Trajectories achived by proposed adaptive closed-loop control in FPL

Fig.3(b) and learning rate for BP rule is 0.01. Fig.3(c) indicates the mean squares of the observed error and the feedback error by FEL. Since the feedback controller prepared and worked as the reference model, the observed error is enough small in initial iterations. However, the feedback error monotonously decreased, while the observed error was not. It means that the neural network imitates the feedback controller through learning. It was found by computer simulation that the required iterations and the calculation time for achieving convergence of FPL is quite less than that of FEL, although the calculation cost per iteration of FPL is more expensive than that of learning schemes based on BP rule.

6 Conclusions

In the current work, FPL has been simplified and extended to closed-loop control. The modified FPL was applied to adaptive closed-loop control of a 2-link arm by computer simulation, and its learning achieved good performance. We confirmed that both feedforward and feedback controllers can be acquired in the multi-layered neural network by FPL. Although there are many works with regard to supervised learning approach to acquire neural inverse models, they are required to apply BP rule. In consideration of our results, however, adaptive biological motor control might be achieved by inverse models as feedforward and feedback controllers without back-propagated signals.

Supervised learning schemes for acquiring neural inverse models require prior information of controlled objects; FPL and FEL require the approximated gradient of the inverse models, while a scheme of forward and inverse modeling requires the approximated gradient of a controlled object. It is interesting how living things can obtain such prior information.

Acknowledgements

This study was supported by the 21st Century COE Program "Intelligent Human Sensing" from the Japanese Ministry of Education, Culture, Sports, Science and Technology, and by the scholarship from Support Center for Advanced Telecommunications Technology Research, Foundation in Japan.

References

1. M.I. Jordan, and D.E. Rumelhert, "Forward models: Supervised learning with a distal teacher," Cognitive Sciences, 16, pp.307-354, 1992.
2. M. Kawato, "Feedback-error-learning neural network for supervised motor learning," in Advanced Neural Computers, ed. R. Eckmiller, pp.365-372, North Holland, Amsterdam, 1990.
3. D.E. Rumelhart, G.E. Hinton, and R.J. Williams "Learning representations by back-propagation errors," Nature, vol.323, pp.533-536, 1986.
4. K. Nagasawa, N. Fukumura, and Y. Uno, "A forward-propagation rule for acquiring inverse models in multi-layered neural networks," (in Japanese), IEICE Trans., J85-D-II, pp.1066-1074, 2002.
5. Y. Ohama, N. Fukumura, and Y. Uno, "A forward-propagation rule for acquiring neural inverse models using a RLS algorithm," in Neural Information Processing. eds. N. R. Pal et al., pp.585-590, Springer-Verlag, Berlin, 2004.
6. Y. Ohama, N. Fukumura, and Y. Uno, "A forward-propagation learning rule in consideration of the correlation of propagated errors," (in Japanese), IEICE Trans., J88-D-II, pp.218-229, 2005.
7. H. Gomi, and M. Kawato, "Adaptive feedback control models of the vestibulocerebellum and spinocerebellum," Biol. Cybern. 68, pp.105-114, 1992.

Improved, Simpler Neural Controllers for Lamprey Swimming

Leena N. Patel[1], John Hallam[2], and Alan Murray[1]

[1] The University of Edinburgh, Kings Buildings, Mayfield Road, Edinburgh EH9 3JL

[2] University of Southern Denmark, Campusvej 55, DK-5230 Odense M, Denmark

Abstract. Swimming for the lamprey (an eel-like fish) is governed by activity in its spinal neural network, called a central pattern generator (CPG). Simpler, alternative controllers can be evolved which provide improved performance over the biological prototype (modelled by Ekeberg). Results of computational evolutions demonstrate several possible outcomes exist, with reduced connectivity (16 connections instead of 26) and a diminished equation set for describing the model. Furthermore, resulting oscillators operate over a wider frequency range (0.99 - 12.67 Hz), outperforming the biological prototype (frequencies 1.74 - 5.56 Hz). Evolving advanced yet simpler controllers provides solutions which are more attainable in silicon (VLSI), determines the extent to which nature's solutions are unique and generates efficient task-specific versions.

1 Introduction

Neural oscillators in nature perform and control functions such as walking, swimming, breathing and digestion. Inspiration from such biological controllers could provide a degree of intelligent control to mechanical operators, improving their productivity and efficiency. Developing these solutions requires detailed analyses of biological neural systems, including an assessment of how unique nature's configurations are, whether simpler versions perform effectively, and if their operation range can be optimised for similar mechanical engineering tasks. This would ascertain whether years of evolution make nature's solutions optimum or if alternatives are possible. The solution space can be explored using genetic algorithm techniques, searching for suitable substitutes.

This research considers the neural architecture responsible for controlling the lamprey's swimming movements in varying water conditions. In parallel, the system is being be developed to optimise the efficiency of wave power devices operating in irregular seas [1]. To contribute towards this goal, this paper explores the flexibility of the lamprey CPG, specifically assessing whether alternative, simpler configurations within the constraints of Ekeberg's model [2] are feasible. Certain neural parameters are evolved together with synaptic weights within the constraints of the biological prototype. Optimum performance is signified by controllers which maintain low level system complexity, yet cover a wider frequency range.

W. Duch et al. (Eds.): ICANN 2005, LNCS 3697, pp. 445–450, 2005.

2 The Lamprey Central Pattern Generator (CPG)

The lamprey propels itself by propagating an undulatory wave (with increasing amplitude) from head to tail. A central pattern generator (CPG) along its spinal column, comprises several copies of an oscillatory neural network which cause rhythmic activity of motoneurons. These, in turn, alternate motion between the two sides of the lamprey's body. The entire network can be represented as a simplified connectionist model, with non-spiking neurons representing populations of functionally similar neurons. Activity of each neuron class is described by a set of first order differential equations [2]:

$$\dot{\xi}_+ = \frac{1}{\tau_D}(\sum_{i\epsilon\Psi_+} u_i w_i - \xi_+) \tag{1}$$

$$\dot{\xi}_- = \frac{1}{\tau_D}(\sum_{i\epsilon\Psi_-} u_i w_i - \xi_-) \tag{2}$$

$$\dot{\vartheta} = \frac{1}{\tau_A}(u - \vartheta) \tag{3}$$

$$u = 1 - \exp\{(\Theta - \xi_+)\Gamma\} - \xi_- - \mu\vartheta \text{ (if } u > 0) \text{ OR } 0 \text{ (if } u \leq 0) \tag{4}$$

Output u (equation 4), of each neural unit represents the mean firing frequency of the population. Synaptic inputs (excitatory (ξ_+) and inhibitory (ξ_-)) are added seperately (equations 1-2) and are subject to time delays (τ_D). Groups of pre-synaptic excitatory and inhibitory neurons are represented by the terms Ψ_+ and Ψ_- respectively, where w_i denotes the synaptic weight associated with each input. A saturating transfer function is applied to high levels of excitatory input. Finally a leak is included as delayed negative feedback (equation 3). Parameters of threshold (Θ), gain (Γ) and adaptation rate (μ) of equation 4 are tuned to match the response characteristics of the corresponding neuron type based on experimentally established connectivity (see [2]).

Two of 100 replicas of an oscillating segment are displayed in Figure 1, with neural connections in a single segment highlighted. Tonic (i.e. non-oscillating) input to the pattern generator is supplied by the brainstem and controls the frequency of oscillation. These signals connect to all the neurons in the CPG; for reasons of clarity they are not shown in figure 1. Each segment functions as a non-

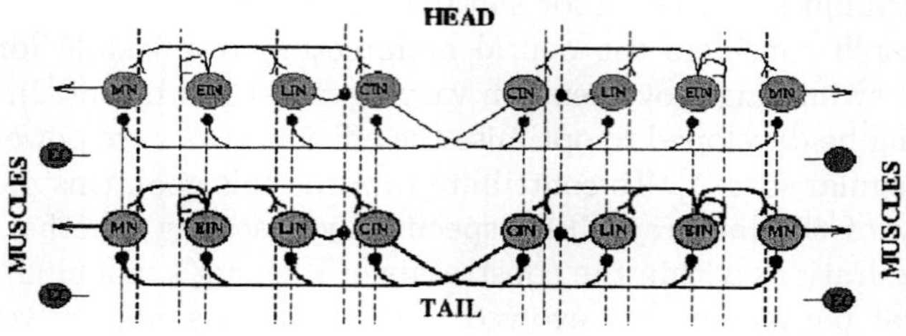

Fig. 1. Connectionist Model of the Lamprey's Spinal CPG

linear oscillator and is coupled to its neighbours through extensions of interneural connections towards the head and the tail (depicted in figure 1 by vertical dotted lines). Output from each segment's motoneurons drive the muscles with a burst activity of 1.74 - 5.56 Hz depending on the level of tonic excitation applied, with higher tonic input resulting in increased oscillation frequency.

The lamprey's CPG is relatively simple and has been isolated *in vivo* to produce detailed cell models, examine electrochemical reactivity to neural stimulation and reproduce the network artificially [2,3]. The simulated CPG is realistic, provides a tool for further exploration of network connectivity and activity, and offers potential for developing systems for more complex control. Alternative controllers have been evolved [4] using genetic algorithms (GAs). Past research manipulated the connection weights and intersegmental extensions. However, optimisation of neural parameters which describe network activity has not been undertaken, nor has reduction and therefore simplification of the equation set, which is especially important for efficient implementation in a real application.

3 Evolving Controllers

Genetic algorithm (GA) techniques are deployed to find acceptable configurations. This method can determine whether the modelled lamprey CPG is unique and optimal for swimming control. Two of three parameters (threshold and gain), which describe the dynamics of each neuron class, are co-evolved with synaptic weights. Adaptation rate, a further neural parameter, is set to zero to eliminate equation 3 (section 2) and thus simplify the system. Each connection is not pre-specified as inhibitory or excitatory and is therefore also evolved.

A real number GA, a variation on the standard binary GA is used. Individual solutions are encoded as chromosomes comprising 39 genes, with each unit corresponding directly to one parameter of the neural configuration. The range of each gene includes the values of the biological CPG. Left-right symmetry is imposed for synaptic weights constituting 24 of the genes. Since, motoneurons only supply output to muscles, connections from them are not evolved. Three chromosome units contain the sign (excitatory or inhibitory) of each neuron group and four genes represent synaptic weights of brainstem input. Finally, eight genes correspond to the equation parameters (threshold and gain).

A random initial population (100 chromosomes) is generated and then operations of selection, variation and rejection are applied to each subsequent generation. Selection involves a fixed number of parents being chosen according to rank-based probability. Fittest individuals are therefore selected more often to create 30 offspring. Variation imposes operations of two-point crossover and mutation on paired chromosomes. Crossover entails swapping parent substrings at two randomly chosen locations. With a 40% probability, each gene is selected exclusively for mutation and a random number $\in$ [-0.5, 0.5] is added to the original value. Every connection is considered independently for pruning and is set to 0 with a probability of 10%. The final population is pruned by eliminating weak connections which do not effect neural activity. Finally, the worst solutions are rejected to maintain a consistent overall population size.

Table 1. Genetic Algorithm Equations for Fitness Evaluation

	Mathematical Definition	Bad	Good	Obj
1	$(zerosL + zerosR)/2$ where $zeros$ = total start/end activity points	3	8	1
2	$\left(\sqrt{\sum_{t=1}^{n}(U_l(t) - \overline{U}_l)^2} + \sqrt{\sum_{t=1}^{n}(U_r(t) - \overline{U}_r)^2}\right)/2n$	0.1	0.5	1
3	$\left(\sum_{cycle=1}^{c} P_l(cycle) + \sum_{cycle=1}^{c} P_r(cycle)\right)/2c$	0.15	0	2
4	$(\lvert P_l(c) - P_l(c-1)\rvert + \lvert P_r(c) - P_r(c-1)\rvert)/2\overline{P}$	0.15	0	2
5	$\left(\frac{\sum_{t \in cycle\_n} \lvert U_l(t) - U_l(t-\overline{P})\rvert}{\sum_{t \in cycle\_n} \lvert U_l(t) - U_l(t-\overline{P})\rvert}\right)/2 + \left(\frac{\sum_{t \in cycle\_n} \lvert U_r(t) - U_r(t-\overline{P})\rvert}{\sum_{t \in cycle\_n} \lvert U_r(t) - U_r(t-\overline{P})\rvert}\right)/2$	0.4	0.05	2
6	$\sum_{t \in cycle\_n} \lvert U_l(t) - U_r(t - \overline{P}/2)\rvert / (\sum_{t \in cycle\_n} \lvert U_l(t) + U_r(t - \overline{P}/2)\rvert$	0.4	0.05	2
7	$\sum_{t=1}^{n} \lvert U_l(t) - U_r(t)\rvert / \sum_{t=1}^{n} \lvert U_l(t) + U_r(t)\rvert$	0	0.8	3
8	$frequency\ range\ (Hz)$	1	12	4
9	$n\_connections/n\_max\_connections$	1	0.3	5

In order to achieve fictive swimming (i.e. oscillatory patterns of activity produced by disconnected pieces of the complete spinal CPG) it is necessary to match certain observed characteristics of the biological oscillating network. These objectives incorporate measurements using the equations in table 1.

Objectives (Obj) include (1) frequency should be controllable by simple tonic excitation from the brainstem (and monotonically increase with the level of excitation), (2) oscillations must be regular with one peak of activity per period and (3) motoneuron activity between the left and right sides of the CPG must alternate. Furthermore, oscillators which operate over a wider frequency range (4) whilst maintaining low connectivity (5) are rewarded. Many of the equations are from Ijspeert's algorithm for evolving synaptic weights [4] but equations 2 and 5 are corrected versions. Procedures analogous to Ijspeert are used to determine the frequency range and combine individual fitness criteria. An additional condition excluded in his work, ensures that the frequency range of the biological model is covered. Also, neural parameters threshold (Θ), gain (Γ) and adaptation rate (μ) are evolved in this work, but fixed to Ekeberg's values in Ijspeert's research. Fitness evaluation is based on motoneural activity as these alone effect muscular activity. In the mathematical definitions of table 1, n is the number of integration steps, c is the number of simulated cycles, U_l and U_r represent left and right motoneuron output respectively, $\overline{P}$ denotes mean period and $P_{l(r)}(j)$ is the period of cycle j for the left (right) motoneuron activity burst.

The outcome of each equation is transformed into a fitness value between 0 and 1 using the function $F(x) = 0.95 * (x - \text{Good})/(\text{Good} - \text{Bad}) + 1$. A result of 1 signifies a good representative, and 0 a poor outcome. Variables 'Good' and 'Bad' depict fitness extremes by which value x is transformed.

4 Results and Discussion

Results of forty experiments, each evolving 500 generations, demonstrate that 40% of the simplified controllers improve performance over that of the prototype

CPG. Table 2 compares the best evolved solution with Ekeberg's biological CPG [2] and Ijspeert's fixed parameter controller [4]. The symmetric weight values are derived by substituting l (left) with r (right) and vice versa in table 2. Fitness of improved controllers (those with greater objective values than the biological prototype's fitness 0.11) range from 0.2 - 0.8. The best overall controller's frequency range is 0.99 - 12.67 Hz (tonic input (ge)= 0.2 - 1.8), substantially greater than the biological frequency range (1.74 - 5.56 Hz, ge=0.2 - 1) and Ijspeert's best controller (1.2 - 8 Hz, ge=0.4 - 2.1). The lowest number of interneuron connections amongst all improved controllers is 16; the highest, 30.

Table 2. Comparing Biological, Fixed Parameter (FP) and Best Evolved Controllers

Run	Obj Val	Frequency Range(Hz) [low - high]	Conns (of 56)	Neural Parameters			Synaptic Weights							
				θ	Γ	μ	from: / to:	EINl	CINl	LINl	EINr	CINr	LINr	BS
bio	0.11	3.82	26	-0.2	1.8	0.3	EINl	0.4	-	-	-	-2.0	-	2.0
		[1.74 - 5.56]		0.5	1.0	0.3	CINl	3.0	-	-1.0	-	-2.0	-	7.0
				8.0	0.5	0	LINl	13.0	-	-	-	-1.0	-	5.0
				0.1	0.3	0	MNl	1.0	-	-	-	-2.0	-	5.0
FP	0.31	6.8	22	-0.2	1.8	0.3	EINl	-0.8	-3.8	-	-0.9	- 0.7	-	0.8
		[1.2 - 8.0]		0.5	1.0	0.3	CINl	-	-	-	-3.5	-3.7	-	13.0
				8.0	0.5	0	LINl	-	-	-	-	-	-	-
				0.1	0.3	0	MNl	-0.4	-3.2	-	-	-	-	3.8
2	0.8	11.68	16	-1	0.7	0	EINl	-	-4.6	-	-	-	-	3.06
		[0.99 - 12.67]		-1	0.48	0	CINl	5.53	-	-	-	-2.9	-	-1.18
				-1	0	0	LINl	-	-	-	-	-	-	-5
				-1	0.27	0	MNl	-	-4.3	-	-	-	-	10.8

Evolved neural parameters and synaptic weights are markedly different from the biological prototype values. Negative synaptic weights denote inhibitory connections, positive weights are excitatory and BS represents brainstem input. It should be noted that although the denominations of EIN, CIN and LIN are kept to distinguish the neuron classes, they lose their functional meaning and even the sign they had in the biological model. Network activity for 3000ms, at the lowest oscillation frequency, is shown in fig. 2 for a) Ekeberg's network and b) the best evolution. Motoneuron burst activity (the dot-dash line), measured for fitness evaluation, is similar for each network. Furthermore, activity is regular and alternates between the left and right sides (top and bottom graphs respectively) as per stipulated conditions for fictive swimming.

5 Conclusion

Experiments evolving Ekeberg style segmental controllers, with added simultaneous adaptation of some neural parameters and connection strengths demonstrate

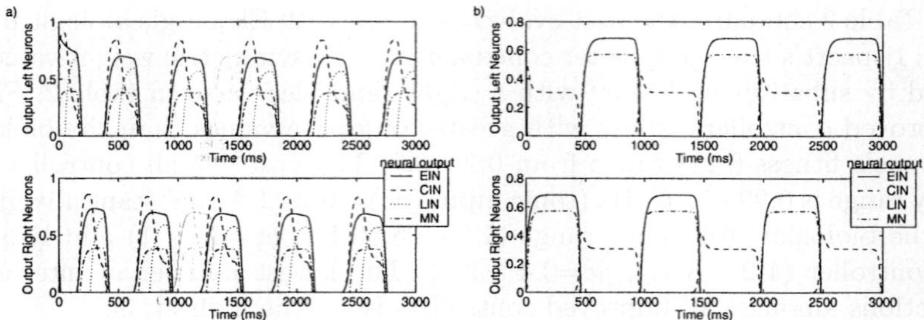

Fig. 2. Network Activity (Lowest Oscillation Frequency) for a) Ekeberg's Biological Controller and b) the Best Evolved CPG

that many effective segmental oscillators can be constructed. The evolved networks can operate over a wider frequency range than their biological prototype [2] and networks with evolution restricted to their weights [4], whilst maintaining low connectivity. Also, since adaptation rate (μ) is set to 0 a simpler model than that used by Ekeberg and Ijspeert is implicitly produced. This property, together with low connectivity, is desirable if an integrated (VLSI) controller is to be constructed. Evolving neural systems according to predefined constraints introduces the potential for developing systems of intelligent control for other non-adaptive operators such as oscillating wave power converters. We are led to speculate that this biological control structure that can be "improved upon" for implementation as an artificial computing system and subsequent use in an application, may not be an isolated example; so can we develop other biologically-inspired structures that lead to new solutions to difficult problems? In summary, by relaxing some of the constraints associated with a biological exemplar, controllers (and potentially other computational structures) can be evolved that capture the strengths of biological "computation" in a simpler, or perhaps more effective manner.

References

1. Mundon, T.R., Murray, A., Hallam, J., Patel, L.: Causal neural control of a latching ocean wave point absorber. Int. Conf. on Artif. Neur. Networks (submitted 2005)
2. Ekeberg, Ö.: A combined neuronal and mechanical model of fish swimming. Biological Cybernetics **69** (1993) 363–374
3. Brodin, L., Grillner, S., Ravainen, C.: N-methyl-*D*-spartate (NMDA), kainate and quisqualate receptors and the generation of fictive locomotion in the lamprey spinal cord. Brain Research **325** (1985) 302–306
4. Ijspeert, A., Hallam, J., Willshaw, D.: Artificial lampreys: Comparing naturally and artificially evolved swimming controllers. 4th Euro. Conf. Artif. Life (1997) 256–265

Supervision of Control Valves in Flotation Circuits Based on Artificial Neural Network

D. Sbarbaro and G. Carvajal

Department of Electrical Engineering, Universidad de Concepción, Concepción, Chile
{dsbarbar, gcarvaja}@udec.cl

Abstract. Flotation circuits play an important role in extracting valuable minerals from the ore. To control this process, the level is used to manipulate either the concentrate or the tailings grade. One of the key elements in controlling the level of a flotation cell is the control valve. The timely detection of any problem in these valves could mean big operational savings. This paper compares two Artificial Neural Network architectures for detecting clogging in control valves. The first one is based on the traditional autoassociative feedforward architecture with a bottleneck layer and the other one is based on discrete principal curves. We show that clogging can can be promptly detected by both methods; however, the second alternative can carry out the detection more efficiently than the first one.

1 Introduction

The flotation of minerals plays an important role in the processing of copper. By using a series of tanks the valuables minerals are extracted from the ore. The tanks are connected in cascade with control valves between them. The concentrate or the tailings grade can be controlled by manipulating the tanks levels; therefore valves are very important for having well performing level control loops. Several authors have proposed fault detection algorithms for control valves based on linear models, parameter estimation and state observers. These approaches require a model of the system and normally they use assumptions concerning the time variations of either signal or parameters[2][3]. In addition, problems concerning the stability of the algorithms must also be addressed during the observer design process. A data driven approach uses the data available to build models of normal operation modes, and then these models can be applied to detect on-line any problem in the incoming data. To this end, ANN have been used in the past to carried out this task. In [4], a three layer bottleneck architecture is proposed to detect changes in the process. This work; however, proposes another alternative based on the use of principal curves, which can be obtained by a Kohonen algorithm. This work is organized as follows: section 2 describes the flotation process model. Section 3 describes two ANN models for addressing the problem of detecting clogging. Section 4 illustrates, by simulation examples, the advantages and limitations of the proposed approaches. Finally some conclusion are given in section 5.

W. Duch et al. (Eds.): ICANN 2005, LNCS 3697, pp. 451–456, 2005.

2 Hydraulic Flotation Process Model

A flotation circuit consists of cascade coupled tanks with control valves after each tank for regulating the levels. Normally all levels are measured. The input signals to the process are the control valve signals and the external inflow. The continuous time model of the level $h(t)$ in the tanks can be described by the following equation:

$$\dot{V}_1(t) = q_{in} - (C_1(u_1) + \Delta C_1)\sqrt{H_1}, \tag{1}$$

$$\dot{V}_i(t) = (C_{i-1}(u_{i-1}) + \Delta C_{i-1})\sqrt{H_{i-1}} - (C_i(u_i) + \Delta C_i)\sqrt{H_i}, \tag{2}$$

where V represents the volume of liquid in the tank, which is a nonlinear function of the height h. The function $C_i(u_i)$ represents the valve opening area of the valve i. The function $H_i(t)$ are defined as:

$$H_i(t) = 2g(h_i - h_{i-1} + \Delta h_i), \tag{3}$$

where g is the acceleration of gravity and Δh_i is the physical height difference between the zero-levels of tanks i and $i+1$. For the last tank

$$H_n(t) = 2g(h_n + \Delta h_n). \tag{4}$$

The valve manufacturer normally provides some nonlinear model for the valve opening as follows:

$$C_i(u)(t) = a_1 d^2 \frac{a_2 u + a_3 u^2}{1 + e^{a_4 u}}, \tag{5}$$

where a_i are constants and d is the diameter of the valve. In addition each level is controlled by single loop PI controllers as depicted in Fig. 1. Clogging manifests

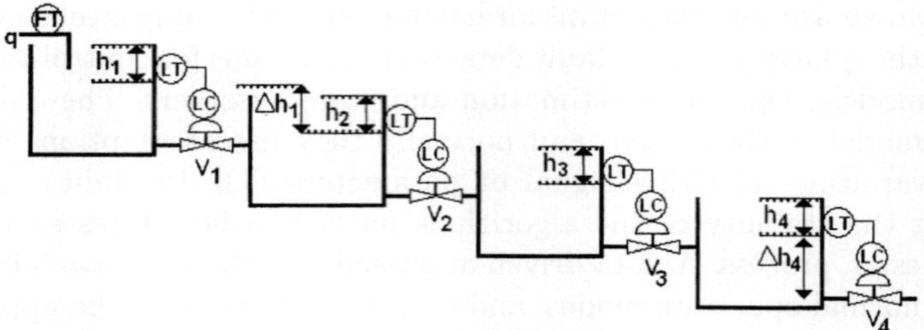

Fig. 1. A conventional flotation circuit

itself as reduction in the effective valve opening. This effect can be represented as an offset added to the position signal. Thus the real opening will be:

$$u = u_c + b, \tag{6}$$

where b is the clogging effect and u_c is the control signal.

3 The ANN Models for Clogging Detection

In industrial processes normally a set of variables are correlated through the natural interactions. These interactions define, for a given operational condition, a principal curve; i.e. a curve that passes through the middle of the data. If there is enough data corresponding to normal operations, it will be possible to build these principal curves. If the variable of interest are stacked in a vector $\mathbf{x}$, then the principal curve will be defined by a function $F(\cdot)$ such that:

$$F(\mathbf{x}) = 0. \tag{7}$$

Then any data lying far from these curves will represent an abnormal operation. In this section we describe two ANN architectures based on the principal curves concept for detecting control valves clogging in a flotation circuit.

3.1 Auto Associative ANN

The traditional approach for dealing with fault detection, using Multilayer Perceptron Network, considers a bottleneck architecture with three hidden layers trained to reproduce the inputs. The fact that the number of units in the hidden layer is smaller than the input, forces the network to rely in the functional dependencies of the variables to reconstruct them. The training is carried out by using standard backpropagation algorithm. Once the network has been trained, the detection can be carried out by comparing each network input and output, and finding the maximum difference:

$$k = \arg\max \|x_i(t) - y_i(t)\| \tag{8}$$

where $\mathbf{y}$ represents the ANN output. If $\|x_k - y_k\| > \delta$ then the output k is faulty [4].

3.2 Discrete Principal Curves and Self-organizing Map Algorithm

In order to train the network to extract the discrete principal curves the following procedure can be followed [1]. The location of units in the feature space are fixed and take values $\mathbf{z} \in \Psi$. Given a training data $\mathbf{x}_i, i = 1, ..., n$ and initial centers $\mathbf{c}_j(0), j = 1, ..., n_u$ repeat the following steps:

1. Projection. For each data point find the closest projected point on the curve:

$$\mathbf{z}_i = \arg\min \|\mathbf{c}_j - \mathbf{x}_i\|^2, \quad i = 1, ..., n \tag{9}$$

2. Determine the conditional expectation using a kernel regression estimate

$$F(\mathbf{z}, \alpha) = \frac{\sum_{i=1}^{n} \mathbf{x}_i K_\alpha(\mathbf{z}, \mathbf{z}_i)}{\sum_{i=1}^{n} K_\alpha(\mathbf{z}, \mathbf{z}_i)} \tag{10}$$

where K_α define the neighborhood function with width parameter α. The principal curve $F(\mathbf{z}, \alpha)$ is then discretized by computing the centers

$$\mathbf{c}_j = F(\mathbf{z}, \alpha), \quad j = 1, ..., n_u \tag{11}$$

3. Decrease α repeat until the empirical risk $R_{emp} = \frac{1}{n}\sum_{i=1}^{N} \|\mathbf{x}_i - F(\mathbf{z}_i, \alpha)\|^2$ reaches some small threshold.

The detection can then be carried out by finding

$$k = \arg\min \|\mathbf{x}(t) - \mathbf{c}_j(t)\|, \quad j = 1, ..., n_u \tag{12}$$

then, if $\|\mathbf{x} - \mathbf{c}_k\| > \delta$ then the vector $\mathbf{x}$ represents a fault.

4 Simulation Results

The equation described in section 2 were simulated for generating data for both normal and faulty conditions. A system with three tank was considered. The clogging was simulated by a adding a constant value to the valves positions, additional noise was also included in all the level signals. For the first valve the variables of vector $\mathbf{x}$ are selected as: $h_1(t), h_2(t), q(t), h_1(t-T)$ and $u_1(t)$. This selection is consistent with the equations describing the hydraulic model;i.e. equation (1). A set of MLPs with one, two and three layers were trained with regularization, and the best results were obtained with the MLP with just three unit in the bottleneck layer are shown in Fig. 2; where the inputs are depicted in blue and the outputs in red, the variable b_1 represents the clogging. The threshold for identifying a fault was selected considering a value bigger than the magnitude of residual error obtained with the training set. As seen in this figure, the difference between the evolution of the fourth input and output of the neural networks, corresponding to the valve opening, does not show up the clogging effect. In order to reach a conclusion concerning the clogging problem, the difference between the input and output vector must be calculated; i.e.$\|\mathbf{x} - \mathbf{y}\|$, Fig. 4 shows this results for three valves, as can be seen this measure is not so effective for detecting problems in the third valve. For the second valve, the required variables are just $h_2(t), h_3(t), u_2(t)$ and $h_2(t-T)$. In this case, a MLP with just three unit in the bottleneck layer is also used. Fig. 3 shows the evolution of the input and output variables, as in valve 1 the fourth pair of input and output, corresponding to the valve opening, does not show any evidence of change. Finally, Fig. 4 summarizes the total distance between the input and

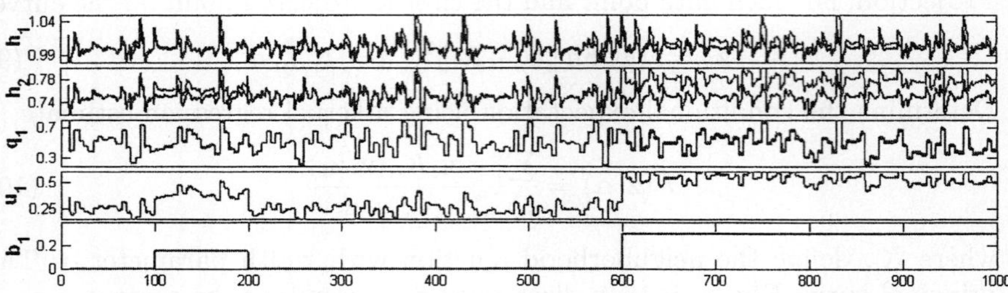

Fig. 2. Clogging detection for valve 1 using autoassociative networks

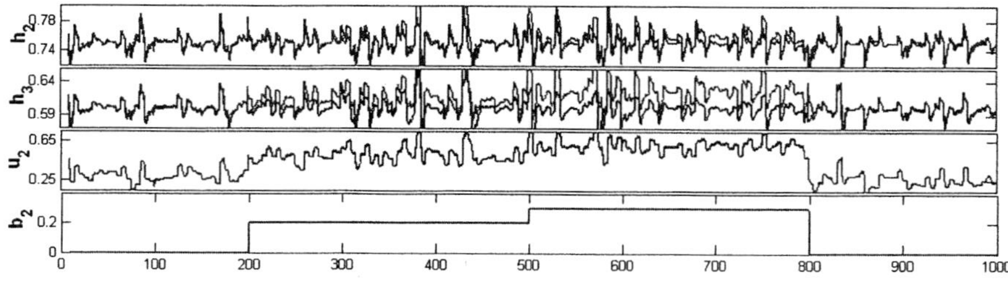

Fig. 3. Clogging detection for valve 2 using autoassociative networks

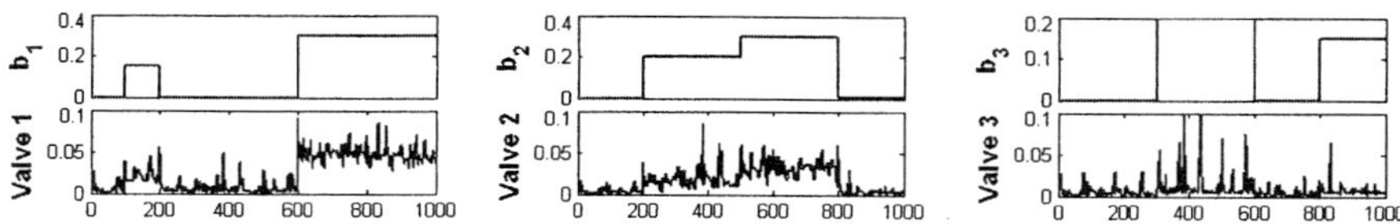

Fig. 4. Summary of results using autoassociative networks

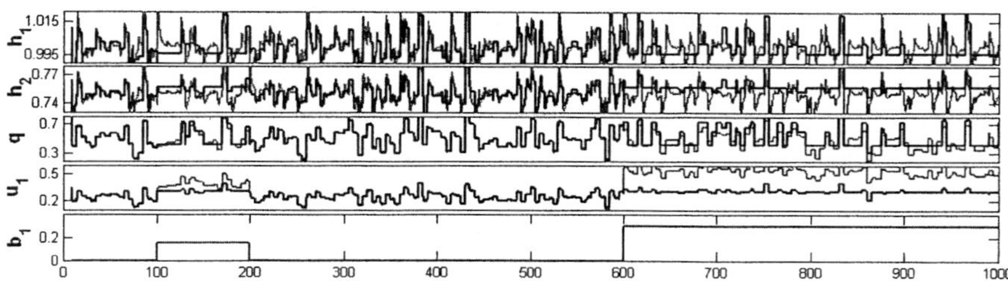

Fig. 5. Clogging detection for valve 1 using a discrete principal curves based approach

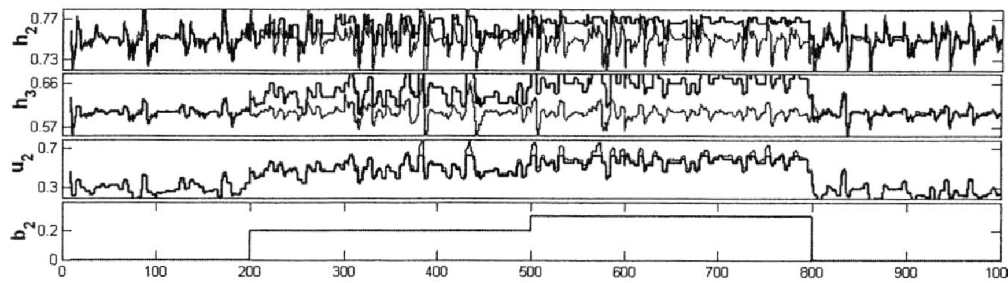

Fig. 6. Clogging detection for valve 2 using a discrete principal curves based approach

output vector, clearly for the first two valves this measure can be used to detect clogging. For calculating the discrete principal curves we have found that a network of 100 units gives a good approximation of the validation data. Fig. 5 shows the evolution of the input variables (red) and the estimated values given by the networks (black); the variable b_1 represents the clogging. As seen in the

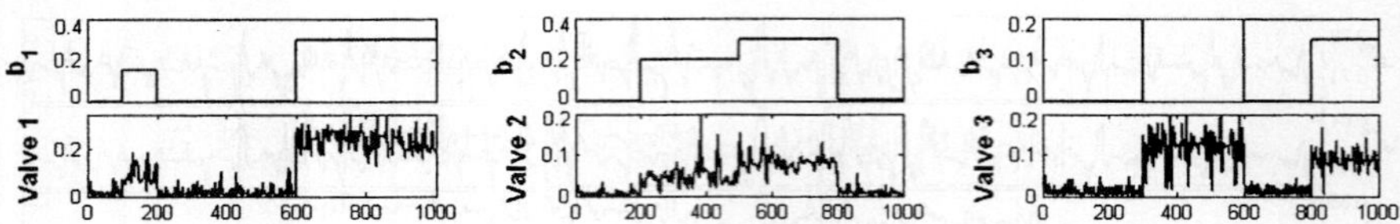

Fig. 7. Summary of results for a discrete principal curves based approach

fourth graph, corresponding to the valve opening, is possible to detect the problem in the valve since the network clearly predict a different opening. However; for the second valve, Fig. 6, the difference between the input and output is not significant. Looking at the total difference, as seen in Fig. 7 is possible to detect the problem in all three valves.

5 Final Remarks

This work has illustrated the use of ANN in the detection of clogging in coupled flotation cells. The use of conventional three bottleneck layer for detecting faults has the disadvantage of requiring a big set of training data and a complex training procedure. In addition, in some cases does not provide consistent results. The discrete principal curves, however, provides a much simpler and efficient alternative. The simulation show that, in this application, it provides more consistent results than the autoassociative network. The next step is to carry out the testing of this strategy with measurement data obtained from some Chilean concentrators.

References

1. Cherkassky, V. and P. Mulier (1997). Learning from data. John Wiley & Sons, New York.
2. Bask, M. and A. Johansson (2003). Model-based supervision of valves in flotation process. In: *CDC 2003,Proceedings of the 42nd IEEE Conference on Decision and Control.* Dec. pp. 744 – 749.
3. Stendlund, B. and A. Medvedev (2001). Supervision of control valves in a series of cascade coupled flotation tanks. In: *4th IFAC Workshop on on-line falut detection and supervision in chemical process industries.* Dec. pp. 320 – 325.
4. Kramer, M.A. (1991). Nonlinear principal Component analysis using autoassociative neural networks. *AIChE J.* **37**(4), 233–243.

Comparison of Volterra Models Extracted from a Neural Network for Nonlinear Systems Modeling

Georgina Stegmayer

Politecnico di Torino, Electronics Deptment, Cso. Duca degli Abruzzi 24, 10129 Turin, Italy
georgina.stegmayer@polito.it
http://www.eln.polito.it

Abstract. In this paper, a Time-Delayed feed-forward Neural Network (NN) is used to make an input-output time-domain characterization of a nonlinear electronic device. The procedure provides also an analytical expression for its behavior, the Volterra Series model, to predict the device response to multiple input power levels. This model, however, can be built to different accuracy degrees, depending on the activation function chosen for the NN used. We compare two Volterra series models extracted from different networks, having hyperbolic tangent and polynomial activation functions. This analysis is applied to the modeling of a Power Amplifier (PA).

1 Introduction

The analysis of electronic systems often requires an analytical model for each nonlinear element (i.e. an equation representing the in/out relationship), that allows to draw conclusions about the system performance. This approach aims to extract a relationship in order to build a model able to generalize the behavior of an electronic component. This procedure is based on the known physical behavior of the modeled device that dictates the equivalent circuit model topology. However, there are difficulties related to the circuit-oriented modeling of microwave electron devices, mainly due to the simultaneous presence of nonlinear phenomena. The process of converting nonlinear device measurements into an equivalent-circuit, and therefore into equations, relies on curve-fitting techniques. However, many of the most common techniques are useful where data trace is well behaved over a defined independent variable range and where behavior of an object is known to follow a specific mathematical model, but problems arise when the object's internal behavior cannot be estimated in advance [1]. In that case, common curve-fitting techniques become useless and a clear need appears for a new procedure, such as the NN approach. In fact, for electronic device modeling, it is receiving increasing attention [2][3] since the training procedure needs only simulation or measurements data. However, an analytical expression is still needed for the analysis and design of electronic devices inside a circuits simulator. In particular, for nonlinear behavior modeling, the analytical Volterra series model has been traditionally used[4].

In a previous work [5] it has been shown that it is possible to build a Volterra Series model for a nonlinear electronic device using a three-layer feed-forward Time-Delayed NN with hyperbolic tangent (*tanh*) activation functions in the hidden layer, trained with

W. Duch et al. (Eds.): ICANN 2005, LNCS 3697, pp. 457–463, 2005.

input-output time-domain device measurements. In this paper a simpler and faster way of obtaining the Volterra series model from a NN is presented, having a simple one-term polynomial activation function in each hidden neuron. The conclusions of the work are drawn from the comparison between these approaches, applied to a practical case study: the modeling of the nonlinear behavior in a PA, according to their estimation accuracy with respect to the original data used for training. The organization of the paper is the following: in the next Section, the neural network model used for the building of the Volterra Series model is introduced. Section 3 presents the Volterra modeling of a nonlinear electronic device (a PA) and how it can be obtained from a NN model is explained in Section 4. Simulation results and conclusions appear on Section 5 and 6.

2 Neural Network Model

The NN used to model the amplifier is a feed-forward Time-Delayed network with three layers, the input time-domain voltage samples and their delayed replies, a nonlinear hidden layer and a linear output. The architecture is shown in Fig. 1, and Eq. 1 shows its corresponding in/out analytical expression, being *f()* a nonlinear activation function for the hidden neurons.

$$Vout(t) = b_0 + \sum_{h=1}^{H} w_h^2 f\left(b_h + \sum_{n=0}^{N} w_{h,n+1}^1 Vin(t - n\Delta) \right) \tag{1}$$

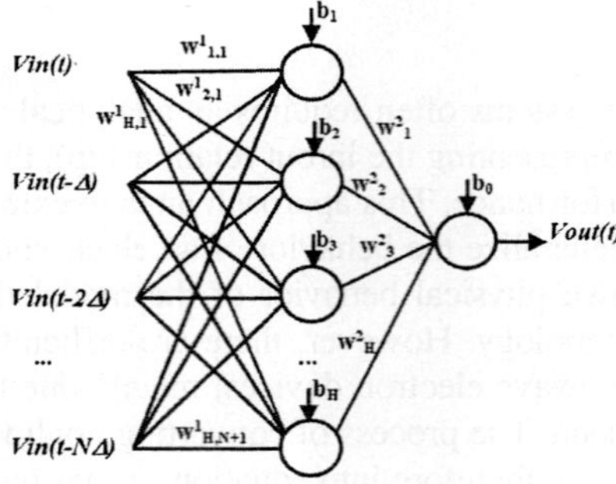

Fig. 1. Time-Delayed NN model for a PA. N: memory deep. H: number of hidden neurons.

This network is trained with PA time-domain measurements. The inputs to the device are sinusoidal voltage waveforms, that are amplified at the output according to the amplifier gain. The input and output waveforms are expressed in terms of their discrete samples in the time domain. An analytical expression for the device under study can be built, as a Volterra series expansion, calculated in function of this NN parameters (connection weights and neuron bias). The series model can be more or less accurate to represent the original device behavior, not only depending on the number of hidden neurons, but also according to the hidden layer nonlinear activation function *f()* chosen and the number of terms included in the series. This is explained in the next Section.

3 Volterra Modeling of a PA

The Volterra approach characterizes a system as a mapping between two function spaces, which represent the input and output spaces of that system [6]. The Volterra series is an extension of the Taylor series representation and it can be represented exactly by a converging infinite series. The series can be described in the time-domain or in the frequency-domain[7]. A Volterra series model in discrete form that represents the nonlinear behavior of a Power Amplifier is presented in Eq. 2. Here, *Vout(t)* is output power of the system at time t; *Vin(t)* is input power at time t and $h_n(k_1, \ldots, k_n)$ is the nth order Volterra kernel. Typically in the series representation, only up to the 3rd order terms are included. The summation interval [0-N] is limited to the practically finite duration of the "memory" effect in the device.

$$Vout(t) = h_0 + \sum_{k=0}^{N} h_1(k)Vin(t-k\Delta) + \sum_{k_1=0}^{N}\sum_{k_2=0}^{N} h_2(k_1,k_2)Vin(t-k_1\Delta)Vin(t-k_2\Delta) + \sum_{k_1=0}^{N}\sum_{k_2=0}^{N}\sum_{k_3=0}^{N} h_3(k_1,k_2,k_3)Vin(t-k_1\Delta)Vin(t-k_2\Delta)Vin(t-k_3\Delta). \tag{2}$$

The Volterra series analysis is well suited to the simulation of nonlinear microwave devices and circuits, in particular in the weakly and mildly nonlinear regime where a few number of kernels (generally up to the 3rd order kernels) are able to capture the device behavior (e.g. for PA distortion analysis) [8]. The Volterra kernels allow the inference of device characteristics of great concern for the microwave designer. However, the number of terms in the kernels of the series increases exponentially with the order of the kernel. Moreover, at microwave frequencies, suitable instrumentation for the measurement of the kernels is still lacking [9]. In spite of this drawback, the Volterra series is used for microwave circuit design, by means of complex and time-consuming analytical or numerical calculation [10] A procedure has been found in [5] that allows generating the Volterra series ad its kernels, for the modeling of a nonlinear electronic device, using the weights and bias values of a Time-Delayed feed-forward NN, after it has been trained with time-domain device measurements. It is briefly explained in the next Section, as it is also presented a new approach for the building of the Volterra series, based on a simpler network model.

4 Volterra Kernels Extraction

The procedure to obtain the Volterra kernels is based on a NN like the one presented in Fig. 1, with *f(x)=tanh(x)* in the hidden layer. This network is trained with time-domain device measurements, up to a predetermined number of epochs or a desired accuracy. Once the network has been trained, and its weights and bias values have been fixed, the network in/out expression (Eq. 1) is developed as a Taylor series around the bias values of the hidden nodes. To do that, the hidden neurons functions derivatives have to be calculated with respect to their bias values. After accommodating the resulting expression in common terms, the Volterra kernels can be easily identified. The general formula that builds the Volterra model using the NN parameters (weights and bias), is shown in Eq. 3, being *d* derivative order. The kernels can be identified as the terms between brackets.

$$Vout(t)=\sum_{d=0}^{\infty}\left[\sum_{h=1}^{H}w_h^2\sum_{n_1=0}^{N}\cdots\sum_{n_n=0}^{N}w_{h,n_1+1}^1\cdots w_{h,n_n+1}^1\left(\frac{\left.\frac{\partial^d f}{\partial x^d}\right|_{x=b_h}}{d!}\right)\right](Vin(t-n_1\Delta)\ldots Vin(t-n_n\Delta))^d \quad (3)$$

The order of the kernel is given by the activation function derivative order. The particular case for the hyperbolic tangent activation function is shown in Eq. 4, where only the most common order kernels (1st and 3rd) are shown. The formulas for the calculation of any order kernels can be found on [5]. Another way of calculating the kernels has been suggested in [11], where it is proposed a new kind of neural network, with a particular topology, having distinct polynomials series with trainable coefficients (i.e. $ax+bx^2+cx^3+\ldots$) as activation functions in the hidden layer. This model requires also a special training algorithm for updating the polynomials coefficients (a,b,c). Later, these coefficients, combined, provide the kernels values.

$$Vout(t)=b_0+\sum_{h=1}^{H}w_h^2\tanh(b_h)+\left[\sum_{h=1}^{H}w_h^2w_{h,1}^1\frac{\partial\tanh}{\partial b_k}\right]Vin(t)+\left[\sum_{h=1}^{H}\sum_{i=1}^{H}w_h^2w_{i,1}^1w_{i,1}^1w_{i,1}^1\left(\frac{\partial^2\tanh}{\partial b_k^{\,2}}/3!\right)\right]Vin(t)^3+\ldots \quad (4)$$

However, a simpler neural network model like the one presented in Fig. 1, with the same order one-term polynomial activation function with coefficient 1 (instead of a polynomial series) in each hidden neuron, using a standard training algorithm for the updating of the network weights and bias, can be used to obtain the same results. The kernels are obtained using the polynomial derivatives, faster and easier to calculate. Once the kernels are found, they are combined with the inputs and the Volterra series model is built. An example is shown in Eq. 5.

$$Vout(t)=b_0+\sum_{h=1}^{H}w_h^2(b_h)^3+\left[\sum_{h=1}^{H}w_h^2w_{h,1}^1 3(b_h)^2\right]Vin(t)+\left[\sum_{h=1}^{H}\sum_{i=1}^{H}w_h^2w_{i,1}^1w_{i,1}^1w_{i,1}^1 1\right]Vin(t)^3+\ldots \quad (5)$$

Concerning this new method, two important observations have to be made. The first one is that, having in mind the objective of simplifying the Volterra kernel calculus and therefore introducing the new cubic method, the universal approximation property of the MLP model is not valid anymore [12] and therefore the model becomes a polynomial one. Second, the order and the type of polynomial function chosen will certainly influence the accuracy of kernels estimation and the type of system that could be approximated through the Volterra Series model. In other words, the order of the polynomial function will influence on the order of the Volterra kernels that could be calculated and added to the series, i.e. with quadratic functions the 3rd order kernels could not be obtained. Therefore, a trade-off has to be made between simplicity and accuracy, at the moment of choosing the type of activation function for the hidden neurons of the network model, from where the Volterra model will be built. The hyperbolic tangent activation function still remains as the preferred choice when the order of the system that has to be modeled is unknown in advance. However, if there is some information about the system under study and its order (i.e., PAs present at most a 3rd order degree nonlinearity) and the main objective is to accelerate the calculus, the use of the corresponding polynomial would yield a better result. A case study that confirms this assumption is presented in the next Section, for the modeling of a PA nonlinear response to several simultaneous input power levels.

5 Model Training and Results

The data for the training of the neural models were obtained in the laboratory from a Cernex 2266 PA, with 1-2 GHz bandwidth and 29 dB gain. The data used to train the NN are eight different power levels used altogether for training the NN model of Fig. 1, having N=4 (five inputs in total). Several networks configuration were tried, finally choosing a topology with ten hidden neurons (H=10), which provided a mean square error (MSE) for the network of approximately 1e-06 after 50 epochs. Two different nonlinear activation functions were tried for the hidden layer: hyperbolic tangent (Eq. 4) and cubic (Eq. 5). After the training phase, the network parameters (weights and bias values) have been used to build a Volterra Series model for the amplifier, including up to the 3rd order kernels in the approximations. From each different NN model, a Volterra series approximation was built. The results are presented in the next figures. We have also calculated the mean squared approximation error (MSAE) between the original device behavior *Vout* and each obtained Volterra series approximations, over the number of samples, using the formula of Eq. 6, being *P* the total number of samples (pairs input/output) used for training.

$$MSAE = \frac{1}{P}\sum_{k=1}^{P}\left(Vout(t) - Volterra(t)\right)^2 \tag{6}$$

The approximation obtained with a Volterra series approximation built from a NN model having hyperbolic tangent (*tanh*) activation function in each hidden neuron is shown in Fig. 2 (left). The MSAE for this Volterra approximation with regard to the original behavior is 0.0011. Concerning a cubic model (right), its MSAE is equal to 0.0121. Analyzing the errors, the differences among the approximations is of roughly one order of magnitude, having the network with *tanh* the lowest error. However, for this problem, the cubic NN model is faster than the other one, reaching the 1e-05 MSE only after 6 epochs (Fig. 3). Taking this into account and looking for a compromise, we can claim that for the device under study, the new cubic NN and its corresponding Volterra model are preferable.

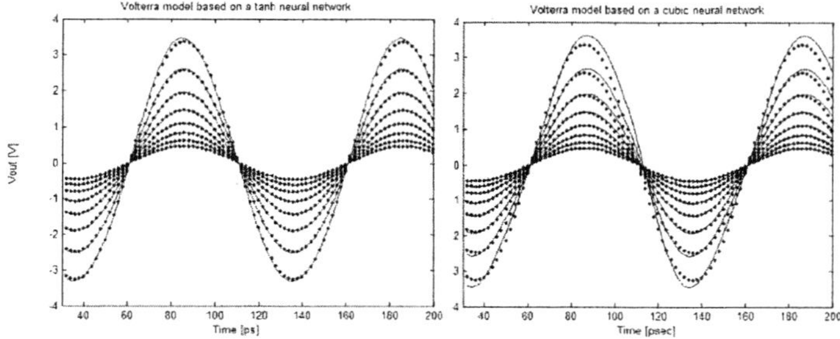

Fig. 2. Results after training the NN with the original data (doted line) and building afterwards the Volterra series model (solid line) from a tanh_NN (left) and a cubic_NN (right)

Finally, a natural question that could arise is why to build and use an approximation of the PA behavior (the Volterra model) that has an error twice bigger than the neural network model itself? The answer is that for simulation purposes, to be able to

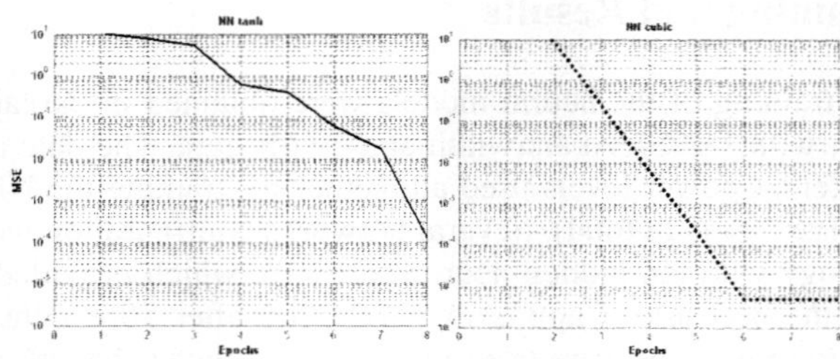

Fig. 3. Speed in NN training: tanh model (left) vs. cubic model (right)

put the device model into a circuits simulator and to interconnect it with other devices models to make systems simulations, an analytical formula is most of the time required. The NN is a valuable tool that saves time and complexity when building the analytical Volterra model, which is a particular suitable model traditionally used for nonlinear device behavior representation, and already available as a component (whose kernels, however, have to be assigned a value) inside most of today circuits simulators.

6 Conclusions

In this paper we have presented a new method for the building of the Volterra series model of a nonlinear PA, using a simple polynomial-based NN trained with data from different power levels simultaneously. The Volterra model, however, can be built to different accuracy degrees, depending on the activation function chosen. That is why two NNs have been tested, having hyperbolic tangent and cubic functions, for the building and comparison between their corresponding Volterra models, according to their estimation accuracy with respect to the original data. Even though the new method, that proposes the use of simpler polynomial activation functions in the hidden layer to further simplify the procedure, the old one assures good approximation accuracy in the case of a nonlinear system that has to be modeled, where system order is unknown in advance.

Acknowledgments

The author thanks Prof. Orengo (Univ.Rome II) for the PA measurements used in this work.

References

1. T.R. Turlington, Behavioral Modeling of Nonlinear RF and Microwave devices, Ed. Artech House, 2000.
2. M.R.G. Meireles, P.E.M. Almeida and M.G. Simoes, "A comprehensive review for industrial applicability of Artificial Neural Networks", IEEE Trans. Industrial Electronics, no. 3, pp. 585-601, 2003.
3. Q.J. Zhang, K.C. Gupta and V.K. Devabhaktuni, "Artificial Neural Networks for RF and Microwave Design – From Theory to practice", IEEE Trans. on MTT, no. 4, pp. 1339-1350, 2003.
4. S.A. Maas, Nonlinear Microwave Circuits, Ed. Artech House, 1988.
5. G. Stegmayer, M. Pirola, G. Orengo, O. Chiotti, "Towards a Volterra series representation from a Neural Network model", WSEAS Trans. on Systems, no. 2, pp 432-437, 2004.
6. W. Rough, Nonlinear System Theory. The Volterra/Wiener Approach, Johns Hopkins University Press, 1981.
7. S. Boyd, L.O. Chua and C.A. Desder, "Analytical Foundations of Volterra Series", IMA Journal of Mathematical Control & Information, vol. 1, pp. 243-282, 1984.
8. D. Weiner and G. Naditch, "A scattering variable approach to the Volterra analysis of nonlinear systems", IEEE Trans. on MTT, no. 24, pp. 422–433, 1976.
9. S. Boyd, Y.S. Tang, and L. 0. Chua, "Measuring Volterra kernels", IEEE Trans. on Circuits and Systems, no. 8, pp. 571-577, 1983.
10. F. Filicori, G. Vannini and V.A. Monaco, "A nonlinear integral model of electron devices for HB circuit analysis", IEEE Trans. on MTT, no. 7, pp. 1456-1465, 1992.
11. M. Iatrou, T.W. Berger and V.Z. Marmarelis, "Modeling of Nonlinear Nonstationary Dynamic Systems with a Novel Class of Artificial Neural Networks", IEEE Trans. on Neural Networks, no. 2, pp. 327-339, 1999.
12. A. Pinkus, "Approximation Theory of the MLP model in neural networks", Acta Numerica, pp. 143-195, 1999.

Identification of Frequency-Domain Volterra Model Using Neural Networks

Georgina Stegmayer[1] and Omar Chiotti[2]

[1] Politecnico di Torino, Electronics Deptment, Cso. Duca degli Abruzzi 24, 10129 Turin, Italy
georgina.stegmayer@polito.it
http://www.eln.polito.it

[2] G.I.D.S.A.T.D., Universidad Tecnológica Nacional, Lavaise 610, 3000 Santa Fe, Argentina
gidsatd@frsf.utn.edu.ar
http://www.frsf.utn.edu.ar/investigacion/grupos/gidsatd/

Abstract. In this paper, a new method is introduced for the identification of a Volterra model for the representation of a nonlinear electronic device in the frequency domain. The Volterra model is a numerical series with some particular terms named kernels. Our proposal is the use of feedforward neural networks (FNN) for the modeling of the nonlinearities in the device behavior, and a special procedure which uses the neural networks parameters for the kernels identification. The proposed procedure has been tested with simulation data from a class "A" Power Amplifier (PA) which validate our approach.

1 Introduction

The classical modeling of electronic devices consists in building empirical models, which are electrical circuits schematics containing capacitors, resistors, transmission lines, among other electronic components representations. The elements with nonlinear behavior are typically defined though analytical functions, and when microwave applications are considered, they are more conveniently defined in the frequency domain. The main problem with this approach is that the model can have hundred of parameters to be tuned to make it work properly. On the other hand, behavioral models propose to characterize a nonlinear system in terms of in/out scattered waves, using relatively simple mathematical expressions. The modeled device is considered as a "black-box", no knowledge of the internal structure is required. The modeling information about its behavior is completely contained in the external response (measurement data) of the device, which help estimating the model parameters [1].

A truncated Volterra series has been successfully used to derive some behavioral models for PA in recent years. It has been traditionally the most general and rigorous modeling approach for systems characterized by nonlinear dynamic phenomena [2]. Black-box models relying on the Volterra series and on directly measured data enable to forget the circuit topology [3]. Concerning a single port (or single in/out) device, the terms "nonlinear" and "dynamic" imply that the output, i.e. the current in the case of a PA, at any time instant, is nonlinearly dependent not only on the applied input at the same instant, but also on its past values, that represent the "memory" effect associated with dynamic phenomena in the device (i.e. charge-storage effects) [4]. This

W. Duch et al. (Eds.): ICANN 2005, LNCS 3697, pp. 465–471, 2005.

makes the modeling of such a system a difficult task. Furthermore, the high computational complexity of the standard methods for Volterra modeling [5][6] are limited in many practical situations. In this paper we present a new method for the identification of the terms of the Volterra series model using a FNN. With the proposed procedure they can be obtained using standard device measurements in the frequency domain in a simple and straightforward way, saving time to the design engineer at the moment of modeling and simulating a nonlinear device. We have tested our approach with simulation data from a class "A" PA with 3rd order nonlinearities and memory effects, which validate our proposal. The organization of the paper is the following: in the next Section, the nonlinear behavior of electronic devices, in particular in the frequency domain, is introduced. Section 3 presents the Volterra modeling of a nonlinear electronic device. The new procedure to obtain the Volterra kernels from a neural network appears in Section 4. Section 5 shows some results obtained from simulations. Finally, the conclusions can be found on Section 6.

2 Electronic Device Nonlinear Behavior

Typically, a black-box representation of a device consists in an abstract block. In general, the incident and reflected power waves are related to the inputs (voltages) and outputs (currents) of the device (i.e. PA) as shows Fig. 1. Here a typical 2-port device (two inputs/outputs) is presented together with its scattering matrix parameters [S], which are a widely known set of parameters that characterize the device and that relate its inputs (a_1 and a_2) and outputs (b_1 and b_2) in a linear way. The S_{ij} parameter describes the influence of the incident wave at port *j* on the resulting wave at port *i*.

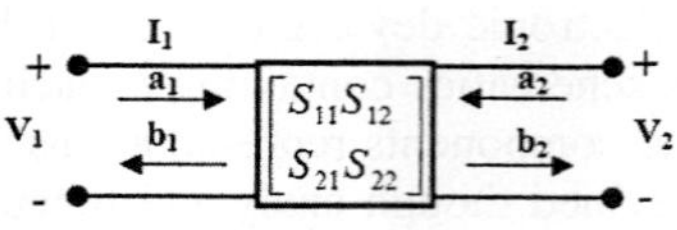

Fig. 1. Representation of a nonlinear two-port device, with the scattering matrix parameters *[S]*. *I* represents current, *V* represents driving voltage and *a* and *b* are the incident and reflected power waves, respectively.

Let us first consider a 2-port device with linear behavior. If both ports are excited by incident waves (a_1, a_2) at frequency f_O (also named the fundamental) the reflected waves (b_1, b_2) will contain a component at that frequency, specified by the S-parameters according to Eqs. (1) and (2).

$$b_1(f_O) = S_{11}(f_O)a_1(f_O) + S_{12}(f_O)a_2(f_O) \quad (1)$$

$$b_2(f_O) = S_{21}(f_O)a_1(f_O) + S_{22}(f_O)a_2(f_O) \quad (2)$$

Linear systems do not generate new frequencies (the frequency content at the output is identical to that of the input, although it can be modified in amplitude and phase). This is represented in Fig. 2. On the left side, the input/output of a linear device is shown, in both time and frequency domain, where can be clearly seen that the input and output frequency is the same and new frequency components are not created. However, on the right side, a nonlinear behavior is shown, where some frequency components apart from f_0 (i.e. f_1, f_2, f_3) may interact nonlinearly to produce frequency components in the output that may not be present in the input signal. This interaction may produce some nonlinear phenomena, such as intermodulation ($f_1 \pm f_2 \pm f_3$) and 3rd harmonic terms ($3f_1$, $3f_2$, $3f_3$), which are of particular interest to the electronic engineer, because they allow to obtain some parameters that describe the device performance [7].

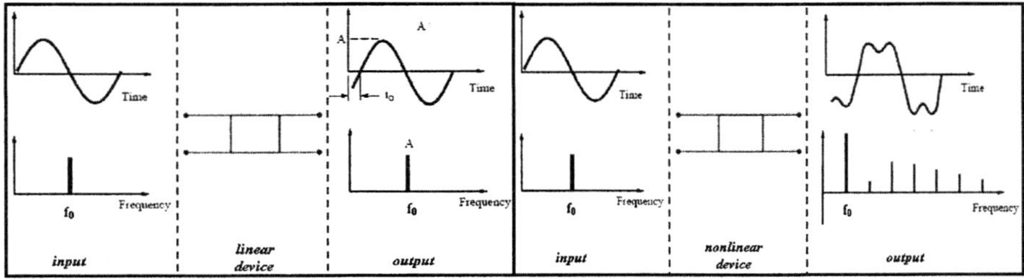

Fig. 2. Left side: representation of a device with linear behavior. Input and output frequencies are the same, no additional frequencies are created. The output frequency may only change in amplitude and phase. Right side: representation of a nonlinear device. Output frequency may suffer a frequency shift and additional frequencies can be created.

3 Volterra Series Model of a Nonlinear Device

The Volterra approach characterizes a system as a mapping between two function spaces, the input and output spaces of that system. The Volterra model is an extension of the Taylor series representation to cover dynamic systems [8]. The series can be described in the time-domain or in the frequency-domain. In the discrete-frequency domain, the series takes the form of Eq. (3), where X(f) is the Fourier transform of the input signal x(t) at the frequency f. The term H_n is the "kernel" which describes the contribution of the nth degree of nonlinearity to the system. This way, H_1 represents the linear transfer function and H_2 and H_3 are the quadratic and cubic transfer functions of the system [9].

$$Y(f) = H_1(f)X(f) + \sum_{f_1}^{\infty}\sum_{\substack{f_2 \\ f_1+f_2=f}}^{\infty} H_2(f_1,f_2)X(f_1)X(f_2) + \tag{3}$$

$$\sum_{f_1}^{\infty}\sum_{f_2}^{\infty}\sum_{\substack{f_3 \\ f_1+f_2+f_3=f}}^{\infty} H_3(f_1,f_2,f_3)X(f_1)X(f_2)X(f_3) + \ldots + \sum_{f_1}^{\infty}\ldots\sum_{\substack{f_n \\ f_1+\ldots+f_n=f}}^{\infty} H_n(f_1,\ldots,f_n)X(f_1)\ldots X(f_n)$$

An example of this model for a nonlinear device is shown below. Eq. (4) shows the linear part of the generated signal at port 1 (b_1) and Eq. (5) models the cubic component that appears in this port as a consequence of the device nonlinear behavior (combination of the input signals). Equivalent equations are valid for port 2 (b_2).

$$b_1(f_0) = H_1(f_0)a_1(f_0) + H_1(f_0)a_2(f_0) \tag{4}$$

$$b_1(3f_0) = H_3(f_0,f_0,f_0)a_1(f_0)a_1(f_0)a_1(f_0) + 3H_3(f_0,f_0,f_0)a_1(f_0)a_1(f_0)a_2(f_0) + 3H_3(f_0,f_0,f_0)a_1(f_0)a_2(f_0)a_2(f_0) + H_3(f_0,f_0,f_0)a_2(f_0)a_2(f_0)a_2(f_0) \quad (5)$$

Application of Volterra system theory has an important role in nonlinear system analysis and identification, due in part to the fact that Volterra series a firm mathematical foundation and nonlinear behavior can be described with reasonably accuracy by a truncated version of the series, which reduces the complexity of the problem and requires a limited amount of knowledge of higher order statistics or higher order spectra [10]. In nonlinear microwave analysis the tool for excellence has been the Volterra-series analysis. The kernels allow the calculation of device parameters of great concern for the microwave designer, i.e. in the case of a PA, among others, nonlinear gain, 3rd order harmonics and intermodulation. However, kernels calculation, analytical expression or measurement can be a very complex and time-consuming task [11]. There have been some approaches to help the calculation of the kernels, in particular in the frequency domain. For example in [7][12] special fitting functions are proposed, using then optimization procedures to find the kernels values in a global manner. Our proposal is a modular approach, simpler and more straightforward, and it is an adaptation of a previous work performed for the kernels identification in the time domain [13]. We use FNNs to do the fitting of standard frequency domain measurements, one for each nonlinear part of the system, and after that, a simple procedure permits obtaining the kernels values directly from the network parameters, just combining its weights and bias values. This procedure is explained in the next Section.

4 Identification Procedure Using Neural Networks

The proposed approach involves the use of FNNs like the one presented in Fig. 3, having H hidden neurons with a generic activation function (*af*) and bias values. The choice of the function will depend on the type of nonlinearity that will be modeled by the network. The inputs to the model are the inputs to the device, i.e. a_1 and a_2, measured at the fundamental frequency f_0, and the output is the output value b_1 measured also at f_0. The output neuron is lineal. The training set is built with these measurements performed in all the frequency range or work interval of the device, where the S-parameters have the same value as long as the device bias point does not change [11].

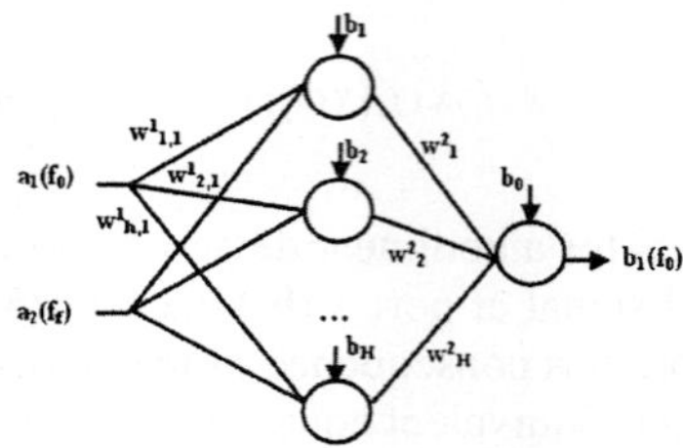

Fig. 3. Feedforward Neural Network model used in the Volterra kernels identification procedure. The inputs/output of the network are the inputs/output measured at the fundamental frequency f_0, in port 1.

First of all we focus our attention on port 1 of the 2-port device of Fig. 1 (equivalent reasoning applies for port 2). Several FNNs like Fig. 3 will be used to model the 1st and 3rd order nonlinearities of output b_1. In general, the network output is Eq. (6).

$$b_1(f_0)=\sum_{h=1}^{H} w_h^2 af\left(w_{h,1}^1 a_1(f_0)+w_{h,2}^1 a_2(f_0)+b_h\right) \tag{6}$$

If the network output is developed as a Taylor series around the bias values of the hidden neurons, and the terms are re-ordered according to derivative order and common terms, Eq. (7) yields, where due to space restrictions only the 1st order terms have been included. The Volterra kernels for b_1 are easily identified as the terms between brackets. Actually this is the general procedure, but for the linear behavior in particular, the activation functions of the hidden neurons are linear and the bias values take zero values. Taking this into account, Eq. (8) and Eq. (9) are obtained for the 1st order kernels at port 1. Comparing Eq. (7) with the in/out relationship of this port (Eq. (1)), becomes clear the fact that the 1st order kernels H_1 happen to be the lineal scattering parameters. The simulations presented in the next Section validate our proposed approach, and therefore we use it for the 3rd order nonlinearity.

$$b_1(f_0)=b_0+\sum_{h=1}^{H} w_h^2 af(b_h)+\left[\sum_{h=1}^{H} w_h^2 w_{h,1}^1\left(\left.\frac{\partial af}{\partial x}\right|_{x=b_h}\right)\right]a_1(f_0)+\left[\sum_{h=1}^{H} w_h^2 w_{h,2}^1\left(\left.\frac{\partial af}{\partial x}\right|_{x=b_h}\right)\right]a_2(f_0)+\ldots \tag{7}$$

$$H_1(a_1)=\sum_{h=1}^{H} w_h^2 w_{h,1}^1 \tag{8}$$

$$H_1(a_2)=\sum_{h=1}^{H} w_h^2 w_{h,2}^1 \tag{9}$$

We apply the same procedure to the identification of higher order kernels, in particular the 3rd order Volterra kernels. But now the FNN used includes bias values in the neurons and cubic ($af = x^3$) in the hidden neurons, and it is trained with the same inputs, but with the output response generated at the 3rd harmonic ($b_1(3f_0)$). The procedure is applied and the formulas obtained this time for the 3rd order Volterra kernels are Eq. (10) to Eq. (13).

$$H_3(a_1,a_1,a_1)=\frac{\sum_{h=1}^{H} w_h^2 w_{h,1}^1 w_{h,1}^1 w_{h,1}^1\left(\left.\frac{\partial^3(x^3)}{\partial x^3}\right|_{x=b_h}\right)}{3!} \tag{10}$$

$$H_3(a_1,a_1,a_2)=\frac{\sum_{h=1}^{H} w_h^2 w_{h,1}^1 w_{h,1}^1 w_{h,2}^1\left(\left.\frac{\partial^3(x^3)}{\partial x^3}\right|_{x=b_h}\right)}{3!} \tag{11}$$

$$H_3(a_2,a_2,a_1)=\frac{\sum_{h=1}^{H} w_h^2 w_{h,2}^1 w_{h,2}^1 w_{h,1}^1\left(\left.\frac{\partial^3(x^3)}{\partial x^3}\right|_{x=b_h}\right)}{3!} \tag{12}$$

$$H_3(a_2,a_2,a_2)=\frac{\sum_{h=1}^{H} w_h^2 w_{h,2}^1 w_{h,2}^1 w_{h,2}^1\left(\left.\frac{\partial^3(x^3)}{\partial x^3}\right|_{x=b_h}\right)}{3!} \tag{13}$$

5 Case of Study and Simulation Results

The simulation data used for the network training were obtained from a class A PA at 1 Ghz. The absolute errors between its scattering parameters the corresponding 1st order Volterra kernels obtained from the FNN at the port 1 are: [Re{S_{11}}=0.99999, $H_1(a_1)$] = 4e-07 and [Re{S_{12}}=0.00001, $H_1(a_2)$] = 3e-07. Concerning port 2, the errors are: [Re{S_{21}}=0.00001, $H_1(a_1)$] = 2e-07 and [Re{S_{22}}=-0.73351, $H_1(a_2)$] = 8e-08. As can be seen, the errors are very low and the procedure is validated. In the case of the 3rd order Volterra kernels estimation, the values are not previously known, so the only way of testing our approach is using the kernels to build a Volterra series model that includes the kernels identified from the network, and compare it with the original nonlinear behavior. The results are shown in Fig. 4, showing an excellent approximation result and validating our proposal once more.

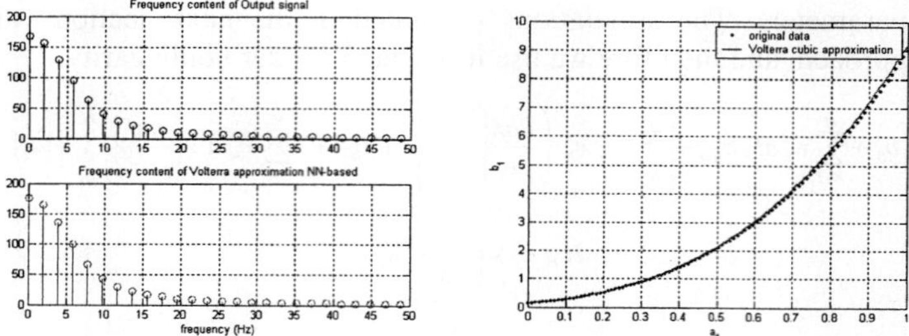

Fig. 4. Simulation results with the 3rd order kernels. Left side: frequency domain plot comparing the original data and the Volterra model. Right side: a_1 vs. b_1 plot comparing original data (dotted line) and the Volterra model (full line). The mean square error between them is of 0.0099.

6 Conclusions

In this work we have presented a new modular method for the identification of the Volterra kernels in the frequency domain. We propose the use of FNNs and a special procedure for the kernels identification, using the neural network parameters. The proposed procedure has been tested with simulation data from a class "A" Power Amplifier (PA), which have validated our approach for the 1st and 3rd order nonlinearities identification.

References

1. Zhu, A., Brazil, T.J.: Behavioral modeling of RF power amplifiers based on pruned Volterra series. IEEE Microwave and Wireless components letters, 14 (2004) 563-565
2. Weiner, D., Naditch, G.: A scattering variable approach to the Volterra analysis of nonlinear systems, IEEE Trans. on Microwave Theory and Techniques 24 (1976), 422–433

3. Harkouss, Y., Rousset, J., Chehade, H., Ngoya, E., Barataud, D., Teyssier J.P.: Modeling microwave devices and circuits for telecommunications system design. Proc. IEEE World Congress on Computational Intelligence, 1 (1998) 128-133
4. Filicori, F., Vannini, G., Monaco, V.A.: A Nonlinear Integral Model of Electron Devices for HB Circuit Analysis. IEEE Trans. on Microwave Theory and Techniques, 40 (1992) 1456-1464
5. Kashiwagi, H., Harada, H., Rong, L.: Identification of Volterra kernels of nonlinear systems by separating overlapped kernel slices. Society Instrument Control Engineers of Japan, 2 (2002) 707-712
6. Bauer, A., Schwarz, W.: Circuit analysis and optimization with automatically derived Volterra kernels. IEEE Int. Symposium on Circuits and Systems, (2000) I-491 - I-494
7. Wang, T., Brazil, T.J.: A Volterra-mapping-based S-parameter behavioral model for nonlinear RF and microwave circuits and systems. IEEE Int. Microwave Theory and Techniques, 2 (1999) 783-786
8. Schetzen, M.: The Volterra and Wiener Theories of Nonlinear Systems. Ed. John Wiley & Sons Inc. (1980)
9. Rough, W.: Nonlinear System Theory. The Volterra/Wiener Approach. Ed. Johns Hopkins University Press (1981)
10. Nam, S.W., Powers, E.J.: Application of Higher-Order Spectral Analysis To Cubically-Nonlinear System Identification. IEEE Trans. on Signal Processing, 42 (1994) 1746-1765
11. Maas, S.A.: Nonlinear Microwave Circuits. Ed. Artech House (1988)
12. Verbeyst, F., Vanden Bossche, M.: The Volterra input-output map of a high-frequency amplifier as a practical alternative to load-pull measurements. IEEE Trans. on Instrumentation and Measurement, 44 (1995) 662-665
13. Stegmayer, G.: Volterra series and Neural Networks to model an electronic device nonlinear behavior. Proc. IEEE Int. Joint Conference on Neural Networks, 4 (2004) 2907-2910

Hierarchical Clustering for Efficient Memory Allocation in CMAC Neural Network

Sintiani D. Teddy[1] and Edmund M.-K. Lai[1]

School of Computer Engineering,
Nanyang Technological University, Singapore, 639798
sdt@pmail.ntu.edu.sg, asmklai@ntu.edu.sg

Abstract. CMAC Neural Network is a popular choice for control applications. One of the main problems with CMAC is that the memory needed for the network grows exponentially with each addition of input variable. In this paper, we present a new CMAC architecture with more effective allocation of the available memory space. The proposed architecture employs hierarchical clustering to perform adaptive quantization of the input space by capturing the degree of variation in the output target function to be learned. We showed through a car maneuvering control application that using this new architecture, the memory requirement can be reduced significantly compared with conventional CMAC while maintaining the desired performance quality.

1 Introduction

The Cerebellar Model Articulation Controller (CMAC) neural network was proposed by Albus [1] as an associative memory neural network that models the mechanisms of the human cerebellum. Since then, CMAC has become a popular choice for real-time control and optimization [2] such as the modeling and control of robotic manipulators [3]. It has also been applied to various signal processing and pattern-recognition applications [4,5].

CMAC learning is based on the principle that similar inputs should produce similar outputs, while inputs that are located distantly in the input space should produce nearly independent outputs. CMAC is an associative memory which stores information locally and behaves as a dynamic look-up table, in which its contents are indexed by the inputs to the network. The advantages of CMAC are simple computation, fast training, local generalization and ease of hardware implementation.

Unfortunately, the look-up table behavior of CMAC also implies that the size of the network increases exponentially as the number of input variables. This causes problems, especially when there are uneven degree of variations in the target function to be learned, where uniform quantization of input space will result in suboptimal space utilization.

It is therefore necessary to find a mechanism for efficient memory space allocation by allocating more storage space in the range of input space which holds more information. Some previously published works have tackled this problem by

W. Duch et al. (Eds.): ICANN 2005, LNCS 3697, pp. 473–478, 2005.

introducing non-uniform quantization of the input space to CMAC [6,7,8]. However, the examples used to illustrate the performance are single input variable cases. Moreover, there is a compromise between the computational complexity and the required memory space.

In this paper, we propose a CMAC architecture for reducing the memory requirements. It makes use of adaptive quantization based on hierarchical clustering technique, which we refer to as Hierarchical-Clustering based Adaptive Quantization Cerebellar Model Arithmetic Computer (HCAQ-CMAC). The proposed architecture is tested on an automated car maneuver control application. The experimental results show a significant improvement on the memory utilization.

2 CMAC Network

The CMAC behaves like a memory, where a particular input to output mapping acts as the address decoder. Each possible input vector selects a unique set of cells, the weighted sum of which is the output of the network for that particular input combinations.

An example of the cell memory allocation is depicted in Figure 2 for a 2-dimensional input. From this point of view, CMAC can be considered a memory in which the memory cells are uniformly distributed along the input dimensions. Each of the input dimension can be considered as being *quantized* into discrete steps or quantization levels. The input value will first be quantized into one of the levels, and the result will be the index which is used to access the memory locations. The idea of HCAQ-CMAC is based on this observation.

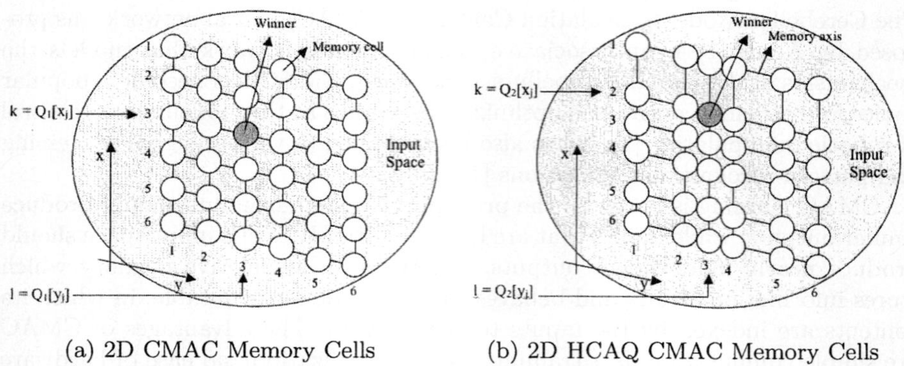

(a) 2D CMAC Memory Cells

(b) 2D HCAQ CMAC Memory Cells

Fig. 1. Comparison of CMAC and HCAQ-CMAC Memory Surface

3 The HCAQ-CMAC

Figure 2 shows an example of a two-dimensional HCAQ-CMAC network. In HCAQ-CMAC, the memory cells are distributed in a non-uniform way according to the degree of variations of the target function to be learned. This is in contrast to Figure 2 where the cells are distributed uniformly.

The idea is to perform a non-uniform quantization of the input variables to obtain a more efficient coverage of the overall input space. The more changes observed in the region of an input variable, the more memory space will be allocated to that particular region along that axis. This results in a finer quantization level inside the input range for which "high frequency of activities" are observed. This implies that more memory cells are allocated to the range of input which holds more information than the rest of the input space.

3.1 Adaptive Quantization

The idea of adaptive quantization is to capture the input distribution and output variation. For this purpose, hierarchical clustering technique is employed. The clustering method is applied separately on each individual input dimension. For each input dimension, we start off by having each individual training data sample as a cluster. In each iteration, the two nearest clusters with the smallest merging cost function are merged to form a single cluster. The cost function is defined as the distance between the mean output value of the two clusters, expressed mathematically as

$$f(M,N) = \frac{\sum_{i \in M} X_i}{n_M} + \frac{\sum_{j \in N} X_j}{n_N} \tag{1}$$

where M and N are two different clusters, and X_i is the ith output value contained in a cluster.

The clusters-merging iteration is continued until the number of clusters in that input dimension reaches the predefined memory size. This step is effectively clustering the nearest data points having similar output together, and allocating more storage cells into those densely populated areas which contain a high degree of variation in the target output.

3.2 Memory Allocation

Following the adaptive quantization, is the memory allocation, in which each of the individual cluster is allocated a memory axes along its particular dimension. The result of the memory allocation is an adaptively quantized CMAC associative neural network, as in the example depicted in Fig. 1 for 2D input case.

3.3 Network 1-Point Training and Neighborhood Retrieval

The learning equation employed is the Widrow-Hoff learning equation, modified for 1-point update and neighborhood retrieval HCAQ-CMAC:

$$\mathbf{Z}^i_{x_j,y_j} = \frac{1}{S_N}\left[\sum_{k \in K, l \in L} \mathbf{W}^i_{k,l}\right] \tag{2}$$

$$K = \{Q[x_j] - NR_x \leq k \leq Q[x_j] + NR_x\} \tag{3}$$

$$L = \{Q[y_j] - NR_y \leq k \leq Q[y_j] + NR_y\} \tag{4}$$

$$\mathbf{W}^{i+1}_{Q[x_j],Q[y_j]} = \mathbf{W}^{i}_{Q[x_j],Q[y_j]} + \alpha \left[\mathbf{W}^{i}_{Q[x_j],Q[y_j]} - \mathbf{D}_{x_j,y_j}\right] \tag{5}$$

Here, i is the iteration number, $\mathbf{V_j} = (x_j, y_j)$ is the two dimensional input to a 2D HCAQ-CMAC, $Q[\cdot]$ is the quantization function, $\mathbf{Z}^i_{x_j,y_j}$ is the output of the network for input $\mathbf{V}_j$, S_N is the number of elements inside the neighborhood of the current input, N is the neighborhood constant, R_x and R_y are both the input space range for input dimension x and y respectively, and $\mathbf{W}_{k,l}$ is the HCAQ-CMAC memory cell at index (k, l). Neighborhood retrieval is employed to smoothen the output of HCAQ-CMAC so that fluctuations of the retrieved output are reduced.

4 Experiments and Results

We demonstrate the performance of the proposed HCAQ-CMAC for a multi-input experiment. In particular, the HCAQ-CMAC network is used as a car a automatic steering controller. The car simulator was developed in [9] and [10]. It consists of a vehicle model together with a 3D virtual driving environment. The simulated car is equipped with 8 directional sensors in the 8 different directions of the car, as shown in Figure 4. The sensor readings were taken at every simulation time interval. The inputs to the HCAQ-CMAC network are the 4 front sensor values: FLSTB (Front Left Sensor to Barrier), FRSTB (Front Right Sensor to Barrier), SFLSTB (Side Front Left Sensor to Barrier), SFTSTB (Side Front Right Sensor to Barrier). The output of the network controls the steering angle. The car is driven along a path in a multi-lane circuit shown in Figure 2. Training data are obtained by sampling human drivers' steering control actions for the specified track. A 4-dimensional HCAQ-CMAC is trained on the steering angle response to data from the four front car sensors. Over 100 seconds of driving data are collected for training. The auto-driving performances are compared with those obtained using a standard CMAC network, employing the same learning

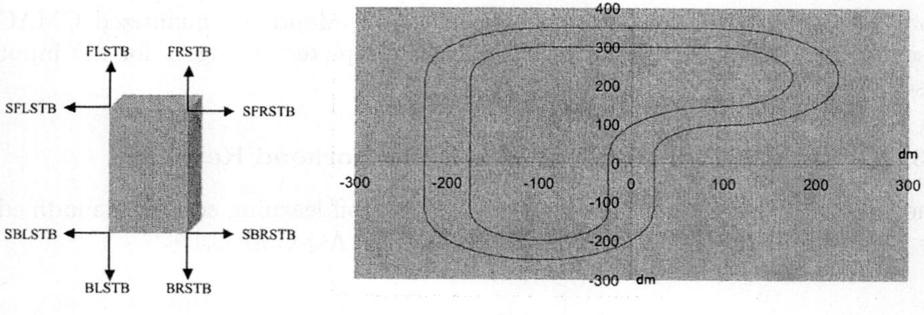

(a) Car sensors' placement (b) Driving Track

Fig. 2. Simulation Environment

Table 1. Comparison of Results from CMAC and HCAQ-CMAC Testing

	CMAC		HCAQ-CMAC	
Memory size per dimension	8	10	5	6
Neighborhood size	0.2	0.2	0.1	0.1
Training				
Learning constant	0.1	0.1	0.1	0.1
Final epoch training error	28.5546	30.104	22.6618	22.6607
Training time	9031 ms	14453 ms	3438 ms	4000 ms
Testing				
Average deviation from centre line	0.3499 m	0.2068 m	0.2216 m	0.2058 m
Average deviation of car orientation	0.7272 rads	0.7594 rads	0.6891 rads	0.6969 rads

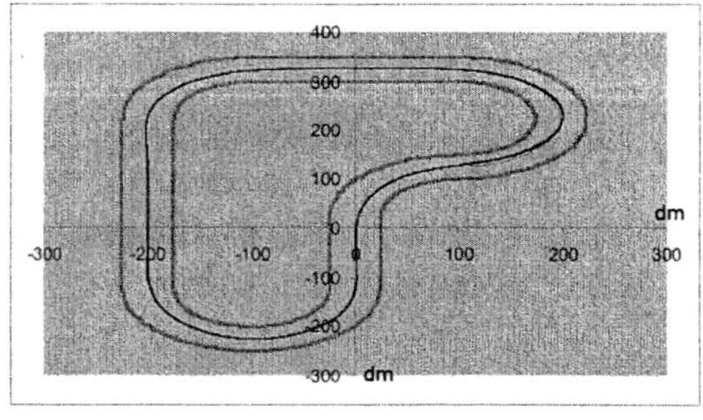

(a) HCAQ CMAC Driving Path (size = 6)

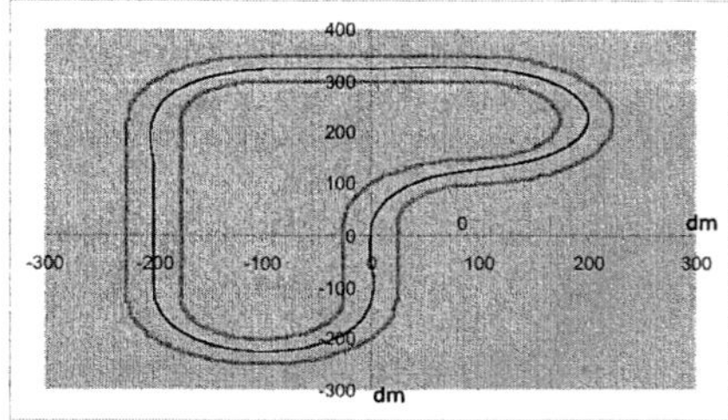

(b) CMAC Driving Path (size = 10)

Fig. 3. Driving paths comparison

function and parameters. The results are tabulated in Table 1. Figure 3 gives a visualization of the tack path obtained using $6 \times 6 \times 6 \times 6$ HCAQ-CMAC as compared to the path obtained using $10 \times 10 \times 10 \times 10$ CMAC. It is observed that using a HCAQ-CMAC whose size is only 60% of the original CMAC network (in each dimension), driving qualities are very similar. This significant improvement on memory allocation will not only reduce memory requirement of a CMAC network, but will also reduce the network training time.

5 Conclusions

We have presented the HCAQ-CMAC as an enhancement to the original CMAC architecture. HCAQ-CMAC improves memory utilization of CMAC by allocating more memory cells in the region where rapid changes in the output of the target function are observed. The performance has been evaluated on multiple-input application – an automated car maneuver control. Simulation results show that significant reduction in memory size can be achieved in HCAQ-CMAC, while still maintaining comparable quality of performance compared to the standard CMAC network. Further research in this direction will include a more detailed study of the computational complexity of the proposed approach and apply it to other application areas.

References

1. Albus, J.S.: A new approach to manipulator control: The cerebellar model articulation controller (CMAC). J. Dynamic Syst., Measurement, Contr., Trans. ASME (1975) 220–227
2. Yamamoto, T., Kaneda, M.: Intelligent controller using CMACs with self-organized structure and its application for a process system. IEICE Trans. Fundamentals **E82-A** (1999) 856–860
3. Commuri, S., Jagannathan, S., Lewis, F.L.: CMAC neural network control of robot manipulators. J. Robot Syst. **14** (1997) 465–482
4. Wahab, A., Tan, E.C., Abut, H.: HCMAC amplitude spectral subtraction for noise cancellation. Intl. Conf. Neural Inform. Processing (2001)
5. Huang, K.L., Hsieh, S.C., Fu, H.C.: Cascade-CMAC neural network applications on the color scanner to printer calibration. Intl. Conf. Neural Networks **1** (1997) 10–15
6. Moody, J.: Fast-learning in multi-resolution hierarchies. In: Adv. Neural Infor. Processing Syst. Volume 14. Morgan Kauffman Publishers (1989) 29–38
7. Menozzi, A., Chow, M.: On the training of a multi-resolution CMAC neural network. 23rd. Intl. Conf. Ind. Electron. Contr. Instrum. **3** (1997) 1130–1135
8. Yeh, M.F., Lu, H.C.: On-line adaptive quantization input space in cmac neural network. IEEE Intl. Conf. Syst., Man, Cybern. **4** (2002)
9. Pasquier, M., Quek, C., Toh, M.: Fuzzylot: A self-organizing fuzzy neural rule-based pilot system for automated vehicle. Neural Networks **14** (2001) 1099–1112
10. Ang, K.K., Quek, C.: An improved mcmac with momentum neighborhood and average trapezoidal output. IEEE Transactions on Systems, Man and Cybernetics Part B **30** (2000) 491–500

Knowledge Extraction from Unsupervised Multi-topographic Neural Network Models

Shadi Al Shehabi and Jean-Charles Lamirel

Loria, Campus Scientifique, BP 239
54506 Vandoeuvre-lès-Nancy Cedex, France
{Shadi.Al-Shehabi, Jean-Charles.Lamirel}@loria.fr

Abstract. This paper presents a new approach whose aim is to extent the scope of numerical models by providing them with knowledge extraction capabilities. The basic model which is considered in this paper is a multi-topographic neural network model. One of the most powerful features of this model is its generalization mechanism that allows rule extraction to be performed. The extraction of association rules is itself based on original quality measures which evaluate to what extent a numerical classification model behaves as a natural symbolic classifier such as a Galois lattice. A first experimental illustration of rule extraction on documentary data constituted by a set of patents issued form a patent database is presented.

1 Introduction

Data mining or knowledge discovery in database (KDD) refers to the non-trivial process of discovering interesting, implicit, and previously unknown knowledge from large databases [4]. Such a task implies to be able to perform analyses on high-dimensional input data. The most popular models used in KDD are the symbolic models. Unfortunately, these models suffer of very serious limitations. Rule generation is a highly time-consuming process that generates a huge number of rules, including a large ratio of redundant rules. Hence, this prohibits any kind of rule computation and selection as soon as data are numerous and they are represented in very high-dimensional description space. This latter situation is very often encountered with documentary data. To cope with these problems, preliminary KDD trials using numerical models have been made. An algorithm for knowledge extraction from self-organizing network is proposed in [3]. This approach is based on a supervised generalized relevance learning vector quantization (GRLVQ) which is used for extracting decision trees. The different paths of the generated trees are then used for denoting rules. Nevertheless, the main defect of this method is to necessitate training data. On our own side, we have proposed a hybrid classification method for matching an explicative structure issued from a symbolic classification to an unsupervised numerical self-organizing map (SOM) [7]. SOM map and Galois lattice are generated on the same data. The cosine projection is then used for associating lattice concepts to the SOM classes. Concepts properties act as explanation for the SOM classes. Furthermore, lattice pruning combined with migration of the associated SOM classes towards the top of the pruned lattice is used to generate explanation of increasing scope on the

W. Duch et al. (Eds.): ICANN 2005, LNCS 3697, pp. 479 – 484, 2005.

SOM map. Association rules can also be produced in such a way. Although it establishes interesting links between numerical and symbolic worlds this approach necessitates the time-consuming computation of a whole Galois lattice. In a parallel way, in order to enhance both the quality and the granularity of the data analysis and to reduce the noise which is inevitably generated in an overall classification approach, we have introduced the MultiSOM model [6]. This model represents a significant extension of the SOM model, in which each viewpoint is represented by a single SOM map. The conservation of an overall view of the analysis is achieved through the use of a communication mechanism between the maps, which is itself based on Bayesian inference [10]. The advantage of the multi-viewpoint analysis provided by MultiSOM as compared to the global analysis provided by SOM [5] has been clearly demonstrated for precise mining tasks like patent analysis [8]. Another important mechanism provided by the MultiSOM model is its on-line generalization mechanism that can be used to tune the level of precision of the analysis. Furthermore, we have proposed in [1] to use the neural gas (NG) model as a basis for extending the MultiSOM model to a MultiGAS model. Hence, NG model [11] is known as more efficient and homogeneous than SOM model for classification tasks where explicit visualization of the data analysis results is not required.

In this paper we propose a new approach for knowledge extraction that consists in using our MultiGAS model as a front-end for unsupervised extraction of association rules. In our approach we specifically exploit the generalization mechanism of the model. We also make use of our own recall and precision measures that derive from the Galois lattice theory and from Information Retrieval (IR) domain [9]. The first section of the paper presents the symbolic approach for rules extraction. The second section presents the rule extraction principles based on the MultiGAS model. The experiment presented in the last section shows how our method can be used both for controlling the rules inflation that is inherent to symbolic methods and for extracting the most significant rules.

2 The Symbolic Model and Association Rules Extraction

The symbolic approach to Database Contents Analysis is mostly based on the Galois lattice model (see [2] and [12]). A Galois lattice, *L(D,P)*, is a conceptual hierarchy built on a set of data *D* which are described by a set of properties *P* also called the intention (Intent) of the concept of the lattice. A class of the hierarchy, also called "formal concept", is defined as a pair *C=(d,p)* where *d* denotes the extension (Extent) of the concept, i.e. a subset of *D*, and *p* denotes the intention of the concept, i.e. a subset of *P*. The lattice structure implies that it exists a partial order on a lattice such that:

$$\forall\, C_1, C_2 \in L,\ C_1 \le C_2 \Leftrightarrow \text{Extent}(C_1) \subseteq \text{Extent}(C_2) \Leftrightarrow \text{Intent}(C_1) \supseteq \text{Intent}(C_2)$$

Association rules are one of the basic types of knowledge extraction from large databases. Given a database, the problem of mining association rules consists in generating all association rules that have certain user-specified minimum support and confidence. An association rule is an expression $A \rightarrow B$ where *A* and *B* are conjunctions of properties. It means that if an individual data possesses all the properties of A then he necessarily possesses all the properties of *B*. The support of the rule is $supp(A \cup B)$, and the confidence: $Conf = supp(A \cup B)/supp(A)$. An approach proposed by [12] shows that a subset of association rules can be obtained following the direct links of heritage between the concepts in the Galois lattice. Even if no satisfactory solution regarding rule computation time have been given, some attempt to solve the rule selection problem by combining rules evaluation measures is also proposed in [2].

3 MultiGAS Model for Rule Extraction

A reliable unsupervised neural model, like a gas, represents a natural candidate to cope with the related problems of rule inflation and rule selection that are inherent to symbolic methods. Hence, its synthesis capabilities that can be used both for reducing the number of rules and for extracting the most significant ones. We will rely on our own class quality criteria for extracting rules from the classes of the original gas and its generalizations, that is the *Precision* and *Recall* measures based on the properties of class members, which are defined in [9]. The *Precision criterion* measures in which proportion the content of the classes generated by a classification method is homogeneous. The greater the *Precision*, the nearer the intensions of the data belonging to the same classes will be one with respect to the other, and consequently, the more homogenous will be the classes. In a complementary way, the *Recall criterion* measures the exhaustiveness of the content of said classes, evaluating to what extent single properties are associated with single classes. We have demonstrated in [9] that if both values of *Recall* and *Precision* reach the unity value, the peculiar set of classes represents a Galois lattice. A class belongs to the peculiar set of classes of a given classification if it possesses peculiar properties. Finally, a property is considered as peculiar for a given class if it is maximized by the class members. As compared to classical inertia measures, averaged measures of *Recall* and *Precision* present the main advantages to be independent of the classification method. They can thus be used both for comparing classification methods and for optimizing the results of a method relatively to a given dataset. In this paper we will focus on peculiar properties of the classes and on local measures of *Precision* and *Recall* associated to single classes. Hence, as soon as these informations can be fruitfully exploited for generating explanations on the contents of individual classes, they also represent a sound basis for extracting rules from these latter classes. The general form of the extraction algorithm follows:

Let C being a class, P_C being the set of properties associated to the members of C, and P_C^* being the set of peculiar properties of C, with $P_C^* \subseteq P_C$:

$\forall\, p_1, p_2 \in P_C^*$
1) ***If*** (Rec(p_1, p_2) = Prec(p_1, p_2) = 1) ***Then*** there is an equivalence rule: $p_1 \leftrightarrow p_2$
2) ***ElseIf*** (Rec(p_1, p_2) = Prec(p_2) = 1) ***Then*** there is an association rule: $p_1 \rightarrow p_2$
3) ***ElseIf*** (Rec(p_1, p_2) = 1) ***Then***
    If (Extent(p_1) $\subset$ Extent(p_2)) ***Then***: $p_1 \rightarrow p_2$
    If (Extent(p_2) $\subset$ Extent(p_1)) ***Then***: $p_2 \rightarrow p_1$
    If (Extent(p_1) $\equiv$ Extent(p_2)) ***Then***: $p_1 \leftrightarrow p_2$

$\forall p_1 \in P_C^*,\ \forall p_2 \in P_C - P_C^*$
4) ***If*** (Rec(p_1) = 1) If (Extent(p_1) $\subset$ Extent(p_2)) ***Then***: $p_1 \rightarrow p_2$ (*)

where Prec and Rec represent the local *Precision* and *Recall* measures, respectively.

The optional step 4) (*) can be used for increasing the number of extracted rules. In this step the constraint of peculiarity is relaxed for the most general property.

The gas generalization principle consists in summarizing the contents of an original gas by progressively reducing its number of neurons. A triangle-based strategy for gas generalization has been successfully tested in [1]. Its main advantage is to produce homogeneous gas generalization levels while ensuring the conservation of the topographic properties of the gas codebook vectors on each level. A basic rule extraction strategy consists in applying the above described extraction algorithm both on an original gas and on its generalizations. The expected result of this strategy is to be able to control the rule number and the rule quality by the choice of a proper generalization level.

4 Experimental Results

Our test database is a database of 1000 patents that has been used in some of our preceding experiments [8]. For the viewpoint-oriented approach the structure of the patents has been parsed in order to extract four different subfields corresponding to fours different viewpoints: Use, Advantages, Titles and Patentees. As it is full text, the content of the textual fields of the patents associated with the different viewpoints is parsed by a lexicographic analyzer in order to extract viewpoint specific indexes. Only, the Use viewpoint will be considered in our experiment. This viewpoint generates itself a description space of size 234. Our experiment is initiated with an optimal gas generated thanks to an optimization algorithm based on the quality criteria [9]:

- Original gas of 100 neurons (optimal) is firstly generated for the Use viewpoint.
- Generalized gases of 79, 62, 50, 40, 31, 26, 16 and 11 neurons are generated for this latter viewpoint by applying the generalization mechanism to the 100 neurons original gas.

Our experiment consists in extracting rules from the single Use viewpoint. Both the original gas and its generalizations are used for extracting the rules. The algorithm is used once without its optional step, and a second time including this step (for more details, see algorithm). The results are presented at figure 1. Some examples of extracted rules are given hereafter:

Bearing of outdoor machines $\leftrightarrow$ *Printing machines* (supp = 2, conf = 100%)
Refrigerator oil $\rightarrow$ *Gear oil* (supp = 3, conf = 100%)

A global summary of the results is given in table 1. The table includes a comparison of our extraction algorithm with a standard symbolic rule extraction method concerning the amount of extracted rules. When our extraction algorithm is used with its optional step, it is able to extract the same number of rules as a classical symbolic model that basically uses a combinatory approach. Indeed, table 1 shows that all the rules of confidence 100% (i.e. 536) are also extracted by the combination of gas levels. Moreover, a significant amount of rule can be extracted from any single level of the gas (see fig. 1b). Even if, in this case, no rule selection is performed, the main advantage of this version of the algorithm, as compared to a classical symbolic method, is the computation time. Indeed, as soon as our algorithm is class-based, the computation time it significantly reduced. Moreover, the lower the generalization level, the more specialized will be the classes, and hence, the lower will be the combinatory effect during computation. Another interesting result is the behavior of our extraction algorithm when it is used without its optional step. The fig. 1a shows that, in this case, a rule selection process that depends of the generalization level is performed: the higher will be the generalization level, the more rules will be extracted. We have already done some extension of our algorithm in order to search for partial rules. Complementary results showed us that, even if this extension is used, no partial rules will be extracted in the low levels of generalization when no optional step is used. This tends to prove that the standard version of our algorithm is able to naturally perform rule selection.

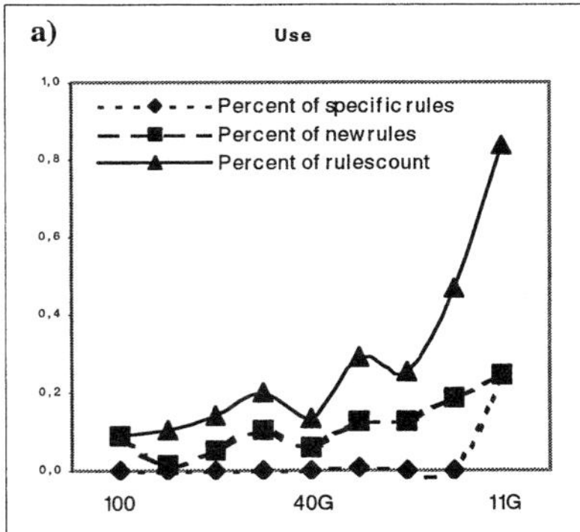

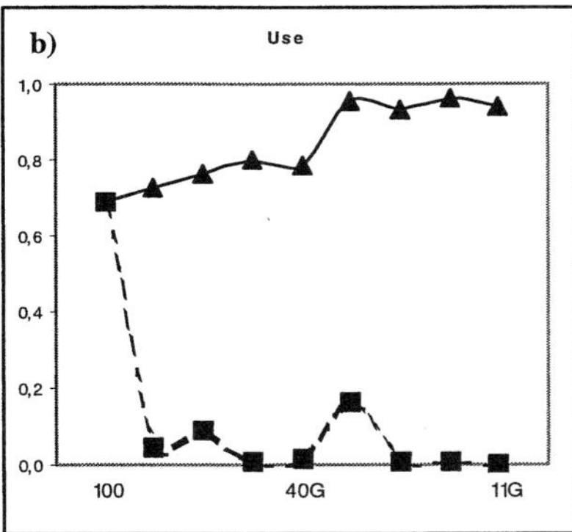

Fig. 1. Rule extraction curves for Use viewpoint. a) extraction algorithm without optional step. b) the same with optional step. New rules: rules that are found in a given level but not in the preceding ones. Specific rules: rules which are found only in a given level. Rules count: is the total number of rules that are extracted from all levels. ((xG): represents a level of generalization of x neurons).

Table 1. Summary of results. The table presents a basic comparison between the standard symbolic rule extraction method and the MultiGAS-based rule extraction method. The global rule count defined for the symbolic model includes the count of partial rules (confidence<100%) and the count of total rules (confidence=100%). The rules generated by the MultiGAS model on the 9 levels are only total rules. The peculiar rule count is obtained with the standard version of the extraction algorithm. The extended rule count is obtained with the extended version of the extraction algorithm including the optional step.

		Use
Symbolic model	Total rule count	536
	Average confidence	100%
	Global rule count	2238
	Average confidence	59%
MultiGAS model (9 levels)	Peculiar rule count	251
	Average confidence	100%
	Extended rule count	536
	Average confidence	100%

5 Conclusion

In this paper we have proposed a new approach for knowledge extraction based on a MultiGAS model. Our approach makes use of original measures of recall and precision for extracting rules from gases. It takes benefit of the generalization mechanism that is embedded in the MultiGAS model. Even if complementary experiments must be done, our first results are very promising. They tend to prove that a neural model, as soon as it is elaborated enough, represents a natural candidate to cope with the related problems of rule inflation, rule selection and computation time that are inherent to symbolic models. One of our perspectives is to adapt this model to the multi-viewpoint context of the MultiGAS model that represents itself a powerful context for knowledge extraction. Furthermore, we plan to test our model on a reference dataset on genome. Indeed, these dataset has been extensively used for experiments of rule extraction and selection with symbolic methods [2].

References

1. S. Al Shehabi, J.C. Lamirel. Multi-Topographic Neural Network Communication and Generalization for Multi-Viewpoint Analysis. International Joint Conference on Neural Networks - IJCNN'05. (Montréal, Québec, Canada). 2005.
2. H. Cherfi. Étude et réalisation d'un système d'extraction de connaissances à partir de textes. Thèse de l'Université de Nancy 1, Henri Poincaré, 2004.
3. B. Hammer, A. Rechtien, M. Strickert, T. Villmann, Rule extraction from self-organizing maps, in: J.R.Dorronsoro (Ed.), Artificial Neural Networks -- ICANN 2002, Springer, 877-882, 2002.
4. J. Han, M. Kamber, A. K. H. Tung. (2001). Spatial clustering methods in data mining: A survey, H. Miller and J. Han (eds.), Geographic Data Mining and Knowledge Discovery, Taylor and Francis.
5. T. Kohonen, Self-Organizing Maps. 3rd ed. Springer Verlag, Berlin, 2001.
6. J.C. Lamirel, Application d'une approche symbolico-connexionniste pour la conception d'un système documentaire hautement interactif. Thèse de l'Université de Nancy 1, Henri Poincaré, 1995.
7. J.C. Lamirel, Y. Toussaint, S. Al Shehabi. A Hybrid Classification Method for Database Contents Analysis. In The 16th International FLAIRS Conference - FLAIRS 2003. (St. Augustine, Florida). 2003.
8. J.C. Lamirel, S. Al Shehabi, M. Hoffmann, C. Francois. Intelligent patent analysis through the use of a neural network : experiment of multi-viewpoint analysis with the MultiSOM model. Proceeding of ACL, Sapporo, Japan. 2003.
9. J.C. Lamirel, S. Al Shehabi, C. Francois, M. Hoffmann. New classification quality estimators for analysis of documentary information: application to web mapping. Scientometrics, vol. 60, no. 3, pp. 445-462, Feb2004.
10. J.C. Lamirel, S. Al Shehabi, C. François, X. Polanco. Using a compound approach based on elaborated neural network for Webometrics: an example issued from the EICSTES Project. Scientometrics, Vol. 61, No. 3 (2004), pp. 427-441.
11. T. Martinetz, K. Schulten. A "neural-gas" network learns topologies. In T. Kohonen, K. Mäkisara, O. Simula, and J. Kangas, editors, Artificial neural networks, North-Holland, Amsterdam, 1991, pp. 397-402.
12. A. Simon and A. Napoli. Building Viewpoints in an Object-based Representation System for Knowledge Discovery in Databases, Proceedings of IRI'99, Atlanta, Geogia, S. Rubin editor, The International Society for Computers and Their Applications, ISCA, pages 04-108, 1999.

Current Trends on Knowledge Extraction and Neural Networks

David A. Elizondo and Mario A. Góngora

School of Computing, Faculty of Computing Sciences and Engineering,
De Montfort University, Leicester, UK
{elizondo, mgongora}@dmu.ac.uk

Abstract. The extraction of knowledge from trained neural networks provides a way for explaining the functioning of a neural network. This is important for artificial networks to gain a wider degree of acceptance. An increasing amount of research has been carried out to develop mechanisms, procedures and techniques for extracting knowledge from trained neural networks. This publication presents some of the current research trends on extracting knowledge from trained neural networks.

1 Introduction

Techniques of artificial neural networks (ANN) have been applied with success to problem domains including classification and continuous function approximation ([1]). However, it is important to understand the process by which ANN arrive at a given result and to provide ways of extracting the knowledge embedded in the network ([2]).

Knowledge extraction (KE) from trained neural networks provides a way for explaining the functioning of a neural network. This is important for artificial networks to gain a wider degree of acceptance. A considerable amount of research has been carried out to develop mechanisms, procedures and techniques for extracting knowledge from trained neural networks. This publication presents some of the current research trends on extracting knowledge from trained neural networks.

This paper is divided into six sections. In the second section some of the current trends for extracting knowledge from neural networks are given. Three separate sections, describing knowledge extraction methods based on these trends, are provided. Finally some conclusions are given in section six.

2 Trends on ANN KE Methods

The knowledge embedded in a trained neural network is represented in terms of the topology, the activation functions, and the weights. Several criteria can be used to classify the different methods for extracting knowledge from neural networks. A variety of methods for dealing with knowledge extraction from neural networks have been proposed, and there seems to be an increase in interest in this area. This survey presents a compilation and an overview of these methods. In this paper, three classes of methods for extracting and representing knowledge from trained neural networks are discussed, the symbolic, the fuzzy logic, and the application dependent methods.

W. Duch et al. (Eds.): ICANN 2005, LNCS 3697, pp. 485–490, 2005.

– **Symbolic Based**. These methods produce rules in the form of logical If-then statements. These rules are expressed as Horn clauses and can also be used for refining the neural network.
– **Fuzzy Logic**. These methods combine fuzzy logic with neural networks in order to facilitate the extraction of knowledge.
– **Application Dependent**. These methods are designed using approaches which are tailored to specific problems or application areas.

In the following, trends on the current methods developed for knowledge extraction are presented. They are arranged into three sections following the trends described above.

3 Symbolic Based KE Methods

The use of symbolic knowledge is a popular trend on knowledge extraction from neural networks. This provides a way of using problem domain knowledge to build the topology, train the network, and extract knowledge from a trained network. An example of this trend is the work developed by [3]. The authors propose a method for extracting knowledge from a pruned fuzzy ARTMAP based neural network (Adaptive Resonance Theory). They used weight quantization to transform real valued weights into feature values (i.e. into fuzzy characteristics for logical rules). This enables their system to provide semantic rules and their interpretation to users. The quantization is done by dividing the range [0 1] into Q intervals and assigning a quantization point to the lower bound of each interval.

A method for extracting symbolic knowledge from neural networks based on linear functions is presented on [4]. The authors show how the knowledge of a Recursive Deterministic Perceptron (RDP) network [5] can always be expressed, transparently, as a finite union of open polytopes which correspond to the decision regions of the RDP. The authors also discuss the combination of the decision regions of RDP models by using boolean operations.

An alternative to the If-Then-Else symbolic rules is proposed by [6]. The paper proposes rules of the form M-of-N which the authors argue maybe more suitable for some applications because it provides better understanding of the problem domain. These rules have the form: 'If M of N conditions $a_1, a_2, ..., a_m$ are true, then the conclusion b is true'.

A symbolic KE method based on a decomposition approach is presented in [7]. Clusters of the hidden unit activation levels are produced and rules are generated in terms of these clusters. Rules for the clusters in terms of the inputs are also generated. Finally, these rules are merged together to create a system that can explain how the system reaches a given answer. Their approach is able to separate core global knowledge, from generic local knowledge, thus extracting additional characteristics that the neural network correlatess during training. The system has been adjusted by using some previous knowledge of the applicability of it.

A method for rule extraction from ANN using a novel gradient-based method with input data dimensionality reduction is given in [8]. The method is ontogenic as it takes place while the neural network is being trained. Rules extracted by their method have hyper-rectangular decision boundaries. The rules are extracted based on the training result of an RBF neural network using gradient descendent theory. Their method leads to smaller rule sets with increased levels of accuracy with respect to other methods and it does not require the transformation, like many methods, from continuous attributes to discrete ones.

A rule-extraction system for ANN using genetic programming is given in [9]. The KE system is independent of the architecture of the neural network architecture and the training method. This method produces a set of if-then rules by means of evolutionary computing, by encoding the rule tree as a chromosome that is evolved to solve a fitness function which relates the training input data set to the output set that the trained network has.

A similar approach for extracting knowledge from neural networks is used by [10]. The authors use the trained network's hidden layer activation functions to relate them to the inputs using a clustering fitness function for the genetic algorithm, rules are generated using each identified cluster. Although their method is specific to software quality applications, it does not need to incorporate any application specific constrains to the search.

Many works including [9,10], show that evolutionary programming based techniques are the main tools used for architecture and training independent symbolic based KE methods.

4 Fuzzy Logic KE Based Methods

An increasing trend for extracting knowledge from trained neural networks involves the use of Fuzzy Logic. These methods extract knowledge from neural networks as a set of fuzzy rules. Gorban ([11]) proposes a method for extracting fuzzy rules from pruned multilayer perceptron trained neural networks. The type and number of rules depend on the type of transfer function of each neuron, and the type of input/output neuron (discrete/continuous). For discrete neurons, the authors propose an automatic way of generating a verbal description of the network.

A neural network model for handling business rules is proposed by [12]. The topology of the neural network is based on both training samples and hidden rules extracted from trained neural networks. The authors use a knowledge based descriptive neural network (DNN) that incorporates embedded business rules extracted from previously trained neural networks. These rules are in the form of fuzzy descriptors. Three steps are involved in the construction of a DNN network: build a neural network forecasting model, extract rules from the trained neural network, and incorporate hidden forecasting rules extracted in the previous step into the DNN.

A neurofuzzy model to create an algorithm for KE from observed finite data sets is proposed by [13]. This is another ontogenic method which works on the basis that the knowledge is being extracted/identified as the training of the neural network happens. In this way it achieves a one-to-one mapping between the rules-base and the model features.

A KE method based on high correlation rules using a fuzzy inference is presented in [14]. Although the method is very effective in extracting the knowledge, it is ontogenic in nature and therefore, has to be applied at training time.

A structural based learning method to extract knowledge from neural networks is described in [15]. The rules obtained are in the form of fuzzy rules. The extraction is correlated to the trained ANN but is actually training dependant. Their method is four fold. To start with, the training data set is fuzzified. With this transformed data set, a back propagation neural network is constructed. The next step is to create an importance index matrix. The final step consists on creating a weighed set of fuzzy rules.

5 Application Dependent KE Based Methods

An application dependent method for extracting knowledge from neural networks trained to model river flow is presented in [16]. The authors explain the internal behaviour of the neural network in terms of the ranges of the outputs of the hidden neurons. They try to map these outputs to portions of the activation function. This mapping is then labelled according to the river flow (high, medium, and low magnitude).

The use of an application specific problem to explore various data mining methods is discussed in [17]. This approach includes the training of a ANN and followed by the extraction of the rules from its knowledge. From five different methods tested in their application, they found that the rule extraction from ANN was second best for accurate knowledge representation, thus proving its potential as a knowledge explicit data mining tool.

A neural network solution for solving an application specific problem is given in [18]. At the same time, a method to create a set of if-then rules representing the knowledge of the ANN is provided. The method for extracting the rules is non-ontogenic implying that the rules are extracted after the ANN was trained but with knowledge about its training. Even though the proposed method is very efficient it is only useful for their application and type of ANN.

A method for extracting knowledge based on pruning and the progressive use of a simpler transfer functions is discussed in [19]. Optimal brain damage is used for pruning the topology of the network. Three steps are used by the authors to extract the rules. The first step consists on considering the first layer of weights as simple perceptrons enabling the extraction of hyperplanes. The second step aims at relating the binary activation of the hidden units with the activation of the output units corresponding to the decision of the network. The third and final step consists on integrating steps one and two to formulate knowledge usefull for the experts. Using this method, the authors were able to find new rules, which the NN had correlated during training. These rules were significantly useful to radar experts.

6 Conclusions

Knowledge extraction is important for artificial networks to gain a wider degree of acceptance. Therefore, this domain has become a mayor field of research since it validates the use of neural networks for applications where reasons or explanations on why or how a result has been achieved are important. Many problems which require the process of reaching a solution to be either accountable, or transparent and interpretable, have found new use for neural networks. This is subject to providing a process for extracting knowledge in the development. In addition, the fact that neural networks are powerfull tools for finding knowledge hidden in data makes KE a natural and sound extension to neural network based data analysis, making this knowledge available in an explicit way.

Several of the most commonly used methods for extracting knowledge from neural networks have been presented. The different methods were divided into three classes of methods, the symbolic, the fuzzy logic, and the application dependent methods.

The symbolic based methods provide the simpler, yet most general and versatile approaches for KE from NNs. The use of Evolutionary Computing techniques in this area enables the extraction of knowledge from ready trained networks, as well as providing the most potential for an architecture and application independent method. Although in principle it provides the highest potential for clarity in the extracted knowledge, most applications still provide very cryptic and large sets of convoluted if-then-rules.

The Fuzzy Logic methods extract if-then type rules either during or after training the neural network. Although in principle very similar to the symbolic methods, they clearly focus to the Fuzzy Logic approach to rule description, thus making their results more human-readable. Unfortunately most applications are still restricted to data-set and architecture dependent methods.

The Application Dependent approaches, as expected, produce the most specific and immediately useful results in the form the knowledge is presented. This advantage is over weighed by the fact that always, for any new application, the whole process from the stage of method design, has to be performed. Still, we envisage that the use of evolutionary computing based methods, which are becoming more general with time, might enable the application specific methods to be generalised to the symbolic methods with good human-readability. This will provide the most useful form of KE for NNs.

References

1. Elizondo, D.A., McClendon, R.W., Hoogenboom, G.: Neural network models for predicting flowering and physiological maturity of soybean. Transactions of the ASAE **37** (1994) 981–988
2. Shellhammer, I., Diederich, J., Towsey, J., Brugman, C.: Knowledge extraction and recurrent neural networks: An analysis of an elman network trained on a natural language learning task. In: Proceedings of the Joint Conference on New Methods in Language Processing and Computational Natural Language Learning, Association for Computational Linguistics (1998) 73–78
3. Chiang Tan, S., Peng Lim, C.: Application of an adaptive neural network with symbolic rule extraction to fault detection and diagnosis in a power generation plant. IEEE Transactions on Energy Conversion **19** (2004) 369–377
4. Tajine, M., Elizondo, D.: The recursive deterministic perceptron neural network. Neural Networks **11** (1998) 1571–1588
5. Tajine, M., Elizondo, D.: Growing methods for constructing recursive deterministic perceptron neural networks and knowledge extraction. Artificial Intelligence **102** (1998) 295–322
6. Browne, A., Hudson, B., Whitley, D., Ford, M., Picton, P., Kazemian, H.: Knowledge extraction from neural networks. In: Proceedings of the IEEE Industrial Electronics Society 29th Annual Conference (IECON-03), IEEE (2003) 1909–1913

7. Setiono, R., Pan, S.L., Hsieh, M.H., Azcarraga, A.P.: Separating core and noncore knowledge: An application of neural network rule extraction to a cross-national study of brand image perception. IEEE Transactions on Systems, Man and Cybernetics-Part C: Applications and Reviews **PP** (2005) 1–11
8. Fu, X., Wang, L.: Rule extraction using a novel gradient-based method and data dimensionality reduction. In: Proceedings of the International Joint Conference on Neural Networks (IJCNN). Volume 2. (2002) 1275–1280
9. Dorado, J., Rabunal, J.R., Rivero, D., Santos, A., Pazos, A.: Automatic recurrent ann rule extraction with genetic programming. In: Proceedings of the International Joint Conference on Neural Networks (IJCNN). Volume 2. (2002) 1552–1557
10. Qi, W., Bo, Y., Jie, Z.: Extracted rules from software quality prediction model based on neural networks. In: International Conference on Tools with Artificial Intelligence. (2004) 191–195
11. Gorban, A.N., Mirkes, E.M., Tsaregorodtsev, V.G.: Generation of explicit knowledge from empirical data through pruning of trainable neural networks. In: Proceedings of the International Joint Conference on Neural Networks (IJCNN) San Diego, IEEE Neural Networks Council; Edward Brothers (1999)
12. Yao, J.: Knowledge based descriptive neural networks. In: Proceedings of the 9th International Conference on Rough Sets, Fuzzy Sets, Data Mining and Granular Computing. Number 2639 (2003) 430–436
13. Hong, X., Harris, C.J.: A neurofuzzy network knowledge extraction and extended gram-schmidt algorithm for model subspace decomposition. IEEE Transactions on Fuzzy Systems **11** (2003) 538–541
14. Iyatomi, H.: Knowledge extraction from scenery images and recognition using fuzzy inference neural networks. Electronics and Communication in Japan, Part 2 **86** (2003) 82–90
15. Fan, T.G., Wang, X.Z.: A new approach to weighted fuzzy production rule extraction from neural networks. In: Proceedings of the International Conference on Machine Learning and Cybernetics. Volume 6. (2004) 3348–3351
16. Sudheer, K.P., Jain, A.: Explaining the internal behaviour of artificial neural network river flow models. Hydrological Processes **18** (2004) 833–844
17. Mitsdorffer, R., Diederich, J., Tan, C.: Rule extraction from technology ipos in the us stock market. In: Proceedings of the 9th International Conference on Neural Information Processing (ICONIP). Volume 5. (2002) 2338–2334
18. Yue, L., Hui, L., Bofend, Z., Gengfend, W.: Extraction of if-then rules from trained neural networks and its application to eartchquake prediction. In: Proceedings of the 3rd IEEE International COnference on Cognitive Informatics. (2004) 109–115
19. Remm, J.F., Aexandre, F.: Knowledge extraction using artificial neural networks: application to radar traget identification. Signal Processing **82** (2002) 117–120

Prediction of *Yeast* Protein–Protein Interactions by Neural Feature Association Rule[1]

Jae-Hong Eom and Byoung-Tak Zhang

Biointelligence Lab., School of Computer Science and Engineering,
Seoul National University,
Seoul 151-744, South Korea
{jheom, btzhang}@bi.snu.ac.kr

Abstract. In this paper, we present an association rule based protein interaction prediction method. We use neural network to cluster protein interaction data and feature selection method to reduce protein feature dimension. After this model training, association rules for protein interaction prediction are generated by decoding a set of learned weights of trained neural network and association rule mining. For model training, the initial network model was constructed with existing protein interaction data in terms of their functional categories and interactions. The protein interaction data of *Yeast* (*S.cerevisiae)* from MIPS and SGD are used. The prediction performance was compared with traditional simple association rule mining method. According to the experimental results, proposed method shows about 96.1% accuracy compared to simple association mining approach which achieved about 91.4%.

1 Introduction

A variety of attempts have been tried to predict protein functions and interactions with various data such as gene expression, protein–protein interaction (PPI) data, and literature analysis. Analysis of gene expression data through clustering also adopted to predict functions of un-annotated proteins based on the idea that genes with similar functions are likely to be co-expressed [1]. Park *et al.* [2] analyzed interactions between protein domains in terms of the interactions between structural families of evolutionarily related domains. Iossifov *et al.* [3] and Ng *et al.* [4] inferred new interaction from existing interaction data.

In this paper, we propose an adaptive neural network (ANN) based feature association mining method for PPI prediction. We used additional association rules for PPI prediction. These are generated by decoding a set of learned weights of adaptive neural network. We assumed that these association rules decoded from neural network (NN) would make the whole prediction procedure more robust to unexpected error factors by accounting relatively robust characteristic of NNs.

---

[1] This research was supported by the NRL Program of the Korea Ministry of Science and by the BK21-IT Program from the Ministry of Education and Human Resources Development of Korea. The ICT at Seoul National University provided research facilities for this study.

W. Duch et al. (Eds.): ICANN 2005, LNCS 3697, pp. 491 – 496, 2005.

Basically, we use ART-1 version of adaptive resonance theory [5] as an ANN clustering model to construct prediction model. The ART-1 [6] is a modified version of ART [7] for clustering binary vectors. Here, we assume again PPI of yeast as feature–to–feature association of each interacting proteins. We also use the same approach of Rangarajan *et al.* [8] for clustering model design and the same feature selection filter of Yu *et al.* [9] to reduce computational complexity.

This paper is organized as follows. In Section 2, we introduce feature selection filter concept and overall architecture of ART-1 based protein interaction clustering model. In Section 3, we present detailed NN training method with PPI data and the decoding method of association rule. In Section 4, we present the representation scheme of PPI for the NN input and association mining method and experimental results. Finally, concluding remarks and future works are given in Section 5.

2 Feature Dimension Reduction and Protein Cluster Learning

Feature Dimension Reduction by Feature Selection

Here, we consider each PPI as feature to feature associations. We constructed massive feature sets for each protein and interacting protein pairs from public protein databases as the same manner of Eom *et al.* [10]. However, there are also many features which have no information of its association with other proteins. Therefore, feature selection may be needed in advance of clustering PPIs. Especially, this feature selection is necessary when dealing with such high dimensional data. So, to filter out these features we used entropy and information gain based measure, *symmetrical uncertainty*, as a measure of feature correlation and which is defined in the work of Press *et al.* [11]. The overall filtering procedures are described in the paper of Eom *et al.* [10]

Clustering Protein Interactions

We use ART-1 NN to group the class of PPIs by their 13 functional classes and the class of interacting counter parts. In our ART-1 based clustering, each PPI is represented by a prototype vector that is generalized representation of the set of features of each interacting proteins. The degree of similarity between the members of each cluster can be controlled by changing the vigilance parameter ρ of Eom *et al.* [12]. The more detailed overall procedures for clustering PPIs with the ART-1 based clustering model is described in our previous work by Eom *et al.* [12]. The set of weights of trained NN were decoded as the form of association rule with the weight-to-rule decoding procedures described in Eom *et al.* [13] to enrich the protein features.

3 Rule Extraction from Trained Neural Network

Learning Feature Association with Neural Network

A supervised ANN uses a set of training examples or records include N attributes. Each attribute, A_n (n = 1, 2, … , N), can be encoded into a fixed length binary substring $\{x_1 \ldots x_i \ldots x_{m(n)}\}$, where $m(n)$ is the number of possible values for an attribute A_n. The element $x_i = 1$ if its corresponding attribute value exists and 0 otherwise. So, the proposed number of input nodes, $\boldsymbol{I}$, in the input layer of ANN can be given by

$$I = \sum_{n=1}^{N} m(n) \tag{1}$$

The input attributes vectors, $\boldsymbol{X}_m$, to the input layer can be rewritten as $\boldsymbol{X}_m = \{x_1 \ldots x_i \ldots x_I\}_m$, $m = (1,2,\ldots, M)$. The M is the total number of input training patterns. The output class vector, $\boldsymbol{C}_k(k = 1, 2, \ldots, K)$, can be encoded as a bit vector of a fixed length K as $C_k\{\psi_1 \ldots \psi_k \ldots \psi_K\}$. Here, K is the number of different possible classes. If the output vector belongs to class$_k$ then the element ψ_k is equal to 1 and 0 otherwise. Therefore, the proposed number of output nodes in the output layer of ANN is K. Then, the input and the output nodes of the ANN are determined and the structure of the ANN. The ANN is trained on the encoded vectors of the input attributes and the corresponding vectors of the output classes. The training of ANN is processed until the convergence rate between the actual and the desired output will be achieved.

After training the ANN, two groups of weights can be obtained. The first group, $(\mathrm{WG1})_{i,j}$, includes the weights between the input node i and the hidden node j. The second group, $(\mathrm{WG2})_{j,k}$, includes the weights between the hidden node j and the output node k. A sigmoid is used for the activation function of the hidden and output nodes of the ANN. The total input to the j–th hidden node, IHN_j and the output of the j–th hidden node, OHN_j, and the total input to the k–th output node, ION_k, are

$$IHN_j = \sum_{i=1}^{I} x_i (WG1)_{i,j}\,,\; OHN_j = \frac{1}{1+e^{-\left[\sum_{i=1}^{I} x_i (WG1)_{i,j}\right]}}\,,\; ION_k = \sum_{j=1}^{J} (WG2)_{j,k} \frac{1}{1+e^{-\left[\sum_{i=1}^{I} x_i (WG1)_{i,j}\right]}} \tag{2}$$

So, the final value of the k–th output node, ψ_k, is given by

$$\psi_k = \left\{ \frac{1}{1+e^{-\left[\sum_{j=1}^{J} WG2_{j,k}\left(\frac{1}{1+e^{-\left[\sum_{i=1}^{I} x_i (WG1)_{i,j}\right]}}\right)\right]}} \right\} \tag{3}$$

The function, $\psi_k = f(x_i, (\mathrm{WG1})_{i,j}, (\mathrm{WG2})_{j,k})$ is an exponential function in x_i since $(\mathrm{WG1})_{i,j}$, $(\mathrm{WG2})_{j,k}$ are constants. Its maximum output value is equal to one.

Association Rule Construction from Trained Neural Network with GA

To extract relations (rules) among the input attributes, $\boldsymbol{X}_m$ relating to a specific $class_k$ one must find the input vector, which maximizes ψ_k. This is an optimization problem and can be stated as $\psi_k(x_i)$ by considering binary data feature vector $\boldsymbol{x}$. In $\psi_k(x_i)$, x_i are binary values (0 or 1).

Since the objective function $\psi_k(x_i)$ is nonlinear and the constraints are binary so, it is a nonlinear integer optimization problem. Thus the genetic algorithm (GA) can be used to solve this optimization problem by maximizing the objective function $\psi_k(x_i)$. In this paper, we used conventional generational-GA procedures with this objective function $\psi_k(x_i)$ to find the best chromosome which provided as an input of NN and produce best network output (highest interaction prediction accuracy).

After we obtain the best chromosomes which produces best network output, we decode these chromosome into the form of association rule (we call this association rule as '*neural feature association rule*', since it's extracted from trained NN). To extract a rule for $class_k$ from the best chromosome selected by GA procedures, we decoded it with several procedures presented in our technical report [13]. The basic approach of above procedures and notations are borrowed from the work of Elalfi *et al.* [14].

4 Experimental Results

Protein Interaction as Binary Feature Vector
An interaction is represented as a pair of two proteins that directly binds to each other. This protein interaction is represented by binary feature vector of interacting proteins and their associations. These interaction representation processes and the processing steps are described in the work of Eom *et al.* [10].

Data Sets
Each *Yeast* proteins have various functions or characteristics which are called 'feature.' Here, set of features of each protein are collected from public genome databases as the same manner of Eom *et al.* [10]. Table 1 shows the statistics of each interaction data source and the number of features before and after the feature filtering.

Table 1. The statistics for the dataset

Data Source	# of interactions	# of initial features	# of filtered features
MIPS	10,641		
YPD	2,952		
SGD	1,482	6,232 (total)	1,293 (total)
Y2H (Ito *et al.*)	957		
Y2H (Uetz *et al.*)	5,086		

Table 2. Accuracy of the proposed methods. The effect of the FDRF-based feature selection and NN-based are shown in terms of prediction accuracy.

Prediction method	Number of interactions			Accuracy (|P|/|T|)
	Training set Size	Test set (T)	Predicted correctly (P)	
Asc. (△)	4,628	463	423	91.4 %
FDRF + Asc. (▽)	4,628	463	439	94.8 %
Asc. + N-Asc. (◇)	4,628	463	432	93.3 %
FDRF + Asc. + N-Asc. (☆)	4,628	463	445	96.1 %

Experiment Procedures
First, we predicted the classes of new PPIs with NN for their 13 functional categories obtained from MIPS [11]. The accuracy of class prediction is measured whether the predicted class of interaction is correctly corresponds to the class of MIPS. After this, we constructed feature association rule from this trained NN.

Next, we trained another NN with PPI data represented as binary feature vector according to the method of Eom *et al.* [10]. After the model training, we extracted again feature association rule from the model with the procedure of Eom *et al.* [13]. Then we predicted test PPI set with these two set of association rules and measured the prediction accuracy of each approaches. Results are measured with 10-fold cross-validation.

Results
Table 2 show the interaction prediction performance of various combination of associantion mining, feature filtering, and exploitation of rules derived from NN.

In Table 2, Simple association mining approach (△) achieved the lowest performance. The number of total feature used in this approach was 6,232. This is quite high feature dimension. We can guess that it may includes lots of non-informative and redundant features and these features may affect the prediction accuracy in negative way by interfering correct rule mining. This assumption confirmed by investigating the result of second approach, FDRF + Asc. (▽), association mining with non-informative and redundant feature filtering. This feature filtering approach improved overall prediction performance about 3.4% than the first approach. But the third approach, Asc. + N-Asc. (◇), prediction with the rules from association rule mining and the rule derived from trained NN only improved overall prediction performance about 1.9% than the first approach. This result can be explained again with the feature dimension problem. In this third approach, there also exist redundant and non-informative garbage features which decrease the prediction performance. But in this approach, eventhough there still lots of garbage features, the over all performance improved about 1.9%. This is the effect of the rule exploitation derived from trained NN. This inference can be confirmed again by investigating the result of fourth approach, FDRF + Asc. + N-Asc (☆), prediction with the rule from association mining and the rule derived from trained NN along with feature filtering. Non-informative and redundant features are filtered out in this approach. Consequently, his approach improved over all prediction accuracy up to 4.7%.

Thus, we can say that both the information theory based feature filtering and the exploitation of the rule derived from trained NN and conventional association rule mining methods are helpful for improving overall performance of feature-to-feature association-based PPI prediction. By considering these experimental results, the proposed approaches in this paper will be useful as a data preprocessing and prediction methods especially when we handle the data which have many features.

5 Conclusions

In this paper, we presented NN based protein interaction learning and association rule mining method from feature set and trained NN model for PPI prediction task. The proposed method (combination of all methods) achieved the improvement of accuracy about 4.7%. The experimental results of various approaches suggest that the NN based feature association learning model could be used for more detailed investigation of the PPIs when the proposed model can learn effectively the hidden patterns of the data which have many features and implicit association of these features. From the result of Section 4, we can conclude that the proposed method is suitable for efficient analysis of PPIs through its hidden feature association learning.

However, current public interaction data have many false positives and some interactions of these false positives are corrected as true positives by recent researches through reinvestigation with new experimental approaches. Thus, the study on the new method for adapting these changes in data set which is related to false positive screening remains as future works. Also, consideration of more biological features such as pseudo amino acid composition or protein localization facts will be helpful for improving overall performance.

References

1. Eisen, M.B., *et al.*: Cluster analysis and display of genome–wide expression patterns. *Proc. Natl. Acad. Sci. USA* **95** (1998) 14863–68.
2. Park, J., *et al.*: Mapping protein family interactions: intramolecular and intermolecular protein family interaction repertoires in the PDB and yeast. *J. Mol. Biol.* **307** (2001) 929–39.
3. Iossifov, I., *et al.* Probabilistic inference of molecular networks from noisy data sources. *Bioinformatics* **20**(8) (2004) 1205–13.
4. Ng, S.K., *et al.* Integrative approach for computationally inferring protein domain interactions. *Bioinformatics* **19**(8) (2003) 923–29.
5. Carpenter, G.A. and Grossberg, S.: A massively parallel architecture for a self-organizing neural pattern recognition machine. *Computer Vision, Graphics and Image Processing* **37** (1987) 54–115.
6. Barbara, M.: ART1 and pattern clustering. In proceedings of the 1988 connectionist models summer 1988. Morgan Kaufmann (1988) 174–85.
7. Heins, L.G. and Tauritz, D.R.: Adaptive resonance theory (ART): an introduction. *Internal Report* 95-35, Dept. of Computer Science, Leiden University, Netherlands, (1995) 174–85.
8. Rangarajan, *et al.*: Adaptive neural network clustering of web users. *IEEE Computer* **37**(4) (2004) 34–40.
9. Yu, L. and Liu, H.: Feature selection for high dimensional data: a fast correlation-based filter solution. In *Proceeding of ICML-03* (2003) 856–863.
10. Eom, J.-H., *et al.*: Prediction of implicit protein–protein interaction by optimal associative feature mining. *Lecture Notes in Computer Science* (*IDEAL'04*) **3177**, (2004) 85–91.
11. Press, W.H., *et al.*: Numerical recipes in C. *Cambridge University Press.* (1988)
12. Eom, J.-H., *et al.*: Adaptive Neural Network based Clustering of Yeast Protein-Protein Interactions. *Lecture Notes in Computer Science* (*CIT'04*) **3356**, (2004) 49–57.
13. Eom, J.-H.: Prediction of Yeast Protein–Protein Interactions by Neural Association Rule. *Internal Report* 04-03, Dept. of Computer Sci.&Eng., Seoul National University, Republic of Korea, (2004) 1–12.
14. Elalfi, A.E., *et al.*: Extracting rules from trained neural network using GA for managing E-business. *Applied Soft Computing* **4**, (2004) 65–77.

A Novel Method for Extracting Knowledge from Neural Networks with Evolving SQL Queries

Mario A. Góngora, Tim Watson, and David A. Elizondo

School of Computing,
Faculty of Computing Sciences and Engineering,
De Montfort University,
Leicester, UK
{mgongora, tw, elizondo}@dmu.ac.uk

Abstract. While artificial neural networks (ANNs) are undoubtedly powerful classifiers their results are sometimes treated with suspicion. This is because their decisions are not open to inspection – the knowledge they contain is hidden. In this paper we describe a method for extracting and representing the knowledge within an ANN. Mappings between inputs and output classifications are stored in a table and, for each classification, Structured Query Language (SQL) queries are evolved using a genetic algorithm. Each evolved query is a simple, human-readable representation of the knowledge used by the ANN to decide on the classification based on the inputs. This method can also be used to show how the knowledge within an ANN develops as it is trained, and can help to identify problems that are particularly hard, or easy, for ANNs to classify.

1 Introduction

Techniques of ANNs have been applied with success to problem domains including classification and continuous function approximation. However, sometimes it is important to understand the process by which ANNs arrive at a given result and to provide ways of extracting the knowledge embedded in the network. There are many techniques that have been researched in the past years to extract the knowledge from ANNs [1], some of which aim to create rules to explain their behaviour; despite the successful results from many of these works, the IF-THEN rules or other logic equations that represent the behaviour of the ANN are still too cryptic to be understood easily.

Another critical aspect in knowledge extraction is the capacity for a method to be able to work after the ANNs have been trained and to be independent of their architecture. Many authors, including [2], point out that the data mining tools that have been shown to work systematically for these constraints are those based on evolutionary programming techniques. These techniques have been applied to the extraction of knowledge from ANNs in different ways, mainly by developing symbolic systems to get different forms of IF-THEN rules to represent the knowledge in the networks.

In this paper we propose a method to create a more clear, human-readable and accountable representation of the knowledge embedded in an ANN based on our past work in the area of data mining. Since our work in 1995 [3], where we successfully used evolutionary computing to evolve SQL queries for data mining applications, to

W. Duch et al. (Eds.): ICANN 2005, LNCS 3697, pp. 497–502, 2005.

more recent approaches to our initial work by other authors [4], the use of evolutionary computation has proven to be a very powerful technique to search for efficient SQL queries to extract rules for data mining in large sets of data. An additional advantage of this approach, apart from its mere functionality, is that the resulting queries for data classification resemble more of a natural language form than complex sets of IF-THEN rules which use mainly logic relational operators and numbers. Data in a database can be arranged in many forms including conveniently semantically labelled categories. The SQL queries use these labels and organisation to form rules that look more natural, and need not use statements based only on number-ranges.

In the rest of this paper we will present our proposed novel approach in which the data classification process of ANNs will be transferred to an organised database and then, using evolutionary computing as a search method, SQL queries will be created to explain the processing model of the network.

This paper is divided into five sections. In the second section, an overview of the use of SQL queries for knowledge representation is presented and the issue of using evolutionary techniques for searching efficient queries is discussed. Our novel method for knowledge extraction is based on previous work in this area. In the third section we discuss the issue of data representation in ANNs. The proposed evolutionary programming method for evolving SQL queries towards the extraction of knowledge from ANNs is presented in section four. Finally, some conclusions and our proposed future research are given in section five.

2 Evolving SQL Queries as Representation of Knowledge

As shown in our previous work [3] and validated as still a current technique by works such as Salim's [4], SQL queries are a powerful method for data mining. When data is structured carefully in a database it presents an underlying organisation relevant to its relation in the database.

Databases provide organisation and conveniently named categories. They also present relations between the data. Part of these relations can be provided manually when this information is known or a level of initial analysis is available. Other relations are intrinsic to the data and have to be extracted using data mining techniques.

When the data mining methodology involves creating SQL queries, natural language explanations of the relations of the data can be created. A means to extract knowledge involves evolving database queries which are rewarded when the correct related data is selected by such queries. This means that each SQL statement represents the knowledge on how the classification of the data is made. Thus, the structure of the query itself explains the knowledge about the data with respect to the analysis being made.

This represents the aims of most current data mining systems which are based on symbolic AI techniques; and using database queries for data mining brings us to tackling the problem of having to analyse very large amounts of data, which we solve by combining the database organisation, the SQL as a powerful tool to explore it, and evolutionary computing techniques to search for efficient and effective queries. One of the main reasons for basing a data mining system on evolutionary computing is that the properties of such a system include both robustness of solutions and scalability.

An additional advantage of evolving the queries to represent knowledge is that a population of SQL queries can be partitioned and the subpopulations evolved in parallel, also the database can be similarly partitioned with subpopulations seeing different subsets of the data (having windows of data sections), with a further corresponding increase in parallelism. This form of representation of knowledge is an advantage that we have explored, both in terms of search efficiency (and ultimately speed of computing) and in terms of data organisation relevant to the representation of the information. Robust solutions can be produced even if they only see a local portion of the data which can then be evolved further after merging successful knowledge clusters and reselecting the window on the database. Thus, an evolutionary system can make the transition from small clean datasets with a clearly organised knowledge representation, to large datasets and sometimes noisy characteristics (like the ones found in knowledge representation inside trained ANNs), without a degradation in performance.

Another critically useful advantage of using SQL queries as representation of knowledge is that it enables the system to be easily used within the majority of commercial relational database application, and more importantly, it allows the relations to be expressed as a set of first order logic expressions. This increases the power of the search as queries can include operations such as EXISTS which cannot be represented in zero order logic. Also SQL queries are easy for people to understand so the results of our knowledge extraction method are naturally presented in a more suitable form.

3 Data Representation from Trained ANNs

Symbolic AI methods are a preferred form of knowledge representation from ANNs [5,6,7,8]. These methods have shown not only that evolutionary computing based techniques are the most powerful way to search for the knowledge, but that even having this powerful search tool, the organisation of the data is also critical to be able to achieve correct knowledge extraction.

Data organisation is critical and in [8] the authors use a clustering genetic algorithm to organise the training data as a first step to applying a knowledge extraction (KE) process during the training of the network. This is a step that has to occur naturally when organising the data for the ANN in a database. The underlying organisation in tables and columns in an appropriately designed database presents a first step towards a correct representation of the KE of the ANN: this includes the data collection process which Yang et al. have found to be a critical step [8] .

In the proposed method presented in this paper the knowledge embedded in a trained ANN can be represented with respect to the data in different ways. One way relies on the training data being available; in this case it is possible to use this data, organised in a database, to create a set of inputs and outputs, equivalent to a fully known sample of the knowledge of the network. Another way relates to the case when the network is already trained and validated. In this case extended data can be included in the database; the coherent way to include this data necessitates new inputs, the organisation of which could be determined. This data is included alongside the outputs which result from the ANN being used to classify these new, previously unseen inputs. A database organised in this way would hold a collection of data representing the knowledge embedded in an ANN.

At this stage the issue of data dimensionality arises. As seen in works such as [6,7], researchers use tools to reduce the dimensionality of the overall data, by removing redundant data, to have more efficient datasets. In [6,7] the authors use genetic algorithms to organise the data and minimise the redundancy in it, for a further step of training and knowledge extraction. Their method, although dependent on the training algorithm and the architecture of the ANN, shows how critical it is to have the correct data representation.

On this basis our data representation, as organised either from the original data only or in an extended data set, by no means corresponds to having redundant data. The difference between the original data only (the training set) and the extended data, is that in the later, the results which correspond to the ANN classification process are included. Data is available from the correctly classified samples out of the training set, as well as from the previously unseen data used to extend the database.

4 Searching for the ANN Knowledge

Our evolutionary algorithm adapted from [3] is to be used to find SQL queries that correctly extract the data from the tables using criteria which match the results obtained with the ANN.

The chromosome structure will be taken from [3] where we have implemented the ability to evolve tree structures that are needed to represent the structure of database queries. Although representation techniques exist for evolving trees using traditional genetic algorithms [9], when dealing with database queries, the ability to evolve the inherent complexity of queries, as well as their constituent elements, is of great value. Consequently we use a more general Genetic Programming based approach which is best suited for the structure of the queries; so we use dynamic trees to represent them.

A standard SQL query has the form:

SELECT attributes FROM tables WHERE logical condition;

Note that the logical condition is in first order logic.

In the initial phase, we have configured the search for queries from a single table. In this phase we also select a constant subset of table attributes (specified by either a configuration file, or interactively by the user). Consequently the system only needs representations of the tree that represents the logical condition of the query.

The fitness of each chromosome (set of SQL queries) will be measured against how effective it is at extracting the correct records from the database. In practical terms we propose two elements of the fitness measurement, relative to the results obtained from the analysed ANN:

- In the first instance, the fitness will be relative to how many correct records are selected by the query in comparison to how many are wrongly selected. This factor will relate to false positives.
- The other factor will be relative to how many correctly classified records were selected out of the total records from its class. This factor will relate to false negatives, relating to the class elements that are left out (not selected) by the query.

Being able to manipulate the levels of importance of these two fitness methods makes the system versatile and more tailored for critical domains such as medical applications. Many authors, including [2], have found that a problem with the use of ANNs in medical applications is that it is difficult to analyse the accountability of results, in particular the relations between false positives and negatives in classification processes. Our method, by using evolutionary computing techniques coupled with SQL format, presents a very powerful case for the use of an ANN as its process can now be more accountable.

The termination criteria of the search for queries present a particular problem. There are currently four possible scenarios that we are taking into account:

- If the fitness was measured using the real outputs of a known data set (i.e. the known answers for the dataset for training and validation) the termination criteria can include an error margin equal to the accuracy of the ANN being tested. It could be argued that if the genetic algorithm finds a perfect representation of the knowledge of the ANN, the output error of the evolved queries would represent the error of the predicted values produced by the network. This is not completely true since there is a possibility that the same type of error as the ANN produced was achieved by the genetic search but in another region of the solution space. This would mean that the real error relative to the knowledge representation of the ANN could be as much as twice as that presented by the fitness measure. On the other hand, an exact match (maximum fitness) would not be a suitable termination criterion since it will guarantee that there is at least the same error of knowledge representation of the ANN as the network itself has with respect to the classification of the data.
- Using this same type of data, the fitness can be measured relative to the results predicted by the ANN rather than from the observed values from the training data set. In this case, the error of the ANN will be included in the data rather than in the error allowance given to the evolutionary search. An exact match (maximum fitness) can then be used safely since it will mean that the knowledge representation is that of the ANN and nothing else.
- The third scenario is to measure the fitness relative to a mixture of known data with known results and new data with the result values as predicted by the ANN. This will provide a compromise between needing a maximum fitness achieved by the search (zero error), or using the maximum error as obtained by the ANN. On the other hand, there is still work to be done to make an error accountable to either to the ANN or the genetic search; provided that it is possible to calculate this in a deterministic way.
- Another situation is to use only data with unknown outputs to measure the fitness, in which case we are definitely unable to tell the origin of the error. This scenario will be used and interesting comparison to the results obtained by applying the 2nd scenario (known data with ANN outputs).

5 Conclusions

Knowledge extraction is important for ANNs to gain a wider degree of acceptance. Therefore, this domain has become a major field of research since it validates the use of

ANNs for applications where reasons or explanations on why or how a result has been achieved are important.

In this paper, a novel method for KE from ANNs is proposed. The processing of the data done by a trained ANN is transformed into a set of tables organised into a relational database. SQL queries from these tables are then created using evolutionary techniques. These SQL queries can then be used to explain how the neural network reaches a particular answer. Since the standard SQL queries are of first order logic form, they are more human-readable representations of the knowledge embedded in ANNs than the IF-THEN rules that symbolic based KE methods use. Details of the evolutionary programming techniques used to evolve the SQL queries are provided. These include the structure of the chromosome and their fitness function. Some discussion concerning the termination criteria for the evolution of queries has been presented.

Future work will involve the implementation and thorough testing of the proposed method for KE from ANNs using machine learning benchmarks and real world problems including the classification of satellite images, and of medical images. Additionally we will explore the use of this method to produce a sequence of SQL queries that show how the knowledge in an ANN develops. Also, potential "fuzzification" of the approach can be sought by using the SQL statement "SOUNDSLIKE". This may be an interesting avenue of investigation.

References

1. Proceedings of the International Conference in Neural Networks (Special Session on Knowledge Extraction and Neural Networks). (2005)
2. Dorado, J., Rabunal, J.R., Rivero, D., Santos, A., Pazos, A.: Automatic recurrent ann rule extraction with genetic programming. In: Proceedings of the International Joint Conference on Neural Networks (IJCNN). Volume 2. (2002) 1552–1557
3. Watson, T., Rakowski, T.: Data mining with an evolving population of database queries. In: Proceedings of MENDEL '95 - The International Conference on Genetic Algorithms. (1995) 169–174
4. Salim, M., Yao, X.: Evolving sql queries for data mining. In Science, L.N.C., ed.: Proceedings of the 3rd International Conference on Intelligent Data Engineering and Automated Learning (IDEAL'02). Volume 2412. (2002) 62–67
5. Duch, W., Adamczak, R., Grabczewski, K.: A new methodology of extraction, optimization, and application of crisp and fuzzy logical rules. IEEE Transactions on Neural Networks **11** (2000)
6. Fu, X., Wang, L.: Rule extraction using a novel gradient-based method and data dimensionality reduction. In: Proceedings of the International Joint Conference on Neural Networks (IJCNN). Volume 2. (2002) 1275–1280
7. Fu, X., Wang, L.: Rule extraction from an rbf classifier based on class-dependent features. In: Proceedings of the 2002 Congress on Evolutionary Computation (CEC). Volume 2. (2002) 1916–1921
8. Wang, Q., Bo, Y., Zhu, J.: Extract rules from software quality prediction model based on neural network. In: Proceedings of the International Joint Conference on Tools with Artificial Intelligence (ICTAI). (2004) 191–195
9. Watson, T.: A new representation technique for genetic algorithms. In: Proceedings of the Seventh Irish Annual Conference on Artificial Intelligence and Cognitive Science. (1994) 233–246

CrySSMEx, a Novel Rule Extractor for Recurrent Neural Networks: Overview and Case Study

Henrik Jacobsson and Tom Ziemke

University of Skövde, School of Humanities and Informatics,
P.O. Box 408, SE-541 28 Skövde, Sweden
{henrik.jacobsson, tom.ziemke}@his.se

Abstract. In this paper, it will be shown that it is feasible to extract finite state machines in a domain of, for rule extraction, previously unencountered complexity. The algorithm used is called the Crystallizing Substochastic Sequential Machine Extractor, or CrySSMEx. It extracts the machine from sequence data generated from the RNN in interaction with its domain. CrySSMEx is parameter free, deterministic and generates a sequence of increasingly deterministic extracted stochastic models until a fully deterministic machine is found.

1 Introduction and Background

The problem of extracting rules, or finite state machines, from recurrent neural networks (RNN Rule Extraction, or RNN-RE) has occupied a number of researchers on and off during the last 15 years. The achievements of this research have recently been compiled into a review [1] which identified four common ingredients of RNN-RE algorithms:

1. *quantization* of the continuous state space of the RNN, resulting in a discrete set of states,
2. state and output *generation* (and observation) by feeding the RNN input,
3. rule *construction* based on the observed state transitions,
4. rule set *minimization.*

These four constituents are often quite distinguishable in the algorithms. For example, in the most commonly used algorithm [2] (1) an equidistant grid partitioning of the state space was used for quantization, (2) states were generated by a breadth-first search, (3) the rules were constructed by transforming the transitions in the quantized space into a deterministic finite automata, and (4) the rules were minimized using a standard minimization algorithm. In another example [3] (1) a self-organising map was used to quantize, (2) states were generated by observing the network in interaction with its domain, and (3) stochastic rules were induced from these observations (no minimization in this case).

As pointed out in [1], none of the previously tested quantization functions have been tailor-made to comply with the specific demands of quantizing the state space of a dynamic system, where the state is recursively enfolded onto

W. Duch et al. (Eds.): ICANN 2005, LNCS 3697, pp. 503–508, 2005.

itself in interaction with a domain. The used quantizers all build (roughly) on the assumption that spatial neighbours should be merged and spatially separated points kept apart. The problem with this approach is that in the RNN, states that are very similar, spatially, may be very different, functionally in the RNN.

We came to the conclusion that the main problem of earlier solutions is in fact the lack of integration of the above presented constituents. More specifically, the quantizer should take into account the dynamics of the RNN through closer integration with the other constituents, so that the state space is quantized based on its functional context as set of a states of a dynamic system in interaction with a domain rather than ordinary points of a Euclidean space.

This realization was the ground for the development of a novel algorithm named `CrySSMEx`[1] (Crystallizing Substochastic Sequential Machine Extractor, see Algorithm 1) which builds on a novel hierarchical quantization algorithm (named Crystalline Vector Quantizer, CVQ) which can merge and split states based on their dynamical properties in the RNN. By the introduction of this algorithm, a novel form of state machine, a substochastic sequential machines (SSM) is also introduced. SSMs can take into account that some data may be missing in the data collected from the RNN. `CrySSMEx` is parameter free and generates a list of SSMs with monotonously increasing determinism.

2 Experiments

The main purpose of the experiments in this paper is simply to show that it is possible to extract a deterministic finite automata from networks in a challenging domain. That it is possible, in theory, to extract finite machines if the RNN is robustly mimicking a regular language recognizer has already been shown [4]. But previous RNN-RE techniques have predominantly been used on quite simple regular binary language classification tasks with relatively few states. The selected domain for this paper is the prediction of the $\mathbf{a}^n\mathbf{b}^n$-language which has been studied extensively in the RNN domain [5,6,7,8,9,10,11,12]. The network which has been chosen for analysis with `CrySSMEx` is a simple recurrent network (SRN) trained using a genetic algorithm with a fitness proportional to how many of the predictable symbols that were correctly predicted. Strings from the $\mathbf{a}^n\mathbf{b}^n$-language with $1 \leq n \leq 10$ were generated and augmented in random order both during training and during generation of data (Ω) for `CrySSMEx` to analyse. See [12] for more details on the training (the SRN in question here is actually one of those behind the statistics in that paper) and for a discussion on the importance of random string order for the analysis of RNNs in the $\mathbf{a}^n\mathbf{b}^n$-domain.

To generate data, the RNN was exposed to a sequence of 5500 symbols (corresponding to 50 strings of each length). The resulting extracted deterministic

[1] Unfortunately, the constituents of the algorithm are quite complex and there is no room for these details here. An open source distribution and an article on `CrySSMEx` are under preparation at the time of submission of this paper. The purpose of this paper is not to present the algorithm as such, but to present the underlying principle of functionally based quantization and to acknowledge the possibility of extracting rules from domains of a complexity previously not considered in RNN-RE contexts.

```
CrySSMEx(Ω, Λ_i, Λ_o)
Input: Time series data from the RNN, Ω, an input quantization
        function, Λ_i, and an output quantization function, Λ_o.
Output: A deterministic machine M mimicking the RNN.
begin
  Let M be the stochastic machine based on Ω resulting from an
  unquantized state space (i.e. only one state);
  repeat
    Select data relevant for splitting indeterministic states;
    Split quanta in state quantizer according to split data;
    Create M using new state quantizer, Λ_i and Λ_o for quantization;
    if M has equivalent states then
      Merge equivalent states;
    end
  until M is deterministic;
  return M;
end
```

Algorithm 1: A simplified description of the main loop of CrySSMEx. M is created from the observed RNN input, output and state contained in Ω by quantization of input, output and state space, of which the latter is optimized.

machine is shown in Figure 1 together with the two first indeterministic (and stochastic) machines. CrySSMEx always start with an initial machine of only one state, in which only the conditional distribution of output symbols given input symbol is modelled. The quantization of the state space of the RNN was then refined such that it contained two states from which the output symbols could be uniquely determined given the input symbol. The algorithm continued to select states that gave rise to indeterminism and split them until a machine with only deterministic states were reached. The whole procedure took ten iterations in the main loop of Algorithm 1.

The extracted deterministic SSM and its stochastic predecessors can all parse the same sequence the RNN was tested on. To parse with an SSM, an initial state must be chosen. If nothing is known about the initial state of the RNN, the initial state of the SSM be a uniform distribution over the states (i.e. that all states are equally probable). When the SSM is then fed input symbols from the domain the stochastic state of the SSM "crystallizes", i.e. the SSM becomes more "certain" about what the state of the underlying RNN would have been given the same input history. In other words, the entropy of the state distribution decreases given "evidence" from the input sequence. In the final, deterministic SSM of Figure 1, the machine eventually narrows down the number of possible occupied states to one.

A second reason for the word "crystalline" in this work is due to how the CVQ divides the state space gradually, in a process that visually resembles crystallization . In the $\mathbf{a}^n\mathbf{b}^n$-RNN of the experiment, the CVQ divides the state space as shown in Figure 2.

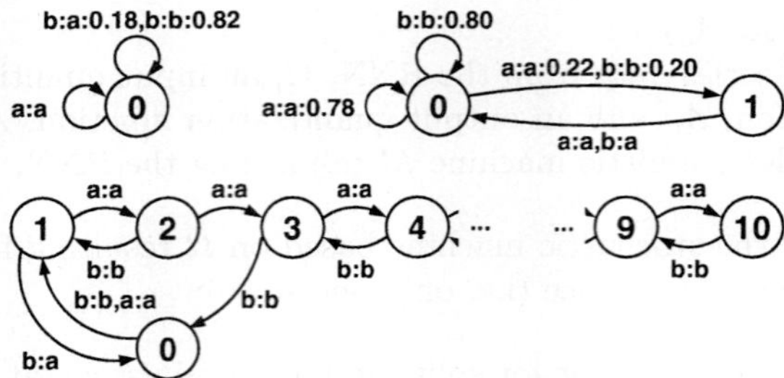

Fig. 1. The two first machines and the last (at iteration ten) in the sequence of machines extracted by `CrySSMEx`. A transition label **x**:**y**:*p* is to be read as a transition with **x** as input and **y** as output and *p* as the probability of this transition. For example, a transition label "**a**:**b**:0.75" from state 0 to 1 would correspond to that the conditional probability that the next state will be 1 and the output symbol **b**, given that the prior state was 0 and the input symbol **a**, is 0.75. Where *p* is 1.0, the probability is omitted from the label. For pairs of transitions between the same pair of states, the labels have in some cases been merged and comma separated to save space.

Apart from extracting rules from RNNs in the domain where $1 \leq n \leq 10$ (in which the RNN is trained to perfection) SSMs have also been extracted from the same RNNs with longer strings. The results varies with what kind of error the RNN makes for longer strings. If it makes no error, it is still trivial for `CrySSMEx` to extract the rules. But if the network cannot predict correctly when exposed to longer strings, extraction is either still trivial, or virtually impossible. In some cases the algorithm had to be aborted when the SSM grew indefinitely (having some 1000 states). Similar results are reached when testing `CrySSMEx` on chaotic systems[2]. The "grammar of mistakes" can obviously be of staggering complexity and this can probably be explained by the near chaotic dynamics of successful RNNs in the $\mathbf{a}^n\mathbf{b}^n$-domain [8,9]. Fortunately, all SSMs that are extracted during the search for a deterministic model are in themselves also models of the RNN, and can parse the input sequence just as the final model can. And the more computational resources invested in the iterations of `CrySSMEx`, the more exact will the extracted model be with respect to the underlying RNN, within the domain. There is however a possibility of data starvation if an SSM of a thousand states is extracted from data of just a few thousand time steps.

3 Discussion

In this paper a simple demonstration has been given to show that it is possible to extract machines from an RNN trained on a task that requires it to embed its memory in deeply recursive manner. Apart from the experiment presented here, `CrySSMEx` has been tested on, for example, autonomous chaotic dynamic

[2] For currently unknown reasons, a similar problem also sometimes occur if the sample set, Ω, is too small.

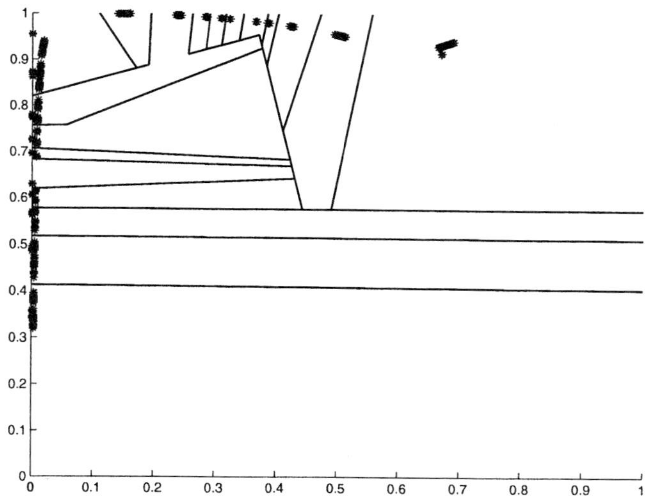

Fig. 2. The state space of the RNN as divided by the quantizer generated by CrySSMEx in order to describe the RNN as a finite deterministic system. The actual states occupied by the RNN when it is predicting $\mathbf{a}^n\mathbf{b}^n$-sequences are also plotted. Note how some disparate points sometimes belong to the same regions while some quite nearby clusters of points are separated.

systems, RNNs (with 10^3 state nodes) with small random weights and RNNs trained on regular languages. The results are overall very promising in that SSMs are extracted reliably and efficiently from all successfully trained networks.

There are of course many open issues and possible enhancements. For example, in the current implementation, CrySSMEx requires a symbolic input domain. If this was not the case, CrySSMEx could be used on systems with continuous input, e.g. a robotic controller reacting to sequences of sensory data. There are also a number of experiments that needs to be done to compare CrySSMEx to earlier approaches more directly, e.g. how many model vectors would typically be required if k-means is used (as in [13]) instead of the CVQ as quantizer in the $\mathbf{a}^n\mathbf{b}^n$-domain.

Extracted rules give us a unique window into the underlying system in that we can, in qualitatively new ways, analyse the rules in place of the RNNs. The next step is to let extracted rules be employed in our ambition to understand these networks. For example, it should be possible to query the rules concerning under what exact conditions an RNN commits mistakes. Such information would be ideal for planning retraining of the RNN. The extracted rules can also be used to test whether RNNs are equivalent to each other, something which is not easily derivable from the weights alone. So, if there is a population of networks, trained on the same domain or not, it would be possible to divide them into families of equivalent networks and describe exactly what distinguishes these families from each other. Similar distinctions have previously been made from a dynamic systems theory standpoint [6].

In future work, we would like to cooperate with a number of researchers who together could provide us with access to a large corpus of recurrent neural networks, of different architectures and trained on different data domains, to be analysed with CrySSMEx. For the future of the research area as a whole, in order to facilitate cooperation, exchange of experimental setups, and validation of results between individual researchers, it would furthermore be desirable to establish a database of benchmark problems and trained (recurrent) neural networks, similar to the benchmark data repositories that are commonly used in other areas of machine learning research, but with the trained models as the main focus.

References

1. Jacobsson, H.: Rule extraction from recurrent neural networks: A taxonomy and review. Neural Computation **17** (2005) 1223–1263
2. Giles, C.L., Miller, C.B., Chen, D., Chen, H.H., Sun, G.Z.: Learning and extracting finite state automata with second-order recurrent neural networks. Neural Computation **4** (1992) 393–405
3. Tiňo, P., Vojtek, V.: Extracting stochastic machines from recurrent neural networks trained on complex symbolic sequences. Neural Network World **8** (1998) 517–530
4. Casey, M.: The dynamics of discrete-time computation, with application to recurrent neural networks and finite state machine extraction. Neural Computation **8** (1996) 1135–1178
5. Wiles, J., Elman, J.L.: Learning to count without a counter: A case study of dynamics and activation landscapes in recurrent neural networks. In: Proceedings of the Seventeenth Annual Conference of the Cognitive Science Society, Cambridge MA: MIT Press (1995) 482–487
6. Tonkes, B., Blair, A., Wiles, J.: Inductive bias in context-free language learning. In: Proceedings of the Ninth Australian Conference on Neural Networks. (1998)
7. Rodriguez, P., Wiles, J., Elman, J.L.: A recurrent network that learns to count. Connection Science **11** (1999) 5–40
8. Tonkes, B., Wiles, J.: Learning a context-free task with a recurrent neural network: An analysis of stability. In Heath, R., Hayes, B., Heathcote, A., Hooker, C., eds.: Dynamical Cognitive Science: Proceedings of the Fourth Biennial Conference of the Australasian Cognitive Science Society. (1999)
9. Bodén, M., Wiles, J., Tonkes, B., Blair, A.: Learning to predict a context-free language: Analysis of dynamics in recurrent hidden units. In: Proceedings of ICANN 99, Edinburgh, IEEE (1999) 359–364
10. Bodén, M., Wiles, J.: Context-free and context-sensitive dynamics in recurrent neural networks. Connection Science **12** (2000) 196–210
11. Gers, F.A., Schmidhuber, J.: LSTM recurrent networks learn simple context free and context sensitive languages. IEEE Transactions on Neural Networks **12** (2001) 1333–1340
12. Jacobsson, H., Ziemke, T.: Improving procedures for evaluation of connectionist context-free language predictors. IEEE Transactions on Neural Networks **14** (2003) 963–966
13. Zeng, Z., Goodman, R.M., Smyth, P.: Learning finite state machines with self-clustering recurrent networks. Neural Computation **5** (1993) 976–990

Computational Neurogenetic Modeling: Integration of Spiking Neural Networks, Gene Networks, and Signal Processing Techniques

Nikola Kasabov, Lubica Benuskova, and Simei Gomes Wysoski

Knowledge Engineering and Discovery Research Institute,
Auckland University of Technology, 581-585 Great South Rd,
Auckland, New Zealand
{nkasabov, lbenusko, swysoski}@aut.ac.nz
http:www.kedri.info

Abstract. The paper presents a theory and a new generic computational model of a biologically plausible artificial neural network (ANN), the dynamics of which is influenced by the dynamics of internal gene regulatory network (GRN). We call this model a "computational neurogenetic model" (CNGM) and this new area of research Computational Neurogenetics. We aim at developing a novel computational modeling paradigm that can potentially bring original insights into how genes and their interactions influence the function of brain neural networks in normal and diseased states. In the proposed model, FFT and spectral characteristics of the ANN output are analyzed and compared with the brain EEG signal. The model includes a large set of biologically plausible parameters and interactions related to genes/proteins and spiking neuronal activities. These parameters are optimized, based on targeted EEG data, using genetic algorithm (GA). Open questions and future directions are outlined.

1 Introduction

We introduce a novel computational approach to brain neural network modeling that integrates ANN with an internal dynamic GRN. Interaction of genes in model neurons affects the dynamics of the whole ANN through neuronal parameters, which are no longer constant, but change as a function of gene expression. Through optimization of the GRN, initial gene/protein expression values and ANN parameters, particular target states of the neural network operation can be achieved. It is illustrated by means of a simple neurogenetic model of a spiking neural network (SNN). The behavior of SNN is evaluated by means of the local field potential (LFP), thus making it possible to attempt modeling the role of genes in different brain states, where EEG data is available to test the model. We use the standard FFT signal processing technique to evaluate the SNN output and compare with real human EEG data. For the objective of this work, we consider the time-frequency resolution reached with the FFT to be sufficient. However, should higher accuracy be critical, Wavelet Transform, which considers both time and frequency resolution, could be used instead. Broader theoretical and biological background of CNGM construction is given in [1]. A simpler linear

W. Duch et al. (Eds.): ICANN 2005, LNCS 3697, pp. 509 – 514, 2005.

version of an internal GRN with preliminary results on epilepsy modeling can be found in [2]. In this paper we (1) introduce and simulate a more realistic nonlinear model of GRN, (2) present a list of real proteins/genes that are involved in CNGM, (3) compare the CNGM performance to real human EEG data using the same signal processing technique, (4) suggest an optimization procedure to obtain a CNGM with parameters leading to modeling of the real EEG signal.

2 A General CNGM and Its Optimization via Evolution

In general, we consider two sets of genes – a set G_{gen} that relates to general cell functions and a set G_{spec} that defines specific neuronal information-processing functions (receptors, ion channels, etc.). The two sets form together a set $\mathbf{G}=\{G_1, G_2, \ldots, G_n\}$. We assume that the expression level of each gene $g_j(t+\Delta t')$ is a nonlinear function of expression levels of all the genes in **G**, inspired by discrete models from [3], [4]:

$$g_j(t+\Delta t') = \sigma\left(\sum_{k=1}^{n} w_{jk} g_k(t)\right) \tag{1}$$

We work with normalized gene expression values in the interval (0, 1). The coefficients $w_{ij} \in (-5,5)$ are the elements of the square matrix **W** of gene interaction weights. Initial values of gene expressions are small random values, i.e. $g_j(0) \in (0, 0.1)$.

In the current model we assume that: (1) one protein is coded by one gene; (2) relationship between the protein level and the gene expression level is nonlinear; (3) protein levels lie between the minimal and maximal values. Thus, the protein level $p_j(t+\Delta t)$ is expressed by

$$p_j(t+\Delta t) = (p_j^{max} - p_j^{min})\sigma\left(\sum_{k=1}^{n} w_{jk} g_k(t)\right) + p_j^{min} \tag{2}$$

The delay $\Delta t < \Delta t'$ corresponds to the delay caused by the gene transcription, mRNA translation into proteins and posttranslational protein modifications [9]. Delay $\Delta t'$ includes also the delay caused by gene transcription regulation by transcription factors.

The GRN model from equations (1) and (2) is a general one and can be integrated with any ANN model into a CNGM. Unfortunately the model requires many parameters to be either known in advance or optimized during a model simulation. In the presented experiments we have made several simplifying assumptions:

1. Each neuron has the same GRN, i.e. the same genes and the same interaction gene matrix **W**.
2. Each GRN starts from the same initial values of gene expressions.
3. There is no feedback from neuronal activity or any other external factors to gene expression levels or protein levels.
4. Delays Δt are the same for all proteins and reflect equal time points of gathering protein expression data.

We have integrated the above GRN model with the SNN illustrated in Fig. 1. Our spiking neuron model is based on the Spike Response Model [5], with excitation and inhibition having both fast and slow components [6], [7] both expressed as double exponentials with amplitudes and the rise and decay time constants.

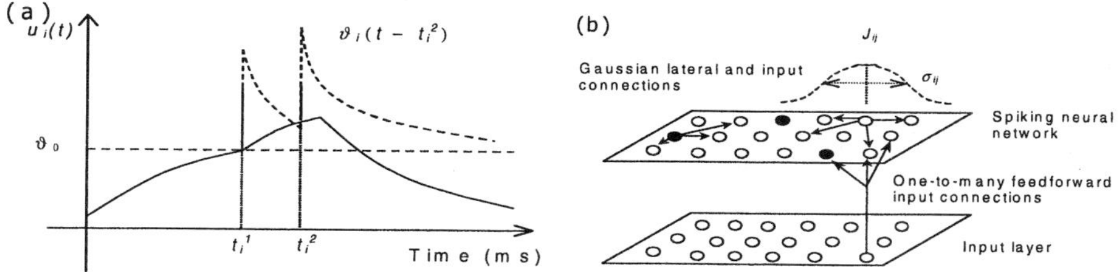

Fig. 1. (a) Spiking neuron model. When the membrane potential $u_i(t)$ of the ith spiking neuron reaches the firing threshold $\vartheta_i(t)$ at time t^k_i, the neuron fires an output spike. $\vartheta_i(t)$ rises after each output spike and decays back to the resting value ϑ_0. (b) The SNN architecture. About 10–20% of $N = 120$ neurons are inhibitory neurons that are randomly positioned on the grid (filled circles). External input is random with average frequency between 10–20 Hz.

Table 1. Neuronal parameters and their corresponding proteins (receptors/ion channels)

Neuron's parameter P_j	Relevant protein p_j
Amplitude and time constants of:	
Fast excitation	AMPAR
Slow excitation	NMDAR
Fast inhibition	GABRA
Slow inhibition	GABRB
Firing threshold and its decay time constant	SCN and/or KCN and/or CLC

Neuronal parameters and their correspondence to particular proteins are summarized in Table 1[1]. Several parameters (amplitude, time constants) are linked to one protein. However their initial values in equation (3) will be different. Relevant protein levels are directly related to neuronal parameter values P_J such that

$$P_j(t) = P_j(0)p_j(t) \tag{3}$$

where $P_j(0)$ is the initial value of the neuronal parameter at time $t = 0$. Moreover, besides the genes coding for the proteins mentioned above, we include in our GRN nine more genes that are not directly linked to neuronal information-processing parameters. These genes are: c-jun, mGLuR3, Jerky, BDNF, FGF-2, IGF-I, GALR1, NOS, S100beta. We have included them for later modeling of some diseases.

We want to achieve a desired SNN output through optimization of the model 294 parameters (we are optimizing also the connectivity and input frequency to the SNN). We evaluate the LFP of the SNN, defined as LFP = $(1/N)\Sigma\, u_i(t)$, by means of FFT in order to compare the SNN output with the EEG signal analyzed in the same way. It has been shown that brain LFPs in principle have the same spectral characteristics as EEG [8]. Because the updating time for SNN dynamics is inherently 1ms, just for computational reasons, we will employ the delays Δt in equation (2) being equal to

[1] Abbreviations: AMPAR = (amino- methylisoxazole- propionic acid) AMPA receptor, NMDAR = (*N*-methyl-D-aspartate acid) NMDA receptor, GABRA = (gamma-aminobutyric acid) GABA receptor A, GABRB = GABA receptor B, SCN = Sodium voltage-gated channel, KCN = kalium (potassium) voltage-gated channel, CLC = chloride channel.

just 1s instead of minutes or tens of minutes [9]. In order to find an optimal GRN within the SNN model so that the frequency characteristics of the LFP of the SNN model are similar to the brain EEG characteristics, we use the following procedure:

1. Generate a population of CNGMs, each with randomly generated values of coefficients for the GRN matrix W, initial gene expression values $g(0)$, initial values of SNN parameters $P(0)$, and different connectivity;
2. Run each SNN over a period of time T and record the LFP;
3. Calculate the spectral characteristics of the LFP using FFT;
4. Compare the spectral characteristics of SNN LFP to the characteristics of the target EEG signal. Evaluate the closeness of the LFP signal for each SNN to the target EEG signal characteristics. Proceed further according to the standard GA algorithm to possibly find a SNN model that matches the EEG spectral characteristics better than previous solutions;
5. Repeat steps 1 to 4 until the desired GRN and SNN model behavior is obtained;
6. Analyze the GRN and the SNN parameters for significant gene patterns that cause the SNN model behavior.

3 Simulation and Results

First, we present the results of analysis performed on real human interictal EEG data obtained with permission from [10]. Fig. 2 shows the brain EEG signal, its FFT power spectrum and evolution of the relative intensity ratios (RIRs) for different clinically relevant sub-bands over time. These sub-bands are: delta (0.5-3.5 Hz), theta (3.5-7.5 Hz), alpha (7.5-12.5 Hz), beta 1 (12.5-18 Hz), beta 2 (18-30 Hz), gamma (above 30 Hz). Each point depicts the RIR over the previous 1s.

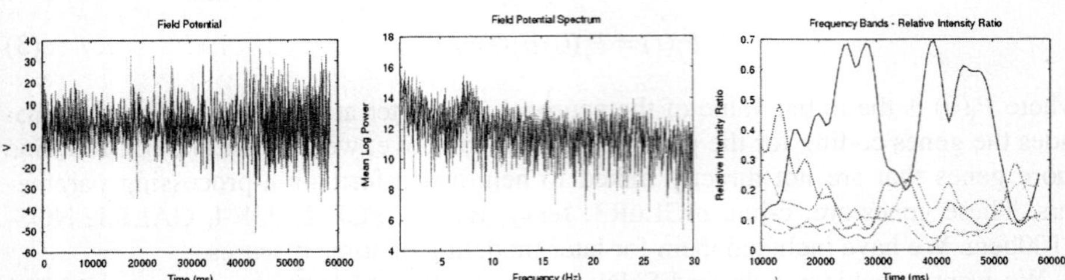

Fig. 2. a) Human interictal EEG Signal; b) classical FFT analysis of the EEG signal, sampling rate is 256 Hz; c) temporal evolution of RIRs for the clinically relevant frequency sub-bands for the EEG signal. The dominant sub-band is delta (0.5-3.5 Hz).

We calculated the average RIRs over the whole time of simulation (i.e., T = 1 min) and used this vector of values as a fitness function for our GA. After 50 generations with 6 solutions in each population we obtained the following result for the best solution, illustrated in Fig. 3. Solutions for reproduction were being chosen according to the roulette rule and the crossover between parameter values was performed as an arithmetic average of the parent values. We performed the same FFT analysis as for the real EEG data with the Min/Max frequency = 0.1 / 50 Hz. This particular SNN had an evolved GRN with only 5 genes out of 16 changing periodically their expression values (s100beta, GABRB, GABRA, mGLuR3, c-jun) and all other genes having constant expression values (see e.g. Fig. 4) with, either minimal or maximal.

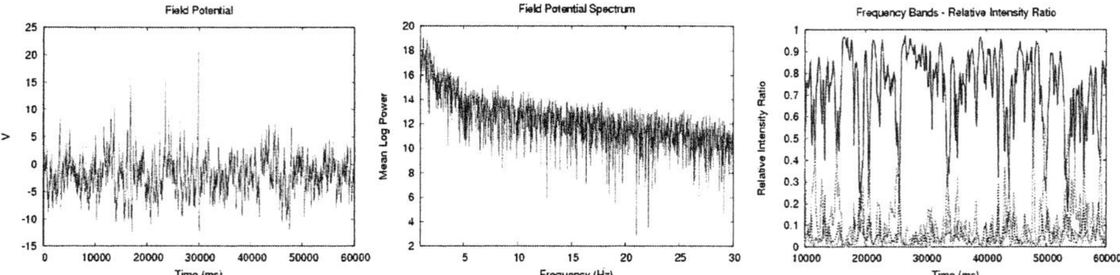

Fig. 3. a) Local field potential of the SNN with GRN; b) classical FFT analysis of the SNN LFP, sampling rate is 1000 Hz; c) temporal evolution of RIRs for clinically relevant frequency sub-bands for the LFP. The dominant sub-band is again delta (0.5-3.5 Hz).

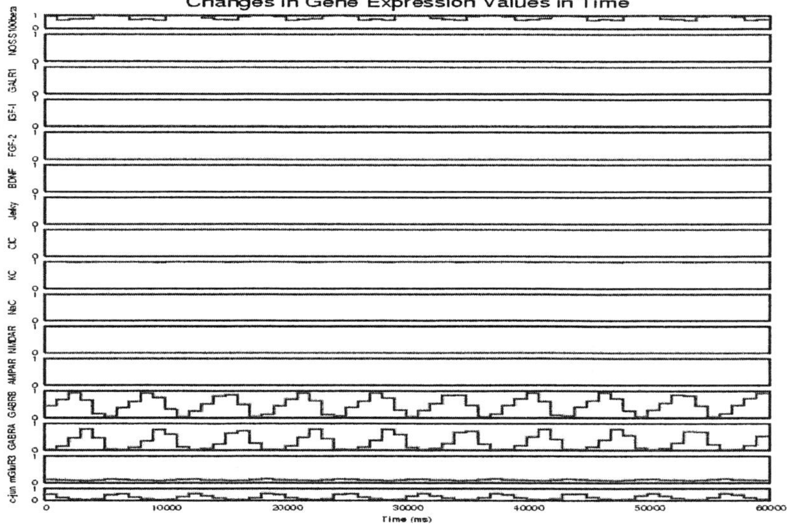

Fig. 4. Changes in gene expression values over time in the model GRN

4 Discussion

Our preliminary results show that the same signal processing techniques can be used for the analysis of both the simulated LFP of the SNN CNGM and the real EEG data to yield conclusions about the SNN behavior and to evaluate the CNGM at a gross level. With respect to our neurogenetic approach we must emphasize that it is still in an early developmental stage and the experiments assume many simplifications. In particular, we would have to deal with the delays in equation (2) more realistically to be able to draw any conclusions about real data and real GRNs. The LFP obtained from our simplified model SNN is of course not exactly the same as the real EEG, which is a sum of many LFPs. However LFP's spectral characteristics are very similar to the real EEG data, even in this preliminary example. Based on our preliminary experimentation, we have come to the conclusion that many gene dynamics, i.e. many interaction matrices **W**s that produce various gene dynamics (e.g., constant, periodic, quasiperiodic, chaotic) can lead to very similar SNN LFPs. In our future work, we want to explore statistics of plausible **W**s more thoroughly and compare it with biological data to draw any conclusions about underlying GRNs. Further research questions are: How many GRNs would lead to similar LFPs and what do they have in common? How to use CNGM to model gene mutation effects? How to use CNGM to predict drug effects? And finally, how to use CNGM for the improvement of individual brain functions, such as memory and learning?

Acknowledgments. Supported by the NERF grant AUTX02001 funded by FRST, and by KEDRI funds.

References

1. Kasabov, N., Benuskova, L.: Computational Neurogenetics. Journal of Computational and Theoretical Nanoscience, Vol. 1. (2004) 47-61
2. Kasabov, N., Benuskova, L., Wysoski, S.G.: Computational neurogenetic modelling: gene networks within neural networks. Proc. IEEE Intl. Joint Conf. Neural Networks, Vol. 2. IEEE Press, Budapest (2004) 1203-1208
3. Weaver, D.C., Workman, C.T., Stormo, G.D.: Modeling regulatory networks with weight matrices. Proc. Pacific Symp. Biocomputing - Hawai, Vol. 4. World Scientific Publ. Co, Singapore (1999) 112-123
4. Wessels, L.F.A., vanSomeren, E.P., Reinders, M.J.T.: A comparison of genetic network models. Proc. Pacific Symp. Biocomputing, Vol. 6. (2001) 508-519
5. Gerstner, W., Kistler, W.M.: Spiking Neuron Models. Cambridge Univ. Press, Cambridge-MA (2002)
6. Destexhe, A.: Spike-and-wave oscillations based on the properties of $GABA_B$ receptors. J. Neuroscience, Vol. 18. (1998) 9099-9111
7. Kleppe, I.C., Robinson, H.P.C.: Determining the activation time course of synaptic AMPA receptors from openings of colocalized NMDA receptors. Biophys. J., Vol. 77. (1999) 1418-1427
8. Destexhe, A. Contreras, D., Steriade, M.: Spatiotemporal analysis of local field potentials and unit discharges in cat cerebral cortex during natural wake and sleep states. J. Neuroscience, Vol. 19. (1999) 4595-4608
9. Lodish, H. Berk, A., Zipursky, S.L., Matsudaira, P., Baltimore, D., Darnell, J.: Molecular Cell Biology. 4th edn. W.H. Freeman & Co, New York (2000)
10. Quiroga, R.Q.: Dataset #3: Tonic-clonic (Grand Mal) seizures. www.vis.caltech.edu/~rodri/ data.htm (1998)

Information Visualization for Knowledge Extraction in Neural Networks

Liz Stuart[1], Davide Marocco[2], and Angelo Cangelosi[1]

[1] School of Computing Communication and Electronics,
University of Plymouth, Drake Circus, PL4 8AA, Plymouth, UK
{lstuart, acangelosi}@plymouth.ac.uk
[2] Institute of Cognitive Science and Technologies,
National Research Council, viale Marx 14, 00137, Rome, Italy
davide.marocco@istc.cnr.it

Abstract. In this paper, a user-centred innovative method of knowledge extraction in neural networks is described. This is based on information visualization techniques and tools for artificial and natural neural systems. Two case studies are presented. The first demonstrates the use of various information visualization methods for the identification of neuronal structure (e.g. groups of neurons that fire synchronously) in spiking neural networks. The second study applies similar techniques to the study of embodied cognitive robots in order to identify the complex organization of behaviour in the robot's neural controller.

1 Introduction

Knowledge extraction using neural networks typically involves the use of analytical methods for the automatic identification of information relevant to specific research goals. For example, the utilization of a finite union of open polytopes permits the transparent expression of knowledge embedded in recursive determinist perceptron [1]. In this paper, a complementary method for knowledge extraction from artificial and natural neural networks is presented. This is characterized by the active role of the researcher in the exploration of the neural network representation and the search for in-depth knowledge. This method is based on recent research and tools for information visualization in neural systems.

Information visualization [2], [3] is one of the fields of computer science that deals with the innovative representation of vast quantities of data. Consequently, it exploits advances in interactive computer graphics hardware, mass storage, and data visualization in order to visualize information. One of the fundamental principles of this field is the role of the investigator interacting with the data being analyzed and their ability to steer the exploration, in order to achieve greater insight. Thus, the investigator needs to be able to navigate throughout the whole dataset, in order to identify and explore specific subsets of interest. However, when visualizing large datasets, the issue of efficient navigation is amplified. It is important that the user is able to move to points of interest quickly without becoming disoriented within the dataset. Thus, it can be beneficial to restrict the mechanisms by which the user can navigate within the dataset. Therefore, the user can be constrained to follow

W. Duch et al. (Eds.): ICANN 2005, LNCS 3697, pp. 515 – 520, 2005.

predetermined paths throughout the data space. In addition to navigation functionality, the investigator should also have control over the data representation itself. Thus, in order to truly steer the analysis, the investigator should be able to manage and overview the whole dataset, they should be able to filter and manipulate the data, select non-sequential subsets of interest and ultimately "drill-down" to inspect the actual data values that underpin the data represented. This follows the design principles of the much cited "information seeking mantra" coined by Shneiderman.

Recent research has focused on the development of new information visualization techniques and tools for neural networks. This includes the visualization of information from neuroscience research and the analysis on neural activity in embodied cognitive systems. In this paper, two such case studies are described. The first study demonstrates the use of information visualization methods for the identification of structure (e.g. groups of neurons that fire synchronously) in spiking neural networks. The second study applies the same techniques to embodied cognitive robots in order to identify the complex organization of sensorimotor behaviour and its management by a neural controller.

2 VISA: Information Visualization for Spiking Neural Networks

The key to numerous issues within the field of neuroscience is linked with the theoretical understanding of vast quantities of experimental neural data. In particular, investigation of information processing in the nervous system is associated with the analysis of this vast resource of neural data, namely, simultaneously recorded multi-dimensional spike train data.

Much of the research focus in this area is focused around the principle of synchronization of neural activity [4], [5]. However, the experimental evidence that is currently available requires further, in-depth analysis in order to extract the knowledge inherent in these datasets. It is clear that analysis of neural data such as multi-dimensional spike trains using traditional tools like raster plots and cross-correlograms increases in complexity as the size of the datasets increase. Therefore, new methods of analyzing this data, designed specifically for large datasets, are required.

Traditionally, analysis of multi-dimensional spike train data has not supported real-time user interaction. In 1996, Shneiderman [2] identified user interaction as a primary essential component of information visualization representations. Shneiderman also introduced the "information–seeking mantra" that highlighted user requirements in this area. It specified that users should have the capability to overview data in order to see the whole dataset in a single display. The mantra also recommended that users needed to be able to zoom in on "interesting" areas of their datasets and to filter out parts of the datasets not required for the current investigation. Finally, the mantra specified that users needed to be able to get details-on-demand, access to the fundamental data items which were used to create the visual representation.

Based on this mantra, a toolbox of interactive methods for exploring neural data was developed by Stuart and collaborators [6] as part a research project called VISA,

Visualization of Inter-Spike Associations. This toolbox facilitates zooming, filtering and manipulation of neural data and supports the use of multiple views of the same data, as well as real time interaction. One of the key visualization tools currently included within VISA is subsequently described.

2.1 Traditional Parallel Coordinates

Interest in parallel coordinates was rejuvenated by Inselberg et al. [7]. The true value of parallel co-ordinates is their ability to represent vast quantities of multi-variate data in a simply 2-d representation. Traditional presentation of parallel coordinates denotes a series of data points as vertical axis coordinate values distributed along a horizontal axis.

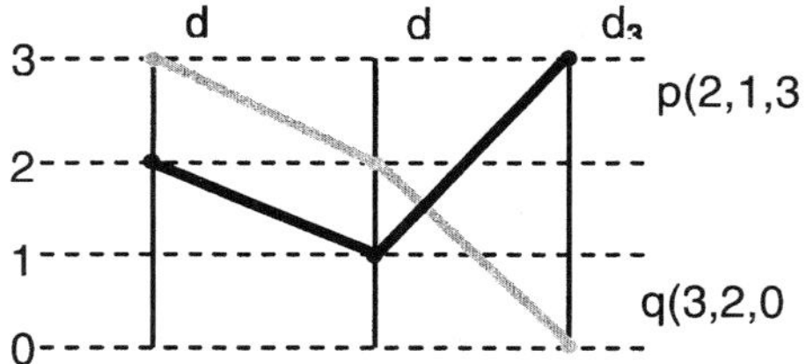

Fig. 1. Representation of the points p(2,1,3) and q(3,2,0) using parallel coordinates

Thus, a specific point in n-dimensional Euclidean space is represented by n vertical axes values distributed along the horizontal axis. To illustrate this, consider the two 3-dimensional points p(2,1,3) and q(3,2,0). Refer to Figure 1 for an illustration of these two points in 3-dimensional space represented in parallel coordinates.

2.2 The Use of Parallel Coordinates in VISA

Parallel coordinates are traditionally used to identify correlations between variables and to convey aggregation information. Although the VISA parallel coordinate tool [8] is based on the original method, it has been adapted for use in the analysis of spiking activity in neural networks.

In brief, the original multi-dimensional spike train dataset is transformed using a well known and respected method of analysis called the gravity transform algorithm defined by Gerstein et al [9]. Thus, the data under investigation is now the position of n particles in n-dimensional space where n is the number of neurons initially specified. The traditional means of displaying output from the Gravity Transform does not scale up easily for large numbers of particles. Thus, parallel coordinates were introduced to represent the n-dimensional positions of the n particles. Therefore, all n particle parallel coordinates were displayed simultaneously on the display to represent the position of all n particles at a specific point in time. Subsequently, animation was used to represent the movement of the particles over time. Numerous trials have been carried out using this technique to support the identification of neuronal assemblies from multi-dimensional spike train datasets. The success of these trials is well documented [8,10]. Furthermore, this work continues as the parallel coordinate tool undergoes further adaptation [11].

3 Sensrimotor Knowledge Integration in the Neural Controller of Cognitive Robots

Research in cognitive systems, including natural (animals and humans) and artificial (agents, robots) systems, supports increased understanding of the relationship between cognitive, neural, social and evolutionary factors. Various researchers are working on the design of cognitive robots that have sensorimotor capabilities, to interact with their environment, and cognitive and linguistic skills to build internal representations of their physical social environment and talk about them [12]. The complex patterns of interaction between the different sensorimotor and cognitive capabilities require the development of new methodologies to increase understanding of the cognitive systems. Information visualization has a role to play in the analysis of these complex sets of behavioural and cognitive data [13].

A new study based on the application of parallel coordinates to the visualization of activity in the neural controllers of linguistic cognitive robots [14] is presented. The primary aim is to investigate the interaction between sensory neurons and the internal categorical and linguistic representations in simulated robotic agents that are able to interact with two different objects (i.e. touch a sphere and avoid a cube). These agents are also able to develop, through evolution, a shared lexicon to name the two objects.

The cognitive robotic model consists of a 3-segment arm with 6 degrees of freedom (DOF). Each segment consists of a basic structure of two cylindrical bodies and two joints. This structure is replicated three times, with the final segment being shorter to represent the hand of the robot (with no fingers). The controller of each individual robot consists of an artificial neural network with 11 sensory neurons connected to 3 hidden neurons. These connect to 8 output neurons. The first 9 sensory neurons encode the angular position (normalized between 0.0 and 1.0) of the 6 DOFs and the state of three contact sensors located in the three corresponding arm segments. The other 2 sensory neurons receive their input from the other agents (name of objects). The first 6 motor neurons control the actuators of the corresponding joints. The motor is activated so that it is able to apply a force proportional to the difference between the current and the desired position of the joint. The last 2 output neurons encode the signal to be communicated to the other agents. Agents are evolved for their ability to interact with the objects, using a genetic algorithm.

To investigate the neural control strategies for the robot's sensorimotor and cognitive behaviour, a dataset from two experimental trials was created. In the first trial, the activity of all the input, hidden and output units were recorded whilst the robot interacted with a sphere. The second trial records the neural activity during the interaction of the robot with the cube. Each trial produced a dataset of 16 variables consisting of the 4 linguistic units, the 9 input units (3 touch sensors and 6 proprioceptive), and the 3 hidden nodes. In addition, each trial lasted for 150 cycles of actions (i.e. neural networks activations), thus producing 150 x 16 data points. All activation values (2 x 16 x 150) were plotted in the parallel coordinate representation. This was produced by a modified version of the VISA software.

This parallel coordinate tool produces a dynamic and interactive representation of the whole dataset. For example, it is possible to search for the different patterns of activation that distinguish the two behavioural tasks. Initially, the whole dataset, 150 x 16 cycles, was plotted using a different colour to represent each of the two tasks. During experimentation, the activation cycles were identified and eliminated. Note: these were the lines (parallel coordinates) from the two tasks which were coincident (i.e. the same unit has the same activation value in both tasks). This process of eliminating redundant (less informative) cycles gradually revealed the small number of critical cycles in the display. These are the lines (parallel coordinates) from each task that have distinct activation patterns.

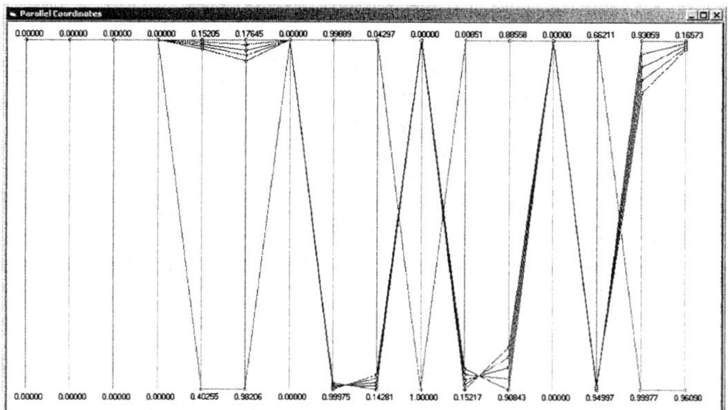

Fig. 2. Parallel coordinate display for cognitive robotics. See text for explanation.

Figure 2 shows a snapshot of the dynamic information visualization process with the data from the robotic experiment. The display shown solely depicts the six critical cycles of interaction (cycles 65-70 for each interaction) when the robot arm makes contact with the two objects (light gray for spheres, dark for cubes). This display clearly shows, with a single view, which input and hidden units are involved values in the two tasks. During interaction with the sphere, the units active solely for the touching behaviour are the first 2 proprioceptive sensors (vertical axes 5 and 6), the touch sensor of the second arm segment (axis 10) and the 3rd hidden unit (the final vertical axes). Instead, the units that specialize for the cube avoiding behaviour are the pairs of proprioceptive sensors for the 2nd and 3rd segment (respectively units 8-9 and 11-12) and the 1st hidden units (vertical axis 14).

4 Conclusions

The two studies presented in this paper demonstrate the usefulness of information visualization methods for knowledge extraction in various types of neural network. In the first study, the parallel coordinate method included in the VISA software project has supported the identification of common activity in networks of spiking neurons. In the second study, research on cognitive robots and the further adapted parallel coordinate tool was used to identify the input and output units involved in the neural control of the sensorimotor behaviour in robots.

The work presented here is an innovative approach to knowledge extraction in artificial neural networks. Instead of relying solely on formal methods for knowledge elicitation, this approach is based on an active exploration and visualization of neural network data. This design is largely based on Shneiderman's information–seeking mantra, where users have the capability to "overview data"; to "zoom and filter" data and also to obtain "details-on-demand".

Future research is looking at the development of new information visualization tools for neural networks and cognitive robotics research as well as the further use of current tools. For example, extension of the parallel coordinate tool is required in order to support greater investigation of the various interrelationships between the neural network units. To achieve this, the tool will require the additional functionality to enable the user to move variables (the vertical axes) along the horizontal axis. This will enable users to analyse the precise relationships between any two variables which are not adjacent by default in the original display. In addition, the user interface will be further developed. It is important that the user is able to specify the colours used to represent both the lines (parallel coordinates) corresponding to correct and incorrect behaviours, and to highlight differences in the pattern of the neural network activation.

References

[1] Tajine, M. & Elizondo, D. (1998), Growing methods for constructing Recursive Deterministic Perceptron neural networks and knowledge extraction. *Artificial Intelligence*, 102, 295-322.

[2] Shneiderman, B. (1996), The eyes have it: A task by data type taxonomy of information visualizations, *Proc. IEEE Symposium on Visual Languages '96*, IEEE, 336-343.

[3] Spence R. Information Visualization. Addison-Wesley: Harlow, UK, 2001.

[4] Borisyuk RM & Borisyuk GN (1997), Information coding on the basis of synchronisation of neuronal activity, *BioSystems*, 40, 3-10.

[5] Fries P, Neuenschwander S, et al. (2001), Rapid feature selective neuronal synchronization through correlated latency shifting, *Nature Neuroscience*, 4(2), 194-200.

[6] http://www.plymouth.ac.uk/infovis

[7] Inselberg A & Dimsdale B (1990), Parallel Coordinates: A tool for visualising multidimensional geometry, *Proc. Visualization'90*, 361-378.

[8] Stuart L., Walter M. and Borisyuk R. (2002), Visualization of synchronous firing in multi-dimensional spike trains, *BioSystems* 67:265-279.

[9] Gerstein G.L. & Aertsen A.M. (1985), Representation of cooperative firing activity among simultaneously recorded neurons, *Journal of Neurophysiology*, 54(6), 1513-1528.

[10] Stuart, L., Walter, M. and R. Borisyuk (2001), Visualization of multi-dimensional Spike Trains, *Proc. 4th International workshop Neural Coding'2001*, 47-48.

[11] Barlow, N. & Stuart, L. (2004), Animator: A Tool for the Animation of Parallel Coordinates, *Proc. IEEE Intl. Conference on Information Visualization*, IV04, 725-730.

[12] Cangelosi, A. & Parisi, D., (Eds.) (2002), *Simulating the Evolution of Language*, Springer.

[13] Smith, T., Bullock, S. & Bird, J. (2002), Beyond fitness: Visualising evolution - Workshop overview, *ALife VIII*, Sydney.

[14] Marocco, D., Cangelosi, A. & Nolfi S. (2003), The emergence of communication in evolutionary robots. *Philosophical Transactions of the Royal Society of London – A*, 361, 2397-2421.

Combining GAs and RBF Neural Networks for Fuzzy Rule Extraction from Numerical Data

Manolis Wallace[1,2] and Nicolas Tsapatsoulis[2,3]

[1] University of Indianapolis, Athens Campus, 9 Ipitou Str., Syntagma, 105 57 Athens, Greece
wallace@uindy.gr

[2] School of Electrical and Computer Engineering, National Technical University of Athens, 9 Iroon Polytechniou Str., Zographou, 157 73, Athens, Greece
{wallace, ntsap}@image.ntua.gr

[3] Dept. of Computer Science, University of Cyprus, 75 Kallipoleos Str., P.O. Box 20537, CY-1678, Nicosia, Cyprus
nicolast@ucy.ac.cy

Abstract. The idea of using RBF neural networks for fuzzy rule extraction from numerical data is not new. The structure of this kind of architectures, which supports clustering of data samples, is favorable for considering clusters as if-then rules. However, in order for real if-then rules to be derived, proper antecedent parts for each cluster need to be constructed by selecting the appropriate subspace of input space that best matches each cluster's properties. In this paper we address the problem of antecedent part construction by (a) initializing the hidden layer of an RBF-Resource Allocating Network using an unsupervised clustering technique whose metric is based on input dimensions that best relate the data samples in a cluster, and (b) by pruning input connections to hidden nodes in a per node basis, using an innovative Genetic Algorithm optimization scheme.

1 Introduction

Extracting if-then rules from numerical data using an RBF neural network can be achieved in the following framework: (a) The RBF-hidden nodes combine inputs in an AND form creating the antecedent part of the rule; that is rule antecedents are considered the input to hidden connections, (b) output nodes combine the outputs of the hidden nodes in an OR form; that is the rule consequents are the hidden to output connections, (c) knowledge in the form of if-then rules can be derived from clustering numeric data, (d) fuzziness is achieved both in the hidden and the output nodes forcing the activation and final output to be in the interval [0 1] instead of having crisp values.

Although the above framework seems reasonable there are two important problems: (i) All inputs are used in the antecedent part of the rule; this leads to inefficiency in creating real linguistic rules especially in cases where the input dimension is relatively high and (ii) the classic clustering approach used in RBF neural networks does not account for specifying different weights for the various input dimensions. While for the second problem one could consider the use of a different metric for creating a more

W. Duch et al. (Eds.): ICANN 2005, LNCS 3697, pp. 521 – 526, 2005.

"rule-like" clustering, the first problem is not that easy to solve. In this paper we address both problems by: (1) using Genetic Algorithms for selecting the appropriate inputs for each hidden node separately; this is radically different from the classic combination of RBF and GAs which focus in feature selection for the whole network [1][2][3] and (2) applying an unsupervised clustering technique, that is based on a data dependent metric, for initializing the parameters of the hidden nodes in the RBF network. In the proposed method clusters are created by data samples that are based on these dimensions that relate them best. This is clearly a more "rule-like" approach that the classic unsupervised clustering methods. The modifications proposed above are applied on an modified Resource Allocation Network consisting of RBF hidden nodes to account for learning required for rule extraction from numerical data.

2 Preliminaries

One approach for extracting rules from numerical data is to apply a supervised training procedure on a domain ***D*** from which the learner has access to a representative example set ***E*** of pairs { $\underline{x}$, $\underline{d}(\underline{x})$ } (numerical data), $\underline{x} \in \Re^n$, $\underline{d} \in \Re^m$. By the end of learning, performed on set ***E***, a set of parameters ***G*** (typically represented as matrices) that model the function $g : \Re^n \rightarrow \Re^p$, is available so that $\|G(\underline{x}) - g(\underline{x})\| < \varepsilon$, ε>0.

In the proposed method the set of parameters consists of four matrices corresponding to the mean vectors (matrix **M**) and spreads (matrix) of the hidden RBF nodes, to the association of the hidden nodes to output classes (matrix **W**), and to the association of input dimensions to hidden nodes (matrix **A**), i.e., which subspace of input space need to be considered for each hidden node for maximum performance in clustering. The values of matrices **M**, and **W** are estimated during the formal training of the RBF network (see Section 2.2) while the matrix **A** is computed by applying GAs optimization to the (already) trained RBF network (see Section 3).

2.1 Unsupervised Clustering of High Dimensional Data

In this work, we extend the classic agglomerative clustering algorithm in order to incorporate soft feature selection in the inter cluster distance estimation process, thus providing an output that is more effective (better results) and more efficient (faster convergence) for initializing the network. Let c_1 and c_2 be two clusters of data samples. Let also r_i, $i \in \{1..F\}$ be a distance metric defined in space $\mathbb{R}^{S_i} \subseteq \mathbb{R}^S$, F the count of distinct metrics that may be defined among a pair of clusters, S the count of features for the data samples and S_i the count of features considered by the i-th sample-to-sample distance metric. A distance metric between the two clusters, when considering the i-th sample-to-sample distance metric, is given by $f_i(c_1,c_2) = \sqrt[\kappa]{\frac{\sum_{a \in c_1, b \in c_2} \left(r_i(\underline{a_i}, \underline{b_i})\right)^{\kappa}}{|c_1| \cdot |c_2|}}$ where $\underline{a_i}$, $\underline{b_i}$ are the positions of data samples a and b

in feature space $\mathbb{R}^{S_i}$, $|c_1|$, $|c_2|$ are the cardinalities of clusters c_1 and c_2 respectively and $\kappa \in \mathbb{R}$ is a constant. The "context" is a selection of features that should be considered when calculating an overall distance value; we define it as a vector $\underline{ctx} \in \mathbb{R}_+^F$ with $\sum_{i=1}^{F} ctx_i = 1$. Given a context, the overall distance between clusters c_1 and c_2 is calculated as $f^*(c_1,c_2) = \sum_{i=1}^{F} (ctx_i(c_1,c_2))^{\lambda} \cdot f_i(c_1,c_2)$. In the non-trivial cases the optimal context is provided by $ctx_i(c_1,c_2) = ctx_F(c_1,c_2) \cdot \left(\frac{f_F(c_1,c_2)}{f_i(c_1,c_2)} \right)^{\frac{1}{\lambda-1}}$, $\forall i \in \{1..F-1\}$ and $ctx_F(c_1,c_2) = \frac{1}{\sum_{i=1}^{F} \left(\frac{f_F(c_1,c_2)}{f_i(c_1,c_2)} \right)^{\frac{1}{\lambda-1}}}$. For the sake of space, the proof for this is omitted and the reader is directed to [4] for more details on the proposed extension to the agglomeration process.

2.2 RBF Network Initialization and Training

Learning is incorporated into the network using the gradient descent method, while a squared error criterion is used for network training. The squared error $e(t)$ at iteration t is computed in the standard way: $e(t) = \frac{1}{2}\sum_{k=1}^{p} (d_k(t) - y_k(t))^2$ where $d_k(t)$ is the desired output and $y_k(t)$ is the output of neuron k given by $y_k(t) = \frac{1-e^{2z_k}}{1+e^{2z_k}}$, $z_k = (\underline{w_k})^T \cdot \underline{\phi(t)}$ where $\underline{w_k} = [w_{k1}, w_{k2} \dots w_{kq(t)}]^T$ are the weights connecting the RBF hidden neurons with the output neurons (note that these parameters are constrained to have binary values so as to better accommodate the extraction of if-then rules) and $\underline{\phi(t)}$ is the output of the hidden layer. For the sake of space, the reader is directed to [6] for more details on the training of the RBF network.

3 Derivation of the Antecedent Part of If-Then Rules Using Genetic Algorithms

The last step for the creation of if-then rules is the derivation of the antecedent part through the estimation of matrix **A.** The initialization of the hidden layer based on the results of the clustering method, described in Section 3, supports the construction of clusters with similarities in subspaces of the input space. However, when the training of the RBF network concludes all inputs are connected to all hidden neurons, thus all values of **A** matrix are set to one. In order to construct a proper antecedent part several input connections to hidden neurons need to removed. Moreover, these connections need, in general, to be different for each hidden node so as allow us to consider each hidden neuron as an if-then rule. In order to accommodate the above requirement a genetic algorithm optimization procedure is followed as described below.

Let ϕ_i be the activation of *i*-th hidden neuron and **I** be the set of data samples of training set **E** (which in addition to **I** contains the corresponding target vectors). Let also S_i be the subset of **I** ($S_i \subset I$) such that every data sample belonging to it ($\forall \underline{x} \in S_i$) activates the most the *i*-th hidden neuron. The aim of the training is to find a string that optimizes the activation ϕ_i over set S_i. For this purpose a genetic algorithm (GA) optimization scheme is used. We utilize a "per rule" feature selection methodology for the construction of the antecedent part of the rules. The coding that has been selected models the presence or absence of the corresponding input dimension in the antecedent part of a rule as 1 or 0 respectively. The fitness function F that is used is given by $F(S_i) = \frac{1}{card(S_i)} \sum_{\underline{x} \in S_i} \phi_i(\underline{x})$ where *card*(S_i) is the cardinality of set S_i and $\phi_i(\underline{x})$ is the activation of the *i*-th hidden neuron when fed by the input vector $\underline{x}$. The objective is to find the binary string that maximizes the fitness function $F(S_i)$. The realization of the genetic operators reproduction, mutation and crossover is as follows: ***Reproduction***. The fitness function $F(S_i)$ is used in the classical "roulette" wheel reproduction operator that gives higher probability of reproduction to the strings with better fitness according to the following procedure: i) an order number, q, is assigned to the population strings. That is q ranges from 1 to Np, where Np is the size of population, ii) the sum of fitness values (Fsum) of all strings in the population is calculated, iii) the interval [0 Fsum] is divided into Np sub-intervals each of one being, iv) a random real number R0 lying in the interval [0 Fsum] is selected, v) the string having the same order number as the subinterval of R0 is selected and vi) steps (4) and (5) are repeated Np times in order to produce the intermediate population to which the other genetic operators will be applied. ***Crossover***. Given two strings of length k (parents) an integer number $r \in \aleph_k$ is randomly selected. The two strings retain their gene values up to gene r and interchange the values of the remaining genes creating two new strings (offspring). ***Mutation***. This operator is applied to each gene of a string and it alters its content, with a small probability.

4 Experimental Results

The *iris* data were used to validate the proposed method as far as the rule extraction efficiency and the classification performance is concerned. When trained with the iris data the proposed combination of RBF-RAN and GAs creates three rules and achieves an overall classification performance of 96.7%. The estimated matrices are given below:

$$\mathbf{M}=\begin{bmatrix}0.66 & 0.59 & 0.50\\ 0.30 & 0.28 & 0.34\\ 0.56 & 0.43 & 0.15\\ 0.20 & 0.13 & 0.03\end{bmatrix},\ \mathbf{\Sigma}=\begin{bmatrix}0.13 & 0.10 & 0.07\\ 0.06 & 0.06 & 0.08\\ 0.11 & 0.10 & 0.04\\ 0.05 & 0.04 & 0.02\end{bmatrix}\ \mathbf{A}=\begin{bmatrix}0 & 0 & 1 & 1\\ 1 & 0 & 0 & 1\\ 0 & 1 & 1 & 0\end{bmatrix},\ \mathbf{W}=\begin{bmatrix}1 & 0 & 0\\ 0 & 1 & 0\\ 0 & 0 & 1\end{bmatrix}$$

We should note that the pruning of input to hidden nodes connections due to the GA optimization has no degradation effect on the classification performance which remains 96.7%. Comparisons with other methods (with the help of [5]), as far is the classification performance (resubstitution accuracy), are given in Table I. We observe that the proposed method (RBF-GAs) outperforms all listed soft computing techniques, with the exception of FuGeNeSys, creating as few as three rules. More results

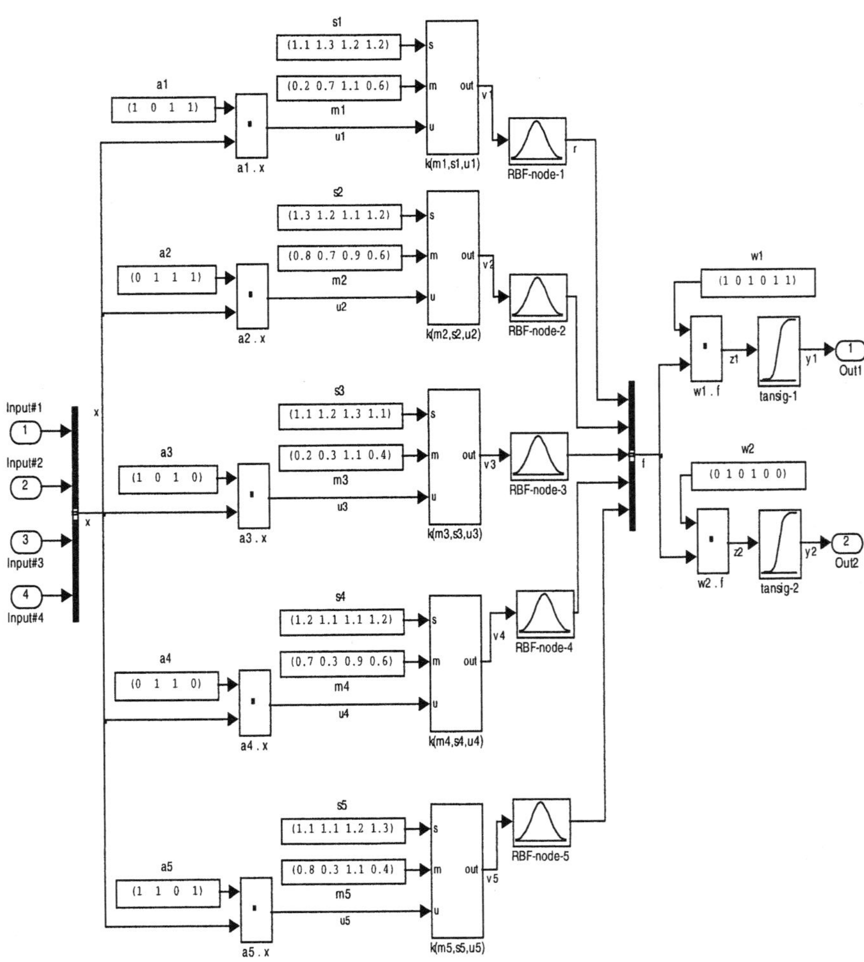

Fig. 1. The proposed RBF architecture for rule extraction

Table 1. Comparison of RBF-GAs with other techniques for Iris Data Classification

Method	Rules	Resubstitution Accuracy (%)
FuGeNeSys	5	100
NEFCLASS	7	96.7
ReFuNN	9	95.3
EFuNN	17	95.3
FuNe-I	7	96.0
RBF-Gas	3	96.7

from the application of the proposed methodology are available and equally promising, but are omitted for the sake of space.

5 Conclusions

In this paper we propose an innovative hybrid architecture, which combines resource allocating properties, novel clustering techniques for multi-dimensional problems and evolutionary weight connection purging in order to incorporate (a) fast classifying capabilities, (b) expert knowledge modeling, (c) knowledge extraction from numerical data. The proposed approach embeds rule-based knowledge directly into its architecture while its resource allocating structure enables new rules to be created. The latter is very important for two reasons: (a) there are several domains in which no estimation about the number of rules that are required to solve a particular problem is available, (b) rules can be created to model a changing of a context.

References

1. J. Yang, V. Honavar, "Feature Subset Selection Using A Genetic Algorithm," ACM computing 1991
2. D. Addison, S. Wermter, G. Arevian, "A Comparison of Feature Extraction and Selection Techniques," *Proceedings of the International Conference on Artificial Neural Networks (ICANN'03),* Istanbul, Turkey, Supplementary Proceedings pp. 212-215, June 2003.
3. Growing Compact RBF Networks Using a Genetic Algorithm," VII Brazilian Symposium on Neural Networks (SBRN'02) November 2002
4. M. Wallace, S. Kollias, "Robust, Generalized, Quick and Efficient Agglomerative Clustering" Proceedings of 6th International Conference on Enterprise Information Systems (ICEIS), Porto, Portugal, April 2004.
5. L. I. Kuncheva and J. C. Bezdek, "Nearest prototype classification: Clustering, genetic algorithms, or random search?" *IEEE Trans. Syst.*
6. Wallace M., Tsapatsoulis N., Kollias S. "`Intelligent Initialization of Resource Allocating RBF Networks" Neural Networks, 2005

Neural Network Algorithm for Events Forecasting and Its Application to Space Physics Data

S.A. Dolenko, Yu. V. Orlov, I.G. Persiantsev, and Ju. S. Shugai

D.V.Skobeltsyn Research Institute of Nuclear Physics,
M.V.Lomonosov Moscow State University,
SINP MSU, Vorobjovy Gory, Moscow, 119992, Russia
Dolenko@srd.sinp.msu.ru

Abstract. Many practical tasks require discovering interconnections between the behavior of a complex object and events initiated by this behavior or correlating with it. In such cases it is supposed that emergence of an event is preceded by some phenomenon – a combination of values of the features describing the object, in a known range of time delays. Recently the authors suggested a neural network based method of analysis of such objects. In this paper, the results of experiments on real-world data are presented. The method aims at revealing morphological and dynamical features causing the event or preceding its emergence.

1 Introduction

The task of search of correlations in multi-dimensional time series as one of the tasks of spatiotemporal image analysis is very topical [1]. A significant condition of such problems being solved at present by various methods is zero delay between the cause (or precursor) of the event and the event itself (or this delay is known and fixed). One more important feature is the opportunity to perform "active investigation" of dependences, i.e. to vary input conditions and to record the response. At the same time, only the acceptable delay range is given in many problems. Thus, search for temporal correlations should be performed in a sufficiently wide time range. One more obstacle is inability to influence the object of investigation ("passive observation").

A demonstrative example is the problem of forecasting geomagnetic storms by revealing those phenomena on the Sun surface that initiate the storm or precede its emergence. Similar tasks also exist in other areas – medicine, seismology, finance etc. In [2,3], the authors have suggested a method for analysis of multi-dimensional time series with the purpose of forecasting occurrence of some *events* and finding their precursors – *phenomena*, i.e. some unknown combinations of the values of the features describing the object. The algorithm is based on the use of a committee of neural networks (NN) trained on different segments of the analyzed time series. A significant feature of the developed algorithm is the possibility of searching non-linear interconnections between the event and the phenomenon.

W. Duch et al. (Eds.): ICANN 2005, LNCS 3697, pp. 527 – 532, 2005.

2 Description of the Algorithm

Let us assume that in the analyzed multi-dimensional time series, possible range of delays between an event and its precursor (from T_{min} to T_{max}, Fig.1) is set *a priori.* This range determines *search interval* $\Delta T = T_{max} - T_{min}$, during which one should reveal a phenomenon (i.e., a combination of the input features) that has initiated the event. Let us further assume that for the event initiation, existence of phenomenon during some specific time interval (*initiation interval* T_{init}) is required. The initiation interval is estimated from *a priori* considerations as well; it should exceed the duration of the sought phenomenon. Finally, assume that the length of the event is much smaller than the search interval. The goal is to find the phenomenon inside the search interval that is the one most probably responsible for the event initiation, and to determine the delay between the event initiation and the event itself.

To accomplish this task, the analyzed search interval is split into overlapping *segments* with length equal to initiation interval T_{init}. Relative position of neighboring segments is the same and it is characterized by the *overlapping interval.* A separate neural network (NN), which is trained to forecast the event based on the features within this segment, is built for each segment. The number of inputs for each NN is equal to the product of the segment length and the dimensionality of the input time series. The number of NN is calculated from the lengths of the search interval and the overlapping interval, so that the search interval is always covered by overlapping data segments, each of which is used as the source of input data for its corresponding NN.

During training, the search interval is shifted along the time axis. When the right border of the search interval is situated at the minimum possible delay between the phenomenon and the moment of the event emergence, the desired output of the NN for each of the segments is set to 1. In such case, the input variables for one or several neighboring NNs with overlapping segments contain the phenomenon that has initiated the event. As the delay between the phenomenon and the event is assumed to be constant, for each event the phenomenon falls into the segments belonging to the same NNs.

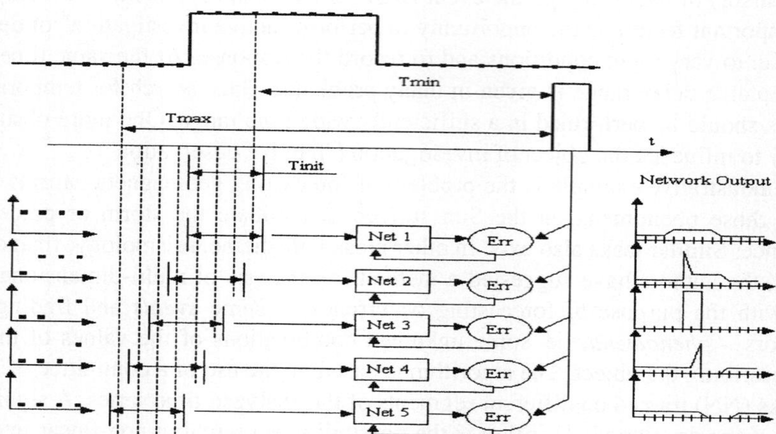

Fig. 1. A 1-dimensional illustration of the algorithm. The event is initiated by 0-1 transition.

In all other cases, the desired output is set to zero. Thus, different NNs of the committee, having the same desired output, are fed each with is own segment of the analyzed multi-dimensional time series as the input data. In this situation those of the networks whose segments happen to contain the precursor (or at least its part), will demonstrate more efficient learning than the networks whose input data is in no way connected with the event (most probably the latter will not learn at all).

When the training is over, one may make a conclusion that the sought phenomenon (the precursor of the event) falls into the segment of the network that gives the most exact prediction of the event (preferably on independent data) as the result of the training. Shifting the search interval along the time axis and applying the set of the NNs to corresponding segments of the analyzed time series, one can forecast the occurrence of an event. The delay between the phenomenon and the event is thus determined with the precision not worse than the length of the initiation interval. Further refinement of the delay value may be achieved by change of the overlapping interval of the segments corresponding to different networks.

The described algorithm has been applied to a set of model tasks. The results of model experiments demonstrate efficiency and high potential of the suggested approach [4,5].

3 Results of Experiments with Space Physics Data

To solve real-world problems, a modification of the algorithm was developed, which forecasts continuous value rather than a binary one. Using a committee of neural networks trained on different segments of time series makes it possible (as in binary version of the algorithm) to discover nonlinear interrelations between the values of the forecasted variable and the input features, as well as to determine which time interval is the most appropriate one for phenomena search.

3.1 Forecasting Geomagnetic Dst index

This experiment was devoted to forecasting hourly values of the geomagnetic storm index Dst [6]. The values of the Dst index were provided by WDC-C2 KYOTO [7]. It is known that the development of a geomagnetic storm depends primarily on two solar wind parameters: B_z-component of the interplanetary magnetic field (IMF) and solar wind velocity V. The parameters of the solar wind were recorded by the satellite ACE [8], located in the point of gravitational equilibrium between the Sun and the Earth. The characteristic duration of the disturbance process registered by the ACE satellite is about several hours.

The values of the solar wind velocity and of the B_z-component of the IMF for 2000 - 2003 years were used as input data. The data was split into training set (2000 - 2001), test set (2002) and examination set (2003). The search interval was 24 hours, the initiation interval was 8 hours, and t he overlapping interval was 4 hours. A committee of 5 neural networks (three-layer perceptrons with 10 hidden neurons) was used.

The algorithm discovered the correct range of possible delays between the phenomenon (being the precursor of geomagnetic storm) and the event (i.e. storm itself).

Namely, it was automatically determined that the values of the solar wind parameters leading to a geomagnetic storm are registered by the satellite at 12 hours or less before the beginning of a geomagnetic storm. According to the results obtained on examination set (Fig.2), correlation coefficient between actual and forecasted values of Dst-index for the best network in the committee is 0.7. This may indicate that the sought phenomenon (the precursor of the event) may be described more adequately with an extended set of input features. Nevertheless, taking into consideration the complexity of the task, these results demonstrate the effectiveness of the approach.

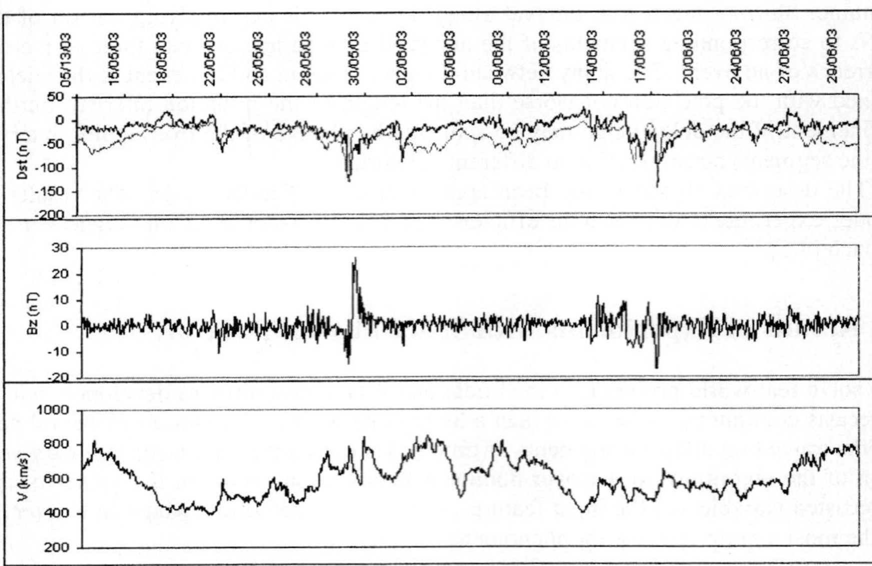

Fig. 2. Results of forecasting Dst index on examination set (May 13 2003 to June 30 2003). Top to bottom: Dst index of geomagnetic activity (dark curve – actual values, light curve – forecasted values); Bz-component of interplanetary magnetic field; solar wind velocity V.

In further experiments, hourly values of the solar wind velocity, of the B_x, B_y, B_z - components of the IMF, and of the solar wind density for 2000 - 2003 years were used as the input data. The algorithm efficiency increased on the higher dimensional data (the correlation coefficient for the best NN approached 0.8).

During this experiment, an attempt was made to identify the role of each solar wind parameter by calculating its contribution factor. The contribution factors were based on the analysis of the weights of perceptron (NN). Such factors are a rough measure of the significance of each input variable relatively to other input variables in the same NN. The higher is the value of the contribution factor, the more the corresponding variable is contributing to the prediction. The B_z - component of the IMF is the most important parameter. Influence of solar wind velocity and density is practically equal and exceeds that of B_x, B_y - components of the IMF. The obtained estimations of significance of the solar wind parameters are in partial agreement with those reported in [6].

3.2 Forecasting Solar Wind Velocity

Periodic increase of solar wind velocity may be correlated with emergence of the so-called large coronal holes (structures on the Sun surface) next to central Solar meridian. A specialized algorithm for automatic detection of coronal holes on the Sun images was used for processing daily snapshots of the Sun, made by the telescope EIT (Extreme ultraviolet Imaging Telescope) from the satellite SOHO (Solar & Heliospheric Observatory) at 284Å wavelength, for the 1997 - 2004 years. The database contains daily data about total area of coronal holes in the central region of the Sun image. With these data, the values of solar wind velocity were forecasted using the suggested algorithm. The parameters of the solar wind were taken from the satellite ACE [8].

Training set contained data from January 1, 2002 to June 30, 2003, test set contained data from July 1, 2003 to December 31, 2003, and examination set contained data from January 1, 2004 to September 1, 2004. Search interval was set to 10 days, initiation interval to 3 days, and overlapping interval to 2 days.

The algorithm determined automatically the range of delays between the moment of passing of the central Sun meridian by equatorial coronal hole and the moment of increasing of the solar wind velocity near the Earth orbit. Namely, it was found that the delay corresponding to the maximum of multiple determination coefficient (R-squared) was in the range of 1 to 3 days. The results on examination set are presented at Fig.3. Correlation coefficient between actual and forecasted values of solar wind velocity for the best network in the committee is 0.58.

The results obtained in this experiment were compared to the results obtained by alternative methods for 1997 - 2004 years. Correlation coefficient between solar wind velocity and total area of coronal holes (shifted for the period of 1 to 6 days) was calculated. It was determined that the delay corresponding to the correlation coefficient maximum, varies from year to year, and is 3-4 days on average, what was in good agreement with the results obtained using the developed algorithm.

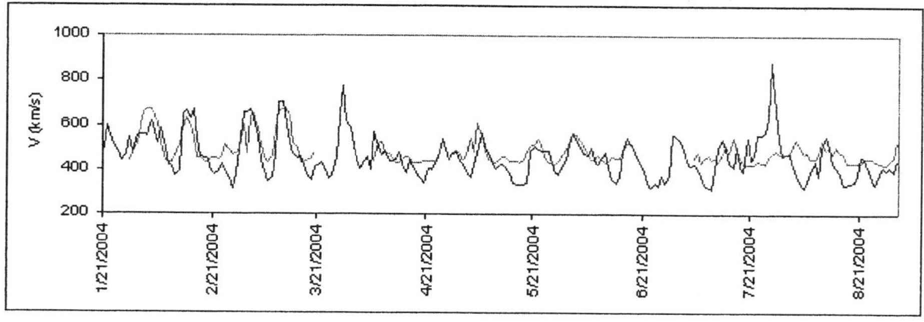

Fig. 3. Results of forecasting solar wind velocity on examination set (January 1 2004 to September 1 2004). Dark curve – actual values, light curve – forecasted values.

4 Conclusion

Recently, the authors have suggested a method for analysis of multi-dimensional time series with the purpose of forecasting occurrence of some events and finding their precursors – phenomena, i.e. some unknown combinations of the values of the features describing the object. The algorithm is based on the use of a committee of neural networks trained on different segments of the analyzed time series. A significant feature of the developed algorithm is the possibility of searching non-linear interconnections between the event and the phenomenon. The approach may be used both for forecasting binary and continuous values.

The suggested approach was also applied to real world problems. First, geomagnetic storm index Dst was forecasted using B_z-component of the interplanetary magnetic field and solar wind velocity. Second, the values of solar wind velocity were forecasted using daily data about total area of coronal holes in the central region of the Sun image. These data were obtained using a specialized algorithm processing daily snapshots of the Sun. In both experiments, the algorithm correctly determined the range of delays between the precursor and the event, and correlation coefficient between actual and forecasted values on independent data set was 0.7 and 0.58 for the first and second task, respectively.

The results obtained in experiments on model and real world problems demonstrate the perspective of the approach. At the next step of the algorithm development, a set of input features describing a phenomenon should be refined automatically once a range of delays containing precursor is found. Gradual reduction of the input data dimensionality along with increasing the precision of delay determination will increase the algorithm efficiency on high dimensional data.

This work has been performed under financial support of the following institutions: Russian Foundation for Basic Research (RFBR), project no. 04-01-00506.

References

1. Cabrera, J.B.D., Mehra, K.R.: Extracting precursor rules from time series – A Classical Statistical Viewpoint. In: Proc. 2nd SIAM Int. Conf. on Data Mining (SDM-2002). April 11-13, 2002, Hyatt Regency, Crystal City at Ronald Reagan National Airport, Arlington, VA
2. Dolenko, S.A., Orlov, Yu.V., Persiantsev, I.G., Shugai, Ju.S.: Discovering temporal correlations by neural networks. Pattern Recognition and Image Analysis 13 (2003) 17-20
3. Orlov, Yu.V. et al. Nuclear Instruments and Methods in Physics Research Section A (NIMA A) 502 (2003) 532-534
4. Dolenko, S.A. et al: A Search for Correlations in Time Series by Using Neural Networks. Pattern Recognition and Image Analysis 13 (2003) 441-446
5. Shugai, Ju.S. et al. Neural network algorithm for events forecasting in multi-dimensional time series and its application for analysis of data in space physics. In: Proc. 7th Int. Conf. on Pattern Recognition and Image Analysis (PRIA-7-2004), October 18-23, 2004, St.Petersburg, Russia, 3 (2004) 908-911
6. Watanabe, Sh. et al. J. Communications Research Laboratory 49 (2002) 69-85
7. http://swdcwww.kugi.kyoto-u.ac.jp/dstdir/dst1/final.html
8. http://www.srl.caltech.edu/ACE

Counterpropagation with Delays with Applications in Time Series Prediction

Carmen Fierascu

Institute of Scientific Computing, University of Salzburg, Austria
cfierasc@cosy.sbg.ac.at
wwww.sbc.ac.at

Abstract. The paper presents a method for time series prediction using a complete counterpropagation network with delay kernels. Our network takes advantage of the clustering and mapping capability of the original CPN combined with dynamical elements and become able to discover and approximate the strongest topological and temporal relationships among the fields in the data. Experimental results using two chaotic time series and a set of astrophysical data validate the performance of the proposed method.

1 Introduction

Over the last two decades, a wide variety of new techniques for analyzing and manipulating time series have been proposed. Neural Networks have become one of the most popular analytical tool in providing solutions to difficult problems. The complexity of useful neural network structures varies from the simple adaline to multi-dimensional arrays with full interconnection utilizing thousands of neurons and multiple feed-back paths. Many neural networks models for time series prediction have been reported: standard feedforward networks (MLP), Elman networks, RBF networks, TDNN, Finite Impulse Response (FIR) networks, with the latter having the distinction of being the winning entry in the *Santa Fe Institute Time Series Prediction Competition* [10].

Kohonen's Self-Organizing Map (SOM) [1] has also found practical application in temporal sequence processing. It is a difficult problem to implement selective responses to dynamical phenomena into the simple SOM, but several extensions using short-term memories and feedback loops have been proven as workable.

The counterpropagation network (CPN) was developed from the instar-outstar model as a network that self-organizes itself to implement an approximation to a function. CPN was successfully used for function approximation, pattern recognition, statistical analysis and data compression [3]. In our previous work [4], we extended the approximation capabilities of the original forward-only CPN by the implementation of delay kernels in its competitive layer and applied the new method to the prediction of currency exchange rates.

In this work we further extend the new method, by implementing delay kernels in the complete version of the original CPN. The new complete counterpropagation network with delays (CCPND) will be validated first by approximating the complex form of the radial velocity function of a spiral galaxy. As a second task, the prediction of chaotic time series will be approached.

W. Duch et al. (Eds.): ICANN 2005, LNCS 3697, pp. 533–539, 2005.

2 Counterpropagation Network with Delays

The CPN architecture was synthesized by Hecht-Nielsen from the Self-Organizing Map of Kohonen and Grossberg's outstar structure [2]. The basic idea is that, during adaptation, a given set of vector pairs $(x_1, y_1), (x_2, y_2), \ldots, (x_L, y_L)$ are presented to the network at the two opposite input layers and then propagate through the network in a counterflow manner to give as return the output vectors **x'** and **y'** as approximations of the input vectors. A schematic representation is shown in Fig.1. The role of SOM is

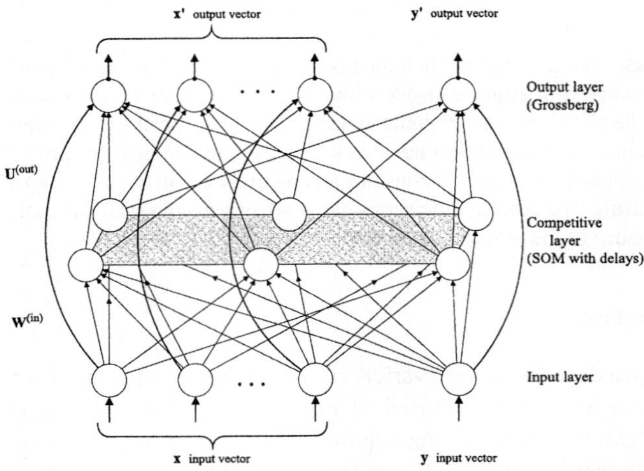

Fig. 1. Architecture of the counterpropagation network with delays

to perform an initial clustering of the input data and to attain a spatial ordering of the map, in the sense that similar input patterns will produce response in neighboring units. Thus, the asymptotic values of the weight vectors will define the vector quantization of the pattern space. But the original SOM idea was based on matching of static signal patterns only. Bringing temporal dependence into the model would extend its capabilities, making it better suited to a variety of tasks such as speech recognition, system identification, time series prediction and other problems where dynamical systems are involved. Therefore we expanded the SOM original learning rule by adding a delay kernel [4], defined as a locally integrable function, at least piece-wise continuous $k(t)$, with a finite L_1 norm and further assume that the best matching units and its' neighbors will change their weights according to the rule

$$\frac{dw(t)}{dt} = -\alpha(t)[w(t) * k(t) - x], \quad 0 < \alpha(t) < 1 \tag{1}$$

where $*$ denotes the convolution and $\alpha(t)$ is the learning rate, that decreases exponentially as learning proceeds. By taking the discrete form of equation (1) and inserting the neighborhood function $h_{U_{BMU(x)}}(t)$, we obtain the following learning rule:

$$w_j(t+1) = w_j(t) - \alpha(t)h_{U_{BMU(x)}}(t)[w_j(0)k(t)+ \\ +w_j(1)k(t-1) + \ldots + w_j(t-1)k(1) - x(t)] \quad (2)$$

where $j = 1, \ldots, N$ with N the dimension of the input space and $\alpha(t)$ is the learning rate which decreases as the learning proceeds. Considering the equation (2) as governing the dynamics of the units in the competitive layer of our CPND, the weights will be allowed to depend also on their past values.

In our previous work [4] we have introduced the forward-only version of the CPND. The CCPND is obtained in a similar way. Both **x** and **y** input vectors are fully connected to the competitive layer and the associated weight vectors w_x and w_y contribute jointly to the calculation of the winner unit. As with the forward network, only the winning unit and its neighbors are allowed to learn for a given input. The weights updating rule is the same as in equation (2) with j covering the extended input space. After the competitive layer has stabilized (the **x-y** vectors have clustered), the w_j vectors are frozen and the Grossberg layer begins to learn.

Like the input layer, the output layer is split into two corresponding parts. The **y'** units have weight vectors $\mathbf{u}_{yi}$ and the **x'** units have weight vectors $\mathbf{u}_{xi}$. The learning laws are:

$$u_{yij}(t+1) = u_{yij}(t) + \beta[y_i - u_{yij}(t)]v_j, \quad 0 < \beta \ll 1 \quad (3)$$

with outputs

$$y'_i = \sum_{j=1}^{N} u_{yij}(t+1)v_j \quad (4)$$

where β is a constant learning rate, v_j is the output signal of the competitive layer and i indexes the component of the desired vector. The equations for u_{xij} and corresponding outputs x'_i have similar forms.

3 Case Studies

In order to evaluate the performance of the proposed network, we used a set of radial velocity measurements of a spiral galaxy and the popular Mackey-Glass and laser generated data.

Multiple CPND architectures for each data set were tested according to the following procedure: First, the training data is normalized to zero mean and variance 1. The last part of observations is reserved for out-of-sample testing periods. The training patterns are obtained by sliding a window through the series. The selection of window widths was based on trial and error along with various heuristics i.e. considering the average mutual information of the given data [5].The order of the input vector has to be large enough to resolve ambiguities, but also cope with "the curse of dimensionality" [7].

Each a CPND and a CCPND are trained using the past few values as **x** vector and the next value as **y** vector. For the neighboring function $h_{U_{BMU(x)}}(t)$ from equation 2, the "Winner Takes All" principle has proven as the best choice. After the network is trained, a long-term iterated prediction is achieved by taking the estimate $\hat{y}$ and feeding it back as input to the network. The normalized root mean square error (NRMSE) is used as the performance measure:

$$NRMSE = \frac{\sqrt{\frac{1}{P}\sum_{p=1}^{P}(y_p - \hat{y}_p)^2}}{\sigma} \tag{5}$$

where P is the number of elements in the test set and σ the standard deviation of the target series. If $NRMSE \approx 0$, the prediction is almost perfect, whereas a $NRMSE$ value equal to 1 is equivalent to using the average as the predictor. Since training is based on only single step predictions, the accuracy of the long iterated predictions cannot be guaranteed. For comparison purposes, we use a MLP with one hidden layer and hyperbolic tangent transfer function, trained with the backpropagation learning algorithm. The architectures (nr. of neurons per layer), the number of training epoches and simulation results are shown in table 1.

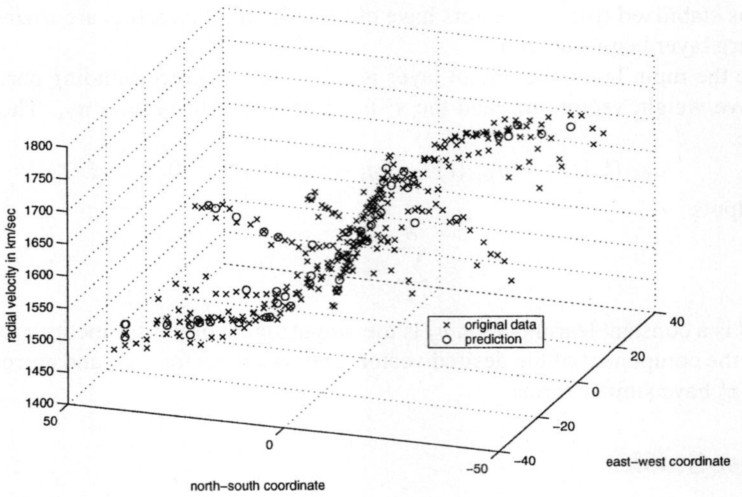

Fig. 2. The distribution of radial velocity points of the galaxy NGC7531 versus position on the celestial sphere

3.1 NGC 7531 Galaxy Data

The galaxy NGC 7531 is an example of a nonbarred spiral in the Southern Hemisphere, possessing a very bright inner ring. R. Buta has explored the properties of all types of ringed galaxies, inspired by the belief that the ringlike patterns could be useful probes of the galaxy dynamics and structure [8].

The dynamics of a galaxy is governed by cooperative gravitational interactions of stars, gas and dark matter. The dominating motion of the disk stars is differential rotation around the center of the galaxy at an angular velocity depending on the Galactocentric distance [9]. Other systematic motions are superimposed on the rotation, as those induced by spiral density waves. Apart from the systematic motion, the stars are also involved in random motions (residual velocities). Conclusively, the rules governing the motion of a spiral galaxy are nonlinear and very complex. The measured data is noisy and could be incomplete. An estimation of the radial velocity for regions where direct measurements are not possible, could be an auxiliary tool for further analysis, since the rotation curve provides the most direct method of measuring the mass distribution of a galaxy.

The variables used for simulation are: the **east-west** coordinate of the velocity measurement, the **north-south** coordinate and the radial **velocity** measured in km/sec.

From the total amount of 323 measurements, 50 points were reserved alternatively for test purposes. The original data and a set of 50 predictions are shown in the Fig.2. The CPND has given slightly better predictions. Consequently, the inclusion of the velocity in the training of the competitive layer doesn't improve the generalization capability of the network.

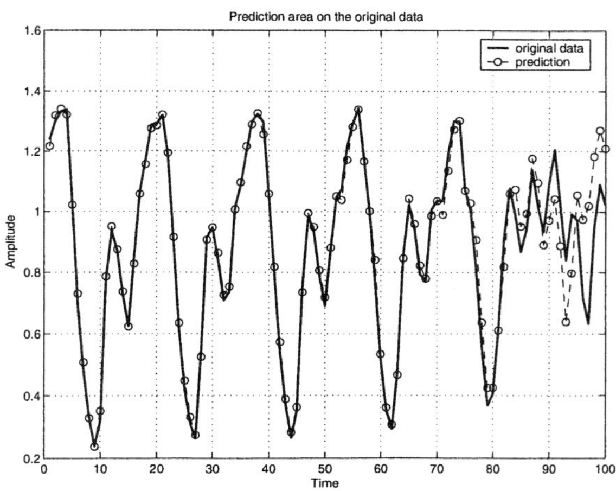

Fig. 3. Sequence of 100 iterative estimated samples of the Mackey-Glass data

3.2 Mackey-Glass Time Series

We consider the well known Mackey-Glass delay-differential equation $dx(t)/dt = -0,1x(t) + [0,2\,x(t-\Delta)]/[1+x(t-\Delta)^{10}]$ which is a model of dynamic respiratory and hematopoiectic diseases [11]. The delay was set to $\Delta = 30$, with initial conditions $x(t) = 0.9$ for $0 \leq t \leq \Delta$ and sampling rate $\tau = 6$. The generated set contains 1500 points of the series. 1400 points are used for training and the next 100 points to validate the iterative prediction capacity of the two networks. This time the CCPND has slightly outperformed the CPND. The prediction area is shown in Fig. 3.

3.3 Laser Generated Data

The data was recorded from a Far-Infrared-Laser in chaotic state and was distributed as part of the *Santa Fe Competition* [10]. The time series shows a cross-cut through periodic to chaotic intensity pulsations of the NH3 laser. Since the size of the provided

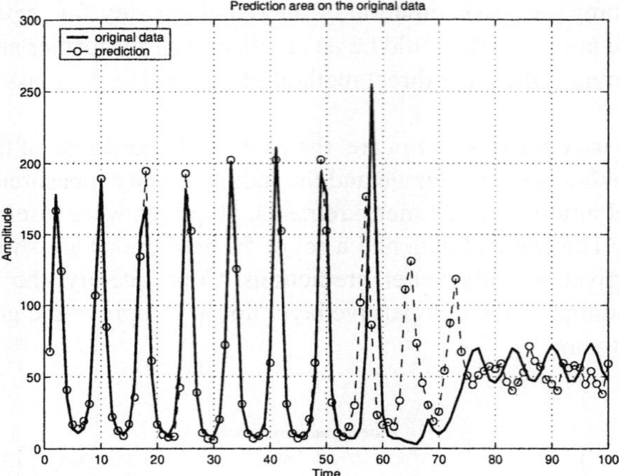

Fig. 4. Sequence of 100 iterative estimated samples of the Laser data

Table 1. Simulation results

Model	Data	Architecture			Training epoches		NRMSE
		Input	Instars/Hidden	Output	Instars	Outstars	
CPND	Galaxy	2	75	1	150	150	0,125
CCPND	Galaxy	3	75	1	150	150	0,157
MLP	Galaxy	2	18	1	20000		0,138
CPND	Mackey-Glass	5	600	1	200	200	0,281
CCPND	Mackey-Glass	6	600	1	200	200	0,268
MLP	Mackey-Glass	5	24	1	10000		1,078
CPND	Laser	4	300	1	200	200	0,744
CCPND	Laser	5	300	1	200	200	0,726
MLP	Laser	4	28	1	10000		1,044

data set was 1000 points, only two collapses in the amplitude are shown. A prediction of the next collapse based on so few instances is a difficult task. None of our networks, trained for one-step ahead prediction has been able to predict the intensity collapse at the point 1065, based only on the first 1000 data points, although the predictions up to the point 1050 were almost perfect. Alternatively we also tested networks trained for a multi-steps *direct prediction*, but these alone gave even worse results. We have adopted the method used by Tim Sauer in his competition entry [10] and combined both approaches in order to optimize the prediction. With each of our models we have trained two nets accordingly. In the simulation phase, a weighted combination of both outputs is used to build the prediction, which is subsequently fed back in both networks' inputs. The prediction accuracy has been improved by 20%.

4 Conclusion

This paper presented an extended version of the counterpropagation neural network with delays, as described in [4]. The method was validated by using examples drawn from the *Santa Fe Institute Time Series Prediction Competition*, computer generated chaos and astrophysical data. The simulations results illustrate the iterative prediction potential of the presented method, compared with those produced by a MLP network. The proposed technique needs to be further consolidated, both in theoretical and practical terms, in order to be become a viable alternative to classical supervised models.

References

1. Kohonen, T.: Self-Organizing Maps. Springer-Verlag. Berlin. (2001)
2. Hecht-Nielsen, R.: Countepropagation networks. Applied Optics. Vol. 26. (1987) 4979–4984
3. Hecht-Nielsen, R.: Applications of Countepropagation Networks. Neural Networks. Vol. 1. (1988) 131–139
4. Fierascu, C. and Badea, C. L.: Counterpropagation with delays for Financial Prediction. Proceedings of IEEE International Joint Conference on Neural Networks. Budapest. (2004)
5. Abarbanel, H.D.I.: Analysis of Observed Chaotic Data. Springer-Verlag. New York. (1996)
6. Casdagli, M.: Nonlinear prediction of Chaotic Time Series. Physica D. Vol.35. (1989) 335–356
7. Hastie, T., Thibshirani, R. and Friedman, J.: The Elements of Statistical Learning. Springer-Verlag. New York. (2001)
8. Buta, R.: The Structure and Dynamics of Ringed Galaxies. III: Surface Photometry and Kinematics of the Ringed Nonbarred Spiral NGC7531. The Astrophysical J. Supplement Ser. 64 (1987) 1–37
9. Binney, J. and Tremaine, S.: Galactic Dynamics. Princeton University Press. New Jersey. (1987)
10. Weigend, A. S. and Gershenfeld, N. A.: Time Series Prediction: Forecasting the Future and Understanding the Past. Addison-Wesley. Reading. MA. (1993)
11. Wan, Eric A.: Finite Impulse Response Neural Networks with Applications in Time Series Prediction. A Dissertation. Stanford University. (1993)

Bispectrum-Based Statistical Tests for VAD

J.M. Górriz[1], J. Ramírez[1], C.G. Puntonet[2], F. Theis[3], E.W. Lang[3]

[1] Dept. Signal theory, University of Granada
[2] Dept. Architecture and Computer Tech., University of Granada,
Fuentenueva s/n, 18071 Granada, Spain
gorriz@ugr.es
Institute of Biophysics, University of Regensburg,
Universitätsstraße 31, D-93040 Regensburg, Germany
fabian.theis@mathematik.uni-regensburg.de

Abstract. In this paper we propose a voice activity detection (VAD) algorithm for improving speech recognition performance in noisy environments. The approach is based on statistical tests applied to multiple observation window based on the determination of the speech/non-speech bispectra by means of third order auto-cumulants. This algorithm differs from many others in the way the decision rule is formulated (detection tests) and the domain used in this approach (bispectrum). It is shown that application of statistical detection test leads to a better separation of the speech and noise distributions, thus allowing a more effective discrimination and a tradeoff between complexity and performance. The experimental analysis carried out on the AURORA databases and tasks provides an extensive performance evaluation together with an exhaustive comparison to the standard VADs such as ITU G.729, GSM AMR and ETSI AFE for distributed speech recognition (DSR), and other recently reported VADs. Clear improvements in Speech Recognition are obtained when the proposed VAD is used as a part of a ASR system.

1 Introduction

Speech/non-speech detection is an unsolved problem in speech processing and affects numerous applications including robust speech recognition [11], discontinuous transmission , real-time speech transmission on the Internet or combined noise reduction and echo cancellation schemes in the context of telephony [1,5][3]. The speech/non-speech classification task is not as trivial as it appears, and most of the VAD algorithms fail when the level of background noise increases. During the last decade, numerous researchers have developed different strategies for detecting speech on a noisy signal [12] and have evaluated the influence of the VAD effectiveness on the performance of speech processing systems. Most of them have focussed on the development of robust algorithms with special attention on the derivation and study of noise robust features and decision rules [14,7,8]. The different approaches include those based on energy thresholds, pitch detection, spectrum analysis, zero-crossing rate, periodicity measure, higher order statistics in the LPC residual domain or combinations of different features [8,5,1,10,14].

W. Duch et al. (Eds.): ICANN 2005, LNCS 3697, pp. 541–546, 2005.

This paper explores a new alternative towards improving speech detection robustness in adverse environments and the performance of speech recognition systems. The proposed VAD proposes a noise reduction block that precedes the VAD, and uses Bispectra of third order cumulants to formulate a robust decision rule. The rest of the paper is organized as follows. Section 2 reviews the theoretical background on Bispectra analysis and shows the proposed signal model. Section 2.1 introduces the Statistical Tests based on Biespectra employed on multiple observation in order to build a robust decision rule. Section 3 describes the experimental framework considered for the evaluation of the proposed algorithm. Finally, section 4 summarizes the conclusions of this work.

2 Model Assumptions

Let $\{x(t)\}$ denote the discrete time measurements at the sensor. Consider the set of stochastic variables y_k, $k = 0, \pm 1 \ldots \pm M$ obtained from the shift of the input signal $\{x(t)\}$:

$$y_k(t) = x(t+k) \tag{1}$$

where k is the differential delay (or advance) between the samples. This provides a new set of $2 \cdot M + 1$ vector variables $\mathbf{y}_j = \{y_j(t_1), \ldots, y_j(t_N)\}$ by selecting $i = 1 \ldots N$ samples of the input signal. It can be represented using an associated Toeplitz matrix. Using this model the speech-non speech detection can be described by using two essential hypothesis(re-ordering indexes):

$$H_o = \left(\mathbf{y}_0 = n_0; \mathbf{y}_{\pm 1} = n_{\pm 1}; \ldots; \mathbf{y}_{\pm M} = n_{\pm M}\right) \tag{2}$$

$$H_1 = \left(\mathbf{y}_0 = s_0 + n_0; \mathbf{y}_{\pm 1} = s_{\pm 1} + n_{\pm 1}; \ldots; \mathbf{y}_{\pm M} = s_{\pm M} + n_{\pm M}\right) \tag{3}$$

where s_k's/n_k's are the speech/non-speech (any kind of additive background noise i.e. gaussian) signals, related themselves with some differential delay (or advance). All the process involved are assumed to be jointly stationary and zero-mean. Consider the third order cumulant function $C_{y_k y_l}$ defined as $C_{y_k y_l} \equiv E[\mathbf{y}_0 \mathbf{y}_k \mathbf{y}_l]$, and the two-dimensional discrete Fourier transform (DFT) of $C_{\mathbf{y}_k \mathbf{y}_l}$, the bispectrum function:

$$\mathcal{C}_{\mathbf{y}_k \mathbf{y}_l}(\omega_1, \omega_2) = \sum_{k=-\infty}^{\infty} \sum_{l=-\infty}^{\infty} C_{\mathbf{y}_k \mathbf{y}_l} \cdot \exp(-j(\omega_1 k + \omega_2 l)) \tag{4}$$

Sampling the equation 4 and assuming a finite number of samples, the biespectrum estimate can be written as:

$$\hat{\mathcal{C}}_{\mathbf{y}_k \mathbf{y}_l}(n, m) = \sum_{k=-M}^{M} \sum_{l=-M}^{M} C_{\mathbf{y}_k \mathbf{y}_l} \cdot w(k,l) \cdot \exp(-j(\omega_n k + \omega_m l)) \tag{5}$$

where $\omega_{n,m} = \frac{2\pi}{M}(n, m)$ with $n, m = -M, \ldots, M$ are the sampling frequencies, $w(k,l)$ is the window function (to get smooth estimates) and $C_{\mathbf{y}_k \mathbf{y}_l} = \frac{1}{N} \sum_{i=0}^{N-1} y_0(t_i) y_k(t_i) y_l(t_i) = \frac{1}{N} \mathbf{y}_0 \mathbf{y}_k \mathbf{y}_l |_{t_0}$.

2.1 Detection Tests for Voice Activity

The decision of our algorithm is based on statistical tests including the Generalized Likelihood ratio tests (GLRT) [13] and the Central χ^2-distributed test statistic under H_O [4] applied to a multiple observation window. This imposes an M-frame delay which is not relevant in the great majority of the real applications, i.e. speech recognition. We will call them GLRT and χ^2 tests.

GRLT: Consider the complete domain in biespectrum frequency for $0 \leq \omega_{n,m} \leq 2\pi$ and define P uniformly distributed points in this grid (m, n), called coarse grid. Define the fine grid of L points as the L nearest frequency pairs to coarse grid points. We have that $2M + 1 = P \cdot L$. If we reorder the components of the set of L Bispectrum estimates $\hat{\mathcal{C}}(n_l, m_l)$ where $l = 1, \ldots, L$, on the fine grid around the bifrequency pair into a L vector β_{ml} where $m = 1, \ldots P$ indexes the coarse grid [13] and define P-vectors $\phi_i(\beta_{1i}, \ldots, \beta_{Pi})$, $i = 1, \ldots L$; the generalized likelihood ratio test for the above discussed hypothesis testing problem:

$$H_0 : \mu = \mu_n \quad against \quad H_1 : \eta \equiv \mu^T \sigma^{-1} \mu > \mu_n^T \sigma_n^{-1} \mu_n \tag{6}$$

where $\mu = 1/L \sum_{i=1}^{L} \phi_i$ and $\sigma = 1/L \sum_{i=1}^{L} (\phi_i - \mu)(\phi_i - \mu)^T$ are the maximum likelihood gaussian estimates of vector $\mathcal{C} = (\mathcal{C}_{\mathbf{y}_k \mathbf{y}_l}(m_1, n_1) \ldots \mathcal{C}_{\mathbf{y}_k \mathbf{y}_l}(m_P, n_P))$, leads to the activity voice detection if:

$$\eta > \eta_0 \tag{7}$$

where η_0 is a constant determined by a certain significance level, i.e. the probability of false alarm. Note that:

1. We have supposed independence between signal s_k and additive noise n_k [1] thus:
$$\mu = \mu_n + \mu_s; \qquad \sigma = \sigma_n + \sigma_s \tag{8}$$
2. The right hand side of H_1 hypothesis must be estimated in each frame (it's a-priori unknown). In our algorithm the approach is based on the information in the previous non-speech detected intervals.

The statistic considered here η is distributed as a central $F_{2P,2(L-P)}$ under the null hypothesis. Therefore a Neyman-Pearson test can be designed for a significance level α.

χ^2 Tests: In this section we consider the χ^2_{2L} distributed test statistic[4]:

$$\eta = \sum_{m,n} 2M^{-1} |\Gamma_{\mathbf{y}_k \mathbf{y}_l}(m, n)|^2 \tag{9}$$

where $\Gamma_{\mathbf{y}_k \mathbf{y}_l}(m, n) = \frac{|\hat{\mathcal{C}}_{\mathbf{y}_k \mathbf{y}_l}(n,m)|}{[S_{\mathbf{y}_0}(m) S_{\mathbf{y}_k}(n) S_{\mathbf{y}_l(m+n)}]^{0.5}}$ which is asymptotically distributed as $\chi^2_{2L}(0)$ where L denotes the number of points in interior of the principal domain. The Neyman-Pearson test for a significant level (false-alarm probability) α turns out to be:

[1] Observe that now we do not assume that n_k $k = 0 \ldots \pm M$ are gaussian.

$$H_1 \quad if \quad \eta > \eta_\alpha \tag{10}$$

where η_α is determined from tables of the central χ^2 distribution. Note that the denominator of $\Gamma_{\mathbf{y}_k\mathbf{y}_l}(m,n)$ is unknown a priori so they must be estimated as the bispectrum function (that is calculate $\hat{C}_{\mathbf{y}_k\mathbf{y}_l}(n,m)$). This requires a larger data set as we mentioned above in this section.

3 Experimental Framework

The ROC curves are frequently used to completely describe the VAD error rate. Only the AURORA subset of the original Spanish SpeechDat-Car (SDC) database [9] was used in this analysis for space reasons. The files are categorized into three noisy conditions: quiet, low noisy and highly noisy conditions, which represent different driving conditions with average SNR values between 25dB, and 5dB. The non-speech hit rate (HR0) and the false alarm rate (FAR0= 100-HR1) were determined in each noise condition. These noisy signals represent the most probable application scenarios for telecommunication terminals (suburban train, babble, car, exhibition hall, restaurant, street, airport and train station). Table 1 shows the averaged ROC curves of frequently referred algorithms [14,7,8,12] for recordings from the distant microphone in quiet, low and high noisy conditions. The working points of the G.729, AMR and AFE VADs are also included. If we compare the two test discussed above we can conclude that GRLT prevails over χ^2 tests. The ROC curves of the two proposed tests are obtained varying the confidence level α (we actually vary the parameter η_α). The results show improvements in detection accuracy over standard VADs and similarities over a representative set VAD algorithms [14,7,8,12] in high noise scenario. The benefits are especially important over G.729 and over the Li's algorithm. On average, it improves Marzinzik's VAD that tracks the power spectral envelopes, and the Sohn's VAD which applies a single observation likelihood ratio test on the voice-pause distributions.

Performance of ASR systems working over wireless networks and noisy environments normally decreases and non-efficient speech/non-speech detection appears to be an important degradation source [6]. Although the discrimination analysis or the ROC curves are effective to evaluate a given algorithm, this section evaluates the VAD according to the goal for which it was developed by assessing the influence of the VAD over the performance of a speech recognition system. The reference framework considered for these experiments was the ETSI AURORA project for DSR [2]. The recognizer is based on the HTK (Hidden Markov Model Toolkit) software package [15].

Table 2 shows the recognition performance for the Spanish SDC databases for the different training/test mismatch conditions (HM, high mismatch, MM: medium mismatch and WM: well matched) when WF and FD are performed on the base system [2]. Again, the VAD outperforms all the algorithms used for reference, yielding relevant improvements in speech recognition. Note that the SDC databases used in the AURORA 3 experiments have longer non-speech periods than the AURORA 2 database and then, the effectiveness of the VAD

Table 1. Average speech/non-speech hit rates for SNRs between $25dB$ and $5dB$. Comparison of the proposed BSVAD to standard and recently reported VADs.

(%)	G.729	AMR1	AMR2	AFE (WF)	AFE (FD)
HR0	55.798	51.565	57.627	69.07	33.987
HR1	88.065	98.257	97.618	85.437	99.750
(%)	Woo	Li	Marzinzik	Sohn	χ^2/GLRT
HR0	62.17	57.03	51.21	66.200	66.520/68.048
HR1	94.53	88.323	94.273	88.614	85.192/90.536

Table 2. Average Word Accuracy (%) for the Spanish SDC databases and tasks

		Base	Woo	Li	Marzinzik	Sohn	G.729	AMR1	AMR2	AFE	**GLRT**
Sp.	WM	92.94	95.35	91.82	94.29	96.07	88.62	94.65	95.67	95.28	96.28
	MM	83.31	89.30	77.45	89.81	91.64	72.84	80.59	90.91	90.23	92.41
	HM	51.55	83.64	78.52	79.43	84.03	65.50	62.41	85.77	77.53	86.70
	Ave.	**75.93**	89.43	82.60	87.84	90.58	75.65	74.33	90.78	87.68	**91.80**

results more important for the speech recognition system. This fact can be clearly shown when comparing the performance of the proposed VAD to Marzinzik's VAD. The word accuracy of both VADs is quite similar for the AURORA 2 task. However, the proposed VAD yields a significant performance improvement over Marzinzik's VAD for the SDC databases.

4 Conclusions

This paper presented a new VAD for improving speech detection robustness in noisy environments. The approach is based on higher order spectra analysis employing noise reduction techniques and statistic tests for the formulation of the decision rule. The VAD performs an advanced detection using the estimated components of the Bispectrum function and robust statistical tests GLRT and χ^2 over the set of vector variables $\mathbf{y}_k$. As a result, it leads to clear improvements in speech/non-speech discrimination especially when the SNR drops. With this and other innovations, the proposed algorithm outperformed G.729, AMR and AFE standard VADs. It also will improve the recognition rate when it was considered as part of a complete speech recognition system. The major benefit of the proposed algorithm is robusteness and simplicity of the decision rule as well as the potential inclusion of the recently reported approaches for endpoint detection.

References

1. ETSI. Voice activity detector (VAD) for Adaptive Multi-Rate (AMR) speech traffic channels. *ETSI EN 301 708 Recommendation*, 1999.
2. ETSI. Speech processing, transmission and quality aspects (stq); distributed speech recognition; front-end feature extraction algorithm; compression algorithms. *ETSI ES 201 108 Recommendation*, 2000.

3. S. Gustafsson, R. Martin, P. Jax, and P. Vary. A psychoacoustic approach to combined acoustic echo cancellation and noise reduction. *IEEE Transactions on Speech and Audio Processing*, 10(5):245–256, 2002.
4. J.R. Hinich. Testing for gaussianity and linearity of a stationary time series. *Journal of Time Series Analisys*, 3:169–176, 1982.
5. ITU. A silence compression scheme for G.729 optimized for terminals conforming to recommendation V.70. *ITU-T Recommendation G.729-Annex B*, 1996.
6. L. Karray and A. Martin. Towards improving speech detection robustness for speech recognition in adverse environments. *Speech Communitation*, (3):261–276, 2003.
7. Q. Li, J. Zheng, A. Tsai, and Q. Zhou. Robust endpoint detection and energy normalization for real-time speech and speaker recognition. *IEEE Transactions on Speech and Audio Processing*, 10(3):146–157, 2002.
8. M. Marzinzik and B. Kollmeier. Speech pause detection for noise spectrum estimation by tracking power envelope dynamics. *IEEE Transactions on Speech and Audio Processing*, 10(6):341–351, 2002.
9. A. Moreno, L. Borge, D. Christoph, R. Gael, C. Khalid, E. Stephan, and A. Jeffrey. SpeechDat-Car: A Large Speech Database for Automotive Environments. In *Proceedings of the II LREC Conference*, 2000.
10. E. Nemer, R. Goubran, and S. Mahmoud. Robust voice activity detection using higher-order statistics in the lpc residual domain. *IEEE Trans. Speech and Audio Processing*, 9(3):217–231, 2001.
11. J. Ramírez, J.C. Segura, C. Benítez, A. delaTorre, and A. Rubio. An effective subband osf-based vad with noise reduction for robust speech recognition. *In press IEEE Transactions on Speech and Audio Processing*, X(X):X–X, 2004.
12. J. Sohn, N. S. Kim, and W. Sung. A statistical model-based voice activity detection. *IEEE Signal Processing Letters*, 16(1):1–3, 1999.
13. T. Subba-Rao. A test for linearity of stationary time series. *Journal of Time Series Analisys*, 1:145–158, 1982.
14. K. Woo, T. Yang, K. Park, and C. Lee. Robust voice activity detection algorithm for estimating noise spectrum. *Electronics Letters*, 36(2):180–181, 2000.
15. S. Young, J. Odell, D. Ollason, V. Valtchev, and P. Woodland. *The HTK Book*. Cambridge University, 1997.

Back-Propagation as Reinforcement in Prediction Tasks

André Grüning

Cognitive Neuroscience Sector,
S.I.S.S.A., via Beirut 4, 34014 Trieste, Italy
gruening@sissa.it

Abstract. The back-propagation (BP) training scheme is widely used for training network models in cognitive science besides its well known technical and biological short-comings. In this paper we contribute to making the BP training scheme more acceptable from a biological point of view in cognitively motivated prediction tasks overcoming one of its major drawbacks.

Traditionally, recurrent neural networks in symbolic time series prediction (e. g. language) are trained with gradient decent based learning algorithms, notably with back-propagation (BP) through time. A major drawback for the biological plausibility of BP is that it is a supervised scheme in which a teacher has to provide a fully specified target answer. Yet, agents in natural environments often receive a summary feed-back about the degree of success or failure only, a view adopted in reinforcement learning schemes.

In this work we show that for simple recurrent networks in prediction tasks for which there is a probability interpretation of the network's output vector, Elman BP can be reimplemented as a reinforcement learning scheme for which the expected weight updates agree with the ones from traditional Elman BP, using ideas from the AGREL learning scheme (van Ooyen and Roelfsema 2003) for feed-forward networks.

Reinforcement learning where the teacher gives only feed-back about success or failure of an answer is thought to be biologically more plausible than supervised learning since a fully specified correct answer might not always be available to the learner or even the teacher ([1], especially for biological plausibility [2]).

In this article we extent the ideas of the AGREL scheme [3] about how to implement (BP) in (FF) networks for classification tasks to encompass Elman (BP) for (SRN) in prediction tasks [4]. The results have relevance especially for the cognitive science community for which (SRN) models have become an important tool [5], since they improve the standing of (SRN) with respect to biological and cognitive plausibility.

1 SRNs and Elman (BP)

A (SRN) (also called Elman network) in its simplest form resembles a 3-layer (FF) network, but in addition the hidden layer is self-recurrent [4]. It is a spe-

W. Duch et al. (Eds.): ICANN 2005, LNCS 3697, pp. 547–552, 2005.

cial case of a general (RNN) and could thus be trained with full (BPTT) and (BPTT)(n) [6]. However, instead of regarding the hidden layer as self-recurrent, one introduces a so-called *context layer* into which the activities of the hidden neurons are stored in each time step and which acts as an additional input to the hidden layer in the next time step and thus effects its recurrency. Regarding the forward-propagation of activity through the (SRN) these two views are equivalent.

For the back-propagation of error, the (SRN) is now viewed as a (FF) network with an additional set of inputs from the context layer. Hence standard (BP) in conjunction with copying the hidden layer into the context layer can be used for training [4]. This scheme is called *Elman (BP)* and has found wide application especially in linguistically motivated prediction task (for an overview see [7]).

Since as described above, with Elman (BP) training (SRN) can be reduced to training layered (FF) networks and since the step back to a (SRN) with context units is obvious, it is sufficient to formulate our reinforcement implementation of Elman (BP) for layered (FF) networks.

2 Recapitulation: Standard (BP)

We state some important formulae of (BP) for layered (FF) networks first since the reinforcement implementation will be based on them. In order to keep notation simple, we will only deal with networks that are strictly layered, i. e. there are connections only between subsequent layers. Generalisations where neurons receive input from other downstream layers are of course possible. Furthermore we refrain from explicitly introducing a bias term, since its effect can easily be achieved by a unit with constant activation 1 in each layer.

Let us assume we have a network with $p+1$ layers in total where the counting starts from 0 for the input layer. Let $y_i^r, r > 0$ denote the output of neuron i in layer r, while $y^0 = (y_i^0)_i$ denotes the input vector. Let us further assume that the set of different input patterns y^0 is classified into classes c and that the target vector t^c only depends on the class of a particular input vector y^0 (or rather the class of a sequence of input vectors in the case of a (SRN)).

Then, for each input y^0 the layers are updated consecutively from 1 to p, $f : x \mapsto \frac{1}{1+e^{-x}}$ is the activation function of the neuron that maps the net input a_i^r of neuron i in layer r to its output y_i^r. Finally the output is read off from y^p and the error E_c against the target vector t^c is computed as follows, $0 < r \leq p$:

$$y_i^r = f(a_i^r), \quad a_i^r = \sum_j w_{ij}^{r,r-1} y_j^{r-1}, \quad E_c(y^p) = \frac{1}{2}\sum_i (t_i^c - y_i^p)^2. \qquad (1)$$

For the update of the weights we need to know how much each single weight $w_{ij}^{r,r-1}$ contributes to the overall error E_c. For input y^0, E_c is an implicit function of all weights w. It helps in book-keeping of each weights' contribution $\frac{\partial E}{\partial w_{ij}^{r,r-1}}$ to the error by first calculating the contribution $\Delta_i^r := \frac{\partial E}{\partial a_i^r}$ of each neuron's net input to the error. The Δs can be recursively computed layer-wise starting from the output layer p and from them the weight updates δw with learning rate ϵ as:

$$\Delta_i^p := (y_i^p - t_i^c) f'(a_i^p), \tag{2}$$

$$\Delta_j^r := f'(a_j^r) \sum_i \Delta_i^{r+1} w_{ij}^{r+1,r}, \quad 0 < r < p. \tag{3}$$

$$\delta w_{ij}^{r,r-1} = -\epsilon \frac{\partial E}{\partial w_{ij}^{r,r-1}} = -\epsilon \frac{\partial E}{\partial a_i^r} \frac{\partial a_i^r}{\partial w_{ij}^{r,r-1}} = -\epsilon \Delta_i^r y_j^{r-1}, \tag{4}$$

3 Prediction Task for (SRN)

Linguistically and cognitively inspired prediction learning means the following [4]: a sequence of unarily encoded symbols is input to the network one symbol at a time. The task is to predict the next symbol of the sequence. A context c for a prediction task would be given by a whole sequence of input symbols allowing for the same possible continuation(s). However it does not determine the next symbol with certainty but rather defines a distribution p_c accounting for linguistic variation. Thus the network has to learn the distribution of next symbols. Cognitively it is implausible to take the precalculated distribution as the target. Instead training is done against a target vector t^c where only one entry t_j^c drawn according to the appropriate distribution is one (all others zero), i. e. $p(t_j^c = 1) = p_c(j)$. For the error contributions Δ^p it follows, taking the expectation value over all possible targets t^c in context c:

$$< \Delta_i^p >_c = < (y_i^p - t_i^c) f'(a_i^p) >_c = (y_i^p - < t_i^c >_c) f'(a_i^p) = (y_i^p - p_c(i)) f'(a_i^p), \tag{5}$$

i. e. the expectation value of Δ_i^p in the output layer p coincides with the Δ derived from training against the distribution of target vectors. The same is true for all other Δs recursively computed from this due to the linearity of (3) in the Δs.

4 Reinforcement Learning

In prediction learning, the output's activation y_i^p corresponds to its symbol's estimated probability $p_{y^p}(i)$. Let us introduce a reinforcement scheme as follows: assume the network is only allowed to select one answer k as a response to its current input context c. It selects this answer according to the distribution $p_{y^p}(k) = y_k^p / |y^p|$: the neuron y_k^p corresponding to symbol k is called the winning neuron.

This answer is then compared to the target k_c drawn from the target distribution p_c and the network receives a reward $r = 1$ only when $k = k_c$, and $r = 0$ otherwise. Thus the objectively expected reward in input context c after selecting neuron k is $< r >_{c,k} = p_c(k)$. The network compares the received reward r

to the subjectively expected reward, namely the activation y_k^p of the winning neuron. The relative difference

$$\delta := \frac{y_k^p - r}{p_{y^p}(k)} \tag{6}$$

is made globally available to all neurons in the network. From δ we then compute the error signals Δ^p for the output layer. Since attention is concentrated on the winning output k, only its Δ_k^p is different from zero, and we set $\Delta_i^p = 0$ for $i \neq k$ and

$$\Delta_k^p = \delta f'(a_k^p). \tag{7}$$

The other Δs and the weight updates δw can be recursively computed as before in (3) and (4). The expectation value of Δ_k^p in context c and with k as the winning unit calculates:

$$< \Delta_k^p >_{c,k} = f'(a_k^p) \frac{y_k^p - p_c(k)}{p_{y^p}(k)} \tag{8}$$

since t^c is drawn independently from k and $< r >_{c,k} = p_c(k)$. Compared to Elman (BP) an error for a certain output is calculated only when it is the winning unit, it is updated less frequently (by a factor $p_y(k)$). But its Δ is larger by $1/p_y(k)$ to compensate for this. Weighting the Δs with the probability $p_y(k)$ that k gets selected as the winning unit in context c, we get

$$< \Delta_k^p >_c = p_y(k) f'(a_k^p) \frac{y_k^p - p_c(k)}{p_y(k)} = f'(a_k^p)(y_k^p - p_c(k)) \tag{9}$$

and this agrees with (2) and (5), keeping in mind that $p_c(k)$ in our scheme would be the target entry t_i^c in the standard (BP) scheme. By linearity the expectation values of all other Δs and the δw in this scheme coincide as well with their counterparts in standard (BP). When we update weights after each input presentation, the average of the Δs over several trials in context c will differ from the expectation value, but this is not a more severe moving target problem than encountered anyway in prediction learning in (5), and can also be dealt with by keeping ϵ low [6].

5 Discussion

We note that the order in which weights receive error signals form the outputs is altered: in standard (BP) weights receive error signals from all output neurons in each time step, while in this reinforcement scheme they receive a signal only from a single output, but with a greater amplitude. No major differences in the course of learning are expected due to this changed order. In fact, it is known that training against the actual successors in the prediction task instead of against their probability distribution leads to faster and more reliable learning because higher error signals enable the network to leave local minima faster. A similar effect can be expected here.

The crucial facts why we can reimplement (BP) as a reinforcement scheme and thus replace a fully specified target vector with a single evaluative feedback are: (i) a probability interpretation of the output vector (and the target). This allows us to regard the network as directing its attention to a single output which is stochastically selected and subsequently to relate the evaluative feedback to this single output, (ii) the error contribution of each single neuron in the hidden layer(s) to the total error in (BP) is a linear superposition of the individual contributions of the neurons in the output layer. This enables us to concentrate on the contribution from a single output in each time step and still arrive at an expected weight update equal to the original scheme.

Obviously the ideas laid out in this paper are applicable to all kinds of (SRN), multilayered or not, and more general (RNN) as long as their recurrence can be treated in the sense of introducing context units with immutable copy weights, and they ought also to be applicable to other networks and gradient based learning algorithms that fulfil the two above conditions (e. g. LSTM [8]).

As regards the biological plausibility, we list the following in favour of this (BP)-as-reinforcement scheme: (i) We have replaced a fully specified target with a single evaluative feed-back. (ii) The relative difference δ of actual and expected reward can be realised as a prediction error neuron [9,2] whose activation is made globally available to all neurons in the network, e. g. by diffusion of a messenger such as dopamine [3]. (iii) Attentive concentration on the winning neuron is physiological plausible ([3] and references therein). (iv) Using the same sets of weights both for forward-propagation of activity and back-propagation of error is made plausible by introducing a second set of weights w' used only for the back-propagation of the Δs in (3) and updated with the same – mutatis mutandis – equation as in (4). This finds its functional equivalent in the ample evidence for recurrent connections in the brain. However we would consider this assumption as the scheme's weakest point that will need further elaboration. (v) Above and beyond [3]'s scheme, the only additional requirement for Elman (BP) as reinforcement has been that the activation of the hidden layer is retrievable in the next time step. This assumption is not implausible either in view of the ample recurrent connections in the brain which locally might well recycle activation from a neighbouring cell for some time from which this activation can be reconstructed. (vi) Finally we need to discuss how $p_{y^p}(k)$ can be derived from y^p_k in a plausible way, technically its is just an addition of the y^ps and dividing each output by this sum. Evidence for pools of neurons in visual cortex doing precisely a divisive normalisation is summarised in [10].[1]

Thus all quantities δ, y^r_j and Δ^{r+1}_i can be made locally available for the weight change at each synapse $w^{r+1,r}_{ij}$. While our reinforcement scheme naturally extends even to fully recurrent networks trained with (BP) through time (again it is mainly a question of the linearity of the Δs), there its application is of course less plausible since we would need to have access to previous activities of neurons for more than one time step.

[1] An anonymous referee pointed this reference out to me.

6 Conclusion

Im sum, we have found a reinforcement learning scheme that behaves essentially like the standard (BP) scheme. It is biologically more plausible by using a success/failure signal instead of a precise target. Essential in transforming (BP) into a reinforcement scheme was (i) that the (BP) error signal for the complete target is a linear superposition of the error for each single output neuron, and (ii) the probabilistic nature of the task: select one possible output randomly and direct the network's attention towards it until it is rewarded. Furthermore we have briefly discussed the physiological or biological plausibility of other ingredients in the (BP)-as-reinforcement scheme. It seems that there is good evidence for all of them at least in some parts of the brain. Enhanced biological plausibly for (BP) thus gives (SRN) usage in cognitive science a stronger standing.

References

1. Sutton, R.S., Barto, A.G.: Reinforcement learning: An Indroduction. Bradford Books, MIT Press, Cambridge (2002)
2. Wörgötter, F., Porr, B.: Temporal sequence learning, prediction, and control – a review of different models and their relation to biological mechanisms. Neural Computation **17** (2005) 245–319
3. van Ooyen, A., Roelfsema, P.R.: A biologically plausible implementation of error-backpropagation for classification tasks. In Kaynak, O., Alpaydin, E., Oja, E., Xu, L., eds.: Artificial Neural Networks and Neural Information Processing – Supplementary Proceedings ICANN/ICONIP, Istanbul (2003) 442–444
4. Elman, J.L.: Finding structure in time. Cognitive Science **14** (1990) 179–211
5. Ellis, R., Humphreys, G.: Connectionist Psychology. Psychology Press, Hove, East Sussex (1999)
6. Williams, R.J., Peng, J.: An efficient gradient-based algorithm for on-line training of recurrent network trajectories. Neural Computation **2** (1990) 490–501
7. Christiansen, M.H., Chater, N.: Toward a connectionist model of recursion in human linguistic performance. Cognitive Science **23** (1999) 157–205
8. Hochreiter, S., Schmidhuber, J.: Long short-term memory. Neural Computation **9** (1997) 1735–1780
9. Schultz, W.: Predictive reward signal of dopaminic neurons. J. Neurophysiol. **80** (1998) 1–27
10. Carandini, M., Heeger, D.J.: Summation and division by neurons in primate visual cortex. Sience **264** (1994) 1333–1336

Mutual Information and k-Nearest Neighbors Approximator for Time Series Prediction

Antti Sorjamaa, Jin Hao, and Amaury Lendasse†

Neural Network Research Centre, Helsinki University of Technology,
P.O. Box 5400, 02150 Espoo, Finland
{asorjama, jhao, lendasse}@cis.hut.fi

Abstract. This paper presents a method that combines Mutual Information and k-Nearest Neighbors approximator for time series prediction. Mutual Information is used for input selection. K-Nearest Neighbors approximator is used to improve the input selection and to provide a simple but accurate prediction method. Due to its simplicity the method is repeated to build a large number of models that are used for long-term prediction of time series. The Santa Fe A time series is used as an example.

Keywords: Time Series, Input Selection, Mutual Information, k-NN.

1 Introduction

In any function approximation, system identification, classification or prediction task one usually wants to find the best possible model and the best possible parameters to have a good performance. Selected model must be generalizing enough still preserving accuracy and reliability without unnecessary complexity, which increases computational load and thus calculation time. Optimal parameters must be determined for every model to be able to rank the models according to their performances.

In this paper we use Mutual Information (MI), described in Section 2, to select the inputs for direct long-term prediction of a time series. Leave-one-out (LOO) method, described in Section 3, is used to select the correct parameter for MI. Both MI and LOO rely on the k-Nearest Neighbors (k-NN) method, which is described in Section 4. Section 5 gives information about the time series prediction problem and finally the obtained experimental results, conclusions and further work are presented in Sections 6 and 7.

2 Mutual Information for Input Selection

Input selection is one of the most important issues in machine learning, especially when the number of observations is relatively small compared to the number of inputs. In practice, the necessary size of the dataset increases dramatically with the

† Part the work of A. Sorjamaa, J. Hao and A. Lendasse is supported by the project of New Information Processing Principles, 44886, of the Academy of Finland.

W. Duch et al. (Eds.): ICANN 2005, LNCS 3697, pp. 553 – 558, 2005.

number of observations (curse of dimensionality). To circumvent this, one should first select the best inputs or regressors in the sense that they contain the necessary information. Then, it would be possible to capture and reconstruct the underlying relationship between input-output data pairs. Within this respect, some approaches have been proposed [1-3]. Some of them deal with the problem of feature selection as a generalization error estimation problem. These approaches are very time consuming and may take several weeks. However, there are other approaches [4-5], which select a priori inputs based only on the dataset, as presented in this paper.

In this paper, the Mutual Information (MI) is used as a criterion to select the best input variables (from a set of possible variables) for the long-term prediction purpose.

The MI between two variables, let say X and Y, is the amount of information obtained from X in the presence of Y, and vice versa. MI can be used for evaluating the dependencies between random variables, and has been applied for Feature Selection and Blind Source Separation [6].

Let's consider two random variables; the MI between them would be

$$I(X,Y) = H(X) + H(Y) - H(X,Y) \ , \tag{1}$$

where $H(.)$ computes the Shannon's entropy. Equation (1) leads to complicated integrations, so some approaches have been proposed to evaluate them numerically. In this paper, a recent estimator based on l-NN statistics is used [7] (l is used instead of k here to avoid confusion with the k appearing in section 4). The novelty of this approach consists in its ability to estimate the MI between two variables of any dimensional spaces. The basic idea is to estimate $H(.)$ from the average distance to the l nearest neighbors. MI is derived from equation (1) and is estimated as

$$I(X,Y) = \psi(l) - 1/l - \langle \psi(n_x) + \psi(n_y) \rangle + \psi(N) \ , \tag{2}$$

where N is the size of the dataset, l is the number of nearest neighbors and $\psi(x)$ is the digamma function,

$$\psi(x) = \Gamma(x) - 1 - d\Gamma(x)/dx, \text{ which satisfies } \psi(x+1) = \psi(x) + 1/x \ , \tag{3}$$

$\psi(1) \approx -0.5772156$ and <...> denotes averages of n_x and n_y over all $1 \leq i \leq N$ and over all realizations of the random samples. $n_x(i)$ and $n_y(i)$ are the number of points in the region $\|x_i - x_j\| \leq \varepsilon_x(i)/2$ and $\|y_i - y_j\| \leq \varepsilon_y(i)/2$, $\varepsilon_x(i)$ and $\varepsilon_y(i)$ are the edge lengths of the smallest rectangle around point i containing l nearest neighbors. Software for calculating the MI based on this method can be downloaded from [8].

3 Leave-One-Out

Leave-one-out [4] is a special case of k-fold cross-validation resampling method. In k-fold cross-validation the training data is divided into k approximately equal sized sets. LOO procedure is the same as k-fold cross-validation with k equal to the size of the training set N. For each model to be tested, LOO procedure is used to calculate the generalization error estimate by removing each data point at a time from the training set, building a model with the rest of the training data and calculating the validation error with the one taken out. This procedure is done for every data point in the training set and the estimate of the generalization error is calculated as a mean of all k, or N, validation errors (4).

$$\hat{E}_{gen}(q) = \frac{\sum_{i=1}^{N}\left(h^{q}(x_i, \theta_i^{*}(q)) - y_i\right)^2}{N} , \tag{4}$$

where x_i is the ith input vector from the training set, y_i is the corresponding output, h^q denotes the qth tested model and $\theta_\iota*(q)$ includes the model parameters without using (x_i, y_i) in the training. Finally, as a result from the LOO procedure, we select the model that gives us the smallest generalization error estimate.

4 *k*-Nearest-Neighbors Approximator

K-Nearest Neighbors approximation method is a very simple, but powerful method. It has been used in many different applications and particularly in classification tasks [9]. The key idea behind the *k*-NN is that similar input data vectors have similar output values. One has to look for a certain number of nearest neighbors, according to Euclidean distance [9], and their corresponding output values to get the output approximation. We can calculate the estimation of the outputs by using the average of the outputs of the neighbors in the neighborhood. If the pairs (x_i, y_i) represent the data with x_i as an n-dimensional input and y_i as a scalar output value, *k*-NN approximation is

$$\hat{y}_i = \frac{\sum_{j=1}^{k} y_{P(j)}}{k} , \tag{5}$$

where $\hat{y}_i$ represents the output estimation, $P(j)$ is the index number of the jth nearest neighbor of the input x_i and k is the number of neighbors that are used. We use the same neighborhood size for every data point, so we use a global k, which must be determined. In our experiments, different k values are tested and the one which gives the minimum LOO error is selected.

5 Time Series Prediction

Time series prediction can be considered as a modeling problem [10]: a model is built between the inputs and the outputs. Then, it is used to predict the future values based on previous values. In this paper we use direct forecast to perform the long-term prediction. In order to predict the values of a time series, M different models are built,

$$\hat{y}(t+m) = f_m(y(t-1), y(t-2), \ldots, y(t-n)) , \tag{6}$$

with m = 0,1,...M-1, M is the maximum horizon of prediction and f_m is the model related to time step m The input variables on the right-hand part of (6) form the regressor, where n is the regressor size.

6 Experimental Results

The dataset used in the experiments is the Santa Fe A Laser Data. To test the influence of the size of dataset on the selection of parameter l in MI calculation, two steps are followed.

The first experiment is done with 10 000 data, 9900 from which is used for training and the rest 100 for testing. In order to apply the prediction model from equation (6), we set the maximum time horizon M to 100 and the regressor size n to 10.

For time step one (m = 0 in equation (6)), MI is used to select the best inputs. The estimation of MI based on equation (2) is calculated with different number of neighbors, with $l = 1,\ldots 10$. All the 2^n-1 combinations of inputs are tested; the one that gives maximum MI is selected.

Finally, k-NN and LOO are used to select the l, which minimizes the LOO error, presented in Fig. 1.a.

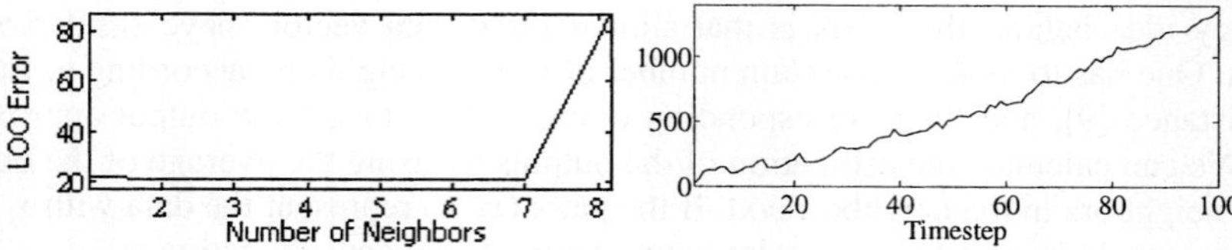

Fig. 1a. The LOO error according to different l in the MI method. **Fig. 1b.** The LOO errors according to 100 time steps.

Based on this result, l = 5 is chosen for the MI estimation. The input selection results for the first 50 time steps are listed in Table 1. After the input selection k-NN and LOO are used with the selected inputs for each time step and the graph of the resulting learning LOO errors are shown in Fig. 1.b.

In Fig. 2, the inputs selected with MI are used to predict 100 time steps. For each time step, f_m in equation (6) is performed using the k-NN method, and the k in equation (5) is determined by the LOO method. The real prediction error (MSE) is then calculated. For this experiment, the MSE is 2.24.

Table 1. Selected Inputs for each prediction with 10 000 data. The rows of the table represents $y(t+m)$, m = 0,…,49, from left to right; the columns represents $y(t-n)$, n = 1,…,10, from top to bottom. The cross mark means for one $y(t+m)$, the related input $y(t-n)$ is selected.

X	X	X				X	X	X	X		X		X	X	X	X	X	X	X	X	X	X	X	X	X	X	X	X	X	X	X	X	X	X	X	X	X	X	X	X	X	X	X	X	X	X	X	X	X
X	X				X	X	X	X	X	X	X	X	X	X	X	X	X	X	X	X	X	X	X	X	X	X	X	X	X	X	X	X	X	X	X	X	X	X	X	X	X	X	X	X	X	X	X	X	X
		X	X	X	X				X	X	X	X	X	X		X	X	X	X	X	X	X	X	X	X	X	X	X	X	X	X	X	X	X	X	X	X	X	X	X	X	X	X	X	X	X	X	X	X
	X	X	X	X					X	X	X	X			X	X	X	X	X	X	X	X	X	X	X	X	X	X	X	X	X	X	X	X	X	X	X	X	X	X	X	X	X	X	X	X	X	X	X
	X	X	X	X				X	X	X	X	X		X	X	X	X	X	X	X	X	X	X	X	X	X	X	X	X	X	X	X	X	X	X	X	X	X	X	X	X	X	X	X	X	X	X	X	X
	X	X	X					X	X	X	X			X	X	X	X	X	X		X	X	X	X	X	X	X	X	X	X	X	X	X	X	X	X	X	X	X	X	X	X	X	X	X	X	X	X	X
X	X						X	X	X	X				X	X	X	X			X	X	X	X	X	X	X	X	X	X	X	X	X	X	X	X	X	X	X	X	X	X	X	X	X	X	X	X	X	X
X						X	X	X	X				X	X	X	X				X	X	X	X	X	X	X	X	X	X	X	X	X	X	X	X	X	X	X	X	X	X	X	X	X	X	X	X	X	X
					X	X	X	X			X	X	X	X	X	X	X	X	X		X	X	X		X		X	X	X				X	X		X	X		X	X	X	X	X	X	X		X		
			X	X	X				X	X	X	X	X	X	X	X	X	X	X	X	X	X	X	X	X	X	X	X	X	X	X	X	X	X	X	X	X	X	X	X	X	X	X	X	X	X	X	X	X

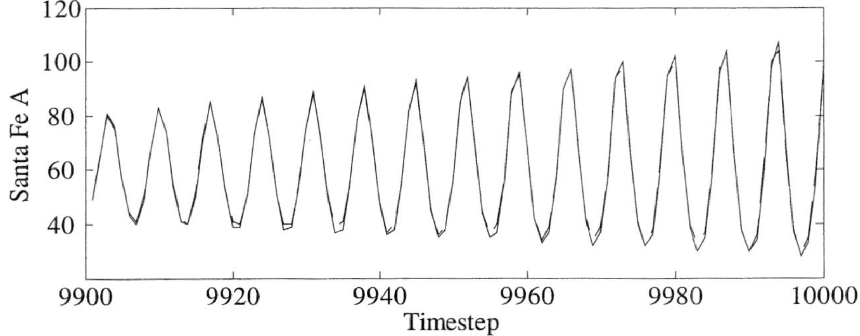

Fig. 2. 100 predictions (solid line) and the real values (dashed line)

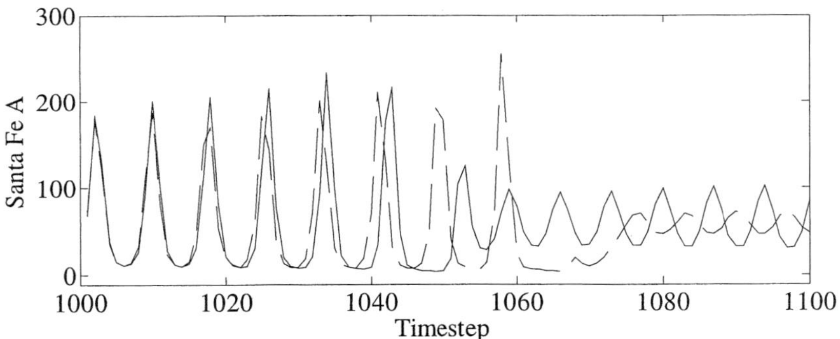

Fig. 3. 100 predictions (solid line) and the real values (dashed line)

In the second experiment, first 1000 data points are used for training and the next 100 points for testing. The procedure follows the first experiment. Based on the LOO error according to different l in the estimation of MI, $l = 2$ is chosen.

The prediction using k-NN and LOO based on the selected inputs by MI is plotted in Fig. 3.

7 Conclusions and Further Work

In this paper, MI is used to select the inputs for time series prediction problem. It has been illustrated with the experiments that the k-NN approximator and LOO method can be used to tune the main parameter of the MI estimator.

k-NN has also been used as an approximation model itself. Although Fig. 3 shows that after step 50, the jump of Santa Fe A Laser Data is not predicted correctly, the results are accurate in other parts. It is also possible to use another regression model to improve the quality of the predictions (Multilayer Perceptrons, Radial Basis Function Networks, Support Vector Machines, etc.). However, the advantage of the k-NN approximators is that it is possible to build a large number of models to perform a direct prediction of a time series in a quite reasonable time.

In the future, we will study different algorithms for estimating the MI and their possible implementations to input selection problems. On the other hand, the implementation of input selection methods directly to *k*-NN approach will also be studied.

References

1. Kwak, N., Chong-Ho, Ch.: Input feature selection for classification problems. Neural Networks, IEEE Transactions, Vol. 13, Issue 1 (2002) 143–159.
2. Zongker, D., Jain, A.: Algorithms for feature selection: An evaluation Pattern Recognition. Proceedings of the 13th International Conference, Vol. 2, 25-29 (1996) 18-22.
3. Xing, E.P., Jordan, M.I., Karp, R.M.: Feature Selection for High-Dimensional Genomic Microarray Data. Proc. of the Eighteenth International Conference in Machine Learning, ICML2001 (2001).
4. Kohavi, R.: A study of Cross-Validation and Bootstrap for Accuracy Estimation and Model Selection. Proc. of the 14th Int. Joint Conf. on A.I., Vol. 2, Canada (1995).
5. Jones, A., J.: New Tools in Non-linear Modeling and Prediction. Computational Management Science, Vol. 1, Issue 2 (2004) 109-149.
6. Yang, H., H., Amari, S.: Adaptive online learning algorithms for blind separation: Maximum entropy and minimum mutual information, Neural Comput., vol. 9 (1997) 1457-1482.
7. Alexander, K., Harald, S., Peter, G.: Estimating Mutual Information. John-von-Neumann Institute for Computing, Germany, D-52425. (2004).
8. URL: http://tinyurl.com/bj73w.
9. Bishop C.M.: Neural Networks for Pattern Recognition. Oxford University Press (1995).
10. Xiaoyu, L., Bing, W., K., Simon, Y., F.: Time Series Prediction Based on Fuzzy Principles. Department of Electrical & Computer Engineering. FAMU-FSU College of Engineering, Florida State University. Tallahassee, FL 32310.

Some Issues About the Generalization of Neural Networks for Time Series Prediction

Wen Wang[1,2], Pieter H.A.J.M. Van Gelder[2], and J.K. Vrijling[2]

[1] Faculty of Water Resources and Environment, Hohai University, Nanjing, 210098, China
[2] Faculty of Civil Engineering & Geosciences, Section of Hydraulic Engineering, Delft University of Technology. P.O.Box 5048, 2600 GA Delft, Netherlands

Abstract. Some issues about the generalization of ANN training are investigated through experiments with several synthetic time series and real world time series. One commonly accepted view is that when the ratio of the training sample size to the number of weights is larger than 30, the overfitting will not occur. However, it is found that even with the ratio higher than 30, overfitting still exists. In cross-validated early stopping, the ratio of cross-validation data size to training data size has no significant impact on the testing error. For stationary time series, 10% may be a practical choice. Both Bayesian regularization method and the cross-validated early stopping method are helpful when the ratio of training sample size to the number of weights is less than 20. However, the performance of early stopping is highly variable. Bayesian method outperforms the early stopping method in most cases, and in some cases even outperforms no-stop training when the training data set is large.

1 Introduction

ANNs are prone to either underfitting or overfitting (Sarle, 2002). A network that is not sufficiently complex can fail to detect fully the signal in a complicated data set, leading to underfitting. A network that is too complex may fit the noise, not just the signal, leading to overfitting, which may result in predictions far beyond the range of the training data. Therefore, one critical issue in constructing a neural network is generalization, namely, the capacity of an ANN to make predictions for cases that are unseen in the training set. Two commonly used techniques for generalization are cross-validated early stopping (e.g., Amari et al., 1997; Prechelt, 1998) and the regularization (or weight decay) technique (e.g., Mackay, 1991; Neal, 1996).

In cross-validated early stopping, the available data are usually split into two subsets: training and cross validation (referred to as CV hereafter) sets. The training set is used for updating the network weights and biases. The CV set is used to monitor the error variation during the training process. When the validation error increases for a specified number of iterations, the training is stopped.

Large weights can cause excessive variance of the output (Geman et al., 1992). A traditional way of dealing with the negative effect of large weights is regularization. The idea of regularization is to make the network response smoother through modification in the objective function by adding a penalty term that consists of the mean square of all network coefficients. Mackay (1991) proposed a technique, called

W. Duch et al. (Eds.): ICANN 2005, LNCS 3697, pp. 559 – 564, 2005.

Bayesian regularization, which automatically sets the optimal performance function to achieve the best generalization based on Bayesian inference techniques.

In this paper, we will discuss three issues about the generalization of networks: (1) How many data are demanded to avoid overfitting; (2) How to split the training samples in cross-validated early stopping; (3) Which generalization technique is better for time series prediction, Bayessian regularization or cross-validated early stopping?

2 Experiments and Result Analyses

2.1 Data

Seven data sets are used in this study, including three synthetic data sets and seven observed data sets. Three synthetic time series are as following: (1) Henon map (Henon, 1976) chaotic series; (2) The discretized chaotic Mackey-Glass flow series (Mackey and Glass, 1977); (3) A stochastic time series generated with an ANN model with a structure 5-3-1. 2% Gaussian noises are added to the two synthetic chaotic time series. The four observed real-world time series include: (1) The monthly sunspot number series (1749.1 ~ 2004.12); (2) The yearly sunspot number series (1700 to 2004); (3) Monthly Southern Oscillation index (SOI) series (1933.1 ~ 2004.12); (4) and (5) daily and monthly streamflow series of the Rhine River at Lobith, the Netherlands (1901.1 ~ 1996.12); (6) and (7) daily and monthly streamflow series of the Danube River at Achleiten, Austria (1901.1 ~ 1990.12).

De Oliveira et al. (2000) suggest to use *m*:2*m*:*m*:1 structure to model chaotic series. Follow their suggestion, we use 6:12:6:1 for Henon series as well as the discretized Mackey-Glass series. ANNs of 2-4-1 (Foresee and Hagan, 1997) and 18-6-1 (Conway, 1998) are used for yearly and monthly sunspot series. With trial and error procedure, the chosen ANN structure is 4-3-1 for the SOI series and the two monthly flow series, 23-12-1 for daily flow of Danube, and 16-8-1 for daily flow of Rhine.

The ANNs are constructed with Matlab Neural network toolbox. In all ANNs, tansig transfer function is used in the hidden layer. To avoid of the problem of sensitivity to initial weights, simple ensemble technique is applied. That is, for each network, we run 10 times with different initial weights, then choose five ones, which have best training performance, and take the average of the outputs of the five networks.

2.2 How Many Data Are Demanded to Avoid Overfitting?

Amari et al. (1997) show that, when the ratio (referred to as R hereafter) of the training sample size to the number of weights is larger than 30, no overtraining is observed. This view is accepted by many researchers as a guideline for training ANNs (e.g., Sarle, 2002).

Is there such a clear cut-off value of R? We make experiments for three synthetic series with different values of R ranging from 5 to 50. To avoid the possible impact of nonstationarity, real world data are not applied here. We use the last 1000 points of each synthetic series as the test data, while the training data vary according to the value of R. Networks are trained with Levenberg-Marquardt backpropagation algorithm and the training epoch is 1000. The variations in root mean squared error (RMSE) of training data and test data with different values of R are plotted in Fig. 1.

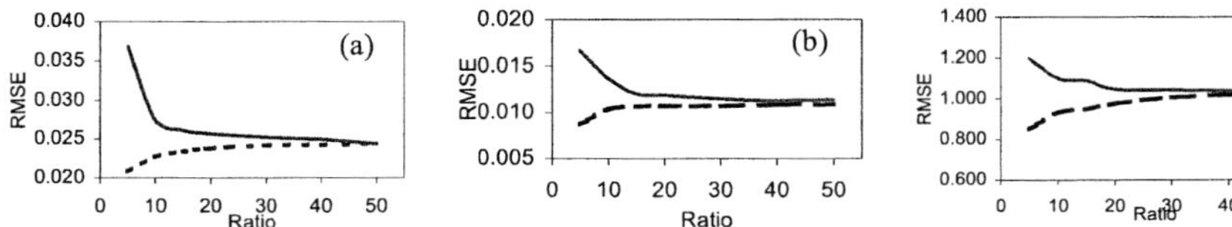

Fig. 1. RMSE of training data and test data with different values of R for (a) Henon series; (b) Mackey-Glass series and (c) ANN series. (solid line: test error; dash line: training error).

From Fig. 1, we see that with the increase of R, the training error grows, whereas, the testing error decreases with the increase of R, which indicates that the intensity of overfitting decrease. However, there is no clear cut-off value of R, above which overfitting vanishes. When R=30, as suggested by Amari et al. (1997), overfitting is still observed for all three fitted networks. When the ratio R is as high as 50, the overfitting basically disappears. However, for the Mackay-Glass series and simulated ANN series, the test error is still slightly higher than, albeit very close to, the training error, which indicates the existence of slight overfitting.

2.3 How to Split the Training Samples in Cross-Validated Early Stopping?

One important issue with regard to cross-validated early stopping is in what ratio to split the total training samples into training set and CV set. Amari et al. (1997) suggest that the average generalization error is minimized asymptotically when the rate of CV set to total training sample is:

$$r_{opt} = \frac{\sqrt{2k-1}-1}{2(k-1)} \tag{1}$$

where k is the total number of weights in the ANN.

In this section we investigate whether there is such an optimal ratio r. We calculate the training error and testing error with different values of r, ranging from 0.02 to 0.3, for different cases where the training data size varies with different values of R. The results for R = 10 are plotted in Fig. 2.

The networks for Henon series and Mackey-Glass series have 169 weights, and the network for the synthetic ANN series has 22 weights. Therefore, according to the optimal ratio proposed by Amari et al. (1997), as shown in Equation (1), for the former two networks, $r_{opt} \cong 0.077$; for the later one, $r_{opt} \cong 0.132$. However, there is no clear evidence of the existence of such optimal ratios from the visual inspection of Fig. 2 as well as the results for other experimental results when R is 5, 15 and 20.

Sarle (2002) comments that the results of Amari et al. (1997) contain serious errors that completely invalidate the results. From the experiments in this study, it seems that this comment is substantiated. In practice, many researchers use a large part, such as 1/3 (e.g., Prechelt, 1998), of training samples as CV set. However, according to this experiment, the ratio of CV set to training set seems to be not very important for early stopping. 10% could be a practical choice when the time series is stationary.

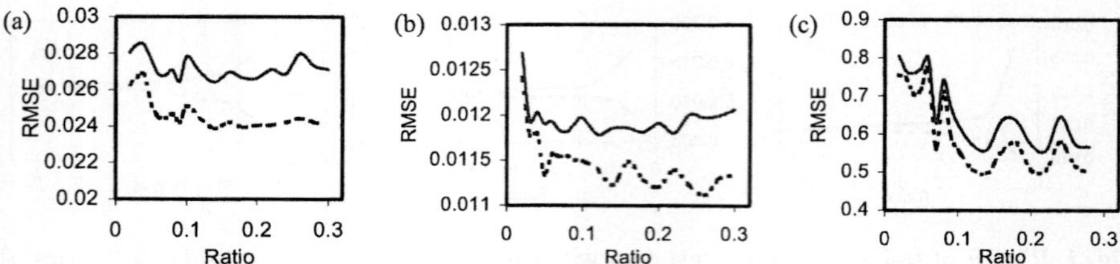

Fig. 2. RMSE of training set, CV set and test data with different CV to training ratio r for (a) Henon series; (b) Mackey-Glass series and (c) ANN series when R = 10. (solid line: test error; dash line: training error).

2.4 Which Technique Works Better, Bayesian Regularization or Cross-Validated Early Stopping?

Now we investigate the performance of Bayesian regularization and cross-validated early stopping technique for one-step ahead time series prediction. As a benchmark, no-stop training is also applied, in which training is stopped after 1000 epochs. Root mean errors (RMSE) of one-step ahead predictions for test data with these three approaches are shown in Table 1. The performance comparison shows that:

(1) Bayesian regularization outperforms CV in most cases, except for the cases of Mackay-Glass series and monthly sunspot series.

(2) When training sample size is small (R < 20), generally, both Bayesian regularization and CV early stopping outperform no-stop training. But these techniques do not always work. For several cases they fail even when $R \leq 10$.

(3) With the increase of the ratio of training data size to the number of weights, the overfitting problem with no-stop training is alleviated. Consequently, no-stop training outperforms or is at least equivalent to CV early stopping in most cases (except for Mackay-Glass series) when $R \geq 20$. That means, cross-validated early stopping does not improve the generalization error when $R \geq 20$ for most cases, even though overfitting still exists. In contrast, Bayesian regularization still outperforms no-stop training in about half of all cases even when $R \geq 30$.

(4) An advantage of CV early stopping is its fastness compared with Bayesian regularization and no-stop training, especially when the network is complicated. Whereas the time-costness is a major problem with Bayesian regularization, especially when the training size is big and the network is complicated.

(5) The performance of CV early stopping is highly variable, indicating that it is much less reliable than the other two techniques, whatever is the training data size. This is because it often ends up the training process too early due to local minimum of error function for the CV data set. Therefore, care must be taken when using CV early stopping in real world application despite of its fastness, unless the speed is of the most importance. It's better to check the error func-

tion surface of CV data to see if there are local minima before we use the CV data for early stopping.

(6) Comparatively, the performance of Bayesian generalization is highly stable, especially compared with CV early stopping. In many cases (e.g., the simulated ANN series), the 10 runs with different initial weights give almost the same result.

Table 1. Compare the performance of Bayesian regularization, CV early stopping and no-stop training according to root mean squared errors (RMSE) of one-step ahead predictions

Series	*R*	Bayes	CV	NST	Series	*R*	Bayes	CV	NST
Henon	5	0.02778	0.02790	0.03684	Rhine	6.3	840	867	886
	10	0.02642	0.02655	0.02745	monthly	12.6	839	935	858
	20	0.02508	0.02575	0.02565		18.9	848	862	867
	30	0.02491	0.02545	0.02520		25.3	850	897	853
	40	0.02468	0.02531	0.02497		31.6	850	898	854
	50	0.02468	0.02482	0.02439		37.9	855	876	855
Mackay-	5	0.01300	0.01253	0.01666	Danube	6.3	380	395	387
Glass	10	0.01198	0.01186	0.01362	monthly	12.6	377	396	376
	20	0.01144	0.01142	0.01186		18.9	369	386	369
	30	0.01140	0.01131	0.01146		25.3	369	384	364
	40	0.01130	0.01130	0.01128		31.6	367	388	361
	50	0.01133	0.01129	0.01134		37.9	367	380	362
ANN	5	1.080	1.098	1.199	Rhine	4.9	227	291	230
	10	1.055	1.072	1.102	daily	10.0	185	277	220
	20	1.045	1.055	1.048		15.0	188	508	218
	30	1.038	1.062	1.044		25.1	168	276	217
	40	1.026	1.044	1.040		37.7	185	280	211
	50	1.017	1.046	1.026		50.3	186	281	217
Sunspot	5	20.9	19.3	19.4	Danube	4.8	218	174	169
monthly	10	19.7	18.1	18.2	daily	9.6	184	178	162
	15	19.5	18.0	17.8		14.5	177	175	163
	20	18.4	17.6	17.6		24.2	172	175	164
SOI	5	1.327	1.409	1.468		36.3	163	170	159
monthly	10	1.357	1.559	1.461		48.5	162	171	159
	15	1.315	1.373	1.354	Sunspot	5	20.9	20.9	20.5
	20	1.333	1.436	1.448	Yearly	10	18.9	21.4	19.6
	30	1.313	1.393	1.363		15	18.8	21.0	19.4

Note: *R* refers to the ratio of the training data size to the number of weights; NST refers to no-stop training with the ANN networks trained 1000 epochs.

3 Conclusions

Some issues about the generalization of ANN training are investigated through experiments with several synthetic data sets and real world time series data. First issue is how many data are demanded to avoid overfitting? It is found that even with the ratio higher than 30, overfitting still would occur, although not significantly. The second issue is how many data should be used as cross-validation data? It is found that the ratio of cross-validation set to training set has no significant impacts on the testing error. For stationary time series, 10% could be a practical choice. The third issue is which method is better for time series prediction, the Bayesian regularization method or the cross-validated early stopping method. The results show that both methods are helpful when the ratio of training sample size to the number of weights is less than 20. But these methods do not always work. For some cases they fail even when the ratio is less than 10. Especially, the performance of CV early stopping is highly variable. Bayesian method outperforms the CV method in most cases, and it even outperforms no-stop training in some cases when the ratio of training sample size to the number of weights is above 30.

References

1. Amari, S.; Murata, N.; Muller, K.-R.; Finke, M.; Yang, H.H. Asymptotic statistical theory of overtraining and cross-validation. IEEE Transactions on Neural Networks, 8(5), 985 – 996, 1997.
2. Conway, A.J.: Time series, neural networks and the future of the Sun. New Astronomy Reviews, 42, 343–394, 1998.
3. De Oliveira , K.A., Vannucci, A., da Silva, E.C. Using artificial neural networks to forecast chaotic time series. Physica A 284, 393-404, 2000.
4. Foresee, F. D., and Hagan M. T. Gauss-Newton approximation to Bayesian learning. Proceeding of the 1997 International Joint Conference on Neural Networks, 1930-1935, 1997.
5. Geman, S., Bienenstock, E. and Doursat, R. Neural Networks and the Bias/Variance Dilemma, Neural Computation, 4, 1-58, 1992.
6. Henon, M. A two-dimensional mapping with a strange attractor. Commun. Math. Phys., 50(1), 69-77, 1976.
7. Mackay, D. Bayesian methods for adaptive models. PhD Thesis, California Institute of Technology, 1991
8. Mackey, M. C. and L. Glass. Oscillations and chaos in physiological control systems. Science 197: 287-289, 1977.
9. Neal, R. M. Bayesian Learning for Neural Networks, New York: Springer-Verlag, 1996.
10. Prechelt, L. Early stopping – But when? In: Orr, G.B., and Mueller, K.-R., eds., Neural Networks: Tricks of the Trade. Berlin: Springer, 55-69, 1998.
11. Sarle, W.S. Neural Network FAQ, part 3 of 7: Generalization. URL: ftp://ftp.sas.com/pub/neural/FAQ3.html, 2002.

Multi-step-ahead Prediction Based on B-Spline Interpolation and Adaptive Time-Delay Neural Network

Jing-Xin Xie[1,2], Chun-Tian Cheng[3], Bin Yu[1], and Qing-Rui Zhang[4]

[1] School of Electronic and Information Engineering,
Dalian University of Technology, Dalian, 116024, P.R. China
xjxie@student.dlut.edu.cn
[2] College of Mechanical and Electronic Engineering, Hebei Agricultural University,
Baoding 071001, P.R. China
[3] Department of Civil Engineering, Dalian University of Technology,
Dalian, 116024, P.R. China
ctcheng@dlut.edu.cn
[4] School of Environmental and Biological Science and Technology,
Dalian University of Technology, Dalian, 116024, P.R. China

Abstract. The availability of accurate empirical models for multi-step-ahead (MS) prediction is desirable in many areas. Motivated by B-spline interpolation and adaptive time-delay neural network (ATNN) which have proven successful in addressing different complicated problems, we aim at investigating the applicability of ATNN for MS prediction and propose a hybrid model SATNN. The annual sunspots and Mackey-Glass equation considered as benchmark chaotic nonlinear systems were selected to test our model. Validation studies indicated that the proposed model is quite effective in MS prediction, especially for single factor time series.

1 Introduction

Multi-step-ahead (MS) is a classical model predictive algorithm with which at any given time the process outputs can predict time series values of many time-steps into the future. Neural networks for MS prediction were reported by Schenker *et al.*(1995), Prasad *et al.* (1998), Parlos *et al.*(2000) and Bone *et al.*(2002). Among all the proposed methods to deal with the problem, the recurrent neural network was proven to be able to improve MS-based prediction (Parlos *et al.*,2000; Bone *et al.*, 2002). Training of a recurrent neural network, however, is usually very time consuming and a single recurrent neural network might lack in robustness (Ahmad *et al.*, 2002). Then, we began to investigate the capacity of TDNN and ATNN base on feedforward network which is easy to implement compared with recurrent neural network. Time-Delay neural network (TDNN) and it adaptive version of TDNN, adaptive time-delay neural network (ATNN) have been successfully applied in many areas. The current and delayed (or past) observations of the measured system input and output are utilized as inputs to the network in the case of single stage MS prediction(Parlos, et.al.,2000). This especially makes it possible that prediction accuracy deteriorated very quickly with increased p. The lack of rigorous proofs regarding the security of

W. Duch et al. (Eds.): ICANN 2005, LNCS 3697, pp. 565 – 570, 2005.

prediction values necessitates the use of measured relevant data with convincing algorithms. In our paper, the purpose for amplifying history data in every step urges us to find a proper method and pay more attention to interpolation for discrete sequences. Interpolation kernels based on B-splines have attracted recent interest because of their potential for efficient implementation and for the avoidance of erroneous `artefacts' arising from oscillations common to a local or running polynomial fit (Unser, et. al.,1993a,b).

In this paper we propose a three-stage prediction model dynamic spline interpolation with ATNN (SATNN). Via interpolation units and dynamic compounding units and ATNN, respectively, the multi-step-ahead prediction can be obtained from single time series. The effectiveness of the model is demonstrated by the application to annual sunspots and Mackey-Glass equation, and comparison is made with a traditional MS ANN model based on TDNN.

The remainder of this paper is organized as follows. Section 2 presents the SATNN model as the resolution for multi-step ahead forecasting, and then gives the model structure and algorithms. Section 3 presents prediction results and discussion. Section 4 presents the conclusion.

2 Model Architecture and Algorithm

2.1 Model Architecture

Given the time series $\{X \mid x_i, 1 \le i \le n\}$, the three-stage architecture is summarized in Fig.1. In the first stage, G_s are B-spline interpolation generator with parameter q equal to time window of delayed input signals of ATNN in the third stage, which is obtained by spectral estimate with MEM1 (Maximum Entropy Method 1). $\{SI_1, SI_2, \ldots, SI_q\}$ are B-spline interpolation digital filter (Unser, 1999). Among them, SI_1 is a simple linear function generating the same data set as X expressed as $\{X_{ijl} \mid x_{ijl} = x_i, i=1, j \in [1,n], l=1\}$. These interpolation units are employed to interpolate the original signal into the smoothed signals $\{X_{ijl} \mid i \in [1,q], j \in [1,n], l=[1,q], l \le i\}$ with various sampling frequencies $\{1/d_1, 1/d_2, \ldots, 1/d_q \mid d1 = D\}$ where D is original sampling period. It is interesting to note that $\{X_{ijl} \mid i=[1,q], j \in [1,n], l=1, l \le i\} = X$. In the second stage, several time series $\{X_{ijl} \mid i \in [1,q], j \in [1,n], l=[1,q], l \le i\}$ are extracted by moving controller directed by controlling signal $\{c_{ti} \mid 1 \le i \le q\}$ generated by C to construct a new time sequence $\{X_t' \mid x'_{t-i}, 0 \le i \le J, J = [q(q+1)]/2\}$ (see Eq.2) via linear integrating unit Σ. In the third stage, $\{N_{ij} \mid i=1, j=J\}$ and N constitute a ATNN which is feedforward network. N denotes one nonlinear hidden layers and output layer. The error between the output and measured value has remarkable impact on whole network.

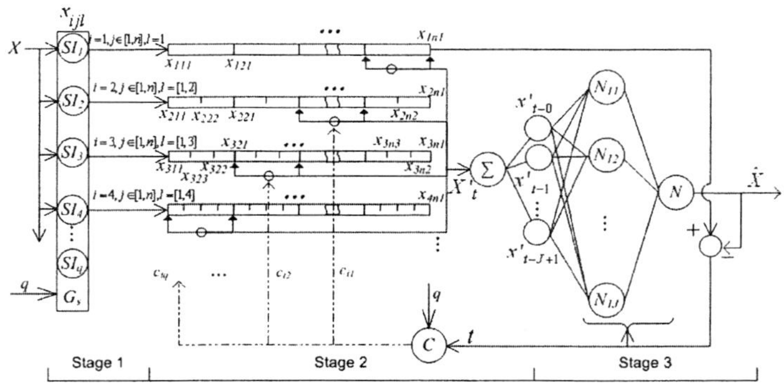

Fig. 1. The three-stage architecture for MS prediction

2.2 Algorithm

In the first stage the time series is interpolated with different units within the original sampling period D by B-spline interpolation transform. B-splines of order n are piecewise polynomial functions of degree n (in our model n=3). The B-spline interpolation equation of the proposed model can be represented as Eq.1

$$\begin{cases} S_3(k') = \sum_{k=1}^{n} c_3(k)\beta_3(k'-k) & k' \in \left\{\frac{1}{r}, \frac{2}{r}, \ldots \frac{r-1}{r} \middle| r \in [2,q]\right\} \\ \beta_3(x) = \beta_0 * \beta_0 * \beta_0 * \beta_0(x) \\ \beta_0(x) = \begin{cases} 1, & -\frac{1}{2} \le x \le \frac{1}{2} \\ 0, & otherwise \end{cases} \end{cases} \tag{1}$$

where k' is the interpolation points , n is the length of original sequence(the same as the umber of data), operator * is convolution, q is the number of B-spline interpolation units which is equivalent to input dimension of ATNN in stage 3.Our aim is to predict the pth sample ahead, $\hat{x}_{t+p}$, of the series. When p>1, the prediction precise begins to depend more and more on the previous forecasting values. Several derivative sequences coming from original observations are used to generate dynamic sequence X' with length J. The procedure can be expressed as follow:

$$\begin{cases} x'_{t-0}=x_{1t1} \\ x'_{t-1}=x_{2t1}, \quad x'_{t-2}=x_{2t2} \\ \vdots \\ \cdots \quad , \ldots, \quad x'_{t-J+1}=x_{q(t-q+1)q} \end{cases} \qquad J=\frac{q(q+1)}{2} \quad ,t\in[p+1,n] \tag{2}$$

where t denotes the current time, J is the length of X'. Note that the interval is $t \in [p+1, n]$, as can be employed to train the neural network. It is reasonable to assume that X' is a dynamic sequence when t different. After interpolation and reconstruction, however, time index in the original has lost meaning, and the natural index in the input of ATNN X' would be paid more attention to. In last stage, t denotes the sequence

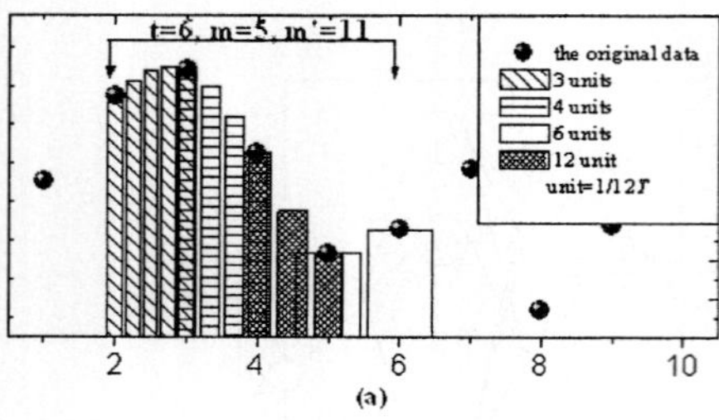

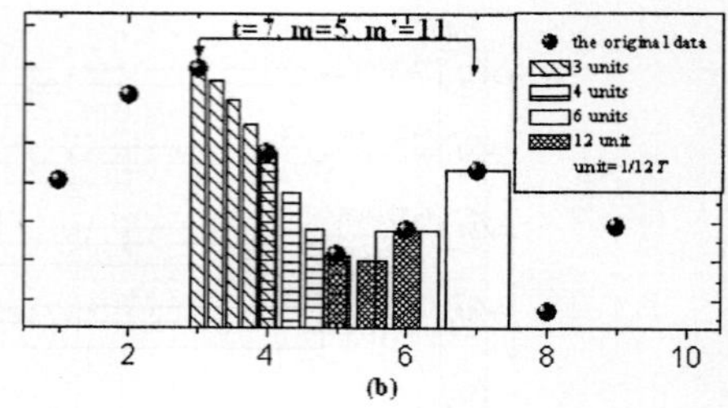

Fig. 2. The dynamic input data set of ATNN, *X'*(assume *q*=4) is restructured When t=6 (a), and t=7 (b) respectively

number of *X'*, and τ denotes the reduce number from current point(for simplification assuming *q*=4, a dynamic sequence *X'* can be obtained as Fig.2).

In this study, ATNN consists of *L* layers with N^L neurons in the *l*th layer. The input-output mapping of the corresponding dynamic neuron of ATNN is governed by

$$y(t) = \sigma\left(\sum_{i=1}^{M} \omega_i x_i\left(t-\tau_i\right)\right) \quad 0 \le \tau_i \le J-1 \tag{3}$$

where ω_i are the neuron weights, τ_i are the delays, and $\sigma(\cdot)$ is a nonlinear activation function. Note that in above equation, the output of the neuron at time t depends on the previous values of the inputs which results in a dynamic behavior. This dynamic behavior will be subsequently modified in appropriate ways to represent different classes of nonlinear systems. Regarding MS prediction, the procedure of input-output can be described as Eq.4. Given the time series data set $\left\{X'_t \mid x'_{t-i}, 0 \le i \le J-1\right\}$ generated in stage 2 where *t* is current time, the new predictions are based on observations, and a group of previous ones, where the quantities with a "hat"

$$\begin{cases} \hat{x}'_{t+p} = F(\hat{x}'_{t+p-1}, \hat{x}'_{t+p-2}, \ldots, \hat{x}'_{t+1}, x'_t, \ldots, x'_{t+p-\tau}) & \tau > p, 0 \le \tau \le J-1 \\ \hat{x}'_{t+p} = F(\hat{x}'_{t+p-1}, \hat{x}'_{t+p-2}, \ldots, \hat{x}'_{t+p-\tau}) & \tau \le p, 0 \le \tau \le J-1 \end{cases} \tag{4}$$

represent estimates of the actual states and outputs, and the others without a "hat" represent observations. *p* is the umber of steps ahead. It should be noticed that the net is trained in a feedforward manner and used as a ATNN model to generate the prediction.The typical ATNN neuron governing equations are developed as follows

$$\begin{cases} net_j^l(t) = \sum_{i=1}^{N^{l-1}} w_{ji}^l o_i^{l-1}(t-\tau_{ji}^l) \\ o_j^l(t) = \sigma^l\left(net_j^l(t)\right) \end{cases}, \quad 0 \le \tau_{ji}^l \le J-1 \tag{5}$$

The output of the *j* th neuron in the *l*th layer at time *t* is denoted by $o_j^l(t)$. The first equation depicts the governing algorithm of original typical multilayer adaptive time-delay, in which the weight and associated delay connecting the *j*th neuron in the *lth* layer to the *i* th neuron in the (*l-1*)th layer are denoted by w_{ji}^l and τ_{ji}^l, respectively. τ_{ji}^l

values form 1 to J . $\left\{o_i^1(t)|x'_{t-i}, 0 \le i \le J-1\right\}$ is the output of the ith neuron in first layer, and $\hat{x}'_{t+1} = \hat{x}_{t+1}$ is the prediction value of x_{t+1}.

The accumulation of the errors in the recursive predictions renders higher difficulty to achieve accurate long range predictions than accurate one-step-ahead predictions. The spline interpolation technology, however, can dynamically enlarge the real within inputs and enforce the robustness of net.

3 Empirical Results and Discussion

In the paper, classical chaotic benchmark time series, the annual sunspots and Mackey-Glass series, are chosen to train and test model for multi-step-ahead forecasting.

- The sunspots of years 1700 through 1959 were chosen to be the training set, and 15-step-ahead forecasting of the sunspots of years 1960 through 1974 was utilized.
- In the discrete-time case (considered here) the series arises from the following delay-difference Eq.6

$$x(t+1) - x(t) = a\frac{x(t-\tau)}{1+x^{10}(t-\tau)} - bx(t) \cdot \tag{6}$$

where both t and τ are integers.The training data are generated using the parameters $a = 0.2, b = 0.1$, $\tau = 17$ and the sampling rate is 5 (only the sample x_0, x_5, x_{10}... are considered). Then the first 225 time steps are used for training, the next 15 time steps for testing.

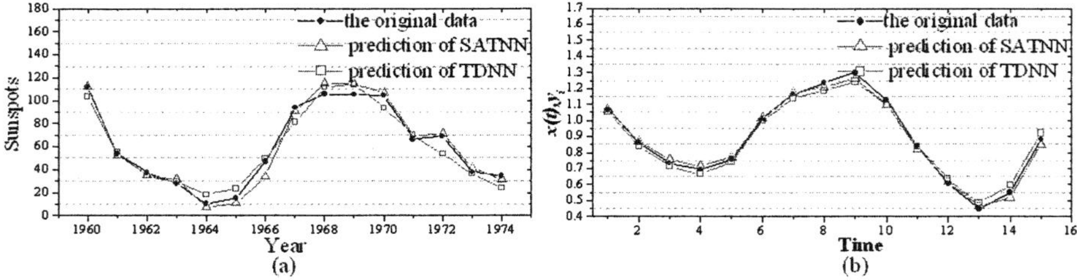

Fig. 3. Prediction results of TDNN and SATNN for (a)annual sunspots;(b) Macky-Glass

In identification, prediction and recognition problems the NMSE is widely used as an evaluation yardstick. We therefore use the NMSE to evaluate the performance of the structures proposed. Meanwhile, TDNN and our model are all trained and tested using time-delay technique. The comparison between them for 15-step-ahead forecasting is given in Fig.3a, b.Though at some data points, the SATNN model gives worse predictions than TDNN, its forecasting capability is improved in all. The results are given in Table 1.

Table 1. The RMSE results for forecasting accuracy measures

	Sunspots$_{1960\text{-}1974}$		Mackey-Glass	
	TDNN	SATNN	TDNN	SATNN
NMSE	0.097857	0.073691	0.00932	0.00753

4 Conclusion

Time series analysis and forecasting is an active research area over the past few decades. Inspired by many technologies such as B-spline interpolation, time-delay and ATNN, we propose a hybrid model for MS forecasting on single factor time series. The model integrates three-stage networks together. The results obtained through the sunspot and the Mackey-Glass chaotic time series substantiate our approach. For MS forecasting based on time-delay problems, the net have both dynamic and correlation structures, and can be extended to other professional areas as well.

Acknowledgments

This research was supported by the National Natural Science Foundation of China (No. 50479055).

References

1. Ahmad, Z., and Zhang, J. : Improving long range prediction for nonlinear process modeling through combining multiple neural network, Proceeding of the 2002 IEEE Interational Conference on Control Applications, (2002) 966-971
2. Bone, R., Crucianu, M.: An evaluation of constructive algorithms for recurrent networks on multi-step-ahead prediction, ICONIP'02 Proceedings of the 9th International Conference on Neural Information Processing, Vol.2 (2002) 547-551
3. Parlos, A. G., Rais, O. T., Atiya, A. F. : Multi-step-ahead prediction using dynamic recurrent neural networks, Neural Network, Vol.13(2000)765-786
4. Prasad, G., Swidenbank, E., and Hogg, B. W.: A neural net model-based multivariable long-range predictive control strategy applied in thermal power plant control, IEEE Transactions on Energy Conversion, Vol.13 (1998) 176-182
5. Schenker, B., and Agarwal, M.: Long-range prediction for poorly-known systems, International Journal of Control, Vol.62 (1995) 227-238
6. Unser, M.: Splines: a perfect fit for signal and image processing, IEEE Signal Processing Magazine, Vol.16 (1999) 22-38
7. Unser, M., Aldroubi, A., and Eden, M. : B-spline signal processing: Part I-Theory, IEEE Signal Processing Magazine, Vol.41 (1993a) 821-833
8. Unser, M., Aldroubi, A., and Eden, M.: Spline Signal Processing: Part II-Efficient Design and Applications, IEEE Signal Processing Magazine, Vol.41 (1993b) 834-848

Training of Support Vector Machines with Mahalanobis Kernels

Shigeo Abe

Graduate School of Science and Technology,
Kobe University, Rokkodai, Nada, Kobe, Japan
abe@eedept.kobe-u.ac.jp
http://www2.eedept.kobe-u.ac.jp/~abe

Abstract. Radial basis function (RBF) kernels are widely used for support vector machines. But for model selection, we need to optimize the kernel parameter and the margin parameter by time-consuming cross validation. To solve this problem, in this paper we propose using Mahalanobis kernels, which are generalized RBF kernels. We determine the covariance matrix for the Mahalanobis kernel using the training data corresponding to the associated classes. Model selection is done by line search. Namely, first the margin parameter is optimized and then the Mahalanobis kernel parameter is optimized. According to the computer experiments for two-class problems, a Mahalanobis kernel with a diagonal covariance matrix shows better generalization ability than a Mahalanobis kernel with a full covariance matrix, and a Mahalanobis kernel optimized by line search shows comparable performance with that with an RBF kernel optimized by grid search.

1 Introduction

Support vector machines have been used for various applications as a powerful tool for pattern classification. One of the advantages of support vector machines is that we can improve generalization ability by proper selection of kernels. In most cases polynomial kernels and radial basis function network (RBF) kernels are used. Mahalanobis kernels [1], which exploit the data distribution information more than RBF kernels do, are expected to ease model selection but how to set the covariance matrix is a difficult problem. Friedrichs and Igel [2] used evolution strategies to tune the parameters obtained by grid search but it is time consuming.

In this paper, we propose model selection for Mahalanobis kernels. Namely, using the data belonging to the two classes, we calculate the covariance matrix for the Mahalanobis kernel. We then optimize the margin parameter and the kernel parameter that scales the Mahalanobis distance by line search: after optimizing the margin parameter by cross validation, we optimize the kernel parameter. We show the usefulness of Mahalanobis kernels over RBF kernels using two-class data sets.

W. Duch et al. (Eds.): ICANN 2005, LNCS 3697, pp. 571–576, 2005.

In Section 2, we discuss Mahalanobis kernels, and in Section 3 we discuss model selection. Finally in Section 4, we compare performance of Mahalanobis kernels with RBF kernels.

2 Mahalanobis Kernels

First we explain the Mahalanobis distance between a datum and the center vector of a cluster. Let the set of M m-dimensional data be $\{\mathbf{x}_1, \ldots, \mathbf{x}_M\}$ for the cluster. Then the center vector and the covariance matrix of the data are given, respectively, by

$$\mathbf{c} = \frac{1}{M} \sum_{i=1}^{M} \mathbf{x}_i, \tag{1}$$

$$Q = \frac{1}{M} \sum_{i=1}^{M} (\mathbf{x}_i - \mathbf{c})\,(\mathbf{x}_i - \mathbf{c})^T. \tag{2}$$

The Mahalanobis distance of $\mathbf{x}$ is given by

$$d(\mathbf{x}) = \sqrt{(\mathbf{x} - \mathbf{c})^T Q^{-1}\,(\mathbf{x} - \mathbf{c})}. \tag{3}$$

Because the Mahalanobis distance is normalized by the covariance matrix, it is linear translation invariant [3]. This is especially important because we need not worry about the scales of input variables.

Another interesting characteristic is that the average of the square of Mahalanobis distances is m [3]:

$$\frac{1}{M} \sum_{i=1}^{M} (\mathbf{x}_i - \mathbf{c})^T Q^{-1}\,(\mathbf{x}_i - \mathbf{c}) = m. \tag{4}$$

Based on the definition of the Mahalanobis distance, we define the Mahalanobis kernel by

$$H(\mathbf{x}, \mathbf{x}') = \exp\left(-(\mathbf{x} - \mathbf{x}')^T A\,(\mathbf{x} - \mathbf{x}')\right), \tag{5}$$

where A is a positive definite matrix. Here, the Mahalanobis distance is calculated between $\mathbf{x}$ and $\mathbf{x}'$, not between $\mathbf{x}$ and $\mathbf{c}$. The Mahalanobis kernel is an extension of the RBF kernel. Namely, by setting

$$A = \gamma\, I, \tag{6}$$

where $\gamma(> 0)$ is a parameter for slope control and I is the $m \times m$ unit matrix, we obtain the RBF kernel:

$$\exp(-\gamma \|\mathbf{x} - \mathbf{x}'\|^2). \tag{7}$$

For a two-class problem, the Mahalanobis kernel is used for the data belonging to one of the two classes. Assuming that $X = \{\mathbf{x}_1, \ldots, \mathbf{x}_M\}$ is the set of data belonging to one of the two classes, we calculate the center and the covariance matrix by (1) and (2), respectively.

Then we approximate the Mahalanobis kernel by

$$H(\mathbf{x}, \mathbf{x}') = \exp\left(-\frac{\delta}{m}\,(\mathbf{x} - \mathbf{x}')^T Q^{-1}\,(\mathbf{x} - \mathbf{x}')\right), \tag{8}$$

where $\delta\,(> 0)$ is the scaling factor to control the Mahalanobis distance.

From (4), by dividing the square of the Mahalanobis distance by m, it is normalized to 1 irrespective of the number of input variables. Although (8) is an approximation of the Mahalanobis kernel, this may enable to limit the search of the optimal δ value in a small range.

If we use the full covariance matrix, it will be time-consuming for a large number of input variables. Thus we consider two cases: Mahalanobis kernels with diagonal covariance matrices and Mahalanobis kernels with full covariance matrices. Hereafter we call the former diagonal Mahalanobis kernels and the latter non-diagonal Mahalanobis kernels.

3 Model Selection

To maximize the generalization ability of the support vector machine we need to optimize the parameters by model selection. The most reliable method is cross validation. In the following, we discuss model selection for RBF kernels and Mahalanobis kernels by cross validation.

3.1 RBF Kernels

For RBF kernels, we need to determine the values of γ and C by grid search. To set the proper search range of γ, it is better to normalize the input ranges into $[0, 1]$. Thus, because the maximum value of $\|\mathbf{x} - \mathbf{x}'\|^2$ is m, we use the following RBF kernels instead of (7) [4]:

$$\exp\left(-\frac{\gamma}{m}\|\mathbf{x} - \mathbf{x}'\|^2\right). \tag{9}$$

However, because RBF kernels are not scale invariant, the range of $[0, 1]$ may not be optimal.

3.2 Mahalanobis Kernels

For Mahalanobis kernels, we need to determine the values of δ and C. But because Mahalanobis kernels given by (8) are determined according to the data distribution and normalized by m, the initial value of $\delta = 1$ is a good selection. Thus, we can carry out model selection by line search not by grid search. Namely, the model selection is done as follows:

1. Set $\delta = 1$ and determine the value of C by cross validation. We call this the first stage.
2. Setting the value of C as that determined by the first stage, determine the value of δ by cross validation. We call this the second stage.

Because $\delta = 1$ is a good initial value, we may search the optimal value around 1, e.g., $[0.1, 2]$.

In addition, because Mahalanobis kernels are normalized by the covariance matrix, it is scale invariant. Therefore, the scale transformation of input variables does not affect the classification performance of the support vector machine.

4 Performance Evaluation

We compared the generalization ability of Mahalanobis kernels and RBF kernels using two-class data sets used in [5].[1] Each problem has 100 or 20 training data sets and their corresponding test data sets. Because there is not much difference of generalization abilities between L1 and L2 support vector machines, we used L1 support vector machines. We determined the optimal values of γ and C for RBF kernels and those of δ and C for Mahalanobis kernels by 5-fold cross validation. Because the input ranges of the data sets were not normalized, we normalized them to $[0, 1]$

For RBF kernels for a value of γ in $\{0.1, 0.5, 1, 5, 10, 15\}$ we performed cross validation of the first five training data sets changing $C = [1, 10, 50, 100, 500, 1000, 2000, 3000, 5000, 8000, 10000, 50000, 100000]$, selected the optimal γ that showed the minimum average error rate for the five validation data sets, and selected the median of the best value of C for the optimal γ. Then, for the optimal values of γ and C, we trained the support vector machine for 100 or 20 training data sets and calculated the average recognition error and the standard deviation for the test data sets.

Similarly for Mahalanobis kernels, at the first stage we determined the optimal value of C by cross validation for the first five training data sets. Then, at the second stage we performed cross validation with the determined value of C, changing $\delta = [0.1, 0.2, \ldots, 1.9, 2]$. As a reference we also performed the grid search of optimum δ for $\delta = [0.1, 0.5, 1.0, 1.5, 2.0]$ and C.

If the recognition rate of the validation set took the maximum value for different values of C, we took the smallest value as the optimal value.

Table 1 lists the parameters obtained by the preceding procedure. Here, we do not include parameters for Mahalanobis kernels obtained by grid search. From the table, it is seen that the values of C for the Mahalanobis kernels are equal to or smaller than those for RBF kernels. In addition, for the image and thyroid data sets, the values for non-diagonal Mahalanobis kernels are smaller than for the diagonal Mahalanobis kernels. This means that the support vector machines with RBF kernels are the most difficult to fit to the data, whereas those with non-diagonal Mahalanobis kernels are the easiest.

[1] http://ida.first.fraunhofer.de/projects/bench/benchmarks.htm

Table 1. Parameter setting

Data	RBF		Diagonal		Non-diagonal	
	γ	C	C	δ	C	δ
Banana	15	100	50	0.8	50	0.9
B. Cancer	1	10	1	0.6	1	0.8
Diabetes	10	1	1	0.5	1	0.2
German	5	1	1	1.7	1	0.9
Heart	0.1	50	1	0.2	1	0.1
Image	10	1000	500	0.7	100	1
Ringnorm	15	1	1	1.5	1	1.3
F. Solar	1	1	1	0.1	1	0.1
Splice	10	10	10	0.8	10	0.5
Thyroid	5	1000	50	0.4	10	0.9
Titanic	10	10	10	0.7	10	0.6
Twonorm	1	1	1	0.9	1	0.2
Waveform	5	10	1	0.6	1	0.4

Table 2. Comparison of average error rates and standard deviations

Data	RBF	Diagonal-1	Diagonal-2	Diagonal	Non-diagonal
Banana	10.5±0.5	10.5±0.4	**10.4**±0.5	**10.4**±0.5	10.5±0.5
B. Cancer	**25.6**±4.4	25.9±4.2	**25.6**±4.4	25.9±4.2	26.1±4.4
Diabetes	23.4±1.7	24.7±1.9	23.7±1.7	23.7±1.7	**23.3**±1.8
German	23.8±2.1	**23.4**±2.1	23.9±2.1	23.7±1.7	23.7±2.2
Heart	16.1±3.1	17.2±3.2	15.7±3.2	**15.6**±3.4	17.2±4.0
Image	**2.8±0.5**	3.1±0.6	3.0±0.5	3.0±0.6	3.2±0.6
Ringnorm	2.6±0.4	1.8±0.2	1.7±0.1	**1.6±0.1**	1.8±0.1
F. Solar	**32.3**±1.8	34.1±2.0	32.5±1.7	32.8±1.7	32.5±1.7
Splice	10.8±0.7	**10.7**±0.7	10.8±0.6	10.8±0.7	13.0±0.6
Thyroid	**4.1**±2.3	4.2±2.0	**4.1**±2.3	4.2±2.3	6.9±2.8
Titanic	**22.5±1.0**	**22.5±1.0**	**22.5±1.0**	**22.5±1.0**	22.6±1.0
Twonorm	**2.4±0.1**	2.7±0.2	2.7±0.2	2.7±0.1	2.8±0.2
Waveform	10.3±0.4	**9.9±0.4**	**9.9**±0.5	10.5±0.4	15.6±1.2

Table 2 lists the average classification errors and the standard deviations with the ± symbol. The "Diagonal-1" and "Diagonal-2" columns list the values for the first and second stages, respectively, and the "Diagonal" column lists the values by the grid search. Performance of RBF kernels with the input range of $[0, 1]$ is different from that with the original input range given in [5]. Except for the ringnorm data set, the performance with the input range of $[0, 1]$ performed better. If we use the original input range for the ringnorm data set, the performance is 1.7±0.1, which is equivalent to that of the second stage using the diagonal Mahalanobis kernel (Diagonal-2). But for Mahalanobis kernels, performance does not change for the change of the input range.

The best performance in the row is shown in boldface. Except for the diabetes, heart, and f. solar data sets, the recognition performance of diagonal Mahalanobis kernels with $\delta = 1$ (Diagonal-1) was comparable with that of the RBF kernels. For these data sets by optimizing the value of δ, performance of the diagonal Mahalanobis kernels (Diagonal-2) was improved and comparable with that of RBF kernels. There is not much difference between Diagonal-2 and Diagonal. But performance of non-diagonal Mahalanobis kernels was not so good. The full covariance matrix might cause overfitting.

5 Conclusions

We discussed how to train support vector machines with Mahalanobis kernels for pattern classification problems. We calculate the covariance matrix using the training data and determine the optimum values of the margin parameter and the kernel parameter by line search. The computer experiments showed that the performance of the Mahalanobis kernels by line search of the optimal margin and kernel parameters was comparable to that of RBF kernels by grid search of the optimal parameters.

References

1. R. Herbrich. *Learning Kernel Classifiers: Theory and Algorithms.* MIT Press, Cambridge, MA, 2002.
2. F. Friedrichs and C. Igel. Evolutionary tuning of multiple SVM parameters. *Proc. ESANN 2004*, pp. 519–524, 2004.
3. S. Abe. *Pattern Classification: Neuro-Fuzzy Methods and Their Comparison.* Springer-Verlag, London, 2001.
4. S. Abe. *Support Vector Machines for Pattern Classification.* Springer-Verlag, New York, 2005.
5. K.-R. Müller, S. Mika, G. Rätsch, K. Tsuda, and B. Schölkopf. An introduction to kernel-based learning algorithms. *IEEE Trans. Neural Networks*, 12(2):181–201, 2001.

Smooth Bayesian Kernel Machines

Rutger W. ter Borg[1] and Léon J.M. Rothkrantz[2]

[1] Nuon NV, Applied Research & Technology,
Spaklerweg 20, 1096 BA Amsterdam, the Netherlands
rutger@terborg.net
[2] Delft University of Technology,
Mekelweg 4, 2628 CD Delft, the Netherlands
l.j.m.rothkrantz@ewi.tudelft.nl

Abstract. In this paper, we consider the possibility of obtaining a kernel machine that is sparse in feature space and smooth in output space. Smooth in output space implies that the underlying function is supposed to have continuous derivatives up to some order. Smoothness is achieved by applying a roughness penalty, a concept from the area of functional data analysis. Sparseness is taken care of by automatic relevance determination. Both are combined in a Bayesian model, which has been implemented and tested. Test results are presented in the paper.

1 Introduction

In tasks such as time series modelling and system control engineering, representing observed data per se does not draw the primary interest, but rather how and how rapidly a system responds to certain events, i.e. the behaviour of derivatives of functions describing such a system.

Kernel machines have become a popular tool to model measured data with. Emerged by the combination of several disciplines, they have in common that they combine the kernel trick [1] and the principle of parsimony [2,3]. The latter is brought forth by obtaining a sparse model that utilises a small subset of the data to represent a function with. However, as yet no special attention has been paid to the smoothness of that function[1], i.e., it should have continuous derivatives up to some order. In case of regression, we want the resulting function to be smooth, and in case of a classification problem, the resulting decision boundary should be smooth.

The remainder of this paper is organised as follows. In section 2, we introduce derivative kernels, and kernel roughness penalties. We note that through penalised regularisation, roughness penalties do not lead to sparseness. In section 3, we introduce a novel Bayesian prior, its related model, and update equations. Section 4 shows experimental results, and section 5 concludes the paper.

[1] The smooth support vector machine [4] entails a reformulation of the quadratic program of the support vector machine.

W. Duch et al. (Eds.): ICANN 2005, LNCS 3697, pp. 577–582, 2005.

2 Smooth Functional Representations

In supervised learning, we consider a data set $\mathcal{D} = (\mathbf{x}_1, y_1), \ldots, (\mathbf{x}_N, y_N)$ containing N input-output pairs $(\mathbf{x}_i, y_i) \in \mathcal{X} \times \mathcal{Y}$, with $\mathcal{X}$ typically containing multidimensional vectors in $\mathbb{R}^M$, and $\mathcal{Y}$ representing either classes in case of classification, or scalars in $\mathbb{R}$ in case of regression. Wahba [5] shows that we can represent our data $\mathcal{D}$ using a linear model of the form

$$y(\mathbf{x}) = w_0 + \sum_{i=1}^{N} w_i k(\mathbf{x}, \mathbf{x}_i) \tag{1}$$

with bias w_0, and parameters $w_1, \ldots, w_N$, and a kernel function k which, under certain conditions, defines an inner product in feature space, $k(\mathbf{x}_i, \mathbf{x}_j) = \Phi(\mathbf{x}_i)^{\mathrm{T}} \Phi(\mathbf{x}_j)$.

In the field of functional data analysis, smoothness is favoured explicitly by applying a roughness penalty [6], a penalty on the degree of curvature of one or more derivatives of $y(\mathbf{x})$. Fortunately, as $y(\mathbf{x})$ in (1) is linear, we can get to its derivatives by establishing the derivatives of the kernel functions $k(\mathbf{x}, \mathbf{x}_i)$.

2.1 Derivative Kernels

For the n-th power of a vector $\mathbf{x}$, we define $\mathbf{x}^n = (\mathbf{x}^{\mathrm{T}}\mathbf{x})^{n/2}$ for n even, and $\mathbf{x}^n = \mathbf{x}(\mathbf{x}^{\mathrm{T}}\mathbf{x})^{(n-1)/2}$ for n odd. Using this definition, the dimension of the n-th order derivative of a kernel $D^n k(\mathbf{x}, \mathbf{x}_i) = \partial^n k(\mathbf{x}, \mathbf{x}_i)/\partial \mathbf{x}^n$ is either 1 for n is even, or M for n being an odd number, where M is the dimension of underlying vectors $\mathbf{x}$ and $\mathbf{x}_i$. Because a kernel is a mapping $\mathbb{R}^M \times \mathbb{R}^M \to \mathbb{R}$, derivative kernels exist for n even or for $M = 1$ (or both). We discuss derivatives of the commonly used Gaussian kernel and derivatives of the polynomial kernel.

Derivatives of the Gaussian kernel $k(\mathbf{x}, \mathbf{x}_i) = \exp(-\frac{1}{2}\sigma^{-2} \|\mathbf{x} - \mathbf{x}_i\|_2^2)$ are identified by Rodrigues' formula for Hermite polynomials

$$H_n(u) = (-1)^n \exp(u^2) D^n \exp(-u^2). \tag{2}$$

By substituting $(\sqrt{2}\sigma)^{-1}(\mathbf{x} - \mathbf{x}_i)$ for u in (2), and keeping track of additional terms of $(\sqrt{2}\sigma)^{-1}$, we arrive at the convenient compact form of the derivatives of Gaussian kernels

$$D^n k(\mathbf{x}, \mathbf{x}_i) = (-\sqrt{2}\sigma)^{-n} H_n((\sqrt{2}\sigma)^{-1}(\mathbf{x} - \mathbf{x}_i)) \exp(-\tfrac{1}{2}\sigma^{-2} \|\mathbf{x} - \mathbf{x}_i\|_2^2).$$

Derivatives of the polynomial kernel $k(\mathbf{x}, \mathbf{x}_i) = \left(\gamma \mathbf{x}^{\mathrm{T}} \mathbf{x}_i + \lambda\right)^d$ are found to be

$$D^n k(\mathbf{x}, \mathbf{x}_i) = \frac{d!}{(d-n)!} \gamma^n \mathbf{x}_i^n \left(\gamma \mathbf{x}^{\mathrm{T}} \mathbf{x}_i + \lambda\right)^{d-n}$$

which is valid in this instantiation as long as $d \geq n$.

2.2 Kernel Roughness Penalties

We use a kernel matrix $\mathbf{K}$ with entries $K_{ij} = k(\mathbf{x}_i, \mathbf{x}_j)$ for inputs $\mathbf{x}_1, \mathbf{x}_2, \ldots, \mathbf{x}_N \in \mathcal{X}$, and a design matrix $\mathbf{H} = [\mathbf{1}\ \mathbf{K}]$ that consists of the combination of $[1, 1, \ldots, 1]^{\mathrm{T}}$ and a kernel matrix. For example, in a regularisation setting, applying a roughness penalty can be accomplished by penalising the summed curvature of the second order derivatives

$$\min \|\mathbf{y} - \mathbf{H}\mathbf{w}\|_2^2 + \lambda \left\|D^2\mathbf{H}\mathbf{w}\right\|_2^2 \tag{3}$$

with $D^2\mathbf{H} = [\mathbf{0}\ D^2\mathbf{K}]$ being a design matrix of a second order derivative kernel. For $\lambda = 0$, this imposes an ordinary least squares, and for $\lambda \to \infty$, the result will approximate a straight line. Besides a penalty on the second order derivative as shown in (3), more generally, a linear differential operator

$$Ly(\mathbf{x}) = \sum_n c_n D^n y(\mathbf{x}) \tag{4}$$

can be defined [7], and used to form a weighted sum of penalties on derivatives of $y(\mathbf{x})$. If applied in penalised regularisation, L takes the shape

$$\min \|\mathbf{y} - \mathbf{H}\mathbf{w}\|_2^2 + \lambda \|\mathbf{L}\mathbf{w}\|_2^2 \tag{5}$$

which is solved by $\mathbf{w} = (\mathbf{H}^{\mathrm{T}}\mathbf{H} + \lambda\mathbf{L}^{\mathrm{T}}\mathbf{L})^{-1}\mathbf{y}^{\mathrm{T}}\mathbf{H}$. Despite reasonable conditions of matrices $\mathbf{H}^{\mathrm{T}}\mathbf{H}$ and $\mathbf{L}^{\mathrm{T}}\mathbf{L}$, in practice, this system of equations is already singular in case of duplicate inputs. In the field of functional data analysis, this is addressed by pre-processing, such as starting with the dominant frequencies [8]. Also, the selection of parameter λ is done manually by methods such as cross-validation.

Although the resulting functions are smooth, regularisation as in (5) does not promote sparseness. We would like to have a representation that is both smooth and sparse, and in addition, an automatic choice of parameter λ.

3 Smooth Relevance Vector Machine

We combine the idea of regularisation by a roughness penalty with a Bayesian sparseness inducing prior. We presume that the outputs are corrupted by Gaussian noise $p(\mathbf{y}|\mathbf{w}, \sigma^2) = \mathcal{N}(\mathbf{y}|\mathbf{H}\mathbf{w}, \sigma^2\mathbf{I})$. To promote both smoothness and sparseness, we propose a zero-mean Gaussian prior over the weights, and a covariance matrix consisting of two terms

$$p(\mathbf{w}|\boldsymbol{\alpha}, \lambda) = \mathcal{N}(\mathbf{w}|\mathbf{0}, (\lambda\mathbf{L}^{\mathrm{T}}\mathbf{L} + \mathbf{A})^{-1}). \tag{6}$$

The first term $\lambda\mathbf{L}^{\mathrm{T}}\mathbf{L}$ is the smoothness promoting part, formed by a roughness penalty matrix defined by a linear differential operator L as in (4). The second term is a diagonal matrix $\mathbf{A} = \mathrm{diag}(\alpha_0, \ldots, \alpha_{\mathrm{N}})$ containing one hyperparameter α_i per weight w_i, as in automatic relevance determination (ARD) [9], as applied in the relevance vector machine (RVM) [10].

By applying Bayes' rule we obtain the posterior distributions

$$p(\mathbf{w}|\mathbf{y}, \boldsymbol{\alpha}, \lambda, \sigma^2) = \mathcal{N}(\mathbf{w}|\boldsymbol{\mu}, \boldsymbol{\Sigma}),$$
$$p(\mathbf{y}|\boldsymbol{\alpha}, \lambda, \sigma^2) = \mathcal{N}(\mathbf{y}|\mathbf{0}, \sigma^2\mathbf{I} + \mathbf{H}\mathbf{S}\mathbf{H}^\mathrm{T}).$$

with $\boldsymbol{\Sigma} = (\sigma^{-2}\mathbf{H}^\mathrm{T}\mathbf{H} + \mathbf{S}^{-1})^{-1}$, $\mathbf{S} = (\lambda\mathbf{L}^\mathrm{T}\mathbf{L} + \mathbf{A})^{-1}$ and $\boldsymbol{\mu} = \sigma^{-2}\boldsymbol{\Sigma}\mathbf{H}^\mathrm{T}\mathbf{y}$.

3.1 Update Rules

To obtain (locally) optimal values for $\boldsymbol{\alpha}$, λ and σ^2, we will follow a type-II maximum likelihood approach similar to that of the RVM [10]. The update equations for the automatic relevance determination hyper-parameters α_i are given by

$$\alpha_i^{t+1} = \frac{\alpha_i^t(S_{ii} - \Sigma_{ii})}{\mu_i^2}. \tag{7}$$

Note that in the case of $\lambda = 0$, then $S_{ii} = (\alpha_i^{-1})^t$, making (7) identical to the update equation of the original RVM. The roughness-penalty parameter λ is updated by

$$\lambda^{t+1} = \frac{(\mathrm{Tr}(\mathbf{S}\lambda^t\mathbf{L}^\mathrm{T}\mathbf{L}) - \mathrm{Tr}(\boldsymbol{\Sigma}\lambda^t\mathbf{L}^\mathrm{T}\mathbf{L}))}{\boldsymbol{\mu}^\mathrm{T}\mathbf{L}^\mathrm{T}\mathbf{L}\boldsymbol{\mu}}, \tag{8}$$

and the variance estimate by

$$(\sigma^2)^{t+1} = \frac{\|\mathbf{y} - \mathbf{H}\boldsymbol{\mu}\|^2}{N - \mathrm{Tr}(\boldsymbol{\Sigma}(\sigma^{-2})^t\mathbf{H}^\mathrm{T}\mathbf{H})}. \tag{9}$$

In practice, we obtain traces of $\mathbf{SA}$ and $\boldsymbol{\Sigma}\mathbf{A}$ by evaluating (7). We fully compute $\mathrm{Tr}(\boldsymbol{\Sigma}\lambda^t\mathbf{L}^\mathrm{T}\mathbf{L})$, and obtain the other trace in (8) by $\mathrm{Tr}(\mathbf{S}\lambda^t\mathbf{L}^\mathrm{T}\mathbf{L}) = N - \mathrm{Tr}(\mathbf{SA})$, and that of (9) by using $\mathrm{Tr}(\boldsymbol{\Sigma}(\sigma^{-2})^t\mathbf{H}^\mathrm{T}\mathbf{H}) = N - \mathrm{Tr}(\boldsymbol{\Sigma}\mathbf{A}) - \mathrm{Tr}(\boldsymbol{\Sigma}\lambda^t\mathbf{L}^\mathrm{T}\mathbf{L})$.

4 Experimental Results

To measure the effectiveness of the proposed Bayesian model, we have performed a series of experiments. Because we need availability of the derivatives of the underlying function in order to be able to measure and compare performance, we have chosen the well-known sinc benchmark.

The data for one experiment consist of 50 points, with x_i uniformly distributed over $[-10, 10]$, and $y_i \sim \mathcal{N}(\mathrm{sinc}(x_i), 0.1)$. We trained a series of kernel machines using these data with our implementation[2]. For our method, we have chosen to penalise the 10th order derivative only, which should effectuate smoothness up to the 8th order [8]. A Gaussian kernel is used by all methods, parametrised by $\sigma = 1.6$. We repeated this experiment 1000 times. Table 1 shows the mean and standard deviation of the results, subsequently of the number of basis vectors, and of the root-mean-square of errors of zeroth, second and fourth order derivatives.

[2] The Kernel-Machine Library is available at `http://www.terborg.net`.

Table 1. Comparative performance

Method	BVs	D^0	D^2	D^4
RVM (old) [10]	6.5 ± 1.0	0.047 ± 0.010	0.054 ± 0.014	0.106 ± 0.030
RVM (new) [11]	6.0 ± 1.1	0.049 ± 0.010	0.056 ± 0.013	0.108 ± 0.026
Figueiredo [12]	6.3 ± 1.3	0.053 ± 0.012	0.067 ± 0.022	0.136 ± 0.051
KRLS [13]	17.0 ± 0.0	0.054 ± 0.012	0.077 ± 0.017	0.242 ± 0.066
SOG-SVR [14]	17.0 ± 0.0	0.062 ± 0.010	0.050 ± 0.008	0.062 ± 0.012
AO-SVR [15]	20.7 ± 3.2	0.057 ± 0.011	0.112 ± 0.029	0.374 ± 0.115
SRVM [this paper]	16.3 ± 3.4	0.039 ± 0.009	0.030 ± 0.009	0.033 ± 0.013

As shown in table 1, smoothness requires notably more basis vectors (BVs) than common sparse Bayesian models. On the other hand, the basis vectors seem to be spent wisely, because the error on all derivative orders is consistently lower than that of the compared methods. Figure 1 illustrates a function that is smooth in output space. Although at first sight the fit of the function itself is not dramatically different from others (left), the gain in quality of fit is clearly visible at a higher order derivative (right).

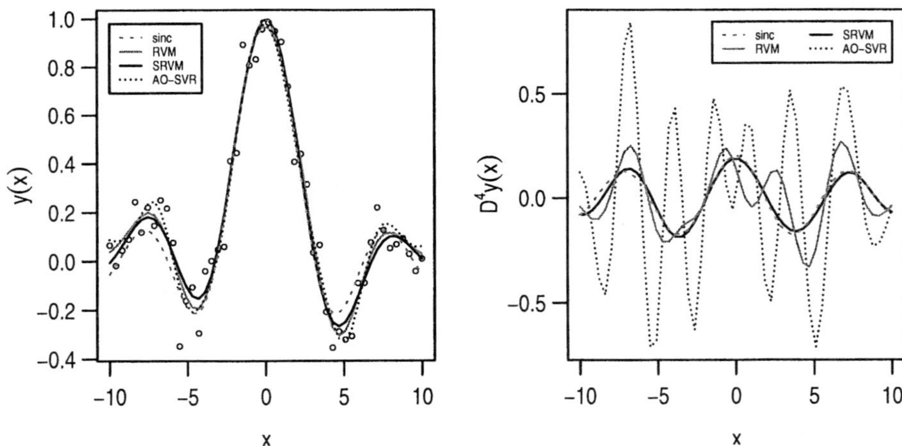

Fig. 1. Input data to a single experiment, the underlying sinc(x) function, and the resulting RVM, SRVM, and AO-SVR (left). The 4th order derivative of sinc(x), RVM, SRVM, and AO-SVR (right).

5 Conclusions

In this paper we introduced a concept from functional data analysis to promote smoothness in output space. We successfully established derivative kernels, kernel roughness penalties, and a Bayesian model to combine kernel roughness penalties with automatic relevance determination. Experiments elicited that smoothness in output space increased the quality of fit on all measured derivative orders. The required number of basis vectors is not the lowest, but the quality of the approximation to the underlying function is empirically better.

Further studies should include the handling of an automatic selection of several c_n in (4), i.e., a penalty on more than one derivative order. An interesting open theoretical issue is the maximum derivative order on which a signal may be expected, given a data set.

References

1. Aizerman, M., Braverman, E., Rozonoèr, L.: Theoretical foundations of the potential function method in pattern recognition learning. Automation and Remote Control **25** (1964) 821–837
2. Schölkopf, B., Smola, A.: Learning with Kernels. Adaptive Computation and Machine Learning. The MIT Press, Cambridge, Massachusetts, USA (2002)
3. Vapnik, V.: The Nature of Statistical Learning Theory. Statistics for Engineering and Information Science. Springer-Verlag, New York (1995)
4. Lee, Y., Mangasarian, O.: SSVM: A smooth support vector machine for classification. Computational Optimization and Applications **20** (2001) 5–22
5. Wahba, G.: Spline Models for Observational Data. Volume 59 of CBMS-NSF Regional Conference Series in Applied Mathematics. Society for Industrial and Applied Mathematics (1990)
6. Green, P., Silverman, B.: Nonparametric Regression and Generalized Linear Models: A Roughness Penalty Approach. Chapman & Hall (1993)
7. Ramsay, J., Silverman, B.: Applied Functional Data Analysis. Springer-Verlag (2002)
8. Ramsay, J., Silverman, B.: Functional Data Analysis. Springer Series in Statistics. Springer-Verlag, New York (1997)
9. MacKay, D.: A practical bayesian framework for backprop networks. Neural Computation **4** (1992) 448–472
10. Tipping, M.: Sparse bayesian learning and the relevance vector machine. Journal of Machine Learning Research **1** (2001) 211–244
11. Tipping, M., Faul, A.: Fast marginal likelihood maximisation for sparse bayesian models. In Bishop, C., Frey, B., eds.: Proceedings of the Ninth International Workshop on Artificial Intelligence and Statistics, January 3–6 2003, Key West, Florida. (2003)
12. Figueiredo, M.: Adaptive sparseness for supervised learning. IEEE Transactions on Pattern Analysis and Machine Intelligence **25** (2003) 1150–1159
13. Engel, Y., Mannor, S., Meir, R.: The kernel recursive least squares algorithm. ICNC03 001, Interdisciplinary Center for Neural Computation, Hebrew University, Jerusalem, Israel (2003)
14. Engel, Y., Mannor, S., Meir, R.: Sparse online greedy support vector regression. In Elomaa, T., Mannila, H., Toivonen, H., eds.: Machine Learning: ECML 2002. Volume 2430 of Lecture Notes in Computer Science., Berlin, Springer-Verlag (2002)
15. Ma, J., Theiler, J., Perkins, S.: Accurate on-line support vector regression. Neural Computation **15** (2003) 2683–2704

A New Kernel-Based Algorithm for Online Clustering

Habiboulaye Amadou Boubacar[1,2] and Stéphane Lecoeuche[1,2]

[1] Laboratoire Automatique, & Génie Informatique et Signal,
Université des Sciences et Technologies de Lille, P2, 59655 Villeneuve d'Ascq, France
[2] Département Génie Informatique et Productique,
Ecole des Mines de Douai, 941, rue Charles Bourseul, BP838, 59508 Douai, France
{amadou, lecoeuche}@ensm-douai.fr

Abstract. This paper presents a kernel-based clustering algorithm called SAKM (Self-Adaptive Kernel Machine) that is developed to learn continuously evolving clusters from non-stationary data. Dedicated to online clustering in multi-class environment, this algorithm is based on an unsupervised learning process with self-adaptive abilities. This process is achieved through three main stages: clusters creation (with an initialization procedure), online clusters adaptation and clusters fusion. Thanks to a new specific kernel-induced similarity measure, the SAKM algorithm is attractive to be very computationally efficient in online applications. At the end, some experiments illustrate the capacities of our algorithm in non-stationary environment.

1 Introduction

Clustering methods group data by using distributions models according to various criteria (distance, membership function...). Numerous techniques have been developed for clustering data in a static environment [1]. However, in many real-life applications, non-stationary (i.e. time-varying) data are generally common and data distributions undergo variations. Therefore, the challenges of online clustering require unsupervised and recursive learning rules that are useful to incorporate new information and to take into account model evolutions over time. Also, the algorithms must be computationally efficient and have to provide good convergence properties in order to be applied to real problems.

So far, several algorithms that have been proposed for online clustering of non-stationary data were generally developed with neural network techniques [2], [3]. More recently, Lecoeuche and Lurette [4] have proposed a new neural architecture (AUDyC: Auto-Adaptive and Dynamical Clustering) that is developed with recursive learning rules. However, this algorithm leads to overfitting and becomes limited in high-dimensional space. Moreover, the convergence bounds of these NN algorithms are not theoretical proved in the context of online clustering.

During recent years, Support Vector Machines (SVM) and related kernel methods have proven to be successful in many applications of pattern recognition [5], [6]. They provide both theoretical and experimental attractions [7], [8], [9]. But, few SVM algorithms exist with recursive learning rules. To online adapt a single-class density support, Gretton and Desobry present an incremental algorithm [10] that provides the

W. Duch et al. (Eds.): ICANN 2005, LNCS 3697, pp. 583 – 588, 2005.

Using the Hinge loss function ξ [11], the derivatives of the *Instantaneous Risk* in the gradient equation (7) provides the iterative update rule of the kernel expansion. Also, the offset ρ^t of the boundary function f^t changes according to data drifts and will be recovered with the hyperplane equation : $f^t(SV) = F^t(SV) - \rho^t = 0$. So, the update equations in the *adaptation stage* (*case* 2, Table 1) are:

$$\begin{cases} \left.\begin{cases} \alpha_{i,m}^{t+1} = (1-\eta)\alpha_{i,m}^t, \quad i<t \\ \alpha_{i,m}^{t+1} = 0 \;[\text{resp. } \eta],\; f_m^t(X_t) \geq 0 \;[\text{resp.} < 0] \end{cases}\right\}, \text{ then } \alpha_{i,m}^{t+1} \leftarrow \alpha_{i,m}^{t+1} \Big/ \sum_i \alpha_{i,m}^{t+1} \\ \rho_m^{t+1} = \sum_{i=\max(1,t-\tau)}^{t} \alpha_{i,m}^{t+1} \kappa(SV_{c,m}^t, SV_{i,m}^t) \end{cases} \tag{9}$$

$\alpha_{i,m}$ is normalised and c is the SV median index chosen for a better estimation. To limit the amount of computation, the kernel expansion is truncated to τ terms by using a sliding window. If the new data is got inside a cluster, no computation is done ; in opposite case, the cluster boundary would change.

Stage 3: In *case 3,* the acquired data is shared by two or more clusters. When the number of these ambiguous data exceeds a threshold A , those clusters are merged :

$$C_{merg} = \left\{ X \in \left(\bigcup_{win} C_{win} \right) \Big/ f_{win}(X) \geq 0 \right\}. \tag{10}$$

f_{merg} is computed with data available in these clusters by using the SAKM update rule. In assumption that two different clusters are disjoints, Fusion preserves the continuity of cluster region and then avoids overlapping. Sets Ω and $\Im$ are modified:

$$\Omega = \left(\Omega \bigcup_{win} \{C_{win}\} \right) \cup \{C_{merg}\} \;\rightarrow\; \Im = \left(\Im \bigcup_{win} \{f_{win}\} \right) \cup \{f_{merg}\}. \tag{11}$$

To reach robustness in non-stationary area, an *elimination stage* is added in [13].

2.3 Performances Analysis

In this section, we point out a brief convergence study of the SAKM update rule in the context of an online estimation of a singe-class distribution. Let $\chi_L = \{X_1, ..., X_t, ..., X_L\}$ be a training data set and $\{f^1, ..., f^t, ..., f^L\}$ boundary functions sequentially learnt on χ_L. Suppose f to be the best learning function on χ_L.

Proposition: Assume that the loss function ξ is convex and Lipschitzian. Assume that $\kappa(\bullet,\bullet)$ is bounded on the training set χ_L. The expectation of instantaneous risk $\bar{E}_{\chi_L}\left[R_{inst}(f^t, X^t)\right]$ of function f^t obtained by using gradient descent in RKHS, converges to the regularised risk of f on set χ_L so that:

$$\overline{E}_{\chi_L}\left[R_{inst}(f^t,X_t)\right]=\frac{1}{L}\sum_{t=1}^{L}R_{inst}(f^t,X_t)\le R_{reg}(f,\chi)+B(L,\delta). \tag{12}$$

with the probability at least $1-\delta$ over random draws of χ_s. The term B is depending of the dataset length L. The proposition (**12**) rises from the NORMA convergence theorem proved in [11]. Hence, using the SAKM update rule, the Vapnik regularised risk will be close to the instantaneous risk with high probability. So, by generalization in multi-class environment, the SAKM algorithm leads to good convergence property. Many tests carried out show that the SAKM update rule is more convenient in learning real drifting targets compare with the NORMA and Gentile' ALMA [13].

3 Experiments

Simulation 1. Data are created from four evolving Gaussian densities(means and dispersals changes as in the experiment [4]). The SAKM algorithm initializes clusters models from the first data acquisitions (Fig. 1.b). Then, models clusters are sequentially update according to their drifting distributions (Fig. 1.c to Fig. 1.f).

Simulation 2. This simulation presents the problem of the clusters fusion (Fig. 2.a). The SAKM algorithm initialises the two clusters (Fig. 2.b), updates iteratively their densities supports according to data evolutions and provides their optimal models (Fig. 2.b to Fig. 2.e). When the two clusters are close enough, the algorithm gives a good fusion result by overcoming overlapping drawbacks (Fig. 2.f).

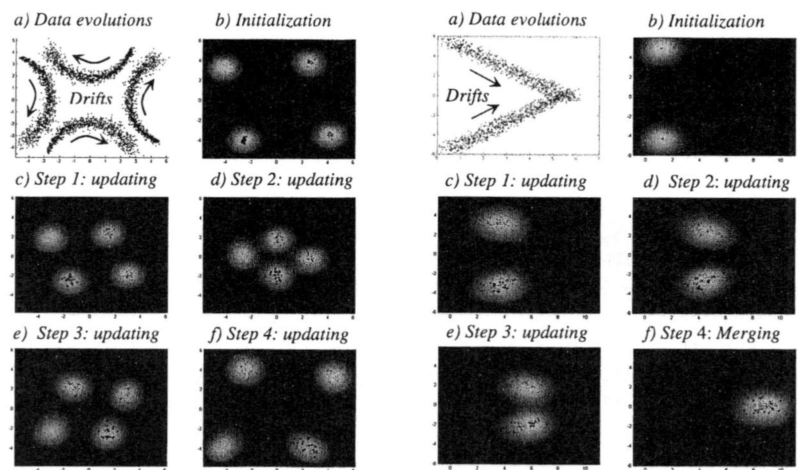

Fig. 1. Online clustering of 4 evolving clusters (real drifting targets) by using SAKM in non-stationary environment

Fig. 2. Fusion of two clusters. The SAKM algorithm takes into account distributions' variations and correctly merges clusters

Parameters settling: $\lambda = 0.8$, η=0.1, $\nu = 0.3$, $\tau = 30$, $\varepsilon_{th} = 0.7$

4 Conclusion

We propose a new online algorithm based on SVM & kernel methods for clustering non-stationary data in multi-class environment. Based on a novel kernel-induced similarity measure in RKHS, the SAKM algorithm is set with an unsupervised learning process using a simple and fast incremental learning rule. The SAKM update rule tracks iteratively the minimisation of Instantaneous Risk in order to avoid overfitting problems. Using density supports in RKHS, the algorithm provides a good reliability to define suitably the clusters structure. After a brief appreciation of the SAKM convergence properties and its attractive validations on artificial data drawing real-drifting targets, experiments have been carried out on real data.

References

1. Bishop C. M.: Neural Networks for Pattern Recognition. Clarendon Press, Oxford, (1995)
2. Eltoft T.: A new Neural Network for Cluster-Detection-and-Labelling. IEEE Trans. on Neural Nets. 9 (1998) 1021-1035
3. Deng, D., Kasabov, N.: On-line pattern analysis by evolving self organizing maps. Neurocomputing. 51 (2003) 87-103
4. Lecoeuche, S., Lurette, C.: Auto-adaptive and Dynamical Clustering Neural Network. Proceedings of ICANN/ ICONIP'03, vol. 2714. Istambul Turkey, (2003) 350-358.
5. Schölkopf, B., Smola, A.: Learning with Kernels. MIT Press, Cambridge MA (2002).
6. Bruges, C.: A tutorial on Support Vector Machines for Pattern Recognition. Data Mining and Knwoledge Discovery. 2 (1998) 121-167.
7. Schölkopf, B. Smola, A., Williamson, R., Bartlett, P.: New Support Vector algorithms. Neural Computation. 12 (2000) 1207-1245
8. Schölkopf, B., Platt, J., Taylor, J. S., Smola, A.: Estimating the Support of a High-Dimensional Distribution. Neural Computation. 13 (2001) 1443-1471
9. Cheong, S., Oh, S., Lee, S.: Support Vector Machines with Binary Tree Architecture for Multi-Class Classification. Neural Information Processing, 2 (2004) 47-51
10. Gretton, A., Desobry, F.: Online one-class nu-svm, an application to signal segmentation. IEEE Proceedings ICASSP'03, vol. 2. Hong Kong (2003) 709-12
11. Kivinen, J., Smola, A., Williamson, R.: Online Learning with Kernel. IEEE Trans. Signal Processing. 52 (2004) 2165-2176
12. Vapnik, V.: An Overview of Statistical Learning Theory. IEEE Trans. on Neural Nets. 10 (1999) 988-999
13. Amadou B., H., Lecoeuche, S., Maouche, S. Self-Adaptive Kernel Machine : Online Clustering in RKHS. IEEE, IJCNN05, Montréal, July 31-August 4 (2005). To be published

The LCCP for Optimizing Kernel Parameters for SVM

Sabri Boughorbel[1], Jean Philippe Tarel[2], and Nozha Boujemaa[1]

[1] IMEDIA Group, INRIA Rocquencourt, 78153 Le Chesnay, France
[2] DESE, LCPC, 58 Bd Lefebvre, 75015 Paris, France

Abstract. Tuning hyper-parameters is a necessary step to improve learning algorithm performances. For Support Vector Machine classifiers, adjusting kernel parameters increases drastically the recognition accuracy. Basically, cross-validation is performed by sweeping exhaustively the parameter space. The complexity of such grid search is exponential with respect to the number of optimized parameters. Recently, a gradient descent approach has been introduced in [1] which reduces drastically the search steps of the optimal parameters. In this paper, we define the LCCP (Log Convex Concave Procedure) optimization scheme derived from the CCCP (Convex ConCave Procedure) for optimizing kernel parameters by minimizing the radius-margin bound. To apply the LCCP, we prove, for a particular choice of kernel, that the radius is log convex and the margin is log concave. The LCCP is more efficient than gradient descent technique since it insures that the radius margin bound decreases monotonically and converges to a local minimum without searching the size step. Experimentations with standard data sets are provided and discussed.

1 Introduction

Support Vector Machine (SVM) [2] is one of the most successful algorithms of machine learning. SVM is flexible since various kernels can be plugged for different data representations. Besides RBF and Polynomial kernels only few other kernels have been used. An interesting and important issue for kernel design consists of assigning, for instance, different scales for each feature component. This is refereed as adaptive metrics [3]. On the other hand, the classical method for tuning the learning algorithm parameters is to select parameters that minimize an estimation or a bound on the generalization error such as cross validation or the radius margin [2]. The latter has been shown to be a simple and predictive enough "estimator" of the generalization error. In this paper, we define the LCCP for optimizing kernel parameters by minimizing the radius margin bound. The LCCP is the direct application of the CCCP [4] to our optimization case.

2 The Log Convex Concave Procedure (LCCP)

The convex concave procedure (CCCP) has been recently introduced [4] for optimizing a function that can be written as a sum of convex and concave functions.

W. Duch et al. (Eds.): ICANN 2005, LNCS 3697, pp. 589–594, 2005.

The advantage of the CCCP compared with gradient descent techniques is that it insures the monotonic decrease of the objective function without searching the size step. In the following, we summarize the main results of the CCCP optimization framework.

Theorem 1. *[4]*

- *Let $E(\boldsymbol{\theta})$ be an objective function with bounded Hessian $\frac{\partial^2 E(\boldsymbol{\theta})}{\partial \boldsymbol{\theta}^2}$. Thus, we can always decompose it into the sum of convex and concave functions.*
- *We consider the minimization problem of a function $E(\boldsymbol{\theta})$ of form $E(\boldsymbol{\theta}) = E_{vex}(\boldsymbol{\theta}) + E_{cave}(\boldsymbol{\theta})$ where E_{vex} is convex and E_{cave} is concave. Then the discrete iterative CCCP algorithm: $\boldsymbol{\theta}_p \to \boldsymbol{\theta}_{p+1}$ given by $\nabla E_{vex}(\boldsymbol{\theta}_{p+1}) = -\nabla E_{cave}(\boldsymbol{\theta}_p)$ decreases monotonically the objective function $E(\boldsymbol{\theta})$ and hence converges to a minimum or a saddle point of $E(\boldsymbol{\theta})$.*
- *The update rule for $\boldsymbol{\theta}_{p+1}$ can be formulated as a minimization of a convex function $\boldsymbol{\theta}_{p+1} = \arg\min_{\boldsymbol{\theta}} E_{p+1}(\boldsymbol{\theta})$ where the convex function $E_{p+1}(\boldsymbol{\theta})$ is defined by*

$$E_{p+1}(\boldsymbol{\theta}) = E_{vex}(\boldsymbol{\theta}) + \boldsymbol{\theta}^\top \nabla E_{cave}(\boldsymbol{\theta_p}).$$

We define the LCCP by applying the CCCP to the case of the minimization of a positive function $J(\boldsymbol{\theta})$ that can be written as a product of log convex and log concave functions $J(\boldsymbol{\theta}) = J_{lvex}(\boldsymbol{\theta})J_{lcave}(\boldsymbol{\theta})$ where $J_{lvex}(\boldsymbol{\theta}) > 0$ is log convex and $J_{lcave}(\boldsymbol{\theta}) > 0$ is log concave. In $\log(J(\boldsymbol{\theta})) = \log(J_{lvex}(\boldsymbol{\theta})) + \log(J_{lcave}(\boldsymbol{\theta}))$, we set $E(\boldsymbol{\theta}) = \log(J(\boldsymbol{\theta}))$, $E_{vex}(\boldsymbol{\theta}) = \log(J_{lvex}(\boldsymbol{\theta}))$ and $E_{cave}(\boldsymbol{\theta}) = \log(J_{lcave}(\boldsymbol{\theta}))$. Hence, we obtain $E(\boldsymbol{\theta}) = E_{vex}(\boldsymbol{\theta}) + E_{cave}(\boldsymbol{\theta})$ where $E_{vex}(\boldsymbol{\theta})$ is convex and $E_{cave}(\boldsymbol{\theta})$ is concave. Moreover, the minima location of $E(\boldsymbol{\theta})$ and $J(\boldsymbol{\theta})$ are the same since the log function is strictly increasing.

3 Parameters Selection Procedure

The optimization of SVM parameters can be performed by minimizing an estimator of the generalization error. The simplest strategy consists in performing an exhaustive search over all possible parameters. When the number of parameters is high, such a technique becomes intractable. In [1], gradient descent framework is introduced for kernel parameter's optimization. Powerful results on the differentiation of various error estimators and generalization bounds are provided. Based of this work, we apply the LCCP framework for optimizing multiple kernel parameters by minimizing the radius margin bound [2]. Indeed, for good choice of kernels, the optimizing problem can be expressed under the condition of LCCP, in particular for the multi-parameters L_1-distance kernel.

3.1 Distance Kernel

In [1], tests with multiple parameters for polynomial and RBF kernels have been successfully carried without over-fitting. From the L_1-distance kernel:

$$K_{L_1}(\boldsymbol{x}, \boldsymbol{x}') = -\sum_{k=1}^{n} |x^k - x'^k|, \tag{1}$$

where $\boldsymbol{x}$ and $\boldsymbol{x}'$ are in $\mathbb{R}^n$ with components x^k and x'^k, we propose its following multiple parameters extension:

$$K_{L_1,\boldsymbol{\theta}}(\boldsymbol{x}, \boldsymbol{x}') = -\sum_{k=1}^{n} \frac{|x^k - x'^k|}{\theta^k}, \tag{2}$$

where $\boldsymbol{\theta}$ is in $\mathbb{R}^{+^n}$ with components θ^k. This kernel is conditionally positive definite, see [5]. We prove that it is possible to use the LCCP for minimizing radius-margin bound $R^2\|\boldsymbol{w}\|^2$, with respect to $\boldsymbol{\theta}$. To do so, we prove the log convexity of the radius R^2 and the log concavity of $\|\boldsymbol{w}\|^2$. Another proof may be used for another kernel. More precisely, for R^2, we will prove that it can be written as a sum of log convex functions. For $\|\boldsymbol{w}\|^2$, it is sufficient to prove that it is concave since the concavity implies the log concavity.

3.2 The Log Convexity of R^2

First, we recall from [6] a useful result on convex functions that we need in the proof of the log convexity of the radius R^2.

Lemma 1. *If for each $\boldsymbol{y} \in \mathcal{A}$, $f(\boldsymbol{x}, \boldsymbol{y})$ is convex in $\boldsymbol{x}$, then the function g, defined as $g(\boldsymbol{x}) = \max_{\boldsymbol{y}\in\mathcal{A}} f(\boldsymbol{x}, \boldsymbol{y})$ is convex in $\boldsymbol{x}$.*

This result can be easily extended to the case of log convex functions. The radius R^2 can be written for the kernel (2) as the following:

$$R^2(\boldsymbol{\theta}) = \max_{\boldsymbol{\beta}\in\mathcal{B}} J_{R^2}(\boldsymbol{\beta}, \boldsymbol{\theta}), \tag{3}$$

where $\mathcal{B} = \{\beta_i \geq 0, \sum_{i=1}^{\ell} \beta_i = 1\}$ and J_{R^2} is the following function:

$$J_{R^2}(\boldsymbol{\beta}, \boldsymbol{\theta}) = -\sum_{i=1}^{\ell} \beta_i \sum_{k=1}^{n} F_{a_{ii}^k}(\boldsymbol{\theta}) + \sum_{i,j=1}^{\ell} \beta_i\beta_j \sum_{k=1}^{n} F_{a_{ij}^k}(\boldsymbol{\theta}), \tag{4}$$

with $F_{a_{ij}^k}(\boldsymbol{\theta}) = f_{a_{ij}^k}(\theta^k) = \frac{a_{ij}^k}{\theta^k}$ and $a_{ij}^k = |x_i^k - x_j^k|$. Since $a_{ii}^k = 0$, the first sum in J_{R^2} is zero. Next, we prove that F is log convex. To do so, it is necessary and sufficient [6] to prove that $\boldsymbol{\nabla}^2 F(\boldsymbol{\theta})F(\boldsymbol{\theta}) - \boldsymbol{\nabla} F(\boldsymbol{\theta})\boldsymbol{\nabla} F(\boldsymbol{\theta})^\top$ is a positive definite matrix. By computing the gradient $\boldsymbol{\nabla} F$ and the Hessian $\boldsymbol{\nabla}^2 F$, it turns out that the obtained matrix is diagonal. Thus the necessary and sufficient condition for the log convexity becomes $f''_{a_{ij}^k}(\theta^k) f_{a_{ij}^k}(\theta^k) - f'^2_{a_{ij}^k}(\theta^k) \geq 0$. We have:

$$f_a(t) = \frac{a}{t},\ f'_a(t) = -\frac{a}{t^2},\ f''_a(t) = \frac{2a}{t^3},\ f''_a(t) f_a(t) - {f'_a}^2(t) = \frac{a^2}{t^4} \geq 0.$$

So J_{R^2} is log convex with respect to $\boldsymbol{\theta}$, as a sum of log convex functions [6]. Lemma 1 implies that R^2 is log convex.

3.3 Log Concavity of $\|w\|^2$

A similar result to Lemma 1, for the concave case, can be derived [6]:

Lemma 2. *Assume that $\mathcal{A}$ is a convex set, if $f(\boldsymbol{x}, \boldsymbol{y})$ is concave in $(\boldsymbol{x}, \boldsymbol{y})$, then the function g, defined by $g(\boldsymbol{x}) = \max_{\boldsymbol{y} \in \mathcal{A}} f(\boldsymbol{x}, \boldsymbol{y})$ is concave in $\boldsymbol{x}$.*

We also need two extra lemmas, which are proved with details in [5]:

Lemma 3. *We define the function f by*

$$f(\boldsymbol{a}, t) = \frac{1}{t}\boldsymbol{a}^\top \boldsymbol{K}\, \boldsymbol{a},\ \boldsymbol{a} \in \mathbb{R}^\ell,\ t \in \mathbb{R}_+,\ \boldsymbol{K} \in \mathbb{R}^{\ell \times \ell}.$$

If $\boldsymbol{K}$ is a positive definite matrix then f is convex in $(\boldsymbol{a}, t)$.

Lemma 4. *We define the function g for $\boldsymbol{t} \in \mathbb{R}^n, \boldsymbol{a} \in \mathbb{R}^\ell$*

$$g(\boldsymbol{a}, \boldsymbol{t}) = \sum_{k=1}^{n} f_k(\boldsymbol{a}, t^k).$$

If each f_k is convex in $(\boldsymbol{a}, t^k)$, then g is convex in $(\boldsymbol{a}, \boldsymbol{t})$.

The expression of $\|\boldsymbol{w}\|^2$ is the following:

$$\|\boldsymbol{w}\|^2(\boldsymbol{\theta}) = \max_{\boldsymbol{\alpha} \in \Lambda} J_{\|\boldsymbol{w}\|^2}(\boldsymbol{\alpha}, \boldsymbol{\theta}),$$

where $\Lambda = \{\alpha_i \geq 0, \sum_{i=1}^{\ell} \alpha_i y_i = 0\}$ and

$$J_{\|\boldsymbol{w}\|^2}(\boldsymbol{\alpha}, \boldsymbol{\theta}) = 2\sum_{i=1}^{\ell} \alpha_i - \sum_{i,j=1}^{\ell} \alpha_i \alpha_j y_i y_j K_{\boldsymbol{\theta}}(\boldsymbol{x}_i, \boldsymbol{x}_j).$$

It is obvious that Λ is a convex set. The first term in $J_{\|\boldsymbol{w}\|^2}$ is linear with respect to $\boldsymbol{\alpha}$, thus it does not affect the convex or concave nature of $J_{\|\boldsymbol{w}\|^2}$. We thus only focus on:

$$J'_{\|\boldsymbol{w}\|^2}(\boldsymbol{\alpha}, \boldsymbol{\theta}) = -\sum_{k=1}^{n} \frac{1}{\theta^k} \sum_{i,j=1}^{\ell} \alpha_i \alpha_j y_i y_j K_k(\boldsymbol{x}_i, \boldsymbol{x}_j)$$

where $K_k(\boldsymbol{x}_i, \boldsymbol{x}_j) = -|x_i^k - x_j^k|$ is conditionally positive definite. We introduce the kernel $\tilde{K}_k$ defined by $\tilde{K}_k(\boldsymbol{x}, \boldsymbol{x}') = K_k(\boldsymbol{x}, \boldsymbol{x}') - K_k(\boldsymbol{x}, \boldsymbol{x}_0) - K_k(\boldsymbol{x}', \boldsymbol{x}_0) + K_k(\boldsymbol{x}_0, \boldsymbol{x}_0)$ where $\boldsymbol{x}_0$ is chosen arbitrary. It is known that $\tilde{K}_k$ is positive definite and that it can be substituted to K_k in the dual SVM problem, see [5]. Similarly, we can substitute K_k by $\tilde{K}_k$ in $J'_{\|\boldsymbol{w}\|^2}$ according to the constraint $\sum_{i=1}^{\ell} \alpha_i y_i = 0$ and rewrite it as $J'_{\|\boldsymbol{w}\|^2}(\boldsymbol{\alpha}, \boldsymbol{\theta}) = -\sum_{k=1}^{n} \frac{1}{\theta^k} \boldsymbol{\alpha_y}^\top \tilde{\boldsymbol{K}}_k\ \boldsymbol{\alpha_y} = -\sum_{k=1}^{n} f_k(\boldsymbol{\alpha}, \theta^k)$, where $\tilde{\boldsymbol{K}}_k$ is the Gram matrix of $\tilde{K}_k$, $\boldsymbol{\alpha_y}$ denotes the vector $[\alpha_1 y_1 \ldots \alpha_\ell y_\ell]^\top$, and $f_k(\boldsymbol{\alpha}, \theta^k) = \frac{1}{\theta^k} \boldsymbol{\alpha}^\top \tilde{\boldsymbol{K}}_{\boldsymbol{y},k}\, \boldsymbol{\alpha}$ with $\left[\tilde{\boldsymbol{K}}_{\boldsymbol{y},k}\right]^{ij} = y_i y_j \left[\tilde{\boldsymbol{K}}_k\right]^{ij}$. We have that $\tilde{\boldsymbol{K}}_{\boldsymbol{y},p}$

is positive definite. Therefore, lemma 3 implies that f_k is convex in $(\boldsymbol{\alpha}, \theta^k)$ and lemma 4 implies that the sum over f_k is convex in $(\boldsymbol{\alpha}, \boldsymbol{\theta})$. Therefore, we have the concavity of $J'_{\|\boldsymbol{w}\|^2}$ in $(\boldsymbol{\alpha}, \boldsymbol{\theta})$ and lemma 2 implies the concavity of $\|\boldsymbol{w}\|^2$ with respect to $\boldsymbol{\theta}$. The log concavity is always obtained when the concavity of positive functions is insured [6]. The conditions of LCCP are all fulfilled. We can thus apply it for the optimization of the L_1-distance kernel parameters.

4 Experiments

Fig. 1 shows the variation of $\log(R^2)$, $\log(\|\boldsymbol{w}\|^2)$ and $\log(R^2\|\boldsymbol{w}\|^2)$ with respect to θ_1 and θ_2 for the kernel (2). It illustrates the log convexity of R^2 and the log concavity of $\|\boldsymbol{w}^2\|$.

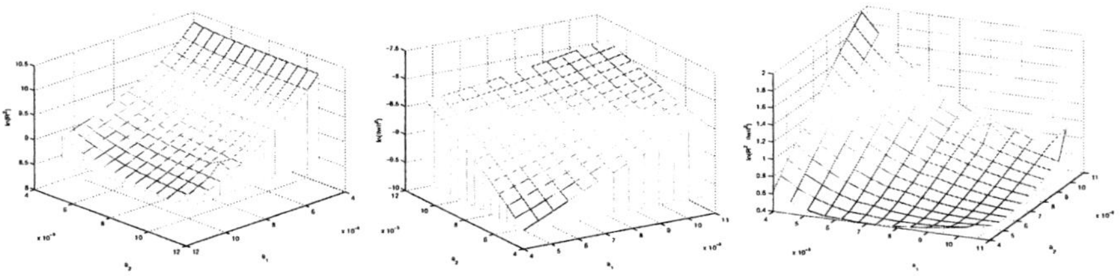

Fig. 1. The left, middle and right figures plot respectively $\log(R^2)$, $\log(\|\boldsymbol{w}\|^2)$ and $\log(R^2\|\boldsymbol{w}\|^2)$ with respect to the L_1-distance kernel parameters (θ^1, θ^2) on banana dataset which is a set of 2D points [7].

Table 1. Test error's comparison of the single parameter L_1 distance kernel (1), L_2 distance kernel (5) and L_1 distance kernel with multiple parameters. n denotes the number of parameters for multi-parameter's kernel. LCCP is used for optimizing the radius margin bound.

	Thyroid	Titanic	Heart	Breast-cancer
K_{L_1} (1)	**5.77%**	22.68%	20.65%	28.97%
K_{L_2} (5)	11.21%	22.56%	18.23%	29.77%
$K_{L_1,\boldsymbol{\theta}}$ (2)	6.20%	**22.08%**	**17.34%**	**27.12%**
n	5	3	13	9

In order to evaluate the performance of the LCCP for optimizing multiple parameters, we performed experiments on datasets obtained from [7]. We compare the L_1-distance kernel without parameters (1), the L_2-distance kernel:

$$K_{L_2}(\boldsymbol{x}, \boldsymbol{x}') = -\sum_{k=1}^{n}(x^k - x'^k)^2, \tag{5}$$

and the L_1-distance kernel with multiple parameters (2). Initial starting point is set to 1 for all θ_i, as in [1]. The stopping criterion is $|log(E_{p+1}(\theta_{p+1})) -$

$log(E_p(\theta_p))| < \epsilon$. The data sets contain 100 realizations of training and test examples. For each realization, we optimize the kernel parameters on the training sample using the LCCP. The obtained parameters are used to estimate the generalization error on the test sample by a 5-fold cross-validation. Tab. 1 summarizes average test errors for different data sets. The L_2-distance kernel is equivalent to the linear kernel when used within SVM. We observe that the L_1-distance kernel performs better or similarly than L_2-distance kernel except on heart dataset. Tab. 1 shows that the use of multiple parameters in L_1-distance kernel allows us most of the time to decrease the test error, despite the weightening of each dataset. This shows clearly the interest of the introduction of multiple parameters in kernels.

5 Conclusion

In this paper, we propose an original way for optimizing of kernel multiple parameters by minimizing the radius margin bound, we named LCCP. The LCCP is derived directly from CCCP optimizing framework. The LCCP approach is more efficient than the gradient descent technique since it converges to a local minimum without searching the size step. We prove that, for the multi-parameters L_1-distance kernel, the radius margin fulfills the conditions for application of the LCCP. Comparison on standard data set leads to improved recognition performance compared to single parameter L_1-distance kernel. The multi-parameters L_1-distance kernel is only one example of kernel which fulfills the conditions for application of the LCCP, but other exists. The formal definition of the set of all kernels that fulfills these conditions is the subject of our future researches.

References

1. O. Chapelle, V. Vapnik, O. Bousquet, and S. Mukherjee, "Choosing multiple parameters for support vector machines," *Machine Learning*, vol. 46, no. 1-3, pp. 131–159, 2002.
2. V. Vapnik, *The Nature of Statistical Learning Theory*, Springer Verlag, 2nd edition, New York, 1999.
3. K. Tsuda, "Optimal hyperplane classifier with adaptive norm," Technical report tr-99-9, ETL, 1999.
4. Yuille A.L. and Rangarajan A., "The concave-convex procedure," *Neural Computation*, vol. 15, no. 4, pp. 915–936, 2003.
5. S. Boughorbel, *Kernels for Image Classification with SVM*, Ph.D. thesis, submited to University of Paris Sud, Orsay, 2005.
6. S. Boyd and L. Vandenberghe, *Convex Optimization*, Cambridge University Press, Cambridge, 2004.
7. G. Ratsch, T. Onoda, and K. R. Muller, "Soft margins for adaboost," *Machine Learning*, vol. 42, no. 3, pp. 287–320, 2001.

The GCS Kernel for SVM-Based Image Recognition

Sabri Boughorbel[1], Jean-Philippe Tarel[2]
François Fleuret[3], and Nozha Boujemaa[1]

[1] IMEDIA Group, INRIA Rocquencourt, 78153 Le Chesnay, France
[2] DESE, LCPC, 58 Bd Lefebvre, 75015 Paris, France
[3] CVLAB, EPFL, 1015 Lausanne, Switzerland

Abstract. In this paper, we present a new compactly supported kernel for SVM based image recognition. This kernel which we called Geometric Compactly Supported (GCS) can be viewed as a generalization of spherical kernels to higher dimensions. The construction of the GCS kernel is based on a geometric approach using the intersection volume of two n-dimensional balls. The compactness property of the GCS kernel leads to a sparse Gram matrix which enhances computation efficiency by using sparse linear algebra algorithms. Comparisons of the GCS kernel performance, for image recognition task, with other known kernels prove the interest of this new kernel.

1 Introduction

Support Vector Machine is one of the successful kernel methods that has been derived from statistical learning theory. Incremental version of SVM has been introduced in [1] allowing faster on-line learning. We focus in this paper on how to improve the computational efficiency of SVM training using compactly supported kernels. We propose a new compactly supported kernel. In §2, we introduce and derive the new kernel, we named Geometric Compactly Supported (GCS) kernel. We provide experimental results proving that GCS kernel leads to good accuracy for image recognition task while being computationally efficient.

2 CS Kernels

A kernel $\varphi(x, y)$ is said to be compactly supported (CS) whenever it vanishes from a certain cut-off distance $2r$ between x and y.

$$\varphi(x, y) = \begin{cases} \varphi(x, y) \text{ if } \|x - y\| < 2r \\ 0 \text{ if } \|x - y\| \geq 2r \end{cases}$$

The main advantage of CS kernels is that their Gram matrices $[\varphi(x_i, x_j)]_{i,j}$ are sparse. If such matrices are associated with a linear system they can be solved efficiently using sparse linear algebra methods. Genton was the first to point

W. Duch et al. (Eds.): ICANN 2005, LNCS 3697, pp. 595–600, 2005.

Table 1. Examples of compactly supported kernels in $\mathbb{R}$, $\mathbb{R}^2$ and $\mathbb{R}^3$

Triangular $K_T(x,y)$	$1-\frac{\|x-y\|}{2r}$, $\|x-y\|<2r$ positive definite in $\mathbb{R}$
Circular	$\arccos(\frac{\\|x-y\\|}{2r})-\frac{\\|x-y\\|}{2r}\sqrt{1-(\frac{\\|x-y\\|}{2r})^2}$, $\\|x-y\\|<2r$ positive definite in $\mathbb{R}^2$
Spherical	$1-\frac{3}{2}\frac{\\|x-y\\|}{2r}-\frac{1}{2}(\frac{\\|x-y\\|}{2r})^3$, $\\|x-y\\|<2r$ positive definite in $\mathbb{R}^3$

out the possible gain in efficiency provided by CS kernels with machine learning techniques [2].

Triangular, circular and spherical kernels, see Tab. 1 for definitions, which are used in geostatistic applications, have been also studied in the context of machine learning [2]. However, the use of these kernels is limited to dimensions from one to three, since they are not positive definite for higher dimensions. Indeed, we provide next a counterexample given in [3] proving that the triangular kernel $K_T(x,y)$ is not positive definite for features living in $\mathbb{R}^2$. Let's take x, $y \in \mathbb{R}^2$, thus $|x-y|$ is replaced by $\|x-y\|$ (the L-2 norm of $\mathbb{R}^2$) in definition of $K_T(x,y)$, see Tab. 1. By choosing $x_{i,j} \in \mathbb{R}^2$ from a 8×8 square grid of spacing $\sqrt{2}r$, and $c_{i,j}$ alternatively $+1$ and -1, we have:

$$\sum_{j_1,j_2=1}^{8} \sum_{k_1,k_2=1}^{8} c_{j_1,j_2} c_{k_1,k_2} K_T(x_{j_1,j_2}, x_{k_1,k_2}) = -1.6081 < 0$$

Therefore K_T is not positive definite on $\mathbb{R}^2$. A few attempts have been carried out to derive compactly support (CS) kernels for high dimensions [2]. In [4], experimentations using a CS kernel are described proving that such kernel does not give usually good performances in the context of non-linear regression (SVR) of functions. Notice that just truncating positive definite kernels does not generally lead to positive definite kernels. We now introduce a new compactly supported kernel that can be viewed as an extension of triangular, circular and spherical kernels to higher dimensions.

3 The GCS Kernel

The derivation of the new Geometric Compactly Supported (GCS) kernel is based on the intersection of two n-dimensional balls. Basically, we use the fact that the intersection volume of two n-dimensional balls leads to a compactly supported and positive definite kernel.

The properties of positiveness and compactness of the GCS kernel are presented in the following proposition.

Proposition 1. *Let x and $y \in \mathbb{R}^n$, $\Psi_n(x,y)$ denotes the intersection volume of two balls having the same radius r, centered in x and y. Thus $\Psi_n(x,y)$ is compactly supported and positive definite kernel.*

Proof. Intersection volume $\Psi_n(x,y)$ can be written as the following integral:

$$\Psi_n(x,y) = \int_{a\in\mathbb{R}^n} \mathbb{1}_{\{\|x-a\|\le r\}}\mathbb{1}_{\{\|y-a\|\le r\}} da \tag{1}$$
$$= \int_{a\in\mathbb{R}^n} \underbrace{f_{a,r}(x) f_{a,r}(y)}_{\text{positive definite}} da$$

where $f_{a,r}(z) = \mathbb{1}_{\{\|z-a\|\le r\}}(z)$. Thus, $\Psi_n(x,y)$ is a positive definite kernel as a mixture of positive definite kernels. Each time the feature point $a \in \mathbb{R}^n$ is in the intersection of the two balls of centers x and y, the function under the integral equals to one, otherwise it equals to zero. As a consequence, the summation over $a \in \mathbb{R}^n$ gives the intersection volume of the two balls. Whenever $\|x-y\| > 2r$, the balls intersection is empty, so $\Psi_n(x,y) = 0$. Therefore, Ψ_n is a CS and positive definite kernel.

To simplify notations, we omitted the radius hyper-parameter r in $\Psi_n(x,y)$. Next, we derive a more explicit formula for the GCS kernel which allows to compute the kernel in a fast recursive way. The volume $\mathcal{V}_n(r)$ of a n-dimensional ball with radius r can be calculated recursively as follows:

$$\mathcal{V}_n(r) = \int_0^r \mathcal{V}_{n-1}(\sqrt{r^2-t^2})dt, \text{ for } n \ge 2 \tag{2}$$

Thus, a general formula of the n-dimensional ball volume can be written as follows:

$$\mathcal{V}_n(r) = \begin{cases} \frac{1}{(\frac{n}{2})!}\pi^{\frac{n}{2}} r^n & \text{if } n \text{ is even} \\ \frac{2^{\frac{n+1}{2}}}{n!!}\pi^{\frac{n-1}{2}} r^n & \text{if } n \text{ is odd} \end{cases}$$

where $n!! = n(n-2)(n-4)\dots 1$ is the double factorial when n is odd. The same recursive approach as in (2) can be used for the derivation of the intersection volume $\Psi_n(x,y)$. We can see in Fig. 1 that the median plane of x and y is a symmetric plane for the intersection volume, so $\Psi_n(x,y)$ can be written, similarly to (2), but now integrating from $\frac{\|x-y\|}{2}$ to r rather than from 0 to r:

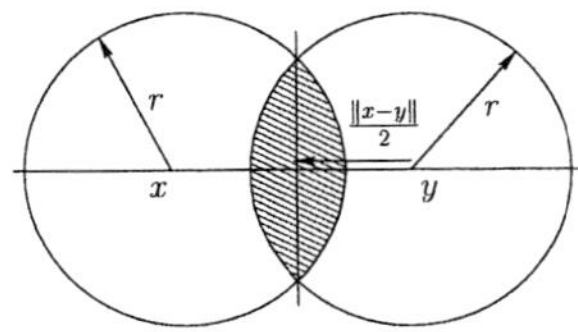

Fig. 1. Intersection volume calculus (case $n = 2$)

$$\begin{aligned}\Psi_n(x,y) &= 2\int_{\frac{\|x-y\|}{2}}^{r} \mathcal{V}_{n-1}(\sqrt{r^2-t^2})dt \\ &= A_n(r)\int_{\frac{\|x-y\|}{2}}^{r} \frac{1}{r}\left(1-\left(\frac{t}{r}\right)^2\right)^{\frac{n-1}{2}} dt \\ &= A_n(r)\Phi_n(x,y)\end{aligned}$$

where $A_n(r)$ is only a normalization factor, which is independent of x and y. SVM decision function is invariant to the product of any positive constant with the kernel. Thus, in the following, we get rid of A_n, and we denote by $\Phi_n(x,y)$ the remaining term which defines the GCS kernel $K_{GCS}(x,y)$.

By variable changing $t = r\sin\theta$, $\Phi_n(x,y)$ can be written as follows:

$$\Phi_n(x,y) = \int_{\arcsin(\frac{\|x-y\|}{2r})}^{\frac{\pi}{2}} (\cos\theta)^n d\theta \qquad (3)$$

By integrating by part, we are able to derive a recursive computation of $\Phi_n(x,y)$. For a dimension n, we define $\varphi_{n,k}(x,y)$ as the value of $\Phi_n(x,y)$ at kth iteration. As a consequence, we have $\Phi_n(x,y) = \varphi_{n,n}(x,y)$. We prove easily that for $\|x - y\| < 2r$:

$$\begin{cases}\varphi_{n,k}(x,y) = \frac{k-1}{k}\varphi_{n,k-2}(x,y) - \frac{1}{k}\frac{\|x-y\|}{2r}\left(1-\left(\frac{\|x-y\|}{2r}\right)^2\right)^{\frac{k-1}{2}} \\ \varphi_{n,2}(x,y) = \arccos\left(\frac{\|x-y\|}{2r}\right) - \frac{\|x-y\|}{2r}\sqrt{1-\left(\frac{\|x-y\|}{2r}\right)^2} \\ \varphi_{n,1}(x,y) = 1 - \frac{\|x-y\|}{2r}\end{cases} \qquad (4)$$

For $\|x - y\| \geq 2r$, $\varphi_{n,k}(x,y) = 0$. Notice that Φ_1 on $\mathbb{R}$ is the triangular kernel, Φ_2 on $\mathbb{R}^2$ is the circular kernel and Φ_3 on $\mathbb{R}^3$ is the spherical kernel. Thus, it can be viewed as the generalization of the spherical kernel to higher dimensions. The GCS kernel Φ_n is positive definite on $\mathbb{R}^n$ as proved before. Nevertheless $\varphi_{n,k}$, $1 \leq k \leq n-1$ are not positive definite kernels on $\mathbb{R}^n$. Indeed, the counterexample given in beginning of §. 2, tells us that $\varphi_{2,1}$, which is the triangular kernel for data living in $\mathbb{R}^2$, is not positive definite. Functions $\varphi_{n,k}$ are only intermediaries functions useful to compute the GCS kernel Φ_n recursively.

4 Complexity Reduction

One of the interest of GCS kernel is the reduction of algorithmic complexity. Actually, we can take advantages of the Gram matrix sparsity to enhance training computation stage. Fig. 2-a shows the sparsity of the GCS Gram matrix. Fig .2-b presents the rearrangement of the Gram matrix using a Cuthill-McKee permutation which leads to a banded matrix [5]. Fig. 2-c presents computational savings for the SVM training stage: it plots the complexity of the quadratic term of the SVM dual problem with respect to the size of training sample n. The usual computation leads to a quadratic complexity of $O(n^2)$, for each iteration. The use of a sparse Gram matrix leads to a better complexity of $O(nB(n))$, where $B(n) < n$ is the bandwidth of the rearranged Gram matrix. Moreover, after rearranging the Gram matrix, only the banded Gram matrix is kept in memory.

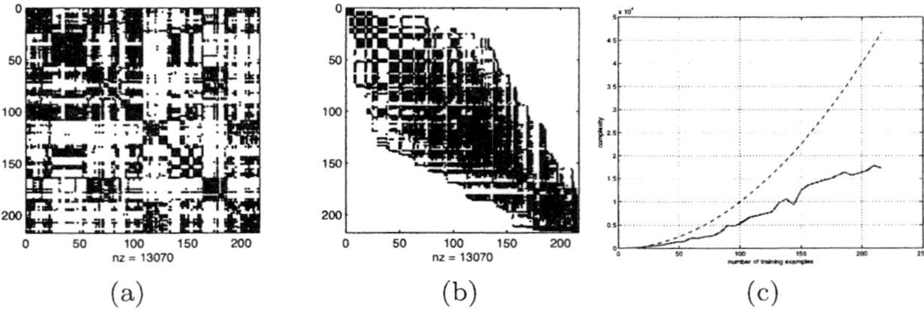

(a) (b) (c)

Fig. 2. (a)-(b) Symmetric reverse Cuthill-McKee permutation of sparse GCS Gram Matrix for radius hyperparameter $r = 0.4$. (c) Complexity of quadratic term of the SVM dual problem using banded matrix representation with respect to the size of training sample.

5 Experiments

Figure 3 shows some images from Corel database that we used for experiments. This database gathers 3200 images in 6 different classes. Images are represented by 64-bin RGB color histogram. We compare 4 kernels namely: Laplace kernel $K_{Lapl}(x, y) = exp(-\frac{\|x^a - y^a\|}{\sigma})$, Polynomial kernel $K_{\text{Poly}}(x, y) = (1 + \langle x^a \cdot y^a \rangle)^d$,

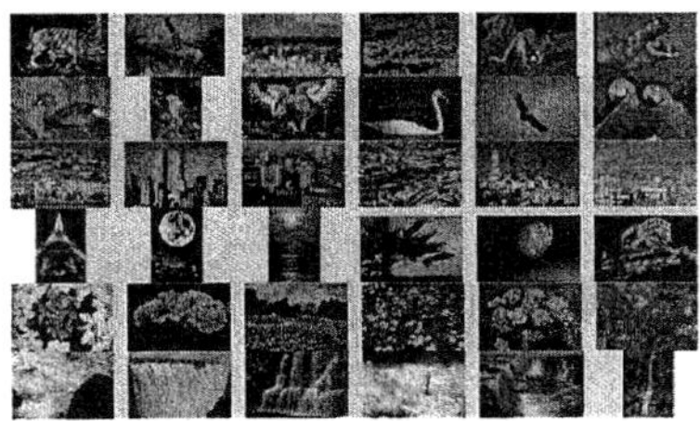

Fig. 3. Image examples from the 6 classes used for experiments

Table 2. Validation and test errors comparisons for the different kernels on Corel database

	valid. err.	test err.
K_{Lapl}	25.32±0.19	25.34±0.42
K_{GCS}	25.19±0.30	**25.07±0.64**
K_{Poly}	27.81±0.17	27.92±0.43
K_{CS}	26.12± 0.37	25.30±0.44

K_{GCS} defined by (4) and $K_{CS} = K_{GCS}K_{Lapl}$. The parameter a applies a non-linear remapping of feature space which is shown to improve drastically performances for image recognition task. We set $a = 0.25$ as in [6]. For K_{GCS}, we tune the radius r. For K_{CS}, we set the radius to $r = 4$ such that we obtain Gram

Table 3. Class-confusion matrix obtained for kernel K_{CS} on Corel database

	Animals	Birds	Buildings	Night scenes	Roses	Water scenes
Animals	536	104	80	69	128	99
Birds	25	62	2	17	35	24
Buildings	37	12	291	95	25	27
Night scenes	13	8	28	82	12	25
Roses	140	189	34	53	542	56
Water scenes	34	36	29	45	27	172

matrix sparsity of 90%, then we tune the σ of inside K_{Lapl}. Table 2 shows that K_{GCS} and K_{CS} yield to similar results to K_{Lapl} which known as the best kernel in the state of the art. Optimal radius of the GCS kernel does not give sparse enough Gram matrices, however combined with K_{Lapl}, K_{CS} has a sparsity of 90%. Table 3 shows the class-confusion matrix obtained with K_{CS}, values on the diagonal gives the number of correctly classified images.

6 Conclusion

In this paper, we have presented a new compactly supported kernel namely the GCS kernel. The construction of this kernel is based on a geometric approach using the intersection volume of two n-dimensional balls. Hence, the GCS kernel can be viewed as the generalization of spherical kernels to higher dimensions. It yields to good recognition performance when the radius is tuned similar to that of Laplace kernel, however optimal radius does not lead to sparse Gram matrix. To recover sparsity, we combine the GCS kernel and the Laplace kernel to obtain an efficient kernel with a highly sparse Gram matrix of 90% of zeros.

References

1. C. Gentile, "A new approximate maximal margin classification algorithm," *Journal of Machine Learning Research*, vol. 2, pp. 213–242, 2001.
2. M. Genton, "Classes of kernels for machine learning: a statistics perspective," *Journal of Machine Learning Research*, vol. 2, pp. 299–312, 2002.
3. N. Cressie, *Statistics for Spatial Data*, New York, Wiley, 1993.
4. B. Hamers, J. A.K. Suykens, and B. De Moor, "Compactly supported rbf kernels for sparsifying the gram matrix in ls-svm regression models," in *International Conference on Artificial Neural Networks (ICANN'02)*, Madrid Spain, 2002, pp. 720–726.
5. J. Gilbert, C. Moler, and R. Schreiber, "Sparse matrices in matlab: design and implementation," *SIAM Journal on Matrix Analysis*, vol. 13, no. 1, pp. 333–356, 1992.
6. O. Chapelle, P. Haffner, and V. Vapnik, "Svms for histogram-based image classification," *IEEE Transactions on Neural Networks*, vol. 9, 1999.

Informational Energy Kernel for LVQ

Angel Caţaron[1] and Răzvan Andonie[2]

[1] Transylvania University of Brasov, Romania
[2] Central Washington University, Ellensburg, USA

Abstract. We describe a kernel method which uses the maximization of Onicescu's informational energy as a criteria for computing the relevances of input features. This adaptive relevance determination is used in combination with the neural-gas and the generalized relevance LVQ algorithms. Our quadratic optimization function, as an L^2 type method, leads to linear gradient and thus easier computation. We obtain an approximation formula similar to the mutual information based method, but in a more simple way.

1 Introduction

Relevance LVQ (RLVQ) [2] uses a weighted distance function for the LVQ classification. A modification of RLVQ has been proposed by Hammer *et al.* [3], Generalized RLVQ (GRLVQ), which obeys a stochastic gradient descent on an energy function.

The neural-gas (NG) algorithm [4] represents a neural model which is applied to the task of vector quantization by using a neighborhood cooperation scheme. The NG network uses an adaptation rule similar to the Kohonen feature map. It replaces the Euclidian distance with the neighborhood ranking of the reference vectors for a given input vector. The Supervised Relevance Neural Gas (SRNG) algorithm [1] combines the NG and the GRLVQ. The idea was to incorporate neighborhood cooperation of NG into the GRLVQ to speedup the convergence and make initialization less crucial.

In our previous work we have introduced two LVQ classificators based on Onicescu's informational energy (IE): the Energy RLVQ (ERLVQ) [5] and the Energy GRLVQ (EGRLVQ) [6]. We have obtained incremental learning algorithms for feature ranking and supervised classification. The sensible part of such an approach is the mutual information estimation, which poses great difficulties as it requires the knowledge on the underlying probability density functions of the data space and the integration on these functions [13]. Our technique proved to be an efficient solution to this problem.

In this paper, we describe the Energy SRNG (ESRNG) classificator, a kernel method which uses the maximization of the IE as a criteria for computing the relevances of input features. This adaptive relevance determination is used in combination with the SRNG model, providing an alternative way for determining the relevances. After introducing the SRNG notations and the relevance determination using IE, we define the ESRNG algorithm and compare it to other algorithms of this family.

W. Duch et al. (Eds.): ICANN 2005, LNCS 3697, pp. 601–606, 2005.

2 SRNG

Assume that a clustering of data into M classes, $c_1, \ldots, c_M$, is implemented and a set of training data is available: $X = \{(\boldsymbol{x}_i, c_i) \subset \mathbb{R}^n \times \{1, \ldots, M\} \mid i = 1, \ldots, N\}$. The training vectors $\boldsymbol{x}_i$ have n components $[x_{i1}, \ldots, x_{in}]$. A subset of reference vectors from $\mathbb{R}^n$ are assigned to each class. Denote the set of all reference vectors by $W = \{\boldsymbol{w}_1, \ldots, \boldsymbol{w}_K\}$. The components of a vector $\boldsymbol{w}_j$ are $[w_{j1}, \ldots, w_{jn}]$.

The NG algorithm optimizes a cost function which uses the rank $r_j(\boldsymbol{x}_i, W)$ of the reference vector $\boldsymbol{w}_j$ for a given input $\boldsymbol{x}_i$ [1], [4]:

$$C_{NG} = \frac{1}{C(\gamma)} \sum_{\boldsymbol{w}_j \in W} \sum_{\boldsymbol{x}_i \in X} h_\gamma(r_j(\boldsymbol{x}_i, W)) \|\boldsymbol{x}_i - \boldsymbol{w}_j\|^2,$$

where $h_\gamma(r_j(\boldsymbol{x}_i, W)) = e^{-r_j(\boldsymbol{x}_i, W)/\gamma}$, $C(\gamma) = \sum_{r=0}^{K-1} h_\gamma(r)$, and γ is a parameter which gives the neighborhood range. The rank $r_j(\boldsymbol{x}_i, W)$ of the reference vector $\boldsymbol{w}_j$ for the input vector $\boldsymbol{x}_i$ is the number of reference vectors that are in the relation $\|\boldsymbol{x}_i - \boldsymbol{w}_k\| \leq \|\boldsymbol{x}_i - \boldsymbol{w}_j\|$, where $j, k \in \{1, \ldots, K\}$ and $j \neq k$. The neighborhood ranking of the reference vectors is updated each time a training vector is applied to the input of the neural network.

The GRLVQ algorithm uses a squared weighted distance between an input vector $\boldsymbol{x}_i$ and a reference vector $\boldsymbol{w}_j$, $D_{ij}^2 = \sum_{k=1}^{n} \lambda_k (x_{ik} - w_{jk})^2$, where $\boldsymbol{\lambda} = [\lambda_1, \ldots, \lambda_n]$ is the relevance vector, with $\lambda_i \geq 0$, $i = 1, \ldots, n$, $\sum_{i=1}^{n} \lambda_i = 1$. The Supervised Relevance NG (SRNG) can be obtained [1] by including the NG idea in the GRLVQ algorithm. The cost function optimized by this algorithm is:

$$C_{SRNG} = \sum_{\boldsymbol{x}_i \in X} \sum_{\boldsymbol{w}_j \in W^{\boldsymbol{x}_i}} \frac{h_\gamma(r_j(\boldsymbol{x}_i, W^{\boldsymbol{x}_i})) f(\mu_\lambda(\boldsymbol{x}_i, \boldsymbol{w}_j))}{C(\gamma, K^{\boldsymbol{x}_i})},$$

with $\mu_\lambda(\boldsymbol{x}_i, \boldsymbol{w}_j) = \frac{|\boldsymbol{x}_i - \boldsymbol{w}_j|_\lambda^2 - D_{ik}}{|\boldsymbol{x}_i - \boldsymbol{w}_j|_\lambda^2 + D_{ik}}$. D_{ik} is the weighted distance between $\boldsymbol{x}_i$ and the closest reference vector that does not belong to $W^{\boldsymbol{x}_i}$, a subset of W which contains the reference vectors from the same class with $\boldsymbol{x}_i$. $K^{\boldsymbol{x}_i}$ is the cardinality of $W^{\boldsymbol{x}_i}$. According to this cost function, all reference vectors from $W^{\boldsymbol{x}_i}$ and the closest reference vector that does not belong to this set are updated by [1]:

$$\Delta \boldsymbol{w}_j = \eta \boldsymbol{\lambda} \boldsymbol{I} \frac{\partial f}{\partial \mu} \frac{D_{ik}}{(|\boldsymbol{x}_i - \boldsymbol{w}_j|_\lambda^2 + D_{ik})^2} (\boldsymbol{x}_i - \boldsymbol{w}_j) \frac{r_j(\boldsymbol{x}_i, W^{\boldsymbol{x}_i})}{C(\gamma, K^{\boldsymbol{x}_i})} \tag{1}$$

where $\boldsymbol{w}_j$ is the closest reference vector from $\boldsymbol{x}_i$ that does not belong to $W^{\boldsymbol{x}_i}$, and

$$\Delta \boldsymbol{w}_k = - \sum_{\boldsymbol{w}_j \in W^{\boldsymbol{x}_i}} \eta_1 \boldsymbol{\lambda} \boldsymbol{I} \frac{\partial f}{\partial \mu} \frac{|\boldsymbol{x}_i - \boldsymbol{w}_j|_\lambda^2}{(|\boldsymbol{x}_i - \boldsymbol{w}_j|_\lambda^2 + D_{ik})^2} (\boldsymbol{x}_i - \boldsymbol{w}_k) \frac{r_j(\boldsymbol{x}_i, W^{\boldsymbol{x}_i})}{C(\gamma, K^{\boldsymbol{x}_i})} \tag{2}$$

for all reference vectors from $W^{\boldsymbol{x}_i}$. In these relations, η and η_1 are two positive constants. We used the sigmoid function $f(\mu) = \frac{1}{1+e^{-\mu\epsilon}}$ for which $\frac{\partial f}{\partial \mu} = f(\mu)(1 - f(\mu))$, with ϵ a positive constant.

3 Relevance Determination Using Informational Energy

Onicescu's IE [7], [8] is defined by: $E(Y) = \int_{-\infty}^{+\infty} p^2(\boldsymbol{y})d\boldsymbol{y}$, where Y is a continuous random variable with probability density function $p(\boldsymbol{y})$. The conditional information energy between Y and a discrete random variable C is: $E(Y|C) = \int_{\boldsymbol{y}} \sum_{p=1}^{M} p(c_p)p^2(\boldsymbol{y}|c_p)d\boldsymbol{y}$.

The unilateral dependence measure $o(Y,X) = E(Y|X) - E(Y)$, defined in [9], quantifies the amount of information contained in random variable X about random variable Y.

The ESRNG algorithm uses a vector of relevances obtained by maximizing $o(Y,X)$ with an ascending gradient method [6]. A transformation which makes the connection between the input vector and the class represented by the reference vector $\boldsymbol{w}_j$ is employed: $\boldsymbol{y}_i = \boldsymbol{\lambda I}(\boldsymbol{x}_i - \boldsymbol{w}_j)$. In this equation, $\boldsymbol{x}_i, i = 1, \ldots, N$, is the set of training vectors that belong to one of the $c_1, c_2, \ldots, c_M$ classes; $\boldsymbol{w}_j, j = 1, \ldots, P$, are the reference vectors of the classes; $\boldsymbol{\lambda}$ is the vector of relevances; $\boldsymbol{I}$ is the unity matrix. The values $\boldsymbol{y}_i, i = 1, \ldots, N$, are samples of the random variable Y.

We obtain the relevance values by an iteratively updating approach:

$$\boldsymbol{\lambda}^{(t+1)} = \boldsymbol{\lambda}^{(t)} + \alpha \sum_{i=1}^{N} \frac{\partial o(Y,C)}{\partial \boldsymbol{y}_i} \boldsymbol{I}\left(\boldsymbol{x}_i - \boldsymbol{w}_j\right).$$

Considering the M class labels as samples of a discrete random variable denoted by C, we have: $o(Y,C) = E(Y|C) - E(Y)$. The conditional information energy can be reformulated as a dependence of the squared mutual probability density $E(Y|C) = \sum_{p=1}^{M} p(c_p) \int_{\boldsymbol{y}} p^2(\boldsymbol{y}|c_p)d\boldsymbol{y} = \sum_{p=1}^{M} \frac{1}{p(c_p)} \int_{\boldsymbol{y}} p^2(\boldsymbol{y}, c_p)d\boldsymbol{y}$.

This allows us to write $o(Y,C) = \sum_{p=1}^{M} \frac{1}{p(c_p)} \int_{\boldsymbol{y}} p^2(\boldsymbol{y}, c_p)d\boldsymbol{y} - \int_{\boldsymbol{y}} p^2(\boldsymbol{y})d\boldsymbol{y}$, which can easily estimated by using the Parzen windows with the Gaussian kernel $G(\boldsymbol{y} - \boldsymbol{y}_i, \sigma) = \frac{1}{\sqrt{2\pi\sigma}} \cdot e^{-\frac{\|\boldsymbol{y}-\boldsymbol{y}_i\|^2}{2\sigma}}$.

The probability density $p(\boldsymbol{y})$ can be expressed [10] as $p(\boldsymbol{y}) = \frac{1}{N}\sum_{i=1}^{N} G(\boldsymbol{y} - \boldsymbol{y}_i, \sigma^2)$. We can write: $\int_{\boldsymbol{y}} p^2(\boldsymbol{y}, c_p)d\boldsymbol{y} = \frac{1}{N^2}\sum_{k=1}^{N_p}\sum_{l=1}^{N_p} G(\boldsymbol{y}_{pk} - \boldsymbol{y}_{pl}, 2\sigma^2)$ and $\int_{\boldsymbol{y}} p^2(\boldsymbol{y})d\boldsymbol{y} = \frac{1}{N^2}\sum_{k=1}^{N}\sum_{l=1}^{N} G(\boldsymbol{y}_k - \boldsymbol{y}_l, 2\sigma^2)$, where $\boldsymbol{y}_{pk}$, $\boldsymbol{y}_{pl}$ are two training samples from class p, and $\boldsymbol{y}_k$, $\boldsymbol{y}_l$ are two training samples from any class. N_p is the number of the training samples from the class p.

We obtain

$$o(Y,C) = \frac{1}{N}\left(\sum_{p=1}^{M} \frac{1}{N_p}\right) \sum_{k=1}^{N_p}\sum_{l=1}^{N_p} G(\boldsymbol{y}_{pk} - \boldsymbol{y}_{pl}, 2\sigma^2 \boldsymbol{I}) - \\ - \frac{1}{N^2}\sum_{k=1}^{N}\sum_{l=1}^{N} G(\boldsymbol{y}_k - \boldsymbol{y}_l, 2\sigma^2 \boldsymbol{I}).$$

We use two consecutive samples $\boldsymbol{y}_1$ and $\boldsymbol{y}_2$ as classes representatives. This expression can only be evaluated when the two training vectors belong to different classes. In this case, we obtain:

$$o(Y, C) = G(0, 2\sigma^2) - \frac{1}{2}G(\boldsymbol{y}_1 - \boldsymbol{y}_2, 2\sigma^2).$$

4 The ESRNG as a Kernel Based Algorithm

When $\boldsymbol{y}_1 \neq \boldsymbol{y}_2$, we have $\|\boldsymbol{y}_1 - \boldsymbol{y}_2\|^2 > 0$ and $G(0, 2\sigma^2) > G(\boldsymbol{y}_1 - \boldsymbol{y}_2, 2\sigma^2)$. This means $o(Y, C) > 0$ for all input vectors. Hence, this is a positive defined kernel.

The squared weighted distance between an input vector $\boldsymbol{x}_i$ and a reference vector $\boldsymbol{w}_j$ $D_{ij}^2 = \sum_{k=1}^{n} \lambda_k (x_{ik} - w_{jk})^2$ requires that $\lambda_k \geq 0$ for all $k = 1, \ldots, n$. In the case when at least one relevance value is negative, this condition can be realized by transforming the relevance vectors with $\lambda_k = \frac{e^{\lambda_k}}{\sum_{i=1}^{n} e^{\lambda_i}} + \epsilon$ or by scaling the relevance components $\lambda_k = \lambda_k + \min_{i=1,\ldots,n} \lambda_i + \epsilon$, where ϵ is a positive constant. We usually apply a transform of the relevance vector in order to keep its component's values in a reasonable domain.

Finally, we obtain:

$$\boldsymbol{\lambda}^{(t+1)} = \boldsymbol{\lambda}^{(t)} - \alpha \frac{1}{4\sigma^2} G(\mathbf{y}_1 - \boldsymbol{y}_2, 2\sigma^2 \boldsymbol{I})(\boldsymbol{y}_2 - \boldsymbol{y}_1)\boldsymbol{I}(\boldsymbol{x}_1 - \boldsymbol{w}_{j(1)} - \boldsymbol{x}_2 + \boldsymbol{w}_{j(2)})(3)$$

where $\boldsymbol{w}_{j(1)}$ and $\boldsymbol{w}_{j(2)}$ are the closest prototypes from the input vectors $\boldsymbol{x}_1$ and $\boldsymbol{x}_2$, respectively.

The ESRNG algorithm adapts the reference vectors for as least as possible quantization error on all feature vectors. After initializing the relevance vector $\lambda_k = 1/n$, $k = 1, \ldots, n$, the codebook vectors, η, α, and σ, the following procedure updates incrementally the codebook vectors, the relevances and the feature ranks, for a given input $\mathbf{x}_i$:

1. Update the codebook vectors using the SRNG relations (1) and (2).
2. Update the relevances according to our formula (3) and transform them.
3. Update the overall rank of each feature as an average over all previous steps.

Since we also obtain a ranking of the input vectors' components, this algorithm can be used not only in classification tasks, but also in feature selection.

The weighted Euclidean metric we use allows for a direct interpretation as kernelized NG if the relevances are fixed [1]. In this case, the relevances should not be updated after processing each input pattern. This may be achieved if we allow a preprocessing of the patterns, where the relevances are computed first.

5 Experiments

The classification results obtained by ESRGN, applied on three well known datasets (Iris, Ionosphere, and Vowel Recognition [11]), are compared in Table 1 with other experiments performed under similar conditions.

We used 6 reference vectors to classify the 150 vectors from the Iris database. The third component was ranked as most important and the least important was the second component, while the recognition rate was 97.33%. The 351 instances of the Ionosphere dataset were split into two subsets. For the first training 200 samples we used 8 reference vectors. The remaining 151 samples were used in the classification tests. We obtained a recognition rate of 94.40%. For the Vowel recognition database (Deterding data) we trained 59 reference vectors and we obtained a recognition accuracy of 47.61%. The second feature was found as most important, whereas the 7-th and 10-th features were ranked as the least important.

Figure 1 shows the average values of the feature relevances obtained with ESRNG experiments.

Table 1. Comparative recognition rates for the test data

	Iris	Vowel	Ionosphere
LVQ	91.33%	44.80%	90.06%
RLVQ	95.33%	46.32%	92.71%
GRLVQ	96.66%	46.96%	93.37%
SRNG	96.66%	47.61%	94.03%
ERLVQ	97.33%	47.18%	94.03%
EGRLVQ	97.33%	47.18%	94.40%
ESRNG	97.33%	47.61%	94.40%

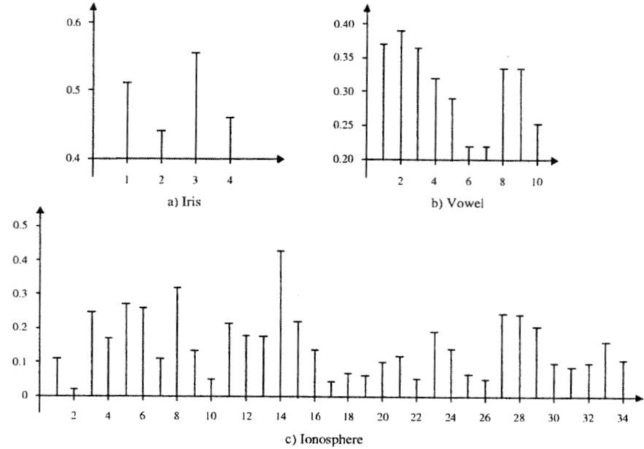

Fig. 1. The average values of the feature relevances obtained with ESRNG experiments

6 Conclusions

Our contribution is an information theory approximation of the relevances in the supervized NG algorithm. This method proves to be computationally effective and leads to good recognition rates.

Jenssen *et al.* [12] have recently proved that information theoretic learning based on Parzen windows density estimation is similar to kernel-based learning. Since the distance we use allows for a direct interpretation as kernelized NG, in our future work we will attempt to combine these two results.

References

1. Hammer, B. , Strickert, M. , Villmann, T.: Supervised neural gas with general similarity measure. Neural Process. Lett. **21** 1 (2005) 21-44
2. Bojer, T., Hammer, B., Schunk, D., von Toschanowitz, K.T.: Relevance Determination in Learning Vector Quantization. Proceedings of the European Symposium on Artificial Neural Networks (ESANN 2001) (2001) 271–276
3. Hammer, B., Villmann, T.: Generalized Relevance Learning Vector Quantization. Neural Networks **15** (2002) 1059–1068
4. Martinetz, T. M. , Berkovich, S. G. , Schulten, K. J.: Neural-gas network for vector quantization and its application to time-series prediction. IEEE Transactions on Neural Networks **4**(1993) 558-569
5. Andonie, R., Cataron, A.: An informational energy LVQ approach for feature ranking. Proceedings of the European Symposium on Artificial Neural Networks (ESANN 2004) (2004) 471–476
6. Cataron, A., Andonie R.: Energy generalized LVQ with relevance factors. Proceedings of the IEEE International Joint Conference on Neural Networks IJCNN 2004, Budapest, Hungary, July 26-29 (2004) 1421-1426
7. Onicescu, O.: Theorie de l'information. Energie informationelle. C. R. Acad. Sci. Paris, Ser. A-B **263** (1966) 841-842
8. Guiasu, S.: Information theory with applications. McGraw Hill New York (1977).
9. Andonie, R., Petrescu, F.: Interacting systems and informational energy. Foundation of Control Engineering **11** (1986) 53-59
10. Principe, J. C., Xu, D., Fisher III, J. W.: Information-theoretic learning. In Unsupervised Adaptive Filtering, S. Haykin, Wiley, New York (2000)
11. Blacke, K., Keogh, E., Merz, C. J.: UCI Repository of Machine Learning Databases. [Online]. Available: http://www.ics.uci. edu/~mlearn/MLSummary.html (1998)
12. Jenssen, R., Erdogmus, D., Principe, J.C., Eltoft, T.: Towards a unification of information theoretic learning and kernel methods. IEEE Workshop on Machine Learning for Signal Processing, Sao Luis, Brazil, (2004)
13. Chow, T. W. S., Huang D.: Estimating optimal feature subsets using efficient estimation of high-dimensional mutual information. IEEE Transactions on Neural Networks **16**(2005) 213–224

Reducing the Effect of Out-Voting Problem in Ensemble Based Incremental Support Vector Machines

Zeki Erdem[1,4], Robi Polikar[2], Fikret Gurgen[3], and Nejat Yumusak[4]

[1] TUBITAK Marmara Research Center, Information Technologies Institute, 41470 Gebze - Kocaeli, Turkey
zeki.erdem@bte.mam.gov.tr
[2] Rowan University, Electrical and Computer Engineering Department, 210 Mullica Hill Rd., Glassboro, NJ 08028, USA
polikar@rowan.edu
[3] Bogazici University, Computer Engineering Department, Bebek, 80815 Istanbul, Turkey
gurgen@boun.edu.tr
[4] Sakarya University, Computer Engineering Department, Esentepe, 54187 Sakarya, Turkey
nyumusak@sakarya.edu.tr

Abstract. Although Support Vector Machines (SVMs) have been successfully applied to solve a large number of classification and regression problems, they suffer from the *catastrophic forgetting* phenomenon. In our previous work, integrating the SVM classifiers into an ensemble framework using Learn++ (SVMLearn++) [1], we have shown that the SVM classifiers can in fact be equipped with the incremental learning capability. However, Learn++ suffers from an inherent *out-voting* problem: when asked to learn new classes, an unnecessarily large number of classifiers are generated to learn the new classes. In this paper, we propose a new ensemble based incremental learning approach using SVMs that is based on the incremental Learn++.MT algorithm. Experiments on the real-world and benchmark datasets show that the proposed approach can reduce the number of SVM classifiers generated, thus reduces the effect of *out-voting* problem. It also provides performance improvements over previous approach.

1 Introduction

As with any type of classifier, the performance and accuracy of SVM classifiers rely on the availability of a representative set of training dataset. In many practical applications, however, acquisition of such a representative dataset is expensive and time consuming. Consequently, it is not uncommon for the entire data to be obtained in installments, over a period of time. Such scenarios require a classifier to be trained and incrementally updated as new data become available, where the classifier needs to learn the novel information provided by the new data without forgetting the knowledge previously acquired from the data seen earlier. We note that a commonly used procedure for learning from additional data, training with the combined old and new data, is not only a suboptimal approach (as it causes catastrophic forgetting), but it

W. Duch et al. (Eds.): ICANN 2005, LNCS 3697, pp. 607 – 612, 2005.

may not even be feasible, if the previously used data are lost, corrupted, prohibitively large, or otherwise unavailable. Incremental learning is the solution to such scenarios, which can be defined as the process of extracting new information without losing prior knowledge from an additional dataset that later becomes available. Various definitions, interpretations, and new guidelines of incremental learning can be found in [2] and references within.

Since SVMs are stable classifiers that use the global learning technique, they are prone to *catastrophic forgetting* phenomenon (also called unlearning) [3] which can be defined as the inability of the system to learn new patterns without forgetting previously learned ones. To overcome some drawbacks, various methods have been proposed for incremental SVM learning in the literature [4, 5]. In this work, we consider the incremental SVM approach based on incremental learning paradigm referenced within [2] and propose an ensemble based incremental SVM construction to solve the catastrophic forgetting problem and out-voting problem by reducing the number of SVM classifiers generated in ensemble.

2 Ensemble of SVM Classifiers

Learn++ uses weighted majority voting, where each classifier receives a voting weight based on its training performance [2]. This works well in practice even for incremental learning problems. However, if the incremental learning problem involves introduction of new classes, then the voting scheme proves to be unfair towards the newly introduced class: since none of the previously generated classifiers can pick the new class, a relatively large number of new classifiers need to be generated that recognize the new class, so that their total weight can out-vote the first batch of classifiers on instances coming from this new class. This in turn populates the ensemble with an unnecessarily large number of classifiers. The Learn++.MT algorithm, explained below, is specifically proposed to address this issue of classifier proliferation [6]. For any given test instance, it compares the class predictions of each classifier and cross-references them with the classes on which they were trained. Essentially, if a subsequent ensemble overwhelmingly chooses a class it has seen before, then the voting weights of those classifiers that have not seen that class are proportionally reduced.

For each dataset (D_k), the inputs to the algorithm are (i) a sequence of *m* training data instances x_i along with their correct labels y_i, (ii) a classification algorithm, and (iii) an integer T_k specifying the maximum number of classifiers to be generated using that database. If the algorithm is seeing its first database ($k=1$), a data distribution (D_t), from which training instances will be drawn, is initialized to be uniform, making the probability of any instance being selected equal. If $k>1$ then a distribution initialization sequence initializes the data distribution. The algorithm adds T_k classifiers to the ensemble starting at $t=eT_k+1$, where eT_k denotes the current number of classifiers in the ensemble. For each iteration *t*, the instance weights, w_t, from the previous iteration are first normalized to create a data distribution D_t. A classifier, h_t, is generated from a subset of D_k that is drawn from D_t. The error, ε_t, of h_t is then calculated; if $\varepsilon_t > ½$, the algorithm deems the current classifier, h_t, to be weak, discards it, and returns and redraws a training dataset, otherwise, calculates the normalized classification error, $\beta_t = \varepsilon_t/(1-\varepsilon_t)$, since for $0 < \varepsilon_t < ½$, $0 < \beta_t < 1$. The class labels of the training instances used to generate this classifier are then stored. The *dynamic weight voting* (*DWV*) algorithm is called to obtain the composite classifier, H_t, of the ensemble. H_t represents the ensemble decision of the first t hypotheses generated thus far. The error of the composite classifier, E_t is then computed and normalized. The instance weights w_t are finally updated according to the performance of H_t such that the weights of instances correctly classified by H_t are reduced and those that are misclassified are effectively increased. This ensures that the ensemble focus on those regions of the feature space that are not yet learned, performing the incremental learning [6].

Given a set of training samples x_i, i=1,…,m, where $\boldsymbol{x}_i \in \boldsymbol{R}^n$ is input patterns, y_i, (i=1,..,m), is the class labels, the SVM classifier function is formulated in terms of kernels functions, such as radial basis function and polynomial:

$$h(x) = sign\left(\sum_{i=1}^{m} \alpha_i y_i K(x_i, x) - b\right). \tag{1}$$

where b is the bias and α_i are the coefficients that are maximized by Lagrangian [7,8]. The final composite SVM classifier is obtained using the *DWV* algorithm for Learn++.MT algorithm, as follows [6]:

$$H_{final}(x_i) = \arg\max_{c} \sum_{t:h_t(x_i)=c} W_t \tag{2}$$

Where c = 1,2,…C is classes, and W_t =log($1/\beta_t$) is the SVMs classifier weights.

3 Simulation Results

Proposed incremental learning approach of SVM ensemble using Learn++.MT has been tested on several datasets. We use the SVMLearn++.MT notation for proposed approach for consistency. Due to space limitations, we present results on one benchmark dataset and one real-world application as explained following sections. We used the LIBSVM library [9] as SVM solver. The Gaussian kernel functions were used in our experiments. We utilized the cross-validation technique with 5-folds to jointly select the SVM parameters, which are the regularization constant C and the RBF width σ.

3.1 Optical Character Recognition Dataset

The Optical Character Recognition dataset is a benchmark dataset from UCI machine learning repository. The OCR dataset features 10 classes (digits 0 ~ 9) with 64 attributes. The dataset was divided into four sets, to create three training subset (**DS1~3**) and a test subset (**Test**), whose distribution can be seen in Table 1. We evaluated the incremental learning capability and also the performance of SVMLearn++ and SVMLearn++.MT on a fixed number of classifiers to allow for a fair comparison. Each algorithm was used to generate seven classifiers with the addition of each dataset, giving a total of 21 classifiers in three training sessions. The data distribution was deliberately made rather challenging, specifically designed to test the ability of proposed approach to learn *multiple* new classes at once with each additional dataset while retaining the knowledge of previously learned classes. In this incremental learning problem, instances from only six of the ten classes are employed in each subsequent dataset resulting in a rather difficult problem.

Table 1. OCR data distribution

Class	C1	C2	C3	C4	C5	C6	C7	C8	C9	C10
DS1	250	250	250	0	0	250	250	250	0	0
DS2	150	0	150	250	0	150	0	150	250	0
DS3	0	150	0	150	400	0	150	0	150	400
Test	110	114	111	114	113	111	111	113	110	112

Results from this test are shown in Tables 2 and 3. Each row shows class-by-class generalization performance of the ensemble on the test data after being trained with dataset **DS$_k$**, k=1,2,3. The last two columns are the average overall generalization performance (**Gen.**) over 20 simulation trials (on the entire test data which includes instances from all ten classes), and the standard deviation (**Std.**) of the generalization performances.

Table 2. SVMLearn++ with RBF kernel ($\sigma = 0.1$, $C = 1$) results on OCR dataset

	C1	C2	C3	C4	C5	C6	C7	C8	C9	C10	Gen.	Std.
DS1	99%	100%	100%	-	-	98%	100%	99%	-	-	60%	0.04%
DS2	99%	73%	100%	44%	-	98%	68%	99%	47%	-	63%	1.54%
DS3	99%	100%	100%	93%	14%	97%	100%	99%	90%	13%	80%	4.17%

Table 3. SVMLearn++.MT with RBF kernel ($\sigma = 0.1$, $C = 1$) results on OCR dataset

	C1	C2	C3	C4	C5	C6	C7	C8	C9	C10	Gen.	Std.
DS1	99%	100%	100%	-	-	98%	100%	99%	-	-	59%	0.05%
DS2	99%	34%	99%	97%	-	93%	20%	99%	60%	-	59%	0.43%
DS3	99%	98%	95%	97%	89%	53%	100%	52%	95%	90%	85%	0.56%

SVMLearn++ was able to learn, the new classes 4 and 9, only poorly after they were introduced in **DS2** but able to learn them rather well, when further trained with these classes in **DS3**. Similarly, it performs rather poorly on classes 5 and 10 after they are first introduced in **DS3**, though it is reasonable to expect that it would do well on these classes with additional training. We note however, SVMLearn++.MT was able to learn new class quite well in first attempt. Finally, recall that the generalization performance of the algorithm is computed on the entire test data which included instances from all classes. This is the reason that the generalization performance is only around 59% after the first training session, since the algorithm has seen only six of the ten classes in the test data. Both SVMLearn++ and SVMLearn++.MT exhibit the ability of incremental learning and an overall increase of generalization performance as new datasets are observed. However, SVMLearn++.MT is able to learn better than SVMLearn++ as shown in Table 2 and 3.

3.2 Volatile Organic Compounds Dataset

The Volatile Organic Compounds (VOC) dataset is a real world dataset that consist of 5 classes (toluene, xylene, hectane, octane and ketone) with 6 attributes coming from six (quartz crystal microbalance type) chemical gas sensors. The dataset was split into three training and a test dataset. The distribution of the data is given in Table 4, where a new class was introduced with each dataset.

Table 4. VOC data distribution

Class	C1	C2	C3	C4	C5
DS1	20	0	20	0	40
DS2	10	25	10	0	10
DS3	10	15	10	40	10
Test	24	24	24	40	52

In this experiment, both algorithms were incrementally trained with three subsequent training datasets. Each algorithm was employed to create as many classifiers as necessary to obtain their maximum performance. As shown in Tables 5 and 6, based on an average of 30 trials, SVMLearn++ generated a total of 33 classifiers to achieve its best performance; however SVMLearn++.MT not only produced a 5% better generalization performance with only 10 classifiers, but it also provided a significantly more stable improvement as seen from the reduced standard deviation.

Table 5. SVMLearn++ with RBF kernel (σ= 3, C =100) results on VOC dataset

	C1	C2	C3	C4	C5	Gen.	Std.
DS1(5)	91%	-	95%	-	99%	58%	1.62%
DS2(10)	97%	91%	81%	-	95%	70%	1.84%
DS3(18)	93%	99%	94%	68%	76%	83%	8.19%

Table 6. SVMLearn++.MT with RBF kernel (σ= 3, C =100) results on VOC dataset

	C1	C2	C3	C4	C5	Gen.	Std.
DS1(6)	93%	-	89%	-	99%	58%	1.67%
DS2(2)	96%	93%	88%	-	95%	70%	1.45%
DS3(2)	95%	94%	100%	99%	73%	88%	1.37%

4 Conclusions

In this paper, we presented a new ensemble based incremental SVM learning algorithm, SVMLearn++.MT, using Learn++.MT. SVMLearn++.MT with RBF kernel functions has been tested on one real world dataset and one benchmark dataset. The results show that while SVM classifier can be equipped with the incremental learning capability, dealing with *catastrophic forgetting* problem, SVMLearn++.MT reduces the effect of *out-voting* problem, and also provides performance improvements over SVMLearn++.

It is also worth noting that, SVMLearn++.MT is more robust than SVMLearn++. One of the reasons why SVMLearn++ is having difficulty in learning a new class when first presented is due to difficulty in choosing the strength of the base classifiers. If we choose too weak classifiers, the algorithm is unable to learn. If we choose too strong classifiers, the training data are learned very well, resulting in very low β values which then causes very high voting weights, and hence even a more difficult *out-voting* problem. Since the SVM classifiers are strong classifiers, we have shown that SVMLearn++.MT, by significantly reducing the effect of the *out-voting* problem, improves the robustness of the algorithm, as the new algorithm is substantially more resistant to more drastic variations in the SVM classifier architecture and parameters (regularization constant C and kernel parameters).

Acknowledgements

This work is supported in part by the National Science Foundation under Grant No. ECS-0239090, "CAREER: An Ensemble of Classifiers Approach for Incremental Learning". Z.E. would like to thank Mr. Michael Muhlbaier and Mr. Apostolos Topalis graduate students at Rowan University, NJ, for their invaluable suggestions and assistance.

References

1. Z. Erdem, R. Polikar, F. Gurgen, N. Yumusak, "Ensemble of SVMs Classifier for Incremental Learning", Proc. of 6th Int. Workshop on Multiple Classifier Systems (MCS 2005), Springer-Verlag LNCS, Vol:3541, pp:246-256, Seaside, CA, USA, 13-15 June 2005.
2. R. Polikar, L. Udpa, S. Udpa, V. Honavar. "Learn++: An incremental learning algorithm for supervised neural networks." IEEE Transactions on Systems, Man, and Cybernetics. Part C: Applications and Reviews 31.4 (2001):497-508.
3. N. Kasabov, "Evolving Connectionist Systems: Methods and Applications in Bioinformatics, Brain Study and Intelligent Machines", Springer Verlag, 2002.
4. L. Ralaivola, F. d'Alché-Buc, "Incremental Support Vector Machine Learning: a Local Approach", In Proceedings of ICANN'01, Vienna, Austria, (2001)
5. C. P. Diehl and G. Cauwenberghs, "SVM Incremental Learning, Adaptation and Optimization", Proc. IEEE Int. Joint Conf. Neural Networks (IJCNN 2003), Portland OR, (2003).
6. M. Muhlbaier, A. Topalis, R. Polikar, Learn++.MT: A New Approach to Incremental Learning, 5th Int. Workshop on Multiple Classifier Systems (MCS 2004), Springer LINS vol. 3077 , pp. 52-61, Cagliari, Italy, June 2004.
7. V. Vapnik, Statistical Learning Theory. New York: Wiley, 1998.
8. N. Cristianini, J. Shawe-Taylor, An Introduction to Support Vector Machines and Other Kernel-based Learning Methods, Cambridge University Press, 2000.
9. C.-C. Chang, C.-J. Lin, "LIBSVM: A library for support vector machines", http://www.csie.ntu.edu.tw/~cjlin/libsvm

A Comparison of Different Initialization Strategies to Reduce the Training Time of Support Vector Machines

Ariel García-Gamboa, Neil Hernández-Gress, Miguel González-Mendoza, Rodolfo Ibarra-Orozco, and Jaime Mora-Vargas

ITESM-CEM, Carretera Lago de Guadalupe Km 3.5, Atizapán de Zaragoza, Estado de México c.p. 52926, México
{ariel.garcia, ngress, mgonza, rodolfo.ibarra, jmora}@itesm.mx

Abstract. This paper presents a comparison of different initialization algorithms joint with decomposition methods, in order to reduce the training time of Support Vector Machines (SVMs). Training a SVM involves the solution of a quadratic optimization problem (QP).The QP problem is very resource consuming (computational time and computational memory), because the quadratic form is dense and the memory requirements grow square the number of data points. The SVM-QP problem can be solved by several optimization strategies but, for large scale applications, they must be combined with decomposition algorithms that breaks up the entire SVM-QP problem into a series of smaller ones. The support vectors found in the training of SVMs represent a small subgroup of the training patterns. Some algorithms are used to initilizate the SVMs, making a fast approximation of the points standing for support vectors, to train the SVM only with those data. Combination of these initializations algorithms and the decomposition approach, coupled with an QP solver specially arranged for the SVM-QP problem, are compared using some well-known benchmarks in order to show their capabilities.

1 Introduction

Support Vector Machine (SVM) is a well known technique for solving classification, regression and density estimation [1] problems. This learning technique provides a convergence to a globally optimal solution and, for several problems, it has shown better generalization capabilities than other learning techniques. Training a SVM involves the solution of a quadratic programming (QP) problem. Solving this QP problem, special training patterns contained in the original training database (X), called support vectors (SV), are identified. Since the number of variables in the QP problem is equal to the number of data patterns, the complexity of the problem grows exponentially with the number of training patterns, and thus, find out a solution for large scale applications use prohibitive computational time. Several researches [1] use the fact that, given a training data set, the QP optimization problem will provide the same result if the entire data

W. Duch et al. (Eds.): ICANN 2005, LNCS 3697, pp. 613–618, 2005.

set or a reduced one (having only support vectors) is used. This paper compares different initialization methodologies in order to find the best one. The initialization algorithm find a reduced set of vectors containing candidate examples to become SV and, then, the training of the SVM is performed only using the reduced training set. The compared methods make use of an initialization by means of different algorithms and the result is used to train a better-posed problem by traditional QP solver. The paper is organized as follows: Section two describes the optimization problem generated by a SVM. Section three describes the different initialization algorithms. Section four describes the way in which initialization strategies are integrated to the decomposition method. Section five compares performance, in terms of time, of the proposed initialization using seven different benchmarks. Finally, results are discussed.

2 Support Vector Machines

For classification tasks, the main idea can be stated as follows: given a training data set ($\mathbf{X}$) characterized by patterns $x_i \in \Re^n$, $i = 1, \ldots, n$ belonging two possible classes $y_i \in \{1, -1\}$, there exist a solution represented by the following optimization problem:

$$\underset{\alpha}{Maximize} \quad L_D(\alpha) = \sum_{i=1}^{l} \alpha_i \alpha_j y_i y_j k(\mathbf{x_i}, \mathbf{x_j}) \tag{1}$$

$$s.t. \quad \sum_{i=1}^{n} y_i \alpha_i = 0, \quad 0 \leq \alpha \leq C \tag{2}$$

where α_i are the Lagrange multipliers introduced to transform the original formulation of the problem with linear inequality constraints into the above representation, [1]. The parameter C controls the misclassification level on the training data and therefore the margin. The $k(\mathbf{x_i}, \mathbf{x_j})$ term represents the so called kernel trick and is used to project data into a Hilbert space F of higher dimension using simple functions for the computation of dot products of the input patterns: $k(\mathbf{x_i}, \mathbf{x_j}) = \phi(\mathbf{x_i})^T \phi(\mathbf{x_j})$, $i, j = 1, \ldots, n$. Once one has the solution, the decision function is defined as:

$$f(\mathbf{x}) = sign\Big(\sum_{i=1}^{l} \alpha_i y_i k(\mathbf{x_i}, \mathbf{x}) + b\Big) \tag{3}$$

The solution to the problem formulated in (2) is a vector $\alpha_i^* \geq 0$ for which the α_i strictly greater than zero are the support vectors. Geometrically, these vectors are at the margin defined by the separator hyperplane. There are different algorithms for the resolution of the QP problem but, for our purposes, a fine tuning dual active set method properly arranged for large datasets is used [4].

3 Initialization Algorithms

3.1 Barycentric Correction Procedure

Barycentric Correction Procedure (BCP) is an algorithm based on geometrical characteristics for training a threshold unit [2]. It is very efficient training linearly separable problems and it was proven that the algorithm rapidly converges towards a solution [3]. The algorithm defines a hyperplane $\mathbf{w}^T\mathbf{x} + \theta$ dividing the input space for each class. Thus, we can define: $I_1 = 1, \ldots, N_1$ and $I_0 = 1, \ldots, N_0$ where N_1 represents the number of patterns of target 1 and N_0 the number of patterns of target -1. Also, let $(b = b_1, b_0)$ be the barycenters of data points belongin to class $\{+1, -1\}$ respectively, and weighted by the positive coefficients $\alpha = \alpha_1, \ldots, \alpha_{N_1}$ and $\mu = \mu_1, \ldots, \mu_{N_0}$ referred as *weighting coefficients* [3]:

$$b_1 = \frac{\sum_{i \in I_1} \alpha_i x_i}{\sum_{i \in I_1} \alpha_i} \qquad b_0 = \frac{\sum_{j \in I_0} \alpha_j x_j}{\sum_{j \in I_0} \mu_j} \tag{4}$$

The weight vector $\mathbf{w}$ is defined as a vector difference $\mathbf{w}= b_1 - b_0$. At each iteration, barycenter moves towards misclassified patterns. Increasing the value of particular barycenter implies hyperplane moves on that direction. For computing the bias term θ, let's define $\vartheta : \Re^n \rightarrow \Re$ such that $\vartheta(x) = -\mathbf{w}{\cdot}x$ The bias term is calculated as follows: $\theta = \frac{\max \vartheta_1 + \min \vartheta_0}{2}$. Assuming the existence of $J_1 \in I_1$ and $J_0 \in I_0$ that refer to misclassified examples of target $\{+1, -1\}$, barycenter modifications are calculated by:

$$\forall_i \in J_1, \;\; \alpha(new)_i = \alpha(old)_i + \beta_i \quad \text{and} \quad \forall_j \in J_0, \;\; \mu(new)_j = \mu(old)_j + \delta_j \tag{5}$$

Where $\beta = \max\left\{\beta_{min}, min[\beta_{min}, \frac{N_1}{N_0}]\right\}$ and $\delta = \max\left\{\delta_{min}, min[\delta_{min}, \frac{N_0}{N_1}]\right\}$. According to [3], β_{min} and δ_{min} can be set to 1 and β_{max} and δ_{max} set to 30.

3.2 Perceptron Algorithm

The Perceptron algorithm [5] is the first approach method to deal with linearly separable problems. It is an incremental algorithm which starts with a weight vector $\mathbf{w} = 0$ and, at each iteration, small modifications to $\mathbf{w}$ are computed until a solution is reached. Convergence is ensured in a finite number of iterations for linearly separable problems. In this research, we make some modifications to the original algorithm to treat with non-linearly separable problems.

Algorithm 1. Perceptron algorithm
1. Initialize the weight vector $\mathbf{w}$, b and choose a learning steep η 2. While there exist $i : i \in N$ such that $f(\mathbf{x_i}) \neq y_i$ a. Update $\mathbf{w}$ and b according to: $\delta w = w_{old} + (\eta/2)(y_i \cdot \mathbf{x_i})$ and $\delta b = b_{old} + (\eta/2)(y_i)$

Kernel Perceptron Extension. The kernel Perceptron algorithm [8] deals with non-linearly separable datasets. Basically the algorithm defines the dual function: $f(\mathbf{x}) = \sum_{i=1}^{n} \gamma_i y_i \{\phi(\mathbf{x}_i^T) \cdot \phi(\mathbf{x})\} + b$, where γ is the set of dual variables to be updated and the dot product $(\phi(\mathbf{x}_i^T) \cdot \phi(\mathbf{x}))$ is replaced for the kernel function $k(\mathbf{x}_i, \mathbf{x})$.

space), while the solution of the SVM is reached in a high dimensional space (in feature space). The $KernPerceptron + QP$ strategy does not have this problem, becuase the Kernel Perceptron algorithm looks for the solution in the same high dimensional space than the SVM algorithm.

Table 1. Time performance of the initialization strategies using the UCI data

	N	$BCP + QP$	$Percep + QP$	$Kpercep + QP$	$KSK + QP$	$Random + QP$
iri	150	**0.014 sec**	0.0673 sec	**0.0160** sec	0.5391 sec	0.016 sec
son	208	**0.181 sec**	0.3588 sec	**0.1231** sec	2.9990 sec	0.391 sec
pid	768	**4.703 sec**	5.4085 sec	**2.9060** sec	10.6013 sec	17.203 sec
tic	958	**3.102 sec**	4.4576 sec	**1.7350** sec	16.6742 sec	15.781 sec
pho	1027	**3.262 sec**	3.7297 sec	**3.5150** sec	21.3721 sec	19.406 sec
adu	5000	**23.38 sec**	29.2773 sec	-	284.96 sec	355.797 sec
shu	13633	**33.77 sec**	72.6192 sec	**22.328** sec	87.532 sec	4256.757 sec

6 Conclusions

Support Vector Machines is a promising methodology used in different research areas. Moreover, the optimization of the SVM is a delicate problem due to computational and memory requirements. This research is focused in comparing different initialization strategies with a decomposition method in order to select one that improves the training time of SVMs. The comparison of the different proposals shows on one hand, that using kernel Perceptron as initialization strategy has a good performance and on the other hand, BCP shows a better behaviour although the limitations that could have training data sets with a high non linearity degree. Additionally, these two approaches ensure a better performance training large scale datasets, in comparison with the Random-QP algorithm.

References

1. Vapnik, V.: Computational Learning Theory. John Wiley & Sons (1998)
2. Poulard, H., Stève, D.: Barycentric Correction Procedure: A fast method of learning threshold unit. World Congress on Neuronal Networks **1** (1995) 710–713
3. Poulard, H., Stève, D.: A convergence theorem for Barycentric Correction Procedure. Technical Report 95180 of LAAS-CNRS (1995)
4. Gonzlez-Mendoza, M.: Quadratic Optimization fine tuning for the learning phase of SVM. ISSADS 2005 **11** (2005).
5. Rossemblat, F.: The Perceptron: a probabilistic model for information storage and organization in the brain. Psychological review (1958) 386–408.
6. Kozinec, B. N.: Recurrent algorithm for separating convex hulls of two sets. Learning algorithms in pattern recognition (1973) 43–50.
7. Vojtech, F.: Generalization of the Schlesinger-Kozinecs algorithm for Support Vector Machines. Centre of machine perception, CTU Prague (2002).
8. Kowalczyk, A.: Maximal margin Perceptron. MT Press, Cambridge (1999).
9. Keogh, E., Blake, C. and Merz, C.J.: UCI repository of machine learning databases. http://kdd.ics.uci.edu (1998).

A Hierarchical Support Vector Machine Based Solution for Off-line Inverse Modeling in Intelligent Robotics Applications

D.A. Karras

Chalkis Institute of Technology, Dept. Automation and Hellenic Open University., Rodu 2, Ano Iliupolis, Athens 16342, Greece
dakarras@teihal.gr, dakarras@ieee.org

Abstract. A novel approach is presented for continuous function approximation using a two-stage neural network model involving Support Vector Machines (SVM) and an adaptive unsupervised Neural Network to be applied to real functions of many variables. It involves an adaptive Kohonen feature map (SOFM) in the first stage which aims at quantizing the input variable space into smaller regions representative of the input space probability distribution and preserving its original topology, while rapidly increasing, on the other hand, cluster distances. During convergence phase of the map a group of Support Vector Machines, associated with its codebook vectors, is simultaneously trained in an online fashion so that each SVM learns to respond when the input data belong to the topological space represented by its corresponding codebook vector. The proposed methodology is applied, with promising results, to the design of a neural-adaptive controller, by involving the computer-torque approach, which combines the proposed two-stage neural network model with a servo PD feedback controller. The results achieved by the suggested SVM approach are favorably compared to the ones obtained if the role of SVMs is undertaken, instead, by Radial Basis Functions (RBF).

1 Introduction

It is known that Artificial Neural Networks (ANNs) and especially Multilayer Perceptrons (MLP) and Radial Basis Functions (RBF) have the theoretical ability to approximate arbitrary nonlinear mappings[1]. Moreover, since ANNs can have multi-inputs and multi-outputs, they can be naturally used for control of multivariable systems. Although MLPs and RBFs have been successfully employed in function approximation tasks, on the other hand, several drawbacks have been revealed in their application. In order to overcome them more powerful ANN models have emerged, namely, Support Vector Machines (SVMs). SVMs, introduced by Vapnik in 1992 [1], have recently started to be involved in many different classification tasks with success. Few research efforts, however, have employed them in nonlinear regression tasks. One of the goals of the herein study was to evaluate the SVM for Nonlinear Regression approach in such tasks, in comparison with the RBF techniques. The results herein obtained justify that the SVM approach could widely and successfully be used in function approximation/regression tasks involved in intelligent control.

W. Duch et al. (Eds.): ICANN 2005, LNCS 3697, pp. 619 – 624, 2005.

Robot manipulators have become increasingly important in the field of flexible automation but they are subject to structured and/or unstructured uncertainties. ANN models and especially MLPs and RBFs have been used for the construction of Neural - Adaptive Controllers to cope with both types of uncertainty [2] following different approaches. The research line herein followed for designing a nonlinear compensator using ANNs is based on the computed torque method proposed in [3], where the ANNs were used to compensate for nonlinearities of the robotic manipulator rather than to learn its inverse dynamics. Another method for the direct control of robot manipulators using ANNs was proposed in [4]. Here, the control system consists of an inverse model of the robot dynamics which produces the forces/torques to be applied to the robot, given desired positions, velocities and accelerations and a neural controller generating a correcting signal. Another approach is to combine an ANN model with a servo PD feedback controller [5]. This approach, illustrated in Figure 1, provides on - line learning of the inverse dynamics of the robot. In this scheme the manipulator's inverse - dynamics model is replaced by generic neural network models, one per joint, each neural network model adaptively approximating the corresponding joint's inverse dyamics.

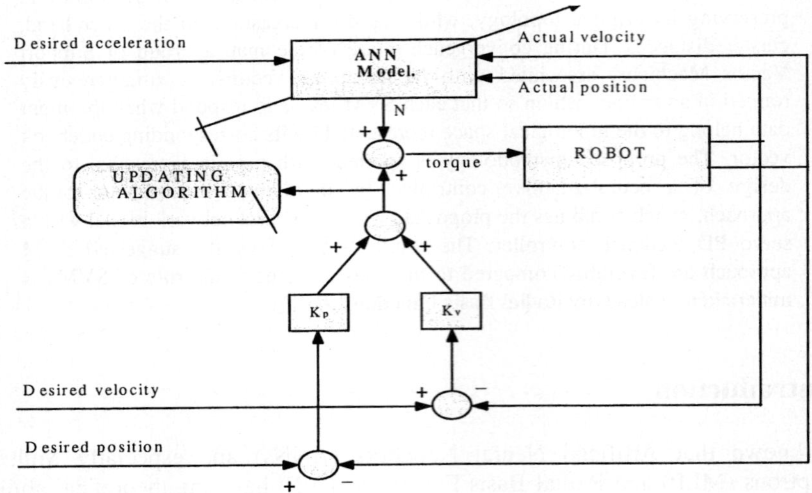

Fig. 1. Neuro-controller based computed torque architecture

2 Hierarchical SVM in the Computed Torque Method

It is herein proposed that ANN performance in control tasks could be improved through a two-stage process employing an adaptive regularized SOFM and a group of SVMs associated with its codebook vectors. The SOFM algorithm could be applied to the available training input vectors of the function approximation problem in order to quantize the space defined by the input variables in smaller topology preserving regions by involving suitable codebook vectors, whose distribution is representative of the joint probability distribution of these input variables. This unsupervised process results in determining the associated codebook vectors, whose number has been predefined by the user. During convergence of this process and at its (t) iteration, the codebook vectors associated to the number NE(t) of neighbouring output units at instance t are updated in accordance with the classical SOFM algorithm. Then, the weights of their corresponding NE(t) group of SVMs are, also, updated. Thus, each such SVM is iteratively being specialized, simultaneously with its associated codebook vector, to respond when the input vector belongs to the topological subspace represented, at the end of the process, by this codebook vector.

In the proposed variation of the SOFM algorithm we attempt to increase the distances between different clusters of the input space probability distribution so as to facilitate convergence of the map and achieve higher resolution of SOFM input space in order to feed the second stage SVMs with groups of input vectors of better quality, in terms of homogeneity. It is herein suggested that such a goal could be attained through adapting not only the winning neuron and its neighboring neurons NE(t) weights but, also, loosing neurons weights. Loosing neurons weights are adapted in a manner similar to that of LVQ, by increasing the distance between these weights vectors and the corresponding input data vectors. More specifically, all neurons outside the winning neuron neighborhood NE(t) at iteration t, are updated by the following formula: $\mathbf{W}_j(t+1) = \mathbf{W}_j(t) - [b(t) \exp\|(\mathbf{X} - \mathbf{W}_j(t))\|] (\mathbf{X} - \mathbf{W}_j(t))$, for all codebook-vectors j not belonging in NE(t) at the t iteration of the algorithm, when an input vector **X** is applied to the SOFM model. All codebook-vectors i belonging in NE(t) are updated as in the conventional SOFM. The term $[b(t) \exp\|(\mathbf{X} - \mathbf{W}_j(t))\|]$ is similar to Kohonen's conventional learning rate but now this parameter depends not only on time but, also, on distance due to $\exp\|(\mathbf{X} - \mathbf{W}_j(t))\|$. That is, the larger the distance between input vector **X** and codebook-vector j the larger the learning parameter so that the updated codebook-vector j increases dramatically its distance from input vector **X**, in order to have faster convergence. This is the reason we call this new SOFM update scheme adaptive SOFM.

With regards to the second stage involved group of SVMs, the task of nonlinear regression could be defined as follows. Let f(X) be a multidimensional scalar valued function to be approximated. Then, a suitable regression model to be considered is: D = f(X) + n, where X is the input vector, n is a random variable representing the noise and d denoting a random variable representing the outcome of the regression process. Given, also, the training sample set {(Xi, Di)} (i=1,..,N) then, the SVM training can be formulated as next outlined:

Find the Lagrange Multipliers {λi} (i=1, ..,N) and {λ'i} (i=1, ..,N) that maximize the objective function,

$Q(\lambda_i, \lambda'_i) = \Sigma_{i=1..N} D_i (\lambda_i - \lambda'_i) - e\, \Sigma_{i=1..N} (\lambda_i + \lambda'_i) - \frac{1}{2} \Sigma_{i=1..N} \Sigma_{j=1..N} (\lambda_i - \lambda'_i)(\lambda_j - \lambda'_j) K(\mathbf{X}_i, \mathbf{X}_j)$ subject to the constraints: $\Sigma_{i=1..N} (\lambda_i - \lambda'_i) = 0$ and $0 <= \lambda_i <= C$, $0 <= \lambda'_i <= C$ for i=1..N, where C is a user defined constant.

In the above definition, $K(\mathbf{X}_i, \mathbf{X}_j)$ are the kernel functions. In the problem at hand we have employed the radial basis kernel $K(\mathbf{X}, \mathbf{X}_j) = \exp(-1/2\sigma^2 \|\mathbf{X} - \mathbf{X}_j\|^2)$. Taking into account all the previous definitions we can then, fully determine the approximating function as $F(\mathbf{X}) = \Sigma_{i=1..N} (\lambda_i - \lambda'_i) K(\mathbf{X}, \mathbf{X}_i)$.

This paper investigates the generalization accuracy of the suggested system in estimating a robotic manipulator's inverse dynamics model in combination with a servo PD feedback controller as in fig. 1. Assuming that structural uncertainty occurs in our n-link robot manipulator, the correct model of its inverse dynamics is given by the differential equation in vectorial form [3],

$$\mathbf{T} = \mathbf{R}^{-1}(\mathbf{q}, \dot{\mathbf{q}}, \ddot{\mathbf{q}}) \tag{1}$$

where, $\mathbf{T}$ is the joint torque, $\mathbf{R}^{-1}$ is a nonlinear mapping from the joint coordinate space to the joint torque space and $\mathbf{q}$, $\dot{\mathbf{q}}$, $\ddot{\mathbf{q}}$ are the robot arm motion parameters: joint trajectory, velocity and acceleration variables. The correct model of its direct dynamics is [3],

$$\ddot{\mathbf{q}} = \mathbf{R}(\mathbf{q}, \dot{\mathbf{q}}, \mathbf{T}) \tag{2}$$

Robot dynamics, however, cannot be modelled exactly. An estimated model $\hat{\mathbf{R}}^{-1}$ is used to predict the feedforward torques and a servo-feedback control scheme is involved to improve robustness. To this end, the approach adopted here involves a PD servo-controller to compensate for the linear changes and the two-stage ANN based controller to compensate for the intrinsic nonlinearities encountered due to parameters uncertainties.Therefore, the supposed correct inverse dynamics model of the robot arm of figure 1 is defined as follows,

$$\mathbf{T} = \hat{\mathbf{R}}^{-1}(\mathbf{q}, \dot{\mathbf{q}}, \ddot{\mathbf{q}}) + \mathbf{T}_{pd} \tag{3}$$

$$\mathbf{T}_{pd} = \mathbf{K}_p(\mathbf{q}_d - \mathbf{q}) + \mathbf{K}_p(\dot{\mathbf{q}}_d - \dot{\mathbf{q}}) \tag{4}$$

where, $\mathbf{T}_{pd}$ is the joint torque estimated by the PD-controller, $\mathbf{T}$ is the total joint torque, $\mathbf{q}_d$ and $\dot{\mathbf{q}}_d$ are the desired trajectory and velocity curves and finally, **Kp**, **Kv** are the gains of the PD controller. From equations (3-4) it is clear that the proposed ANN model should be trained to predict joint torque $\mathbf{N} = \mathbf{T} - \mathbf{T}_{pd}$ provided, the curves $\mathbf{q}$, $\dot{\mathbf{q}}$, $\ddot{\mathbf{q}}_d$ of actual trajectory, velocity and desired acceleration are given as inputs. It is a common practice to use $\ddot{\mathbf{q}}_d$ instead of differentiating the velocity $\dot{\mathbf{q}}$ to get $\ddot{\mathbf{q}}$. In a simulated version of the system, it is clear that equispaced samples of these three curves should be given as inputs to the ANN model. The algorithm of the simulation of the above control scheme can be depicted as follows : At every instance t the desired curves $\mathbf{q}_d$, $\dot{\mathbf{q}}_d$, and $\ddot{\mathbf{q}}_d$ as well as the curves of the actual trajectory and velocity ($\mathbf{q}(t)$ and $\dot{\mathbf{q}}(t)$) are given. Applying them as inputs to equation (1) when the true masses are involved we have the desired joint torque T = T(true). Then, if the false link masses are used in equation (4) a $\hat{\mathbf{T}}$ = Tpd(false) joint torque is computed. Subsequently, this false torque is applied to equation (4) involving the true parameters in order to derive the instance t+1 curves of the actual trajectory and velocity ($\mathbf{q}(t+1)$ and $\dot{\mathbf{q}}(t+1)$) of the robot arm. And so on. Therefore, we can take equispaced samples of the curves $\mathbf{q}(t)$, $\dot{\mathbf{q}}(t)$ as well as $\ddot{\mathbf{q}}_d(t)$ to form the suggested ANN's input patterns. In addition, we can take N= T (= T(true)) - $\hat{\mathbf{T}}$ (= Tpd(false)) to form its desired output. By varying link masses within given intervals, we can obtain a large set of training, validation and test set patterns.

3 Experimental Results and Discussion

A simple two-link planar elbow arm was used to test the performance of the proposed two-stage methodology comparing the application of SVMs and RBFs in the second stage of the proposed approach. The manipulator was modeled as a two rigid links of lengths l1 = 1m and l2 = 1m with point masses m1 =0.8kg and m2 = 2.3kg at the distal end of the links corresponding to the false PD model, while the true masses varied within the 10% confidence interval of these values. Twelve variations have been considered for each such link mass. The simulation was carried out invoking a fourth order Runge-Kutta algorithm, with step size h= 0.01 The desired position trajectory has the components

$$\theta_1 = g_1 \sin(2\pi t / T) \text{ and } \theta_2 = g_2 \sin(2\pi t / T) \tag{5}$$

with period T = 2s and amplitudes $g_i = 0.1$ rad. For good tracking the time constant of the closed-loop system was selected as 0.1s. For critical damping this means that for the PD Outer-Loop design $\mathbf{K}_v = diag\{k_v\}$, $\mathbf{K}_p = diag\{k_p\}$ with $k_p = 100$ and $k_v = 20$.

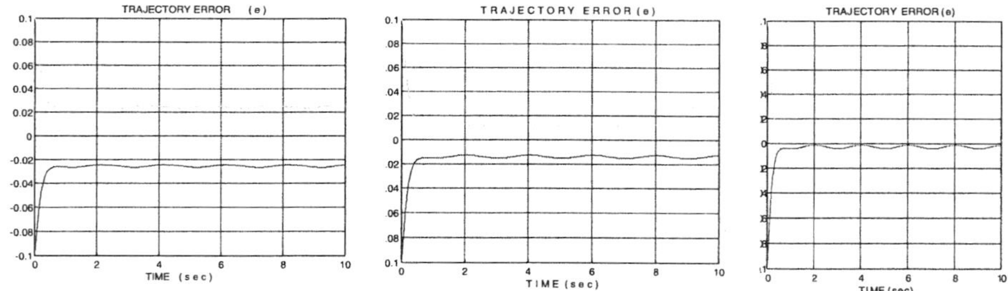

Fig. 2. (a) Trajectory tracking errors for the two joints when the true values of masses are ±10% of the false masses. (b) The same experiment using the correcting torque (N) obtained after using the two stage methodology involving RBFs in the second stage. (c) The same experiment using the correcting torque (N) obtained after using the proposed two stage methodology, i.e involving SVMs in the second stage.

Application of equations (1), (3), (4) employing these definitions leads, for every one of the 144 total pair masses variations, to obtaining the curves of desired acceleration, actual trajectory and actual velocity along with the associated torque N to be modelled by the proposed two-stage ANN. Each such curve has been sampled into 30 points, from which, using sliding windows of length l_w, we have formed training and test patterns for the ANN as follows. Each ANN input pattern contains l_w points for the desired acceleration, the corresponding l_w points for the actual trajectory and finally, the l_w corresponding points for the actual velocity. The desired output value is the associated torque N of the l_w -th sample of the sliding window under consideration. In our simulations $l_w = 6$ and we have used 2000 patterns for ANN training and the rest 1600 patterns for testing. Finally, the resulting network was combined with the PD robotic controller to test the performance of the method. Figure 2(a) shows the trajectory tracking error when the true masses are well apart from the false masses. In figures 2(b)/2(c) the same experiment is repeated but this time an additional correcting torque, produced by the ANN involving RBFs/SVMs in the second stage, is applied to improve the performance of the controller. This experiment is trajectory dependent. Further study is needed to make the method trajectory independent.

References

1. Haykin S., "Neural Networks: A comprehensive foundation", Second Edition, Prentice Hall, 1999.
2. M. Kawato et al., "Hierarchical neural network model for voluntary movement with applicationn to robotics", *IEEE Contr. Syst. Mag.*, 8 - 16
3. A. M. S. Zalzala and A. S. Morris, "A neural network approach to adaptive robot control", *Int. J, Neural Networks*, 2, pp 17-35, 1989.
4. A. Y. H. Zomaya, M. E. Suddaby and A. S. Morris, "Direct neuro - adaptive control of robot manipulators", *Proc. IEEE Int. Conf. On Robotics and Automation*, Nice, France, May 1992, pp. 1902 - 1907.
5. S. Khemaissia and A. S. Morris, "Neuro-adaptive control of robotic manipulators", *Robotica*, 11, pp. 465-473.

LS-SVM Hyperparameter Selection with a Nonparametric Noise Estimator

Amaury Lendasse[1], Yongnan Ji[1], Nima Reyhani[1], and Michel Verleysen[2]

[1] Neural Network Research Centre,
Helsinki University of Technology, P.O. Box 5400,
02150 Espoo, Finland
{lendasse, yji, nreyhani}@.hut.fi
[2] Machine Learning Group,
Université catholique de Louvain, DICE, 3 place du Levant, 1348
Louvain-la-Neuve, Belgique
verleysen@dice.ucl.ac.be

Abstract. This paper presents a new method for the selection of the two hyperparameters of Least Squares Support Vector Machine (LS-SVM) approximators with Gaussian Kernels. The two hyperparameters are the width σ of the Gaussian kernels and the regularization parameter λ. For different values of σ, a Nonparametric Noise Estimator (NNE) is introduced to estimate the variance of the noise on the outputs. The NNE allows the determination of the best λ for each given σ. A Leave-one-out methodology is then applied to select the best σ. Therefore, this method transforms the double optimization problem into a single optimization one. The method is tested on 2 problems: a toy example and the Pumadyn regression Benchmark.

Keywords: Least Squares Support Vector Machines, Leave-one-out, Noise Estimation, Regression.

1 Introduction

The selection of hyperparameters is a important issue in the fields of Artificial Neural Networks, Machine Learning and System Identification. Many resampling techniques have been successfully used as Leave-One-Out (LOO), Bootstrap and Cross-Validation [1, 2].

Least Squares Support Vector Machines with Gaussian kernels are efficient regression models [3]. For example, they do not suffer from the problem of local minima. Unfortunately, two hyperparameters have to be tuned, for example using LOO [4]. The two hyperparameters are the width σ of the Gaussian kernels and the regularization parameter λ. This problem leads to a grid search that is highly time consuming. In this paper, we propose the use of Nonparametric Noise Estimator (NNE) in order to select the regularization parameter as a function of the width σ.

The paper is organised as follows: LS-SVM are introduced in Section 2, NNE in Section 3 and the methodology in Section 4. In Section 5, the method is successfully tested on 2 problems: a toy example and the Pumadyn regression Benchmark.

W. Duch et al. (Eds.): ICANN 2005, LNCS 3697, pp. 625 – 630, 2005.

2 Least Squares Support Vector Machines

LS-SVM are regularized supervised approximators, which has been proved to be efficient for function approximation. Only solving linear equation is needed in the optimization process, which not only simplifies the process, but also avoids the problem of local minima in SVM. In this section, a short summary of the LS-SVM model is given. The LS-SVM model [4, 5] is defined in its primal weight space by,

$$\hat{y}(x) = \omega^T \varphi(x) + b \quad (1)$$

where $\varphi(x)$ is a function which maps the input space into a higher dimensional feature space, x is the M-dimensional vector of inputs x_j, and ω and b the parameters of the model. Given N input-output learning pairs $(x^i, y^i) \in R^M xR$, Least Squares Support Vector Machines for function estimation formulate the following optimization:

$$\min_{\omega,b,e} J(\omega, e) = \frac{1}{2}\omega^T \omega + \gamma \frac{1}{2}\sum_{i=1}^{N} e_i^2 \text{ subject to } y^i = \omega^T \varphi(x^i) + b + e^i, i = 1, \ldots, N \quad (2)$$

The parameter set θ consists of vector ω and scalar b. Solving this optimization problem in dual space leads to finding the α_i and b coefficients in the following solution:

$$h(x) = \sum_{i=1}^{N} \alpha_i K(x, x^i) + b \quad (3)$$

Function $K(x, x^i)$ is the kernel defined as the dot product between the $\varphi(x)^T$ and $\varphi(x)$ mappings. The meta-parameters of the LS-SVM model are the width of the Gaussian kernels (taken identical for all kernels) and the γ regularization factor. The training method for the estimation of ω and b can be found in [4].

3 Nonlinear Noise Estimator

The problem of function approximation consists in the determination of the relationship between a set x of inputs and one single output y. Given N inputs-output pairs $(x^i, y^i) \in R^M xR$, the relationship between x_i and y_i can be expressed as $y_i = f(x_i) + \varepsilon_I$,.where f is the unknown relationship and ε_i the noise. Any estimation of model f based on a finite number N of learning data goes through a compromise between a low learning error (small bias) and a smooth model (small variance). In the case of LS-SVM, this compromise is implemented through the choice of an adequate value of γ. If the value of γ is set too large, the model will overfit the data, including the noise. Still, the value of γ should be set as large as possible; a too small value of γ would simply mean that the model does not fit the learning data! It is therefore suggested to select the largest value of γ so that the learning error does goes below the level of noise. Indeed it is unreasonable to expect that a model could lead to an error that is lower than the level of noise; if it was the case, the model would be in overfitting region.

Selecting γ then means first to estimate the learning error of the model in function of γ, and secondly to estimate the variance of the noise. Of course, the noise estimator should not use the model itself, but only the data at disposal; it should be nonparametric.

An approach called "Delta Test" has been proposed for estimating the variance of the noise on the output [6]. It is based on the similarity of the noise behaviour between two closed data points. As the distance δ between two close points x and x' goes to zero, the average MSE between the corresponding outputs tends to var(ε) [7]:

$$E\left\langle \frac{1}{2}(y'-y)^2 \middle| \left|x'-x\right| < \delta \right\rangle \rightarrow \mathrm{var}(\varepsilon) \quad as \quad \delta \rightarrow 0 \tag{6}$$

Despite this approach seems to be promising for noise estimation purposes, it fails when the size of the data set is small with respect to the complexity of underlying function and noise distribution. Jones *et al.* [6] improved the Delta test using the k-nearest neighbour distances between data in the input space and corresponding data in the output space. This leads to an approach called here Nonparametric Noise Estimator (NNE). Referring to [6], the estimate of noise variance is the intercept of the linear regression line which is drawn between the average of the k nearest distances in the inputs space and the corresponding average of the k nearest distances in the output space (see equation 7 below). A proof of NNE (which is also called Gamma Test in some papers) can be found in [7] and is based on a generalization of Chybechov inequality and the property of k-nearest neighbor structures. Moreover, it has been shown that NNE is useful too for evaluating the nonlinear correlation between two random variables, or input and output pairs realizations. In the proof, the following conditions are necessary:

- the first and second partial derivatives of the underlying function exist;
- the first to the fourth moments of the noise distribution exist;
- the noise is independent from the input.

Using this three conditions, the variance of noise is given by the intercept with the vertical line $\delta(k)$=0, of the regression line between $\gamma(k)$ and $\delta(k)$, where $1 \leq k \leq p$ and

$$\delta(k) = \frac{1}{N}\sum_{i=1}^{N}\left|x_{NN(x_i,k)} - x_i\right|^2 \text{ and } \gamma(k) = \frac{1}{2N}\sum_{i=1}^{N}\left|y_{NN(x_i,k)} - y_i\right|^2 . \tag{7}$$

In (7), $NN(x_i,k)$ is the index of the kth neighbour of x_i. According to [6], p=10 is used in experiments presented in section 5.

This noise variance estimator based on [6] is similar to the variogram based estimator detailed in [8]. However, it differs from the fact that Jones' estimator only uses the k nearest neighbours of the data points. This reduces the computation time and makes the estimator efficient when the number of data points is large enough by concentrating on small values of $\delta(k)$.

4 Methodology

The goal of the presented methodology is to transform the double optimization of γ and σ in LS-SVM into a simple optimization procedure. The double optimization of the metaparameters using LOO presented in [3, 4] is very efficient but is highly time consuming.

Our methodology can be expressed as the following:

1) A range of σ is selected.
2) For each σ, the Nonparametric Noise Estimate is performed.
3) A bisection method is used to estimate the value of γ such that the training error of the LS-SVM is equal to the value of the Nonparametric Noise Estimate. The training error is strictly decreasing with respect to γ and then the solution is unique and its computation is very fast. Taking the largest γ value such that the training error does not exceed the noise variance leads to the more accurate mode without overfitting.
4) The LOO error (LOO MSE) is estimated for each value of σ.
5) The value of σ and corresponding γ minimizing the LOO error are selected.

5 Experiment

5.1 Toy Example

A toy example with 1000 samples is build using the following function:

$$y = \sin(x) + \sin(5x) + \sin(15x) + \varepsilon \tag{8}$$

with ε an uniform noise in [-0.5, 0.5]. The function is represented in Fig.2 The real value of the variance of the noise is 0.0822 and the estimate obtained with the NNE is also 0.0822. The methodology presented in section 4 is applied. The range of σ is between 0.01 and 0.4 by step of 0.005. For each value of σ, γ is calculated using the estimate of the NNE (see Fig. 1. a).

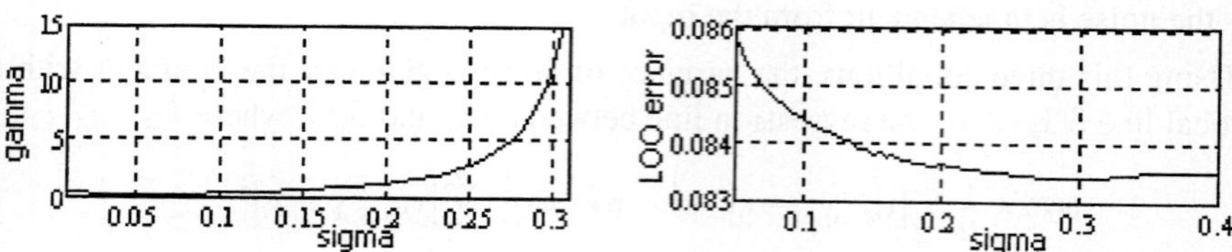

Fig. 1. Toy example results. **a** - γ with respect to σ. **b** – LOO error with respect to σ.

For each value of σ (using the corresponding γ), the LOO error is computed (see Fig. 1. b). The optimum is obtained for $\sigma = 0.295$ and the corresponding $\gamma = 9.727$. The approximation obtained the selected LS-SVM is represented in Fig. 2.

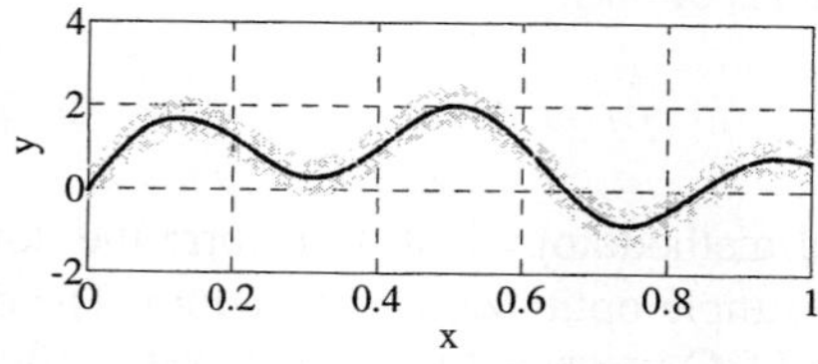

Fig. 2. The toy example and the approximation after the selection of the hyperparameters

5.2 Pumadyn Benchmark

The pumadyn datasets [9] are a family of datasets synthetically generated from a realistic simulation of the dynamics of a Puma robot arm. The tasks associated with these datasets consist of predicting the angular acceleration of one of the links of the robot arm given the angular positions, velocities, torques, and in some cases, other dynamic parameters of the robot arm. The dataset contains 8192 samples, 8 inputs and one output. The methodology presented in section 4 is applied. The range of σ is between 5 and 110 by step of 5. For each value of σ, γ is calculated using the estimate of the NNE (see Fig. 3. a).

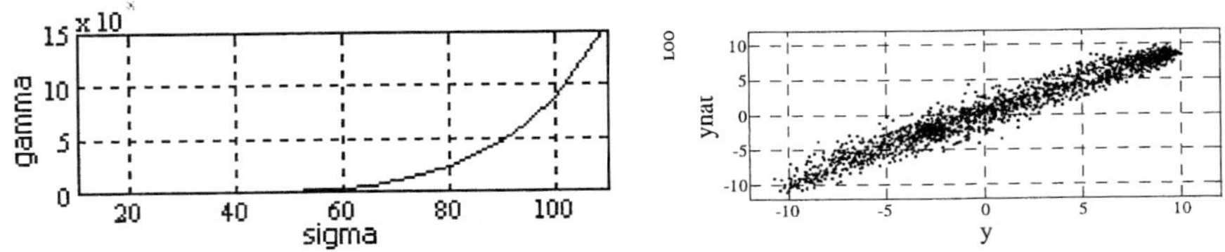

Fig. 3. a γ with respect to σ. **b** – LOO error with respect to σ.

For each value of σ (using the corresponding γ), the LOO error is computed; a smooth slope similar to the one in Fig. 1. b. is obtained. Its minimum is found for σ = 95 and the corresponding γ = 6.4749e+008. The approximation with respect to the target value y is represented in see Fig. 3. b. The LOO error that is obtained is 1.81.

6 Conclusion and Further Work

In this paper, a Nonparametric Noise Estimator has been introduced for the selection of the hyperparameters of LS-SVM. The proposed methodology transforms the double optimization problem of the selection of the hyperparameters into a single optimization one, therefore reducing drastically the computation time for similar results.

The method has been illustrated on two examples and gives accurate approximations. Further work includes the test of other methods for Nonparametric Noise Estimation (see for example [8]) and their embedding into the same methodology to select hyperparameters in LS-SVM and other learning schemes.

Acknowledgements

LS-SVMlab [3, 4] has been used for the optimization of the LS-SVM models and to perform the Leave-one-Out procedures. Part of work of A. Lendasse, Y.N. Ji and N. Reyhani is supported by the project of New Information Processing Principles, 44886, of the Academy of Finland. M. Verleysen is a Senior Research Associate of the Belgian F.N.R.S. (National Fund For Scientific Research).

References

1. Bishop, C. M., *Neural Networks for Pattern Recognition*. New York: Oxford, 1995.
2. Lendasse A., Wertz V., Verleysen M.: Model selection with cross-validations and bootstraps – Application to time series prediction with RBFN models. In: Artificial Neural Networks and Neural Information Processing – ICANN/ICONIP (2003), Kaynak O., Alpaydin E., Oja E., Xu L. (eds): Springer-Verlag Lecture Notes in Computer Science 2714, Berlin (2003) 573-580.
3. Suykens, J., A., Van Gestel, K., T., De Brabanter, J., De.Moor, B., Vandewalle, J.: Least Squares Support Vector Machines. World Scientific, Singapore, ISBN 981-238-151-1 (2002).
4. http://www.esat.kuleuven.ac.be/sista/lssvmlab/
5. Suykens, J., A., De brabanter, K., J., Lukas, L., Vandewalle, J.: Weighted least squares support vector machines: robustness and sparse approximation. Neurocomputing, Special Issue on fundamental and information processing aspects of neurocomputing.
6. Jones, A. J. New Tools in Non-linear Modeling and Prediction. Computational Management Science, Vol. 1, Issue 2, p.p. 109-149, 2004.
7. Evans, D. and Jones, A. J., A proof of the Gamma test, Proc. Roy. Soc. Lond. A, Vol. 458, pp. 1-41, 2002.
8. Pelckmans K., De Brabanter J., Suykens J.A.K., De Moor B., Variogram based noise variance estimation and its use in Kernel Based Regression, in *Proc. of the IEEE Workshop on Neural Networks for Signal Processing*, pp. 199-208.
9. Corke, P. I. (1996). A Robotics Toolbox for MATLAB. *IEEE Robotics and Automation Magazine,* **3** (1): 24-32.

Building Smooth Neighbourhood Kernels via Functional Data Analysis

Alberto Muñoz[1] and Javier M. Moguerza[2]

[1] University Carlos III, c/ Madrid 126, 28903 Getafe, Spain
alberto.munoz@uc3m.es
[2] University Rey Juan Carlos, c/ Tulipán s/n, 28933 Móstoles, Spain
javier.moguerza@urjc.es

Abstract. In this paper we afford the problem of estimating high density regions from univariate or multivariate data samples. To be more precise, we propose a method based on the use of functional data analysis techniques for the construction of smooth kernel functions oriented to solve the One-Class problem. The proposed kernels increase the precision of One-Class estimation procedures. The advantages of this new point of view are shown using data sets drawn from representative density functions.

1 Introduction

The task of estimating high density regions from data samples arises explicitly in a number of works involving interesting problems such as outlier detection or cluster analysis. One-Class Support Vector Machines (SVM) [3] and Support Neighbour Machines (SNM) [1] are designed to solve this problem with tractable computational complexity. We refer to [3] and references therein for a complete description of the problem and its ramifications.

The concrete problem to solve is the estimation of minimum volume sets of the form $S_\alpha(f) = \{x | f(x) \geq \alpha\}$, such that $P(S_\alpha(f)) = \nu$, where f is the density function and $0 < \nu < 1$. The goal is to obtain some decision function $h(x)$ which solves this problem, that is, $h(x) = +1$ if $x \in S_\alpha(f)$ and $h(x) = -1$ otherwise.

The strategy of One-Class SVM is to map the data points into a feature space determined by a kernel function, and to separate them from the origin with maximum margin. In order to build a separating hyperplane between the origin and the mapped points, the quadratic One-Class SVM method solves the following problem:

$$\begin{aligned} \min_{w,\rho,\xi} \ & \frac{1}{2}\|w\|^2 - \rho + \frac{1}{\nu n}\sum_{i=1}^{n} \xi_i \\ \text{s.t.} \ & w^T \phi(x_i) \geq \rho - \xi_i \,, \\ & \xi_i \geq 0, \qquad\qquad i = 1, \ldots, n \,, \end{aligned} \tag{1}$$

where ϕ is the mapping defining the kernel function, ρ represents the decision value which determines if a given point belongs to the support of the distribution,

W. Duch et al. (Eds.): ICANN 2005, LNCS 3697, pp. 631–636, 2005.

ξ_i are slack variables, and $\nu \in [0,1]$ is an a priori fixed constant. The decision function will be $h(x) = sign(w^{*T}\phi(x) - \rho^*)$, where w^* and ρ^* are the values of w and ρ at the solution of problem (1) (see [3] for details). In the following we will show how to build smooth ϕ functions for the construction of $h(x)$.

The rest of the paper is organized as follows. Section 2 introduces a new kind of kernels. In Section 3 the smoothing methodology is shown. Section 4 concludes.

2 Neighbourhood Measures and Kernels

There are data analysis problems where the knowledge of an accurate estimator of the density function $f(x)$ is sufficient to solve them, for instance, mode estimation, or the present task of estimating $S_\alpha(f)$. However, density estimation is far from trivial [4,3]. The next definition is introduced to relax the density estimation problem: the task of estimating the density function at each data point is replaced by a simpler measure that asymptotically preserves the order induced by f.

Definition 1 (Neighbourhood Measures). *Consider a random variable X with density function $f(x)$ defined on $\mathbb{R}^d$. Let S_n denote the set of random independent identically distributed (iid) samples of size n (drawn from f). The elements of S_n take the form $s_n = (x_1, \cdots, x_n)$, where $x_i \in \mathbb{R}^d$. Let $M : \mathbb{R}^d \times S_n \longrightarrow \mathbb{R}$ be a real-valued function defined for all $n \in \mathbb{N}$. (a) If $f(x) < f(y)$ implies $\lim_{n\to\infty} P(M(x,s_n) > M(y,s_n)) = 1$, then M is a* ***sparsity measure****. (b) If $f(x) < f(y)$ implies $\lim_{n\to\infty} P(M(x,s_n) < M(y,s_n)) = 1$, then M is a* ***concentration measure****.*

Example 1. $M(x,s_n) \propto 1/\hat{f}(x,s_n)$, where $\hat{f}$ can be any consistent non- parametric density estimator, is a sparsity measure; while $M(x,s_n) \propto \hat{f}(x,s_n)$ is a concentration measure. A commonly used estimator is the kernel density one $\hat{f}(x,s_n) = \frac{1}{nh^d}\sum_{i=1}^{n} K(\frac{\|x-x_i\|}{h})$.

Example 2. Consider the distance from a point x to its kth-nearest neighbour in s_n, $x^{(k)}$: $M(x,s_n) = d_k(x,s_n) = d(x,x^{(k)})$: it is a sparsity measure. Note that d_k is neither a density estimator nor is it one-to-one related to a density estimator. Thus, the definition of 'sparsity measure' is not trivial. Another valid choice is given by the average distance over all the k nearest neighbours: $M(x,s_n) = \bar{d}_k = \frac{1}{k}\sum_{j=1}^{k} d_j = \frac{1}{k}\sum_{j=1}^{k} d(x,x^{(j)})$. Extensions to other centrality measures, such as trimmed-means are straightforward.

In the case of SNM (see [1]) neighbourhood measures can be used directly to build the decision function $h(x)$. For One-Class SVM, a particular class of neighbourhood measures has to be defined.

Definition 2 (Positive and Negative Neighbourhood Measures). *$MP(x,s_n)$ is said to be a* ***positive sparsity (concentration) measure*** *if*

$MP(x, s_n)$ is a sparsity (concentration) measure and $MP(x, s_n) \geq 0$. $MN(x, s_n)$ is said to be a ***negative sparsity (concentration) measure*** *if $-MN(x, s_n)$ is a positive concentration (sparsity) measure.*

Given that negative neighbourhood measures are in one-to-one correspondence to positive neighbourhood measures, only positive neighbourhood measures need to be considered. The following classes of kernels can be defined using positive neighbourhood measures.

Definition 3 (Neighbourhood Kernels). *Consider the mapping $\phi : \mathbb{R}^d \rightarrow \mathbb{R}^+$ defined by $\phi(x) = MP(x, s_n)$, where $MP(x, s_n)$ is a positive neighbourhood measure. The function $K(x, y) = \phi(x)\phi(y)$ is called a* ***neighbourhood kernel****. If $MP(x, s_n)$ is a positive sparsity (concentration) measure, $K(x, y)$ is a* ***sparsity (concentration) kernel****.*

In [1] it is shown that, using concentration kernels, the One-Class SVM method proposed in [3] will detect asymptotically the desired high density regions.

3 Building Smooth Neighbourhood Measures

Neighbourhood measures are not necessarily smooth functions. Consider, for instance, the neighbourhood measure defined in Example 2, $M(x, s_n) = d(x, x^{(k)})$, the distance to the kth-nearest neighbour. Figure 1 (left) shows the sparsity measure plotted for a normal sample of size $n = 100$, $k = 40$, apparently non-smooth. Figure 1 (right) shows the kernel estimator of the density function, a regularized (therefore smooth) concentration measure. Note that, independently of smoothness, both graphs show a (positive) bias in the mode estimation (it should be zero).

Figure 2 (left) shows the corresponding measures plotted for a sample of size $n = 2000$, $k = 437$. It is apparent that the sparsity measure becomes smooth

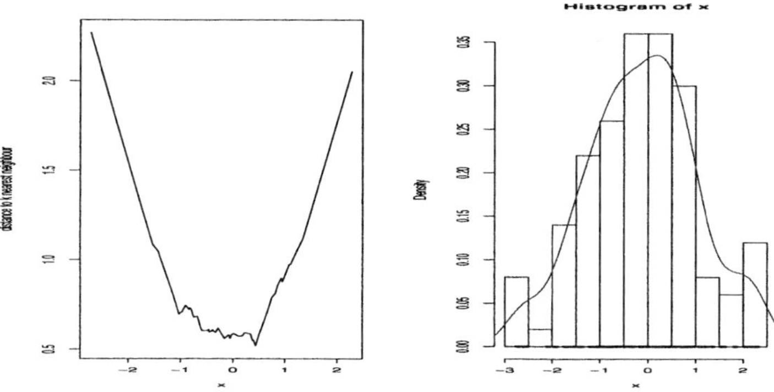

Fig. 1. Left. Sparsity measure. Right. Kernel density estimator. Normal distribution, sample size $n = 100$, $k = 40$.

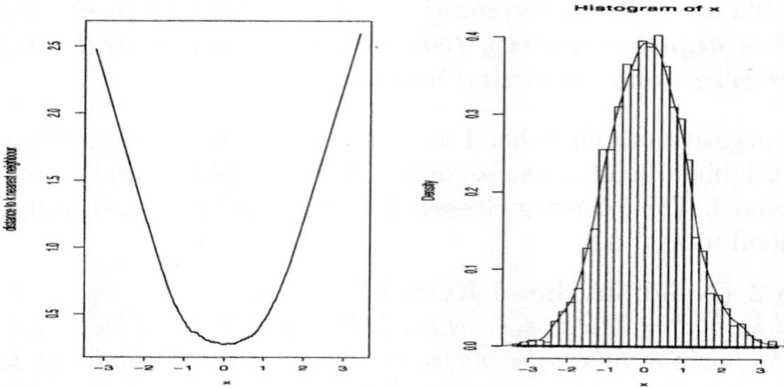

Fig. 2. Left. Sparsity measure. Right. Kernel density estimator. Normal distribution, sample size $n = 2000$, $k = 437$.

as $n \to \infty$, $k \to \infty$ and $k/n \to 0$, and the bias disappears in both graphs. In both cases $k = n^{4/(d+4)}$, where d is the space dimension. This value is known to be proportional to the (asymptotically) optimal value [4] for density estimation tasks.

For small data samples the direct use of the neighbourhood measure could cause problems for those points lying near the decision surface. A way to overcome this difficulty is to apply regularization theory (see [5]). The common approach constructs the regularized curve by minimizing the functional:

$$\min_{f \in H_K} \frac{1}{n} \sum_{i=1}^{n} L(y_i, f(x_i)) + \lambda \|f\|_K^2 , \tag{2}$$

where H_K is a Reproducing Kernel Hilbert Space (RKHS) with kernel K, L is a loss function, $\|f\|_K^2$ is the square of the norm of f in the RKHS, and λ is a positive real constant (usually fixed by cross-validation). Now consider a linear differential operator D, and choose K as Green's function for the operator D^*D, where D^* is the adjoint operator of D [2]. It is easy to show that $\|f\|_K^2 = \|Df\|_{L_2}^2$, where in the right hand side of the equality the norm is taken in L_2. Thus the second term in (2) imposes smoothing conditions on the solution f.

The regularization process tries to find a good estimation of the limit case $(n \to \infty, k \to \infty, k/n \to 0)$ from a finite sample. The straightforward solution to the previous regularization problem is to directly build a smooth curve from the sample $\{x_i, y_i = M(x_i, s_n)\}$. However, regularization solely based on a single sample could suffer from bias, inherited from the non-regularized curve, particularly if the sample is small (see for instance Figure 1, left).

An alternative approach uses a particular application of the technique known as Functional Data Analysis (FDA) [2]. This application consists in using a family of curves to build a regularized average curve. The point in this approach is that the regularized average curve will be less dependant of the sample than the regularized curve arising from a single sample. Next we describe the method for the measure $M(x, s_n; k) = d(x, x^{(k)})$, where k has been included as an explicit argument for the sake of clarity. Consider the family of curves induced by the data sets $M(x, s_n; k) = d(x, x^{(k)})$ for $k \in I$, where I is a predefined discrete set. Then the average curve is calculated and subsequently regularized by performing a standard SVM regression with exponential kernel based on (2). Figure 3 shows an example for a normal distribution with $n = 200$ and $k \in [60, 80]$. On the left, the average curve and the M curves surrounding it are shown. On the right, the average curve has been replaced by its regularized version. The choice of I depends on the data set. In this case I has been chosen as a set of values around $k = n^{4/(d+4)}$, where d is the space dimension. As mentioned before, this value is known to be proportional to the (asymptotically) optimal value [4]. For other neighbourhood measures similar ideas can be applied, where the role of k will be played by the corresponding parameters of the selected measure.

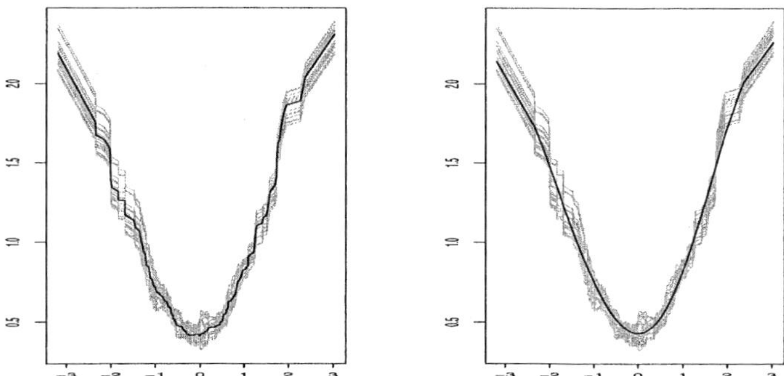

Fig. 3. Left. M_k neighbourhood curves for various k and its average. Right. The same curves together with the regularized average curve.

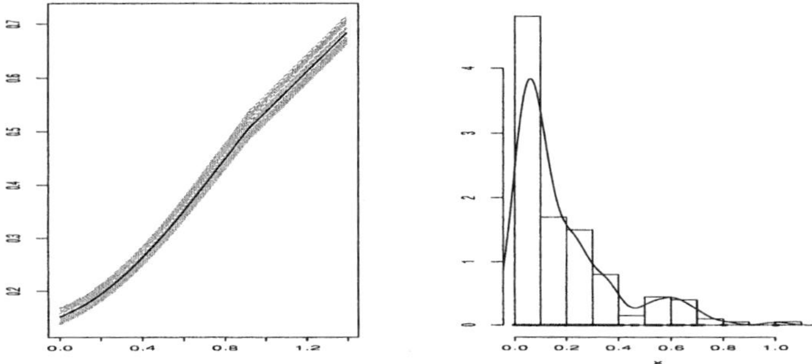

Fig. 4. Left. Regularized neighbourhood curves and average curve. Right. Kernel density estimator. $\gamma(1, 5)$, $n = 200$.

Regularization demonstrates nice properties in some problematic cases. For instance, if the mode of the distribution is at the boundary of the data domain, accurate estimation of the support becomes a difficult task. Figure 4 (left) shows the average curve of the functional data arising from a gamma distribution $\gamma(1,5)$ with $n = 200$ data points, surrounded by the regularization of each functional sample. Again, $k \in [60,80]$. Figure 4 (right) shows the kernel density estimator for the same distribution. Note that for the kernel estimator there is a clear bias in the mode estimation (the true mode is zero), while the regularized curve attains its minimum at zero. In this way, when the regularized curve is used within the One-Class method, the nearest point to the mode (zero) is always included in the estimated support of the distribution. It is apparent that the regularized kernel introduced here provide improved results agains non-regularized (standard) kernels using One-Class methods.

4 Conclusions

We have proposed a method based on the use of functional data analysis techniques for the construction of smooth kernel functions oriented to solve the One-Class problem. The method makes use of neighbourhood measures. These measures asymptotically preserve the order induced by the density function f. In this way the complexity of solving a pure density estimation problem is avoided. The regularized kernels obtained make One-Class estimation procedures become more accurate. In particular, this methodology is specially useful when the sample size is small.

Acknowledgments. This work was partially supported by spanish grants SEJ2004-03303, 06/HSE/0181/2004, TIC2003-05982-C05-05 (MCyT) and PPR-2004-05 (URJC).

References

1. J.M. Moguerza and A. Muñoz. *Solving the One-Class Problem Using Neighbourhood Measures.* Lecture Notes in Computer Science, 3138:680-688, 2004.
2. J.O. Ramsay and B.W. Silverman. *Functional Data Analysis.* Springer, 1997.
3. B. Schölkopf, J.C. Platt, J. Shawe-Taylor, A.J. Smola and R.C. Williamson. *Estimating the Support of a High Dimensional Distribution.* Neural Computation, 13(7):1443-1471, 2001.
4. B.W. Silverman. *Density Estimation for Statistics and Data Analysis.* Chapman and Hall, 1990.
5. A.N. Tikhonov and V.Y. Arsenin. *Solutions of ill-posed problems.* John Wiley & Sons, New York, 1977.

Recognition of Heartbeats Using Support Vector Machine Networks – A Comparative Study

Stanisław Osowski[1,2], Tran Haoi Linh[3], and Tomasz Markiewicz[1]

[1] Warsaw University of Technology,
[2] Military University of Technology, Warsaw, Poland
{sto, Markiewt}@iem.pw.edu.pl
[3] Hanoi University of Technology, Hanoi, Vietnam
evnbk@hn.vnn.vn

Abstract. The paper presents the comparison of performance of the individual and ensemble of SVM classifiers for the recognition of abnormal heartbeats on the basis of the registered ECG waveforms. The recognition system applies two different Support Vector Machine based classifiers and the ensemble systems composed of the individual classifiers combined together in different way to obtain the best possible performance on the ECG data. The results of numerical experiments using the data of MIT BIH Arrhythmia Database have confirmed the superior performance of the proposed solution

1 Introduction

The paper is concerned with the application of the Support vector Machine (SVM) classifiers for the recognition of heart rhythms on the basis of the registered ECG waveforms. The recognition system will be composed of the SVM used as the basic recognizing and classifying system and two different preprocessing stages resulting in different feature sets, composing the input signals to the classifier.

We will compare the performance of the heartbeat recognition system relying on the individual classifiers acting independently and on the ensemble of classifiers combined together to form the final recognizing system. The SVM networks have been used as the individual classifiers applied in the recognition of the ECG beats. The input vector **x** for these classifiers has been formed by the features following from the application of Hermite basis function expansion and of the higher order statistics (HOS) of the QRS complex of ECG. The individual classifiers will be combined together to form the ensemble network by applying the concept of voting. Three different methods of voting will be considered. The results of the numerical experiments for the recognition of 13 types of heart rhythms are presented and discussed.

2 The Individual SVM Classifiers for Heart Rhythm Recognition

To solve the problem of heartbeat recognition on the basis of the registered ECG waveform we have to generate the appropriate features describing the registered ECG waveform and apply them as the input signals to the classifier [1,2,3,4]. We have employed here the SVM classifier and two types of pre-processing of the data.

W. Duch et al. (Eds.): ICANN 2005, LNCS 3697, pp. 637 – 642, 2005.

2.1 Feature Extraction

The recognition of the patterns of the heart rhythms needs generation of the features, well characterizing patterns in a way enabling differentiation among different types of them. This is very important demand, since we observe great variation of signals among samples of the same type of beats, as is shown for example in the MIT BIH Arrhythmia Database [5].

In our work we have applied two methods of feature extraction: Hermite basis function expansion of the QRS complex and characterization of QRS by the cumulants of the second, third and fourth orders, well satisfying these general requirements [3]. In the numerical calculations, we presented the QRS segment of the ECG signal by 91 data points around the R peak (45 points before and 45 ones after).

In Hermite method we represent the QRS by the coefficients c_n of Hermite basis function expansion. If the analysed waveform is denoted by x(t) the expansion is defined by [2]

$$x(t) = \sum_{n=0}^{N-1} c_n \phi_n(t,\sigma) \tag{1}$$

where c_n are the expansion coefficients, σ - the width parameter and $\phi_n(t,\sigma)$ - the Hermite basis functions of nth order for n= 0, 1, 2, ..., N-1. After some preliminary experiments we have applied 15 Hermite coefficients for ECG data representation.

In higher order statistics (HOS) approach [6] we represent the QRS complex by the values of the cumulants of the 2nd, 3rd and 4th orders [3], each calculated at five points distributed evenly within the QRS length (for the 3rd and 4th order cumulants the diagonal slices have been calculated). The application of HOS description reduces the variance of the registered ECG signals of each type and makes the recognition problem easier. For 91-element vector representation of the QRS complex the cumulants corresponding to the time lags of 15, 30, 45, 60 and 75 have been chosen.

Additionally we have added two temporal features: one corresponding to the instantaneous RR interval of the beat and the second representing the average RR interval of 10 preceding beats. In this way each beat, irrespective of its description, has been represented here by the 17-element feature vector, with the first 15 elements corresponding to either the higher order statistics or Hermite characterization of QRS complex, and the last two - the temporal features of the actual QRS signal.

2.2 Support Vector Machine Classifier

The Support Vector Machine (SVM) applied as the classifier is a linear machine [7] working in the high dimensional feature space formed by the linear or non-linear mapping of the n-dimensional input vector **x** into a *K*-dimensional feature space, usually of K>n through the use of a function $\varphi(\mathbf{x})$. The SVM network separates the data into two classes with maximum margin of separation. The hyperplane equation separating two different classes is given by $y(\mathbf{x}) = \mathbf{w}^T \boldsymbol{\varphi}(\mathbf{x}) = \sum_{j=1}^{K} w_j \varphi_j(\mathbf{x}) + w_0 = 0$, where $\boldsymbol{\varphi}(\mathbf{x}) = [\varphi_0(\mathbf{x}), \varphi_1(\mathbf{x}), ..., \varphi_K(\mathbf{x})]^T$ with $\varphi_0(\mathbf{x}) = 1$ and **w** – the weight vector of the network. The learning and testing modes of work are performed in SVM using so-called kernel functions, satisfying the Mercer conditions [7,8]. The kernel $K(\mathbf{x},\mathbf{x}_i)$ is defined as the inner product of the vectors $\boldsymbol{\varphi}(\mathbf{x}_i)$ and $\boldsymbol{\varphi}(\mathbf{x})$, i.e., $K(\mathbf{x}_i,\mathbf{x}) = \boldsymbol{\varphi}^T(\mathbf{x}_i)\boldsymbol{\varphi}(\mathbf{x})$. The primal learning problem of SVM, formulated as the task of separating learning vectors $\mathbf{x}_i$ (i=1, 2, ..., p) into two classes of the destination values either $d_i=1$ (one class) or $d_i=-1$ (the opposite class), with the maximal separation margin is transformed to the so called dual problem of maximization of the function $Q(\boldsymbol{\alpha})$ with respect to the Lagrange multipliers α_i forming vector $\boldsymbol{\alpha}$ [7,8]. Dual problem solution results in the optimal values of the Lagrange multipliers, on the basis of which the output signal $y(\mathbf{x})$ of the SVM is determined as [7]

$$y(\mathbf{x}) = \sum_{i=1}^{N_S} \alpha_{si} d_i K(\mathbf{x}_{si}, \mathbf{x}) + w_o \tag{2}$$

If $y(\mathbf{x})>0$ the feature vector $\mathbf{x}$ belongs to the particular class and if $y(\mathbf{x})<0$ to the opposite one. The recognition of more classes is done in SVM by applying either "one against one" or "one against all" methods [8]. We have applied "one against one" approach, in which the SVM networks are trained to recognize between all combinations of two classes of data. For M classes we have to train M(M-1)/2 individual SVM networks. In the retrieval mode the vector $\mathbf{x}$ belongs to the class of the highest number of winnings in all combinations of classes.

3 The Integration Systems

Let us assume that there exist M channels of individual classifiers combined into one classifying system by the integrating part of the network. The measured signals of the ECG form the n-dimensional vector $\mathbf{x}_{in}$. This vector is transformed into different feature vectors $\mathbf{x}_i$ by the appropriate preprocessing blocks, forming the inputs to the classifiers. Each classifier has N binary outputs corresponding to N classes and the output signals of each classifier form the vector $\mathbf{y}_j$, for j=1, 2, …, M. The results of classifications of different classifiers may be combined together using different methods of integration. We will compare here 3 methods: the weighted voting (WV), Kulback-Leibler (K-L) and the modified Bayes approach (MB) [9].

The weighted voting combines the results $\mathbf{y}_i$ of M classifiers through the integrating matrix $\mathbf{W}$ to form one output vector $\mathbf{z}$ of the classifying system. The result of integration of all classifiers can be expressed by the relation

$$\mathbf{z}=\mathbf{W}\mathbf{y} \tag{3}$$

where $\mathbf{y} = \left[\mathbf{y}_1^T, \mathbf{y}_2^T, ..., \mathbf{y}_M^T\right]^T$. The position of the highest value element of $\mathbf{z}$ indicates the membership to the appropriate class. In adjusting the values of elements of the matrix $\mathbf{W}$ we have applied the minimization of the sum of squared error of the whole ensemble of the classifiers, measured on the learning data set [3]. This minimization leads to the solution expressed through the Moore-Penrose pseudoinverse in the following form $\mathbf{W} = \mathbf{D}\mathbf{Y}^+$, where $\mathbf{Y}$ is the $NM \times p$ matrix composed of p vectors $\mathbf{y}$ corresponding to p results of individual M classifications for learning data and $\mathbf{D}$ is the appropriate $N \times p$ matrix formed by the destination vectors associated with each learning pair of data.

In Kullback-Leibler divergence [9] we calculate the ensemble probability μ_j supporting jth class given the actual input **x**, as the normalized arithmetic mean

$$\mu_j = \frac{1}{M}\sum_{i=1}^{M} d_{ij} \tag{4}$$

where d_{ij} means the probability of indicating jth class by ith classifier for the data of this class. This probability is determined in the testing mode for each classifier as the ratio of the number of victories of jth class to all possible indications in one against one mode of operation of SVM.

In modified naive Bayes combination [9] the ensemble probability μ_j for jth class is determined on the basis of results of testing the networks on learning data and is given in the form

$$\mu_j = \prod_{i=1}^{M} \frac{cm_{js_i}^{(i)} + 1/N}{n_j + 1} \tag{6}$$

where n_j is the number of elements in training set for class j and $cm_{js_i}^{(i)}$ is the element of the confusion matrix generated for learning data of ith classifier. The (j,s)th entry of the confusion matrix is the number of elements of the data set whose true class label was j and were assigned by ith classifier to sth class.

4 The Results of Numerical Experiments

The experiments have been performed for 13 types of heartbeats contained in MIT BIH Arrhythmia Database [5]. In this data base there are ECG waveforms of 12 types of abnormal beats: left bundle branch block (L), right bundle branch block (R), atrial premature beat (A), aberrated atrial premature beat (a), nodal (junctional) premature beat (J), ventricular premature beat (V), fusion of ventricular and normal beat (F), ventricular flutter wave (I), nodal (junctional) escape beat (j), ventricular escape beat (E), supraventricular premature beat (S) and fusion of paced and normal beat (f) and the waveforms corresponding to the normal sinus rhythm (N). We have used 12785 heart rhythms (6690 for learning and 6095 for testing). The SVM networks of radial Gaussian kernels have been applied and one against one strategy. The optimal values of regularization constant C and parameter σ of Gaussian function have been adjusted using cross validation approach for the learning data (C=100, σ=0.5). They have been set constant for all SVM networks.

Table 1 presents the comparison of total results achieved by individual SVM based classifiers (Hermite and HOS preprocesing), as well as by the ensembles of both classifiers for the learning and testing data (not used in learning). The rows labelled as WV, K-L and MB denote the results of integration of classifier results. The misclassification rate has been calculated as the mean of errors of each class recognition. There is a visible improvement of the results of testing after application of ensemble of

Table 1. The comparison of the average misclassification rate of testing different SVM solutions for the recognition of 13 types of the heartbeats

Classification method	Total number of learning errors	Average relative learning error	Total number of testing errors	Average relative testing error
Hermite	103	3.02%	172	5.23%
HOS	173	5.19%	216	6.28%
WV	74	1.83%	159	4.09%
K-L	108	2.27%	147	3.77%
MB	73	1.80%	171	4.39%

Table 2. The confusion matrix of the ensemble classifier system for 13 types of rhythms of the testing data

Heart type	A	a	f	E	F	j	I	J	L	N	R	S	V	Total
A	399	3	1	0	1	5	1	0	1	16	0	0	6	433
a	2	57	0	0	1	0	0	0	0	2	0	0	3	65
f	0	0	195	0	0	0	0	1	6	1	0	0	0	203
E	0	0	0	48	0	0	0	0	0	0	0	0	0	48
F	0	0	0	0	358	0	0	0	0	2	1	0	5	366
j	2	0	1	1	1	95	0	0	0	5	0	0	0	105
I	0	0	0	0	0	0	193	0	1	1	0	1	2	198
J	0	0	0	0	1	0	0	36	0	3	0	0	0	40
L	1	0	1	0	0	0	0	0	488	5	0	0	5	500
N	10	3	1	0	1	4	0	1	3	1959	2	0	1	1985
R	1	0	0	1	0	0	0	1	0	1	395	0	1	400
S	1	0	0	0	0	1	6	0	0	0	2	511	0	521
V	2	1	1	0	7	0	0	0	1	5	0	0	1214	1231
Total	418	64	200	50	370	105	200	39	500	2000	400	512	1237	6095

classifiers. For the best Kulback-Leibler (K-L) integration it is more than 20% in relation to the best individual classifier. Table 2 presents the confusion matrix of classification for the best ensemble system.

The diagonal entries of this matrix represent right recognition of the beat type and the off diagonal – the misclassifications. Each column presents how the beats of particular type have been classified. The row indicates which beats have been classified as the type mentioned in this row.

In practice the most dangerous case is when the ill person is diagnosed as the healthy one (false negative diagnosed patient). To deal with such case we have introduced the parameter DUV (dangerous uncertainty value), defined as the ratio of the number of all false negative diagnosed patients to the total number of misclassifications. Table 3 presents the values of this parameter for the individual SVM classifiers and for the integrated system. There is an evident improvement of the quality of classification, both in learning and in the testing mode.

Table 3. The comparison of DUV values for different SVM based classifiers

Quality measure	Learning data			Testing data		
	HOS	HER	Integration	HOS	HER	Integration
DUV	31.79%	38.83%	27.03%	18.52%	23.84%	17.68%

5 Conclusions

The paper has presented the application of the single SVM based classifier and the ensemble of classifiers for the recognition of heartbeats on the basis of ECG waveforms. The numerical results performed on the MIT BIH AD examples have shown that the SVM networks combined into the ensemble system composed of individual classifiers brings in significant improvement of the classification results, especially reduction of the most dangerous misclassification cases. The experimental results have shown that instead of designing one high performance classifier we can build a number of classifiers, each of possibly not superb performance and as a results we get the classifying system of significantly higher quality.

References

1. Hu, Y. H., Palreddy, S., Tompkins, W.: Patient adaptable ECG beat classifier using a mixture of experts approach. IEEE Tr. Biomed. Eng. Vol. 44 (1997) 891 - 900
2. Lagerholm, M., Peterson, C., Braccini, G., Edenbrandt, L., Sornmo, L.: Clustering ECG complexes using Hermite functions and self-organizing maps. IEEE Tr. Biomed. Eng. Vol. 47 (2000) 838-847
3. Osowski, S., Tran Hoai, L., Markiewicz, T.: Support Vector Machine based expert system for reliable heartbeat recognition. IEEE Tr. on Biomed. Eng. Vol. 51 (2004) 582-589
4. de Chazal, P., O'Dwyer, M., Reilly, R. B.: Automatic classification of heartbeats using ECG morphology and heartbeat interval features. IEEE Tr. on Biomed. Eng. Vol. 51 (2004) 1196-1206
5. Mark, R., Moody, G.: MIT-BIH arrhythmia database directory. MIT (1988)
6. Nikias, C., Petropulu, A.: Higher order spectral analysis. Prentice Hall, N. J. (1993)
7. Vapnik, V.: Statistical learning theory. Wiley, N.Y. (1998)
8. Hsu, C. W., Lin, C. J.: A comparison methods for multi class support vector machines. IEEE Trans. Neural Networks Vol. 13 (2002) 415-425
9. Kuncheva, L.: Combining pattern classifiers: methods and algorithms, Wiley, N. J. (2004)

Componentwise Support Vector Machines for Structure Detection

K. Pelckmans, J.A.K. Suykens, and B. De Moor

K.U.Leuven ESAT-SCD/SISTA, Kasteelpark Arenberg 10, B-3001 Leuven, Belgium
{kristiaan.pelckmans, johan.suykens}@esat.kuleuven.ac.be
http://www.esat.kuleuven.ac.be/sista/lssvmlab

Abstract. This paper extends recent advances in Support Vector Machines and kernel machines in estimating additive models for classification from observed multivariate input/output data. Specifically, we address the question how to obtain predictive models which gives insight into the structure of the dataset. This contribution extends the framework of structure detection as introduced in recent publications by the authors towards estimation of componentwise Support Vector Machines (cSVMs). The result is applied to a benchmark classification task where the input variables all take binary values.

1 Introduction

The theory, methodology and application of Support Vector Machines (SVMs) has gained a mature status in the last decade, see e.g. [16,3,12,13]. This work extends recent advances on primal-dual kernel machines for learning classification rules based on additive models [7] where the primal-dual optimization point of view [2] (as exploited by SVMs [16] and LS-SVMs [14,13]) is seen to provide an efficient implementation [9]. Although relations exist with results on ANOVA kernels [16,6], the optimization framework established a solid foundation for extensions towards structure detection similar to LASSO [15] and bridge regression [1] in the context of regression as elaborated in [9,10]. The key idea was to employ a measure of maximal variation (as defined in the sequel) for the goal of regularization. Extensions towards handling missing values amongst the observed inputs were described in the context of cSVMs in [8].

This paper is organized as follows. Section 2 gives the main result of componentwise SVMs equipped with a measure of maximal variation. Section 3 then illustrates the concept on a UCI benchmark prediction task.

2 Componentwise Support Vector Machines

Given a set of observed input/output data-samples $\mathscr{D} = \{(x_i, y_i)\}_{i=1}^N \subset \mathbb{R}^D \times \mathbb{D}$ where $\mathbb{D} = \{-1, +1\}$. Learning a decision rule then amounts to identifying a function $f : \mathbb{R} \to \mathbb{D}$ such that any data-sample $(x_*, y_*) \in \mathbb{R}^D \times \mathbb{D}$ sampled from the same distribution $P_{XY} : \mathbb{R}^D \times \mathbb{D} \to [0, 1]$ underlying the dataset $\mathscr{D}$ deviates minimal from the prediction using the model f. More formally, let F denote the class of admissible functions f. Learning amounts to the task of approximating the minimizer $f^* = \arg\min_{f \in F} \int (y - f(x))^2 dP_{XY}$. The framework of SVMs as given in [16] is adopted.

W. Duch et al. (Eds.): ICANN 2005, LNCS 3697, pp. 643–648, 2005.

Definition 1. *(Additive Classifier) Let* $x \in \mathbb{R}^D$ *be a point with components* $x \doteq \left(x^{(1)}, \ldots, x^{(P)}\right)$. *Additive classifiers then take a componentwise form [7] defined as*

$$\operatorname{sign}[f(x)] = \operatorname{sign}\left[\sum_{p=1}^{P} f_p\left(x^{(p)}\right) + b\right], \tag{1}$$

with sufficiently smooth mappings $f_p : \mathbb{R}^{D_p} \to \mathbb{R}$ *such that the decision boundary is described as in [16,12]*

$$\mathscr{H}_f = \left\{ x_0 \in \mathbb{R}^D \;\middle|\; \sum_{p=1}^{P} f_p\left(x_0^{(p)}\right) + b = 0,\; x_0 \in \mathbb{R}^P \right\}. \tag{2}$$

It is well-known [16] that the distance of any point x the hyper-plane $\mathscr{H}_{f_p}$ is given as

$$d\left(x, \mathscr{H}_f\right) = \frac{|f(x)|}{\|f'(x)\|} \geq \frac{y_i \left(\sum_{p=1}^{P} f_p\left(x^{(p)}\right) + b\right)}{\sum_{p=1}^{P} \|f^{(p)'}(x^{(p)})\|}, \tag{3}$$

as $\left\|\sum_{p=1}^{P} f^{(p)'}\left(x^{(p)}\right)\right\| \leq \sum_{p=1}^{P} \left\|f^{(p)'}\left(x^{(p)}\right)\right\|$ due to the triangle inequality. The optimal separating hyper-plane can be expressed as the model (3) solving

$$\max_{M \geq 0, f_p, b} M \quad \text{s.t.} \quad d(x_i, \mathscr{H}_{f_p}) \geq M. \tag{4}$$

After the change of variables in the function f such that $M \sum_{p=1}^{P} \|f^{(p)'}\| = 1$ and the application of the lower-bound (3), one can write alternatively

$$(\hat{f}, \hat{b}) = \arg\min_{f,b} \mathscr{J}(f) = \sum_{p=1}^{P} \left\|f^{(p)'}\right\|$$
$$\text{s.t.} \quad y_i \left(\sum_{p=1}^{P} f_p\left(x_i^{(p)}\right) + b\right) \geq 1, \; \forall i = 1, \ldots, N. \tag{5}$$

Then the size of the margin is given as $M = 1/\sum_{p=1}^{P} \|f^{(p)'}\|$.

2.1 Structure Detection and Maximal Variation

Structure detection as in the case of LASSO and bridge regression in the case of linear parametric models becomes hard to incorporate into non-parametric and kernel methods. A possible approach then is to employ a measure of the contribution of any component which is not expressed directly in terms of the parameters. The following measure was proposed.

Definition 2. *(Maximal Variation) The maximal variation of a function* $f_p : \mathbb{R}^{D_p} \to \mathbb{R}$ *is defined as*

$$\mathscr{M}_p = \max_{x^{(p)} \in \mathbb{R}^{D_p}} \left| f_p\left(x^{(p)}\right) \right|, \tag{6}$$

for all $x^{(p)} \in \mathbb{R}^{D_p}$. *The empirical maximal variation can be defined as*

$$\hat{\mathscr{M}}_p = \max_{x_i^{(p)} \in \mathscr{D}} \left| f_p\left(x_i^{(p)}\right) \right|, \tag{7}$$

with $x_i^{(p)}$ denoting the p-th component of the i-th sample of the training set $\mathscr{D}$.

Adopting this definition, it becomes clear that when a certain component f_p finally has a maximal variation $\mathscr{M}_p$ equal to zero, the corresponding variables do not contribute to the learned classifier and may be omitted for the sake of prediction.

2.2 Componentwise Primal-Dual Kernel Classifiers

Consider the model

$$f(x) = \sum_{p=1}^{P} w_p^T \varphi_p\left(x^{(p)}\right) + b, \tag{8}$$

where $\varphi_p(\cdot) : \mathbb{R}^{D_p} \to \mathbb{R}^{n_h}$ denote the potentially infinite dimensional feature map and $w_p \in \mathbb{R}^{n_h}$ is the unknown parameter of the pth component for all $p = 1, \ldots, P$. The following regularized cost-function is considered:

$$\min_{w,b,e,t} \mathscr{J}_{\gamma,C}(w,t) = \gamma \sum_{p=1}^{P} t_p + \frac{1}{2} \sum_{p=1}^{P} w_p^T w_p$$

$$\text{s.t.} \begin{cases} y_i \left(\sum_{p=1}^{P} w_p^T \varphi_p\left(x_i^{(p)}\right) + b \right) \geq 1 - e_i & \forall i = 1, \ldots, N \\ \sum_{i=1}^{N} e_i \leq C, \;\; e_i \geq 0 & \forall i = 1, \ldots, N \\ -t_p \leq w_p^T \varphi_p\left(x_i^{(p)}\right) \leq t_p & \forall i = 1, \ldots, N, p = 1, \ldots, P. \end{cases} \tag{9}$$

The dual problem is given in the following Lemma.

Lemma 1. *(Dual of Componentwise SVM with Maximal Variation) Given the primal problem (9), the dual solution is*

$$\max_{\alpha_i, \rho_{ip}^+, \rho_{ip}^-, \lambda} -\frac{1}{2} \sum_{i,j=1}^{N} \left(\alpha_i y_i + \rho_{ip}^+ - \rho_{ip}^- \right) \left(\alpha_j y_j + \rho_{jp}^+ - \rho_{jp}^- \right) \tilde{\Omega}_{ij}^P + \sum_{i=1}^{N} \alpha_i - \lambda C$$

$$s.t. \begin{cases} \sum_{i=1}^{N} y_i \alpha_i = 0 & \\ 0 \leq \alpha_i \leq \lambda & \forall i = 1, \ldots, N \\ \gamma = \sum_{i=1}^{N} (\rho_{ip}^+ + \rho_{ip}^-) & \forall p = 1, \ldots, P \\ \rho_{ip}^+, \rho_{ip}^- \geq 0 & \forall i = 1, \ldots, N, \forall p = 1, \ldots, P, \end{cases} \tag{10}$$

where $\tilde{\Omega}_{ij}^P = \sum_{p=1}^{P} \tilde{K}_p\left(x_i^{(p)}, x_j^{(p)}\right)$ for all $i, j = 1, \ldots, N$ and where $\tilde{K}_p\left(x_i^{(p)}, x_j^{(p)}\right) = K_p\left(x_i^{(p)}, x_j^{(p)}\right)$. The resulting nonlinear classifier evaluated on a new data point $x_* = \left(x_*^{(1)}, \ldots, x_*^{(P)}\right)$ takes the form

$$\text{sign}\left[\sum_{p=1}^{P} \sum_{i=1}^{N} \alpha_i^{(p)} K_p\left(x_i^{(p)}, x_*^{(p)}\right) + b \right], \tag{11}$$

where $\hat{\alpha}_i^{(p)} = \left(\hat{\alpha}_i y_i + \hat{\rho}_{ip}^+ - \hat{\rho}_{ip}^-\right)$ *for all* $i = 1,\ldots,N$ *and* $p = 1,\ldots,P$ *follow from the unique solution to (10).*

Proof. The dual solution is given after construction of the Lagrangian

$$\mathscr{L}_{\gamma,C}(w_p,b,e_i,t_p;\alpha_i,v_i,\rho_{ip}^+,\rho_{ip}^-,\lambda) = \mathscr{J}_{\gamma,C}(w_p,t_p) - \sum_{i=1}^N v_i e_i$$
$$-\sum_{i=1}^N \alpha_i \left(y_i\left(\sum_{p=i}^P w_p\varphi_p\left(x_i^{(p)}\right)+b\right)-1+e_i\right)+\lambda\left(\sum_{i=1}^N e_i - C\right)$$
$$-\sum_{i,p}\rho_{ip}^+\left(t_p + w_p^T\varphi_p\left(x_i^{(p)}\right)\right) - \sum_{i,p}\rho_{ip}^-\left(t_p - w_p^T\varphi_p\left(x_i^{(p)}\right)\right), \quad (12)$$

with positive multipliers $0 \leq \alpha_i, v_i, \rho_{ip}^+, \rho_{ip}^-$ and $\lambda \geq 0$. The solution is given by the saddle point of the Lagrangian [2]

$$\max_{\alpha_i,v_i,\rho_{ip}^+,\rho_{ip}^-,\lambda} \; \min_{w_p,b,e_i,t_p} \; \mathscr{L}_{\gamma,C}. \quad (13)$$

By taking the first order conditions $\frac{\partial \mathscr{L}_{\gamma,C}}{\partial w_p} = 0$, $\frac{\partial \mathscr{L}_{\gamma,C}}{\partial b} = 0$, $\frac{\partial \mathscr{L}_{\gamma,C}}{\partial e_i} = 0$ and $\frac{\partial \mathscr{L}_{\gamma,C}}{\partial t_p} = 0$, one obtains the (in)equalities $w_p = \sum_{i=1}^N \left(\alpha_i y_i + \rho_{ip}^+ - \rho_{ip}^-\right)\varphi_p(x_i^{(p)})$, $\sum_{i=1}^N \alpha_i y_i = 0$, $0 \leq \alpha_i \leq \lambda$ and $\gamma = \sum_{i=1}^P \left(\rho_{ip}^+ + \rho_{ip}^-\right)$. By application of the kernel trick $K_p(x_i,x_j) = \varphi_p(x_i^{(p)})^T\varphi_p(x_j^{(p)})$, the solution to (8) is found by solving the dual problem The primal variables b, e_i and t_p can be recovered from the complementary slackness conditions. □

3 Example: Learning Logical Rules Using Componentwise SVMs

In order to explore the capabilities of the presented method, the 1984 United States Congressional voting records database available on the UCI benchmark repository is used. This data set includes votes for each of the U.S. House of Representatives Congressmen on the 16 key votes identified by the CQA. A 90%- 95% performance was reported in [11]. Here we explore the capabilities of the described framework to learn a parsimonious decision system from the observed data.

We first elaborate on the issue how to handle the special structure of the input data. Inspired by the method of first order inductive logic programming, one may let the variable x be mapped on the truth-value as $T(x) = TRUE$ if $x = +1$ and $T(x) = FALSE$ otherwise. Then the following relation holds

$$T(y) = OR(T(x^1),\ldots,T(x^P)) \Leftrightarrow y = \operatorname{sign}\left(\sum_{p=1}^P x^p + P - 1\right), \quad (14)$$

where *OR* is the logical OR operation. This motivates the use of the additive model (3). Furthermore note that $\neg T(x) = T(-x)$ holds where $\neg$ denotes the logical negation. The *AND* operator is induced by the use of the following kernel.

Definition 3. *(Logic 'AND' Kernel) Let π_p be a nonempty set of indices $1,\dots,D$ for all $p = 1,\dots,P$ and let D_p denotes the number of different indices in π_p. Let the feature space mapping of the pth component be defined as*

$$\varphi_p(x) = I_{AND}\left(x^{\pi_p(1)},\dots,x^{\pi_p(D_p)}\right), \tag{15}$$

where the indicator function $I_{AND}(\cdot)$ is $+1$ if all arguments equal $+1$ and -1 otherwise. The corresponding Mercer kernel becomes

$$K_p(x_i,x_j) = I_{AND}\left(x_i^{\pi_p(1)},\dots,x_i^{\pi_p(D_p)}\right) I_{AND}\left(x_j^{\pi_p(1)},\dots,x_j^{\pi_p(D_p)}\right). \tag{16}$$

For this example, a maximal order of $D_p = 2$ is used in order to keep the computations tractable. The datset was divided in disjunct training dataset ($N = 250$), validation set (of size 100) and test set (of size 85). The pameters γ and C where tuned minimizing the misclassification rate on the validation set. A Monte Carlo simulation was conducting resulting in mean testset performance of 96.24% with a one sigma bound as 96.24% $\pm$ 1.2%. The vote of congressmen pro or contra the democration candidate is in most predictors of the sample proportional to their vote pro the resolutions of (3) cost-sharing of water-projects (4) inversely proportional to the adoption of the budget-resolution and (4) proportional to the vote concerning immigration.

4 Conclusions

This paper studied the estimation of additive classifiers by componentwise SVMs. The measure of maximal variation was used to perform structure detection. An example was elaborated where the task is to infer logical rules from binary observations by use of the additive model structure and the use of a logical AND kernel.

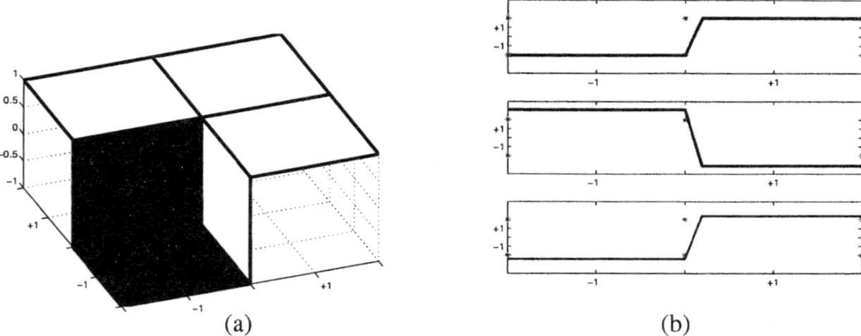

Fig. 1. **(a)** Indicator function of the function $y = \text{sign}\left(x^1 + x^2 + 1\right)$ and **(b)** The tuned predictor is a function of the votes on vote (3), (4) and (11)

Acknowledgments. This research work was carried out at the ESAT laboratory of the KUL. Research Council KU Leuven: Concerted Research Action GOA-Mefisto

666, GOA-Ambiorics IDO, several PhD/postdoc & fellow grants; Flemish Government: Fund for Scientific Research Flanders (several PhD/postdoc grants, projects G.0407.02, G.0256.97, G.0115.01, G.0240.99, G.0197.02, G.0499.04, G.0211.05, G.0080.01, research communities ICCoS, ANMMM), AWI (Bil. Int. Collaboration Hungary/Poland), IWT (Soft4s, STWW-Genprom, GBOU-McKnow, Eureka-Impact, Eureka-FLiTE, several PhD grants); Belgian Federal Government: DWTC IUAP IV-02 (1996-2001) and IUAP V-10-29 (2002-2006) (2002-2006), Program Sustainable Development PODO-II (CP/40); Direct contract research: Verhaert, Electrabel, Elia, Data4s, IPCOS. JS is an associate professor and BDM is a full professor at K.U.Leuven Belgium, respectively.

References

1. A. Antoniadis and J. Fan. Regularized wavelet approximations (with discussion). *Jour. of the Am. Stat. Ass.*, 96:939-967, 2001.
2. S. Boyd and L. Vandenberghe. *Convex Optimization.* Cambridge University Press, 2004.
3. N. Cristianini and J. Shawe-Taylor. *An Introduction to Support Vector Machines.* Cambridge University Press, 2000.
4. L.E. Frank and J.H. Friedman. A statistical view of some chemometric regression tools. *Technometrics*, 35:109-148, 1993.
5. W.J. Fu. Penalized regression: the bridge versus the LASSO. *Journal of Computational and Graphical Statistics,* 7:397-416, 1998.
6. S. R. Gunn and J. S. Kandola. Structural modelling with sparse kernels. *Machine Learning,* 48(1):137-163, 2002.
7. T. Hastie and R. Tibshirani. *Generalized additive models.* Chapman and Hall, 1990.
8. K. Pelckmans, J. De Brabanter, J.A.K. Suykens, and B. De Moor. Maximal variation and missing values for componentwise support vector machines. Technical report, SCD - ESAT - KULeuven, Leuven, 2005.
9. K. Pelckmans, I. Goethals, J. De Brabanter, J.A.K. Suykens, and B. De Moor. Componentwise least squares support vector machines. Chapter in *Support Vector Machines: Theory and Applications,* L. Wang (Ed.), Springer, 2004, In press.
10. K. Pelckmans, J.A.K. Suykens, and B. De Moor. Building sparse representations and structure determination on LS-SVM substrates. *Neurocomputing, in press,* 2005.
11. J.C. Schlimmer. *Concept acquisition through representational adjustment.* PhD thesis, Department of Information and Computer Science, University of California, Irvine, CA, 1987.
12. B. Schölkopf and A. Smola. *Learning with Kernels.* MIT Press, Cambridge, MA, 2002.
13. J.A.K. Suykens, T. Van Gestel, J. De Brabanter, B. De Moor, and J. Vandewalle. *Least Squares Support Vector Machines.* World Scientific, Singapore, 2002.
14. J.A.K. Suykens and J. Vandewalle. Least squares support vector machine classifiers. *Neural Processing Letters,* 9(3):293300, 1999.
15. R.J. Tibshirani. Regression shrinkage and selection via the LASSO. *Journal of the Royal Statistical Society,* 58:267-288, 1996.
16. V.N. Vapnik. *Statistical Learning Theory.* Wiley and Sons, 1998.

Memory in Backpropagation-Decorrelation O(N) Efficient Online Recurrent Learning

Jochen J. Steil

Bielefeld University, Neuroinformatics Group, Faculty of Technology,
Universittsstr. 25, D-33615 Bielefeld, Germany
jsteil@techfak.uni-bielefeld.de
www.jsteil.de

Abstract. We consider regularization methods to improve the recently introduced backpropagation-decorrelation (BPDC) online algorithm for O(N) training of fully recurrent networks. While BPDC combines one-step error backpropagation and the usage of temporal memory of a network dynamics by means of decorrelation of activations, it is an online algorithm using only instantaneous states and errors. As enhancement we propose several ways to introduce memory in the algorithm for regularization. Simulation results of standard tasks show that different such strategies cause different effects either improving training performance at the cost of overfitting or degrading training errors.

1 Introduction

Recently, recurrent neural networks have matured into a fundamental tool for trajectory learning or time-series prediction and generation and increasingly find application in domains such as speech recognition, adaptive control, or biological modeling. In particular for discrete time standard and recursive networks approximation capabilities and suitable training algorithms have been intensively investigated [1]. The main drawbacks for their more widespread application are the difficulties to chose a suitable architecture among the many possible connecting schemes which reach from partially recurrent or cascade correlation networks to fully connected RNN's. Also the usually high degree of architecture specialization with respect to applications and the complexity of respective specialized training algorithms for these architectures prohibits simple transfer of methods among problem domains. Finally there is much ongoing research to devise efficient recurrent learning schemes often employing regularization techniques [2]. However, most of the efficient existing algorithms are quite complex and most online techniques typically need proper adjustment of learning rates and time-constants.

In [3], we proposed a new efficient online technique, the BPDC rule, combining three principles: (i) one-step back propagation of errors; (ii) the usage of the temporal memory in the network dynamics based on decorrelation of the activations, and (iii) the employment of a non-adaptive reservoir of hidden neurons to reduce complexity of training. The BPDC rule roots in a combination of recent ideas to differentiate the error function with respect to the states in order to obtain a "virtual teacher" target, with respect to which the weight changes are computed [4,5]. Further, under the notion "echo state network" [6] and "liquid state machine" [7] non-adaptive recurrent networks as a

W. Duch et al. (Eds.): ICANN 2005, LNCS 3697, pp. 649–654, 2005.

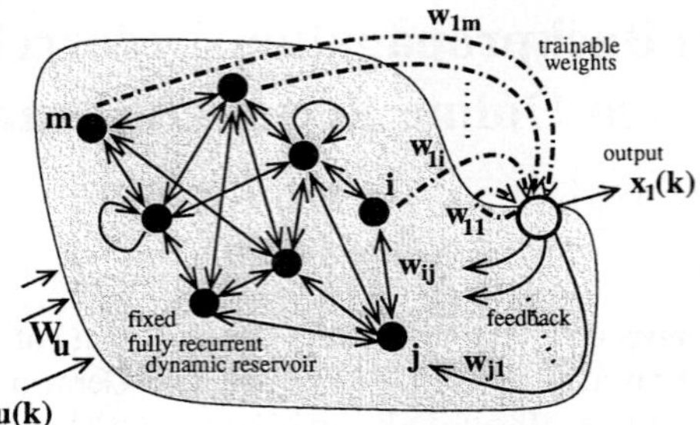

Fig. 1. The BackPropagation-DeCorrelation (BPDC) rule adapts only the output weights of a fully recurrent network with fixed internal reservoir and output-to-reservoir connections

kind of dynamic reservoir to store information about the temporal behavior of inputs have been proposed, which allow to effectively learn simple readout functions.

In the BPDC network, the output weights also implement a linear readout function while at the same time the output neurons provide full feedback into the reservoir, see Fig. 1. In [3,8] it has been shown that BPDC performs well on a number of standard tasks and in [8] there has also been derived an stability condition, which can easily be monitored online. While BPDC can learn very fast and with large learning rates, in this contribution we show that regularization can improve BPDC performance because the generalization error tends to increase already after few epochs like shown in Fig. 2 a). Section 2 introduces the BPDC approach and its regularization, Section 3 gives simulation results on three different nonlinear prediction tasks, and Section 4 concludes.

2 The BPCD-Rule

We consider fully connected recurrent networks

$$\mathbf{x}(k+1) = (1-\Delta t)\mathbf{x}(k) + \Delta t \mathsf{W}\varphi(\mathbf{x}(k)) + \Delta t \mathsf{W}_u \mathbf{u}(k), \tag{1}$$

where $x_i, i = 1, \ldots, N$ are the states, $\mathsf{W} \in \mathbb{R}^{N\times N}$ is the weight matrix, W_u the input weight matrix. Let $k = \hat{k}\Delta t, \hat{k} \in N_+$ be a discretized time variable and with a slight abuse of notation interprete time arguments and indices $(k+1)$ as $((k+1)\Delta t)$ for simplicity. For small Δt we obtain an approximation of the continuous time dynamics $dx/dt = -x + W\varphi(x)$ and for $\Delta t = 1$ the standard discrete dynamics. We assume that φ is a standard sigmoidal differentiable activation function with $\varphi' \leq 1$ and is applied component wise to the vector $\mathbf{x}$. We further assume that W, W_u are initialized with small random values in a certain weight initialization interval $[-a, a]$. Denote by $O \subset \{1, .., N\}$ the set of indices s of N_O output neurons (i.e. x_s output $\Rightarrow s \in O$) and let for a single output neuron w.r. $O = \{1\}$ such that x_1 is the respective output of the network like shown in Fig. 1, which is a linear combination of the internal neuron outputs. In [3], the *Backpropagation-Decorrelation* rule

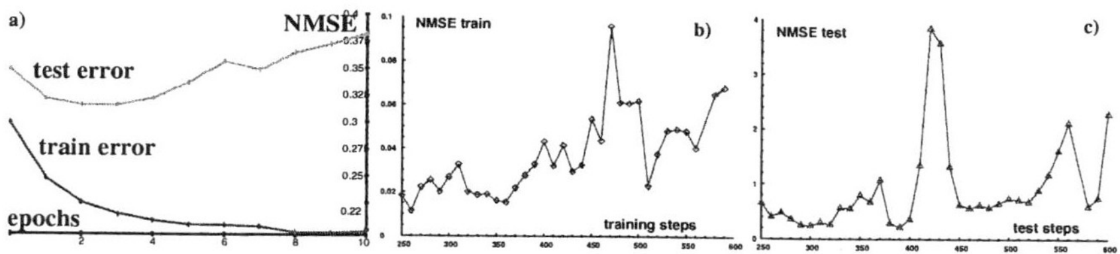

Fig. 2. a) Typical error curves for train/test error motivating need for regularization. Sensitivity of b) training and c) test NMSE to number of training data for the Mackey-Glass data (40 neurons, 50 runs of 10 epochs, weights in $[-0.2, 0.2]$, η =0.5, $\varepsilon = 0.002$, timestep 0.2).

$$\Delta w_{ij}(k+1) = \frac{\eta}{\Delta t} \frac{\varphi(x_j(k))}{\sum_s \varphi(x_s(k))^2 + \varepsilon} \gamma_i(k+1), \tag{2}$$

$$\text{where} \quad \gamma_i(k+1) = \sum_{s \in 0} \Big((1-\Delta t)\delta_{is} + \Delta t w_{is} \varphi'(x_s(k)) \Big) e_s(k) - e_i(k+1),$$

has been introduced, where η is the learning rate, $\varepsilon > 0$ a small regularization constant, and $e_s(k)$ are the non-zero error components for $s \in O$ at time $k : e_s(k) = x_s(k) - y_s(k)$ with respect to the teaching signal $y_s(k)$. The γ_i propagate a mixture of the current errors $e_i(k+1)$ and the errors in the last time step $e_s(k)$ weighted by a typical back-propagation term involving φ'.

The term $\varphi(x_j(k))/(\sum_s \varphi(x_s(k))^2 + \varepsilon)$ can be further interpreted as enforcing an approximative decorrelation of the neurons' outputs $\varphi(x_j(k))$ over time: denoting $\mathsf{f}_k = (\varphi(x_1(k)), \varphi(x_2(k)), \ldots, \varphi(x_n(k)))^T$ the vector of outputs and $C(k) = \varepsilon I + \mathsf{f}_k \mathsf{f}_k^T$ the regularized instantaneous correlation matrix of the outputs, we obtain

$$\frac{\mathsf{f}_k}{\sum \varphi(x_s(k))^2 + \epsilon} = \mathsf{f}_k \left[\frac{1}{\epsilon} - \frac{1}{\epsilon^2} \frac{\mathsf{f}_k^T \mathsf{f}_k}{1 + \frac{1}{\epsilon} \|\mathsf{f}_k\|^2} \right] = \left[\frac{1}{\epsilon} \mathrm{I} - \frac{[\frac{1}{\epsilon}\mathrm{I}\mathsf{f}_k][\frac{1}{\epsilon}\mathrm{I}\mathsf{f}_k]^T}{1 + \mathsf{f}_k^T \frac{1}{\epsilon} \mathrm{I} \mathsf{f}_k} \right] \mathsf{f}_k = C(k)^{-1} \mathsf{f}_k$$

$$\Rightarrow \Delta w_{ij}(k+1) = \frac{\eta}{\Delta t} [C(k)^{-1} \mathsf{f}_k]_j \gamma_i(k+1) \tag{3}$$

The BPDC rule in the form (3) is a modification and simplification of an online learning approach introduced by Atiya & Parlos [4] (AP learning), which can be given as ([3])

$$\begin{aligned} \Delta w_{ij}^{AP}(k+1) &= \frac{\eta}{\Delta t} \left[C_k^{-1} \mathsf{f}_k \right]_j \gamma_i(k+1) + \frac{\eta}{\Delta t} (C_k^{-1} - C_{k-1}^{-1}) \sum_{r=0}^{k-1} [\mathsf{f}_r]_j \gamma_i(r+1) \\ &= \frac{\eta}{\Delta t} \left[C_k^{-1} \mathsf{f}_k \right]_j \gamma_i(k+1) + \frac{\eta}{\Delta t} \left(C_k^{-1} C_{k-1} - I \right) \Delta w_{ij}^{AP_{batch}}(k) \end{aligned} \tag{4}$$

where $C_k = \sum_{r=0}^{k} \mathsf{f}_k \mathsf{f}_k^T$ is the full correlation matrix of the neurons' outputs accumulated over time and $\Delta w_{ij}^{AP_{batch}}$ is the accumulated Atiya-Parlos learning update step until time step k. BPDC in (3) is the first term in (4) using $C(k)$ instead of C_k The restriction of learning in BPDC to the output weights, i.e. only $i \equiv 0$ in (3) is now motivated by the observation that in AP recurrent learning the hidden neurons update slowly and in a highly coupled way [5]. These simplifications lead to the O(N) efficiency of BPDC.

2.1 Memory Issues

AP recurrent learning employs memory in the update in two ways: (i) by accumulating the full correlation matrix and (ii) by using a momentum term providing some memory about the previous update history. The O(N) efficiency of BPDC originates in the simplifications in (3) in comparison to (4), which skip both memory terms of the APRL. Further BPDC (as well as AP learning) does not follow the standard gradient of the quadratic error [5], such that the online rule itself can not be interpreted in the usual way as stochastic and regularizing approximation of the true gradient function. Lacking any memory in the algorithm itself BPDC thus is fully dependent on the memory contained in the dynamics of the network. This has the advantage that BPDC learns extremely fast with very few data and is highly capably of tracking quickly changing signals. On the other hand , BPDC generalization becomes sensitive to the length of the training sequence, because it is most sensitive to the last data provided. Fig. 2 b), c) shows this dependency for the training error of the Mackey-Glass data. Further BPDC can show strong overfitting behavior as demonstrated in Fig. 2 a).

This motivates to investigate different ways to reintroduce some regularizing memory in the BPDC algorithm without sacrificing the O(N) complexity: the use of a simple momentum term mixing the actual update with the last one $\Delta w(k+1)^{mom} = \Delta w(k+1) + \alpha\Delta w(k)$; a local batch version, which accumulates $\sum_{k+1}^{k+p} \Delta w_{ij}$ for p steps and then performs an update; the third way is motivated more directly from the BPDC-formula (2.1) using[1] (interprete $(k-s)$ as $((k-s)\Delta t)$):

$$\Delta w_{ij}(k+1) = \frac{\eta}{\Delta t} \frac{\sum_{s=0}^{p} \varphi(x_j(k-s))}{\|\varphi(\mathbf{x}(k))\|^2 + \varepsilon} \gamma_i(k+1) \tag{5}$$

This is an elegant way to accumulate a weighted sum of the instantaneous decorrelation factors, because (neglecting ε) we have

$$\frac{\sum_{s=0}^{p} \varphi(x_j(k-s))}{\|\varphi(\mathbf{x}(k))\|^2} = \sum_{s=0}^{p} C^{-1}(k-s)\varphi(x_j(k-s)) \frac{\|\varphi(\mathbf{x}(k-s))\|^2}{\|\varphi(\mathbf{x}(k))\|^2}.$$

3 Simulation Results

The different strategies to introduce memory are investigated on three standard datasets[2], for which reference results are also available in [3],[4]. We compute the NMSE defined as $E[(x_1(k) - y(k))^2]/\sigma^2$, where σ^2 is the variance of the input signal. For comparison we use a common parameter setting for all networks and all data: $\varepsilon = 0.002$ (is not critical), the initialization range $[-0.2, 0.2]$ provides sufficient dynamics to enable fast learning. As learning rate we use $\mu = 0.2, \mu = 0.2/p$ in the batch mode and the correlation mode of equation (5). This ensures that for the accumulated updates the steps size does not become too large. Preliminary experiments have shown

[1] This idea is due to M. Wardermann, personal communication.

[2] The exact data used and full tables of results for all combinations of memory parameters can be obtained from www.jsteil.de/BPDC.html

Table 1. Selected Average test/training NMSE for 50 runs of 50 epochs, one line at end of training, second line best results within 50 epochs

data/net	no reg.	mom. .8	batch 3	batch 10	corr. 1	corr. 2
Las/40	.548/.656	.490/.248	.368/.431	.275/.324	1.53/1.57	2.35/3.37
	.254/.271	.259/.263	.245/.273	.263/.305	.305/.297	.766/.665
Las/75	1.05/1.42	1.07/1.528	.156/.185	.175/.196	1.18/1.52	1.64/2.13
	.157/.248	.156/.247	.150/.181	.174/.196	.200/.299	.375/.391
Las/100	.320/.485	.311/.499	.141/.168	.164/.179	.537/.671	.924/1.19
	.143/.230	.141/.220	.139/.168	.163/.179	.204/.277	.312/.359
10th/40	.183/.361	.187/.364	.244/.368	.291/.507	.198/.381	.316/.516
	.183/.342	.186/.339	.243/.363	.291/.503	.193/.352	.210/.382
10th/75	.153/.309	.152/ .310	.169/.247	.213/.332	.163/.296	.350/.461
	.153/.277	.151/ .280	.169/.237	.213/.324	.161/.275	.178/.288
10th/100	.155/.323	.161/.302	.186/.282	.233/.328	.175/.319	.251/.394
	.159/.287	.159/.285	.185/.271	.232/.326	.173/.299	.192/.312
MG/40	.029/.212	.030/.208	.052/.280	.070/.265	.033/.210	.036/.211
	.286/.181	.029/.181	.0516/.238	.070/.240	.031/.182	.035/.183
MG/75	.033/.254	.033/.248	.053/.290	.074/.268	.037/.254	.041/.252
	.030/.214	.031/.205	052/.245	.073/.245	.0344/.208	.038/.214
MG/100	.034/.302	.032/.313	.054/.297	.079/.269	.036/.303	.04/.313
	.031/.256	.028/.248	.051/.259	.077/.258	.032/.253	.035/.258

that neither small learning rates nor a small initialization rate provides the same regularization and degrades performance. We use three datasets:

Example 1 (Santa Fee Laser Data)): We predict the $y(k+1)$ based on a time window of 15 past values $y(k-i), i=0..15$ using the first 1000 points for training and the next 2000 for test.

Example 2 (Tenth-order system (10th)): The following hard problem [4]predicts

$$y(k+1) = 0.3y(k) + 0.05y(k)\Big[\sum_{i=0}^{9} y(k-i)\Big] + 1.5u(k-9)u(k) + 0.1.$$

based on $u(k), u(k-9)$ as input fed to network, # training/test data $= 500$.

Example 3 (Mackey-Glass (MG)): As higher order benchmark we use the well known Mackey-Glass system with standard parameters

$$\dot{y}(t) = -0.1y(t) + \frac{0.2y(t-17)}{1+y(t-17)^{10}}.$$

Network inputs are $y(k), y(k-6), y(k-12), y(k-18)$ and the target $y(k+84)$, # training/test data $= 500$. To disregard initial transients, the training error is taken only after 200 steps.

Table 1 shows the reference performance without additional memory and selected results for the different memory methods. The classical momentum term does not give significant results for any of the problems. The batch update can improve performance for both training and test data for the Laser data, where it also keeps the error at the end of training close to the best result achieved overall. The role of the accumulated decorrelation factor p. For $p=1$, 2 the effect depends on the problem: for the Laser data and the tenth order data it disturbs learning while for the Mackey-Glass data the training error can decrease $p=1$,2 while the test error increases, a sign of overfitting. We suspect that this difference is caused by the different character of the time-series; the 10th-order system is driven by random input and the Laser data have large and fast changing oscillations such that the correlation memory of a time-step before can be totally misleading and even reinforce a already too large step. The Mackey-Glass data are much more smooth and therefore can profit from taking into account the last state as well.

4 Conclusion

We have introduced a new effective and simple learning rule for online adaptation of recurrent networks, which combines backpropagation, virtual teacher forcing, and decorrelation. In its derivation it is stripped all elements of memory in the algorithm itself and therefore in its basic form has to rely on the implicit memory in the network. This leads to fast learning and online tracking but as well to a high sensitivity to the last training data shown such that a careful choice of the training sequence is essential. We have further identified two ways to introduce memory, which have different effects: a local batch version which can regularize learning in particular for undersized networks, while an extended correlation factor can improve approximation at the cost of over-fitting and a decay in test performance. In control experiments we have also tried to combine these two mechanisms, however, with no significant other result. The different effects these mechanism have with respect to the different problems also show that an optimization of parameters may be difficult. Nevertheless the current work offers simple and computationally feasible extensions of the BPDC algorithms which are worth being explored if an optimized performance is desired.

References

1. Kolen, Kremer, eds.: A Field Guide to Dynamical Recurrent Networks. Wiley (2001)
2. Hammer, B., Steil, J.J.: Tutorial: Perspectives on learning with recurrent neural networks. In: Proc. ESANN. (2002) 357–369
3. Steil, J.J.: Backpropagation-decorrelation: online recurrent learning with O(N) complexity. In: Proc. IJCNN. (2004) 843–848
4. Atiya, A.B., Parlos, A.G.: New results on recurrent network training: Unifying the algorithms and accelerating convergence. IEEE Trans. Neural Networks **11** (2000) 697–709
5. Schiller, U.D., Steil, J.J.: Analyzing the weight dynamics of recurrent learning algorithms. Neurocomputing **vol. 63C** (2005) 5–23
6. Jaeger, H.: Adaptive nonlinear system identification with echo state networks. In: NIPS 15. (2003) 593–600
7. Natschläger, T., Maass, W., Markram, H.: The "liquid computer": A novel strategy for real-time computing on time series. TELEMATIK **8** (2002) 39–43
8. Steil, J.J.: Stability of backpropagation-decorrelation efficient O(N) recurrent learning. In: Proc. ESANN, d-facto publications (2005)

Incremental Rule Pruning for Fuzzy ARTMAP Neural Network

A. Andrés-Andrés, E. Gómez-Sánchez, and M.L. Bote-Lorenzo

School of Telecommunications Engineering, University of Valladolid,
47011 Valladolid, Spain
{aandand@ribera., edugom@, migbot@}tel.uva.es

Abstract. Fuzzy ARTMAP is capable of incrementally learning interpretable rules. To remove unused or inaccurate rules, a rule pruning method has been proposed in the literature. This paper addresses its limitations when incremental learning is used, and modifies it so that it does not need to store previously learnt samples. Experiments show a better performance, especially in concept drift problems.

1 Introduction

Fuzzy ARTMAP [1] is the most popular supervised architecture based on the Adaptive Resonance Theory (ART). One of its most appealing features is its capability for incremental learning: it can learn new patterns without forgetting previous knowledge, and without the need to present previously learned patterns again [2]. This is useful in a number of cases: if a model has to be first trained on a few available samples, and then improved with fresh data as they are collected; if a dataset is too large and sweeping over it is computationally very costly; or if data distribution varies with time.

Another significant feature of fuzzy ARTMAP is that, for classification tasks, it finds recognition categories of input patterns, associating each of them with the predicted class label. These associations can be translated into IF-THEN rules, simple to understand if not too many recognition categories are created during training (a phenomenon called *category proliferation*). To reduce category proliferation, some authors have devised alternative learning algorithms [3], that reduce the number of generated rules but loose the incremental learning capabilities. Others [2] proposed rule pruning after learning is complete, preserving the properties of the original training algorithm. However, this pruning mechanism needs to store all previously presented patterns in order to compute usage and accuracy indices that determine which rules should be pruned. Moreover, computing these indices every time a new pattern is presented is computationally demanding. Finally, in an adaptive setting, recently generated rules will have low usage, thus being likely to be pruned, hindering learning.

This paper addresses the limitations of the rule pruning mechanism proposed in [2] for incremental learning settings. A new method is proposed that does not use previously presented patterns, thus reducing computational demands.

W. Duch et al. (Eds.): ICANN 2005, LNCS 3697, pp. 655–660, 2005.

In addition, a novelty index is computed for each rule, in order to avoid early pruning. These modifications are illustrated with an experiment where a network must be trained with an initial set of samples, and then incrementally trained with recent ones as they become available. In addition, the proposed method will also be tested in scenarios where the concepts being learned drift with time [4], thus making the network obsolete and calling for incremental learning.

After this statement of the problem, section 2 briefly recalls the rule pruning method proposed in [2], and details the new method. Section 3 will illustrate this method experimentally, and section 4 will conclude.

2 Incremental Rule Pruning for Fuzzy ARTMAP

Due to space constraints, fuzzy ARTMAP cannot be explained in detail, and the reader is referred to [1,2]. However, as long as this work is concerned, it suffices to know that, when used for classification tasks, fuzzy ARTMAP partitions the input space with (hyper)boxes, associating a classification label to each one. The size of the boxes is determined by the distribution of data samples with the match tracking algorithm [2]. During the test, a pattern *selects* the closest box if it is outside all existing boxes, or the smallest hyperbox that contains it, and the associated classification label is predicted. Thus, each of these recognition categories and their associated labels can be seen as a set of IF-THEN rules.

In order to produce a small set of rules, [2] proposes to divide labelled samples into a training and a validation set. After fuzzy ARTMAP is trained on the whole training set, a confidence factor CF_j is computed for each rule:

$$CF_j = \gamma_U U_j + \gamma_A A_j \ . \tag{1}$$

where U_j is the *usage* of rule j, A_j its accuracy, and γ_U, γ_A are weighting factors that meet $\gamma_U + \gamma_A = 1$. The usage of rule j, provided that it predicts label k, is the number of *training samples* used to learn this rule (C_j), divided by the maximum C_J used to learn any rule J that predicts the same classification label:

$$U_j = C_j / \max\{C_J : \text{rule } J \text{ predicts label } k\} \ . \tag{2}$$

The accuracy of rule j, provided that it predicts label k, is the number of *validation samples* that selected this rule and were correctly classified (P_j), divided by the maximum P_J of any rule J that predicts the same classification label:

$$A_j = P_j / \max\{P_J : \text{rule } J \text{ predicts label } k\} \ . \tag{3}$$

The scaling ensures that $U_j \in [0,1]$, $A_j \in [0,1]$ and thus $CF_j \in [0,1]$. In addition, at least one rule has $U_j = 1$, and at least one rule (but not necessarily the same) has $A_j = 1$. Threshold pruning can be carried out by pruning those rules with $CF_j < \tau$, where $\tau \in [0,1]$, i.e. those that are infrequent and inaccurate (in a balance that depends on γ, and to an extent determined by τ).

This method has several limitations when using incremental learning, if rule pruning has to be carried out frequently. First, all training and validation samples have to be retained in order to recompute the usage and accuracy. Moreover, if any of the existing rules is modified, or a new one created, all indices should be recomputed, because samples that previously selected another rule may select the one recently modified or created, and viceversa. This also applies after pruning a rule. As the computational cost of evaluating usage and accuracy increases with the number of samples, rule pruning becomes eventually unfeasible.

Moreover, incremental learning can be applied to a problem where concepts drift with time (e.g. the preferences of a web portal user). In this scenario, initial data becomes obsolete after sources drift, as well as rules that were learnt using those data. However, a rule pruning mechanism that uses old data to compute the confidence factors would retain old rules while pruning recent ones.

This paper proposes two modifications to the rule pruning method: computing the confidence factor without the need to store previous patterns, and introducing a novelty index in the confidence factor to avoid early rule pruning.

2.1 Computing CF_j Without Storing Previous Patterns

If training samples are discarded once they are used, the count C_j of samples that selected rule j, needed in eq. (2) cannot be recomputed. However, an estimated count, $\hat{C}_j$, can be incremented after training with one sample if it selects rule j, as explained in Fig. 1. Nevertheless, since rules evolve during training, a sample that chooses rule j when it is learnt, may select a different rule, $j\prime$, if presented after the network is trained. Here, $\hat{C}_j$ is incremented instead of $\hat{C}_{j\prime}$. This might fake the real usage of a rule that was initially quite specific (hence with precedence over more general rules) but grew later becoming more general. This way, a rule that is sparsely used might remain after pruning takes place. However, this effect disappears after a reasonable amount of samples.

To compute the accuracy index, A_j, eq. (3) could be used, with an estimated $\hat{P}_j$. Each $\hat{P}_j$ can be expressed as P_j^C/P_j^T, where P_j^T is the total number of validation samples that selected category j, and P_j^C and the number of them that were correctly labelled. These two counts can be incremented after each validation sample is presented (see Fig. 1). Nevertheless, after pruning rule j, if the P_j^T samples that selected this rule were recorded they could have been represented again to contribute to the accuracy indices of some other rules. In our approach they are not recorded, and thus the quantities P_j^T and P_j^C could be added to other $P_{j\prime}^T$ and $P_{j\prime}^C$. However, $j\prime$ should predict the same class label that j, otherwise $P_{j\prime}^C$ would not make sense as a number of *correctly* labelled samples. But it may well be the case that the samples (if recorded), would select rule $j\prime$ with different class label, and the P_j^C should be used to increment the number of *wrong* predictions of $j\prime$. In summary, there is not enough information on how to assign these counts of samples, and it seems better to discard them.

- Present a *training* sample
- It selects rule j
  - Update $\hat{C}_j := \hat{C}_j + 1$
  - Compute U_j with $\hat{C}_j$ in eq. (2)

- Present a *validation* sample
- It selects rule j that predicts label k
  - Update $P_j^T := P_j^T + 1$
  - Only if k is the *correct* label
    - Update $P_j^C := P_j^C + 1$
  - Update $\hat{P}_j := P_j^C / P_j^T$
  - Compute A_j with $\hat{P}_j$ in eq. (3)

Fig. 1. Pseudo-code to compute of $\hat{C}_j$ and $\hat{P}_j$ without storing previous patterns

2.2 A Novelty Index to Prevent Early Pruning

If rule pruning is done frequently during incremental learning, recent rules will have very small usage and are likely to be pruned, making the network quite unstable. To avoid this, a novelty index is introduced in eq. (1), as follows:

$$CF_j = \gamma_U U_j + \gamma_A A_j + \gamma_N N_j \ . \tag{4}$$

where γ_U, γ_A, γ_N are weighting factors (and in general $\gamma_U + \gamma_A + \gamma_N \geq 1$), and N_j is the novelty index defined by

$$N_j = 1 - \frac{\text{Age}_j}{K} \ . \tag{5}$$

where Age_j is the number of training samples presented to the network after rule j was created, and K is a constant, as a rule of thumb, two or three times the number of patterns presented before a significant change of the data sources. This equation gives a chance to each rule to be selected by some training patterns.

3 Experimental Study

As stated in the introduction, incremental learning makes sense if a network needs to be trained on a few initial samples, and then its knowledge incremented with the use of more recent samples, or if data sources vary with time.

To asses rule pruning when incremental learning is used in order to improve the knowledge of the network with **time-invariant data sources**, 12 initial samples were generated from two gaussian sources with $\mu_1 = (0.25, 0.5)$ and $\mu_2 = (0.75, 0.5)$, and $\sigma = 0.20$. After that, sets of 12 training and 4 validation samples were generated, the networks updated, and rule pruning carried out. Parameters were heuristically selected to be $\gamma_U = 0.5$, $\gamma_A = 0.5$, $\gamma_N = 0$ (novelty is not relevant since data sources do not vary) and $\tau = 0.5$. The experiment was completed after the presentation of 1000 samples in total. Fig. 2a shows that, if the network is not pruned it ends up with a larger set of rules and a larger error. This is because rule pruning removes inaccurate rules in the boundary between the two gaussian sources. Moreover, the rule pruning method proposed here yields the same result as the original from [2]. After some more samples, the count indices computed without storing and processing all previously learnt patterns approach those of the original method, and the selection of rules to

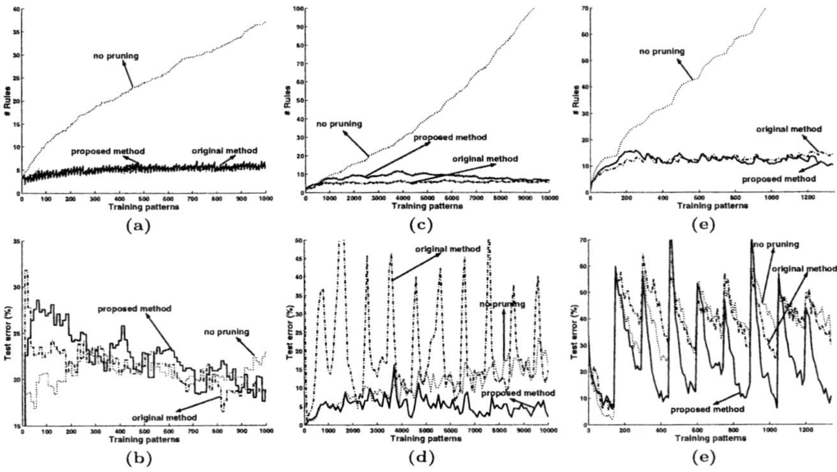

Fig. 2. **(a,c,e)** Number of rules in the network and **(b,d,f)** test error, for the classification of samples from **(a,b)** two fixed gaussian sources, **(c,d)** two drifting gaussian sources that start as in the previous case, and **(e,f)** noisy Stagger concepts

prune is similar. Significantly, the simulation with our method took 1/2 the time and 1/20 the memory of the original method.

Two experiments have been proposed to evaluate the usefulness of rule pruning on Fuzzy ARTMAP when it is used to carry out incremental learning of a problem where **time varying data sources**. In both, pruning parameters were $\gamma_U = 0.3$, $\gamma_A = 0.3$, $\gamma_N = 0.7$, $\tau = 0.5$ and K in eq. (5) was set to 250. In the first experiment, patterns come from two gaussian sources as in previous example (but with $\sigma = 0.1$). However, after 100 patterns have been generated, the sources are changed by drifting their means clockwise by $\pi/10$. The experiment concludes when five turns are completed. Fig. 2c shows how pruning is necessary since, as old rules become incorrect, new rules are created being more specific, and hence many rules become necessary. Pruning removes obsolete rules, and lets new rules become more general, which in general results into a smaller and more accurate set of rules. Furthermore, as shown in Fig. 2d, the original pruning method has a higher classification error, since recent, good rules may be removed due to the use of obsolete samples to compute the confidence factors.

The second experiment is a variation of the Stagger Concepts [4]. An object is described by size, shape and color, with three values each. Initially, some of the objects are labelled in one class (say "objects I like"), as shown in Fig. 3a. After 150 samples are presented, the classification drifts to Fig. 3b, and 150 later to Fig. 3c, looping through them three times. To represent the problem numerically, we could state that *color* can take values 0 (*red*), 0.5 (*green*) and 1 (*blue*). However, in many real problems there is noise and natural overlap between classes. In our experiments, once the feature value is selected (say *color* is *green*), gaussian noise ($\sigma = 0.1$) is added to its numerical representation (i.e. *color*$=0.5+\mathcal{N}(\prime, \sigma)$).

		S	M	L
R	△	N	N	N
	○	N	N	N
	□	N	N	N
G	△	N	N	N
	○	N	N	N
	□	N	N	N
B	△	Y	N	N
	○	Y	N	N
	□	Y	N	N

(a)

		S	M	L
R	△	Y	Y	Y
	○	Y	Y	Y
	□	Y	Y	Y
G	△	N	N	N
	○	Y	Y	Y
	□	N	N	N
B	△	N	N	N
	○	Y	Y	Y
	□	N	N	N

(b)

		S	M	L
R	△	N	Y	Y
	○	N	Y	Y
	□	N	Y	Y
G	△	N	Y	Y
	○	N	Y	Y
	□	N	Y	Y
B	△	N	Y	Y
	○	N	Y	Y
	□	N	Y	Y

(c)

Shape	Size	Color	Label
○, △	S	G,B	N
○, △	S,M	G	Y
○, □	M,L	R	Y
○, △	S	R	N
○, △	M	B	Y
△	M,L	B	Y
□	S	R,G,B	N
○, □	L	G,B	Y
△	L	B	Y
□	M	G,B	Y

(d)

Fig. 3. (a,b,c) The three possible classifications in the Stagger concepts, using shape (○, □ or △), size (S, M or L) and color (R, G or B). **(d)** Rules after pruning with the proposed method, for case (c).

Fig. 2e shows how the number of rules grows indefinitely if no pruning is applied. Furthermore, the proposed method shows smaller classification error for the reasons mentioned above. Fig. 3d shows the set of rules that remained after having successively learnt the classifications in Figs. 3a, 3b and 3b, and having pruned several times. Though suboptimal, these rules are easy to understand and yield a good classification performance.

4 Conclusions

A new rule pruning method for fuzzy ARTMAP has been proposed, with advantages for incremental learning: it does not store all previous patterns, and it does not use obsolete information to determine which rules to prune. Experimentally, it performed with less memory and computation, yielding similar results to the original method with invariant data sources, and better otherwise.

Acknowledgement

This work has been partially funded by Spanish project TIC2002-04258-C03-02.

References

1. Carpenter, G.A., Grossberg, S., Markuzon, N., Reynolds, J.H., Rosen, D.B.: Fuzzy ARTMAP: A neural network architecture for incremental supervised learning of analog multidimensional maps. IEEE Trans. Neural Networks, **3** (1992) 698–713
2. Carpenter, G.A., Tan, H.A.: Rule extraction: From neural architecture to symbolic representation. Connection Science **7** (1995) 3–27
3. Gómez-Sánchez, E., Dimitriadis, Y.A., Cano-Izquierdo, J.M., and López-Coronado, J.: μARTMAP: use of mutual information for category reduction in fuzzy ARTMAP. IEEE Trans. Neural Networks **13** (2002) 58–69
4. Schimmer, J., Granger. R.: Beyond incremental procesing: Tracking concept drift. In *Proc. Fifth Nat. Conf. Artificial Intelligence*, Philadelphia, PA (1996) 502–507

An Inductive Learning Algorithm with a Partial Completeness and Consistence via a Modified Set Covering Problem

Janusz Kacprzyk and Grażyna Szkatuła

Systems Research Institute, Polish Academy of Sciences,
ul. Newelska 6, 01-447 Warsaw, Poland
{kacprzyk, szkatulg}@ibspan.waw.pl

Abstract. We present an inductive learning algorithm that allows for a partial completeness and consistence, i.e. that derives classification rules correctly describing, e.g, most of the examples belonging to a class and not describing most of the examples not belonging to this class. The problem is represented as a modification of the set covering problem that is solved by a greedy algorithm. The approach is illustrated on some medical data.

1 Introduction

In inductive learning (from examples) we traditionally seek a classification rule satisfying *all* positive and *no* negative examples which is often strict and unrealistic. Here we assume: (1) *a partial completeness*, and (2) *a partial consistency*, i.e. that it is sufficient to describe – respectively – e.g., *most* of the positive and, e.g., *almost none* of the negative examples. We also add: (3) *convergence*, i.e. that the rule must be derived in a *finite* number of steps, and (4) that the rule is of a *minimal "length"*.

Examples are described (cf. Michalski [14]) by a set of K "attribute - value" pairs $e = \wedge_{j=1}^{K}[a_j \# v_j]$; a_j denotes attribute j with value v_j and # is a relation (=, <,…).

We propose a modified inductive learning procedure based on Michalski's [14] star-type methodology, related to our previous work (cf. Kacprzyk and Szkatuła [5 – 13]). The problem is represented as a modified set covering problem solved by a greedy algorithm. Medical data are employed for testing.

2 A Softened Problem Formulation of Inductive Learning

Sets of examples U and attributes $A = \{a_1,...,a_K\}$ are finite. $V_{a_j} = \{v_{i_1}^{a_j},...,v_{i_j}^{a_j}\}$ is a domain of a_j, $j = 1,...,K$, $V = \bigcup_{j=1,...,K} V_{a_j}$. $f : U \times A \to V$, $f(e, a_j) \in V_{a_j}$, $\forall a_j \in A$, $\forall e \in U$. Each $e \in U$, with K attributes, $A = \{a_1,...,a_K\}$, is written $e = \wedge_{j=1}^{K}[a_j = v_i^{a_j}]$, where $v_i^{a_j} = f(e, a_j) \in V_{a_j}$ denotes attribute a_j taking on value $v_i^{a_j}$ for example e. An e in (1) is composed of K "attribute-value" pairs, denoted $s_j = [a_j = v_i^{a_j}]$ (selectors). The conjunction of $l \le K$ "attribute-value" pairs, i.e.

W. Duch et al. (Eds.): ICANN 2005, LNCS 3697, pp. 661 – 666, 2005.

$$C^I = \underset{j\in I}{\wedge} s_j = \underset{j\in I}{\wedge} [a_j = v_i^{a_j}] = [a_{j_1} = v_i^{a_{j_1}}] \wedge \ldots \wedge [a_{j_l} = v_i^{a_{j_l}}] \tag{1}$$

where $I = \{j_1, j_2, \ldots, j_l\} \subseteq \{1, \ldots, K\}$ is called a *complex.*

Let us have example e and a complex $C^I = [a_{j_1} = v_i^{a_{j_1}}] \wedge \ldots \wedge [a_{j_l} = v_i^{a_{j_l}}]$ corresponding to the set of indices $I = \{j_1, \ldots, j_l\} \subseteq \{1, \ldots, K\}$; $\{j_1, \ldots, j_l\}$ is equivalent to a vector $x = [x_j]^T$, $j = 1, \ldots, K$, such that $x_j = 1$ if $s_j = [a_j = v_i^{a_j}]$ occurs in C^I, and 0 otherwise. C^I *covers* e if $f(C^I, a_j) = f(e, a_j), \forall j \in I$. Now, a_d is a decision attribute and $V_{a_d} = \{v_{i_1}^{a_d}, \ldots, v_{i_d}^{a_d}\}$ is a domain of a_d. Each $e \in U$ is described by $\{a_1, a_2, \ldots, a_K\} \cup \{a_d\}$. So, a_d determines a partition $\{Y_{v_{i_1}^{a_d}}, Y_{v_{i_2}^{a_d}}, \ldots, Y_{v_{i_d}^{a_d}}\}$ of U, where $Y_{v_{i_t}^{a_d}} = \{e \in U : f(e, a_d) = v_{i_t}^{a_d}\}$, $v_{i_t}^{a_d} \in V_{a_d}$ for $t = 1, \ldots d$. Set $Y_{v_{i_t}^{a_d}}$ is called the t-th *decision class* (for $v_{i_t}^{a_d} \in V_{a_d}$), $Y_{v_{i_1}^{a_d}} \cup \ldots \cup Y_{v_{i_d}^{a_d}} = U$, $Y_{v_i^{a_d}} \cap Y_{v_j^{a_d}} = \varnothing$ *for* $i \neq j$.

Suppose that we have a set of *positive* and *negative examples* for a class $Y_{v_{i_t}^{a_d}}$

$$S_P(Y_{v_{i_t}^{a_d}}) = \{e \in U : f(e, a_d) = v_{i_t}^{a_d}\} \tag{2}$$

$$S_N(Y_{v_{i_t}^{a_d}}) = \{e \in U : f(e, a_d) \neq v_{i_t}^{a_d} \text{ and } \forall e' \in S_P(Y_{v_{i_t}^{a_d}})\ \exists a_j \in P,\ f(e, a_j) \neq f(e', a_j)\}$$

and $S_P(Y_{v_{i_t}^{a_d}}) \cap S_N(Y_{v_{i_t}^{a_d}}) = \varnothing$ and $S_P(Y_{v_{i_t}^{a_d}}) \neq \varnothing$, $S_N(Y_{v_{i_t}^{a_d}}) \neq \varnothing$.

The descriptions of $Y_{v_{i_t}^{a_d}}$ can be given as "IF *certain conditions are fulfilled* THEN *membership in a definite class takes place*". The rule $rul(P, v_{i_t}^{a_d})$: "IF C^I THEN $[a_d = v_{i_t}^{a_d}]$" is called an "*elementary*" *rule* for class $Y_{v_{i_t}^{a_d}}$, $v_{i_t}^{a_d} \in V_{a_d}$, $I = \{j_1, \ldots, j_l\} \subseteq \{1, \ldots, K\}$, $P = \{a_{j_1}, \ldots, a_{j_l}\} \subseteq A \setminus \{a_d\}$, where C^I is a description of example in terms of attributes $a_j, j \in I$, and this example belongs to $Y_{v_{i_t}^{a_d}}$.

The *strength of a rule* is defined in the following manner:

$$q(rul(P, v_{i_t}^{a_d})) = \frac{card(\{e : e \in [C^I] \text{ and } f(e, a_d) = v_{i_t}^{a_d}\})}{card(\{e : e \in U\})}. \tag{3}$$

We consider the classification rules:

$$\text{IF } C^{I_1} \cup \ldots \cup C^{I_L} \text{ THEN } [a_d = v_{i_t}^{a_d}] \tag{4}$$

with: $I_1,\ldots,I_L \subseteq \{1,\ldots,K\}$, $C^{I_l} = \underset{j\in I_l}{\wedge} [a_j = v_i^{a_j}]$, $l = 1,\ldots,L$.

Let us have P positive examples, $e^m \in S_P(Y_{v_{i_t}^{a_d}})$, $m = 1,\ldots,P$, and N negative examples, $e^n \in S_N(Y_{v_{i_t}^{a_d}})$, $n = 1,\ldots,N$. For each a_j, each possible value occurs at some intensity (frequency). If it occurs more frequently in the positive and less frequently in the negative examples, then it is somehow typical and should appear in the rule sought. So, we introduce the function, for each a_j, $j = 1,\ldots,K$ and $v \in V_{a_j}$

$$g_j(v) = \frac{1}{P}\sum_{m=1}^{P} \delta(e^m, v) - \frac{1}{N}\sum_{n=1}^{N} \delta(e^n, v) \tag{5}$$

where: $\delta(e^m, v) = \begin{cases} 1 & for\ v_i^{a_j} = v \\ 0 & otherwise \end{cases}$, and: $e^m \in S_P$, $v_i^{a_j} = f(e^m, a_j) \in V_{a_j}$; and analogously for $\delta(e^n, v)$. So, we may expresses to which degree the particular $v \in V_{a_j}$ of a_j occurs more often in the positive than negative examples; the normalized $g_j(v)$ is used as a weight of $v \in V_{a_j}$ (cf. Kacprzyk and Szkatuła [10].

Example e_W with weights is $e_W = \overset{K}{\underset{j=1}{\wedge}} [a_j = v_i^{a_j}; g_j(v_i^{a_j})]$, i.e. is a conjunction of weighted selectors, $s_j^W = [a_j = v_i^{a_j}; g_j(v_i^{a_j})]$, that is: $C_W^I = \underset{j\in I\subseteq\{1,\ldots,K\}}{\wedge} s_j^W$, and is called a *weighted complex*. Notice that for C_W^I x has the elements $x_j = 1$ for $j \in I$, while, for $j \in \{1,2,\ldots,K\}\setminus I$, $x_j = 0$. For C_W^I its *weighted length* is:

$$d_W(C_W^I) =$$

$$= \sum_{j\in I}(1 - g_j(v_i^{a_j}))\cdot x_j + \sum_{j\in\{1,2,\ldots,K\}\setminus I}(1 - g_j(v_i^{a_j}))\cdot x_j = \sum_{j=1}^{K}(1 - g_j(v_i^{a_j}))\cdot x_j \tag{6}$$

which reflects a higher relevance of those values of attributes which occur more often in the positive than in negative examples.

The length of $R_W = C_W^{I_1} \cup \ldots \cup C_W^{I_L}$ is $d_{R_W}(C_W^{I_1} \cup \ldots \cup C_W^{I_L}) = \max_{l=1,\ldots,L} d_W(C_W^{I_l})$, and we look for an optimal classification rule $R_W^* = C_W^{I_1^*} \cup \ldots \cup C_W^{I_L^*}$ such that

$$\min_{I_1,\ldots,I_L} d_{R_W}(C_W^{I_1} \cup \ldots \cup C_W^{I_L}) \tag{7}$$

As the (exact) solution of (7) is very difficult, an auxiliary problem is solved (cf. Kacprzyk and Szkatuła [11]) whose solution is in general very close but much easier, i.e. an $R_W^* = C_W^{I_1*} \cup ... \cup C_W^{I_L*}$ is sough such that $\min_{I_1} d_W(C_W^{I_1}),..., \min_{I_L} d_W(C_W^{I_L})$.

3 Formulation as a Modified Set Covering Problem

For $e^p \in S_P$, and all the negative examples $e^{P+n} \in S_N$, $n = 1,..., N$, we construct a 0-1 matrix $Z_{N*K} = [z_{nj}]$, $j = 1,..., K$, $z_{nj} = \begin{cases} 1 & for\ f(e^p, a_j) = f(e^{P+n}, a_j) \\ 0 & for\ f(e^p, a_j) \neq f(e^{P+n}, a_j) \end{cases}$ whose rows correspond to the consecutive $e^{P+n} \in S_N$, $n = 1,..., N$ and columns to attributes $a_1,..., a_K$; $z_{nj} = 1$ if a_j has different values in the positive and negative examples, i.e. $f(e^p, a_j) \neq f(e^{P+n}, a_j)$, and $z_{nj} = 0$ otherwise. There are no rows with all 0s since the sets of positive and negative examples are disjoint (and non-empty). So, for any positive and negative example there is always at least one attribute with a different value in these examples.

Consider now the following inequality: $\sum_{j=1}^{K} z_{nj} x_j \geq \gamma_n$, $n = 1,..., N$, where $\gamma = [\gamma_1,..., \gamma_N]^T$ is a 0-1 vector, and $x_j \in \{0,1\}$, for $j = 1,..., K$. Any vector x satisfying $Zx \geq \gamma$ determines uniquely a complex describing at least one example from the set of positive ones, and does not describe most of the negative examples. If x does not describe the n-th negative example, then $\gamma_n = 1$; and $\gamma_n = 0$ otherwise.

So, the problem is, using the above inequality:

$$\min_{x:Zx \geq \gamma} \sum_{j=1}^{K} (1 - g_j(v_i^{a_j})) \cdot x_j \tag{8}$$

and, in a simplified form: $\min_{x:Z^1 x \geq \gamma} d_W(C_W^{I_1}),..., \min_{x:Z^L x \geq \gamma} d_W(C_W^{I_L})$ in which each minimization with respect to x is equivalent to the determination of a 0-1 vector x^* which uniquely determines the complex of the shortest weighted length. On the other hand, the satisfaction of $Zx \geq \Lambda$ (Λ is a unit vector) guarantees that such a complex would not describe all negative examples. If rules should describe *almost none* of the negative examples, the problem can be written as a modification of the set covering problem

$$\min_{x, \gamma} \sum_{j=1}^{K} c_j x_j \tag{9}$$

subject to: $\sum_{j=1}^{K} z_{1j} x_j \geq \gamma_1$,..., $\sum_{j=1}^{K} z_{Nj} x_j \geq \gamma_N$, with an additional constraint: $\sum_{n=1}^{N} \gamma_n \geq N - rel$

where $c_j = (1 - g_j(v_i^{a_j}))$, $z_{nj} \in \{0,1\}$, $x_j \in \{0,1\}$, $j = 1,...,K$, $rel \geq 0$, $\gamma = [\gamma_1,...,\gamma_N]^T$, $\gamma_n \in \{0,1\}$.

This is as the original set covering problem except for that no more then *rel* rows are uncovered. Then, no more then *rel* rows can be deleted though we may loose some information, and this reduction cannot always be applied. In the set covering problem (cf. [1-4]) there is only constraint, and $\gamma = [\gamma_1,...,\gamma_N]^T$ is a unit vector. Problem (11) is that of covering at least *N-rel* rows of an *N*-row, *K*-column, zero-one matrix (z_{nj}) by a subset of the columns at minimal cost c_j. We define $x_j = 1$ if column *j* with cost $c_j > 0$ is in the solution, and $x_j = 0$ otherwise. Then, most rows (at least *N-rel* rows) are covered by at least one column. It always has a feasible solution (*x* of *K* element), due to the required disjointness of the sets of positive and negative examples and the way the matrix *Z* was constructed.

We seek a 0-1 vector *x* at the minimum cost and a 0-1 vector $\gamma = [\gamma_1,...,\gamma_N]^T$ that determines the covered rows, $\gamma_n = 1$ if row *n* is covered by *x*, and $\gamma_n = 0$, otherwise. By assumption, at least *N-rel* rows must be covered by *x*. Then, an "*elementary" rule* for $Y_{v_{i_t}^{a_d}}$, $v_{i_t}^{a_d} \in V_{a_d}$, may not describe at least $(100 / N \sum_{n=1}^{N} \gamma_n)$% negative examples.

The set covering problem is a well-known NP-complete combinatorial optimization problem. Many optimal and faster heuristic algorithms exist, cf. Balas and Padberg [1], Beasley [2], Christofides [3], presented a genetic algorithm, with modified operations,too. One can also use here a greedy algorithm (cf. Chvatal [4]) and we use it here.

4 An Example Using Heart Disease Data

We have 90 examples, ill or healthy, 60 are a training set and 30 are for testing. The following 12 blood factors (attributes) are measured: *lk1* - blood viscosity for coagulation quickness 230/s, *lk2* - blood viscosity for coagulation quickness 23/s, *lk3* - blood viscosity for coagulation quickness 15/s, *lp1* - plasma viscosity for coagulation quickness 230/s, *lp2* - plasma viscosity for coagulation quickness 23/s, *agr* - aggregation level of red blood cells, *fil* - blood cells capacity to change shape, *fib* - fibrin level in plasma, *ht* - hematocrit value, *sas* - sial acid rate in blood serum, *sak* - sial acid rate in blood cells, *ph* - acidity of blood.

We seek classification rules into: *class 1*: patients have no coronary heart disease, *class 2*: patients have a coronary heart disease. Some results are shown below:

$A_{learning}$ %, by assumption	Number of iterations for class 1/2	Number of selectors in rule for class 1/2	$A_{learning}$ %, by assumption	$A_{learning}$ %, attained
100%	16/19	43/55	100%	90.0%
at least 97%	13/17	26/33	at least 97%	96.7%

and the results are encouraging, in fact comparable to the use of a genetic algorithm (cf. Kacprzyk and Szkatuła [13]).

5 Concluding Remarks

We proposed a improved inductive learning algorithm allowing for a partial completeness and consistence that is based on a set covering problem formulation solved by a greedy algorithm. Results seem to be very encouraging.

References

1. Balas E., Padberg M.W.: Set partitioning - A survey. In: N. Christofides (ed.) *Combinatorial Optimisation*, Wiley, New York (1979).
2. Beasley J.E.: A genetic algorithm for the set covering problem. *European Journal of Operational Research* 94 (1996) 392-404.
3. Christofides N., Korman S.: A computational survey of methods for the set covering problem. *Management Sci.* 21 (1975) 591-599.
4. Chvatal V.: A greedy heuristic for the set-covering problem. *Maths. of Oper. Res.* 4 (3) (1979) 233-235.
5. Kacprzyk J., Szkatuła G.: Machine learning from examples under errors in data, Proceedings of IPMU'1994 – 5th Int'l Conf. on Information Processing and Management of Uncertainty in Knowledge-Based Systems Paris France, Vol.2 (1994) 1047-1051
6. Kacprzyk J., Szkatuła G.: Machine learning from examples under errors in data. In B. Bouchon-Meunier, R.R. Yager and L.A. Zadeh (eds.): Fuzzy Logic and Soft Computing, World Scientific, Singapore, (1995) 31-36.
7. Kacprzyk J., Szkatuła G.: An algorithm for learning from erroneous and incorrigible examples, *Int. J. of Intelligent Syst.* 11 (1996) 565-582.
8. Kacprzyk J., Szkatuła G.: An improved inductive learning algorithm with a preanalysis of data", In Z.W. Ras, A. Skowron (eds.): Foundations of Intelligent Systems (Proc. of 10th ISMIS'97 Symposium, Charlotte, NC, USA), LNCS, Springer, Berlin (1997) 157-166.
9. Kacprzyk J., Szkatuła G.: IP1 - An Improved Inductive Learning Procedure with a Preprocessing of Data. In L. Xu, L.W. Chan, I. King and A. Fu (eds.): Intelligent Data Engineering and Learning. Perspectives on Financial Engineering and Data Mining (Proc. of IDEAL'98, Hong Kong), Springer, Hong Kong (1998) 385-392, 1998.
10. Kacprzyk J., Szkatuła G.: An inductive learning algorithm with a preanalysis of data. *Int. J. of Knowledge - Based Intelligent Engineering Systems*, vol. 3 (1999) 135-146.
11. Kacprzyk J., Szkatuła G.: An integer programming approach to inductive learning using genetic algorithm. In: Proceedings of CEC'02 - The 2002 Congress on Evolutionary Computation, CEC'02, Honolulu, Hawaii (2002) 181-186.
12. Kacprzyk J., Szkatuła G.: An integer programming approach to inductive learning using genetic and greedy algorithms, In L.C. Jain and J. Kacprzyk (eds.): New Learning Paradigms in Soft Computing, Physica-Verlag, Heidelberg and New York (2002) 322-366.
13. Kacprzyk J., Szkatuła G.: A softened formulation of inductive learning and its use for coronary disease data. In: M.-S. Hacid, N.V. Murray, Z.W. Raś and S. Tsumoto (eds.) Foundations of Intelligent Systems (Proc. of 15th ISMIS'2005 Symaposium, Saratoga Springs, NY, USA), LNCS, Springer, Berlin, (2005), 200-209.
14. Michalski R.S.: A theory and methodology of inductive learning. In: R. Michalski, J. Carbonell and T.M. Mitchell (Eds.), Machine Learning. Tioga Press (1983).

A Neural Network for Text Representation

Mikaela Keller and Samy Bengio

IDIAP Research Institute, 1920 Martigny, Switzerland*
{mkeller, bengio}@idiap.ch

Abstract. Text categorization and retrieval tasks are often based on a good representation of textual data. Departing from the classical *vector space model*, several probabilistic models have been proposed recently, such as PLSA. In this paper, we propose the use of a neural network based, non-probabilistic, solution, which captures jointly a rich representation of words and documents. Experiments performed on two information retrieval tasks using the TDT2 database and the TREC-8 and 9 sets of queries yielded a better performance for the proposed neural network model, as compared to PLSA and the classical TFIDF representations.

1 Introduction

The success of several real-life applications involving tasks such as text categorization and document retrieval is often based on a good representation of textual data. The most basic but nevertheless widely used technique is the *vector space model* (VSM) [1] (also often called *bag-of-words*), which makes the assumption that the precise order of the words is uninformative.

Such representation neglects potential semantic links between words. In order to take them into account, several more recent models have been proposed in the literature, mostly based on a probabilistic approach, including the Probabilistic Latent Semantic Analysis (PLSA) [2]. They in general factor the joint or conditional probability of words and documents by assuming that the choice of a word during the generation of a document is independent of the document given some hidden variable, often called *topic* or *aspect*.

In this paper, we would like to argue that while the basic idea behind probabilistic models is appealing (trying to extract higher level concepts, such as topics, from raw texts), there is no need to constrain the model to be probabilistic. Indeed, most of the applications relying on text representation do not really need precise probabilities. It was recently argued [3] that in such a case, one should probably favor so-called *energy-based models*, which associate an unnormalized energy to each target configuration, instead of a proper probability, and then simply compare energies of competing solutions in order to take a final

* This work was supported in part by the Swiss NSF through the NCCR on IM2 and in part by the European PASCAL Network of Excellence, IST-2002-506778, through the Swiss OFES.

W. Duch et al. (Eds.): ICANN 2005, LNCS 3697, pp. 667–672, 2005.

decision. It is argued that this scheme enables the use of architectures and loss functions that would not be possible with probabilistic models.

We thus propose here a neural network based representation that can be trained on a large corpus of documents, and which associates a high score to pairs of word-document that appear in the corpus and a low score otherwise. The model, which automatically induces a rich and compact representation of words and documents, can then be used for several text-related applications such as information retrieval and text categorization.

The outline of the paper is as follows. Section 2 briefly summarizes current state-of-the-art techniques used for text representation, mostly based on probabilistic models. Then, Section 3 presents our proposed neural network based model. This is followed in Section 4 by some experiments on two real information retrieval tasks. Finally, Section 5 concludes the paper.

2 Related Work

In most Textual Information Access applications, documents are represented within the *Vector Space Model* (VSM) [4]. In this model, each document d is represented as a vector $(\alpha_1, ..., \alpha_M)$, where α_j is a function of the frequency of the jth word w_j in a chosen dictionary of size M. To be more concrete, let us consider the Document Retrieval task, - used in the experimental section - and how VSM is implemented there. In a Document Retrieval task, a user formulates a query q addressed to a database, and the database documents d are then ranked according to their *Relevance Status Value*, $RSV(q,d)$, which is defined so that documents relevant to q should have higher values than non-relevant ones. In the VSM, $RSV(q,d)$ is defined as the scalar product of the query's and document's representations: $RSV(q,d) = \sum_{j=1}^{M} \alpha_j^q \cdot \alpha_j^d$, where α_j^d (resp. α_j^q) is the weight in the document (resp. query) representation of the jth dictionary word. A simple way to implement this value function is to choose α_j^q as a binary weight stating the presence or absence of the word in the query, and d_j as the well-known TFIDF weight [5]:

$$\alpha_j^d = tf_j(d) \cdot \log(\frac{N}{df_j}) ,$$

where $tf_j(d)$ corresponds to the number of occurrences of w_j in d, N is the number of documents in the database and df_j stands for the number of documents the term w_j appears in. It is designed to give more importance to terms frequent in the document while penalizing words appearing in too many documents.

In the VSM, two documents (or a query and a document) are considered similar if they are composed of the same words. However, a property of most human languages is that a certain topic can be expressed with different words (*synonyms*). Moreover, a given word can often be used in totally different contexts (*polyseme*). The VSM does not model these links between words. Several attempts have been proposed to take that into account, among which the use of the so-called Latent Semantic Analysis (LSA)[6]. LSA tries to link words together according to their co-occurrences in a database of documents by performing a Singular Value Decomposition. The more recent Probabilistic Latent Semantic Analysis (PLSA) model [2] seeks a generative model for word/document co-occurrences. It makes the assumption that each word w_j in a given document d_δ, is generated from a latent aspect t taking values among $\{1, \ldots, K\}$, K being a chosen hyper-parameter, and δ a variable picking one document among the others in the database. The joint probability of a word w_j and a document d_δ is then:

$$P(d_\delta, w_j) = P(\delta) \sum_{k=1}^{K} P(t = k|d_\delta) P(w_j|t = k) \; . \tag{1}$$

PLSA can then be used to replace the original VSM document representation by a representation in a low-dimensional "latent" space. In [2], the components of the document in the low-dimensional space are $P(t = k|d), \forall k$; for each unseen document or query these are computed by maximizing the log likelihood of (1) with $P(w_j|t = k)$ fixed. Successful Document Retrieval experiments have been reported in [2], for which documents were ranked according to a combination of their cosine similarities with the query in the latent space and in the VSM.

Several other probabilistic models have also been proposed lately, including a hierarchical version of PLSA [7], Latent Dirichlet Allocation [8], multinomial Principal Component Analysis [9], Theme Topic Mixture Model [10], etc. Kernel methods pursuing the same goal have been also proposed [11].

3 Proposed Model

As seen so far, most recent research work have concentrated on novel probabilistic models for document representation. But is the probabilistic framework really necessary at all? In the resolution of a machine learning task, such a probabilistic framework appears necessary in two cases: either some probabilities are involved in the final decision, or probabilities are to be used as a tool for exploring the solution space. The tasks related to Information Access do not necessarily belong to the first case; for example in a Document Retrieval task, what we seek is a ranking of RSV, for which a probabilistic setting is not particularly needed. Regarding the second case, as it has been suggested in [3], while probabilities are a useful tool, they establish constraints, *eg* of normalization or cost function to be minimized, which are not always justified and are difficult to deal with. In addition, in a non-probabilistic framework, a lot of powerful tools allowing different kinds of exploration are available, among which the well-established margin and kernel concepts as well as the stochastic approximation.

The model we propose in this paper is designed to take advantage of the huge amount of unlabeled textual documents, using them as a clue *per se* to the links between words. The basic idea is to train a Neural Network using couples *(word, document)* as inputs and the absence or presence of the word in the document as targets.

A similar approach has been first proposed successfully in the context of statistical language modeling under the name of *Neural Probabilistic Language*

Model (NPLM) [12], which learns a distributed representation for each word alongside with the probability of word sequences in this representation.

Here we adapt the same idea and call our model *Neural Network for Text Representation* (NNTR). As illustrated in Figure 1, there are two input vectors in an NNTR: the first one is a word w_j represented by a one-hot encoding, and the second one is a document d_i represented as a VSM with TFIDF weighting. The output is a score which target is high if w_j is *in the context of* d_i, and low otherwise. As depicted in Figure 1, the word (resp. document) vector is first passed through a Multi-Layer Perceptron MLP_W (resp. MLP_D) that extracts a richer and more distributed representation of words (resp. documents); these two representations are then concatenated and transformed non-linearly in order to obtain the target score using MLP_T, as summarized in (2):

$$\text{NNTR}(w_j, d_i) = \text{MLP}_T\left\{[\text{MLP}_W(w_j), \text{MLP}_D(d_j)]\right\} . \tag{2}$$

All the parameters of the model are trained jointly on a text corpus, assigning high scores to pairs (w_j, d_i) corresponding to documents d_i containing words w_j and low scores for all the other pairs.

A naive criterion would be to maximize the likelihood of the correct class, however, doing so would give the same weight to each seen example. Note that our data presents a huge imbalance between the number of *positive* pairs and the number of *negative* pairs (each document only contains a fraction of words of the vocabulary). Thus, the model would quickly be biased towards answering negatively and would then have difficulties in learning anything else. Another kind of imbalance specific to our data is that, among the positive examples, a few words tend to appear really often, while a lot appear only in few documents, which would bias the model to give lower probabilities to pairs with infrequent words independently of the document.

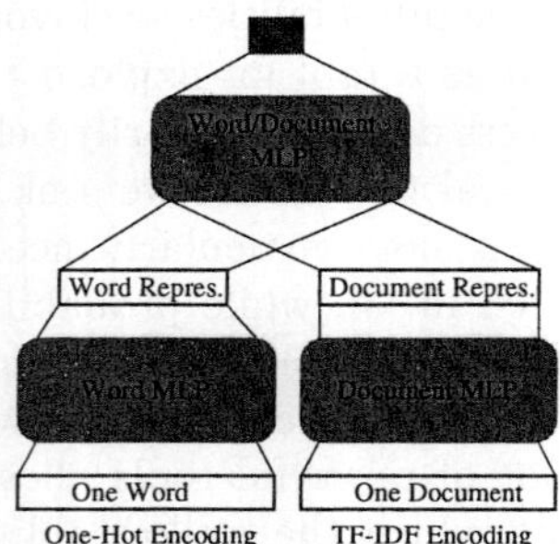

Fig. 1. The NNTR model

Task	TFIDF	PLSA	NNTR
Retrieval	0.170	0.199	**0.215**
Filtering	0.185	0.189	**0.192**

Fig. 2. Compared mean Averaged Precisions (the higher the better) for Document Retrieval and Batch Document Filtering tasks

The approach we thus propose does not try to obtain probabilities but simply unnormalized scores. Thanks to this additional freedom, we can help the optimization process by weighting the training examples in order to balance the total number of positive examples (words w_j that are indeed in documents d_i) with the total number of negative examples. We can also balance each positive example independently of its *document frequency* (the number of documents into which the word appears). The criterion we thus optimize can be expressed as follows:

$$\mathcal{C} = \frac{1}{L^-}\sum_{l=1}^{L^-} Q(x_l, -1) + \frac{1}{M}\sum_{j=1}^{M}\frac{1}{\mathrm{df}_j}\sum_{i=1}^{\mathrm{df}_j} Q((d_i, w_j), 1), \qquad (3)$$

where L^- is the number of negative examples, M the number of words in the dictionary extracted from the training set, df_j the number of documents the word w_j appears in, and $Q(x, y)$ the cost function for an example $x = (d, w)$ and its target y. Note that using this weighting technique we do not need to present the whole negative example set but a sub-sampling of it at each iteration, which makes stochastic gradient descent training much faster. Furthermore, we use a margin-based cost function $Q(x, y) = |1 - y \cdot \mathrm{NNTR(x)}|_+$, as proposed in [13], where $|z|_+ = \max(0, z)$.

4 Experiments

TDT2 is a database of transcripted broadcast news in American English. For this experiment we used 24 823 documents from a manually produced transcription and segmentation, referred to in the following as TDT2-clean. Two sets of 50 queries for documents of TDT2, called TREC-8 and TREC-9 were collected during TREC SDR evaluation. In this **Document Retrieval** classical setting the database documents are available as development data as well as the TREC-8 queries and their corresponding relevance judgements, while the TREC-9 queries are for evaluation. Using TDT2-clean, we trained a PLSA model with 1000 aspects and a NNTR with the following architecture: the word and document sub-MLPs had 25 438 inputs each (corresponding to the size of the training set vocabulary), no hidden unit, and 10 outputs each; the joint word-document MLP_T had 20 inputs, 25 hidden units, and one output unit. Similarly to [2], the relevance of a document d to a query q was computed as $\lambda \cdot RSV_{tfidf}(q, d) + (1 - \lambda) \cdot RSV_{model}(q, d)$, where $RSV_{tfidf}(q, d)$ is a normalized version of the scalar product described in Sect. 2, $RSV_{model}(q, d)$ in the case of PLSA is the cosine similarity and in the case of NNTR the normalized sum, over words w of q, of $\mathrm{NNTR}(w, q)$. All hyper-parameters of the compared models, including λ, were tuned by cross-validation using TREC-8 queries, and we report the mean averaged precision of each model for TREC-9 queries in Figure 2.

Another Document Retrieval setting, described in [14], is the **Batch Filtering** task. In this application, the targeted documents are not immediately available. Thus, we trained our models using a parallel corpus, which we called TDT2-par, of 28 843 documents from other medias covering the same period of news as TDT2-clean. Using TDT2-par, we trained a PLSA model with 500 aspects and a NNTR with the word and document sub-MLPs having 63 736 inputs each, no hidden unit, and 10 outputs each, and the joint word-document MLP_T having 20 inputs, 10 hidden units, and one output unit, with all hyper-parameters tuned using cross-validation over the training set. Note however that in that case there were no data available to tune λ, and it was thus set to 0.5 arbitrarily for both models. Figure 2 reports the results in terms of mean averaged precision for the TREC-9 queries.

5 Conclusion and Discussion

In this paper, we have proposed a novel, non-probabilistic, text representation model, which yields rich internal representations of both words and documents. It has been applied to two text-related tasks, namely **document retrieval** and **batch filtering**. In both cases, the proposed neural network yielded a better mean averaged precision than the well-known PLSA model. Several extensions of NNTR are currently investigated, including representing a full query/document relation instead of a word/document relation, with shared parameters between all word sub-MLPs of the query.

References

1. Sebastiani, F.: Machine learning in automated text categorization. ACM Computing Surveys **34** (2002) 1–47
2. Hofmann, T.: Unsupervised learning by Probabilistic Latent Semantic Analysis. Machine Learning **42** (2001) 177–196
3. LeCun, Y., Huang, F.J.: Loss functions for discriminative training of energy-based models. In: Proc. of AIStats. (2005)
4. Salton, G., Wong, A., Yang, C.: A Vector Space Model for Automatic Indexing. Communication of the ACM **18** (1975)
5. Salton, G., Buckley, C.: Term-weighting approaches in automatic text retrieval. Information Processing and Management **24** (1988) 513–523
6. Deerwester, S.C., Dumais, S.T., Landauer, T.K., Furnas, G.W., Harshman, R.A.: Indexing by Latent Semantic Analysis. Journal of the American Society of Information Science **41** (1990) 391–407
7. Gaussier, E., Goutte, C., Popat, K., Chen, F.: A hierarchical model for clustering and categorising documents. In: Advances in Information Retrieval – Proceedings of the 24th BCS-IRSG European Colloquium on IR Research (ECIR-02), Glasgow. (2002) 229–247
8. Blei, D., Ng, A., Jordan, M.: Latent Dirichlet Allocation. JMLR **3** (2003) 993–1022
9. Buntine, W.: Variational extensions to em and multinomial pca. In: ECML. (2002) 23–34
10. Keller, M., Bengio, S.: Theme topic mixture model: A graphical model for document representation. In: PASCAL Workshop on Learning Methods for Text Understanding and Mining. (2004)
11. Cristianini, N., Shawe-Taylor, J., Lodhi, H.: Latent semantic kernels. J. Intell. Inf. Syst. **18** (2002) 127–152
12. Bengio, Y., Ducharme, R., Vincent, P., Gauvin, C.: A Neural Probabilistic Language Model. JMLR **3** (2003) 1137–1155
13. Collobert, R., Bengio, S.: Links between perceptrons, MLPs and SVMs. In: Proceedings of ICML. (2004)
14. Lewis, D.D.: The trec-4 filtering track. In: TREC. (1995)

A Fuzzy Approach to Some Set Approximation Operations*

Anna Maria Radzikowska

Faculty of Mathematics and Information Science,
Warsaw University of Technology,
Plac Politechniki 1, 00–661 Warsaw, Poland
System Research Institute, Polish Academy of Science,
Newelska 6, 01–447 Warsaw, Poland
annrad@mini.pw.edu.pl

Abstract. In many real–life problems we deal with a set of objects together with their properties. Due to incompleteness and/or imprecision of available data, the true knowledge about subsets of objects can be determined approximately. In this paper we present a fuzzy generalisation of two relation–based operations suitable for set approximations. The first approach is based on relationships between objects and their properties, while the second set approximation operations are based on similarities between objects. Some properties of these operations are presented.

1 Introduction and Motivation

In many applications the available information has the form of a set of objects (examples) and a set of properties of these object. Relationships between objects and their properties can be naturally modelled by a binary relation connecting objects with their properties. For a subset X of objects, which might be viewed as an expert decision, the real knowledge about X is actually twofold: selected elements of the set X itself and properties of these (and other) objects. Since explicit information is usually insufficient to precisely describe objects of X in terms of their properties, some approximation techniques are often applied. The theory of rough sets provides methods for set approximation basing on similarities between objects – relationships among objects determined by properties of these objects (see e.g. [5],[7],[8]).

A more general approach has been recently proposed by Düntsch et al. ([1],[2]). Sets of objects are approximated by means of specific operations based only on object–property relations themselves, without referring to similarities between objects.

Both these approaches were developed under the assumption that the available information, although incomplete, is given in a precise way. However, it

* The work was carried out in the framework of COST Action 274/TARSKI on *Theory and Applications of Relational Structures as Knowledge Instruments* (*www.tarski.org*).

W. Duch et al. (Eds.): ICANN 2005, LNCS 3697, pp. 673–678, 2005.

is often more meaningful to know *to what extent* an object has some property than to know that it has (or does not have) this property. When imprecision of data is admitted, it usually cannot be adequately represented and analysed by means of standard methods based on classical structures. A natural solution seems to be a fuzzy generalisation ([16]) of the respective methods. Hybrid fuzzy–rough approaches were widely discussed in the literature (see, e.g., [3],[6], [10],[11],[12],[15]).

In this paper we present a fuzzy generalisation of set approximation operations proposed in [1] and [2]. An arbitrary continuous triangular norm and its residuum are taken as basic fuzzy logical connectives. Some properties of these operations are presented. We show that in general these operations give a better set approximation than fuzzy rough set–style methods. It will be also shown that under some assumptions both these techniques coincide.

2 Preliminaries

Fuzzy Logical Operations. Fuzzy logical operations are generalisations of classical logical connectives. Triangular norms ([13]), or t–norms for short, are fuzzy generalisations of classical conjunction. Recall that a *t–norm* is any associative and commutative mapping $\otimes : [0,1]^2 \to [0,1]$, non–decreasing in both arguments and satisfying the boundary condition $x \otimes 1 = 1 \otimes x = x$ for every $x \in [0,1]$. Typical examples of t–norms are the Zadeh's triangular norm $x \otimes_Z y = min(x,y)$, and the Łukasiewicz triangular norm $x \otimes_L y = \max(0, x+y-1)$.

A *fuzzy implication* ([4]) is any mapping $\Rightarrow: [0,1]^2 \to [0,1]$, non-increasing in the 1st and non–decreasing in the 2nd argument, and satisfying $1 \Rightarrow 1 = 0 \Rightarrow 0 = 0 \Rightarrow 1 = 0$ and $1 \Rightarrow 0 = 0$. For a continuous t–norm $\otimes$, a *residual implication determined by* $\otimes$ (the *residuum of* $\otimes$) is a fuzzy implication $\otimes\!\!\rightarrow$ defined by: $x \otimes\!\!\rightarrow y = \sup\{z \in [0,1] : x \otimes z \leqslant y\}$ for all $x, y \in [0,1]$. The Gödel implication $x \otimes\!\!\rightarrow_G y = 1$ iff $x \leqslant y$ and $x \otimes\!\!\rightarrow_G y = y$ otherwise, is the residuum of $\otimes_Z$, while the Łukasiewicz implication $x \otimes\!\!\rightarrow_L y = \min(1, 1-x+y)$ is the residuum of $\otimes_L$.

Fuzzy Sets and Fuzzy Relations. Given a non–empty domain U, a *fuzzy set in* U is a mapping $X : U \to [0,1]$. For any $u \in U$, $X(u)$ is the degree to which u belongs to X. The family of all fuzzy sets in U will be written $\mathcal{F}(U)$. For $X, Y \in \mathcal{F}(U)$, we will write $X \subseteq Y$ iff $X(u) \leqslant Y(u)$ for every $u \in U$.

For two non–empty domains U and V, a *fuzzy relation from* U *to* V is a fuzzy set in $U \times V$, i.e. this is any mapping $R : U \times V \to [0,1]$. For any $u \in U$ and for any $v \in V$, $R(u,v)$ is the degree to which u is R–related with v. The family of all fuzzy relations from U to V will be denoted by $\mathcal{R}(U,V)$. For $R \in \mathcal{R}(U,V)$, the *converse* fuzzy relation $R^{\smile} \in \mathcal{R}(V,U)$ is given by $R^{\smile}(v,u) = R(u,v)$ for all $u \in U$ and $v \in V$. For $u \in U$ (resp. $v \in V$) we write uR (resp. Rv) to denote the fuzzy sets in V (resp. in U) given by: $(uR)(z) = R(u,z)$, $z \in V$ (resp. $(Rv)(z) = R(z,v)$, $z \in U$). If $U = V$, then R is called a fuzzy relation on U. A fuzzy relation R on U is *reflexive* iff for any $u \in U$, $R(u,u) = 1$, *symmetric* iff $R(u,v) = R(v,u)$ for all $u, v \in U$, and $\otimes$*–transitive*, where $\otimes$ is a t–norm, iff $R(u,v) \otimes R(v,z) \leqslant R(u,z)$

for all $u, v, z \in U$. A fuzzy relation R is called a $\otimes$*–equivalence* relation iff it is reflexive, symmetric and $\otimes$–transitive.

Recall ([9]) two specific fuzzy relations on $\mathcal{F}(U)$. Let $\otimes$ be a t–norm and let $\Rightarrow$ be a fuzzy implication. For two fuzzy sets $X, Y \in \mathcal{F}(U)$ a *fuzzy inclusion* $inc_{\Rightarrow}$ and a *fuzzy compatibility* $com_{\otimes}$ are defined by: $inc_{\Rightarrow}(X,Y) = \inf_{u\in U}(X(u) \Rightarrow Y(u))$ and $com_{\otimes}(X,Y) = \sup_{u\in U}(X(u) \otimes Y(u))$, respectively. Observe that $inc_{\Rightarrow}(X,Y)$ is the degree to which X is included in Y, whereas $com_{\otimes}(X,Y)$ is the degree to which X and Y overlap. For a residuum $\overset{\otimes}{\rightarrow}$ of a continuous t–norm $\otimes$, a fuzzy inclusion will be written $inc_{\otimes}$.

3 Approximation Operations

Let U be a non–empty set of *objects*, V be a non–empty set of their *properties*, and let $R : U \times V \rightarrow [0,1]$ be a fuzzy relation representing relationships between objects and their properties. For any $u \in U$ and for any $v \in V$, $R(u,v)$ is the degree to which the object u has the property v. Moreover, let a continuous t–norm $\otimes$ be given. A system $\Sigma = (U, V, R, \otimes)$ is called a *fuzzy information structure.*

Given a fuzzy information structure $\Sigma = (U, V, R, \otimes)$, let us define two operations from $\mathcal{F}(V)$ to $\mathcal{F}(U)$ as follows: for any $P \in \mathcal{F}(V)$ and for any $u \in U$,

$$[R]_{\Sigma} P(u) = inc_{\otimes}(uR, P) \tag{1}$$

$$\langle R\rangle_{\Sigma} P(u) = com_{\otimes}(uR, P). \tag{2}$$

Following the terminology from modal logics the operations (1) and (2) are called the *necessity* and the *possibility* operator, respectively. $[R]_{\Sigma} P(u)$ is the degree to which all properties characterizing the object u are in P, whereas $\langle R\rangle_{\Sigma} P(u)$ is the degree to which some property from P characterizes u. Note that for any $X \in \mathcal{F}(U)$ and for any property $v \in V$, $[R^{\smile}]_{\Sigma} X(v) = inc_{\otimes}(Rv, X)$ is the degree to which all objects characterized by v are in X. Similarly, $\langle R^{\smile}\rangle_{\Sigma} X(v) = com_{\otimes}(Rv, X)$ is the degree to which X contains an object which has the property v.

Basing on the operations (1) and (2), let us define now the following two fuzzy operations $\Delta_{\Sigma}, \nabla_{\Sigma} : \mathcal{F}(U) \rightarrow \mathcal{F}(U)$ as follows: for any $X \in \mathcal{F}(U)$,

$$\Delta_{\Sigma}(X) = \langle R\rangle_{\Sigma} [R^{\smile}]_{\Sigma} X \tag{3}$$

$$\nabla_{\Sigma}(X) = [R]_{\Sigma} \langle R^{\smile}\rangle_{\Sigma} X. \tag{4}$$

Note that for every object $u \in U$, $\Delta_{\Sigma}(X)(u) = \sup_{v\in V}(R(u,v) \otimes inc_{\otimes}(Rv, X))$. Intuitively, this is the degree to which some property of u characterizes only (some) objects from X. Also, $\nabla_{\Sigma}(X)(u) = \inf_{v\in V}(R(u,v) \overset{\otimes}{\rightarrow} com_{\otimes}(Rv, X))$, for any $u \in U$. Intuitively, this is the degree to which every property of the object u characterizes some object from X.

The operators (3) and (4) have the following properties.

Proposition 1. *For every $\Sigma = (U, V, R, \otimes)$ and for all $X, Y \in \mathcal{F}(U)$,*

(i) *Approximation property:* $\Delta_{\Sigma}(X) \subseteq X \subseteq \nabla_{\Sigma}(X)$

(ii) *Monotonicity:* $X \subseteq Y$ *implies* $\Delta_{\Sigma}(X) \subseteq \Delta_{\Sigma}(Y)$ *and* $\nabla_{\Sigma}(X) \subseteq \nabla_{\Sigma}(Y)$. ■

Due to the approximation property, the operations (3) and (4) are useful for set approximations. These operations are respectively called a *fuzzy lower* and *fuzzy upper bound of* X *in* Σ. Consequently, $\nabla_\Sigma X(u)$ (resp. $\Delta_\Sigma X(u)$) might be interpreted as the degrees to which u *certainly* (resp. *possibly*) belongs to X. Note also that monotonicity is the natural property of approximation operations: the larger the approximated set is, the larger its lower (upper) bound should be.

Example 1 (Knowledge assessment) Let U be a set of *problems*, V be a set of *skills* (*abilities*) necessary to solve these problems, and let $R \in \mathcal{R}(U,V)$ be a fuzzy relation from U to V such that for any problem $u \in U$ and for any $v \in V$, $R(u,v)$ represents the degree to which the skill v is necessary to solve the problem u. Let $T \in \mathcal{F}(U)$ be a result of some test an agent solved. This set reflects a state of the agent's knowledge. Applying the operators Δ_Σ and ∇_Σ to T, we get the following approximation of the agent's knowledge: $\Delta_\Sigma(T) \subseteq T \subseteq \nabla_\Sigma(T)$. Intuitively, for any problem $p \in U$, $\Delta_\Sigma T(p)$ (resp. $\nabla_\Sigma T(p)$) is the degree to which the agent is *certainly* (resp. *possibly*) capable to solve p, i.e. his true state of knowledge allows to solve p. Basing on characterizations of particular problems and the result of a test the agent solved, we have the assessment of the true knowledge of the agent.

Table 1.

R	*EN*	*CP*	*DB*	T	$\Delta_\Sigma(T)$	$\nabla_\Sigma(T)$
p_1	0.9	0.1	0.4	0.9	0.8	1.0
p_2	1.0	0.2	0.4	1.0	0.9	1.0
p_3	0.2	1.0	0.1	0.2	0.2	0.2
p_4	0.1	0.8	0.2	0.3	0.0	0.4
p_5	0.3	0.5	0.7	0.6	0.5	0.7
p_6	0.6	0.3	0.7	0.5	0.5	0.7

Consider, for example, a structure $\Sigma=(U,V,R,\otimes_L)$, where $U=\{p_1,\ldots,p_6\}$ and V consists of three abilities necessary to solve these problems: *EN* (English speaking), *CP* (Computer Programming), and *DB* (expertise in Data Bases). In Table 1 a fuzzy relation R is given, together with the test result T. Using the Łukasiewicz triangular norm $\otimes_L$ and its residuum $\otimes\!\!\rightarrow_L$, by simple calculations we get the lower $\Delta_\Sigma(T)$ and the upper $\nabla_\Sigma(T)$ bound of T (see Table 1). Note that the problem p_4 was solved to the degree 0.3, whereas the agent knowledge wrt this problem was assessed between 0.0 and 0.4. Therefore, the agent could obtain a weaker evaluation for his solution of p_4. Also, his solution of p_6 was evaluated to 0.5, which coincides with the (fuzzy) lower approximation of the test. Since the upper approximation for p_6 is 0.7, the agent's real knowledge allows him to get a better result for the solution of p_6. □

4 Similarity–Based Set Approximations

In this section we assume that $U=V$. Then we actually deal with a set U of objects and a fuzzy relation R on U representing relationships among objects.

This relation, called a *similarity* relation, is usually determined by properties of objects. As before, let $\otimes$ be a continuous t–norm. A system $\Sigma=(U,R,\otimes)$ is called a *fuzzy approximation space*. Clearly, this is a specific case of fuzzy information structure.

In rough set theory ([7],[8],[5]) similarity relations (originally referred to as indiscernibility relations) are assumed to be equivalence relations (i.e. reflexive, symmetric and transitive), but in recent research ([14]) only reflexivity and symmetry are required properties of these relations.

In fuzzy rough set theory ([3],[10],[11],[12],[15]) any set $X \in \mathcal{F}(U)$ of objects is approximated by means of similarity classes of the relation R: for any $X \in \mathcal{F}(U)$, the *fuzzy lower* and the *fuzzy upper rough approximation of* X *in* $\Sigma=(U,R,\otimes)$ are respectively defined by: $\underline{\Sigma}(X)(u)=[R]_{\Sigma}X$ and $\overline{\Sigma}(X)(u)=\langle R\rangle_{\Sigma}X$. For any $X \in \mathcal{F}(U)$ and for any $u \in U$, $\underline{\Sigma}(X)(u)$ (resp. $\overline{\Sigma}(X)(u)$) is the degree to which all (resp. some) elements similar to u are in X.

The following proposition presents some connections between fuzzy rough approximation operations and the operations (3)–(4).

Proposition 2. *For every* $\Sigma=(U,R,\otimes)$ *and for every* $X \in \mathcal{F}(U)$,

(i) *if* R *is reflexive, then* $\underline{\Sigma}(X) \subseteq \Delta_{\Sigma}(X) \subseteq X \subseteq \nabla_{\Sigma}(X) \subseteq \overline{\Sigma}(X)$

(ii) *if* R *is a* $\otimes$*–equivalence relation, then* $\Delta_{\Sigma}(X)=\underline{\Sigma}(X)$ *and* $\nabla_{\Sigma}(X)=\overline{\Sigma}(X)$. ■

The above proposition says that only for (at least) reflexive fuzzy relation R we get a sensible approximation of any set $X \in \mathcal{F}(U)$ of objects. Also, in view of Proposition 1(**i**), the operations ∇_{Σ} and Δ_{Σ} give an approximation of X for an arbitrary fuzzy relation, regardless of its properties. Moreover, the pair $(\Delta_{\Sigma},\nabla_{\Sigma})$ of fuzzy operators gives a tighter set approximation than the pair $(\underline{\Sigma},\overline{\Sigma})$. However, from **(iii)** it follows that for a $\otimes$–equivalence relation, the operators ∇_{Σ} and $\overline{\Sigma}$, as well as Δ_{Σ} and $\underline{\Sigma}$, coincide.

5 Conclusions

In this paper we have presented a fuzzy generalisation of two relation–based approximation operations. We have taken an arbitrary continuous t–norm and its residuum as fuzzy generalisation of classical conjunction and implication. The first approach allows for (fuzzy) set approximation on the basis of a heterogeneous fuzzy relation connecting objects and their properties. The second approach uses similarities of objects, so it is based on homogeneous binary fuzzy relations among objects. Some basic properties of these operations have been shown. In particular, it has turned out that the operations Δ_{Σ} and ∇_{Σ} give tighter set approximations that fuzzy–rough approximation operations. Moreover, these operations do not depend on properties of relations, whereas in the fuzzy–rough approach similarity relations should be at least reflexive in order to obtain a sensible set approximation. Finally, basic properties of the operations Δ_{Σ} and ∇_{Σ} are straightforward generalisations of the properties of the respective operations based on classical structures.

References

1. Düntsch I. and Gediga G. (2002), Approximation operators in qualitative data analysis, in *Theory and Application of Relational Structures as Knowledge Instruments*, de Swart H., Orłowska E., Schmidt G., Roubens M. (eds), Lecture Notes in Computer Science 2929, Springer–Verlag, 214–230.
2. Düntsch I. and Gediga G. (2002), Modal–like operators in qualitative data analysis, in Proceedings of the 2nd IEEE International Conference on Data Mining ICDM-2002, 155–162.
3. Dubois D., Prade H. (1990), Rough Fuzzy Sets and Fuzzy Rough Sets, Int. J. of General Systems 17(2–3), 191–209.
4. Klir G. J., Yuan B. (1995), *Fuzzy Logic: Theory and Applications*, Prentice–Hall, Englewood Cliffs, NJ.
5. Orłowska E. (ed) (1998), *Incomplete Information: Rough Set Analysis*, Physica–Verlag.
6. Pal S. K., Skowron A. (1999), Rough Fuzzy Hybridization: A New Trend in Decision Making, Springer–Verlag.
7. Pawlak Z. (1982), Rough sets, Int. Journal of Computer and Information Science 11(5), 341–356.
8. Pawlak Z. (1991), *Rough Sets – Theoretical Aspects of Reasoning about Data*, Kluwer Academic Publishers.
9. Radzikowska A. M., Kerre E. E. (2002), A fuzzy generalisation of information relations, in *Beyond Two: Theory and Applications of Multiple–Valued Logics*, Orłowska E. and Fitting M. (eds), Springer–Velag, 287–312.
10. Radzikowska A. M., Kerre E. E. (2002), A comparative study of fuzzy rough sets, Fuzzy Sets and Systems 126, 137–155.
11. Radzikowska A. M., Kerre E. E. (2004), On L–valued fuzzy rough sets, Artificial Intelligence and Soft Computing – ICAISC 2004, Rutkowski L., Siekmann J., Tadeusiewicz R., Zadeh L. A. (eds), Lecture Notes in Computer Science 3070,Springer–Verlag, 526–531.
12. Radzikowska A. M., Kerre E. E. (2003), Fuzzy rough sets based on residuated lattices, in *Transactions on Rough Sets II*, Peters J.F., Skowron A., Dubois D., Grzymała-Busse J., Inuiguchi M., Polkowski L. (eds), Lecture Notes in Computer Science 3135, Springer–Verlag, 278–297.
13. Schweizer B. and Sklar A. (1983), *Probabilistic Metric Spaces*, North Holland, Amsterdam.
14. Słowiński R. and Vanderpooten D. (1997), Similarity relation as a basis for rough approximations, in *Advances in Machine Intelligence and Soft Computing*, vol.4, Duke University Press, Durham, NC, 17–33.
15. Thiele H. (1993), On the definition of modal operators in fuzzy logic, Proceedings of ISMVL'93, 62–67.
16. Zadeh L. A. (1965), Fuzzy Sets, Information and Control 8, 338–358.

Connectionist Modeling of Linguistic Quantifiers

Rohana K. Rajapakse[1], Angelo Cangelosi[1], Kenny R. Coventry[2], Steve Newstead[2], and Alison Bacon[2]

[1] Adaptive Behaviour & Cognition Group, School of Computing Comms and Electronics,
[2] Centre for Thinking & Language, School of Psychology,
University of Plymouth, Drake Circus,
PL4 8AA, Plymouth, UK
{rrajaakse, acangelosi kcoventry, snewstead, abacon}@plymouth.ac.uk
http://www.tech.plym.ac.uk/soc/research/ABC

Abstract. This paper presents a new connectionist model of the grounding of linguistic quantifiers in perception that takes into consideration the contextual factors affecting the use of vague quantifiers. A preliminary validation of the model is presented through the training and testing of the model with experimental data on the rating of quantifiers. The model is able to perform the "psychological" counting of objects (fish) in visual scenes and to select the quantifier that best describes the scene, as in psychological experiments.

1 Introduction

The selection and use of vague linguistic quantifiers, such as *a few, few, several, many, lots of* is greatly influenced by the communicative aim of the speaker. For example, in the two sentences "A few people went to the cinema. They liked the movie" and "Few people went to the cinema. They preferred the restaurant", the selection of the quantifier *a few* vs. *few* indicates differences in the focus of attention signaled by the quantifier. *A few* is chosen to put emphasis on those that actually went to the cinema; *Few*, instead, shifts the attention to the people that did not go. Understanding the meaning of such terms is important as they are among the set of closed class terms which are generally regarded as having the role of acting as organizing structure for further conceptual material. Although some researchers have proposed that quantifiers can be mapped directly to numbers on a scale (e.g. [3]), there is compelling evidence that the comprehension and production of quantifiers can be affected by a range of factors which go beyond the number of objects present. These include contextual factors (e.g. the relative size of the objects involved in the scene, the expected frequency of those objects based on prior experience, the functionality present in the scene - e.g. [15]) and communicative factors (e.g. to control the pattern of inference as in the example above - see [14]).

The existence of these contextual and communicative effects highlights the fact that language cannot be treated as an abstract, self-referential symbolic system. On the contrary, the understanding of language strongly depends on its grounding in the individual's interaction with the world. The importance of grounding language in perception, action and cognition has recently received substantial support both from

W. Duch et al. (Eds.): ICANN 2005, LNCS 3697, pp. 679 – 684, 2005.

cognitive psychology experiments (e.g. [2,9,11]) and computational models (e.g. [4,12]). In particular, connectionist systems are being increasingly used as the basis for modeling grounding [6]. They permit a straightforward way to link perceptual stimuli to internal categorical representations, upon which semantic and linguistic representations are anchored. The use of connectionist components within embodied cognitive systems, such as agents and robots, also permit the link between language and sensorimotor stimuli.

Various connectionist models of quantification and number learning have been proposed, although none has directly focuses on linguistic quantifiers. Two main directions of research can be identified in the neural network literature. A first set of models focuses on learning number sequences. They are able to process the objects in the input scene sequentially to reproduce the sequences and/or compute distances between two numbers. For example, Rodriguez et al. [18] modeled the learning of sequences of letters. The identification of the correct number of presentations of a first letter permits the prediction of the presentation of a second letter. Ma and Hirai [13] simulated the production of the number word sequence as observed from children.

The second approach includes models that learn to identify the number of objects in the input visual scene. Dehaene & Changeux [10] developed a numerosity detection system comprised of three modules: an input retina, an intermediate topological map of object locations, and a map of numerosity detectors). This was able to replicate the distance effect in counting, by which performance improves with increasing numerical distance between two discriminated quantities. Peterson and Simon [16] presented a connectionist model of subitizing, the phenomenon by which subjects appear to produce immediate quantification judgments (normally up to 4 objects) without the need to do sequential counting. The simulation results suggested that subitizing emerges through experience, rather than being the result of a limited representational capacity of the architecture. Similar results were found in the model by Ahmad et al. [1], which uses Kohonen's SOM networks. This model also used a recurrent backpropagation network for articulating the numerosity of individual objects in the collection and a static backpropagation network for the next object pointing task. This model is distinct in the way it is trained to count by decomposing the counting task into that of number word update\storage from the next-object pointing task.

In this paper we present a new connectionist model of the grounding of linguistic quantifiers in perception that takes into consideration the contextual factors affecting the use of vague quantifiers. The model is able to perform both the "psychological" counting of objects (fish) in visual scenes and to select the quantifier that best describes the scene. A preliminary validation of the model will be presented through the training and testing of the model with experimental data on the rating of quantifiers.

2 Architecture of the Model

The computational model consists of a hybrid artificial vision-connectionist architecture (Figure 1). The model has four main modules: (1) Vision Module, (2) Compression Networks, (3) Quantification Network, and (4) Dual-Route Network. This architecture is partially based on a previous model on the grounding of spatial language [5,8]. The overall idea is to ground the connectionist and linguistic representation of quantification judgments directly in input visual stimuli.

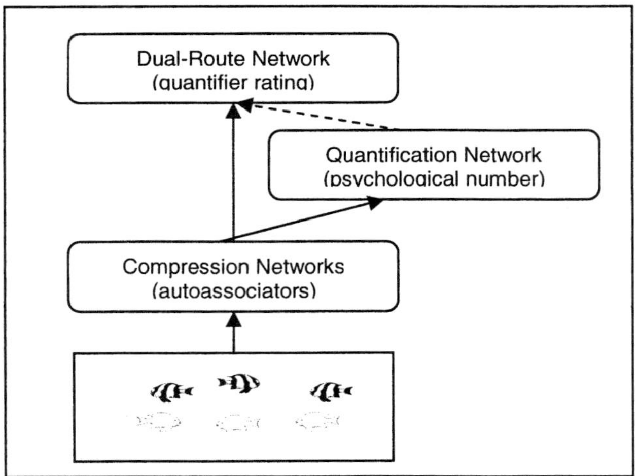

Fig. 1. Modular architecture of the model

The vision module processes visual scenes involving varying quantities of two types of objects: striped fish and white fish. It uses a series of Ullman-type vision routines to identify the constituent objects in the scene. The input to the Vision module consists of static images with the two kinds of fish. The system must pay attention to striped fish, whilst white fish are only used as distracters (or vice versa). The input images are processed at a variety of spatial scales and resolutions for object features yielding a visual buffer. The processing of each image results in two retinotopically organized arrays of 30x40 activations (one per fish type). The output of the vision module represents data of isotropic receptive fields.

The Compression Networks is needed to convert the output data from the vision module into compressed neural representation of the input scene. This is to reduce the complexity of the vision module output. Two separate auto-associative networks are used, respectively for each of the object types in the scene (stripy fish and white fish). Both networks have 1200 input and output units, and 30 hidden units. The activation values of the hidden units will be utilized by the following networks to make quantification and linguistic judgments. The compression network for each type of fish will learn to autoassociate all the stimili with varying number of fish.

The Quantification Network is a feedforward multi-layer perceptron trained to reproduce the quantification judgments of the number of fish made by subjects during experiments on psychological counting [7]. Previous simulations only focused on the counting of the striped fish, those the subjects are asked to consider when making quantification decisions [17]. In the current, updated version, the same network has to count both sets of fish. The network has 60 input units (30 per compressed fish type), 50 hidden nodes, and 2 output nodes. Each output node has a modified activation function that produce activation values in the range 0 to 20, to include the actual range of 0 to 18 stripy fish used in the stimulus set.

The fourth module consists of a Dual-Route neural network. This architecture combines visual and linguistic information for both linguistic production and comprehension tasks [6]. This is the core linguistic component of the model, as it integrates visual and linguistic knowledge to produce a description of the visual scene. The network receives, in input, information on the scene through the activation values of the compression (and quantification) networks' hidden units. It will then produce, in output, judgments regarding the appropriate ratings for the quantifier terms describing the visual scene. The activation values of the linguistic output nodes correspond to rating values given by subjects for the five quantifiers considered: a few, few, several, many, lots.

In the current version of the model, only the hidden representations of the compression networks are used. The dual-route network will require 60 input visual units and 5 input linguistic nodes, one for each quantifier. The linguistic units correspond to the 5 vague quantifiers *a few*, *few*, *several*, *many* and *lots*. The 60 visual nodes corresponded to the 30 hidden units of the two compression networks (of stripy fish and white fish). The output layer has the same number and type of units as those in the input layer. After training with data from psycholinguistic experiments, the network will be capable of producing two different outputs: (1) acceptability ratings for quantifiers given only the vision inputs (language production) and (2) imaginary output pictures, given only a description of the scene in terms of quantifiers (comprehension). Results of the simulation on the production route (predicted ratings for the quantifiers) will be compared to the actual ratings of experiments with human subjects.

3 Simulation Results

The model uses as input stimuli to the vision module 216 scenes used in quantification experiments with subjects [7]. In the experimental design, the number of fish (of both types) is varied from zero to 18 fish per scene, with incremental steps of 3. The fish are arranged in random locations, with equal spacing between them (two levels of inter-fish distances are used). In addition, two levels of grouping of fish from the same type (grouped or mixed) are used, with another factor regarding the two levels of the position of the grouped stimuli (top or bottom of the image).

The 216 scenes are first presented to the vision module. Its output is then used to train the autoassociative networks of the Compression module. For the training, 195 scenes are used as training stimuli and 21 as generalization test stimuli. The learning rate is 0.01 and momentum 0.8. The networks are trained for 2000 epochs. The autoassociative network is able to learn both training stimuli (average RMS error of 0.019 and 0.014 for stripy and fish data respectively) and novel generalisation stimuli (average RMS error of 0.080 and 0.070 for stripy and white fish data). This permits a significant reduction of complexity of the 1200 output values of the visual module into only 30 compressed hidden activation values.

Results with the Quantification network [17] have shown that networks are able to reproduce the production of "psychological numbers" produced by subjects [7] and that they use similar mechanisms in producing such judgements.

New simulations have focused on the training of the dual route network. The same stimulus set is used to collect data on the use of the five vague quantifiers *a few*, *few*, *several*, *many* and *lots*. In this psycholinguistic experiment, subjects are asked to rate the use of vague quantifiers by using a 9-point Likert scale for the appropriateness of sentences like "There are a few stripy fish". The average rating data of the subjects are converted into presentation frequencies for the training of the dual route network, as in Cangelosi et al. [6]. For the training, 195 scenes are used as training stimuli and 21 as generalization test stimuli. The learning rate is 0.001 and momentum 0.8. The networks are trained for 1000 epochs. The autoassociative network is able to learn both training stimuli (average RMS error of 0.051) and novel generalisation stimuli (average RMS error of 0.084).

Simulation No.	Learning rate	Random novel set	Hidden nodes	Training error	Novel error
1	0.001	TrnTst1	30	0.055079	0.086295
2	0.001	TrnTst2	30	0.055002	0.070183
3	0.001	TrnTst3	30	0.043332	0.095066
			average	0.051138	0.083848

4 Conclusions

This paper reports some preliminary simulation results on a new artificial vision/connectionist model of the grounding of linguistic quantifiers in perception. Experimental data are used for the training and testing of the dual-route neural network that selects the linguistic quantifiers that best describe the scene.

Future simulation experiments will focus on the analyses of the specific effects that the various contextual factors manipulated in the experiment (e.g. number, spacing and grouping of fish) produce in the selection of vague quantifiers. In addition, future simulations will address the contribution of explicit numerical judgments, as in quantification experiments [7], in the use of linguistic quantifiers.

The model presented here proposes a new approach to the study of linguistic quantifiers, by grounding quantification judgments and vague quantifier directly in perception. From the semantic point of view, these terms have the virtue of relating in some way to visual scenes being described. Hence, it will be possible to offer more precise semantic definitions of these, as opposed to many other expressions, because the definitions can be grounded in perceptual representations.

Acknowledgements

This research was supported by the UK Engineering and Physical Research Sciences Council (EPSRC Grant GR/S26569).

References

1. Ahmad, K., Casey, M.C., Bale, T: (2002). Connectionist Simulation of Quantification Skills. Connection Science , 14 (2002), 1739-1754
2. Barsalou, L. W.: Perceptual symbol systems. Behavioral and Brain Sciences, 22 (1999), 577-660.
3. Bass, B.M, Cascio, W.F., O'Connor, E.J.: Magnitude estimations of expressions of frequency and amount. Journal of Applied Psychology, 59 (1974), 313-320.
4. Cangelosi, A., Bugmann, G., Borisyuk, R. (eds.): Modeling Language, Cognition and Action: Proceedings of the 9th Neural Computation and Psychology Workshop. World Scientific, Singapore (2005)
5. Cangelosi, A., Coventry, K.R., Rajapakse, R., Bacon, A., Newstead S.N.: Grounding language into perception: A connectionist model of spatial terms and vague quantifiers. In: [4].
6. Cangelosi, A., Greco, A., Harnad S.: From robotic toil to symbolic theft: Grounding transfer from entry-level to higher-level categories. Connection Science, 12 (2000) 143-162
7. Coventry, K.R., Cangelosi, A., Newstead, S.N., Bacon, A., Rajapakse R.: Vague quantifiers and visual attention: Grounding number in perception. XXVII Annual Meeting of the Cognitive Science Society (2005)
8. Coventry, K. R., Cangelosi, A., Rajapakse, R., Bacon, A., Newstead, S., Joyce, D., Richards, L. V:. Spatial prepositions and vague quantifiers: Implementing the functional geometric framework. In C. Freksa et al. (eds.), Spatial Cognition, Volume IV. Reasoning, action and interaction, pp 98-110. Lecture notes in Computer Science. Springer (2005)
9. Coventry, K.R., Garrod, S.C.: Saying, Seeing and Acting. The Psychological Semantics of Spatial Prepositions. Essays in Cognitive Psychology Series. Psychology Press. Hove and New York (2004)
10. Dehaene, S., Changeux, J.P.: Development of Elementary Numerical Abilities: A Neuronal Model. Journal of Cognitive Neuroscience, 5 (1993), 390-407
11. Glenberg, A.M., Kaschak, M.: Grounding language in action. Psychonomic Bulletin and Review, 9 (2002) 558-565
12. Harnad S. (1990). The symbol grounding problem. Physica D, 42: 335-346
13. Ma, Q., Hirai, Y.: Modeling the Acquisition of Counting with an Associative Network. Biological Cybernetics, 61 (1989) 271-278
14. Moxey, L.M., Sanford, A.J.: Communicating Quantities. A Psychological Perspective. Lawrence Erlbaum Associates; Hove, East Sussex (1993)
15. Newstead, S.N., Coventry, K.R.: The role of expectancy and functionality in the interpretation of quantifiers. European Journal of Cognitive Psychology, 12(2) (2000), 243–259
16. Peterson, S.A., Simon, T.J.: Computational Evidence for the Subitizing Phenomenon as an Emergent Property of the Human Cognitive Architecture. Cognitive Science, 24 (2000) 93-122
17. Rajapakse, R.K., Cangelosi, A., Coventry, K., Newstead, S., Bacon A.: Grounding linguistic quantifiers in perception: Experiments on numerosity judgments. 2nd Language & Technology Conference: Human Language Technologies as a Challenge for Computer Science and Linguistics. April 21-23, 2005, Poznañ, Poland (submitted).
18. Rodriguez, P., Wiles, J., Elman, J.L.: A Recurrent Neural Network that Learns to Count. Connection Science, 11 (1999), 5-40

Fuzzy Rule Extraction Using Recombined RecBF for Very-Imbalanced Datasets

Vicenç Soler, Jordi Roig, and Marta Prim

Dept. Microelectronics and Electronic Systems, Edifici Q, Campus UAB, 08193 Bellaterra, Spain
{Vicenc.Soler, Jordi.Roig, Marta.Prim}@uab.es

Abstract. An introduction to how to use RecBF to work with very-imbalanced datasets is described. In this paper, given a very-imbalanced dataset obtained from medicine, a set of Membership Functions (MF) and Fuzzy Rules are extracted. The core of this method is a recombination of the Membership Functions given by the RecBF algorithm which provides a better generalization than the original one. The results thus obtained can be interpreted as sets of low number of rules and MF.

1 Introduction

Classical learning methods are always based on balanced datasets, where the data of different classes are divided proportionally. Neural Networks, Classical Extraction of Rules, etc. are some examples.

However, the very imbalanced datasets are present in many real-world domains, for example, in diagnosis of diseases, text classification, etc. The solution consists on methods specialized in imbalanced datasets [1][2]. These methods work assuming that the dataset is imbalanced and either they try to balance the dataset or they try to adapt/build a new method.

In this paper, we will work with a very-imbalanced dataset. Very-imbalanced datasets are datasets with very low proportion of data (i.e., for a 2-class system, 1000 instances of class-1 and 20 instances of class-2).

This dataset belongs to the problem of Down syndrome. Down syndrome detection during the second trimester gestation is a difficult problem to solve because the data is highly unbalanced. Current methods are statistically based and are only successful in a maximum of 70% of the cases [3].

To achieve the goal of this paper, Recombined RecBF and Genetic Algorithms will be used.

RecBF network [4] is a variation of RBF networks, in which the representation is a set of hyper-rectangles belonging to the different classes of the system. Every dimension of the hyper-rectangles represents a membership function. Finally, a network is built, representing, on every neuron, the Membership Function (MF) found.

Genetic Algorithms (GA) are methods based on principles of natural selection and evolution for global searches. Given a problem, GAs run repeatedly by using the three

W. Duch et al. (Eds.): ICANN 2005, LNCS 3697, pp. 685–690, 2005.

fundamentals operators: reproduction, crossover and mutation. These operators, combined randomly, are based on a fitness function evolution to find a better solution in the searching space. Chromosomes represent the individuals of the GA and a chromosome is divided in genes. GAs are used to find solutions to problems with a large set of possible solutions and they have the advantage of only requiring information concerning the quality of the solution. This fact makes GA a very good method to solve complex problems, [5][6][7][8].

This paper is divided in four parts. In the first one, the use of the RecBF and the method used to recombine are explained. In the second one, the GA is described. In the third part, the results will be shown and, in the last part, the conclusions of this paper are described.

2 Part I: MFs from RecBF

RecBFN is a good method to obtain MFs. In this paper, we only use RecBF as a method to obtain MFs, without its neural network. The hyper-rectangles defining the MF are created by means of splitting a bigger one (shrinking). This splitting can be either done in all dimensions or only in one (which looses less volume in its splitting). In this article, the latter is used, to avoid granulation of the MFs of the minor-class and reduce the number of total MFs. This will make the problem simpler and more understandable.

In this case, MFs and the rules are generated according to the given dataset, although this dataset is very-imbalanced. That is, some rules/MFs are specialized for these few cases, as happen in classical learning methods.

To solve this problem, we introduce in this paper a method to combine these MFs to create new ones.

This method consists on applying to the maximum variability for our rules/MF, i.e similar competition between the patterns to match the rules, without taking into account which class they belong to.

A Fuzzy Rule is a set of MFs, each belonging to a variable. The value of every rule *i* is represented by the equation (1):

$$\mu_i(x) = \prod_j \mu_{MFij}(x) \tag{1}$$

If the result of this formula for every pattern is >0, then any pattern can introduce variability to (1).

Another factor which plays an important role in the final fuzzy calculation (1) is the area covered by every MF, that is, how wide is the area where $\mu_{MFij}(x) = 1$ and where $0 < \mu_{MFij}(x) < 1$. Where $\mu_{MFij}(x)$ is 1, this area marks the alpha-cut as 1 (maximum area for the centroid defuzzyfication method). If this area is $0 < \mu_{MFij}(x) < 1$, marks the minimum of all the alpha-cuts. So, always a fuzzy rule is calculated by the minimum alpha-cut.

The solution is to use maximum variability for the MFs corresponding to the very-imbalanced classes. The RecBF returns the MFs as trapezoids, and classifies them by classes.

The method is described in the following 4 steps. From now on, we will use the term major-class to refer to the class with the most quantity of patterns and the term minor-class to refer to the class with the lowest quantity of patterns.

1. Train the RecBF with ordered patterns, first major-class and then minor-class. The results are much better with ordered patterns than with non-ordered patterns. The results are MFs.
2. Take only the MFs belonging to the major-class which include, at least, the 10% of patterns. Thus, we avoid having specialized cases and exceptions to general rules [9][10].
3. Reorder and transform the MFs of major-class into new trapezoids, and those of minor-class into triangles, just calculating the centre as the average of the points *b* and *c*. (Figure-1). Every MF has to have the points *a* and *d* (Figure-2) as the minimum and maximum values of its variable.
4. Apply a Genetic Algorithm to find the Fuzzy Rules.

Since the shrinking method in RecBF algorithm is only in one dimension (to avoid granulation of MFs), superposed MFs are given as result of RecBF algorithm (Figure-2). If a set of rules is tried to be obtained from them, much noise is introduced into the system, because of the system has much more probability to not obtain good rules which match with test patterns. In this case, MFs are transformed, splitting them by the b and c points of every MF. Figure-3 shows the result obtained from the original (Figure2).

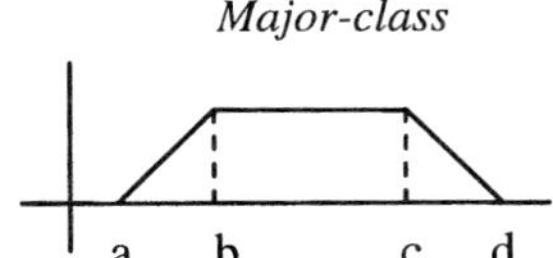

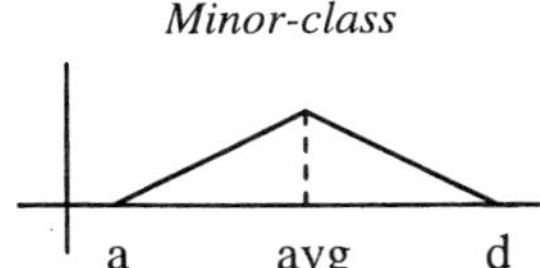

Fig. 1.

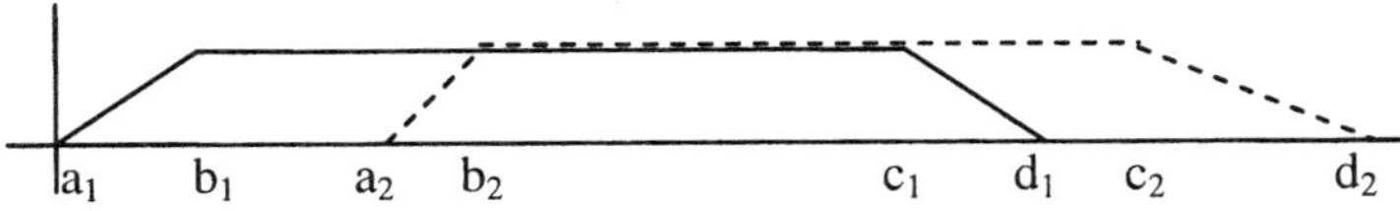

Fig. 2.

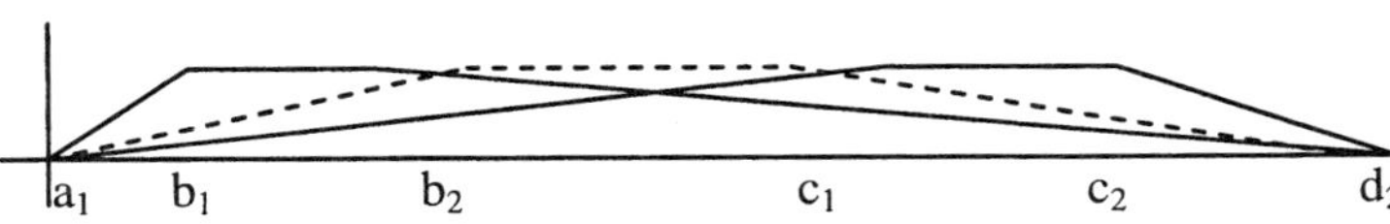

Fig. 3.

3 Part II: The GA

The codification of one chromosome of our GA is expressed in the following lines:

$$(x_{1,1}, \ldots, x_{1,n}, x_{2,1}, \ldots, x_{2,n}, x_{m,1}, \ldots, x_{m,n})$$

where n is the number of variables (input variables plus output variables) and m is the number of rules. $x_{i,j}$ is the value a gene can take, which is an integer value compressed in the interval [0 , $n\_fuzzysets_j$] where $n\_fuzzysets_j$ is the number of MFs of the jth variable. If a $x_{i,j}$ has value 0 it express that this variable is not present in the rule. Every $x_{i,1}, \ldots, x_{i,n}$ corresponds to a rule of the system.

If the set of rules is set to m, the system is able to find a set of rules less than m, just putting 0 in the output fuzzy set of the rule.

The initial population is either taken randomly or by an initial set of rules. Every gene of a chromosome is generated randomly in the interval [0, $n\_fuzzysets_j$], but some rules can be fixed for all the simulation or just given as an initial set of rules. If a fuzzy set is 0, it means that the variable is not taken into account in the rule.

The GA has to find a set of rules which prioritizes matching the minor-class patterns over the major-class ones. To do this, the fitness calculation of the GA for a class c is:

$$\text{Fitness (c)} = \frac{\text{total number of patterns matched (c)}}{\text{total number of patterns (c)}} \quad (2)$$

4 Results

The presented method has been applied to the problem of Down's syndrome Detection. This is a very good example of very-imbalanced problem. In this case, to compare results, the same dataset can be presented with either 5 variables or 3 variables, thanks to an applied reduction of variables. The characteristics of these datasets are: 2 output classes, continuous input data in 3060 patterns belonging to the major-class and just 11 belonging to the minor-class.

This dataset has been tested in Neural Networks (Backpropagation, BayesNN), classical methods of Fuzzy Rule Extraction [11] and other methods, like decision trees. The results were negative, either because they did were not generalized or because the minor-class patterns were ignored (they tried always to match the major-class patterns without taking into account the minor-class patterns).

The training was always done with the worst case (for minor-class), that is 1000 or 1500 patterns of major-class and 5 for the minor-class. If we tried to train with less than 5 patterns, the results were not acceptable. If we tried with 6 patterns, the results were much better than with 5.

The main problem is thus to choose the training dataset. Depending on which patterns are selected, the results are different, even if the patterns chosen from the minor-class are the same.

The results obtained are the following:

1. The lower the input variables are in the dataset, the better the method generalizes. A difference of one magnitude order is presented.
2. The generalization never reached the whole matching of the minor-class patterns. At least, one could not be matched. Anyhow, the best results reached more than 85% of the input patterns.
3. The number of MFs is always small, smaller than 20. Normally it is between 5 and 15. That makes the resulting fuzzy set understandable and allows GA to reach the solution in just a few steps. In addition, the number of MFs found for minor-class is less than half of patterns trained (in this case, 5 patterns trained and 2 or 3 MFs of that class, as maximum).
4. With the same training dataset, different quality solutions are obtained, but there is always a good combination of parameters that gives a good solution. Consequently, every dataset has one good solution.
5. The perfect matching is never obtained, but the minor-class patterns are always matched.

5 Conclusions

This study reflects the application of a new method for very-imbalanced datasets, working on a 2-class dataset. The method can generalize, at least, half of the patterns, given only 5 patterns of minor-class. Other experiments, with more test data, have shown that about 85-87% of patterns are matched for major-class, even if other 5000 patterns (from the same medical problem) not shown before are tested, with 12-14% of false positives.

The set of rules generated is small - between 4 and 6, with few MFs as well- and it is easy to understand for the medical staff.

Optimal results were obtained in 90% of the cases. On the other hand, since pattern matching was never 100% obtained, it is necessary to try different combinations of patterns in order to improve the generalization. However, these results improve the current ones (given by a statistical method), which are 70% of good classified patterns with 10% of false positives.

References

1. G. Wu, E. Y. Chang. "Class-boundary alignment for imbalanced dataset learning". Workshop on Learning from Imbalanced Datasets (ICML'03), Washington DC (USA, 2003).
2. N. Japkowicz and S. Stephen, "The Class Imbalance Problem: A Systematic Study", *Intelligent Data Analysis,* Volume 6, Number 5, pp. 429-450, November 2002.
3. J. Sabrià, "Screening bioquímico del segundo trimestre. Nuestra experiencia", Progresos en Diagnóstico Prenatal, vol. 10, nº 4, pag.147-153. 1998.
4. M. Berthold and K.P. Huber, "Constructing Fuzzy Graphs from Examples", Intelligent Data Analysis, 3, pp. 37-53, 1999.

5. T.P. Wu and S.M. Chen, "A New Method for Constructing MFs and Fuzzy Rules from Training Examples", IEEE Transactions on Systems, Man and Cybernetics-Part B: Cybernetics, Vol. 29, No.1, pp. 25-40, Feb. 1999.
6. L.X. Wang and J.M. Mendel, "Generating Fuzzy Rules by Learning from Examples", IEEE Transactions on Systems, Man and Cybernetics, Vol. 22, No.6, pp. 1414-1427, Nov./Dec.1992.
7. S.K. Pal, S. Bandyopadhyay and A. Murphy, "Genetic Algorithms for Generation of Class Boundaries", IEEE Transactions on Systems, Man and Cybernetics-Part B: Cybernetics, Vol. 28, No.6, pp. 816-828, Dec. 1998.
8. R. Setiono and W.K. Leow, "FERNN: An Algorithm for Fast Extraction of Rules from Neural Networks", Technical Report, National University of Singapore.
9. V. Soler, J. Roig, M. Prim, "*Finding Exceptions to Rules in Fuzzy Rule Extraction*", KES 2002, Knowledge-based Intelligent Information Engineering Systems, Part 2, pp.1115-1119, 2002.
10. V. Soler, J. Roig, M. Prim, "*A Study of GA Convergence Problems in the Iris Data Set*", Third International NAISO Symposium on Engineering of Intelligent Systems, 2002.
11. F. Herrera, M.Lozano, J.L. Verdegay, "A Learning Process for Fuzzy Control Rules using Genetic Algorithms", Technical Report #DECSAI-95108, Feb 1995.

An Iterative Artificial Neural Network for High Dimensional Data Analysis

Armando Vieira

Physics Dept. ISEP, Rua S. Tome 4200 Porto, Portugal
asv@isep.ipp.pt

Abstract. We present an Iterative Artificial Neural Network (IANN) in which computation is performed through a set of successive layers sharing the same weights. This network requires fewer weights while it can handle high-dimensional inputs. IANN is applied, with good results, to a time series prediction and two classification problems.

1 Introduction

Artificial Neural Networks (ANN) are universal computing machines capable to approximate an arbitrary function. In a multilayer perceptron computation is performed through a nonlinear projection over successive layers of processing units called neurons. In general large set of adjustable weights connecting the nodes are required.

In difficult problems with high dimensional data, large networks may be necessary. In these situations, training is difficult and may become an ill-conditioned optimization problem. Furthermore, such large networks have a considerable risk of overfitting. To alleviate these difficulties we can either prune the network to decrease its complexity or reduce the dimensionality of the problem using feature extraction techniques. However, these techniques discharge some information, while some problems are intrinsically high dimensional.

In bioinformatics these cases abound, for example, gene identification or prediction of secondary structure of proteins. In the latest case long-range interactions between amino acids are common, due to protein folding, and to surpass accuracies of 80% much larger sequences have to be consider. In time series prediction of systems with stiff dynamics, several time scales may be present and to take into account long range correlations a large time window is necessary leading to very large networks.

We present a new neural network architecture, mainly to deal with high dimensional data. Although related to recursive networks [1], this network, called Iterated Artificial Neural Network (IANN), differs from them in several aspects. In an IANN computation is performed iteratively through successive layers using the same set of connection weights. IANN is therefore more compact than traditional perceptrons.

The paper is organized as follows. Section 2 explains the IANN, section 3 presents some applications to time series prediction, a benchmark classification problem and a bankruptcy prediction problem. Section 4 presents the conclusions.

W. Duch et al. (Eds.): ICANN 2005, LNCS 3697, pp. 691 – 696, 2005.

2 Iterative Artificial Neural Networks (IANN)

Many complex functions can be decomposed in a n-fold repetition of a simple process [2]. Mathematically this is equivalent to the problem of computing iterative or functional roots. Iterative functions are very difficult to handle by analytical methods but they play a key role in nature. It is known from chaos theory that relatively simple rules, iterated many times, can produce complex mappings. For instance, the logistic equation $x_{t+1} = \lambda x_t (1 - x_t)$ generates chaotic sequences for $\lambda > 3.57$, and the famous Mandelbrot-set results from this equation for a complex-value λ.

The use of iterative processes to approximate complex functions has, however, several difficulties. In most cases the system is driven to equilibrium points or is trapped on attraction basins. This represents a computational limitation since the same fixed point is reached irrespectively of the inputs presented to the system.

With these considerations in mind, we developed the Iterative Artificial Neural Network - figure 1. The IANN is composed of a sequence of identical fully connected layers. The input is presented to layer 1 and passed to the following layer using a set of weights connecting node i of layer t+1 to node j of layer t

$$w_{ij} = p_i x_j^0 , \tag{1}$$

where x_j^0 is the element j of input vector $\vec{x}^0$, and the array $\vec{p}$ contains the adjustable weights. Note that the weights are the input vector scaled by the vector $\vec{p}$. We used this definition in order preserve information of the initial state of the system, $\vec{x}^0$. Each layer has the same dimension, of the inputs, N. Note that, in contrast to recurrent neural networks, the weights are the same for *all* layers.

Each neuron has a nonlinear transference function, and the input is propagated through successive layers following the rule:

$$\boldsymbol{x}^{t+1} = f(\sum_{j=1}^{N} w_{ij} x_j^t) , \tag{2}$$

where $\boldsymbol{x}^{t+1}$ is the output of layer t+1, N the dimension of the input, and f a nonlinear function. For this work the sigmoid was used:

$$f = \frac{1}{1+e^{-\eta(x+\theta)}} , \tag{3}$$

with a bias term θ and a parameter η to adjust the slope of the sigmoid. A different bias is used for each neuron. The output is obtained through a readout neuron connected to the last layer:

$$y = f(\sum_{i=1}^{N} w_i^f x_i^T) . \tag{4}$$

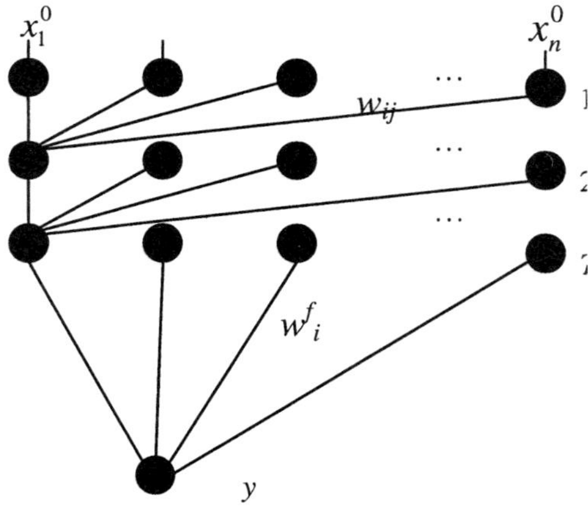

Fig. 1. A Iterative Artificial Neural Network architecture with a single output

The total number of training parameters is $3N+1$, which is linear with the number of input features, N. The number of layers T is not specified *apriori* but is adapted to the complexity of the problem. The convergence criterion must be verified:

$$\left| x^{T+1} - x^{T} \right| < \varepsilon \,, \tag{5}$$

for all x, being ε a small quantity.

Training the IANN consists in computing the weights that minimize the quadratic error:

$$E = (y - o)^2 \,, \tag{7}$$

where y is the actual output of the network and o the desired output.

The backpropagation algorithm cannot be used since the network is heterogenous and has iterative characteristics. The parameters $\boldsymbol{p}$, $\boldsymbol{\theta}$, $\boldsymbol{w}^f$ and η were optimized using a Genetic Algorithm with elitism. A mutation rate of 0.002, crossover probability of 60% and elitism ratio of 20% were used.

3 Applications

The IANN was applied to several problems and its performance compared with tradition neural networks and other machine learning approaches.

3.1 Sunspot Prediction

For this classical time-series prediction we used four time delayed windows of size k = 5, 10, 15 and 20 and one step ahead prediction. From the 280 points, the initial two hundreds were used for training and the remaining reserved for test. We compare our results with predictions made by a MLP optimized for the problem - Table 1. For the

MLP five runs were evaluated using different weights initializations. For the IANN, five runs were also computed using different initial populations for the GA. The smallest error were obtained by a MLP with k = 10 using 5 hidden neurons, which corresponds to 70 weights. While the performance of IANN is very close to the MLP, it does not degrade as much when k is increased, reaching almost the same level of accuracy as the best MLP but using less parameters. For this problem $T = 9$ and $\eta = 0.73$ were selected.

Table 1. MSE and standard deviation for the sunspot problem

k	MLP	IANN
5	15.1 ± 0.3	15.9 ± 0.5
10	13.2 ± 0.5	15.3 ± 1.1
15	13.4 ± 0.7	13.8 ± 0.7
20	14.3 ± 0.4	13.3 ± 0.9

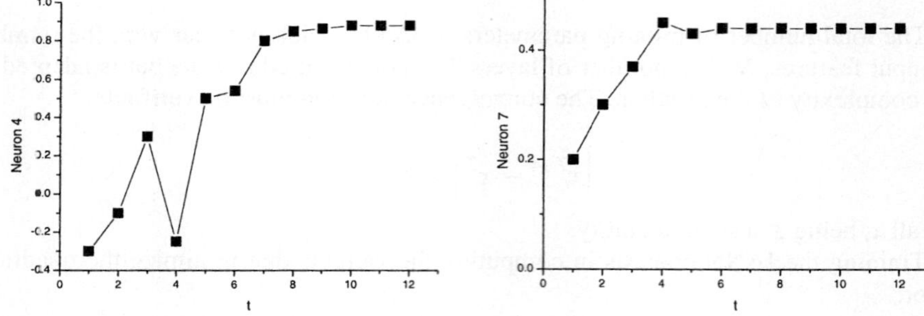

Fig. 2. Activity of two neurons over successive iterations for the sunspot problem with k = 15

To test the robustness of the predictor a 10% gaussian noise was added to the signal. While the accuracy of the best MLP degrade from 13.1% to 15.6 % the accuracy of the best IANN only decreased from 13.2% to 13.4%. Figure 2 show the activity of some neurons as a function of iteration t for k = 20. Note that some neurons take more time to stabilize that others.

3.2 The Ionosphere Problem

The IANN was applied to the ionosphere dataset to test its capabilities as a classifier. This dataset (see http://www.ics.uci.edu/~mlearn), containing 351 examples, was chosen due to its high dimensionality: 33 attributes and two classes. Five fold cross validation was used to measure the efficiency of the classifier. For this problem the values chosen were $T = 6$ and $\eta = 1.17$. This small T is an indication that the problem is more linear than the sunspot (T = 10), thus requiring less iterations to reach convergence.

Table 2 show our results compared with MLP, Genetic Programming (GP) [3] and Support Vector Machines (SVM) [4]. In this problem SVM achieved the best results but IANN clearly outperforms MLP and even GP.

Table 2. Efficiency of several classifiers in the ionosphere dataset

Method	Test Error (%)
SVM	5.9
GP	9.8
MLP	11.4 ± 1.1
IANN	8.3 ± 1.4

3.3 Bankruptcy Prediction

Finally, we apply IANN to a important problem largely discussed in the literature: bankruptcy prediction. This problem consists in discriminating between healthy and distressed companies based on the record of several financial indicators from previous years [5, 6].

The sample has 583 financial distressed French companies firms, most of them of small and medium size, with 35 to 400 employees. To test IANN we used a balanced dataset containing the same number of healthy and distressed companies. We discriminate the accuracy of the classifiers for type I, type II error and the overall misclassification. Table 3 summarizes the results with data from 1999, one-year prior to the announcement of bankruptcy. We used $T = 12$ and $\eta = 1.30$ and results were obtained using ten-fold cross validation.

Our method surpass discriminant analysis (MDA) and MLP, both in the overall accuracy, and, more important, on type I error which is the term with higher costs for banks and insurance companies. For details see Ref [7].

Table 3. Average generalization errors for the bankruptcy problem

Model	Error I	Error II	Total
MDA	26.4	21.0	23.7
SVM	17.6	12.2	14.8
MLP	25.7	13.1	19.4
IANN	19.1	13.3	16.2

4 Conclusions

Iterative Neural Networks is a promising approach for high dimensional data analyse. It is robust, relatively simple to implement and it can handle many features, even if they are irrelevant for the solution. Applications in bioinformatics, like protein-protein interactions and secondary protein structure prediction will be addressed in a future work.

References

1. Kenji D., Recurrent Networks: Supervised Learning, In: Michael A. Arbib (ed): The Handbook of Brain Theory and Neural Networks, MIT Press (1995) 796-800
2. Kindermann L.: Computing Iterative Roots with Neural Networks, V Int. Conf. on Neural Information Processing, (1998) Kitakyushu.
3. Zhou C., Xiao W., Tirpak T. M., and Nelson P. C.: Evolving Accurate and Compact Classification Rules With Gene Expression Programming, IEEE Trans. Evol. Comp. **7** (2003) 519
4. Vieira A., Castillo P.A. and Merelo J.J.: Comparison of HLVQ and GProp in the problem of bankruptcy prediction, IWANN03 - International Workshop on Artificial Neural Networks, J. Mira (Ed.), LNCS 2687, Springer-Verlag (2003) pp. 655-662
5. Meyer D., Leisch F., Hornik K.: "Benchmarking Support Vector Machines", Report n. 78, Vienna University of Economics and Business (2002).
6. Zhang G., Hu M. Y., Patuwo B. E., Indro D. C.: Artificial neural networks in bankruptcy prediction: General framework and cross-validation analysis, Europ. J. Op. Research. **116** (1999) 16
7. Vieira A., Castillo P.A. and Merelo J.J.: Comparison of HLVQ and GProp in the problem of bankruptcy prediction, IWANN03 - International Workshop on Artificial Neural Networks, J. Mira (Ed.), LNCS 2687, Springer-Verlag (2003) pp. 655-662
8. Vieira A. S., Ribeiro B., Mukkamala S., Neves J. C., Sung A. H.: On the Performance of Learning Machines for Bankruptcy Detection, Proc. IEEE International Conference on Computational Cybernetics, Vienna (2004) 223 – 227

Towards Human Friendly Data Mining: Linguistic Data Summaries and Their Protoforms

Sławomir Zadrożny[1], Janusz Kacprzyk[1,2] and Magdalena Gola[1]

[1] Systems Research Institute, Polish Academy of Sciences,
ul. Newelska 6, 01–447 Warsaw, Poland
[2] Warsaw School of Information Technology (WIT),
ul. Newelska 6, 01–447 Warsaw, Poland
{kacprzyk, zadrozny}@ibspan.waw.pl

Abstract. We show how linguistic database summaries can provide tools for human friendly data mining. The relevance of Zadeh's concept of a protoform is indicated. We present the use of our fuzzy databse querying interface for an effective and efficient mining of such linguistic data summaries. We outline an implementation for a computer retailer involving both data from an internal database of the company and data downloaded from external databases via the Internet.

1 Introduction

Data mining is meant here as a process to obtain from a very large, uncomprehensible to a human being data set some simpler, condensed form that would "subsume" the contents and essence of that data set, and is comprehensible to a human being. Needless to say, that since for a human being the only fully natural way of communication and articulation is natural language, then human consistent and friendly data mining would ideally involve natural language as much as possible. A promising approach, used here, is based on the concept of a *linguistic data(base) summary* proposed by Yager [10] and further developed mainly by Kacprzyk and Yager [1], and Kacprzyk, Yager and Zadrożny [2]. The essence of linguistic data summaries is that a set of data, e.g., on employees, with (numeric) data on their age, salaries, etc., can be summarized linguistically with respect to a selected attribute(s) like age and salaries, by linguistically quantified propositions like *most young and highly qualified employees are very well paid.*

We present such linguistic summaries, indicate the use of Zadehs protoform of a fuzzy linguistic summary (cf. Zadeh [12]) providing a generalization, portability and scalability, and then some approaches to mining of linguistic summaries, notably along Kacprzyk and Zadrożnys [4,7] interactive approach via Kacprzyk and Zadrożny's [3,5] FQUERY for Access, a fuzzy querying add-on to Microsoft Access©. By relating types of linguistic summaries to fuzzy queries, with various known and sought elements, we obtain a hierarchy of protoforms of linguistic

W. Duch et al. (Eds.): ICANN 2005, LNCS 3697, pp. 697–702, 2005.

data summaries. We will show an implementation of the data summarization system proposed for the derivation of linguistic data summaries in a sales database of a computer retailer.

2 Linguistic Data Summaries via Fuzzy Logic with Linguistic Quantifiers

In Yagers approach (cf. Yager [10], Kacprzyk and Yager [1], and Kacprzyk, Yager and Zadrożny [2]) we have: (1) $Y = \{y_1, \ldots, y_n\}$ is a set of objects (records) in a database, e.g., the set of workers, and (2) $A = \{A_1, \ldots, A_m\}$ is a set of attributes characterizing objects from Y, e.g., salary, age, etc. in a database of workers, and $A_j(y_i)$ denotes a value of attribute A_j for object y_i.

A linguistic summary of a data set D consists of:

- a summarizer S, i.e. an attribute together with a linguistic value (fuzzy predicate) defined on the domain of attribute A_j (e.g. 'low salary' for attribute 'salary');
- a quantity in agreement Q, i.e. a linguistic quantifier (e.g. most);
- truth (validity) T of the summary, i.e. a number from the interval $[0, 1]$ assessing the truth (validity) of the summary (e.g. 0.7); usually, only summaries with a high value of T are interesting;
- optionally, a qualifier R, i.e. another attribute together with a linguistic value (fuzzy predicate) defined on the domain of attribute A_k determining a (fuzzy subset) of Y (e.g. 'young' for attribute 'age').

Thus, linguistic summaries may be exemplified by

$$T(\text{most of employees earn low salary}) = 0.7 \tag{1}$$

$$T(\text{most of young employees earn low salary}) = 0.7 \tag{2}$$

and their foundation is Zadeh's [11] *linguistically quantified proposition* corresponding to either, for (1) and (2):

$$Qy\text{'s are } S \tag{3}$$

$$QRy\text{'s are } S \tag{4}$$

The T, i.e., the truth value of (3) or (4), may be calculated by using either original Zadehs calculus of linguistically quantified statements (cf. [11]), or other interpretations of linguistic quantifiers.

Using Zadeh's [11] fuzzy calculus of linguistically quantified propositions, a (proportional, nondecreasing) linguistic quantifier Q is assumed to be a fuzzy set in $[0, 1]$ and then truth(Qy's are S) $= \mu_Q[\frac{1}{n}\sum_{i=1}^{n} \mu_S(y_i)]$ or truth(QRy's are S) $= \mu_Q[\frac{\sum_{i=1}^{n}(\mu_R(y_i) \wedge \mu_S(y_i))}{\sum_{i=1}^{n} \mu_R(y_i)}]$, respectively. Clearly, the fuzzy predicates S and R need not be in such a simplified, atomic form referring to one attribute.

Recently, Zadeh [12] introduced the concept of a *protoform* defined as a more or less abstract prototype (template) of a linguistically quantified proposition.

The most abstract protoforms correspond to (3) and (4), while (1) and (2) are examples of fully instantiated protoforms. Thus, protoforms form a hierarchy, where higher/lower levels correspond to more/less abstract protoforms. Going down this hierarchy one has to instantiate particular components of (3) and (4), i.e., Q, and S and R. A protoform may provide a guiding paradigm for a user interface for the mining of linguistic summaries. Basically, the more abstract protoform the less should be assumed about summaries sought, i.e., the wider range of summaries is expected by the user, between: (A) a totally abstract protoform is specified, i.e., (4), and (B) all elements of a protoform are totally specified as given linguistic terms; in the former case (more intersting but more complicated) the system has to construct all possible summaries (with all possible linguistic components and their combinations) for the context of a given database and present to the user those verifying the validity to a degree higher than some threshold. In the second case, the whole summary is specified by the user and the system has only to verify its validity. In Table 1 basic types of protoforms/linguistic summaries are shown, of a more and more abstract form. Each of fuzzy predicates S and R may be defined by listing its atomic fuzzy

Table 1. Classification of protoforms/linguistic summaries

Type	Protoform	Given	Sought
0	QRy's are S	All	validity T
1	Qy's are S	S	Q
2	QRy's are S	S and R	Q
3	Qy's are S	Q and structure of S	linguistic values in S
4	QRy's are S	Q, R and structure of S	linguistic values in S
5	QRy's are S	Nothing	S, R and Q

predicates (pairs of "attribute/linguistic value") and structure, i.e., how these atomic predicates are combined. In Table 1 S (or R) corresponds to the full description of both the atomic fuzzy predicates (referred to as linguistic values, for short) as well as the structure. The higher the type of a summary, the more interesting but more difficult to mine it is.

3 Mining of Linguistic Data Summaries

In the process of mining of linguistic summaries, at the one extreme, the system may be responsible for both the construction and verification of summaries (which corresponds to Type 5 protoforms/summaries in Table 1). At the other extreme, the user proposes a summary and the system only verifies its validity (which corresponds to Type 0 protoforms/summaries in Table 1). The former approach seems to be more attractive and in the spirit of data mining meant as the discovery of interesting, unknown regularities in data. On the other hand, the latter approach, obviously secures a better interpretability of the results. Thus, we will discuss now the possibility to employ a flexible querying interface for the purposes of linguistic summarization of data, and indicate the implementability of a more automatic approach.

In Kacprzyk and Zadrożnys [4,7] approach, the interactivity, i.e. a user assistance, in the mining of linguistic summaries is a key point, and is in the definition of summarizers (indication of attributes and their combinations). This proceeds via a user interface of a fuzzy querying add-on. In Kacprzyk and Zadrożny [3,5,8], a conventional database management system is used with a fuzzy querying tool, FQUERY for Access. For example, an SQL query searching for *troublesome orders* may take the following WHERE clause:

WHERE *Most* of the conditions are met out of
PRICE*ORDERED-AMOUNT IS *Low*
DISCOUNT IS *High*
ORDERED-AMOUNT IS *Much Greater Than* ON-STOCK

Obviously, the condition of such a fuzzy query directly corresponds to summarizer S in a linguistic summary. Moreover, the elements of a dictionary (of terms) are perfect building blocks of such a summary. Thus, the derivation of a linguistic summary of type (3) may proceed in an interactive (user-assisted) way as follows: (1) the user formulates a set of linguistic summaries of interest (relevance) using the fuzzy querying add-on, (2) the system retrieves records from the database and calculates the validity of each summary adopted, and (3) a most appropriate linguistic summary is chosen.

Referring to Table 1, we can observe that Type 0 as well as Type 1 linguistic summaries may be easily produced by a simple extension of FQUERY for Access. For Type 3 summaries, a query/summarizer S consists of only one simple condition built of the attribute whose typical (exceptional) value is sought. Type 5 summaries represent the most general form considered: fuzzy rules describing dependencies between specific values of particular attributes. The summaries of Type 1 and 3 have been implemented as an extension to Kacprzyk and Zadrożnys [6] FQUERY for Access.

Since the discovery of Type 5 rules is difficult, some simplifications about the structure of fuzzy predicates and/or quantifier are needed, for instance to obtain association rules – cf. Kacprzyk and Zadrożny [9] for details.

4 Examples of an Implementation

We will briefly present an implementation for deriving linguistic database summaries for a sales database of a computer retailer. First, suppose that we are interested in a relation between the commission and the type of goods sold. The best linguistic summaries obtained are as shown in Table 2.

Next, let us show in Table 3 some of the obtained linguistic summaries expressing relations between the indicated attributes.

Notice that the linguistic summaries obtained do provide much of relevant and useful information, and can help the decision maker make decisions. It should

Table 2. Linguistic summaries expressing relations between the group of products and commission

Summary
About 1/3 of sales of network elements is with a high commission
About 1/2 of sales of computers is with a medium commission
Much sales of accessories is with a high commission
Much sales of components is with a low commission
About 1/2 of sales of software is with a low commission
About 1/3 of sales of computers is with a low commission
A few sales of components is without commission
A few sales of computers is with a high commission
Very few sales of printers is with a high commission

Table 3. Linguistic summaries expressing relations between the attributes: size of customer, regularity of customer (purchasing frequency), date of sale, time of sale, commission, group of product and day of sale

Summary
Much sales on Saturday is about noon with a low commission
Much sales on Saturday is about noon for bigger customers
Much sales on Saturday is about noon
Much sales on Saturday is about noon for regular customers
A few sales for regular customers is with a low commission
A few sales for small customers is with a low commission
A few sales for one-time customers is with a low commission
Much sales for small customers is for nonregular customers

be stressed that in the construction of the data mining paradigm presented we do not want to replace the decision maker but just to provide him or her with a help (support).

The system for deriving linguistic summaries developed and implemented for a computer retailer has been found useful by the user who has indicated its human friendliness, and ease of calibration and adaptation to new tasks (summaries involving new attributes of interest) and users (of a variable preparation, knowledge, flexibility, etc.). However, after some time of intensive use, the user has expressed his intention to go beyond data from the own database of a company, and use some external data We have extended the class of linguistic summaries handled by the system to include those that take into account data easily (freely) available from Internet sources, more specifically data on weather conditions as, first, they have an impact on the operation, and are easily and inexpensively available from the Internet, for instance to obtain relations between group of products, time of sale, temperature, precipitacion, and type of customers, the best linguistic summaries (of both our "internal" data from the sales database, and external meteorological data from an Internet service). Notice that the use of external data gives a new quality to possible linguistic summaries.

5 Concluding Remarks

We briefly showed the use of linguistic data(base) summaries, and their protoforms, handled by a fuzzy logic based calculus of a linguistically quantified propositions as a promising tools to obtain a greater human friendliness and consistency i datamining, mainly by a more explicit use of a natural language. We presented the use of fuzzy querying for an effcient implementation, and showed some practical application.

References

1. J. Kacprzyk and R.R. Yager. Linguistic summaries of data using fuzzy logic. International Journal of General Systems, 30, 33 - 154, 2001.
2. J. Kacprzyk, R.R. Yager and S. Zadrożny. Fuzzy linguistic summaries of databases for an efficient business data analysis and decision support. In W. Abramowicz and J. Zurada (Eds.): Knowledge Discovery for Business Information Systems, pp. 129-152, Kluwer, Boston, 2001.
3. J. Kacprzyk and S. Zadrożny. FQUERY for Access: fuzzy querying for a Windows-based DBMS. In P. Bosc and J. Kacprzyk (Eds.): Fuzziness in Database Management Systems, pp. 415-433, Springer-Verlag, Heidelberg, 1995.
4. J. Kacprzyk and S. Zadrożny. Data Mining via Linguistic Summaries of Data: An Interactive Approach. In T. Yamakawa and G. Matsumoto (Eds.): Methodologies for the Conception, Design and Application of Soft Computing. Proc. of IIZUKA98, pp. 668 - 671, Iizuka, Japan, 1998.
5. J. Kacprzyk and S. Zadrożny. The paradigm of computing with words in intelligent database querying. In L.A. Zadeh and J. Kacprzyk (Eds.): Computing with Words in Information/Intelligent Systems. Part 2. Foundations, pp. 382 - 398, Springer–Verlag, Heidelberg and New York, 1999.
6. J. Kacprzyk J. and S. Zadrożny. On combining intelligent querying and data mining using fuzzy logic concepts. In G. Bordogna and G. Pasi (Eds.): Recent Research Issues on the Management of Fuzziness in Databases, pp. 67 - 81, Springer–Verlag, Heidelberg and New York, 2000.
7. J. Kacprzyk and S. Zadrożny. Data mining via linguistic summaries of databases: an interactive approach. In L. Ding (Ed.): A New Paradigm of Knowledge Engineering by Soft Computing, pp. 325-345, World Scientific, Singapore, 2001.
8. J. Kacprzyk and S. Zadrożny. Computing with words in intelligent database querying: standalone and Internet-based applications. Information Sciences, 134, 71 - 109, 2001.
9. J. Kacprzyk, S. Zadrożny. Linguistic database summaries and their protoforms: towards natural language based knowledge discovery tools. Information Sciences, 173, 281 - 304, 2005.
10. R.R. Yager R.R.: On linguistic summaries of data. In W. Frawley and G. Piatetsky-Shapiro (Eds.): Knowledge Discovery in Databases. AAAI/MIT Press, pp. 347 - 363, 1991.
11. L.A. Zadeh. A computational approach to fuzzy quantifiers in natural languages. Computers and Mathematics with Applications. 9, 149 - 184, 1983.
12. L.A. Zadeh. A prototype-centered approach to adding deduction capabilities to search engines - the concept of a protoform. BISC Seminar, 2002, University of California, Berkeley, 2002.

Localization of Abnormal EEG Sources Incorporating Constrained BSS

Mohamed Amin Latif, Saeid Sanei, and Jonathon A. Chambers

Center for Digital Signal Processing, School of Engineering, Cardiff University, Cardiff CF24 0YF, UK

Abstract. An effective method has been developed to solve the localization problem of the brain sources. A priori knowledge about normal source locations has been effectively exploited in estimating the rotation matrix, which inherently permutes the estimated separating matrix in the blind source separation (BSS) algorithm. An important application of this method is to localize the Focal epilepsy sources, which causes changes in attention, movement and behavior. Here, an effective and simple technique for both separation and localization of the EEG sources has been developed incorporating BSS. The criterion is subject to having some of the sources known. The constraint is then incorporated into the separation objective function using Lagrange multipliers whereby changing it to an unconstrained problem.

1 Introduction

Electroencephalogram (EEG) signals are the major source of information for diagnosis of anatomical, pathological, physiological, and functional abnormalities. These signals include normal and abnormal rhythms within the frequency range of 0.3 to more than 40 Hz. This range is divided into five main subbands of $0.3-3.5$ Hz (Delta), $3.5-7.5$ Hz (Theta), $7.5-13$ Hz (Alpha), $13-30$ Hz (Beta), and more than 30 Hz (Gamma). Although the nature of the mixing medium is not completely known. A reasonable assumption is that the EEG mixtures are isotropically propagated, linearly mixed and are considered stationary within a short interval of about ten seconds. Localization of abnormal sources within the brain has been a serious and important problem within both neurophysiology and signal processing communities. A number of methods for localization of EEG sources has been investigated by researchers. Among them the methods based on dipole assumption of the sources have been very well established. MUSIC and its extension RAP-MUSIC have been extended to estimation of the source locations. However, the accuracy of such algorithms is dependent on the number of both sources and sensors [1]. Other techniques solely based on independent component analysis (ICA) cannot ensure a unique solution to the problem [2]. In this paper we show that the ICA based algorithm can be modified in order to have a unique solution to the localization algorithm. We are therefore considering an inverse solution to the EEG recordings. Here it is assumed that the EEG sources are independent. In addition the locations of the normal brain rhythms

W. Duch et al. (Eds.): ICANN 2005, LNCS 3697, pp. 703–708, 2005.

are known (obtained from EEG recordings filtered within specific bands). The instantaneous BSS formulation is as follows. Denote the time varying observed signals by $\mathbf{X} = [x_1(t), x_2(t), \ldots, x_n(t)]^T$ where $\mathbf{X} \in \mathbb{R}^n$ and the unknown independent sources by $\mathbf{S} = [s_1(t), s_2(t), \ldots, s_m(t)]^T$ where $\mathbf{S} \in \mathbb{R}^m$.

$$\mathbf{x} = A\mathbf{s} + \mathbf{v} \tag{1}$$

and

$$\mathbf{y} = W\mathbf{x} \tag{2}$$

here $\mathbf{v} \in \mathbb{R}^n$ is assumed to be a white zero mean Gaussian noise vector, $\mathbf{A} \in \mathbb{R}^{n\times m}$ and $\mathbf{W} \in \mathbb{R}^{m\times n}$ are unknown constant mixing and unmixing matrices respectively, and $(.)^T$ is vector transpose. The mixture is assumed to be overdetermined (valid for usual cases), i.e. $m \leq n$. $\mathbf{y} = [y_1(t), y_2(t), \ldots, y_m(t)]^T$, where $\mathbf{y} \in \mathbb{R}^m$ is the output vector. The separation matrix, W, can be found by finding the global minima (or maxima) of a cost function $J_M(W)$, which provides a measure of independency of the estimated sources.

Using ICA we can separate the signals into their independent components. The number of outputs may be approximated by one of the methods described in [3]. However, the separation is subject to the scaling and permutation of the sources i.e.

$$\mathbf{A} = \mathbf{DRW}^{-1} \tag{3}$$

Where $\mathbf{D}$ and $\mathbf{R}$ are the scaling and permutation matrices respectively. The effect of $\mathbf{D}$ can be constrained by the size of the head and it can be generally disabled by normalization of the estimated separating matrix after each iteration. However, without solving the permutation problem there won't be any solution to the estimation of $\mathbf{A}$. This means there will be no clue to find a unique solution to the localization problem. However, a priori information about the locations of some of the sources, say $k < m$, leads us to a more accurate estimation of $\mathbf{A}$ and, as a result, the locations of other sources.

Alpha rhythm occurs in fully alert and awake subjects. Therefore it is convenient to choose the location of the Alpha generator as the reference point.

2 Source Localization

Figure 1 shows part of the scalp including three electrodes and the two sources located within the brain as an example. Assuming the head as a homogenous medium, the link weights are inversely proportional to attenuation of the signals crossing the brain tissues i.e. $a_{ij} = \Gamma_{ij}^{-1}$. To formulate the problem consider k out of m sources are known. This means that the scaled values of the k columns of $\mathbf{A}$ are known. Moreover since the order of the electrodes is conventional we know exactly which columns are known. In the example of Figure 1 we may expand the BSS system as

$$\mathbf{X} = \begin{bmatrix} x_1 \\ x_2 \\ x_3 \end{bmatrix} = \mathbf{AS} = \begin{bmatrix} a_{11} & a_{12} \\ a_{21} & a_{22} \\ a_{31} & a_{32} \end{bmatrix} \cdot \begin{bmatrix} s_1 \\ s_2 \end{bmatrix} \tag{4}$$

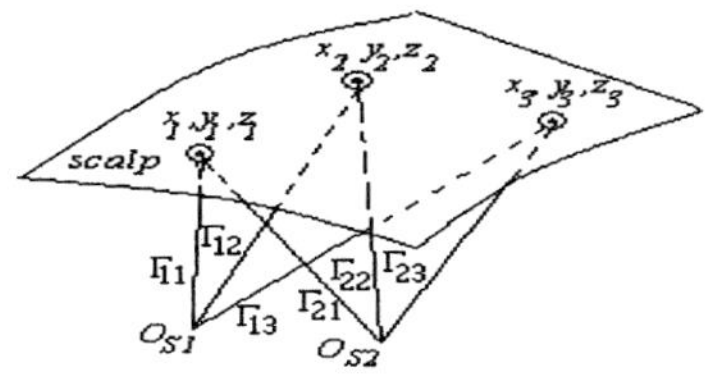

Fig. 1. Part of the scalp including three electrodes, and locations of the sources, (assuming the head is homogenous)

if source s_1 is known we define a new matrix as $\tilde{\mathbf{A}}$ such that only those of its columns related to the known sources and the unknown source can be presented by "a_{ij}" and "$\hat{a}_{ij}$" respectively, another words we can assume the column related to the known sources as nonzero, and unknown sources are arbitrary values i.e. in this case

$$\tilde{A} = \begin{bmatrix} a_{11} \ \hat{a}_{12} \\ a_{21} \ \hat{a}_{22} \\ a_{31} \ \hat{a}_{32} \end{bmatrix} \tag{5}$$

Now, during the separation process we may simultaneously try to minimize the following constraint

$$J_c = \|\tilde{\mathbf{A}} - \tilde{\mathbf{P}}\|_F^2 \tag{6}$$

where $\|.\|_F^2$ is the Frobenius norm (Euclidean norm may also be used) and $\tilde{\mathbf{P}}$ is

$$\tilde{\mathbf{P}} = \{\tilde{P}_{ij}\}, \quad i = 1,2,3 \ \ and \ \ j = 1,2 \tag{7}$$

Considering

$$\mathbf{P} = \mathbf{DRW}^{-1} = \{P_{ij}\}, \ i = 1,2,3 \ \ and \ \ j = 1,2 \tag{8}$$

then $\tilde{\mathbf{P}}$ and $\mathbf{P}$ will correspond to each other through

$$\tilde{\mathbf{P}}_{ij} = \begin{cases} P_{ij}, if & a_{ij} \neq 0; \\ 0, \quad if & a_{ij} = 0. \end{cases} \tag{9}$$

This constraint is then incorporated into the main BSS cost function to effectively estimate the rotation matrix iteratively. Such a strategy ensures the solutions to both the estimation of the separation matrix $\mathbf{W}$ based on the main BSS cost function, and the recovery of the mixing matrix $\mathbf{A}$ based on estimation of the permutation matrix. In order to exactly locate the source the non-homogeneity of the head region has to be encountered. A novel method is described in section 4 to solve this problem.

3 Constrained Problem

EEG signals are statistically non-stationary. They are affected by the human internal signals, noise of the measurement system, environment noise and interference from the adjacent electrode signals. In this work, we assume that the effects of system noise and other human internal signals are filtered out. The effective bandwidth for EEGs is from 0.3 to 40 Hz. To decorrelate the electrode sources, adaptive filtering has been traditionally used. Since each electrode signal is in fact a combination of more than one nearby sources, blind separation of these signals appears to be more favorable. Since the signals are not stationary, an accurate separation technique is hard to achieve. A number of recently developed techniques such as time-lagged second-order blind identification (SOBI) or in [5] can better cope with nonstationarity of the data. On the other hand the signals may be considered stationary within short segments of about 10 seconds (or about 2000 samples). Here, a method based on the SOBI algorithm has been considered. Since our main objective is localization of the sources having a priori information about locations of some of the sources, the separation method is of less concern. By estimating the inverse of the mixing matrix $\mathbf{A}$ we may be able to localize the location of the sources. Although solution to equation (6), given the separating matrix $\mathbf{W}$, is a linear programming problem only in places where the number of unknown source locations is equal to the number of mixtures a unique solution can result. In general cases this may not be true. Therefore to find $\mathbf{W}$ and $\mathbf{R}$ we can incorporate a constant into the cost function and solve the following unconstrained problem:

$$\mathbf{J}(\mathbf{W}) = \mathbf{J}_m(\mathbf{W}) + \lambda \mathbf{J}_c(\mathbf{W}) \tag{10}$$

where $\mathbf{J}_m(\mathbf{W})$ is the main BSS cost function, $\mathbf{J}_c(\mathbf{W})$ is the constraint defined by equation (6), and λ is the lagrange multiplier. The effect of Lagrange multiplier here is to incorporate the constraint into the main cost function whereby changing the constrained problem into an unconstrained one. An efficient algorithm to minimize equation (10) ensures the best solution to the problem especially when the number of unknown source locations is close to the number of mixtures.

4 Non-homogeneity Problem

With some indeterminacy in the result we can approximate the location of the sources within the brain. Unlike the methods in [2] and [3], which consider the sources as magnetic dipoles, we simply consider them as the sources of isotropic signal propagations. Therefore the head (mixing media) model only mix and attenuates the signals. The attenuation directly corresponds to the distance and the resistance between the sources and the fixed electrodes. Based on the method developed in section 2 we need to have the location of some of the sources known. This is not a difficult problem in the context of EEGs. Normal Alpha rhythms have a fixed source location at the contralateral central sulcus. These sources generate reference signals within a small frequency band of $7 - 13$ Hz in a fully awake person. Since we can measure both the link weights and the energy of the mixtures within the selected bands we will be able to estimate the nonhomogenity by finding a relationship between $\tilde{A}$, found through measurement of the geometrical locations and $\mathbf{A}_g$, found through measurement of the energy of the signal(s) of the known source(s). The energy within Alpha band is obtained by carefully bandpass filtering the EEGs around the peak in Alpha wave. These amplitudes are then inverted to give the entries of k columns of $\tilde{\mathbf{A}}$. On the other hand the geometrical location of the known sources can be roughly determined off-line (call it $\tilde{\mathbf{A}}_g$). The rest of $(m-k)$ columns are set to zero for both $\tilde{\mathbf{A}}$ and $\tilde{\mathbf{A}}_g$.

In a spherical model of the head we may consider three main layers; brain, skull, and scalp for which the thickness is known. The conductivity of the skull is about 10 to 100 times less than those of brain and scalp. In our proposed algorithm although there won't be any change in the overall BSS algorithm but to incorporate the non-homogeneity into account f has to be completely identified for all the sources. Here, we consider $f(\tilde{\mathbf{A}}_g) = \psi.\tilde{\mathbf{A}}_g$, where ψ is a weighing $n \times m$ matrix and "." refers to element-wise multiplication. Having more than one known source locations, in order to extend the above nonlinear map to all the estimated source locations a simple means of extrapolation of the columns of the estimated mixing matrix will be adequate.

5 The Experiments

The proposed algorithm was implemented for separation and localization of the EEG sources [6]. In order to separate the signals detected from EEG recordings obtained from head of a patient the cost function based on SOBI was used. Furthermore to localize the positions of the sources a constraint was incorporated into the main cost function using a Lagrange multiplier. For one source a minimum of three electrodes can be used for correlation measure. A matrix of three signals containing three synthetic sinusoidal sources were generated and the resulting separated signals are shown in Figure 2(b). A random mixing matrix $\mathbf{A}$ representing the weight links is generated and the sensor signals were mixed [7]. In estimation the constraint on the cost function

$$\mathbf{W}_{t+1} = \mathbf{W}_t + \mu_1 \nabla_{\mathbf{W}} \mathbf{J} \tag{11}$$

where $\mathbf{J}$ is main cost function, $\nabla_{\mathbf{W}} \mathbf{J}$ is the gradient of $\mathbf{J}$ with respect to $\mathbf{W}$ also:

$$\mathbf{J}_c = \|\tilde{\mathbf{A}} - \mathbf{R}\mathbf{W}^{-1}\|_F^2 \tag{12}$$

where $\tilde{A}$ is the estimated mixing matrix, which refers to the position of the sources , $\mathbf{W}^{-1}$ is pseudo inverse for $m \neq n$, and μ is learning rate.

Furthermore the permutation matrix $\mathbf{R}$ is updated through

$$\mathbf{R}_{t+1} = \mathbf{R}_t + \mu_2 \nabla_{\mathbf{R}} \mathbf{J}_c \tag{13}$$

where $\nabla_{\mathbf{R}} \mathbf{J}_c$ is the gradient of $\mathbf{J}_c$ with respect to rotation matrix $\mathbf{R}$ as:

$$\nabla_{\mathbf{R}} \mathbf{J}_c = 2\mathbf{W}^{-1}(\tilde{\mathbf{A}} - \mathbf{R}\mathbf{W}) \tag{14}$$

After estimating $\mathbf{W}$ in each iteration the Rotation matrix $\mathbf{R}$ is also iteratively calculated. Hence the solution to the Mixing matrix can be calculated.The separation plot is shown in Figure 2(b).

Hence knowing the rotation and the unmixing matrix the mixing matrix which refers to the weight links of sources can be obtained, as shown in Figure 2(c).

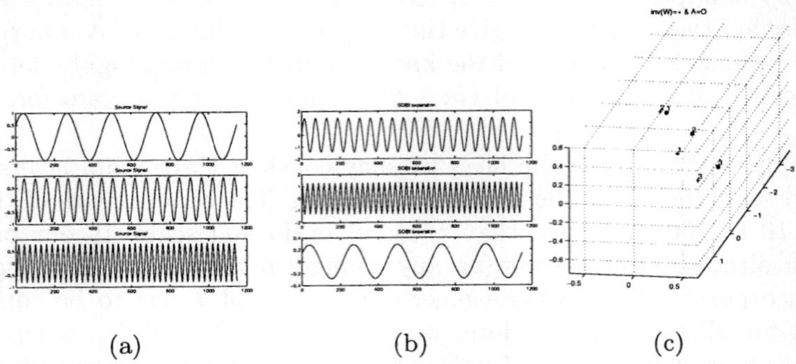

(a) (b) (c)

Fig. 2. (a)The modelled sinusoidal sources (b)Separated sources using SOBI and (c) The points represented by "O" refer to **A** *and* $\mathbf{RW}^{-1}$ *represented by* "+"

6 Results and Conclusions

The proposed algorithm was implemented for separation and localization of abnormal sources. The solution gives the approximate locations of all the sources. In order to mitigate the ambiguities we need to use a priori information about the known sources. This is done in two steps: The estimated sources are scaled to make energy of the known sources equal to those extracted by bandpass filtering. The Rotation matrix, R, is estimated by minimizing $\|\mathbf{A}_{es}\text{-}\mathbf{RW}^{-1}\|_F^2$. Moreover, the computational cost is much lower.

References

1. Gutiérrez, D., Nehorai, A.,: Estimating Brain Conductivities and Dipole Source Signals with EEG Arrays, *IEEE transactions on biomedical engineering, vol 51 no 12 (2004)*.
2. Mosher, J.C., and Leahy, R. M.,: Source Localization using recursivley applied and projected (RAP) MUSIC, *IEEE transactions on signal processing (1999) 332-340*.
3. Leahy, R. M. et al,: A study of dipole localization accuracy for MEG and EEG using a human skull phantom, *Electroencephalgraphy and Clinical Neurophysiology, vol 107 (1998) 159-173*.
4. Pascual-Marqui, R. D., Michel, C. M., and Lehmann, D.,: Low resolution electromagnetic tomography: a new method for localizing electrical activity in the brain,*International Journal of Psychophysiology, vol. 18, (1995) 49-65*.
5. Wang, W., Chambers, J. A., and Sanei, S.: A joint diagonalization method for convolutive blind separation of nonstatioanry sources in the frequency domain, *Proc of ICA2003 Nara Japan (2003) 939-944*.
6. Marcel Joho Heinz Mathis,: Joint diagonalization of correlation matrices by using gradeint method with application to BSS,*SAM2002 August 4-6 Rosslyn VA USA (2002) 273-277*.
7. Belouchrani, A., Abed-Meriam, K., Cardoso, J.F., Moulines, E.: A blind source separation techinque using second order statisitcs , *IEEE trans. on signal processing Vol XX No Y (1997)*.

Myocardial Blood Flow Quantification in Dynamic PET: An Ensemble ICA Approach

Byeong Il Lee[1], Jae Sung Lee[1], Dong Soo Lee[1], and Seungjin Choi[2]

[1] Department of Nuclear Medicine,
Seoul National University Hospital, Korea
dewpapa@hanmail.net, jaes@snuvh.snu.ac.kr, dsl@plaza.snu.ac.kr
[2] Department of Computer Science,
Pohang University of Science and Technology,
San 31 Hyoja-dong, Nam-gu, Pohang 790-784, Korea
seungjin@postech.ac.kr

Abstract. Linear models such as factor analysis, independent component analysis (ICA), and nonnegative matrix factorization (NMF) were successfully applied to dynamic myocardial $H_2^{15}O$ PET image data, showing that meaningful factor images and appropriate time activity curves were estimated for the quantification of myocardial blood flow. In this paper we apply the ensemble ICA to dynamic myocardial $H_2^{15}O$ PET image data. The benefit of the ensemble ICA (or Bayesian ICA) in such a task is to decompose the image data into a linear sum of independent components as in ICA, with imposing the nonnegativity constraints on basis vectors as well as encoding variables, through the rectified Gaussian prior. We show that major cardiac components are separated successfully by the ensemble ICA method and blood flow could be estimated in 15 patients. Mean myocardial blood flow was 1.2 ± 0.40 ml/min/g in rest, 1.85 ± 1.12 ml/min/g in stress state. Blood flow values obtained by an operator in two different occasion were highly correlated (r=0.99). In myocardium component images, the image contrast between left ventricle and myocardium was 1:2.7 in average.

1 Introduction

Linear model-based methods, including factor analysis, independent component analysis (ICA), nonnegative matrix factorization (NMF), were shown to useful in analyzing dynamic positron emission tomography (PET) image data, demonstrating that meaningful factor images and appropriate time activity curves could be extracted [1,2,3]. In the application of such linear models to PET, a main focus was to extract left ventricle input function [1,2], which is an essential part for the calculation of myocardial blood flow (MBF) in the tracer kinetics model of dynamic $H_2^{15}O$ cardiac PET. However, the extraction of the input function is a difficult task, because of the partial volume effect resulting from the limitation of system resolution and the spill-over of left ventricle, right ventricle, and myocardium by the motion of heart. Consequently, a new method

W. Duch et al. (Eds.): ICANN 2005, LNCS 3697, pp. 709–714, 2005.

for the input function extraction is required to estimate the blood flow more accurately. $H_2^{15}O$ dynamic cardiac PET has been used for the quantification of MBF as an ideal blood flow tracer [4,5,6]. The half life of $H_2^{15}O$ is about 2 minutes, which makes repetitive and short interval estimation of MBF possible. In this paper, we apply the ensemble ICA [7] to $H_2^{15}O$ dynamic cardiac PET image data. In the ensemble learning [8], the inference is carried out by averaging over the posterior distribution of the parameters. The main benefit of the ensemble ICA over the conventional ICA or NMF, is to decompose the image data into a linear sum of independent components as in ICA, with imposing the nonnegativity constraints on basis vectors as well as encoding variables, through the rectified Gaussian prior. We evaluate the ensemble ICA for the quantification of regional myocardial blood flow (rMBF) after segmentation of left ventricle, right ventricle, and myocardium images.

2 PET Image Acquisition and Processing

PET images were acquired from ECAT EXACT47 (Siemense-CTI, Knoxville, USA) in Seoul National University Hospital. Totally 24 frames 47 transaxial images were acquired; 12 frames for 5 seconds, 9 frames for 10 seconds, and 3 frames for 30 seconds. After bolus injection of $H_2^{15}O$ (555-740 MBq), adenosine stress was carried out during 7 minutes. $H_2^{15}O$ was injected after 3 minutes during stress, and then dynamic PET images were acquired during 4 minutes continuously. Images were reconstructed using FBP (image matrix = 128 128, magnification factor = 1.5). Twenty patients were investigated using $H_2^{15}O$ dynamic myocardial PET. Patients were underwent gated 99mTc-MIBI myocardial perfusion SPECT for the suspicious coronary artery disease. Rest image and adenosine stress images were acquired. All frame data was reoriented to short axis and two plans were summed in order to extract myocardium component automatically using ensemble ICA. Nine region of interest (ROI) were drawn on left ventricle and myocardium (1 apex, 4 middle wall, 4 basal wall) to take out the time-activity curve of dynamic PET image. Using input function and time-activity curve of each region, rMBF was calculated. The values of rMBF were compared with angiography and gated myocardial perfusion SPECT. Regional perfusion was relocated to 9 regions used in dynamic PET analysis.

3 Factor Image Extraction Using Ensemble ICA

$H_2^{15}O$ PET images are converted to vector sequences $\mathcal{D} = \{x_t \in \mathbb{R}^m\}$. ICA assumes that data vectors x_t are generated by

$$x_t = As_t + \epsilon_t, \tag{1}$$

where $s_t \in \mathbb{R}^n$ correspond to factor images (independent components), column vectors of the matrix $A \in \mathbb{R}^{m \times n}$ represent time activity curves, and $\epsilon_t \in \mathbb{R}^m$ reflect the model uncertainty which is assumed to be Gaussian.

In the context of $H_2^{15}O$ PET images, independent components are expected to images corresponding to left ventricle, right ventricle, myocardium, and background, which reasonable satisfy spatial independence. In such a case, basis vectors (corresponding to the column vectors of A) represent the time activity curves which reflect the time-varying influence in PET images [9]. The standard ICA, including mutual information minimization, maximum likelihood estimation (MLE), output entropy maximization, and so on (see [10] for recent review), incorporates with the prior probability of parameters in a limited way and neglects the uncertainty term in (1). That is, in the standard ICA, parameters were inferred by maximizing the likelihood in the limit of zero noise.

On the other hand, NMF [11] also considers the linear model (1) but infers parameters with constraining both A and s_t to be nonnegative, whereas ICA incorporates with independence conditions for s_t. Inference in NMF can also be illustrated in the framework of maximum likelihood estimation with assuming Poisson distribution for ϵ_t. Application of NMF to dynamic PET can be found in [3].

Here we use the ensemble ICA [7] to extract factor images in $H_2^{15}O$ PET. In the Bayesian framework, the posterior probability of parameters θ, given a set of data points $\mathcal{D}$, is described by

$$P(\theta|\mathcal{D},\mathcal{H}) = \frac{P(\mathcal{D}|\theta,\mathcal{H})P(\theta|\mathcal{H})}{P(\mathcal{D}|\mathcal{H})}, \tag{2}$$

where $\mathcal{H}$ represents a model. In the ensemble learning, the inference is performed by averaging over the posterior distribution, so that the inference is sensitive to regions where the probability mass is large, in contrast to ML or MAP where the inference is sensitive to regions where the probability density is large. In practice, exact inference is often intractable. The ensemble learning approximation finds an approximate a posterior distribution Q for the model parameters by minimizing the Kullback-Leibler divergence between the approximate posterior Q and the true posterior

$$\begin{aligned} KL[Q||P] &= \left\langle \log\left[\frac{Q(\theta)}{P(\theta|\mathcal{D},\mathcal{H})}\right]\right\rangle_Q \\ &= \left\langle \log\left[\frac{Q(\theta)}{P(\mathcal{D},\theta|\mathcal{H})}\right]\right\rangle_Q + \log P(\mathcal{D}|\mathcal{H}). \end{aligned} \tag{3}$$

where $\langle\cdot\rangle_Q$ denotes the statistical expectation under the approximate distribution Q.

The following objective function $\mathcal{J}$ was considered in [7]

$$\begin{aligned} \mathcal{J} &= KL[Q||P] - \log P(\mathcal{D}|\mathcal{H}) \\ &= \left\langle \log\left[\frac{Q(\theta)}{P(\mathcal{D},\theta|\mathcal{H})}\right]\right\rangle_Q \\ &\geq -\log P(\mathcal{D}|\mathcal{H}). \end{aligned} \tag{4}$$

The minimization of the objective function $\mathcal{J}$ in (4) is equivalent to maximizing a bound on the log-evidence $\log P(\mathcal{D}|\mathcal{H})$.

The main benefit of the ensemble ICA is to decompose the PET images as a linear combination of factor images with encoding variables being statistically independent as in ICA, with imposing nonnegativity constraints on A and s_t through rectified Gaussian prior. In other words, the ensemble ICA leads us to incorporate with both independence and nonnegativity constraints in the context of the linear model (1). Empirical results in Sec. 4 demonstrate that the ensemble ICA, indeed, works well in the task of analyzing $H_2^{15}O$ PET image data.

4 Quantification Results of rMBF

The rMBF from $H_2^{15}O$ dynamic myocardial PET were compared with the results of perfusion SPECT. Image contrast between myocardium and left ventricle were estimated in segmented myocardial independent component images (see Fig. 1). Image contrast of myocardium was $1 : 2.97$ (LV:myocardium) in the rest image and was $1 : 2.56$ in stress image of separated independent component images (see Fig. 2). The number of subjects with the image contrast under 2.0 was 6 and the highest value of image contrast was 4.63. Blood flow obtained from PET was 1.2 ± 0.40 ml/min/g in rest state, 1.85 ± 1.12 ml/min/g in stress state. Reproducibility of myocardial blood flow of 15 subjects PET image data which were acquired twice for each region was high. ($r = 0.99$, $P < 0.0001$) Myocardial perfusion was quantified by autoQuant program. Uptake value of normal segments group were 67.613.3in stress (reversibility score = 1.9), while that of stenotic group were 71.9±9.8in rest and 69.1±12.8There was no significant difference between normal group and stenotic group in terms of reversibility score. The rMBF of reversible segments were 0.98 ± 0.30 ml/min/g in rest, 1.78 ± 0.76 ml/min/g in stress, and blood flow reserve was 0.80 ± 0.69 ml/min/g. The rMBF of persistent segments in myocardial perfusion SPECT was 1.10 ± 0.40 ml/min/g in rest, 2.06 ± 1.35 ml/min/g in stress, and the blood flow reserve was 0.95 ± 1.32 ml/min/g (see Fig. 3).

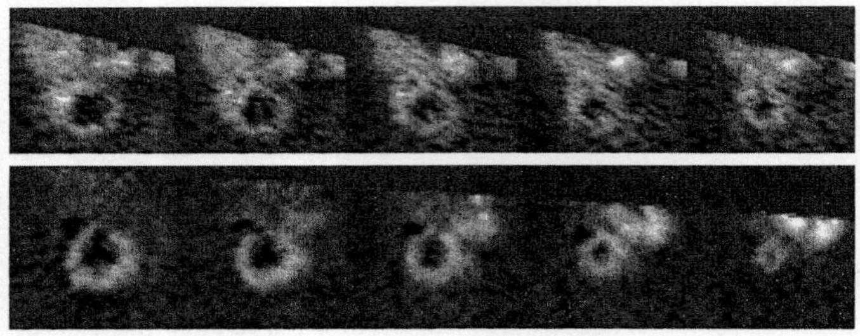

Fig. 1. Segmented component images using the ensemble ICA method

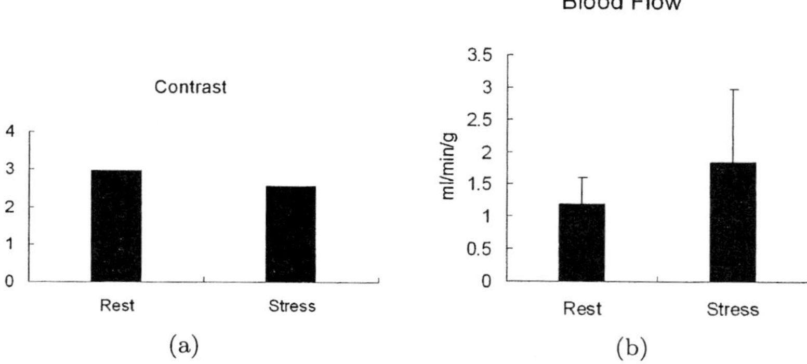

(a) (b)

Fig. 2. (a) Image contrast of myocardium was improved in rest image than stress image; (b) Estimated myocardial blood flow values using $H_2^{15}O$ PET

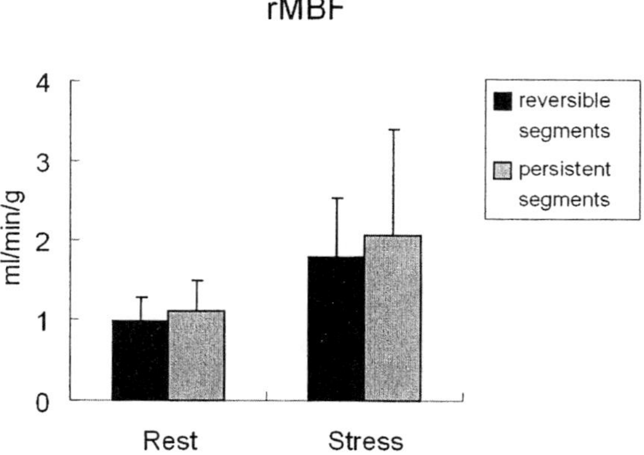

Fig. 3. Regional myocardial blood flow of reversible segments and persistent segments in perfusion SPECT

5 Discussions

The standard ICA had difficulty in extracting appropriate factor images in our clinical data, because of the difference of injection dose according to weight and low sensitivity of hardware system. Recently, left ventricle and myocardium image were visualized through the NMF method in clinical study, and the MBF of patient could be estimated using the NMF [12] The nonnegativity is a natural constraint in medical imaging such as PET. The ensemble ICA is a technique which incorporates with both independence and nonnegativity constraints. In our study, we have observed that the ensemble ICA had a a merit of improved image contrast and quality for ROI processing, compared to the NMF method. The rMBF was estimated using the ensemble ICA in $H_2^{15}O$ dynamic myocardial PET. Reproducibility of measurement and image contrast were good enough to segment myocardium. We expect that dynamic myocardial PET analysis using the ensemble ICA can be used to assess the absolute myocardial blood flow in clinical situations.

Acknowledgments. This work was supported by KOSEF 2000-2-20500-009-5, and Basic Research Fund in POSTECH.

References

1. Ahn, J.Y., Lee, D.S., Lee, J.S., Kim, S.K., Cheon, G.J., Yeo, J.S.: Quantification of regional myocardial blood flow using dynamic $H_2^{15}O$ PET and factor analysis. J. Nucl. Med. **42** (2001) 782–787
2. Lee, J.S., Lee, D.S., Ahn, J.Y., Cheon, G.J., Yeo, J.S.: Blind separation of cardiac components and extraction of input function from $H_2^{15}O$ dynamic myocardial PET using independent component analysis. J. Nucl. Med. **42** (2001) 938–943
3. Lee, J.S., Lee, D.D., Choi, S., Lee, D.S.: Application of non-negative matrix factorization to dynamic positron emission tomography. In: Proc. ICA, San Diego, California (2001) 629–632
4. Iida, H., Tamura, Y., Kitamura, K., Eberl, P.M.B.S., Ono, Y.: Histochemical correlates of ^{15}O-water-perfusable tissue fraction in experimental canine studies of old myocardial infarction. J. Nucl. Med. **41** (2000) 1737–1745
5. Iida, H., Kanno, I., Takahashi, A.: Measurement of absolute myocardial blood flow with $H_2^{15}O$ and dynamic positron-emission tomography strategy for quantification in relation to the partial volume effect. Circulation **78** (1998) 104–115
6. Kaufmann, P.A., Gnecchi-Ruscone, T., Yap, J.T., Rimoldi, O., Camici, P.G.: Assessment of the reproducibility of baseline and hyperemic myocardial blood flow measurements with ^{15}O-labeled water and pet. J. Nucl. Med. **40** (1999) 1848–1856
7. Miskin, J.W., MacKay, D.J.C.: Ensemble learning for blind source separation. In Roberts, S., Everson, R., eds.: Independent Component Analysis: Principles and Practice. Cambridge University Press (2001) 209–233
8. Miskin, J.W., MacKay, D.J.C.: Application of ensemble learning to infra-red imaging. In: Proc. ICA. (2000) 399–404
9. Schafers, K.P., Spinks, T.J., Camici, P.G., Rhodes, P.M.B.C.G., Law, M.P.: Absolute quantification of myocardial blood flow with $H_2^{15}O$ and 3-dimensional PET: an experimental validation. J. Nucl. Med. **43** (2002) 1031–1040
10. Choi, S., Cichocki, A., Park, H.M., Lee, S.Y.: Blind source separation and independent component analysis: A review. Neural Information Processing - Letters and Review **6** (2005) 1–56
11. Lee, D.D., Seung, H.S.: Learning the parts of objects by non-negative matrix factorization. Nature **401** (1999) 788–791
12. Schaefer, W.M., Nowak, B., Kaiser, H.J., Block, K.C.K.S., Dahl, J.V.: Comparison of microsphere-equivalent blood flow (^{15}O-water PET) and relative perfusion (99m Tc-Tetrofosmin SPECT) in myocardium showing metabolism-perfusion mismatc. J. Nucl. Med. **44** (2003) 33–39

Data Fusion for Modern Engineering Applications: An Overview

Danilo P. Mandic[1], Dragan Obradovic[2], Anthony Kuh[3], Tülay Adali[4], Udo Trutschel[5], Martin Golz[6], Philippe De Wilde[1], Javier Barria[1], Anthony Constantinides[1], and Jonathon Chambers[7]

[1] Imperial College London, UK
[2] Siemens AG, Munich, Germany
[3] University of Hawaii, USA
[4] UMBC, USA
[5] Circadian Technologies, USA
[6] University of Schmalkalden Germany
[7] Cardiff University, UK

{d.mandic, pdewilde, j.barria, a.consantinides}@imperial.ac.uk
dragan.obradovic@siemens.com, kuh@spectra.eng.hawaii.edu
adali@umbc.edu, utrutschel@circadian.com
golz@informatik.fh-schmalkalden.de, chambersj@cf.ac.uk

Abstract. An overview of data fusion approaches is provided from the signal processing viewpoint. The general concept of data fusion is introduced, together with the related architectures, algorithms and performance aspects. Benefits of such an approach are highlighted and potential applications are identified. Case studies illustrate the merits of applying data fusion concepts in real world applications.

1 Introduction

The data fusion approach combines data from multiple sensors (and associated databases if appropriate) to achieve improved accuracies and more specific inferences that could not be achieved by the use of only a single sensor [1]. This concept is hardly new:- living organisms have the capability to use multiple senses to learn about the environment. The brain then **fuses** all this available information to perform a decision task.

One of the first definitions of data fusion came form the North American Joint Directors of Laboratories (JDL) [2,3], who define data fusion as a:- *multilevel, multifaceted process dealing with the automatic detection, association, correlation, estimation and combination of data from single and multiple sources.*

Data fusion principles apply to many domains, and have been (often implicitly) at the core of modern applications in the diverse areas spanning engineering, computing, and biomedicine. The recent interest in the theory and taxonomy of multisensor data fusion has been reflected by a number of special issues of leading international journals and conferences, which have been dedicated to this

W. Duch et al. (Eds.): ICANN 2005, LNCS 3697, pp. 715–721, 2005.

area (e.g. Proc. of the IEEE in 1997 [1] and 2003 [4], JMLR in 2003 [5], and IEEE TNN 2002 [6]).

There has been a somewhat conflicting use of terminology within the data–sensor–information fusion community. People working at the sensor level view data fusion as basically operating with raw data which have undergone at the most only some preliminary processing [7]. Others, like JDL, have a more general view which includes both raw and processed data – in short, all the inputs to some higher level decision making/classifying stages.

Our aim in this paper is therefore to provide a systematic overview of the existing data fusion philosophy and methods for engineering applications.

2 Data Fusion Principles

When approaching a problem from the data fusion viewpoint, we differentiate between the following levels of abstraction:

- **Observation/measurement space** contains vectors of measurement functions which can be univariate, multivariate, and/or multidimensional, depending on temporal, spatial or other independent variables. It may be possible to build a state–space model, or to assess the data modality [8,9];
- **Transform domain representations**, which seek features from time and/or frequency models (fast Fourier transform (FFT), (nonlinear) autoregressive (N)ARMA models [10], wavelet), blind processing (independent component analysis (ICA), blind source separation (BSS) [11]), particle/Kalman filter [12], kernels and support vector machines (SVM) [25], kernel ICA);
- **Decision space**, where the classes within the data fusion model (and the corresponding basins of attraction from the measurement space) are mapped into the relevant probabilities of the occurrence of an event.

Similarly, authors distinguish between the Data, Information, and Knowledge semantic levels [14] (Figure 1). This simple taxonomy has been very useful in the diverse applications of data fusion, such as in:- i) *transportation*, aviation, intelligent car traffic and motorways management; ii) *multimedia communications*, audio–visual fusion for teleconferencing; iii) *robotics*, 3-D vision; iv) *wearable computing*, monitoring the disabled and elderly.

2.1 Models of Data Fusion

Data fusion is based on the manipulation of multiple measurements, where *classifiers* operate on *features* extracted from the real world *measurements*; an overview

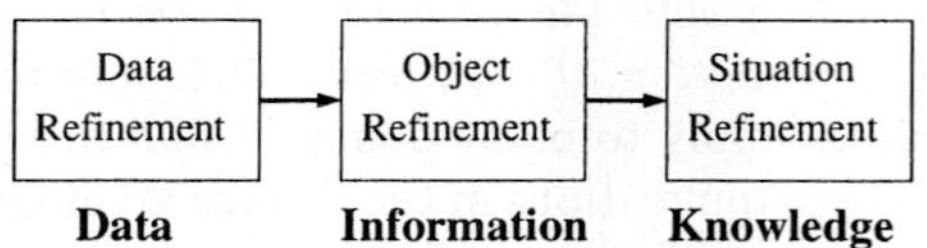

Fig. 1. General data fusion concept

of the ways for combining classifiers can be found in [15]. Authors distinguish between the two fusion classes:-

i) **Data fusion**, where the classifier operates on either the *raw data* or *features* extracted directly from the measurements;
ii) **Decision fusion**, where the decisions from the *individual classifiers* for different data channels are *combined.*

The choice depends on the statistical relationship between the data channels, mutual entropy, or joint Gaussianity [16], and to this end coupling of mathematical modelling and information processing is under investigation [17]. The main issues are signal nonlinearity (with associated non–Gaussianity), nonstationary, intermittent data natures and noises. This makes it very difficult to perform estimation by standard methods since no assumption on the data model and distribution can be ascertained. In some applications, such as functional Magnetic Resonance Imaging (fMRI), there is even no "ground truth", to rely upon. Multisensor practical systems therefore aim at providing higher accuracy and improved robustness against uncertainty and sensor malfunction [18], and also for the information extracted from different sources to be integrated into a single signal or quantity.

Signal processing algorithms for "sensor" or "data fusion" can be based on [19]:-

– *Probabilistic models:* Bayesian reasoning, evidence theory, robust statistics;
– *Least squares:* Kalman filtering, regularization, set membership;
– *Intelligent fusion:* Fuzzy logic, neural networks, genetic algorithms.

One of the first proposed data fusion models was the "waterfall model" (Figure 2), developed for the UK Defence Evaluation Research Agency (DERA) [3].

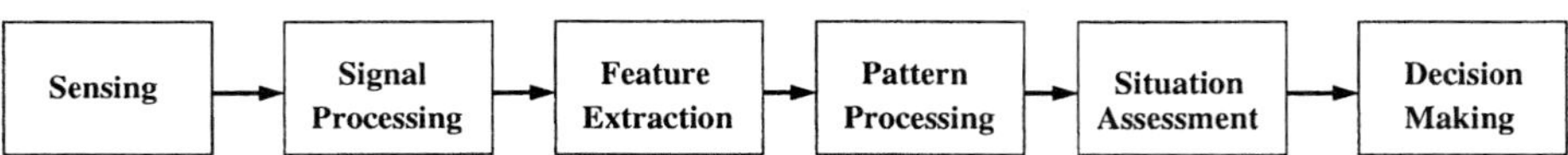

Fig. 2. The Waterfall model

2.2 Data Fusion and Sufficient Information

We can think of the heterogeneous sensors monitoring a certain process as being "windows" into the phenomenon under observation. Sensors can either have their own window, or the windows "overlap" in space or time. This way, the information obtained can be thought of as "decomposed" or "fragmented" by the sensors, which is sometimes called sensor fission [7], and is related to so–called *sufficient information* (whether the character and number of sensors can indeed describe the phenomenon). This is analogous to the notion of *embedology*, where we wish to model the nonlinear dynamics of a multidimensional process based on its time delay representation [8]. The information fragments coming from sensors are exposed to spectral shaping, saturation, and noise; *data fusion* aims at retrieving the "interesting" characteristics of the phenomenon.

3 Architectures and Performance Aspects

Combining multi–sensor data in the data fusion framework has the potential of faster and cheaper processing and new interfaces, together with reducing overall uncertainty (increase in reliability). Such data can be combined in various ways, for instance by:- i) linear combiner, ii) combination of posteriors (weights, model significance), iii) product of posteriors (independent information). Based on the different ways of combining information and different semantic levels, we differentiate between the following data fusion architectures, shown in Figure 3:-

- **Centralised:** simple algorithms, but inflexible to sensor changes;
- **Hierarchical:** collaborative processing, two way communication;
- **Decentralised:** robust to sensor changes and failures, complex algorithms.

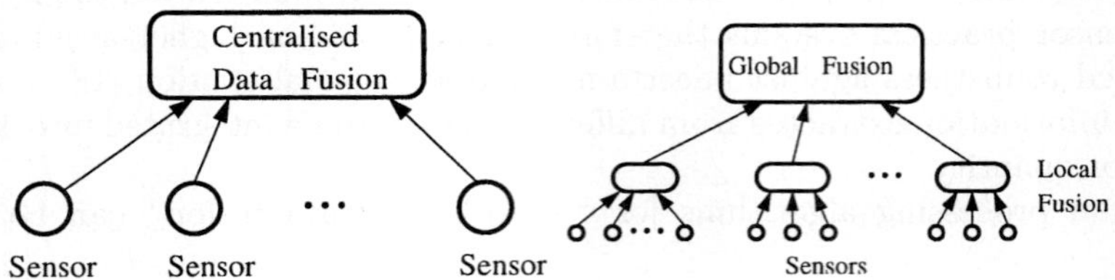

Fig. 3. Centralised and hierarchical data fusion

This synergy [20] of information fragments offers some advantages over standard algorithms, such as:-

- Improved confidence due to complementary and redundant information;
- Robustness and reliability in adverse conditions (smoke, noise, occlusion);
- Increased coverage in space and time; dimensionality of the data space;
- Better discrimination between hypotheses due to more complete information;
- System being operational even if one or several sensors are malfunctioning;
- Possible solution to the vast amount available information.

The paradigm of *optimal fusion* in this sense is to *minimise* the probability of unacceptable error.

Based on the taxonomy presented in Section 2, depending on the stage at which fusion takes place, data fusion is often categorized as the *low*– (LLF), *intermediate*– (ILF) or *high*–level (HLF) fusion, where:-

- LLF (*data fusion*) combines raw data sources to provide better information;
- ILF (*feature fusion*) combines features that come from heterogeneous or homogeneous raw data. The aim is to find *relevant* features amongst various features coming from different methods (FFT, discrete cosine transform (DCT), wavelet, delay vector variance (DVV) [9]);
- HLF (*decision fusion*), combines decisions or confidence levels coming from several experts (*hard* and *soft* fusion).

In practice, any combination of these three levels can be employed, for instance [7]: *Data in – Data out, Data in – Feature out, Feature in – Feature out, Feature in – Decision out, Decision in - Decision out.*

4 Data Alignment and Fusion of Attributes

Depending on where the fusion process occurs, open literature differentiates between the *temporal*, *spatial*, and *transform domain* fusion. Notice, however, that the latter two can be considered as examples of the low– or intermediate–level fusion. Temporal fusion is different in the sense that it may occur at any level:- inputs from one sensor taken at different instants are combined.

The information entering a fusion process should be aligned, a difficult problem for which there is no general supporting theory. Alignment should be applied to both homogeneous (commensurate) and heterogeneous (non–commensurate) information, which may require conversion or transformation of observations [13].

The concept of alignment assumes "common language" between the inputs, for instance:- i) standardisation of measurement units; ii) sensor calibration; or iii) corrections for different illuminants and shading [21]. Alignment may operate at any of the three semantic levels: *measurements*, *attributes*, and *rules*, with possible crossings between levels [21]. For instance, for aligned and associated sources of information, fusion of attributes concatenates attributes of the same object, derived from different representations of the object. Fusion of representations performs meta–operations, it is applicable to any representation, and can be combined with other types of fusion.

Data fusion also applies to cyberspace, where intrusion detection (ID) systems fuse data from heterogeneous distributed network sensors to create "situational awareness" [14], such as the detection of network anomalies and virus attack. Information of interest are the identity, threat, rate of attack, and target of intruders [22].

Performance aspects of a fusion system [20] are domain–specific:-

- Detection performance and characteristics (false alarm rate);
- Spatial/temporal resolution and ability to distinguish between signals;
- Spatial and temporal coverage (span or viewfield of a sensor);
- Detection/tracking mode (scanning, tracking, multiple target tracking);
- Measurement accuracy and dimensionality.

5 Case Studies

We next provide three case studies to illustrate the data fusion concept:- the examples in car navigation, sleep science and multimedia.

Car navigation systems perform three main tasks: positioning, routing and navigation (guidance). The car position is calculated from several information sources including on–board odometers and gyroscopes, the global positioning system (GPS) and digital maps. On–board sensors measure acceleration and

angular rates, for which the short–term precision is high, but the accumulated errors grow with time, producing a poor long–term position estimate.
On the other hand, the GPS exhibits excellent overall performance, but its accuracy is highly sensitive to factors such as "blind" areas (tunnels, garages) and the number of "visible" satellites. One way to circumvent these sensor limitations would be to exploit the potential a combination of the short–term accuracy of on–board sensors and long–term accuracy of the GPS system.
This has been achieved in the Siemens car navigation system [23], where the *fusion* of the information from vehicles' internal sensors and the GPS position reading provides 80% improved navigation accuracy within the given time interval as compared to the estimate based on the on–board sensors only.

Awareness/fatigue modelling is important in the detection of sleep stages and also for the detection of microsleep for drowsy drivers. The observed signals are the electroencephalogram (EEG), electro-oculogram (EOG), and respiratory signal. There are also several sources of artifacts, such as the eye blink artifact in EEG. Although it is possible to detect sleep stages or microsleep events using only one sensor modality (typically EEG), the classification accuracy is not sufficient to warant real world applications, and the data fusion approach is one viable solutions which combines the EEG and EOG features. In addition, in order to achieve high detection and classification rates, the temporal fusion over the observation windows is even more important than feature selection. For sleep stage detection, feature fusion can be performed using the DVV method [9], which gives features related to the signal nonlinearity [24]. Such a fusion of EEG and EOG features provides $\approx$ 99% accuracy in training and $\approx$ 90 % accuracy on test data. Similarly, the feature fusion of EEG and EOG channels significantly improves the detection of microsleep [26].

Video assisted speech separation, where the task is to integrate complementary audio and visual modalities to enhance speech separation. Rather than using independence criteria suggested in most BSS systems, visual features from a video signal are used as additional information to optimise separation. The Bayesian framework can be applied for feature fusion, where the mel–frequency cepstrum and "active apperance model" provide audio and video features. This way a performance improvement of several dB can be achieved [27].

6 Conclusions

Data fusion provides a theoretical, computational, and implementational framework for combining data and knowledge from different sources with the aim of maximising the useful information content. In this way, reliability and discrimination capability are improved while the amount of required data is minimised. Through the three overlapping stages: preprocessing, data alignment, and decision making, the performance of a system is improved. Data fusion spans disciplines such as signal detection, pattern recognition, and tracking, with applications in domains such as military, robotics, medicine, and space research. This paper sumarises some of the recent developments in data fusion, and gives an overview of concepts, architectures and potential benefits of using this approach.

References

1. Hall, D.L., Llinas, J.: An introd. to multis. data fus.. Proc. IEEE **85** (1997) 6–23
2. White, Jr., F.E.: Joint directors of laboratories data fusion subpanel report. In: Proceedings of the Joint Service Data Fusion Symposium, DFS–90. (1990) 496–484
3. Worden, K., Dulieu-Barton, J.M.: An overview of intelligent fault detection in systems and structures. Structural Health Monitoring **3** (2004) 85–98
4. Chong, C.Y., Kumar, S.P.: Sensor networks: evolution, opportunities, and challenges. Proceedings of the IEEE **91** (2003) 1247–1256
5. Dybowski, R. et al., Eds: Journal of Machine Learning Research: Special issue on the fusion of domain knowledge with data for decision support (July 2003)
6. Adali, T. et al. Eds: IEEE Transactions on Neural Networks: Special issue on intelligent multimedia processing (July 2002)
7. Dasarathy, B.V.: Sensor fusion potential exploitation – Innovative architectures and illustrative applications. Proceedings of the IEEE **85** (1997) 24–38
8. Kantz, H., Schreiber, T.: Nonlinear TSE. Cambridge University Press (2004)
9. Gautama, T. et al.: A novel method for determining the nature of time series. IEEE Transactions on Biomedical Engineering **51** (2004) 728–736
10. Mandic, D.P., Chambers, J.A.: RNNs for Prediction. Wiley (2001)
11. Cichocki, A., Amari, S.I.: Adaptive Blind Signal and Image Proc. Wiley (2002)
12. Deco, G., Obradovic, D.: An I.T. Approach to Neural Computing. Springer (1997)
13. G. Wald: (http://www.data-fusion.org/article.php?sid=70)
14. Bass, T.: Intrus. detect. and multis. data fusion. Comm. ASM **43** (2000) 99–105
15. Tax, D.M. et al: Combining multiple classifiers. Pat. Rec. **33** (2000) 1475–1485
16. Brooks, R.R., Ramanathan, P., Sayeed, A.M.: Distributed target classification and tracking in sensor networks. Proceedings of the IEEE **91** (2003) 1162–1171
17. Coatrieux, J.L.: A look at integrative science: Biosignal processing and modelling. IEEE Engineering in Medicine and Biology Magazine **23** (2004) 9–12
18. Zhao, F. et al.: Collaborative signal and information processing: An information–directed approach. Proceedings of the IEEE **91** (2003) 1199–1209
19. Sasiadek, J.Z.: Sensor fusion. Annual Reviews in Control **26** (2002) 203–228
20. Waltz, E., Llinas, J.: Multisensor Data Fusion. Artech House (1990)
21. Pau, L. F. Sensor Data Fusion. Jnl. of Intel. and Robot. Sys. **1** (1988) 103–116
22. Alarcon, V., Barria J.: Anom. det. in com. net. IEE Proc. Com. **148** (2001) 355–362
23. Obradovic, D. et al.: Sensor Fusion In Siemens Car Navigation System. Proc. of MLSP 2004, (2004) 655–664
24. Mandic, D. P.: (http://www.commsp.ee.ic.ac.uk/~mandic)
25. Zhu, C., Kuh, A.: Sensor Network Loc. Using Pat. Rec. Proc. of HISC, (2005)
26. Sommer, D. et al.: Appl. LVQ to detect drivers dozing–off. Proc. EUNITE, (2002)
27. Wang, W. et al.: Video Assisted Speech Source Sep. Proc. ICASSP (2005) 425–427

Modified Cost Functions for Modelling Air Quality Time Series by Using Neural Networks

Giuseppe Nunnari and Flavio Cannavó

Dipartimento di Ingegneria Elettrica, Elettronica e dei Sistemi,
Universitá di Catania, Viale A.Doria 6, 95125 Catania, Italy
{gnunnari, fcannavo}@diees.unict.it

Abstract. In this paper a new Backpropagation algorithm appropriately studied for modelling air pollution time series is proposed. The underlying idea is that of modifying the error definition in order to improve the capability of the model to forecast episodes of poor air quality. Five different expressions of error definition are proposed and their cumulative performances are rigorously evaluated in the framework of a real case study which refers to the modelling of 1 hour average daily maximum Ozone concentration recorded in the industrial area of Melilli (Siracusa, Italy). Furthermore, two new performance indices to evaluate the model prediction capabilities referred to as Probability Index and Global Index respectively, are introduced. Results indicate that the traditional and the proposed version of Backpropagation perform quite similarly in terms of the Global Index which gives a cumulative evaluation of the model. However the latter algorithm performs better in terms of the percentage of exceedences correctly forecast. Finally a criterion to make the choice among various air quality prediction models is proposed.

1 Introduction

Non linear regression techniques based on Multilayer Perceptron (MLP) neural networks have drawn the attention of several scientists involved in the stochastic modelling of pollutant time series. Several authors have shown that MLP works better than traditional linear regression techniques [1] and also many other non-linear techniques as short-term predictors of pollutant concentrations at a point [2], [3], [4]. The Backpropagation algorithm [5], which is the basic approach to training a supervised Multilayer Perceptron (MLP) neural networks, is based on the minimisation of the traditional average squared error cost function defined as follows

$$J_0 = \frac{1}{2N}\sum_{p=1}^{N} E_p = \frac{1}{2N}\sum_{p=1}^{N}(T_p - Y_p)^2 \tag{1}$$

In (1) T_p and Y_p represent the target and actual model output value respectively. However, it is easy to understand that this assumption is not the most appropriate when dealing with pollution time series containing a relatively small

W. Duch et al. (Eds.): ICANN 2005, LNCS 3697, pp. 723–728, 2005.

number of episodes of poor air quality. The drawback that arises considering the cost function (1) in a similar case is due to the fact there is distinction between targets above or below a given threshold. Hence the learning algorithm will give to the exceeding episodes the same weight as the remaining events. The immediate consequence is that although one of the main targets of the models is the prediction of episodes of poor air quality, these events may not be relevant during the model identification process. The idea underlying this work is to modify expression (1) in order to weigh exceeding events more appropriately.

2 Modified Cost Functions

In this paper five different cost functions (i.e. error definitions) are considered, as expressed in (2) to (6) respectively

$$J_1 = \frac{\sum_{p=1}^{N}(T_p - Y_p)^2(T_p - M)^2}{2N} \tag{2}$$

$$J_2 = \frac{\sum_{p=1}^{N}(T_p - Y_p)^2[(T_p - M)^2 + (Y_p - M)^2]}{2N} \tag{3}$$

$$J_3 = \frac{\sum_{p=1}^{N}(T_p - Y_p)^2 e^{-(\frac{Y_p}{T}-1)(\frac{T_p}{T}-1)}}{2N} \tag{4}$$

$$J_4 = \frac{\sum_{p=1}^{N} e^{-(Y_p-T)(T_p-T)(T_p-Y_p)^2}}{N} \tag{5}$$

$$J_5 = \begin{cases} \frac{\sum_{p=1}^{N}(T_p-Y_p)^2}{2N} & T_p <= T \\ \frac{\sum_{p=1}^{N}2(T_p-Y_p)^2}{2N} & T_p > T \end{cases} \tag{6}$$

In expressions (2) and (3) M is a constant value and T is the threshold (e.g. 180 $\mu g/m^3$ for ozone daily maximum concentration). We will assume that M is the average value of the pollutant time series, i.e.

$$M = \frac{1}{N}\sum_{p=1}^{N} T_p \tag{7}$$

Furthermore we will indicate the algorithms corresponding to cost functions (2) to (6) as BP1 to BP5 respectively.

3 The Improved Backpropagation Algorithm

As is known the Backpropagation is a recursive algorithm to update the weights of Multilayer Perceptron (MLP) neural networks, based on the deepest-descent formula:

$$\Delta w_{ij}^{p} = -\epsilon \frac{\partial E_p}{\partial w_{ij}^{p}} = -\epsilon \delta_{i,p}^{(S)} O_{j,p}^{(S-1)} \tag{8}$$

where ϵ and w_{ij} are the learning velocity and the weight of the interconnections between the i-th neuron of the layer $S-1$ and the j-th neuron of the layer S. $\delta_{i,p}^{(S)}$ is the local gradient of the i-th neuron in the layer (S) and $O_{j,p}^{(S-1)}$ is the output of the j-th neuron in the layer $(S-1)$. In view of implementing modified versions of the BP algorithm we observe that different cost functions will affect the local gradient of the output layer neurons, only. We have computed $\delta_{i,p}^{(n)}$ for the five different cost functions given in (2) to (6) and the results are listed below.

Cost function J1:

$$\delta_{i,p}^{(n)} = (T_p - Y_p)(T_p - M)^2 \dot{f}(Net_i^p) \tag{9}$$

Cost function J2:

$$\begin{aligned} \delta_{i,p}^{(n)} &= (T_p - Y_p)[(T_p - M)^2 + (Y_p - M)^2 - \\ &-(T_p - Y_p)(Y_p - M)]\dot{f}(Net_i^p) \end{aligned} \tag{10}$$

Cost function J3:

$$\begin{aligned} \delta_{i,p}^{(n)} &= (T_p - Y_p)e^{-(\frac{Y_p}{T}-1)(\frac{T_p}{T}-1)}[2 + \\ &+ \frac{(T_p - Y_p)(T_p - T)}{T^2}]\dot{f}(Net_i^p) \end{aligned} \tag{11}$$

Cost function J4:

$$\begin{aligned} \delta_{i,p}^{(n)} &= e^{-(Y_p - T)(T_p - T)(T_p - Y_p)^2}(T_p - T) \\ &(T_p - Y_p)[(T_p - Y_p) - 2(Y_p - T)]\dot{f}(Net_i^p) \end{aligned} \tag{12}$$

Cost function J5:

$$\delta_{i,p}^{(n)} = \begin{cases} (T_p - Y_p)\dot{f}(Net_i^p) & T_p < T \\ 2(T_p - Y_p)\dot{f}(Net_i^p) & T_p > T \end{cases} \tag{13}$$

4 Modelling Daily Maximum Ozone Concentrations at Melilli (SR)

The backpropagation algorithm proposed in this paper was considered to model 1 hour average daily maximum concentrations (DMAX) of Ozone (O3) recorded at various recording stations located in the industrial area of Siracusa (Italy). To compare the different algorithms 10 trials have been carried out changing the test set. The process being modelled was assumed to be stationary during the time interval considered.

4.1 Performance Indices

In order to evaluate the capabilities of the training algorithms to predict exceedences of the attention level the indices defined in (14)-(18) were computed.

$$SP = \frac{N_p}{N_o}; \;\; SR = \frac{N_p}{N_f}; \;\; FA = 1 - SR; \tag{14}$$

$$SI = (\frac{N_p}{N_o} + \frac{N + N_p - N_o - N_f}{N - N_o} - 1) \tag{15}$$

In expressions (14)-(18) N_o is the total number of observed exceedences of a given threshold, N_p is the number of correctly predicted exceedences, N_f is the total number of forecast exceedences and N the total number of data points. The meaning of the indices defined above is the following. SP indicates the percentage of exceedences correctly forecast, FA is the percentage of false alarms, SR gives the percentage of predicted exceedences which actually occurred and, finally, SI is the success index which gives a cumulative evaluation of how well the exceedences are predicted. Details about the measuring of the aforementioned performance indices can be found in [6].

$$PI = (1 - \frac{N_o + N_f - 2N_p}{N}) \tag{16}$$

$$PI = P(O, Y) + P(\overline{O}, \overline{Y}) \tag{17}$$

Unfortunately the SI index does not express a probability of success in strictly probabilistic sense. To overcome this drawback we propose here a new index referred to as PI (Probability Index) expressed by (16). It is easy to demonstrate that PI can also be represented as indicated by (17). The right term of expression (17) represents the sum of two probabilities: $P(O, Y)$ which gives the probability that an observed exceedence will be correctly predicted by the model and $P(\overline{O}, \overline{Y})$ which represents the probability that non exceeding values will also correctly forecast. In (17), the argument O represents a boolean variable defined as following:

$$O = \begin{cases} True & \text{when the pollutant time series to} \\ & \text{be modelled exibits an} \\ & \text{exceedence (e.g. } O3_{MAX} > T) \\ False & \text{otherwise} \end{cases}$$

The arguments Y and $\overline{Y}$ have the same meaning as O and $\overline{O}$ but refer to the estimated values (i.e. the output of the prediction model). Furthermore, in this paper we introduce another new index, referred to as GI (Global Index) expressed by (18) which gives a measure of the success of the forecasting model independent on the number of samples (N) in the modelled time series.

$$GI = (\frac{N_p}{N_o + N_f - N_p}) \tag{18}$$

It is to stress that both PI and GI assume values in the $[0, 1]$ interval. For a good prediction model PI and GI should approach 1.

4.2 Experimental Framework

To evaluate the peculiarities of adopting the modified cost functions (2) to (6) in comparison with the traditional MSE given in (1), a software tool was coded which implements the modified back propagation algorithms as described in the previous section. All these algorithms were considered to train the model. In order to obtain a measure of the generalization capabilities not affected by a particular training and testing set, the learning phase was organized as follows. The available data set spanning for 1995 to 1998 was divided into ten overlapping data sets, each containing one year data (i.e. 365 samples of daily maximum ozone concentration). For each backpropagation algorithm ten different trials were performed. During each trial 9 of the 10 data set were used for training, and the remaining one for testing. This should guarantee a non biased evaluation of the performance (i.e. the set of indices is representative of the generalization capabilities of the neural model). During all the experiments the number of learning cycles, hidden neurons (in the unique hidden layer considered) and the learning velocity were considered constant in order to assure a more objective inter-comparison exercise. In particular the number of hidden neurons was set to 6, the learning velocity ϵ to 0.1 and the number of learning cycles was set to 10000. Results in terms of averaged values of the performance indices over ten trials for each algorithm are summarized in Fig. 1. In particular Fig. 1a gives the SP, FA and SI indices and Fig. 1b gives the introduced set of indices (PI and GI). From Fig. 1 it appears that all the modified backpropagation algorithms (except BP3) perform better than the traditional BP in terms of SP and SI. In particular SP is about 0.60 for BP, 0.90 for BP1, 0.78 for BP2, 0.84 for BP4 and 0.70 for BP5. However this result is accompanied by a larger number of false alarms. This agrees with the fact that the PI and the GI are almost constant for all the considered algorithms. In other words, the proposed backpropagation algorithms do not perform globally better than the traditional BP but if the modeller is interested in maximizing the performance in terms of percentage of exceedences correctly forecasted it is quite evident that a benefit can be obtained from adopting one of the introduced algorithms. The price to pay is an increased

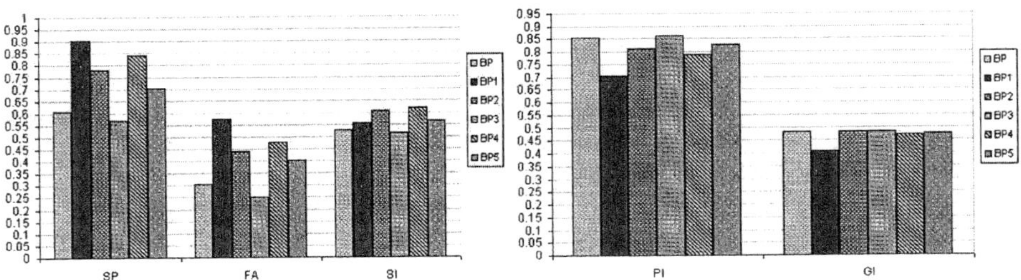

Fig. 1. Comparison among all proposed algorithms in terms of SP, FA, SI (a) and PI, GI (b)

level of false alarms which is usually acceptable provided that it is lower than a prefixed threshold (say 0.40). Fig. 1a shows that BP5, among the inter-compared algorithms, is the best compromise between a high level of SP (0.70) and an acceptable level of FA (0.40) whilst the traditional BP exhibits $SP = 0.60$ and $FA = 0.30$. It is interesting to stress here that the choice of $BP5$ is also confirmed by the following reasoning carried out in terms of PI and SP. Indeed Fig. 1b shows that the best three models in terms of PI are BP ($PI = 0.854$), $BP3$ ($PI = 0.863$) and $BP5$ ($PI = 0.827$) since they exhibit almost the same value. However $BP5$ is the best with respect to BP and $BP3$ in terms of SP . Hence we may suggest this criterion to make the choice among various air quality prediction models.

5 Conclusions

In this paper a novel backpropagation algorithm to improve the capabilities of the traditional backpropagation algorithm used to predict episodes of poor air quality has been proposed. The rigorous intercomparison, performed in the framework of the described case study show that eventhough the traditional and the proposed algorithms perform quite similarly in terms of success index and global index, the latter algorithms perform better in terms of the percentage of exceedences correctly forecasted. The price to pay for this is a slight increase in the percentage of false alarms.

References

1. G. Nunnari and A. Nucifora and C. Randieri: The application of neural techniques to the modelling of time-series of atmospheric pollution data. Ecological Modelling, (1998), vol. 111, pages 187–205
2. A.B. Chelani and R.C.V. Chalapati and K.M. Phadke and M.Z. Hasan: Prediction of Sulphure Dioxide Concentration Using Artificial Neural Network. .Environmental Modelling and Software,(2002), vol. 17, pages 161–168
3. U. Schlink and S. Dorling and E. Pelikan and G. Nunnari and G. Cawley and H. Junninen and A. Greig and R. Foxall and K. Eben and T. Chatterton and J. Vondrcek and M. Richter and M. Dostal and L. Bertucco and M. Kolehmainen and M. Doyle: A rigorous inter-comparison of ground-level ozone predictions. Atmospheric Environment, (2003), vol. 37, pages 3237–3253
4. S.R. Dorling and R.J. Foxall and P.D. Mandic and G.C. Cawley: Maximum likelihood cost functions for neural networks models of air quality data. Atmospheric Environment, (2003), vol. 37, pages 3435–3443
5. D.E. Rumelhart and G.E. Hinton and L. McLelland: Parallel Distributed Processing Exploration in the Microstructure of Cognition. MIT Press, (1986), vol. 1, Cambridge, pages 45–76
6. R. M. Van Aalst and F. A. A. M. De Leeuw: National Ozone Forecasting System and International Data Exchange in Northwest Europe. European Topic Centre on Air Quality, (1997)

Troubleshooting in GSM Mobile Telecommunication Networks Based on Domain Model and Sensory Information

Dragan Obradovic and Ruxandra Lupas Scheiterer

Siemens AG, Corporate Technology, Information and Communications,
81730 Munich, Germany
{dragan.obradovic, ruxandra.scheiterer}@siemens.com

Abstract. Mobile cellular telecommunication networks are complex dynamic systems whose troubleshooting presents formidable challenges. Typically, the network performance analysis is carried out on a network cell basis and it is based on the traffic information obtained from various sensors such as the number of requested calls, number of dropped calls, number of handovers, etc. This paper presents a novel troubleshooting system, which provides likelihood of different user-specified root causes of performance degradation based on the observed sensory information and the underlying domain model. This domain model has a form of a Causal Network whose structure is appropriately chosen. The novelty of the herein presented approach is that the domain model is initially based on expert knowledge and later on refined via supervised learning with the data gathered during system operation.

1 Introduction

Detecting and explaining faulty states in complex telecommunication systems such as GSM networks are challenging tasks. The mobile networks are hierarchical cell-based systems with complex dynamics influenced by the stochastic user demand, the network operating characteristics such as hand-over algorithms, carrier frequency, etc., and the non-stationary influence of the environment (interference, channel properties). Once the quality of service (QoS) is unacceptably low in any part of the network, it is necessary to perform troubleshooting, i.e. to identify and remedy the problem causes in order to bring the QoS to the required level.

Troubleshooting in telecommunication networks is typically carried out by human experts. The experts analyze system parameters and available measurements and, based on their prior experience, try to identify the possible causes of problems. Usually, this expert knowledge is formalized as a set of simple rules describing the analysis of individual potential network problems completely neglecting their possible simultaneous occurrence.

Due to the increasing complexity of communication networks and the increasing cost of human expert labor, it is necessary to develop a semi-automatic system for troubleshooting in communication networks. This system is intended to provide decision support to the expert in highly developed areas, or to allow deployment of

W. Duch et al. (Eds.): ICANN 2005, LNCS 3697, pp. 729 – 734, 2005.

networks in areas where human expertise is scarce. The system presented in this paper will use a domain model describing the interaction of different network variables with the goal of estimating the likelihood of different possible problem causes.

2 Bayesian Network Model of the GSM Domain

Bayesian Networks are acyclic directed graphs whose nodes represent random variables and whose edges indicate causal relationships [1]. A Bayesian Network is completely specified only when, in addition to the nodes and edges, the underlying conditional and marginal probabilities are also known. Bayesian Networks provide a rigorous and efficient framework for inference, i.e. for calculating probabilities of non-observable variables given a set of observations of related observable variables. Hence, a Bayesian Network can be seen as a data fusion paradigm which is based on the probabilistic model of the analyzed domain.

In order to specify a Bayesian Network model of the GSM domain suitable for troubleshooting, we have to select random variables, i.e. Bayesian Network nodes, indicate their causal relationship, and provide the underlying conditional and marginal probabilities. Our model has the following characteristics:

i. The networks variables, i.e. nodes, belong to three distinctive groups: Factors, Root Causes, and Symptoms. Factors are variables representing the GSM network characteristics at the cell basis, such as the carrier frequency or cell type. Root causes are the network problems whose occurrence we would like to investigate. And symptoms are either basic traffic measurements or their functions, typically called Key Performance Indicators (KPI). The KPI are used since they are typically used by the operators and since they reduce the redundancy of the basic traffic measurements. The justification of using the three-layered structure is two-fold: It reflects the composition of the GSM network meaning that the network characteristics (cell type, planning features...) and customer behavior (mobility...) are the factors that contribute to network problems, which are then observed through KPI, i.e. network measurements. In addition, this structure enables a meaningful and comprehensive collection of human expert knowledge. Lack of a clear model structure or its too high complexity (i.e. arbitrary number of layers and their interconnectivity) would make the knowledge collection process practically impossible.

ii. Factors are defined as discrete random variables that can have more than two states. Factors are variables that describe the network settings (e.g. cell size: pico, micro, and macro) and user behavior characteristics and, therefore, there is no reason to assume that they should be limited to binary variables. Furthermore, they are in most cases discrete per definition.

iii. Root causes are defined as two-state discrete variables with states YES (present) and NO (absent). In the first approximation we are primarily interested in the probability of the presence/absence of a particular problem. A further subdivision of the original problem into different problem types, when meaningful, is possible and can be always achieved by substitution of the original problem by a set of sub-problems. An alternative approach is to have an additional probabilistic model of the problem that is activated only when the probability of its state YES is greater than a given threshold. The additional model will then use extra measurements (possibly per request) in order to provide additional information about the problem (its sub-types). The latter case effectively describes a hierarchical model with possibly different sets of measurements (KPIs).

iv. Symptoms are defined as discrete random variable that can have more than two states and are functions of measured variables (KPIs). Typically these functions are the ones human experts have evolved to best capture the complex correlations of the numerous elementary counters and root causes. The measured traffic variables are continuous variables that we have to discretize. The number of discrete intervals has to take into account that a single KPI can be correlated with several different problems where each problem determines specific regions (intervals) of interest. Hence, the final discretization of the given KPI can result in more than two states (discretization intervals).

Once the nodes are defined, the edges and the corresponding probabilities have to be specified. They are either learned from the data describing results of the human troubleshooting process, or obtained by interviewing experts. Our approach uses both, initially relying on expert knowledge and then refining through data gathered during operation of the troubleshooting module.

3 Probabilistic Model Generation

The generation of the GSM domain model is carried out in three steps:

First, extraction and formalization of human expert knowledge about the technical domain of interest with the "domain expert" tool, which we have developed for this purpose [2], [3]. The experts interact with the tool's graphical user interface where they specify relevant system variables and potential system failures/problems and their causal relationships in the form of "factor-problem-symptom". The obtained information is saved in an appropriate database, in our case in Microsoft Access.

Second, automatic generation of the probabilistic model from the knowledge database by the so-called "knowledge compiler" program developed for this purpose [4]. The resulting model has the form of a Bayesian Network.

Third, refinement, i.e. the on-line learning of the expert model based on the gathered data during operation. The initial model based on the expert knowledge contains inaccuracies stemming from the subjectivity of the available human knowledge and the assumptions made for the purpose of automatic generation of the corresponding probabilistic model. In order to improve the initial model, a learning algorithm [5] is applied to update the model parameters (conditional probabilities) based on the data gathered during system operation.

Once the expert knowledge is gathered in the database and the probabilistic domain model is generated, it can be used for actual troubleshooting. The resulting domain model is fed together with the available process information and measurements into the "Inference Engine", a program that computes the probabilities of different possible system problems. Here one of the advantages of using a Bayesian probabilistic domain model becomes apparent, namely that it can work with incomplete information, using the available data in an incremental way. At each moment the instantaneous probabilities of the Root Causes reflect the total available knowledge about the problem, and denote the probability of the respective Root Cause given the available known measurements and factor values. If enough information is presented to the probabilistic model, the probability of the actual problem will be sufficiently increased in comparison to the probabilities of other alternative problems, to trigger new steps, such as issuing a report to the operator, recommending the necessary steps for circumventing the problem, etc.

In the case when certain system measurements are not readily available, our inference program processes the existing ones and makes a recommendation about which additional measurements are needed to properly identify the cause of the system problem. This recommendation is a result of a tradeoff between the benefit of the new measurement for the problem identification and the cost of its collection.

The process of the domain model generation and its usage in troubleshooting is depicted in the following figure.

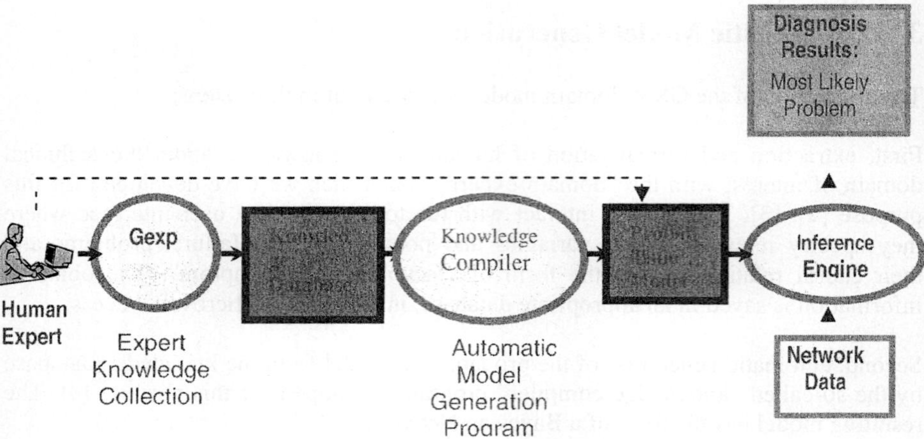

Fig. 1. The operator knowledge is collected with the Gexp tool and translated into a probabilistic model by the knowledge compiler. The inference engine delivers the probabilities of different modeled root causes based on the observed data.

Besides the here considered application to troubleshooting in mobile telephony, our approach is equally applicable in any technical or complex non-technical domain (as for example the medical or biological disciplines, e.g. [4]), where for diagnostic purposes and troubleshooting the user relies on the valuable knowledge and experience of domain experts.

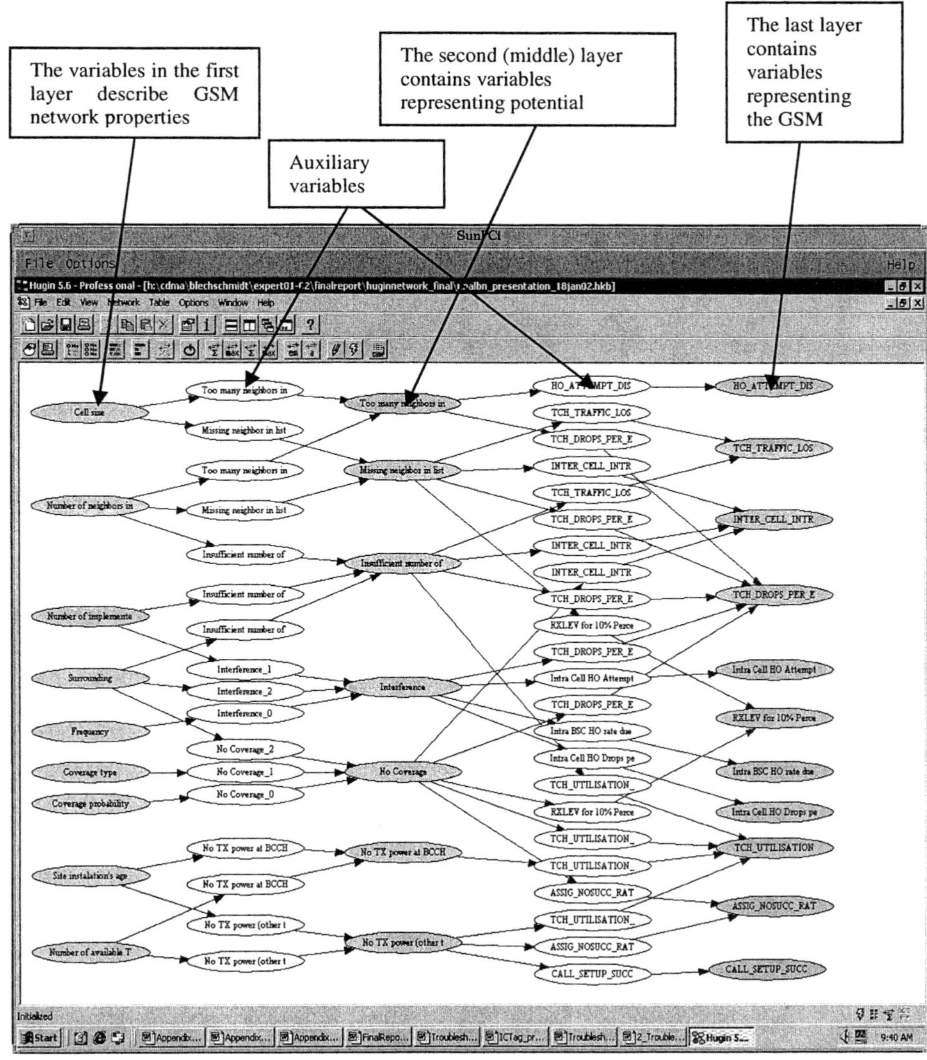

Fig. 2. Bayesian Network model generated for troubleshooting in GSM networks. The Bayesian Network consists of layers with different groups of variables. The GSM network properties are in the leftmost layer, the potential failures are in the middle layer, while the measurements are in the rightmost layer. The remaining two layers contain auxiliary variables that can be masked and not shown. The auxiliary variables are generated by the knowledge compiler and reflect the built-in assumptions about some entries of the conditional probability tables [6].

4 Implementation and Testing Issues

The Bayesian Network modeling tool was tested at one of the major mobile phone providers in Germany. The interviews with experts resulted in the description of ten typical root-cause problems in GSM mobile telecommunication networks. The generated Bayesian Networks was fed daily with the mobile traffic data of the previous day. The results of the Root-Cause Analysis were made available to the operators who were asked to evaluate them. Every time an expert would disagree with the results of the Troubleshooting system, he would have to report his findings, i.e. the root-causes he had detected. This information was then used to refine the Bayesian Network model by adapting its conditional probabilities. The overall performance of the tool was very satisfactory in detecting the modeled problem causes but cannot be currently presented to the proprietary restrictions.

During the development of the modeling tool, we noticed that a typical problem in interviewing experts lies in the direction of human reasoning and its relationship to causality. In explaining root-causes experts follow the forward cause-consequence relation i.e. given a cause they estimate the conditional probabilities of the possible consequences. On the other hand, during the troubleshooting stage the experts work backwards, quantifying the probabilities of different causes given the symptoms. The information about different conditional probabilities has to be optimally combined by the knowledge compiler during the generation of a Bayesian Network model. Ref. [6] discusses the involved aspects and shows how to combine the expert domain description with the requirements for generating a practical Bayesian Network.

References

1. J. Pearl: Probabilistic Reasoning in Intelligent Systems, Morgan Kaufmann Publishers, San Francisco, USA, (1998).
2. D. Obradovic et al: Detecting probability of occurrence of specific state of radio communications system using Bayes network probability theory model with at least one accessible state per process variable, patent EP1385348-A1, Siemens AG, Munich, Germany, (2004)
3. J. Horn, M. Pellegrino and R. Lupas Scheiterer: Method for creating a knowledge based causal or Bayesian network by carefully structuring base knowledge data so that the causal network can be automatically generated by a computer, patent WO2003005297-A2, Siemens AG, Munich, Germany, (2004)
4. J. Horn, T. Birkhölzer, O. Hogl, M. Pellegrino, R. Lupas Scheiterer, K. -U. Schmidt, V. Tresp: Knowledge Acquisition and Automated Generation of Bayesian Networks for a Medical Dialogue and Advisory System,, in S. Quaglini, P. Barahona, S. Andreassen, Eds, Artificial Intelligence in Medicine, Springer-Verlag, (2001), 199-202.
5. D. Heckerman: A Tutorial on Learning with Bayesian Networks. In Learning in Graphical Models, M. Jordan, ed. MIT Press, Cambridge, MA, (1999).
6. R. Lupas Scheiterer, D. Obradovic: Bayesian Network Modeling Aspects Resulting from Applications in Medical Diagnosis and GSM Troubleshooting, submitted to ICANN 2005, Krakow, Poland, (2005).

Energy of Brain Potentials Evoked During Visual Stimulus: A New Biometric?

Ramaswamy Palaniappan[1] and Danilo P. Mandic[2]

[1] Department of Computer Science, University of Essex, Colchester, United Kingdom
rpalan@essex.ac.uk
[2] Department of Electrical and Electronic Engineering, Imperial College London, United Kingdom
d.mandic@imperial.ac.uk

Abstract. We further explore the possibility of using the energy of brain potentials evoked during processing of visual stimuli (VS) as a new biometric tool, where biometric features representing the energy of high frequency electroencephalogram (EEG) spectra are used in the person identification paradigm. For convenience and ease of processing of cognitive processing, in the experiments, simple black and white drawings of common objects are used as VS. In the classification stage, the Elman neural network is employed to classify the generated EEG features. The high recognition rate of 99.62% on an ensemble of 800 raw EEG signals indicates the potential of the proposed method.

1 Introduction

Over the last decade or so, there has been ongoing research into the possibility of employing some alternative biometrics for identifying individuals, instead of the standard one based on fingerprints [1]. These include techniques which focus on:- face [2], palm print [3], hand geometry [4], heart signal [5], iris [6], odor [7], and brain signals [8-10]. The brain *fingerprints* have also been studied for aiding criminal investigations [11]. Methods based on the use of brain electrical signals (electroencephalogram (EEG)) as a biometric are relatively recent compared to the other established biometric tools. Paranjape *et al* [8] achieved the classification accuracy ranging between 49% and 85%, by using autoregressive (AR) modelling of EEG in combination with discriminant analysis. Poulus *et al* looked at the problem of distinguishing between individuals, based on a set of EEG recordings [9]. Their objective was to find an individual as distinct from other individuals as possible [9]. Their method was based on AR modelling of EEG signals and Linear Vector Quantisation neural network (NN), with 72-80% of classification success. However, this method was not tested on the task of recognition of individual subjects.

In this paper, we provide further perspective on the possibility of EEG based person identification. This is an extension of our approach proposed in [10], where features computed from 61 EEG channels, were used for person identification. The fact that it is virtually impossible that different persons will have similar activity in all parts of the brain, and that brain responses cannot be faked, makes this method suitable for the use in biometric applications. The extracted EEG biometric features are processed with the Elman NN (ENN) to classify (that is recognise) different persons.

W. Duch et al. (Eds.): ICANN 2005, LNCS 3697, pp. 735 – 740, 2005.

Despite its relative simplicity and high success ratio, the method proposed in [10] suffered from several drawbacks, such as a decrease in the accuracy with the increase in the size of the recorded dataset and number of individuals to classify, and a relatively narrow frequency range used in the processing of signals. This paper therefore introduces several improvements in order to increase the success rate of person identification. This is supported by a comprehensive analysis and experimentation tackling all the aspects of the method, such as pre-filtering, usable frequency range, postprocessing and dimensionality reduction.

2 Data and Experiment

EEG biometrics as a data fusion problem: The processing of EEG recordings coming from multiple electrodes may be considered as a multi-channel signal processing problem. Notice, however, that the electrodes on the scalp of a subject are located so as to record the electrical activity of different brain areas. These areas in the cortex are responsible for a variety of cognitive and motor tasks, and the brain electrical activity recorded from these spatially distributed electrodes reflects the nature of the task being processed. For instance, the P3 area is responsible for decision making processes arising from visual stimuli [12]. Therefore, the processing of multi-channel EEG recordings represents a data fusion problem, since we combine the data coming from different information processing mechanisms within the brain.
Data used: We use a non-invasive technique based on the EEG signals recorded from the scalp. EEG signals are potentials exhibited by neuronal excitations in the cortex [13], and were recorded with the subjects observing drawings of common black and white objects.
Data processing: To obtain high frequency EEG signals in the gamma band range (30-70 Hz), filtering was performed, and the energy of these signals was used as a set of features (after some preprocessing) to be classified by the Elman neural network (ENN) [14] trained by the resilient backpropagation (RB) algorithm. This frequency band was suggested by other studies [15, 16] which have, for instance, successfully used gamma band spectral features to differentiate between alcoholics and non-alcoholics. Gamma band is suitable as EEG oscillations in this frequency band are believed to be involved in feature binding process during visual perception [16].
Data acquisition: The subjects (totalling 40) were seated in a reclining chair located in a sound attenuated RF shielded room. Measurements were taken from 61 channels placed on the subject's scalp, sampled at 256 Hz. The electrode positions were according to the extension of Standard Electrode Position Nomenclature, recommended by the American Encephalographic Association.
Visual stimuli: The EEG signals were recorded from subjects while being exposed to a stimulus, which consist of drawings of objects chosen from Snodgrass and Vanderwart picture set [17]. These pictures represent common black and white objects, such as, for instance, airplane, banana, and ball. These were chosen according to a set of rules that provides consistency of pictorial contents. They have been standardised based in the variables of central relevance to memory and cognitive processing, for instance, objects can be named (definite verbal labels).

Mental task: The subjects were asked to remember or recognise the stimulus. Stimulus duration of every picture was 300 ms with an inter-trial interval of 5.1 s. All the stimuli were shown using a display located 1 meter away from the subjects. One-second EEG measurements after each stimulus onset were stored. Figure 1 illustrates the stimulus presentation. This data set used is a subset of a larger experiment designed to study the short-term memory [18].
Artifact removal: EEG signals contaminated with eye blink artifacts were not considered in the classification, and were detected using a 100 μV threshold. This is a common threshold value in EEG studies, and is used since blinking produces 100-200 μV potential lasting 250 milliseconds [19]. A total of 40 artifact free trials were considered for every subject, to make a total 1600 EEG data sets.

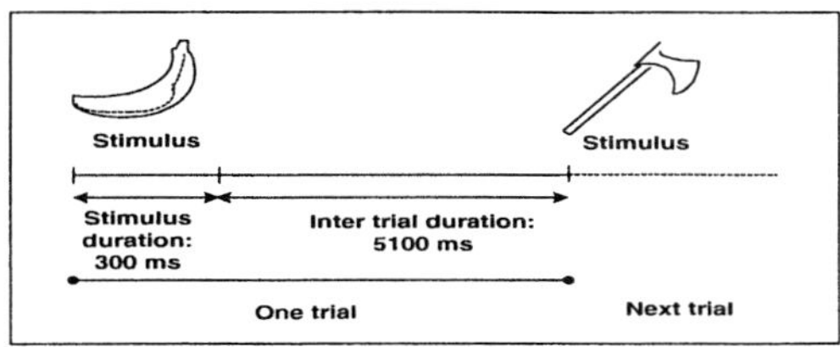

Fig. 1. Example of visual stimulus presentation

3 Method

Original method: In the original method proposed in [10], the EEG signals were filtered using a forward and reverse Butterworth band-pass digital filter, to obtain zero phase distortion. The 3-dB pass-band was chosen to be between 30 and 50 Hz, whereas the stop-band was fixed at 28 and 52 Hz. A model order of 14 was used to attain a minimum stopband attenuation of 20 dB. To form the EEG features, the energy of the EEG signal from each channel was computed and normalised according to the total energy from all 61 channels. These 61 EEG features were then classified by a multi-layer perceptron (MLP) NN trained by a standard backpropagation algorithm [20]. The training was conducted until the average error fell below 0.01.
Improved method: In the proposed method in this paper, several improvements were made for every aspect of the method:-

i) EEG signals were filtered (in the forward and reverse direction) using Elliptic filter as this required a lower order as compared to the Butterworth filter;

ii) frequency range was changed from 30-70 Hz because of the reported existence of gamma band oscillations in this range [21];

iii) order 5 was sufficient to obtain a 3-dB passband of 30-70 Hz with minimum stopband attenuation of 30 dB below 25 Hz and above 75 Hz. The ripple in the passband was kept below 0.1 dB;

iv) energy of the filtered EEG signal from each channel were computed and normalized with the total energy from 61 channels to form the EEG features;

v) EEG features were normalised to unit variance and zero mean;
vi) principal component analysis (PCA) was used to reduce the feature set by selecting the more discriminating features.

The PCA setting: The standard PCA method [21] was used, where variable **z** represents the extracted signal, for which its covariance matrix is computed as

$$\mathbf{R} = \mathbf{E}(\mathbf{z}\mathbf{z}^{\mathbf{T}}) \;, \tag{1}$$

where **E**(.) is the mathematical expectation operator. Next, matrices **V** and **D** were computed, where **V** is the orthogonal matrix of eigenvectors of **R** and **D** is the diagonal matrix of its eigenvalues, that is $D=diag(d_1,...,d_n)$. The features with reduced dimensionality **y** were then found as

$$\mathbf{y} = \mathbf{V}_r^{\mathbf{T}}\mathbf{z}^{\mathbf{T}} \;, \tag{2}$$

where $\mathbf{V_r}$ denotes the reduced eigenvector matrix corresponding to the selected principal components (PCs). In our work, the PCs that contribute to 99.9% of the total variance were selected, which amounted to 52 features. These features were normalised to the range [-1,1], using maximum and minimum values of each feature, with the idea to improve the NN training.

Neural network classification: ENN was used for feature processing. The ENN is effectively a MLP in which the hidden layer outputs are delayed and fed back into the network, thereby providing a state feedback. A three-layer network was used here, with the hyperbolic tangent activation function in its hidden layer, and a sigmoid activation function in its output layer. As compared to the standard MLP network, only one hidden layer, but with more hidden neurons is needed for the function approximation task. Network weights and biases were initialised following the Nguyen-Widrow algorithm [22]. This algorithm distributes the active region of each neuron in the layer evenly across the layer's input space, which is advantages to speed up training and to efficiently use the available neurons. After some preliminary simulations, the resilient-backpropagation (RB) algorithm [23] was used to train the ENN, and the training was conducted until the mean-square error fell below a threshold of 0.0001.

4 Results and Discussion

Table 1 shows the classification results based on the original and improved methods. For both the methods, 800 EEG patterns (20 from each subject) were used to train the NNs, while the previously unseen 800 EEG patterns were used to test the classification performance (%).

In the original method [10], the numbers of hidden units were varied from 10 to 50 in steps of 10 but using the current dataset, the classification performances were less than 93% (except for 50 hidden units) using these numbers of hidden units and are therefore not reported here. The poorer classification performance is most likely due to the increase in the complexity of the dataset, due to an increase in the number of subjects and patterns. In addition, to ensure a fairer comparison with the proposed improved method, the numbers of hidden units were varied from 50 to 300 in steps of 50. The ENN required a higher number of hidden units due to the increased state feedback (outputs of hidden neurons). The number of epochs (iterations), training and testing times (in seconds) for 800 EEG patterns are also shown in the Table. We chose to compare the execution times, rather than the number of operations needed, which is done for convenience. Since the algorithms were run on the same code, using Matlab, this still provides a fair comparison of the complexity of the algorithms. From the Tables, it can be seen that the proposed improved method gives better classification performance in addition to fewer training epochs and a decrease in the training duration. This applies for all the cases and sizes of hidden units.

Table 1. Classification results using the original and improved methods

Original method					Improved method				
Hidden units	Epochs	Train time (s)	Testing time(s)	%	Hidden units	Epochs	Train time (s)	Testing time(s)	%
50	191	51.85	0.19	95.50	50	40	4.26	0.06	98.75
100	161	54.52	0.20	95.50	100	32	6.81	0.10	99.12
150	217	96.75	0.25	95.87	150	25	8.29	0.14	99.62
200	151	83.59	0.29	96.13	200	32	14.12	0.19	99.00
250	185	122.21	0.35	95.37	250	28	17.29	0.26	99.00
300	157	120.67	0.39	95.75	300	31	24.48	0.33	99.00
Average	177.0	88.27	0.28	95.69	Average	31.3	12.54	0.18	99.08

5 Conclusion

In this paper, we embark upon the results from [10] and propose an improved method for employing EEG features as a biometric to identify individuals. This is achieved in a data fusion setting, where the energy of high frequency EEG signals has been used as a classification criterion, and has obtained when subjects were seeing a common black and white line drawing of common objects. The features have undergone several pre- and post-processing operations, to be used as features for classification by Elman neural network. The results obtained have shown the potential in applications such a stand alone individual identification system or as a part of a multi-modal individual identification system.

Acknowledgement

We thank Prof. Henri Begleiter at the Neurodynamics Laboratory at the State University of New York Health Centre at Brooklyn, USA who generated the raw EEG data and Mr. Paul Conlon, of Sasco Hill Research, USA for assisting us with the database.

References

1. Pankanti, S, Bolle, R.M., Jain, A.: Biometrics: The Future of Identification. Special issue of IEEE Comp. on Biometrics (2000) 46-49
2. Samal, A., Iyengar, P.: Automatic recognition and analysis of human faces and facial expressions: A survey. Pattern Recognition, Vol. 25, No. 1 (1992) 65-77

3. Duta, N., Jain, A.K., Mardia, K.V.: Matching of Palmprints. Pattern Recognition Letters, Vol. 23, No. 4 (2002) 477-485
4. Jain, A.K., Ross, A., Pankanti, S.: A Prototype Hand Geometry-based Verification System. Proc. Int. Conf. on Audio & Video-Based Biometric Person Identification (1999) 166-171
5. Biel, L., Pettersson, O., Philipson, L., Wide, P.: ECG Analysis: A New Approach in Human Identification. IEEE Trans. Instrument & Measurement, Vol. 50, No.3 (2001) 808-812
6. Daugman,J.:Recognizing Persons by Their Iris Patterns.In:Jain,A.K.,Bolle,R.,Pankanti, S.,(eds.):Biometrics:Personal Identification in Networked Society.Kluwer Academic (1999)
7. Korotkaya, Z.: Biometric Person Authentication: Odor. Available online: http://www.it.lut.fi/kurssit/03-04/010970000/seminars/Korotkaya.pdf (2003)
8. Paranjape, R.B., Mahovsky, J., Benedicenti, L., Koles, Z.: The Electroencephalogram as a Biometric. Proc. Canadian Conf. on Elect. & Comp. Eng., Vol. 2 (2001) 1363-1366
9. Poulos, M., Rangoussi, M., Chrissikopoulos, V., Evangelou, A.: Person Identification Based on Parametric Processing of the EEG. Proc. IEEE Int. Conf. on Electronics, Circuits, and Systems, Vol. 1 (1999) 283-286
10. Palaniappan, R.: A New Method to Identify Individuals Using VEP Signals and Neural Network. IEE Proc. - Science, Measurement and Tech., Vol. 151, No. 1 (2004) 16-20
11. Farwell, L. A. Smith, S. S.: Using Brain MERMER Testing to Detect Concealed Knowledge Despite Efforts to Conceal. J. of Forensic Sciences, Vol. 46 (2001) 1-9
12. Polich, J.: P300 in Clinical applications: Meaning, method, and measurement. In: Niedermeyer, E., da Silva, F.L., (eds.): Electroencephalography Basic Principles, Clinical Applications, and Related Fields. William and Wilkins, Baltimore (1993) 1005-1018
13. Misulis K.E.: Spehlmann's Evoked Potential Primer: Visual, Auditory and Somatosensory Evoked Potentials in Clinical Diagnosis. Butterworth-Heinemann (1994)
14. Elman, J. L.: Finding structure in time. Cognitive Science, Vol. 14 (1990) 179-211
15. Palaniappan, R., Raveendran, P., Omatu, S.: EEG Optimal Channel Selection Using Genetic Algorithm for Neural Network Classification of Alcoholics. IEEE Trans. Neural Networks, Vol. 13, No. 2 (2002) 486-491
16. Basar, E., Eroglu, C.B., Demiralp, T., Schurman, M.: Time and Frequency Analysis of the Brain's Distributed Gamma-Band System. IEEE Eng. in Med. & Bio. Mag. (1995) 400-410
17. Snodgrass, J.G., Vanderwart, M.: A Standardized Set of 260 Pictures: Norms for Name Agreement, Image Agreement, Familiarity, and Visual Complexity. J. of Exp. Psychology: Human Learning and Memory, Vol. 6, No. 2 (1980) 174-215
18. Zhang, X.L., Begleiter, H., Porjesz, B., Wang, W., Litke, A.: Event related potentials during object recognition tasks. Brain Research Bulletin, Vol. 38, No. 6 (1995) 531-538
19. Kriss, A.: Recording Technique. In: Halliday, A.M. (ed.): Evoked Potentials in Clinical Testing. Churchill Livingstone (1993)
20. Rumelhart, D.E., McCelland, J.L.: Parallel Distributed Processing: Exploration in the Microstructure of Cognition. MIT Press, Cambridge, MA, Vol. 1 (1986)
21. Jolliffe, I.T.: Principal Component Analysis. Springer-Verlag (1986)
22. Nguyen, D., Widrow, B.: Improving the learning speed of 2-layer neural networks by choosing initial values of the adaptive weights. Proc. Int. J. Conf. on Neural Networks, Vol. 3 (1990) 21-26
23. Riedmiller, M., Braun, H: A direct adaptive method for faster backpropagation learning: The RPROP algorithm. Proc. IEEE Int. Conf. on Neural Networks (1993) 586-591

Communicative Interactivity – A Multimodal Communicative Situation Classification Approach

Tomasz M. Rutkowski[1,*] and Danilo Mandic[2]

[1] Academic Center for Computing and Media Studies, Kyoto University, Kyoto, Japan
[2] Department of Electrical and Electronic Engineering, Imperial College London, United Kingdom
tomek@mm.media.kyoto-u.ac.jp
d.mandic@imperial.ac.uk

Abstract. The problem of modality detection in so called Communicative Interactivity is addressed. Multiple audio and video recordings of human communication are analyzed within this framework, based on fusion of the extracted features. At the decision level, Support Vector Machines (SVM) are utilized to segregate between the communication modalities. The proposed approach is verified through simulations on real world recordings.

1 Introduction

Multimodal interfaces are an emerging interdisciplinary discipline which involves different modalities of a generic communication process, such as speech, vision, gestures, and haptic feedback. The main goal is to enable better understanding and hence more convenient, intuitive, and efficient interaction between humans and computers. Further requirements are that the users ought to interact with such technology in a natural way, without the need for special skills. Emerging work on communicative activity monitoring addresses the problem of automatic activity evaluation in audio and visual channels for distance learning applications [1]. These approaches, however, are limited in that they focus only on separated activity in communicative interaction evaluation, without considering other aspects of the behavioral interdependence in communication.

In this work, we present an attempt to combine knowledge from human communication theory and signal/image processing in order to provide an intelligent way to evaluate communicative situations among humans. This analysis will be used in later stages for implementation of communicative interaction models in human–machine interfaces. For the purpose of evaluation of multimodal interaction, in order to classify them according to "communicative intelligibility" (a measure of potential affordance [2] (usability) of the analyzed interaction), it is necessary to first identify certain illustrative communicative situations from the recorded multimedia streams. We next provide some theoretical background,

* Currently at: Brain Science Institute RIKEN, Saitama, Japan.

W. Duch et al. (Eds.): ICANN 2005, LNCS 3697, pp. 741–746, 2005.

which is followed by a proposal of a multidimensional interaction evaluation engine, supported with some experimental results.

2 Multimodal Features

The underlying aim of this study is to identify those audio–visual features of the (human) communication process that can be tracked and which, from an information processing point of view, are sufficient to create and recreate the climate of a meeting ("communicative interactivity"). This communicative interactivity analysis provides a theoretical, computational and implementational framework in order to characterize the human–like communicative behavior.

The analysis of spoken communication is an already mature field, and following the above arguments, our approach will focus on the dynamical analysis of non–spoken components of the communication. In the proposed model of communicative interactivity, based on interactive (social) features of captured situations, two sensory modalities (visual and auditory) of communicative situations are utilized.

The working hypothesis underlying our approach is therefore that observations of the non–verbal communication dynamics contain sufficient information to allow us to estimate the climate of a situation, that is, the communicative interactivity. To that end, the multimodal information about the communication environment must be first separated into the *communication related* and *environmental* components. Notice the analogy to the Wold decomposition theorem which states that every signal can be decomposed into its deterministic and stochastic part. In this way, the audio and video streams can be separated into the information of interest and background noise [3,4].

Audio Features: Since the audio signal carries much redundant information, techniques used in speech recognition, such as the mel-frequency cepstrum coefficients (MFCC) [5] can be used for feature extraction and compression. In our work we found that the first 24 MFCC coefficients were enough as audio fetaures.

Video Features: We desire to obtain video features that carry information about the communication–related motion, and are also compatible with the audio features. Two modalities: the search for faces and moving contours are combined to detect communicating humans in video. This is achieved as follows: for two consecutive video frames $\mathbf{f}_{[h\times w]}(t-1)$ and $\mathbf{f}_{[h\times w]}(t)$, the temporal gradient is expressed as a smoothed difference between the images convoluted with a two-dimensional Gaussian filter $\mathbf{g}$ with the adjusted standard deviation σ. The pixel $\mathbf{G}_{[h\times w]}(t,n,m)$ of the gradient matrix at time t is calculated as follows:

$$\mathbf{G}_{[h\times w]}(t,n,m) = \left|\sum_{i=1}^{x}\sum_{j=1}^{y}\mathbf{d}_{[h\times w]}(t,n-i,m-j)\mathbf{g}_{[x\times y]}(\sigma,i,j)\right|, \qquad (1)$$

where $\mathbf{d}_{[h\times w]}(t)$ is the difference between consecutive frames of the size $h\times w$ pixels: $\mathbf{d}_{[h\times w]}(t) = \mathbf{f}_{[h\times w]}(t) - \mathbf{f}_{[h\times w]}(t-1)$. The absolute value is taken to remove the gradient directional information and to enhance the movement capture.

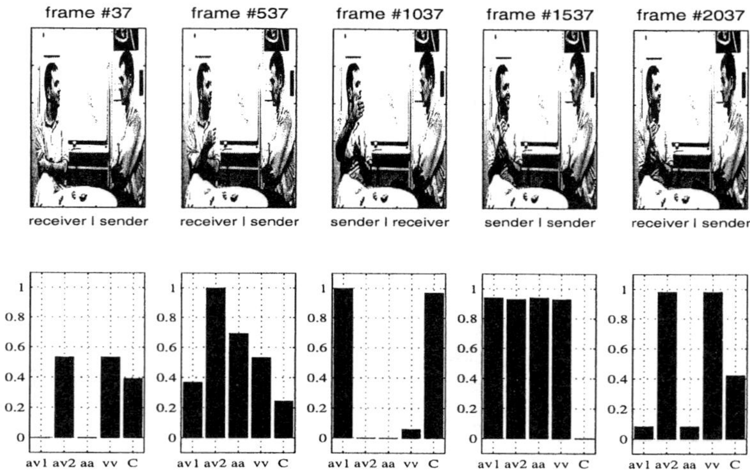

Fig. 1. All mutual information tracks together with efficiency estimate (av1; av2; aa; vv; C; stand respectively for $I_{A_1V_1}; I_{A_2V_2}; I_{A_1A_2}; I_{V_1V_2}; C(t);$)

For the face tracking, the features used were: i) skin color (in three– dimensional color space domain, albeit very sensitive to illumination variations) [6], ii) moving face pattern recognition with patterns obtained using non-negative matrix factorization method [7]. Features obtained in this way (details in [8]) are more localized and correspond to the intuitively perceived parts of faces (face, eyes, nose, and mouth contours).

So extracted facial features, together with the audio synchronized information, permit us to classify a participant in communication as being engaged in *talking*, *listening*, and *responding*. To enable compatibility with the extracted audio features and for information compression, the two–dimensional discrete cosine transformation (DCT) is used, where most of the energy within the processed image is contained in a few uncorrelated DCT coefficients (24 DCT coefficients in our approach [8,4]).

3 Evaluation of Communicative Interactivity

Communicative interactivity evaluation assesses the behavior of the participants in the communication from the audio–visual channel, and reflects their ability to "properly" interact in the course of conversation. This is quantified by synchronization and interaction measures [4]. In [8] the *communication efficiency* is defined as *a measure that characterizes the behavioral coordination of communicators.* Here, a measure of the communication efficiency is proposed as a combination of four estimates of mutual information [9]: i) two visual (V_i), ii) two audio (A_i), and iii) two pairs of audiovisual features ($A_i; V_i$).

Presence of communication is judged based on mutual information(s) for selected regions of interest (ROI) [10], as

$$\begin{aligned} I_{A_iV_i} &= H(A_i) + H(V_i) - H(A_i, V_i) \\ &= \frac{1}{2}\log(2\pi e)^n |R_{A_i}| + \frac{1}{2}\log(2\pi e)^m |R_{V_i}| - \frac{1}{2}\log(2\pi e)^{n+m} |\mathbf{R}_{A_iV_i}| \\ &= \frac{1}{2}\log\frac{|R_{A_i}||R_{V_i}|}{|R_{A_iV_i}|}, \end{aligned} \quad (2)$$

where $i = 1, 2$ and R_{A_i}, R_{V_i}, $R_{A_iV_i}$ stand for empirical estimates of the corresponding covariance matrices of the feature vectors [8] (computed recursively).

Simultaneous activity estimates in the same modes (audio and video respectively) are calculated for video and audio streams, respectively, as:

$$I_{V_1V_2} = \frac{1}{2}\log\frac{|R_{V_1}||R_{V_2}|}{|R_{V_1V_2}|} \quad \text{and} \quad I_{A_1A_2} = \frac{1}{2}\log\frac{|R_{A_1}||R_{A_2}|}{|R_{A_1A_2}|}, \quad (3)$$

where $R_{A_1A_2}$ and $R_{V_1V_2}$ are the empirical estimates of the corresponding covariance matrices for unimodal feature sets representing different communicator activities. Quantities $I_{A_1V_1}$ and $I_{A_2V_2}$ evaluate the local synchronicity between the audio (speech) and visual (mostly facial movements) flows and it is expected that the sender should exhibit the higher synchronicity, reflecting the higher activity. Quantities $I_{V_1V_2}$ and $I_{A_1A_2}$ are related to the possible crosstalks in same modalities (audio–audio, video–video). The latter is also useful to detect the possible activity overlapping, which can impair the quality of the observed communication. A combined measure of temporal communication efficiency can be calculated as (see Figure 1. for dynamic track of multimodal features):

$$C(t) = \left(1 - \frac{I_{V_1V_2}(t) + I_{A_1A_2}(t)}{2}\right) |I_{A_1V_1}(t) - I_{A_2V_2}(t)|, \quad (4)$$

The communicator's role, that is, (*sender* or *receiver*) can be estimated by monitoring the behaviour of audiovisual features over time. An indication of higher synchronization across the audio and video features characterizes the active member - the sender, while the lower one indicates the receiver. This synchronized audiovisual behavior of the sender and the unsynchronized one of the receiver characterizes an efficient communication [8,4].

The pair of the mutual information estimates for the local synchronization of the senders and the receivers in Equation (2) is used to give clues about concurrent individual activities during the communication event, while the unimodal cross-activities estimates in Equations (3), are used to evaluate the interlaced activities for a further classification. Intuitively, the efficient sender–receiver interaction involves *action* and *feedback*. The interrelation between the actions of a sender and feedback of a receiver is therefore monitored, whereby the audiovisual synchronicity is used to determining the roles.

In our approach, the interactions between individual participants in communication are modelled within the *data fusion* framework, based on features coming simultaneously from both the audio and video. A multistage and a multisensory classification engine [11] based on the support vector machine (SVM) approach is used at the *decision making* level of the data fusion framework, where the *one-versus-rest-fashion* approach is used to identify the phases during ongoing communication (based on the mutual information estimates from Equations (2) and (3)).

At the Decision Level. SVMs are particularly suited when *sender–receiver* or *receiver–sender* situations are to be discriminated from the *noncommunicative* or *multi–sender* cases. In this work, a kernel based on a radial basis function (RBF) is utilized [12], and is given by

$$K(\mathbf{x}, \mathbf{x}_i) = e^{-\gamma||\mathbf{x}-\mathbf{x}_i||^2} \tag{5}$$

where γ is a kernel–function width parameter.

Using the above concept, an arbitrary temporal multimodal mutual information estimates from Equations (2) and (3) together with efficiency from Equation (4) can be categorized online into four categories: i) for the *noncommunicative* case with no interaction (no communication or a single participant); ii) for the *sender–receiver* case; iii) for the *receiver–sender* case; iv) for the *sender–sender* case. The categories i) and iv) are somehow ambiguous due to the lack of clear separation boundaries, and they are treated by separately trained SVM classifiers.

4 Experimental Results and Conclusions

The experiments, where the participants in communication were engaged in a face–to–face conversation, were conducted to validate the proposed approach. The proposed approach was able to evaluate the communication interactivity level with similar performance to that of subjective evaluations of human experts for the analyzed videos as summarized in Table 1. This way, the proposed *data fusion* approach for the evaluation of communicative interaction represent a step forward in the modeling of communication situation, as compared to the existing audio– and video–only approaches [1]. The experiments have clearly shown the possibility to estimate the interactivity level, based on the behavioral analysis of the participants in communication. The mutual information based feature extraction of multimodal audio and video data streams makes it possible to detect the presence participants and to classify them according their role. Despite some difference between the conclusions of a seven human experts and the proposed method, our results show strong correlation between the two. In

Table 1. Comparison of objective and subjective (seven experts) communication interactivity evaluations (the score around 100% would suggest fully interactive event, while lower one characterizes overlapped discourse between communicators)

Case	Objective (proposed method)	Subjective (human experts)	Method's error
teacher & student #1	51%	50%	1%
teacher & student #2	63%	60%	3%
friends #2	75%	80%	−5%

fact, the human judgement is also highly subjective, therefore further studies will have a larger population of human experts to balance their opinion.

Acknowledgements

Authors would like to thank Prof. Toyoaki Nishida, Prof. Michihiko Minoh, and Prof. Koh Kakusho of Kyoto University for their support and fruitful discussions in frame of the project *Intelligent Media Technology to Support Natural Human Communication*, where the presented approach was developed.

References

1. Chen, M.: Visualizing the pulse of a classroom. In: Proceedings of the Eleventh ACM International Conference on Multimedia, ACM Press (2003) 555–561
2. Gibson, J.J.: The theory of affordances. In Shaw, R., Bransford, J., eds.: Perceiving, Acting and Knowing. Erlbaum, Hillsdale, NJ (1977)
3. Rutkowski, T.M., Yokoo, M., Mandic, D., Yagi, K., Kameda, Y., Kakusho, K., Minoh, M.: Identification and tracking of active speaker's position in noisy environments. In: Proceedings of International Workshop on Acoustic Echo and Noise Control (IWAENC2003), Kyoto, Japan (2003) 283–286
4. Rutkowski, T.M., Kakusho, K., Kryssanov, V.V., Minoh, M.: Evaluation of the communication atmosphere. In Negoita, M., Howlett, R.J., Jain, L.C., eds.: Proceedings of 8th International Conference on Knowledge-Based Intelligent Information and Engineering Systems KES 2004, Part I. Volume 3215 of Lecture Notes in Computer Science., Wellington, New Zealand, Springer-Verlag Heidelberg (2004) 364–370
5. Becchetti, C., Ricotti, L.P.: Speech Recognition. John Wiley & Sons, Inc., Great Britain (1999)
6. Sigan, L., Sclaroff, S., Athitsos, V.: Estimtion and prediction of evolving color distributions for skin segmentation under varying illumination. In: Proceedings of the IEEE Conference on Computer Vision and Pattern Recognition, (CVPR). Volume 2., IEEE (2000) 152–159
7. Lee, D.D., Seung, H.S.: Learning the parts of objects by non-negative matrix factorization. Nature **401** (1999) 788–791
8. Rutkowski, T.M., Seki, S., Yamakata, Y., Kakusho, K., Minoh, M.: Toward the human communication efficiency monitoring from captured audio and video media in real environments. In Palade, V., Howlett, R.J., Jain, L.C., eds.: Proceedings of 7th International Conference on Knowledge-Based Intelligent Information and Engineering Systems KES 2003, Part II. Volume 2774 of Lecture Notes in Computer Science., Oxford, UK, Springer-Verlag Heidelberg (2003) 1093–1100
9. Shannon, C., Weaver, W.: The Mathematical Theory of Communication. University of Illinois Press, Urbana (1949)
10. Hyvarinen, A., Karhunen, J., Oja, E.: Independent Component Analysis. John Wiley & Sons (2001)
11. Hsu, C.W., Lin, C.J.: A comparison of methods for multi-class support vector machines. IEEE Transactions on Neural Networks **13** (2002) 415–425
12. Cherkassky, V., Mulier, F.: Learning from Data. Adaptive and Learning Systems for Signal Processing, Communication, and Control. John Wiley & Sons, Inc., USA (1998)

Bayesian Network Modeling Aspects Resulting from Applications in Medical Diagnostics and GSM Troubleshooting

Ruxandra Lupas Scheiterer and Dragan Obradovic

Siemens AG, Corporate Technology, Information and Communications

Abstract. This paper addresses issues in constructing a Bayesian Network (BN) domain model for diagnostic purposes from expert knowledge. The novelty of this paper is the approach for structured generation of a model that incorporates the unstructured multifaceted and possibly conflicting probabilistic information provided by the experts.

1 Introduction

Diagnostic systems are prime examples for data fusion, since they rely on a multitude of measurements, composite indicators, and human models of the domain. These models describe causal relationships between problems and observables, and are typically obtained by interviewing domain experts [1], [2] . We examine modeling aspects in building a Bayesian Network[1] for diagnosis assistance. Our considerations apply to any domain governed by cause and effect principles.

2 Forward and Backward Modeling

2.1 2 Problems

Consider a single binary symptom or indicator K that is causally dependent on two binary problems P_1 and P_2, as shown in Fig. 1. The table lists the conditional probabilities $P(K|P_1, P_2)$ for all 2^3 values of the three random variables, for example $P(K = 1|P_1 = 0, P_2 = 0) = l$, the so-called "leak" probability, which is the probability that the effect is present even though none of the causes within the considered domain is present. In order to determine a causal model, [3], the probabilities of the root nodes and the conditional probability tables of any child node given its parents have to be specified. In this case this means specifying the probability distributions of the random variables P_1, P_2 and $K|P_1, P_2$, that is, the following 6 numbers, the other being determined by the requirement that the probability of disjoint mutually exhaustive events sums to one (we abbreviate $P(R) = P(R = 1)$, $P(\overline{R}) = P(R = 0)$):

$$\begin{array}{lll} P(P_1), & P(K|P_1,\overline{P_2}), = p_1, & P(K|P_1,P_2), = p \text{ (free parameter)} \\ P(P_2), & P(K|\overline{P_1},P_2), = p_2, & P(K|\overline{P_1},\overline{P_2}), = 0 \text{ or } l \text{ (leak, free parameter)} \end{array}$$

[1] Bayesian or Causal Networks are directed acyclic graphs whose edges indicate causality and whose structure captures the joint probability functions of the node variables.

W. Duch et al. (Eds.): ICANN 2005, LNCS 3697, pp. 747–752, 2005.

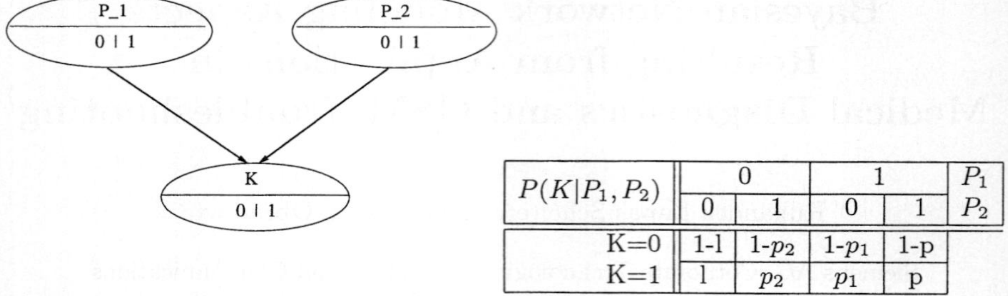

$P(K\|P_1,P_2)$	0		1		P_1
	0	1	0	1	P_2
K=0	1-l	1-p_2	1-p_1	1-p	
K=1	l	p_2	p_1	p	

Fig. 1. Cause-to-effect model of simple domain with two problems and a single common symptom

The expert can readily estimate the values of the marginal probabilities $P(P_1)$, $P(P_2)$ which are the probabilities that problem i is present, as well as the values p_1, p_2 that the symptom is present given that exactly one of the alternative causes is present. But it is prohibitive to expect him to estimate the value of p, because two (or more) causes being present at the same time is such a rare event that he has no intuition about this estimate.

He has no problem though with the values of the *backward* probabilities $P(K)$, $P(P_1|K)$ and $P(P_2|K)$, since this is how he reasons when doing troubleshooting. The first value is the probability that the symptom K has a conspicuous value, and the other numbers are the probabilities that problem P_i is present given that the symptom K has a conspicuous value.

When constructing the model from probability estimates provided by the expert or in the presence of incomplete data, it is desirable to use exactly those probabilities that the expert can specify with the highest confidence. These are:

$$\begin{array}{llll} P(P_1), & P(K|P_1,\overline{P_2}), = p_1 & P(K), & P(P_1|K) \\ P(P_2), & P(K|\overline{P_1},P_2), = p_2 & & P(P_2|K) \end{array}$$

that is, a mixture of forward and backward probabilities. Thinking of the freedom left in the model after specification of the forward probabilities (first 4) we see that *we would like to fix 3 backward probabilities, having only two free parameters, p and l.* This leaves as the only choice a causal model that exhibits probabilities with the "closest" approximation to the desired ones, for example in the minimum mean square error sense (MMSE), but of course any suitable non-quadratic cost function is possible . There are several ways to do the proposed MMSE approximation of the model to the provided expert probabilities.

1. Fix $P(P_1), P(P_2), p_1, p_2$, optimize only over the free parameters p, l, such that the three backward probabilities $P(K)$, $P(P_1|K)$, $P(P_2|K)$ are approximated as closely as possible. That is, define the cost function:

$$\begin{aligned} C_0 = & \left[P(K)_{(P(P_1),P(P_2),p_1,p_2,p,l)} - P^t(K)\right]^2 \\ & + \left[P(P_1|K)_{(P(P_1),P(P_2),p_1,p_2,p,l)} - P^t(P_1|K)\right]^2 \\ & + \left[P(P_2|K)_{(P(P_1),P(P_2),p_1,p_2,p,l)} - P^t(P_2|K)\right]^2 \end{aligned}$$

where in each paranthesis the second term is the target value specified by the expert, and the first term is the value resulting from fixing the functional arguments given in subscript parantheses.

2. Alternatively allow the forward probabilities to deviate from their specified values by a small amount ϵ. That is, define

$$C = C_0 + [P(P_1) - P^t(P_1)]^2 + [P(P_2) - P^t(P_2)]^2 + [p_1 - p_1^t]^2 + [p_2 - p_2^t]^2 \quad (1)$$

where in each summand the first term is the variable, and the second value is the target value specified by the expert. Hence the desired MMSE optimum of the cost function C over the variable space is the solution of the minimization:

$$\min_{p,l \in [0,1]} C \quad \text{subj.to} \quad \begin{matrix} |P(P_1) - P^t(P_1)| < \epsilon_1 \\ |P(P_2) - P^t(P_2)| < \epsilon_2 \\ |p_1 - p_1^t| < \epsilon_3 \\ |p_2 - p_2^t| < \epsilon_4 \end{matrix} \quad \text{s.t.} \quad \begin{matrix} P(K) \in [0,1], \\ P(P_1|K) \in [0,1] \\ P(P_2|K) \in [0,1] \end{matrix} \quad (2)$$

The second minimization includes the first as a special case and will therefore result in a lower or equal minimum at the expense of increased computational effort. It is possible that the expert finds it easy to specify the leak value $l = P(K|\overline{P_1}, \overline{P_2})$. Then there is only one free parameter left, p, with obvious modifications.

2.2 Effect of a Leak from Outside the Domain

The effect K results from an effect KD within the domain at hand and possibly an effect L from outside this domain. The two are combined into K according to the same law governing the P_i combination to K. The following equation holds for the probabilities of K, KD and L:

$$P(\overline{K}) = P(\overline{KD})P(\overline{L}) \quad (3)$$

From this equation we see that by addition of a leak to our model $P(\overline{K})$ can not increase, hence $P(K)$ cannot decrease. Hence adding a leak to the model makes sense if and only if the marginal probability $P(K)$ resulting from inputting the forward probabilitites into the BN is *smaller* than the probability $P(K)$ specified by the expert or resulting from the data. If this is not the case, then the probability $P(K)$ in the Bayesian Network has to be decreased, for example by introduction of an inhibitor [6].

2.3 n Problems

In the case of n problems the probability table of the binary random variable $K|P_1, ..., P_n$ has 2^n entries that have to be specified. The expert finds it feasible to specify for each symptom $2n+2$ probabilities: n forward values of "probability that the symptom is present given that exactly one problem is present", the value of the leak "probability that the symptom is present given that no problem is present", $P(K)$, probability of the symptom being present, and n backward probabilities $P(P_i|K), i = 1, ..., n$, the probabilities that the problem is present given the symptom present. For $n = 2$ the model is under-dimensioned, and can only approximate the expert estimates in for example a mean-square sense. Equality is given for 3 problems, $n = 3$, when $2^n = 2n + 2$, as after constructing the model there are exactly 4 free parameters to capture the 4 specified backward probabilities. For $n > 3$ the model is over-dimensioned compared to what an expert can reasonably specify, and has to be reduced in complexity.

2.4 Noisy-OR, or Reducing Complexity via Proxy Modeling

The classical way to reduce complexity in Bayesian Networks is to model the n-way interaction as "noisy-OR", or as "noisy" versions of AND, MAX, MIN, ADDER [4], SUM or ELENI [5], which in the "noisy-OR" case means associating with each cause a random inhibitory mechanism that prevents the cause from producing the effect, and linking the single noisy causes with an OR function. This model has n unknowns in the $P(K|P_i)$ probability table, one for each inhibitory mechanism, and thus reduces the assessmemt burden from 2^n to n. Note that if a simple OR function would be used to link the causes, instead of a "noisy-OR" function, there would be no parameter whatsoever available to tune the model to the probabilities found in the domain at hand.

The way complexity is reduced in proxy modeling, such as in HealthMan [2], is by introduction of intermediate "noisy-OR" nodes between problem and indicator, Fig. 2. The figure looks analogously with n problems. The significance

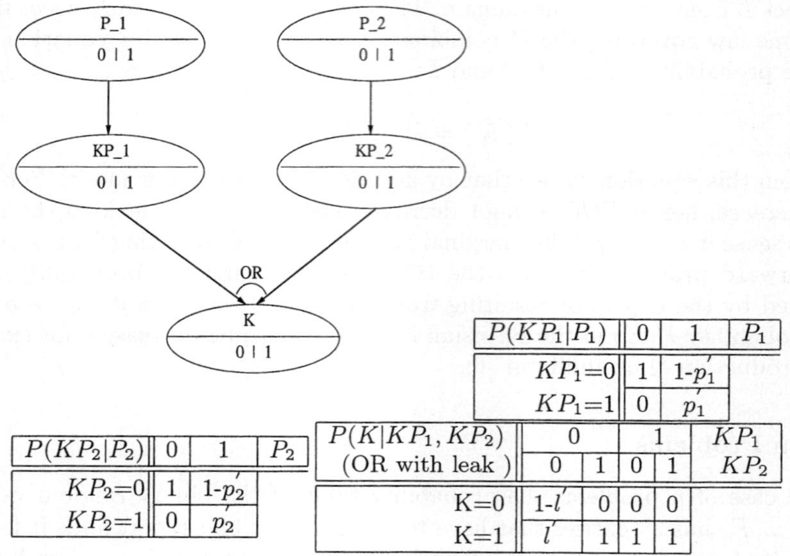

$P(KP_1|P_1)$	0	1	P_1
KP_1=0	1	1-p_1'	
KP_1=1	0	p_1'	

$P(KP_2|P_2)$	0	1	P_2
KP_2=0	1	1-p_2'	
KP_2=1	0	p_2'	

$P(K|KP_1, KP_2)$ (OR with leak)	0		1		KP_1
	0	1	0	1	KP_2
K=0	1-l'	0	0	0	
K=1	l'	1	1	1	

Fig. 2. Causal model of two problems with proxy nodes between problem and indicator layers

of the proxy nodes KP_i, "K due to P_i", is that they capture the event that the indicator K is due to precisely the problem P_i, which is seen from the probability table: $P(KP_i = 1|P_i = 0) = 0$, combined with the following OR-combination of the proxies. That is, given that the P_i is not present, a possible $K = 1$ value cannot be due to P_i.

The following relations hold between the probabilities in the proxy model and those in the direct causal model specified by the expert: 1) The leaks in the two models are identical. 2) The $p_i^{'}$ in the proxy model are related to the p_i of the direct model via: $p_i = p_i^{'} + l * (1 - p_i^{'})$. 3) Thus, in the absence of a leak, the $p_i^{'}$ in the proxy model are identical to the p_i in the direct model. 4) Beside the $p_i^{'}$, in the proxy model there is no free parameter, except for the leak, which is a free parameter only if the expert feels there is a leak, but is not comfortable in estimating it. $P(K|P_1, P_2)$ is not a free parameter, as in the direct model, but fixed by specification of p_1, p_2, l.

So, after specifying the forward probabilities, "noisy-OR" or proxy modeling has only one free parameter, which is often not sufficient to accomodate the backward probabilities the expert can provide. Thus an important issue in reducing complexity is to reduce it to exactly the right amount to incorporate the probabilities that are easy to obtain.

3 The Minimum Forward and Backward Model

Let us call n problem nodes with a common proxy and no leak, having m free parameters, an *n-m-cluster*. Then the *3-6-cluster* shown in Fig.3, has exactly the amount of free parameters necessary to model the 3 forward and 3 backward probabilities resulting from the 3 problem nodes. Other than the simple one problem proxy block (1-1-cluster), we have a range of BN building blocks to choose from, e.g. the 2-3-cluster, the 3-7-cluster (3-6 with leak or inh.) or the 3-6-cluster. To incorporate the $n + 1$ forward and $n + 1$ backward probabilities:

1. Choice one is to lump sets of 3 problems into 3-6-clusters[2], choice two (e.g. due to semantics or to the specific number of problems) into 2-3-clusters, then merge these into the end-of-domain node KD.
2. Be prepared to add a leak respectively inhibitor [6].
3. Compute the values of the free parameters from the given backward probabilities $P(P_i|K)$ and P(K). If there are less unknowns than equations, then via MMSE over the free parameter values between 0 and 1 (i.e. probabilities).

This model has to be tested against its alternatives, a) the simple proxy model and b) the proxy model optimized via MMSE as shown in 2.1, generalized to n problems, in order to avoid overfitting to the prior knowledge. Incorporating mechanisms for learning from the incoming data reduces the dependence of the quality of the prior knowledge, and provides a desirable adaptive component.

[2] The importance of semantically meaningful grouping is a consideration that has to be further investigated in the domain at hand. The resulting architecture should be meaningful to the human expert.

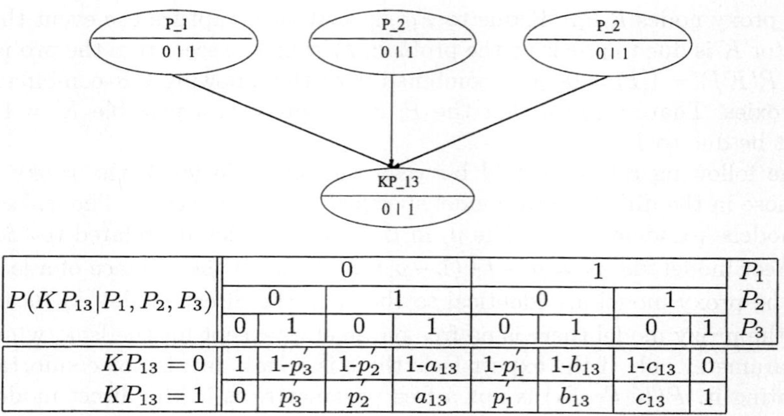

$P(KP_{13}\|P_1,P_2,P_3)$	0				1				P_1
	0		1		0		1		P_2
	0	1	0	1	0	1	0	1	P_3
$KP_{13}=0$	1	$1\text{-}p'_3$	$1\text{-}p'_2$	$1\text{-}a_{13}$	$1\text{-}p'_1$	$1\text{-}b_{13}$	$1\text{-}c_{13}$	0	
$KP_{13}=1$	0	p'_3	p'_2	a_{13}	p'_1	b_{13}	c_{13}	1	

Fig. 3. Definition of a 3-6-cluster

4 Conclusion

It was shown that a typical problem in expert interviewing is the presence of partial conditional probabilities in two directions: from root-causes to symptoms and vice-versa. A suitable parametric model was introduced which can accomodate both directions. The discussed examples were binary, however the applications in [1], [2] deal with mixtures of binary and multivariate variables and the considerations presented here are amenable to multivariate extensions.

References

1. D. Obradovic, R. Lupas Scheiterer, *Troubleshooting in GSM Mobile Telecommunication Networks based on Domain Model and Sensory Information*, ICANN 2005.
2. J. Horn, T. Birkhlzer, O. Hogl, M. Pellegrino, R. Lupas Scheiterer, K.-U. Schmidt, V. Tresp, *Knowledge Acquisition and Automated Generation of Bayesian Networks*, Proc. AIME '01, Cascais, Portugal, July 2001, pp. 35-39.
3. F. V. Jensen, *An Introduction to Bayesian Networks*, UCL Press, 1996.
4. D. Heckerman, *Causal independence for knowledge acquisition and inference*, Proceedings of the 9th Conference on Uncertainty in Artificial Intelligence, Washington D.C., Morgan Kaufmann, San Mateo, Calif., 1993, pp. 122-127.
5. R. Lupas Scheiterer, *HealthMan Bayesian Network Description: Disease to Symptom Layers, Multi-Valued Syptoms, Multi-valued Diseases*, CT IC 4 Internal Report, October 1999.
6. R. Lupas Scheiterer, *BN Modeling Aspects Resulting from the HealthMan and GSM Troubleshooting Applications*, CT IC 4 Internal Report, April 2003.

Fusion of State Space and Frequency-Domain Features for Improved Microsleep Detection

David Sommer[1], Mo Chen[2], Martin Golz[1], Udo Trutschel[3], and Danilo Mandic[2]

[1] Univ. of Appl. Sciences Schmalkalden, Dptm. of CS, 98574 Schmalkalden, Germany
[2] Dptm.of Electr. & Electr. Engin, Imperial College, London, SW7 2BT, UK
[3] Circadian Technologies, Inc. 24 Hartwell Avenue, Lexington, MA 02421, USA

Abstract. A novel approach for Microsleep Event detection is presented. This is achieved based on multisensor electroencephalogram (EEG) and electrooculogram (EOG) measurements recorded during an overnight driving simulation task. First, using video clips of the driving, clear Microsleep (MSE) and Non-Microsleep (NMSE) events were identified. Next, segments of EEG and EOG of the selected events were analyzed and features were extracted using Power Spectral Density and Delay Vector Variance. The so obtained features are used in several combinations for MSE detection and classification by means of populations of Learning Vector Quantization (LVQ) networks. Best classification results, with test errors down to 13%, were obtained by a combination of all the recorded EEG and EOG channels, all features, and with feature relevance adaptation using Genetic Algorithms.

1 Introduction

One of the main problems associated with data fusion for real-world applications is related to combining the information coming from heterogeneous sensors, acquired at different sampling rates and at different time scales. Data/sensor fusion approaches dealing with combining data from homogeneous sensors are normally based either in the time domain, or in some transform domain, for instance on features coming from the frequency representation of signals, their time-frequency, or state-space features [1].

Notice that in this framework we deal with multivariate and multimodal processes, for which either there are no precise mathematical relationships, or if they exist they are too complex. Such is the case with the detection of lapses of attention in car drivers, due to fatigue and drowsiness, the so-called Microsleep Event. Their robust detection is a major challenge. Recent developments in this field have shown that most promising approaches for this purpose are based on a fusion of multiple electrophysiological signals coming from different sources together with Artificial Neural Networks in the detection and prediction [2-4].

In general, there are two standard approaches to combine multiple electroencephalogram (EEG) and electrooculogram (EOG) signals. In the first approach, called **Raw Data Fusion** the sensor data are merged without prior preprocessing or dimensionality reduction. Despite its simplicity, the major disadvantage here is the potentially vast amount of data to be handled. In the second approach, the so-called **Feature Fusion**, features extracted from signals coming from different sources and/or

W. Duch et al. (Eds.): ICANN 2005, LNCS 3697, pp. 753 – 759, 2005.

extracted by different methods are fused. In our investigations, the frequency domain features obtained from the Power Spectral Density (PSD) [4] and state space features obtained from the Delay Vector Variance (DVV) [5, 6], are combined in order to show whether such a combination of different features shows improvement in MSE detection over the standard approaches using only one of the signals and one class of features. The motivation for such an approach is as follows: PSD estimation is a "linear" frequency domain method which can be conveniently performed using the periodogram. This has been shown to perform particularly well in applications related to EEG signal processing, [2-4]. It is, however, natural to ask ourselves whether such an approach, based solely on the second order statistics conveys enough information to provide fast and reliable detection of such a complex event as the MSE.

On the other hand, the recently introduced DVV approach [5, 6] is a method based on the local predictability in the state space. The virtue of the DVV approach is that it can show both qualitatively and quantitatively whether the linear, nonlinear, deterministic or stochastic nature of a signal has undergone a modality change or not. This way, the DVV methodology represents a complement to the widely used linear PSD estimation. Notice that the estimation of nonlinearity associated with the DVV method is intimately related to non-Gaussianity, and we also set ourselves to investigate whether this additional information, which cannot be estimated by PSD, contributes to the discrimination ability, and if so, to estimate its importance level, as compared to the PSD based discrimination.

The purpose of this paper is therefore to provide a theoretical and computational framework for the combination of the two classes of features (PSD and DVV) and to show whether such a combination has the potential benefits for multivariate and multimodal signals over standard approaches. This is illustrated on a practical problem of detection of MSE in car drivers. Reliable methods to detect MSE in continuously recorded signals will be an important milestone in the development of drowsiness warning systems in real car cockpits. At present, however, achieving highly reliable MSE detection [3] is still a major issue to be resolved.

2 Data Fusion Architecture

To achieve the detection of MSE in real-world car driving situations, both estimated feature sets are merged by an adaptive feature weighting system (Fig. 1). The error of the training set is used to optimize the parameters of feature extraction based on PSD and DVV and also serves as fitness function in a genetic algorithm that examines the relevance of the different features employed. Subsequent multiple hold-out validation of LVQ networks yields the mean test set error for the evaluation of MSE detection ability. Consequently, test set errors were not used, directly and indirectly, for any step of optimization.

Fig. 1 shows the block diagram of the proposed data fusion system, which allows for the solution of the extensive data management problem in real time processing of multivariate and multimodal signals. Fusion based on features provides a significant advantage by means of a reduction in the dimensionality of the space in which the information to be processed resides. There is a trade-off associated with this strategy, since in principle, feature fusion may not be as accurate as raw data fusion because portions of raw signal information could have been eliminated.

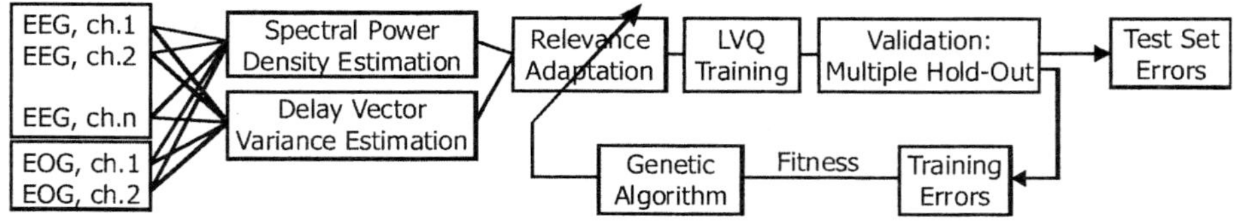

Fig. 1. Proposed Microsleep detection system based on feature fusion

In general, it is not known a priori which features within the different sets of features (EEG-FFT, EEG-DVV, EOG-FFT, EOG-DVV) are best suited for detection of MSE. It is intuitively clear that the obtained features differ in their importance level with respect to the classification accuracy. We therefore combine all different feature sets obtained from EEG and EOG by means of PSD and DVV. To prove whether our hypothesis that a combination of features coming from two different sources will indeed improve classification accuracy, we propose to use Genetic Algorithms (GA) to determine a scaling factor for every single feature coming from the four different sets. The scaling factors are used as gene expressions and the training error rate as fitness function. The sensitive adaptation of scaling factors by GA leads to a weighted Euclidean metric in the feature space, and can be interpreted as relevance factors [12]. For the purpose of comparison, the classification task is also performed without application of the relevance adaptation step (Fig. 3).

3 Experimental Setup

Our experimental setup was similar to the one presented in [4]. Seven EEG channels from different scalp positions and two EOG-signals (vertical, horizontal) were recorded from 23 young subjects (age range: 19 - 35 years) during driving simulation sessions lasting for 35 minutes. These sessions were repeated every hour between 1 a.m. and 8 a.m. This way, the likelihood of the occurrence of MSE was gradually increasing due to at least 16 hours without sleep prior to the experiment.

MSE are typically characterized by driving errors, prolonged eye lid closures or nodding-off. Towards automatic detection, two experts performed the initial MSE scoring, whereby three video cameras were utilized to record i) drivers portrait, ii) right eye region and iii) driving scene. For further processing, only clear-cut cases, where all the experts agreed on the MSE, were taken into account. Despite providing enough test data to tune our algorithms, the human experts could not detect some of the typical attention lapses, such as the one with open eyes and stare gaze. The number of MSE varied amongst the subjects and was increasing with time of day for all subjects. In all 3,573 MSE (per subject: mean number 162±91, range 11-399) and 6,409 NMSE (per subject: mean number 291±89, range 45-442) were scored. This clearly highlights the need for an automated data fusion based MSE detection system, which would not only detect the MSE also recognized by human experts, but would also offer a possibility to detect the critical MSE cases which are not recognizable by human experts.

4 Feature Extraction

In our experimental set-up, we varied two preprocessing parameters, the segment length and temporal offset. Evaluating test errors of our processing cue (Fig. 1) without relevance adaptation yields an optimal offset of 3 and 1sec and optimal segment length of 8 and 4 sec for the PSD and DVV, respectively. This means that EEG and EOG segments are beginning 3/1 sec (PSD/DVV) before and are finishing 5/3 sec after the onset of (N)MSE.

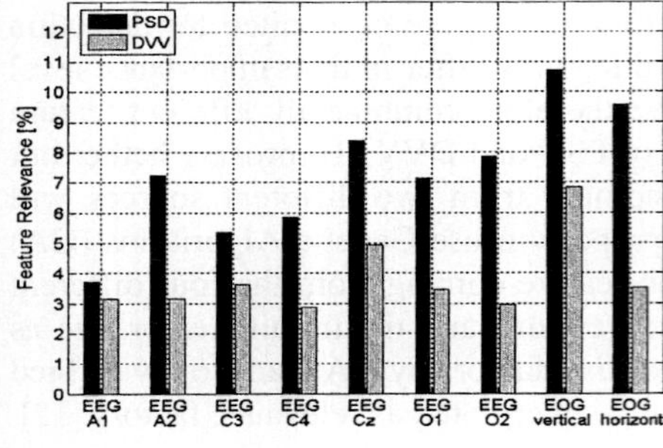

Fig. 2. Feature relevances for MSE detection estimated using GA
Fig. 2a. (left top): Normalized feature relevances for different data channels of EEG and EOG
Fig. 2b. (left bottom): PSD feature relevances over all EEG and EOG signals
Fig. 2c. (right bottom): DVV feature relevances over all EEG and EOG signals

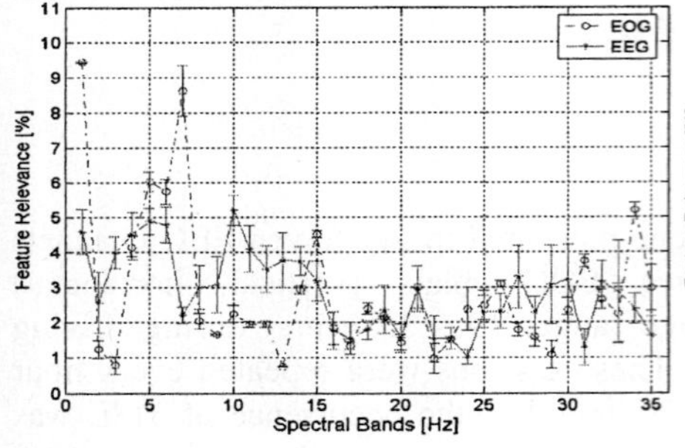

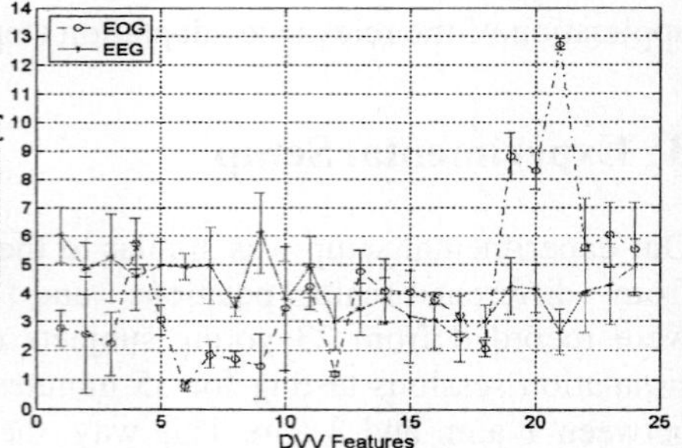

The preprocessing involves linear trend removal and applying the Hanning window to the data segments. PSD estimation was performed by the discrete Fourier transform. The so calculated PSD coefficients were averaged within 1.0 Hz wide bands. Further improvements in classification were achieved by applying a monotonic continuous transformation log (x) to the PSD [7].

The linear feature extraction method PSD was accompanied by a feature extraction method originating from the nonlinear dynamics, the DVV. The DVV features were calculated with the embedding dimension (m=3). Basically, they are variances of distances between delay vectors calculated on original and on surrogate data; further details are presented elsewhere [5, 6]. In contrast to PSD features, classification results did not improve by applying log(x) to DVV features.

Before examining the MSE detection performance in such a feature fusion setting, we perform a rigorous analysis of the feature relevance for the different EEG and EOG signals. This was achieved by means of GA. The relevance scores for the single EEG and EOG signals (Fig 2a) were calculated and normalized using the sum over all

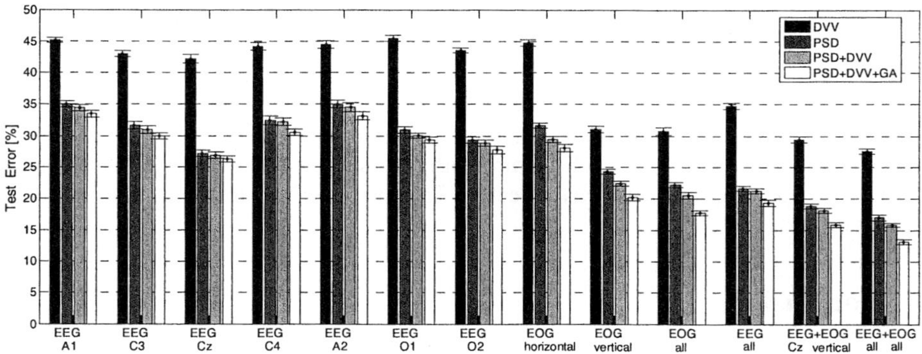

Fig. 3. Mean values and standard deviations of test errors for different single signals (first 9 groups) and for different combination of fused signals (last 4 groups)

feature relevance coefficients (35 for PSD, 24 for DVV). The normalized relevances for each PSD feature (Fig. 2b) and each DVV feature (Fig. 2c) were determined by averaging relevances of all single EEG / EOG signals.

5 Discriminant Analysis

Feature sets extracted by both methods (PSD and DVV) and of each of 7 EEG and of 2 EOG signals are merged in their multiple combinations both without and with an adaptive feature scaling system (GA). For each feature vector a label "MSE" or "NMSE" was assigned, thus introducing a two-class classification setting. Networks utilizing the OLVQ1 learning rule were used for analysis. Multiple hold-out validation [8] of the LVQ networks yields the mean test set error depicted in Fig 3. The test error rate was estimated as the ratio between the number of false classifications and the number of all classifications.

The error bars in Fig. 3 represent the standard deviation, which is caused by different initializations of LVQ networks and by the nature of the training progress due to randomly applied input vectors. To avoid the possibility of excellent results for some arbitrary settings, we repeated random partitioning 50 times, following the paradigm of multiple hold-out validation. For each partition, training and testing were repeated 25 times with different weight matrix initializations. The LVQ network was trained and tested by different selections of signals. Every signal was first selected alone for both training and testing (Fig. 3, first 9 groups). The feature extraction methods, PSD and DVV, were applied individually and in combination. The best single channel detection result was achieved with a combination of PSD, DVV and GA for the EOG channel 'vertical' followed by the EEG channel 'Cz'.

In an earlier work we pronounce that a combination of EEG and EOG measures should be most successful in predicting MSE [4]. Our results (Fig. 3, right) lend further support to this statement, independent from the feature extraction method used. Our simulations on DVV and PSD features achieved mean test error rates of 28 % and 17 % respectively. We judge the standard deviations of 1.4 % as moderate. The fusion of DVV and PSD features from all signals, which yields 531 features (7 EEG + 2 EOG signals) x (35 PSD + 24 DVV features), gained only a small improvement in the test error rates, namely from 17 % to 16 %. This result is not satisfactory enough and can be corrected by applying a GA where the DVV and PSD features had to compete with each other regarding their relevance for the MSE detection. After multiplying each feature with the estimated relevance factor obtained by GA, the training of LVQ was repeated. This way the best test error rate of 13 % was achieved.

6 Conclusions

We have presented an adaptive system for the analysis of Microsleep events (MSE), where several combinations of feature fusion were used for MSE detection and classification by means of populations of Learning Vector Quantization (LVQ) networks. Best results, with test errors down to 13 %, were obtained by a combination of all the recorded EEG and EOG channels, all features, and with feature relevance adaptation using Genetic Algorithms (GA).

Due to their complementing abilities to represent the linear and nonlinear nature of the EEG and EOG signals [13], simple feature extraction methods, PSD and DVV, were applied before and during an onset of a MSE. The results showed PSD to be more effective as a feature extraction method. This was also confirmed by our feature relevance results using GA, which detects features that were most relevant for the MSE detection. The relevances of the PSD features were similar to other findings [2-4], but for the understanding of the DVV feature relevance more research is needed. Furthermore, there are large inter-individual differences of the EEG- and EOG- characteristic [9, 10]. It would be interesting to ascertain whether the found feature relevance distribution can be confirmed or whether the DVV features play more significant role in certain cases. In general, there are strong indications that the role of the DVV features as compared to PSD features increases for the EOG signals. Another issue to be investigated is the fusion of EEG- / EOG- features and other oculomotoric features such as pupillography [11] using a greater variety of feature extraction methods. This is likely to improve and stabilize the discrimination of MSE, an issue of important real world applications.

References

1. Hall, D., Llinas, J.; An introduction to multisensor data fusion, Proceedings of the IEEE, 85; (1997)
2. Polychronopoulos, A., Amditis, A., Bekiaris, E.; Information data flow in AWAKE multi-sensor driver monitoring system, Proc. IEEE Intell. Vehicles Symp. (2004), 902-906
3. Sagberg, F., Jackson, P., Krüger, H-P., Muzet, A., Williams, A.J.; Fatigue, sleepiness and reduced alertness as risk factors in driving, Project Report, Transport RTD (2004)
4. Sommer, D.; Hink, T.; Golz, M.; Application of Learning Vector Quantization to detect drivers dozing-off, European Symposium on Intelligent Technologies, Hybrid Systems and their implementation on Smart Adaptive Systems (2002), 119-123

5. Gautama, T., Van Hulle, M.M., Mandic, D.P.; On the characterization of deterministic/stochastic and linear/nonlinear nature of time series, Technical Report (2004)
6. Gautama, T., Mandic, D.P., & Van Hulle, M.M. (2004). A Novel Method for Determining the Nature of Time Series. *IEEE Trans. Biomedical Engineering*, **51**(5), 728-736
7. Gasser, T., Bächer, P., Möcks, J.; Transformations Towards the Normal Distribution of Broad Band Spectral Parameters of the EEG, Electroenceph. clin. Neurophysiol. 53, (1982), 119-124
8. Devroye, L., Gyorfi, L. & Lugosi, G.; A probabilistic theory of pattern recognition; Springer, New York; 1996
9. Golz, M., Sommer, D., Seyfarth, A., Trutschel, U., Moore-Ede, M.; Application of vector-based neural networks for the recognition of beginning Microsleep episodes with an eyetracking system. In: Kuncheva, L. I. (ed.) Computational Intelligence: Methods & Applications, (2001), 130-134
10. Jung, T.-P., Stensmo, M., Sejnowski, T., Makeig, S.; Estimating Alertness from the EEG Power Spectrum. IEEE Transactions on Biomedical Engineering 44, (1997), 60-69.
11. Wilhelm, B., Giedke, H., Lüdtke, H., Bittner, E., Hofmann, A., Wilhelm, H.; Daytime variations in central nervous system activation measured by a pupillographic sleepiness test. J Sleep Research 10, (2001), 1-8
12. Hammer, B., Villmann, T.; Generalized relevance learning vector quantization. Neural Networks, 15, (2002), 1059–1068
13. Fell, J., Röschke, J., Mann, K., Schaffner, C.; Discrimination of sleep stages: a comparison between spectral and nonlinear EEG measures. Electroencephalogr. Clin. Neurophysiol. 98 (1996), 401-10

Combining Measurement Quality into Monitoring Trends in Foliar Nutrient Concentrations

Mika Sulkava[1], Pasi Rautio[2], and Jaakko Hollmén[1]

[1] Helsinki University of Technology,
Laboratory of Computer and Information Science,
P.O. Box 5400, FI-02015 HUT, Finland
{Mika.Sulkava, Jaakko.Hollmen}@hut.fi
[2] The Finnish Forest Research Institute, Parkano Research Station,
Kaironiementie 54, FI-39700 Parkano, Finland
Pasi.Rautio@metla.fi

Abstract. Quality of measurements is an important factor affecting the reliability of analyses in environmental sciences. In this paper we combine foliar measurement data from Finland and results of multiple measurement quality tests from different sources in order to study the effect of measurement quality on the reliability of foliar nutrient analysis. In particular, we study the use of weighted linear regression models in detecting trends in foliar time series data and show that the development of measurement quality has a clear effect on the significance of results.

1 Introduction

Analyzing chemical characteristics in samples collected from different components of ecosystems (e.g. biological samples, soil, water, etc.) are key methods in environmental monitoring. Chemical analyses are, however, prone to many errors which has brought up concern about the reliability of the analyses. Great improvements in laboratory quality have been achieved in the past two decades due to, for example, use of international reference material and interlaboratory comparisons (so-called ring tests). Despite the general improvement in laboratory quality some of the latest ring tests still reveal problems in quality [1].

The aim of this paper is to study how large of an impact laboratory quality has on detecting changes in environment. We use data from the ring tests, where laboratories analyzing foliar samples have been surveyed (see e.g. [1]). Further we link the data of conifer foliar concentrations measured in samples collected from 36 Finnish ICP Forests Level I plots (International Co-operative Programme on Assessment and Monitoring of Air Pollution Effects on Forests, see http://www.icp-forests.org/) and the results obtained by the Finnish laboratory in internal quality tests and the above mentioned ring tests.

Both theoretical computations and real-world data were used to study the effect of changing data quality on trend detection. Foliar nutrient data from Finland were analyzed using weighted regression. In our previous research the

W. Duch et al. (Eds.): ICANN 2005, LNCS 3697, pp. 761–767, 2005.

use of sparse linear models for finding other linear dependencies in the data have been briefly discussed in [6].

2 Data

2.1 Foliar Nutrient Data

Annual nutrient concentration data of conifer needles (Norway spruce [Picea abies (L.) Karsten] and Scots pine [Pinus sylvestris L.]) collected from 36 Finnish ICP Forests Level I stands were available for years 1987–2002. Foliage from 20 pine stands and 16 spruce stands around the country were collected yearly in October or November. Concentrations of 12 elements were measured from the needles, but in this study we focus on two elements: nitrogen (N) and sulfur (S). For details concerning the sampling procedure, see [5]. A more comprehensive characterization of the data using nutrition profiles is presented in [3].

2.2 Laboratory Quality Data

The quality of measurements of the laboratory analyzing ICP Forests foliar samples in Finland was studied in national calibration tests and international interlaboratory tests. The test data can be used to estimate the accuracy and precision of the nutrient measurement data. Between 1987 and 1994 the quality of measurements was surveyed in laboratory comparisons arranged by IUFRO (International Union of Forest Research Organizations). Since 1993 the measurement quality was tested in seven ICP Forests biennial ring tests (see e.g. [1]). In addition to interlaboratory tests, starting from 1995 the quality of the Finnish laboratory was measured in repeated measurements of certified reference samples (CRM 101). Before 1995 the methods were more varied. The quality control of the laboratory is discussed in more detail in [3].

3 Methods

3.1 Weighted Regression

If the precision of the observations is not constant, fitting an ordinary least squares linear regression model in order to analyze the data is not well justified, because homoscedasticity[1] is one of the basic assumptions of the model. Instead, weighted regression [4] is an effective method with heteroscedastic data. The regression model with heteroscedastic data can be expressed as follows:

$$Y_i = \beta_0 + \beta_1 X_i + \epsilon_i, \quad i = 1, \ldots, n \tag{1}$$

where β_0 and β_1 are regression coefficients ($\boldsymbol{\beta} = [\beta_0 \; \beta_1]^T$), X_i are known constants and error terms ϵ_i are independent $N(0, \sigma_i^2)$. If X_i denote time steps, the model assumes that there is a linear trend in the time series data.

[1] Homoscedasticity = property of having equal variances.

The weight w_i is defined as the inverse of the noise variance and thus, the method gives weights to observations according to their uncertainty

$$w_i = \frac{1}{\sigma_i^2}. \tag{2}$$

For example, completely uncertain ($\sigma_i = \infty$) measurements are eliminated from the model. **Y** and **X** the dependent and independent variables expressed in vector and matrix terms and **W** is a diagonal matrix containing the weights w_i

$$\mathbf{Y} = \begin{bmatrix} Y_1 \\ Y_2 \\ \vdots \\ Y_n \end{bmatrix}, \quad \mathbf{X} = \begin{bmatrix} 1 & X_1 \\ 1 & X_2 \\ \vdots & \vdots \\ 1 & X_n \end{bmatrix}, \quad \mathbf{W} = \begin{bmatrix} w_1 & 0 & \cdots & 0 \\ 0 & w_2 & \cdots & 0 \\ \vdots & \vdots & \ddots & \vdots \\ 0 & 0 & \cdots & w_n \end{bmatrix} \tag{3}$$

The maximum likelihood estimators of the regression coefficients are

$$\hat{\beta} = (\mathbf{X}^T\mathbf{W}\mathbf{X})^{-1}\mathbf{X}^T\mathbf{W}\mathbf{Y}. \tag{4}$$

The statistical significance of $\hat{\beta}_j \neq 0$ can be evaluated using the F-test [2].

3.2 Parameter Estimation

The error in the measurement of laboratory j in year i is assumed to be normally distributed with standard deviation (precision) σ_{ij} and mean (accuracy) μ_{ij}. The ICP Forests ring tests and a part of the IUFRO tests contained repeated measurements of the same sample. This makes it possible to estimate both the accuracy and precision of a tested laboratory. The estimated accuracy is the average deviation of n_{ij} repetitions from the average of all laboratories

$$\hat{\mu}_{ij} = \frac{1}{n_{ij}} \sum_{k=1}^{n_{ij}} Z_{ijk} - \hat{\mu}_i, \tag{5}$$

where Z_{ijk} is the value of kth repetitive measurement and

$$\hat{\mu}_i = \frac{\sum_{j=1}^{m_i} \sum_{k=1}^{n_{ij}} Z_{ijk}}{\sum_{j=1}^{m_i} n_{ij}}. \tag{6}$$

Above, m_i is the number of laboratories with acceptable results (i.e. laboratories fulfilling the quality requirements set by ICP Forests [1]). The estimate of the precision of a laboratory is the unbiased estimate for the standard deviation

$$\hat{\sigma}_{ij} = \sqrt{\frac{1}{n_{ij}-1} \sum_{k=1}^{n_{ij}} (Z_{ijk} - \hat{\mu}_{ij})^2}. \tag{7}$$

In case there are no repetitions, i.e. $n_{ij} = 1$, the precision is estimated to be the standard deviation of all acceptable laboratories

$$\hat{\sigma}_{ij} = \sqrt{\frac{1}{m_i - 1} \sum_{j=1}^{m_i} (Z_{ij1} - \hat{\mu}_i)^2}. \tag{8}$$

The minimum relative standard deviation (RSD) of the methods was determined to be 0.7% for N and 1.5% for S between 1987 and 2000. The precision estimated using Equation 7 may be lower than these minimum values, because n_{ij} is too small to make reliable estimates. If this is the case, the minimum RSD is used instead of the estimated value.

4 Experiments

4.1 Theoretical Computations

The effect of measurement quality (i.e. accuracy and precision of measurements) on trend detection was studied. Here we assume that there is a linear trend (see Equation 1) in the time series data (e.g. decreasing foliar sulfur concentrations in the course of time) and that this trend will also continue in the future.

The development of quality in different laboratories was visually inspected. According to that two simple scenarios roughly corresponding to typical development of real measurement precision were constructed. Either the precision does not change with time or the precision changes linearly from initial precision level c in a time steps to level b and then stays constant

$$\sigma_i = \begin{cases} \frac{b-c}{a} X_i + c & \text{if } X_i \leq a \\ b & \text{if } X_i > a \end{cases} \tag{9}$$

Trend detection with weighted regression was studied using the scenarios explained above. The hypothesis H_1 that there exists either an increasing or a decreasing trend in the data was tested against the null hypothesis H_0 that there is no trend. That is, H_0: $\beta_1 = 0$, H_1: $\beta_1 \neq 0$. The different parameter values, i.e. a, b, c, β_1, and n were varied and the p-value was calculated using the F-test. Significance level 0.05 was used to reject the null hypothesis.

The results for linearly changing precision are shown in Figure 1. The time needed to detect a trend with improvement in precision ($b < c$) can be seen in Figure 1 above the diagonals of the subfigures. The initial precision c, final precision b, and parameter a notably affect the time needed for detecting a trend. For example, when we look at the subfigure in the second column and top row in Figure 1, we can see (in the center of the subfigure) that the time needed to detect a trend with slope $\beta_1 = 0.1$ is greater than 10, if the precision is constant 0.5. However, if the final precision b is improved to 0.25 or 0.05 in two time steps ($a = 2$), the time n needed to detect the trend decreases to $8 < n \leq 10$ or $4 < n \leq 6$, respectively. We also studied the effect of an exponential change in precision, and found that the results are very similar to linear change in precision.

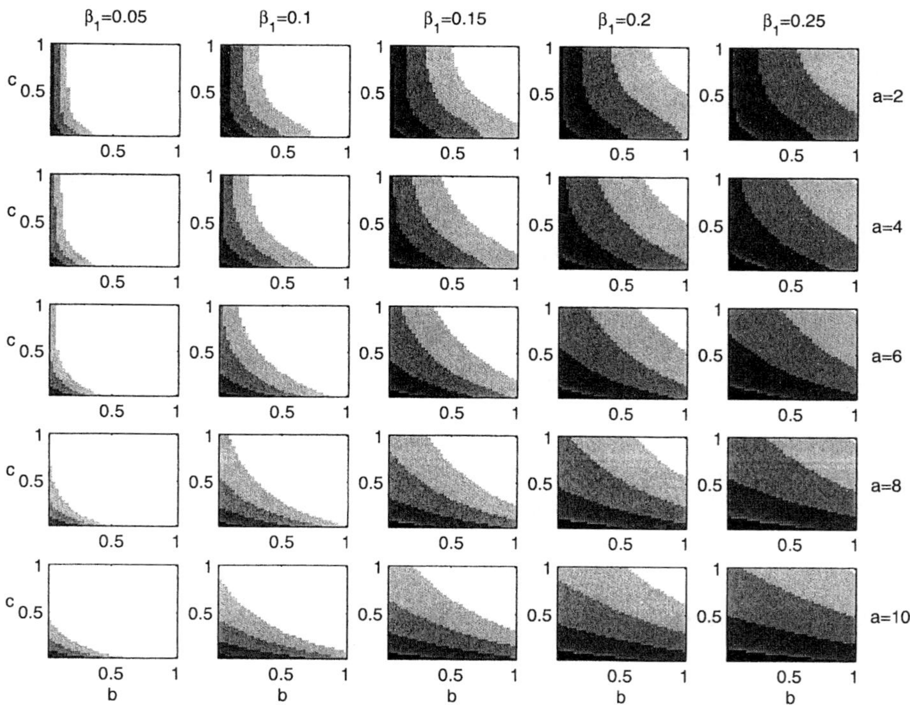

Fig. 1. Trend detection for linearly changing precision with different parameter values. The colors indicate the length n of time series that is needed to detect a trend with significance level 0.05. Black: $n \leq 4$, dark gray: $4 < n \leq 6$, medium gray: $6 < n \leq 8$, light gray: $8 < n \leq 10$, white: $n > 10$. In a subfigure the abscissa represents the final precision b and ordinate the initial precision c. Different columns of subfigures ($\beta_1 = 0.05, \ldots, 0.25$) represent the slope of the trend. Different rows of subfigures ($a = 2, \ldots, 10$) represent the speed of change in measurement precision.

4.2 Trends in Measurement Data

Quality data from all different sources was combined and the accuracy and precision of the Finnish laboratory was estimated for years 1988–2003 using methods described in Section 3.2. Measuring of the needles is done the year after the sample collection and therefore, quality data of year $t + 1$ was used for needles of year t. Because weighted regression requires that the noise in the measurements has zero mean, the accuracies μ_{ij} were subtracted from the measurements.

First, the average N and S foliar concentration data of pine between 1987–2002 were studied. Using weighted regression, a significant ($p < 0.05$) decreasing trend was found in S concentration and an insignificant ($p \geq 0.05$) weakly increasing trend in N concentration. The trend in average N concentration would become significant in three years if the data quality stays the same as in year

2003. The trend could be detected the following year, if the precision would be improved to $b = 0.06$.

It was also studied if there are significant trends in N and S concentrations in pine needles collected from the 20 individual pine stands between years 1987 and 2000. Out of 20 time series in 3 a significant trend was found in the N data and in 7 in the S data. In both cases there were three time series, where too many missing values made fitting a regression model unsubstantial.

We also experimented how long we would have to continue measuring in the stands, where a significant trend was not found, assuming that the trend continues, to be able to tell that the trend is significant. If the precision of measurements stays the same as in year 2001, the trends would become significant in 1–146 years depending on the estimated slope and number of measurements. However, if the precision of measurements is improved to approximately 0.5% RSD, the time needed decreases clearly. If the standard deviation of N measurements decreases linearly in $a = 3$ years to value $b = 0.06$, the time needed to detect a trend decreases on average 46%. Similarly, if the standard deviation of S measurements decreases linearly in $a = 3$ years to value $b = 0.005$, the time needed to detect a trend decreases on average 15%.

5 Conclusions

The results show that measurement precision strongly affects trend detection. Improving data quality can decrease clearly the time needed for finding statistically significant trends in environmental monitoring. Even though the Finnish laboratory analyzing the foliar samples has always fulfilled the quality demanded by the ICP Forests programme for both nitrogen (less than 10% deviation from the mean values of all labs) and sulfur (less than 20% deviation) it can still take many years to detect a possible ongoing trend with this measurement precision. For a laboratory not meeting the criteria set by ICP Forest programme (e.g. showing deviation greater than 20% in case of sulfur) it can take years or even decades to detect possible ongoing changes in the state of the environment. In all our results from theoretical computations and real world data clearly highlight the importance of quality in laboratory analyses.

Acknowledgements

We would like to thank Dr. Sebastiaan Luyssaert for collaboration in setting the objectives of the research and valuable discussions in early phase of the work.

References

1. Alfred Fürst. 7th needle/leaf interlaboratory comparison test 2004/2005. Technical report, United Nations Economic Commission for Europe, European Union, 2005.
2. Bent Jørgensen. *The theory of linear models.* Chapman & Hall, 1993.

3. Sebastiaan Luyssaert, Mika Sulkava, Hannu Raitio, and Jaakko Hollmén. Evaluation of forest nutrition based on large-scale foliar surveys: are nutrition profiles the way of the future? *Journal of Environmental Monitoring*, 6(2):160–167, February 2004.
4. John Neter, Michael H. Kutner, Christopher J. Nachtsheim, and William Wasserman. *Applied Linear Statistical Models.* McGraw-Hill, 4th edition, 1996.
5. Klaus Stefan, Alfred Fürst, Robert Hacker, and Ulrich Bartels. Forest foliar condition in Europe - results of large-scale foliar chemistry surveys 1995. Technical report, European Commission, United Nations Economic Commission for Europe, Brussels, Geneva, 1997.
6. Mika Sulkava, Jarkko Tikka, and Jaakko Hollmén. Sparse regression for analyzing the development of foliar nutrient concentrations in coniferous trees. In Sašo Džeroski, Bernard Ženko, and Marko Debeljak, editors, *Proceedings of the Fourth International Workshop on Environmental Applications of Machine Learning (EAML 2004)*, pages 57–58, Bled, Slovenia, September/October 2004.

3. Sebastian Luyssaert, Mika Sulkava, Hannu Raitio, and Jaakko Hollmén. Evaluation of forest nutrition based on large-scale foliar surveys: are nutrition profiles the way of the future? *Journal of Environmental Monitoring*, 6(2):160–167, February 2004.
4. John Neter, Michael H. Kutner, Christopher J. Nachtsheim, and William Wasserman. *Applied Linear Statistical Models*. McGraw-Hill, 4th edition, 1996.
5. Klaus Stefan, Alfred Fürst, Robert Hacker, and Ulrich Bartels. Forest foliar condition in Europe – results of large-scale foliar chemistry surveys 1995. Technical report, European Commission, United Nations Economic Commission for Europe, Brussels, Geneva, 1997.
6. Mika Sulkava, Jarkko Tikka, and Jaakko Hollmén. Sparse regression for analyzing the development of foliar nutrient concentrations in coniferous trees. In Sašo Džeroski, Bernard Ženko, and Marko Debeljak, editors, *Proceedings of the Fourth International Workshop on Environmental Applications of Machine Learning (EAML 2004)*, pages 57–58, Bled, Slovenia, September/October 2004.

A Fast and Efficient Method for Compressing fMRI Data Sets

Fabian J. Theis[1,2] and Toshihisa Tanaka[1]

[1] Department of Electrical and Electronic Engineering, Tokyo University of Agriculture and Technology, Tokyo 184-8588, Japan
[2] Institute of Biophysics, University of Regensburg, 93040 Regensburg, Germany

Abstract. We present a new lossless compression method named FTT-coder, which compresses images and 3d sequences collected during a typical functional MRI experiment. The large data sets involved in this popular medical application necessitate novel compression algorithms to take into account the structure of the recorded data as well as the experimental conditions, which include the 4d recordings, the used stimulus protocol and marked regions of interest (ROI). We propose to use simple temporal transformations and entropy coding with context modeling to encode the 4d scans after preprocessing with the ROI masking. Experiments confirm the superior performance of FTTcoder in contrast to previously proposed algorithms both in terms of speed and compression.

1 Introduction

Modern imaging techniques become increasingly important and common in medical diagnostics; such techniques include X-ray *computerized tomography (CT)* and *magnetic resonance imaging (MRI)*. Although much of this paper can be applied to more general (medical) image series, we will in the following focus on the latter. MRI visualizes three-dimensional objects by measuring magnetic properties within small volume elements called *voxels*. Even single scans of CT and MRI are already large in size (up to several megabytes). More recently, it has become popular to also study time series of MRI scans, during which the subject performs various functional tasks, hence the term *functional MRI (fMRI)*. These data sets measure up to several hundred megabytes, and efficient storage methods have to be applied. As even small deviations within adjacent scans may already contain important information (which can for example be revealed by blind source separation [1]), lossy image and time series compression is out of question, and lossless (or at most near lossless) techniques have to be used. Moreover efficiency not so much in compression rate but also in speed is essential, as fast and preferably sequential, single-pass processing enables close to real-time data preprocessing and analysis.

The field of medical image and volume compression is still rather young, with popular algorithms being direct generalizations of image compression techniques such as for example 3d context modeling [2]. Image sequence coding methods

W. Duch et al. (Eds.): ICANN 2005, LNCS 3697, pp. 769–777, 2005.

have also recently gained some attention, for instance by compressing sequences using integer wavelet transform with subsequent entropy coding [3]. However specific application to fMRI are rare, and to our knowledge only some simple algorithms exist at present: SmallTime [4] performs a very fast but rather inefficient fMRI compression by simply taking difference images between adjacent recordings and then storing the typically 8-bit difference image in contrast to the 16-bit recorded image. Adamson employs the efficient *LOCO-I (low complexity lossless compression for images)* algorithm [5] to compress MRI and fMRI recordings in sequence [6]. This enables high data throughput with acceptable compression efficiency, and we will generalize his proposal using *JPEG-LS('Joint Photographic Experts Group'-lossless)* with added preprocessing in the following. Lossy fMRI compression has been proposed in [7] but has not been adopted by the community, most probably due to the above mentioned sensitivity of MRI recordings to small deviations.

Our algorithm processes the time series, interpreted as recordings from multiple sources by also taking into account additional recordings such as a *region-of-interest (ROI)* or mask selection. This information, stemming from multiple sources, is packed into a single data stream allowing for efficient and fast storage and recovery, outperforming present fMRI compression algorithms considerably both in speed and compression ratio.

2 Lossless Image Compression Using JPEG-LS

Lossless compression should be used for a variety of applications, particularly those involving medical imaging such as CT and (f)MRI. For these applications, ISO/IEC provided a lossless algorithm in the JPEG international standard, but, unlike the baseline JPEG (lossy), this lossless version is poor in compression performance. Instead, a new lossless image compression standard called *JPEG-LS* is provided [8]. This compression algorithm draws heavily from the LOCO-I method developed by Weinberger at al. [5] and aims at improving compression performance with low complexity.

JPEG-LS is an image compression algorithm involving prediction and entropy coding. The coder effectively uses four neighboring pixels. Let $x[n_1, n_2]$ be a value of the current pixel. The neighborhood consists of the four samples: $x[n_1, n_2 - 1]$, $x[n_1 - 1, n_2]$, $x[n_1 - 1, n_2 - 1]$, and $x[n_1 - 1, n_2 + 1]$, respectively denoted by x_a, x_b, x_c, and x_d, which are exploited for modeling contexts of an entropy coder as well as determining a 'mode'. JPEG-LS switches between two modes: normal and run modes, depending on the neighborhood. At the current pixel, if one of gradients defined as $\Delta_1 = x_d - x_b$, $\Delta_2 = x_b - x_c$, and $\Delta_3 = x_c - x_a$ are non-zero, then the JPEG-LS coder is in normal mode.

In normal mode, prediction of x denoted by μ_x is obtained by a non-linear function of the four neighbor pixels. Then, the residual that is actually coded in JPEG-LS is given by $e_x = s_x(x - \mu_x) - \beta_x$, where s_x represents the sign and β_x the bias compensating term, which is needed to make the probability distribution of residuals unbiased symmetric. The JPEG-LS coder represents the mapped

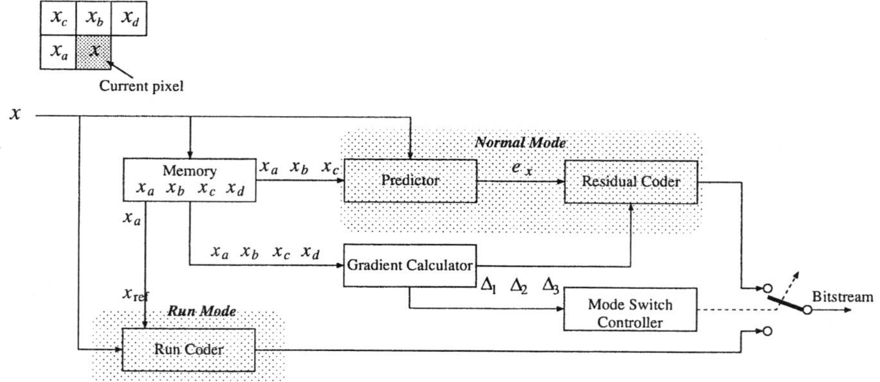

Fig. 1. Schematic JPEG-LS compression

residuals by using an adaptive Golomb code [9] with context modeling. This residual coding is context adaptive, where the context used for the current sample is identified by a context quantization function of the three gradients. Then, context-dependent Golomb and bias parameters are estimated sample by sample.

If all the gradients are identical, then the JPEG-LS coder moves to the run mode. The assumption here is that x and possibly a large number of consecutive samples are all likely to have the same value as x_a. The number of samples which are all the same as x_a in a scanning direction is called the run-length. This run-length is coded by using Golomb code again. But, here a so-called MELCODE [5], which is specialized for encoding the run-length, is utilized in JPEG-LS. This compression scheme is visualized in figure 1; for details, see [5] for example.

3 Compressing Information from Multiple Sensors

The goal of the proposed compression algorithm is to fuse the data set, acquired by multiple scans over time, into a single easy to store file of decreased size as quickly as possible. Furthermore, additional information such as masked voxels or stimulus components corresponding to the protocol used in the fMRI recordings can be merged with the data stream.

3.1 fMRI Compression

The MRI measurements $\mathbf{x}(t) \in \{0, 1, \ldots, \alpha\}^{w \times h \times d}$ are taken for time points $t = 1, \ldots, T$ with a temporal resolution in the range of seconds and size $T \approx 100$. The measured data for each time point is a three-dimensional data structure; each scan is of size $w \times h \times d$ and voxels take integer values between 0 and α.

Our compression algorithm is simple in concept — just apply the efficient and fast context-based JPEG-LS onto each slice image, but after some intelligent preprocessing. A desirable image property enabling high compression rates is simple structure together with low α. This property can be achieved in our case of biomedical time series by temporal operations. Different methods and filters are possible such as integer wavelet transformation using for example a simple Haar wavelet, or discrete cosine transform. For fMRI it is advisable to employ a structurally simple, preferably even linear transformation that uses a rather small time window — this would increase coding and decoding speed and keep the memory consumption low albeit at the possible loss of some compression efficiency. Hence we decided to fully encode the first scan $\mathbf{x}(1)$, and then only encode the difference images $\Delta\mathbf{x}(t) := \mathbf{x}(t) - \mathbf{x}(t-1)$ (after possible translation to have a zero-valued minimum).

Another property of fMRI data is that typical scans are taken of restricted regions that do not fully fill out the whole scan volume — for example fMRI is a common tool for brain imaging, and the non-brain volume or at least voxels outside of the head can be easily identified and masked out, see figure 2(a). After preprocessing by motion alignment [10] the scans are temporally aligned. Hence we can assume to have a single time-independent mask $\mathbf{y} \in \{0,1\}^{w\times h\times d}$. The mask is assumed to be given by the user, and to be binary with ones indicating the ROI. Context-based coding of the full volumes is not as efficient, so the additional ROI data can be used to enhance compression: only encode the non-masked voxels of each line, and during reconstruction recover the full line by adding zeros (or mask values) at the masked voxels. Then of course the mask $\mathbf{y}$ has to be stored in the compressed file. For this we use run-length encoding, which works well due to the often simple structure of the mask. Given a sufficiently large number of scans additional masking turns out to be of significantly higher efficiency.

The resulting algorithm is single-run, as there is no need of returning to previous scans if at most two scans are held in memory at the same time (which is acceptable, typical sizes are up to 2MB per scan). So memory efficiency is also provided, and the algorithm is fast as confirmed by the experiments later. We call the resulting compression algorithm *FTTcoder (fMRI temporal transformation coder)*, and refer to algorithms 1 and 2 for details.

3.2 Fusing the Stimulus

A peculiarity of functional MRI data sets is the presence of a *stimulus protocol*. It describes the functional task used in the experiment. Typically either a simple *block design* or a so-called *event-based protocol* is used. The former describes a periodic on-off stimulus, whereas the latter consists of activities interspersed with a varying period of non-activity, which can also depend on the subject's interaction. In this section, we will fuse the additional before-hand knowledge of the stimulus with the data to achieve more efficient compression ratios; for simplicity, we will use a block design protocol.

In addition to the observed data set $\mathbf{x}(t)$ assume that the binary stimulus protocol $\sigma(t) \in \{0,1\}$ is given for $t = 1,\ldots,T$. A simple model for incorporating the stimulus with the MRI data can be built by assuming that an underly-

Data: scan sizes $w \times h \times d$, scan range $[0, \alpha] \subset \mathbb{N}_0$, T fMRI scans $\mathbf{x}(1), \ldots, \mathbf{x}(T) \in [0, \alpha]^{w \times h \times d}$, optional common mask $\mathbf{y} \in \{0, 1\}^{w \times h \times d}$
Result: compressed bit stream b

store sizes h, w, d, T and range α in b
if *mask is used* **then** run-length encode and store binary $\mathbf{y}$ in b
for $t \leftarrow 1, \ldots, T$ **do**
  if $t = 1$ **then** $\mathbf{z} \leftarrow \mathbf{x}(1)$ **else** $\mathbf{z} \leftarrow \mathbf{x}(t) - \mathbf{x}(t-1)$
  if necessary translate $\mathbf{z}$ and determine and store new range $[0, \alpha']$ by calculating minima and maxima of $\mathbf{z}$
  for $j \leftarrow 1, \ldots, h,\ k \leftarrow 1, \ldots, d$ **do**
    determine masked current and previous lines:
    $\mathbf{l} \leftarrow \{\mathbf{z}(i,j,k) | i \text{ with } \mathbf{y}(i,j,k) = 1\}$
    $\mathbf{pl} \leftarrow \{\mathbf{z}(i,j-1,k) \text{ resp. } \mathbf{z}(i,j,k-1) | i \text{ with } \mathbf{y}(i,j,k) = 1\}$
    context encode line $\mathbf{l}$ using context $(\mathbf{l}, \mathbf{pl})$ to stream b by JPEG-LS
  end
end

Algorithm 1. Compression algorithm FTTcoder

Data: fMRI-compressed bit stream b
Result: fMRI scans $\mathbf{x}(1), \ldots, \mathbf{x}(T) \in [0, \alpha]^{w \times h \times d}$

read scan sizes $h \times w \times d$, number of scans T and maximal range α from b
run-length decode optional binary mask $\mathbf{y} \in \{0, 1\}^{w \times h \times d}$ from b
for $t \leftarrow 1, \ldots, T$ **do**
  for $j \leftarrow 1, \ldots, h,\ k \leftarrow 1, \ldots, d$ **do**
    determine masked previous line:
    $\mathbf{pl} \leftarrow \{\mathbf{z}(i,j-1,k) \text{ resp. } \mathbf{z}(i,j,k-1) | i \text{ with } \mathbf{y}(i,j,k) = 1\}$
    context decode line $\mathbf{l}$ using context $\mathbf{pl}$ from stream b by JPEG-LS
    recover unmasked line: $\mathbf{z}(\{i | \mathbf{y}(i,j,k) = 1\}, j, k) \leftarrow \mathbf{l}$
  end
  if necessary translate $\mathbf{z}$
  if $t = 1$ **then** $\mathbf{x}(1) \leftarrow \mathbf{z}$ **else** $\mathbf{x}(t) \leftarrow \mathbf{x}(t-1) + \mathbf{z}$
end

Algorithm 2. Decompression algorithm inverting FTTcoder

ing stimulus-independent activity is additively overlayed by the stimulus-related brain activity at time instants t where $\sigma(t) = 1$. If $\mathbf{x}^{(0)}, \mathbf{x}^{(1)} \in \mathbb{R}^{w \times h \times d}$ denote the stimulus-independent and stimulus-dependent data component respectively, then according to the model the data time series $\mathbf{x}(t)$ can be written as

$$\mathbf{x}(t) = \mathbf{x}^{(0)} + \sigma(t)\mathbf{x}^{(1)} + \mathbf{e}(t), \tag{1}$$

where $\mathbf{e}(t)$ denotes the model error at time instant t. The model is fulfilled well if $\mathbf{e}(t)$ is small for all t, and compression can be improved in this case. Please note that of course more advanced models (including convolutions induced by the BOLD effect, using blind separation for additional component identification etc.) are possible and used in the analysis of fMRI data sets, but our goal is to keep the algorithm simple, fast and efficient; especially the single-run property must not be destroyed, so more complex models might be difficult to include.

Let $\Delta\sigma(t) := \sigma(t) - \sigma(t-1)$, $t = 2, \ldots, T$, and denote time instants t with $\Delta\sigma(t) \neq 0$ as *stimulus jumps*. The compression performance can be increased by reducing the range and the deviation of the difference image $\Delta\mathbf{x}(t)$ to be compressed. However at stimulus jumps, the differences can be expected to be larger than at other time instants. By using model (1), we get

$$\Delta\mathbf{x}(t) = \mathbf{x}^{(0)} + \sigma(t)\mathbf{x}^{(1)} + \mathbf{e}(t) - \mathbf{x}^{(0)} - \sigma(t-1)\mathbf{x}^{(1)} - \mathbf{e}(t-1) = \Delta\sigma(t)\mathbf{x}^{(1)} + \Delta\mathbf{e}(t)$$

with $\Delta\mathbf{e}(t) := \mathbf{e}(t) - \mathbf{e}(t-1)$. So $\Delta\mathbf{x}(t)$ is larger at stimulus jumps.

For compression, we now propose to estimate the stimulus component $\mathbf{x}^{(1)}$ by the normalized difference image $\hat{\mathbf{x}}^{(1)} := \Delta\sigma(t_0)\Delta\mathbf{x}(t_0)$ (which equals $\mathbf{x}^{(1)} \pm \Delta\mathbf{e}(t)$ and hence approximately $\mathbf{x}^{(1)}$) at the first jump t_0. Subsequently, instead of encoding $\Delta\mathbf{x}(t)$, we compress

$$\mathbf{z}(t) := \Delta\mathbf{x}(t) - \Delta\sigma(t)\hat{\mathbf{x}}^{(1)} = \Delta\mathbf{x}(t) - \Delta\sigma(t)\Delta\sigma(t_0)\Delta\mathbf{x}(t_0), \quad (2)$$

which equals $\Delta\sigma(t)\left(\mathbf{x}^{(1)} - \hat{\mathbf{x}}^{(1)}\right) + \Delta\mathbf{e}(t)$ and can therefore expected to be small. The stimulus can be included in the compressed stream using run-length encoding.

For decompression, this can easily be inverted by restoring the stimulus approximation $\hat{\mathbf{x}}^{(1)}$ at the first jump, and then reconstructing subsequent $\mathbf{x}(t)$ from the decompressed frame $\mathbf{z}(t)$ by

$$\mathbf{x}(t) := \mathbf{x}(t-1) + \mathbf{z}(t) + \Delta\sigma(t)\hat{\mathbf{x}}^{(1)}. \quad (3)$$

4 Results

We demonstrate the performance of FTTcoder (freely available for download at `http://fabian.theis.name/`) when applied to real data sets. For this we use two data sets. The first one is a two-dimensional slice with $w = h = 128$, $\alpha = 2048$ and $T = 98$ scans. The data set has been masked with a rather large mask, see figure 2(a) for the (masked) first scan of the series. Physically, the data set is represented by a large file containing a concatenation of all scans. The second data set is given by 240 analyze files with scan size $64 \times 64 \times 12$. The data has not been masked, and the largest possible $\alpha = 65536$ is used for the 16bit range.

For illustration, FTTcoder performance is compared with two previously proposed fMRI compression schemes as well as three general algorithms namely direct file copy, zip and the efficient bzip2. For the latter three algorithms, the plain data itself is directly compressed, not the temporally preprocessed one. The two fMRI compression utilities are iterative LOCO-I compression of images [6], which due to lack of code we emulate by using FTTcoder without temporal differences and masking, and direct storage by mapping 16bit values into 8bit values if possible (SmallTime, [4]). We compare both compressed file size and

Table 1. Algorithm performance; speed was measured for combined compression/decompression using the mean over 100 runs (first data set) respectively 10 runs (second data set)

data set	original size	algorithm	compression	speed
2d data set	3137 kB	FTTcoder (mask)	245 kB (7.8%)	0.182s
2d data set	3137 kB	FTTcoder (no mask)	261 kB (8.3%)	0.185s
2d data set	3137 kB	FTTcoder (no diffs)	350 kB (11.2%)	0.166s
2d data set	3137 kB	FTTcoder (no mask&diffs)	366 kB (11.7%)	0.162s
2d data set	3137 kB	SmallTime	1589 kB (50.7%)	1.3s
2d data set	3137 kB	bzip2	394 kB (12.5%)	0.514s
2d data set	3137 kB	zip	562 kB (17.9%)	0.255s
2d data set	3137 kB	file copy	3137 kB (100%)	0.192s
analyze data set	22.6 MB	FTTcoder (no mask)	14.1 MB (62.4%)	5.16s
analyze data set	22.6 MB	FTTcoder (no mask&diffs)	17.0 MB (75.2%)	5.9s
analyze data set	22.6 MB	SmallTime	runtime error	-
analyze data set	22.6 MB	tar and bzip2	15.6 MB (69.0%)	14.6s
analyze data set	22.6 MB	tar and gzip	18.0 MB (79.7%)	12.5s
analyze data set	22.6 MB	file copy	22.6 MB (100%)	14.3s

speed. The experiments have been made on a Pentium M 2.0GHz using cygwin. The results are shown in table 1.

Clearly FTTcoder outperforms the other algorithms both in compression ratio as well as in speed — the latter is at first a bit astonishing when comparing against file copy, but this is due to the fact that the much smaller compressed file takes less time to be stored on hard disc than the file copy of the larger one. Apparently due to its implementation, SmallTime is considerably slower than the other algorithms and also less efficient, although we note that the given data set contains a large number of non-ROI voxels, which SmallTime does not compress efficiently. SmallTime was unable to compress the second data set, which we believe is due to the fact that almost all differences were 16bit. Direct application of LOCO-I performs comparably well in terms of speed as FTTcoder, but the compression rate is considerably lower. In practical applications, FTTcoder is about 3 times as fast as traditional zip algorithms, and considerably more efficient.

We finally compare the proposed extended stimulus-based compression algorithm from section 3.2 with plain FTTcoder. The algorithm performance depends greatly on how well the model (1) is fulfilled i.e. how large the error terms $\mathbf{e}(t)$ are. In the following, we construct a toy data set by choosing $\mathbf{x}^{(0)}$ to be the brain slice from figure 2(a) with an additive stimulus component $\mathbf{x}^{(1)}$ constructed by setting pixels randomly to ± 1 within a fixed rectangle in the brain part, see figure 2(b). Within the brain, white Gaussian noise is added with varying SNR in $[-1.5dB, \infty dB]$. A stimulus of 6 off, 6 on periods of total length $T = 98$ was used. We compare the file sizes of the compressed data sets. Figure 2(c) shows the ratios. Clearly, the stimulus based algorithm considerably outperforms the normal one in the no- and low-noise cases, but the performance decreases with increas-

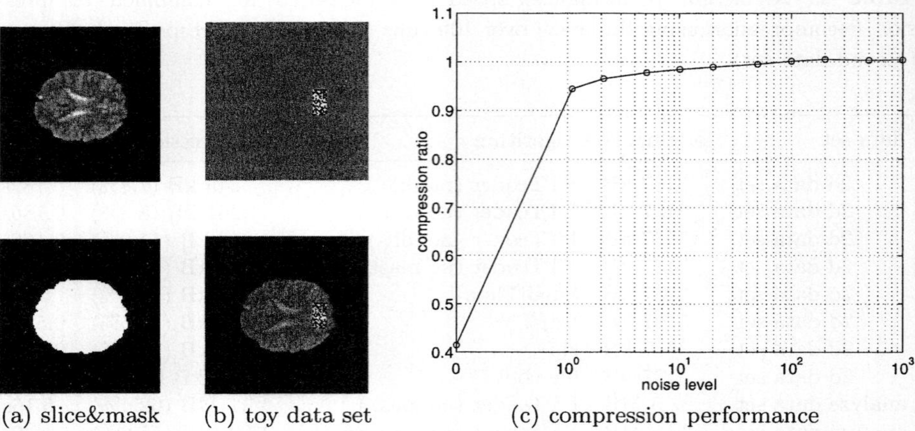

(a) slice&mask (b) toy data set (c) compression performance

Fig. 2. In (a) a single fMRI slice together with a selected ROI/mask differentiating brain from non-brain voxels is presented. (b) shows the toy stimulus component $\mathbf{x}^{(1)}$ together with a slice image at active stimulus and noise level 100. Figure (c) compares the fMRI compression based on stimulus-fusion versus normal compression by plotting the ratio of the stimulus-compressed file size and the non-stimulus-compressed size.

ing noise. Similar performance is achieved starting at SNRs of around $8dB$. We conclude that depending on how well the stimulus model is fulfilled, fusing the additional stimulus component with the data may increase compression ratios. In the future, we will study more advanced temporal preprocessing.

5 Conclusion

We have proposed a novel lossless compression scheme for sequences of medical images, focusing on fMRI recordings. The algorithm is based on JPEG-LS and turns to be more efficient in both speed and compression rate than more generic compression algorithms. It is well known that even for data sets with a small number of two-dimensional slices and a high slice distance there is a significant gain in compression ratio by compressing the 3d data spatially in contrast to compressing the two-dimensional slices separately [2], and we currently work on generalizing JPEG-LS to 3d contexts and employing this for fMRI compression.

References

1. Keck, I., Theis, F., Gruber, P., Lang, E., Specht, K., Puntonet, C.: 3D spatial analysis of fMRI data - a comparison of ICA and GLM analysis on a word perception task. In: Proc. IJCNN 2004, Budapest, Hungary (2004) 2495–2500
2. Klappenecker, A., May, F., Beth, T.: Lossless compression of 3D MRI and CT data. In Laine, A., Unser, M., Aldroubi, A., eds.: Proc. SPIE, Wavelet Applications in Signal and Imaging Processing VI. Volume 3458. (1998) 140–149

3. M.Wu, Forchhammer, S.: Medical image sequence coding. In: Proc. DSAGM, Copenhagen, Denmark (2004) 77–87
4. Cohen, M.: A data compression method for image time series. Human Brain Mapping **12** (2001) 2024
5. Weinberger, M., Seroussi, G., Sapiro, G.: The LOCO-I lossless image compression algorithm: Principles and standardization into JPEG-LS. IEEE Transactions on Image Processing **9** (2000) 1309–1324
6. Adamson, C.: Lossless compression of magnetic resonance imaging data. Master's thesis, Monash University, Melbourne, Australia (2002)
7. Taswell, C.: Wavelet transform compression of functional magnetic resonance image sequences. In: Proc. SIP, Las Vegas, USA (1998) 725–728
8. Taubman, D.S., Marcellin, M.W.: JPEG 2000 Image Compression Fundamentals, Standards and Practice. Kluwer Academic Publishers, Massachusetts (2002)
9. Golomb, S.: Run-length encodings. IEEE Transactions on Information Theory **12** (1966) 399–401
10. Woods, R., Cherry, S., Mazziotta, J.: Rapid automated algorithm for aligning and reslicing pet images. Journal of Computer Assisted Tomography **16** (1992) 620–633

Non-linear Predictive Models for Speech Processing

M. Chetouani[1], Amir Hussain[2], M. Faundez-Zanuy[3], and B. Gas[1]

[1] Laboratoire des Instruments et Systèmes d'Ile-De-France, Université Paris VI, Paris, France
[2] Dept. of Computing Science and Mathematics, University of Stirling, Scotland, U.K
[3] Escola Universitària Politècnica de Mataró, Barcelona, Spain

Abstract. This paper aims to provide an overview of the emerging area of non-linear predictive modelling for speech processing. Traditional predictors are linear based models related to the speech production model. However, non-linear phenomena involved in the production process justify the use of non-linear models. This paper investigates certain statistical and signal processing perspectives and reviews a number of non-linear models including their structure and key parameters (such as prediction context).

1 Introduction

Linear prediction has been intensively used in speech processing: coding, synthesis and recognition. For instance, the Linear Predictive Coding (LPC) [1] forms the basis of several speech compression techniques. The key idea is the estimation of the current speech sample $\widehat{y}_k$ using a linear combination of P past speech samples or the prediction context $\{x_{k-1}, x_{k-2}, \ldots, x_{k-P}\}$:

$$\widehat{x}_k = \sum_{i=1}^{P} a_i x_{k-i} + e(k) \tag{1}$$

Where $e(k)$ is the prediction error. The LPC model is based on an Auto-Regressive (AR) model which can be related, under linear assumptions, to the vocal tract [1]. The a_i coefficients and the predictor order P are related to the different resonators forming the vocal tract model.

Even if there is some relationship relation between the two models (i.e. LPC and vocal tract), the basic linear AR models do not offer a sufficiently efficient modeling of the vocal tract. For this purpose, several alternative models have been proposed. The Perceptual Linear Predictive (PLP) coding method [2] is an example of human auditory knowledge integration for the improvement of AR models' computation. Other strategies have been also investigated, Ref. [3] gives an overview of them. In this paper, we deal with non-linear speech processing for the improvement of predictors.

W. Duch et al. (Eds.): ICANN 2005, LNCS 3697, pp. 779–784, 2005.

Recently, several authors have shown that non-linear processing can improve the performance of speech predictors in terms of prediction gains. Thyssen [4] compared prediction gains of linear and non-linear predictors. Results showed a better fitting using non-linear models. The use of non-linear models for speech samples prediction is motivated by physiological reasons [5]. Other reasons also have been advanced including statistical ones which are further explored in this contribution.

The paper is organized as follows: Section 2 introduces the principle of prediction. Then, the extension of predictors to the non-linear domain is described, followed by a description of non-linear models. Finally some concluding remarks are presented.

2 Statistical Aspects

2.1 Regression

From a statistic point of view, prediction can be related to the regression process which aims to estimate a function f:

$$\mathbf{Y} = \mathrm{f}(\mathbf{X}) + \varepsilon \tag{2}$$

where $\mathbf{X}$ and $\mathbf{Y}$ are two random variables. $\mathbf{X}$ denotes the random variable representing the prediction context: $\{x_{k-1}, x_{k-2}, \ldots, x_{k-P}\}$. $\mathbf{Y}$ is also a random variable representing the distribution of the predicted sample ($\widehat{x}_k$). ε is a noise modeled by a random variable with zero mean and it is independent of the variables $\mathbf{X}$ and $\mathbf{Y}$.

The function f represents the relationship existing between the two variables ($\mathbf{X}$ and $\mathbf{Y}$), or, in other words, the process which generates the time series. For instance in Linear Predictive Coding (LPC), the coefficients are related to the speech production model. However, the linear predictor is known to be non-efficient for this kind of a model. The optimal predictor f can be computed in a statistical way from the equation 2:

$$E[\mathbf{Y}|\mathbf{X}] = E[\mathrm{f}(\mathbf{X}) + \varepsilon|\mathbf{X}] = f(\mathbf{X}) \tag{3}$$

According to this result, the predictor appears as the conditional expectation of $\mathbf{Y}$ given $\mathbf{X}$. The major obstacle is that, in practice, we never know the probability joint distribution of the couple ($\mathbf{X}$,$\mathbf{Y}$) and thus the conditional expectation $E[\mathbf{Y}|\mathbf{X}] = \mathbf{E}[\widehat{\mathbf{x}}_\mathbf{k}|\{\mathbf{x}_{\mathbf{k-1}}, \mathbf{x}_{\mathbf{k-2}}, \ldots, \mathbf{x}_{\mathbf{k-P}}\}]$.

2.2 Predictor Adequacy

The approximation of the optimal predictor is done by the computation of a predictor h with a fixed structure (Finite Impulse Response filter, polynomial, neural networks, etc.). The models are commonly computed by the minimization of quadratic prediction error. The measure of the prediction precision is made using the overall risk [6]:

$$R(\mathrm{h}) = \int \left(\widehat{\mathbf{x}} - \mathrm{h}(\mathbf{x})\right)^2 dF(\mathbf{X}, \mathbf{Y}) \tag{4}$$

For an efficient estimation of h, the overal risk has to be minimized. On can show that under the same assumptions of the equation 2 (zero mean and standard deviation σ), the overall risk becomes:

$$R(\mathrm{h}) = \sigma^2 + \int \left(\mathrm{h}(\mathbf{x}) - \mathrm{f}(\mathbf{x})\right)^2 p(\mathbf{X}) d\mathbf{X} \tag{5}$$

This result shows that the prediction precision depends on:

- The noise deviation σ^2.
- The capacity of the used predictor h to approximate the optimal predictor f: $\left(\mathrm{h}(\mathbf{x}) - \mathrm{f}(\mathbf{x})\right)^2$.

This theoretical result is important because it highlights the relationship between prediction error minimization and model approximation.

In other terms, in speech processing, the choice of the predictor h should allow an efficient approximation of the f (speech production model). Non-linear speech processing is one of the solutions. In the next section, we describe the main reasons from a signal processing point of view.

3 Extension to the Non-linear Domain

3.1 Non-linear Aspects of Speech Signal Samples

The speech signal is special since it is produced by human beings involving, during its production, several non-linear phenomena [5]. Consequently, such signals are characterized by non-linear methods exploiting for example, chaos based assumptions [7] and dynamical theories [8]. These methods aim to model the signals' non-linearities.

Another point of view is to characterize the speech signals' distribution. Such studies have shown that the distributions are not Gaussian, but rather Laplacian or Gamma based [9].

3.2 A Non-linear Solution

Section 2 shows that the prediction strategy has to take into account the characteristics of the speech generating process. And it is known that non-linear models are efficient alternative models for speech processing because they can take into account the speech samples' distribution [10]. Indeed, basic AR models are based on second order statistics (auto-covariance or auto-correlation) which (optimally) characterize only Gaussian processes.

Higher Order Statistics (HOS) have been investigated in speech processing [11]. They provide an adaptive mathematical framework for non-linear characterization. However, the estimation is not reliable for non-stationary signals because of the amount of data needed for an efficient computation of the HOS.

Non-linear models can be more efficient for speech prediction due to their modeling capacities. However theoretical and practical aspects limit their utilization. In the next section, we describe characteristics of non-linear models in speech prediction.

4 Non-linear Predictive Models

4.1 Non-linear Function

Basic Non-linear Predictive models use non-linear functions for prediction:

$$\widehat{x}_k = F(x_{k-1}, x_{k-2}, \ldots, x_{k-P}) \tag{6}$$

The difference between the models relies on the deployed function F. One possible classification of the predictors involves classifying them as parametric, nonparametric or semiparametric [12]. Parametric models require a fixed structure which means that this structure is known and obviously adapted to the problem. Non-Linear Auto-Regressive (NLAR) are examples of parametric models. Contrary to these models, nonparametric models do not require a function definition. They can be based on look-up tables which is the basis for codebook prediction [13]. Semiparametric models allow a combination of the two previous ones.

4.2 Global and Local Prediction

Solutions for Non-linear Prediction. Most commonly used non-linear functions are based on polynomial approaches such as Volterra filters [4] or Neural Networks with ridge/gaussian functions [10]. The advantage of Volterra filters is that they can be computed using traditional signal processing algorithms in the same way as linear models. They have the attractive property of being linear in their parameters. However the number of parameters grows can grow exponentially with increasing order of the input. Neural networks can help limit this number but learning is often more time consuming.

Both of the above models can be viewed as global non-linear predictor models in the sense that all the speech signals are processed in a same way. Due to the non-stationary characteristics of speech signals, other types of non-linear functions have been investigated including the so called, local non-linear functions. The key idea here is to adapt the modeling and make it more flexible: for instance, in Threshold Auto-Regressive (TAR) models [14], speech signals are divided into different parts, and the linear AR model is then fitted for each partition. Similar ideas are also used in Code-Excited Linear Predictive (CELP) for speech coding. An excitation codebook is used for speech synthesis using linear predictive filters. The major problem with local non-linear predictors lies in the initial partitioning process, for which certain learning based approaches have been proposed and these are reviewed next.

Local Predictive Models. The partitioning problem for Local Linear Predictive models can be solved using learning approaches. For example, the Hierarchical Mixtures of Experts [15] can allow an optimal partition with the help of the (Expectation Maximisation) EM algorithm [6]. State based models can also be used, such as the well-known Hidden Markov Models (HMM) which provide a general and elegant framework [16]. Moreover, HMMs coupled with Multi-Layered Perceptron (MLP) neural networks have also been used for non-stationary time series prediction, including speech signals [17]. Hidden Control Neural models [18] are another example of connectionist based state models.

4.3 Predictor Structure

Prediction Context. The input window size or the prediction context (cf. equation 6) is a key parameter in predictive models. The dimension P should be sufficient to capture the dynamical characteristics of the speech signals. This parameter is usually determined using the embedding theorems of Takens [19]. The determination of the input window size P is a major problem in non-linear prediction [10]. It has to be determined for each application and various trade-offs may be required: for example, in [20] the input window size is optimized for classification but not for prediction.

Type of Predictor Architecture. Traditional predictors are generally based on a feedforward architecture. Even if the prediction context is theoretically sufficient, a lack of modelling efficiency can be due to the architecture. Recurrent models have been recently shown to offer a better solution compared to feed-forward models [12], but stability and optimization are important problems in these recurrent structures which need further investigation.

5 Learning Based Predictor Models

Most of the non-linear predictors reported to-date are machine learning based. The learning scheme can be very different: for example, feedforward Neural Networks cz trained using the well known backpropagation algorithm or its variants, and recurrent models can also be trained using several methods [12]. Statistical methods such as HMMs are usually trained using the EM algorithm. Predictive machine learning can be improved using statistical and kernel methods as shown in [21].

In all machine learning based methods, the generalization capability of the predictors is an important issue. Several methods including Bagging, Boostrap and Regularization have been successfully used to-date [10].

6 Conclusions and Perspectives

This paper briefly reviewed the current state of the art in non-linear predictive modeling for speech processing, focusing in particular, on the statistical and signal processing perspectives. We showed that non-linear predictors are mainly needed due to their statistical (i.e. non-gaussian) modeling requirements. The main structures of the non-linear models (including choice of non-linear functions, type of architecture, etc.) and their key parameters (such as the prediction context) were highlighted.

References

1. L. Rabiner and B.J. Juand, "Fundamentals of speech processing", *Prentice-Hall*, (1993).
2. H. Hermansky, "Perceptual linear predictive (plp) analysis of speech", *The Journal of the Acoustical Society of America*, pp. 1738-1752, (1990).
3. W. B. Kleijn, "Signal Processing Representations of Speech", *IEICE Trans. Inf. and Syst.*, **E86-D**, 3, pp. 359-376, March (2003).
4. J. Thyssen, H. Nielsen and S.D. Hansen, "Non-linearities short-term prediction in speech coding", *Proc. ICASSP*,**1**, pp. 185-188 (1994).
5. H. Teager and S. Teager, "Evidence for nonlinear sound production mechanisms in the vocal tract", *Proc. NATO ASI on Speech production and Speech Modeling*, **II**, pp. 241-261 (1989).
6. Richard O. Duda and Peter E. Hart and David G. Stork, "Pattern Classification", *Wiley-Interscience Publication*, (2001).
7. V. Pitsikalis and I. Kokkonos and P. Maragos, "Nonlinear analysis of speech signals: generalized dimensions and Lyapunov exponents", *Proc. EUROSPEECH*, pp. 817-820 (2003).
8. A. C. Lindgren, M. T. Johnson and R. J. Povinelli, "Speech Recognition using Reconstructed Phase Space Features", *ICASSP*, **1**,pp. 61-63 (2003).
9. S. Gazor and W. Zhang, "Speech probability distribution", *IEEE Signal Processing Letters*, 10(7), pp. 204-207, July 2003.
10. J.C. Principe, "Going Beyond linea, gaussian and stationary time series modeling", *International Summer Schoool on Neural Nets "E.R. Caianiello", IX Course: Nonlinear Speech Processing: Algorithms and Applications*, (2004).
11. J. Soraghan and A. Hussain and A. Alkulaibi abd T.S. Durrani, "Higher Order Statistics based non-linear speech analysis", *Journal of Control and Intelligent Systems, ACTA Press*, **30**,1, pp. 11-18 (2002).
12. D. P. Mandic and A. Chambers, "Recurrent Neural Networks for Prediction: Learning Algorithms, Architectures and Stability", *Wiley*, (2001).
13. S. Wang and E. Paksoy and A. Gersho, "Performance of nonlinear prediction of speech", *ICSLP*,**1**, 29–32 (1990).
14. H. Tong, "Nonlinear Time Series Analysis: A Dynamical System Approach", *Oxford University Press*, (1990).
15. S.R. Waterhouse and A.J. Robinson, "Non-linear Prediction of Acoustic Vectors Using Hierarchical Mixtures of Experts", *Tesauro*, pp.824-842 (1995).
16. P. Gallinari, "An MLP/HMM hybrid model using nonlinear predictors", *Adaptive Processing of Sequences and Data Structures, C. Lee Giles and Mori (Eds), Springer*, pp.418-434 (1998).
17. Y.J. Chung and C. K. Un, "Predictive Models for Sequence Modelling, Application to Speech and Character Recognition", *Speech Communication* , 19(4):307-316 (1996).
18. E. Levin, "Hidden control neural architecture modeling nonlinear time varying and its applications", *IEEE Trans. Neural Networks*, 4:109-116 January (1996).
19. H. Takens, "On the numerical determination of the dimension of an attractor", *Dynamical Systems and Turbulence*,(1981).
20. M. Chetouani, "Codage neuro-prédictif pour l'extraction de caractéristiques de signaux de parole", PhD Thesis, Université Paris VI, 2004.
21. J. H. Friedman, "Recent advances in Predictive (Machine) Learning", *PHYSTAT2003*, 2003.

Predictive Speech Coding Improvements Based on Speaker Recognition Strategies

Marcos Faundez-Zanuy

Escola Universitària Politècnica de Mataró,
Universitat Politècnica de Catalunya, Barcelona Spain
faundez@eupmt.es
http://www.eupmt.es/veu

Abstract. This paper compares the speech coder and speaker recognizer applications, showing some parallelism between them. In this paper, some approaches used for speaker recognition are applied to speech coding based on neural networks, in order to improve the prediction accuracy. Experimental results show an improvement in Segmental SNR (SEGSNR) up to 1.7 dB.

1 Introduction

We can establish a parallelism between speech coding and speech/speaker recognition. This is stated in the literature as classification versus regression [1], defined as:

Regression: It is the process of investigating the relationship between a dependent (or response) variable Y and independent (or predictor) variables $X_1, \ldots, X_P$; a regression function expresses the expected value of Y in terms of $X_1, \ldots, X_P$ and model parameters.

Discrimination: The problem itself is one in which we are attempting to predict the values of one variable (the class variable) given measurements made on a set of independent variables (the pattern vector $\boldsymbol{x}$). In this case, the response variable is categorical.

When comparing speech coding with speech/speaker recognition, we can identify the following similarities:

1.1 Problem Statement

Without loss of generality we will assume that the speech signal $x(t)$ is normalized in order to achieve maximum absolute value equal to 1.

For speaker recognition applications, with a closed set of N users inside the database, the problem is: given a vector of samples $\left[x(1), x(2), \cdots, x(L)\right]$, try to guess to whom speaker $speaker_i$ it belongs, with $speaker_i \in \left[1, 2, \cdots, N\right]$. In order to achieve statistical consistency, this task is performed using several hundreds of vectors, and some kind of parameterization is performed over the signal samples, like a bank of filters, cepstral analysis, etc. For speech coding applications, the problem is: given a vector of previous samples $\left[x(1), x(2), \cdots, x(L)\right]$, try to guess which is the next sample: $x(L+1)$, $\quad x(L+1) \in \left[-1, 1\right]$. Thus, the problem statement is the same with the

W. Duch et al. (Eds.): ICANN 2005, LNCS 3697, pp. 785 – 791, 2005.

exception that the former corresponds to a discrete set of output values, and the latter corresponds to a continuum set of output values.

Taking into account this fact, the "speech predictor" can be seen as a "classifier".

1.2 Signal Parameterization

Although strong efforts have been done in speaker recognition for obtaining a good parameterization of the speech samples in order to improve the results and to reduce the computational burden, this step is ignored in speech coding, where the parameterized signal is the own signal without any processing.

1.3 Model Computation

For speaker recognition applications, some kind of model must be computed. Usually it is computed using some training material different from the test signals to be classified. This model can be as simple as the whole set of input vectors (Nearest Neighbor model [2]) or the result of some reduction applied on them (Codebook obtained with Vector Quantization [3], Gaussian Mixture Model; etc.), being the most popular the GMM.

For predictive speech coding applications, the model is usually a LPC (Linear predictive Coding) model, but it can also be a codebook, Volterra series, neural networks, etc. These last kind of predictors belong to the nonlinear prediction approaches, that can outperform the classical linear ones [4]. If the neural network is a Radial Basis Function [5], the similarity is considerable with the GMM model of the speaker recognition applications.

1.4 Decision

For speaker recognition applications, the decision (classification) is done taking into account the fitness of the test observation to the previous computed models, and some decision rule (for instance, maximum likelihood). On the other hand, in predictive speech coders, the predicted sample is the output of the predictor given one input vector. While the speech prediction is the result of one single predictor, in speaker recognition it is often used a combination of several classifiers performing their task over the same observations or sometimes even different (multimodal biometrics). This strategy is known as data fusion [6] and committee machines [7], and it can be considered that it relies on the principle of divide and conquer, where a complex computational task is split into several simpler tasks plus the combination of them. A committee machine is a combination of experts by means of the fusion of knowledge acquired by experts in order to arrive to an overall decision that it is supposedly superior to that attainable by any one of them acting alone.

2 Proposed Improvements

2.1 Contribution on "Signal Parameterization"

One drawback of the classical LPC predictors is that there is just one parameter that can be set up: the prediction order (or vector dimension of the input vectors). Using a nonlinear predictor there is more feasibility and better results, because linear models are optimal just for Gaussian signals, that it is not the case of speech signals. Another advantage of nonlinear models for (instance neural networks) is that they can integrate different kinds of information. They can use "hints" in order to improve the accuracy, etc. For this reason we propose, in addition to the speech samples, the use of delta parameters. This kind of information has been certainly useful for speaker recognition [3], where it is found that instantaneous and transitional representations are relatively uncorrelated, thus providing complementary information for speaker recognition. The computation of the transitional information is as simple as the first order finite difference. This transitional information is also known as delta parameters.

2.2 Contributions on "Decision"

The combination of several predictors is similar to the Committee machines strategy [7]. If the combination of experts were replaced by a single neural network with a large number of adjustable parameters, the training time for such a large network is likely to be longer than for the case of a set of experts trained in parallel. The expectation is that the differently trained experts converge to different local minima on the error surface, and overall performance is improved by combining the outputs of each predictor. Different predictors can be combined, being the result of the same architecture, with the same training vectors and algorithm, but different initialization weights and biases. In speaker recognition, this strategy provides improved security [8-10].

3 Experimental Results

The experiments have been done using the same database of our previous work on nonlinear speech coding [8]. We have encoded eight sentences uttered by eight different speakers (4 males and 4 females).

3.1 Predictor Based on Radial Basis Functions

The RBF network consists on a Radial Basis layer of S neurons and an output linear layer. The output of ith Radial Basis neuron is $R_i = radbas\left(\left\|\vec{w}_i - \vec{x}\right\| \times b_i\right)$, where $\vec{x}$ is the L dimensional input vector, b_i is the scalar bias or spread (σ) of the Gaussian, $\vec{w}_i$ is the L dimensional weight vector of the Radial Basis neuron i, also known as center, and the transfer function is $radbas[n] = e^{-n^2}$.

In our case, the output is one neuron. We have used two different training algorithms:

RBF-1:
The variance is manually setup in advance. Thus, it is one parameter to fix. The algorithm iteratively creates a radial basis network one neuron at a time. Neurons are added to the network until the maximum number of neurons has been reached. At each iteration, the input vector that results in lowering the network error the most, is used to create a radial basis neuron.

RBF-2:
The weights are all initialized with a zero mean, unit variance normal distribution, with the exception of the variances, which are set to one. The centers are determined by fitting a Gaussian mixture model with circular covariances using the EM algorithm. (The mixture model is initialized using a small number of iterations of the K-means algorithm). The variances are set to the largest squared distance between centers. The hidden to output weights that give rise to the least squares solution are determined using the pseudo-inverse.

3.2 Predictive Speech Coding

We have used an ADPCM scheme with an adaptive scalar quantizer based on multipliers [8]. The number of quantization bits is variable between Nq=2 and Nq=5, that correspond to 16 kbps and 4 0kbps (the sampling rate of the speech signal is 8 kHz). We have used a prediction order L=10

3.3 Experiments with Algorithm RBF-1

In order to setup the RBF architecture, we have studied the relevance of two parameters: spread (variance) and number of neurons. First, we have evaluated the SEGSNR for an ADPCM speech coder with RBF prediction and adaptive quantizer of 4 bits, as function of the spread of the Gaussian functions.

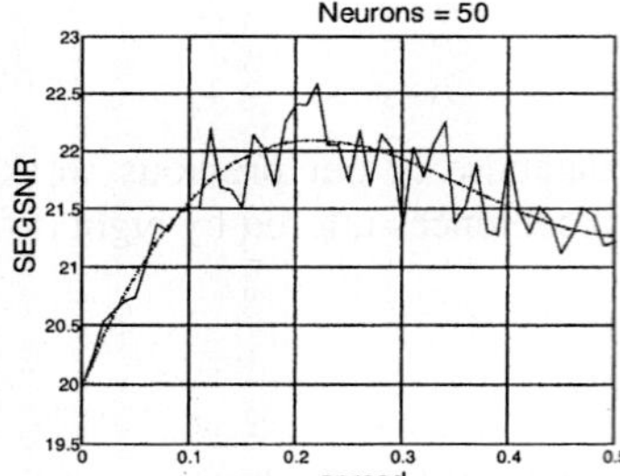

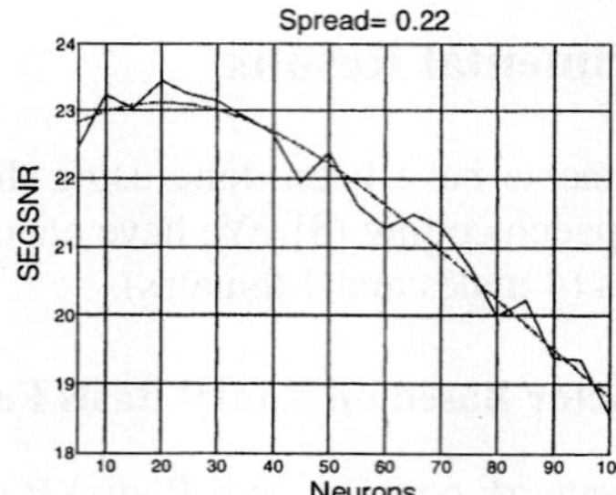

Fig. 1. RBF-1, left: SEGSNR vs spread for 50 neurons; right: SEGSNR vs number of neurons for spread=0.22

Figure 1, on the left, shows the results using one sentence, for spread values ranging 0.011 to 0.5 with an step of 0.01 and S=50 neurons. It also shows a polynomial interpolation of third order, with the aim to smooth the results. Based on this plot, we have chosen a spread value of 0.22. Using this value, we have evaluated the relevance of the number of neurons. Figure 3, on the center, shows the results using the same sentence and a number of neurons ranging from 5 to 100 with a step of 5. This plot also shows an interpolation using a third order polynomial. Using this plot we have chosen an RBF architecture with S=20 neurons. If the number of neurons (and/ or the spread of the guassians) is increased, there is an overfit (over parameterization that implies a memorization of the data and a loose of the generalization capability).
The use of delta parameters can be seen in figure 2.

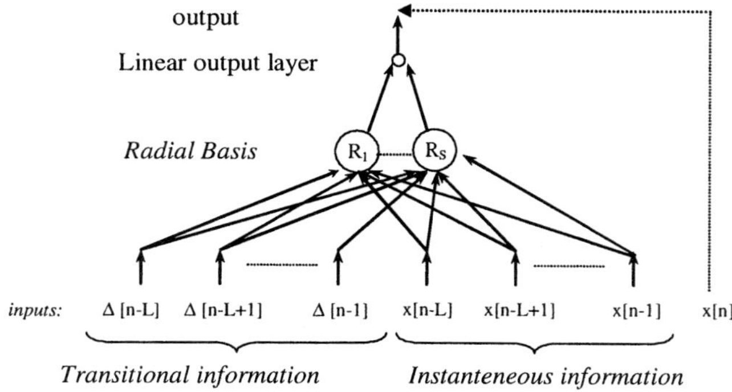

Fig. 2. RBF scheme using delta parameters

Figure 3 shows the results in the same conditions than figure 1, but adding delta parameters. Comparing figures 1 and 3 it is clear that the use of delta information can improve the SEGSNR, even without an increase of the number of centers.

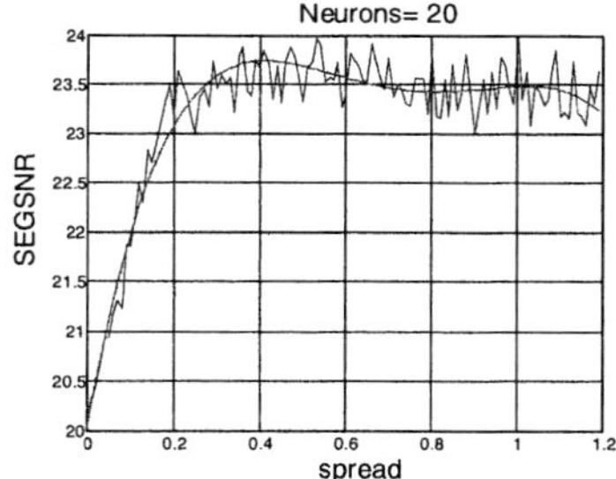

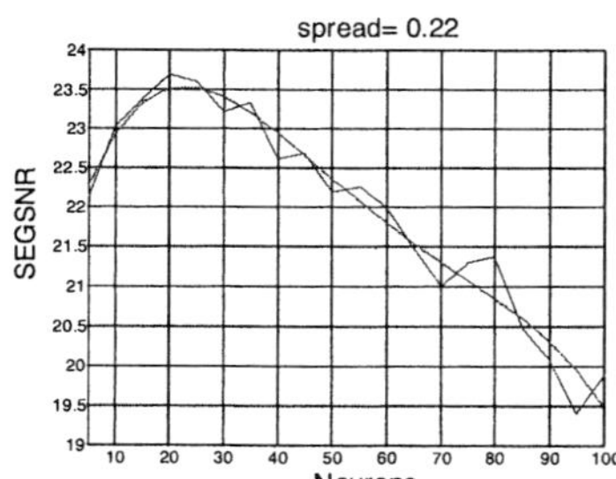

Fig. 3. RBF-1, delta parameters left: SEGSNR vs spread for 20 neurons; right: SEGSNR vs neurons for spread=0.22

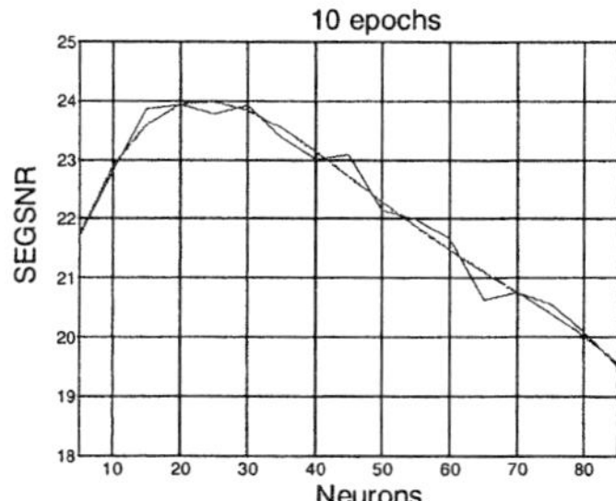

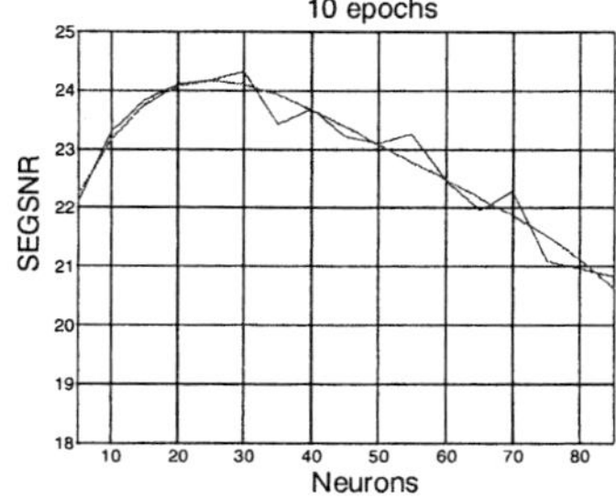

Fig. 4. SEGSNR vs number of neurons for RBF-2 trained with 10 epochs. Left: without delta parameters, right: with delta parameters.

Table 1. SEGSNR for ADPCM with several predictors

	RBF-1 (spread=0.22)				RBF-1 (spread=0.4)				RBF-2			
Parameterization →	$\vec{x}$		$\vec{x}+\vec{\Delta}$		$\vec{x}$		$\vec{x}+\vec{\Delta}$		$\vec{x}$		$\vec{x}+\vec{\Delta}$	
Nq	m	σ	m	σ	m	σ	m	σ	m	σ	m	σ
2	11.65	7.63	10.94	7.21	12.05	8.88	12.21	8.36	13.75	5.96	14.05	5.78
3	18.40	6.56	18.99	6.13	19.18	7.94	19.17	7.71	20.23	6.41	20.60	6.22
4	23.69	6.12	23.40	6.05	24.33	7.20	24.47	6.99	25.35	6.57	25.69	6.63
5	28.22	6.34	28.13	6.14	29.16	7.28	29.29	6.95	30.22	6.90	30.42	6.86

3.4 Experiments with Algorithm RBF-2

Comparing figures 1 and 3 it can be seen that the system with delta parameters achieves and improvement without an increase of the number of neurons (centers), for the whole range of spreads and number of neurons. Anyway, the algorithm needs the setup of the spread. In order to avoid this, we can use the algorithm RBF-2 described before, that overcomes this drawback.

Figure 4 shows the obtained results for RBF-2 and the architecture of figure 2.

Table 1 shows the results for RBF-1 (with spread=0.4), RBF-2 and a committee RBF-1+RBF-2 with and without delta parameters.

Table 2. SEGSNR for ADPCM with several combined predictors

	Combined RBF-1 spread=0.22+RBF2				Combined RBF-1 spread=0.4+RBF2			
Parameterization →	$\vec{x}$		$\vec{x}+\vec{\Delta}$		$\vec{x}$		$\vec{x}+\vec{\Delta}$	
Nq	m	σ	m	σ	m	σ	m	σ
2	13.46	6.58	13.66	6.12	13.57	7.27	13.88	6.90
3	19.97	6.13	20.05	5.80	20.21	7.02	20.60	6.56
4	25.19	6.18	25.15	6.00	25.52	6.87	25.69	6.80
5	29.91	6.43	29.87	6.35	30.22	7.03	30.42	6.92

Acknowledgement

This work has been supported by FEDER and the Spanish grant MCYT TIC2003-08382-C05-02.

References

1. Webb A. "Statistical Pattern Recognition" 2nd edition. John Wiley and sons, 2002
2. Higgins A. L., Bahler L. G. & Porter J. E. "Voice identification using nearest-neighbor distance measure". Proc. ICASSP 1993, Vol. II, pp. 375-378

3. F. K. Soong and A. E. Rosenberg "On the use of instantaneous and transitional spectral information in speaker recognition". IEEE Trans. On ASSP, Vol. 36, Nº 6, pp.871-879, June 1988
4. Faundez-Zanuy M., McLaughlin S., Esposito A., Hussain A., Schoentgen J., Kubin G., Kleijn W. B. & Maragos P.. "Nonlinear speech processing: overview and applications". Control and intelligent systems, Vol. 30 Nº 1, pp.1-10, 2002. ACTA Press.
5. Birgmeier M., "Nonlinear prediction of speech signals using radial basis function networks". Proc. of EUSIPCO 1996, vol. 1, pp. 459-462.
6. Faundez-Zanuy M., "Data fusion in biometrics" IEEE Aerospace and Electronic Systems Magazine, Vol.20 nº 1, pp.34-38, January 2005.
7. Haykin S., "Neural nets. A comprehensive foundation", 2on ed. Prentice Hall 1999
8. Faundez-Zanuy M., Vallverdú F., Monte E., "Nonlinear prediction with neural nets in adpcm" IEEE International Conference on Acoustics, Speech, and Signal Processing, ICASSP 1998, Vol I, pp.345-348.Seattle
9. Faundez-Zanuy M., "On the vulnerability of biometric security systems". IEEE Aerospace and Electronic Systems Magazine. Vol.19 nº 6, pp.3-8, June 2004.
10. Faundez-Zanuy M. and E. Monte-Moreno, "State-of-the-art in speaker recognition". IEEE Aerospace and Electronic Systems Magazine. Vol.20 nº 5, pp 7-12, May 2005.

Predictive Kohonen Map for Speech Features Extraction

Bruno Gas, Mohamed Chetouani, Jean-Luc Zarader, and Christophe Charbuillet

Laboratoire des Instruments et Systèmes d'Ile de France, Université Paris VI, France
Bruno.Gas@upmc.fr
http://www.lisif.jussieu.fr

Abstract. Some well known theoretical results concerning the universal approximation property of MLP neural networks with one hidden layer have shown that for any function f from $[0,1]^n$ to $\Re$, only the output layer weights depend on f. We use this result to propose a network architecture called the *predictive Kohonen map* allowing to design a new speech features extractor. We give experimental results of this approach on a phonemes recognition task.

1 Introduction

Most of the speech recognition systems require in the very first stage to model the short-term spectrum of the signal (typically windows from 10 to 20 ms). MFCC parameters (Mel Frequency Cepstrum Coding) are for a long time used because of their robustness and of the quality of their statistical distribution. Authors as Hermansky [1] however pointed out the importance to revisit the stage of feature extraction. He proposed to use the more recent perceptual auditive models such as the PLP and RASTA-PLP. Instead of using directly the short-term spectrum as for MFCC, one can approximate it by parametric approaches like it is done in the well-known LPC (Linera Predictive Coding). Usually these approximations are based on linear assumptions of the speech production model (i.e. vocal tract).

Discriminative Models. One drawback of the NPC parameters, inherited from LPC parameters, is their lack of discrimination. In fact, they are more adapted to speech coding and synthesis applications. Juang and Katigiri (1992) showed that a reinforcement of the discriminant property can be obtained by adapting the features extraction to the classification task. For example, Biem and Katagiri [6] proposed to estimate the optimal spectral width of the MFCC filters bank during the classifier training stage. Similar ideas have been used to make improvements of the NPC coder. Two new versions of the coder were thus proposed (DFE-NPC and LVQ-NPC). They were tested on phonemes recognition [7] and speaker recognition [8].

Unsupervised Models. Some applications (for example the segmentation of unknown speakers in radio broadcast news) do not provide classes membership

W. Duch et al. (Eds.): ICANN 2005, LNCS 3697, pp. 793–798, 2005.

information (the speakers). An alternative consists in using unsupervised algorithms. We propose in this article a new unsupervised version of the coder called SOM-NPC. The output layer cells are organized according to a topological map called *topological predictive map*. We show by experiments that a specialization of the output layer weights is obtained by self-organization, according to the membership class of the input signals.

2 SOM-NPC Parameters

In 1957, Kolmogorov proved with its superposition theorem (13th Hilbert problem refutation) that every continuous function f from $\mathcal{E}^n$ to $\Re$ defined on the n-dimensional Euclidean unit cube $\mathcal{E}^n$ and with range on the real line $\Re$ can be represented as a sum of continuous functions:

$$f(x_1, \dots, x_n) = \sum_{q=1}^{2n+1} \phi_q(\sum_{p=1}^{n} \psi_{pq}(x_p)) \tag{1}$$

Hecht-Nielsen [9] recognized that this specific format of Kolmogorov's superpositions can be interpreted as a feedforward neural network with a hidden layer that computes the variables $y_q = \sum_{p=1}^{n} \psi_{pq}(x_p)$. This suggestion, has been criticized by Poggio and Girosi [10] for several reasons, one being that applying Kolmogorov's theorem would require the learning of nonparametric activation functions. However, others similar result have been obtained by the use of functional analysis theorems [11]. What makes Hecht-Nielsen's network particularly attractive for us is that the hidden layers are fixed independently of any function f, so that in theory this part of the neural network is trained once for n (It was demonstrated by Kurkova (1992), Sprecher (1993) and Katssura (1994) and others that there are universal hidden layers that are independant even of n). The NPC features extractor is builded from this principle : only the output layer weights are the feature vector. The remaining problem is then to estimate the hidden layer weights. Four estimation methods have been already proposed which are the NPC, NPC-2, DFE-NPC and LVQ-NPC. The proposed one here has the advantage of being unsupervised and clearly puts in obviousness the output weights specialization.

2.1 SOM-NPC Coder Definition

Following now the Lapedes and Farber [2] model, one can see the NPC encoder as a layered neural network trained to predict time series. For a given signal frame m generated by an unknown non linear operator f, it is trained from examples of pairs of $\mathbf{x}_k = [y_{k-1}, y_{k-2}, \dots, y_{k-\lambda}]^\top$ input vectors and y_k output samples, while minimizing the mean square error:

$$Q_m(\Omega, \mathbf{a}) = \frac{1}{2} \sum_{k}^{K} (y_k - F_{\Omega,\mathbf{a}}(\mathbf{x}_k))^2 \tag{2}$$

where $F_{\Omega,\mathbf{a}}$ is the non linear λ dimensional function realized by the neural network with parameters noted Ω (first layer weights) and $\mathbf{a} = [a_1, \dots, a_N]^\top$ (output layer weights) including sigmoidal node functions. More precisely, $F_{\Omega,\mathbf{a}}$ can be viewed as the composition of two functions G_Ω (corresponding to the network first layer) and $H_\mathbf{a}$ (corresponding to the network output layer) such that:

$$F_{\Omega,\mathbf{a}}(\mathbf{x}_k) = \sum_i a_i \sigma[\sum_j \omega_{ij} y_{k-j}] = G_\Omega \circ H_\mathbf{a}(\mathbf{x}_k) \tag{3}$$

The NPC coding needs two computing stages. 1) the *parameters adjustment stage* which consists in the learning of the weights of the first layer Ω once a time; 2) the *features extraction stage* which occurs at every signal frame coding: only the $\mathbf{a}$ weights are learned while the hidden layer weights (issued from the first stage) remain fixed. The prediction error which must be minimized over all the sample vectors $\mathbf{x}_k$ of the frame m is then given by : $Q_m(\mathbf{a}) = \sum_k (y_k - H_\mathbf{a}(\mathbf{z}_k))^2$ with $\mathbf{z}_k = G_\Omega(\mathbf{x}_k)$, using a standard multidimensional optimisation method, e.g. steepest descent (error back propagation).

2.2 NPC Distance

The first stage (first layer weights learning), which is unsupervised in our case, is done by defining a set of predictive output cells organized on a 2 dimension map. Because the comparison between patterns from the input signals space and vectors from the second layer weights space is not immediate, we need to define a specific distance. The *NPC distance* between two signal frames l and m is defined as the Itakura's distance measure was in the framework of linear prediction techniques [7]:

$$d_\Omega^{NPC}(l,m) = \log \frac{Q_m(\mathbf{a}_l)}{Q_m(\mathbf{a}_m)} \tag{4}$$

(4) gives the ratio of the frame m prediction error using the frame l NPC parameters $\mathbf{a}_l$ and the same frame prediction error, but using the frame m NPC parameters $\mathbf{a}_m$. When applying the m signal frame to the NPC (for a given Ω) with its adapted coding coefficients $\mathbf{a}_m$, the output residual error $Q_m(\Omega, \mathbf{a}_m)$ is minimal. On the other hand, when applying the same signal to the NPC with the adapted coding coefficients $\mathbf{a}_l$ of the l signal frame, the residual error $Q_m(\Omega, \mathbf{a}_l)$ is not minimal and one obtains $Q_m(\Omega, \mathbf{a}_l) \geq Q_m(\Omega, \mathbf{a}_m)$. For $l = m$, one has $d_\Omega(l,m) = 0$. Let us note that $d_\Omega^{NPC}(l,m)$ is a not a true distance since it is not symetrical.

2.3 First Layer Weight Training

One define a network structure with L output cells on a 2D map with a local neighborhood V^σ (fig. 1). The learning algorithm is that of a traditional Kohonen map which one would have replaced the Euclidean distance in the input space by the NPC distance in the input signal space. The algorithm is as follows:

For all the training frames m :

1) finding the winner neuron l^* of the map such that :

$$l^* = \arg \min_{l=1,\ldots,L} \{\log \frac{Q_m(\mathbf{a}_l)}{Q_m(\mathbf{a}_m)}\} = \arg \min_{l=1,\ldots,L} \{Q_m(\mathbf{a}_l)\} \quad (5)$$

2) updating the winner neuron and its neighbors weights such as to minimize the $d_{\Omega}^{NPC}(l^*, m)$ distance (this is equivalent to minimize the square prediction error) :

$$Q_m(\mathbf{a}_{1,\ldots,L}) = \sum_{l}^{L} \sum_{k(m)} (y_k - G_{\Omega} \circ H_{\mathbf{a}_l}(\mathbf{x}_k))^2 V^{\sigma}(l^*, l) \quad (6)$$

were $V^{\sigma}(l, l^*) = e^{-\frac{d(l,l^*)}{2\sigma}}$ is the neighborhood function (a gaussian low in our case, $d(l, l^*)$ being the length of the shortest way between l and l^* in the map and σ the standard deviation). σ is a decreasing function of the learning time such that $\sigma(q) = [\frac{\sigma_f}{\sigma_i}]^{\frac{1}{N}} \sigma(q-1)$ where σ_i and σ_f are the initial and the final imposed values of the standard deviation and N the learning iteration number.

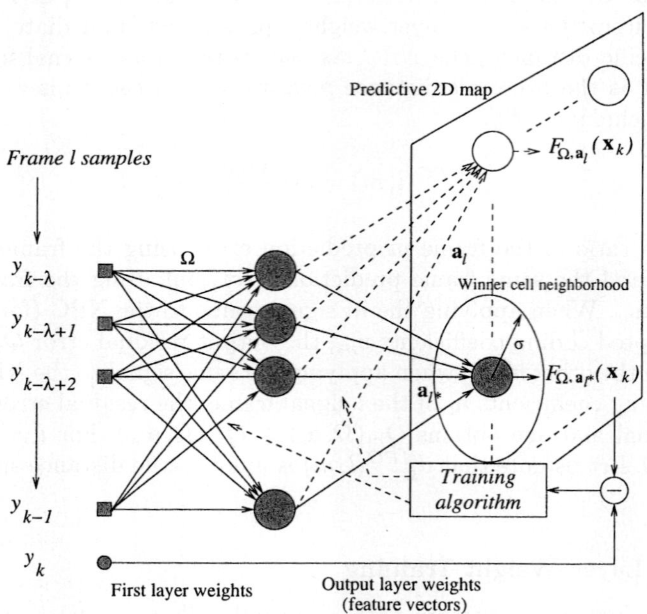

Fig. 1. NPC-K coder

2.4 Experimental Results

We built three phoneme bases each extracted from the Darpa-TIMIT speech database. The first base groups four classes of voiced phonemes (vowels) very commonly used: */aa/*, */ae/*, */ey/* and */ow/*. the second and the third bases group two series of phonemes : */b/*,*/d/*,*/g/* (voiced plosives) and */p/*,*/t/*,*/k/* (unvoiced plosives). Those phonemes are frequently used and simultaneously difficult to process. We used the two first Dialect Regions : *DR1* (see table 1) for the training set of both the SOM-NPC first layer estimation and the MLP weights estimation, *DR2* for the test set.

We trained 3 SOM-NPC coders of 16 inputs, 16 hidden cells and $8 \times 8 = 64$ predictive cells. After 50 training epochs (for example each epoch means 11701 frames presented to the network for the first vowels base) we then obtained the map cells labelling in table 2. A map cell is labelled according to the most frequently winner classe. The coder can be then used as a phonemes classifier. The number of cells sharing the same label depends on the signal class complexity but also of the ratio of the corresponding frames used for the training (see the

Table 1. Phoneme training bases

	vowels				voiced plosives			unvoiced plosives		
frames	11701				883			3223		
phones	*/aa/*	*/ae/*	*/ey/*	*/ow/*	*/b/*	*/d/*	*/g/*	*/p/*	*/t/*	*/k/*
frames	2924	4600	2161	2016	258	312	313	623	1100	1510
%	24%	39%	18%	17%	29%	35%	35%	19%	34%	46%
cells (/64)	13	32	13	6	14	32	18	19	34	46
%	20.3%	50%	20.3%	9.3%	22%	50%	28%	15.6%	53.1%	31.2%

Table 2. Map cells labelling for the 3 phonemes bases and phonemes recognition rates

d d d d d g g g	q q q q q q q t	ow ow aa ae ae ae ae ey
d d d d d b g g	q q q q q q p t	ow aa aa aa ae ae ae ey
d d d d d b g g	q q p t t t t t	ow aa aa aa ae ae ey ey
d d d d g b g g	q p p p t t t t	ow ae aa aa ae ae ey ey
g d d d d d g g	p q p p t t t t	aa ae ae ae ae ae ae ey
d d d b d d g g	p q t t t t t t	aa aa ae ae ae ae ey ey
d b d b b b b b	p q q t t t t t	ow ae ae ae ae ae ey ey
d b b b g g g b	t t t t t t t t	aa ey ey ae ae ae ae ae

classifier	data set	recognition rate		
		vowels	voiced plosives	unvoiced plosives
NPC-K map	training set	64%	66%	76%
NPC-K map	test set	59%	63%	69%
MLP	training set	64%	88%	86%
MLP	test set	56%	64%	77%

table 1). Once the first stage was ended, we computed the SOM-NPC parameters of the *DR1* and *DR2* frames. The *DR1* features were used to train a 2 layers MLP as a phoneme classifier (50000 training iterations). We reported on table 2 the recognition rates obtained on the three bases from both the coder and the MLP classifier. These results and the visible differentiation of the output cells on the 2D map show that the output layer weights carry really important features related to the modelized short-term spectrum.

3 Conclusions

We have proposed a predictive self-organizing map architecture which ensure the unsupervised training of a NPC coder under the assumption that only the second layer weights carry the modelized signal features. Phoneme feature extraction experiments given in this article have shown an interesting self-organizing process of the output cells which seems to confirm the initial assumptions. Our current works are devoted to the study of an adaptative neighborhood function. We are also focusing on a non deterministic reading of the predictive map mainly because the higher levels of speech systems usually need class probability estimation.

References

1. Hermansky, H.: Should recogizers have ears ? Speech Communication **25** (1998) 3–27
2. Lapedes A., Farber R.: Nonlinear signal processing using neural networks: Prediction and system modelling. Internal Report, Los Alamos National Laboratory (1987)
3. Thyssen J., Nielsen H., Hansen S.D.: Non-linear short-term prediction in speech coding. Proc. of Int. Conf. on Signal and Speech Processing **1** (1994) 185–188
4. Reichl W., Harengel S., Wolferstetter F., Ruske, G.: Neural networks for nonlinear discriminant analysis in continuous speech recognition. Eurospeech (1995) 537–540
5. Hunt M.J., Lefebvre C.: A comparison of several acoustic representations for speech recognition with degraded and undegraded speech. Int. Conf. on Speech and Signal Processing **2** (1989) 262–265"
6. Biem A., Katagiri S.: Filter bank design based on Discriminative Feature Extraction. Proc. of Int. Conf. on Signal and Speech Processing **1** (1994) 485–488
7. Gas, B. and Zarader, J.L. and Chavy, C., and Chetouani, M.: Discriminant neural predictive coding applied to phoneme recognition. Neurocomputing, **56** (2004) 141–166
8. Chetouani, M. and Faundez-Zanuy, M. and Gas, B. and Zarader, J.L.: A New Non-linear speaker parameterization algorithm for speaker identification. Proc. of ISCA Tutorial and Research Workshop on Speaker and Recognition Langage Workshop (2004) 309–314
9. Hecht-Nielsen R.: Kolmogorov's mapping neural network existence theorem. Proc. of Int. Conf. on Neural Networks (1987) 11–13
10. Girosi F., Poggio T.: Representation properties of networks: Kolmogorov's theorem is irrelevant. Neural Computation **1**(4) (1989) 465–469
11. Hornik, K.: Multilayer feedforward networks are universal approximators. Neural Networks, **2** (1089) 359–366

Bidirectional LSTM Networks for Improved Phoneme Classification and Recognition

Alex Graves[1], Santiago Fernández[1], and Jürgen Schmidhuber[1,2]

[1] IDSIA, Galleria 2, 6928 Manno-Lugano, Switzerland
{alex, santiago, juergen}@idsia.ch
[2] TU Munich, Boltzmannstr. 3, 85748 Garching, Munich, Germany

Abstract. In this paper, we carry out two experiments on the TIMIT speech corpus with bidirectional and unidirectional Long Short Term Memory (LSTM) networks. In the first experiment (framewise phoneme classification) we find that bidirectional LSTM outperforms both unidirectional LSTM and conventional Recurrent Neural Networks (RNNs). In the second (phoneme recognition) we find that a hybrid BLSTM-HMM system improves on an equivalent traditional HMM system, as well as unidirectional LSTM-HMM.

1 Introduction

Because the human articulatory system blurs together adjacent sounds in order to produce them rapidly and smoothly (a process known as co-articulation), contextual information is important to many tasks in speech processing. For example, when classifying a frame of speech data, it helps to look at the frames after it as well as those before — especially if it occurs near the end of a word or segment. In general, recurrent neural networks (RNNs) are well suited to such tasks, where the range of contextual effects is not known in advance. However they do have some limitations: firstly, since they process inputs in temporal order, their outputs tend to be mostly based on *previous* context; secondly they have trouble learning time-dependencies more than a few timesteps long [8]. An elegant solution to the first problem is provided by bidirectional networks [11,1]. In this model, the input is presented forwards and backwards to two separate recurrent nets, both of which are connected to the same output layer. For the second problem, an alternative RNN architecture, LSTM, has been shown to be capable of learning long time-dependencies (see Section 2).

In this paper, we extend our previous work on bidirectional LSTM (BLSTM) [7] with experiments on both framewise phoneme classification and phoneme recognition. For phoneme recognition we use the hybrid approach, combining Hidden Markov Models (HMMs) and RNNs in an iterative training procedure (see Section 3). This gives us an insight into the likely impact of bidirectional training on speech recognition, and also allows us to compare our results directly with a traditional HMM system.

2 LSTM

LSTM [9,6] is an RNN architecture designed to deal with long time-dependencies. It was motivated by an analysis of error flow in existing RNNs [8], which found that long

W. Duch et al. (Eds.): ICANN 2005, LNCS 3697, pp. 799–804, 2005.

time lags were inaccessible to existing architectures, because the backpropagated error either blows up or decays exponentially.

An LSTM hidden layer consists of a set of recurrently connected blocks, known as memory blocks . These blocks can be thought of a differentiable version of the memory chips in a digital computer. Each of them contains one or more recurrently connected memory cells and three multiplicative units - the input, output and forget gates - that provide continuous analogues of write, read and reset operations for the cells. More precisely, the input to the cells is multiplied by the activation of the input gate, the output to the net is multiplied by the output gate, and the previous cell values are multiplied by the forget gate. The net can only interact with the cells via the gates.

Some modifications of the original LSTM training algorithm were required for bidirectional LSTM. See [7] for full details and pseudocode.

3 Hybrid LSTM-HMM Phoneme Recognition

Hybrid artificial neural net (ANN)/HMM systems are extensively documented in the literature (see, e.g. [3]). The hybrid approach benefits, on the one hand, from the use of neural networks as estimators of the acoustic probabilities and, on the other hand, from access to higher-level linguistic knowledge, in a unified mathematical framework.

The parameters of the HMM are typically estimated by Viterbi training [10], which also provides new targets (in the form of a new segmentation of the speech signal) to re-train the network. This process is repeated until convergence. Alternatively, Bourlard *et al.* developed an algorithm to increase iteratively the global posterior probability of word sequences [2]. The REMAP algorithm, which is similar to the Expectation-Maximization algorithm, estimates local posterior probabilities that are used as targets to train the network.

In this paper, we implement a hybrid LSTM/HMM system based on Viterbi training compare it to traditional HMMs on the task of phoneme recognition.

4 Experiments

All experiments were carried out on the TIMIT database [5]. TIMIT contain sentences of prompted English speech, accompanied by full phonetic transcripts. It has a lexicon of 61 distinct phonemes. The training and test sets contain 4620 and 1680 utterances respectively. For all experiments we used 5% (184) of the training utterances as a validation set and trained on the rest.

We preprocessed all the audio data into frames using 12 Mel-Frequency Cepstrum Coefficients (MFCCs) from 26 filter-bank channels. We also extracted the log-energy and the first order derivatives of it and the other coefficients, giving a vector of 26 coefficients per frame in total.

4.1 Experiment 1: Framewise Phoneme Classification

Our first experimental task was the classification of frames of speech data into phonemes. The targets were the hand labelled transcriptions provided with the data,

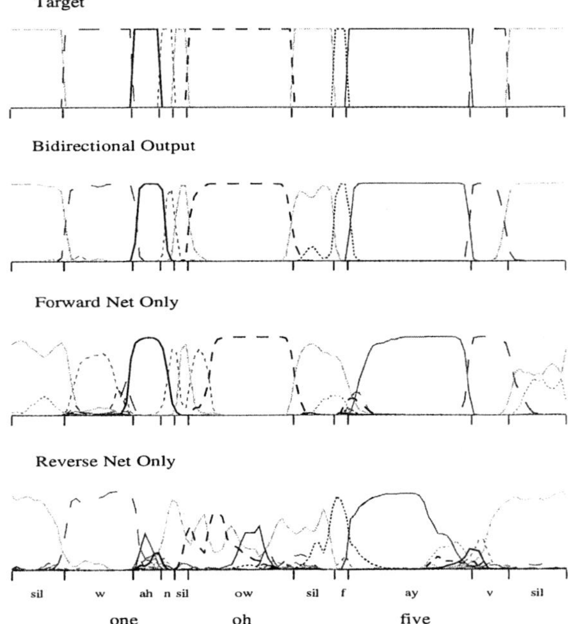

Fig. 1. A bidirectional LSTM net classifying the utterance "one oh five" from the Numbers95 corpus. The different lines represent the activations (or targets) of different output nodes. The bidirectional output combines the predictions of the forward and reverse subnets; it closely matches the target, indicating accurate classification. To see how the subnets work together, their contributions to the output are plotted separately ("Forward Net Only" and "Reverse Net Only"). As we would expect, the forward net is more accurate. However there are places where its substitutions ('w'), insertions (at the start of 'ow') and deletions ('f') are corrected by the reverse net. In addition, both are needed to accurately locate phoneme boundaries, with the reverse net tending to find the starts and the forward net tending to find the ends ('ay' is a good example of this).

and the recorded scores were the percentage of frames in the training and test sets for which the output classification coincided with the target.

We evaluated the following architectures on this task: bidirectional LSTM (BLSTM), unidirectional LSTM (LSTM), bidirectional standard RNN (BRNN), and unidirectional RNN (RNN). For some of the unidirectional nets a delay of 4 timesteps was introduced between the target and the current input — i.e. the net always tried to predict the phoneme of 4 timesteps ago. For BLSTM we also experimented with duration weighted error, where the error injected on each frame is scaled by the duration of the current phoneme.

We used standard RNN topologies for all experiments, with one recurrently connected hidden layer and no direct connections between the input and output layers. The LSTM (BLSTM) hidden layers contained 140 (93) blocks of one cell in each, and the RNN (BRNN) hidden layers contained 275 (185) units. This gave approximately 100,000 weights for each network.

All LSTM blocks had the following activation functions: logistic sigmoids in the range $[-2, 2]$ for the input and output squashing functions of the cell , and in the range $[0, 1]$ for the gates. The non-LSTM net had logistic sigmoid activations in the range $[0, 1]$ in the hidden layer.

All nets were trained with gradient descent (error gradient calculated with Backpropagation Through Time), using a learning rate of 10^{-5} and a momentum of 0.9. At the end of each utterance, weight updates were carried out and network activations were reset to 0.

As is standard for 1 of K classification, the output layers had softmax activations, and the cross entropy objective function was used for training. There were 61 output nodes, one for each phonemes At each frame, the output activations were interpreted as the posterior probabilities of the respective phonemes, given the input signal. The phoneme with highest probability was recorded as the network's classification for that frame.

4.2 Experiment 2: Phoneme Recognition

A traditional HMM was developed with the HTK Speech Recognition Toolkit (http://htk.eng.cam.ac.uk/). Both context independent (mono-phone) and context dependent (tri-phone) models were trained and tested. Both were left-to-right models with three states. Models representing silence (h#, pau, epi) included two extra transitions: from the first to the final state and vice versa, in order to make them more robust. Observation probabilities were modelled by eight Gaussian mixtures.

Sixty-one context-independent models and 5491 tied context-dependent models were used. Context-dependent models for which the left/right context coincide with the central phone were included since they appear in the TIMIT transcription (e.g. "my eyes" is transcribed as /m ay ay z/). During recognition, only sequences of context-dependent models with matching context were allowed.

In order to make a fair comparison of the acoustic modelling capabilities of the traditional and hybrid LSTM/HMM, no linguistic information or probabilities of partial phone sequences were included in the system.

For the hybrid LSTM/HMM system, the following networks (trained in the previous experiment) were used: LSTM with no frame delay, BLSTM and BLSTM trained with weighted error. 61 models of one state each with a self-transition and an exit transition probability were trained using Viterbi-based forced-alignment. Initial estimation of transition and prior probabilities was done using the correct transcription for the training set. Network output probabilities were divided by prior probabilities to obtain likelihoods for the HMM. The system was trained until no improvement was observed or the segmentation of the signal did not change. Due to time limitations, the networks were not re-trained to convergence.

Since the output of both HMM-based systems is a string of phones, a dynamic programming-based string alignment procedure (HTK's HResults tool) was used to compare the output of the system with the correct transcription of the utterance. The accuracy of the system is measured not only by the number of hits, but also takes into account the number of insertions in the output string (accuracy = ((Hits - Insertions) /

Total number of labels) x 100%). For both the traditional and hybrid system, an insertion penalty was estimated and applied during recognition.

5 Results

From Table 1, we can see that bidirectional nets outperformed unidirectional ones in framewise classification. From Table 2 we can also see that for BLSTM this advantage carried over into phoneme recognition.

Overall, the hybrid systems outperformed the equivalent HMM systems on phoneme recognition. Also, for the context dependent HMM, they did so with far fewer trainable parameters.

The LSTM nets were 8 to 10 times faster to train than the standard RNNs, as well as slightly more accurate. They were also considerably more prone to overfitting, as can be seen from the greater difference between their training and test set scores in Table 1. The highest classification score we recorded on the TIMIT training set with a bidirectional LSTM net was 86.4% — almost 17% better than we managed on the test set. This degree of overfitting is remarkable given the high proportion of training frames to weights (20 to 1, for unidirectional LSTM). Clearly, better generalisation would be desirable.

Using duration weighted error slightly decreased the classification performance of BLSTM, but increased its recognition accuracy. This is what we would expect, since its effect is to make short phones as significant to training as longer ones [4].

Table 1. Framewise Phoneme Classification

Network	Training Set	Test Set	Epochs
BLSTM	77.4%	69.8%	21
BRNN	76.0%	69.0%	170
BLSTM Weighted Error	75.7%	68.9%	15
LSTM (4 frame delay)	77.5%	65.5%	33
RNN (4 frame delay)	70.8%	65.1%	144
LSTM (0 frame delay)	70.9%	64.6%	15
RNN (0 frame delay)	69.9%	64.5%	120

Table 2. Phoneme Recognition Accuracy for Traditional HMM and Hybrid LSTM/HMM

System	Number of parameters	Accuracy
Context-independent HMM	80 K	53.7 %
Context-dependent HMM	>600 K	64.4 %
LSTM/HMM	100 K	60.4 %
BLSTM/HMM	100 K	65.7 %
Weighted error BLSTM/HMM	100 K	66.9 %

6 Conclusion

In this paper, we found that bidirectional recurrent neural nets outperformed unidirectional ones in framewise phoneme classification. We also found that LSTM networks were faster and more accurate than conventional RNNs at the same task. Furthermore, we observed that the advantage of bidirectional training carried over into phoneme recognition with hybrid HMM/LSTM systems. With these systems, we recorded better phoneme accuracy than with equivalent traditional HMMs, and did so with fewer parameters. Lastly we improved the phoneme recognition score of BLSTM by using a duration weighted error function.

Acknowledgments

The authors would like to thank Nicole Beringer for her expert advice on linguistics and speech recognition. This work was supported by SNF, grant number 200020-100249.

References

1. P. Baldi, S. Brunak, P. Frasconi, G. Soda, and G. Pollastri. Exploiting the past and the future in protein secondary structure prediction. *BIOINF: Bioinformatics*, 15, 1999.
2. H. Bourlard, Y. Konig, and N. Morgan. REMAP: Recursive estimation and maximization of a posteriori probabilities in connectionist speech recognition. In *Proceedings of Europeech'95*, Madrid, 1995.
3. H.A. Bourlard and N. Morgan. *Connnectionist Speech Recognition: A Hybrid Approach.* Kluwer Academic Publishers, 1994.
4. R. Chen and L. Jamieson. Experiments on the implementation of recurrent neural networks for speech phone recognition. In *Proceedings of the Thirtieth Annual Asilomar Conference on Signals, Systems and Computers*, pages 779–782, 1996.
5. J. S. Garofolo, L. F. Lamel, W. M. Fisher, J. G. Fiscus, D. S. Pallett, , and N. L. Dahlgren. Darpa timit acoustic phonetic continuous speech corpus cdrom, 1993.
6. F. Gers, N. Schraudolph, and J. Schmidhuber. Learning precise timing with LSTM recurrent networks. *Journal of Machine Learning Research*, 3:115–143, 2002.
7. A. Graves and J. Schmidhuber. Framewise phoneme classification with bidirectional lstm and other neural network architectures. *Neural Networks*, August 2005. In press.
8. S. Hochreiter, Y. Bengio, P. Frasconi, and J. Schmidhuber. Gradient flow in recurrent nets: the difficulty of learning long-term dependencies. In S. C. Kremer and J. F. Kolen, editors, *A Field Guide to Dynamical Recurrent Neural Networks*. IEEE Press, 2001.
9. S. Hochreiter and J. Schmidhuber. Long Short-Term Memory. *Neural Computation*, 9(8):1735–1780, 1997.
10. A. J. Robinson. An application of recurrent nets to phone probability estimation. *IEEE Transactions on Neural Networks*, 5(2):298–305, March 1994.
11. M. Schuster and K. K. Paliwal. Bidirectional recurrent neural networks. *IEEE Transactions on Signal Processing*, 45:2673–2681, November 1997.

Improvement in Language Detection by Neural Discrimination in Comparison with Predictive Models

Sébastien Herry

Human Interaction Technologies, Advance Software Department,
Thales Research & Technology, France
sebastien.herry@thalesgroup.com

Abstract. In this paper, we present a new method of language detection. This method is based on language pair discrimination using neural networks as classifier of acoustic features. No acoustic decomposition of the speech signal is needed. We present an improvement of our method applied to the detection of English for a signal duration of less than 3 seconds (Call Friend corpus), as well as a comparison with a neural predictive model. The obtained results highlight scores ranging from 74.7% to 76.9% according to the method used.

1 Introduction

Language detection is the process which decides if a language is spoken or not in a speech stream. It is a part of the Language Identification (LID) which determines the language spoken from a set of given language (English, Farsi, French, German…).

LID takes benefit of the interest for multilingual systems, which target the international call center for example. The techniques usually used in LID research are based on spectral parameters distribution modeling and/or language modeling, by n-gram for the phoneme series. The phonotactic approach is mostly used, observing the phoneme series to establish a statistical model, like the model proposed by Zissman [1]. It is based on the Parallel Phone Recognition followed by Language Modeling (PPRLM), which needs Acoustic-Phonetic Decoders (APD). This imposes a heavy constraint, because APDs need phonetically labeled corpus which are only available for few languages. Techniques have been developed which bypass this problem, for example P. Torres [2] replaces phonemes by automatically created acoustic units. J. Farinas [3] uses also a creation of pseudo-phonemes, and in the past Y. Muthusamy [4] used a technique to generate automatically phonetic macro-class. Unlike the precedents authors, W. Wu & C. Kwasny [5] model the speech signal with recurrent neural networks without acoustic units. They make language identification English and French, from audio files of 12.5s duration, by acoustic vector series classification.

We propose in this paper a method using the concept of acoustic vector series classification too. It is only based on acoustic discrimination of language between them, with neural networks. We present an application of detection of English against 10 other languages, for phone speech signal of 3 seconds duration. The main objective of our method is to obtain a technique that allows to reduce the speech duration required for identification in order to give an answer more quickly. As 3 seconds is the shortest duration where some results have been published, we chose this duration for our tests

W. Duch et al. (Eds.): ICANN 2005, LNCS 3697, pp. 805 – 810, 2005.

for comparison reasons. The detection is based on acoustic discrimination between languages. We start from the hypothesis that language information for a large part is present in the spectrum. We use several neural networks, each of them discriminates a couple of languages. The result of this discrimination is merged by language to provide first signal detection. The continuation of this paper is cut in five parts, the second part describes the corpus used, the third one explains the detection method for a language, the fourth presents the predictive model, the fifth shows results obtained, and finally we conclude.

2 The Data Description

We use the Call Friend (CF) corpus [6]. This corpus is composed of 12 languages and 3 dialects but we use only 11 languages to be compliant with previous research (English, Farsi, French, German, Hindi, Japanese, Korean, Mandarin, Spanish, Tamil and Vietnamese). For each of them 120 speakers have spoken with a friend, their subjects were unbounded. Our objective is to work on sentences of less than 3 seconds duration, thus we perform a transformation of sentences. All 11 languages sentences have been divided into sentence segments of 3s duration. Next, we distribute these parts among 3 subsets : learning, development for tuning the system and test. The ratio of these subsets are respectively 3/5, 1/5, 1/5. Since the Call Friend corpus contains many hours of speech for each language, each part of the subset is significant.

3 English Detection Process

Our first goal is to detect the English language among 11 languages. To do that we use neural nets, each of them identifies English versus an other language for example English versus French. Thus we are able to discriminate English from the remainder of corpus, by merging discrimination coming from all the networks. Acoustic vectors are then processed by the 10 networks, each of them giving as output a discriminating signal for a language pair. These signals, computed for all the acoustic vectors extracted from 3s speech duration are then used to calculate an average. Thus we obtain 10 average values (each of them corresponding to a network discriminating a language pair) which are merged to give a global English detection signal.

3.1 The Front End Processing

We used a detector of speech and we perform a normalization of speech before the Mel Frequency Cepstral Coefficient (MFCC) processing to generate acoustic vectors of dimension 36 ($12+12\Delta+12\Delta\Delta$). This front end is basically use in language recognition. We use the MFCC processing on 30ms of signal with an overlap of 50%.

3.2 Discriminating Neural Nets

At this step we train a neural net for each pair of language: (English vs. another one) on MFCC acoustic vectors. Each of the networks are multi-layer perceptron (MLP) with sigmoidal activation functions. They are composed with 36 inputs, 72 hidden cells and two outputs (one for each language). The learning algorithm used was the stochastic back propagation algorithm with mean square criterion, to increase the learning speed [7]. The development subset is used to prevent overlearning.

3.3 Merging Local Discrimination for English Detection

At this stage of the model, we discriminate only pair of language. To detect English we proceed to a merging between outputs of networks. The process is represented on Fig. 1. In the figures and tables below, we name each language by its two first letters (EN : English).

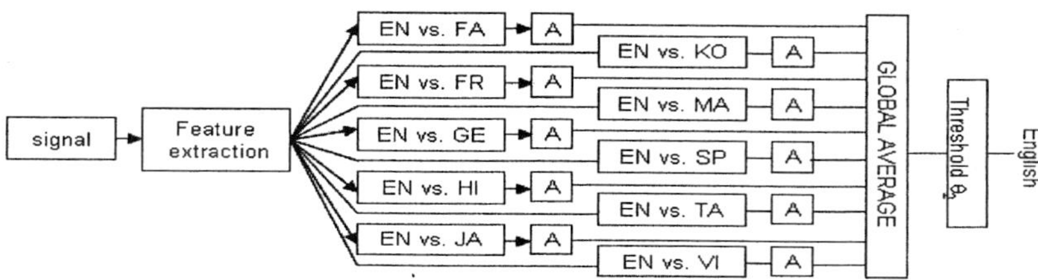

Fig. 1. Scheme of English detector

Following the first pair of language detection step, the goal of this merging is to emphasize the detection of English from each networks in the time. Because each network gives a detection every 30ms, while sentence duration is 3s, the merging is done in two steps: "step 2" for the time (output networks are merged to compute the average over the 3s duration signal) and "step 3" to convert language pair identification into English detection .

Since first step, the process converts the speech signal into acoustic vectors with the processing MFCC. These acoustic vectors are evaluated by each neural networks. Series of acoustic vectors produces series of neural networks outputs. We have improved the 2nd step, by summing the image of the neural network output by rejection graph of this neural net computed on evaluation corpus, that allows to take into account all outputs. Let $O_{EN}^{EN-FA}(q)$ be the image of English output of the network: English versus Farsi (En vs. FA), when presenting the acoustic vector q as input. We compute the output of the English detector as:

$$\begin{cases} A_{EN} = \dfrac{1}{q \times |p|} \sum_p \sum_q O_{EN}^{EN-p}(q); \quad p = \{fa, fr, ge, hi, ja, ko, ma, sp, ta, vi\}. \\ A_{other} = \dfrac{1}{q \times |p|} \sum_p \sum_q O_{p}^{EN-p}(q); \quad p = \{fa, fr, ge, hi, ja, ko, ma, sp, ta, vi\}. \\ Detection \;\; of \;\; English \;\; if: \quad A_{EN} + \theta > A_{other}. \end{cases} \tag{1}$$

Where θ is a threshold determined to induce the best score on the learning corpus.

4 Predictive Model

In order to compare the predictive model to our model with the intension to improve the results through time, we have trained MLP to predict the next acoustic vector, with a prediction order of two. The acoustic vector for this model is composed by 12 MFCC coefficients for 30ms of signal. Thus we create for each language an MLP comprising 24 inputs, 25 hidden cells, and 12 outputs. We trained the networks with back propagation algorithm and with the Modelisation Error Ration (MER) criterion [8] (F in eq. (5)). The coefficient α is used for French language modeling and the coefficient (1- α) is applied to the discriminated language (English). The development subset is used to stop the learning.

$$C = F\left(\alpha_{\in fr} + (1-\alpha)_{\in en\cdot}\right) \tag{2}$$

Each network learns one language and at the same time performs a discrimination with another one. Thus we need two networks to replace a discriminating network, one learns English and discriminates French and the second one learns French and discriminates English (Fig. 2). Through time, we sum the error of prediction of each network. We choose the network having the smallest error of prediction to identify the language. We use the Euclidean distance to compute the prediction error of networks. The next section compares them with our model.

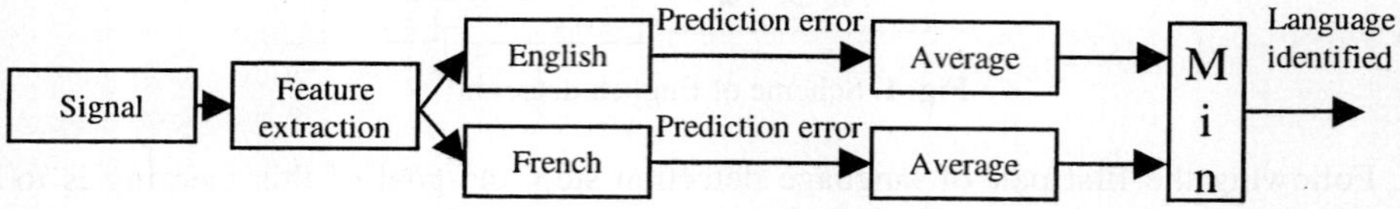

Fig. 2. Scheme of English and French detector with prediction

5 Experimentation

We present in this section the experimental results obtained on pair language identification with our model and with the predictive model. Table 1 presents the scores of identification obtained on the test corpus for each pair of language. The learning is done on acoustic vectors which represents a duration of 30ms, currently used to detect phoneme. For instance, we obtained on the first network (see the EN vs. FA column of table 1) an English identification score of 62.5% (62.5% of the 30ms English frames have been recognized as English language frames) and a Farsi identification score of 62.3%. Thus the average identification score of English/Farsi detector was 62.4%. The results involve that the average identification score is 66.2%. These results are encouraging because they were obtained from decisions based on frames of only 30ms duration. They are improved up to 76.9% by considering sentences of 3s duration and using the whole model defined by eq. (2) and (3).

Table 1. Scores of language pair identification in percent

EN vs. FA		EN vs. FR		EN vs. GE		EN vs. HI		EN vs. JA		EN vs. KO		EN vs. MA		EN vs. SP		EN vs. TA		EN vs. VI	
EN	FA	EN	FR	EN	GE	EN	HI	EN	JA	EN	KO	EN	MA	EN	SP	EN	TA	EN	VI
62,5	62,3	74,7	74,6	67,3	67,6	67,6	68,1	68,5	68,6	63,6	63,2	62,7	62,9	66,4	65,9	67,1	67,9	61,8	61,7
62,4		74,6		67,5		67,8		68,6		63,4		62,8		66,2		67,5		61,7	

5.1 Results of English Detection Through Time

We have performed a test to compare the two temporal fusion techniques (CF corpus). The first temporal fusion technique is the average of network outputs. The second is the average of network output images by rejection graph. We obtained a gain of 2.3% on sentences of 3s duration for the English detection against the 10 other languages as the Table 2 shows. We have placed in this Table, the results of output average technique developed with OGI corpus [9],[10], to carry out the comparison with preceding work.

Table 2. Scores of English in percent

	image of rejection graph with CF		Neural output with CF		Neural output with OGI	
	EN	OTHERS	EN	OTHERS	EN	OTHERS
EN	**82,9%**	17,1%	**85,3%**	14,7%	**71,6%**	28,4%
OTHERS	29,1%	**70,9%**	36,0%	**64,0%**	25,4%	**74,6%**
GLOBAL	**76,9%**		**74,7%**		**73,1%**	

If we compare the results obtained with the OGI corpus, we have improved our results by 3.8%. This increase is due to a better front-end processing and to the use of rejection graph. The corpus is different but its languages are matching those of the OGI corpus previously used. One of the others significant characteristic of the model is its response time which we present with the Fig. 3 as those of the predictive network and the discriminating network.

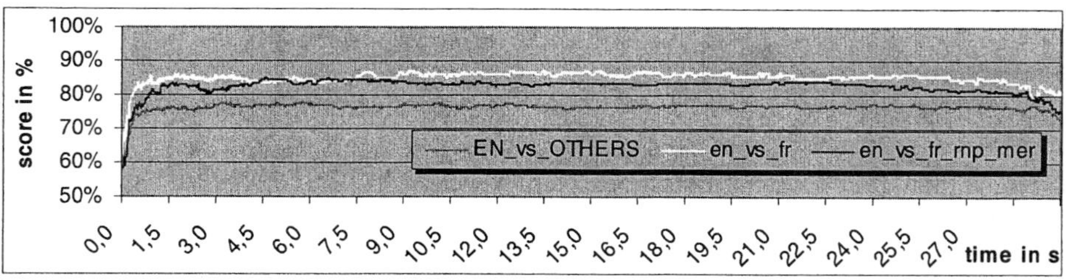

Fig. 3. Response time for: the detection of English (En_vs_OTHERS), one of the neural network use in the model (en_vs_fr) and for the predictive model (en_vs_fr_rnp_mer)

Those graphs show us that the response time is very short, less than 3s, at 1s we are near to the maximum score of detection, that's very interesting because it allows us to reduce the time needed to detect without accuracy loss. We also see that the graph of the predictive model is slightly below the graph of the discriminating network, not increasing our scores, implying that the information type modeled by the two methods is the same.

6 Conclusion

We have proposed an improvement of the method based on neural networks to detect one language among several others and tested it on English detection. The results obtained with our design allow us to detect language very fast in approximately 1s. The scores English of detection is less than the best detection systems which are around 81% [11], [12] of detection for 3s of speech. Nevertheless the work presented here is the first stage of the complete model [9]. The second stage improves the first with a gain about 4.5% then we hope obtain a final detection rate at least 81.4%. This score is near from the best systems but with a speech duration reduced from 3s to 1s. Moreover, The modeling technique includes two interesting properties: the first is the real time operation on a P4 1.7Ghz, and the second is the representation of every language with a speech corpus without a phonetic labeling. Unfortunately the predictive model cannot be use to improve the model as seen in the previous section.

References

1. Zissman, M. A.: Comparison of Four Approaches to Automatic Language Identification of Telephone Speech. IEEE Trans. Speech and Audio Proc, Vol. 4. (1996) 31-44.
2. Torres-Carasquillo, P.A., Reynolds, D.A., Deller Jr., J.R.: Language identification using Gaussian Mixture Model Tokenization. ICASSP, Vol. 1. Olando, Floride, USA, (2002) 757-760.
3. Farinas, J., Pellegrino, F .: Automatic Rhythm Modeling for Language Identification. Proc. of Eurospeech, Vol. 4. Scandinavia, Aalborg, September (2001) 2539-2542.
4. Muthusamy, Y. K.: A Segmental Approach to Automatic language Identification. Oregon Graduate Institute of Science and Technology, October (1993).
5. Weilan, W., Kwasny, S.C., Kalman, B.L., Maynard Engebretson, E.: Identifying Language from Raw Speech, An application of Recurrent Neural networks. 5th Midwest Artificial Intelligence and Cognitive Science Conference. (1993) 53-57.
6. CallFriend Corpus, Linguistic Data Consortium, (1996) http://www.ldc.upenn/ldc/about/ callfriend.html.
7. Bottou, L., Murata, N.: Stochastic Approximations and Efficient Learning. The Handbook of Brain Theory and Neural Networks, Second edition, (M. A. Arbib, ed.), (Cambridge, MA), (2002).
8. Chetouani, M., Gas, B., Zarader, J.L.: Maximization of the modelisation error ratio for neural predictive coding. NOLISP (2003).
9. Herry, S., Gas, B., Sedogbo, C., Zarader, J.L.: Language detection by neural discrimination. the Proc. of ICSLP, Vol 2. Jeju, Korea, (2004) 1561-1564.
10. Muthusamy, Y.K., Cole, R.A., Oshika, B.T.: The OGI Multilingual Telephone speech Corpus. Proc. of ICSLP. Banff, (1992) 895-898.
11. Martin, A., Przybocki, M.: NIST 2003 Language Recognition Evaluation. the Proc. of Eurospeech. Switzerland, Geneva, September (2004)
12. Gauvain, J.L., Messaoudi, A., Schwenk, H.: Language Recognition using Phone Lattices. Proc. ICSLP. Jeju, Korea (2004)

Learning Ontology Alignments Using Recursive Neural Networks

Alexandros Chortaras*, Giorgos Stamou, and Andreas Stafylopatis

School of Electrical and Computer Engineering,
National Technical University of Athens,
Zografou 157 80, Athens, Greece
{achort,gstam}@softlab.ntua.gr, andreas@cs.ntua.gr

Abstract. The Semantic Web is based on technologies that make the content of the Web machine-understandable. In that framework, ontological knowledge representation has become an important tool for the analysis and understanding of multimedia information. Because of the distributed nature of the Semantic Web however, ontologies describing similar fields of knowledge are being developed and the data coming from similar but non-identical ontologies can be combined only if a semantic mapping between them is first established. This has lead to the development of several ontology alignment tools. We propose an automatic ontology alignment method based on the recursive neural network model that uses ontology instances to learn similarities between ontology concepts. Recursive neural networks are an extension of common neural networks, designed to process efficiently structured data. Since ontologies are a structured data representation, the model is inherently suitable for use with ontologies.

1 Introduction

The purpose of the Semantic Web is to introduce structure and semantic content in the huge amount of unstructured or semi-structured information available in the Web. The central notion behind the Semantic Web is that of ontologies, which describe the concepts and the concept relations in a particular field of knowledge. The data associated with an ontology acquire a semantic meaning that facilitates their machine interpretation and makes them reusable by different systems. However, the distributed development of domain-specific ontologies introduces a new problem: in the Semantic Web many independently developed ontologies, describing the same or very similar fields of knowledge, will coexist. These ontologies will not be identical and will present from minor differences, such as different naming conventions, to higher level differences in their structure and in the way they represent knowledge. Moreover, legacy ontologies will have to be used in combination with new ones.

For this reason, before being able to combine similar ontologies, a semantic and structural mapping between them has to be established. The process of establishing such a mapping is called *ontology alignment*. It will become increasingly significant as the Semantic Web evolves, it is already an active research area and several auto-

* The author is funded by the Alexander S. Onassis Public Benefit Foundation.

W. Duch et al. (Eds.): ICANN 2005, LNCS 3697, pp. 811 – 816, 2005.

matic or semi-automatic ontology alignment tools have been proposed (e.g. [5, 6, 7]). Most of the tools rely on heuristics that detect some sort of similarity in the description of the concepts and the structure of the ontology graphs, by using e.g. string and graph matching techniques. They usually work at the terminological level of the ontologies without taking into account their instances. A different method is proposed in [2], where a machine learning methodology is used. The approach is *extensional*, i.e. it exploits the information contained in the ontology instances. For all concepts in the ontologies to be aligned, a naïve Bayes classifier is built. The instances of each concept are then presented to the classifiers of the other ontology and, depending on the degree of overlap of the classifications, a similarity measure for each concept pair is computed. The classifiers do not take into account the structure of the ontologies; this is considered at a subsequent stage by integrating a relaxation labelling technique.

The method we propose follows a similar machine learning approach, but takes directly into account the structure of the ontologies, by relying on the use of recursive neural networks [3], which are a powerful tool for the processing of structured data.

The rest of the paper is organized as follows: section 2 discusses the basic ideas underlying the recursive neural network model, section 3 describes the details of our method and presents a simple example, and section 4 discusses future work and concludes.

2 Recursive Neural Networks

The recursive neural network model was proposed in [3, 8] as an extension to the recurrent neural networks, and is capable of efficiently processing structured data. The data are represented as labelled directed ordered acyclic graphs (DOAGs), on whose structure a neural network (*encoding neural network*) is repeatedly unfolded. Because the representation of the data as DOAGs is in many applications too restrictive, some extensions to the initial model have been proposed, that generalize the type of graphs on which it can be applied. For example, in [4] the graphs are allowed to have labels attached also to their edges. This extension, which we use in our method, lifts a significant constraint of the initial model that required the graphs to have a maximum, a priori known out-degree as well as an ordering on their out-going edges.

In the model the data are represented as directed acyclic graphs, each node v of which is assigned a label $\mathbf{L}_v \in \mathbb{R}^m$ and each edge connecting nodes v and w a label $\mathbf{L}_{(v,w)} \in \mathbb{R}^k$. To each node v an encoding neural network is attached, that computes a *state vector* $\mathbf{X}_v \in \mathbb{R}^n$ for v. Let $\mathrm{ch}(v)$ be the set of children of v, p the cardinality of $\mathrm{ch}(v)$ and $\mathrm{ch}_i(v)$ the i-th child of v. The input to the encoding neural network of v is a) a vector, function of the state vectors $\mathbf{X}_{\mathrm{ch}_1(v)}, \ldots \mathbf{X}_{\mathrm{ch}_p(v)}$ of the node's children and of the corresponding edge labels $\mathbf{L}_{(v,\mathrm{ch}_i(v))}$, and b) the label $\mathbf{L}_v$ of v. The encoding neural network is usually an MLP and is identical for all the nodes of the graph. The output of the recursive neural network is obtained at the graph's *super-node*, a node from which a path to all other nodes of the graph exists. The output is computed by a common neural network applied on the state vector of the super-node.

The strength of the model is that the state of each node, calculated by the encoding neural network and encoded in the state vector, is computed as a function not only of the label of the node, but also of the states of all its children. On their turn, the states of the children depend recursively on the states of their respective children. As a result, the state vector of each node encodes both the structure and the label content of the sub-graph that stems from the node. Thus, if a recursive neural network classifier is trained appropriately on data represented as graphs of different structures then it will be able to identify data similar both in their content and structure.

3 Neural Extensional Ontology Alignment

Our method is based on the fact that an ontology can be considered as a graph, with the ontology concepts (relations) corresponding to the nodes (edges) of the graph. The graph is directed because the ontology relations are in general not symmetric, and each node (edge) has a label consisting of the name and the attributes of the corresponding concept (relation). This holds at the terminological level of the ontology.

At the instance level, there are instances belonging to concepts and pairs of instances connected by relation instances. Thus, if we consider an instance I of a concept C, then by following the relation instances stemming at I, we obtain a tree that consists of I in its root (*root instance*), which is connected with nodes that correspond to the instances of the ontology concepts with which I is related. The tree can grow up to n levels, by recursively following the relation instances of the new nodes that are added to the tree. We call this graph an *instance tree* of level n of concept C for the instance I. Our tool computes similarities between the concepts of two ontologies by training and applying a recursive neural network classifier on such instance trees. In detail, the method has as follows:

Let O_2 be the ontology whose concepts we want to map to the concepts of a similar ontology O_1, and C_{1i}, $i = 1,\dots p_1$, C_{2i}, $i = 1,\dots p_2$ be the concepts in O_1 and O_2 respectively. We first decompose the graph of O_1 (at the terminological level) into a set of p_1 sub-graphs, in particular into trees that we call *concept trees*, one for each concept in O_1. Each tree is constructed by setting as its root the concept that it corresponds to and its children are taken to be the concepts that are directly connected to it with a relation in the ontology. As in the case of the instance trees we define a maximum level up to which the tree can grow by recursively following the relations defined for the children that are added to it. In the concept trees we ignore the direction of the relations as well as the edges corresponding to hierarchical (is-a) relations.

A problem that arises while constructing the concept trees is that the graph of the ontology in general contains cycles. This obstacle can be overcome by following a methodology like the one used in [1] and appropriately expanding the graph into an equivalent tree, by traversing it in a specific order and breaking the circles by duplicating the nodes that form them. An example of a concept tree is shown in Fig. 1.

Once the p_1 concept trees have been constructed, they are used as templates for the generation of the instance trees, by assigning a unique instance to each one of their nodes. Each concept tree will in general give birth to several instance trees, not only because of the several instances available for its root concept (root instances), but also because the ontology relations may have cardinalities higher than one. In this case, each edge in a concept tree will correspond to several instance pairs. In fact, for a particular instance in the role of the root instance, the total number of instance trees that can be produced is equal to the product of the number of instances that can be assigned to each node in the corresponding concept tree. In the instance trees we assume that all edges have the same label, i.e. that all the relations are equivalent and that all edges have direction towards the root of the tree.

The instance trees are used as training data in order to build a recursive neural network classifier. The desired output for each instance tree is the concept in O_1 its root belongs to. Assuming a two layer perceptron with a linear output activation function as the encoding neural network, the state vector of node v is computed as:

$$\mathbf{X}_v = \sigma\left(\mathbf{A} \cdot \overline{\mathbf{X}}_v + \mathbf{B} \cdot \mathbf{L}_v + \mathbf{C}\right) \tag{1}$$

where $\sigma(\cdot)$ is a sigmoid function and $\mathbf{A} \in \mathbb{R}^{q \times n}$, $\mathbf{B} \in \mathbb{R}^{q \times m}$, $\mathbf{C} \in \mathbb{R}^{q}$ are parameters to be learned and q is the number of hidden neurons. $\mathbf{L}_v$ is the label of the node and $\overline{\mathbf{X}}_v$ a vector, function of the state vectors of its children. A similar equation holds for the neural network that computes the output of the super-node. Since for simplicity we have considered all edge labels in the graphs to be the same, we can write for $\overline{\mathbf{X}}_v$:

$$\overline{\mathbf{X}}_v = \frac{1}{|\mathrm{ch}(v)|} \sum_{i=1}^{|\mathrm{ch}(v)|} \mathbf{D} \cdot \mathbf{X}_{\mathrm{ch}_i(v)} \tag{2}$$

where $\mathbf{D} \in \mathbb{R}^{n \times n}$ is a matrix of parameters. The parameters can be learned by the back-propagation through structure algorithm [3].

What remains to be defined are the labels $\mathbf{L}_v$, which must be descriptive of the instances contents. In the simplest case we can consider as label space the space of the attribute values of all instances in O_1 and use as label for each instance its term frequency vector in this space.

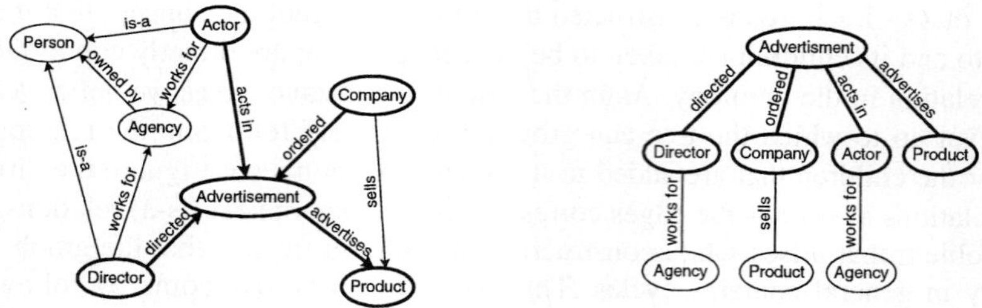

Fig. 1. Left: A graph representing an ontology of the advertising domain. The sub-graph that corresponds to the concept *Advertisement* and includes the concepts related to it with direct relations is marked with thick lines. Right: The corresponding concept-tree of level two.

After the classifier for the concepts of O_1 has been trained, the same procedure is followed for O_2, from which p_2 concept trees and the corresponding instance trees are generated. The instance trees of O_2 are presented to the classifier, which classifies them to one of the concepts of O_1. Let $t^{C_{2i}}$ be the number of instance trees of O_2 belonging to concept C_{2i} and $t^{C_{2i},C_{1j}}$ those of them that have been assigned by the classifier to concept C_{1j}. Given that in general several instance trees correspond to the same root instance, the values of $t^{C_{2i}}$ and $t^{C_{2i},C_{1j}}$ are normalized with respect to the number of instance trees that correspond to the same root instance. Then, assuming that the instances that we use are a representative sample of the instance space of the two ontologies, we estimate the conditional probabilities:

$$\hat{P}\left(C_{1j} \mid C_{2i}\right) = \frac{t^{C_{2i},C_{1j}}}{t^{C_{2i}}} \quad \forall i, j \tag{3}$$

that an instance of O_2 belonging to concept C_{2i} is also an instance of C_{1j} of O_1. We use this probability estimate as the similarity measure $s\left(C_{2i}, C_{1j}\right)$ of C_{2i} with C_{1j}. Other similar similarity measures can also be computed, like the Jaccard coefficient. The final output of the tool is a set of similarity pairs for all concepts O_2 and O_1.

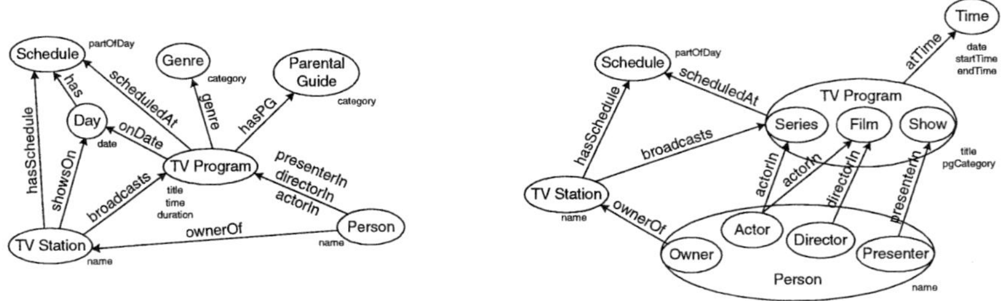

Fig. 2. Two ontologies for the TV programs domain. Left: ontology A, Right: ontology B.

Table 1. Computed similarities for the concepts of ontology B with the concepts of ontology A

B \ A	Parental Guide	TV Program	Day	Schedule	Person	TV Station	Genre
Actor	0.00	0.06	0.05	0.09	**0.59**	0.20	0.02
Presenter	0.01	0.06	0.07	0.02	**0.64**	0.15	0.04
Director	0.00	0.14	0.05	0.03	**0.48**	0.29	0.02
Owner	0.00	0.03	0.11	0.02	**0.68**	0.16	0.00
Series	**0.49**	0.22	0.05	0.02	0.09	0.08	0.05
Film	**0.31**	**0.30**	0.04	0.07	0.11	0.14	0.01
Show	**0.42**	0.19	0.01	0.01	0.10	0.17	0.10
Schedule	0.02	0.02	0.16	**0.65**	0.14	0.01	0.02
Time	0.08	**0.36**	**0.26**	0.05	0.15	0.04	0.06
Station	0.02	0.13	0.15	0.02	0.16	**0.50**	0.01

The proposed method has been implemented and tested on small-scale datasets with promising initial results. As an example, the similarities of the concepts of the two ontologies of Fig. 2 computed by our method are presented in Table 1. The recursive neural network classifier has been trained with instance trees of ontology B of level 2, using as label space the stemmed words of the attribute values, excluding proper names and numbers. The dataset was taken from [9]. The results are intuitively correct; it is however worth noticing that the classifier does not produce very high similarity scores. This reflects the fact that some of the distinctive attributes of the domain concepts are distributed over different ontology concepts in the two ontologies. The recursive neural network classifier takes thus into account such inter-concept dependencies. Moreover, the classifier performs well with those instances, the concept to which they belong can correctly be determined only if the information about the instances with which they are related is also considered. For each instance this information is provided to the classifier through the corresponding instance tree.

4 Conclusions

We described a machine learning ontology alignment tool based on the use of recursive neural networks. Our method exploits the ability of recursive neural networks to efficiently process structured data and builds a classifier which is used to estimate a distribution-based similarity measure between the concepts of two ontologies. Our research is ongoing and we are at the stage of configuring and evaluating our method, having some promising initial results. There are several points where the suggested method may be improved. Particularly important is the definition of the label space of the instance trees. Currently, we use the attribute values, but it is desirable to reduce the label space dimensionality by extracting more general labels. For this purpose the Wordnet ontology could e.g. be used to map the individual attribute values to more general features, moving in this way the attribute values closer to the abstract attributes they represent and improving the generalization properties of the classifier.

References

1. Bianchini, M., Gori, M., Scarselli, F.: Recursive Processing of Directed Cyclic Graphs, In: Proc. IEEE Int. Conf. Neural Networks, (2002) 154-159.
2. Doan, A., Madhavan, J., Domingos, P., Halevy A.: Ontology Matching: A Machine Learning Approach, In: Staab S., Studer, R., (eds.): Handbook on Ontologies in Information Systems, Springer-Velag (2004) 397-416.
3. Frasconi, P., Gori, M., Sperduti M.: A General Framework for Adaptive Processing of Data Structures, In: IEEE Trans. Neural Networks, Vol. 9:5 (1997) 768-786.
4. Gori, M., Maggini, M., Sarti, L.: A Recursive Neural Network Model for Processing Directed Acyclic Graphs with Labeled Edges, In: Proc. Int. J. Conf. Neural Networks, Vol. 2 (2003) 1351-1355.
5. Melnik, S., Garcia-Molina, H., Rahm, E.: Similarity Flooding: A Versatile Graph Matching Algorithm, Extended Technical Report (2001).
6. Noy, N., Musen, M. A.: PROMPT: Algorithm and Tool for Automated Ontology Merging and Alignment: In Proc. 7th Nat. Conf. Artificial Intelligence (2000).
7. Noy, N. Musen, M. A.: Anchor PROMPT: Using Non-Local Context for Semantic Mapping, In Proc. Int. Conf. Artificial Intelligence (2001).
8. Sperduti, A., Starita A.: Supervised Neural Network for the Classification of Structures, In: IEEE Trans. Neural Networks, Vol. 8 (1997) 429-459.
9. Web-based Knowledge Representation Repositories, http://wbkr.cs.vu.nl.

Minimizing Uncertainty in Semantic Identification When Computing Resources Are Limited

Manolis Falelakis[1], Christos Diou[1],
Manolis Wallace[2], and Anastasios Delopoulos[1]

[1] Department of Electrical and Computer Engineering,
Aristotle University of Thessaloniki - Greece
{fmanf, diou}@olympus.ee.auth.gr,adelo@eng.auth.gr
[2] Department of Computer Science,
University of Indianapolis, Athens Campus - Greece
wallace@uindy.gr

Abstract. In this paper we examine the problem of automatic semantic identification of entities in multimedia documents from a computing point of view. Specifically, we identify as main points to consider the storage of the required knowledge and the computational complexity of the handling of the knowledge as well as of the actual identification process. In order to tackle the above we utilize (i) a sparse representation model for storage, (ii) a novel transitive closure algorithm for handling and (iii) a novel approach to identification that allows for the specification of computational boundaries.

1 Introduction

During the last years the scientific community has realized that semantic analysis and interpretation not only requires explicit knowledge, but also cannot be achieved solely through raw media processing. For this purpose, multimedia research has now shifted from the query by example approach, where the aim was to provide meaningful handling and access services directly from the low level processing of media, to a two step approach including (i) the identification of high level entities in raw media and (ii) the utilization of this *semantic indexing* towards the offering of more efficient multimedia services. Research efforts in standardizing the metadata representations of multimedia documents have led to the MPEG-7 standard which, however, does not suggest methods for automatic extraction of high level information.

In well-structured specific domains (e.g., sports and news broadcasting), domain-specific features that facilitate the modelling of higher level semantics can be extracted (see e.g., [6]). Typically, a priori knowledge representation models are used as a knowledge base that assists semantic-based classification and clustering [9]. In [7], for example, the task of bridging the gap between low-level representation and high-level semantics is formulated as a probabilistic pattern recognition problem.

W. Duch et al. (Eds.): ICANN 2005, LNCS 3697, pp. 817–822, 2005.

In the proposed paper we aim to deal with the identification of semantic entities in raw media, focusing on issues related to efficient operation under computing resources limitations. The aim is to automatically configure the identification process so as to achieve optimal results, i.e. maximize the relevance of the retrieved multimedia documents when the amount of available computing resources is constrained. Constraints are directly related to hard bounds regarding physical memory, processing power and time availability.

These considerations turn to be important due to two major reasons. The first is that the knowledge base involved in semantic identification procedures naturally contains a vast amount of items so that even simple operations on its content may lead to overwhelming the existing computational power and/or memory. The second reason is that although retrieval is performed on the basis of semantic entities, identification of the latter necessarily resorts to quantification of large numbers of measurable features (syntactic entities) i.e., requires execution of multiple signal and image processing algorithms. The latter may sum up to an execution cost that is prohibitive especially for real-time or online scenarios.

2 Knowledge Representation

In order to extract high level (semantic) information from multimedia, low level (directly measurable) features have to be evaluated and combined with the use of a properly structured knowledge base [10]. This structure is often called "semantic encyclopedia" and consists of relationships either among semantic entities or between semantic entities and low level features [2]. For example semantic entity "planet" can be related with semantic entity "star" while feature "blue color" can be related with semantic entity "sea". In most cases such relationships are valid *up to a certain degree*, that is, there is an inherent uncertainty and/or degree of validity associated with them. This makes representation of the aforementioned relationships by using fuzzy relations a natural choice. Considering the set of both semantic and syntactic entities as our universe of discourse, the semantic encyclopedia can be modelled as a large fuzzy relation describing the degrees of association among the elements of this universe. Using such an encyclopedia, it is possible to build systems for automatic or semi-automatic identification of semantic entities in raw media, thus contributing to the bridging of the semantic gap [5].

Even in the case of semantic encyclopedias that are limited to specific thematic categories (e.g., sports, politics, etc) the number of included "terms" and syntactic features may easily reach the order of tens of thousands (in [2], for example, the universe of discourse contains definitions for 70000 semantic entities). This alone is prohibitive for the complete representation of the semantic relations. On the other hand, classical linked list sparse array representations, as well as hash table approaches, are both inadequate to handle such sizes of data, the former due to the $O(n)$ access time and the latter due to the augmented requirements in physical memory. In this work we utilize a novel sparse fuzzy binary relation model that is based on pairs of AVL trees and provides for both space efficient storage and time efficient access.

Specifically, the representation model proposed in order to overcome these limitations is as follows: a binary relation is represented using two *AVL trees*; an AVL tree is a binary, balanced and ordered tree that allows for access, insertion and deletion of a node in $O(\log m)$ time, where m is the count of nodes in the tree [1]. If $n \log n$ nodes exist in the tree, as will be the case for the typical sparse relation, then the access, insertion and deletion complexity is again $O(\log n)$ since $n < n \log n < n^2 \Rightarrow O(\log n) \leq O(\log(n \log n)) \leq O(\log n^2) = O(\log n)$.

In both trees, both row index i and column index j are utilized to sort the nodes lexicographically; however, the first tree, the row-tree, is sorted according to index i, and in case of common row positions i, column position j is utilized, and vice versa for the second tree, the column-tree. The resulting vectors can then be represented as AVL trees.

Furthermore, most of the semantic relations that participate in the semantic encyclopedia are of a transitive form. For example, a texture feature may be associated with entity "skin" and "skin" with semantic entity "human", which should imply that the specific feature is also associated with entity "human". In order to populate the semantic encyclopedia with the links that can be inferred in this way and to allow for more efficient time wise access to such implied links, a transitive closure of these relations needs to be acquired. As conventional transitive closure algorithms either have a high complexity ($O(n^3)$ or higher) or cannot handle archimedean t-norms and asymmetrical relations, such as the ones typically included in a semantic encyclopedia, this task is not trivial.

Based on the proposed representation model, a novel transitive closure algorithm that is targeted especially to generalized, sparse fuzzy binary relations and has an almost linear complexity can be utilized [8]. This algorithm has the added advantage of allowing updates to the relation with trivial computational burden while at the same time maintaining the property of transitivity.

3 Semantic Identification

The methodologies referred to up to this point guarantee that the encyclopedia will be well structured, consistent and informative enough (due to its transitive closure characteristic) and represented in a compact manner. The remaining of the paper refers to its effective use in identifying semantic entities and retrieving the corresponding multimedia documents. The core idea is that the analysis of the raw media by applying signal and image processing algorithms yields quantification of existence of low level features as a first stage of the retrieval procedure. The second step is the assessment of the degree up to which certain semantic entities are identified within the given multimedia documents by exploiting the relations of the semantic encyclopedia. We choose to model this identification procedure as a fuzzy inference mechanism. Considering, though, the count of semantic entities in the universe of discourse, as well as the count of distinct features that may be extracted and evaluated, it is easy to see that this process quickly becomes inapplicable in real life scenarios.

The way followed in this paper in order to tackle semantic identification, without suffering the expense of immense processing power, is to partially identify an entity i.e., to evaluate only a subset of the involved syntactic features and produce an estimation based on this imperfect and incomplete input. The challenge is to automatically select this subset that includes characteristics providing the highest possible validity regarding the computed result, combined with minimum complexity requirements. Utilization of fuzzy logic theory is shown to provide methods for quantifying both validity and complexity of the involved algorithms.

Evaluation of a Syntactic Entity Y_i participating in a detailed definition is equivalent to running its corresponding algorithm τ and computing the membership degree μ_{Y_i} up to which the document under examination assumes property Y_i. In a similar manner, a metric is defined that denotes the degree up to which a Semantic Entity exists in a document and is called *Certainty* of the identification. Given a detailed definition of a Semantic Entity E_k in the form

$$E_k = F_{1k}/S_1 + F_{2k}/S_2 + \ldots + F_{nk}/S_n,$$

and the membership degrees μ_{Y_i} of the Syntactic Entities Y_i in a specific document, Certainty that E_k exists in that document is defined as

$$\mu_{E_k} \triangleq \mathcal{U}_i(\mathcal{I}(F_{Y_i E_k}, \mu_{Y_i}))$$

where the operators $\mathcal{U}$ and $\mathcal{I}$ denote fuzzy union and intersection operators respectively.

The maximum possible value of μ_{E_k} is assigned the term *Validity* of the definition and is equal to

$$\mathcal{V}(E_k) \triangleq \mathcal{U}_i(F_{Y_i E_k}),$$

attained for $\mu_{Y_i} = 1$ for all Y_i in the scope of E_k and the use of the identity $I(a, 1) = a$ (true for every t-norm $\mathcal{I}$).

Validity denotes the maximum amount of information that a definition can provide and is used extensively in the identification design process. We must note that Validity is independent of the data set under examination and can be computed *prior* to the identification. Validity is therefore a property of the definition itself.

Another characteristic of a definition is the computational complexity associated with the algorithms corresponding to its Syntactic Entities. We assign a computational cost $c(t)$ to every syntactic feature t that is essentially equal to the cost of it's corresponding algorithm τ. Hence, we may now define *Complexity* of a definition as

$$\mathcal{C}(E_k) = \sum_i c(t_i)$$

where t_i are the syntactic features required to evaluate the properties Y_i of the definition E_k. Notice that this value will normally depend on the size of the input data, as will the values $c(t_i)$. At least, though, worst or average case expressions of $c(t_i)$ can be considered as independent of the actual content of the examined data sets. In this perspective $\mathcal{C}(E_k)$ is also computable prior to identification.

Based on these definitions, and following the approach of [3] and [4], the task of optimal semantic identification given some hard complexity boundaries is reduced to a dynamic programming optimization problem. Thus, semantic identification can be performed in real-time, while the uncertainty in the output is guaranteed to be minimized.

4 Experimental Results

In this section we provide some brief, yet indicative, results acquired through the application of the proposed methodologies.

Representation. A knowledge of 70000 semantic entities has been developed [2]. Although loading a fuzzy binary relation defined on this set requires more than 50GB of main memory, assuming a double precision number format, the knowledge base is loaded in less that 100MB of memory using the proposed representation model.

Handling. Transitive closure of the above-mentioned relation is calculated to require more than 5 days of computing time. Using the proposed representation model and applying the proposed transitive closure algorithm, the time required for the complete transitive closure is approximately 20 seconds on the same computer. For the case of simple update of an already transitive relation, the processing time is less than a millisecond.

Identification. Figure 1 presents the validity achieved when applying the prosed approach on a random data set with different complexity thresholds. The non linear character of the graph shows the benefit of the optimized selection of the part of the definition to evaluate in each case. Note that for a threshold $C_T = 12$ the Validity is $V \approx 0.9$ while the total Complexity of the definition is $C_t = 132$ for the uniformly distributed values.

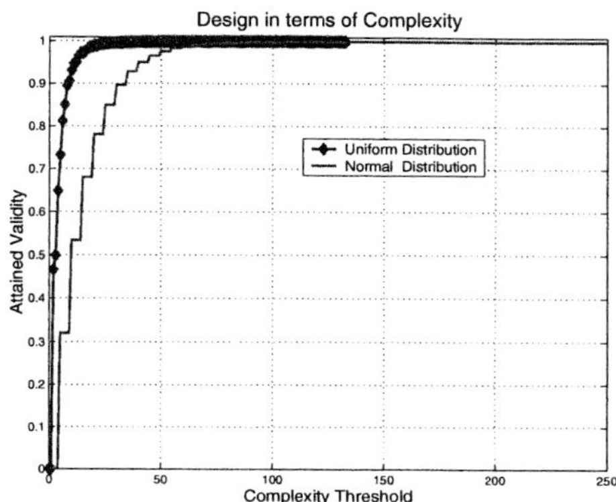

Fig. 1. Achieved validity with respect to the complexity threshold

5 Conclusions

In this paper we made a first attempt to provide an integrated solution to computing problems related to semantic identification of entities in multimedia streams. Problems addresses include the storage requirements of the semantic knowledge, the computational complexity for the handling of the knowledge, as well as the computational complexity of the identification process itself.

The above have been tackled through a novel representation model and matching algorithms for transitive closure and semantic identification. The experimental results verify the efficiency and prospect of the proposed approaches.

References

1. Adelson-Velskii G.M., Landis E.M., "An algorithm for the organization of information" Doklady Akademia Nauk SSSR, Vol 146, pp. 263-266, 1962; English translation in Soviet Math, Vol 3, pp. 1259-1263, 1962.
2. Y. Avrithis, G. Stamou, M. Wallace, F. Marques, P. Salembier, X. Giro, W. Haas, H. Vallant and M. Zufferey, "Unified Access to Heterogeneous Audiovisual Archives", Journal of Universal Computer Science Vol 9(6), pp. 510-519, 2003.
3. M. Falelakis and C. Diou and A. Valsamidis and A. Delopoulos, "Complexity Control in Semantic Identification", IEEE International Conference on Fuzzy Systems, Reno, Nevada, USA, May 2005.
4. M. Falelakis and C. Diou and A. Valsamidis and A. Delopoulos, "Dynamic Semantic Identification with Complexity Constraints as a Knapsack Problem", IEEE International Conference on Fuzzy Systems, Reno, Nevada, USA, May 2005.
5. X. Giro and F. Marques, "Semantic Entity Detection using Description Graphs", International Workshop on Image Analysis for Multimedia Interactive Services (WIAMIS), London, England, April 2003.
6. W. Al-Khatib and Y.F. Day and A. Ghafoor and P.B. Berra, "Semantic Modeling and Knowledge Representation in Multimedia Databases", IEEE Transactions on Knowledge and Data Engineering, Vol 11(1), pp. 64-80, 1999.
7. M. Ramesh Naphade and I.V. Kozintsev and T.S. Huang, "A factor graph Framework for Semantic Video Indexing", IEEE Trans. on Circuits and Systems for Video Technology, Vol 12(1), pp. 40-52, 2002.
8. M. Wallace, Y.Avrithis and S. Kollias, " Computationally efficient sup-t transitive closure for sparse fuzzy binary relations", Fuzzy Sets and Systems, in press.
9. A. Yoshitaka and S. Kishida and M. Hirakawa and T. Ichikawa, "Knowledge-assisted content based retrieval for multimedia databases" IEEE Multimedia, Vol. 1(4), pp. 12-21, 1994.
10. R. Zhao and W.I. Grosky, "Narrowing the Semantic Gap-Improved Text-Based Web Document Retrieval Using Visual Features", IEEE Transactions on Multimedia, Vol 4(2), 2002.

Automated Extraction of Object- and Event-Metadata from Gesture Video Using a Bayesian Network

Dimitrios I. Kosmopoulos

National Centre for Scientific Research "Demokritos",
Institute of Informatics & Telecommunications,
15310, Aghia Paraskevi, Greece
dkosmo@iit.demokritos.gr

Abstract. In this work a method for metadata extraction from sign language videos is proposed, by employing high level domain knowledge. The metadata concern the depicted objects of the head and the right/left hand and the occlusion events, which are essential for interpretation and therefore for subsequent higher level semantic indexing. The occlusions between hands, head and hands and body and hands, can easily confuse metadata extraction and can consequently lead to wrong gesture interpretation. Therefore, a Bayesian network is employed to bridge the gap between the high level knowledge about the valid spatiotemporal configurations of the human body and the metadata extractor. The approach is applied here in sign-language videos, but it can be generalized to video indexing based on gestures.

1 Introduction

The extraction of mid- and high-level semantics from video content is important for tasks as video indexing and retrieval, video summarization and non-linear content organization. This applies also to videos depicting gestures, since they constitute a very useful source of semantic information for multimedia content analysis. The automated extraction of metadata, e.g., according to MPEG-7 or extensions of it, is a prerequisite for the above tasks. However, automated extraction of metadata regarding gesture is lagging behind processing of other modalities such as speech. Apart from the variability of spatiotemporal gesture patterns and coarticualtion effects (merging of gestures) that are responsible for this slow progress, occlusions introduce additional complexity. Failure to produce correct metadata as a result of using conventional extractors can lead to wrong semantics. Such metadata may concern association of color regions to the objects in this context, which are the head, the left and right hand or the visual *objects* that result from their mutual occlusions, or the *appearance* and *disappearance* or *occlusion* events for the *head* and *left/right hand.*

Here the occlusion problem is handled through the analysis of temporally structured events by combining the two-dimensional visual features and the high – level knowledge about the human gestures and the related body configurations. A Bayesian network is employed for probabilistic knowledge modeling. Inferencing over this network serves the purpose of bridging the semantic gap in a top-down fashion.

W. Duch et al. (Eds.): ICANN 2005, LNCS 3697, pp. 823 – 828, 2005.

The rest of the paper is organized as follows: in the next section the research context concerning semantics extraction from gesture videos is briefly discussed; in section 3 the structure of the proposed semantic model through a Bayesian network is presented; in section 4 the calculation of the evidence nodes and the experimental results are presented; finally section 6 summarizes the results and suggests future directions.

2 Related Work

The extraction of semantics from gesture videos has attracted the interest of many researchers in the past. A big portion of them concerns sign language and gestures for human-computer interaction. Gesture recognition methods can be used for extraction of semantic metadata, with the Hidden Markov Models being the most remarkable (e.g., [4], [5]). Other approaches include, neural networks, principal component analysis, motion history images, and their comparative features are discussed extensively in surveys such as [7], [8]. Trajectory-based techniques were presented recently [6]. Although these techniques may provide significant results for gesture recognition tasks, they require accurate hand and (sometimes) head segmentation. Occlusions are not handled at all or become resolved through stereoscopic camera configurations, e.g., in [1], which concerns very limited content. Another approach for occlusion handling is the employment of a 3D hand model, e.g., [9], however optimization of models for articulated objects with so many degrees of freedom, such as hands, is a very challenging task. In works similar to [2] the optical flow constraint is used, however they assume that the sampling rate is constantly high and that the movement is smooth in terms of shape and position.

The above approaches use low-level features to infer higher-level semantics but they don't address the inverse information flow. Reasoning about low- and mid-level features using high-level knowledge (thus enabling a closed-loop semantic extraction process) would be a major step for bridging the semantic gap and could be complementary to the aforementioned methods. An approach that has motivated the present work is given in [3], where a Bayesian network is used to model known spatiotemporal configurations of the human body.

3 Semantic Model

The aim of this work is to facilitate mid-level gesture metadata extraction using high level knowledge. It extends the work presented in [3] mainly in three ways: (a) A different probabilistic network structure is applied that allows explicit modeling of all possible states (occlusions) as well as their types, while not assuming known head position. (b) The modeled temporal relations are minimal in order to cope with motion discontinuities. (c) The observation variables provide a detailed skin region representation, including internal structure, based on Zernike moments. In [3] the regions are represented only by the area the aspect ratio and orientation, which are insufficient for modeling the complex hand shapes appearing in sign language gestures.

In this work a Bayesian network is used as a semantic model due to its ability to capture uncertainty of the domain knowledge. This is not possible when using a fixed set of rules, which necessitate a manual threshold definition. Here, provided a set of variables **x** (network nodes), we seek to find the instance vector $\mathbf{x}_m$ of those variables that maximizes their joint distribution, given some evidence **e** coming from image measurements (associated with other network nodes). In other words:

$$\mathbf{x}_m = \{\mathbf{x}_0 : \forall\, \mathbf{x},\ P(\mathbf{x}_0|\mathbf{e}) > P(\mathbf{x} \mid \mathbf{e})\} \quad (1)$$

The value of $\mathbf{x}_m$ provides the current gesture state, i.e., position of head and hands in the image as well as current occlusions.

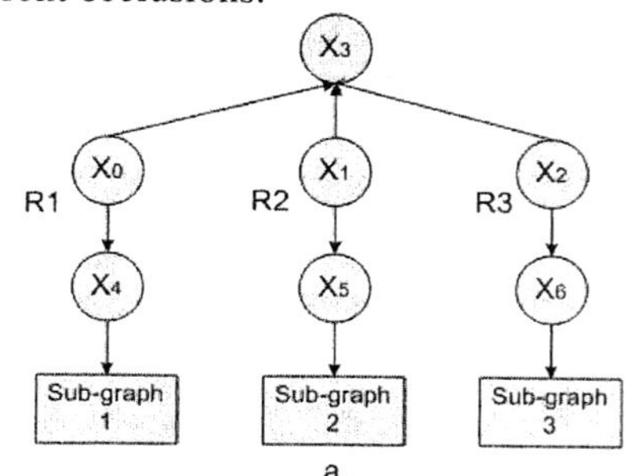

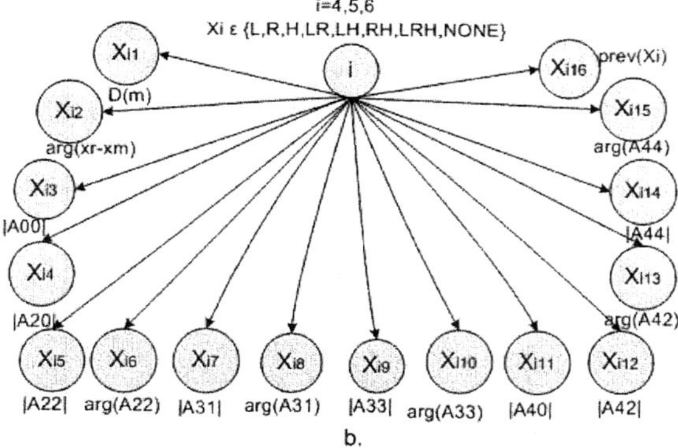

Fig. 1. (a) Bayesian network representing the semantic model and (b) the subgraphs

The network encoding the high level semantics of the gesture metadata extraction task is presented in Fig.1(a-b). More specifically:

- X_0, X_1, X_2: Variables corresponding to the three (maximum) skin regions expressing the probability that one of them corresponds to left hand, right hand, head, mutual occlusion of hands, head – left hand occlusion, head-right hand occlusion, occlusion of head by both hands or noise. The corresponding values belong to the set **A**=$\{L, R, H, LR, LH, RH, LRH, N\}$.
- X_3: Binary variable depending on X_0, X_1, X_2 and used for excluding the non acceptable associations of regions to visual objects. It is only true when the following are the values for X_0, X_1, X_2: (L,R,H), (L,H,N), (L,RH,N), (R,H,N), (R,LH,N), (H,N,N), (H,LR,N), (LH,N,N), (RH,N,N), (LRH,N,N), at any possible order.
- X_4, X_5, X_6: Auxiliary variables used to decouple X_0, X_1, X_2, X_3 from the sub-graphs presented in detail in Fig.1b. for simplifing the network, e.g., during training and

execution sub-graphs 1-3 are treated independently using the same learning and inference engine. Their values coincide with those of their parent nodes.

The aforementioned sub-graphs are identical, each associated with a skin region and used for capturing visual evidence. Their root nodes are the X_i (i=7,8,9).

The child nodes in the sub-graphs provide evidence for inferring the system state:

- X_{i1} X_{i2}: The distance and angle of the current region center of gravity from the common center of gravity of all regions.
- X_{i16}:The region label in the previous time instance. Models motion continuity.
- $X_{i3}, X_{i4}, X_{i5}, X_{i7}, X_{i9}, X_{i11}, X_{i12}, X_{i14}$: Correspond to the Euclidean norm of the Zernike moments up to 4th order of the i-region around its center of gravity ($|A_{11}|=0$).
- $X_{i6}, X_{i8}, X_{i10}, X_{i13}, X_{i15}$: Correspond to non-zero orientations of the above moments.

The complex Zernike moments have been selected due to noise resiliency, reduced information redundancy (orthogonality) and reconstruction capability. Their definition in polar coordinates (unit circle) for order p (p-q = even and $0 \leq q \leq m$) is [11]:

$$A_{pq} = \frac{p+1}{\pi} \int_0^1 \int_{-\pi}^{\pi} R_{pq}(r) \cdot e^{-jq\theta} f(r,\theta) \cdot r \cdot dr d\theta \quad , \qquad R_{pq}(r) = \sum_{s=0}^{\frac{p-q}{2}} (-1)^s \cdot \frac{(m-s)!}{s! \cdot (\frac{m+n}{2} - s)!(\frac{m-n}{2} - s)!} r^{m-2s} \qquad (2)$$

Restating the goal expressed at the beginning of this section, we aim to find the value $\mathbf{x}_m$ of the vector (X_0, X_1, X_2) that maximizes the joint distribution P(**x**|**e**) where **e** is the evidence provided by the measured variables and $\mathbf{e}=\{\mathbf{x}_{i,j} : i \in \{4,5,6\}, j \in \{1,2,3,..,15\}\}$. $\mathbf{x}_m$ provides the association of skin color regions in the image space to context objects, in the "world" space. The objects appearance or disappearance (whenever the vector elements change) signifies the visual events that are extracted as metadata. For inferring the current state, algorithms such as the junction tree [12] can be employed.

The efficiency of the followed approach does not depend on the vocabulary size. If new gestures need to be analysed they will have to be trained. This will modify the probability distributions of the network variables but the system complexity will not be affected. Virtual reality systems will be able to facilitate this procedure.

4 Experimental Results

To measure evidence we first locate the target's face in the image (Haar face detection) and we use a part of the face region for probabilistic skin color modeling. The target is assumed to be dressed and to face the camera with the upper body part within the image. For this reduced skin color modeling problem, a single multivariate Gaussian model for color is used. Robust estimation functions [10] are employed to remove outliers in the face region due to noise, eyes, hair, glasses etc. Using the extracted color model, the skin regions are segmented, keeping a maximum of three regions (the biggest ones). The final mask is applied to the intensity image to obtain a masked gray-level image including only these regions. The Zernike moment norms are normalized with regard to the initial face area to decrease influence of non-uniform body dimensions or distance from camera. For simplicity all continuous variables are discretized.

The system was trained for several thousand frames using a vocabulary of 15 words (including many occlusion types) and was tested on more than 1500 frames with promising results, even at lower space and time resolution (Table 1). The latter case implies discontinuous motion, and the present approach has an advantage compared to algorithms based on optical flow due to the lack of strict temporal constraints. Furthermore, it has been shown that probabilistic approaches like the one presented here have been proved superior compared to particle filtering and namely the Condensation algorithm, as regards tolerance to occlusions and performance, provided that proper training has been previously performed [3]. After experimental comparison to the work in [3], using a similar training procedure, the results don't differ significantly in full spatial and time resolution but the current method is superior in discontinuous motion (approximately 25% less errors). This is well explained by the fact that only one network variable (per skin region) models the temporal relations (X_{i16}) in comparison to 12 such variables used in [3]. Typical examples for the current method are displayed in Fig. 2, where the occlusion types are visible.

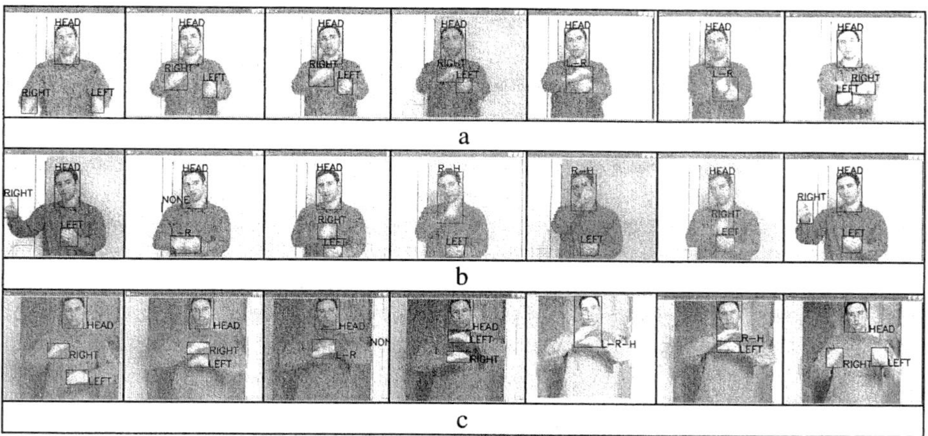

Fig. 2. The extraction of semantics from skin regions in sign language videos for the words: (a) change (b) correct, (c) achieve. The identified skin regions are enclosed in rectangles, annotated with the region type. Changes in the skin region states signify visual events.

Table 1. Classification results of skin regions. The three numbers in each cell (a-b-c) provide the results for the following cases regarding the same gestures (a) full spatial and time resolution, i.e., 640x480 - 15fps (b) 320x240 – 15fps (c) 640x480 – 7.5fps.

		Recognised as							
		L	**R**	**H**	**LR**	**LH**	**RH**	**LRH**	**NONE**
Actual	**L**	1224-1190-633	33-72-11	9-42-10	0-0-0	0-0-0	0-0-0	0-0-0	45-7-3
	R	71-96-20	1168-1096-601	0-20-7	0-0-0	0-0-0	0-0-0	0-0-0	3-30-0
	H	9-22-6	0-63-0	2415-2321-1206	0-21-0	0-0-0	0-0-0	0-0-0	36-33-18
	LR	3-23-12	0-19-9	33-42-11	414-378-180	0-0-0	0-0-0	0-0-0	45-60-33
	LH	3-0-3	0-0-1	6-13-5	0-0-0	105-101-49	0-0-0	0-0-0	12-12-15
	RH	0-6-0	0-0-0	22-36-15	0-0-0	0-0-0	134-117-45	0-0-0	9-6-0
	LRH	0-0-0	0-0-0	9-9-6	0-0-0	0-0-0	0-0-0	55-47-27	2-10-0
	NONE	1-6-24	7-7-11	4-6-20	0-3-0	0-0-0	0-0-0	0-0-0	1425-1415-615

5 Conclusions

A method for extraction of gesture-related metadata from video based on a Bayesian network has been presented. Its ability to overcome occlusions was shown, which is difficult to achieve with conventional methods. Furthermore, it is more resilient to image discontinuities than the optical-flow based methods or other probabilistic methods that use temporal relations extensively. Higher order Zernike moments (or temporal variables) can be easily included for more detailed area representation, but the application context has to be considered in that case for a performance-effectiveness trade-off. Tests with bigger vocabularies will verify scalability.

Future work includes the integration of the network in a closed-loop gesture recognition scheme, e.g., in combination to a HMM, for more focused low level feature extraction. Furthermore, virtual reality systems will be evaluated for automating the training procedure. Within the scope of research is also the development of invariant measures, to minimize the effect of different camera viewpoints as well as human body variations. These measures will be integrated as evidence network nodes.

References

1. C. Vogler, D. Metaxas, A Framework for Recognizing the Simultaneous Aspects of american Sign Language, Computer Vision and Image Understanding, 81 (2001) 358-384
2. N. Tanibata, N. Shimada, Y. Shirai, "Extraction of Hand Features for Recognition of Sign Language Words", International Conference on Vision Interface, 391-398, 2002
3. S. Gong, J. Ng, J. Sherrah, On the semantics of visual behavior, structured events and trajectories of human action, Image and Vision Computing, 20 (2002) 873-888
4. A. D. Wilson, A. Bobick, Parametric Hidden Markov Models Approach for Gesture Recognition, IEEE Transactions on Pattern Analysis and Machine Intelligence, 21(12) (1997) 1325 – 1337
5. Hyeon-Kyu Lee, Kim, J.H., An HMM-based threshold model approach for gesture recognition, IEEE Transactions on Pattern Analysis and Machine Intelligence, 21(10) (1999) 961 – 973
6. C. Rao, A. Yilmaz, M. Shah, View-Invariant representation and recognition of actions, International Journal of Computer Vision, 50(2) (2002) 203-226
7. V. Pavlovic, R. Sharma, T. S. Huang, Visual Interpretation of Hand Gestures for Human-Computer Interaction: a Review, IEEE Transactions on Pattern Analysis and Machine Intelligence, 19 (1997) 677 – 695
8. Y. Wu, T. Huang, Vision-based gesture recognition: a Review, Gesture-Based Communication in Human-Computer Interaction (A. Braffort, R. Gherbi, S. Gibet, J. Richardson, D. Teil, Eds.), 1739, Lecture Notes in Artificial Intelligence, (1999) 103-115
9. J. Rehg, T. Kanade, DigitEyes: Vision-Based Human Hand Tracking. Tech. Rep. CMU-CS -93-220, School of Comp. Science, Carnegie Mellon University, Pittsburgh., 1993
10. F. R. Hampel, E. M. Ronchetti, P.J. Rousseauw, W. A. Stahel, Robust Statistics : The Approach Based on Influence Functions, Wiley, New York, 1986
11. R. Mukundan, K. R. Ramakrishnan, Moment Functions in Image Analysis: Theory and Applications, World Scientific, Singapore, 1998.
12. F. V. Jensen, Bayesian Networks and Decision Graphs,Springer Verlag, New York, 2001
13. Michael Isard and Andrew Blake, CONDENSATION conditional density propagation for visual tracking, Int. J. Computer Vision, 29 (1), (1998) 5-28

f-SWRL: A Fuzzy Extension of SWRL

Jeff Z. Pan[2], Giorgos Stamou[1], Vassilis Tzouvaras[1], and Ian Horrocks[2]

[1] Department of Electrical and Computer Engineering, National Technical University of Athens, Zographou 15780, Greece
[2] School of Computer Science, The University of Manchester, Manchester, M13 9PL, UK

Abstract. In an attempt to extend existing knowledge representation systems to deal with the imperfect nature of real world information involved in several applications like multimedia analysis and understanding, the AI community has devoted considerable attention to the representation and management of uncertainty, imprecision and vague knowledge. Moreover, a lot of work has been carried out on the development of reasoning engines that can interpret imprecise knowledge. The need to deal with imperfect and imprecise information is likely to be common in the context of multimedia and the (Semantic) Web. In anticipation of such requirements, this paper presents a proposal for fuzzy extensions of SWRL, which is a rule extension to OWL DL.

1 Introduction

According to widely known proposals for a Semantic Web architecture, Description Logics (DLs)-based ontologies will play a key role in the Semantic Web [5]. This has led to considerable efforts to developing a suitable ontology language, culminating in the design of the OWL Web Ontology Language [2], which is now a W3C recommendation. SWRL (Semantic Web Rule Language) [3] is proposed as a well known Horn clause rules extension to OWL DL.[1]

Experience in using ontologies and rules in applications has shown that in many cases we would like to extend their representational and reasoning capabilities to deal with vague or imprecise knowledge. For example, multimedia applications have highlighted the need to extend representation languages with capabilities which allow for the treatment of the inherent imprecision in multimedia object representation, matching, detection and retrieval. Unfortunately, neither OWL nor SWRL provides such capabilities.

In order to capture imprecision in rules, we propose a fuzzy extension of SWRL, called f-SWRL. In f-SWRL, fuzzy individual axioms can include a specification of the "degree" (a truth value between 0 and 1) of confidence with which one can assert that an individual (resp. pair of individuals) is an instance of a given class (resp. property); and atoms in f-SWRL rules can include a "weight" (a truth value between 0 and 1) that represents the "importance" of the atom in

[1] OWL DL is a key sub-language of OWL.

W. Duch et al. (Eds.): ICANN 2005, LNCS 3697, pp. 829–834, 2005.

a rule. For example, the following fuzzy rule asserts that being healthy is more important than being rich to determine if one is happy:

$$\mathsf{Rich}(?p) * 0.5 \wedge \mathsf{Healthy}(?p) * 0.9 \rightarrow \mathsf{Happy}(?p),$$

where Rich, Healthy and Happy are classes, and 0.5 and 0.9 are the weights for the atoms Rich(?p) and Healthy(?p), respectively. A detailed motivating use case for fuzzy rules can be found in [11].

In this paper, we will present the syntax and semantics of f-SWRL. We will use standard Description Logics [1] notations in the syntax of f-SWRL, while the model-theoretic semantics of f-SWRL is based on the theory of fuzzy sets [14]. To the best of our knowledge, this is the first paper describing a fuzzy extension of the SWRL language.

2 Preliminaries

2.1 SWRL

SWRL is proposed by the Joint US/EU ad hoc Agent Markup Language Committee.[2] It extends OWL DL by introducing *rule axioms*, or simply *rules*, which have the form:

$$\mathsf{antecedent} \rightarrow \mathsf{consequent},$$

where both antecedent and consequent are conjunctions of atoms written $a_1 \wedge \ldots \wedge a_n$. Atoms in rules can be of the form C(x), P(x,y), Q(x,z), sameAs(x,y) or differentFrom(x,y), where C is an OWL DL description, P is an OWL DL *individual-valued* property, Q is an OWL DL *data-valued* property, x,y are either *individual-valued* variables or OWL individuals, and z is either a *data-valued* variable or an OWL data literal. An OWL data literal is either a typed literal or a plain literal; see [2,6] for details. Variables are indicated using the standard convention of prefixing them with a question mark (e.g., $?x$). For example, the following rule asserts that one's parents' brothers are one's uncles:

$$parent(?x, ?p) \wedge brother(?p, ?u) \rightarrow uncle(?x, ?u), \quad (1)$$

where *parent*, *brother* and *uncle* are all *individual-valued* properties.

The reader is referred to [3] for full details of the model-theoretic semantics and abstract syntax of SWRL.

2.2 Fuzzy Sets

While in classical set theory any element belongs or not to a set, in fuzzy set theory [14] this is a matter of degree. More formally, let X be a collection of elements (the universe of discourse) with cardinality m, i.e $X = \{x_1, x_2, \ldots, x_m\}$. A fuzzy subset A of X, is defined by a membership function $\mu_A(x)$, or simply $A(x)$, $x \in X$. This membership function assigns any $x \in X$ to a value between 0

[2] See `http://www.daml.org/committee/` for the members of the Joint Committee.

and 1 that represents the degree in which this element belongs to X. The *support*, *Supp(A)*, of A is the crisp set $Supp(A) = \{x \in X \mid A(x) \neq 0\}$.

Using the above idea, the most important operations defined on crisp sets and relations (complement, union, intersection etc) are extended in order to cover fuzzy sets and fuzzy relations. The complement $\neg A$ of a fuzzy set A is given by $(\neg A)(x) = c(A(x))$ for any $x \in X$. The intersection of two fuzzy sets A and B is given by $(A \cap B)(x) = t[A(x), B(x)]$, where t is a triangular norm (t-norm). The union of two fuzzy sets A and B is given by $(A \cup B)(x) = u[A(x), B(x)]$, where u is a triangular conorm (u-norm). A binary fuzzy relation R over two countable crisp sets X and Y is a function $R : X \times Y \rightarrow [0, 1]$. The composition of two fuzzy relation $R_1 : X \times Y \rightarrow [0, 1]$ and $R_2 : Y \times Z \rightarrow [0, 1]$ is given by $[R_1 \circ^t R_2] = \sup_{y \in Y} t[R_1(x, y), R_2(y, z)]$. The reader is referred to [4] for details of fuzzy logics and their applications.

3 f-SWRL

Fuzzy rules are of the form antecedent → consequent, where atoms in both the antecedent and consequent can have weights, i.e., numbers between 0 and 1. More specifically, atoms can be of the forms C(x)$*w$, P(x,y)$*w$, sameAs(x,y)$*w$ or differentFrom(x,y)$*w$, where $w \in [0, 1]$ is the weight of an atom,[3] and omitting a weight is equivalent to specifying a value of 1. For instance, the following fuzzy rule axiom asserts that if a man has his eyebrows raised enough and his mouth open then he is happy, and that the condition that he has his eyebrows raised is a bit more important than the condition that he has his mouth open.

$$\mathsf{EyebrowsRaised}(?a) * 0.9 \wedge \mathsf{MouthOpen}(?a) * 0.8 \rightarrow \mathsf{Happy}(?a), \tag{2}$$

In this example, EyebrowsRaised, MouthOpen and Happy are classes, ?a is a *individual-valued* variable, and 0.9 and 0.8 are the weights of the atoms EyebrowsRaised(?a) and MouthOpen(?a), respectively.

In this paper, we only consider *atomic* fuzzy rules, i.e., rules with only one atom in the consequent. The weight of an atom in a consequent, therefore, can be seen as indicating the weight that is given to the rule axiom in determining the degree with which the consequent holds. Consider, for example, the following two fuzzy rules:

$$\mathsf{parent}(?x, ?p) \wedge \mathsf{Happy}(?p) \rightarrow \mathsf{Happy}(?x) * 0.8 \tag{3}$$

$$\mathsf{brother}(?x, ?b) \wedge \mathsf{Happy}(?b) \rightarrow \mathsf{Happy}(?x) * 0.4, \tag{4}$$

which share Happy($?x$) in the consequent. Since $0.8 > 0.4$, more weight is given to rule (3) than to rule (4) when determining the degree to which an individual is Happy.

In what follows, we formally introduce the syntax and model-theoretic semantics of fuzzy SWRL.

---

[3] To simplify the presentation, we will not cover datatype property atoms in this paper.

3.1 Syntax

In this section, we present the syntax of fuzzy SWRL. Due to space limitation, we use DL syntax (see the following definition) instead of the XML, RDF or abstract syntax of SWRL.

Definition 1. *Let* a,b *be individual URIrefs, C, D OWL class descriptions, r, s OWL role descriptions, r_1, r_2 role URIrefs, $m_1, m_2, w, w_1, \ldots, w_n \in [0,1]$, $\vec{v}$, $\vec{v}_1$ $, \ldots, \vec{v}_n$ are (unary or binary) tuples of variables and/or individual URIrefs, $a_1(\vec{v_1}), \ldots, a_n(\vec{v_n})$ and $c(\vec{v})$ are of the forms $C(x)$, $r(x,y)$, $sameAs(x,y)$ or $differentFrom(x,y)$, where x, y are* individual-valued *variables or individual URIrefs.*

An f-SWRL ontology can have the following kinds of axioms:

- *class axioms: $C \sqsubseteq D$ (class inclusion axioms);*
- *property axioms: $r \sqsubseteq s$ (property inclusion axioms),* $\mathsf{Func}(r_1)$ *(functional property axioms),* $\mathsf{Trans}(r_2)$ *(transitive property axioms);*
- *individual axioms: $(\mathtt{a} : C) \geq m_1$ (fuzzy class assertions), $(\langle \mathtt{a}, \mathtt{b} \rangle : r) \geq m_2$ (fuzzy property assertions), $\mathtt{a} = \mathtt{b}$ (individual equality axioms) and $\mathtt{a} \neq \mathtt{b}$ (individual inequality axioms);*
- *rule axioms: $a_1(\vec{v_1}) * w_1 \wedge \cdots \wedge a_n(\vec{v_n}) * w_n \rightarrow c(\vec{v}) * w$ (fuzzy rule axioms).*

Omitting a degree or a weight is equivalent to specifying the value of 1. ◇

According to the above definition, f-SWRL extends SWRL with fuzzy class assertions, fuzzy property assertions and fuzzy rule axioms.

3.2 Model-Theoretic Semantics

In this section, we give a model-theoretic semantics for fuzzy SWRL. Although many f-SWRL axioms share the same syntax as their counterparts in SWRL, such as concept inclusion axioms, they have different semantics because we use fuzzy interpretations in the model-theoretic semantics of f-SWRL.

Definition 2. *A fuzzy interpretation is a pair $\mathcal{I} = \langle \Delta^{\mathcal{I}}, \cdot^{\mathcal{I}} \rangle$, where the domain $\Delta^{\mathcal{I}}$ is a non-empty set and $\cdot^{\mathcal{I}}$ is a fuzzy interpretation function, which maps*

1. *individual names and* individual-valued *variables to elements of $\Delta^{\mathcal{I}}$,*
2. *a class description C to a membership function $C^{\mathcal{I}} : \Delta^{\mathcal{I}} \rightarrow [0,1]$,*
3. *an* individual-valued *property name R to a membership function $R^{\mathcal{I}} : \Delta^{\mathcal{I}} \times \Delta^{\mathcal{I}} \rightarrow [0,1]$,*
4. *the built-in property sameAs to a membership function*

$$sameAs^{\mathcal{I}}(x,y) = \begin{cases} 1 \text{ if } x^{\mathcal{I}} = y^{\mathcal{I}} \\ 0 \text{ otherwise,} \end{cases}$$

5. *the built-in property differentFrom to a membership function*

$$differentFrom^{\mathcal{I}}(x,y) = \begin{cases} 1 \text{ if } x^{\mathcal{I}} \neq y^{\mathcal{I}} \\ 0 \text{ otherwise.} \end{cases}$$

A fuzzy interpretation $\mathcal{I}$ *satisfies a class inclusion axiom* $C \sqsubseteq D$, *written* $\mathcal{I} \models C \sqsubseteq D$, *if* $\forall o \in \Delta^{\mathcal{I}}, C^{\mathcal{I}}(o) \leq D^{\mathcal{I}}(o)$.

A fuzzy interpretation $\mathcal{I}$ *satisfies a property inclusion axiom* $r \sqsubseteq s$, *written* $\mathcal{I} \models r \sqsubseteq s$, *if* $\forall o, q \in \Delta^{\mathcal{I}}, r^{\mathcal{I}}(o,q) \leq s^{\mathcal{I}}(o,q)$. $\mathcal{I}$ *satisfies a functional property axiom* $\mathsf{Func}(r_1)$, *written* $\mathcal{I} \models \mathsf{Func}(r_1)$, *if* $\forall o, q \in \Delta^{\mathcal{I}}, \mid Supp[r_1^{\mathcal{I}}(o,q)] \mid \leq 1$. $\mathcal{I}$ *satisfies a transitive property axiom* $\mathsf{Trans}(r_2)$, *written* $\mathcal{I} \models \mathsf{Trans}(r_2)$, *if* $\forall o, q \in \Delta^{\mathcal{I}}, r_2^{\mathcal{I}}(o,q) = \sup_{p \in \Delta^{\mathcal{I}}} t[r_2^{\mathcal{I}}(o,p), r_2^{\mathcal{I}}(p,q)]$, *where* t *is a triangular norm.*

A fuzzy interpretation $\mathcal{I}$ *satisfies a fuzzy class assertion* $(\mathtt{a} : C) \geq m$, *written* $\mathcal{I} \models (\mathtt{a} : C) \geq m$, *if* $C^{\mathcal{I}}(\mathtt{a}) \geq m$. $\mathcal{I}$ *satisfies a fuzzy property assertion* $(\langle \mathtt{a}, \mathtt{b} \rangle : r) \geq m_2$, *written* $\mathcal{I} \models (\langle \mathtt{a}, \mathtt{b} \rangle : r) \geq m_2$, *if* $r^{\mathcal{I}}(\mathtt{a}, \mathtt{b}) \geq m$. $\mathcal{I}$ *satisfies an individual equality axiom* $\mathtt{a} = \mathtt{b}$, *written* $\mathcal{I} \models \mathtt{a} = \mathtt{b}$, *if* $\mathtt{a}^{\mathcal{I}} = \mathtt{b}^{\mathcal{I}}$. $\mathcal{I}$ *satisfies an individual inequality axiom* $\mathtt{a} \neq \mathtt{b}$, *written* $\mathcal{I} \models \mathtt{a} \neq \mathtt{b}$, *if* $\mathtt{a}^{\mathcal{I}} \neq \mathtt{b}^{\mathcal{I}}$.

A fuzzy interpretation $\mathcal{I}$ *satisfies a fuzzy rule axiom* $a_1(\vec{v_1}) * w_1 \wedge \cdots \wedge a_n(\vec{v_n}) * w_n \rightarrow c(\vec{v}) * w$, *written* $\mathcal{I} \models a_1(\vec{v_1}) * w_1 \wedge \cdots \wedge a_n(\vec{v_n}) * w_n \rightarrow c(\vec{v}) * w$, *if* $t(t(a_1^{\mathcal{I}}(\vec{v_1}^{\mathcal{I}}), w_1), \ldots, t(a_n^{\mathcal{I}}(\vec{v_n}^{\mathcal{I}}), w_n)) \leq t(c^{\mathcal{I}}(\vec{v}^{\mathcal{I}}), w)$, *where* t *is a triangular norm.* ◇

Let us take the rule (2) as an example to illustrate the above semantics. Assuming that EyebrowsRaised, MouthOpen and Happy are class URIrefs, then given a fuzzy interpretation $\mathcal{I} = \langle \Delta^{\mathcal{I}}, \cdot^{\mathcal{I}} \rangle$, the rule (2) is satisfied by $\mathcal{I}$ iff for all $a \in \Delta^{\mathcal{I}}$, we have $t(t(\mathsf{EyebrowsRaised}^{\mathcal{I}}(a), 0.9), t(\mathsf{MouthOpen}^{\mathcal{I}}(a), 0.8)) \leq t(\mathsf{Happy}^{\mathcal{I}}(a), 1)$.

Note that in SWRL the class assertion $\mathtt{Tom} : \mathsf{Happy}$ is equivalent to the rule axiom $\rightarrow \mathsf{Happy}(\mathtt{Tom})$. In f-SWRL, we have the following equivalence between the f-SWRL individual axiom and rule axiom: The fuzzy assertion $(\mathtt{Tom} : \mathsf{Happy}) \geq 0.8$ is equivalent to the rule axiom $\top(\mathtt{Tom}) * 0.8 \rightarrow \mathsf{Happy}(\mathtt{Tom})$. According to the above semantics, we have: $t(\top^{\mathcal{I}}(\mathtt{Tom}), 0.8) \leq \mathsf{Happy}^{\mathcal{I}}(\mathtt{Tom})$. From a semantics point of view an individual always belong to a degree of 1 to the top concept, so we have: $t(1, 0.8) \leq \mathsf{Happy}^{\mathcal{I}}(\mathtt{Tom})$. Due to the boundary condition of t-norms, we have $\mathsf{Happy}^{\mathcal{I}}(\mathtt{Tom}) \geq 0.8$.This suggests that fuzzy assertion can be represented by fuzzy rule axioms.

4 Discussion

Several ways of extending Description Logics and logic programming with the theory of fuzzy logic have been proposed [13,10,8,9,7,12]; however, we have not seen any publications on fuzzy extensions of SWRL. We believe that the combination of Semantics Web ontology and rules languages provides a powerful and flexible knowledge representation formalism, and that f-SWRL is of great interest to the ontology community as well as to communities in which ontologies with vague information can be applied, such as multimedia and the Semantic Web.

Our future work includes logical properties and computational aspect of f-SWRL. Another interesting direction is to extend f-SWRL to support datatype groups [5], which allows the use of customised datatypes and datatype predicates in ontologies.

References

1. F. Baader, D. L. McGuiness, D. Nardi, and P. Patel-Schneider, editors. *Description Logic Handbook: Theory, implementation and applications.* Cambridge University Press, 2002.
2. Sean Bechhofer, Frank van Harmelen, Jim Hendler, Ian Horrocks, Deborah L. McGuinness, Peter F. Patel-Schneider, Lynn Andrea Stein; Mike Dean, and Guus Schreiber (editors). OWL Web Ontology Language Reference. Technical report, W3C, February 10 2004. `http://www.w3.org/TR/2004/REC-owl-ref-20040210/`.
3. Ian Horrocks, Peter F. Patel-Schneider, Harold Boley, Said Tabet, Benjamin Grosof, and Mike Dean. SWRL: A Semantic Web Rule Language — Combining OWL and RuleML. W3C Member Submission, `http://www.w3.org/Submission/SWRL/`, May 2004.
4. G. J. Klir and B. Yuan. *Fuzzy Sets and Fuzzy Logic: Theory and Applications.* Prentice-Hall, 1995.
5. Jeff Z. Pan. *Description Logics: Reasoning Support for the Semantic Web.* PhD thesis, School of Computer Science, The University of Manchester, Oxford Rd, Manchester M13 9PL, UK, Sept 2004.
6. Jeff Z. Pan and Ian Horrocks. OWL-Eu: Adding Customised Datatypes into OWL. In *Proc. of Second European Semantic Web Conference (ESWC 2005)*, 2005. To appear.
7. G. Stoilos, G. Stamou, V. Tzouvaras, J.Z. Pan, and I. Horrocks. A fuzzy description logic for multimedia knowledge representation. Proc. of the International Workshop on Multimedia and the Semantic Web, 2005.
8. U. Straccia. Reasoning within fuzzy description logics. *Journal of Artificial Intelligence*, 14:137–166, 2001.
9. Umberto Straccia. Towards a fuzzy description logic for the semantic web (preliminary report). In *2nd European Semantic Web Conference (ESWC-05)*, Lecture Notes in Computer Science, Crete, 2005. Springer Verlag.
10. C. Tresp and R. Molitor. A description logic for vague knowledge. In *In proc of the 13th European Conf. on Artificial Intelligence (ECAI-98)*, 1998.
11. Vassilis Tzouvaras and Giorgos Stamou. A use case of fuzzy *swrl*. Technical report, 2004. http://image.ntua.gr/~tzouvaras/usecase.pdf.
12. P. Vojtás. Fuzzy logic programming. *Fuzzy Sets and Systems*, 124:361–370, 2001.
13. J. Yen. Generalising term subsumption languages to fuzzy logic. In *In Proc of the 12th Int. Joint Conf on Artificial Intelligence (IJCAI-91)*, pages 472–477, 1991.
14. L. A. Zadeh. Fuzzy sets. *Information and Control*, 8:338–353, 1965.

An Analytic Distance Metric for Gaussian Mixture Models with Application in Image Retrieval

G. Sfikas, C. Constantinopoulos*, A. Likas, and N.P. Galatsanos

Department of Computer Science,
University of Ioannina
Ioannina, Greece GR 45110
{sfikas, ccostas, arly, galatsanos}@cs.uoi.gr

Abstract. In this paper we propose a new distance metric for probability density functions (PDF). The main advantage of this metric is that unlike the popular Kullback-Liebler (KL) divergence it can be computed in closed form when the PDFs are modeled as Gaussian Mixtures (GM). The application in mind for this metric is histogram based image retrieval. We experimentally show that in an image retrieval scenario the proposed metric provides as good results as the KL divergence at a fraction of the computational cost. This metric is also compared to a Bhattacharyya-based distance metric that can be computed in closed form for GMs and is found to produce better results.

1 Introduction

The increasing supply of cheap storage space in the past few years has led to multimedia databases with ever-increasing size. In this paper we consider the case of content-based image retrieval (CBIR) [3]. That means that the query is made using a sample image, and we would like the CBIR system to give us the images that resemble the most our sample-query. A common approach to CBIR is through the computation of image *feature histograms* that are subsequently modeled using probability density function (PDF) models. Then, the PDF corresponding to each image in the database is compared with that of the query image, and the images closest to the query are returned to the user as the query result. The final step suggests that we must use some distance metric to compare PDFs. There is no universally accepted such distance metric; two commonly used metrics for measuring PDF distances is the *Kullback-Liebler* divergence and the *Bhattacharyya* distance [4]. In this paper, we explore a new distance metric that leads to an analytical formula in the case where the probability density functions correspond to Gaussian Mixtures.

It is obvious that the distance metric we choose to employ is of major importance for the performance of the CBIR system. It is evident that the query results are explicitly affected by the metric used. Also, a computationally demanding metric can slow

* This research was funded by the program "Heraklitos" of the Operational Program for Education and Initial Vocational Training of the Hellenic Ministry of Education under the 3rd Community Support Framework and the European Social Fund.

W. Duch et al. (Eds.): ICANN 2005, LNCS 3697, pp. 835 – 840, 2005.

down considerably the whole retrieval process, since the sample image must be compared with every image in the database.

2 GMM Modeling and PDF Distance Metrics

At first, we need as we noted to construct a *feature histogram* for each image in the database, as shown in [2] for color features. There are a number of reasons, though, that feature histograms are not the best choice in the context of image retrieval and it is preferable to model the feature data using parametric probability density function models, like for example Gaussian mixture models (GMM).

Consider the case where we choose color as the appropriate feature and construct color histograms. It is well-known that color histograms are sensitive to noise interference like lighting intensity changes or quantization errors ("binning problem"). Also, the number of bins in a histogram grows exponentially with the number of feature components ("curse of dimensionality"). These problems, which apply in feature histograms in general, can be solved by modeling the histogram using a probability density function model.

A good way to model probability density functions (PDF) is assuming that the target distribution is a Finite Mixture Model [1]. A commonly used type of mixture model is the Gaussian Mixture Model (GMM). This model represents a PDF as

$$p(x)=\sum_{j=1}^{K}\pi_j N\left(x:\mu_j,\Sigma_j\right) \tag{1}$$

where K stands for the number of Gaussian kernels mixed, π_j are the mixing weights and μ_j,Σ_j are the mean vector and the covariance matrix of Gaussian kernel *j*. GMMs can be trained easily with an algorithm such as EM (Expectation – Maximization) [1].

So we come to the point where the sample image used for the query and the images in the database have their feature histogram and Gaussian Mixture Model been generated. The final step is to compare the GMM of the sample image with the GMMs of the stored images in order to decide which images are the closest to the sample. Therefore, we need a way to calculate a distance metric between PDFs.

A common way to measure the distance between two PDFs *p(x)* and *p'(x)*, is the Kullback-Liebler divergence [4]:

$$KL(p\,\|\,p')=\int p(x)\ln\frac{p(x)}{p'(x)}dx\cdot$$

Notice that $KL(p\,\|\,p')$ is not necessarily equal to $KL(p'\,\|\,p)$. Thus, it is more reasonable to use a symmetric version of the Kullback-Liebler divergence:

$$SKL(p,p')=\left|\frac{1}{2}\int p(x)\ln\frac{p(x)}{p'(x)}dx+\frac{1}{2}\int p'(x)\ln\frac{p'(x)}{p(x)}dx\right| \tag{2}$$

where SKL stands for Symmetric Kullback-Liebler. The absolute value is taken in order for the metric to have distance properties. Since the SKL metric cannot be computed in closed form, we have to resort to a Monte-Carlo approximation based on the formula $\int f(x)p(x)dx \to (N)^{-1}\sum_{i=1}^{N} f(x_i)$ as $N \to \infty$, where the samples x_i are assumed to be drawn from $p(x)$. Thus, SKL can be computed as:

$$SKL_{MK}(p,p') = \left| \frac{1}{2N}\sum_{x \sim p} \ln p(x) - \frac{1}{2N}\sum_{x \sim p} \ln p'(x) + \frac{1}{2N}\sum_{x \sim p'} \ln p(x) - \frac{1}{2N}\sum_{x \sim p'} \ln p'(x) \right|$$

where *N* is the number of data samples generated from the *p(x)* and *p'(x)*. Note that the above formula can be very computationally demanding, since it consists of sums over *4xN* elements – *N* must be large if we want to get an accurate result. Also, when the dimensionality of the *x* vectors is high, things get worse, since *N* must be even larger.

3 The PDF Distance Metric

We can take advantage of the fact that the PDFs we need to compare are Gaussian Mixtures, not *any* distributions. A GMM can be described only by the mean and co-variance of its Gaussian kernels, plus the mixing weights. This suggests that we might construct a distance metric using the values μ, Σ, π for each one of the two distributions compared, thus creating a fast to compute metric.

The metric we considered in its general form is the following [5]:

$$C2(p,p') = -\log\left[\frac{2\int p(x)p'(x)dx}{\int p^2(x) + p'^2(x)dx}\right] \tag{3}$$

This metric is zero when $p(x)$ and $p'(x)$ are equal and is symmetric and positive. In the case where the PDFs compared are GM, eq. (3) yields

$$C2(p,p') = -\log\left[\frac{2\sum_{i,j}\pi_i\pi_j'\sqrt{\frac{|V_{ij}|}{e^{k_{ij}}|\Sigma_i||\Sigma_j'|}}}{\sum_{i,j}\left\{\pi_i\pi_j\sqrt{\frac{|V_{ij}|}{e^{k_{ij}}|\Sigma_i||\Sigma_j|}}\right\} + \sum_{i,j}\left\{\pi'_i\pi'_j\sqrt{\frac{|V_{ij}|}{e^{k_{ij}}|\Sigma'_i||\Sigma'_j|}}\right\}}\right] \tag{4}$$

where

$$V_{ij} = \left(\Sigma_i^{-1} + \Sigma'^{-1}_j\right)^{-1},$$

$$k_{ij} = \mu_i^T\Sigma_i^{-1}(\mu_i - \mu'_j) + \mu'^T_j\Sigma'^{-1}_j(\mu'_j - \mu_i),$$

π, π' the mixing weights, i and j are indexes on the gaussian kernels, and, finally, μ, Σ and μ', Σ' are mean and covariance matrices for the kernels of the Gaussian mixtures $p(x)$ and $p'(x)$ respectively.

4 Numerical Experiments

To test the effectiveness of the above distance metric we consider an image database consisting of pictures that can be classified in 5 categories, according to their theme. These are: 1) Pictures of cherry trees ("Cherries"), 2) Pictures of bushes and trees in general ("Arborgreens"), 3) Pictures of a seaside village in a rainy day ("Cannon-beach"), 4) Pictures in a university campus (outdoor) in Fall ("Campus in Fall") and 5) Shots of a rugby game ("Football"). Forty 700x500 images per class were considered.

We have generated a Gaussian mixture model for each of the images, using color (RGB space) as the feature vector. The number of the Gaussian components for every GMM was empirically chosen to be five and the Gaussian mixture models were trained using the EM algorithm. In the case of an actual image retrieval query, we would need to compare the GMM of the sample image with every other model in the database. Instead, in this experiment we compare the models of every image with one another, once for each of three distance metrics, which are 1) Symmetric Kullback-Liebler (with 4096 samples per image), 2) a Bhattacharyya based distance for GMMs and 3) the proposed C2 distance. The times required to compute all distances among the five sets are 154,39 sec., 674,28 sec. and 33161,62 sec for Bhattacharyya-based

Table 1. Average distance among classes for three distance metrics

	Cherr	**Arbor**	**Football**	**Cann**	**Campus**
Cherries	1	1,12	1,12	1,43	1,67
Arbor	2,84	1	1,87	2,57	2,94
Football	4,96	3,26	1	6,98	3,87
Cann	1,88	1,32	2,07	1	2,35
Campus	2,94	2,03	1,54	3,15	1

(a) Average SKL distances

	Cherr	**Arbor**	**Football**	**Cann**	**Campus**
Cherr	1	1,55	1,08	1,28	1,91
Arbor	1,89	1,05	1	2,78	1,92
Football	1,66	1,25	1	2,34	1,76
Cann	1	1,77	1,19	1,02	1,87
Campus	1,65	1,36	1	2,08	1,36

(b) Average Bhattacharyya-based distances

	Cherr	**Arbor**	**Football**	**Cann**	**Campus**
Cherr	1	1,64	1,69	1,45	1,5
Arbor	1,75	1	1,91	2,15	1,42
Football	2,28	2,43	1	2,67	1,82
Cann	1,12	1,56	1,52	1	1,5
Campus	1,57	1,39	1,4	2,02	1

(c) Average C2 based distances

distance, C2, and Symmetric Kullback-Liebler metrics respectively. The computations were performed in Matlab on a Pentium 2.4 GHz PC.

Note that the Bhattacharyya-based distance that was used is

$$BhGMM(p, p') = \sum_{i=1}^{N} \sum_{j=1}^{M} \pi_i \pi'_j B(p_i, p'_j),$$

where p, p' are Gaussian mixture models consisting of N and M kernels respectively, p_i, p'_j denote the kernel parameters and π_i, π'_j are the mixing weights. B denotes the Bhattacharyya distance between two Gaussian kernels, defined as [4]:

$$B(p, p') = \frac{1}{8}(\mu - \mu')^{T} \left(\frac{\Sigma + \Sigma'}{2} \right)^{-1} (\mu - \mu') + \frac{1}{2} \ln \left[\frac{\left| \frac{\Sigma + \Sigma'}{2} \right|}{\sqrt{|\Sigma||\Sigma'|}} \right]$$

where μ, Σ and μ', Σ' stand for the means and covariance matrices of Gaussian kernels p, p' respectively.

In Table 1 we provide for each metric the resulting distances among image classes normalized so that the minimum distance value over each line is 1. These distances are the means over each of the image categories. For example, by distance of group 'Cherry' to group 'Campus in fall', we mean the average distance of every image in 'Cherry' to every image in 'Campus in fall'. An issue to check out in this Table is the distance of an image group with itself (i.e the diagonal elements); if it is comparatively small, then the metric works well. In other words, the more ones in the diagonal the better the metric is. Notice that while C2 is about four times slower than the Bhattacharyya-based distance, it provides better results.

Table 2. Average between-class distances between original and sub-sampled images

	Cherr	**Arbor**	**Foot**	**Cann**	**Camp**
S-Cher	3,6e16	2,8e19	1	1,21	5,7e19
S-Arbo	2,14	1,21	1	2,87	2,14
S-Foot	1,86	1,38	1	2,45	1,94
S-Cann	1	2,12	1,15	1	2,86
S-Camp	1,8	1,56	1	2,12	1,7

(a) Average Bhattacharyya-based distances

	Cherr	**Arbor**	**Foot**	**Cann**	**Camp**
S-Cher	1	1,56	1,66	1,39	1,52
S-Arbo	1,5	1	1,62	1,68	1,27
S-Foot	2,17	2,41	1	2,22	1,82
S-Cann	1,23	1,74	1,67	1	1,69
S-Camp	1,47	1,39	1,31	1,67	1

(b) Average C2 distances

The Symmetric Kullback – Liebler (SKL) distance provides good results, however it is very slow to compute even when only 4096 (about 1/85 of the total) pixels per image are used.

To test the robustness of the metrics, we have conducted a second set of experiments. That is, we produced a sub-sampled copy of each of the original images, which has only half the width and height of the original. Then, based on the RGB values of the sub-images the GM models have been computed. Then, the distances of the GM models of the sub-sampled images were compared to those of the full images.

We have conducted the above test for the Bhattacharyya and C2 metrics, computing average distances as in the 'non-subsampled' scenario. This time, we compare each original image category with each sub-sampled image category. The distances computed are shown in Table 2. (Note that the S- prefix is used for the sub-sampled images).

5 Conclusions – Future Work

We have experimented with a new distance metric for PDFs that seems to work well for image retrieval when the images are modeled using GMMs. The metric is fast to compute, since it has a closed form when a GM model is used for the PDF, it also provides as good separation between different classes of images, similar to that produced by symmetric KL divergence which was computed using Monte-Carlo. Furthermore, in an initial test it also seems to be robust. We also compared this metric with a Bhattacharyya-based metric which, although it is fast to compute, it does not provide as good results in terms of class separation. In the future we plan test this metric with more features (edge, texture) and with a larger image database. Also we plan to test the accuracy of the SKL metric as the number of samples used in the Monte-Carlo approximation is reduced.

References

1. G. McLachlan, D. Peel: "*Finite Mixture Models*", Wiley 2000
2. J.Han, K.Ma: "Fuzzy Color Histogram and its use in Color Image Retrieval", *IEEE Trans. on Image Processing*, vol. 11, No. 8, August 2003.
3. A. Del Bimbo: "*Visual information Retrieval*", Morgan Kaufmann publishers, San Francisco, 1999.
4. K. Fukunaga: "Introduction to Statistical Pattern Recognition", Academic Press 1990.
5. Surajit Ray, "Distance-Based Model Selection with Application to the Analysis of Gene Expression Data", Ms Thesis, Dept. of Statistics, Pennsylvania State University, 2003.

Content-Based Retrieval of Web Pages and Other Hierarchical Objects with Self-organizing Maps

Mats Sjöberg and Jorma Laaksonen

Laboratory of Computer and Information Science*,
Helsinki University of Technology,
P.O.Box 5400, 02015 HUT, Finland
{mats.sjoberg, jorma.laaksonen}@hut.fi
http://www.cis.hut.fi/picsom/

Abstract. We propose a content-based information retrieval (CBIR) method that models known relationships between multimedia objects as a hierarchical tree-structure incorporating additional implicit semantic information. The objects are indexed based on their contents by mapping automatically extracted low-level features to a set of Self-Organized Maps (SOMs). The retrieval result is formed by estimating the relevance of each object by using the SOMs and relevance sharing in the hierarchical object structure. We demonstrate the usefulness of this approach with a small-scale experiment by using our PicSOM CBIR system.

1 Introduction

Large multi-modal databases, with objects of many different domains and formats, are becoming more common. Multimedia databases containing texts, images, videos and sounds require sophisticated search algorithms. Such algorithms should take into account all available semantic information including the actual contents of the database objects as well as their relationships to other objects.

We propose a content-based information retrieval (CBIR) method that models known relationships between objects in a hierarchical parent-child tree structure. We have used the CBIR system PicSOM [1] as a framework for our research and extended it to incorporate hierarchical object relationships. By mapping low-level features of the database objects, such as colour or word-frequency, to a set of Self-Organized Maps (SOMs) [2] we can index the objects based on their contents. The known relationships of objects, such as being a part of another object (eg. image attachment of an e-mail) or appearing near each other (eg. two images in the same web page) are modelled as object trees.

Section 2 presents the hierarchical object concept in more detail, Section 3 reviews the PicSOM CBIR system. Sections 4 and 5 discuss an experiment using a small set of web pages. Finally, conclusions are drawn in Section 6.

* Supported by the Academy of Finland in the projects *Neural methods in information retrieval based on automatic content analysis and relevance feedback* and *New information processing principles*, a part of the Finnish Centre of Excellence Programme.

W. Duch et al. (Eds.): ICANN 2005, LNCS 3697, pp. 841–846, 2005.

2 Multi-part Hierarchical Objects

A recent review of image retrieval from the World Wide Web [3] shows that most systems use only information of the images themselves, or in combination with textual data from the enclosing web pages. ImageRover [4] for example combines the visual and textual features into one unified vector. WebMars [5] is the only system in the review that allows multimodal browsing and it uses a hierarchical object model. Other systems include AMORE [6] that uses multiple media types in a single retrieval framework, and the Informedia project [7] that seeks to provide full content search and browsing of video clips by integrating speech, closed captioning and image recognition.

In this work, a *multi-part object* is a hierarchical object structure, organised in a tree-like manner modelling relationships between the objects. A hierarchical object tree can consist of objects of many different types and can, in principle, be of any depth. The trees are usually formed from natural relationships, like the child objects being parts of the parent object in their original context. For example an e-mail message as a parent object can consist of attachments as child objects. Likewise, an image can be parent of its segments.

As an example of the forming of a multi-part object, a web page with link information and embedded images is shown on the left in Fig. 1. The different parts have been enumerated and marked with a red rectangle. On the right we see the multi-part object tree structure created from this web page. The URL of the web page itself and links to other pages, images and other objects, are collected into one common "links" object, while the images and textual content of the web page are stored as objects by themselves.

The relevance of each object in a multi-part tree can be considered to be a property of not only the object itself, but to some extent also of its parents, children and siblings in the tree structure. We call this idea *relevance sharing*, which means that the relevance assessments originally received from user feedback will be transfered from the object to its parents, children and siblings. For example, if an e-mail message is considered relevant in a certain query, its attachments will also get increased relevance values. As a result of the *relevance propagation* performed by the PicSOM system, e-mail messages with similar attachments will then later get a share of that relevance.

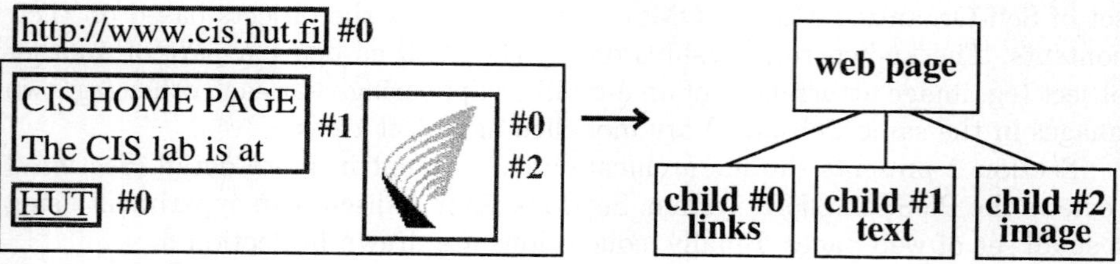

Fig. 1. A web page (left) with its corresponding multi-part object tree (right)

3 PicSOM CBIR System

The content-based information retrieval system PicSOM [1] has been used as a framework for the research described in this paper. PicSOM uses several Self-Organizing Maps (SOMs) [2] in parallel to index and determine the similarity and relevance of database objects for retrieval. These parallel SOMs have been trained with different data sets acquired by using different feature extraction algorithms on the objects in the database. This results in each SOM arranging the objects differently, according to the corresponding feature.

3.1 Relevance Feedback

Query by example (QBE) is the main operating principle in PicSOM, meaning that the user is presented with a set of objects of the desired target type, from which he selects the relevant ones. This *relevance feedback* information [8] is returned to the PicSOM system which expands it from parent objects to children, and from children to parents, and possibly also to siblings, depending on the types of the objects. This relevance sharing stage was added to the baseline PicSOM system for this work to gain an advantage from the dependencies between the objects.

For each object type, all relevant-marked objects in the database of that type get a positive weight inversely proportional to the total number of relevant objects of the given type. Similarly the non-relevant objects get a negative weight inversely proportional to their total number. The grand total of all weights is thus always zero for a specific type of objects. On each SOM, these values are summed into the best-matching units (BMUs) of the objects, which results in sparse value fields on the map surfaces.

After that the value fields on the maps are low-pass filtered or “blurred” to spread the relevance information between neighbouring units. This produces to each map unit a *qualification value*, which is given to all objects that are mapped to that unit (i.e. have it as the BMU). Map areas with a mixed distribution of positive and negative values will even out in the blurring, and get a low average qualification value. Conversely in an area with a high density of mostly positive values, the units will reinforce each other and spread the positive values to their neighbours. This automatically weights the maps according to relevance and coherence with the user’s opinion.

The next processing stage is to combine the qualification values gained from each map to the corresponding objects. These values are again summed between parents and children of the object trees. The final stage is to select a specific number of objects of the desired target type with the highest qualification values. These will be shown to the user in the next query round.

4 Data and Features

We collected a set of web pages from the intranet of our institution. This resulted in a database of over 7000 web pages and almost 2900 images. In the hierarchical model each web page forms a tree with the page itself as parent and the embedded text, images and links as children as illustrated in Fig. 1.

Two ground truth classes containing images as the target type were selected manually. *Tourist* class (907 images, *a priori* 31%) was from a conference or vacation and mainly outdoor tourist-type photography with attractions like monuments and buildings. *Face* images (253, 8.6%) were such that the main target was a human head.

4.1 Visual Features

From the images we extracted MPEG-7 still image descriptors, using the MPEG-7 Experimentation Model (XM) Reference Software [9]. As colour descriptors we used *Colour Layout* (dimension: 12) and *Scalable Colour* (256), as shape descriptor *Region-based Shape* (35), and as texture descriptor *Edge Histogram* (80), all calculated from the entire image area.

4.2 Weblink Feature

In our experiments we used a *weblink* feature calculated from all the URLs related to a web page. Each distinct URL can be regarded as one dimension in a very high-dimensional binary space of all valid URLs. Our weblink feature extraction algorithm is based on the idea initially presented in [10], and uses the *Secure Hash Algorithm* (SHA-1) [11] for performing *random mapping* to combine and to reduce the dimensionality of such vectors. Random mapping replaces an orthogonal base with a new base of lower dimensionality that is almost orthogonal. SHA-1 produces a condensed and nearly unique representation of a text string or message, called a *message digest*.

In the weblink feature extraction algorithm, each of the URLs related to a web page is recursively pruned into shorter URLs: first the original URL, then the web page directory and each higher level directory, and finally the bare domain part. We calculate an SHA-1 message digest for each of these generated URLs and form a 1024-dimensional binary random projection vector for each by looking at the first 32 bits of the digest. These bits are interpreted as four 8-bit indices into separate ranges of a 1024-dimensional vector where the corresponding components are set to unity and the others to zero. These vectors are then weighted, summed and normalised to unit length. Finally, the link feature vector for the web page is given as the sum of the normalised per URL vectors. The dimensionality is fixed at 1024 which is computationally sound.

4.3 Text Feature

We extracted the text from the HTML files and calculated a character *trigram* feature. For each character trigram in the text we calculate an SHA-1 message digest and form a 1024-dimensional vector from the first 32 bits in the same manner as with the weblink feature. The final feature vector is the sum of all these vectors from each trigram of the text document, normalised by their number. The text feature can thus be regarded as a random-projected histogram of character trigrams.

5 Experiments and Results

We trained a total of six SOMs, one for each of the weblink and character trigram features and one for each four MPEG-7 features. Every feature vector was used 100 times to train the corresponding SOM of size 256×256 map units.

The experiments were run in four ways: using only the MPEG-7 image features and combining MPEG-7 with weblink, trigram or both. Each query was initialised with one image of the pre-selected ground truth class. 50 query rounds were performed with 20 returned images in each round, where the relevance of each image could be automatically determined by using the ground truth data. The experiment was repeated so that each ground truth image was used once as the initialiser and the results were then averaged over all experiments.

Fig. 2 shows the recall–precision graphs where the precision has been normalised relative to the *a priori* of the class. In all plots the precision initially increases and then begins to decline when a clear majority of the relevant images has been found. The additional non-visual features can be seen to increase the precision of the retrieval in all the three combinations. In those cases the recall level where the precision starts to decline is also substantially higher. Using the non-visual features seems to bring the final recall very close to unity already when only one third of the images has been retrieved.

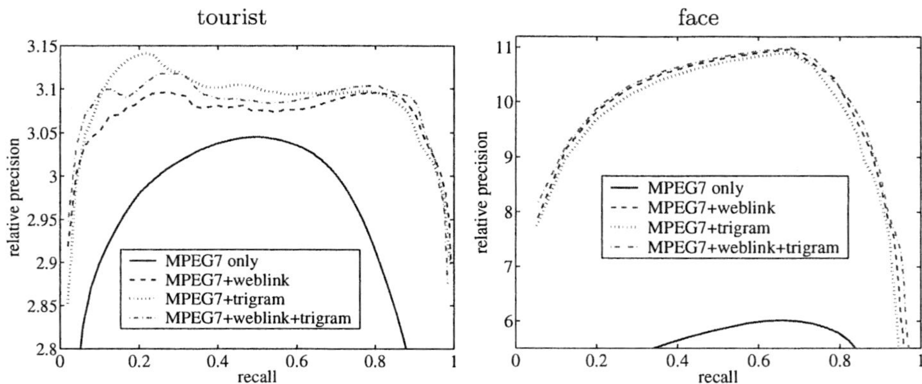

Fig. 2. Recall–relative precision graphs for classes tourist and face

6 Conclusions

In this paper, we have studied the use of hierarchical object trees to represent relationships between multimedia objects in a content-based information retrieval system. We have demonstrated that such structures can improve the performance of the system by implementing parent-child relevance sharing in the object trees. The novel idea is that while we are searching for a certain object type, for example images, other related object types in the database, like web page links, can implicitly contribute to the retrieval. This technique can be used whenever one has a database with multiple object types with hierarchical interrelations.

Our approach can be seen to complement the *semantic web* paradigm [12], where semantic information is explicitly embedded in web documents. The hierarchical object structures used in our work incorporate certain forms of semantic knowledge in an automated way, which will reduce the required manual annotation work. On the other hand, future developments in our system could utilise semantic web information as an additional feature in the hierarchical structure.

References

1. Laaksonen, J., Koskela, M., Oja, E.: PicSOM—Self-organizing image retrieval with MPEG-7 content descriptions. IEEE Transactions on Neural Networks, Special Issue on Intelligent Multimedia Processing **13** (2002) 841–853
2. Kohonen, T.: Self-Organizing Maps. Third edn. Volume 30 of Springer Series in Information Sciences. Springer-Verlag (2001)
3. Kherfi, M., Ziou, D.: Image retrieval from the world wide web: Issues, techniques and systems. ACM Computing Surveys **36** (2004) 35–67
4. M. La Cascia, S. Sethi, S.S.: Combining textual and visual cues for content-based image retrieval on the world wide web. IEEE Workshop on Content-Based Access of Image and Video Libraries (1998) 24–28
5. Ortega-Binderberger, M., Mehrotra, S., Chakrabarti, K., Porkaew, K.: Webmars: A multimedia search engine. In: Proceedings of the SPIE Electronic Imaging 2000: Internet Imaging, San Jose, CA (2000)
6. Mukherjea, S., Hirata, K., Hara, Y.: Amore: A world wide web image retrieval engine. World Wide Web **2** (1999) 115–132
7. Wactlar, H.D., Kanade, T., Smith, M.A., Stevens, S.M.: Intelligent access to digital video: Informedia project. IEEE Computer **29** (1996) 46–52
8. Salton, G., McGill, M.J.: Introduction to Modern Information Retrieval. Computer Science Series. McGraw-Hill (1983)
9. MPEG: MPEG-7 visual part of the eXperimentation Model (version 9.0) (2001) ISO/IEC JTC1/SC29/WG11 N3914.
10. Laakso, S., Laaksonen, J., Koskela, M., Oja, E.: Self-organizing maps of web link information. In Allinson, N., Yin, H., Allinson, L., Slack, J., eds.: Advances in Self-Organising Maps, Lincoln, England, Springer (2001) 146–151
11. FIPS: Secure hash standard (1995) PUB 180-1, `http://www.itl.nist.gov/fipspubs/fip180-1.htm`.
12. Berners-Lee, T., Hendler, J., Lassila, O.: The semantic web. Scientific American **284** (2001) 28–37

Fusing MPEG-7 Visual Descriptors for Image Classification

Evaggelos Spyrou[1], Hervé Le Borgne[2], Theofilos Mailis[1], Eddie Cooke[2], Yannis Avrithis[1], and Noel O'Connor[2]

[1] Image, Video and Multimedia Systems Laboratory, National Technical University of Athens, 9 Iroon Polytechniou Str, 157 73 Athens, Greece
espyrou@image.ece.ntua.gr
http://www.image.ece.ntua.gr/~espyrou/

[2] Center for Digital Video Processing, Dublin City University, Collins Ave., Ireland

Abstract. This paper proposes three content-based image classification techniques based on fusing various low-level MPEG-7 visual descriptors. Fusion is necessary as descriptors would be otherwise incompatible and inappropriate to directly include e.g. in a Euclidean distance. Three approaches are described: A "merging" fusion combined with an SVM classifier, a back-propagation fusion combined with a KNN classifier and a Fuzzy-ART neurofuzzy network. In the latter case, fuzzy rules can be extracted in an effort to bridge the "semantic gap" between the low-level descriptors and the high-level semantics of an image. All networks were evaluated using content from the repository of the aceMedia project[1] and more specifically in a *beach/urban* scene classification problem.

1 Introduction

Content-based image retrieval (CBIR) consists of locating an image or a set of images from a large multimedia database. Such a task can not be performed by simply manually associating words to each image , firstly because it would be a very tedious task with the exponential increasing quantity of digital images in all sort of databases (web, personal database from digital camera, professional databases and so on) and secondly because "images are beyond words" [1], that is to say their content can not be fully described by a list of words. Thus an extraction of visual information directly from the images is required, and is usually called *low-level features extraction.*

Unfortunately, bridging the gap between the target semantic classes and the available low-level visual descriptors is an unsolved problem. Hence it is crucial to select an appropriate set of visual descriptors that capture the particular properties of a specific domain and the distinctive characteristics of each image class. For instance, local color descriptors and global color histograms are used

[1] This work was supported by the EU project aceMedia "Integrating knowledge, semantics and content for user centered intelligent media services" (FP6-001765). Hervé Le Borgne and Noël O'Connor acknowledge Enterprise Ireland for its support through the Ulysse-funded project ReSEND FR/2005/56.

W. Duch et al. (Eds.): ICANN 2005, LNCS 3697, pp. 847–852, 2005.

in indoor/outdoor classification [2] to detect e.g. vegetation (green) or sea (blue). Edge direction histograms are employed for city/landscape classification [3] since city images typically contain horizontal and vertical edges. Additionally, motion descriptors are also used for sports video shot classification [4].

Nonetheless, the second crucial problem is to combine the low-level descriptors in such a way that the results obtained with individual descriptors are improved. The combination of features is performed before or at the same time as the estimation of the distances between images (*early fusion*)or directly at the matching scores (*late fusion*) [5].

In this work, fusion of several MPEG-7 descriptors is approached using three different machine learning techniques. A SVM is used with a "merging" descriptors' fusion, a Back-Propagation neural network is trained to estimate the distance between two images based on their low-level descriptors and a KNN Classifier is applied to evaluate the results. Finally in order to extract fuzzy rules and bridge low-level features with the semantics of images, a Falcon-ART Neurofuzzy Network is used.

Section 2 gives a brief description of the scope of the MPEG-7 standard and presents the three low-level MPEG-7 descriptors used in this work. Section 3 presents the three different techniques that aim at image classification using these descriptors.Section 4 describes the procedure followed to train the machine learning systems along with the classification results. Finally conclusions are drawn in section 5.

2 Feature Extraction

In order to provide standardized descriptions of audio-visual (AV) content, MPEG-7 standard [6] specifies a set of descriptors, each defining the syntax and the semantics of an elementary visual low-level feature *e.g.*, color, shape. In this work, the problem of image classification is based on the use of three MPEG-7 visual descriptors which are extracted using the aceToolbox, developed within the aceMedia project[7][2] and is based on the architecture of the MPEG-7 eXperimentation Model [8]. A brief overview of each descriptor is presented below, while more details can be found in [9].

Color Layout Descriptor. (CLD) is a compact and resolution-invariant MPEG-7 visual descriptor defined in the YCbCr color space and designed to capture the spatial distribution of color in an image or an arbitrary-shaped region. The feature extraction process consists of four stages.

Scalable Color Descriptor. (SCD) is a Haar-transform based encoding scheme that measures color distribution over an entire image, in the HSV color space, quantized uniformly to 256 bins. To reduce the large size of this representation, the histograms are encoded using a Haar transform.

Edge Histogram Descriptor. (EHD) captures the spatial distribution of edges. Four directions of edges ($0\,°$, $45\,°$, $90\,°$, $135\,°$) are detected in addition

[2] http://www.acemedia.org

to non-directional ones. The input image is divided in 16 non-overlapping blocks and a block-based extraction scheme is applied to extract the five types of edges and calculate their relative populations.

3 Image Classification Based on MPEG-7 Visual Descriptors

Several distance functions, MPEG-7 standardized or not, can be used when a single descriptor is considered. However, in order to handle all the above descriptors at the same time for tasks like similarity/distance estimation, feature vector formalization or training of classifiers, it is necessary to fuse the individual, incompatible elements of the descriptors, with different weights on each.

Three methods are considered for this purpose, combined with appropriate classification techniques. *Merging fusion* combines the three descriptors using a Support Vector Machine for the classification, *Back-propagation fusion* produces a "matrix of distances" among all images to be used with a K-Nearest Neighbor Classifier. Finally, a *Fuzzy-ART neurofuzzy network* is used not only for classification but also to extract semantic fuzzy rules.

3.1 Merging Fusion/SVM Classifier

In the first fusion strategy, all three descriptors are merged into a unique vector, thus is called *merging fusion*. If $D_{SCD}, D_{CLD}, D_{EHD}$ are the three descriptors referenced before then the merged descriptor is equal to:

$$D_{merged} = [D_{SCD}|D_{CLD}|D_{EHD}]$$

All features must have more or less the same numerical values to avoid scale effects. In our case, the MPEG-7 descriptors are already scaled to integer values of equivalent magnitude. A Support Vector Machine [10] was used to evaluate this fusion.

3.2 KNN Classification Using Back-Propagation Fusion

The second method is based on a back-propagation feed-forward neural network with a single hidden layer. Its input consists of the low-level descriptions of two images and its output is the normalized estimation of their distance. The network is trained under the assumption that the distance of two images belonging in the same class is 0, otherwise, it is 1. These distances are used as input of a K-Nearest Neighbor (KNN) classifier that assigns to an image the same label as the majority of its K nearest neighbors.

A problem that occurs is that the distance between descriptors belonging to the same image is estimated rather as a very small number than zero. However,it is a priori set to zero. Moreover, even for a well-trained network, the output would be slightly different depending on the row they are presented, thus the distance matrix would not respect the symmetry property needed by the KNN classifier. To overcome this, we used only the distances either of the upper or of the lower triangular matrix, or replacing a distance by the average of the two corresponding outputs of the neural network.

Another approach to efficiently fuse the different visual descriptors uses pre-calculated distance matrices for individual visual descriptors assigning weights on each one, to produce a weighted sum. This time, the input of the network consists of the three distances and results to a distance matrix which is used again as the input of a KNN classifier.

3.3 Classification Using a Falcon-ART Neurofuzzy Network

Image classification using a neural network or a SVM fails to provide semantic interpretation of the underlying mechanism that realizes the classification. In order to extract semantic information, a neurofuzzy network can be applied. To achieve this, we used the Falcon-ART network [11].

The training of the network is done in two phases, the "structure learning phase", where the Fuzzy-ART algorithm is used to create the structure of the network, and the "parameters learning stage", where the parameters of the network are improved according to the back-propagation algorithm.

The input of the network is a merged descriptor according to the process of section 3.1. After training, the network's response is the class that the input belongs. Hence, the way that the low-level features of the image determine the class to which it belongs becomes more obvious and can be described in natural language.

In order to have a description close to human perception for the rules of the Falcon-Art algorithm, each dimension of an image descriptor was divided into three equal parts, each one corresponding to *low*, *medium*, *high* values; each hyperbox created by the Falcon-ART then leads to a rule that uses these values. We present an example of such a rule, when classification considers only the EHD descriptor. The subimages are grouped to those describing the upper, middle and lower, parts of the image and a qualitative value (*low*, *medium* or *high*) is estimated for each type of edges. Thus, a fuzzy rule can be stated as:

> IF the number of $0°$ edges on the *upper* part of the image is *low* AND the number of $45°$ edges on the *upper* part of the image is *medium* AND ... AND the number of non-directional edges on the *lower* part of the image is *high*, THEN the image belongs to *Beach*

4 Experimental Results

The image database used for the experiments is part of the aceMedia content repository [3] and more specifically of the Personal Content Services database. It consists of 767 high quality images divided in two classes *beach* and *urban*. All the results are presented in table 1. 40 images from the *beach* category and 20

[3] http://driveacemedia.alinari.it/

Fig. 1. Representative Images - 1-3:Beach Images, 4-6: Urban Images

Table 1. Classification rate using several approaches on different MPEG-7 descriptors: edge histogram (EH), color layout (CL) and scalable color (SC)

Classification	EH	CL	SC	EH+CL	EH+SC	CL+SC	EH+CL+SC
Merging/linear SVM	79.5%	82.3%	83.6%	87.1%	88.7%	86.9%	89.0%
Back-Prop.L2 dist./KNN	-%	-%	-%	88.97%	89.25%	88.54%	93.49%
Back-Prop./KNN.	81.9%	87.13%	85.86%	67.04%	90.1%	91.37%	86.28%
Falcon-ART	81.4%	84.7%	83.67%	82.4%	83.6%	86.3%	87.7%

Table 2. Fuzzy Rules created by the Falcon-ART, trained with the EH descriptor

part of image	edge type	Rule 1	Rule 2	Rule 3	Rule 4	Rule 5
upper	0 °	M	L	M-L	M-L	L
	45 °	M	L	M-L	M	M
	90 °	M	L	M	M	M
	135 °	H	M	M	M	M
	nondir. °	M	M	M	M	M
center	0 °	M	L	M	M	M
	45 °	M	M-L	H	M	H
	90 °	M	M	M	M	H
	135 °	H	M	M-L	H	H
	nondir. °	H	M	M	H	M
lower	0 °	M	L	L	M	L
	45 °	M	M	H	H	H
	90 °	H	M	M	H	H
	135 °	M	M-L	M	H	M
	nondir. °	M	M	M-L	H	M
class		*urban*	*beach*	*urban*	*beach*	*beach*

from the *urban* were selected and used as training dataset. The remaining 707 (406 from *beach* and 301 from *urban*) images were used for evaluation.

SVM Classifier using Merging Fusion: The merged vectors were directly used as input of a SVM classifier with a polynomial kernel of degree one (*i.e* a linear kernel). Results with polynomial kernels of higher degree (up to 5) give similar results. While individual features lead to classification results from 79.5% to 83.6%, the merging of two of them improve the classification results from 86.9% to 88.7%, and reaches 89% with the merging of the three.

Back-Propagation Fusion of Merged Descriptors: The distance between two images was determined manually and was set to 0 for images belonging to the same category and to 1 otherwise. The symmetric distance matrices were used with the KNN classifier, as described in section 3. Best performance was achieved using all the descriptors and the distances between the images. In this case the success rate was 93.49%. All the results are shown on table 1.

Falcon-ART Neurofuzzy Network: The same 60 images' merged descriptions were presented randomly at the Falcon-ART neurofuzzy network. In the case of the EHD descriptor, the Falcon-ART has created 5 fuzzy rules which are presented in detail in table 2. The success rate was 95.8% on the training set and 87.7% on the test set, with the Fuzzy-ART algorithm creating 8 hyperboxes (rules) and the Falcon-ART neurofuzzy network being trained for 275 epochs.

5 Conclusion and Future Works

All methods were applied successfully to the problem of image classification using three MPEG-7 descriptors. Back-propagation fusion showed the best results followed by the merging fusion using the SVM. The Falcon-ART provided a linguistic description of the underlying classification mechanism. Future work will aim to use more MPEG-7 descriptors. Additionally, these classification strategies may be extended in matching the segments of an image with predefined object models with possible applications in image segmentation.

References

1. Smeulders, A.W.M., Worring, M., Santini, S., Gupta, A., Jain, R.: Content-based image retrieval at the end of the early years. IEEE t. PAMI **22** (2000) 1349–1380
2. Szummer, M., Picard, R.: Indoor-outdoor image classification. In: IEEE international workshop on content-based access of images and video databases,. (1998)
3. Vailaya, A., Jain, A., Zhang, H.J.: On image classification: City images vs. landscapes. Pattern Recognition **31** (1998) 1921–1936
4. D.H. Wang, Q. Tian, S.G., Sung, W.K.: News sports video shot classification with sports play field and motion features. ICIP04 (2004) 2247–2250
5. Mc Donald, K., Smeaton, A.: A comparison of score, rank and probability-based fusion methods for video shot retrieval. In: CIVR. (2005) Singapore.
6. Chang, S.F., Sikora, T., Puri, A.: Overview of the mpeg-7 standard. IEEE trans. on Circuits and Systems for Video Technology **11** (2001) 688–695
7. Kompatsiaris, I., Avrithis, Y., Hobson, P., Strinzis, M.: Integrating knowledge, semantics and content for user-centred intelligent media services: the acemedia project, Proc. of WIAMIS 04, Portugal, April 21-23, 2004. (2004)
8. MPEG-7: Visual experimentation model (xm) version 10.0. ISO/IEC/JTC1/SC29/WG11, Doc. N4062 (2001)
9. Manjunath, B., Ohm, J.R., Vasudevan, V.V., Yamada, A.: Color and texture descriptors. IEEE trans. on Circuits and Systems for Video Technology **11** (2001) 703–715
10. Vapnik, V.: The Nature of Statistical Learning Theory. NY:Springer-Verlag (1995)
11. Lin, C.T., Lee, C.S.G.: Neural-network-based fuzzy logic control and decision system. IEEE trans. Comput. **40** (1991) 1320–1336

The Method of Inflection Errors Correction in Texts Composed in Polish Language – A Concept

Tomasz Kapłon and Jacek Mazurkiewicz

Wroclaw University of Technology, Institute of Engineering Cybernetics,
ul. Janiszewskiego 11/17, 50-372 Wroclaw, Poland
{tomasz.kaplon, jacek.mazurkiewicz}@pwr.wroc.pl

Abstract. The idea of verification the inflection correctness of sentences composed in polish language is presented in this paper. The idea and its realization is based on the formal model of modified link grammar, grammatical rules of polish language and neural network as a classification tool of individual words for the sake of gender, number, person, mood and case. The crucial to determine the inflection correctness is the third of mentioned items. The proposition of the artificial neural classifier is presented. Finally, the application and expect results are discussed.

1 Introduction

The problem of automatic verification and correction grammatical errors in text files created by OCR systems or speech-to-text transcription systems is still important in natural language processing. There are many applications which try to solve this problem in practice. Unfortunately, all of them are related to texts composed in English so they use English grammar rules only. The best systems have effectiveness at level of 97.8%. It is very good result in comparison with human (99.7%). We try to propose the idea of such system focused on Polish grammar. Taking into consideration the specific for Polish language: inflection, changeable word order and fact that in living language most of people speak not so clear but nevertheless understandable (because of semantic and intentional understanding), is very possible that efficiency in this case would be much lower using the methods applied in known English-dedicated systems.

Except the process of voice or image (scanned documents, handwriting) recognition and transformation them into text, verification of inflection correctness is the next important problem to get errorless text and realize e.g. speech-to-text transformation correctly and effectively.

Using the modified link grammar (MLG) [5] the correctness of number and relations among lexical-marked groups of words in sentence is verified. In the MLG the inflection of individual words is not considered, so the sentence could be recognized as a syntactic correctly but in fact it wouldn't be true, because one or more neighboring words would be incorrect with relation to the other.

This is the reason why it is necessary to have the knowledge at least about gender, number and case of each word in sentence. Having such knowledge and verification rules – based on rules of grammar – is possible to mark and to correct inflection mistakes.

W. Duch et al. (Eds.): ICANN 2005, LNCS 3697, pp. 853 – 858, 2005.

2 Aim and Main Idea

The conception presented in this paper can be put into practice in automatic verification and correction grammatical errors in text files from OCR or speech-to-text transcription systems. Preliminary, no context, correction would be realized with the aid of dictionary. It is helpful but insufficient. Especially in polish language (and the other inflection languages), when the most grammatical errors are caused by using incorrect word endings. So, indispensable is a formal method (algorithm) of correction the inflection errors basis on the grammatical context of words in sentence.

Actual correction is realized using rules with conditions taken from neural classifier. The classifier provided information about gender, number (singular or plural), person and case of each known word. The knowledge serves as the condition in verification and correction rules.

3 Verification and Correction Process

3.1 Preparation

Each entered sentence is subjected to syntactic analysis, which is based on modified link grammar. This step is essential in order to facilitate further inflection verification and correction. During analysis the syntactic incorrect sentences are marked. The sentences marked as a syntactic incorrect are treated as the exceptions and are not further considered. As these sentences do not meet syntactic rules, an analysis by means of verification rules is difficult and requires additional semantic analysis. Moreover, it needs to follow the semantic context in previous and next sentences what is extremely hard.

Actual verification is realized by means of rules *if then*, which describes – formally – inflection relations between words marked by links and determined in syntactic analysis. The conditions of verification rules are taken from the neural classifier database. Based on effects of training, the neural classifier creates for each word (N_n) its individual set of features (*data*) (Fig. 1). For nouns, data-set consists information about gender, number, person, case and for verbs: person, aspect, tense, mood, voice.

Since as only last two letters in word are crucial to describe the difference of each inflection form of word we define the ending as at most two last letters of the word. It causes simplification of stem-ending separation algorithm and what is much useful, to code the ending are needed only ten bits – five for each letter.

3.2 Correction and Verification

The correction process (Fig.1) starts with the text consisting only sentences without syntax errors – "clear" text. Next, from each sentence S_s pairs of neighboring words (L_{l-1}, L_l) are taken. Each word in pair is divided into stem and ending (according assumption that endings consist of at most last two letters of each word). Next, the database of the neural classifier is searching for the compatible couple, stem and ending of word L_{l-1} and L_l. If exists any compatible word N_n, the set of its features (*data*) is placed into actual verification rule. Then the rule is fired. If conclusion is true, neighboring words are used in proper inflection form. In case of discrepancy between N_n and L_l ending, the sentences is marked as grammatically incorrect (inflection error) and follows the search for the correct form in classifier database. If exist, the incorrect form is replaced. The process is repeated until the sentences S_s is over. When the process successfully ends – each word in pair have proper ending – the sentence is including back to the text. Otherwise correction is impossible.

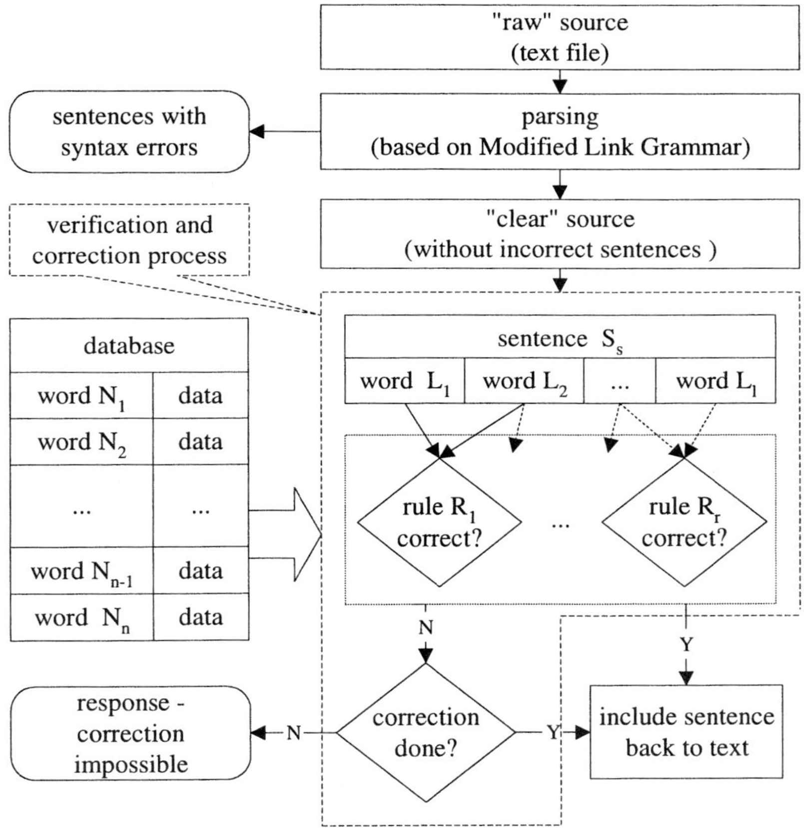

Fig. 1. Verification and correction process

4 Neural Network as Word Classifier

We propose to use artificial neural network as a word classifier. The classifier is focused on two parts of speech: nouns and verbs. The classification of nouns means that the following features of declension ought to be pointed: gender, number (singular or plural), person and case. For verbs – the following features of conjugation are the most important: person, aspect, tense, mood and voice (active or passive). The each classified word is described by the set of features, some of them are correlated, and some of them are completely independent. The rules which drive the word to the proper class during declension or conjugation process are quite complex in polish language and it is rather hard to describe them in clear algorithmic way.

This is the reason why we use artificial neural network to realize the classification – the rules are elaborated during learning process based on set of examples.

4.1 Details

The input vector for neural network is created based on word endings separated from stems. The number of different endings which are characteristic for nouns and verbs in polish is rather large, but it is possible to reduce them using the language specific word dependences. In our opinion the set of 50 input learning vectors is enough representatives to realize correct weights calculation. There is no chance to create the single neural network which is able to describe all features, which characterize a single noun or a single verb [2, 3].

The classification of word is realized using the set of neural networks. The single neural network gives an answer related to the single feature. The post processing methods combine these elementary answers to general answer about the class of input word. Such approach seems to be little bit complicated, but it gives a chance to tune the sensitivity of single noun or verb feature – neural networks are trained independently. On the other hand it is possible to implement the set of neural networks in very effective way using parallel processing methods.

The endings used as neural network inputs are binary coded. The answer of the neural network – related to the single feature - is presented using one-of-*N* code, where *N* is the range of feature – number of cases for example. This coding technique is justifiable, because the single word is described by only single case, single gender, etc.

4.2 Neural Network Organization

We propose multilayer perceptrons as a neural networks responsible for word classification. The nets are equipped by single input layer, single hidden layer and single output layer. The number of neurons in input layer equals to the number of bits used for ending coding. During the experiment we used 10 input neurons – 5 bits for the first and 5 bits for the second character of single ending. The hidden layer is composed of ten (the number matched experimentally) neurons and the number of output neurons equals to the range of tested feature. The each multilayer perceptron is trained with use of backpropagation algorithm with momentum and variable learning rate [3]. Mean square error measure (MSE) is used as the base of network energy function. Unipolar sigmoid function:

$$f(u_i) = \frac{1}{1 + exp(-\beta u_i)} \tag{1}$$

where: u_i is weighted sum of i-th perceptron input values,
β is parameter defining “steepness” of activation function *f*,

is chosen for the nonlinear activation element, so all neuron output values fall in range (0, 1). The post processing method converts the output into binary values using experimentally chosen thresholds [4]. This way we propose the most standard neuron classifier with many different and efficient learning algorithms. We can use one of them instead of backpropagation if the results of training are not sufficient.

5 Experiments and Initial Results

We carried out some experiments using NeuroSolutions system. Both neural nets were trained with use of backpropagation algorithm with constant learning rate 0.25. The results of two experiments are presented below.

The first one concerned recognition of noun case. The neural net has in input layer 10 neurons (each ending included two letter and is coded by ten-bits sequences – five for each letter – where, 00000|00000 – no letters, 00001|00000 – a|-, 00010|00000 – b|-, ...,). The output layer consists of 6 neurons (six bits sequence in 1 of 6 code describes each case. 000001 – nominative, 000010 – genitive, ...). Hidden layer is made of 10 neurons (matched experimentally). The training set: 100 words – 20 words, each in three genders and two numbers. The test set: 100 words – 20 different words, each in three genders and two numbers.

The level of nouns recognition correctness: 65% genitive, 66% accusative, 77% dative, 79% instrumental, 86%, locative 84% and 87% nominative.

The level of recognition correctness of nouns is not adequate to expected but the reaction of neural net is correct and is similar to human. The improvement of recognition is possible to achieve increasing number of words in training sets and the neural net modifications – e.g. in hidden layer.

The second experiment concerned recognition the person of verb. The input and hidden layer as above. The output layer consists of 8 neurons. The learning set consists of 90 verbs – 30 to each tenses. The test set included 90 different verbs.

The results of recognition where definitely better and achieved average level 92% of correctness and are convergent to our expectations. The endings of verbs in polish language reveals high regularity what was confirmed by the results of experiment and what points the proposed neural nets is well organized.

6 Conclusions

The presented method of automatic verification and correction grammatical errors in text files is possible to use in speech-to-text transcription systems, scanned documents or handwritten recognition systems. We think the conception will find its first application in text files processing. The preparation of some testing files in order to specify neural network classifier organization, number and forms of verification rules shows the limitations of conception and applied methods.

We expect effectiveness at the level of human but in this moment, due to lack of enough number of experiments, the factual progress is unknown. However, the initial results confirmed the validity of assumption, the neural classifier could be useful in effective inflection error correction in text composed in polish language.

The idea of verification the inflection correctness of sentences was introduced. The pattern of the verification and correction process was presented. The organization of neural classifier, expecting results and application of system was discussed.

References

1. Bischop, Ch.: Neural Networks for Pattern Recognition. Clarendon Press Oxford (1996)
2. Hecht-Nielsen, R.: Neurocomputing, Addison-Wesley, (1990)
3. Hertz, J., Korgh, A., Palmer R.: Introduction to the Theory of Neural Computation, Addison-Wesley, (1991)
4. Kung, S. Y.: Digital neural networks, PTR Prentice Hall, (1993)
5. Mierzwa, J.: The Formal Model of Decomposition Natural Language Sentences into Group of Words in order to Computer Inference, Raport PRE, No 39, Wroclaw University of Technology Press, (PhD Project), (2001). (in polish)

Coexistence of Fuzzy and Crisp Concepts in Document Maps

Mieczysław A. Kłopotek, Sławomir T. Wierzchoń,
Krzysztof Ciesielski, Michał Dramiński, and Dariusz Czerski

Institute of Computer Science, Polish Academy of Sciences,
ul. Ordona 21, 01-237 Warszawa, Poland
{kciesiel, dcz, mdramins, klopotek, stw}@ipipan.waw.pl

Abstract. SOM document-map based search engines require initial document clustering in order to present results in a meaningful way. This paper[1] reports on our ongoing research in applications of Bayesian Networks for document map creation at various stages of document processing. Modifications are proposed to original algorithms based on our experience of superiority of crisp edge point between classes/groups of documents.

1 Introduction

Nowadays, human beings searching for information are overwhelmed by information rain. Even small businesses collect huge amount of written documents both on their internal activities, advertisement materials and market situation. Search engines retrieve ever growing number of documents in response to typical requests. Publicly available libraries contain papers on any subject. Multimedia encyclopedias run a race in providing a potential customer with more and more information. Present legal documents - treaties, constitutions, law books - are too elaborate for an ordinary man in the street to read. So there is an urgent need for such a presentation of a document collection to its potential user that one can grasp it as comprehensively as possible.

Within a broad stream of various novel approaches, one can encounter the well known WebSOM project, producing two-dimensional maps of documents [7,8]. A pixel on such a map represents a cluster of documents. The document clusters are arranged on a 2-dimensional map in such a way that the clusters closer on the map contain documents more similar in content.

At the heart of the process is of course the issue of document clustering. In fact, to be efficient and reproducible, the clustering process must be multi-stage one: clustering for identification of major topics, initial document grouping, WebSOM-like clustering on document groups, fuzzy cell clusters extraction and labeling [2,3,4].

[1] Research partially supported under KBN research grant 4 T11C 026 25 "Maps and intelligent navigation in WWW using Bayesian networks and artificial immune systems".

W. Duch et al. (Eds.): ICANN 2005, LNCS 3697, pp. 859–864, 2005.

These steps are at the heart of a BEATCA system created in our research group[2] for intelligent navigation in document maps.

In a separate paper [3], we described a document clustering method via fuzzy-set approach and immune-system-like clustering.

This paper focuses on another research path, based on Bayesian networks. In the subsequent sections we introduce the concept of Bayesian networks and then step by step explain, how Bayesian networks may be (and in our system are) applied in document clustering. Bayesian networks have been applied to document processing in the past. Modifications are proposed to original algorithms based on our experience of superiority of crisp edge point between classes/groups of documents.

2 Bayesian Networks

Bayesian networks (BN) encode efficiently properties of probability distributions. Their usage is spread among many disciplines. A Bayesian network is an acyclic directed graph (dag) nodes of which are labeled with variables and conditional probability tables of the node variable given its parents in the graph. The joint probability distribution is then expressed by the formula:

$$P(x_1, ., x_n) = \prod_{i=1n} P(x_i|\pi(x_i)) \qquad (1)$$

where $\pi(X_i)$ is the set of parents of the variable (node) X_i. On the one hand, BNs allow for efficient reasoning, and on the other many algorithms for learning BNs from empirical data have been developed.

A well-known problem with Bayesian networks is the practical limitation for the number of variables for which a Bayesian network can be learned in reasonable time [6]. Reasonable execution times are known for a few classes of BN only, including "naive Bayes" classifier (a BN with decision node connected to all other variables, which are not connected themselves), the Chow-Liu-tree structured BN, the TAN classifier (Tree Augmented Naive Bayes Network, a combination of naive Bayes classifier with Chow/Liu tree-like Bayesian network).

In our approach we substitute Chow/Liu algorithm for learning tree-like BN with the ETC algorithm [6] learning BNs in tens of thousands of variables.

3 Bayesian Networks in Document Processing

Bayesian networks are exploited in BEATCA system for: rough clustering of documents - topic extraction, expansion, keyword extraction, relevance feedback and document classification.

Subsequently we briefly point at our experience with using crisp and fuzzy aspects of reasoning with Bayesian networks.

---

[2] http://www.ipipan.waw.pl/ klopotek/mak/current_research/KBN2003/KBN2003Translation.htm

3.1 PLSA Technique for Document Clustering

SOM document-map based search engines require initial document clustering in order to present results in a meaningful way. Latent semantic indexing based methods appear to be promising for this purpose. We have investigated empirically one of them, the PLSA [5].

The so-called Probabilistic Latent Semantic Analysis (PLSA) assumes a model of document generation based on two assumptions: the first is that a document is a bag of words, the second is that the words are inserted into a document based on a probability distribution which depends on the topic to which the document belongs. This results in the concept of (fuzzy) "degrees of membership" of a document to a concept. The document clustering obtained is just a non-disjoint one.

3.2 Our Modifications to PLSA Technique

In the BEATCA system, we wanted to use the PLSA approach "as is" for creation of the initial broad topic identification. We applied it for grouping into 3 or 4 clusters that would initiate some "fix points" of the document map. However, the technique proved to be unable to distinguish linearly separable (carefully chosen) sets of documents. The reason for the problem is probably the too high number of adjustable variables (degrees of freedom). Beside this we complained about long computation times and instability of the derived clusters.

We feel that the inability to discriminate linearly separable (in terms of words) sets of documents is an important drawback of the technique, because human beings appear to think using categories such as coincidence of terms. Therefore we have investigated an alternative approach to PLSA, substituting the "fuzzy" boundaries between clusters with crisp ones. One can say that in our experiment "Naive Bayes" was at work. The linear separability was achieved for the artificial sets of documents, and experiments with the original Syskill and Webert data [9] generated meaningful clusters.

So our next step was to extend this "Naive Bayes" version of PLSA by substituting the Naive Bayesian network structure with a TAN structure. The TAN here is learnt with the ETC algorithm from [6], as the typical Chiow/Liu algorithm based approach fails due to the dictionary size (and hence the number of attributes). The approach has been implemented and is now subject of intense testing.

3.3 PHITS Techniques

Not only words, but also links may be a useful source of clustering information when topics are concerned. PHITS algorithm [5] does the same with document link information as the PLSA with document textual content. From mathematical point of view, PHITS is identical to PLSA, with one distinction: instead of modeling the citations contained within a document (corresponding to PLSA's modeling of terms in a document), PHITs models "in-links", the citations to a document. It substitutes a citation-source probability estimate for PLSA's term probability estimate. On the Web and in other document collections, usually both links and terms could or should be used for document clustering. The mathematical similarity of PLSA and PHITS creates a joint clustering algorithm [5].

3.4 Our Considerations of PHITS

The link-based clustering is applicable and valuable only in case of a really large body of documents. Hence we consider a mixed approach to the problem of link exploration. We extend the contents of a document with the contents of documents it is pointed to (anchor related information). In this way we may overcome the loss of information resulting from poor knowledge of related pages by those that create the links. Most important here is the possibility of straight application of BNs to combined link and text analysis.

3.5 Query Expansion

The query may not fully express users information needs. A document may not contain all the words that should or might be contained (may be reasonable for the document content). Hence it is worth effort to extend users query by terms that are likely to correspond to user queries BN approach: attach terms from the "neighborhood" in a BN of presence/absence of terms in the documents (approach of e.g. Acid et al system [1]). Theoretically, it is possible to compute the probability of each term given the user query terms. So one may consider the user query as a vector of terms weighed via these conditional probabilities and seek documents similar to this vector. However, experiments show that more human-acceptable results are obtained if we take only the most probable terms and ignore the other ones.

It seems that unsharp relations among terms maintained by a Bayesian network prove to be helpful to identify documents only slightly missing the query formulation but still relevant to the user.

3.6 Relevance Feedback

A user, after obtaining the results of a query, may precise the query by pointing at documents that were relevant to his query and the ones that were not. For relevant and non-relevant documents separate Bayesian networks over the union of all terms occurring in both document sets are constructed. Further documents are estimated for relevance by computation of $P(document|relevant) - P(document|irrelevant)$. The same is done for irrelevant document BN to calculate a substitute for $P(document|irrelevant)$. Weighting is applied if $P(relevant)$ and $P(irrelevant)$ may be significantly different. This is a standard technique.

Note that a sharp boundary is drawn by the user between relevant and irrelevant documents. Apparently, there is something significant for humans drawing sharp boundaries. But at the same time the system itself works with a kind of "grades of membership", when trying to accommodate to user's way of thinking.

3.7 Document Classification

Document classification is needed at various stages of a search engine. First of all, as a complement for the clustering task. Usually, only a subset of documents is used to identify clusters in the data. The remaining documents need to be classified into emerging clusters.

In our crawler, that should focus on some document categories, a classifier is needed to predict which path for collecting documents seems to be the most promising one at the given state of search.

Note that the class assignment may be unsharp *to some extent*; the documents will not be classified into this or that particular class, but will be considered to be more or less likely belonging to some classes. So we get a graded membership, which may be used for ranking the documents, which is in fact the way we try to use BN for crawler navigation.

4 Conclusions

The case studies associated with the development of our search engine show an interesting mixture of situations where the "fuzziness" of concepts is used and situations where sharp boundaries have to be drawn. Bayesian networks, with their probabilistic nature, represent unsharp measures of membership in various contexts. The membership degrees are used for ranking of documents at some stage, while subsequently they are thresholded to induce sharp memberships, in order to provide convincing, human-interpretable results. Therefore, crisp and fuzzy concepts have to coexist and be interchangeably used depending on the stage of processing. When such switching should take place is not only a philosophical question, but also a practical one.

References

1. de Campos, L.M., Fernndez, J.M., & Huete, J.F.: Query expansion in information retrieval systems using a Bayesian network-based thesaurus. In Proc.14th Conference on Uncertainty inAI, Madison, July 1998 (pp. 53-60).
2. K. Ciesielski, M. Dramiński, M. Kłopotek, M. Kujawiak, S. Wierzchoń: On some clustering algorithms. To appear in Proc. Intelligent Information Processing and Web Mining, Gdansk 2005.
3. M. Kłopotek, M. Dramiński, K. Ciesielski, M. Kujawiak, S.T. Wierzchoń: Mining document maps. Proc. Statistical Approaches to Web Miningof PKDD'04, M. Gori, M. Celi, M. Nanni eds., Pisa, Italy, September 20-24, pp.87-98
4. K. Ciesielski, M. Dramiński, M. Kłopotek, M. Kujawiak, S. Wierzchoń: Mapping document collections in non-standard geometries. B.De Beats, R. De Caluwe, G. de Tre, J. Fodor, J. Kacprzyk, S. Zadrony (eds): Current Issues in Data and Knowledge Engineering. EXIT Publ., Warszawa 2004.. pp.122-132.
5. D.Cohn, T.Hofmann, *The missing link - a probabilistic model of document content and hypertext connectivity*, in T.K.Leen et. al (eds), Advances in Neural Information Processing Systems, Vol. 10, 2001

6. M.A.Kłopotek: A New Bayesian Tree Learning Method with Reduced Time and Space Complexity. Fundamenta Informaticae, 49(no 4)2002, IOS Press, pp. 349-367
7. T. Kohonen, *Self-Organizing Maps*, Springer Series in Information Sciences, vol. 30, Springer, Berlin, Heidelberg, New York, 2001
8. K.Lagus, *Text Mining with WebSOM*, PhD Thesis, HUT, Helsinki, 2000
9. Syskill & Webert http://kdd.ics.uci.edu/databases/SyskillWebert/SyskillWebert.html

Information Retrieval Based on a Neural-Network System with Multi-stable Neurons

Yukihiro Tsuboshita and Hiroshi Okamoto

Corporate Research Laboratory, Fuji Xerox Co., Ltd.
430 Sakai, Nakai-machi, Ashigarakami-gun, Kanagawa 259-0157, Japan
{Yukihiro.Tsuboshita, hiroshi.okamoto}@fujixerox.co.jp

Abstract. Neurophysiological findings of graded persistent activity suggest that memory retrieval in the brain is described by dynamical systems with continuous attractors. It has recently been shown that robust graded persistent activity is generated in single cells. Multiple levels of stable activity at a single cell can be replicated by a model neuron with multiple hysteretic compartments. Here we propose a framework to simply calculate the dynamical behavior of a network of multi-stable neurons. We applied this framework to spreading activation for document retrieval. Our method shows higher performance of retrieval than other spreading activation methods. The present study thus presents novel and useful information-processing algorithm inferred from neuroscience.

1 Introduction

It is generally accepted that memory retrieval in the brain is organized by persistent activation of an ensemble of neurons. In computational neuroscience, such activation is considered to emerge as a discrete attractor of a dynamical system describing a neural network in which multiple distributed patterns are embedded by recurrent connections [1] [2] (Fig. 1a).

Recent neurophysiological findings of graded persistent activity, however, dispose us to reconsider this traditional view. The firing rate of neurons recorded from the prefrontal cortex of the monkey performing vibrotactile discrimination tasks varied, during the delay period between the base and comparison stimuli, as a monotonic function of the base stimulus frequency [3]. The firing rate of neurons in the oculomotor system of the goldfish during fixations was associated with the history of spontaneous saccadic steps [4]. These phenomena cannot simply be described by dynamical systems with discrete attractors. They are more likely to be described by dynamical systems with attractors that continuously depend on the initial state (Fig. 1b).

Our previous study revealed unique functional properties of a neural-network system with continuous attractors, context-dependent retrieval of information. These were illustrated by the use of a typical document-processing task, keyword extraction from documents [5]. We considered a network of terms (words) appearing in a set of documents; the link between two terms was defined by their co-occurrence. It was shown that a continuous attractor led by activation propagation in this network gave appropriate keywords for a document (i.e. context) represented by the initial state of the network-activation pattern. To endow the network system with continuous

W. Duch et al. (Eds.): ICANN 2005, LNCS 3697, pp. 865 – 872, 2005.

attractors, we assumed that each neuron had two stable states; in one state ('on' state) a neuron is active and in the other state ('off' state) it is inactive. It was originally proposed by Koulakov et al. [6] that a network of bistable neurons could produce robust graded persistent activity.

The problem accompanying a neural-network system with bistable neurons is that the output from each neuron is necessarily dichotomous (this is also the case for recurrent networks of single-stable neurons with a sigmoidal input/output relation). This prevents the network system to deal with refined information about a term. For instance, information about a term in a given document is represented not only by its presence or absence but also by the degree of its importance, which should be represented by an analog value.

Recent remarkable findings by Egorov et al. [7] will shed a hint on this problem. They demonstrated in vitro experiment that individual neurons in the layer V of the rat entorhinal cortex (EC) responded to consecutive stimuli with graded change in the firing frequency that remained stable after each stimulus presentation. This was observed under pharmacological blockade of synaptic transmission. These observations clearly show that graded persistent activity can emerge at a single-cell level.

The purpose of this study is to demonstrate the functional advantage of a neural-network system with neurons endowed with mechanisms to generate graded persistent activity at a single-cell level. When applied to a document retrieval task, the system shows high performance compared with a standard linear spreading activation method, as well as our previous neural-network system with bistable neurons.

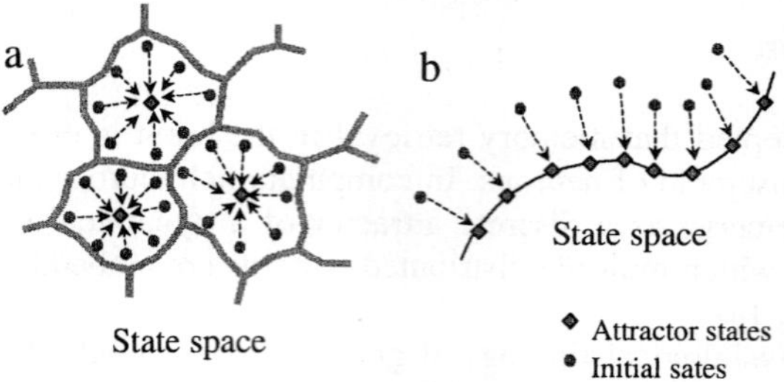

Fig. 1. a, discrete attractors. b, continuous attractor.

2 Model

To replicate robust graded persistent activity by single-cell level properties (i.e., without exploiting network level properties), model neurons with multiple bistable (hysteretic) compartments piled up in such a way as illustrated in Fig.2a were proposed[8] [9] [10]. As input I increases, transitions of the compartments from the 'off' to 'on' states occur in ascending order. As I decreases, transitions of the compartments from the 'on' to 'off' states occur in descending order. Each state of a cell appearing in such ascending or descending processes is stable because of the hysteretic characteristics of the compartments. This means that a neuron, as a whole, is multi-stable. Therefore, if the activity of a neuron is proportional to the number of compartments in the 'on' state, multiple levels of activity can become stable. Thus, robust graded persistent activity can be gained only by using single-cell level properties.

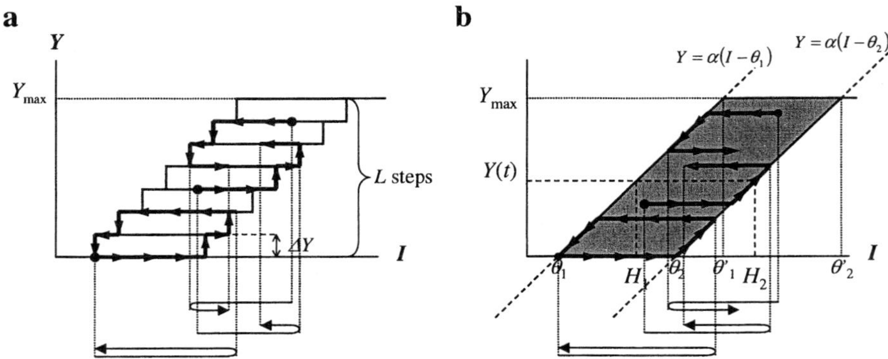

Fig. 2. Input/output relation of a multiple hysteretic neuron. a, relation for finite L. b, relation in the continuous limit: $L \to \infty$ and $\Delta Y \to 0$ while Y_{max} being fixed

Now we consider a network consisting of such multi-stable neurons. To describe a single neuron with L hysteretic compartments, we need at least L variables. Calculating the dynamics of a network consisting of N such multi-stable neurons, therefore, confronts computational overload (at least, $L \times N$ variables are required to be calculated). To solve this problem, we take the continuous limit: $L \to \infty$ and $\Delta Y \to 0$ with Y_{max} being fixed (Fig.2b). In the continuous limit, for given I and Y at time step t, Y at $t+1$ is defined by the rule:

I) If $\theta_2' < I$, $Y(t+1) = Y_{max}$.

II) If $\theta_1 < I < \theta_2'$,
- if $H_2 < I$, $Y(t+1) = \alpha(I - \theta_2)$;
- if $I < H_1$, $Y(t+1) = \alpha(I - \theta_1)$;
- else $Y(t+1) = Y(t)$.

III) If $I < \theta_1$, $Y(t+1) = 0$.

Here, θ_1, θ_2, θ_1', θ_2', α, H_1 and H_2 are given as in Fig. 2b. Thus, computational load is dramatically reduced.

Let Y_i be the output from neuron i and T_{ij} be the strength of the link from neuron j to neuron i. The input to neuron i is hence given by $I_i = \sum_{j=1}^{N} T_{ij} Y_j$. The network dynamics follows the standard asynchronous neural-network dynamics [1]:

1) Select one (say, neuron i) randomly from N neurons;
2) calculate I_i;
3) update Y_i according to the rule I) - III);
4) repeat 1) - 3) until the network state reaches equilibrium.

Here we simply verify that graded outputs from individual neurons are realized in a continuous attractor of our system. Let Φ symbolize the input/output relation defined by the rule I)-III); i.e., $Y_i = \Phi(I_i)$. Consider a state in which a cluster of neurons, say C, has persistent activity sustained by reverberating activation within these neurons. We further assume that, for neuron i ($\in C$) with fairly large I_i, Y_i is proportional to $T_i = \sum_{j\in C} T_{ij}$: i.e. $Y_i = \gamma T_i$ with γ being a constant. Hence we have $I_i = \sum_{j=1}^{N} T_{ij}Y_j = \sum_{j\in C} T_{ij}Y_j = \gamma \sum_{j\in C} T_{ij}T_j$. By mean field approximation, $\sum_{j\in C} T_{ij}T_j = \sum_{j\in C} T_{ij}m = mT_i$ with $m = \sum_{i=1}^{N} T_i / N$. Thus we obtain $Y_i = \Phi(m\gamma T_i)$. Fig. 3 intuitively shows that $Y_i = \gamma T_i$ ($i \in C$) is gained in a continuous attractor, for which multi-stability of a single neuron (indicated by the grey region) is crucial.

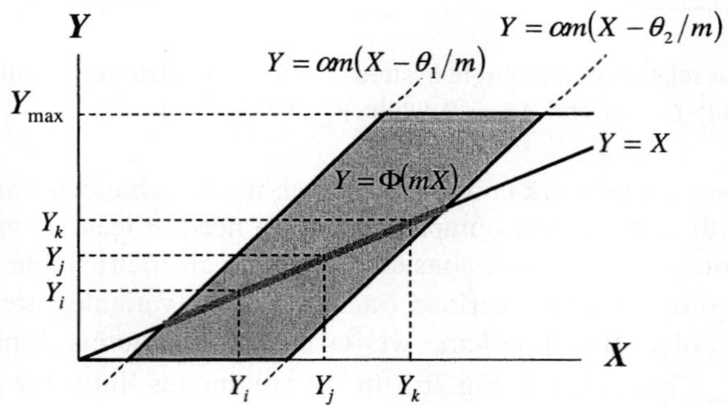

Fig. 3. Graded outputs from individual neurons, $Y_i=\gamma T_i$, $Y_j=\gamma T_j$, $Y_k=\gamma T_k$, ... , are given in a continuous attractor

3 Results and Discussions

The effect of a single-cell level of graded persistent activity upon a network-level of information processing was examined by applying the neural-network system with multi-stable neurons, as formulated in the preceding section, to spreading activation for document retrieval. Spreading activation refers to a process to improve the vector representation of a given query. We expected that graded representation of output from individual neurons operated in favor of spreading activation.

In conventional document retrieval, texts are converted to vectors in the space spanned by terms. Let $\vec{D}^{(p)} = (w_{1p}, w_{2p}, \cdots, w_{Np})$ be the vector representation of document p with w_{ip} being the relative importance of term i in document p; $\vec{Q}^{(r)} = (q_{1r}, q_{2r}, \cdots, q_{Nr})$ be the vector representation of query r with q_{ir} being the relative importance of term i in query r. The relevance of document p to query r is estimated by the degree of similarity (e.g. cosine) between $\vec{D}^{(p)}$ and $\vec{Q}^{(r)}$.

One can consider a network of terms. The weight of the link between a pair of terms represents the level of association between them, which is usually defined by the frequency of 'co-occurrence'. Each term has 'activity' that varies as spreading activation evolves in the network. For query $\vec{Q}^{(r)}$, the initial value of activity of term i is defined by q_{ir}. The modified query is given by an attractor state of spreading activation; $\vec{Q}^{*(r)} = (q^*_{1r}, q^*_{2r}, \cdots, q^*_{Nr})$. In $\vec{Q}^{*(r)}$, terms that are closely relevant to the underlying meaning of the query but not highly rated in $\vec{Q}^{(r)}$ will be associated with high values; on the other hand, terms that are associated with high values in $\vec{Q}^{(r)}$ but actually have little relevance to the query will be low rated. Thus, $\vec{Q}^{*(r)}$ will more adequately represent the underlying meaning of query r. Therefore, $\vec{Q}^{*(r)}$ is more appropriate for estimating the relevance of a given document to query r than $\vec{Q}^{(r)}$. Fig. 4 illustrates the process of document retrieval by spreading activation.

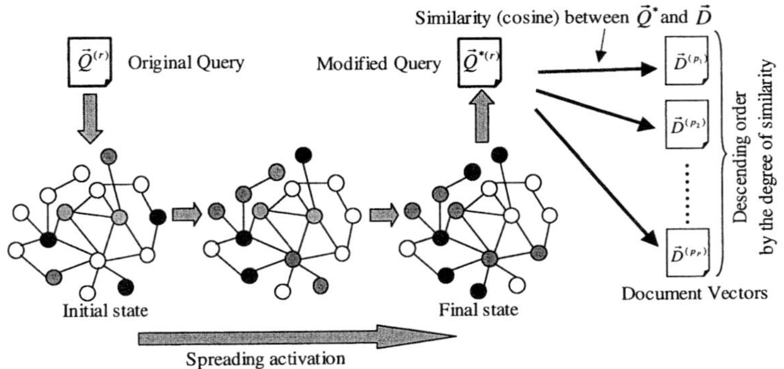

Fig. 4. Outline of document retrieval by spreading activation

We examined three kinds of spreading activation dynamics:

i) *Liner spreading activation model (LSA)*
LSA is a standard spreading activation method widely used in information retrieval, described by the linear dynamics [11] [12] [13]: $\vec{Q}^*(t) = \vec{Q} + \mathbf{M}\vec{Q}^*(t-1)$ with $\mathbf{M} = (1-\gamma)\mathbf{I} + \alpha\mathbf{T}$ where $\vec{Q}^*$ describes activities of terms at time t; $\vec{Q}$, which represents activities of terms in the original query, acts as constant inputs to terms; $\mathbf{T} = (t_{ij})$ is the matrix defined by co-occurrence of terms (the covariance-learning rule, see later).

ii) *Bistable neuron model (BSN)*
Query vectors are modified by the dynamics of a network of bistable neurons proposed in our previous study [5].

iii) *Multi-stable neuron model (MSN)*
Query vectors are modified by the dynamics of a network of multi-stable neurons whose input/output relation is described by the rule illustrated in Fig. 2.

We compared the average precisions of document retrieval given by the above three methods. The average precision (its mathematical definition is given in Fig. 5, legend) is a measure to evaluate both recall and precision by a single value. Medline1033 [14], a data set consisting of 1,033 medical paper abstracts with 30 text queries, was used as a test collection. The relative importance of term i in document p, w_{ip}, is defined by TFIDF (Term Frequency Inverse Document Frequency) [15]; $w_{ip} = \log(1+a_{ip})\log(P/b_i)$ where a_{ip} represents the frequency of term i in document p, b_i is the number of documents containing term i, and P is the number of documents. The weight of the link between terms i and j, T_{ij}, is defined by the covariance learning rule that extracts statistically significant co-occurrence of these terms: $T_{ij} = \sum_{p=1}^{P}(w_{ip}-m_i)(w_{jp}-m_j) \Big/ \sqrt{\sum_{p=1}^{P}(w_{ip}-m_i)^2}\sqrt{\sum_{p=1}^{P}(w_{jp}-m_j)^2}$ with $m_i = \sum_{p=1}^{P} w_{ip}/P$. In addition, we also examined Vector Space Model (VSM) as a baseline, in which the original query vectors ($\vec{Q}$) were used.

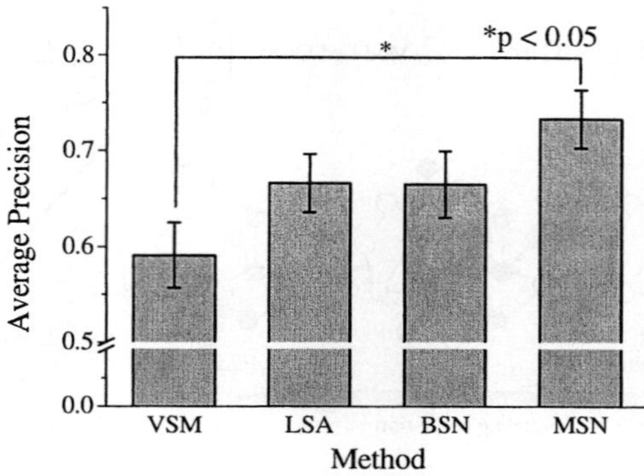

Fig. 5. Mean values of average precision for all queries. Average precision [15] is defined by $v = (1/R)\sum_{p=1}^{P} z_p\left(1+\sum_{q=1}^{p-1} z_q\right)/p$ where P is the number of output documents, R is the number of relevant documents, and z_p is 1 if document p is relevant, 0 otherwise. The bar chart indicates mean values of average precisions for 26 queries and the error bars indicate standard errors of the mean.

Fig. 5 shows the result of our examination. There were significant difference only between MSN and the baseline method, VSM ($p < 0.05$, ANOVA). MSN performed best among three spreading activation methods. Each of LSA, BSN and MSN shows higher performance than VSM, but MSN is the best. Actually, difference from VSN is statistically significant only for MSN. The superiority of MSN to BSN and LSA can be explained as follows. In BSN, output from each neuron is dichotonomous, refined information about the relative importance of terms, which should be represented by analog values, is lost. In LSA, query is expressed by constant external input ($\vec{Q}$). Therefore, if a query contains irrelevant terms, they lastingly affect search processes. On the other hand, in MSN, query is expressed by an initial state of the network activation pattern. Irrelevant terms in a query will be reduced during spreading activation.

A novel information-retrieval method by spreading activation has been inferred from possible neural mechanisms for graded persistent activity. In this method, for a given query encoded in the initial state of the activation pattern of a network of multi-stable neurons, information is retrieved as a continuous attractor attained by spreading activation in this network. The method, when applied to query extension for document retrieval, shows higher performance than other spreading activation methods such as LSA. Although comparison with query extension using other methods than spreading activation [16] still remains to be addressed, we believe that our method presents a novel and useful approach for document processing.

In cognitive psychology, it has been proposed that long-term memory is organized in the network structure [12] [13], which resembles the network of terms examined in the present study. The performance of our method demonstrated by the use of the network of terms suggests that the continuous-attractor dynamics of a network of multiple hysteretic neurons might describe essential features of real brain processes of short-term memory retrieval from long-term memory.

Acknowledgements

This work was partly supported by JSPS, KAKENHI (16500190).

References

[1] Hopfield, J. J., "Neural networks and physical systems with emergent collective computational abilities", Proc. Natl. Acad. Sci. USA 79, 2554-2558 (1982).

[2] Durstewitz, D., Seamans, J. K. & Sejnowski, T. J., "Neurocomputational model of working memory", Nature neurosci. 3 supplement, 1184-1191 (2000).

[3] Romo, R., Brody, C. D., Hernandez, A. & Lemus, L., "Neuronal correlates of parametric working memory in the prefrontal cortex", Nature 399, 470-473 (1999).

[4] Aksay, E., Gamkrelidze, G., Seung, H. S., Baker, R. & Tank, D. W., "In vivo intracellular recording and perturbation of persistent activity in a neural integrator", Nature neurosci. 4, 184-193 (2001).

[5] Tsuboshita, Y., Okamoto, H., "Extracting information in a graded manner from a neural-network system with continuous attractors", IJCNN 2004 (2004).

[6] Koulakov, A. A., Raghavachari, S., Kepecs, A. & Lisman, J. E., "Model for a robust neural integrator", Nature neurosci 5, 775-710 (2002).

[7] Egorov, A. V., Hamam, B. N., Fransen, E., Hasslmo, M. E. & Alonso, A. A., "Graded persistent activity in entorhinal cortex neurons", Nature 420, 173-178 (2002).

[8] Goldman MS, Levine JH, Major G, Tank DW, Seung HS., "Robust persistent neural activity in a model integrator with multiple hysteretic dendrites per neuron", Cereb. Cortex 13, 1185-1195 (2003).

[9] Loewentein, Y. & Sompolinsky, H., "Temporal integration by calcium dynamics in a model neuron", Nature neurosci. 6, 961-967 (2003).

[10] Teramae, J. & Fukai, T. "A cellular mechanism for graded persistent activity in a model neuron and its implications in working memory", J. Comput. Neurosci. 18, 105-121 (2005).

[11] Anderson, J., R., Pirolli, P., L., "Spread of Activation", Journal of Experimental Psychology: Learning, Memory, and Cognition, Vol. 10, No. 4, 791-798 (1984).

[12] Quillian, M. R., *Semantic memory*: in Semantic Information Processing (Minsky, M. eds, MIT press, Cambridge MA), 227-270 (1968).

[13] Collins, A. M. & Loftus, E. F., "A spreading activation theory of semantic processing", Psychological Review 82, 407-428 (1975).
[14] Available at ftp://ftp.cs.cornell.edu/pub/smart/med/
[15] Salton, G., &. McGill, M. J., Introduction to Modern Information Retrieval, McGraw-Hill (1983).
[16] Ricardo Baeza-Yates and Berthier Ribeiro-Neto. Modern Information Retrieval. ACM Press / Addison-Wesley (1999).

Neural Coding Model of Associative Ontology with Up/Down State and Morphoelectrotonic Transform

Norifumi Watanabe and Shun Ishizaki

Keio University Graduate School of Media and Governance,
5233, Endo, Fujisawa, Kanagawa, 252-8520 Japan
(norifumi, ishizaki)@sfc.keio.ac.jp

Abstract. We propose a new coding model to the associative ontology that based on result of association experiment to person. The semantic network with the semantic distance on the words is constructed on the neural network and the association relation is expressed by using the up and down states. The associative words are changing depending on the context and the words with the polysemy and the homonym solve vagueness in self organization by using the up and down states. In addition, the relation of new words is computed depending on the context by morphoelectrotonic transform theory. In view of these facts, the simulation model of dynamic cell assembly on neural network depending on the context and word sense disambiguation is constructed.

1 Introduction

The language function in the brain is supported by a complex, exquisite system. Therefore, even if the activity of the individual neuron related to the language function is clarified, the mechanism of the entire system is not understood. If only the activity of the entire system is examined from the outside the mechanism that supports it is not understood. In this research, the neuron that shows the word by the up and down state and the morphoelectrotonic transform on neurophysiology is assembled, and the meaning of word sense ambiguity is specified from the relation of the network in which it gathered.

Elman 's Simple recurrent network(SRN) model targets the words generated with simple context-free grammar including the relative clause generation. When the word was input one by one, study that forecast a approaching word next was learned [1]. However, only the grammatical relations between individual words were able to be learned, and semantic relations between words were not acquired in Elman 's model. The semantic network where semantic relations between words had been shown as a distance was implemented on the neural network in this model. The sentence with the ambiguity resolves vagueness by the specify the meaning of the word with the cell assembly.

The relations between words use the association ontology constructed based on stimulus word-association word relation by the association experiment in the Ishizaki laboratory [2]. We proposes the method for computing the relations between concepts based on the morphoelectrotonic transform theory from the firing rate between

W. Duch et al. (Eds.): ICANN 2005, LNCS 3697, pp. 873 – 879, 2005.

neuron. Dynamical cell assembly is symbolized by the mechanism of the up and down state and the morphoelectrotonic transform on the network, and the method for clarifying the word sense disambiguation is described.

2 Up and Down States of Neurons

When the activity of the cerebral cortex was observed, the experiment result that there were two states that were called "Up state" and "Down state" in the state of the neuron under the threshold of the fire was announced [3]. Up state indicates the case where there is a state of the membrane potential in the place immediately before the spike's fire. Down state shows the state of the low membrane potential called the state of geostationary. The state that changed at the average cycle of about one second was discovered between these two states (Fig.1).

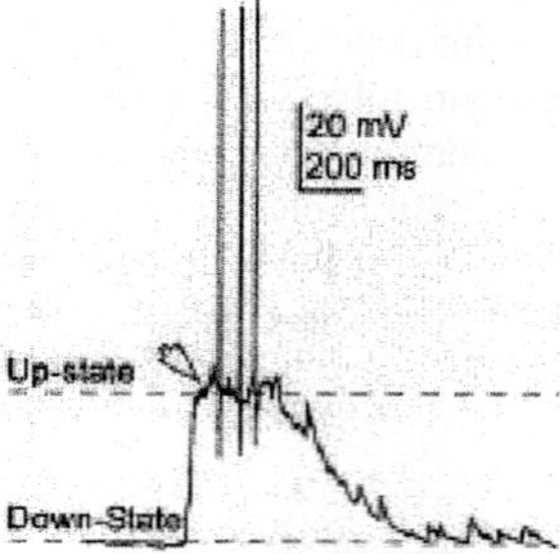

Fig. 1. Spontaneous, suprathreshold up state. The fast transition to the up state, delays in burst firing onset during the up state, and relatively slow return to the down state [3].

The model such as neurons that fire easily according to the input and neurons that do not fire easily can be constructed by using up and down state. For example, two states of this fire are built into the recurrent neural network and the model where the neuron group has working as the integrator by strength of the input is proposed [4]. As a result, the neuron that fire sequentially by the firing pattern of the input is dynamically changed, and grouping of the neuron can be changed dynamically.

3 The Morphoelectrotonic Transform

When the potential injected into a membrane exceeds a constant threshold, a morphoelectrotonic potential corresponding to the input stimulation is generated. For example, this morphoelectrotonic potential is observed in neocortical pyramidal neuron dendrites of rat olfactory cortex [5]. The potential decreases in proportion to the length of neural axon and dendrite because of the resistance of the membrane and the cytoplasm. The morphoelectrotonic transform measures the distance between two neurons using the attenuation rate of the current that flows between them [6]. A morphoelectrotonic transform is expressed by a distance which is calculated by using the amount of the morphoelectrotonic potential when it decreases to 1/e (e: common logarithm).

$$L_{ij} = \log\left(\frac{V_i}{V_j}\right) \tag{1}$$

L_{ij}: attenuation function, V_i: input voltage, V_j: measured voltage

The distance between neurons changes dynamically by using morphoelectrotonic transform theory, and it is proposed that the model into which the neuron group that fires according to the input is changed can be constructed. For example, the model is proposed as a research of the perceptual reconstruction into which the object recollected according to the input is changed by learning the distance between neurons [7]. It is possible that dynamically constructing the neural network based on the relation of the input words by using morphoelectrotonic transform.

4 Neural Cording Model

We explains the coding model by which the associative ontology is implemented on the neural network based on the theory of neurophysiology. The associative ontology is what structurizes and computerizes the association data by the association experiment (in Japanese) executed in the Ishizaki laboratory. It has the network structure with which a related concept is connected directly and the distance between concepts is requested quantitatively. It is calculated based on a large amount of data of about 42,000 difference words, of which association words are 160,000 in a present version. It is implemented on the neural network by setting this association ontology as follows.

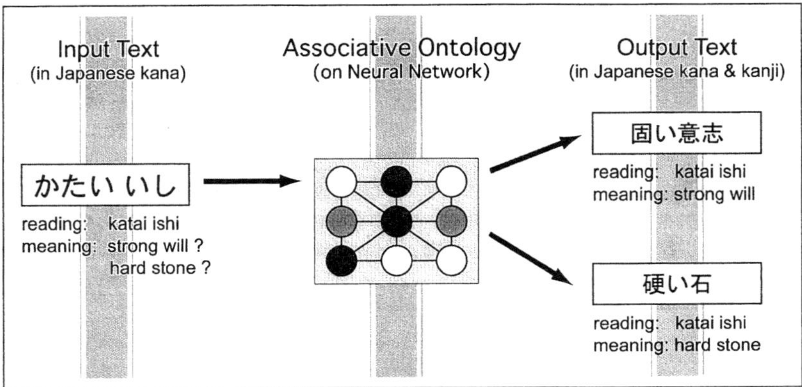

Fig. 2. Outline of neural coding model of association ontology. When the text of the Japanese kana is input to the associative ontology, it converts it into the text that contains the Japanese kana and kanji with a correct meaning relation.

1. Nodes of the association ontology as for the neuron.
2. Words with the association relation as for an excitatory connection.
3. Homonym as for an inhibitory connection.

The input signal to the model is vague sentences that are including the polysemy and the homonym (Fig.2). The meaning of the word greatly changes into the sentence to which the voice is input in the natural language depending on the context. And, we recognize it specifying the meaning from the relation of the word before and after. For

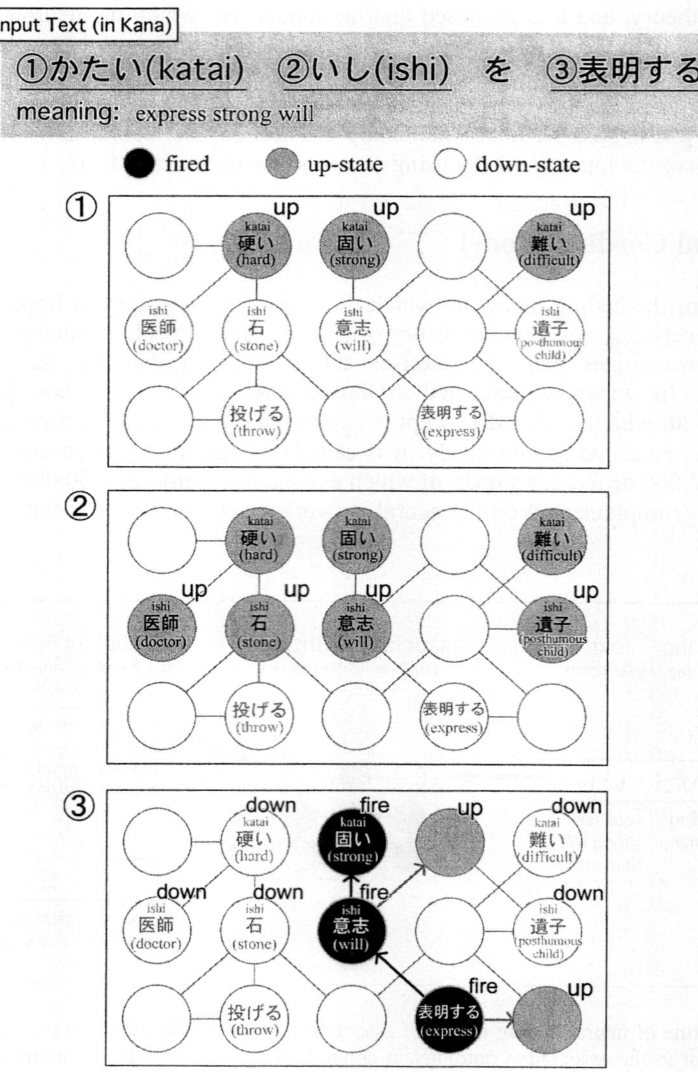

Fig. 3. Formation process of cell assembly with up and down states

example, "Pen" is specified by having the meaning such as instrument and building, and containing the following contexts.

- The pen and pencil are sold in the shop.(instrument)
- A man who had committed the crime was released from the pen.(building)

The neurons with the relation between the word and the homonym are put into the up state when the word into which the expression changes depending on the context is input. Next, when the word that was able to specify the meaning without a homonym and polysemy is input, the neuron corresponding to the word fires. The neuron which has the associative relation with firing neuron in the up state neurons fires in self organization for excitatory connection. Another neuron which has the homonym relation with firing neuron is the down state by inhibitory connection. Vagueness is resolved according to the context input, and this model corresponds to the input of a consecutive sentence.

In addition, the relation of the word newly extracted is learned from the resolution of vagueness to which the up and down states is taken by using distance information on the association ontology. The distance of the neurons with a new relation by using (Exp.1) is requested by assuming the reciprocal of the distance between the firing neurons to be a size of the firing potential. A dynamic model by which the relation of new words is extracted can be automatically constructed with this algorithm.

A Japanese sentence is used as input and output data of this model. Two kinds of characters like the Japanese kana and kanji, etc. are changes into the literation by the meaning. In this research, the polysemy and the homonym of the Japanese kana are resolved by the input context, and the model that literates a specific Japanese kanji is constructed.

4.1 Cell Assembly with Up and Down States

We explains the process in cell assembly by the sentence that continues to the phrase of "katai ishi". The sentence of "katai ishi (wo) hyoumeisuru" is input. Each part is sequentially input to the associative ontology (Fig.3).

The neurons with the same reading as "katai" are the up states because it cannot specify the meaning at this stage when "katai" is input (1). Next, the neurons with the same reading as well as (1) are the up states because it cannot specify the meaning even here when "ishi" was input (2). When the part of "hyoumeisuru" is input, the meaning can be specified (Because another doesn't have the meant neuron "hyoumeisuru"). Therefore, when the "express" neuron fires, an excitatory signal is sent to the related neurons. When the neuron that receives an excitability signal is the up state, the neuron fires and "will" neuron fires like (3). It sends an excitatory signal to the neurons with the relation in the chain reaction by "will" neuron's firing. The firing "will" neuron sends the homonym inhibitory signal at the same time, and puts them into the down states. The homonyms similarly are the down states for "strong" neuron that fires in the chain reaction (3).

4.2 Learning of Distance by Morphoelectrotonic Transform

This simulation is a result of inputting 100 documents that show " katai ishi (wo) hyoumeisuru " (Fig.4). The document is extracted by inputting the key word by the search engine. In this simulation, "express" and "strong" neuron were connected from the input sentence, and association distance 0.6 was studied according to the morphoelectrotonic transform.

There is a famous Elman network as a network model to the appearance word is calculated by the context dependent. But in this network, it is a problem that firing in steady order.In this network, it is not necessary to adjust the weight of individual neuron, and if average strength is decided, the neuron group is integrated even if neuron connects at random.

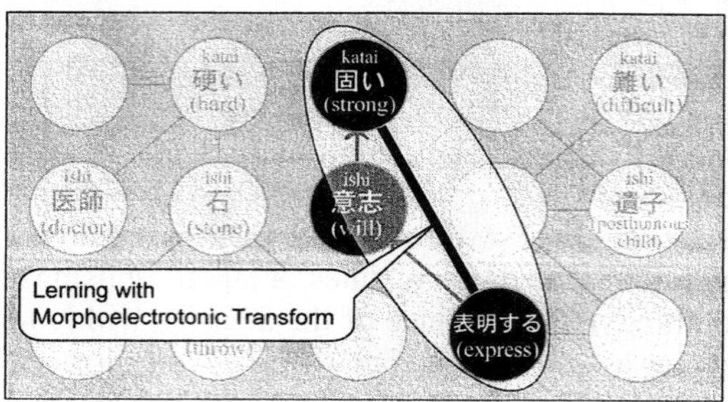

Fig. 4. Learning of relation of extracted new word

Table 1. Associative distance after relation of word of "express strong will" is extracted. The number in parentheses is a value before learning.

Associative distance	strong	will	express
strong		2	0
will	**3.6**(3)		4
express	**0.6**	**5.6**(5)	

5 Conclusion

This neural coding model that is able to treat various meanings was constructed by the using up and down state. By morphoelectrotonic transform, it was possible to extract automatically from the large-scale text data the relation of a new word without the association experiment. Large-scale learning of the associative distance and the evaluation are scheduled by the association concept dictionary after this.

References

1. Elman, J. L. Distributed representations, simple recurrent networks, and grammatical structure. Machine Learning, vol. 7, pp. 195-225, 1991.
2. Jun Okamoto, Shun Ishizaki. Associative concept dictionary construction and its comparison with electronic concept dictionaries. PACLING2003, pp. 1-7, August. 2003.
3. Kerr, J.N. and Plenz, D. Dendritic calcium encodes striatal neuron output during Up-states. J. Neurosci. 22, pp. 1499-1512, 2002.

4. Kitano K, Cateau H, Kaneda K, Nambu A, Takada M, Fukai T. Two-State Membrane Potential Transitions of Striatal Spiny Neurons as Evidenced by Numerical Simulations and Electrophysiological Recordings in Awake Monkeys. J. Neurosci. 22:RC230, pp. 1-6, 2002.
5. WR. Chen, J. Midtgaard, GM. Shephered. Forward and Backward Propagation of Dendritic Impulses and Their Synaptic Control in Mitral Cells. Science, vol. 278, pp. 463-467, 1997
6. AM. Zador, H. Agmor-Snir and I. Segev. The morphoelectrotonic transform: A Grafical Approch to Dendritic Function. J.Neuroscience, vol. 15, pp. 1669-1682, 1995
7. Norifumi Watanabe and Shun Ishizaki. Neural Coding Model of Perceptual Reconstruction Using the Morphoelectrotonic Transform Theory. In Proceeding of the International Joint Conference on Neural Networks, IJCNN2004, pp. 327-332, July. 2004.

Robust Structural Modeling and Outlier Detection with GMDH-Type Polynomial Neural Networks

Tatyana Aksenova[1,2], Vladimir Volkovich[3], and Alessandro E.P. Villa[1]

[1] Inserm U318, Laboratory of Neurobiophysics, University Joseph Fourier, Grenoble, France
{Tatyana.Aksyonova, Alessandro.Villa}@ujf-grenoble.fr
http://www.nhrg.org/
[2] Institute of Applied System Analysis, Prospekt Peremogy, 37, Kyiv 03056, Ukraine
[3] International Researching-Training Center of Information Technologies, Glushkova 40, 252022, Kyiv, Ukraine
volk@volk.kiev.ua

Abstract. The paper presents a new version of a GMDH type algorithm able to perform an automatic model structure synthesis, robust model parameter estimation and model validation in presence of outliers. This algorithm allows controlling the complexity – number and maximal power of terms – in the models and provides stable results and computational efficiency. The performance of this algorithm is demonstrated on artificial and real data sets. As an example we present an application to the study of the association between clinical symptoms of Parkinsons disease and temporal patterns of neuronal activity recorded in the subthalamic nucleus of human patients.

1 Introduction

Artificial Neural Networks (ANN) have been successfully applied in many fields to model complex non–linear relationships. ANNs may be viewed as the universal approximators but the main disadvantage of this approach is that detected dependencies are hidden within the neural network structure. Conversely, Group Method of Data Handling (GMDH) [1] are aimed to identify the functional structure of a model hidden in the empirical data. The main idea of GMDH is the use of feedforward networks based on short-term polynomial transfer function whose coefficients are obtained using regression technique combined with the emulation of the self-organizing activity for the neural network (NN) structural learning. In order to reduce the sensitivity of GMDH to outliers a Robust Polynomial Neural Network (RPNN) approach was recently developed [2]. This paper presents a new version of RPNN using new robust criteria for model selection and measures of goodness of fit and a demonstration of its performance on artificial and real data sets.

W. Duch et al. (Eds.): ICANN 2005, LNCS 3697, pp. 881–886, 2005.

2 GMDH Approach

The GMDH approach for complex system modeling and identification is based on given multi-unit–single-output data invented by Ivakhnenko [1]. Traditional GMDH is a multi-layered perceptron type NN formed by neurons whose transfer function g, $g = a + bw_i + cw_j + dw_iw_j + ew_i^2 + fw_j^2$ is a short-term polynomial of two variables w_i, w_j. The GMDH training algorithm is based on an evolutionary principle. The algorithm begins with regression-type data, the observations of vector of independent variables $x = (x_1, \ldots, x_m)^T$ and one dependent variable y. The data set is subdivided into training and test sets. At the 1st layer all possible combinations of two inputs generate the first population of neurons according to the transfer function g. The size of the population at the 1st layer is equal to C_m^2. The coefficients of the polynomials of g are estimated by Least Square fitting using the training set. The best neurons are selected by evaluating the performance on the test set according to a criterion value. The outputs of selected neurons of the first layer are treated as the inputs to the neurons of the 2nd layer, and so on for the next layers. The size of the population of the successive layers become equal to C_f^2. The process is terminated if there is no improvement of the performance according to the criterion. The GMDH model can be computed by tracing back the path of the polynomials. The composition of quadratic polynomials of g forms a high-order regression polynomial known as the Ivakhnenko polynomial. Notice that the degree of the polynomial doubles at each layer and the number of terms in the polynomial increases.

3 Robust Polynomial Neural Networks

Basically RPNN are described as follows (see [2,3] for more details). Let $\mathbf{x} = (x_1, \ldots, x_m)^T$ be the vector of input variables and let y be the output variable that is a function of a subset of input variables $y = u(x_{i1}, x_{i2}, \ldots, x_{ip})$. Let $\mathbf{X} = (x_{ij})$ be a $[m \times n]$ matrix and $\mathbf{Y} = (y_1, \ldots, y_n)^T$ the vector of observations. The random errors ξ of observations are assumed to be uncorrelated, identically distributed with finite variance $\mathbf{Y} = E(\mathbf{Y}|\mathbf{X}) + \xi$. The goal of the method is to find a subset of variables $x_{i1}, \ldots, x_{ik}$ and a model belonging to the class of polynomial that minimizes some criteria values (CR). Thus, model identification means both structure synthesis and parameters estimation. The main modifications according to the original GMDH are the following:

1. An expanded vector of initial variables $\mathbf{x} = (x_1, \ldots, x_m, x_{m+1}, x_{m+2})^T$, $x_{m+1} = 1$, $x_{m+2} = 0$ is available at each layer;

2. The following nonlinear transfer function which generates the class of polynomials is used [3]:

$$g(w_i, w_j, w_k) = aw_i + bw_jw_k, \qquad i, j, k = 1 \ldots m \quad . \tag{1}$$

Triplets of inputs are considered instead of pairs. The coincidences of indexes lead up to triple the number of connections. Neurons with one or two inputs as well as several transform functions (including the linear one) are generated according to Eq. 1 and additional variables $x_{m+1} = 1$ and $x_{m+2} = 0$. Notice that only two coefficients a and b are estimated. In traditional GMDH the Mean Least Square (MLS) method is used. Thus the second order matrices are only inverted. This provides fast learning of NN. The number of neurons at each layer of the net that depends on the form of the transfer function g and the number f of output variables which were selected from previous layer equals C^3_{m+2+f} .

3. Each term $x_i^{q1}, \ldots, x_j^{q2}$ in the equation is coded as a product of the appropriate powers of a prime numbers, i.e. the polynomial is coded by a vector of Gedels numbers [3]. Because of the one-to-one correspondence between the terms of the polynomials and their Gedels numbers this coding scheme can be used to transfer the results of the ANN to the parametric form of equation.

4. The polynomials of high power are unstable and sensitive to outliers. Therefore, a twice-hierarchical ANN structure based on the polynomial complexity control [2] was proposed to increase the stability and computational efficiency of GMDH. This structure allows the convergence of the MLS coefficients, as proven mathematically for algorithm with linear transform [4]. The vector $(p, c)^T$, where c is the number of terms and p is the power of the polynomial is considered as the polynomial complexity. Gedels coding scheme allows to calculate the number of terms for each intermediate model that equals to the number of non zero element of its vector of Gedels numbers. The power of intermediate model $g(w_i, w_j, w_l) = aw_i + bw_j w_l$ is controlled by the condition $p(g(w_i, w_j, w_l)) = maxp(w_i), p(w_j), p(w_l)$ where $p(w_i), p(w_j), p(w_l)$ are the power of inputs w_i, w_j, w_l. This allows to control the complexity by restricting the class of the models by $p(w_i) < p_{max}$ and $c < c_{max}$. The RPNN are twice-multilayered since multilayered neurons are connected into a multilayered net. The *external iterative procedure* controls the complexity of the models, i.e. the number of the terms and the power of the polynomials in the intermediate models. The best models form the initial set for the next iterative procedure. The *internal iterative procedure* realizes a search for optimal models given the fixed complexity and discard models that are out of the specified range. Both external and internal iteration procedures are terminated if there is no improvement of the criterion values CR.

5. Robust M–estimates [5] of the coefficients a and b of the transfer functions $g(w_{j1}, w_{j2}, w_{j3}) = aw_{j1} + bw_{j2}w_{j3}$ were applied instead of MLS estimates.

6. Robust versions of CR are used for model selection:

$$CR1 = \frac{\hat{\sigma}}{n-p} \sum_{i=1}^{n} \rho(r_i/\hat{\sigma}) \quad , \quad CR2 = \hat{\sigma}\frac{n+p}{n-p} \sum_{i=1}^{n} \rho(r_i/\hat{\sigma}) \quad . \tag{2}$$

If the data is splitted into training and test sets A and B, then the robust version of regularity criterion AR used in GMDH [1] is implemented. The parameters $\hat{a}_A$, $\hat{b}_A$, and the variance $\hat{\sigma}_A$ estimated on the set A are used to calculate the residuals r_i for the set B. Then, the regularity criterion AR is expressed by $AR = \sigma_A^2 \sum_{i \in B} \rho(r_i/\hat{\sigma}_A)$.

7. The ρ-test (robust variant of F-criteria) and R_n^2-test [6] are applied to the final models for the canonical hypothesis $H_0 : \beta_i = 0$, β is the vector of parameters of the resulted model, to avoid the appearance of spurious terms. Robust correlation and robust deviation for both training set A and test set B are used as measures of goodness of fit.

8. The residuals r_i, of the final models are used for the outlier detection: $outlier = 0, if|r_i| \leq k\hat{\sigma}$, otherwise $outlier = 1$.

4 Validation on an Artificial Data Set

Let us consider the vector of $m = 5$ input variables $\mathbf{x} = (x_1, \ldots, x_5)^T$, and the fourth power polynomial $y = 10.0 + 1.0 \cdot x_1 \cdot x_5^3 + \xi$ generally used for testing GMDH. The matrix $\mathbf{X} = (x_{ij})$ $[5 \times 15]$, $n = 15$ was generated at random with a uniform distribution on the interval $[1, 10]$. Random values ξ were generated according to the model of outliers $P_\delta(\xi) = (1 - \delta)\phi(\xi) + \delta h(\xi)$. Here $\phi(\xi)$ is the Normal distribution density $N(0, \sigma_b)$; $h(\xi)$ is the distribution density of the outliers $N(0, \sigma_{out})$; δ is the level of the outliers. Twenty realizations were considered for the following combinations of parameters: (A) $\sigma_b = 10$, $\delta = 0$; (B) $\sigma_b = 10$, $\delta = 0.2$, $\delta_{out} = 1000$; (C) $\sigma_b = 10$, $\delta = 0.2$, $\delta_{out} = 2000$; (D) $\sigma_b = 10$, $\delta = 0.2$, $\delta_{out} = 3000$. Structural indexes $StrInd$ were determined for each term of the equation: $StrInd = 1$ if the term was present in the synthesized equation and $StrInd = 0$ otherwise. The mean value of the structural indexes corresponds to the frequency of the appearance of the term in the resulted equations over all computational experiments. Table 1 shows that RPNN provides the best structural synthesis irrespective of the increasing variance of the outliers. Table 2 summarizes the results of the coefficients estimation with RPNN and PNN (MLS). Table 3 presents the quality of approximation with the measures of goodness of fit of the calculated model according to the exact model $y_{exact} = 10.0 + 1.0 \cdot x_1 \cdot x_5^3$, μ_{exact} is the mean value:

$$MSD = \frac{1}{n}\sum(y_{calc} - y_{exact})^2, R^2 = \frac{\sum(y_{exact} - \mu_{exact})^2 - \sum(y_{calc} - y_{exact})^2}{\sum(y_{exact} - \mu_{exact})^2}. \tag{3}$$

Table 1. Mean values of structural indexes $StrIn$ for the constant, monomial $x_1 \cdot x_5^3$ and additional terms with significant (signif.) and non-significant (n.s.) coefficients. δ: level of the outliers, σ_b, σ_{out}: : SD of basic Normal distribution and of the outliers.

		MLS		**RPNN**			**PNN with MLS**		
δ		$\delta = 0$		$\delta = 0.2$			$\delta = 0.2$		
σ_b		0	10	$\sigma_b = 10$			$\sigma_b = 10$		
σ_{out}		$--$	$--$	1000	2000	3000	1000	2000	3000
const		1.00	0.76	0.47	0.53	0.47	0.07	0.21	0.07
$x_1x_5^3$		1.00	1.00	1.00	1.00	1.00	0.80	0.50	0.33
add	*n.s.*	0.00	0.65	0.60	0.53	0.60	0.40	0.64	0.33
	significant	0.00	0.00	0.00	0.00	0.00	1.33	0.71	1.60

Table 2. Coefficients estimated in case the terms were present in the resulted equation

		MLS		RPNN			PNN with MLS		
δ		$\delta = 0$		$\delta = 0.2$			$\delta = 0.2$		
σ_b		0	10	$\sigma_b = 10$			$\sigma_b = 10$		
σ_{out}		−−	−−	1000	2000	3000	1000	2000	3000
const	*mean*	10.0	11.69	10.86	10.54	10.53	687.00	127.50	247.29
	SD	–	2.59	1.96	1.90	1.88	–	1041.41	–
$x_1x_5^3$	*mean*	1.0	0.997	0.996	0.992	0.996	1.0541	1.1104	1.0673
	SD	–	0.004	0.003	0.003	0.003	0.1516	0.6058	0.2412
σ_b	*mean*	10.0	8.25	14.50	12.53	13.21	304.62	537.20	785219
	SD	–	1.99	6.93	4.90	5.64	197.60	261.58	758152.26

Table 3. The measures of goodness of fit according to the exact data

		MLS	RPNN			PNN with MLS		
δ		$\delta = 0$	$\delta = 0.2$			$\delta = 0.2$		
σ_b		$\sigma_b = 10$	$\sigma_b = 10$			$\sigma_b = 10$		
σ_{out}		−−	1000	2000	3000	1000	2000	3000
MSD	*mean*	5.49	6.13	5.83	5.96	272.43	447.79	711.28
	SD	1.82	3.08	2.97	3.12	139.11	159.32	309.02
R^2	*mean*	0.99996	0.99995	0.99996	0.99995	0.91598	0.78919	0.58605
	SD	0.00002	0.00004	0.00004	0.00004	0.07346	0.16420	0.34052

5 Application to Experimental Data

RPNN was applied to study the association between clinical symptoms of Parkinsons disease and firing patterns in the subthalamic nucleus of patients that underwent surgical operation for Deep Brain Stimulation (DBS) [7]. The set of parameters determined from the neurological examination of the patients (x_i) are based on scores defined by the Unified Parkinsons Disease Rating Scale (x_1=RT: resting tremor; x_2=AT: action tremor, i.e. essential tremor during voluntary movement; x_3=RG: rigidity of upper limbs; x_4=AK: akinesia of upper limbs) and the parameters defining the firing activity (y_i) are obtained from the electrophysiological recordings (y_1=Syn: percentage of pairs of units with synchronous firing; y_2=Bst: % of units with bursting activity; y_3=(1-2): % of units with oscillatory activity [1-2 Hz]; y_4=(4-6): % of units with oscillatory activity [4-6 Hz]; y_5=(8-12): % of units with oscillatory activity [8-12 Hz]; y_6=FR: average firing rate in the subthalamic nucleus of patients operated at the Grenoble University Hospital. We have proceeded by considering the clinical parameter vector X as the independent variable and the neurophysiological vector Y as the dependent variable. However, the relation of causality between clinical and neurophysiological parameters is not known and we have analyzed the data considering also X dependent on Y. Two outliers exceeding 3σ-confidential interval were detected (Fig. 1a). The final result of the analysis allowed to generate a new model of the associations between clinical symptoms and neurophysiological data (Fig. 1b). In this Figure the arrows show the presence of the variables in the models and the sign of the corresponding terms. Notice that only models with criteria values $R^2 > 0.6$ were considered.

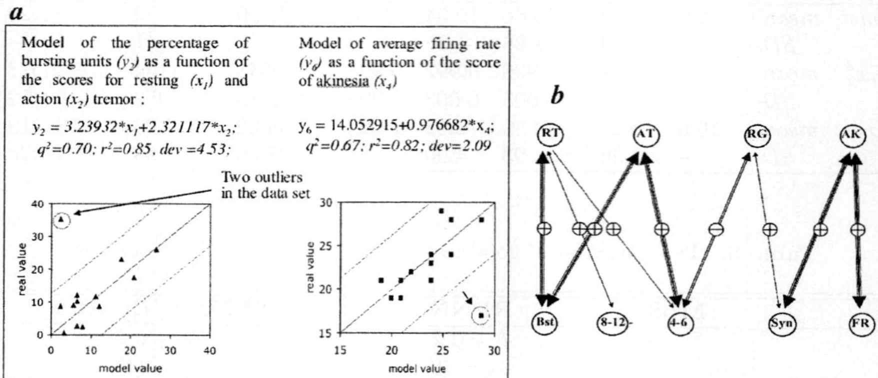

Fig. 1. (a) Examples of polynomial estimates; (b) Results of modeling presented as a scheme of the dependencies

Acknowledgments. The authors thank Pr. A.-L. Benabid, Dr. S. Chabardes and all members of the neurosurgical and neurological teams of Inserm U318 at Grenoble University Hospital for providing the data used for an application of this technique.

References

1. Madala, H., Ivakhnenko, A.: Inductive Learning Algorithms for Complex Systems Modeling. CRC Press Inc., Boca Raton, FL, USA (1994)
2. Aksenova, T.I., Volkovych, V., Tetko, I.: Robust Polynomial Neural Networks in Quantative-Structure Activity Relationship Studies. SAMS **43** (2003) 1331–1341
3. Yurachkovsky, Y.: Restoration of polynomial dependencies using self-organization. Soviet Automatic Control **14** (1981) 17–22
4. Yurachkovsky, Y.: Convergence of multilayer algorithms of the Group Method of Data Handling. Soviet Automatic Control **14** (1981) 29–35
5. Huber, P.: Robust Statistics. John Wiley & Sons Inc. (2003)
6. Hampel, F., Ronchetti, E.M., Rousseeuw, P., Stahel, W.: Robust Statistics: The Approach Based on Influence Function. John Wiley & Sons Inc. (2005)
7. Chibirova, O., Aksenova, T.I., Benabid, A., Chabardes, S., Larouche, S., Rouat, J., Villa, A.: Unsupervised Spike Sorting of extracellular electrophysiological recording in subthalamic nucleus parkinsonian patients. Biosystems **79** (2005) 59–71

A New Probabilistic Neural Network for Fault Detection in MEMS*

Reza Asgary[1] and Karim Mohammadi[2]

IranUniversity of Science and Technology,
Department of Electrical engineering
{ R_Asgary, Mohammadi}@iust.ac.ir

Abstract. Micro Electro Mechanical Systems will soon usher in a new technological renaissance. Learn about the state of the art, from inertial sensors to microfluidic devices [1]. Over the last few years, considerable effort has gone into the study of the failure mechanisms and reliability of MEMS. Although still very incomplete, our knowledge of the reliability issues relevant to MEMS is growing. One of the major problems in MEMS production is fault detection. After fault diagnosis, hardware or software methods can be used to overcome it. Most of MEMS have nonlinear and complex models. So it is difficult or impossible to detect the faults by traditional methods, which are model-based.In this paper, we use Robust Heteroscedastic Probabilistic Neural Network, which is a high capability neural network for fault detection. Least Mean Square algorithm is used to readjust some weights in order to increase fault detection capability.

1 Introduction

Reliability of Micro Electro Mechanical Systems (MEMS) is a very young and fast-changing field. Fabrication of a MEMS System involves many new tools and methods, including design, testing, packaging and reliability issues. Especially the latter is often only the very last step that is considered in the development of new MEMS. The early phases are dominated by considerations of design, functionality and feasibility; not reliability[2].

Only a few fault detection methods have been introduced for fault detection in MEMS. Additionally most of them need a precise model of system[3-5]. In MEMS most of the parts are strictly nonlinear and finding a proper model is difficult or sometimes impossible. The constraints of this kind of model have motivated the development of artificial intelligent approaches[6,7]. Different neural networks have been trained and used for fault detection in MEMS. The Multi Layer Perceptron(MLP), Radial Basis Function (RBF), Probabilistic Neural Network (PNN) and RHPNN has been used. The best results obtained for RHPNN[8].In this paper, we will use a new probabilistic neural network for fault detection in MEMS.

* This work has been partially supported by Iran Telecommunication Research Center.

W. Duch et al. (Eds.): ICANN 2005, LNCS 3697, pp. 887–892, 2005.

2 Robust Heteroscedastic Probabilistic Neural Network

A PNN classifies data by estimating the class conditional probability density functions, because the parameter of a PNN cannot be determined analytically. To do this it requires a training phase, followed by a validation phase, before it can be used in a testing phase. A PNN consists of a set of Gaussian kernel functions. The original PNN uses all the training patterns as the centers of the Gaussian kernel functions and assumes a common variance or covariance, which is named homoscedastic PNN. To avoid using a validation data set and to determine analytically the optimal common variance, a maximum likelihood procedure was applied to PNN training [9]. On the other hand, the Gaussian kernel functions of a heteroscedastic PNN are uncorrelated and separate variance parameters are assumed. This type of PNN is more difficult to train using the ML training algorithm because of numerical difficulties. A robust method has been proposed to solve this numerical problem by using the jackknife, a robust statistical method, hence the term 'robust heteroscedastic probabilistic neural networks' [10]. The RHPNN is a four layer feedforward neural network based on the Parzen window estimator that realizes the Bayes classifier given by

$$g_{Bayes} = arg\left(max\left\{\alpha_j f_j(x)\right\}\right) \tag{1}$$

Where x is a d-dimensional pattern, $g(x)$ is the class index of x, the a priori probability of class $j(1 \leq j \leq k)$ is α_j and the conditional probability density function of class j is f_j. The object of the RHPNN is to estimate the values of f_j. This is done using a mixture of Gaussian kernel functions.

RHPNN has been shown in Fig.1. In this figure two classes are shown. First class is considered for fault free and the second class for faulty patterns. There is only one fault free kernel because with only one Gaussian function all fault free patterns can be shown. There are many different faults and the distances between them are unknown, so in second class, more than one kernel is considered. The optimum number of kernels in second class is the minimum that each kernel has at least one faulty pattern. The first layer of the PNN is the input layer. The second layer is divided into k groups of nodes, one group for each class.

The ith kernel node in the jth group is described by a Gaussian function

$$p_{i,j} = \frac{1}{(2\pi\sigma_{i,j}^2)^{d/2}} \exp\left(-\frac{\|x - c_{i,j}\|^2}{2\sigma_{i,j}^2}\right) \tag{2}$$

Where $c_{i,j}$ is the mean vector and $\sigma_{i,j}^2$ is the variance. The third layer has k nodes; each node estimates f_j, using a mixture of Gaussian kernels

$$f_j(x) = \sum_{i=1}^{M_j} \beta_{i,j} p_{i,j}(x), 1 \leq j \leq k \tag{3}$$

$$\sum_{i=1}^{M_j} \beta_{i,j} = 1, 1 \leq j \leq k \tag{4}$$

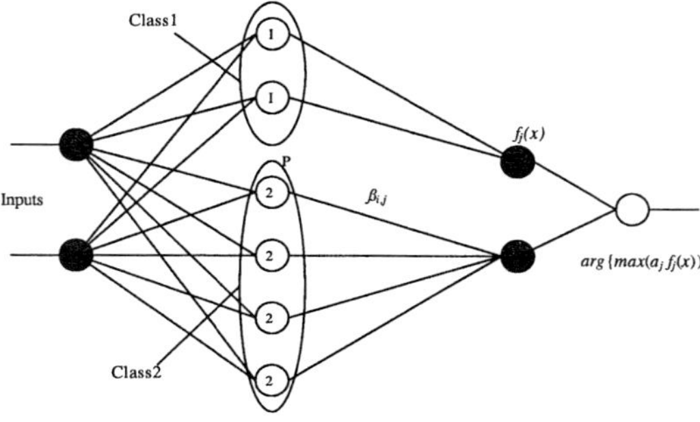

Fig. 1. Four layer feed forward RHPNN

Where M_j is the number of nodes in the jth group in the second layer. The fourth layer of the PNN makes the decision from Eq(1). The PNN is heteroscedastic when each Gaussian kernel has its own variance. The centers, $c_{i,j}$, the variance, $\sigma^2_{i,j}$ and the mixing coefficients, $\beta_{i,j}$ have to be estimated from the training data. One assumption for α_j weights is $\alpha_j = \frac{1}{k}, 1 \leq j \leq k$. The EM algorithm has been used to train homoscedastic PNN's. Each iteration of the algorithm consists of an expectation (E) followed by a maximization process (M). This algorithm converges to the ML estimate. For the heteroscedastic PNN, the EM algorithm frequently fails because of numerical difficulties. These problems have been overcome by using a jackknife, which is a robust statistical method. Suppose the training data is partitioned into k subsets $\{x_n\}_{n=1}^{N} = \{\{x_{n,j}\}_{n=1}^{N_j}\}_{j=1}^{k}$, where $\sum_{j=1}^{k} N_j$ is the total number of samples and N_j is the number of training samples for class j. The training algorithm is now expressed as follows, where $\tilde{\sigma}^2_{m,i}|^k$ and $\tilde{c}_{m,i}|^k$ are the jackknife estimates of the previous values of $\sigma^2_{m,i}$ and $c_{m,i}$ at step t, respectively.

Step 1: Compute weights for $1 \leq m \leq M_i$, $1 \leq n \leq N_i$ and $1 \leq i \leq k$.

$$\omega^{(t)}_{m,i}(x_{n,i}) = \frac{\beta_{m,i} p^{(t)}_{m,i}(x_{n,i})}{\sum_{l=1}^{M_i} \beta_{i,l} p^{(t)}_{l,i}(x_{n,i})} \tag{5}$$

$$p^{(t)}_{l,i}(x_{n,i}) = \frac{1}{(2\pi\tilde{\sigma}^2_{l,i}|^{(t)})^{d/2}} \exp\left(-\frac{\|x_{n,i} - \tilde{c}_{l,i}|^{(t)}\|^2}{2\tilde{\sigma}^2_{l,i}|^{(t)}}\right) \tag{6}$$

Step 2: Update the parameters for $1 \leq m \leq M_i$ and $1 \leq i \leq k$

$$\tilde{c}_{m,i}|^{t+1} = N_i c_{m,i}|^{t+1} - \frac{N_i - 1}{N_i} \sum_{j=1}^{N_i} c_{m,i}|^{t+1}_{-j} \tag{7}$$

$$c_{m,i}|^{t+1} = \frac{\sum_{n=1}^{N_i} \omega^{(t)}_{m,i}(x_{n,i}) x_{n,i}}{\sum_{n=1}^{N_i} \omega^{(t)}_{m,i}(x_{n,i})} \tag{8}$$

$$c_{m,i}|_{-j}^{t+1} = \frac{\sum_{n=1,n\neq j}^{N_i} \omega_{m,i}^{(t)}(x_{n,i})x_{n,i}}{\sum_{n=1,n\neq j}^{N_i} \omega_{m,i}^{(t)}(x_{n,i})}, 1 \leq j \leq N_i \tag{9}$$

$$\tilde{\sigma}_{m,i}^2|^{t+1} = N_i \sigma_{m,i}^2|^{t+1} - \frac{N_i - 1}{N_i} \sum_{j=1}^{N_i} \sigma_{m,i}^2|_{-j}^{t+1} \tag{10}$$

$$\sigma_{m,i}^2|^{t+1} = \frac{\sum_{n=1}^{N_i} \omega_{m,i}^{(t)}(x_{n,i}) \|x_{n,i} - \tilde{c}_{m,i}|^{(t)}\|^2}{d \sum_{n=1}^{N_i} \omega_{m,i}^{(t)}(x_{n,i})} \tag{11}$$

$$\sigma_{m,i}^2|_{-j}^{t+1} = \frac{\sum_{n=1,n\neq j}^{N_i} \omega_{m,i}^{(t)}(x_{n,i}) \|x_{n,i} - \tilde{c}_{m,i}|^{(t)}\|^2}{d \sum_{n=1,n\neq j}^{N_i} \omega_{m,i}^{(t)}(x_{n,i})}, 1 \leq j \leq N_i \tag{12}$$

$$\beta_{m,i}|^{t+1} = \frac{1}{N_i} \sum_{n=1}^{N_i} \omega_{m,i}^{(t)}(x_{n,i}) \tag{13}$$

3 Modified RHPNN

In RHPNN training phase, centers and variances of kernel functions change and finally fix at values to cover faults. Additionally, $\beta_{i,j}$ which are weights between second and third layers change to final values. In Modified Robust Heteroscedastic Probabilistic Neural network (MRHPNN), all $\beta_{i,j}$ weights are readjusted by Least Mean Square, LMS, algorithm. The MRHPNN learning has two phases. First, like RHPNN all training patterns (faulty and fault free) are applied to network and in two recursive steps, network parameters are defined. In this phase all centers and variances and $\beta_{i,j}$s are adjusted. In second phase, all training patterns are applied to network again and for each misclassified pattern, error e_j is defined at third layer output. For each class, this error is defined and back propagates to change all $\beta_{i,j}$:

$$\beta_{i,j}(t+1) = \beta_{i,j}(t) + \eta \sum_{n=1}^{N_{Mis}} e_j(x_n) p_{i,j}^{(t)}(x_n), 1 \leq j \leq k \tag{14}$$

In which total number of misclassified patterns is N_{Mis} and η is a small number. We have used 0.001 for this application. When the first training phase is finished, for each faulty pattern, one or more kernel produce a great probability density function. In other words, when a faulty pattern is given to network, the outputs of one or more kernel in second class (faulty class), is greater than the output of fault free kernel (first cell in second layer). If faulty pattern is far from the fault free kernel center, the output of faulty group kernels is obviously greater than the output of fault free kernel. But for faulty patterns which are near the fault free kernel center, output of fault free kernel is near or sometimes greater than the output of faulty group kernels. In this case, faulty pattern is classified as a fault free pattern. We used second training phase to solve this problem. The difference between fault free class and faulty class outputs is defined as error which back propagate to adjust $\beta_{i,j}$ weights, as Eq(14). After updating all $\beta_{i,j}$, again all the patterns are applied to network, and for each misclassified pattern an error is defined at third layer output. In each step all $\beta_{i,j}$ weights and the number of misclassified weights are saved. The optimum values of $\beta_{i,j}$ weights occur in accordance with the least number of misclassified patterns.

4 Simulation Results

EM3DS is MEMS simulator software, which has been used for fault simulation in RF MEMS. 20 faults and one fault free pattern have been simulated in a RF low pass filter MEMS. These 20 faults consist of both digital and analog faults. Changing substrate resistance, magnetic and electric properties, shorts

Table 1. Fault detection results using RHPNN

	Detected as Faulty	Detected as Fault free	Results for RF Lp filter	Detected as Faulty	Detected as Fault free	Results for RF capacitor
40 faulty pattern	37	3	%92.5	35	5	%87.5
10 fault free pattern	1	9	%90	2	8	%80
Total	(37+9)	/(40+10)=	%92	(35+8)	/(40+10)=	%86

Table 2. Fault detection results using MRHPNN

	Detected as Faulty	Detected as Fault free	Results for RF Lp filter	Detected as Faulty	Detected as Fault free	Results for RF capacitor
40 faulty pattern	38	2	%95	37	3	%92.5
10 fault free pattern	1	9	%90	1	9	%90
Total	(38+9)	/(40+10)=	%94	(37+9)	/(40+10)=	%92

and opens, disconnections, connection between separate parts and some other faults have been simulated by software. The S parameters are calculated and used for training and testing all neural networks. We have used real and imaginary parts of S_{11}, 2 dimensional data, as input to neural networks. For each fault 3 patterns and 7 patterns for fault free case have been simulated, so total number of learning patterns is (20*3+7)=67. The other RF MEMS which is simulated by EM3DS is Inter digital capacitor. We considered 32 digital and analog faults and for each one three pattrens have been simulated. Also, seven fault free patterns have been used. Total number of learning patterns is (32*3+7)=103.

The optimal number of kernels in second layer for RF low pass filter is 6 and for inter digital capacitor is 8. One kernel is belonged to fault free class and the others are belonged to faulty class. All the faulty patterns are labeled with the same number when training a MRHPNN model. During training the MRHPNN is able to cluster the patterns automatically. This is an advantage compared with most of other neural networks. After training neural networks, faulty and fault free patterns have been applied to them. Table1 shows the results of RHPNN fault recognition in RF low pass filter and RF interdigital capacitor. Similarly, Table2 shows MRHPNN results for the same MEMS.

5 Conclusion

MEMS usually have nonlinear and complex models. Most of the times, novel and unknown faults occur in them, too. A powerful recognition method is essential to detect/diagnose the faults. This part can be inserted in MEMS as a Built In Self Test (BIST) mechanism. With respect to nonlinearity and novel faults in MEMS, neural networks are proposed as a BIST mechanism. In this paper, we propose MRHPNN for fault detection in MEMS in which for improving the performance, some of the weights are readjusted by LMS algorithm. Improper η values cause divergence. For every input data space, a different η value should be selected. Extra work is needed to determine proper η. Finding better learning methods for determining centers and variances, is future work.

References

1. B. Murari, "Integrated Nanelectronic Components into Electronic Microsystems", IEEE Trans. on Reliability, vol.52, No.1, 2003, pp.36-44.
2. R. Muller, U. Wagner, W. Bernhard, "Reliability of MEMS-a methodical approach", Proc.11th European symposium on reliability of electron devices, failure physics and analysis, ,2001, pp.1657-62.
3. R. Rosing, A. Richardson, "Test Support Strategies for MEMS", IEEE International Mixed Signal Test Workshop, Whistler, Canada, June 1999.
4. S. Mir, B. charlot, "On the Integration of Design and Test for chips embedding MEMS", IEEE design and Test of Computers, Oct-Dec 1999.
5. TIMA Lab research reports, Http://Tima.imag.fr, 2002-2004.
6. S.H. Yang, B.H. Chen, "Neural Network Based Fault Diagnosis Using Unmeasurable Inputs", Journal of Artificial Intelligence-PERGAMON, Vol.13, 2000, pp.345-356.
7. M.A. El-Gamal, "Genetically Evolved Neural Networks for Fault Classification in Analog Circuits", Journal of Neural Computing & applications, Springer, Vol.11, 2002, pp.112-121.
8. R.Asgary, K.Mohammadi, "Pattern Recognition and Fault Detection in MEMS", Fourth International Conference on Computer Recognition System, Poland, May 2005, pp.877-884.
9. R.L.Streit and T.E.Luginbuhl, "Maximum Likelihood Training of Probabilistic Neural Network", IEEE Trans. Neural Networks, Vol.5, No.5, 1994, pp.764-783.
10. Z. Yang, Zwolinski, "Applying A Robust Heteroscedastic Probabilistic Neural Network to Analog Fault Detection and Classification", IEEE Trans. Computer-Aided Design of Integrated Circuits and Systems, Vol.19, No.1, 2000, pp.142-151.

Analog Fault Detection Using a Neuro Fuzzy Pattern Recognition Method*

Reza Asgary[1] and Karim Mohammadi[2]

IranUniversity of Science and Technology,
Department of Electrical engineering
R_A_Moghadam@yahoo.com, Mohammadi@iust.ac.ir

Abstract. There are different methods for detecting digital faults in electronic and computer systems. But for analog faults, there are some problems. This kind of faults consist of many different and parametric faults, which can not be detected by digital fault detection methods. One of the proposed methods for analog fault detection, is neural networks. Fault detection is actually a pattern recognition task. Faulty and fault free data are different patterns which must be recognized. In this paper we use a probabilistic neural network to recognize different faults(patterns) in analog systems. A fuzzy system is used to improve performance of network. Finally different network results are compared.

1 Introduction

The primary objective of cluster analysis is to partition a given set of data or subjects into clusters. Analytical, statistical and intelligent methods have been proposed to partition input data to clusters(subsets, groups or classes). Analog fault detection is a pattern recognition task. Some methods have been proposed for MEMS fault detection [1-3]. One of the best proposed methods for analog fault detection is neural networks [4]. Analog faults have statistical characteristics, so in this paper we use a Probabilistic Neural Network to recognize analog faults in Micro Electro Mechanical Systems, MEMS. A fuzzy system is used to readjust some weights in order to improve performance.

2 Robust Heteroscedastic Probabilistic Neural Network

A Probabilistic Neural Network (PNN), classifies data by estimating the class conditional probability density functions. To do this it requires a training phase, followed by a validation phase, before it can be used in a testing phase. A PNN consists of a set of Gaussian kernel functions. The original PNN uses all the training patterns as the centers of the Gaussian kernel functions and assumes a common variance or covariance, which is named homoscedastic PNN. To avoid using a validation data set and to determine analytically the optimal common

* This work has been partially supported by Iran Telecommunication Research Center.

W. Duch et al. (Eds.): ICANN 2005, LNCS 3697, pp. 893–898, 2005.

variance, a maximum likelihood procedure was applied to PNN training [5]. On the other hand, the Gaussian kernel functions of a heteroscedastic PNN are uncorrelated and separate variance parameters are assumed. This type of PNN is more difficult to train using the ML training algorithm because of numerical difficulties. A robust method has been proposed to solve this numerical problem by using the jackknife, a robust statistical method, hence the term 'Robust Heteroscedastic Probabilistic Neural Networks' [6]. The RHPNN is a four layer feedforward neural network based on the Parzen window estimator that realizes the Bayes classifier given by

$$g_{Bayes} = arg\left(max\left\{\alpha_j f_j(x)\right\}\right) \tag{1}$$

Where x is a d-dimensional pattern, $g(x)$ is the class index of x, the a priori probability of class $j (1 \leq j \leq k)$ is α_j and the conditional probability density function of class j is f_j. The object of the RHPNN is to estimate the values of f_j. This is done using a mixture of Gaussian kernel functions.

RHPNN has been shown in Fig.1. In this figure two classes are shown. First class is considered for fault free and the second class for faulty patterns. We assume only one fault free kernel because with only one Gaussian function all fault free patterns can be shown. There are many different faults and the distances between them are unknown, so in second class, more than one kernel is considered. The optimum number of kernels in second class is the minimum that each kernel has at least one faulty pattern. The first layer of the PNN is the input layer. The second layer is divided into k groups of nodes, one group for each class.

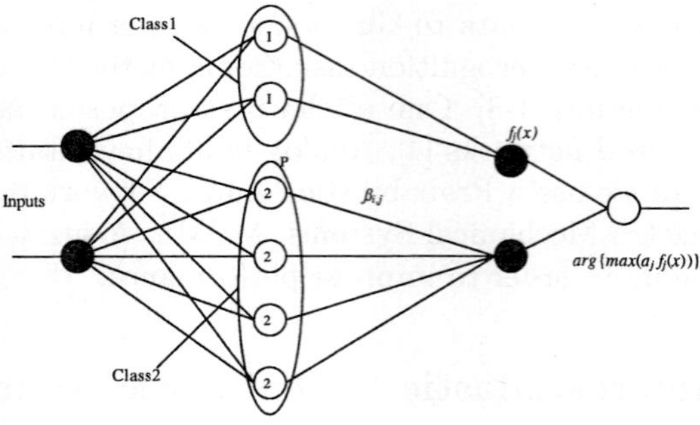

Fig. 1. Four layer feed forward RHPNN

The ith kernel node in the jth group is described by a Gaussian function

$$p_{i,j} = \frac{1}{(2\pi\sigma_{i,j}^2)^{d/2}} \exp\left(-\frac{\|x - c_{i,j}\|^2}{2\sigma_{i,j}^2}\right) \tag{2}$$

Where $c_{i,j}$ is the mean vector and $\sigma_{i,j}^2$ is the variance. The third layer has k nodes; each node estimates f_j, using a mixture of Gaussian kernels

$$f_j(x) = \sum_{i=1}^{M_j} \beta_{i,j} p_{i,j}(x), 1 \leq j \leq k \tag{3}$$

$$\sum_{i=1}^{M_j} \beta_{i,j} = 1, 1 \leq j \leq k \tag{4}$$

Where M_j is the number of nodes in the jth group in the second layer. The fourth layer of the PNN makes the decision from Eq(1). The PNN is heteroscedastic when each Gaussian kernel has its own variance. The centers, $c_{i,j}$, the variance, $\sigma_{i,j}^2$ and the mixing coefficients, $\beta_{i,j}$ have to be estimated from the training data. One assumption for α_j weights is $\alpha_j = \frac{1}{k}, 1 \leq j \leq k$. The EM algorithm has been used to train homoscedastic PNN's. Each iteration of the algorithm consists of an expectation (E) followed by a maximization process (M). This algorithm converges to the ML estimate. For the heteroscedastic PNN, the EM algorithm frequently fails because of numerical difficulties. These problems have been overcome by using a jackknife, which is a robust statistical method. RHPNN training algorithm consists of two recursive steps that finally determines kernel centers, variances and $\beta_{i,j}$ weights. All of the formulas can be found in Ref[6].

3 Fuzzy RHPNN

In RHPNN training phase, centers and variances of kernel functions change and finally fix at values to cover faults. Additionally, $\beta_{i,j}$ which are weights between second and third layers change to final values. In Fuzzy Robust Heteroscedastic Probabilistic Neural network (FRHPNN), all $\beta_{i,j}$ weights are readjusted by a fuzzy system. The FRHPNN learning has two phases. First, like RHPNN all training patterns (faulty and fault free) are applied to network and in two recursive steps, network parameters are defined [6]. In this phase all centers and variances and $\beta_{i,j}$s are adjusted. In second phase, all training patterns are applied to network again and for each misclassified pattern, a fuzzy system make proper changes to all $\beta_{i,j}$ weights.

When the first training phase is finished, for each faulty pattern, one or more kernels in second class, produce a great probability density function. In other words, when a faulty pattern is given to network, the outputs of one or more kernel in second class (faulty class), is greater than the output of fault free kernel (first cell in second layer). If faulty pattern is far from the fault free kernel center, the output of faulty group kernels is obviously greater than the output of fault free kernel. But for faulty patterns which are near the fault free kernel center, output of fault free kernel is near or sometimes greater than the output of faulty group kernels. In this case, faulty pattern is classified as a fault free pattern. We used second training phase to solve this problem.

3.1 Fuzzy System

$\beta_{i,j}$ are the weights connecting second layer to third layer. For each cell in the third layer, inputs are $\beta_{i,j}p_{i,j}(x)$ and output is $f_j(x)$ as shown in Eq(3). for each misclassified pattern, an error is defined. The error is the difference between other cell output and output of the current cell, both in third layer.

For each class and each input pattern x, maximum $\beta_{i,j}p_{i,j}(x)$ is defined as $MAX_k(x)$. All the inputs to third layer cell and also error are divided by $MAX_k(x)$ to be normalized as BP_N and e_N, respectively. Inputs of fuzzy system are e_N and BP_N. For every misclassified input pattern x, and each $\beta_{i,j}$, there are BP_N and e_N as inputs and fuzzy system produces an output. Output of fuzzy system, $\delta F(t)$, is used to update $\beta_{i,j}$ as:

$$\beta_{i,j}(t+1) = \beta_{i,j}(t)(1 + \delta F(t)) \tag{5}$$

After applying all misclassified patterns to change $\beta_{i,j}$ weights, all the input patterns are given to network and new set of misclassified patterns is defined. Again, for all of these misclassified pattern, fuzzy system is used to readjust $\beta_{i,j}$ weights. In each step all $\beta_{i,j}$ weights and the number of misclassified weights are saved. The optimum values of $\beta_{i,j}$ weights occur in accordance with the least number of misclassified patterns. In fig.2 all membership functions are shown. Table1 shows fuzzy rules.

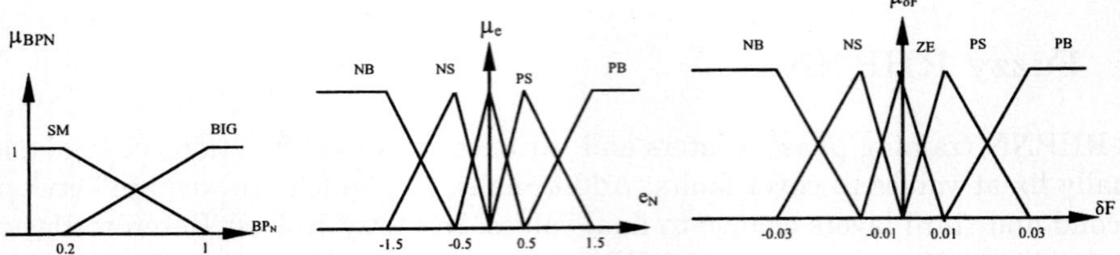

Fig. 2. Input and output membership functions

Table 1. Fuzzy rules

$\overrightarrow{e_N}$ / $\downarrow BP_N$	NB	NS	ZE	PS	PB
SM	NS	NS	ZE	PS	PS
BIG	NB	NS	ZE	PS	PB

4 Simulation Results

EM3DS is MEMS simulator software, which has been used for fault simulation in RF MEMS. 20 faults and one fault free pattern have been simulated in a RF low pass filter MEMS. These 20 faults consist of both digital and analog faults. Changing substrate resistance, magnetic and electric properties, shorts and opens, disconnections, connection between separate parts and some other faults have been simulated by software. The S parameters are calculated and used for training and testing all neural networks. We have used real and imaginary parts of S_{11}, 2 dimensional data, as input to neural networks. For each fault 3 patterns have been simulated, so total number of learning patterns is (20*3+7)=67.

Table 2. Fault detection results using RHPNN

	Detected as Faulty	Detected as Fault free	Results for RF LP filter	Detected as Faulty	Detected as Fault free	Results for RF capacitor
40 faulty pattern	37	3	%92.5	35	5	%87.5
10 fault free pattern	1	9	%90	2	8	%80
Total	(37+9)/	(40+10)=	%92	(35+8)/	(40+10)=	%86

Table 3. Fault detection results using FRHPNN

	Detected as Faulty	Detected as Fault free	Results for RF Lp filter	Detected as Faulty	Detected as Fault free	Results for RF capacitor
40 faulty pattern	38	2	%95	37	3	%92.5
10 fault free pattern	1	9	%90	2	8	%80
Total	(38+9)/	(40+10)=	%94	(37+8)/	(40+10)=	%90

The other RF MEMS which is simulated by EM3DS is Inter digital capacitor. 32 faults and one fault free patterns have been simulated. Total number of learning patterns, in this case, is (32*3+7)=103.

For training FRHPNN, at first all the patterns in the pool are used to find centers, variances and weights. Then network is able to group all the fault free and faulty patterns into n groups. The strategy for selecting the value of n is to ensure each kernel has at least one pattern of a fault free or faulty pattern, falling in it. The optimal n for RF low pass filter is 6 and for inter digital capacitor is 8. One kernel is belonged to fault free class and the others are belonged to faulty class. All the faulty patterns are labeled with the same number when training FRHPNN network. As mentioned, training is done in two phases, then 50 test patterns are given to network. Table2 shows the results of RHPNN fault detection in RF low pass filter and RF interdigital capacitor. Similarly, Table3 shows FRHPNN results for the same MEMS.

5 Conclusion

MEMS usually have nonlinear and complex models. Most of the times, novel and unknown faults occur in them, too. A powerful recognition method is essential to detect/diagnose the faults. This part can be inserted in MEMS as a Built In Self Test (BIST) mechanism. With respect to nonlinearity and novel faults in MEMS, neural networks are proposed as a BIST mechanism. In this paper, we propose FRHPNN for fault detection in MEMS in which for improving the performance, some of the weights are readjusted by a fuzzy system. Simulation results show that using the fuzzy system as second training phase, improves fault detection percentage. Extra work is needed to find better learning methods for determining centers and variances.

References

1. R. Muller, U. Wagner, W. Bernhard, "Reliability of MEMS-a methodical approach", Proc.11th European symposium on reliability of electron devices, failure physics and analysis, ,2001, pp.1657-62.
2. R. Rosing, A. Richardson, "Test Support Strategies for MEMS", IEEE International Mixed Signal Test Workshop, Whistler, Canada, June 1999.
3. TIMA Lab research reports, Http://Tima.imag.fr, 2002-4.
4. R.Asgary and K.Mohammadi, "Pattern Recognition and Fault Detection in MEMS", Fourth International Conference on Computer Recognition System, Poland, May 2005, pp.877-884.
5. R.L.Streit and T.E.Luginbuhl, "Maximum Likelihood Training of Probabilistic Neural Network", IEEE Trans. Neural Networks, Vol.5, No.5, 1994, pp.764-783.
6. Z. Yang, Zwolinski, "Applying A Robust Heteroscedastic Probabilistic Neural Network to Analog Fault Detection and Classification", IEEE Trans. Computer-Aided Design of Integrated Circuits and Systems, Vol.19, No.1, 2000, pp.142-151.

Support Vector Machine for Recognition of Bio-products in Gasoline

Kazimierz Brudzewski, Stanisław Osowski, Tomasz Markiewicz, and Jan Ulaczyk

Warsaw University of Technology,
00-661 Warsaw, Poland
sto@iem.pw.edu.pl

Abstract. The paper presents the application of Support Vector Machine for recognition and classification of the bio-products in the gasoline. We consider the supplement of such bio-products, as ethanol, MTBE, ETBE and benzene. The recognition system contains the measuring part in the form of semiconductor array sensors responding with a signal pattern characteristic for each gasoline blend type. The SVM network working in the classification mode processes these signals and associates them with an appropriate class. It will be shown that the proposed measurement system represents an excellent tool for the recognition of different types of the gasoline blends. The results are compared with application of multilayer perceptron.

1 Introduction

The paper is concerned with the recognition of the bio-products added to the gasoline. The bio-based fuels (bio-fuels) or bio-products such as ethanol, methyl tertiary butyl ether (MTBE), ethyl tertiary butyl ether (ETBE), tertiary amyl methyl ether (TAME) as the supplements to the gasoline may form an alternative solution to the increasing worldwide needs for a fuels in the next decade. The most commonly used alcohol is ethanol, while the most popular ether is MTBE. Other oxidants such as ETBE and TAME have recently started being used on a commercial scale. It is also well known that aromatic compounds, in particular benzene contribute to the reasonable octane properties.

The paper will consider the application of the artificial nose measurement system to the recognition and quantification of these bio-products added to the gasoline. The artificial nose is composed of the array of semiconductor sensors and the postprocessing stage in the form of Support Vector Machine (SVM), used as the calibrator. The results of numerical experiments will be presented and discussed in the paper. They will be compared to the results obtained by using multilayer perceptron.

2 The Description of the System

The recognition of the gasoline blends on the basis of its odour applies the fact that the blends are associated with different aroma resulting from varieties in technologies and their chemical composition. To solve the problem of the analysis of aroma we

W. Duch et al. (Eds.): ICANN 2005, LNCS 3697, pp. 899 – 904, 2005.

will use the electronic nose and Support Vector Machine (SVM). The patterns of signals of the vapour sensitive detectors are processed by the neural system composed of SVM and associated with different types of the gasoline blends.

In the computerized measurement system [2] we have applied the array of tin oxide-based gas sensors of Figaro Engineering Inc. These sensors have been mounted into an optimised test chamber. The chamber is placed in a mass flow controlled measurement system with laminar gas flow and controlled gas temperature conditions. The synthetic air used as the carrier was used for delivering an atmosphere from the 'head-space' of the sample chamber with the testing gasoline sample to sensors. The carrier flow, the temperature, the volume of the gasoline sample as well as the volume of the measuring chamber are kept constant in the whole measurements.

The features used for gasoline data analysis have been extracted from the averaged temporal series of sensor resistances R(j), one for each j-th sensor of the array. To obtain the consistent data for pattern recognition process some form of pre-processing of the data from the sensor array is necessary. We have used the relative variation r(j) of each sensor resistance

$$r(j) = \frac{R(j) - R_0(j)}{R_0(j)} \tag{1}$$

where R(j) is the actual resistance of the j-th sensor in the array and $R_0(j)$ represents the baseline value of resistance. The baselines values of the measured resistance of the sensors in the synthetic air atmosphere have been used as the reference.

All numerical experiments have applied seven tin oxide-based gas sensors (TGS815, TGS821, TGS822, TGS825, TGS824, TGS842, TGS822-modificated) from Figaro Engineering Inc., mounted into an optimised test chamber. The measurements were carried out under the following conditions: carrier flow - 0.2 l/min, gasoline temperature - 25°C, volume of the gasoline sample - 100 ml, the volume of the sample chamber - 200 ml. In the experimental system we have used 8-Channel Analog Input Module Rev.D1 type ADAM-4017 as serial communication interface with computer. The resistance sampling rate used in experiments was 30 times per minute. The measured sensor resistances R(j) have been pre-processed according to the relation (1) described in the previous section, delivering the relative variations of each sensor resistance r(j), for j=1, 2, …,7, used as the features. The feature vector **x** applied to the neural classifier takes the form $\mathbf{x} = [r(1), r(2), ..., r(7)]^T$.

The baseline resistance of the sensors used for generation of features r(j) according to the equation (1), was acquired at stabilized temperature of 25°C of a synthetic air. The baseline value of resistance was obtained by averaging 36 samples of the measured values within 72s. The sampling rate was 30 times per minute. The washing interval at the measurement was typically 10 min.

3 The Classifier Network

We have applied the Support Vector Machine as the basic recognizing and classifying system. The choice was made after checking other possible neural network solution, multilayer perceptron. Basically, the SVM is a linear machine working in the high dimensional feature space formed by the linear or non-linear mapping of the n-dimensional input vector **x** into a K-dimensional feature space usually of K>n through the use of functions $\varphi(\mathbf{x})$ forming the K-dimensional vector $\boldsymbol{\varphi}(\mathbf{x})$. SVM is a kernel type network using kernel function $K(\mathbf{x},\mathbf{x}_i)$ [3],[4]. The kernel $K(\mathbf{x},\mathbf{x}_i)$ is defined as the inner product of the vectors $\boldsymbol{\varphi}(\mathbf{x}_i)$ and $\boldsymbol{\varphi}(\mathbf{x})$, i.e., $K(\mathbf{x},\mathbf{x}_i)=\boldsymbol{\varphi}^T(\mathbf{x}_i)\boldsymbol{\varphi}(\mathbf{x})$. To the most often used kernel functions, satisfying the Mercer conditions belong: linear kernel $K(\mathbf{x},\mathbf{x}_i)=\left(\mathbf{x}^T\cdot\mathbf{x}_i+\gamma\right)$ and radial Gaussian kernel $K(\mathbf{x},\mathbf{x}_i)=\exp\left(-\gamma\|\mathbf{x}-\mathbf{x}_i\|^2\right)$. The learning problem of SVM is formulated as the task of separating learning vectors $\mathbf{x}_i$ (i=1, 2, ..., p) into two classes of the destination values, either $d_i=1$ (one class) or $d_i=-1$ (the opposite class), with the maximal separation margin. After some transformation it is reformulated as the quadratic programming task with respect to Lagrange multipliers [3]. The SVM network separates the data into two classes. The final equation of the hyperplane separating these two different classes is given by

$$y(\mathbf{x})=\sum_{i=1}^{N_{sv}}\alpha_i d_i K(\mathbf{x}_i,\mathbf{x})+w_0 \qquad (2)$$

where α_i are the nonzero Lagrange multipliers, d_i – the destinations associated with the input vectors $\mathbf{x}_i$, w_0 – bias and N_{sv} – the number of support vectors, that is the input vectors $\mathbf{x}_i$ associated with nonzero Lagrange multipliers. If the actual data vector **x** fulfils the condition $y(\mathbf{x})>0$ it will be regarded as the member of one class and when $y(\mathbf{x})<0$ – as the member of the opposite one.

The important role in learning fulfils the regularization constant C determining the balance between the complexity of the network, characterized by the weight vector **w** and the error of classification of learning data. Low values of C mean smaller significance of the learning errors on the adaptation stage and leads to the wider separation margin. For the normalized input signals the value of C is usually much bigger than 1 and adjusted by the cross validation technique.

Although SVM separates the data into two classes only, the recognition of more classes is straightforward by applying either "one against one" or "one against all" methods [8]. In "one against one" approach the SVM networks are trained to recognize between all combinations of two classes of data. For M classes we have to train M(M-1)/2 individual SVM networks. In the retrieval mode the vector **x** belongs to the class of the highest number of winnings in all combinations of classes. In "one against all" we build M SVM networks, each trained on the data opposing one class against the rest. The vector x belongs to the class of the highest decision value $y(\mathbf{x})$.

To compare the performance of SVM classifier we have also employed the multilayer perceptron (MLP) as the alternative solution of the classifying system. MLP network consists of many simple neuron-like processing units of sigmoidal activation function grouped together in layers [5]. The typical network contains one hidden layer followed by the output layer of neurons. Information is processed locally in each unit by computing the dot product between the corresponding input vector and the weight vector of the neuron. Training the network to produce a desired output vector $\mathbf{d}_i$ when presented with an input vector $\mathbf{x}_i$ involves systematically changing the weights of all neurons until the network produces the desired output within a given tolerance (error). This is repeated over the entire training set. Learning is just reduced to a minimization procedure of the error measure over the entire learning set, continued finite number of cycles to prevent over fitting [5]. The most effective learning methods rely on gradient application, where gradient is computed using the back propagation algorithm [5]. The highest efficiency of learning is achieved at application of Levenberg-Marquardt or BFGS algorithms [5].

4 The Results of Numerical Experiments

All numerical experiments have been performed on the pure extracted gasoline enriched by different supplements of various concentration of the blend. The added supplements included ethanol, ETBE, MTBE and benzene. Three different types of blends have been prepared. The first consisted of extracted gasoline and ethanol of different concentration (5%, 10%, 15% and 20% of volume). The second blend family was formed by adding two supplements to the extracted gasoline: MTBE and ETBE of the same proportion: MTBE (3%) and ETBE (97%). Four different blends of different concentrations of supplements (5%, 10%, 15% and 20% of volume) have been created in this way. The third blend family was created by adding benzene as the supplement. The same four concentrations (5%, 10%, 15% and 20% of volume) have been produced. In this way twelve gasoline blend types have been prepared altogether. In total, 432 gasoline samples have been produced for the study, from which 360 have been used as the training data set. The rest (72) available gasoline blend samples have been used as the testing data only.

The difficulty of the recognition task has been assessed by exploring the measured data using Principal Component Analysis (PCA). PCA is described as the linear transformation $\mathbf{y}=\mathbf{W}\mathbf{x}$, mapping the N-dimensional original vector $\mathbf{x}$ into K-dimensional output vector $\mathbf{y}$, where $K<N$. The vector $\mathbf{y}$ preserves most important elements of original information. The matrix $\mathbf{W}$ is the PCA transformation matrix composed of the eigenvectors of the correlation matrix $\mathbf{R}_{xx}$ associated with the set of input vectors $\mathbf{x}_i$. Fig. 1 presents the results of PCA analysis of the data. We have mapped the 7-dimenensional input data on three most important principal components PC1, PC2 and PC3, presenting the distribution of data as a 3D plot. The analysis of distribution of the data samples performed by using PCA tools has convinced us that the clusters corresponding to different supplements of the gasoline are well separated from each other. This gives good perspective for application of linear discriminant classifier, for example the linear kernel SVM network, the simplest possible solution. So the SVM networks of linear kernel have been applied for the recognition and classification. The gasoline blend samples have been split into two groups as mentioned in the previous subsection (360 samples in the learning set and 72 samples for testing only). The total number of learning as well testing data samples was equal (30 for learning and 6 samples for testing) for each category of the gasoline blends.

Two kinds of experiments have been performed. In the first one we have aimed at the recognition of the supplement only, without any attention to the concentration of the supplement. Three classes have been defined in this way. The first one corresponds to the ethanol as the supplement, the second – MTBE and ETBE and the third one – the benzene.

Some introductory experiments using cross-validation of the learning data have allowed to find the optimal value of the regularization coefficient, $C=300$. For the recognition of three classes we have applied the "one against all" approach [8]. For the 3-class recognition problem we had to train 3 of 2-class recognizing SVM networks. The trained SVM networks have been tested on the testing data. The performance of the network was perfect. All samples used in testing have been classified without error (100% of accuracy). The number of support vectors (the different learning vectors **x** chosen by all trained SVM networks) was 13.

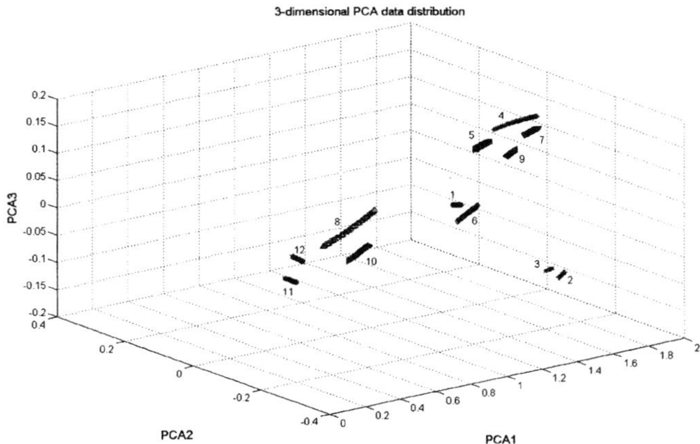

Fig. 1. The 3-dimensional PCA plots of the data samples. The numbers from 1 to 12 are the notations of gasoline blend types.

The problem of 12 class recognition (recognition of the supplement and its concentration) is more difficult and has to be dealt with more attention. The main task was to find the SVM network structures at application of linear kernel functions in order to achieve the best accuracy of classification for the testing data. The experiments have been also performed using cross-validation technique on the learning data to determine the optimal value of the regularization constant C. The optimal results have been obtained at C=2000 and linear kernel. The recognition of the gasoline blends was perfect (100% accuracy).

To compare the quality of solution we have trained the MLP of the same structure (no hidden layer) as SVM. It was also possible to train this network, recognizing the blends perfectly. Since both structures are the same it was interesting to compare their performance. We have done it in "one against all" mode by comparing the distances of the training data **x** from the separating hyperplane in both cases. The wider is the distance, the better the solution and more probable perfect performance at the future measurement data. Fig. 2 presents the results of such comparison for the recognition of the first class from all others. The superiority of SVM over MLP solution is evident. In all cases the distance of **x** to the hyperplane is much bigger for SVM network. The average distance of the vectors **x** to the hyperplane for MLP was equal 0.035, while for SVM network this value was 0.139.

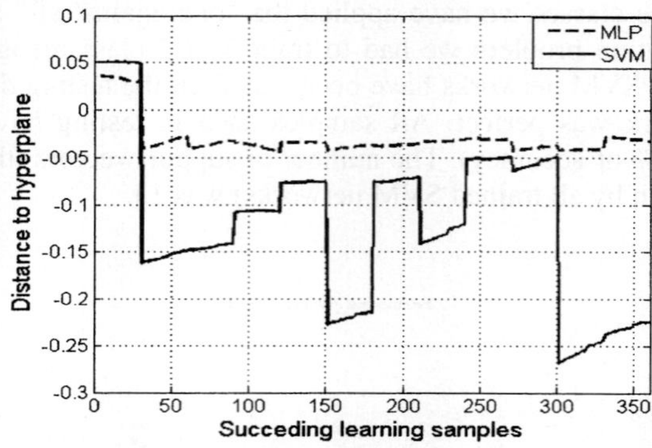

Fig. 2. The distances of learning data to separating hyperplane for MLP and SVM

5 Conclusions

The paper has presented the gasoline blend recognising system based on the application of the electronic nose and linear SVM classifier. The proposed solution has been tested on the samples of the gasoline blends related to different supplements of various concentrations. In all cases we have obtained 100% accuracy of the recognition of the samples. The results confirm that the proposed measurement system applying Support Vector Machine represents an excellent tool for the recognition of the type of the gasoline blend. The comparison to MLP has confirmed the superiority of SVM approach.

References

1. C. W. McCarrick, D. T. Ohmer, L. A. Gilliland, P. A. Edwards, Fuel identification by neural network analysis, Anal.Chem.,68 (1996),4264-4269.
2. K. Brudzewski, S. Osowski, T. Markiewicz, Classification of milk by means of an electronic nose and SVM neural network, Sensors and Actuators B-Chem. 98, pp. 291-298 , 2004
3. V. Vapnik, Statistical learning theory, Wiley, New York, 1998
4. C. Burges, A tutorial on support vector machines for pattern recognition, in: U. Fayyad (Ed.), Knowledge Discovery and Data Mining, Kluwer, 2000, pp.1-43
5. S. Haykin, Neural Networks - A Comprehensive Foundation, Macmillan, New York, 1999
6. O. L. Mangasarian, D. Musicant, Lagrangian SVMs, J. Mach. Learn. Res. 1 (2001) 161-177
7. J. Platt, Fast training of SVM using sequential optimisation, in: B. Scholkopf, C. Burges, A. Smola (Eds.), Advances in Kernel Methods, MIT Press, Cambridge, 1998, pp. 185-208
8. C. W. Hsu and C. J. Lin. A comparison of methods for multi-class support vector machines, IEEE Transactions on Neural Networks, 13, 2002, pp. 415-425

Detecting Compounded Anomalous SNMP Situations Using Cooperative Unsupervised Pattern Recognition

Emilio Corchado, Álvaro Herrero, and José Manuel Sáiz

Department of Civil Engineering, University of Burgos, Spain
escorchado@ubu.es

Abstract. This research employs unsupervised pattern recognition to approach the thorny issue of detecting anomalous network behavior. It applies a connectionist model to identify user behavior patterns and successfully demonstrates that such models respond well to the demands and dynamic features of the problem. It illustrates the effectiveness of neural networks in the field of Intrusion Detection (ID) by exploiting their strong points: recognition, classification and generalization. Its main novelty lies in its connectionist architecture, which up until the present has never been applied to Intrusion Detection Systems (IDS) and network security. The IDS presented in this research is used to analyse network traffic in order to detect anomalous SNMP (Simple Network Management Protocol) traffic patterns. The results also show that the system is capable of detecting independent and compounded anomalous SNMP situations. It is therefore of great assistance to network administrators in deciding whether such anomalous situations represent real intrusions.

1 Introduction

Intrusion Detection Systems (IDS) are tools designed to monitor the events occurring in a computer system or network, analysing them to detect suspicious patterns that may be related to a network or system attack. They have become a necessary additional tool to the security infrastructure as the number of network attacks has risen very sharply over recent years.

There are currently several techniques used to implement IDS. Some are based on the use of expert systems (containing a set of rules that describe attacks), signature verification (where attack scenarios are converted into sequences of audit events), petri nets (where known attacks are presented with graphical petri nets) or state-transition diagrams (representing attacks with a set of goals and transitions). One of the main disadvantages of these techniques is the fact that new attack signatures are not automatically discovered without updating the IDS.

Connectionist models have been identified as very promising methods of addressing the ID problem due to two key features: they are suitable to detect day-0 attacks and they are able to classify patterns (attack classification, alert validation). There have recently been several attempts to apply artificial neural architectures [1, 2] (such as Self-Organising Maps [3, 4] or Elman Network [5]) to the field of network security. This paper presents an IDS based on a neural architecture that has never before been applied to the problem of ID.

W. Duch et al. (Eds.): ICANN 2005, LNCS 3697, pp. 905–910, 2005.

2 The Cooperative Unsupervised IDS Model

Exploratory Projection Pursuit (EPP) [6, 7, 8, 9] is a statistical method for solving the complex problem of identifying structure in high dimensional data. It is based on the projection of the data onto a lower dimensional subspace in which its structure is searched by eye. It is necessary to define an "index" to measure the varying degrees of interest generated by each projection. Subsequently, the data is transformed by maximizing the index and the associated interest. From a statistical point of view the most interesting directions are those that are as non-Gaussian as possible.

The Data Classification and Result Display steps performed by this IDS model are based on the use of a neural EPP model called Cooperative Maximum Likelihood Hebbian Learning (CMLHL) [10, 11, 12]. It was initially applied to the field of Artificial Vision [10, 11] to identify local filters in space and time. Here, we have applied it to the field of Computer Security [2, 13, 14]. It is based on Maximum Likelihood Hebbian Learning (MLHL) [8, 9]. Consider an N-dimensional input vector, $\mathbf{x}$, and an M-dimensional output vector, $\mathbf{y}$, with W_{ij} being the weight linking input j to output i and let η be the learning rate. MLHL can be expressed as:

$$y_i = \sum_{j=1}^{N} W_{ij} x_j, \forall i . \tag{1}$$

The activation (e_j) is fed back through the same weights and subtracted from the input:

$$e_j = x_j - \sum_{i=1}^{M} W_{ij} y_i, \forall j . \tag{2}$$

Weight change:

$$\Delta W_{ij} = \eta . y_i . sign\left(e_j\right) | e_j |^{p-1} . \tag{3}$$

Lateral connections [10, 11] have been derived from the Rectified Gaussian Distribution [15] and applied to the MLHL. The resultant net can find the independent factors of a data set but do so in a way that captures some type of global ordering in the data set. So, the final CMLHL model is as follows:

Feed forward step: Equation (1)

Lateral activation passing: $$y_i\left(t+1\right) = \left[y_i(t) + \tau\left(b - Ay\right)\right]^+ . \tag{4}$$

Feed back step: Equation (2)

Weight change: Equation (3)

Where: η is the learning rate, τ is the "strength" of the lateral connections, b is the bias parameter and p is a parameter related to the energy function [8, 9, 11].

Finally A is a symmetric matrix used to modify the response to the data. Its effect is based on the relation between the distances among the output neurons.

3 Model Structure

The aim of this research is to design a system capable of detecting anomalous situations within a computer network. The information analysed by our system is obtained from the packets that travel along the network, meaning that it is a Network-Based IDS. The data needed to analyse the traffic is contained on the captured packets headers, obtained using a network analyser.

The structure of the IDS model is described as follows:

First step.- Network Traffic Capture: one of the network interfaces is set up in "promiscuous" mode. It captures all the packets travelling along the network.
Second step.- Data Pre-processing: the captured data is pre-processed and used as an input data in the following stage.
Third step.- Data Classification: once the data has been pre-processed, the connectionist model (section 2) analyses the data and identifies anomalous patterns.
Fourth step.- Result Display: the last step is related to the visualization stage. Finally the output is presented to the network administrator.

4 Real Data Sets Containing Compounded and Independent Anomalous SNMP Situations

We have decided to study anomalous SNMP situations because an attack based on this protocol may severely compromise system security [17]. CISCO [18] ranked the top five most vulnerable services in order of importance, and SNMP was one of them. In the short-term, SNMP was oriented to manage nodes in the Internet community [19].

Our efforts have focussed on the study of two of the most dangerous anomalous situations related to SNMP [2, 13, 14]:
SNMP port sweep: it is a scanning of network computers for the SNMP port using sniffing methods. The aim is to make a systematic sweep within a group of hosts to verify if SNMP is active in any port. Both default port numbers (161 and 162) and random port number (3750) are used.

MIB information transfer: the MIB (Management Information Base) can be defined in broad terms as the database used by SNMP to store information about the elements that it controls. This situation is a transfer of some information contained in the SNMP MIB. This kind of transfer is potentially quite a dangerous situation because anybody who possesses some free tools, some basic SNMP knowledge and the community password (in SNMP v. 1 and SNMP v. 2) will be able to access all sorts of interesting and sometimes useful information.

In this work, the IDS analysed three different data sets:

1st Data set (Fig 1): this includes an example of each one of the anomalous situations defined above: an SNMP port sweep and an MIB information transfer. We have

called this a compounded anomalous SNMP situation because it involves simple but different anomalous events that occur at the same time.
2nd Data set (Fig 2.a): this contains an example of an SNMP port sweep situation (an independent anomalous SNMP situation).
3rd Data set (Fig 2.b): an example of an MIB information transfer situation (another independent anomalous SNMP situation).

In addition to the SNMP packets, these data sets contain traffic related to other protocols installed in our network, such as NETBIOS and BOOTPS.

In the Data Pre-processing step, the system performs a data selection from all of the captured information. As a result, all of the above-mentioned data sets contain the following five variables extracted from the packet headers: timestamp (the time when the packet was sent in relation to the first one), protocol (all the protocols contained in the data set have been codified, taking values between 1 and 35), source port (the port number of the source host that sent the packet), destination port (the destination host port number to which the packet is sent) and size (total packet size in Bytes).

5 Results, Conclusions and Future Work

Scatterplot Matrix is used to analyse pairwise relationships between variables in high dimensional data sets. Each factor pair highlights different structures or clusters in the projections of the same data set. It was used to analyse the results obtained from the connectionist IDS model. The system identified (Fig 1.a) the two anomalous situations contained in the real compounded data set. The analysis took account of such aspects as traffic density or "anomalous" traffic directions.

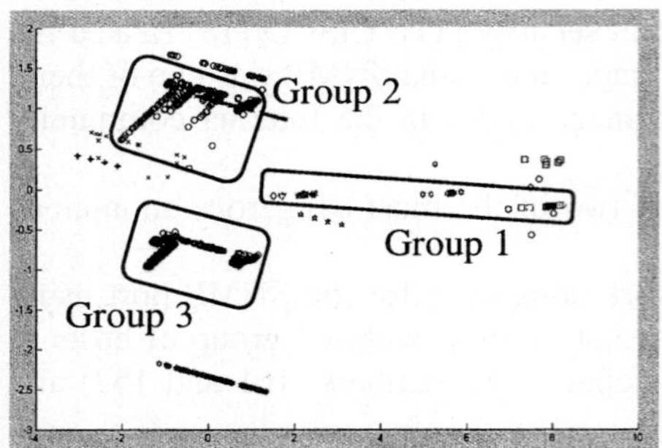

Fig. 1. a. Scatterplot Matrix factor pair 2-1 generated by the model for the 1st data set

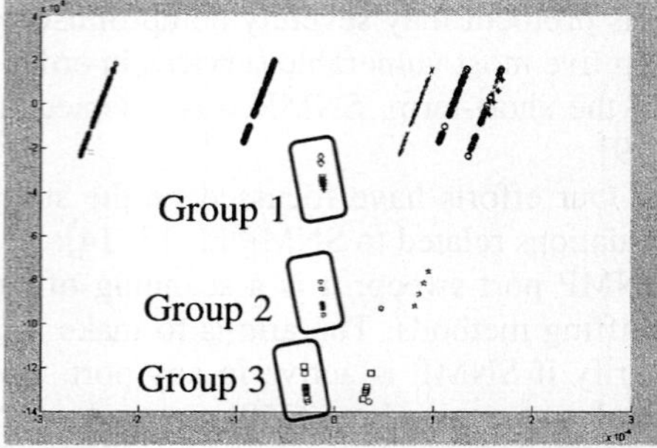

Fig. 1. b. PCA projection for the 1st data set

Factor pair 2-1 (Fig 1.a) contains the best representation of this anomalous situation, where the horizontal axe is related with the time feature and the vertical axe represents a combination of the protocol and size features. There are several issues to highlight about this figure: *Group 1* (Fig 1.a) identifies the sweep by means of normal and abnormal directions. It is clear that packets contained in this group do not progress in the same direction as the rest of packets groups (related to normal situations). On the other hand, *Groups 2* and *3* (Fig. 1.a) bring together packets related to the MIB information transfer. These groups are identified as anomalous due to their high temporal packets concentrations.

We have applied different connectionist methods such as Principal Component Analysis (PCA) [20] (Fig. 1.b) or MLHL to the same data set. CMLHL provides more sparse projections than the others [11]. CMLHL is able of identifying both anomalous situations while PCA (Fig. 1.b) is only able to identify the sweep (*Groups 1, 2 and 3*).

On the other hand, as can be seen in Fig. 2.a and Fig. 2.b, the neural IDS is capable of identifying both anomalous situations independently. The following figures (Fig. 2.a and Fig 2.b) show how the system performs successfully in those cases where there is only one anomalous situation within normal ones (2nd and 3rd Data Sets). In Fig 2.a we have identified the sweep (*Groups 1, 2 and 3*) by means of normal/abnormal direction and in Fig 2.b we have identified the MIB transfer (*Groups 1 and 2*) by means of high temporal concentration of packets.

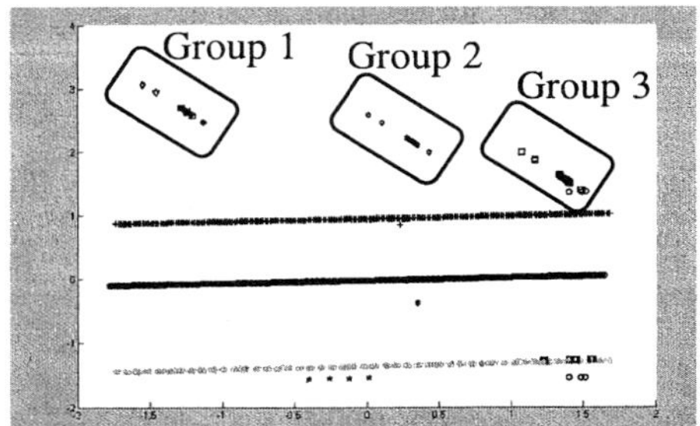

Fig. 2. a. Independent SNMP anomalous situation by a port sweep (2nd data set)

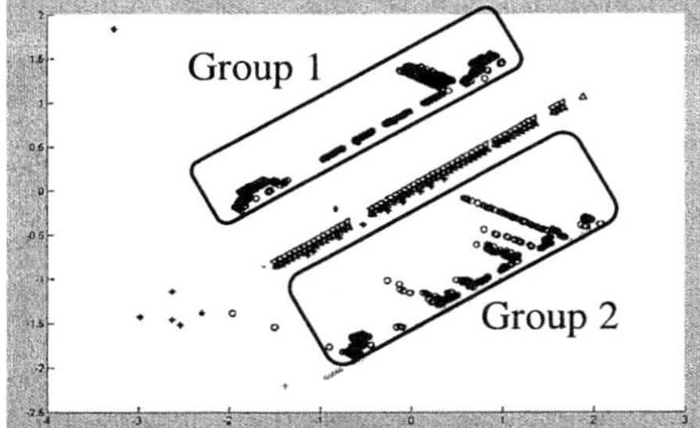

Fig. 2. b. Independent SNMP anomalous situation by a MIB transfer (3rd data set)

This research demonstrates the effectiveness and robustness of this novel IDS due to its capability to identify anomalous situations in two different ways: whether or not they are contained in the same data set. In summary, the connectionist IDS described in this paper is able to identify both independent and compounded anomalous SNMP situations showing its capability for generalization.

The visualization tool used in the Result Display step, shows data projections that highlight anomalous situations sufficiently clearly to alert the network administrator, taking into account such aspects as traffic density or "abnormal" directions.

One of the most common IDS techniques is the one called signature verification [20b]. Most of signature verification systems use pattern matching algorithms based on previously established rules included in a database. To reduce the number of posterior false alarms, this database should be adapted to the work environment by studying the traffic patterns that circulate along the network segment where the IDS is set up. One disadvantage of this method is the high processing time consume. This can be reduced by speeding up the packets analysis [21]. In comparison with this method, the advantages of our novel neural IDS are the following: it does not require any previous knowledge in the form of rules and it is able to detect unknown attacks day-0 ones.

Further work will be focused on the application of GRID [22] computation with more complex data sets and the use of multi-agent distributed systems.

References

1. Debar, H., Becker, M., Siboni, D.: A Neural Network Component for an Intrusion Detection System. IEEE Symposium on Research in Computer Security and Privacy (1992)
2. Corchado, E., Herrero, A., Baruque, B., Sáiz, J.M.: Intrusion Detection System Based on a Cooperative Topology Preserving Method. International Conference on Adaptive and Natural Computing Algorithms. Springer Computer Science. SpringerWienNewYork (2005)
3. Hätönen, K., Höglund, A., Sorvari, A.: A Computer Host-Based User Anomaly Detection System Using the Self-Organizing Map. International Joint Conference of Neural Networks (2000)
4. Zanero, S., Savaresi, S.M.: Unsupervised Learning Techniques for an Intrusion Detection System. ACM Symposium on Applied Computing (2004) 412-419
5. Ghosh, A., Schwartzbard, A., Schatz, A.: Learning Program Behavior Profiles for Intrusion Detection. Workshop on Intrusion Detection and Network Monitoring (1999)
6. Friedman, J., Tukey, J.: A Projection Pursuit Algorithm for Exploratory Data Analysis. IEEE Transaction on Computers 23 (1974) 881-890
7. Hyvärinen, A.: Complexity Pursuit: Separating Interesting Components from Time Series. Neural Computation 13 (2001) 883-898
8. Corchado, E., MacDonald, D., Fyfe, C.: Maximum and Minimum Likelihood Hebbian Learning for Exploratory Projection Pursuit. Data Mining and Knowledge Discovery. Kluwer Academic Publishing 8(3) (2004) 203-225
9. Fyfe, C., Corchado, E.: Maximum Likelihood Hebbian Rules. European Symposium on Artificial Neural Networks (2002)
10. Corchado, E., Han, Y., Fyfe, C.: Structuring Global Responses of Local Filters using Lateral Connections. Journal of Experimental and Theoretical Artificial Intelligence 15(4) (2003) 473-487
11. Corchado, E., Fyfe, C.: Connectionist Techniques for the Identification and Suppression of Interfering Underlying Factors. International Journal of Pattern Recognition and Artificial Intelligence 17(8) (2003) 1447-1466
12. Corchado, E., Corchado, J.M., Sáiz, L., Lara, A.: Constructing a Global and Integral Model of Business Management Using a CBR System. 1st International Conference on Cooperative Design, Visualization and Engineering (2004)
13. Herrero, A., Corchado, E., Sáiz, J.M.: A Cooperative Unsupervised Connectionist Model Applied to Identify Anomalous Massive SNMP Data Sending. 1st International Conference on Natural Computation (2005) ("*In press*")
14. Herrero, A., Corchado, E., Sáiz, J.M.: Identification of Anomalous SNMP Situations Using a Cooperative Connectionist Exploratory Projection Pursuit Model. 6th International Conference on Intelligent Data Engineering and Automated Learning (2005) ("*In press*")
15. Seung, H.S., Socci, N.D., Lee, D.: The Rectified Gaussian Distribution. Advances in Neural Information Processing Systems 10 (1998) 350-356
16. Myerson, J.M.: Identifying Enterprise Network Vulnerabilities. International Journal of Network Management 12 (2002)
17. Cisco Secure Consulting: Vulnerability Statistics Report (2000)
18. Case, J., Fedor, M.S., Schoffstall, M.L., Davin, C.: Simple Network Management (SNMP). RFC-1157 (1990)
19. Oja, E.: Neural Networks, Principal Components and Subspaces. International Journal of Neural Systems 1 (1989) 61-68
20. Aldwairi, M., Conte, T., Franzon, P.: Configurable string matching hardware for speeding up intrusion detection. ACM SIGARCH Computer Architecture News 33(1) (2005)
21. Foster, I., Kesselman, C.: The Grid: Blueprint for a New Computing Infrastructure. 1[t] edn. Morgan Kaufmann Publishers (1998)

Using Multilayer Perceptrons to Align High Range Resolution Radar Signals

R. Gil-Pita, M. Rosa-Zurera, P. Jarabo-Amores, and F. López-Ferreras*

Departamento de Teoría de la Señal y Comunicaciones, Universidad de Alcalá
Ctra. Madrid-Barcelona, km. 33.600, 28805, Alcalá de Henares - Madrid (Spain)
{roberto.gil, manuel.rosa, mpilar.jarabo, francisco.lopez}@uah.es

Abstract. In this paper we propose the use of Multilayer Perceptrons (MLPs) to align High Range Resolution (HRR) radar signals circularly shifted in time. To study the performance, the error of shift estimation is measured for different values of Signal to Noise ratio (SNR). The Zero Phase method is used for comparison purposes. Results show the best performance of the Zero Phase method with completely misaligned patterns, and the best performance of the MLP with low grades of misalignment. Using these results, a new method is proposed. First, the Zero Phase algorithm is used to pre-align the signals. Then, a MLP is trained using the pre-aligned signals in order to get more accuracy on the estimation of the shift. Results show an improvement up to 30%.

1 Introduction

Automatic classification of High Range Resolution (HRR) radar targets is a difficult task. This kind of radar uses broad-band linear frequency modulation or step frequency waveforms to measure range profiles (signatures) of targets [1] in order to increase the resolution in range of the received signal. Due to the inner characteristics of this HRR radar signals, small variations in the distance to the target cause circular shifts of the received signal. Results obtained with most of the classification algorithms are very dependent on shifts over the input signal. So, it is necessary to align the signal previously to any classification technique or preprocessing stage.

In the literature there are descriptions of the methods used to align HRR signals. In [2] a preprocessing method based on the extraction of the position of the main scatterer is proposed. Using the position of the main scatterer as reference, a new set of aligned profiles is obtained. This method is very sensible to the presence of noise. In [3] a non linear classification method based on the comparison of each pattern with all patterns in the training set is presented. This comparison takes into account possible shifts, scalings and DC component of the signals. For each input pattern and training pattern, the optimum shift, scale and direct current (DC) component is estimated trying to minimize a simplified version of

* This work has been supported by the "Consejería de Educación de la Comunidad de Madrid" (SPAIN), under Project 07T/0036/2003 1.

W. Duch et al. (Eds.): ICANN 2005, LNCS 3697, pp. 911–916, 2005.

the square error. A comparative study of different HRR alignment methods is carried out in [4]. The paper concludes that the best absolute alignment method is the Zero Phase method (ZP).

In this paper we propose a novel use of neural networks (NNs). NNs have never been used to align signals before. The capability to learn from the environment in noisy conditions makes them an option in signal alignment problems. So, we propose the use of Multilayer Perceptrons (MLPs) to align HRR radar signals. Results are compared with those obtained using the ZP alignment method. At last, using the obtained results, a new method is proposed, combining the ZP method with a MLP.

2 Materials and Methods

Our objective is to study the capabilities of MLPs to align HRR profiles. For this purpose, a database containing HRR radar profiles of six types of aircrafts has been used. The assumed target position is head-on with an azimuth range of 25° and elevations of -20° to 0°. The database contains 4349 profiles, and the length of each profile is 128. For the experiments, the database has been divided into three subsets: a training set composed of 1450 randomly selected profiles, a validation set composed of other 1449 randomly selected profiles, and a test set, composed of the resting 1450 profiles. The test set is used to assess the classifier's quality after training.

Each profile of the database has been randomly shifted, in order to study the capabilities of the alignment methods. Three experiments have been defined in function of the probability density function used to shift the patterns. In the first experiment, original data have been shifted using an uniform integer random variable from 0 to 127, which represents a complete misalignment of the profiles (100% misalignment), with a standard deviation of the shift of 36.95. In the second experiment, the shift varies from 0 to 63, which corresponds to a partial misalignment (50% misalignment) with a standard error deviation of the shift of 18.48. In the third experiment the shift varies from 0 to 31 (25% misalignment), which supposes a standard deviation of 9.24. So, the performance of the methods can be evaluated in function of the grade of misalignment of the data.

In this paper, the Signal to Noise Ratio (SNR) has been a parameter of the study, varying from 15 dB to 40 dB in steps of 5 dB. Due to the temporal localization of the energy of the signal, the SNR has been defined using the peak energy of the signal (1). A peak SNR value of 15 dB represents a very high amount of noise and a peak SNR of 40 dB implies a very low amount of noise. So, using this range of values we cover a wide range of SNR conditions.

$$SNR = 10 \log \left(\frac{\max\{x[n]\}^2}{\sigma_n^2} \right) dB \tag{1}$$

The performance of the alignment methods is given by the standard deviation of the error in the estimation of the shift of the signal. We have generated a correctly aligned database using the ZP method applied to the profiles without noise, and we have used this database as reference for the experiments. Only integer shifts have been considered in order to obtain the performance.

3 Aligning HRR Radar Signals Using the Zero Phase Algorithm

The ZP method has been previously used for alignment of panoramic images [5]. The basis of this method is the shift property of the Discrete Fourier Transform (DFT). For any function $x[n]$ with DFT $X(k)$, the DFT of $x[((n-m))_N]$ denoted by $Y(k)$ is given by (2):

$$DFT\{x[((n-m))_N]\} = Y(k) = X(k)\exp(-j2\pi km/N) \tag{2}$$

So, for a discrete shift m, and supposing the phase of the original signal $\phi(X(k))$ equal to zero, the phase of the k-th component of the Fourier transform $\phi(Y(k))$ $(k = 0, ..., N-1)$ will be shifted by $-2\pi km/N$. These phase shifts can be used to obtain an estimation of the discrete shift m.

A phase shift of $\phi + 2k\pi$ generates uncertainty, because it is taken like a phase shift of ϕ. This fact makes necessary to measure differences between two consecutive phase shifts. So, if the phase shift for the component k is $\phi(Y(k)) = -2\pi km/N$ and the phase shift for the component $k+1$ is $\phi(Y(k+1)) = -2\pi(k+1)m/N$ then the difference between both phases is:

$$\phi(Y(k)) - \phi(Y(k+1)) = 2\pi m/N \Rightarrow m = \frac{N}{2\pi}(\phi(Y(k)) - \phi(Y(k+1))) \tag{3}$$

And this value is annotated between 0 and 2π, solving the uncertainty. So, in order to study the shift of the signal, it is necessary to study the differences in phase of two consecutive Fourier components. Using $k = 0$, it is only necessary to study one phase, because the phase of the DC component $(k = 0)$ is always zero. Moreover, the assumption of $\phi(X(k))$ equal to zero is more suitable for low frequency values. So, applying (3) with $k = 0$ we obtain (4):

$$\phi(Y(1)) = -2\pi m/N \tag{4}$$

And so, the shifting can be estimated by (5):

$$m = -\frac{N}{2\pi}\phi(Y(1)) \tag{5}$$

Table 1 shows the shifting error standard deviation using the ZP method over the test set for the selected SNR values and the three grades of misalignment. Results demonstrate the accuracy of the method, which obtains error standard deviations lower than the unity with SNRs higher than 25 dB. Otherwise, this accuracy does not depend on the grade of misalignment of the profiles. The ZP method performance is nearly independent on the grade of misalignment of the original data. This fact makes this method unpractical for applications with low grades of misalignment and low values of SNR simultaneously.

Table 1. Error standard deviation aligning the test set with different SNRs, using the ZP method with the three grades of misalignment

SNR:	*15 dB*	*20 dB*	*25 dB*	*30 dB*	*35 dB*	*40 dB*
100% misalignment	8.08	3.11	1.53	0.89	0.62	0.48
50% misalignment	7.68	3.16	1.50	0.91	0.63	0.47
25% misalignment	7.47	3.12	1.55	0.88	0.60	0.48

4 Aligning HRR Radar Signals Using Multilayer Perceptrons

The Perceptron was developed by F. Rosenblatt [6] in the 1950s for optical character recognition. The Perceptron has multiple inputs fully connected to an output layer with multiple outputs. Each output y_j is the result of applying the linear combination of the inputs to a non linear function called activation function. Multilayer Perceptrons (MLPs) extend the Perceptron by cascading one or more extra layers of processing elements. These extra layers are called hidden layers, since their elements are not connected directly to the external world. In order to implement an alignment method, the number of inputs of the network must be equal to the length of the profile (128), and the network must have one output, in order to obtain a shift value to align the signal.

Cybenko's theorem [7] states that any continuous function $f : \mathbb{R}^n \rightarrow \mathbb{R}$ can be approximated with any degree of precision by a network of sigmoid functions. Therefore we choose an MLP with one hidden layer using the hyperbolic tangent sigmoid transfer function given in (6) as the activation function.

$$L(x) = \frac{1 - e^{-2x}}{1 + e^{-2x}} \tag{6}$$

In order to study the influence of the network size on the performance of the MLP-based method, we vary the number of neurons in the hidden layer M from 2 to 60 in increments of 2 ($M = 2, 4, 6, \ldots, 60$). The validation set has been used to obtain the best value of M.

The use of a MLP to align signals has never been studied, and their ability to solve some non linear problems can be very interesting in the field of study. There are several options in order to select the targets for training the NN. In this paper we have selected as targets the shifts needed to align each pattern. These values are normalized in order to use an output in the interval $(-1, 1)$. The MLPs are trained using the gradient descent with momentum and adaptive learning rate backpropagation algorithm. The validation set is used to early stop the training process.

Table 2 presents the results obtained for each SNR value with the profiles corresponding to the three grades of misalignment. Results show a global reduction of the error deviation when the misalignment of the profiles decreases. Comparing this results with those obtained in table 1, we can observe that the ZP method performs better than the MLP-based method, when completely

Table 2. Error standard deviation aligning the test set with different SNRs, using the MLP-based method with the three grades of misalignment

SNR:		*15 dB*	*20 dB*	*25 dB*	*30 dB*	*35 dB*	*40 dB*
100% misalignment	Lowest Error	16.62	11.65	7.88	7.15	6.27	6.80
	Neurons	18	40	38	20	24	10
50% misalignment	Lowest Error	6.61	3.54	2.64	2.50	1.99	1.92
	Neurons	8	6	20	20	20	12
25% misalignment	Lowest Error	3.64	2.15	1.48	1.19	1.12	1.09
	Neurons	6	8	8	12	6	12

misaligned patterns and high SNRs are considered. With partially misaligned patterns, the MLP is the best choice for low SNRs.

5 Aligning Signals Combining MLPs and the Zero Phase Algorithm

Analyzing results observed in tables 1 and 2 we can extract some conclusions. Results obtained using the MLPs are very different for different grades of misalignment, obtaining better results with low grades of misalignment. The error standard deviation obtained with the MLP directly depends on the grade of misalignment of the patterns. On the contrary, the error standard deviation obtained with the ZP method does not depend on the grade of misalignment.

Taking this fact into account, a new combined method is proposed, which consists in a two-stage alignment method. The first stage is a preprocessing one, in which all profiles are aligned using the ZP algorithm. In the second stage, a MLP is trained to realign these pre-aligned patterns, using the reconstructed profiles as inputs of the networks.

Table 3 presents the best results obtained with the completely misaligned profiles for each SNR value, and different number of hidden units. Due to the studied characteristics of the ZP method, results with other grades of misalignment are similar and, therefore, they are not considered. Results show the best performance of the proposed method with low SNR values, compared with the ZP method. This improvement does not depend on the number of hidden units,

Table 3. Error standard deviation aligning the test set with different SNRs, using the proposed MLP-based method with the completely misaligned patterns

SNR:	*15 dB*	*20 dB*	*25 dB*	*30 dB*	*35 dB*	*40 dB*
ZP	8.08	3.11	1.53	0.89	0.62	0.48
ZP + MLP with 2 hidden n.	6.69	2.15	1.14	0.76	0.60	0.49
ZP + MLP with 16 hidden n.	6.66	2.16	1.11	0.81	0.63	0.48
ZP + MLP with 28 hidden n.	6.75	2.15	1.18	0.81	0.64	0.48
Improvement	17.57%	30.87%	27.45%	14.61%	3.23%	0.00%

making two neurons the best choice in all cases. Otherwise, the use of the NN does not supposes an advantage for high SNR values. In these cases the ZP method obtains very low error standard deviations (less than the unity).

6 Conclusions

In this paper we propose the use of MLPs to align HRR radar signals. This kind of signals are presented circularly shifted in time, so they must be previously aligned to any later feature extraction stage. To study the performance of the alignment methods, the error standard deviation of the shift estimation is measured for different values of SNR. The ZP method is used for comparison purposes. Results show the best performance of the ZP method with high grades of misaligment and high values of SNR. Otherwise, the MLP performance depends on the grade of misalignment of the profiles, obtaining better results than those obtained with the ZP method with low misalignment and low SNR values. Using these results, a new method for completely misaligned patterns is proposed, which tries to combine the results obtained with both methods. First, the ZP algorithm is used to pre-align the signals. Then, a MLP is trained with the pre-aligned profiles, in order to get more accuracy on the estimation of the shift. Results show an improvement up to 30% with low SNR values.

As a global conclusion, we can propose a new alignment method which uses a NN to get more accuracy on the alignment of the profiles for low SNR values.

References

1. C.R. Smith, P.M. Goggans, "Radar Target Identification", *IEEE Antennas and Propagation Magazine*, vol. 35, no. 2, pp. 27-37, April 1993.
2. D. F. Fuller, A. J. Terzuoli, P. J. Collins, R. Williams, 1-D feature extraction using a dispersive scattering center parametric model, "Antennas and Propagation Society International Symposium", vol. 2, pp. 1296-1299, June 1998.
3. R. Wu, Q. Gao, J. Liu and H. Gu, "ATR scheme based on 1-D HRR profiles", *Electronic Letters*, vol. 38, no. 24, pp. 1586-1588, November 2002.
4. J. P. Zwart, R. van der Heiden, S. Gelsema, F. Groen, "Fast translation invariant classification of HRR range profiles in a zero phase representation", *IEE Proceedings Radar, Sonar and Navigation*, vol. 150, no. 6, pp. 411-418, December 2003.
5. T. Pajdla and V. Hlavac, Zero phase representation of panoramic images for image based location "Proc. 8th Int. Conf. on Computer analysis of images and patterns", pp. 550-557, 1999.
6. F. Rosenblatt, *Principles of Neurodynamics*, Spartan books, New York, 1962.
7. G. Cybenko, "Approximation by superpositions of a sigmoidal function", *Mathematics of Control, Signals and Systems*, vol. 2, pp. 303-314, 1989.

Approximating the Neyman-Pearson Detector for Swerling I Targets with Low Complexity Neural Networks

D. de la Mata-Moya, P. Jarabo-Amores, M. Rosa-Zurera, F. López-Ferreras, and R. Vicen-Bueno*

Departamento de Teoría de la Señal y Comunicaciones,
Escuela Politécnica Superior, Universidad de Alcalá,
Ctra. Madrid-Barcelona, Km. 33.600, 28805, Alcalá de Henares – Madrid (Spain)
{daid.mata, mpilar.jarabo, manuel.rosa, francisco.lopez, raul.vicen}@uah.es

Abstract. This paper deals with the application of neural networks to approximate the Neyman-Pearson detector. The detection of Swerling I targets in white gaussian noise is considered. For this case, the optimum detector and the optimum decision boundaries are calculated. Results prove that the optimum detector is independent on TSNR, so, under good training conditions, neural network performance should be independent of it. We have demonstrated that the minimum number of hidden units required for enclosing the optimum decision boundaries is three. This result allows to evaluate the influence of the training algorithm. Results demonstrate that the LM algorithm is capable of finding excellent solutions for MLPs with only 4 hidden units, while the BP algorithm best results are obtained with 32 or more hidden units, and are worse than those obtained with the LM algorithm and 4 hidden units.

1 Introduction

This paper deals with the application of neural networks (NNs) to approximate the Neyman-Pearson (NP) detector. This detector maximizes the probability of detection (P_D), while maintaining the probability of false alarm (P_{FA}) lower than or equal to a given value [1]. The problem of detecting radar echoes in additive white gaussian noise (AWGN) is studied.

Ruck et al. [2] and Wan [3] demonstrated that a NN can be used to approximate the optimum bayessian classifier, when trained using the least mean squared-error (LMSE) criterion. NNs have also been applied to approximate the NP detector [4,5,6]. In these works, multi-layer perceptrons (MLPs) with a hidden layer and one output, trained using the standard back-propagation algorithm (BP) are used. In [4] MLPs with ten inputs, five hidden units and one output are proposed for detecting deterministic signals in different environments, but no study is presented about the influence of network size. In [5,6] a trial and error strategy is carried out to find a trade-off

* This work has been supported by the "Consejería de Educación de la Comunidad de Madrid" (SPAIN), uder Project 07T/0036/2003 1.

W. Duch et al. (Eds.): ICANN 2005, LNCS 3697, pp. 917–922, 2005.

solution between performance and complexity. MLPs with sixteen inputs, eight hidden neurons and one output are proposed for detecting non-fluctuating targets in AWGN. But in any case, NN performance is compared to the NP detector one, and the fact that the studies are based on a trial an error process, does not allow us to evaluate the influence of network size and training algorithm efficiency separately.

In this paper, the application of NNs for detecting fluctuating targets is considered. For simulating the target return, the "Swerling 1" model (SWI) has been used [7]. Although this model can be too simple and in many cases do not characterize actual targets properly, it is useful to evaluate the powerful of alternative detection schemes, such us NNs. To evaluate the performance of any detection scheme that approximates the NP detector, the Receiver Operating Characteristic (ROC) curves of both detectors must be compared. For a P_{FA} value, the difference between the corresponding P_Ds must be as low as possible. For the case of study, the NP detector is easily obtained and evaluated. These ROC curves are compared with those estimated for the trained NNs, proving the ability of NNs to approximate the NP detector for the case of study. As no assumption is made about the detection problem during training, the obtained results can be generalized to prove the possibility of using NNs in practical situations where target and interference statistics are unknown and difficult to estimate.

As a previous step, the NP detector and the optimum decision boundaries are calculated. Taking into consideration the behavior of MLPs [8], the minimum number of hidden units that must be used is obtained. The standard BP and the Levenberg-Marquardt (LM) [9] algorithms are used for training the MLPs. Since the minimum number of hidden units is known, the influence of the algorithm can be studied.

2 Optimum Detector

We assume that the scanning radar collects N target echoes in a scan, and each input pattern, $\mathbf{z}=[z_1,z_2,\ldots,z_{2N}]^T$ is composed by the in-phase (the first N samples) and in-quadrature components (the remaining N samples) of each pulse. Under H_0, $\mathbf{z}$ is a vector of zero mean independent gaussian random variables with variance σ_n^2, so $f(\mathbf{z}|H_0)$ is a generalized gaussian of zero mean and covariance matrix $\mathbf{C_0} = \sigma_n^2 \cdot \mathbf{I}$, where $\mathbf{I}$ is the 2N×2N identity matrix. Under H_1, $f(\mathbf{z}|H_1)$ depends on the assumed target model. For Swerling 1 targets [7], the target magnitude is Rayleigh and the one-lag correlation coefficient is equal to unity. Assuming that the phase of radar echoes is an uniform random variable in the interval $[0,2\pi)$, and independent of the magnitude, the quadrature components of each returned pulse are independent gaussian random variables of zero mean and variance σ_s^2. So, under H_1, $\mathbf{z}$ is a vector of zero mean gaussian random variables with variance $\sigma_s^2+\sigma_n^2$, and $f(\mathbf{z}| H_1)$ is a generalized Gaussian of zero mean and covariance matrix C_1 given in (1), where **O**, **U** and **I** are NxN matrixes: **O** is a matrix of zeros, **U** is a unity matrix and **I** is the identity matrix.

$$\mathbf{C}_1 = \begin{pmatrix} \sigma_s^2\mathbf{U}+\sigma_n^2\mathbf{I} & \mathbf{O} \\ \mathbf{O} & \sigma_s^2\mathbf{U}+\sigma_n^2\mathbf{I} \end{pmatrix} \tag{1}$$

Defining the signal-to-noise ratio (SNR) as $SNR=10\log_{10}(snr)=10\log_{10}(\sigma_s^2/\sigma_n^2)$, and assuming $\sigma_n^2=1$, $SNR=10\log_{10}(\sigma_s^2)$.

To calculate the likelihood ratio, a SNR value must be assumed. When designing a NN based detector, this design SNR value is denoted as TSNR (Training Signal-to-Noise Ratio), because it is the SNR of the training set. $\mathbf{C}_1$ can be re-written as in (2), and an expression of the decision rule in the NP sense is presented in (3), where $\mathbf{I}$ is the (2Nx2N) identity matrix, η_{lr} is the detection threshold of the decision rule based on the likelihood ratio, and det($\mathbf{C}_1$) has been substituted by its value: $(1+N\cdot tsnr)^2$.

$$\mathbf{C}_1 = \begin{pmatrix} tsnr\mathbf{U}+\mathbf{I} & \mathbf{O} \\ \mathbf{O} & tsnr\mathbf{U}+\mathbf{I} \end{pmatrix} \tag{2}$$

$$\tfrac{1}{2}\cdot \mathbf{z}^{\mathrm{T}} \cdot \left(\mathbf{I}-\mathbf{C}_1^{-1}\right)\cdot \mathbf{z} \underset{\mathrm{H}_0}{\overset{\mathrm{H}_1}{\gtrless}} \ln\left[\eta_{lr}\left(1+N\cdot tsnr\right)\right] \tag{3}$$

3 Structure of the MLP-Based Neural Detector

If $\mathbf{\Sigma}= \mathbf{I}\text{-}\mathbf{C}_1^{-1}$, $\mathbf{z}^{\mathrm{T}}\mathbf{\Sigma}\mathbf{z}$ is a general quadratic function. When $\mathbf{\Sigma}$ is positive definite, the decision boundaries defined in (3) are hyper-ellipsoids in a space of 2N dimensions. If λ_{Σ} are the eigenvalues of $\mathbf{\Sigma}$, hyper-ellipsoids principal axes are given by the eigenvectors of $\mathbf{\Sigma}$, and its lengths are proportional to $(\lambda_{\Sigma})^{-1/2}$. In the case of study, $\mathbf{C}_1$ has 2N-2 eigenvalues equal to one and 2 eigenvalues equal to 1+Ntsnr, so $\mathbf{C}_1$ is positive definite. As the relation between $\mathbf{C}_1$ and $\mathbf{\Sigma}$ eigenvalues can be expressed as $\lambda_{\Sigma}=1-(1/\lambda_C)$, $\mathbf{\Sigma}$ has only two non zero eigenvalues and, hence, 2N-2 axes tends to infinity. So, the decision regions are open volumes, and the decision boundaries are hyper-cylinders.

In a 2N dimensions space, a hyper-cylinder can be enclosed with three hyper-planes, and this number will tend to infinity when the approximating error tends to zero. The sigmoid activation functions improve this approximation, but for an error being close to zero, the number of hidden units tends to infinity. As to implement an approximation of the optimum decision boundary, at least, the approximated boundary must enclose it, a one hidden layer MLP with a minimum number of three hidden neurons is needed. Clearly, the approximation error will depend on the number of hyper-planes (hidden units) and how the neuron activation functions transform them. Combining expressions (2) and (3), rule (3) is expressed as:

$$\left(\sum_{i=1}^{N} z_i\right)^2 + \left(\sum_{i=(P+1)}^{2N} z_i\right)^2 \underset{\mathrm{H}_0}{\overset{\mathrm{H}_1}{\gtrless}} 2\frac{1+Ptsnr}{tsnr}\ln\left[\eta_{cv}\left(1+Ptsnr\right)\right] \tag{4}$$

This expression reveals that, the distance to the hyper-cylinder axis is a sufficient statistic. So the approximation error will be a function of the hyper-cylinder radius and, because of that, of the desired P_{FA}.

Since $\sigma_n^2=1$, the required detection threshold for a given P_{FA} is independent of TSNR, so the NP detector is independent of this value, and when training a NN, its performance is expected to be independent of TSNR.

4 Design of Experiments and Results

In Air Traffic Control radar, the usual number of collected pulses in a scan is N=8, so MLPs with 16 inputs have been trained. A hard threshold detector has been used: if the NN output is greater than the threshold, H_1 is accepted, in other case, H_0 is accepted. To study the dependence on TSNR, different values have been selected. For each TSNR, separated training and validation sets composed of 50,000 randomly distributed patterns from H_0 and H_1 have been generated. To study the dependence on network size, MLPs with one hidden layer with different number of neurons have been trained. Taking into consideration the conclusion presented in previous section, starting with three hidden units, we have increased this value until no improvement has been observed. In all cases (NN size and TSNR), the log sigmoid transfer function has been used.

NNs have been trained for minimizing the LMSE, using two algorithms: the BP (with momentum and adaptive learning rate), and the LM (with adaptive parameter) [9]. While BP is based on the steepest descent method, the LM is based on the Newton method, and has been designed specifically for minimizing the LMSE.

A cross-validation technique has been used to avoid over-fitting and all NNs have been initialized using the Nguyen-Widrow method [10]. For each case, the training process has been repeated ten times, to check if the performances of the ten trained networks were similar in average.

Since for radar applications only low P_{FA} values are of interest, results are presented for P_{FA} lower than 10^{-4}. These values have been estimated using Importance Sampling techniques (relative error lower than 10% in the presented results) [11,12]. P_D values have been estimated using conventional Montecarlo simulation.

In figure 1, MLPs with 8 hidden units trained using the BP algorithm, show a high dependence on TSNR and poor detection capabilities. MLPs with 32 or more hidden units are needed to make TSNR dependence insignificant.

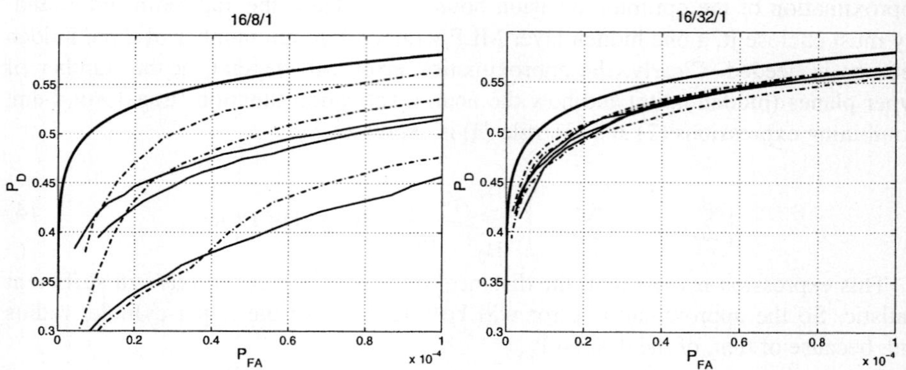

Fig. 1. ROC curves for MLPs trained using the BP algorithm and the LMSE criterion with TSNR= 0, 3, 7, 11, 15 and 19 dB, different number of hidden neurons (8 and 32), and SNR=3dB. The optimum detector ROC curve is drawn up with wider line.

In figure 2, ROC curves for MLPs trained using the LM algorithm are presented. LM not only is faster (training epochs can be reduced by more than an order of magnitude), but it is capable of finding a much better solution than the BP for a MLP with only 4 hidden neurons. This was a expected result, because for NNs which have up to a few hundreds of weights, the LM algorithm is more efficient that the BP with variable learning rate or the conjugate gradient algorithms, being able to converge in many cases when the other two algorithms failed to converge [9]. LM uses more information about the error surface in each iteration to find the minimum, and remain robust even if the needed line searches are only performed to relatively low accuracy.

Apart from the MLP trained with TSNR=19dB, the dependence of network performance on TSNR is insignificant, and the difference between the network ROC curve and the optimum detector one is lower than that observed for the MLPs with 32 hidden units trained using the BP. LM requires the estimation of the inverse of the hessian matrix. For TSNR=19dB, the determinant of the hessian matrix is close to zero and the MLP performance is clearly worse. As the hessian matrix is WxW, for a MLP with W weights, the results obtained for bigger networks are poorer (figure 2, MLPs with 8 hidden units).

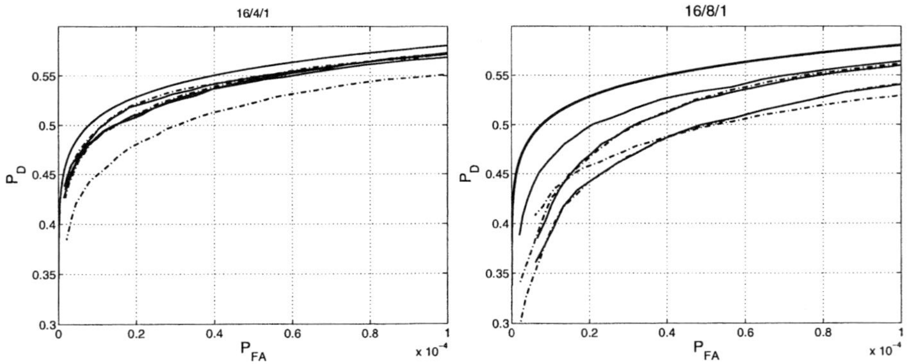

Fig. 2. ROC curves for MLPs trained using the LM algorithm with TSNR= 0, 3, 7, 11, 15 and 19 dB, different number of hidden neurons (4 and 8), and SNR=3dB. The optimum detector ROC curve is drawn up with wider line.

5 Conclusions

The application of NNs to approximate the NP detector for the detection of SWI targets in AWGN is studied. The optimum detector and decision boundaries are calculated, with two objectives: to calculate its ROC curves and to determine the minimum number of hidden units for enclosing the optimum decision boundaries, and implementing an approximation of the optimum detector. We have concluded that the optimum detector is independent on TSNR, so, under good training conditions, NN performance should be independent of this parameter. Comparing the NP detector ROC curves with those estimated for the trained NNs, we have evaluated the implemented approximations, and the influence of the TSNR. Also, we have demonstrated that the minimum number of hidden units required for enclosing the optimum decision boundaries is three. The conventional trial an error process for finding the best network structure has been transformed in a guided one. Results demonstrate that the LM algorithm is capable of finding very good solutions for MLPs with only 4 hidden units, while the BP algorithm best results are obtained with 32 or more hidden units, and are worse than that obtained with LM algorithm and 4 hidden units.

References

1. Van Trees, H.L: Detection, estimation, and modulation theory. Vol. 1. Willey. (1968)
2. Ruck, D.W. et al : The multilayer perceptron as an aproximation to a Bayes optimal discriminant function. IEEE Trans. on Neural Networks. Vol. 1, No. 1 (1990) 296-298.
3. Wan, E.A: Neural network classification: A Bayesian interpretation. IEEE Trans. on Neural Networks. Vol. 1, No. 1, (1990) 303-305.
4. Gandhi, P. P., Ramamurti, V: Neural Networks for Signal Detection in Non-Gaussian Noise. IEEE Trans. on Signal Proc. Vol. 45, No. 11 (1997) 2846-2851.
5. Andina, D, Sanz-gonzalez, J.L: On the problem of binary detection with neural networks. Proc. 38th Midwest Symp. on Circuits and Systems, Vol. 1 (1995)13-16.
6. Andina, D., Sanz-Gonzalez, J.L: Comparison of a neural network detector vs Neyman-Pearson optimal detector. Proc. of ICASSP-96, USA.(1996) 3573-3576.
7. Skolnik, M: Radar Handbook, Second Edition. McGraw-Hill, Inc. USA (1990)
8. Makhoul, J, El-Jaroudi, A., Schwartz, R: Partitioning capabilities of two-layer neural networks. IEEE Trans. on Signal Proc. Vol. 39, No. 6 (1991)1435-1440.
9. Hagan, M.T., Menhaj, M.B: Training feedforward networks with the Marquardt algorithm. IEEE Trans. on Neural Networks, Vol. 5, No. 6 (1994) 989-993.
10. Nguyen, D., Widrow, B: Improving the learning speed of 2-layer neural networks by choosing initial values of the adaptive weights. Proc. of the Int. Joint Conf. on Neural Networks, Vol 3 (1990) 21-26.
11. Grajal, J., Asensio, A: Multiparametric importance sampling for simulation of radar systems. IEEE Trans. on Aerospace and Electronic Systems, Vol. 35, No. 1 (1999) 123-137.
12. Sanz-Gonzalez, J. L, Andina, D: Performance analysis of neural network detectors by importance sampling techniques. Neural Proc. Letters, No. 9, (1999) 257-269.

Completing Hedge Fund Missing Net Asset Values Using Kohonen Maps and Constrained Randomization

Paul Merlin[1] and Bertrand Maillet[2]

[1] A.A.Advisors-QCG (ABN Amro Group), Variances and Paris-1, (TEAM/CNRS and SAMOS/MATISSE), 72 rue Regnault F-75013 Paris
paul.merlin@malix.univ-paris1.fr

[2] A.A.Advisors-QCG (ABN Amro Group), Variances and Paris-1, (TEAM/CNRS), 106 bv de l'hôpital F-75647 Paris cedex 13
bmaillet@univ-paris1.fr

Abstract. Analysis of financial databases is sensitive to missing values (no reported information, provider errors, outlier filters...). Risk analysis and portfolio asset allocation require cylindrical and complete samples. Moreover, return distributions are characterised by non-normalities due to heteroskedasticity, leverage effects, volatility feedbacks and asymmetric local correlations. This makes completion algorithms very useful for portfolio management applications, specifically if they can deal properly with the empirical stylised facts of asset returns. Kohonen maps constitute powerful non-linear financial classification tools (see [3], [4] or [6] for instance), following the approach of Cottrell *et al.* (2003), we use a Kohonen algorithm (see [2]), altogether with the Constrained Randomization Method (see [8]) to deal with mutual fund missing Net Asset Values. The accuracy of rebuilt NAV estimated series is then evaluated according to a comparison between the first moments of the series.

1 Introduction

The presence of missing data in the underlying time series is a recurrent problem for asset allocation and risk measure which require to deal with cylindrical and complete samples. Moreover, many financial databases contain missing values. For common stock returns measured at a low frequency, the Gaussian hypothesis is considered as a fairly good approximation, but financial assets such as options can introduce non-linearities and asymmetries to the portfolio returns. Because of the non-normality, symmetric measures of risk as the standard deviation cannot be applied; they do not distinguish between heavy left tails and heavy right tails. Hedge Fund asset return in this sense seems to be very particular. Several empirical studies conclude that many hedge fund index return distributions are not normal and exhibit negative skewness, positive excess kurtosis, and highly significant positive first-order autocorrelation (see [1] for instance). Thus, for hedge fund asset class, higher moments should be taken into account for the analysis. The importance of higher moments of returns, especially the skewness and kurtosis in evaluating portfolio risk and performance has been

W. Duch et al. (Eds.): ICANN 2005, LNCS 3697, pp. 923 – 928, 2005.

already highlighted by a number of authors, proposing and analyzing the inclusion of higher moments in portfolio theory. For illustration in the following, we extracted from the large HFR™ database, a dataset of hedge fund net asset values composed with 49 funds on a 5-year period of 60 monthly values. Note that, at purpose, no missing values are contained in this database.

2 Classical Self-Organized Maps Algorithm

The SOM algorithm is based on the unsupervised learning principle where the training is entirely data-driven and no information about the input data is required (see [5]). The SOM consist of a network, compound in n neurons, units or code vectors organised on a regular low-dimensional grid. If $I=[1, 2, ..., n]$ is the set of the units, the neighbourhood structure is provided by a neighbourhood function Λ defined on I^2. The network state at time t is given by:

$$\mathbf{m}(t)=[\mathbf{m}_1(t), \mathbf{m}_2(t), ..., \mathbf{m}_T(t)] \tag{1}$$

where $\mathbf{m}_i(t)$ is the T-dimensional weight vector of the unit i.

For a given state $\mathbf{m}$ and input $\mathbf{x}$, the winning unit $i_w(\mathbf{x},\mathbf{m})$ is the unit whose weight $\mathbf{m}_{i_w(\mathbf{x},\mathbf{m})}$ is the closest to the input $\mathbf{x}$.

The SOM algorithm is recursively defined by the following steps:

1. Draw randomly an observation $\mathbf{x}$.
2. Find the winning unit $i_w(\mathbf{x},\mathbf{m})$ also called the Best Matching Unit (noted BMU) such that:

$$BMU_{t+1}=i_w[\mathbf{x}(t+1),\mathbf{m}(t)]=\underset{\mathbf{m}_i, i\in I}{Argmin}\left\{\left\|\mathbf{x}(t+1)-\mathbf{m}_i(t)\right\|\right\} \tag{2}$$

where $\|.\|$ is the Euclidian norm.

3. Once the BMU is found, the weight vectors of the SOM are updated so that the BMU and his neighbours are moved closer to the input vector. The SOM update rule is:

$$\mathbf{m}_i(t+1)=\mathbf{m}_i(t)-\varepsilon_t \Lambda(BMU,i)[\mathbf{m}_i(t)-\mathbf{x}(t+1)], \forall i\in I \tag{3}$$

where ε_t is the adaptation gain parameter, which is]0,1[-valued, generally decreasing with time. The number of neurons taken into account during the weight updates depends on the neighbourhood function Λ that also generally decreases with time (see [5]).

Figure 1 represents the code vectors obtained using the dataset of hedge funds described above.

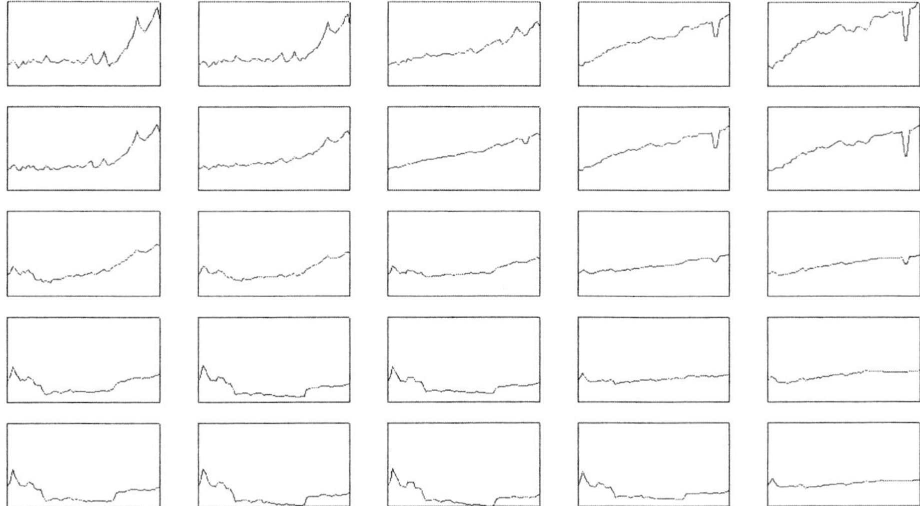

Fig. 1. Representation of Code Vectors on the Kohonen maps

3 Self-Organizing Maps with Partial Data Algorithm

SOM allows for classification of data samples with multiple variables and missing values (see [7]). Cottrell *et al.* (2003) propose an adapted Kohonen algorithm that first clusters the data, and then replaces the missing observations (see [2]). When the SOM algorithm iterates, if a vector $\mathbf{x}$ with missing value(s) is drawn, we consider the subset *NM* of variables which are not missing in vector $\mathbf{x}$. We define a norm on this subset (denotes $\|\cdot\|_M$) that allows us to find the *BMU* (with previous notations):

$$BMU = i_w[\mathbf{x}(t+1), \mathbf{m}(t)] = \underset{i \in I}{Argmin} \{\|\mathbf{x}(t+1), \mathbf{m}(t)\|_M\} \tag{4}$$

with:

$$\|\mathbf{x} - \mathbf{m}_i\|_M = \sum_{k \in NM} (\mathbf{x}_k - \mathbf{m}_{i,k})^2$$

where:

$$\begin{cases} \mathbf{x}_k \text{ for } k = [1, \ldots, T] \text{ denotes the } k^{th} \text{ value of the chosen vector;} \\ \mathbf{m}_{i,k} \text{ for } k = [1, \ldots, T], \text{ for } i = [1, \ldots, n] \text{ is the } k^{th} \text{ value of the } i^{th} \text{ code vector;} \\ NM = \text{is the set of the net asset values } \mathbf{x}_k \text{ that are not missing.} \end{cases}$$

Once the Kohonen algorithm has converged, we got some cluster containing our time series. Cottrell *et al.* (2003) first propose to fill the missing values of time-series by the cross-sectional mean of observed values present in the cluster.

4 Combing Self-Organized Maps and Constrained Randomization for Data Completion

Such an approach, when dealing with financial time series, will affect drastically some important statistical properties of the over-all rebuilt dataset. In particular, higher moments (second, third and fourth centred moments), auto-correlations and the correlations with the other time-series are neglected in the analysis. We propose here to combine the Self-Organizing Maps, adapted to the presence of missing values, and the Constrained Randomization algorithm introduced in [8]. This last computational method - initially presented as a specific reshuffling data sampling technique - allows for the simulation of artificial time-series that fulfil given constraints, but are random in other aspects.

The Figure 2 summarizes the proposed procedure for data completion. The first step starts with computing some empirical features of the data (moments of returns in our present case). Then, in parallel, a SOM is run with the non-missing values in the original dataset. Coordinates of Code Vectors in each of *BMU* are then considered as natural first candidates for missing value completion. The constrained randomization, using as constraints some of the empirical features of the data determined at the first step, can then start. If the candidate meets the constraints, then it takes the place of the missing value into the original data; if not, a standard normal residual is drawn, then added to the previous candidates and the test for the constraints starts again. This process lasts until all constraints are fulfilled and all missing values replaced.

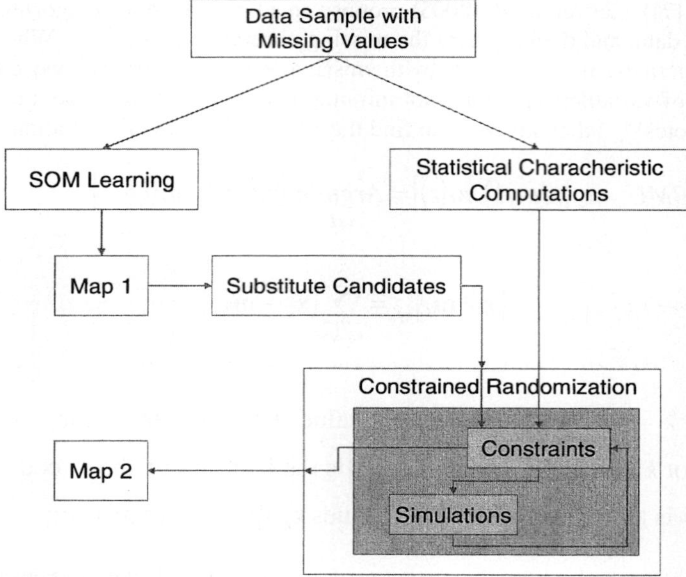

Fig. 2. Representation of the Scheme when Mixing Self-Organizing Maps and Constrained Randomization in Data Completion

5 Empirical Illustrations

Table 1 and Table 2 hereafter summarize the mean properties of the errors in moments when - respectively - using the adapted Kohonen algorithm alone and the two-step procedure presented in this article. As a first remark, we can note that - with no surprise - the addition of a Constrained Randomization procedure allows to recover missing values that more in line with the statistical characterization of the original series, as indicated by the comparison of Table 1 and Table 2. As the second remark, this is true in our example for all (reasonable) level of missing values in the original database. Finally, in this example, the improvement of accuracy regarding the moments is between 11% and 46%, the mean improvement is of order of 27%.

Table 1. Mean Errors on Moments when using the adapted SOM algorithm for Missing Values-fifty draws

	Absolute Error (in %) after Completion *via* Kohonen Maps			
Missing Values (in %)	Mean	Variance	Skewness	Kurtosis
5.00	2.99	6.96	11.55	9.81
10.00	5.38	12.71	20.71	17.71
15.00	7.47	17.82	27.20	24.21
20.00	9.15	23.51	31.90	29.74
25.00	10.85	28.04	36.89	34.40
30.00	13.91	34.04	40.44	40.85
35.00	15.59	38.83	45.62	45.83
40.00	19.57	42.81	45.04	47.97
45.00	21.62	47.01	54.13	52.88
50.00	23.60	54.77	61.67	62.21

Source: HFR™; Monthly Net Asset Values (12/1999-12/2004). Computations from the authors.

Table 2. Mean Errors on Moments when using the adapted SOM algorithm for Missing Values and Constrained Randomization - fifty draws

	Absolute Error (in %) after Completion *via* Kohonen Maps Combined with Constrained Randomization			
Missing Values (in %)	Mean	Variance	Skewness	Kurtosis
5.00	2.65	4.63	6.26	5.59
10.00	4.56	8.95	12.99	10.52
15.00	6.13	13.39	17.72	15.43
20.00	7.26	17.69	21.55	20.13
25.00	8.55	22.08	27.26	24.82
30.00	9.42	26.65	30.79	30.38
35.00	10.43	30.04	34.07	34.62
40.00	12.02	33.48	36.22	36.82
45.00	12.98	37.42	41.45	41.47
50.00	14.43	43.82	48.51	49.56

Source: HFR™; Monthly Net Asset Values (12/1999-12/2004). Computations from the authors.

For a more illustrative example, let us suppose a 17% annualized return fund. We destruct artificially 5%, 20% and 50% of the time series, at a 5% level of missing values, both methodologies get the same result with 0.5 points error on annualized return estimated (the annualized return estimated is between 16.5% and 17.5%). At a 20% level of missing value, the difference between the two methodologies is more observable: 1.5 points for a completion with Kohonen maps *versus* 1 point for a completion with Kohonen Maps combined with Constrained Randomization Method. At a 50% level of missing value, the difference becomes explicit: 4 points for a completion with Kohonen Maps *versus* 2 points for a completion with Kohonen Maps combined with Constrained Randomization Method.

6 Conclusion

The presented method for data completion uses SOM description of the data as the starting point for a constrained randomization. The main interest of the technique can be found in the fact that some of the important empirical features of the input are respected during the rebuilding process of missing observations. Specifically higher moments, whose accuracy of estimations are crucial in some financial applications, are taken into account when substitutions. Moreover, one can easily think about some generalizations of the proposed algorithm, adding for instance some features under studies into the constraints of the so-called Constrained Randomization procedure, such as local correlation structure or tail of the density focuses, depending on what is the final aim of the financial applications (asset allocation or risk management). One may also think about the robustness of the algorithm, namely specifying robust estimators in the constraints and allowing for data resampling when building the Kohonen Maps (see [4]).

References

1. Agarwal, V., Naïk, N.: Multi-period Performance Persistence Analysis of Hedge Funds, Journal of Financial and Quantitative Analysis 35 (2000), 327-342.
2. Cottrell, M., Ibbou, S., Letrémy, P.: *Traitement des données manquantes au moyen de l'algorithme de Kohonen*, in french in Proceedings of the tenth ACSEG Conference (2003), 12 pages.
3. Cottrell, M., de Bodt, E., Grégoire, P.: Financial Application of the Self-Organizing Map, Proceedings of EUFIT'98 (1), Verlag Mainz, (1998), 205-209.
4. De Bodt, E., Cottrell, M.: Bootstrapping Self-Organizing Maps to Assess the Statistical Significance of Local Proximity, European Symposium on Artificial Neural Networks (2000), 245-254.
5. Kohonen, T.: Self-Organizing Maps, Springer, Berlin (1995), 362 pages.
6. Maillet, B., Rousset, P.: Classifying Hedge Funds using Kohonen Map, in Connectionist Approaches in Economics and Management Sciences, Series in Advances in Computational Management Science, Vol. 6, Cottrell-Lesage (Eds), Kluwer Academic Publisher, 2003, 233-259
7. Samad, T., Harp, S.: Self Organization with Partial Data, Network 3, (1992), 205-212.
8. Schreiber, T.: Constrained Randomization of Times Series Data, Physical Review Letter 80 (10) (1998), 2105-2108.

Neural Architecture for Concurrent Map Building and Localization Using Adaptive Appearance Maps*

St. Mueller, A. Koenig, and H.-M. Gross

Department of Neuroinformatics and Cognitive Robotics,
Ilmenau Technical University,
98684 Ilmenau, Germany
Steffen.Mueller@tu-ilmenau.de

Abstract. This paper describes a novel omnivision-based Concurrent Map-building and Localization (CML) approach which is able to localize a mobile robot in complex and dynamic environments. The approach extends or improves known CML techniques in essential aspects. For example, a more flexible model of the environment is used to represent experienced observations. By applying an improved learning regime, observations which are not longer of importance for the localization task are actively forgotten to limit complexity. Furthermore, a generalized scheme for hypotheses fusion is presented that enables the integration of further multi-sensory position estimators.

1 Introduction

Robust self-localization plays a central role in our long-term research project PERSES (PERsonal SErvice System) which aims to develop an interactive mobile shopping assistant which can autonomously guide its user within a home store [1]. To accommodate the challenges that arise from the specifics of this scenario and the characteristics of the operation area, a regularly structured, maze-like and populated environment, we placed special emphasis on vision-based methods for robot navigation. In our previous approach [1], we have employed a static graph representation as map of the environment, which is build up manually. The nodes of the graph are labeled with visual observations extracted from omnidirectional images and corresponding position information. Given this map, localization was realized employing a Particle Filter to estimate the robot's state. The main drawback of this and other appearance-based approaches for localization published in recent years is, however, that localization is only possible in manually mapped areas. Furthermore, the learned map is only valid as far as no important modifications of the operation area occur. Therefore, we developed an alternative technique which is able to perform an omnivision-based Concurrent Map-building and Localization (CML) to overcome this drawback. Inspired by former approaches like [5] but especially the work of Porta and Kroese [2]

* This work is partially supported by TMWFK-Grant # B509-03007 to H.-M. Gross.

W. Duch et al. (Eds.): ICANN 2005, LNCS 3697, pp. 929–934, 2005.

and continuing our former work, we present a neural architecture (see Fig. 1), which is able to track multiple state hypotheses (position and orientation of a mobile robot) in a short-term memory (STM) using odometry data and previous state estimations, while building up a kind of long-term memory (LTM) used for associating omnidirectional views to already observed and learned states. This appearance map afterwards directly influences the tracked state hypotheses in the STM to reduce their uncertainty.

Main advantage of this approach is the advanced learning scheme used in the LTM. The network is able to actively forget information about observations that became irrelevant because of changes in the environment. This guarantees that the complexity remains limited for a given operation area and independent from working time, which is of fundamental importance for a continuous duty.

2 Neural Architecture for Probabilistic Localization

Our architecture consists of three main components, the short-term memory (STM), the long-term memory (LTM), and the fusion subsystem shown in Fig. 1. The STM is responsible for representing the distribution of possible states the robot might currently be in. By placing linear RBF-neurons in the *State Space* S (x, y, ϕ) and summing up their weighted outputs, this structure represents a Mixture of Gaussians (MoG) characterizing one hypothesis for the current

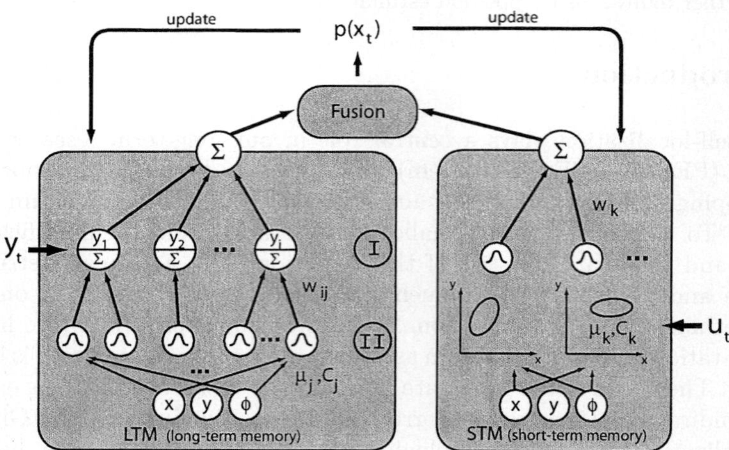

Fig. 1. Architcture of our probabilistic localization system: last known state hypothesis (position x,y and orientation ϕ of the robot) from STM (right) and a hypothesis from LTM (left) resulting from current observation $\boldsymbol{y}_t$ become merged and approximated by a Mixture of Gaussians $p(\boldsymbol{x}_t)$. Afterwards, this resulting distribution (top) is used to adapt STM again and to teach the observation-state associations in LTM. During the next step, hypotheses in STM will be updated using odometry data $\boldsymbol{u}_t$ and a motion model, then the output $p(\boldsymbol{x}_t)$ can be estimated again.

state estimation. The resulting activity h^{STM} at the STM-output node can be determined as follows:

$$h^{STM}(\boldsymbol{x}_t) = \sum_k w_k \phi(\boldsymbol{x}|\boldsymbol{\mu}_k, \boldsymbol{C}_k) \tag{1}$$

whereby ϕ is a Gaussian with mean $\boldsymbol{\mu}_k$ and covariance matrix $\boldsymbol{C}_k$, and w_k is the weight of the respective connection to the output node. The LTM (Fig. 1, left) consists of a layer of nodes representing prototypes $\boldsymbol{y}_i$ of observations and performing a clustering of the *Observation Space*. Each node in this layer I receives the current observation $\boldsymbol{y}_t$ and a weighted sum of activity from layer II, which consists of linear RBF-neurons connected to exactly one prototype node of layer I. Experiments showed that connections to more than one prototype nodes destabilize the state estimation. While layer II nodes are representing positions in *State Space*, layer I combines them considering the similarity between the respective reference observation $\boldsymbol{y}_i$ and the current observation $\boldsymbol{y}_t$, whereby $S(\boldsymbol{y}_t, \boldsymbol{y}_i)$ is a similarity function delivering a maximum (1.0) for identical views and decreasing continuously to zero up to a minimum similarity. As a result, the activity h^{LTM} at the LTM-output node is given by:

$$h^{LTM}(\boldsymbol{x}_t|\boldsymbol{y}_t) = \sum_i (S(\boldsymbol{y}_t, \boldsymbol{y}_i) \cdot \sum_j w_{ij} \phi(\boldsymbol{x}|\boldsymbol{\mu}_j, \boldsymbol{C}_j)) \tag{2}$$

The LTM-output node integrates the activation over all reference nodes, such that the resulting output characterizes the distribution of possible states under the given observation. The sum of activation characterizes the certainty of this hypothesis resulting from more or less similarity between observation $\boldsymbol{y}_t$ and the learned prototypes. Concerning this, the output is not a true probability distribution because weights do not sum up to one.

The last component, that receives the two hypotheses h^{STM} and h^{LTM}, is responsible for their fusion. In this module, a kind of probabilistic inference takes place, which leads to a probability distribution of the robot's current state.

2.1 Fusion of Hypotheses

The fusion module has to evaluate the activity distribution of different sources of information in the *State Space*, in the case shown here of $h^{LTM}(\boldsymbol{x}_t|\boldsymbol{y}_t)$ and $h^{STM}(\boldsymbol{x}_t)$, but hypotheses from further state estimators can be integrated. To simplify the fusion process, inputs are given in form of a weighted sum of Gaussians, whereas different to a mixture probability the sum of the weights w^i needs not to be one. First, in this pool of Gaussians one has to decide which Gaussians are representing the same hypothesis. Therefore, a spatial distance criterion is applied, similar to [2] the Mahalanobis distance is employed. So the inference can realize a logic AND for all the combinations of Gaussians within a maximum spatial distance. This is done by *Covariance Intersection* similar to [2] and [4]. However, in our approach the weights w^i are explicitly considered to take the reliability of the different Gaussians into account. Gaussians that have no corresponding counterpart, are taken into account in form of a logic OR. This way, single hypothesis can be transfered into the resulting set of Gaussians, too. Final step is to normalize the weights such that the weighted sum can be interpreted as a probability distribution $p(\boldsymbol{x}_t)$. Further on, the resulting MoG can be simplified if two or more Gaussians resemble each other. This is done by approximating the overlapping Gaussians by a single one. Also components with too small weights can be removed. At this point, we want to place emphasis on the necessity of the inference realizing an AND. Without the reduction of uncertainty by means of Covariance Intersection, a convergence of the whole model cannot be forced and variances of the participating Gaussians would grow over time.

2.2 Short-Term Memory (STM)

Main part for tracking the state hypotheses is the STM. After computation of the localization distribution $p(\boldsymbol{x}_t)$, the weights and parameters of the RBF nodes in the STM have to be adapted to represent the new hypothesis. This is done by transferring the weights of the MoG $p(\boldsymbol{x}_t)$ to w_k and setting up the mean values $\boldsymbol{\mu}_k$ and covariance matrices $\boldsymbol{C}_k$ according to the MoG components, while the number of nodes is adapted to the number of components in $p(\boldsymbol{x}_t)$. An other kind of STM-update takes place if a motion $\boldsymbol{u}_t$ is measured by odometry. Then a motion model is applied to each partial hypothesis represented by one RBF node. This results in new parameters $\boldsymbol{\mu}_k$ and $\boldsymbol{C}_k$. Concurrently a new visual observation $\boldsymbol{y}_t$ is captured and a new estimation of $p(\boldsymbol{x}_t)$ will be initiated.

2.3 Long-Term Memory (LTM) - Adaptive Environment Model

The LTM is performing a mapping from observations $\boldsymbol{y}_t$ to a distribution of states the robot has already been in while receiving a similar observation. Unlike to our former model [1], this mapping is learned and adapted online while using it for localization. Therefore, pairs of observation $\boldsymbol{y}_t$ and related estimated state hypotheses $p(\boldsymbol{x}_t)$ serve as teach value. To speed up convergence and to reduce faulty entries in the LTM, principles similar to [2] are employed. So an update takes place only if $p(\boldsymbol{x}_t)$ is unimodal. In all other cases, the updates will be delayed until $p(\boldsymbol{x}_t)$ reaches unimodality again. Then disambiguated former positions can be reconstructed by using stored motion information (see [3]). Once given an update request, the structure and parameters of LTM are changed in three steps.

First, the clustering of *Observation Space* in layer I is updated. Therefore, if similarity of $\boldsymbol{y}_t$ to each prototype $\boldsymbol{y}_i$ falls below a threshold, a new node representing the current observation $\boldsymbol{y}_t$ is inserted. During this operation, similarities $S(\boldsymbol{y}_t, \boldsymbol{y}_i)$ of all layer I prototypes have to be computed.

Second step: In this phase, the parameters of the RBF nodes in layer II are updated. For that, first the output (merged hypotheses) $p(\boldsymbol{x}_t)$ is back-propagated to each RBF node by multiplying the weights of the MoG components by $S(\boldsymbol{y}_t, \boldsymbol{y}_i)$ according to that prototype $\boldsymbol{y}_i$ the layer II node is connected to. If this is done, a single Gaussian $\phi(\boldsymbol{x}|\boldsymbol{\mu}_t, \boldsymbol{C}_t)$ with a weight $w_t = S(\boldsymbol{y}_t, \boldsymbol{y}_i)\, w_{p(\boldsymbol{x}_t)}$ is given for updating all layer II nodes that are connected to the prototype node $\boldsymbol{y}_i$. This update is done by introducing a new RBF node, representing the new observation. Finaly, nodes with nearly similar Gaussians become merged, to reduce redundancy.

Third step: In this step the connections w_{ij} from layer II to layer I are adapted. Here, relevanceweights w_{ij} of layer II hypotheses will be increased with a learning rate β if layer I is activated by a high similarity $S(\boldsymbol{y}_t, \boldsymbol{y}_i)$ and layer II is activated by a low spatial distance to the Gaussian in $p(\boldsymbol{x}_t)$.

$$w_{ij} := \beta\, S(\boldsymbol{y}_i, \boldsymbol{y}_t)\, w_{p(x_t)} + (1 - \beta\, S(\boldsymbol{y}_i, \boldsymbol{y}_t)\, w_{p(x_t)})\, w_{ij} \tag{3}$$

To reach a stabilizing behavior (and for solving the kidnapped robot problem while building up the internal representation), on the other hand connections to inactive layer II nodes have to be reduced if the respective layer I node is activated.

$$w_{ij} := (1 - \beta\, S(\boldsymbol{y}_i, \boldsymbol{y}_t)\, w_{p(x_t)})\, w_{ij} \tag{4}$$

So long, only new information were captured and the complexity of the LTM increases continuously. But it is also necessary to delete information, because the operation area is extremely dynamic. So situations that will not be observed again can be forgotten, if there is a new observation at the same position. For this purpose, similarity $D(t, j)$ in *State Space* between $p(\boldsymbol{x}_t)$ and the Gaussian represented by each RBF node has to be evaluated. So connections from activated RBF nodes to deactivated prototype nodes in layer I will be decreased,

$$w_{ij} := (1 - f\,(D(t,j), |\boldsymbol{C}_t|))\; w_{ij} \tag{5}$$

and if these weights reach a lower bound, the respective RBF node can be deleted. If no layer II node is connected any longer to a certain prototype node in layer I, this prototype node is deleted, too. The forgetting function f decreases with growing spatial distance D and growing variance of the new Gaussian, which is contained in the determinant of its covariance matrix. Only by means of this third rule, a limitation of the number of nodes, responsible for a restricted area, can be reached.

3 Experimental Results and Conclusion

First, the algorithm was analyzed in a part of the home store with low changes and dynamic modifications. In these preliminary experiments, a mean localization error of about $0.6m$ in an area of about $25m$ by $10m$ could be reached. Observable was a localization error growing with distance to the initial position. The reason for this behavior is the erroneous odometry data used during the first lap for building the initial model. So the LTM represents correct spatial relations of the world in an internal coordinate system, which typically can be rotated to world space. Binding the model at absolute world coordinates is a general problem of this class of CML approaches.

In our desired application, the main task is not to build a model of a completely unknown area but to continuously adapt the model learned before to a changing environment, so this problem is secondary. Therefore, further long-term experiments were done in the home store. After building up an initial model similar to the first experiment, the representation in LTM was rotated and translated to fit the absolute world coordinates by means of minimizing the error between the true path and the estimation of the localization system. Afterwards the experiment was continued for several days. The result is a model of a $30m$ by $30m$ area that allows a localization with an average absolute error of less than $0.45m$, built up without any a priori information. The long-term experiments also clearly demonstrates the merits of our model. Using a model similar to the one presented in [2], the number of layer I nodes in LTM was growing continuously as long as the environment changed. The method presented here handles the situation by replacing irrelevant prototype views by new ones, finally leading to a limited number of nodes for this restricted operation area. The presented approach, thus realizes an applicable long-term localization in a continuously changing environment based on an adaptive statistical distribution with different time-scales.

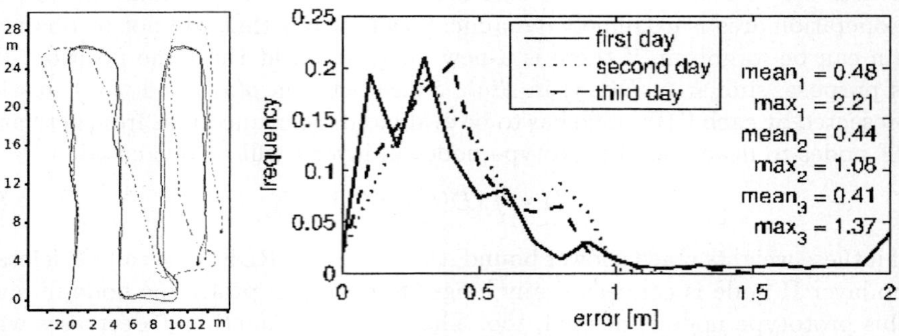

Fig. 2. Results of the long-term experiment in the home store: localization test after three days of operation (left): real path (red/grey), estimated path (blue/solid) and odometric data (green/dotted), histograms of the localization error for three trials (right) the development of means and the rising concentration on small errors visualize the convergence of the approach

References

1. H.-M. Gross, A. Koenig, Chr. Schroeter and H.-J. Boehme, "Omnivision-based Probabilistic Self-localization for a Mobile Shopping Assistant Continued", in: Proc. IEEE-IROS 2003, pp. 1505-1511
2. J. M. Porta and B. J.A. Kroese, "Appearance-based Concurrent Map Building and Localization using a Multi-Hypotheses Tracker", in: Proc. IEEE-IROS 2004, pp. 3424-3429
3. S. H. G. ten Hagen and B. J. A. Kroese, "Trajectory reconstruction for self-localization and map building", in: Proc. IEEE-ICRA 2002, pp. 1796-1801
4. J. K. Uhlmann, S. Julier and M. Csorba,"Nondivergent Simultaneous Map Building and Localization using Covariance Intersection",in Proc. of the SPIE Aerosense Conference, 3087, 4/1997
5. T. Duckett and U. Nehmzow, "Experiments in Evidence-Based Localisation for a Mobile Robot", in AISB-97, Technical Report Series UMCS-97-4-1,1996

New Neural Network Based Mobile Location Estimation in a Metropolitan Area

Javed Muhammad[1], Amir Hussain[1], Alexander Neskovic[2], and Evan Magill[1]

[1] Dept. of Computing Science and Mathematics, University of Stirling, FK9 4LA, Scotland, UK
{ahu, ehm, jmu}@cs.stir.ac.uk
[2] Faculty of Electrical Engineering, University of Belgrade
neshko@etf.bg.ac.yu

Abstract. This paper presents a new neural network based approach to the prediction of mobile locations using signal strength measurements in a simulated metropolitan area. The prediction of a mobile location using propagation path loss (signal strength) is a very difficult and complex task. Several techniques have been proposed recently mostly based on linearized, geometrical and maximum likelihood methods. An alternative approach based on artificial neural networks is proposed in this paper which offers the advantages of increased flexibility to adapt to different environments and high speed parallel processing. The paper first gives an overview of conventional location estimation techniques and the various propagation models reported to-date, and a new signal-strength based neural network technique is then described. A simulated mobile architecture based on the COST-231 Non-line of Sight (NLOS) Walfisch-Ikegami implementation of a metropolitan environment is used to assess the generalization performance of a Multi-Layered Perceptron (MLP) Neural Network based mobile location predictor with promising initial results.

1 Introduction

Location Estimation is the process of localizing an object on the basis of some parameter. This parameter can be proximity to a detector, or some other parameter like radiated energy. The latter parameter is the one of interest in our case. In the particular context of cellular systems, this translates to the localization of the transmitter or the receiver.

Proper location estimation is very important in making many crucial decisions in cellular networks [1]. Handoff management is one such example. When a mobile station enters from the region of service of one base station (BS) to another, a handoff is to be made. The initiation of the handoff process depends on the location of the mobile. A delay in the initiation of handoff will result in very low signal strength or in the adverse case, a call drop. Applications like handoff management don't require very accurate location estimates; all that is required is to determine which cell the mobile is in. But there are applications that ask for a very accurate estimate e.g., intelligent transport systems, fleet management and security applications etc. [1].

Many authors have shown that neural networks provide a good way of approximating non-linear functions [7, 8]. The application of neural networks discussed in this

W. Duch et al. (Eds.): ICANN 2005, LNCS 3697, pp. 935 – 941, 2005.

paper is considered as a function approximation problem consisting of a non-linear mapping of signal strength input (received at several Base Stations) onto a dual output variable representing the mobile location co-ordinates. The signal strength data is generated using a COST 231-Walfisch Ikegami Non-line of Sight (NLOS) model for the metropolitan area.

This paper is organized as follows: section 2 gives an overview of conventional location estimation techniques, and a brief overview of propagation models is presented in section 3, together with a description of the COST-231 Walfisch Ikegami NLOS model used in this paper. The proposed neural network based location estimation model is briefly described in section 4 and preliminary results are presented in section 5 together with a description of the mobile architecture used in the simulation case study. Finally some concluding remarks are given in section 6.

2 Conventional Location Determination Technologies (LDT)

At present conventional LDTs fall into two main classes [1], namely handset-based and network bases LDT's. Currently, GPS based location information services are in commercial use. However, in a city or building where there is often no direct Line of Sight (LoS) between GPS satellite and the terminal, which causes a severe degradation of accuracy. In such cases, location estimation using cellular network systems can offer advantages, and estimating a location using the signal from BS's becomes a highly non-linear problem. Few linearized and geometrical methods have been proposed for calculating the mobile position based on measured signal strengths [12].

Although signal strength based location estimation algorithms may not be the preferred approach at present for providing location services, signal strength is the only common attribute available between various kinds of mobile networks and deserves more attention than received to-date due to its ability to provide network-based mobile location solutions (without the need to modify the handsets). In this paper, we investigate the use of neural networks for mapping the outputs of a selected signal-strength propagation model to predict the mobile location co-ordinates.

3 Brief Review of Propagation Models

Propagation models are used in the field of wireless communications in order to predict signal strength at a signal receiving point, or an entire sector of a wireless system. However, there is no clear cut definition of a propagation model, because there are so many methods of modelling outdoor (as well as indoor) propagation of transmitted and received signals [10]. On the basis of the radio environment, the prediction models can be classified into two main categories, outdoor and indoor propagation models [11]. In this paper we employ the well-known COST 231 model also known as the Walfisch Ikegami model (WIM) which is described in the next section.

3.1 COST-231 Walfisch Ikegami Model

Developing a model for propagation characteristics is an important problem for mobile communications engineering. The problem becomes even more challenging when urban environments are the regions of interest. Many different models have so far been developed to solve this problem.

The Walfisch-Ikegami model (WIM) has been shown to be a good fit to measured propagation data for frequencies in the range of 800 to 2000MHz and the path distances in the range of 0.02 to 5km. The WIM distinguishes between LOS and NLOS propagation situations. For NLOS path situations, the WIM gives the following expression for the path loss in dB [9]

$$L_{NLOS} = \begin{cases} L_{fs} + L_{rts} + L_{mds'} & L_{mds} + L_{rts} \geq 0 \\ L_{fs}, & L_{mds} + L_{rts} < 0 \end{cases}.$$

Where

L_{fs} = Free space loss,

L_{rts} = Roof-to-street diffraction and scatter loss, and

L_{msd} = Multiscreen diffraction loss

Our system assumes a frequency of 1000 MHz, base station antenna height of 30m and mobile antenna height of 1m, building separation of 40m, street width of 20m and angle of incidence of 20°, and a NLOS metropolitan setting.

4 New Signal Strength Based Neural Network Techniques

BS assisted or network based signal strength location estimation techniques are investigated in this work using neural networks (NN), for an urban environment model. In this technique, the signal attenuation is used by the trained neural network model to estimate the distance travelled by electromagnetic waves and an estimate of the location is made. Real data or a realistic propagation model such as the COST 231 WIM can be used to generate training data by analytically calculating the received signal power at various BS's for any given link distance to the mobile. Analytical signal strength based approaches for location determination are further analyzed by Song [3] who demonstrates that multi-path propagation and shadowing effects are the main sources of error in conventional signal strength based location determination techniques, and in this paper we propose the use of neural networks to overcome these problems.

Next we present a brief overview of the multi-layered neural network model used in this work.

4.1 Neural Network Overview

The general structure of a multi-layered perceptron (MLP), also sometimes known as the back propagation network, is illustrated in Figure 1, which can comprise one or more hidden layers [4].

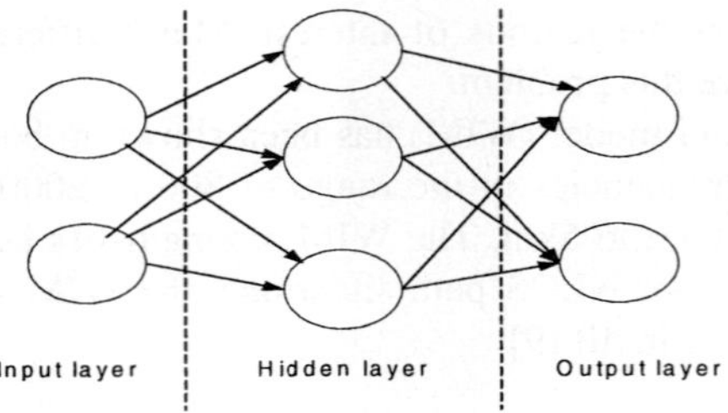

Fig. 1. General architecture of MLP

In the MLP structure illustrated in Figure 1, the output Yi of each neuron of the nth layer is defined by a derivable nonlinear function F:

$$y_1 = F\left(\sum_j w_{ji} y_i\right).$$

Where F is the nonlinear activation function, w_{ji} are the weights of the connection between the neuron N_j and N_i, y_j is the output of the neuron of the $(n-1)^{th}$ layer. In our application, the neural networks are trained with the Levenberg-Marquardt algorithm, which converge faster than the backpropagation algorithm with adaptive learning rates and momentum. The Levenberg-Marquardt rule for updating parameters (weights and biases) is given by [4]:

$$\Delta W = \left(J^T J + \mu I\right)^{-1} J^T e .$$

where e is an error vector, μ is a scalar parameter, W is a matrix of networks weights and J is the Jacobian matrix of the partial derivatives of the error components with respect to the weights.

5 Simulation Results

5.1 Mobile Architecture

The mobile architecture used for the simulations is discussed here. For the sake of simplicity, a square cell of dimensions 3km X 4km is assumed, as shown in Figure 2. Three fixed BSs are used for measuring signal strengths as used in trilateration [13]. The coverage area is divided into grids of dimensions 0.3km X 0.3km for training purposes. The idea is to place the mobile in each of these grid intersections and transmit the signal. All the three BSs measure the received signal strengths from each position of the mobile [2] using the WIM NLOS model described in the previous section 3.1.

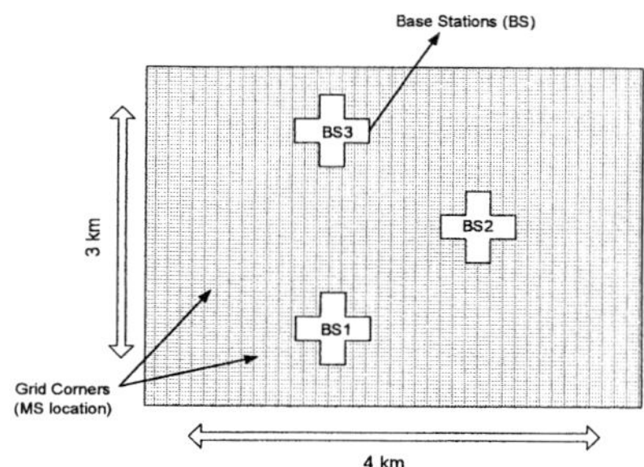

Fig. 2. The square cell used for the simulation of neural network assisted location estimation

The neural net is trained on the generated data using the corresponding mobile location co-ordinates as its target outputs. The origin of coordinates is taken at the left bottom corner and all measurements are taken relative to it. The trained neural network's generalization capability is assessed by testing on data generated with a different grid size (0.1km x 0.1km) to that used for training, as described in the next section.

5.2 Multi-layered Perceptron (MLP) Based Location Estimatiom

For the situation described in Figure 3, the training set consisted of 154 samples of signal strength measurements received at the three fixed BSs and the corresponding mobile location co-ordinates. A two-hidden layered (3-4-8-2) MLP comprising 3 inputs, 2 hidden layers of 4 and 8 nodes, and 2 outputs, was trained using the Levenberg-Marquardt back propagation algorithm [4], and the error was reduced to 9.91947×10^{-7} after 404 epochs, with the result that the net maps any measurement of the training set perfectly to the location of MS for that set. For testing the trained neural network's generalised capability, points other than the training set were generated within the same (3km x 4km) coverage area by dividing the coverage area into

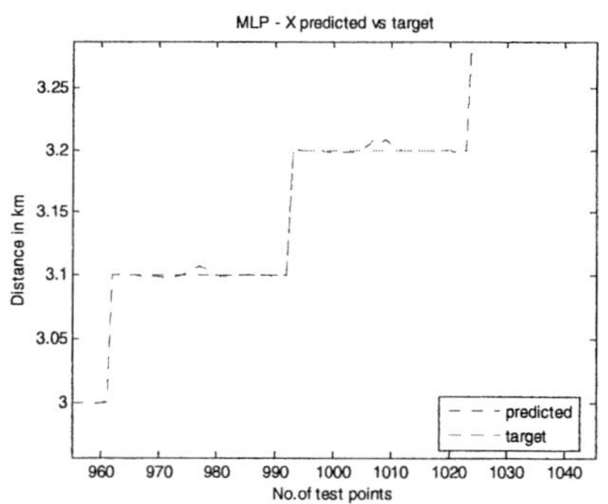

Fig. 3. Sample X-location co-ordinates (MLP predicted vs. target test data)

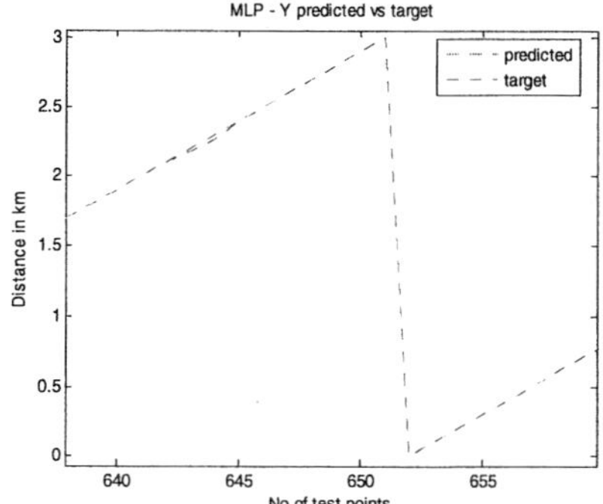

Fig. 4. Sample Y-location co-ordinates (MLP predicted vs. target test data)

grids of dimensions 0.1km X 0.1km (rather than the 0.3km x 0.3km grids used to generate the training data). Sample test results for the MLP location predictor are shown in Figures 3 and 4, for which the mobile was assumed to be at 1271 different points on the test grid. For each of these unseen test points the signal strength received at all the four BSs was calculated using the NLOS WIM model and was fed to the MLP which mapped these to the estimated locations. Figures 3 and 4 show part of the target (test) versus MLP predicted X and Y location coordinates. Note that the MLP predictions on the test data can be further improved by training the net on a larger set of readings (using a smaller grid than 0.3km x 0.3km).

6 Discussion and Conclusions

The motivation behind application of neural networks to solve the location estimation problem is that the neural network technique is adept to the use of intelligence in the cellular system. The greatest benefit is the one time training. In practice however, field collection of the signal strength data is the most laborious part (and downside of the neural network approach in general) but this one-time effort can give location estimates for years until the terrain changes considerably and another training trial is required. Also, the inherent nature of the location estimation problem makes neural nets selection a wise choice for tackling this problem. Modeling the propagation of radio waves by mathematical models is quite complex involving numerous interacting variables. In addition, multipath, diffraction and non line of sight (NLOS) cause problems. Also weather conditions affect the radio wave propagation. These are the types of complex modelling problems neural networks are known to be well suited for [6, 11], and in this research their application is extended to learn the non-linear mapping between the propagation model outputs (path loss) and the corresponding mobile location co-ordinates at various link distances. The preliminary results reported in this paper demonstrate that neural networks can be effectively trained on signal strength measurements obtained using a realistic WIM simulation model for a NLOS metropolitan environment.

Further work will assess the performance of the developed MLP based mobile locator using real field measurements, and comparing with other neural network models (e.g., Radial Basis Function and Recurrent Neural Networks) as well as with other related non-linear function approximation techniques (such as the volterra model).

References

1. Location-based Services, Geo Informatics, April; 2001, http://www.geoinformatics.com.
2. Wamiq M.Ahmed, Amir Hussain and Syed I. Shah, "Location Estimation in Cellular Networks using neural networks", *Proc. International (NAISO-IEEE) Symposium on Info. Science Inovations (ISI'2001)*, Dubai, 19-21 March 2001.
3. Han-Lee Song, "Automatic Vehicle Location in Cellular Communications Systems", *IEEE Transactions on Vehicular Technology*, 43(4), November 1994.
4. Haykin, S, *Neural Networks: A Comprehensive foundation.* Upper Saddle River, NJ: Prentice Hall, 1994.

5. J.Muhammad, A. Hussain and W.M.Ahmed, *"Location Estimation in Cellular Networks Using Neural Networks"*, 1st IEEE-IEE International Workshop on Signal Processing for Wireless Communications (SPWC2003), pages.243-247, London, UK., May 2003.
6. A.Hussain, J.J Soraghan and T.S.Durrani, *A new Adaptive Functional-Link Neural Network Based DFE for Overcoming Co-channel Interference*, IEEE Transactions on Communications, 45(11):1358-1362, 1997.
7. B.E. Gschwendtner and F.M. Landstorfer, *"Adaptive propagation modelling using a Hybrid Neural Technique"*, Electronics Letters, vol. 32, pp. 162-164, Feb.1996
8. P-R. Chang, W-H Yang, "Environment-Adaptation Mobile Radio Propagation Prediction Using Radial basis Function Neural Networks", IEEE Trans. Vech. Technol., vol. 46, no, 1, pp 155-160, Feb.1997
9. J.S. Lee and L.E.Miller, "CDMA Systems Engineering Handbook", pp. 190-199, 1998 (ISBN: 0-89006-990-5)
10. Okumura Propagation Modelling, Tony Ambrosini, Wireless Communications, November 23, 1999
11. A. Neskovic, N. Neskovic and G. Paunovic, *"Modern Approaches in Modelling of Mobile Radio Systems Propagation Environment"*, IEEE communications Surveys and Tutorials, 2000
12. M. Aso, T. Saikawa, T. Hattori, *Maximum Likelihood Location Estimation using Signal Strength and the Mobile Station Velocity in Cellular Systems*, Proc. IEEE Vehicular Technology Conference, 2003.
13. http://electronics.howstuffworks.com/gps1.htm

Lagrange Neural Network for Solving CSP Which Includes Linear Inequality Constraints

Takahiro Nakano[1] and Masahiro Nagamatu[2]

[1] Graduate School of Life Science and Systems Engineering, Kyushu Institute of Technology, Kitakyushu, Japan, 2-4, Hibikino, Wakamatsu, Kitakyushu 808-0196, Japan
nakano-takahiro@edu.brain.kyutech.ac.jp

[2] Graduate School of Life Science and Systems Engineering, Kyushu Institute of Technology, Kitakyushu, Japan 2-4, Hibikino, Wakamatsu, Kitakyushu 808-0196, Japan
nagamatu@brain.kyutech.ac.jp

Abstract. We proposed a neural network called LPPH-CSP (Lagrange Programming neural network with Polarized High-order connections for Constraint Satisfaction Problem) to solve the CSP. The CSP is a problem to find a variable assignment which satisfies all given constraints. Because the CSP has a well defined representation ability, it can represent many problems in AI compactly. From experimental results of LPPH-CSP and GENET which is a famous CSP solver, we confirmed that our method is as efficient as the GENET. In addition, unlike the other conventional CSP solvers which are discrete-valued methods, our method is a continuous-valued method and it can update all variables simultaneously, while the conventional csp solvers cannot find a solution by updating all variables simultaneously Because of the oscilation of the states. Therefore, we can expect the speed-up of LPPH-CSP if it is implemented by the hardware such as FPGA. In this paper, we extend LPPH-CSP to deal with the linear inequality constraints. By using this type of constraint, we can represent various practical problems more briefly. In this paper, we also define the CSP which has an objective function, and we extend LPPH-CSP to solve this problem. In experiment, we apply our method and OPBDP to the warehouse location problem and compare the effectiveness.

1 Introduction

The constraint satisfaction problem (CSP) is a combinatorial problem to find a solution which satisfies all given constraints. Since the CSP is a well-defined problem, it can represent many problems in the field of information science. There are two kinds of methods for solving the CSP, the complete search method [1] and the incomplete search method [2,3]. The complete search method can determine the inconsistency of the given problem, while the incomplete search method can not. If the given problem has solutions, the incomplete search method can find a solution quickly. We proposed a neural network called LPPH-CSP [4,5] to the CSP. In LPPH-CSP dynamics, the variables are applied force so as to decrease

W. Duch et al. (Eds.): ICANN 2005, LNCS 3697, pp. 943–948, 2005.

the energy of the system, i.e., so as to satisfy all constraints, and the strength weight of the force applied by a constraint increases if the constraint is not satisfied. This means that the dynamics of LPPH-CSP changes the energy landscape dynamically, and the trajectory of LPPH-CSP is not trapped by any point which is not the solution of the CSP. LPPH-CSP belongs to the continuous-valued incomplete search method. It can update all variables simultaneously, while many other discrete-valued incomplete methods for the CSP such as the MCHC [2] and the GENET [3] must update variables sequentially. If they update all variables simultaneously, this may cause the network to oscillate between a small number of states indefinitely [3]. So, if we implement LPPH-CSP by VLSI, we can expect significant speed-up for solving the CSP. We implemented LPPH [5] which is a solver for the SAT (SATisfiability problem) on a digital circuit using pulse density modulation. Therefore, we think the hardware implementation of LPPH-CSP is not difficult.

The LPPH-CSP can deal with only logical constraints. However, more general types of constraints are required to represent practical problems. In this paper we extend LPPH-CSP to deal with the linear inequality constraints. By incorporating this type of constraint, we can represent various CSPs more briefly. In this paper we also define the CSP which has an objective function (OCSP) and extend LPPH-CSP to solve it. In experiments, we apply the extended LPPH-CSP to the warehouse location problem (WLP) [6] which is a kind of OCSP and examine the effectiveness of our method by comparison with other existing method.

2 CSP

The CSP is a combinatorial problem to find a solution which satisfies all given constraints. The CSP is defined by a triple (X, D, C).

- $X = \{X_1, X_2, \cdots, X_n\}$ is a finite set of variables.
- $D = \{D_1, D_2, \cdots, D_n\}$ is a finite set of domains. Each domain D_i is a finite set of values and each variable X_i is assigned a value in D_i.
- $C = \{C_1, C_2, \cdots, C_m\}$ is a finite set of constraints.

A solution of the CSP is a value assignment to the variables in X which satisfies all constraints in C. Let x_{ij} be a Boolean variable which represents the variable X_i is assigned the jth value in D_i. x_{ij} is called a VVP (Variable-Value Pair). If x_{ij} is true ($x_{ij} = 1$), the variable X_i is assigned the jth value in D_i. If x_{ij} is false ($x_{ij} = 0$), the variable X_i is not assigned the jth value in D_i. The constraint C_r consists of a set of VVPs. In this paper, we consider the following types of constraints. The CSP can represent briefly in comparison with the SAT by using the following constraints.

- ALT(n, S) [at-least-n-true constraint]
  S is a finite set of VVPs. The ALT constraint requires that at least n of VVPs in S must be true.
- ALF(n, S) [at-least-n-false constraint]
  The ALF constraint requires that at least n of VVPs in S must be false.

- AMT(n, S) [at-most-n-true constraint]
  The AMT constraint requires that at most n of VVPs in S must be true.
- AMF(n, S) [at-most-n-false constraint]
  The AMF constraint requires that at most n of VVPs in S must be false.

The ordinary definition of the CSP includes only binary constraint which requires at least one of the given two VVPs is false, and is represented by ALF($1, S$). By introducing the above four types of constraints, we can represent many combinatorial problems more compactly. These four types of constraints represent that logical relationships between VVPs. In this paper, in addition to the above logical constraints, we will consider the following linear inequality constraint.

$$\sum c_{rij} x_{ij} \leq \sigma,$$

where c_{rij} is a positive or negative coefficient for VVP x_{ij} which appears in the linear inequality constraint C_r, and σ is a constant. By using this type of constraints, we can represent various problems in AI or OR field more briefly.

3 Lagrange Neural Network for CSP

Let VVP x_{ij} represent the degree of certainty that the variable X_i is assigned the jth value of D_i, i.e., x_{ij} has the continuous value between 0 and 1. The dynamics of LPPH-CSP is defined as follows.

$$\frac{dx_{ij}}{dt} = x_{ij}(1 - x_{ij}) \sum_{r=1}^{m} w_r s_{rij}(\boldsymbol{x}), \qquad \text{for all VVP } x_{ij},$$

$$\frac{dw_r}{dt} = h_r(\boldsymbol{x}) - \alpha w_r, \qquad r = 1, 2, \cdots, m,$$

where $s_{rij}(\boldsymbol{x})$ represents a force put on x_{ij} for satisfying constraint C_r, w_r is the weight of constraint C_r, and $h_r(\boldsymbol{x})$ represents the degree of unsatisfaction of constraint C_r. In LPPH-CSP dynamics, each variable changes its value so as to satisfy all constraints, and weight w_r increases, if constraint C_r is not satisfied. The factor $x_{ij}(1 - x_{ij})$ plays a role of keeping x_{ij} between 0 and 1. LPPH-CSP searches a solution of the CSP by numerically solving the above dynamics. This dynamics is a generalization of the dynamics of LPPH for the SAT. In the following we explain how functions s_{rij} and h_r are defined for ALT, ALF, and linear inequality constraints. These functions for AMT and AMF constraints are defined similarly.

3.1 C_r=ALT(n, S)

$$h_r(\boldsymbol{x}) = 1 - \mathrm{NMax}(n, S),$$

$$s_{rij}(\boldsymbol{x}) = \begin{cases} 1 - \mathrm{NMax}(n+1, S), & \text{if } x_{ij} \geq \mathrm{NMax}(n, S), \\ 1 - \mathrm{NMax}(n, S), & \text{otherwise}, \end{cases}$$

where $\mathrm{NMax}(n, S) = n$th maximum value in S.

3.2 C_r=ALF(n, S)

$$h_r(\boldsymbol{x}) = \text{NMin}(n, S),$$
$$s_{rij}(\boldsymbol{x}) = \begin{cases} -\text{NMin}(n+1, S), & \text{if } x_{ij} \geq \text{NMin}(n, S), \\ -\text{NMin}(n, S), & \text{otherwise,} \end{cases}$$

where NMin$(n, S) = n$th minimum value in S.

3.3 $C_r = (\sum c_{rij} x_{ij} \leq \sigma)$

$$h_r(\boldsymbol{x}) = \begin{cases} 0, & \text{if } \sum c_{rij} x_{ij} - \sigma \leq 0, \\ \sum c_{rij} x_{ij} - \sigma, & \text{otherwise,} \end{cases}$$
$$s_{rij}(\boldsymbol{x}) = h_r\left(\boldsymbol{x}^{[ij,0]}\right) - h_r\left(\boldsymbol{x}^{[ij,1]}\right),$$
$$\text{where } x_{kl}^{[ij,0]} = \begin{cases} 0, & \text{if } (k,l) = (i,j), \\ x_{kl}, & \text{otherwise,} \end{cases} \quad \text{and}$$
$$x_{kl}^{[ij,1]} = \begin{cases} 1, & \text{if } (k,l) = (i,j), \\ x_{kl}, & \text{otherwise.} \end{cases}$$

We relax VVPs $\boldsymbol{x}$ from discrete space to continuous space when we apply LPPH-CSP to solve the CSPs. Therefore, the inequality constraints may be satisfied by the VVPs which have continuous values between 0 and 1. However, as mentioned in sec.2, there is the CSP's inherent constraint which means "each variable must have only one value in its domain". This inherent constraint can be represented by ALT$(1, S)$ and AMT$(1, S)$. Accordingly, if all constraints are satisfied, all values of VVPs become 0 or 1 and the inequality constraints never be satisfied by VVPs which have continuous values.

4 Lagrange Neural Network for CSP with Objective Function

As mentioned in Sec.2, the CSP is a problem to find a solution which satisfies all given constraints. However, practical problems in the real world may have an objective function in addition to the constraints. Thus, in this section we will consider the CPS with an objective function (OCSP).

$$\begin{aligned} \text{(OCSP)} \quad & \text{minimize } \ E(\boldsymbol{x}), \\ & \text{subject to } \{C_r | r = 1, 2, .., m\}, \\ & \boldsymbol{x} \in \{0, 1\}^n. \end{aligned}$$

We extend LPPH-CSP dynamics to solve OCSP. The extension LPPH-OCSP is defined as follows.

$$\frac{dx_{ij}}{dt} = x_{ij}(1 - x_{ij}) \left(\sum_{r=1}^{m} w_r s_{rij}(\boldsymbol{x}) - \delta L H_{ij} \frac{\partial F(\boldsymbol{x})}{\partial x_{ij}} \right),$$

$$\frac{dw_r}{dt} = h_r(\boldsymbol{x}) - \alpha w_r,$$

$$\frac{dH_{ij}}{dt} = x_{ij} \frac{\partial F(\boldsymbol{x})}{\partial x_{ij}} - \beta H_{ij},$$

$$\frac{dL}{dt} = \begin{cases} \eta > 0, & \text{if all constraints are satisfied,} \\ 0, & \text{if } L = 1 \text{ and some constraints are not satisfied,} \\ -\epsilon < 0, & \text{if } L > 1 \text{ and some constraints are not satisfied,} \end{cases}$$

where $F(\boldsymbol{x})$ is obtained by normalizing the objective function $E(\boldsymbol{x})$. In this paper, we attempt to apply LPPH-OCSP to the warehouse location problem (WLP) which is a kind of OCSP. In the WLP, $E(\boldsymbol{x})$ is a linear function with positive cofficients. $\partial F(\boldsymbol{x})/\partial x_{ij}$ represents a coefficient of x_{ij} in $F(\boldsymbol{x})$. Therefore, H_{ij} becomes large when x_{ij} has a positive value and a large coefficient in $E(\boldsymbol{x})$. L plays a role to move away from current state which satisfies all constraints to explore new state for minimizing the objective function.

5 Experiments

To investigate the efficiency of LPPH-OCSP, we applied our method to the warehouse location problem (WLP). This problem can be represented by AMT, AMT, and linear inequality constraints and has a linear objective function. The WLP can be formed as 0-1 integer programming. We compared our method with OPBDP [7] which is a 0-1 integer programming solver. Fig.1 and 2 show experimental results for randomly generated WLPs which have 30 stores and 15 candidate locations for warehouses. These graphs show how the best known value of objective function decreases with time for LPPH-OCSP with 3 initial points and OPBDP. The horizontal axis represents the CPU time, and the vertical axis represents the best known value of objective function. For LPPH-OCSP, we used $\alpha = 0.06, \beta = 0.1, \delta = 15, \eta = 10$, and $\epsilon = 0.1$ which were determined by preliminary experiments.

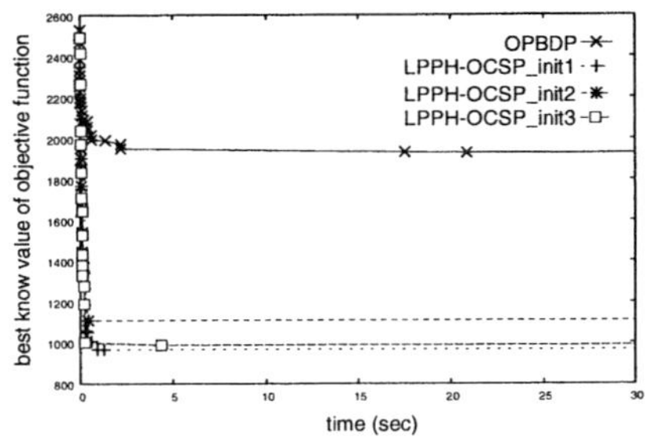

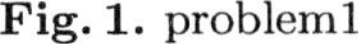
Fig. 1. problem1

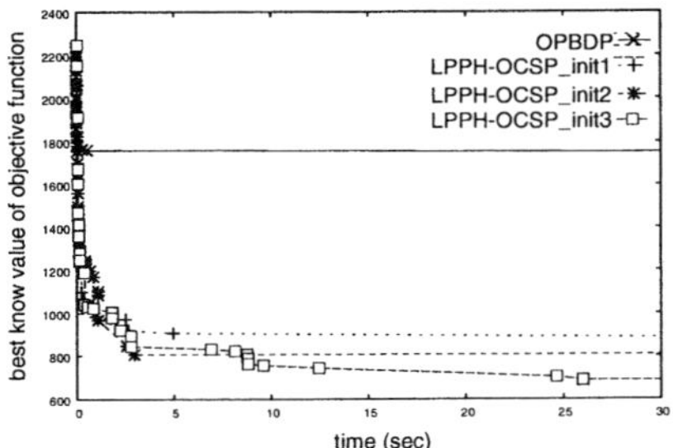

Fig. 2. problem2

6 Conclusion

In this paper, we extend LPPH-CSP to deal with linear inequality constraints. This enables to represent various CSPs more briefly. Furthermore, we define the CSP which has an objective function and propose LPPH-OCSP to solve this problem. In experiments, we apply LPPH-OCSP and OPBDP to the WLP. Experimental results show that our method can find a good near optimal solutions quickly compared to OPBDP. A future direction of this research is to apply LPPH-CSP/LPPH-OCSP various practical problems. In addition, we are also planning to study a hardware implementation of our method.

Acknowledgement. This work was supported by a COE program (Center #J19) granted to Kyushu Institute of Technology by MEXT of Japan.

References

1. Haralick, R., Elliot, G.: Increasing Tree Search Efficiency for Constraint Satisfaction Problems. Artificial Intelligence **14** (1980) 263–313.
2. Minton, S., Johnston, M., Philips, A., Laird, P.: Minimizing Conflicts: A Heuristic Repair Method for Constraint Satisfaction and Scheduling Problems. Artificial Intelligence **58** (1992) 161–205.
3. Davenport, A., Tsang, E., Wang, C. J., Zhu, K.: GENET: A Connectionist Architecture for Solving Constraint Satisfaction Problems by Iterative Improvement. Proceedings of National Conference on Artificial Intelligence (1994) 325–330.
4. Nakano, T., Nagamatu, M.: Solving csp via neural network which can update all neurons simultaneously. In Proceedings of the SCIS & ISIS (2004).
5. Nagamatu, M., Yanaru, T.: On the stability of Lagrange programming neural networks for satisfiability problems of propositional calculus. Neurocomputing **13** (1995) 119–133.
6. Scaparra, M.P., Scutella, M.G: Facilities, Locations, Customers: Building Blocks of Location Models. A Survey. Technical report, University of Pisa (2001).
7. Barth, P.: A Davis-Putnam Based Enumeration Algorithm for Linear Pseudo-Boolean Optimization. Tech. Rep. MPI-I-95-2-003 (1995).

Modelling Engineering Problems Using Dimensional Analysis for Feature Extraction

Noelia Sánchez-Maroño[1], Oscar Fontenla-Romero[1], Enrique Castillo[2], and Amparo Alonso-Betanzos[1]

[1] University of A Coruña, Department of Computer Science, 15071 A Coruña, Spain
{noelia, oscarfon, ciamparo}@udc.es

[2] University of Cantabria, Department of Applied Mathematics and Computer Science, 39005 Santander, Spain
castie@unican.es

Abstract. The performance of a method for the reduction of the input space dimensionality of a physical or engineering problem is analyzed. The results of its application to several engineering problems are compared with those obtained by other well-known methods for the reduction of input space dimensionality, such as Principal Component Analysis and Independent Component Analysis. In order to carry out this study, the features extracted by the three methods were used as inputs to a feed-forward neural network. The advantages of the proposed method are that it presents a computational complexity depending on the number of variables and guarantees dimensional homogeneity in the new space.

1 Introduction

Many scientific disciplines use modelling and simulation processes and techniques in order to obtain a non-linear mapping between the input and the output variables of a given system under study. Neural networks are among the several learning methods used for modelling [1]. In many cases, it is interesting to reduce the dimensionality of the problem, in order to save computational resources such as memory and time, and to avoid the curse of dimensionality. There are two main methods to reduce the dimensionality: feature extraction and feature selection. In feature extraction, which is the method used by our approach, the aim is to find a new set of r dimensions that are the combination of the original n dimensions.

On the other hand, from a physical point of view, a neural network has to satisfy the fundamental prerequisite of dimensional homogeneity when employed as an approximation tool for a given dimensionally homogeneous relation. This fact is not always accomplished, due to the use of standard neural network topologies that do not satisfy the requirements for valid physical relations when used with dimensional variables.

To overcome the inherent disadvantages of a simple mapping between input and output, the proposed method, that employs dimensional analysis and the Π–theorem, yields an improved generalization of a neural network, while

W. Duch et al. (Eds.): ICANN 2005, LNCS 3697, pp. 949–954, 2005.

achieving a dimensionality reduction. This fact is of great interest for complex engineering problems, in which this reduction can mean a significant saving in computational memory and time. Besides, the computational complexity of the method depends on the number of variables and fundamental magnitudes of the problem, and not on the number of data points, as most of the other methods used for dimensionality reduction [2]. In this way, the neural network models used can be more simple, and so more robust. Other authors [3] have used dimensionally homogeneous neural networks, but a formal methodology considering all the possible dimensionless approaches has not been proposed until [4].

The performance of the proposed method for feature extraction is shown over two engineering examples, and its results are compared to those obtained by two of the best known and widely used feature extraction methods, such as Principal Component Analysis (PCA)[5] and Independent Component Analysis (ICA)[6].

2 The *Π*–Theorem

There are some fundamental magnitudes in any physical system, such as length (L), time (T) and mass (M), i.e., some sets of magnitudes such that any other magnitude (called secondary or derived magnitude), can be written in terms of them, using certain formulas (for example, velocity is LT^{-1}). Then, consider a physical problem with $n-1$ input variables, $x_1, \cdots, x_{n-1}$, and one output variable, x_n. This set of variables can be represented using s fundamental magnitudes, $M_1, \cdots, M_s$, i.e., the variables are expressed as:

$$x_j = \prod_{i=1}^{s} M_i^{a_{ij}}; \quad j = 1, 2, \ldots, n, \tag{1}$$

where a_{ij} are the exponents associated with variable j, and the fundamental magnitude i. The elements a_{ij} form the matrix $\mathbf{A}_{s \times n}$ shown below:

	x_1	x_2	...	x_n
M_1	a_{11}	a_{12}	...	a_{1n}
M_2	a_{21}	a_{22}	...	a_{2n}
...	...	...	...	...
M_s	a_{s1}	a_{s2}	...	a_{sn}

In this context, the Buckingham *Π*–Theorem, a fundamental theorem used in dimensional analysis [7], can be applied to know the minimum set of dimensionless variables involved on it. This theorem can be enunciated as follows:

Theorem 1 (The *Π*–Theorem). *If a physical phenomena can be expressed in a given measure-system in which there exist m fundamental magnitudes by means of a function of n parameters* $(x_1, \cdots, x_n)$*, which represent other magnitudes, then, if r is the rank of the matrix* $\mathbf{A}$ *which elements* a_{ij} *are the exponents of the fundamental magnitudes in the corresponding expressions as in (1), then there exist n-r dimensionless monomials by means of which the physical phenomena* *can be represented. These dimensionless monomials are formed by products of powers of such magnitudes, i.e., they are of the form:*

$$\pi_j = \frac{x_j}{\prod_i x_i^{d_{ji}}} \tag{2}$$

where $j \in \{1, \ldots, n-r\}$ and d_{ji} are constants.

It means that any physically meaningful relation $\Phi(x_1, \ldots, x_n) = 0$ is equivalent to a relation of the form $\Psi(\pi_1, \ldots, \pi_{n-r}) = 0$, where $\pi_1, \ldots, \pi_{n-r}$ are dimensionless monomials. The important fact to notice is that a relation of n variables can be rewritten in a new relation involving r fewer variables than the original one, simplifying the theoretical analysis and the experimental design.

3 The Proposed Method for Feature Extraction

Considering a physical problem as the one described in the previous page, the methodology proposed in [4] will be applied as follows:

3.1 Obtaining the Dimensionless Ratios

First, the number of input variables will be reduced. The following algorithm is automatically applied to determine all the sets of dimensionless ratios, ensuring that the number of ratios is less than the number of variables:

1. Write the variables in terms of fundamental magnitudes. The variables are expressed in terms of the fundamental magnitudes using (1), obtaining a Matrix **A** as shown in section 2.
2. Determine the number of dimensionless ratios. The Buckingham Π–Theorem allows to determine the number of dimensionless ratios involved in a given problem. A submatrix **C** of **A** leading to the rank is calculated. The indices of the columns (input variables) of the matrix **A** that form the submatrix **C** compound the set $\mathcal{B}$, analogously, the set $\mathcal{F}$ is formed by indices of rows. It is necessary to choose the variables among the $n-1$ input variables, so that only one of the dimensionless ratios would contain the output variable in order to be able to recover it later.
3. Reduce dimensionality. Build a matrix **B** by removing from **A** the rows not in $\mathcal{F}$ and the columns in $\mathcal{B}$.
4. Change basis. Calculate the matrix $\mathbf{D} = \mathbf{C}^{-1}\mathbf{B}$, that gives the variables in terms of the new basic variables (those in $\mathcal{B}$).
5. Build the dimensionless ratios. Using the Π–theorem, the ratios are selected as:
$$\pi_k = \frac{x_k}{\prod_{\ell \in \mathcal{F}} (X_\ell)^{d_{\ell k}}}; \forall k \notin \mathcal{B} \tag{3}$$
where $d_{\ell k}$ are the elements of matrix **D**.

In step 2 of this algorithm, several submatrices **C** could be selected leading to different sets of dimensionless ratios. So, the algorithm is automatically repeated from step 2 until all the possible sets are obtained.

3.2 Estimating the Dimensionless Output

Once the dimensionless ratios are known, neural networks are employed to estimate the dimensionless ratio which includes the dimensional output, π_q, using all the others ratios as inputs, i.e., the function g' will be estimated as follows:

$$\pi_q = g'(\pi_1, \pi_2, \ldots, \pi_{q-1}). \tag{4}$$

As the ratios are dimensionless, any neural network used will generate a valid physical relation.

3.3 Recovering the Dimensional Output

The last step consists in recovering the original dimensional output X_n, using equation (3), from π_q as:

$$x_n = \pi_q \left(\prod_{\ell \in \mathcal{F}} (x_\ell)^{d_{\ell n}} \right) \tag{5}$$

4 A Comparison with Other Methods for Dimensionality Reduction

The performance of the proposed method (DA) is illustrated by its application to two engineering problems: the estimation of the sinking speed of a ball in a liquid and the optimization design of a vertical breakwater. Then, the results obtained are compared to those derived from the application of PCA and ICA and also with the performance of the same neural network but without previous dimensionality reduction. The software used for implementing the neural network, PCA and the proposed method was Matlab 6.0, while the FastICA algorithm [8] was employed for ICA.

Table 1. All possible sets of dimensionless ratios for the the sinking-speed problem

Approach	Columns	π_1	π_2	π_3
1	$\mathcal{B}=\{1,2,4\}$	$\frac{\rho_{liq}}{\rho_b}$	$\frac{\mu}{D^{3/2}\rho_b g^{1/2}}$	$\frac{v}{D^{1/2}g^{1/2}}$
2	$\mathcal{B}=\{1,2,5\}$	$\frac{\rho_{liq}}{\rho_b}$	$\frac{gD^3\rho_b^2}{\mu}$	$\frac{vD\rho_b}{\mu}$
3	$\mathcal{B}=\{1,3,4\}$	$\frac{\rho_b}{\rho_{liq}}$	$\frac{\mu}{D^{3/2}\rho_{liq} g^{1/2}}$	$\frac{v}{D^{1/2}g^{1/2}}$
4	$\mathcal{B}=\{1,3,5\}$	$\frac{\rho_b}{\rho_{liq}}$	$\frac{gD^3\rho_{liq}^2}{\mu}$	$\frac{vD\rho_{liq}}{\mu}$
5	$\mathcal{B}=\{1,4,5\}$	$\frac{\rho_b D^{3/2}g^{1/2}}{\mu}$	$\frac{\rho_{liq} D^{3/2}g^{1/2}}{\mu}$	$\frac{v}{D^{1/2}g^{1/2}}$
6	$\mathcal{B}=\{2,4,5\}$	$\frac{D\rho_b^{2/3}g^{1/3}}{\mu^{2/3}}$	$\frac{\rho_{liq}}{\rho_b}$	$\frac{v\rho_b^{1/3}}{g^{1/3}\mu^{1/3}}$
7	$\mathcal{B}=\{3,4,5\}$	$\frac{D\rho_{liq}^{2/3}g^{1/3}}{\mu^{2/3}}$	$\frac{\rho_b}{\rho_{liq}}$	$\frac{v\rho_{liq}^{1/3}}{g^{1/3}\mu^{1/3}}$

The sinking-speed problem was presented in [9] and it consists on determining the speed, v, in terms of 5 dimensional input variables: diameter and density of the ball, density and viscosity of the liquid, and, finally, the gravity, which abbreviations are, respectively, D, ρ_b, ρ_{liq},μ and g. These variables involve 3 fundamental magnitudes: length, mass and time. Then, applying the first step of the methodology presented, the 5 input variables are reduced to 2 dimensionless ratios, therefore, the problem is considerably simplified. Besides, 7 different sets of dimensionless ratios are obtained (see Table 1), i.e., the output variable v can be estimated using 7 different approximations. A multilayer perceptron with 7 neurons in the hidden layer, sigmoidal logarithm as transference functions and the Levenberg-Marquardt training algorithm was used to estimate each approximation. The overall number of samples was 12120 and a 10−fold cross-validation was carried out to obtain a more accurate error. As each approximation leads to different performance results, only the best one is shown in the first column of Table 2. For the sake of comparison, the same dimensionality reduction was carried out using ICA and PCA. Then, a neural network with the same architecture was used for estimating the output variable using the transformed inputs. Moreover, the same perceptron was applied over the real inputs, after normalizing them. Their performance results can be checked in Table 2.

Table 2. Mean and standard deviation for the Normalized Mean Squared Error of the test data in the 10-fold cross-validation. DA stands for dimensional analysis approach.

	v	p	p_u
NN	$1.63 \times 10^{-4} \pm 1.38 \times 10^{-4}$	$8.85 \times 10^{-2} \pm 4.96 \times 10^{-2}$	$7.92 \times 10^{-4} \pm 4.07 \times 10^{-4}$
DA	$9.80 \times 10^{-10} \pm 7.20 \times 10^{-10}$	$2.77 \times 10^{-3} \pm 4.07 \times 10^{-3}$	$2.48 \times 10^{-4} \pm 1.87 \times 10^{-4}$
ICA	$7.30 \times 10^{-1} \pm 1.45 \times 10^{-2}$	$2.41 \times 10^{-1} \pm 4.15 \times 10^{-2}$	$4.67 \times 10^{-2} \pm 1.42 \times 10^{-3}$
PCA	$5.01 \times 10^{-1} \pm 1.59 \times 10^{-1}$	$1.23 \times 10^{-1} \pm 4.80 \times 10^{-2}$	$8.14 \times 10^{-3} \pm 2.67 \times 10^{-3}$

The second problem, the design of a vertical breakwater, is a difficult engineering optimization problem that have been studied for years [10]. The aim is to find the optimal cross section that minimizes the construction and maintenance costs during the useful life of the vertical breakwater, while at the same time, satisfies some reliability constraints that guarantee that the work is reasonable safe for each mode of failure. This implies determining the water pressures, p and p_u, produced by the sea waves on the breakwater crownwall in function of 9 variables such as the depth of the water, the height of the wave, etc. The input variables involve 3 fundamental magnitudes. However, most of them implies only one (length) and therefore just a reduction of 2 variables is achieved (from 9 to 7). Again, 7 different approaches are derived after applying the algorithm presented in section 3.1. In order to compare the performance results, ICA and PCA were applied to obtain a similar dimensionality reduction. A neural network with the same characteristics as the one used for the sinking-speed problem was employed for each dimensionless approach, the transformed variables derived from ICA and PCA and the real normalized variables. The number of samples available was 1500 and a 10-fold cross validation was carried out. The performance results are presented in the second and third column of Table 2, although only the dimensionless approach with the best performance is included.

5 Conclusions

A method for dimensionality reduction for physical or engineering problems is presented. The dimensionality reduction of the proposed method is based on the physical dimension of the variables involved in the problem and not on the data; then, it can be applied independently on the number of samples available. This is an important advantage, because in most cases the number of inputs is much lesser than the number of samples. Its performance was illustrated by its application to two physical problems. Also, two standard techniques for dimensionality reduction, ICA and PCA, were applied for solving the same problems and it was demonstrated that the proposed method is more suitable for engineering problems than the other two.

Acknowledgments

This work has been partially funded by the Spanish Ministry of Science and Technology (TIC-2003-00600) with FEDER funds, and by the Xunta de Galicia (PGIDIT04PXIC10502PN).

References

1. Bishop, C.: Neural Networks for Patter Recognition. Oxford University Press (1995)
2. Alpaydin, E.: Introduction to Machine Learning. MIT Press (2004)
3. Brückner, S., Rudolph, S.: Dimensionally homogeneous neural networks for system identification. In: Proc. Int. Workshop on Similarity Methods. (1998) 179–199
4. Alonso-Betanzos, A., Castillo, E., Fontenla-Romero, O., Sánchez-Maroño, N.: Shear strength prediction using dimensional analysis and functional networks. In: Proc. European Symp. on Artificial Neural Networks (ESANN04). (2004) 251–256
5. Diamantaras, K.I., Kung, S.: Principal Component Neural Networks. Theory and Applications. John Wiley and Sons (1996)
6. Hyvärinen, A., Karhunen, J., Oja, E.: Independent Component Analysis. John Wiley & Sons (2001)
7. Buckingham, E.: On physically similar systems: Illustrations of the use of dimensional equations. Phys. Rev. **4** (1914) 345–376
8. FastICA: The fastICA package for Matlab. (2004) Helsinki University of Technology. Laboratory of Computing and Information Science. Neural Networks Research Centre. `http://www.cis.hut.fi/projects/ica/fastica/`.
9. Schneider, G., Korte, D., Rudolph, S.: Neural network correspondencies of engineering principles. In: Proceedings SPIE Conference on Applications and Science of Computational Intelligence III. (2000)
10. Takahashi, S.: Design of vertical breakwaters. Technical Report 34, Port and Harbour Research Institute. Ministry of Transport, Yokosuka, Japan (1996)

Research on Electrotactile Representation Technology Based on Spatiotemporal Dual-Channel

Shuai Liguo, Kuang Yinghui, Xuemei Wang, and Xu Yanfang

Department of Instrument Science and Engineering, Southeast University,
Nanjing 210096 P.R. China
liguo.shuai@cern.ch
kuangyh@seu.edu.cn

Abstract. An electro-tactile representation technology based on spatio-temporal dual-channel is presented and discussed. Both the stimuli current on temporal channel and the signal for tactile element selection on spatial channel are provided by sound waves. Signals on two channels are composed into a file in WAV format by a special wave editor. This WAV format file can be converted to a dimensional wave by the sound card of a computer. When the output of the sound card is connected to the current stimulator, it provides the control signal for tactile stimuli current on temporal channel and tactile element selection signal on spatial channel. The analysis on the model of electrotactile representation shows that the whole tactile perception can be divided into the base volume and the fluctuant amount. To obtain a comfortable electrotactile sensation, a limit to the fluctuant amount is needed.

1 Introduction

Tactile tele-presence plays an important role in master-slave teleoperator system. In 1990, the contrastive experiment[1] by Patrick and his partners showed that tactile feedback can effectively improve remote manipulator's operation to the object and help the user complete the remote task efficiently and accurately. There are many way for direct tactile representation such as pneumatic stimulation[2] ,vibratory stimulation[3], electrical stimulation[4], and functional neuromuscular stimulation[5] etc. Among them, electrotactile is fairly favored by researchers for its small volume, easy to install and to connect to the computer. Electrotactile also has shortcomings of uncomfortable sensation such as adaptation and electrical sharp and burning pain, to solve these problems we put forward a spatiotemporal dual-channel electrotactile representation method on basis of the discovery by Dr. E. Ahissa that the tactile information obtained by the brain is encoded in both spatial channel and temporal channel[6].

2 Scheme

The scheme of spatiotemporal dual-channel eletrotactile representation is shown in Fig.1 (a). The electrotactile data is obtained from the sensors mounted on the fingers of the remote robot, and transmitted to the local computer in real time and stored in a tactile data file. When a tactile representation experiment starts, tactile data are first

W. Duch et al. (Eds.): ICANN 2005, LNCS 3697, pp. 955–961, 2005.

read out from tactile data file, and then composed to a stereo file by a special audio editor together with the scale control signal of expected stimuli current. The audio output of sound card is connected to the tactile current stimulator, the right part with dotted line frame in Fig.1 (a). The output of left channel is a voltage and can be converted to electro-tactile stimuli current by voltage to current convector（VIC）. The output of right channel is connected to the spatial signal process module of the stimulator, which drives the relay group to control the electrical connection (on/off) between the electrode elements on tactile array and the stimuli current node, which can make the user obtain an electrotactile map of the remote tactile information.

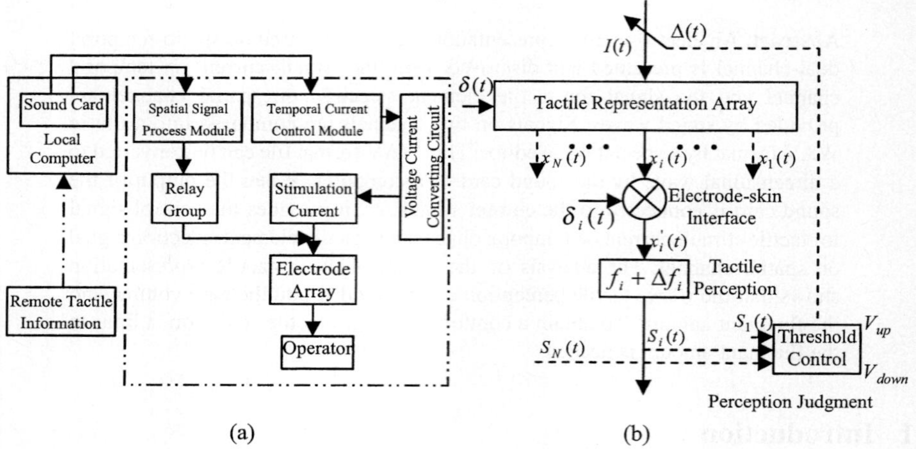

Fig. 1. Tactile tele-presence scheme (a) and electrotactile model (b)

Fig.2 shows a WAV format stereo file, where, (a) is the stimuli current control signal on temporal channel, and (b) is the elements selection signal on spatial channel, Different frequency is mapping different electrodes selection of tactile information.

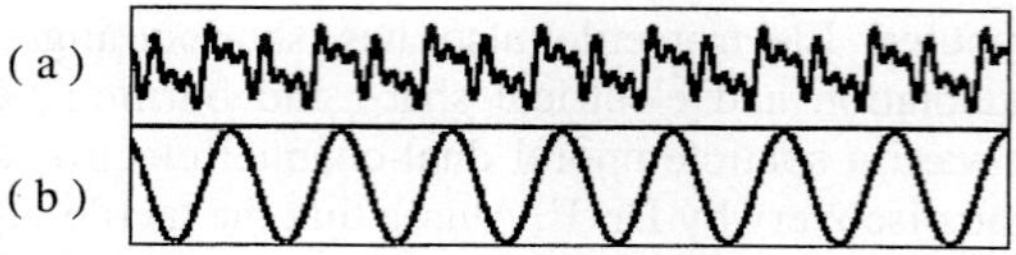

Fig. 2. Wave of a spatiotemporal dual-channel stereo file

3 Model

The model of spatiotemporal dual-channel electrotactile tele-presence is shown in Fig.1 (b). $I(t)$ is the stimuli current, $x_i(t)$ is the actual current on each electrode, where

$i = 1,2,\cdots,N$, $N = K \times L$ is the amount of tactile elements, K and L correspond to the row and column amount of the tactile array. The current on each electrode is $x_i(t) = k_i * I(t)$, k_i is a constant. In Fig.1 (b), electrode i is taken for instance to further discussion. Suppose $\delta_i(t)$ is the influencing factor at time t . $x_i^{'}(t)$ is the current applied onto the skin of the finger. The relationship between $x_i(t)$ and $x_i^{'}(t)$ is:

$$x_i(t) + \delta_i(t) = x_i^{'}(t) \tag{1}$$

$\delta_i(t)$ is related to both the electrode-skin contact resistance and skin characteristics under electrode i. During an eletrotactile experiment, users often coat the electrodes with electric pastern to keep a stable electrode-skin contact status.

In Fig.1 (b), $S_i(t)$ is the tactile perception under stimulation $x_i^{'}(t)$, $f_i + \Delta f_i$ is the current-perception converting function of the tactile receptor, Δf_i is a fluctuant parameter. $S_i(t)$, $x_i^{'}(t)$ and $f_i + \Delta f_i$ can be formulated as follows:

$$S_i(t) = (f_i + \Delta f_i)[x_i^{'}(t)] \tag{2}$$

The whole tactile sensation on operator's finger-tip can be described as:

$$S = \{S_i(t)\} = \{(f_i + \Delta f_i)[x_i^{'}(t)]\} = \{[f_i[k_i * I(t) + \delta(t)] + \Delta f_i[k_i * I(t) + \delta(t)]\} \tag{3}$$

$i = 1、2 \cdots\cdots N$, $t \geq 0$

The last item is a second order infinitesimal and can be removed, so

$$S = \{[f_i[k_i * I(t) + \delta(t)] + \Delta f_i[k_i * I(t)]]\} \tag{4}$$

Since $\delta(t)$ is an electrical interference, $f_i * \delta(t)$ can be adjusted to a constant by improving the electrode-skin contact status, and the equation can be expressed as:

$$S = \{[f_i[k_i * I(t) + \delta(t)] + \Delta f_i[k_i * I(t)]]\} = \{[(S_i^0(t) + \Delta S_i(t)]\} \tag{5}$$

Where , $S_i^0(t) = f_i[k_i * I(t) + \delta(t)]$, can named as base volume of tactile perception which does not change with time, while $\Delta S_i(t) = \Delta f_i[k_i * I(t)]$ is a fluctuant value varying with time. Wide fluctuant range of $\Delta S_i(t)$ will make a worse performance. Suppose $\Delta S_i^M(t)$ is the tolerance limit. if $\Delta S_i(t) \geq \Delta S_i^M(t)$, the user will have uncomfortable sensations as electrical sharp, burning pain, etc..

Theoretically, there should exist an optimal solution set of stimuli current $x_i(t)$ to provide the user a most satisfying tactile perception at a certain time t .

$$U = \{U \mid u_i(t), i = 1、2 \cdots\cdots N\} \tag{6}$$

The actual stimuli current $x_i(t)$ is different from $u_i(t)$, and there is a difference $\zeta_i(t)$ = $x_i(t)$ − $u_i(t)$. Studies on electrotactile is to find the optimal $x_i(t)$ to make $\zeta_i(t) \to 0$ or $\zeta_i(t) \leq \zeta_0(t)$.

$$x_i(t) = u_i(t) + \zeta_i(t) \tag{7}$$

Where, $\zeta_i(t)$ is the difference of the stimuli current, $u_i(t)$ has only a range while its exact value is hard to estimate theoretically or to measure practically.

Although $u_i(t)$ can not be obtained theoretically, it does exist. In Fig.3 (a), I_i stands for the fluctuant range of the stimuli current on electrode i which can provide the user a comfortable electrotactile perception. The intersection of all I_i is $U_1(t)$.

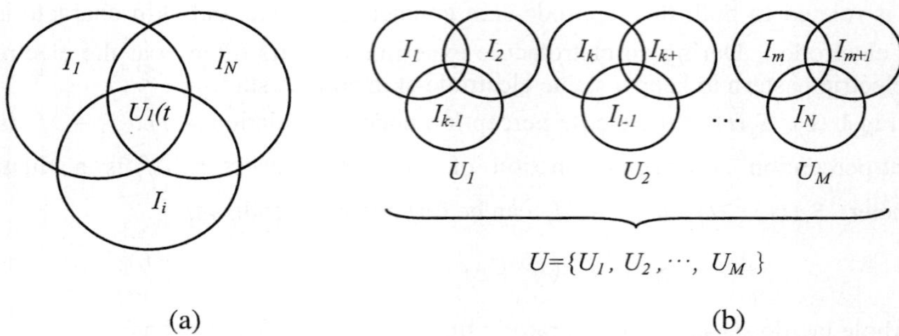

Fig. 3. Current set with a range

$U_1(t)$ might be an empty set theoretically, in this case, the electrode elements should be divided and organized in small groups, and the electrodes in each small group will have the same current set, so the current set of all electrodes should be,

$$U = \{U_1, U_2, \cdots U_M\} \tag{8}$$

In worst case the current on each electrode might be set strictly and separately. The current relationship mentioned above is shown in Fig.3 (b).

4 Frequency Encoding

As the sound card is a standard part of a computer, to simplify the connection between the computer and the stimulator, data in spatial channel are encoded in frequency and composed in WAV format audio file with the stimuli current data.

The maximum BCD (Binary Coded Decimal) code of a digit is 1001(bit3 bit2 bit1 bit0). When bit3 is 1, bit1, bit2 can not be 1 simultaneously, otherwise it would be over 9, so the highest bit is left off. The maximum of each digit is 111(bit2 bit1 bit0), and the upper-limit frequency is 17777Hz, although it can reach 20 kHz. As the lower-limit of audio frequency is 20Hz, bit1 on the ten's place must be fixed to 1. The available bits are: bit0 on ten thousand's place, bit2, bit1, bit0 on thousand's place and hundred's place, bit2 and bit0 on ten's place, and, bit2, bit1, bit0 on unit's place.

The array is as table 1(Left). Each element of tactile array is presented by Dij, i is line number, j is column number. The frequency for each element selection is shown in

Table 1. Collocation of the array (Left) and the frequency of element selection (Right)

Collocation of the array		10,000's place	ten's place		unit's place			line
		Bit0	Bit2	Bit0	Bit2	Bit1	Bit0	
1000's place	Bit2	D_{11}	D_{12}	D_{13}	D_{14}	D_{15}	D_{16}	1
	Bit1	D_{21}	D_{22}	D_{23}	D_{24}	D_{25}	D_{26}	2
	Bit0	D_{31}	D_{32}	D_{33}	D_{34}	D_{35}	D_{36}	3
100's place	Bit2	D_{41}	D_{42}	D_{43}	D_{44}	D_{45}	D_{46}	4
	Bit1	D_{51}	D_{52}	D_{53}	D_{54}	D_{55}	D_{56}	5
	Bit0	D_{61}	D_{62}	D_{63}	D_{64}	D_{65}	D_{66}	6
column		1	2	3	4	5	6	

Frequency of element selection		10,000's place	ten's place		unit's place			line
		Bit0	Bit2	Bit0	Bit2	Bit1	Bit0	
1000's place	Bit2	14***	4*4*	4*1*	4**4	4**2	4**1	1
	Bit1	12***	2*4*	2*1*	2**4	2**2	2**1	2
	Bit0	11***	1*4*	1*1*	1**4	1**2	1**1	3
100's place	Bit2	1*4**	*44*	*41*	*4*4	*4*2	*4*1	4
	Bit1	1*2**	*24*	*21*	*2*4	*2*2	*2*1	5
	Bit0	1*1**	*14*	*11*	*1*4	*1*2	*1*1	6
column		1	2	3	4	5	6	

table 1 (Right), * presents any numerical value. When several elements are selected, the frequency is logic OR of all selected single element, for example, suppose * as 0, the frequency of D11 and D52 is 14***+*24*=1424*Hz, the frequency of D11 and D12 is 14***+4*4*=14*4* Hz, the frequency of D11 and D51 is 14***+1*2**=142** Hz, and the frequency of D11, D52 and D12 is 14***+*24*+4*4*=1424* Hz. Here is a problem that D12 and D51 are selected at the same time when D11 and D52 are selected. Because mistake selections happen only between different lines, and elements on the same line do not interfere with each other, we can avoid mistake selection by selecting different line at different time, that is, to work in time-sharing mode.

5 Experiments

The early tactile array is made of printed circuit with electrodes of square shaped, shown in Fig.4 (a). Purpose for this design is to enlarge the contact area between the electrodes and the finger tip to obtain a larger dynamic range. Experiments show that this kind of electrode is prone to cause electrical sharp or burning pain. The reason is that the acuate rim causes severe charge collection, especially at the sharp angles of the square electrode, thus can easily result in an electrical sharp or burning pain. Our solutions are as follows, first to reduce the charge accumulation to a lower level by using semi spherical electrodes as shown in Fig.4 (b), and then to isolate the charge accumulated rim from the user's finger by covering the electrodes with an 0.3mm-thick insulating film with small pinholes on it for the electrodes to stick out, and the diameter of each pinhole is a little smaller than the diameter of the electrode. Experiments show that this kind of electrode can effectively avoid the electrical sharp or burning pain.

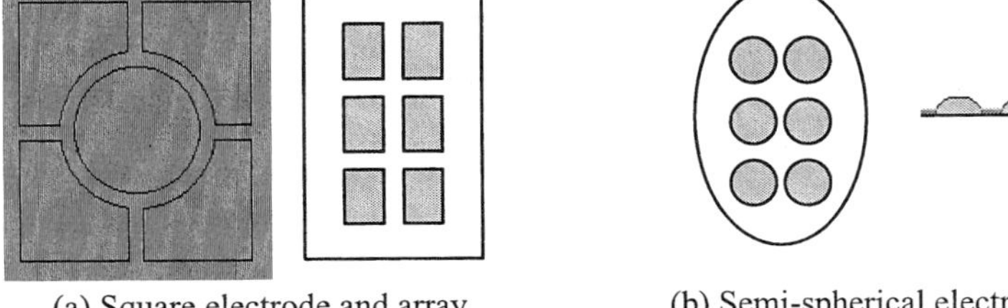

(a) Square electrode and array

(b) Semi-spherical electrode and array

Fig. 4. Eectrotactile electrode and array

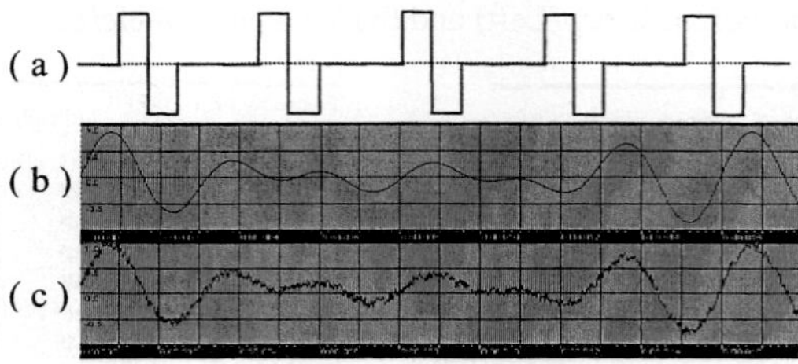

Fig. 5. Waves of stimulating current in tactile representation experiments

The stimuli currents used in tactile representation are shown in Fig.5, where (a) is a rectangle pulse used in early experiments; (b) is a periodic wave alternating with time, the function expression is 0.3*sin(2*pi*t*f)+ 0.3*sin(2*pi*t*f*1.2599)+ 0.3*sin(2*pi*t*f*1.4983) (f=250Hz); (c) is the wave of (b) with a random noise of 0.1*rand (2). Experiments show that wave (a) is apt to bring on tingle and adaptation; while wave (b) doesn't bring on tingle during long time experiment, and can improve adaptation; while wave (c) can provide the user a comfortable tactile sensation.

6 Conclusions

This paper is primary research in electrotactile field. The frequency encoding and the relationship between the tactile tolerance limit and the stimuli current are discussed and analyzed in detail, which might be helpful in further researches on eletrotactile.

Acknowledgements

We are grateful for supports from both National Natural Scientific Foundation of China (60475033) and State Key Laboratory of Intelligent Technology and Systems of Tsinghua University (0415).

References

1. N.J.M. Patrick, T.B. Sheridan, M.J. Massimino & B.A Marcus, Design and Testing of a non-reactive, fingertip, tactile display for interaction with remote environments.Proceedings of SPLE-The International Society for Optical Engineering,1990,v 1387, pp215-222
2. M. Enriquez, O. Afonin, B. Yager, and K. Maclean, A pneumatic tactile alerting system for the driving environment, Proceedings of the 2001 workshop on Percetive user interfaces, November 2001
3. A. M. Okamura, M. W. Hage, M. R. Cutosky & J. T. Dennerlein, Improving reality-based models for vibration feedback. Proc. Haptic Interfaces for Virtual Environment and Teleoperator Systems Symposium, ASME, IMECE-2000, vol.69, no.2, 2000. pp1117-1124.
4. K. A. Kaczmarek, M. E. Tyler and P. Bach-y-Rita, Electrotactile haptic display on the fingertips: Preliminary results, in Proc. 16th Annu. Int. Conf. IEEE Eng. Med. Biol. Soc., Baltimore, IEEE, 1994, pp 940-941

5. K. Yoshida & K. Horch, Closed-loop control of ankle position using muscle afferent feedback with functional neuromuscular stimulation. IEEE Trans. Biomed, 1996, Eng. 43:167-176
6. E. Ahissar, R. Sosnik, S. Haidarliu. Transformation from temporal to rate coding in a somatosensory thalamocortical pathway. Nature406, 2000.6 pp302 – 306

Application of Bayesian MLP Techniques to Predicting Mineralization Potential from Geoscientific Data

Andrew Skabar

Department of Computer Science and Computer Engineering,
La Trobe University, Bundoora, 3083, VIC, Australia
a.skabar@latrobe.edu.au

Abstract. Conventional neural network training methods attempt to find a single set of values for the network weights by minimizing an error function using a gradient descent based technique. In contrast, the Bayesian approach estimates the posterior distribution of weights, and produces predictions by integrating over this distribution. A distinct advantage of the Bayesian approach is that the optimization of parameters such as weight decay regularization coefficients can be performed without use of a cross-validation procedure. In the context of mineral potential mapping, this leads to maps which display far less variability than maps produced using conventional MLP training techniques, the latter which are highly sensitive to factors such as initial weights and cross-validation partitioning.

1 Introduction

Mineral potential mapping is the process of producing a map which ranks areas according to their potential to host deposits of a particular type [1]. More formally, the task can be expressed as follows:

Given:

1. Background information provided by *m* layers of data, each of which represents the value of a distinct geoscientific variable x_i at each pixel *p;*
2. A subset of pixels, each of which is known from historical data to contain one or more deposits of the sought after mineral;

Find:

A function $f(\mathbf{x})$ that assigns to each pixel *p* in the study area a value that represents the probability that pixel *p* is mineralized, given the evidence supplied by the background information.

Thus, assuming that the evidence for a pixel *p* is described by a vector $\mathbf{x} = (x_1, \ldots, x_m)$, the objective is to learn a function f: $\mathbf{X} \rightarrow [0,1]$, where $f(\mathbf{x})$ represents the conditional probability that *p* contains one or more of the known deposits, given the evidence provided by $\mathbf{x}$. The function *f* can then be used to map the probabilities over all pixels.

There are several characteristics of this problem domain that distinguish it from many other domains to which multilayer perceptrons (MLPs) are commonly applied. For example, mineralization is an inherently rare event, and consequently, the number

W. Duch et al. (Eds.): ICANN 2005, LNCS 3697, pp. 963–968, 2005.

of mineralized pixels will be a very small proportion of the total number of pixels in the study area. This can be thought of as a class imbalance problem. A second problem concerns ground truth: while it is known from historical records that a small number of pixels are mineralized, we cannot assume for all other pixels that the lack of a known deposit means that the call is barren; that is, there is still some probability that these pixels are mineralized.

In [2] we have described an approach by which MLPs can be applied to this task. Important features of the approach are that: (i) all examples from the study region are used for training, with known mineralized cells being assigned a target value of 1 and all other cells a target value of 0; (ii) use of cross-entropy (as opposed to quadratic) error reduction ensures that the output of the MLP represents strictly the posterior probability, and (iii) a special cross-validation procedure allows the target values (but not the input vector) of some of the mineralized cells to be held out from training, thus allowing objective estimation of important parameters such as the weight regularization coefficient, early stopping point, and number of hidden layer units. (See [3] for details on the cross-validation procedure).

One of the problems with the approach described above is that the resulting maps are sensitive to the particular cross-validation partitions used. This paper reports on the application of Bayesian MLP methods [4, 5, 6] to mineral potential mapping. Because Bayesian MLP methods do not require a cross-validation procedure for optimization of parameters such as regularization coefficients, the resulting maps are expected to display significantly less variability than those produced using conventional methods.

The paper is structured as follows. Section 2 describes the Bayesian MLP technique used in this research. Section 3 provides empirical results of applying the technique to mapping gold mineralization potential in the Castlemaine region of Victoria, Australia, and compares these results to the conventional MLP training approach. Section 4 concludes the paper.

2 Bayesian Learning for MLPs

In the Bayesian approach, the predicted output corresponding to some input vector $\mathbf{x}^n$ is obtained by performing a weighted sum of the predictions over all possible weight vectors, where the weighting coefficient for a particular weight vector depends on the posterior distribution of $\mathbf{w}$ given data D. Thus,

$$\hat{y}^n = \int f(\mathbf{x}^n, \mathbf{w})\, p(\mathbf{w} \mid D)\, d\mathbf{w} \tag{1}$$

where $f(\mathbf{x}^n, \mathbf{w})$ is the MLP output, and $p(\mathbf{w}|D)$ is the posterior weight distribution. The fact that $p(\mathbf{w}|D)$ is a probability density function allows us to express the integral in Equation 1 as the expected value of $f(\mathbf{x}^n, \mathbf{w})$ over this density:

$$\int f(\mathbf{x}^n, \mathbf{w})\, p(\mathbf{w} \mid D)\, d\mathbf{w} = E_{p(\mathbf{w}|D)}\left[f(\mathbf{x}^n, \mathbf{w}) \right] \simeq \frac{1}{N} \sum_{i=1}^{N} f(\mathbf{x}^n, \mathbf{w}) \tag{2}$$

Thus, the integral can be estimated by drawing N samples from the density $p(\mathbf{w}|D)$, and averaging the predictions due to these samples. This process is known as *Monte Carlo* integration.

The density $p(\mathbf{w}|D)$ can be estimated using the fact that $p(\mathbf{w} \mid D) \propto p(D \mid \mathbf{w}) p(\mathbf{w})$, where $p(\mathbf{w}| D)$ is the likelihood, and $p(\mathbf{w})$ is the prior weight distribution. For the mineral potential mapping problem, the target values are binary, and hence the likelihood can be expressed as

$$p(D \mid \mathbf{w}) = \exp\left(-\left(-\sum_n \left\{t^n \ln f(\mathbf{x}^n, \mathbf{w}) + (1-t)^n \ln(1 - f(\mathbf{x}^n, \mathbf{w}))\right\}\right)\right) \tag{3}$$

where t^n is 1 if pixel n contains a known deposit, and 0 otherwise.

The prior weight distribution, $p(\mathbf{w})$, should reflect any prior knowledge that we have about the complexity of the MLP. To reflect the fact that we want it to be a smooth function, $p(\mathbf{w})$ is commonly assumed to be Gaussian with zero mean and inverse variance α, thus giving preference to weights with smaller magnitudes; *i.e.*,

$$p(\mathbf{w}) = \left(\frac{\alpha}{2\pi}\right)^{m/2} \exp\left(-\frac{\alpha}{2}\sum_{i=1}^{m} w_i^2\right) \tag{4}$$

where m is the number of weights in the network [6]. However, we usually do not know what variance to assume for the prior distribution, and for this reason it is common to set a distribution of values. As α must be positive, a suitable form for its distribution is the gamma distribution [6]. Thus,

$$p(\alpha) = \frac{(a/2\mu)^{a/2}}{\Gamma(a/2)} \alpha^{a/2-1} \exp(-\alpha a / 2\mu) \tag{5}$$

where the a and μ are respectively the shape and mean of the gamma distribution, and are set manually. Note that a single α need not be used for all weights and biases. For example, it is common to use separate values of α for input-hidden-layer weights, input-to-hidden-layer biases, hidden-to-output layer weights, and hidden-to-output layer biases. This is the approach used in this paper. Another common approach is the automatic relevance detection (ARD) approach in which all weights emanating from a common input node share the same α value [4,6].

Because the prior depends on α, Equation 1 should be modified such that it includes the posterior distribution over α parameters:

$$\hat{y}^n = \int f(\mathbf{x}^n, \mathbf{w})\, p(\mathbf{w}, \alpha \mid D)\, d\mathbf{w}\, d\alpha \tag{6}$$

where

$$p(\mathbf{w}, \alpha \mid D) \propto p(D \mid \mathbf{w}) p(\mathbf{w}, \alpha) \tag{7}$$

Monte Carlo integration depends on the ability to obtain samples from the posterior distribution. The objective is to sample preferentially from the region where $p(\mathbf{w},\alpha \mid D)$ is large. The Metropolis algorithm [7] achieves this by generating a sequence of vectors in such a way that each successive vector depends on the previous vector as well as having a random component; i.e., $\mathbf{w}_{\text{new}} = \mathbf{w}_{\text{old}} + \varepsilon$, where ε is a small random vector. Preferential sampling is then achieved using the criterion:

$$\begin{aligned}&\text{if } p(\mathbf{w}_{\text{new}} \mid D) > p(\mathbf{w}_{\text{old}} \mid D) \text{ accept} \\ &\text{if } p(\mathbf{w}_{\text{new}} \mid D) < p(\mathbf{w}_{\text{old}} \mid D) \text{ accept with probability } \frac{p(\mathbf{w}_{\text{new}} \mid D)}{p(\mathbf{w}_{\text{old}} \mid D)}\end{aligned} \tag{8}$$

The difficulty in using the Metropolis algorithm to estimate the integrals for neural networks stems from the strong correlations in the posterior weight distribution; *i.e.*, the great majority of the candidate steps generated in the random walk will be rejected as they lead to a decrease in $p(\mathbf{w}|D)$ [8]. The Hybrid Monte Carlo algorithm [9] reduces the random walk behaviour by using gradient information, which, in the case of MLPs, can be readily calculated. While the Hybrid Monte Carlo algorithm allows for the efficient sampling of parameters (*i.e.*, weights and biases), the posterior distribution for α should also be determined. In this paper we use Neal's (1996) approach, and use Gibbs sampling [10] for the αs.

3 Results

The approach described above has been applied to the production of a mineral potential map showing the favourability for reef gold deposits over the Castlemaine region, Victoria, Australia. Based on a grid-cell resolution of 50m by 50m, the study region was represented by a rectangular grid consisting of 29,046 cells, 148 of which were known from historical records to contain deposits of the sought-after type. A total of 16 input layers describing geophysical, geochemical and geological data were used.

The MLP consisted of 6 hidden layer units, which we know from past work is sufficient to accurately model the posterior probabilities [3,4]. The prior distribution for the hyperparameters (*i.e.*, the weight regularization coefficients for the four weight groupings) was set to a Gamma distribution with mean 0.1 and shape parameter 0.1. For Monte Carlo sampling, a burn–in period of 500 samples was used (allowing the sampling procedure to converge to the target distribution), following which the next 1000 samples were stored. To reduce the chance of any correlations between the samples, every tenth one of these 1000 samples was selected to be used for marginalization. These 100 samples were then used to predict probabilities for each pixel in the study region, and, for each pixel, the mean was calculated.

Figure 1 is the resulting mineral potential map. Points indicate the locations of known deposits; values to the rights of the colour bar indicate the posterior probability of mineralization (prior probability is approximately 0.0051). Visual inspection of the map reveals that the known deposits coincide quite well with regions assigned high potential; however, this relationship can be seen more closely by ranking pixels according to their assigned probability values (highest to lowest), and tallying the number of observed known deposits as the area is traversed from highest probability to lowest probability. This is shown in Figure 2. Note that the dashed line, which represents the cumulative sum of probabilities, lies significantly below the line representing observed deposits. This is an effect of the use of regularization to prevent overfitting. It is the dashed line which provides the best indication of mineralization potential for unknown deposits.

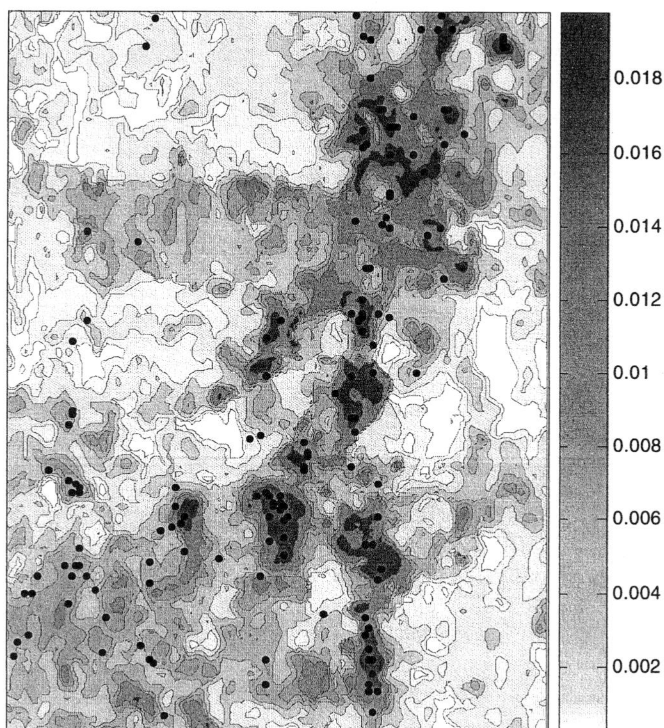

Fig. 1. Contour map showing probabilities. Points indicate location of known deposits. Colour bar indicates posterior probabilities. Contours are placed at probabilities of 0.0001, 0.0005, 0.001, 0.0025, 0.005, 0.0075, 0.01 and 0.02.

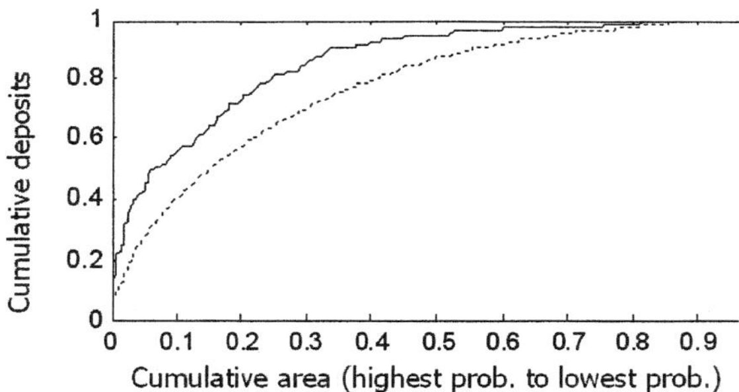

Fig. 2. Cumulative deposits versus cumulative area represented by pixels ranked from highest probability to lowest probability. Solid curve represents prediction on known deposits; dashed curve represents cumulative sum of probabilities

Finally, we compare the map produced using the Bayesian approach with maps produced using the conventional MLP approach. The pairwise correlation between four maps produced using the conventional approach ranged from 0.290 to 0.730, with a mean of 0.429. This variation is due predominantly to the selection of cross-validation partitions, which were chosen randomly, and were different for each run. The pairwise correlation between the Bayesian map and each of the four maps produced using the conventional approach ranges from 0.551 to 0.592, with an average of 0.570, and is consistent with the interpretation of the Bayesian-produced map as being the average of many networks (*i.e.*, weight vectors) drawn from the posterior distribution.

4 Conclusion

The essential difference between conventional maximum likelihood approaches to MLP training and Bayesian MLP techniques is that whereas the former *optimize* over parameters, the latter *integrate* over parameters, thus taking into account the inherent uncertainty in the parameters. In the context of mineral potential mapping, the advantages of the Bayesian approach are that: (i) it determines regularization coefficients automatically, i.e., without having to hold out any data, and thus avoiding a complex and noisy cross-validation procedure; (ii) it reduces variability due to factors such initial weight assignment and cross-validation partitioning. Further advantages of the Bayesian approach, although not discussed in this paper, are that it does not depend on the number of hidden layer units, thus allowing complex models to be formed without the risk of overfitting that occurs in conventional approaches (see [6]), and that it allows estimates of the uncertainties in the predicted probabilities to be easily calculated.

References

[1] Bonham-Carter, G.F., 1994, Geographic information systems for geoscientists: modeling with GIS, Pergamom Press, Oxford.

[2] Skabar, A. 2003 Mineral Potential Mapping using Feed-Forward Neural Networks, *Proc. Int.l Joint Conf. on Neural Networks* (IJCNN). Portland, Oregon, pp. 1814-1819.

[3] Skabar, A., 2004 Optimization of MLP parameters on Mineral Potential Mapping Tasks, *Proceedings of ICOTA: International Conference on Optimization: Techniques and Applications*, 9-11 December 2004, Ballarat, Australia.

[4] MacKay, D.J.C. 1992, A practical Bayesian framework for backpropagation networks. In *Neural Computation*, 4(3), pp. 448-472.

[5] Neal, R.M., 1992, Bayesian Training of Backpropagation Networks by the Hybrid Monte Carlo Method, Technical Report CRG-TR-92-1, Department of Computer Science, University of Toronto.

[6] Neal, R.M., 1996, Bayesian Learning for Neural Networks, New York: Springer-Verlag.

[7] Metropolis, N.A., Rosenbluth, A.W., Rosenbluth, M.N., Teller & Teller, E., A.H., 1953, Equation of State Calculations by Fast Computing Machines. *Journal of Chemical Physics* 21(6), pp. 1087-1092.

[8] Bishop, C., 1995, Neural networks for pattern recognition, Oxford University Press, Oxford.

[9] Duane, S., Kennedy, A.D., Pendleton, B.J. and Roweth, D., 1987, Hybrid Monte Carlo. *Physics Letters B.* 195(2), pp. 216-222.

[10] Geman, S. and Geman, G., 1984, Stochastic Relaxation, Gibbs Distributions and the Bayesian Restoration of Images. *IEEE Transactions on Pattern Analysis and Machine Intelligence*, 6:721-741.

Solving Satisfiability Problem by Parallel Execution of Neural Networks with Biases

Kairong Zhang[1] and Masahiro Nagamatu[2]

[1] Graduate School of Life Science and Systems Engineering, Kyushu Institute of Technology, Kitakyushu, Japan 2-4, Hibikino, Wakamatsu, Kitakyushu 808-0196 Japan
choh-kaie@edu.brain.kyutech.ac.jp

[2] Graduate School of Life Science and Systems Engineering, Kyushu Institute of Technology, Kitakyushu, Japan 2-4, Hibikino, Wakamatsu, Kitakyushu 808-0196 Japan
nagamatu@brain.kyutech.ac.jp

Abstract. We have proposed a neural network named LPPH (Lagrange programming neural network with polarized high-order connections) for solving the SAT (SATisfiability problem of propositional calculus), together with parallel execution of LPPHs to increase efficiency. Experimental results demonstrate a high speedup ratio of this parallel execution. LPPH dynamics has an important parameter named attenuation coefficient which strongly affects LPPH execution speed. We have proposed a method in which LPPHs have different attenuation coefficients generated by a probabilistic generating function. Experimental results show the efficiency of this method. In this paper, to increase the diversity we propose a parallel execution in which LPPHs have mutually different kinds of biases, e.g., positive bias, negative bias, and centripetal bias. Experimental results show the efficiency of this method.

1 Introduction

For the SAT (SATisfiability problem of propositional calculus), we proposed a neural network named LPPH (Lagrange programming neural network with polarized high-order connections) [1], which is based on the Lagrangian method and has the following properties: (1) The solutions of the SAT are the equilibrium points of LPPH and vice versa. (2) When a trajectory of LPPH passes near a solution of the SAT, it converges to the solution.

When a neural network is simulated by software, usually parallel processing is done by, first, dividing the network into parts, then executing each part on a computer individually. This type of parallel processing requires high communication overheads. We proposed a parallel execution of LPPHs, in which plural LPPHs are prepared, and the LPPHs find solutions from different initial states. It is quit different from the previous one. Experimental results show a high speedup ratio is obtained by using this parallel technique of LPPHs.

There is an important parameter, the attenuation coefficient, in LPPH dynamics. And this parameter strongly influences the speed of LPPH execution. Furthermore it is difficult to decide its good value in advance. To overcome this difficulty, we proposed a method to determine the value of the attenuation coefficient of each LPPH using a probabilistic generating function in the parallel execution of LPPHs.

W. Duch et al. (Eds.): ICANN 2005, LNCS 3697, pp. 969 – 974, 2005.

In some cases, solutions of the SAT may have high percentage of 1s, and in some other cases, high percentage of 0s. Suppose, if any information about the percentage of 1s or 0s is known, we can introduce a bias to control the direction of value change of variables in LPPH dynamics, which is expected to find a solution much faster for many cases. But it is not easy to get such information in advance. Furthermore, the speed of LPPH execution depends on not only the percentage but also other reasons, such as some detailed structure of network of LPPH for the problem at hand. It is more difficult to get such information. In parallel execution of LPPHs, even if the above information is not known, we can prepare LPPHs with several kinds of biases, e.g., a bias toward 1(positive bias), a bias toward 0 (negative bias), and a bias toward 0.5 (centripetal bias). If the percentage of 1s is high, LPPHs with a positive bias is expected to find a solution faster than the others. In the opposite case, LPPHs with a negative bias is expected to find a solution faster. Centripetal bias has proposed by us [2], which helps variables to change their values more easily. In this paper, parallel execution with mixed biases is proposed. In this method, LPPHs are divided into four groups. LPPHs in the first group have no bias, the second group positive biases, the third group negative biases, and the fourth group centripetal biases. Experimental results show this method is efficient.

2 LPPH

The SAT is defined as follows:

$$\begin{aligned} \text{(SAT)}\quad & \text{find } \boldsymbol{x}, \\ & \text{such that } \boldsymbol{x} \text{ satisfies } \quad C_r \quad r = 1,2,\cdots m, \\ & \boldsymbol{x} \in \{0,1\}^n, \end{aligned} \tag{1}$$

Where $C_r\,(r=1,2,\cdots,m)$ is a clause (a disjunction of literals). To solve SAT by LPPH, we define the following continuous valued functions. For each $i=1,2,\cdots,n$, and $r=1,2,\cdots,m$, a function $g_{ir}:[0,1]^n \to [0,1]$ is defined as follows:

$$g_{ir}(\boldsymbol{x}) = \begin{cases} x_i & \text{if } x_i \text{ appears in } C_r \text{ as a negative literal,} \\ 1-x_i & \text{if } x_i \text{ appears in } C_r \text{ as a positive literal,} \\ 1 & \text{otherwise.} \end{cases} \tag{2}$$

Corresponding with $C_r\,(r=1,2,\cdots,m)$ a function $h_r:[0,1]^n \to [0,1]$ is defined as follows:

$$h_r(\boldsymbol{x}) = \prod_{i=1}^{n} g_{ir}(\boldsymbol{x}). \tag{3}$$

The dynamics of LPPH is composed of the following differential equations.

$$\begin{aligned}
\frac{dx_i}{dt} &= -x_i(1-x_i)\frac{\partial F(\boldsymbol{x},\boldsymbol{w})}{\partial x_i} \qquad i = 1,2,\cdots,n, \\
\frac{dw_r}{dt} &= -\alpha w_r + h_r(\boldsymbol{x}) \qquad r = 1,2,\cdots,m.
\end{aligned} \tag{4}$$

Where $F(\boldsymbol{x},\boldsymbol{w})$ is defined as follows:

$$F(\boldsymbol{x},\boldsymbol{w}) = \sum_{r=1}^{m} w_r h_r(\boldsymbol{x}), \quad \boldsymbol{x} \in [0,1]^n, \boldsymbol{w} \in (0,\infty)^m \tag{5}$$

By solving the above differential equations numerically, LPPH can find a solution of the SAT. α is a parameter called an attenuation coefficient.

3 Parallel Execution of LPPHs

We have proposed a parallel execution of LPPHs:

(1) Prepare plural LPPHs.
(2) Start the LPPHs simultaneously from different initial points from each other.
(3) When any of the LPPHs finds a solution, halt all LPPHs and return the solution.

It is very easy to realize the parallel execution of LPPHs by hardware. Only we have to do is to prepare plural LPPHs. The total system is very simple and executable at high-speed.

Experimental results are shown in Fig.1. Suppose that P is the number of LPPHs. Let t_j be the execution time of jth neural network for finding a solution. Then, $T_p = \min_j \{t_j \mid 1 \le j \le p\}$ is the execution time of parallel execution. The horizontal axis indicates the number of LPPHs, and the vertical axis indicates the speedup ratio, namely $E(T_p)/E(T_1)$, where $E(T_p)$ and $E(T_1)$ are the CPU time of the parallel execution of LPPHs, respectively. In this experiment, parallel execution of p (p=1, 2, … , 50) LPPHs are used, and randomly generated 3-SAT problems are solved. They are exp-r300 (300 variables and 1275 clauses), exp-r200 (200 variables and 860 clauses), exp-r100 (100 variables and 430 clauses) and exp-r50 (50 variables and 215 clauses). From Fig.1, it is shown that high radio of speedup is obtained. This is remarkable for large and difficult problems, e.g. exp-r300.

4 Parallel Execution of LPPHs with Different Attenuation Coefficients

The optimum value of attenuation coefficient strongly depends on the problems at hand. To resolve this problem, we proposed a parallel execution in which LPPHs have different values of attenuation coefficient from each other. To generate a set of values of attenuation coefficient, a probabilistic generating function is proposed [3]. Experimental results show the parallel execution of LPPHs which uses the generating function is efficient and near optimum for many problems. The function furthermore eases the difficulties to select appropriate value of attenuation coefficient.

5 Parallel Execution of LPPHs with Mixed Biases

In this paper, we assign four kinds of biases to LPPHs. We call this parallel execution of LPPHs with mixed biases. The dynamics of this parallel execution is described as follows:

$$\frac{dx_i}{dt} = -x_i(1-x_i)(\sum_{r=1}^{m} w_r \frac{\partial h_r(\boldsymbol{x})}{\partial x_i} + bias), \quad i = 1,2,\cdots,n, \text{ for each LPPH},$$

$$\frac{dw_r}{dt} = -\alpha w + h_r(x), \quad r = 1,2,\cdots,m, \text{ for each LPPH},$$

$$bias = \begin{cases} 0 & \text{for LPPH } 1 \sim n/4, \\ 1 \times coef\_bias & \text{for LPPH } n/4+1 \sim 2n/4, \\ -1 \times coef\_bias & \text{for LPPH } 2n/4+1 \sim 3n/4, \\ (0.5 - x_i) \times coef\_cb & \text{for LPPH } 3n/4+1 \sim n, \end{cases} \tag{6}$$

$coef\_bias$: strength of positive and negative biases,
$coef\_cb$: strength of centripetal bias.

6 Experiments

In experiments, we set the parameter *coef_bias*=0.1 and *coef_cb*=1.0. Figs.2 and 3 show experimental results comparing the following parallel execution methods. (1) All LPPHs have a same fixed value for attenuation coefficient and no bias (In Figs.2 and 3, results of this method are indicated as "alpha=0.14", etc). (2) Attenuation coefficients are generated by a generating function, and no bias is added. (In Figs.2 and 3, "generating function"). (3) Attenuation coefficients are generated by the same way as (2), and biases are added according to (6) (In Figs.2 and 3, "mixed biases"). In Figs.2 and 3, the horizontal axis indicates the number of LPPHs and the vertical axis (logarithmic) indicates the number of updates. "exp-hm1020" is a Hamilton circuit problem with 100 variables and 762 clauses. It is known that α=0.14 is the optimal value, and α=0.06 is a bad value for exp-r200. From Fig.2, we can see that the result of the parallel execution with mixed biases is near to the result of the optimum value of attenuation coefficient. In Fig.3, α=0.05 (optimal value) and α=0.01 (a bad value) are used as the fixed values. From this experiment, we can see the result of parallel execution with mixed biases is better than that with optimal value, extensively.

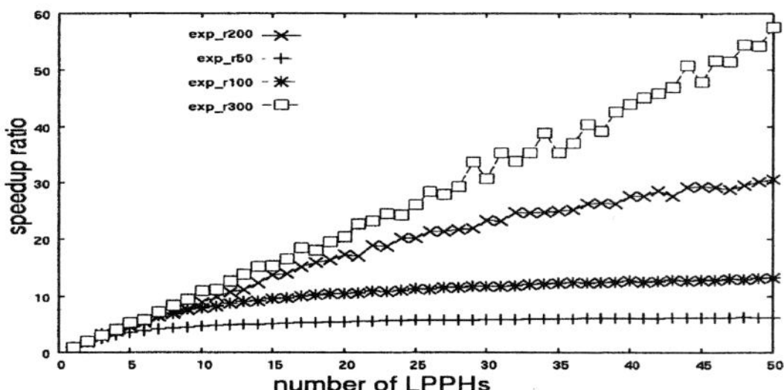

Fig. 1. Speedup ratio for random 3-SAT problem

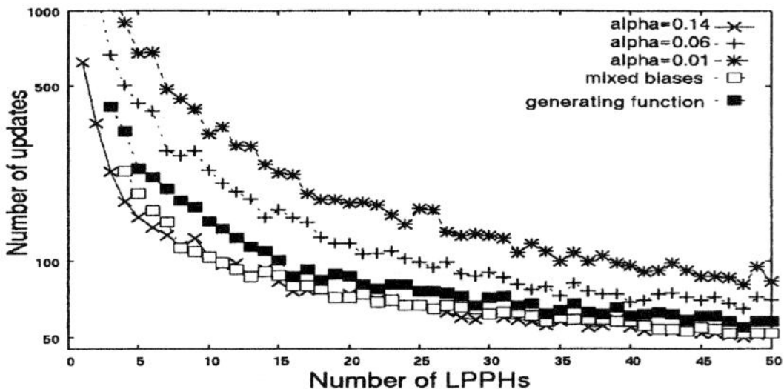

Fig. 2. Comparison of Several parallel executions for exp-r200

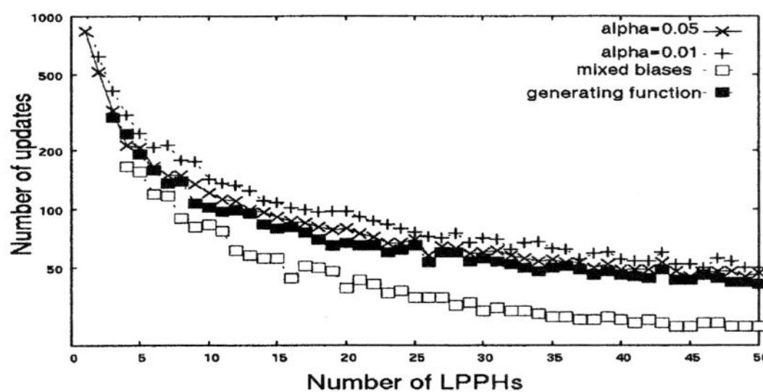

Fig. 3. Comparison of Several parallel executions for exp-hm1020

In general, we can get no information about the percentage of 1s or 0s in a solution of the SAT before solving it. So it cannot be determined whether positive bias or negative bias is good in advance. Centripetal bias is efficient for many problems, but for some problems it is not efficient. We have compared the efficiency of parallel execution with positive bias, negative bias, centripetal bias, and the mixed biases. Experimental results show that the result of parallel execution of LPPHs with mixed biases is near to the best one among the parallel execution of positive bias, negative bias or centripetal bias only.

7 Conclusion

We have proposed a parallel technique named parallel execution of LPPHs, which is easy to realize not only by software but also by hardware, merely preparing plural LPPHs, and executing them simultaneously. Experimental results show a high speedup ratio is obtained especially for difficult problems. It is known that the attenuation coefficient strongly influences execution time of LPPH. And it is also known to be difficult to find a good value of the attenuation coefficient in advance. Experimental results show the dependence of the execution time on the attenuation coefficient can be eased by the parallel execution of LPPHs. Experimental results also show the parallel execution of LPPHs which uses a generating function of attenuation coefficient is efficient and near optimum for many problems. The function furthermore eases the difficulties to select a good value of attenuation coefficient.

In this paper, we propose a parallel execution of LPPHs with mixed biases. From experiments, it is known, the proposed method is efficient. Furthermore for some problems, it is known, the proposed method is better than the parallel execution with optimum value of the attenuation coefficient. However, the proposed method is not efficient for every problem. For some problems, such as an unique-solution random 3-SAT, the parallel execution with mixed biases is not efficient. For our future works, we want to study why the parallel execution with mixed biases is not useful for some problems, such as unique solution random 3-SAT, and find a new efficient bias for parallel execution of LPPHs for all types of problems. Deciding the strength of bias automatically is also our future work.

Acknowledgements

This work was supported by a COE program (Center #J19) granted to Kyushu Institute of Technology by MEXT of Japan.

References

[1] M. Nagamatu and T. Yanaru, "On the stability of Lagrange programming neural networks of satisfiability problems of propositional calculus", Neurocomputing, 13, 119-133, 1995.

[2] M. Nagamatu and M. Hoshiura, "Using Centripetal Force to Solve SAT by Lagrange Programming Neural Network", Proceeding of Knowledge-Based Intelligent Information Engineering System & Allied Technologies (KES'2001) pp.476-480, 2001.

[3] K. Zhang and M. Nagamatu "Parallel Execution of Neural Networks for Solving SAT" JACIII, 2005. (to appear)

Local vs Global Models in Pong

Colin Fyfe

Applied Computational Intelligence Research Unit,
The University of Paisley, Scotland
colin.fyfe@paisley.ac.uk

Abstract. We review two previous simulations in which opponent modelling was performed within the computer game of pong. These results suggested that sums of local models were better than a single global model on this data set. We compare two supervised methods, the multilayered perceptron, which is global, and the radial basis function network which is a sum of local models on this data and again find that the latter gives better performance. Finally we introduce a new topology preserving network which can give very local or more global estimates of results and show that, while the local estimates are more accurate, they result in game play which is less human-like in behaviour.

1 Introduction

One of the most satisfying aspects of most computer games is that there is generally no transitivity in strategies; i.e. we rarely have a situation in which because strategy A beats strategy B and strategy B beats strategy C, this automatically means that strategy A will beat strategy C. Rather we generally have to select the optimal strategy in the context of what strategy the opponent uses.

One of the main disappointments for computer games players is that the computer opponent, the "AI", is often rather unintelligent: it generally has a fixed, finite repertoire of strategies which it uses again and again. Once a human has discovered these strategies, they can be responded to: since there is no overall best strategy, there is always a response which will beat it. Thus, rather quickly, computer games become rather less interesting. However when we play games against other humans, these games can often engage us quite happily for a lifetime.

Thus we have the recent effort to make AIs more human like. Note that this does not necessarily mean that the AI must perform better: it must rather perform in a more human-like manner; in contemporary games, the AI often wins because it is better at micro-management of resources than the human and again playing against such an opponent is not a rewarding experience. In this paper, we discuss a series of experiments in which we compare various artificial neural network methods of modelling human behaviour in the context of the computer game, Pong.

W. Duch et al. (Eds.): ICANN 2005, LNCS 3697, pp. 975–980, 2005.

2 Background

The self-organising map (SOM) [4] quantises a data set in such a way that topology relations are preserved: nearby points in data space are quantised to similar values while distant points in data space are quantised to very different values. The quantised values, known as nodes or neurons, are assumed to lie on a regular grid in some latent space and to have an associated centre, $\mathbf{c}_i$ in data space. Let an input data point be $\mathbf{x}$. We then select the winning neuron as that neuron whose centre is closest to the input.

$$i* = \arg\min_i ||\mathbf{x} - \mathbf{c}_i|| \tag{1}$$

The centres of the winning neurons and those closest to it are then updated according to

$$\Delta\mathbf{c}_i = \eta\Lambda(i, i*)(\mathbf{x} - \mathbf{c}_i) \tag{2}$$

where η is a small learning rate and $\Lambda(i, i*)$ is known as the neighbourhood function and is such that those neurons which are closest to $i*$ are updated with greatest magnitude.

The game of Pong is a computerised table tennis game: each opponent is in control of one bat at either side of the screen and must move the bat vertically in order to intercept a ball which moves back and forth across the playing surface. He who does not intercept the ball is the loser. Of course, it is very easy to program an "AI" to be perfect at this game, however in [6], McGlinchey showed how to make a game against an AI more interesting by creating an AI who learned to behave as a human would: McGlinchey modelled a human player by training a SOM on the data of a game played between two humans. The training data is the ball's position and speed on which (1) is determined while one human's response to this state is also recorded and associated with the winning node: if d is the human player's bat position and the $i*$ neuron currently has an estimate of this as f_{i*}, then $\Delta f_{i*} = \eta(d - f_{i*})$.

The generative topographic mapping (GTM) [1] is a probabilistic model which treats the data as having been generated by a set of latent points. We have a set of K latent points which are mapped through a set of M basis functions and a set of adjustable weights to the data space. The parameters of the combined mapping are adjusted to make the data as likely as possible under this mapping. The GTM is a probabilistic formulation so that if we define $\mathbf{y} = \mathbf{\Phi W} = \mathbf{\Phi}(\mathbf{t})\mathbf{W}$, where $\mathbf{t}$ is the vector of latent points, the probability of the data is determined by the position of the projections of the latent points in data space and so we must adjust this position to increase the likelihood of the data. More formally, let

$$\mathbf{m}_i = \mathbf{\Phi}(\mathbf{t}_i)W \tag{3}$$

be the projections of the latent points into the feature space. Then, if we assume that each of the latent points has equal probability

$$p(\mathbf{x}) = \sum_{i=1}^{K} P(i)p(\mathbf{x}|i) = \sum_{i=1}^{K} \frac{1}{K}\left(\frac{\beta}{2\pi}\right)^{\frac{D}{2}} \exp\left(-\frac{\beta}{2}||\mathbf{m}_i - \mathbf{x}||^2\right) \tag{4}$$

where D is the dimensionality of the data space. i.e. all the data is assumed to be noisy versions of the mapping of the latent points.

In the GTM, the parameters W and β are updated using the EM algorithm though the authors do state that they could use gradient ascent.

In [5], we conjectured that the GTM which models the data in a more global manner would be better at capturing the main points from play than the SOM. Also the SOM prediction of bat position requires to be interpolated between several nodes to give a smooth transition of bat positions while this comes naturally with the GTM since the positions may take any values in the latent space. Therefore, we trained the GTM on the same data as McGlinchey but, while we also mimicked human behaviours, we were somewhat disappointed that the GTM performed less well than the SOM. The SOM-AI could mostly beat the GTM-AI in play. In this paper, we first investigate two supervised methods, one of which performs a global mapping while the other performs a local mapping before introducing a new mapping [2] which can do both simultaneously.

3 Supervised Artificial Neural Networks

The two supervised artificial neural networks which we use are the multilayered perceptron (MLP) trained with backpropagation and the radial basis function network (RBF) [3]. These are both well described in the literature; the only feature in which we are interested is that the MLP is trained in a global manner while the RBF learns a sum of local models. We trained our networks on a data set created by two human players playing a game of pong. During the game, the ball made approximately 80 double traverses of the pitch. This gave us 16733 samples but we cleaned the data by ignoring those samples when the ball was outwith the pitch i.e. in the few samples when one player had missed the ball. This left us with 15442 samples. The input data was the ball's x coordinate, y-coordinate, x-velocity, y-velocity and a parameter which determined the overall speed of the ball. The target data was the player's bat position which had to be learned from the input data.

We will call the left player Stephen and the right Danny. We trained separate artificial neural networks on these data sets i.e. one network is trained on the input data + Stephen's bat position while the other is trained on the input data + Danny's bat position so that there is no interference in the network from the other function which must be learned.

We wish to compare two types of supervised learning networks, the multilayered perceptron (MLP) and the radial basis function network (RBF) on a level playing field: we attempt to level the field by giving each network approximately 100 parameters which will be learned during training.

3.1 Semi-final 1: Radial Basis Networks

We first compare two radial basis networks. 10-fold cross validation suggested that the radial basis network should be trained for 500000 iterations with inverse width parameter =0.0001. We used 100 basis functions since that gives us 100 parameters (weights) which can be adjusted; comparative experiments showed little improvement when we increase the number of basis functions and a small decrease in performance as we decrease the number of basis functions. The learning rate was annealed from 0.1 to 0 during the simulation. When the trained networks played each other rbf_{Danny} beat $rbf_{Stephen}$ by 14 games to 2 on new games (i.e. not data in the training set) over 10000 time instances.

An important point to note is that we select the centres randomly from the data points at the start of the game and do not adjust their positions during the game. The *same* centres are used for both RBF networks.

3.2 Semi-final 2: Multilayered Perceptrons

We repeat the experimental situation with multilayered perceptrons. We note that if we have n inputs, h hidden neurons and 1 output, the number of adjustable parameters for the mlp is $(n+1)h+h+1$[1]: we use 20 hidden neurons so that we have 141 parameters to adjust. Again cross validation showed that 20 hiddden neurons and a slope parameter of 0.000000001 were appropriate. We also allowed 500000 iterations but used a lower initial learning rate (=0.001) since the higher learning rate produced overfitting with this number of iterations. With the same test set as before (i.e. the ball appeared at the same position for the first shot of any rally), we find that $mlp_{Stephen}$ beat mlp_{Danny} by 37 to 19.

3.3 The Final

In the final, we play $mlp_{Stephen}$ against rbf_{Danny} and find that the latter wins by a convincing 10 games to 1. Even if we reduce the number of basis functions to 50, rbf_{Danny} beats $mlp_{Stephen}$ by 18 games to 14.

We also might consider, for example, improving the RBF's performance by selecting the centres only from points close to the position from which the ball must be played. Thus we might select each RBF's centres only from points within 100 units (the pitch is 600 long) from the appropriate side. This does improve performance even more but between shots, the RBF networks output nothing meaning the bat returns to 0 (the bottom of the pitch) after each shot and only comes up the pitch as the ball re-enters the 100 wide strip near where the ball must be played; this is hardly a human-like performance and so will not be considered any further.

3.4 Discussion

The above might seem to suggest that we might always favour the radial basis network over the multilayered perceptron. However, there are advantages and disadvantages to be aware of in both networks. In particular, the MLP performs a global mapping while the RBF is best considered a sum of local mappings. This means that the MLP is better placed to extrapolate (into new regions of data space) than the RBF and also that the RBF can be too data-specific. Ideally we

[1] The extra 1 appears because of the bias term.

wish our machines to be robust and able to perform under new conditions not specifically met during training.

For example, the human players modelled above tried to hit the ball with the edges of their bats, something which the artificial neural networks also learned though this actually has no effect in the game. If we reduce the size of the bat from 64 pixels to 34 pixels, $mlp_{Stephen}$ beats rbf_{Danny} (with 100 basis functions) by 23 games to 16. The RBF has learned to hit the ball with the edge of the bat; presumably the MLP's edge hitting behaviour has been moderated by its neurons learning the global mapping.

4 The Topographic Product of Experts

The Topographic Product of Expers (ToPoE) is discussed in more detail in [2]. We may quickly describe it as a latent variable model which learns using gradient ascent. We envisage that the underlying structure of the data can be represented by K latent points, $t_1, t_2, \cdots, t_K$. To allow local and non-linear modeling, we map those latent points through a set of M basis functions, $f_1(), f_2(), \cdots, f_M()$. This gives us a matrix Φ where $\phi_{kj} = f_j(t_k)$. Thus each row of Φ is the response of the basis functions to one latent point, or alternatively we may state that each column of Φ is the response of one of the basis functions to the set of latent points. One of the functions, $f_j()$, acts as a bias term and is set to one for every input. Typically the others are gaussians centered in the latent space. The output of these functions are then mapped by a set of weights, W, into data space. W is $M \times D$, where D is the dimensionality of the data space and is the sole parameter which we change during training. We use $\mathbf{w}_i$ to represent the ith column of W and Φ_j to represent the row vector of the mapping of the jth latent point. Thus each basis point is mapped to a point in data space, $y_j = \Phi_j W$.

We may update W either in batch mode or with online learning. To change W in online learning, we randomly select a data point, say $\mathbf{x}_i$. We calculate the responsibility of each latent point for this data point using

$$r_{ij} = \frac{exp(-\gamma d_{ij}^2)}{\sum_k exp(-\gamma d_{ik}^2)} \tag{5}$$

where $d_{pq} = ||\mathbf{x}_p - \sum_k (\phi_{qk} . w_k)||$, the euclidean distance between the pth data point and the projection of the qth latent point (through the basis functions and then multiplied by W). If no weights are close to the data point (the denominator is zero), we set $r_{ij} = \frac{1}{K}, \forall j$.

We calculate $p_{kd} = \sum_{m=1}^{M} w_{md}\phi_{km}$, the projection of the kth latent point on the dth dimension in data space and then use this in the update rule

$$\Delta_n w_{md} = \sum_{k=1}^{K} \eta \phi_{km} (x_d - p_{kd}^n) r_{kn} \tag{6}$$

so that we are summing the changes due to each latent point's response to the data points. Note that, for the basic model, we do not change the Φ matrix during training at all.

With this model on the same pong data as before, we may either during testing take the estimate of the bat position from the modal node (that with highest responsibility) or we may incorporate the responsibilities into the estimate. Thus the estimated response to $\mathbf{x}_n$ is either f_{i*} (totally local) or $\sum_i f_i r_{in}$ (somewhat distributed). This is an even more level playing field with which to compare the modelling since both estimates, one very local and one more distributed, are coming from exactly the same model and indeed simulation. Our findings are that, with a 10$\times$ 10 grid of basis functions and a 20$\times$20 grid of latent points, the local method makes only 75.3% of the misses that the more distributed method makes but visually it looks rather un-human-like: the bat movement moves jerkily between positions whereas in the global method the transitions are smooth and human-like.

5 Conclusion

In this paper, we have compared various forms of opponent-modelling for the computer game, Pong. If the aim of the game is to make the AI as good as possible at the game, then local models tend to be better. However, it is very easy with this simple game to make an unbeatable (and hence not very interesting) AI. Rather our aim is firstly to investigate how we can best model human behaviour and only subsequently to investigate how to improve the AI player. In the former task the global models may be better.

The series of simulations in this paper deal with the very simple game of Pong. Our future research will investigate whether our findings are repeatable in more complex environments.

References

1. C. M. Bishop, M. Svensen, and C. K. I. Williams. Gtm: The generative topographic mapping. *Neural Computation*, 1997.
2. C. Fyfe. Topographic product of experts. In *International Conference on Artificial Neural Networks, ICANN2005*, 2005.
3. S. Haykin. *Neural Networks- A Comprehensive Foundation.* Macmillan, 1994.
4. Tuevo Kohonen. *Self-Organising Maps.* Springer, 1995.
5. G. Leen and C. Fyfe. Training an ai player to play pong using the gtm. In *IEEE Symposium on Computational Intelligence and Games*, 2005.
6. S. McGlinchey. Learning of ai players from game observation data. In *Proceedings of Fourth International Conference on Intelligent games and Simulation, ISBN: 90-77381-05-8*, pages 106–110, 2003.

Evolution of Heuristics for Give-Away Checkers

Magdalena Kusiak, Karol Walędzik, and Jacek Mańdziuk

Faculty of Mathematics and Information Science,
Warsaw University of Technology, Plac Politechniki 1, 00-661 Warsaw, Poland

Abstract. The efficacy of two evolutionary approaches to the problem of generation of heuristical linear and non-linear evaluation functions in the game of give-away checkers is tested in the paper. Experimental results show that both tested methods lead to heuristics of reasonable quality and evolutionary algorithms can be successfully applied to heuristic generation in case not enough expert knowledge is available.

1 The Game of Give-Away Checkers (GAC)

The game of US give-away checkers [1] is played according to the same rules as US checkers. The way of determining the winner is the only exception. In order to win in the game of GAC a player has to lose all his/her pieces or be unable to perform a move. The rules of the game are simple and widely known and - at the same time - the results of brute-force algorithm using trivial strategy of losing pieces as quickly as possible are unsatisfactory. One of the reasons for unsuitability of this simple algorithm is the fact that a single piece is barely mobile and has very restricted choice of possible moves. Fig. 1 presents two situations in which white loses despite having only one piece left.

2 The Evaluation Function

In each of heuristical evaluation functions discussed in this paper some or all of the following components (factors) were considered (each of them calculated either separately for each player or as a difference of respective values for both players): numbers of (1) pawns (i.e. pieces other than kings), (2) kings and (3) pieces; numbers of (4) safe (i.e. adjacent to the edge of the board) pawns, (5) safe kings and (6) safe pieces; numbers of (7) moveable (i.e. able to make a move other than capturing – this feature was calculated without considering capturing priority) pawns, (8) moveable kings and (9) moveable pieces; (10) aggregated distance of all pawns to promotion line; (11) number of unoccupied fields on the promotion line. Each heuristic consisted of linear combination of some (or none) of the above parameters and arbitrary number of nonlinear components, each of them of the following form:
IF [**NOT**](*param1* **BETWEEN** minVal1 **AND** maxVal1 **AND/OR** *param2* **BETWEEN** minVal2 **AND** maxVal2 **AND/OR** ...) **THEN**

W. Duch et al. (Eds.): ICANN 2005, LNCS 3697, pp. 981–987, 2005.

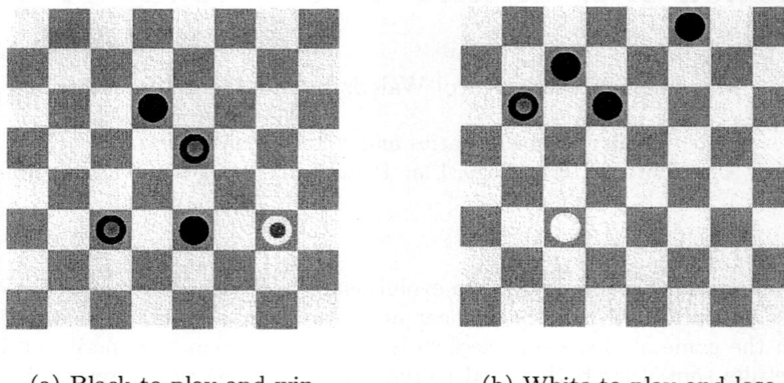

(a) Black to play and win (b) White to play and lose

Fig. 1. Examples of inefficiency of greedy heuristics. White move bottom-up. Open circles denote kings.

$Heuristic\_Result \mathrel{+}= LinearCombinationOfParameters$

Only one logical operator (AND or OR) could be used in each nonlinear component and negation could be applied to the whole condition only.

3 Evolutionary Algorithms

Genetic algorithms were used to generate weights in each of the above *LinearCombinationOfParameters.* All variable parameters of the heuristics were represented as a vector of real numbers and each number was a single gene. Conditions defined in nonlinear components were not modified by the evolutionary process. Two different approaches, described in the following subsections were implemented.

3.1 Heuristic Generator (HG)

One of the problems encountered while designing a genetic algorithm was defining fitness function for the heuristics. The general idea to solve this problem was based on [2]. The game was divided into several stages and the purpose of the first phase of the algorithm was to obtain a heuristic that would be able to assess correctly situations close to the end of the game. In order to achieve this, a number of situations close to the leaves of the game tree were generated and assessed using alpha-beta algorithm without heuristic. If alpha-beta failed to reach a leaf of the tree, the situation was considered to be a draw. Subsequently, each specimen assessed the same situations and its fitness was defined as $[n/\sum(h_i - a_i)^2]$, where n is the number of test positions, h_i - assessment of the i-th test situation by the heuristic specimen and a_i by the alpha-beta algorithm.

Depending on the algorithm settings the fitness of each specimen could be divided by the sum of similarities of all specimens from the population to it. Similarity of two specimens was defined as $exp(-d^2)$, where d is Euclidean distance between their genotypes. This was done as a mean to encourage speciation [3,4], which in turn might lead to improved exploration of the problem space. Since GAC has only three possible results and the values of heuristics belong to a continuous interval, the depth of a leaf in the game tree was taken into consideration when assessing it.

Once the initial stage had ended, some new situations were generated, that were closer to the root of the game tree and worst fitted fraction of the population was replaced by random specimens. The fittest specimen of the previous phase was used by alpha-beta to assess these new situations. The process continued until the beginning of the game was reached.

In each phase a constant fraction of all test boards came from the stage of the game nearest to the beginning. Depending on the settings of the algorithm, the situations closer to the leaves of the game tree were either regenerated and reassessed by the newest heuristic or once generated they were used throughout all subsequent phases. The following genetic operators were used:

Selection. Tournament selection was implemented. Several specimens were randomly chosen from the population. The fittest among them was the winner of the tournament. In order to determine a pair of specimens to crossbreed, two such tournaments were held, and their winners were coupled.

Crossover. Each pair of respective linear combinations contained by a heuristic was crossed over independently. The genotype of each linear combination was randomly divided into two parts and values of each part were inherited from one parent. The value of the gene on which the division was placed was taken from the interval defined by the values of this gene in parent specimens. The descendant replaced the weakest specimen in the population.

Mutation. Three kinds of mutations occurred in population: multiplying a value of a gene by two, dividing it by two or changing its sign. Multiplying or dividing a value by two were twice as probable as changing the sign. Each gene of a specimen mutated independently[1].

3.2 Simple Heuristic Generator (SHG)

The idea of the algorithm was based on a simplistic assumption that results of games played by pairs of specimens define a relation close to partial order. In order to determine the result, two specimens played one or two (with sides swap) games against each other. By default the search depth during the games was set to 3. Basing on the relation described above, it was relatively easy to compare and sort specimens within a small set. Therefore, no fitness function was necessary to carry out tournament selection.

[1] Instead of multiplication/division of gene's value also addition/subtraction within some range was tested, but results were poorer in that case.

The genetic operators used in this algorithm bear great resemblance to those described above. The only significant difference is the necessity to normalize specimens genotypes. Two specimens to crossbreed were chosen by means of tournaments. Additional tournament was held to determine the weakest specimen to be replaced by the descendant.

4 Results

4.1 Heuristic Types

Based on preliminary tests we have decided to inspect four types of heuristics in more detail. Two of them (8F and 10F) were linear and two other (3Ph and 5Ph) consisted only of nonlinear components. Each heuristic was generated twice: once using HG and once with SHG.

8Factors (8F) heuristic was a linear combination of the differences (between the player and his opponent) in the following parameters described in Sect. 2: (1), (2), (4), (5), (7), (8), (10) and (11). For example (2) in the above denotes the following feature: *the number of kings owned by the player minus the number of opponents kings.*

10Factors (10F) heuristic was a linear combination of the differences in the following six parameters described in Sect. 2: (4), (5), (7), (8), (10), (11) and of the four raw values, namely (1) and (2), each of them calculated for both playing sides.

Basing on the analysis of games played it was decided that it would be advantageous to divide the entire game into several disjoint stages and to use different heuristic for each stage. Two crucial moments requiring change of the heuristic were identified. Firstly, presence of kings certainly indicates that the game has entered an advanced stage. Moreover, due to the mobility issues mentioned earlier, end-game positions might also require defining a new heuristic.

3Phase (3Ph) heuristic assigned each situation on the board to one of three disjoint categories: (*a*) **ending**: one of the players has at most three pieces left; (*b*) **kings**: both opponents have more than 3 pieces and at least one player has some kings; (*c*) **beginning**: both opponents have more than 3 pieces and no kings exist. A linear heuristic respective to 8F was assigned to evaluate situations belonging to each category. For example phase (*c*) was encoded in the following pseudo-algorithm:

IF ABS(Players_pieces_count - 10.0) $<$ 6.5 AND ABS(Opponent's_pieces_count - 10.0) $<$ 6.5 AND ABS(Total_kings_count) $<$ 0.5 THEN $C_1 * Diff(1) + C_2 * Diff(4) + C_3 * Diff(7) + C_4 * Diff(10) + C_5 * Diff(11)$, where $C_1, \ldots, C_5$ are evolvable coefficients, and $Diff(n)$ denotes the value equal to the difference of feature n (listed in sect. 2) between player and its opponent.

5Phase (5Ph) heuristic was similar to 3Ph with the only exception being that **kings** category (i.e. (b) in the above) was subdivided into three categories depending on which of the players was in possession of kings. Again, a linear heuristic respective to 8F was assigned to evaluate situations belonging to each of 5 categories.

4.2 Algorithm Settings

In case of HG the depth of the initial situations in the game tree was between 81 and 87. The interval was determined basing on preliminary tests calculating average number of moves necessary to finish a game of GAC performing random moves. The difference in depths between subsequent phases was set to 6 since alpha-beta search with depth limit of 6 was still reasonably fast.

For linear heuristics generation the number of test boards for each phase was 3 000 and reusing test boards was disabled whereas for nonlinear ones the count of the boards was 6 000 and they were reused in different phases. While fewer boards were assessed in each phase during the generation of linear heuristics the total number of situations was greater which resulted in better exploration of the problem space. On the other hand, evaluating as many as 6 000 boards during each phase while generating nonlinear heuristics minimized the chance of considering too few situations belonging to certain categories and propagating the error upwards.

Test populations consisted of 350 specimens. In each phase the weakest 80% of the population were regenerated.

For SHG each test population consisted of $100 - 150$ specimens. Populations had to be smaller because of the way specimens were compared with each other. During all tests alpha-beta search limit in the games played for comparison purposes was set to the depth of 3 and maximum of 150 expanded nodes. Comparisons were symmetrical, i.e. two specimens played two games against each other swapping sides after the first game.

During all tests a newly created specimen replaced the weakest specimen in the population (in SHG the specimen to be replaced was chosen by means of a tournament). However, this only happened if the descendant was fitter than the specimen to be replaced. The potential crossovers to effective crossovers (i.e. the ones in which created specimen was actually added to the new population) ratio was investigated. It turned out that the fraction remained fairly stable throughout the process. About $80\% - 90\%$ of all the crossovers were effective in HG and about 70% in SHG. The stability resulted from the fact that in the initial stages of the algorithm convergence was comparatively quick and therefore descendants tended to be fitter than specimens from previous generations. On the other hand, in the final stages vast majority of the specimens were almost identical and there were virtually no difference in fitness between ancestors and descendants in which case new specimens were preferred and added to the population.

The convergence of the evolution is clearly illustrated by changes in lengths of intervals for different parameters as well as by distinct declines in their variances. For most parameters variances dropped by more than a thousand times in the course of evolution.

It appeared that mutations had no significant influence on the results of evolution. In SHG best specimens were saved every 1 000 crossbreedings and in most runs not a single mutated specimen was logged. In the process of evolving linear heuristics using HG about 1% of the fittest specimens saved turned out to have

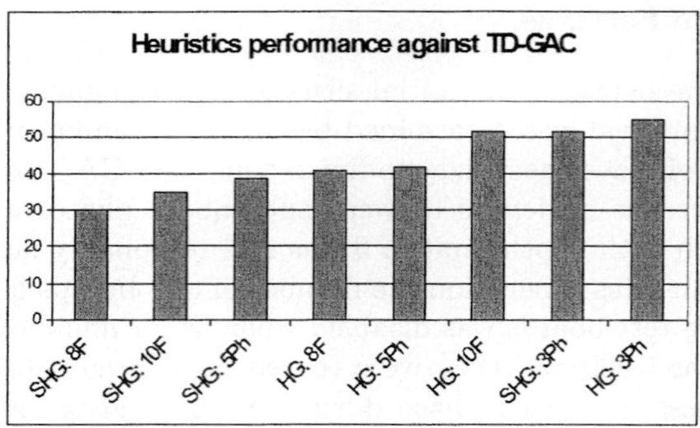

Fig. 2. Performance of the heuristics against TD-GAC

experienced mutation. The fraction was about 10% for nonlinear heuristics. The significant difference may result from the fact that in initial stages of the algorithm some genes were not applicable to the situations evaluated and therefore their mutation had no influence on the overall fitness of the specimen.

4.3 Heuristics' Performance

In order to find out about the quality of different heuristics generated, tests were run during which each heuristic played 40 games (swapping sides after each game) against TD-GAC program [5,6,7], which uses temporal difference algorithm and learns from games played. During the tests alpha-beta search depth was set to 6 in evolved heuristics and was set to 4 in TD-GAC (since TD-GAC heuristic makes use of more sophisticated parameters, including indirect exploration of the game tree one ply further). In order to make a fair comparison of heuristics the learning ability of TD-GAC was temporarily disabled. Please note, that due to some randomness in searching the game tree implemented in alpha-beta, for any particular heuristic games played with TD-GAC were not identical.

The results of comparison presented in Fig. 2 show clearly that the heuristics tended to perform well, taking into account simple parameters they considered. As it can be seen in Fig. 2 nonlinear heuristics (particularly 3Ph) generally performed slightly better than linear ones which could be expected. Worse performance of 5Ph heuristic might stem from its greater complexity which could have hindered evolutionary process.

Additional tests were carried out to measure performance of the alpha-beta algorithm. It turned out that the results depended to a great extent on the players strategies. During quick games with a lot of compulsory capture sequences lower average branching factors were reported and fewer nodes had to be analyzed as well. During games with the search depth of 6 linear heuristics needed approximately 54 − 78ms to assess a situation. The number of nodes analyzed was between 3 700 and 6 700 (at about 75 000 − 85 000 nodes per second). For nonlinear heuristics assessment lasted on average 120 – 280ms. Heuristics analyzed 6 000 – 10 000 nodes with the speed of about 40 000 nodes per second.

Estimations were also made as to pruning efficiency of different heuristics, which was defined as $(1 - n/s)$, where n denotes the number of nodes analyzed and s-theoretical size of the game subtree calculated basing on branching factor reported and search depth. Approximated pruning efficiency turned out to be rather stable for all heuristics ranging from 0.75 to 0.9.

5 Conclusions and Directions for Future Research

The main research goal of this paper concerns the possibility of building efficient heuristic evaluation functions based on evolutionary approach. In particular the efficacy of nonlinear vs. linear heuristics is verified along with comparison of HG and SHG.

Results of games played against TD-GAC support the hypothesis that HG generally outperforms SHG, and hence it can be concluded that due to its non-transitivity a direct assessing method of SHG may not be the appropriate evaluation method.

As it can be expected non-linear heuristics dominated over linear ones. However, based on some other tests (not presented) it is strongly recommended that nonlinear components be defined over **disjoint conditions**. Otherwise, having several overlapping conditions makes it possible to achieve very similar results in many ways, each time with very different values of parameters.

In future we plan to verify other schemes of evolving non-linear evaluation functions in GAC as well as apply these methods to other board games.

References

1. Alemanni, J.B. http://perso.wanadoo.fr/alemanni/ give_away.html (1993)
2. Borkowski, M.: Analysis of algorithms for two-player games. M.Sc. Thesis, Warsaw University of Technology (in Polish) (2000)
3. Goldberg, D.E., Richardson, J.: Genetic algorithms with sharing for multimodal function optimisation. In: Proceedings of the 2nd International Conference on Genetic Algorithms. (1993) 41–49
4. Goldberg, D.E.: Genetic Algorithms in Search, Optimization and Machine Learning. Addison-Wesley Pub. Co. (1989)
5. Osman, D., Mańdziuk, J. http://gac-arena.gt.pl/ (2004)
6. Mańdziuk, J., Osman, D.: Temporal difference approach to playing give-away checkers. In Rutkowski, L., et al., eds.: 7th Int. Conf. on Art. Intell. and Soft Comp. (ICAISC 2004), Zakopane, Poland. Volume 3070 of LNAI., Springer (2004) 909–914
7. Osman, D., Mańdziuk, J.: Comparison of tdleaf(λ) and td(λ) learning in game playing domain. In Pal, N.R., et al., eds.: 11th Int. Conf. on Neural Inf. Proc. (ICONIP 2004), Calcutta, India. Volume 3316 of LNCS., Springer (2004) 549–554

Nonlinear Relational Markov Networks with an Application to the Game of Go

Tapani Raiko

Neural Networks Research Centre, Helsinki University of Technology,
P.O.Box 5400, FI-02015 HUT, Espoo, Finland
Tapani.Raiko@hut.fi

Abstract. It would be useful to have a joint probabilistic model for a general relational database. Objects in a database can be related to each other by indices and they are described by a number of discrete and continuous attributes. Many models have been developed for relational discrete data, and for data with nonlinear dependencies between continuous values. This paper combines two of these methods, relational Markov networks and hierarchical nonlinear factor analysis, resulting in joining nonlinear models in a structure determined by the relations in the data. The experiments on collective regression in the board game go suggest that regression accuracy can be improved by taking into account both relations and nonlinearities.

1 Introduction

Growing amount of data is collected every day in all fields of life. For the purpose of automatic analysis, prediction, denoising, classification etc. of data, a huge number of models have been created. It is natural that a specific model for a specific purpose works often the best, but still, a general method to handle any kind of data would be very useful. For instance, if an artificial brain has a large number of completely separate modules for different tasks, the interaction between the modules becomes difficult. Probabilistic modelling provides a well-grounded framework for data analysis. This paper describes a probabilistic model that can handle data with relations as well as discrete and continuous values with nonlinear dependencies.

Terminology: Using Prolog notation, we write $\mathrm{knows}(\mathrm{alex}, \mathrm{bob})$ for stating a fact that the knows relation holds between the objects alex and bob, that is, Alex knows Bob. The arity of the relation tells how many objects are involved. The knows relation is binary, that is, between two objects, but in general relations can be of any arity. The atom $\mathrm{knows}(\mathrm{alex}, B)$ matches all the instances where the variable B represents an object known by Alex. In this paper, the terms are restricted to constants and variables, that is, compound terms such as $\mathrm{thinks}(A, \mathrm{knows}(B, A))$ are not considered. For every relation that is logically true, there are associated attributes $\mathbf{x}$, say a class label or a vector of real numbers. The attributes $\mathbf{x}(\mathrm{knows}(A, B))$ describe how well A knows B and whether A likes or dislikes B. The attribute vector $\mathbf{x}(\mathrm{con}(A))$ describes what kind of a consumer the person A is. Given a relational database describing relationships between people and their consuming habits, we might study the dependencies that might be found. For instance, some people cloth like their idols, and nonsmokers tend to be

W. Duch et al. (Eds.): ICANN 2005, LNCS 3697, pp. 989–996, 2005.

KNOWS

who	whom	how
Alex	Bob	friend
Bob	Carl	neighbour
Carl	Alex	colleague

CONSUMER

who	smoker	...
Alex	no	...
Bob	no	...
Carl	no	...

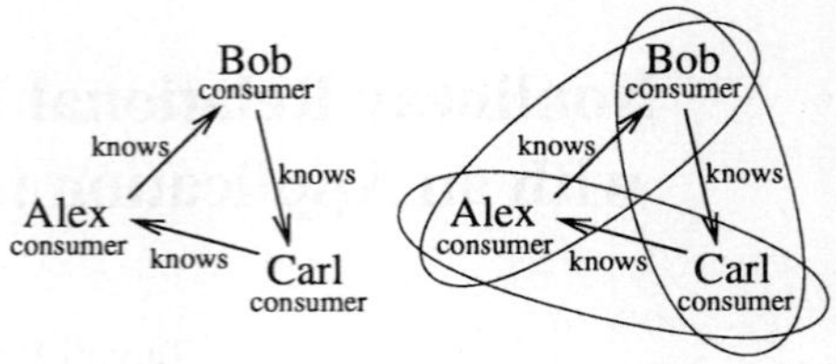

Fig. 1. Consider a relational database describing the relationships and consumer habits of three people. The two tables are shown on the left. On the right, the database is represented graphically, with the occurrences of the template $(\text{con}(A), \text{knows}(A, B), \text{con}(B))$ marked with ovals on the very right.

friends with nonsmokers. The modelling can be done for instance by finding all occurrences of the template $(\text{con}(A), \text{knows}(A, B), \text{con}(B))$ in the data and studying the distribution of the corresponding attributes. The situation is depicted in Figure 1.

Bayesian networks[6] are popular statistical models based on a directed graph. The graph has to be acyclic, which is in line with the idea that the arrows represent causality: an occurrence cannot be its own cause. In relational generalisations of Bayesian networks [7], the graphical structure is determined by the data. Often it can be assumed that the data does not contain cycles, for instance in the case when the direction of the arrows is always from the past to the future. Sometimes the data has cycles, like in Figure 1. Markov networks [6], on the other hand, are based on undirected graphical models. A Markov network does not care whether A caused B or vice versa, it is interested only whether there is a dependency or not.

2 Model Description

This section describes the models that are combined into nonlinear relational Markov networks.

2.1 Hierarchical Nonlinear Factor Analysis (HNFA)

In (linear) factor analysis, continuous valued observation vectors $\mathbf{x}(t)$ are generated from unknown factors (or sources) $\mathbf{s}(t)$, a bias vector $\mathbf{b}$, and noise $\mathbf{n}(t)$ by $\mathbf{x}(t) = \mathbf{A}\mathbf{s}(t) + \mathbf{b} + \mathbf{n}(t)$. The factors and noise are assumed to be Gaussian and independent. The index t may represent time or the object of the observation. The mapping $\mathbf{A}$, the factors, and parameters such as the noise variances are found using Bayesian learning. Factor analysis is close to principal component analysis (PCA). The unknown factors may represent some real phenomena, or they may just be auxillary variables for inducing a dependency between the observations.

Hierarchical nonlinear factor analysis (HNFA) [11] generalises factor analysis by adding more layers of factors that form a multi-layer perceptron type of a network. In this paper, there are two layers of factors $\mathbf{h}$ and $\mathbf{s}$, and the mappings are:

$$\mathbf{h}(t) = \mathbf{B}\mathbf{s}(t) + \mathbf{b} + \mathbf{n}_h(t) \tag{1}$$

$$\mathbf{x}(t) = \mathbf{A}\mathbf{f}[\mathbf{h}(t)] + \mathbf{C}\mathbf{s}(t) + \mathbf{a} + \mathbf{n}_x(t)\,, \tag{2}$$

where the nonlinearity $\mathbf{f}(\xi) = \exp(-\xi^2)$ operates on each element separately. HNFA can easily be implemented using the Bayes Blocks software library [10,12]. The update rules are automatically derived in a manner shortly described below.

The unknown variables $\boldsymbol{\theta}$ (factors, mappings, and the parameters) are learned from data with variational Bayesian learning [4]. A parametric distribution $q(\boldsymbol{\theta})$ over the unknown variables $\boldsymbol{\theta}$ is fitted to the true posterior distribution $p(\boldsymbol{\theta} \mid \boldsymbol{X})$ where the matrix $\boldsymbol{X}$ contains all the observations $\mathbf{x}(t)$. The misfit is measured by Kullback-Leibler divergence $D(\cdot \parallel \cdot)$. An additional term $-\log p(\boldsymbol{X})$ is included to avoid calculation of the model evidence term $p(\boldsymbol{X}) = \int p(\boldsymbol{X}, \boldsymbol{\theta}) d\boldsymbol{\theta}$. The cost function is

$$\mathcal{C} = D(q(\boldsymbol{\theta}) \parallel p(\boldsymbol{\theta}|\boldsymbol{X})) - \log p(\boldsymbol{X}) = \left\langle \log \frac{q(\boldsymbol{\theta})}{p(\boldsymbol{X}, \boldsymbol{\theta})} \right\rangle, \tag{3}$$

where $\langle \cdot \rangle$ denotes the expectation over distribution $q(\boldsymbol{\theta})$. Note that since $D(q \parallel p) \geq 0$, it follows that the cost function provides a lower bound for the model evidence $p(\boldsymbol{X}) \geq \exp(-\mathcal{C})$. The posterior approximation $q(\boldsymbol{\theta})$ is chosen to be Gaussian with a diagonal covariance matrix.

It is possible, though slightly impractical, to model also discrete values in HNFA by using the discrete variable with a soft-max prior [12]. In the binary case, the ith component of $\mathbf{x}(t)$ is left as a latent auxiliary variable, and an observed binary variable $y(t)$ is conditioned by $p(y(t) = 1 \mid x_i(t)) = \frac{\exp x_i(t)}{1+\exp x_i(t)}$. The general discrete case follows analogously requiring more than one auxiliary component of $\mathbf{x}(t)$. The experiments in Section 3 use a thousand copies of a binary variable having the same conditional probability. They can be united into one variable by multiplying its cost by one thousand. Observing 800 ones and 200 zeros corresponds to fixing the variable to a distribution of 0.8 times one and 0.2 times zero.

2.2 Relational Markov Networks (RMN)

A relational Markov network (RMN) [9] is a model for data with relations and discrete attributes. It is specified by a set of clique templates $\mathbf{C}$ and corresponding potentials $\boldsymbol{\Phi}$. Using the example in the introduction, a model can be formed by defining a single clique template $C = (\mathrm{con}(A), \mathrm{knows}(A, B), \mathrm{con}(B))$ and the corresponding potential ϕ_C over $\mathbf{x}(C)$ which is (a subset of) the concatenation of attribute vectors $\mathbf{x}(\mathrm{con}(A))$, $\mathbf{x}(\mathrm{knows}(A, B))$, and $\mathbf{x}(\mathrm{con}(B))$. Given a relational database, the RMN produces an unrolled Markov network over all the attributes $\boldsymbol{X}$. The cliques $c \in C(\mathcal{I})$ instantiated by a template C share the same clique potential ϕ_C. The combined probabilistic model is $p(\boldsymbol{X}) = \frac{1}{Z} \prod_{C \in \mathbf{C}} \prod_{c \in C(\mathcal{I})} \phi_C(\mathbf{x}(c))$, where Z is a normalisation constant and $C(\mathcal{I})$ contains all the instantiations of the template C. In general, a template can be any boolean formula over the relations.

The general inference task is to compute the posterior distribution over all the variables $\boldsymbol{X}$. The network induced by data can be very large and densely connected, so exact inference is often intractable [9]. The belief propagation (BP) algorithm [6] is guaranteed to converge to the correct marginal probabilities only for singly connected Markov networks, but it is used as a good approximation also in the loopy case. The learning task, or the estimation of the potentials $\boldsymbol{\Phi}$ is done using the maximum a posteriori criterion. It requires an iterative algorithm alternating between updating the parameters of the potentials and running the inference algorithm on the unrolled Markov network.

2.3 Nonlinear Relational Markov Networks (NRMN)

In nonlinear relational Markov networks (NRMN), the clique potentials are replaced by a probability density function for continuous values, in this case HNFA[1] The combination of these two methods is not completely straightforward. For instance, marginalisation required by the BP algorithm is often difficult with nonlinear models. Also, algorithmic complexity needs to be considered, since the model will be quite demanding. One of the key points is to use probability densities p in place of potentials Φ. Then, overlapping templates give multiple probability functions for some variable and they are combined using the product-of-experts combination rule described below.

Combination Rules: One of the non-trivialities in making relational extensions of probabilistic models is the so called combination rule [7]. When the structure of the graphical model is determined by the data, one cannot know in advance how many links there are for each node. One solution is to use combination rules such as the noisy-or. Combination rule transforms a number of probability functions into one. Noisy-or does not generalise well to continuous values, but two alternatives are introduced below.

Using a Markov network and the BP algorithm corresponds to using probability densities as potentials and the *maximum entropy* combination rule. The probability densities $p_C(\mathbf{x}(C))$ are combined to form the joint probability distribution by maximising the entropy of $p(\boldsymbol{X})$ given that all instantiations of $p_C(\mathbf{x}(C))$ coalesce with the corresponding marginals of $p(\boldsymbol{X})$. For singly connected networks, this means that the joint distribution is $p(\boldsymbol{X}) = \prod_c p_c(\mathbf{x}(c))/\prod_k p_k(\mathbf{x}(k))$, where k runs over pairs of instantiations of templates and $\mathbf{x}(k)$ contains the shared attributes in those pairs. Marginalisation of nonlinear models cannot usually be done exactly and therefore one should be very careful with the denominator. Also, one should take care in handling loops.

In the *product-of-experts* (PoE) combination rule, the logarithm of the probability density of each variable is the average of the logarithms of the probability functions that the variable is included in: $p(x) \propto \sqrt[n]{\prod_{C\in\mathbf{C}} \prod_{c\in C(\mathcal{I})} p_C(x)}$ for all $x \in \boldsymbol{X}$, where only those n instantiated templates c that contain x, are considered. PoE is easy to implement in the variational Bayesian framework because the term in the cost function (3) can be split into familiar looking terms. Consider the combination of two probability functions $p_1(x)$ and $p_2(x)$ (that are assumed to be independent):

$$\left\langle \log \frac{q(x)}{\sqrt{p_1(x)p_2(x)}} \right\rangle = \frac{\left\langle \log \frac{q(x)}{p_1(x)} \right\rangle + \left\langle \log \frac{q(x)}{p_2(x)} \right\rangle}{2}. \quad (4)$$

A characteristic of PoE is that implicit weighting happens in some sense automatically. When one of the experts gives a distribution with a large variance and another one with a small variance, the combination is close to the one with small variance.

Inference in Loopy Networks: Inferring unobserved attributes in a database is in this case an iterative process which should end up in a cohesive whole. Information can traverse through multiple relations.[2] The basic element in the inference algorithm of

[1] One could also think in terms of e.g. a mixture model.

[2] In mixture of experts (MoE), it is enough when only one of the experts explains the data even if all the other disagree. The ignored experts will not pass information on. This explains why the author did not consider MoE as a combination rule.

Bayes Blocks is the update of the posterior approximation $q(\cdot)$ of a single unknown variable, assuming the rest of the distribution fixed. The update is done such that the cost function (3) is minimised. One should note that when the distribution over the Markov blanket of a variable is fixed, the local update rules apply, regardless of any loops in the network. Therefore the use of local update rules is well founded, that is, local inference in a loopy network does not bring any additional heuristicity to the system. Also, since the inference is based on minimising a cost function, the convergence is guaranteed, unlike in the BP algorithm.

Learning: The learning or parameter estimation problem is to find the probability functions associated with the given clique templates $\mathbf{C}$. Now that we use probability functions instead of potentials, it is possible in some cases to separate the learning problem into parts. For each template $C \in \mathbf{C}$, find the appropriate instantiations $c \in C(\mathcal{I})$ and collect the associated attributes $\mathbf{x}(c)$ into a table. Learn a HNFA model for this table ignoring the underlying relations. This divide-and-conquer strategy makes learning comparatively fast, because all the interaction is avoided. There are some cases that forbid this. If the data contains missing values, they need to be inferred using the method in the previous paragraph. Also, it is possible to train experts cooperatively rather than separately [3].

Clique Templates: In data mining, so called frequent sets are often mined from binary data. Frequent sets are groups of binary variables that get the value 1 together often enough to be called frequent. The generalisation of this concept to continuous values could be called the *interesting sets*. An interesting set contains variables that have such strong mutual dependencies that the whole is considered interesting. The methodology of inductive logic programming could be applied to finding interesting clique templates. The definition of a measure for interestingness is left as future work. Note that the divide-and-conquer strategy described in the previous paragraph becomes even more important if one needs to consider different sets of templates. One can either learn a model for each template separately and then try combinations with the learned models, or try a combination of templates and learn cooperatively the models for them. Naturally the number of templates is much smaller than the number of combinations and thus the first option is computationally much cheaper.

So what are meaningful candidates for clique templates? For instance, the template $(\text{con}(A), \text{con}(B))$ does not make much sense. Variables A and B are not related to each other, so when all pairs are considered, $\text{con}(A)$ and $\text{con}(B)$ are always independent and thus uninteresting by definition. In general, a template is uninteresting, if it can be split into two parts that do not share any variables. When considering large templates, the number of involved attributes grows large as well, which makes learning more involved. An interesting possibility is to make a hierarchical model. When a large template contains others as subtemplates, one can use the factors $\mathbf{s}$ in Eq. (1, 2) of the subtemplate as the attributes for the large template. The factors already capture the internal structure of the subtemplates and thus the probabilistic model of the large template needs only to concentrate on the structure between its subtemplates.

3 Experiments with the Game of Go

Game of Go: Go is an ancient oriental board game. Two players, black and white, alternately place stones on the empty points of the board until they both pass. The standard board is 19 by 19 (i.e. the board has 19 lines by 19 lines), but 13-by-13 and 9-by-9 boards can also be used. The game starts with an empty board and ends when it is divided into black and white areas. The one who has the larger area wins. Stones of one colour form a block when they are 4-connected. Empty points that are 4-connected to a block are called its liberties. When a block loses its last liberty, it is removed from the board. After each move, surrounded opponent blocks are removed and only after that, it is checked whether the block of the played move has liberties or not. There are different rulesets that define more carefully what a "larger area" is, whether suicide is legal or not, and how infinite repetitions are forbidden.

Computer Go: Of all games of skill, go is second only to chess in terms of research and programming efforts spent. While go programs have advanced considerably in the last 10-15 years, they can still be beaten easily by human players of moderate skill. [5] One of the reasons behind the difficulty of static board evaluation is the fact that there are stones on board that will eventually be captured, but not in near future. In many cases experienced go players can classify these dead stones with ease, but using a simple look-ahead to determine the status of stones is not always feasible since it might take dozens of moves to actually capture the stones.

Experiment Setting: The goal of the experiments was to learn to determine the status of the stones without any lookahead. An example situation is given in Figure 2. The data was generated using a go-playing program called Go81 [8] set on level 1 and using randomness to have variability. By playing the game from the current position to the end a thousand times, one gets an estimate of who is going to own each point on the board. Information on the board states was saved to a relational database with two tables for learning an NRMN. The $\mathbf{x}(\text{block}(A))$ contains the colour, the number of liberties, the size, distances from the edges, influence features in the spirit of [1], and finally the count of how many times the block survives in the 1000 possible futures. The $\text{ally}(A, B)$ and $\text{enemy}(A, B)$ contain a measure of strength of the connection between the blocks A and B estimated using similar influence features [1]. Only the pairs with a strong enough influence on each other (> 0.02) were included. One thousand 13-by-13 board positions after playing 2 to 60 moves were used for learning.

Two clique templates, $((\text{block}(A), \text{ally}(A, B), \text{block}(B))$ and an analogous one for enemy, were used. HNFA models were taught with 28 attributes of the two blocks and the pair. The dimensionality of the $\mathbf{s}$ layer was 8. The learning algorithm pruned the dimensionality of the $\mathbf{h}$ layer to 41 for allies and to 47 for enemies. The models were learned for 500 sweeps through the data. A linear factor analysis model was learned with the same data for comparison. A separate collection of 81 board positions with 1576 blocks was used for testing. The status of each block was now hidden from the model and only the other attributes were known. With inference in the network, the status were collectively regressed. As a comparison experiment, the inferences were also done separately, and combined only in the end. Inference required from four to thirty iterations to converge. As a postprocessing step, the regressed survival probabilities $\hat{x}$

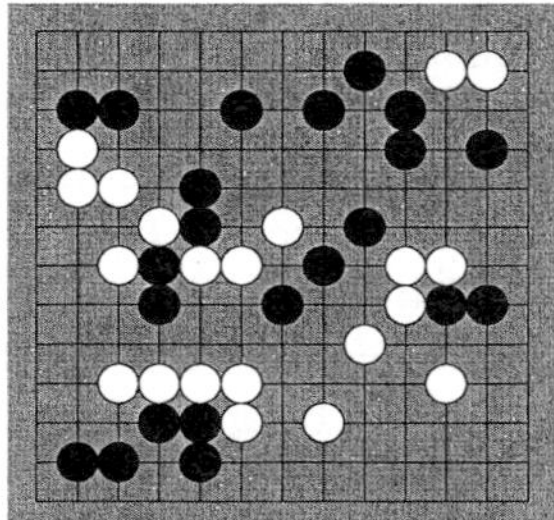 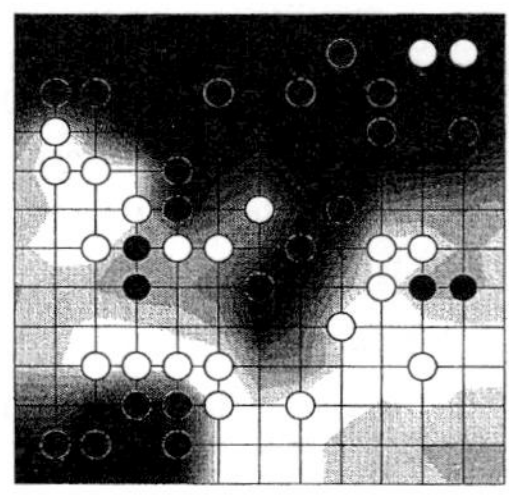 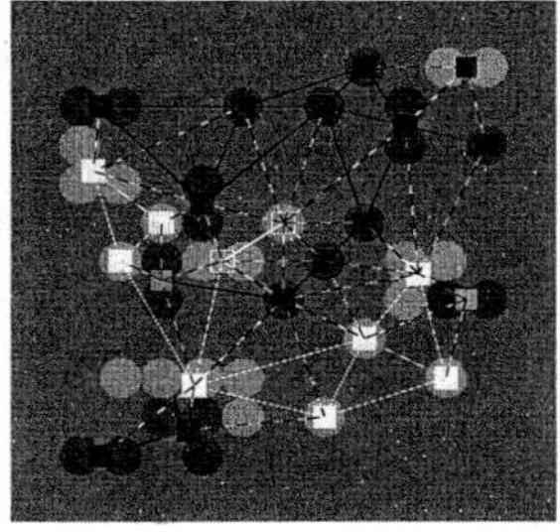

Fig. 2. The leftmost subfigure shows the board of a go game in progress. In the middle, the expected owner of each point is visualised with the shade of grey. For instance, the two white stones in the upper right corner are very likely to be captured. The rightmost subfigure shows the blocks with their expected owner as the colour of the square. Pairs of related blocks are connected with a line which is dashed when the blocks are of opposing colours.

were modified with a simple three-parameter function $\hat{x}_{new} = a\hat{x}^b + c$ and the three parameters that gave the smallest error for each setting, were used.

Results: The table below shows the root mean square (rms) errors for inferring the survival probabilities of the blocks in test cases. They can be compared to the standard deviation 0.2541 of the probabilities.

rms error	Linear	Nonlinear
Separate regression	0.2172	0.2052
Collective regression	0.2171	0.2037

As expected, nonlinear models were better than linear ones and collective regression was better than separate regression.

4 Discussion and Conclusion

A traditional Markov network was applied for statically determining the status of the go board in [2]. Games played by people were used as data. Humans play the game better, but still, this approach has an important downside. The data contains only one possible future for each board position whereas a computer player can produce many possible futures. At the learning stage, all those futures can be used together for the computational price of one. Also, stones that are provably determined to be captured under optimal play (*dead*), might still be useful: By threathening to revive them, the player can gain elsewhere. When data is gathered with unoptimal play, the stones are marked as *not quite dead*, which might be desirable.

NRMN includes a probabilistic model only for the attributes and not for the logical relations. Link uncertainty means that one models the possibility of a certain relation to exist or not. Actually one can model link uncertainty using just the proposed methodology. All the uncertain relations are assumed to be logically true and an additional binary attribute is included to mark whether the link exists or not. One only needs to take into account that when this binary attribute gets the value zero, the dependencies between the other attributes are not modelled. Also, time series data can be represented using relations obs(T) for the observations at time T and ensues($T1, T2$) to denote that the time indices $T1$ and $T2$ are adjacent. These two examples give light to the generality of the proposed method. In [12], HNFA is augmented with a variance model. Modelling variances would be important also in the NRMN setting, because then each expert would produce an estimate of its accuracy and thus implicitly a weight compared to other experts. In [9], relational Markov networks were constructed to be discriminative so that the model is specialised to classification. The same could be applied here.

Conclusion: A model was proposed for data containing both relations and nonlinear dependencies. The model was built by combining two state-of-the art probabilistic models, hierarchical nonlinear factor analysis and relational Markov networks by using the product-of-experts combination rule. Many simplifying assumptions were made, such as diagonality of the posterior covariance matrix, and separate learning of experts. Also, learning the model structure (the set of clique templates) was left as future work. Experiments with the game of go give promise for the proposed methodology.

Acknowledgements: The author thanks Kristian Kersting, Harri Valpola, Markus Harva, and Alexander Ilin for useful discussions and comments. This research has been funded by the Finnish Centre of Excellence Programme (2000-2005) under the project New Information Processing Principles, and by the IST Programme of the European Community, under the PASCAL Network of Excellence, IST-2002-506778. This publication only reflects the author's views.

References

1. B. Bouzy. Mathematical morphology applied to computer go. *IJPRAI*, 17(2), 2003.
2. T. Graepel D. Stern and D. MacKay. Modelling uncertainty in the game of Go. In *Proc. of the Conference on Neural Information Processing Systems*, Vancouver, December 2004.
3. G.E. Hinton. Modelling high-dimensional data by combining simple experts. In *Proc. AAAI-2000*, Austin, Texas.
4. M. Jordan, Z. Ghahramani, T. Jaakkola, and L. Saul. An introduction to variational methods for graphical models. In M. Jordan, editor, *Learning in Graphical Models*, pages 105–161. The MIT Press, Cambridge, MA, USA, 1999.
5. M. Müller. Computer Go. *Special issue on games of Artificial Intelligence Journal*, 2001.
6. J. Pearl. *Probabilistic Reasoning in Intelligent Systems: Networks of Plausible Inference.* Morgan Kaufmann Publishers Inc., San Francisco, CA, USA, 1988.
7. L. De Raedt and K. Kersting. Probabilistic logic learning. *ACM-SIGKDD Explorations, special issue on Multi-Relational Data Mining*, 5(1):31–48, July 2003.
8. T. Raiko. The go-playing program called Go81. In *Proceedings of the Finnish Artificial Intelligence Conference, STeP 2004*, pages 197–206, Helsinki, Finland, 2004.
9. B. Taskar, P. Abbeel, and D. Koller. Discriminative probabilistic models for relational data. In *Proc. Conference on Uncertainty in Artificial Intelligence (UAI02)*, Edmonton, 2002.
10. H. Valpola, A. Honkela, M. Harva, A. Ilin, T. Raiko, and T. Östman. Bayes blocks software library. `http://www.cis.hut.fi/projects/bayes/software/`, 2003.
11. H. Valpola, T. Östman, and J. Karhunen. Nonlinear independent factor analysis by hierarchical models. In *Proc. ICA2003*, pages 257–262, Nara, Japan, 2003.
12. H. Valpola, T. Raiko, and J. Karhunen. Building blocks for hierarchical latent variable models. In *Proc. ICA2001*, pages 710–715, San Diego, USA, 2001.

Flexible Decision Process for Astronauts in Marsbase Simulator

Jean Marc Salotti

Laboratoire de Sciences Cognitives, Bordeaux 2 University
salotti@idc.u-bordeaux2.fr

Abstract. We present a friendly software platform called Marsbase for the simulation of human activities on Mars. The interface allows the user to command every astronaut activity on Mars, typically exploring the region, analyzing rocks or building new facilities. In the AI mode, astronauts' decisions and actions are defined by a competition between different basic processes. Some human factors such as tiredness or perseverance have been implemented.

1 Introduction

A software platform called "Marsbase" has been developed to simulate astronauts' activities on the Martian surface. The simulator has been designed with the help of experts of human-based space exploration programs [10]. It takes into account a large number of objects and enables numerous astronauts' actions. AI modules have also been implemented, allowing an interesting interaction of the user with the system. In Section 2, the main features of the simulator are described. In Section 3, a brief overview of artificial intelligence techniques for simulating virtual characters are discussed. We then present the concepts of the AI modules, which are inspired from Brooks' subsumption architecture [2]. An application is also presented to show how cognitive behaviors can be observed from different implementations of the layers.

2 Marsbase Simulator

2.1 Objects and Transformations

In order to develop a realistic simulator, a large number of objects have been defined, such as different kinds of rocks, chemical elements (water, oxygen, silicon, etc.), industrial products and tools (photovoltaic cells, bricks, mass spectrometers, spacesuits, automatons, etc.), buildings (habitat, greenhouse, chemical unit, nuclear reactor, etc.), vehicles and astronauts. An object is defined by a name, a weight, the consumption or the supply of a given amount of energy; it can be visible (with a picture and a position) or not; it can move with a variable speed; it eventually includes other objects; it is eventually included in a containing object, etc. Astronauts are defined as particular objects that perform specific tasks. They can "transform" objects, move around, carry or put down objects, etc. In addition, they consume oxygen all the time, consume water and food a few times per day and need to sleep sometime.

W. Duch et al. (Eds.): ICANN 2005, LNCS 3697, pp. 997 – 1002, 2005.

Transformations are performed by astronauts, automatons and robots. They have been normalized to enable information processing standardization. They are defined in text files by a list of resources with their corresponding masses, a list of tools (not consumed), a list of products with their corresponding masses, the energy required for the transformation and its duration. Most actions are defined as transformations. For instance, agriculture is a process that requires seeds, fertilizer and water as sources, a greenhouse and a human as tools and produces cereals and fruits. Chemical or industrial transformations as well as constructions of new buildings are defined in a similar way. An icon is associated with every transformation and all of them are accessible in several grids on the action panel. Once an astronaut is selected, the user can start a transformation by a click on the corresponding icon, provided that all required resources and tools are available. Objects and transformations are defined in text-files independently from the code.

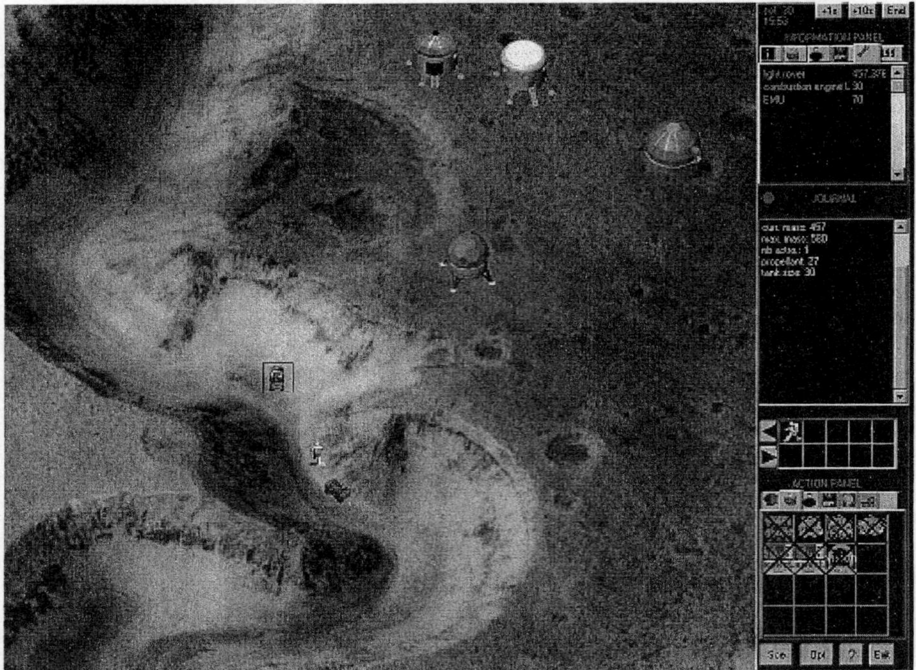

Fig. 1. Gray-scale snapshot of the simulator. On the planetary surface, from top to bottom: A cargo, a habitat, a greenhouse, a nuclear reactor, an astronaut driving a light vehicle (selected), an astronaut walking and another light vehicle. The right panel is divided into 3 parts. At the top, information on the selected object is available. A journal provides information on current events (action completed, oxygen missing, etc.). In the middle, a small grid enables the selection of astronauts and at the bottom, transformations icons are displayed in different grids.

2.2 Interface and Scenarios

The interface of the simulator is user-friendly and is inspired from well-known simulation/strategic games. As can be observed in Fig. 1, a picture of Mars is used to display human-based objects on the planetary surface and a panel allows the user to get information about any object and to command astronauts' tasks by means of the mouse. Astronauts can move with a spacesuit, enter a vehicle or a building, carry objects or start (or stop) a transformation. The time scale of the simulation is 3 minutes for one sol (1 Martian day) and the scenarios can last an arbitrary number of sols. At night, a darker image of the same area is automatically loaded. During the simulation, there is a constant interaction with the user. When a transformation starts, its icon is slowly redrawn below the astronaut to inform the user of its achievement.

A scenario is defined by a list of objects located on the map, for instance a habitat, a pressurized rover, solar panels, interesting rocks etc. Each object may contain a list of smaller objects (oxygen, water, mass spectrometer, etc.) and eventually astronauts.

The objective of the scenario is presented in a text window at the beginning of the simulation. It is usually defined in the scenario as a list of objects in a specific state that should exist at the end. It is also possible to use the simulator without specific objective. Another option is to start with a preparation phase, in which the user has to define the payload of the rocket before it is sent to Mars.

3 Artificial Intelligence

3.1 Virtual Characters and Brooks' Principles

A lot of work has been carried out in the field of artificial intelligence for the simulation of human behaviors in artificial environments, in particular in strategic/tactical games [4], [7], [8]. There are basically two different approaches to the problem of creating virtual characters [5], [6], [7]:

- The first one is top-down and characters are accurately designed but fixed (Oz, Petz, Improv).
- The second is bottom-up and characters are basically designed but evolving (Creatures, Silas, Sims' creatures).

The top-down approach generally requires each behavior to be explicitly defined, whereas the bottom-up approach is based on learning techniques such as neural networks or genetic algorithms that allow adaptive emergent behaviors.
Multi-agents systems usually provide the conceptual framework for the design of reactive or cognitive behaviors [3], [9]. Reactivity is preferred when there is no need to define a complex behavior or when time is too short for a complex decision process. Cognitive agents generally embed the required knowledge and know-how for an accurate and sometimes complex decision process.

Brooks suggested that simple principles should be considered to design virtual characters [2]:

- Complex behaviors emerge from hierarchically organized basic behaviors, which can be performed by finite state machines.

- Time constraints can be added for global coherence.
- Inhibition between different modules provides distributed control.

Brooks' ideas are inspired from cognitive science theories, which state that the attention process in human brains consists in a complex interaction of several hierarchical processes that try to inhibit each other and to monopolize brain resources to achieve predetermined behaviors [1], [2].

4 Marsbase Decision Layers

4.1 Five Important Layers

Our approach is precisely inspired from Brooks' ideas. Five basic astronaut behaviors have been identified:

- Layer 1: The search for oxygen for short term survival. Astronauts have to care about the oxygen level when they are exploring the Martian surface with a spacesuit, walking or driving a non pressurized rover. If the remaining oxygen quantity is less than an adaptive threshold, the astronaut stops all actions and goes immediately towards the nearest safe structure.
- Layer 2: The search for water and food. As it is actually implemented in the simulator, water and food are consumed each day at fixed intervals, provided that a life support system is available and that enough water and food are present. If the right conditions are never encountered after a period of 1 sol (a sol is a Martian day, it is equivalent to 24 hours and 40 minutes), the priority of the astronaut is to look for the nearest place to fulfill his needs.
- Layer 3: The search for a comfortable place to rest when he is tired. In order to take into account tiredness, astronauts working time can never exceed 1 sol and a minimum continuous sleeping time of 8 hours is expected before working again. Therefore, after 1 sol of work, whatever the number of short breaks, the astronaut immediately goes towards the nearest safe structure where he can rest (life support system required).
- Layer 4: Obedience to user's orders. The user can select any astronaut with the mouse and send him anywhere or ask him to work on any transformation. However, the astronaut is not a perfect slave. He usually obeys to orders, but he might decide to do something else, typically in case of a priority action decided in a previous layer. If the user commands the astronaut to go out or to perform a specific task, an obedience mode is activated and layer 5 is not processed. The obedience mode stops when the astronaut is back in the habitat and rested.
- Layer 5: Complex automatic actions. Complex actions require cognitive abilities. It would be very difficult to implement a coherent strategy for the automatic development of a Martian base, building new facilities, extracting ores and setting up new industries, trying to organize the cooperation between astronauts. Though it is a perspective of that work, only simple behaviors have been implemented in the current version, typically exploring the region, analyzing rocks, farming or building new facilities, depending on the scenario.

4.2 Decision Process

Such layers are normally ordered according to a priority criterion such that layer N is processed before layer N+1, and if an action is required, the successive layers are not considered (Brooks' subsumption principle). However, human behaviors are complex and priorities of decisions processes are never a priori fixed. It usually depends on the context, the motivation and global physiological states. We therefore suggest defining the priority of a layer as a subjective variable, which can evolve according to internal parameters. In its current version, only two internal parameters have been implemented to validate the principles.

- An important internal parameter is the tiredness. An astronaut is tired (but still allowed to work) when the duration of its working activity exceeds a given threshold, which depends on the character of the astronaut. In the case of tiredness, the priority of all layers is randomly decreased. If all priority levels falls to 0, there is neither action nor decision, the astronaut simply stops doing things.
- Another important internal parameter is the motivation of the astronaut to complete his current task. If the objective is close to be achieved, the motivation is increased. Humans typically try to finish their action if the remaining time to achieve the objective is relatively low, even in the case of alarming signals like little oxygen left (providing that there is some margin). In order to observe a similar behavior, the remaining time to achieve the objective is computed. If it is a transformation, its end is known and if it is a walk towards a specific place, it is inferred from the speed and the distance to the location. Then, if the time is below a threshold, the motivation to complete that action is set to a high value and the priority is set to the maximum.

5 Experiments and Discussion

Some experiments have been made to evaluate the global decision architecture. In most cases, the hierarchy of the layers is respected and the behavior of the astronauts is logical and corresponds to what is expected.
However, if an astronaut is tired, the observed behavior is unpredictable. As the tiredness increase, humans usually become less efficient and the decision process takes more time. In Marsbase, if appropriate random functions are used, a tired astronaut who is walking around looking for ferrous rocks stops walking several times in a random fashion. When the tiredness reaches its maximum, the astronaut finally goes back to the habitat for sleeping.

In the previous version, in which the hierarchy of the layers was fixed, some players were complaining that astronauts decide to come back to the base while they were very close to take the requested rocks. In a new scenario, astronauts have been programmed to explore the region and collect as much interesting rocks as possible. The astronaut walks towards the closest ones in order to carry them back to the habitat. Now, if the astronaut is very close when the oxygen comes below the threshold that triggers the survival process, the priority of the current task is set to 10, which is greater than the priority of survival for oxygen. He therefore decides to take some samples before coming back safe to the habitat. The new behaviors are now more similar to what is expected from intelligent creatures.

6 Conclusion

Marsbase simulator has been implemented like a strategic game with a strong emphasis on the quality of the interface and the relevance of simulated human activities for the development of a Martian colony. The architecture of artificial intelligence modules is based on Brooks' ideas but the priority of layers evolves according to internal parameters. The competition between processes is probably a key idea to implement human based behaviors, as it simulates in some sense humans' decision making process. If the priority of the processes fluctuates according to internal parameters, the behavior becomes more unpredictable and it is easier to take into consideration physiological or emotional parameters with a global impact. We therefore believe that the principles of our architecture could be extended to other games and simulations.

References

[1] Amir E. and Maynard-Zhang P. Logic-Based Subsumption Architecture, Artificial Intelligence journal, 153 (2004), 167-237.

[2] Brooks R. How to build complete creatures rather than isolated cognitive simulators. In Architectures for Intelligence, K. VanLehn (ed.), Lawrence Erlbaum Assosiates, Hillsdale, NJ (1991), 225-239.

[3] Daniels M. Integrating Simulation Technologies With Swarm. Proceedings of the conference Agent Simulation: Applications, Models, and Tools, Chicago, (Argonne National Laboratory, University of Chicago.), (October 15-16 1999).

[4] Funge J.D. AI for Games and Animation, a Cognitive Modeling Approach. A.K. Peters, Natick, Massachusetts (1999).

[5] Grand S. and Cliff D.. Creatures: Artificial Life Autonomous Software Agents for Home Entertainment. In Proceedings of the First International Conference on Autonomous Agents Marina del Rey, California, United States (1997), 22-29.

[6] Mateas M. An Oz-Centric Review of Interactive Drama and Believable Agents. Technical Report CMU-CS-97-156, School of Computer Science, Carnegie Mellon University, Pittsburgh, PA. (June 1997).

[7] Meyer J.A. and Wilson S. From Animals to Animats: Proceedings of the First International Conference on Simulation of Adaptive Behavior, MIT Press, Cambridge MA (1991).

[8] Rabin S. and coauthors. AI Game Programming Wisdom. Steve Rabin (ed.), Charles River Media (2002).

[9] Wooldridge M. and Jennings N. Agent Theories, Architectures, and Languages: a Survey. In Wooldridge and Jennings (ed.)., Intelligent Agents, 1-22, Berlin: Springer-Verlag (1995), 1-22.

[10] Zubrin R. with Wagner R. The Case for Mars, The Plan to Settle the Planet and Why We Must. Touchstone Ed., N.Y. (1997).

Tolerance of Radial Basis Functions Against Stuck-At-Faults*

Ralf Eickhoff and Ulrich Rückert

Heinz Nixdorf Institute,
System and Circuit Technology,
University of Paderborn, Germany
eickhoff, rueckert@hni.upb.de

Abstract. Neural networks are intended to be used in future nanoelectronic systems since neural architectures seem to be robust against malfunctioning elements and noise in their weights. In this paper we analyze the fault-tolerance of Radial Basis Function networks to Stuck-At-Faults at the trained weights and at the output of neurons. Moreover, we determine upper bounds on the mean square error arising from these faults.

1 Introduction

Neural networks are used as function approximators for continuous functions [1,2]. Especially, Radial Basis Function networks are utilized to perform a local approximation of an unknown function specified by a set of test data. The main reason why neural networks are used for this purpose is the adaptability of the network due to the learning process. Moreover, the networks seem to be fault-tolerant against malfunctioning neurons [2] which can be modeled as Stuck-At faults and to be robust against noise corrupted weights and inputs [3].

Digital and analog implementations of neural networks have always to face malfunctioning elements [4], especially in future nanoelectronic realizations [5]. Moreover, when using analog hardware noise is always present due to thermal or flicker noise [6,7,8] and even if digital hardware is used quantization noise contaminates the weights and inputs [9]. Thus, the artificial neural network structure should handle malfunctioning elements and noise contaminated weights.

In this paper we analyze the Radial Basis Function network with respect to errors based on Stuck-At-Faults. In [10] these properties are demonstrated for sigmoidal feedforward networks. First, a short overview about the analyzed neural network architecture is given. The fault-tolerance against different types of Stuck-At-Faults is analyzed afterwards. Section 4 determines upper bounds on the mean square error for Stuck-At-Faults occuring in the weights and output of neurons and necessary restrictions are introduced leading to bounded errors.

* This work was supported by the Graduate College 776 - Automatic Configuration in Open Systems- funded by the Deutsche Forschungsgemeinschaft.

W. Duch et al. (Eds.): ICANN 2005, LNCS 3697, pp. 1003–1008, 2005.

2 Radial Basis Functions

In this section a short overview about the architecture of a Radial Basis Function network is given. The network consists of an input vector with dimension $\dim \boldsymbol{x} = n$. At a second step m different Basis Functions which have different centers $\boldsymbol{x}_i$ are superposed and denoted by a weight α_i to produce the output.

The Radial Basis Function network (RBF) can be used for local function approximation [11]. Basing on the regularization theory the quadratic error is minimized with respect to a stabilizing term [12]. Due to this stabilizer the interpolation and approximation quality is controlled in order to achieve a smooth approximation. Based on this stabilizer different Basis Functions can be performed for superposition. As a consequence, the network function can be expressed as

$$f_m(\boldsymbol{x}) = \sum_{i=1}^{m} \alpha_i h_i \left(\|\boldsymbol{x} - \boldsymbol{x}_i\| \right) \tag{1}$$

where m denotes the number of superposed Basis Functions.

The function $h_i(z)$ can be any function related to a (radial) regularization stabilizer. Here, the stabilizer leading to a Gaussian function is considered, thus it follows

$$h_i(z) = \exp\left(\frac{-z^2}{2\sigma_i^2} \right) \tag{2}$$

and therefore

$$f_m(\boldsymbol{x}) = \sum_{i=1}^{m} \alpha_i \exp\left(\frac{-\|\boldsymbol{x} - \boldsymbol{x}_i\|^2}{2\sigma_i^2} \right) \tag{3}$$

Moreover, the parameters $\boldsymbol{x}_i$ are the individual centers of each Basis Function, σ_i^2 resembles the variance of each Gaussian function and α_i are the weights from each neuron to the output neuron, which performs a linear superposition of all Basis Functions.

3 Stuck-At-Faults in Radial Basis Function Networks

In future nanoelectronic systems one major problem will be the massive amount of malfunctioning elements [5]. Therefore, fault-tolerant architectures have to be established in order to achieve reliable systems and predictable system behavior. From biology it is well known that the human brain allows the loss of several neurons because of the redundancy in its structure [2,13]. However, it was proven that sigmoidal feedforward neural networks are not fault-tolerant against Stuck-At-Faults [10,14]. Here, the fault-tolerance of an RBF network is analyzed against Stuck-At-Faults at the output weights, at the Basis Function centers and at the variance of the Gaussian Basis Function.

3.1 Stuck-At-Fault at the Output Weight

First, it is assumed that the Stuck-At-Fault occurs in the weights from the neuron to the output α_i. In order to achieve a universal expression it is assumed that the weight is sticking at the value μ. Moreover, with loss of generality only one weight from the whole Basis Functions is imposed by this fault. Thus, it follows for the difference of both network outputs

$$f_m(\boldsymbol{x}) - \hat{f}_m(\boldsymbol{x}) = \sum_{i=1}^{m} \alpha_i \exp\left(-\frac{\|\boldsymbol{x}-\boldsymbol{x}_i\|^2}{2\sigma_i^2}\right) - \sum_{i=1}^{m} \hat{\alpha}_i \exp\left(-\frac{\|\boldsymbol{x}-\boldsymbol{x}_i\|^2}{2\sigma_i^2}\right) \quad (4)$$

$$= \sum_{i=1}^{m} (\alpha_i - \hat{\alpha}_i) \exp\left(-\frac{\|\boldsymbol{x}-\boldsymbol{x}_i\|^2}{2\sigma_i^2}\right) \quad (5)$$

where $\hat{f}_m(\boldsymbol{x})$ denotes the network function due to the Stuck-At-Fault at the k-th neuron. Since nearly all α_i are identical to $\hat{\alpha}_i$ the difference vanishes except of the k-th term where the weight is sticking at the value μ. Therefore, under the assumption that only one weights is imposed by a Stuck-At-Fault (5) leads to

$$f_m(\boldsymbol{x}) - \hat{f}_m(\boldsymbol{x}) = (\alpha_k - \mu) \cdot \exp\left(-\frac{\|\boldsymbol{x}-\boldsymbol{x}_k\|^2}{2\sigma_k^2}\right) \quad (6)$$

3.2 Stuck-At-Fault at an RBF Center

Here, the Stuck-At-Fault occurs in the center of a Basis Function resulting in an unintentional movement of the center. The k-th neuron is interfered by a Stuck-At-Fault at the ν-th entry of the vector $\boldsymbol{x}_k$, and therefore this produces a faulty output behavior of the neural network. The difference between both network responses can be expressed

$$f_m(\boldsymbol{x}) - \hat{f}_m(\boldsymbol{x}) = \sum_{i=1}^{m} \alpha_i \exp\left(-\frac{\|\boldsymbol{x}-\boldsymbol{x}_i\|^2}{2\sigma_i^2}\right) - \sum_{i=1}^{m} \alpha_i \exp\left(-\frac{\|\boldsymbol{x}-\hat{\boldsymbol{x}}_i\|^2}{2\sigma_i^2}\right) \quad (7)$$

$$= \alpha_k \left[\exp\left(-\frac{\|\boldsymbol{x}-\boldsymbol{x}_k\|^2}{2\sigma_k^2}\right) - \exp\left(-\frac{\|\boldsymbol{x}-\hat{\boldsymbol{x}}_k\|^2}{2\sigma_k^2}\right)\right] \quad (8)$$

$$= \alpha_k \exp\left(-\frac{\sum_{j=1, j\neq\nu}^{n} (x_j - x_{kj})^2}{2\sigma_k^2}\right) \times \quad (9)$$

$$\left[\exp\left(-\frac{(x_\nu - x_{k\nu})^2}{2\sigma_k^2}\right) - \exp\left(-\frac{(x_\nu - \mu)^2}{2\sigma_k^2}\right)\right] \quad (10)$$

where x_{kj} denotes the j-th entry in the center of the k-th Basis Function.

3.3 Stuck-At-Fault at the Variance of a Gaussian Basis Function

Now, it is assumed that the variance of a certain Basis Function is affected by a Stuck-At-Fault at μ. Here, the k-th Basis Function is disturbed, leading to

$$f_m(\boldsymbol{x}) - \hat{f}_m(\boldsymbol{x}) = \sum_{i=1}^{m} \alpha_i \exp\left(-\frac{\|\boldsymbol{x}-\boldsymbol{x}_i\|^2}{2\sigma_i^2}\right) - \sum_{i=1}^{m} \alpha_i \exp\left(-\frac{\|\boldsymbol{x}-\boldsymbol{x}_i\|^2}{2\hat{\sigma}_i^2}\right) \tag{11}$$

$$= \alpha_k \cdot \left[\exp\left(-\frac{\|\boldsymbol{x}-\boldsymbol{x}_k\|^2}{2\sigma_k^2}\right) - \exp\left(-\frac{\|\boldsymbol{x}-\boldsymbol{x}_k\|^2}{2\mu^2}\right)\right] \tag{12}$$

4 Bounds on the Mean Square Error

In this section we analyze the fault-tolerance of an RBF network against the Stuck-At-Faults. Hence, for the three different types of Stuck-At-Faults necessary restrictions are introduced to achieve an upper bound on the mean square error of the difference between both network functions. Therefore, the input vector is assumed to be a random variable with a certain distribution function. In the following E denotes the expected value.

Concerning (6), the mean square error due to a Stuck-At-Fault at the output weights is determined by

$$\text{mse}_\alpha = (\alpha_k - \mu)^2 \cdot E\underbrace{\left[\exp\left(-\frac{\|\boldsymbol{x}-\boldsymbol{x}_k\|^2}{\sigma_k^2}\right)\right]}_{\leq 1} \tag{13}$$

Equation (13) can be estimated by the mean value theorem of integral calculus [15] resulting in a mean square error of

$$\text{mse}_\alpha \leq (\alpha_k - \mu)^2 \tag{14}$$

Thus, if the weights of the RBF network are not bounded rather arbitrary the mean square has no upper bound. Moreover, the error is depending on the Stuck-At value. The value of the Stuck-At-Fault is a consequence of the technical implementation. In analog hardware μ can be restricted to any continuous value in a certain interval which is determined by operating conditions [6,8]. In the case of a digital realization μ can only adopt discrete values leading to a quantized error. However, in both implementations μ is restricted by an upper bound and with restricted weights

$$|\alpha_i| \leq B \quad \forall\, i = 1 \ldots n \tag{15}$$

equation (14) can be further evaluated

$$\text{mse}_\alpha \leq (B + |\mu|)^2 \tag{16}$$

In the same way equation (10) and (12) can be determined leading to

$$\begin{aligned} \text{mse}_{\boldsymbol{x}_i} &= \alpha_k^2 E\left[\left(\exp\left(-\frac{\|\boldsymbol{x}-\boldsymbol{x}_i\|^2}{2\sigma_i^2}\right) - \exp\left(-\frac{\|\boldsymbol{x}-\hat{\boldsymbol{x}}_i\|^2}{2\sigma_i^2}\right)\right)^2\right] \\ &\leq \alpha_k^2 \end{aligned} \tag{17}$$

$$\begin{aligned} \text{mse}_\sigma &= \alpha_k^2 E\left[\left(\exp\left(-\frac{\|\boldsymbol{x}-\boldsymbol{x}_i\|^2}{2\sigma_i^2}\right) - \exp\left(-\frac{\|\boldsymbol{x}-\boldsymbol{x}_i\|^2}{2\mu^2}\right)\right)^2\right] \\ &\leq \alpha_k^2 \end{aligned} \tag{18}$$

From equation (17) and (18) can be concluded that both mean square errors have no upper bound since the weights of the neural network can be arbitrary. In contrast to (14) the mean square errors do not depend on the technical realization. If the weights are restricted by an upper bound $|\alpha| \leq B$ both errors are restricted by an upper bound

$$\text{mse} \leq B^2 \tag{19}$$

5 Conclusion

Artificial neural networks are intended to be fault-tolerant against noise contaminated inputs and malfunctioning elements like biological neural networks. However, it was shown in [10,14] that sigmoidal feedforward networks are not fault-tolerant. In this work the fault-tolerance against malfunctioning elements is determined for Radial Basis Function networks. These interferences can be modeled as Stuck-At-Faults at the output weights and at the output behavior of the neurons.

As in the case for multilayer perceptrons the Radial Basis Function network is not immune to malfunctioning elements. If arbitrary weights can be used no upper bound on the mean square error exists. Therefore, a well-defined system behavior due to sticking elements can not be guaranteed. Furthermore, if the error occurs in the output weights of a neuron the mean square error is depending on the sticking value and thus on the technical realization.

The technical implementation of neural networks in analog or digital hardware provides restrictions on the weights which are resulting in fault-tolerant networks. As the weights are bounded by an upper bound (15) the mean square error is restricted (cf. (16) and (19)). In the case of analog hardware the Stuck-At-Fault can be assigned to any continuous value within a certain interval determined by the operating conditions. For digital implementations the Stuck-At-Fault are restricted to '1' and '0' leading to quantized steps of the error.

However, both technical implementations provide upper bounds on the Stuck-At-Faults as was shown in section 4. Therefore, technical realizations of an RBF network are still fault-tolerant against malfunctioning elements. By providing adequate bounds on the weights a reliable network response can be guaranteed.

References

1. Geva, S., Sitte, J.: A constructive method for multivariate function approximation by multilayer perceptrons. IEEE Transactions on Neural Networks **3** (1992) 621–624
2. Haykin, S.: Neural Networks. A Comprehensive Foundation. Second edn. Prentice Hall, New Jersey, USA (1999)
3. Rückert, U., Surmann, H.: Tolerance of a binary associative memory towards stuck-at-faults. In Kohonen, T., ed.: Artificial Neural Networks. Volume 2., Amsterdam, North-Holland (1991) 1195–1198
4. Rückert, U., Kreuzer, I., Tryba, V.: Fault-tolerance of associative memories based on neural networks. In: Proceedings of the International Conference on Computer Technology, Systems and Applications, Hamburg, Germany (1989) 1.52–1.55

5. Beiu, V., Rückert, U., Roy, S., Nyathi, J.: On nanoelectronic architectural challenges and solutions. Fourth IEEE Conference on Nanotechnology (2004)
6. Razavi, B.: Design of Analog CMOS Integrated Circuits. McGraw-Hill (2000)
7. Sitte, J., Körner, T., Rückert, U.: Local cluster neural net: Analog vlsi design. Neurocomputing **19** (1998) 185 – 197
8. Körner, T., Rückert, U., Geva, S., Malmstrom, K., Sitte, J.: Vlsi friendly neural network with localied transfer functions. In: Proceedings of the IEEE International Conference on Neural Networks. Volume 1., Perth, Australia (1995) 169 – 174
9. Widrow, B., Kollár, J.: Quantization Noise. Prentice Hall PTR, New Jersey, USA (2002)
10. Chandra, P., Singh, Y.: Feedforward sigmoidal networks - equicontinuiy and fault-tolerance properties. IEEE Transactions on Neural Networks **15** (2004) 1350–1366
11. Girosi, F., Poggio, T.: Networks and the best approximation property. Biological Cybernetics **63** (1990) 169–176
12. Girosi, F., Jones, M., Poggio, T.: Regularization theory and neural networks architectures. Neural Computation **7** (1995) 219–269
13. Bose, N.K., Liang, P.: Neural network fundamentals with graphs, algorithms, and applications. McGraw-Hill, Inc. (1996)
14. Phatak, D.S., Koren, I.: Complete and partial fault tolerance of feedforward neural nets. IEEE Transactions on Neural Networks **6** (1995) 446–456
15. Bronstein, I.N., Semendyayev, K.A.: Handbook of mathematics (3rd ed.). Springer-Verlag (1997)

The Role of Membrane Threshold and Rate in STDP Silicon Neuron Circuit Simulation

Juan Huo[1] and Alan Murray[2]

School of Electronics and Engineering, The University of Edinburgh, Edinburgh, UK
EH9 3JL
J.Huo@ed.ac.uk, Alan.Murray@ed.ac.uk

Abstract. Spike-timing dependent synaptic plasticity (STDP) circuitry is designed in $0.35\mu m$ CMOS VLSI. By setting different circuit parameters and generating diverse spike inputs, we got different steady weight distributions. Through analysing these simulation results, we show the effect of membrane threshold and input rate in STDP adaptation.

1 Introduction

Synaptic plasticity in biological neurons is widely believed to be important in memory storage and other brain functions [1,2]. Recently, significant interest has centred on synaptic learning rules that rely upon spike timing (Spike Timing Dependent Plasticity, STDP). Such rules are most obviously of interest in the context of vision processing and robotics [3,4].

STDP rules adapt synaptic weights by pairing pre- and postsynaptic action potentials within a time window (Fig.1). A weight is increased when a presynaptic spike precedes a post-synaptic spike and the weight is decreased when the post-synaptic spike arrives first.

Earlier experimental results have shown that on-chip neurons with STDP could detect and amplify spike-timing synchrony and create bimodal weight distributions [5]. However, a recent study indicates that timing is not the only determinant factor of plasticity [6] and the steady weight distribution is affected by many other factors.

We report further experiments with analog circuit in a $0.35\mu m$ process, to explore the role of membrane threshold and spike firing rate in STDP learning and characteristics of this circuit. We show that change of these factors could lead to different weight distribution. Comparing to [5], model used in this paper is simpler, one neuron connected with several learning synapses. The number of input synapses is uncertain and adapted to the number of input spike trains.

2 Method

The circuit is based on that in [5], translated and re-designed from a $0.6\mu m$ process to a $0.35\mu m$ process.

W. Duch et al. (Eds.): ICANN 2005, LNCS 3697, pp. 1009–1014, 2005.

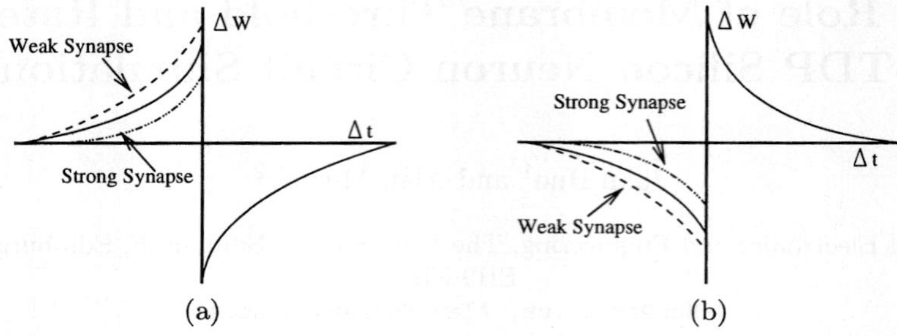

Fig. 1. Learning Window for STDP. $\Delta t = t_{pre} - t_{post}$, ΔW refers to the amount of weight change. **(a)** Weight dependent STDP, within which ΔW depends upon the magnitude of W. **(b)** Change in (capacitive) weight voltage in our STDP circuits. Weight voltage change is inverse to weight change. The closer the value of V_w is to GND, the stronger is the synapse.

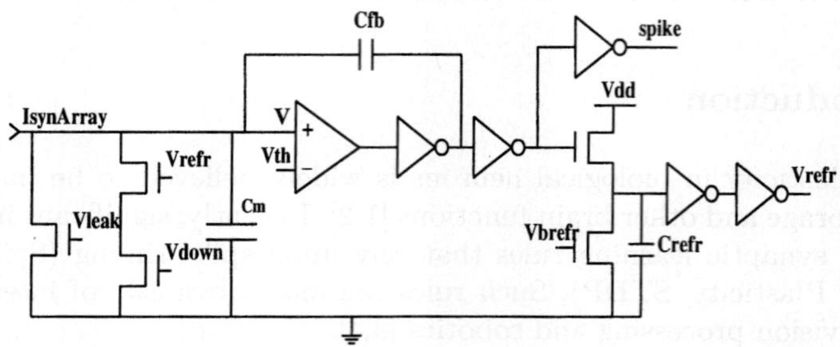

Fig. 2. Neuron circuit. $I_{synArray}$ is the sum of I_{syn} from each synapse circuit

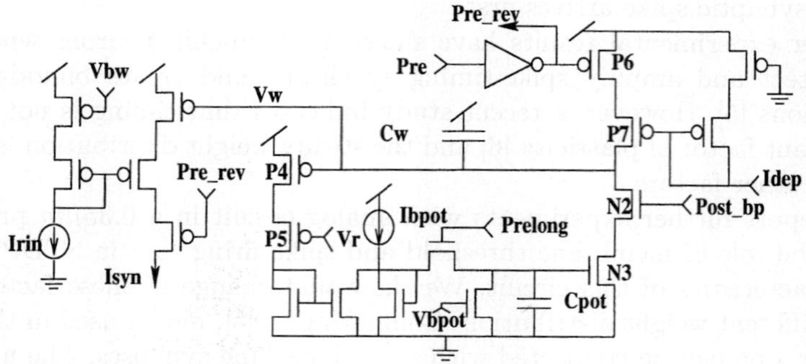

Fig. 3. Synapse and weight change circuit. Cw is discharged through N2 & N3 and charged through P6&P7.

The leaky Integrate & Fire neuron model circuit is shown in Fig. 2. The synaptic current ($I_{synArray}$) is integrated by the membrane capacitor C_m and a postsynaptic spike is generated, when the activity voltage V passes the threshold V_{th}.

Fig. 3 shows the synapse and weight modification circuit. The prominent characteristic of this circuit is weight dependent. Vw is stored on MOS capacitor Cw. When a postsynaptic spike fires shortly after a presynaptic spike, Cw is discharged through N2 and N3. The amount of discharge is determined by P4 and P5 connected with Vw to control the voltage on Cpot. Consequently, when the synaptic weight is increasing, the discharging current (I_{syn}) is proportional to Vw. The decaying current, Idep, which controls the weight depression, comes from a causal circuit switched on by postsynaptic spike. Details of the connection and the rest parts of the circuit could be found in [5].

2.1 Membrane Threshold and Synaptic Current

To understand the effect of the membrane threshold to STDP in our circuit, we generate four input spike trains from inhomogeneous Poisson processes. Two cases are discussed and compared with each other. In case one, we set V_{th}=1V, $I_{syn}=1.5\mu A$. These values are selected from a parameter space, where no bifurcation occurs. In case two, we set V_{th}=600mv, $I_{syn}=10\mu A$, which comes from another parameter space where bifurcation occurs. The other circuit parameters and input presynaptic spike trains in both cases are the same.

2.2 Input Firing Rate

To explore the effects of variable firing rate synthetically, there are two simulations in this section. The first one uses 5 uncorrelated poisson spike trains with rates ranging from 10 to 40Hz. The presynaptic inputs produced by an inhomogeneous poisson processes in the second simulation have different standard deviations (STD) ranging from 0 to 0.55 times the mean rate of 25Hz.

3 Simulation Results

3.1 The Effect of V_{th} and I_{syn}

Since in section 2.1, the input spike trains are independent of each other, the correlation between them is nearly nonexistent. Thus the weight's potentiation or depression is unpredictable. Although the input spike trains are unchanged, in Fig. 4(a) the steady weight distribution appears to be random while in Fig. 4(b) the weight distribution is clearly bimodal. The reason of the difference between these simulations is that in case one, no matter whether the synapse is weak or strong, the input synaptic current is very weak, so integration is the only way to make the membrane potential achieve the threshold voltage. In contrast to case one, in case two the I_{syn} is very high and V_{th} is low when the synapses become strong. Fig. 4(d) shows that postsynaptic firing rate is higher than Fig. 4(c) in the first 3 seconds, during which S4 wins the competition by random chance. Here is

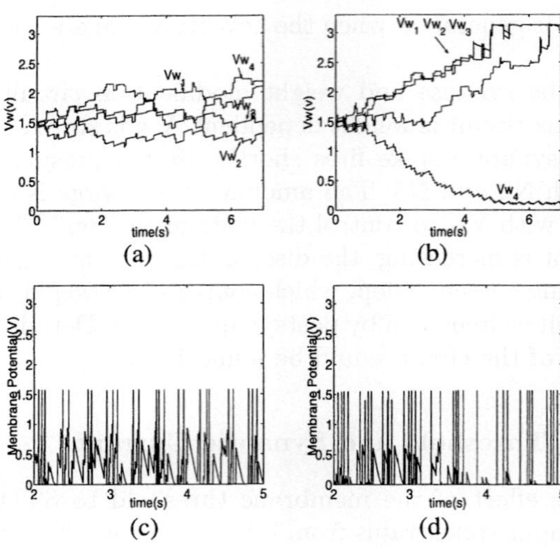

Fig. 4. Simulation results of uncorrelated poisson spike trains. (a) Case one with V_{th}=1V, I_{syn}=1.5μA, the weight distribution is random without obvious tendency. **(b)** Case two with V_{th}=600mv, I_{syn}=10μA, the weight distribution is bimodal. **(c)** The membrane potential and postsynaptic spikes of case one. **(d)** The membrane potential and postsynaptic spikes of case two.

a bifurcation point in the parameter space of V_{th} and I_{syn}, which distinguishes a region of the steady bimodal weight distribution from one of non-bimodal weight distribution. After the bifurcation point, S4 persistently fires a postsynaptic spike by itself. Eventually, the weight of S4 gets the maximum value and the circuit serves like the "winner-take-all" schema.

From the points above, V_{th} and I_{syn} could determine the final weight distribution in the circuit. We could conjecture that the value of V_{th} and I_{syn} even affect the number of synapses whose weights will reach the upper boundary. These values are fixed in the biological neuron model, but in fact, they are experimental inductions from biological experiment with fluctuation depending on the circumstances [7]. Our simulation shows the importance to reconsider the scale of V_{th} and I_{syn} when a neuron is being trained by STDP.

3.2 The Effect of Variable Firing Rates

It can not be directly seen from Fig. 5(a), (b) whether synapses firing at either faster or slower rates are preferentially strengthened by STDP. The results are consistent with [3], within the rate range presented 10 to 40Hz, STDP modification is insensitive to the firing frequency and degree of variability of the presynaptic input.

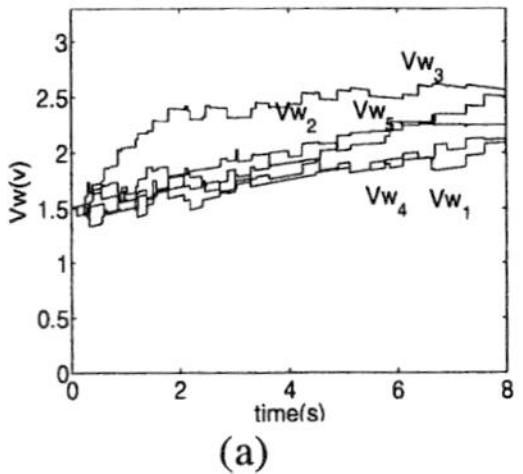
(a)

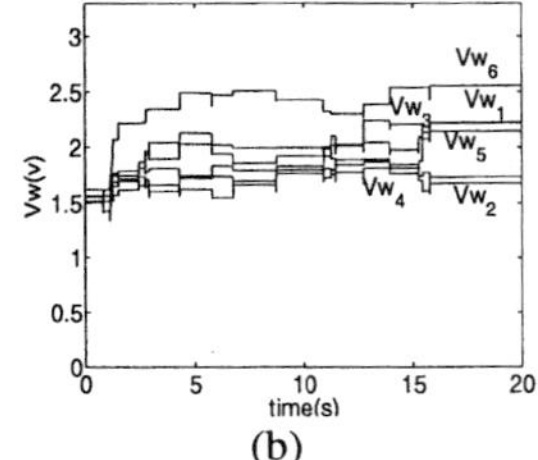
(b)

Fig. 5. Simulation results of firing rates. (a) Result of first simulation, whose presynaptic input have different rates ranging from 10Hz to 40Hz. **(b)** Result of second simulation, whose variable firing rates of presynatpic inputs with standard deviation ranging from 0 to 0.5 times the mean rate.

4 Conclusion

As our circuit is redesigned from [5] under $0.35\mu m$ process, 4 metals and 2 poly. The upper boundary of weight voltage is reduced by 25% and the value of weight modification is decreased proportionally. The simulation results tell us that the membrane threshold, correlation and firing rate jointly determines the steady weight distribution. First, low membrane threshold and high synaptic current could produce weight bifurcation, although the input spike trains are independent. Furthermore, for all input spike trains, the number of strengthened or weakened synapses could be influenced by different vaule of V_{th} and I_{syn}. The similar characters could be found in the relationship between membrane potential and postsynaptic firing rate of the cat visual cortex, which has been investigated in [10]. Changes of the membrane potential threshold could sharp orientation tuning of visual cortex and then modify the postsynaptic spike frequency. As is known, enhanced ability of a given synapse to rapidly evoke a postsynaptic spike will lead to synapse strengthening through STDP.

Second, our STDP circuit is insensitive to firing rate value and variability in the range of 10 to 40Hz, which is consistent with biological findings [11].

The work in this paper helps us to produce circuits with better matching for STDP in the future. With this circuit and the discussion of the weight distribution above, it is possible to organise a dynamic I&F network with different synapses which are sensitive to different input patterns.The network can be potentially used in pattern segmentation and classification tasks. The immediately related new work is to find the mechanism of weight bifurcation presented in section3.1.

Acknowledgement

We thank Adria Bofill, his circuit implemented in $0.6\mu m$ is our foundation. We also appreciate the helpful discussion with Katherine Cameron, Zhijun Yang, Hisham Hamid and Vasin Boonsobhak.

References

1. GG M.C.W. van Rossum, G.Q.Bi& Turrigiano: Stable hebbian learning from spike timing-dependent plasticity. J. Neurosci **20** (2000) 8812–8821.
2. R. G. M. Morris S. J. Martin, P. D. Grimwood: Synaptic plasticity and memory: An evaluation of the hypothesis. Annual Review of Neuroscience **23** (2000) 649–711.
3. Kenneth D. Miller Sen Song and L.F.Abbott: Competitive hebbian learning through spike-timing-dependent synaptic plasticity. Nature Neurosci **3** (2000) 919–926.
4. Ezequiel Di Paolo: Spike-timing dependent plasticity for evolved robots. Adaptive Behavior **10** (2002) 243–263.
5. Adria Bofill-I-Petit and Alan F. Murray: Synchrony detection and amplification by silicon neurons with stdp synapses. IEEE Transactions On Neural Networks **15** (2004) 1296–1304.
6. Misha Tsodyks: Spike-timing-dependent synaptic plasticity - the long road towards understanding neuronal mechanisms of learning and memory. Trends in Neurosciences **25**(12) (2002) 599–600,
7. Chritof Koch: Biophysics of Computation: Information Processing in Single Neurons. Oxford University Press, New York (1999).
8. Aertsen A Kuhn A, Rotter S: Correlated input spike trains and their effects on the response of the leaky integrate-and-fire neuron. Neurocomputing **44-46** (2002) 121–126.
9. G. G. Sjostrom P. J., Turrigiano and Nelson S. B.: Rate, timing, and cooperativity jointly determine cortical synaptic plasticity. Neuron **32** (2001) 1149–1164.
10. Ferster D. Carandini M: Membrane potential and firing rate in cat primary visual cortex. Neuroscience **20** (2000) 470–484.
11. John Curtin: Determinants of Spike Timing-Dependent Synaptic Plasticity. Neuron. **32**(6) (2001) 966–968.

Systolic Realization of Kohonen Neural Network

Jacek Mazurkiewicz

Institute of Engineering Cybernetics, Wroclaw University of Technology,
ul. Janiszewskiego 11/17, 50-372 Wroclaw, Poland
Jacek.Mazurkiewicz@pwr.wroc.pl

Abstract. The paper is focused on partial parallel realization of retrieving phase as well as learning phase of Kohonen neural network algorithms. The method proposed is based on pipelined systolic arrays – an example of SIMD architecture. The discussion is realized based on operations which create the following steps of learning and retrieving algorithms. The data which are transferred among the calculation units are the second criterion of the problem.

1 Introduction

Kohonen maps algorithms were in different way implemented using dedicated neuro-computers [3] [6]. A main problem related to the hardware implementation is focused on necessary modifications of algorithms to fit them to neuro-computer architecture [5] [6] [9]. The experiments were successful [2] [3] and it was possible to use implemented Kohonen maps in serious industrial applications [4] [6]. The paper proposes systolic approach to retrieving phase as well as learning phase of Kohonen neural network algorithms. The method is based on the most classical description of Kohonen algorithms with no modifications. The main goal is to divide the whole algorithm into subtasks. The subtasks can be realized by software or hardware simulator of Kohonen map. The presented method is based on Data Dependences Graphs [7]. Elementary processors which are defined after linear projection of Data Dependence Graphs onto lattice of points can be realized by processes or real device. The details related to processor construction are not available yet.

2 Kohonen Neural Network Algorithms

2.1 Learning Algorithm

The learning algorithm is based on the Grossberg rule [6] [7]. All weights are modified according to the following equation:

$$w_{lij}(k+1) = w_{lij}(k) + \eta(k)\Lambda(i^w, j^w, i, j)(x_l - w_{lij}(k)) \tag{1}$$

where:
k - iteration index, η - learning rate function, x_l - component of input learning vector,
w_{lij} - weight associated with connection from component of input learning vector x_l and neuron indexed by (i, j),

W. Duch et al. (Eds.): ICANN 2005, LNCS 3697, pp. 1015 – 1020, 2005.

Λ - neighborhood function, (i^w, j^w) - indexes related to winner neuron, (i, j) - indexes related to single neuron from Kohonen map.

The learning rate η we assume as a linear decreasing function. Learning rate function is responsible for the number of iterations - it marks the end of learning process. The presented solution is based on the following description of the neighborhood function [1]:

$$\Lambda\left(i^w, j^w, i, j\right) = \begin{cases} 1 & \text{for} \quad r = 0 \\ \dfrac{\sin(ar)}{ar} & \text{for} \quad r \in \left(0, \dfrac{2\pi}{a}\right) \\ 0 & \text{for} \quad \text{other values } r \end{cases} \tag{2}$$

where:
a - neighborhood parameter,
r - distance from winner neuron to each single neuron from Kohonen map, calculated by indexes of neurons as follow:

$$r = \sqrt{(i^w - i)^2 + (j^w - j)^2} \tag{3}$$

The learning procedure is iterative: weights are initialized by random values; position of winner neuron for each learning vector is calculated by ordinary Kohonen retrieving algorithm using random values of weights; weights are modified using Grossberg rule (1); the learning rate is modified, the neighborhood parameter a (2) is modified and if the learning rate is greater than zero weights are modified by the next learning vector, else the learning algorithm stops [7].

2.2 Retrieving Algorithm

During the retrieving phase the Euclidean distance: the weights vector and the output vector is calculated. The winner neuron is characterized by the shortest distance [6] [7]. Each neuron from Kohonen map calculates the output value according to the classical weighted sum:

$$Out(i, j) = \sum_{l=0}^{N-1} x_l w_{lij} \tag{4}$$

where:
$Out(i, j)$ - output value calculated by single neuron of Kohonen map indexed by (i, j).

3 Data Dependence Graphs for Kohonen Neural Network

A Data Dependence Graph is a directed graph that specifies the data dependencies of an algorithm. Nodes of the Data Dependence Graph represent computations and arcs specify the data dependencies between computations [7].

3.1 Data Dependence Graph for Learning Algorithm

For 1-D Kohonen map neurons are placed is single line, each neuron has two neighbours, excluding neurons at the ends of line. For such topology there are $(N \times K)$

weights if we assumed N-element input vector and K neurons which create the Kohonen map. 1-D Kohonen map ought to be described by rectangular Data Dependence Graph (Fig. 2.). Each node of the graph is responsible for single weight calculation. using Grossberg rule (1) (Fig. 1.). The current value of the weight is stored in the local memory of each node. The node decreases the learning rate in automatic way. The size of the graph equals to the size of the weight matrix. Each node of the graph is loaded by two signals. The neighborhood function is calculated using sinus function. We propose to place the values of sinus in a table and store them in a local memory of each node. The neighborhood parameter a (2) is also stored in the local memory and is sequential reduced by negative counter.

3.2 Data Dependence Graph for Retrieving Algorithm

1-D Kohonen map is described by rectangular Data Dependence Graph (Fig. 3.). Each node of the graph calculates the component of the weighted sum (4) (Fig. 1.). The necessary weight value is stored in a local memory of the node. The size of the graph equals to the size of the weight matrix.

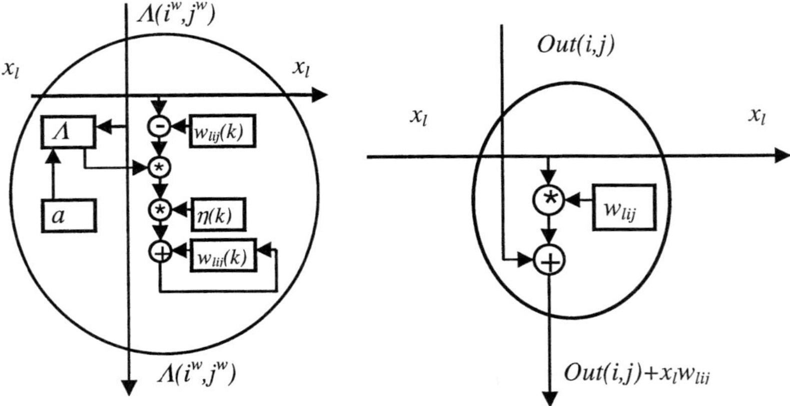

Fig. 1. Single node of Data Dependence Graph for learning algorithm (left side)and single node of Data Dependence Graph for retrieving algorithm (right side)

4 Mapping Data Dependence Graphs onto Systolic Array

The Data Dependence Graphs for retrieving and learning algorithms are local and composed by the same number of nodes. The single neuron operations are described by the column of the graph [7]. Multi-dimensional Kohonen map is described by the set of 1-D Data Dependence Graphs (Fig. 2.) (Fig. 3.). It means that the slabs work in parallel [1] [8]. The graphs can be converted to an universal structure able to implement learning algorithm as well as retrieving algorithm using processors with switched functions (Fig. 4.) [3] [1]. The systolic arrays are the result of the linear projection of Data Dependence Graphs onto lattice of points, known as processor space. The elementary processor combines operations described by nodes taken from single vertical line of the graph [7].

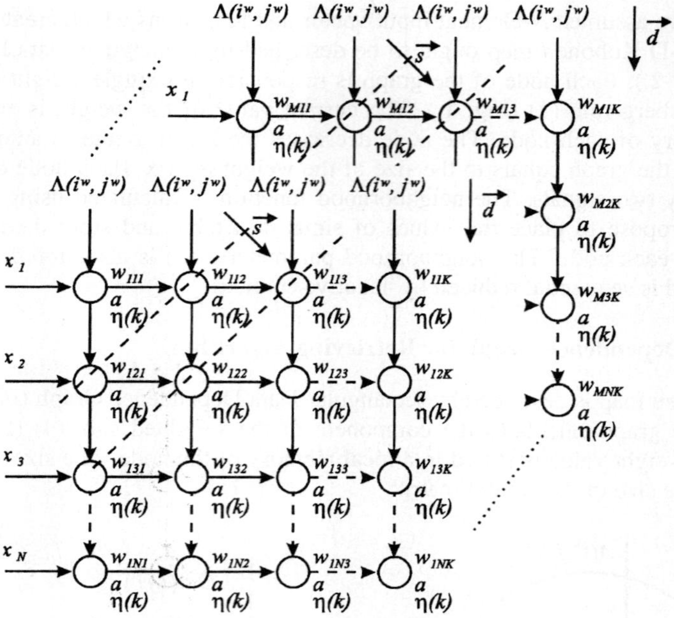

Fig. 2. Data Dependence Graph for learning algorithm of multi-dimensional Kohonen map

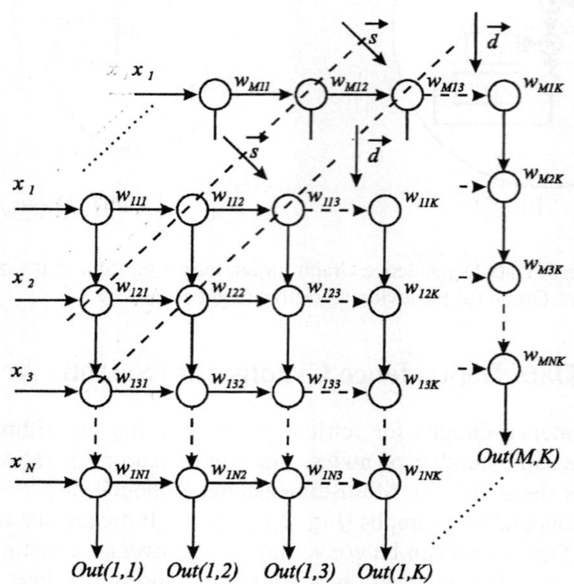

Fig. 3. Data Dependence Graph for retrieving algorithm of multi-dimensional Kohonen map

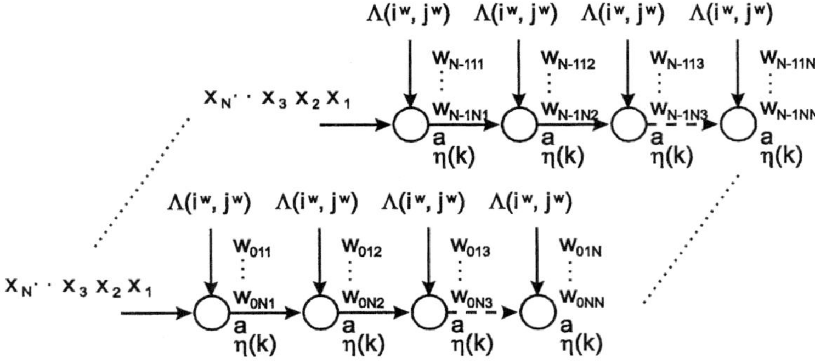

Fig. 4. Systolic array for learning algorithm of multi-dimensional Kohonen neural network

5 Efficiency of Proposed Approach

An efficiency of proposed approach is estimated using the algorithm proposed by Kung [7] and modified for MANTRA computer analysis [6]. The estimation is based on the dimensions and organization of the Data Dependence Graphs. A computation time for retrieving algorithm equals:

$$T = (N + K - 1)\tau \tag{5}$$

where τ - processing time for elementary processor.

The computation time for learning algorithm:

$$T = (N + K - 1)M\eta\tau \tag{6}$$

where M - number of learning vectors.

Speed-up and processor utilization rate are exactly the same for retrieving and learning algorithms - assuming possible sequential computation time:

$$speed - up = \frac{NK}{N + K - 1} \qquad utilization\ rate = \frac{N}{N + K - 1} \tag{7}$$

Of course it is necessary to verify the estimated values of efficiency after the implementation of proposed structures in silicon. The implementation is started but is not finished yet.

6 Conclusion

Summarizing, the paper proposed a methodology for Kohonen neural network simulation based on systolic array structure. The methodology is based on classical and not modified algorithms related to Kohonen maps. It is possible to realize the obtained subtasks by software processes, but also using dedicated neuro-computers like MANTRA [6] or to create your own elementary processors in PLD or FPGA.

References

1. Cabestany J., Ienne P., Moreno J. M., and Madrenas J.: Is There a future for ANN Hardware? Workshop on Mixed Design of Integrated Circuits and Systems, Poland, Lodz (1996)
2. Cornu, T., Ienne, P.: Performance of Digital Neuro-computers. Fourth International Conference on Microelectronics for Neural Networks and Fuzzy Systems, Italy, Turin, (1994), 87-93
3. Ienne, P.: Digital Hardware Architectures for Neural Networks. SPEEDUP Journal, (1995). Vol. 9, No. 1, 18-25
4. Ienne, P., Kuhn, G.: Digital Systems for Neural Networks. Digital Signal Processing Technology, SPIE Optical Engineering, Vol. 57 of Critical Review Series, USA, Orlando, Fla., (1995), 314-345
5. Ienne, P., Thiran, P., Vassilas, N.: Modified Self-Organizing Feature Map Algorithms for Efficient Digital Hardware Implementation. IEEE Transactions on Neural Networks. Vol. 8, No. 2, (1997), 315-330
6. Ienne, P., Viredaz, M.: Implementation of Kohonen's Self-Organizing Maps on MANTRA, Fourth International Conference on Microelectronics for Neural Networks and Fuzzy Systems, Italy, Turin (1994), 273-279
7. Kung, S. Y.: Digital neural networks, PTR Prentice Hall, (1993)
8. Mazurkiewicz, J., Zamojski W.: Systolic-Based Hardware Realization of Hopfield Neural Network, Electrónica e Telecomunicações, Revista do Departamento de Electrónica e Telecomunicações da Universidade de Aveiro, Vol. 3, No. 5, Portugal, Aveiro, (2002) 392-399
9. Vassilas, N., Thiran, P., and Ienne, P.: On modifications of Kohonen's feature map algorithm for an efficient parallel implementation, International Conference on Neural Networks, USA, Washington, D.C., (1996), Vol. II, 932-937

A Real-Time, FPGA Based, Biologically Plausible Neural Network Processor

Martin Pearson[1], Ian Gilhespy[1], Kevin Gurney[2], Chris Melhuish[1], Benjamin Mitchinson[2], Mokhtar Nibouche[1], and Anthony Pipe[1]

[1] Intelligent Autonomous Systems laboratory, University of the West of England
[2] Adaptive Behaviour Research Group, University of Sheffield

Abstract. A real-time, large scale, leaky-integrate-and-fire neural network processor realized using FPGA is presented. This has been designed, as part of a collaborative project, to investigate and implement biologically plausible models of the rodent vibrissae based somatosensory system to control a robot. An emphasis has been made on hard real-time performance of the processor, as it is to be used as part of a feedback control system. This has led to a revision of some of the established modelling protocols used in other hardware spiking neural network processors. The underlying neuron model has the ability to model synaptic noise and inter-neural propagation delays to provide a greater degree of biological plausibility. The processor has been demonstrated modelling real neural circuitry in real-time, independent of the underlying neural network activity.

1 Introduction and Background

The hardware processor detailed in this paper can model large networks of Leaky-Integrate-and-Fire (LIF) neural processing nodes (described in numerous sources, e.g., [1], [2]) which are themselves based on the observed phenomenological function of biological neurons. Additional biologically plausible features of the model used here include synaptic and membrane threshold noise, inter-neural propagation delays, and individual membrane potential and post synaptic current decay constants. The hard-real-time constraints that are encountered in the field of embedded computing is an area which has been applied to artificial neural networks before [3]. However, to the best of our knowledge, this has not been used in relation to 'Spiking' artificial Neural Network (SNN) implementations.

SNNs differ from more conventional rate-coded artificial neural networks in that the information passed between neurons is expressed as temporally separated discrete events, or spikes. SNNs and Pulse Coded Neural Networks (PCNNs) [4] can generate behaviours and reproduce coding schemes closely analogous to biological neural systems [5]. They are consequently used extensively by computational neuroscientists in experiments to model and obtain insights into the operational functionality of the brain. Typically these models are simulated using software simulators, such as [6], compiled to conventional Personal

W. Duch et al. (Eds.): ICANN 2005, LNCS 3697, pp. 1021–1026, 2005.

Computers (PC) or parallel High Performance Computing (HPC) systems such as Beowulf clusters [7]. These simulators utilise the inherent characteristics of biologically plausible neural networks (low average network activity and sparse inter-neural connectivity [2]) to maximise the utility of the processing space and minimise simulation time. This approach is also adopted by dedicated biologically plausible hardware neural network accelerators, for example, [8], [9]. To test neural network models for robustness in real-world control environments, such as mobile robotics, the underlying network processing platform must guarantee hard-real-time performance. The computationally efficient, activity dependant approach to network modelling, as described above, can not guarantee this performance at all levels of network activity. Therefore, a new SNN processing architecture, which trades some network complexity for a guaranteed temporal performance at all levels of SNN activity, is preferential for the stable on-line control of, for example, a mobile robot.

2 The Neuron Model

The neuron model used in this processor is a single point (or single compartmental) model which exhibits class I excitability [10]. The weight of each synapse can be subjected to multiplicative Rayleigh distributed noise. Gaussian distributed noise can also be injected additively to the magnitude of the membrane threshold potential. The noise distributions used here were chosen to best fit the model output to empirical biological data. A variable inter-neural delay is associated with each synapse which is used to model spatially distributed networks. Both the absolute and the relative refractory periods of each neuron are also modelled. The use of floating point arithmetic to represent and manipulate these parameters is not available to FPGA without incurring a substantial cost in silicon real estate. For this reason, fixed point, 16-bit integers have been used to approximate the more accurate representation of these floating point values.

3 The Processor Architecture

The architecture of this design is best described as a Single Instruction path, Multiple Data path (SIMD) array processor, Fig. 1. It consists of an array of Processing Elements (PE) operating concurrently on the same instruction, issued from a central sequencer, using locally stored data. The input stimulus to the processor is ported via 2 input modules which can read in data asynchronously. Similarly there are also 2 output modules. The neurons and synapses are implemented in what we have called Neural Processing Elements (NPE). The SIMD neural processor, detailed in this report, has 10 NPEs, each of which emulates 120 virtual neurons and 912 synapses. The update period of the processor is set at 500μS which is regulated by a real time counter in the sequencer module.

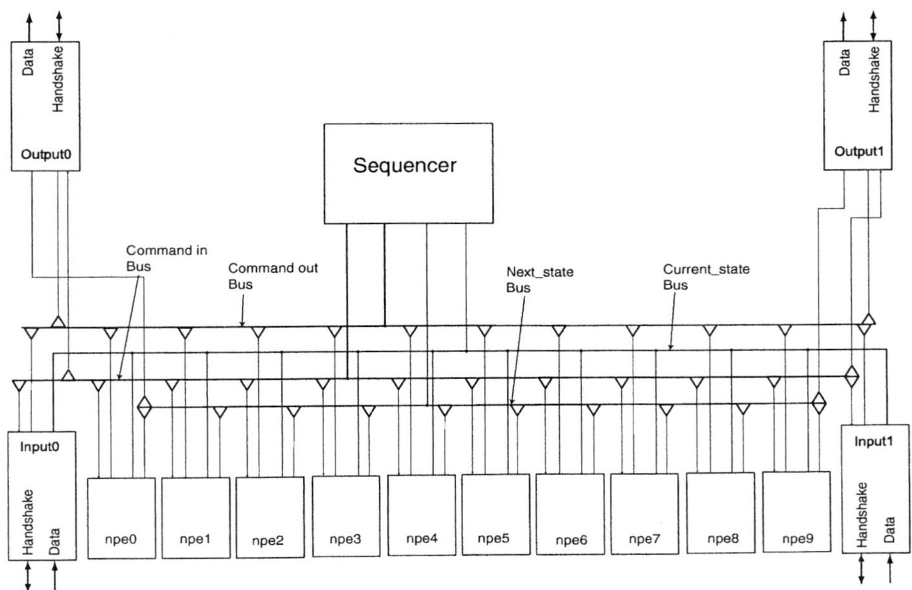

Fig. 1. Block diagram of SIMD neural processor topology

4 Module Specifics

The Sequencer maintains the real time performance and coordinates the activity of all the concurrently operating PEs of the processor. This coordination is performed using a 2 bit control channel and a 16 bit data bus between each of the elements.

The input module has a 64-bit input port (multiplexed onto 384 input channels) which can be connected either to physical pins or an appropriate internal interface using the logic array of the host FPGA. Handshaking lines facilitate asynchronous operation and allow communications across different clock domains. The 384 input channels are passed onto the internal 16 bit data bus of the module and consequently stored in the current state memory.

The output module has a similar architecture and operation to the input module but with an additional block of RAM containing a list of the network outputs.

The Neural Processing Element contains a hardware implementation of a neuron and a single synapse, Fig. 2. The contextual information of 120 'virtual' neurons and 912 synapses are stored locally in 4 banks of RAM. The context for each neuron and synapse are sequentially multiplexed onto the hardware at super-real-time. A copy of the state of the entire network is stored locally in each NPE (as in the output module) which serves as the input stimulus for the virtual neurons/synapses. The updated state of each of the neurons in the NPE are stored in the local next state memory space and is subsequently broadcast to the other PEs.

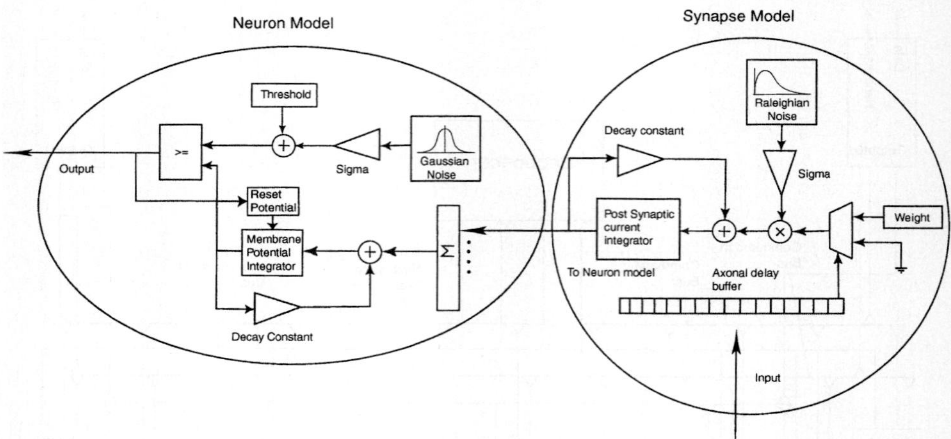

Fig. 2. Block diagram of individual neuron and a single synapse

5 Results

An existing model of part of the Basal Ganglia [11] was used to test the performance of the system. The parameters, generated from a floating point software simulator, were translated into fixed point integers. Tests were conducted, using a C++ coded hardware simulator, to assess the degradation in the network performance compared to the original floating point model. These parameters were then passed into the physical synthesis process of the FPGA design flow and the subsequent bit stream was used to configure the device. The target FPGA was a Xilinx Virtex-II (XC2V1000-4), 1 million gate equivalent, clocked at 50MHz and situated on a Celoxica RC200 development board. The raster plot shown in the top panel of figure 3, is from 1200 hardware implemented neurons over a 400 millisecond time period (800 operational epochs) buffered from the FPGA via an RS232 serial port. The histogram, shown in the lower panel, was generated from the spike event data taken from the raster plot and clearly indicates the periods of peak neural activity during this trial.

In a previous study it was found that floating point SNN simulator software compiled and executed on a Pentium 4 based PC [1] could maintain real-time performance whilst modelling a network of 7000 neurons with an average network activity of 50 spike events per neuron per second (50Hz) and an average divergence of 16 synapses per neuron. This equates to a total of 350,000 spike events per second or 175 per update period (500μS), above which this processing paradigm will require more than 500μS to update the state space of the network. To assess whether a neural network can be modelled in real time a metric which establishes a measure of peak activity, size of the network itself and the required update period was derived. This was referred to as the activity quota of the network; peak number of spike events per neuron per update period. Therefore, a

[1] 3GHz processor, utilising Microsoft Windows XP operating system.

network of 7000 neurons generating a constant spike activity of 175 spike events per update period can be categorised as requiring an activity quota of 0.025 from the underlying processing modality to remain operating in real-time. In figure 3, the histogram shows that the network used in this test, of 1200 neurons, has instances of network activity in excess of 30 spike events per update period. This network therefore requires an activity quota greater than 0.025 to remain operating in real-time. Were the network size to be increased to 7000, and the activity quota remained high as shown in Fig. 3, this network could not be modelled in real-time by the PC. The processor described in this paper has been designed to maintain real-time performance up to a network activity quota of 1, i.e., it is activity independent.

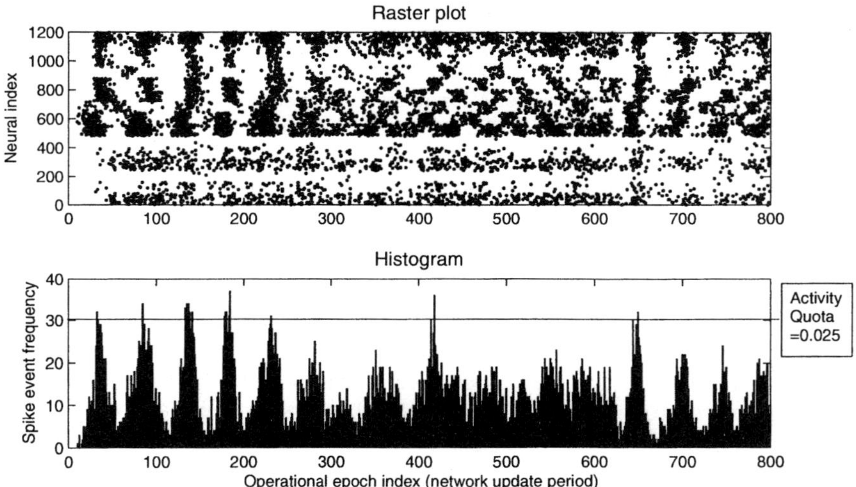

Fig. 3. Raster plot and corresponding activity histogram of 1200 neurons modelling the Basal Ganglia over a 400mS period

6 Discussion and Conclusion

The architecture has been designed to accommodate a substantial size increase in the near future. Further, the hardware has been designed from the outset to cater easily for cascading processor cores, either on the same FPGA or via physical pins to separate devices. Thus, very large networks could be generated. In fact, a matrix of 6 interconnected processor cores could simulate a network of almost 7000 neurons in hard real-time independent of network activity.

The work reported on in this paper has demonstrated that a large SNN model, based closely on the observed behaviour of biological neurons, can be simulated in real-time using a single FPGA. The emphasis on hard real-time performance has resulted in a re-evaluation of some of the existing optimisation techniques which take advantage of the temporal and spatial characteristics of biologically plausible SNNs to provide hardware acceleration for off-line modelling.

Acknowledgments

This work forms part of and is supported by EPSRC Grant No. GR/S19639/01 "whiskerbot project". Acknowledgement is given to all members of the project.

References

1. Gerstner, W., Kistler, W.M.: Spiking Neuron Models: Single Neurons, Populations, Plasticity. Cambridge University Press, Cambridge, MA (2002)
2. Koch, C., Segev, I., eds.: Methods in Neuronal Modeling: From Synapses to Networks. MIT Press, Cambridge, Massachusetts (1989)
3. Goerick, C., Noll, D., Werner, M.: Artificial neural networks in real-time car detection and tracking applications. Pattern Recogn. Lett. **17** (1996) 335–343
4. Eckhorn, R.: Neural mechanisms of scene segmentation: recordings from visual cortex suggest basic circuits for linking field models. IEEE Transactions on Neural Networks **10** (1999) 464–
5. Maass, W.: Computing with spiking neurons. In Maass, W., Bishop, C.M., eds.: Pulsed Neural Networks. MIT Press (Cambridge) (1999) 55–85
6. Beeman, D.: Simulating a neuron soma. In Bower, J.M., Beeman, D., eds.: The Book of GENESIS: Exploring Realistic Neural Models with the GEneral Neural SImulation System (2nd Ed.), New York, Springer-Verlag (1998) 215–224
7. T.L. Sterling, J. Salmon, D.B., Savarese, D.: How to Build a Beowulf: A Guide to the Implementation and Application of PC Clusters. Scientific and engineering computation. MIT Press (1999)
8. Mehrtash, N., Jung, D., Hellmitch, H., Schoenauer, T., Lu, V., Klar, H.: Synaptic plasticity in spiking neural networks (sp^2inn): A system approach. IEEE Transactions on Neural Networks **14** (2003)
9. Schemmel, J., Meier, K., Mueller, E.: A new vlsi model of neural microcircuits including spike time dependent plasticity. In: Proceedings of IJCNN'04, IEEE Press (2004) 1711–1716
10. Izhikevich, E.M.: Which model to use for cortical spiking neurons? IEEE Transactions on neural networks **15** (2004) 1063–1070
11. Gurney, K., Prescott, T., Redgrave, P.: A computational model of action selection in the basal ganglia i: A new functional anatomy. Biological Cybernetics **84** (2001) 401–410

Balancing Guidance Range and Strength Optimizes Self-organization by Silicon Growth Cones

Brian Taba and Kwabena Boahen

University of Pennsylvania, Philadelphia PA 19104, USA
{btaba, boahen}@seas.upenn.edu
http://www.neuroengineering.upenn.edu/boahen

Abstract. We characterize the first hardware implementation of a self-organizing map algorithm based on axon migration. A population of silicon growth cones automatically wires a topographic mapping by migrating toward sources of a diffusible guidance signal that is released by postsynaptic activity. We varied the diffusion radius of this signal, trading strength for range. Best performance is achieved by balancing signal strength against signal range.

1 Introduction

Neuromorphic engineers seek to migrate the computational efficiency of neurobiological systems into engineering applications by building silicon chips that faithfully reproduce neural function. For example, silicon retinae now emulate up to thirteen different cell types to encode distinct stimulus properties in four types of spiking ganglion cell output [1]. This level of detail is possible because the biological retina is relatively accessible experimentally. However, no comparable circuit description exists for higher order processing centers.

Where circuit details are lacking, a viable alternative is to leverage recent rapid progress in developmental neuroscience to design systems that can self-organize their own connectivity. During development, neurons wire themselves into their mature networks by sprouting axonal and dendritic precursors called *neurites*. Each neurite is tipped by a sensory structure called a *growth cone* that uses local chemical cues to guide the elongating neurite. Growth cones move by continually extending and retracting finger-like appendages called *filopodia* whose dynamics are biased by diffusible ligands [2].

Adopting this developmental approach, we previously described the first self-organizing map chip that is based on neurite outgrowth [3]. Our chip's neuromorphic cells are equipped with growth cone circuits that enable them to wire themselves into a mature network automatically, without an explicit blueprint.

Although analogous self-organizing algorithms have been successfully implemented digitally, previous analog implementations required high precision components that are expensive in chip area (e.g., [4],[5]). By contrast, neurobiological

W. Duch et al. (Eds.): ICANN 2005, LNCS 3697, pp. 1027–1034, 2005.

systems achieve tolerable performance with low precision components. By mimicking their growth cones, our neuromorphic approach realizes the low power benefit of analog implementation without the associated cost in chip area.

We previously described the design of the Neurotrope1 chip and its ability to self-organize topography [3]. Here, we extend those results by illustrating how topographic self-organization depends on the range of the diffusible growth cone guidance signal. Section 2 introduces the learning algorithm, Section 3 describes the neuromorphic implementation, and Section 4 presents chip measurements from a topographic self-organization task. We find that performance on this task is optimized by balancing signal strength with signal range.

2 Neurotropic Axon Guidance

In our implementation, an active growth cone's filopodia bind a diffusible guidance signal called *neurotropin* that is released by active target cells (Fig. 1). The growth cone measures the local concentration gradient by comparing the rates at which its filopodia accumulate neurotropin, and maximizes its neurotropin uptake by climbing the measured gradient. Since filopodial neurotropin binding is gated by presynaptic activity and neurotropin release is gated by postsynaptic activity, neurotropic gradient ascent implements a Hebbian update rule under which cells that fire at the same time wire to the same place.

This algorithm for the self-organization of connections from one layer of neurons to another is formally described as follows. Source cells occupy nodes of a regular two-dimensional (2D) lattice, while growth cones and target cells occupy nodes of separate regular 2D lattices that are interleaved. Nodes are indexed by their positions in their respective layers, labeled by Greek letters in the source layer (e.g. $\alpha \in \mathbb{Z}^2$) and by Roman letters in the target layer (e.g. $x, c \in \mathbb{Z}^2$).

Target cell x fires at a rate $a_{\mathrm{TC}}(x)$ that is proportional to its excitation:

$$a_{\mathrm{TC}}(x) = \sum_{\alpha} a_{\mathrm{SC}}(\alpha) A(c(\alpha) - x)$$

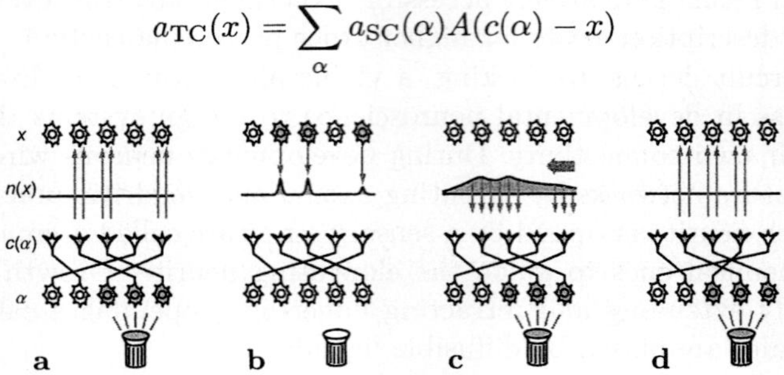

Fig. 1. Neurotropic attraction model. **a.** A stimulus coactivates a contiguous patch of source cells α, which fire spikes down their axons to induce their growth cones $c(\alpha)$ to excite nearby target cells x. **b.** Active target cells release neurotropin $n(x)$ into the extracellular medium. **c.** Neurotropin spreads laterally before being bound by active growth cones, which measure the direction of the local gradient, indicated by the arrow. **d.** Active growth cones climb the gradient by displacing other growth cones.

where $a_{\mathrm{SC}}(\alpha)$ is the activity of source cell α, and $A(c(\alpha) - x)$ is the branch density of the excitatory arbor elaborated by the growth cone-tipped axon trunk projected by source cell α to target site $c(\alpha)$. $A(c(\alpha) - x)$ decreases with target layer separation $\|c(\alpha) - x\|$.

Active target cells release neurotropin, which spreads laterally until consumed by constitutive scavenger processes, establishing a concentration

$$n(x) = \sum_y a_{\mathrm{TC}}(y) N(x - y)$$

that sums contributions from all active target cells y, weighted by a spreading function $N(x - y)$ that decreases with $\|x - y\|$.

An active growth cone located at $c(\alpha)$ uses a local winner-take-all function to identify the node

$$c'(\alpha) = \arg\max_{x \in \mathcal{C}(c(\alpha))} n(x)$$

that contains the most neurotropin during the growth cone's activity, where $\mathcal{C}(x)$ contains x and its nearest neighbors. Upon identifying $c'(\alpha)$, the growth cone swaps nodes with the growth cone currently occupying $c'(\alpha)$, moving the entire axon arbor with it, thereby increasing its neurotropic uptake while maintaining a constant axon density. Growth cones initiate swaps independently, at a rate $\lambda(\alpha) \propto a_{\mathrm{SC}}(\alpha) \max_{x \in \mathcal{C}(c(\alpha))} n(x)$. Swaps are serviced asynchronously, in the order in which they arrive.

Software simulation of similar equations yields self-organized topography when driven with appropriate presynaptic correlations [6]. In this paper, we probe the effect of varying the width of $N(x - y)$ in a hardware implementation.

3 Neuromorphic Implementation

We implemented this model in a full custom VLSI chip that interleaves a 24×40 array of growth cone circuits with a 24×20 array of target cell circuits and a neurotropin spreading network. The Neurotrope1 chip was fabricated through MOSIS using the TSMC 0.35μm process, and is 11.5 mm^2 in area. Axons are implemented as entries in an off-chip lookup table that are updated by Neurotrope1 activity, as described in Subsection 3.1. Subsection 3.2 explains how Neurotrope1 computes these updates from neurotropin in the spreading network and Subsection 3.3 explains how the spreading network circuit shapes $N(x - y)$.

3.1 Axon Remapping

Chips in our system exchange spikes encoded in the address-event representation (AER) [7], an asynchronous protocol that pools spikes from all the cells on the same chip onto a shared data link. AER tags each spike with the address of its originating cell body for transmission off-chip onto the data link. Each spike is filtered through a *forward lookup table* that translates the source layer address of

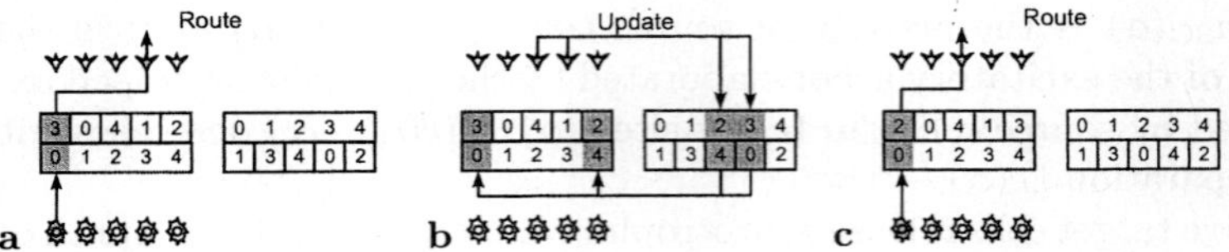

Fig. 2. Axon remapping. **a.** Cell bodies tag their spikes with their source layer addresses, which are decoded through the forward lookup table into the target layer addresses of their growth cones. **b.** Migrating growth cones decode their target layer addresses through the reverse lookup table into the source layer addresses of their cell bodies, which index their entries in the forward lookup table. **c.** Growth cones swap locations by modifying their four entries in the forward and reverse lookup tables.

its origin into the target layer address of its destination (Fig. 2a). The receiving chip uses the delivered address to route the spike to the appropriate target.

An axon is remapped to a new target site simply by updating its entry in the lookup table. Updates are requested by Neurotrope1 growth cone circuits and communicated as address-events to a Ubicom ip2022 microcontroller for processing. Each update request identifies a pair of adjacent growth cones whose target layer addresses are to be swapped. These addresses are translated through a *reverse lookup table* that decodes target layer growth cone addresses into the source layer cell body addresses that index the forward lookup table (Fig. 2b). Axons migrate by modifying their entries in each lookup table (Fig. 2c).

Both lookup tables are stored in a random access memory (RAM) chip (Fig. 3a). The ip2022 processes the axon updates computed by the growth cone circuits and overwrites the appropriate RAM cells. The ip2022 also supports a universal serial bus (USB) over which a computer can write to and read from the RAM. Any AER-compliant device can implement the source cell population; in this paper, we simulate the source layer with a second ip2022.

3.2 Axon Updates

Axon updates are computed by Neurotrope1 using the growth cone circuits described in [3]. Each growth cone occupies one node of the neurotropin spread-

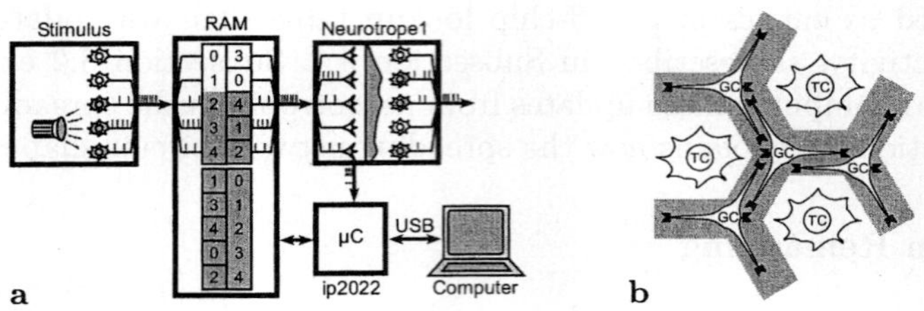

Fig. 3. a. Neurotrope1 system. **b.** Neurotrope1 cell mosaic. The neurotropin spreading network (grey) is interleaved with the array of target cell circuits (TC). Each growth cone circuit (GC) occupies one node of the spreading network and extends filopodia to the three adjacent nodes, expressing neurotropin receptors (black) at all four nodes.

ing network and extends filopodia to the three adjacent nodes, expressing neurotropin receptors at all four sites (Fig. 3b). Neurotropin is represented as charge in the spreading network. During a presynaptic spike, receptor circuits tap charge from their nodes of the spreading network and store it on separate capacitors. The first capacitor voltage to cross a threshold triggers an update request and resets all the growth cone's capacitors. If the receptor at the growth cone body won the race to threshold, no action is required and the request is dropped. If one of the filopodial receptors won, the growth cone transmits a request off-chip to swap places with the growth cone currently occupying the winning node.

Gradient measurements are noisy, so we require the ip2022 to process multiple requests from the same pair of adjacent growth cone circuits before actually executing the swap. We maintain a running count in the RAM of accumulated requests for each pair, and only execute a swap after its count exceeds a preprogrammed threshold. An executed swap resets the counts of five affected growth cone pairings among the two growth cone circuits and their four neighbors. The effect is to screen out spurious update requests and brake growth cone velocity.

3.3 Neurotropin Spreading Circuit

The neurotropin spreading function $N(x-y)$ is shaped by the transistor circuit in Fig. 4a, which implements a neighborhood function similar to [8]. Transistors M1 and M2 are gated by a facilitation circuit that only allows charge to be injected into the spreading network during a burst of postsynaptic activity, since $\tilde{}tb$ requires several consecutive spikes to bring it low enough to open M1, although $\tilde{}ts$ allows individual spikes to open M2. Similarly, a growth cone can only sample charge during bursts of presynaptic activity through transistors M3-5. (The sampled current I_{NT} is limited by bias V_{b}.)

Between spikes, charge spreads laterally through the unnumbered pFETs until shunted to ground through one of the unnumbered nFETs located at each lattice node. The two gate biases V_{spread} and V_{shunt} control the distance to which

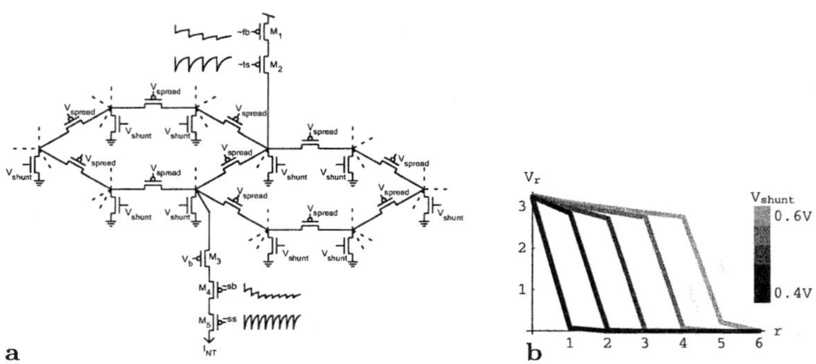

Fig. 4. a. Neurotropin spreading circuit. **b.** SPICE simulation of node voltages V_{r} during a neurotropin release pulse, as a function of distance r from the release site. $V_{\mathrm{spread}} = 2.0V$.

charge can spread from its injection site. V_{shunt} gates the shunt transistor hosted by each lattice node. This transistor enters saturation as the node voltage rises and becomes a constant current sink. Incoming current in excess of this sink charges the node capacitance until the node voltage exceeds V_{spread}, allowing the remaining current to flow outward to as many nodes as needed to sink all the injected current. Larger shunt currents sink the injected current closer to the release site, so increasing V_{shunt} reduces the spreading range (Fig. 4b).

4 Topographic Self-organization

To induce the growth cone population to self-organize a topographic image of the source layer, we drive them with bursts of presynaptic spikes from contiguous patches of coactive source cells. Each patch consists of a randomly selected source cell and its three immediately adjacent neighbors, a presynaptic activity correlation kernel with sufficient structure to instruct topographic ordering [9].

We trace the evolution of the growth cone population by sampling the contents of the forward lookup table every five minutes, an interval long enough to present each of the 24×20 possible patches about once per sample. Starting from a random projection at $n = 0$ (Fig. 5a), small chunks of local topography are visible by $n = 20$. These local crystals eventually merge into the larger, more global clusters observed at $n = 200$. A similar endstate is reached by a perfect initial projection as it relaxes to a more sustainable topographic level (Fig. 5b).

We evaluate performance quantitatively by defining an *order parameter* $\Phi^{(n)}$ whose value measures the relative topography at a source cell's location at time n. One definition of topography is that adjacent source cells extend axons to adjacent target sites, so we choose $\Phi^{(n)}$ for a given source cell to be the average target layer distance separating its growth cone from those of its three nearest source layer neighbors in the nth sample. (In a perfectly topographic projection, $\Phi^{(n)} = 1$.) The population average $\langle \Phi^{(n)} \rangle$ converges to similar values from both random and perfect initial projections (Fig. 6a), so sustainable topography is not limited by the initial conditions, but by some intrinsic property of the system. This intrinsic limit depends on the neurotropin spreading range, which we control with the shunt bias V_{shunt}. At equilibrium, $\langle \Phi^{(n)} \rangle$ is minimized by an intermediate value of V_{shunt} that corresponds to an optimal spreading range (Fig. 6b).

To investigate this optimal spreading range, we examine the probability $P^{(n)}(\Phi)$ that a pair of growth cones projected by adjacent source cells is sep-

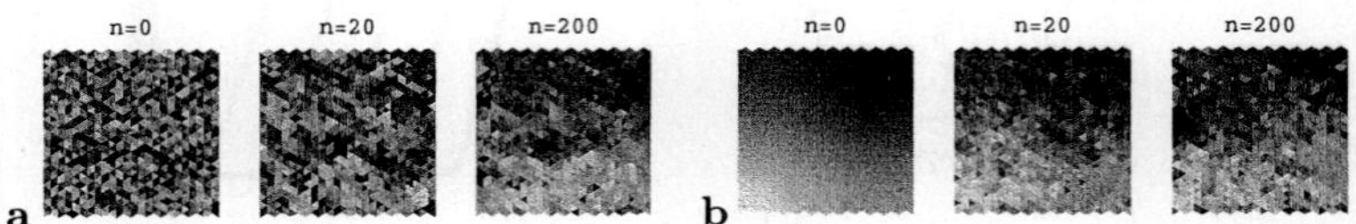

Fig. 5. Topographic self-organization. **a.** Source layer maps generated from random initial projection at time steps n. Source cells are colored by the target layer locations of their growth cones. **b.** Source layer maps generated from perfect initial projection.

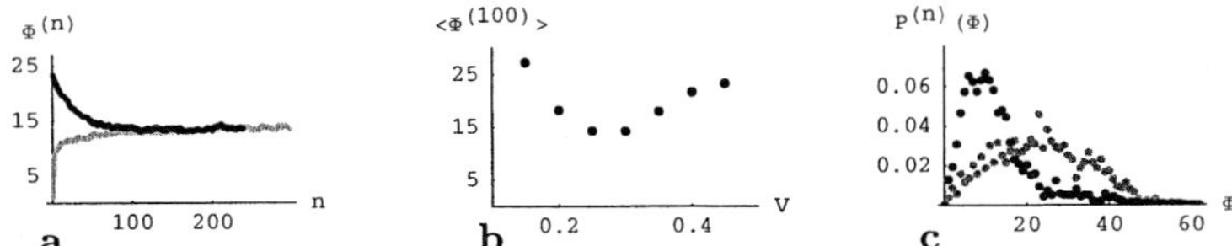

Fig. 6. a. Order parameter Φ evolution from random (black) and perfect (grey) initial projections. **b.** Average equilibrium order parameter $\langle \Phi^{(100)} \rangle$ dependence on neurotropin spreading range, as controlled by V_{shunt}. **c.** Instantaneous order parameter probability $P^{(n)}(\Phi)$ measured during evolution from random initial projection. Grey: $P^{(0)}(\Phi)$; black: $P^{(200)}(\Phi)$.

arated in the target layer by a distance Φ in the nth axon projection sample. We construct $P^{(n)}(\Phi)$ from the relative frequency with which each value of Φ is observed within the nth sample of the population ensemble of growth cone positions downloaded from the RAM. Perfect growth cone guidance elicits a $P^{(\infty)}(\Phi)$ that is 1 at $\Phi = 1$ and 0 elsewhere. Unguided growth cones are distributed in proportion to the number of target sites located a distance Φ from a given attractor (grey in Fig. 6c). This number initially increases as $2\pi\Phi$, but falls to zero at large Φ, since growth cone separations cannot exceed the finite array dimensions. The actual distribution achieved by our physical system lies somewhere between these two extremes (black in Fig. 6c).

The optimal spreading range balances a neurotropin release site's ability to hold growth cones with its ability to attract them (Fig. 7a). For short spreading ranges, $P^{(100)}(\Phi)$ resembles a random distribution except for a small peak at low Φ that captures growth cones that manage to fall within each other's detection horizon. Increasing the spreading range allows growth cones to lure coactive peers from greater distances, siphoning $P^{(100)}(\Phi)$ into a peak at low Φ. However, as the spreading range approaches the array size, the ability of larger attraction basins to rope in more distant growth cones is outweighed by the inability of captured growth cones to localize within the basin. Consequently, for long spreading ranges, $P^{(100)}(\Phi)$ broadens and shifts toward the random distribution.

We dissociate the attraction and confinement aspects of guidance by tracking the topographic evolution of the 25% of growth cones that are furthest from and

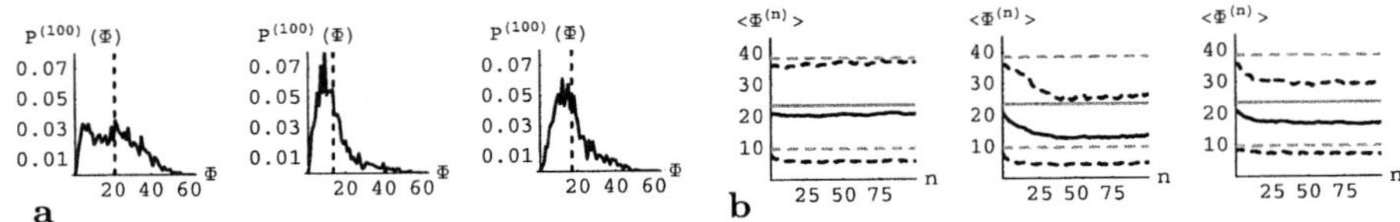

Fig. 7. Optimal neurotropin spreading range. Left: short ($V_{\text{shunt}} = 0.40V$); middle: medium ($0.30V$); right: long ($0.20V$). **a.** Equilibrium distribution $P^{(100)}(\Phi)$. Dashed line indicates $\langle \Phi^{(100)} \rangle$. **b.** Evolution of $\langle \Phi^{(n)} \rangle$ within full growth cone population (black solid line) and growth cone subpopulations with the 25% highest and lowest $\Phi^{(n)}$ values (black dashed lines). Grey lines plot the corresponding $\langle \Phi \rangle$ for a random distribution.

closest to their topographic neighbors (black dashed lines in Fig. 7b). Proximate growth cones do better at short spreading ranges, but distant growth cones do better at long spreading ranges. The optimal spreading range improves the performance of both proximate and distant growth cones.

5 Conclusions

We characterized the first hardware implementation of a self-organizing map algorithm based on axon migration. We varied the range of the neurotropic signal that guides the silicon growth cone population to automatically wire a topographic map when driven by correlated activity. Long neurotropin spreading ranges attract distant growth cones but cannot hold them to their targets, while short neurotropin spreading ranges hold growth cones to nearby targets but cannot rescue distant growth cones. This tradeoff between recovery and confinement implies that future systems should address these aspects separately, perhaps through other developmental mechanisms like synaptogenesis, to consolidate accurately placed growth cones, and pruning, to eliminate outliers.

Acknowledgment

This work was supported by the David and Lucille Packard Foundation and the NSF/BITS program (EIA0130822). B.T. received support from the Dolores Zohrab Liebmann Foundation.

References

1. Zaghloul, K.A., Boahen, K.: Optic nerve signals in a neuromorphic chip I: Outer and inner retina models. IEEE Trans. Bio-Med. Eng. **51**:4 (2004) 657–666
2. Huber, A.B., Kolodkin, A.L., Ginty, D.D., Cloutier, J.-F.: Signaling at the growth cone: ligand-receptor complexes and the control of axon growth and guidance. Annu. Rev. Neurosci. **26** (2003) 509–563
3. Taba, B., Boahen, K.: Topographic map formation by silicon growth cones. Advances in Neural Information Processing Systems **15** (MIT Press, Cambridge, eds. Becker, S., Thrun, S., Obermayer. K.) 1163–1170
4. Melton, M., Phan, T., Reeves, D.S., Van den Bout, D.E.: The TInMANN VLSI chip. IEEE Trans. Neural Networks **3** (1992) 375–384
5. Porrmann, M., Witkowski, U., Ruckert, U.: A massively parallel architecture for self-organizing feature maps. IEEE Trans. Neural Networks **14** (2003) 1110–1121
6. Lam, S.Y.M., Shi, B.E., Boahen, K.A.: Self-organized cortical map formation by guiding connections. Proc. 2005 IEEE Int. Symp. Circ. & Sys. (in press)
7. Boahen, K.: Point-to-point connectivity between neuromorphic chips using address-events. IEEE Trans. Circ. & Sys. II **47** (2000) 416–434
8. Heim, P., Hochet, B., Vittoz, E.A.: Generation of learning neighbourhood in Kohonen feature maps by means of simple nonlinear network. Electronics Letters **27** (1991) 275–277
9. Miller, K.: A model for the development of simple cell receptive fields and the ordered arrangement of orientation columns through activity-dependent competition between on- and off-center inputs. J. Neurosci. **14** (1994) 409–441

Acknowledgments to the Reviewers

We wish to express our sincere gratitude to the following reviewers and advisors for their valuable remarks and sugestions:

C. Aaron
S. Abe
R. Adamczak
F. Aiolli
A. Albrecht
E. Alhoniemi
H. Amin
A. Anastasiadis
R. Andonie
P. Andras
D. Anguita
C. Angulo-Bahon
B. Apolloni
C. Archambeau
M. Atencia
J. Avrithis
Y. Avrithis
G. Bedoya
M. Bianchini
R. Birkenhead
L. Bobrowski
M. Bogdan
S. Bohte
A. Boni
E. Bottino
S. Breutel
J. Bródka
D. Buldain
P. Campadelli
A. Cangelosi
B. Caputo
G. Cawley
L.-W. Chan
M. Chetouani
A. Chung Tsoi
A. Cichocki
E. Corchado
M. Cottrell
S. Coupland
N. Crook
L. Csato
P. Dario
N. Delannay
C. Dimitrakakis
K. Doherty
E. Dominguez Merino
G. Dorffner
W. Duch
D. Elizondo
P. Erdi
M. Faundez-Zanuy
M. Fernandez-Redondo
A. Flanagan
P. Fleury
R. Folland
D. Franois
C. Fyfe
C. Garcia-Osorio
N. Garcia-Pedrajas
B. Gas
S. Gielen
M. Gola
M. Gongora
L. Gonzales Abril
K. Goser
B. Gosselin
K. Grąbczewski
M. Grana
R. Grothmann
M. Grzenda
B. Hammer
R. Haschke
G. Heidemann
J. Henderson
J. Himberg
E. Hines

J. Hollmen
A. Honkela
O. Hryniewicz
D. Huang
T. Huang
M. Huelse
A. Hussain
C. Igel
G. Indiveri
S. Ishii
Y. Ito
N. Jankowski
M. Jirina
R. John
G. Joya
A. Kaban
J. Kacprzyk
M. de Kamps
B. Kappen
J. Karhunen
N. Kasabov
S. Kasderidis
D. Kim
M. Kimura
S. Kollias
J. Korbicz
J. Koronacki
W. Kosiński
M. Koskela
O. Kouropteva
R. Kozma
M. Krawczak
L. Kruś
F. Kurfess
M. Kurzyński
J. Laaksonen
E. Lang
V. Laxmi
E. Leclercq
J. Lee
P. Lehtimäki
K. Leiviskä
A. Lendasse
A. Likas
T. Lourens
C. Lu
B. Macukow
J. Madrenas
M. Maggini
B. Maillet
N. Mammone
D. Mandic
J. Mańdziuk
T. Marcu
U. Markowska-Kaczmar
M. Martin-Merino
T. Martinetz
F. Masulli
J. Meeus
J. Meller
A. Menciassi
A. Micheli
L. Morra
F. Moutarde
N. Mtetwa
R. Muresan
M. Nakayama
N. Nedjah
L. Niklasson
K. Nikolopoulos
W. Nowak
D. Obradovic
E. Oja
F. Okulicka
M. Olteanu
D. Ortiz Boyer
S. Osowski
X. Parra
H. Paugam-Moisy
C. Pedreira
W. Pedrycz
K. Pelckmans
B. Pelletier
J. Pestian
G. Peters
A. Piegat
J. Pizarro Junquera
D. Polani
M. Porrmann
R. R. Poznański

J. Prévotet
C. Puntonet
D. Puzenat
F. Queirolo
T. Raiko
K. Raivio
C. Reyes Garca
M. Rocha
O. Rochel
M. Rodriguez-Alvarez
A. Romariz
F. Rossi
S. Rovetta
D. Rutkowska
L. Rutkowski
J. Rynkiewicz
J. Salojarvi
J. Salotti
J. Sarela
B. Schrauwen
F. Schwenker
U. Seiffert
M. Sfakiotakis
V. Siivola
G. Simon
O. Simula
R. Słowiński
D. Sona
A. Sperduti
A. Stafylopatis
G. Stamou
J. Steil
M. Steuer
M. Strickert
M. Sugiyama
J. Suykens
P. Szczepaniak
R. Tadeusiewicz
R. Tagliaferri
J. G. Taylor
M. Terra
F. Theis
A. Tomé
E. Trentin
D. Tsakiris
V. Tzouvaras
S. Usui
M. Valle
F. van der Velde
T. Van Gestel
S. Van Looy
M. Vannucci
G. Vaucher
J. Venna
J. Verbeek
M. Verleysen
N. Viet
V. Vigneron
A. E. P. Villa
T. Villmann
J. Vitay
M. Wallace
Z. Waszczyszyn
N. Watanabe
T. Wennekers
S. Wermter
W. Wiegerinck
B. Wilamowski
A. Wilbik
K. Wills
P. Wira
A. Wróbel
B. Wyns
S. Xavier de Souza
I. Yaesh
A. Yanez Escolano
Y. Yang
Z. Yang
J. Yearwood
S. Zadrożny
H. Zimmermann
A. Żochowski
J. Żurada

Author Index

Lecture Notes in Computer Science

For information about Vols. 1–3567

please contact your bookseller or Springer

Vol. 3697: W. Duch, J. Kacprzyk, E. Oja, S. Zadrożny (Eds.), Artificial Neural Networks: Formal Models and Their Applications - ICANN 2005, Part II. XXXII, 1045 pages. 2005.

Vol. 3696: W. Duch, J. Kacprzyk, E. Oja, S. Zadrożny (Eds.), Artificial Neural Networks: Biological Inspirations - ICANN 2005, Part I. XXXI, 703 pages. 2005.

Vol. 3687: S. Singh, M. Singh, C. Apte, P. Perner (Eds.), Pattern Recognition and Image Analysis, Part II. XXV, 809 pages. 2005.

Vol. 3686: S. Singh, M. Singh, C. Apte, P. Perner (Eds.), Pattern Recognition and Data Mining, Part I. XXVI, 689 pages. 2005.

Vol. 3674: W. Jonker, M. Petković (Eds.), Secure Data Management. X, 241 pages. 2005.

Vol. 3672: C. Hankin, I. Siveroni (Eds.), Static Analysis. X, 369 pages. 2005.

Vol. 3671: S. Bressan, S. Ceri, E. Hunt, Z.G. Ives, Z. Bellahsène, M. Rys, R. Unland (Eds.), Database and XML Technologies. X, 239 pages. 2005.

Vol. 3670: M. Bravetti, L. Kloul, G. Zavattaro (Eds.), Formal Techniques for Computer Systems and Business Processes. XIII, 349 pages. 2005.

Vol. 3664: C. Türker, M. Agosti, H.-J. Schek (Eds.), Peer-to-Peer, Grid, and Service-Orientation in Digital Library Architectures. X, 261 pages. 2005.

Vol. 3663: W. Kropatsch, R. Sablatnig, A. Hanbury (Eds.), Pattern Recognition. XIV, 512 pages. 2005.

Vol. 3662: C. Baral, G. Greco, N. Leone, G. Terracina (Eds.), Logic Programming and Nonmonotonic Reasoning. XIII, 454 pages. 2005. (Subseries LNAI).

Vol. 3660: M. Beigl, S. Intille, J. Rekimoto, H. Tokuda (Eds.), UbiComp 2005: Ubiquitous Computing. XVII, 394 pages. 2005.

Vol. 3659: J.R. Rao, B. Sunar (Eds.), Cryptographic Hardware and Embedded Systems – CHES 2005. XIV, 458 pages. 2005.

Vol. 3658: V. Matoušek, P. Mautner, T. Pavelka (Eds.), Text, Speech and Dialogue. XV, 460 pages. 2005. (Subseries LNAI).

Vol. 3654: S. Jajodia, D. Wijesekera (Eds.), Data and Applications Security XIX. X, 353 pages. 2005.

Vol. 3653: M. Abadi, L.d. Alfaro (Eds.), CONCUR 2005 – Concurrency Theory. XIV, 578 pages. 2005.

Vol. 3649: W.M.P. van der Aalst, B. Benatallah, F. Casati, F. Curbera (Eds.), Business Process Management. XII, 472 pages. 2005.

Vol. 3648: J.C. Cunha, P.D. Medeiros (Eds.), Euro-Par 2005 Parallel Processing. XXXVI, 1299 pages. 2005.

Vol. 3645: D.-S. Huang, X.-P. Zhang, G.-B. Huang (Eds.), Advances in Intelligent Computing, Part II. XIII, 1010 pages. 2005.

Vol. 3644: D.-S. Huang, X.-P. Zhang, G.-B. Huang (Eds.), Advances in Intelligent Computing, Part I. XXVII, 1101 pages. 2005.

Vol. 3642: D. Ślezak, J. Yao, J.F. Peters, W. Ziarko, X. Hu (Eds.), Rough Sets, Fuzzy Sets, Data Mining, and Granular Computing, Part II. XXIV, 738 pages. 2005. (Subseries LNAI).

Vol. 3641: D. Ślezak, G. Wang, M.S. Szczuka, I. Düntsch, Y. Yao (Eds.), Rough Sets, Fuzzy Sets, Data Mining, and Granular Computing, Part I. XXIV, 742 pages. 2005. (Subseries LNAI).

Vol. 3639: P. Godefroid (Ed.), Model Checking Software. XI, 289 pages. 2005.

Vol. 3638: A. Butz, B. Fisher, A. Krüger, P. Olivier (Eds.), Smart Graphics. XI, 269 pages. 2005.

Vol. 3637: J. M. Moreno, J. Madrenas, J. Cosp (Eds.), Evolvable Systems: From Biology to Hardware. XI, 227 pages. 2005.

Vol. 3636: M.J. Blesa, C. Blum, A. Roli, M. Sampels (Eds.), Hybrid Metaheuristics. XII, 155 pages. 2005.

Vol. 3634: L. Ong (Ed.), Computer Science Logic. XI, 567 pages. 2005.

Vol. 3633: C. Bauzer Medeiros, M. Egenhofer, E. Bertino (Eds.), Advances in Spatial and Temporal Databases. XIII, 433 pages. 2005.

Vol. 3632: R. Nieuwenhuis (Ed.), Automated Deduction – CADE-20. XIII, 459 pages. 2005. (Subseries LNAI).

Vol. 3629: J.L. Fiadeiro, N. Harman, M. Roggenbach, J. Rutten (Eds.), Algebra and Coalgebra in Computer Science. XI, 457 pages. 2005.

Vol. 3628: T. Gschwind, U. Aßmann, O. Nierstrasz (Eds.), Software Composition. X, 199 pages. 2005.

Vol. 3627: C. Jacob, M.L. Pilat, P.J. Bentley, J. Timmis (Eds.), Artificial Immune Systems. XII, 500 pages. 2005.

Vol. 3626: B. Ganter, G. Stumme, R. Wille (Eds.), Formal Concept Analysis. X, 349 pages. 2005. (Subseries LNAI).

Vol. 3625: S. Kramer, B. Pfahringer (Eds.), Inductive Logic Programming. XIII, 427 pages. 2005. (Subseries LNAI).

Vol. 3624: C. Chekuri, K. Jansen, J.D.P. Rolim, L. Trevisan (Eds.), Approximation, Randomization and Combinatorial Optimization. XI, 495 pages. 2005.

Vol. 3623: M. Liśkiewicz, R. Reischuk (Eds.), Fundamentals of Computation Theory. XV, 576 pages. 2005.

Vol. 3621: V. Shoup (Ed.), Advances in Cryptology – CRYPTO 2005. XI, 568 pages. 2005.

Vol. 3620: H. Muñoz-Avila, F. Ricci (Eds.), Case-Based Reasoning Research and Development. XV, 654 pages. 2005. (Subseries LNAI).

Vol. 3619: X. Lu, W. Zhao (Eds.), Networking and Mobile Computing. XXIV, 1299 pages. 2005.

Vol. 3618: J. Jedrzejowicz, A. Szepietowski (Eds.), Mathematical Foundations of Computer Science 2005. XVI, 814 pages. 2005.

Vol. 3617: F. Roli, S. Vitulano (Eds.), Image Analysis and Processing – ICIAP 2005. XXIV, 1219 pages. 2005.

Vol. 3615: B. Ludäscher, L. Raschid (Eds.), Data Integration in the Life Sciences. XII, 344 pages. 2005. (Subseries LNBI).

Vol. 3614: L. Wang, Y. Jin (Eds.), Fuzzy Systems and Knowledge Discovery, Part II. XLI, 1314 pages. 2005. (Subseries LNAI).

Vol. 3613: L. Wang, Y. Jin (Eds.), Fuzzy Systems and Knowledge Discovery, Part I. XLI, 1334 pages. 2005. (Subseries LNAI).

Vol. 3612: L. Wang, K. Chen, Y. S. Ong (Eds.), Advances in Natural Computation, Part III. LXI, 1326 pages. 2005.

Vol. 3611: L. Wang, K. Chen, Y. S. Ong (Eds.), Advances in Natural Computation, Part II. LXI, 1292 pages. 2005.

Vol. 3610: L. Wang, K. Chen, Y. S. Ong (Eds.), Advances in Natural Computation, Part I. LXI, 1302 pages. 2005.

Vol. 3608: F. Dehne, A. López-Ortiz, J.-R. Sack (Eds.), Algorithms and Data Structures. XIV, 446 pages. 2005.

Vol. 3607: J.-D. Zucker, L. Saitta (Eds.), Abstraction, Reformulation and Approximation. XII, 376 pages. 2005. (Subseries LNAI).

Vol. 3606: V. Malyshkin (Ed.), Parallel Computing Technologies. XII, 470 pages. 2005.

Vol. 3604: R. Martin, H. Bez, M. Sabin (Eds.), Mathematics of Surfaces XI. IX, 473 pages. 2005.

Vol. 3603: J. Hurd, T. Melham (Eds.), Theorem Proving in Higher Order Logics. IX, 409 pages. 2005.

Vol. 3602: R. Eigenmann, Z. Li, S.P. Midkiff (Eds.), Languages and Compilers for High Performance Computing. IX, 486 pages. 2005.

Vol. 3599: U. Aßmann, M. Aksit, A. Rensink (Eds.), Model Driven Architecture. X, 235 pages. 2005.

Vol. 3598: H. Murakami, H. Nakashima, H. Tokuda, M. Yasumura, Ubiquitous Computing Systems. XIII, 275 pages. 2005.

Vol. 3597: S. Shimojo, S. Ichii, T.W. Ling, K.-H. Song (Eds.), Web and Communication Technologies and Internet-Related Social Issues - HSI 2005. XIX, 368 pages. 2005.

Vol. 3596: F. Dau, M.-L. Mugnier, G. Stumme (Eds.), Conceptual Structures: Common Semantics for Sharing Knowledge. XI, 467 pages. 2005. (Subseries LNAI).

Vol. 3595: L. Wang (Ed.), Computing and Combinatorics. XVI, 995 pages. 2005.

Vol. 3594: J.C. Setubal, S. Verjovski-Almeida (Eds.), Advances in Bioinformatics and Computational Biology. XIV, 258 pages. 2005. (Subseries LNBI).

Vol. 3593: V. Mařík, R. W. Brennan, M. Pěchouček (Eds.), Holonic and Multi-Agent Systems for Manufacturing. XI, 269 pages. 2005. (Subseries LNAI).

Vol. 3592: S. Katsikas, J. Lopez, G. Pernul (Eds.), Trust, Privacy and Security in Digital Business. XII, 332 pages. 2005.

Vol. 3591: M.A. Wimmer, R. Traunmüller, Å. Grönlund, K.V. Andersen (Eds.), Electronic Government. XIII, 317 pages. 2005.

Vol. 3590: K. Bauknecht, B. Pröll, H. Werthner (Eds.), E-Commerce and Web Technologies. XIV, 380 pages. 2005.

Vol. 3589: A M. Tjoa, J. Trujillo (Eds.), Data Warehousing and Knowledge Discovery. XVI, 538 pages. 2005.

Vol. 3588: K.V. Andersen, J. Debenham, R. Wagner (Eds.), Database and Expert Systems Applications. XX, 955 pages. 2005.

Vol. 3587: P. Perner, A. Imiya (Eds.), Machine Learning and Data Mining in Pattern Recognition. XVII, 695 pages. 2005. (Subseries LNAI).

Vol. 3586: A.P. Black (Ed.), ECOOP 2005 - Object-Oriented Programming. XVII, 631 pages. 2005.

Vol. 3584: X. Li, S. Wang, Z.Y. Dong (Eds.), Advanced Data Mining and Applications. XIX, 835 pages. 2005. (Subseries LNAI).

Vol. 3583: R.W. H. Lau, Q. Li, R. Cheung, W. Liu (Eds.), Advances in Web-Based Learning – ICWL 2005. XIV, 420 pages. 2005.

Vol. 3582: J. Fitzgerald, I.J. Hayes, A. Tarlecki (Eds.), FM 2005: Formal Methods. XIV, 558 pages. 2005.

Vol. 3581: S. Miksch, J. Hunter, E. Keravnou (Eds.), Artificial Intelligence in Medicine. XVII, 547 pages. 2005. (Subseries LNAI).

Vol. 3580: L. Caires, G.F. Italiano, L. Monteiro, C. Palamidessi, M. Yung (Eds.), Automata, Languages and Programming. XXV, 1477 pages. 2005.

Vol. 3579: D. Lowe, M. Gaedke (Eds.), Web Engineering. XXII, 633 pages. 2005.

Vol. 3578: M. Gallagher, J. Hogan, F. Maire (Eds.), Intelligent Data Engineering and Automated Learning - IDEAL 2005. XVI, 599 pages. 2005.

Vol. 3577: R. Falcone, S. Barber, J. Sabater-Mir, M.P. Singh (Eds.), Trusting Agents for Trusting Electronic Societies. VIII, 235 pages. 2005. (Subseries LNAI).

Vol. 3576: K. Etessami, S.K. Rajamani (Eds.), Computer Aided Verification. XV, 564 pages. 2005.

Vol. 3575: S. Wermter, G. Palm, M. Elshaw (Eds.), Biomimetic Neural Learning for Intelligent Robots. IX, 383 pages. 2005. (Subseries LNAI).

Vol. 3574: C. Boyd, J.M. González Nieto (Eds.), Information Security and Privacy. XIII, 586 pages. 2005.

Vol. 3573: S. Etalle (Ed.), Logic Based Program Synthesis and Transformation. VIII, 279 pages. 2005.

Vol. 3572: C. De Felice, A. Restivo (Eds.), Developments in Language Theory. XI, 409 pages. 2005.

Vol. 3571: L. Godo (Ed.), Symbolic and Quantitative Approaches to Reasoning with Uncertainty. XVI, 1028 pages. 2005. (Subseries LNAI).

Vol. 3570: A. S. Patrick, M. Yung (Eds.), Financial Cryptography and Data Security. XII, 376 pages. 2005.

Vol. 3569: F. Bacchus, T. Walsh (Eds.), Theory and Applications of Satisfiability Testing. XII, 492 pages. 2005.

Vol. 3568: W.-K. Leow, M.S. Lew, T.-S. Chua, W.-Y. Ma, L. Chaisorn, E.M. Bakker (Eds.), Image and Video Retrieval. XVII, 672 pages. 2005.